S0-AVI-174

Traffic Engineering

Fourth Edition

Roger P. Roess, Ph.D.
Professor of Transportation Engineering
Polytechnic Institute of New York University

Elena S. Prassas, Ph.D.
Associate Professor of Transportation Engineering
Polytechnic Institute of New York University

William R. McShane, Ph.D., P.E., P.T.O.E.
President, KLD Engineering, P.C.
Professor Emeritus
Polytechnic Institute of New York University

Boston Columbus Indianapolis New York San Francisco Upper Saddle River
Amsterdam Cape Town Dubai London Madrid Milan Munich Paris Montreal Toronto
Delhi Mexico City Sao Paulo Sydney Hong Kong Seoul Singapore Taipei Tokyo

Vice President and Editorial Director, ECS: Marcia J. Horton
Senior Editor: Holly Stark
Editorial Assistant: Keri Rand
Vice President, Production: Vince O'Brien
Marketing Manager: Tim Galligan
Marketing Assistant: Mack Patterson
Senior Managing Editor: Scott Disanno
Production Project Manager: Clare Romeo
Senior Operations Specialist: Alan Fischer
Operations Specialist: Lisa McDowell
Art Director, Interior: Greg Dulles
Art Director, Cover: Kristine Carney
Cover Designer: Bruce Keneslaar
Cover Illustration/Photo(s): Shutterstock
Manager, Rights and Permissions: Zina Arabia
Manager, Visual Research: Beth Brenzel
Image Permission Coordinator: Debbie Latronica
Manager, Cover Visual Research & Permissions: Karen Sanatar
Composition: Integra Software Services Pvt. Ltd.
Full-Service Project Management: Integra Software Services Pvt. Ltd.
Printer/Binder: Hamilton
Typeface: 10/12 Times

Credits and acknowledgments borrowed from other sources and reproduced, with permission, in this textbook appear on appropriate page within text.

Library of Congress Cataloging-in-Publication Data

Roess, Roger P.
Traffic engineering / Roger P. Roess, Elena S. Prassas, William R. McShane. — 4th ed.
p. cm.
Includes bibliographical references and index.
ISBN-13: 978-0-13-613573-9 (alk. paper)
ISBN-10: 0-13-613573-0 (alk. paper)
1. Traffic engineering—United States. I. Prassas, Elena S. II. McShane, William R. III. Title.
HE355.M43 2011
388.3'120973—dc22

2010013599

10 9 8 7 6 5 4 3 2 1

Prentice Hall
is an imprint of

www.pearsonhighered.com

ISBN-10: 0-13-613573-0
ISBN-13: 978-0-13-613573-9

Contents

Preface

The transportation system is often referred to as the nation's "lifeblood circulation system." Our complex system of roads and highways, railroads, airports and airlines, waterways, and urban transit systems provides for the movement of people and goods within and between our densest urban cities and the most remote outposts of the nation. Without the ability to travel and to transport goods, society must be structured around small self-sufficient communities, each of which produces food and material for all of its needs locally and disposes of its wastes in a similar manner. The benefits of economic specialization and mass production are possible only where transportation exists to move needed materials of production to centralized locations, and finished products to widely dispersed consumers.

Traffic engineering deals with one critical element of the transportation system: streets and highways, and their use by vehicles. This vast national system provides mobility and access for individuals in private autos and for goods in trucks of various sizes and forms, and facilitates public transport by supporting buses, bicycles, and pedestrians.

Because the transportation system is such a critical part of our public infrastructure, the traffic engineer is involved in a wide range of issues, often in a very public setting, and must bring a wide range of skills to the table to be effective. Traffic engineers must have an appreciation for and understanding of planning, design, management, construction, operation, control, and system optimization. All of these functions involve traffic engineers at some level.

This text focuses on the key engineering skills required to practice traffic engineering in a modern setting. This is the fourth edition of this textbook. It includes material on the latest standards and criteria of the *Manual on Uniform Traffic Control Devices* (2003 Edition and forthcoming 2010 Edition), the *Policy on Geometric Design of Highways and Streets* (2004 Edition), the *Highway Capacity Manual* (2000 Edition and forthcoming 2010 Edition), and other critical references. It also presents both fundamental theory and a broad range of applications to modern problems.

The text is organized in five major functional parts:

- Part 1—Traffic Components and Characteristics
- Part 2—Traffic Studies and Programs
- Part 3—Freeways and Rural Highways
- Part 4—The Intersection
- Part 5—Arterials, Networks, and Systems

This text can be used for an undergraduate survey course, or for more detailed graduate courses. At Polytechnic Institute of New York University, it is used for two undergraduate courses and a series of three graduate courses.

As in previous editions, the text contains many sample problems and illustrations that can be used in conjunction with course material. A solutions manual is available. The authors hope that practicing professionals and students find this text useful and informative, and they invite comments and/or criticisms that will help them continue to improve the material.

What's New in This Edition

This edition of the textbook adds a significant amount of material, including, but not limited to:

1. New homework problems for most chapters.
2. New chapters on Traffic Flow Theory, Analysis of Arterials in a Multimodal Setting, Critical Movement Analysis of Signalized Intersections, and Traffic Impact Studies.
3. Material from the latest editions of key traffic engineering references, including the *Traffic Engineering Handbook*, the *Manual of Uniform Traffic Control*

Devices, the *Traffic Signal Timing Handbook*, and the *Policy on Geometric Design of Highways and Streets*.

4. Substantial material from forthcoming new editions of the *Highway Capacity Manual* (2010) and *Manual of Uniform Traffic Control Devices* (2010), which were obtained from research documents, draft materials, and other source documents has been included. Since some of this material has not yet been officially adopted, it provides a preview, but not final information on these standard documents.
5. New material on actuated signal systems and timing.
6. New material on coordination of signal systems.
7. Reference links to important Web sites, as well as demonstration solutions using current software packages.

ROGER P. ROESS

ELENA S. PRASSAS

WILLIAM R. MCSHANE

CHAPTER 1

Introduction to Traffic Engineering

1.1 Traffic Engineering as a Profession

The Institute of Transportation Engineers defines traffic engineering as a subset of transportation engineering as follows [*1*]:

> Transportation engineering is the application of technology and scientific principles to the planning, functional design, operation, and management of facilities for any mode of transportation in order to provide for the safe, rapid, comfortable, convenient, economical, and environmentally compatible movement of people and goods.

and:

> Traffic engineering is that phase of transportation engineering which deals with the planning, geometric design and traffic operations of roads, streets, and highways, their networks, terminals, abutting lands, and relationships with other modes of transportation.

These definitions represent a broadening of the profession to include multimodal transportation systems and options, and to include a variety of objectives in addition to the traditional goals of safety and efficiency.

1.1.1 Safety: The Primary Objective

The principal goal of the traffic engineer remains the provision of a safe system for highway traffic. This is no small concern. In recent years, fatalities on U.S. highways have ranged between 40,000 and 43,000 per year. Although this is a reduction from the highs experienced in the 1970s, when highway fatalities reached more than 55,000 per year, it continues to represent a staggering number. Rising fuel prices in 2008 and 2009 have had an impact on both fatalities and vehicle-miles travelled. In 2008, fatalities were reduced to 37,261, the first time the number dipped below 40,000 in many years. Some of this was due to a reduction in vehicle-miles travelled, which dipped under 3.0 trillion miles after two years over this level. It remains to be seen whether this reduction is sustainable or whether fatalities will rise once again when (and if) the fuel cost issues are resolved. One point, however, remains: More Americans have been killed on U.S. highways than in all of the wars in which the nation has participated, including the Civil War.

Although total highway fatalities per year have remained relatively stable over the past two decades, accident rates based on vehicle-miles traveled have consistently declined. That is because U.S. motorists continue to drive more miles each year. With a stable total number of fatalities, the increasing number of annual vehicle-miles traveled produces a declining fatality rate. This trend will also be affected by the decrease in vehicle use in 2008 and 2009.

Improvements in fatality rates reflect a number of trends, many of which traffic engineers have been instrumental in implementing. Stronger efforts to remove dangerous drivers from the road have yielded significant dividends in safety. Driving under the influence (DUI) and driving while intoxicated (DWI) offenses are more strictly enforced, and licenses are suspended or revoked more easily as a result of DUI/DWI convictions, poor accident record, and/or poor violations record. Vehicle design has greatly improved (encouraged by several acts of Congress requiring certain improvements). Today's vehicles feature padded dashboards, collapsible steering columns, seat belts with shoulder harnesses, air bags (some vehicles now have as many as eight), and antilock braking systems. Highway design has improved through the development and use of advanced barrier systems for medians and roadside areas. Traffic control systems communicate better and faster, and surveillance systems can alert authorities to accidents and breakdowns in the system.

Despite this, however, approximately 40,000 people per year still die in traffic accidents. The objective of safe travel is always number one and is never finished for the traffic engineer.

1.1.2 Other Objectives

The definitions of transportation and traffic engineering highlight additional objectives:

- Speed
- Comfort
- Convenience
- Economy
- Environmental compatibility

Most of these are self-evident desires of the traveler. Most of us want our trips to be fast, comfortable, convenient, cheap, and in harmony with the environment. All of these objectives are also relative and must be balanced against each other and against the primary objective of safety.

Although speed of travel is much to be desired, it is limited by transportation technology, human characteristics, and the need to provide safety. Comfort and convenience are generic terms and mean different things to different people. Comfort involves the physical characteristics of vehicles and roadways, and it is influenced by our perception of safety. Convenience relates more to the ease with which trips are made and the ability of transport systems to accommodate all of our travel needs at appropriate times. Economy is also relative. There is little in modern transportation systems that can be termed "cheap." Highway and other transportation systems involve massive construction, maintenance, and operating expenditures, most of which are provided through general and user taxes and fees. Nevertheless, every engineer, regardless of discipline, is called on to provide the best possible systems for the money.

Harmony with the environment is a complex issue that has become more important over time. All transportation systems have some negative impacts on the environment. All produce air and noise pollution in some forms, and all utilize valuable land resources. In many modern cities, transportation systems use as much as 25% of the total land area. "Harmony" is achieved when transportation systems are designed to minimize negative environmental impacts, and where system architecture provides for aesthetically pleasing facilities that "fit in" with their surroundings.

The traffic engineer is tasked with all of these goals and objectives and with making the appropriate trade-offs to optimize both the transportation systems and the use of public funds to build, maintain, and operate them.

1.1.3 Responsibility, Ethics, and Liability in Traffic Engineering

The traffic engineer has a very special relationship with the public at large. Perhaps more than any other type of engineer, the traffic engineer deals with the daily safety of a large segment of the public. Although it can be argued that any engineer who designs a product has this responsibility, few engineers have so many people using their product so routinely and frequently and depending on it so totally. Therefore, the traffic engineer also has a special obligation to employ the available knowledge and state of the art within existing resources to enhance public safety.

The traffic engineer also functions in a world in which a number of key participants do not understand the traffic and transportation issues or how they truly affect a particular project. These include elected and appointed officials with decision-making power, the general public, and other professionals with whom traffic engineers work on an overall project team effort. Because all of us interface regularly with the transportation system, many overestimate their understanding of transportation and traffic issues. The traffic engineer must deal productively with problems associated with naive assumptions, plans and designs that are oblivious to transportation and traffic needs, oversimplified analyses, and understated impacts.

Like all engineers, traffic engineers must understand and comply with professional ethics codes. Primary codes of ethics for traffic engineers are those of the National Society of Professional Engineers and the American Society of Civil Engineers. The most up-to-date versions of each are available online. In general, good professional ethics requires that traffic engineers work only in their areas of expertise; do all work completely and thoroughly; be completely honest with the general public, employers, and clients; comply with all applicable codes and standards; and work to the best of their ability. In traffic engineering, the pressure to understate negative impacts of projects, sometimes brought to bear by clients who wish a project to proceed and employers who wish to keep clients happy, is a particular concern. As in all engineering professions, the pressure to minimize costs must give way to basic needs for safety and reliability.

Experience has shown that the greatest risk to a project is an incomplete analysis. Major projects have been upset because an impact was overlooked or analysis oversimplified. Sophisticated developers and experienced professionals know that the environmental impact process calls for a fair and complete statement of impacts and a *policy decision by the reviewers* on accepting the impacts, given an overall good analysis report. The process does not require zero impacts; it does, however, call for clear and complete disclosure of impacts so that policy makers can make informed decisions. Successful challenges to major projects are almost always based on flawed analysis, not on disagreements with policy makers. Indeed, such disagreements are not a valid basis for a legal challenge to a project. In the case of the Westway Project proposed in the 1970s for the west side of Manhattan, one of the bases for legal challenge was that the impact of project construction on striped bass in the Hudson River had not been properly identified or disclosed. In particular, the project died due to overlooking the impact on the reproductive cycle of striped bass in the Hudson River. Although this topic was not the primary concern of the litigants, it was the legal "hook" that caused the project to be abandoned.

The traffic engineer also has a responsibility to protect the community from liability by good practice. Agencies charged with traffic and transportation responsibilities can be held liable in many areas. These include (but are not limited to) the following:

- Placing control devices that do not conform to applicable standards for their physical design and placement.
- Failure to maintain devices in a manner that ensures their effectiveness; the worst case of this is a "dark" traffic signal in which no indication is given due to bulb or other device failure.
- Failure to apply the most current standards and guidelines in making decisions on traffic control, developing a facility plan or design, or conducting an investigation.
- Implementing traffic regulations (and placing appropriate devices) without the proper legal authority to do so.

A historic standard has been that "due care" be exercised in the preparation of plans and that determinations made in the process be reasonable and "not arbitrary." It is generally recognized that professionals must make value judgments, and the terms *due care* and *not arbitrary* are continually under legal test.

The fundamental ethical issue for traffic engineers is to provide for the public safety through positive programs, good practice, knowledge, and proper procedure. The negative (albeit important) side of this is the avoidance of liability problems.

1.2 Transportation Systems and Their Function

Transportation systems are a major component of the U.S. economy and have an enormous impact on the shape of the society and the efficiency of the economy in general. Table 1.1 illustrates some key statistics for the U.S. highway system for the base year 2007 and two preliminary statistics for 2008.

America moves on its highways. Although public transportation systems are of major importance in large urban areas such as New York, Boston, Chicago, and San Francisco, it is clear that the vast majority of person-travel as well as a large proportion of freight traffic is entirely dependent on the highway system. The system is a major economic force in its own right: Over $90 billion per year is collected by state and federal governments directly from road users in the form of focused user taxes and fees. Such taxes and fees include excise taxes on gasoline and other fuels, registration fees, commercial vehicles fees, and others. Other funds are allocated from federal and state general funds for highway use. As indicated in Table 1.1, by 2007, $161 billion was being collected and spent on highway and traffic improvements by all units of government.

Table 1.1: Important Statistics on U.S. Highways

Statistic	2007	2008
Miles of public roadway	4.05 million	4.04 million
Vehicle-miles traveled	3.05 trillion	2.93 trillion
Total population of United States	301 million	304 million
Licensed drivers	205 million	208 million
Registered vehicles	247 million	248 million
Total receipts: Taxes, fees, tolls, general fund allocations	$161 billion	
Total expenditures: Federal, state, local	$161 billion	
Fatalities	41,059	37,261

The American love affair with the automobile has grown consistently since the 1920s when Henry Ford's Model T made the car accessible to the average wage earner. This growth has survived wars, gasoline embargoes, depressions, recessions, and almost everything else that has happened in society. As seen in Figure 1.1, annual vehicle-miles traveled reached the 1 trillion mark in 1968, the 2 trillion mark in 1987, and the 3 trillion mark in 2005.

This growth pattern is one of the fundamental problems to be faced by traffic engineers. Given the relative maturity of our highway systems and the difficulty faced in trying to add system capacity, particularly in urban areas, the continued growth in vehicle-miles traveled leads directly to increased congestion on our highways. The inability to simply build additional capacity to meet the growing demand creates the need to address alternative modes, fundamental alterations in demand patterns, and management of the system to produce optimal results.

1.2.1 The Nature of Transportation Demand

Transportation demand is directly related to land-use patterns and to available transportation systems and facilities. Figure 1.2 illustrates the fundamental relationship, which is circular and ongoing. Transportation demand is generated by the types, amounts, and intensity of land use, as well as its location. The daily journey to work, for example, is dictated by the locations of the worker's residence and employer and the times that the worker is on duty.

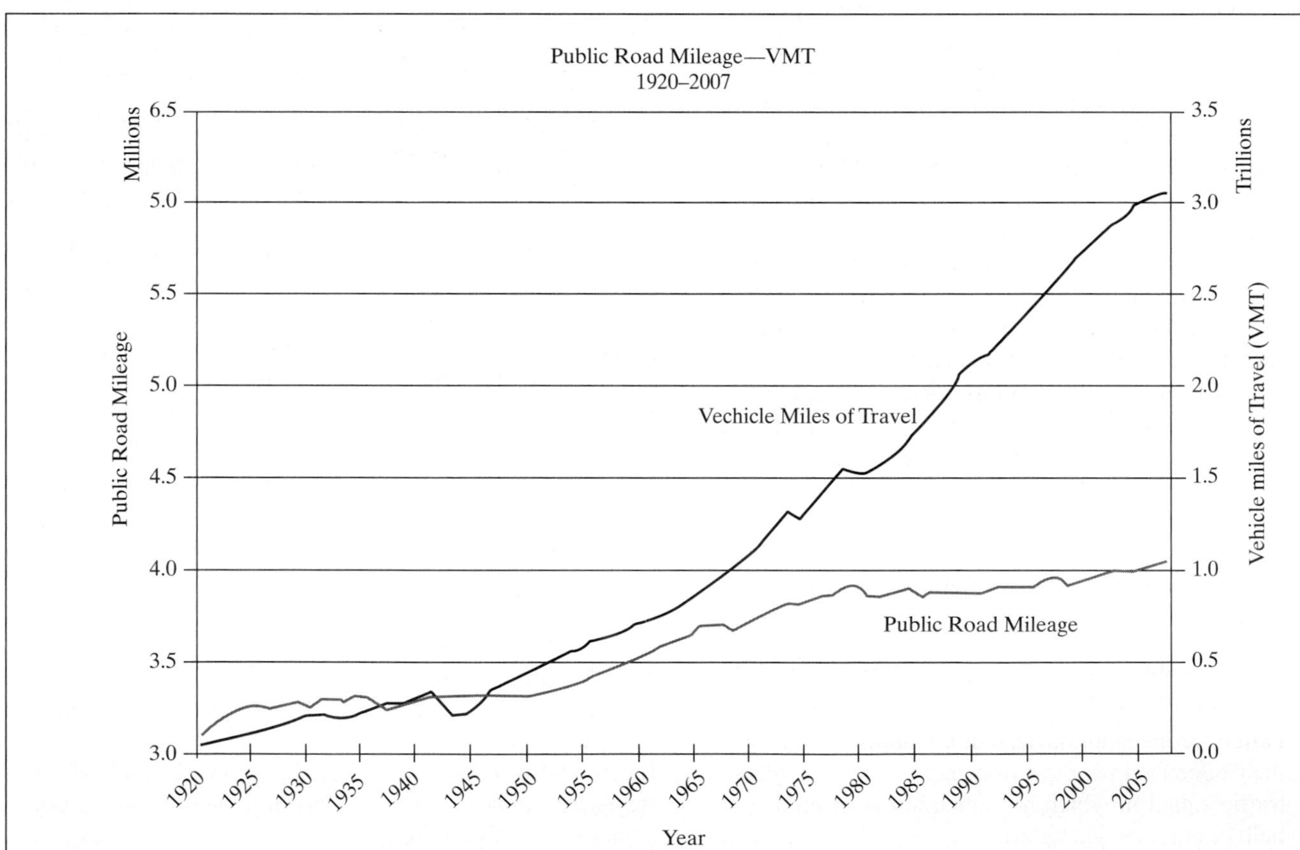

Figure 1.1: Public Highway Mileage and Annual Vehicle-Miles Traveled in the United States, 1920–2007

(*Source: Highway Statistics 2007*, Federal Highway Administration, U.S. Department of Transportation, Washington DC, 2008, Table VMT 421.)

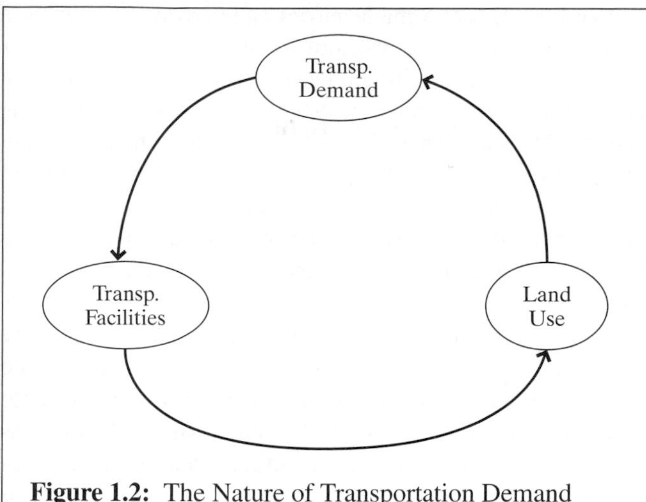

Figure 1.2: The Nature of Transportation Demand

Transportation planners and traffic engineers attempt to provide capacity for observed or predicted travel demand by building transportation systems. The improvement of transportation systems, however, makes the adjacent and nearby lands more accessible and, therefore, more attractive for development. Thus building new transportation facilities leads to further increases in land-use development, which (in turn) results in even higher transportation demands. This circular, self-reinforcing characteristic of traffic demand creates a central dilemma: Building additional transportation capacity invariably leads to incrementally increased travel demands.

In many major cities, this has led to the search for more efficient transportation systems, such as public transit and car-pooling programs. In some of the largest cities, providing additional system capacity on highways is no longer an objective because such systems are already substantially choking in congestion. In these places, the emphasis shifts to improvements within existing highway rights-of-way and to the elimination of bottleneck locations (without adding to overall capacity). Other approaches include staggered work hours and workdays to reduce peak hour demands, and even more radical approaches involve development of satellite centers outside of the central business district (CBD) to spatially disperse highly directional demands into and out of city centers.

Demand, however, is not constrained by capacity in all cities, and the normal process of attempting to accommodate demand as it increases is feasible in these areas. At the same time, the circular nature of the travel/demand relationship will lead to congestion if care is not taken to manage both capacity and demand to keep them within tolerable limits.

It is important that the traffic engineer understand this process. It is complex and cannot be stopped at any moment in time. Demand-prediction techniques (not covered in this text) must start and stop at arbitrary points in time. The real process is ongoing, and as new or improved facilities are provided, travel demand is constantly changing. Plans and proposals must recognize both this reality and the professional's inability to precisely predict its impacts. *A 10-year traffic demand forecast that comes within approximately ±20% of the actual value is considered a significant success.* The essential truth, however, is that traffic engineers cannot simply build their way out of congestion.

If anything, we still tend to underestimate the impact of transportation facilities on land-use development. Often, the increase in demand is hastened by development occurring simply as a result of the planning of a new facility.

One of the classic cases occurred on Long Island, in New York State. As the Long Island Expressway was built, the development of suburban residential communities lurched forward in anticipation. While the expressway's link to Exit 7 was being constructed, new homes were being built at the anticipated Exit 10, even though the facility would not be open to that point for several years. The result was that as the expressway was completed section by section, the 20-year anticipated demand was being achieved within a few years, or even months. This process has been repeated in many cases throughout the nation.

1.2.2 Concepts of Mobility and Accessibility

Transportation systems provide the nation's population with both mobility and accessibility. The two concepts are strongly interrelated but have distinctly different elements. *Mobility* refers to the ability to travel to many different destinations with relative ease, whereas *accessibility* refers to the ability to gain entry to a particular site or area.

Mobility gives travelers a wide range of choices as to where to go to satisfy particular needs, and it provides for efficient trips to get to them. Mobility allows shoppers to choose from among many competing shopping centers and stores. Similarly, mobility provides the traveler with many choices for all kinds of trip purposes, including recreational trips, medical trips, educational trips, and even the commute to work. The range of available choices is enabled by having an effective transportation network that connects to many alternative trip destinations within a reasonable time, with relative ease, and at reasonable cost. Thus mobility provides access to many travel opportunities, and it provides relative speed and convenience for the required trips.

Accessibility is a major factor in the value of land. When land can be accessed by many travelers from many potential origins, it is more desirable for development and, therefore, more valuable. Thus proximity of land to major highways and public transportation facilities is a major factor determining its value.

Mobility and accessibility may also refer to different portions of a typical trip. Mobility focuses on the through portion of trips and is most affected by the effectiveness of through facilities that take a traveler from one general area to another. Accessibility requires the ability to make a transfer from the transportation system to the particular land parcel on which the desired activity is taking place. Accessibility, therefore, relies heavily on transfer facilities, which include parking for vehicles, public transit stops, and loading zones.

As we discuss in Chapter 3, most transportation systems are structured to separate mobility and access functions because the two functions often compete and are not necessarily compatible. In highway systems, mobility is provided by high-type facilities, such as freeways, expressways, and primary and secondary arterials. Accessibility is generally provided by local street networks. Except for limited-access facilities, which serve only through vehicles (mobility), most other classes of highway serve both functions to some degree. Access maneuvers (e.g., parking and unparking a vehicle, vehicles entering and leaving off-street parking via driveways, buses stopping to pick up or discharge passengers, trucks stopped to load and/or unload goods), however, retard the progress of through traffic. High-speed through traffic, in contrast, tends to make such access functions more dangerous.

A good transportation system must provide for both mobility and accessibility and should be designed to separate the functions to the extent possible to ensure both safety and efficiency.

1.2.3 People, Goods, and Vehicles

The most common unit used by the traffic engineer is "vehicles." Highway systems are planned, designed, and operated to move vehicles safely and efficiently from place to place. Yet the movement of vehicles is not the objective; the goal is the movement of the people and goods that occupy vehicles.

Modern traffic engineering now focuses more on people and goods. Although lanes must be added to a freeway to increase its capacity to carry vehicles, its person-capacity can be increased by increasing the average vehicle occupancy. Consider a freeway lane with a capacity of 2,000 vehicles per hour (veh/h). If each vehicle carries one person, the lane has a capacity of 2,000 persons/hour as well. If the average car occupancy is increased to 2.0 persons/vehicle, the capacity in terms of people is doubled to 4,000 persons/hour. If the lane were established as an exclusive bus lane, the vehicle-capacity might be reduced to 1,000 veh/h due to the larger size and poorer operating characteristics of buses as compared with automobiles. However, if each bus carries 50 passengers, the people-capacity of the lane is increased to 50,000 persons/hour.

The efficient movement of goods is also vital to the general economy of the nation. The benefits of centralized and specialized production of various products are possible only if raw materials can be efficiently shipped to manufacturing sites and finished products can be efficiently distributed throughout the nation and the world for consumption. Although long-distance shipment of goods and raw materials is often accomplished by water, rail, or air transportation, the final leg of the trip to deliver a good to the local store or the home of an individual consumer generally takes place on a truck using the highway system. Part of the accessibility function is the provision of facilities that allow trucks to be loaded and unloaded with minimal disruption to through traffic and the accessibility of people to a given site.

The medium of all highway transportation is the vehicle. The design, operation, and control of highway systems relies heavily on the characteristics of the vehicle and of the driver. In the final analysis, however, the objective is to move people and goods, not vehicles.

1.2.4 Transportation Modes

Although the traffic engineer deals primarily with highways and highway vehicles, there are other important transportation systems that must be integrated into a cohesive national, regional, and local transportation network. Table 1.2 provides a comprehensive listing of various transportation modes and their principal uses.

The traffic engineer deals with all of these modes in a number of ways. All over-the-road modes—automobile, bus transit, trucking—are principal users of highway systems. Highway access to rail and air terminals is critical to their effectiveness, as is the design of specific transfer facilities for both people and freight. General access, internal circulation, parking, pedestrian areas, and terminals for both people and freight are all projects requiring the expertise of the traffic engineer. Moreover, the effective integration of multimodal transportation systems is a major goal in maximizing efficiency and minimizing costs associated with all forms of travel.

Table 1.2: Transportation Modes

Mode	Typical Function	Approximate Range of Capacities*
Urban People Transportation Systems		
Automobile	Private personal transportation; available on demand for all trips.	1–6 persons/vehicle; approx. 2,000 veh/h per freeway lane; 400–700 veh/h per arterial lane.
Taxi/For-hire vehicles	Private or shared personal transportation, available by prearrangement or on call.	1–6 persons/vehicle; total capacity limited by availability.
Local bus transit	Public transportation along fixed routes on a fixed schedule; low speed with many stops.	40–70 persons/bus; capacity limited by schedule; usually 100–5,000 persons/h/route.
Express bus transit	Public transportation along fixed routes on a fixed schedule; higher speed with few intermediate stops.	40–50 persons/bus (no standees); capacity limited by schedule.
Para-transit	Public transportation with flexible routing and schedules, usually available on call.	Variable seating capacity depends on vehicle design; total capacity depends on number of available vehicles.
Light Rail	Rail service using one- to two-car units along fixed routes with fixed schedules.	80–120 persons/car; up to 15,000 persons/h/route.
Heavy Rail	Heavy rail vehicles in multicar trains along fixed routes with fixed schedules on fully separated rights-of-way in tunnels, on elevated structures, or on the surface.	150–300 persons/car depending on seating configuration and standees; up to 60,000 persons per track.
Ferry	Waterborne public transportation for people and vehicles along fixed routes on fixed schedules.	Highly variable with ferry and terminal design and schedule.
Intercity People Transportation Systems		
Automobile	Private transportation available on demand for all trip purposes.	Same as urban automobile.
Intercity bus	Public transportation along a fixed intercity route on a fixed (and usually limited) schedule. Provides service to a central terminal location in each city.	40–50 passengers per bus; schedules highly variable.
Railroad	Passenger intercity rail service on fixed routes on a fixed (and usually limited) schedule. Provides service to a central terminal location or locations within each city.	500–1,000 passengers per train, depending on configuration; schedules highly variable.
Air	A variety of air-passenger services from small commuter planes to jumbo jets on fixed routes and fixed schedules.	From 3–4 passengers to 500 passengers per aircraft depending on size and configuration. Schedules depend on destination and are highly variable.
Water	Passenger ship service often associated with onboard vacation packages on fixed routes and schedules.	Ship capacity highly variable from several hundred to 3,500 passengers; schedules often extremely limited.

*Ranges cited represent typical values, not the full range of possibilities.

(*Continued*)

Table 1.2: Transportation Modes (*Continued*)

Mode	Typical Function	Approximate Range of Capacities*
Urban and Intercity Freight Transportation Systems		
Long-haul trucks	Single, double, and triple tractor-trailer combinations and large single-unit trucks provide over-the-road intercity service, by arrangement.	Hauling capacity of all freight modes varies widely with the design of the vehicle (or pipeline) and limitations on fleet size and schedule availability.
Local trucks	Smaller trucks provide distribution of goods and services throughout urban areas.	
Railroad	Intercity haulage of bulk commodities with some local distribution to locations with rail sidings.	
Water	International and intercity haulage of bulk commodities on a variety of container ships and barges.	
Air freight	International and intercity haulage of small and moderately sized parcels and/or time-sensitive and/or high-value commodities where high cost is not a disincentive.	
Pipelines	Continuous flow of fluid or gaseous commodities; intercity and local distribution networks possible.	

*Ranges cited represent typical values, not the full range of possibilities.

1.3 Highway Legislation and History in the United States

The development of highway systems in the United States is strongly tied to federal legislation that supports and regulates much of this activity. Key historical and legislative actions are discussed in the sections that follow.

1.3.1 The National Pike and the States' Rights Issue

Before the 1800s, roads were little more than trails cleared through the wilderness by adventurous travelers and explorers. Private roadways began to appear in the latter part of the 1700s. These roadways ranged in quality and length from cleared trails to plank roadways. They were built by private owners, and fees were charged for their use. At points where fees were to be collected, a barrier usually consisting of a single crossbar was mounted on a swiveling stake, referred to as a "pike." When the fee was collected, the pike would be swiveled or turned, allowing the traveler to proceed. This early process gave birth to the term *turnpike,* often used to describe toll roadways in modern times.

The National Pike

In 1811, the construction of the first national roadway was begun under the direct supervision of the federal government. Known as the "national pike" or the "Cumberland Road," this facility stretched for 800 miles from Cumberland, Maryland, in the east, to Vandalia, Illinois in the west. A combination of unpaved and plank sections, it was finally completed in 1852 at a total cost of $6.8 million. A good deal of the original route is now a portion of U.S. Route 40.

Highways as a States' Right

The course of highway development in the United States, however, was forever changed as a result of an 1832 Supreme Court case brought by the administration of President Andrew Jackson. A major proponent of states' rights, the Jackson Administration petitioned the court claiming that the U.S. Constitution did not specifically define transportation and roadways as federal functions; they were, therefore, the responsibility of the individual states. The Supreme Court upheld this position, and the principal administrative responsibility for transportation and highways was forevermore assigned to state governments.

The Governmental Context

If the planning, design, construction, maintenance, and operation of highway systems is a state responsibility, what is the role of federal agencies—for example, the U.S. Department of Transportation and its components, such as the Federal Highway Administration, the National Highway Safety Administration, and others in these processes?

The federal government asserts its overall control of highway systems through the power of the purse strings. The federal government provides massive funding for the construction, maintenance, and operation of highway and other transportation systems. States are not *required* to follow federal mandates and standards but must do so to qualify for federal funding of projects. Thus the federal government does not force a state to participate in federal-aid transportation programs. If it chooses to participate, however, it must follow federal guidelines and standards. Because no state can afford to give up this massive funding source, the federal government imposes strong control of policy issues and standards.

The federal role in highway systems has four major components:

1. Direct responsibility for highway systems on federally owned lands, such as national parks and Native American reservations.
2. Provision of funding assistance in accord with current federal-aid transportation legislation.
3. Development of planning, design, and other relevant standards and guidelines that must be followed to qualify for receipt of federal-aid transportation funds.
4. Monitoring and enforcing compliance with federal standards and criteria, and the use of federal-aid funds.

State governments have the primary responsibility for the planning, design, construction, maintenance, and operation of highway systems. These functions are generally carried out through a state department of transportation or similar agency. States have:

1. Full responsibility for administration of highway systems.
2. Full responsibility for the planning, design, construction, maintenance, and operation of highway systems in conformance with applicable federal standards and guidelines.
3. The right to delegate responsibilities for local roadway systems to local jurisdictions or agencies.

Local governments have general responsibility for local roadway systems as delegated in state law. In general, local governments are responsible for the planning, design, construction, maintenance, and control of local roadway systems. Often, assistance from state programs and agencies is available to local governments in fulfilling these functions. At intersections of state highways with local roadways, it is generally the state that has the responsibility to control the intersection.

Local organizations for highway functions range from a full highway or transportation department to local police to a single professional traffic or city engineer.

There are also a number of special situations across the United States. In New York State, for example, the state constitution grants "home rule" powers to any municipality with a population in excess of 1 million people. Under this provision, New York City has full jurisdiction over all highways within its borders, including those on the state highway system.

1.3.2 Key Legislative Milestones

Federal-Aid Highway Act of 1916

The Federal-Aid Highway Act of 1916 was the first allocation of federal-aid highway funds for highway construction by the states. It established the "A-B-C System" of primary, secondary, and tertiary federal-aid highways, and provided 50% of the funding for construction of highways in this system. Revenues for federal aid were taken from the federal general fund, and the act was renewed every two to five years (with increasing amounts dedicated). No major changes in funding formulas were forthcoming for a period of 40 years.

Federal-Aid Highway Act of 1934

In addition to renewing funding for the A-B-C System, this act authorized states to use up to 1.5% of federal-aid funds for planning studies and other investigations. It represented the entry of the federal government into highway planning.

Federal-Aid Highway Act of 1944

This act contained the initial authorization of what became the National System of Interstate and Defense Highways. No appropriation of funds occurred, however, and the system was not initiated for another 12 years.

Federal-Aid Highway Act of 1956

The authorization and appropriation of funds for the implementation of the National System of Interstate and Defense Highways occurred in 1956. The act also set the federal share of the cost of the Interstate System at 90%, the first major change in funding formulas since 1916. Because of the major impact on the amounts of federal funds to be spent, the act also created the *Highway Trust Fund* and enacted a series of road-user taxes to provide it with revenues. These taxes included excise taxes on motor fuels, vehicle purchases, motor oil, and replacement parts. Most of these taxes, except for the federal fuel tax, were dropped during the Nixon Administration. The monies housed in the Highway Trust Fund may be disbursed only for purposes authorized by the current federal-aid highway act.

Federal-Aid Highway Act of 1970

Also known as the Highway Safety Act of 1970, this legislation increased the federal subsidy of non-Interstate highway projects to 70% and required all states to implement highway safety agencies and programs.

Federal-Aid Highway Act of 1983

This act contained the "Interstate trade-in" provision that allows states to "trade in" federal-aid funds designated for urban interstate projects for alternative transit systems. This historic provision was the first to allow road-user taxes to be used to pay for public transit improvements.

ISTEA and TEA-21

The single largest overhaul of federal-aid highway programs occurred with the passage of the Intermodal Surface Transportation Efficiency Act (ISTEA) in 1991 and its successor, the Transportation Equity Act for the 21st Century (TEA-21) in 1998.

Most importantly, these acts combined federal-aid programs for all modes of transportation and greatly liberalized the ability of state and local governments to make decisions on modal allocations. These are the key provisions of ISTEA:

1. Greatly increased local options in the use of federal-aid transportation funds.
2. Increased the importance and funding to Metropolitan Planning Organizations (MPOs) and required that each state maintain a state transportation improvement plan (STIP).
3. Tied federal-aid transportation funding to compliance with the Clean Air Act and its amendments.
4. Authorized $38 billion for a 155,000-mile National Highway System.
5. Authorized an additional $7.2 billion to complete the Interstate System and $17 billion to maintain it as part of the National Highway System.
6. Extended 90% federal funding of Interstate-eligible projects.
7. Combined all other federal-aid systems into a single surface transportation system with 80% federal funding.
8. Allowed (for the first time) the use of federal-aid funds in the construction of toll roads.

TEA-21 followed in kind, increasing funding levels, further liberalizing local options for allocation of funds, further encouraging intermodality and integration of transportation systems, and continuing the link between compliance with clean-air standards and federal transportation funding.

The creation of the National Highway System (NHS) answered a key question that had been debated for years: What comes after the Interstate System? The new, expanded NHS is not limited to freeway facilities and is over three times the size of the Interstate System, which becomes part of the NHS.

SAFETY-LU

President Bush signed the most expensive transportation funding act into law on August 10, 2005. The act was a mile wide, and more than four years late, with intervening highway funding being accomplished through annual continuation legislation that kept TEA-21 in effect.

The Safe, Accountable, Flexible and Efficient Transportation Equity Act—A Legacy for Users (SAFETY-LU) has been both praised and criticized. Although it retains most of the programs of ISTEA and TEA-21, and expands the funding for most of them, the act also adds many new programs and provisions, leading some lawmakers and politicians to label it "the most pork-filled legislation in U.S. history." Table 1.3 provides a simple listing of the programs covered under this legislation. The program, which authorizes over $248 billion in expenditures, includes many programs that represent items of special interest inserted by members of Congress.

Table 1.3: Programs Covered by SAFETY-LU*

Program	Amount
Interstate Maintenance Program	$25.1
National Highway System	$30.5
Surface Transportation System	$32.4
Congestion Mitigation/Air Quality Improvement Program	$8.5
Highway Safety Improvement Program	$5.1
Appalachian Development/Highway System Program	$2.4
Recreational Trails Program	$0.4
Federal Lands Highway Program	$4.5
National Corridor Infrastructure Improvement Program	$1.9
Coordinated Border Infrastructure Program	$0.8
National Scenic Byways Program	$0.2
Construction of Ferry Boats/Terminals	$0.3
Puerto Rico Highway Program	$0.7
Projects of National and Regional Significance Program	$1.8
High-Priority Projects Program	$14.8
Safe Routes to School Program	$0.61
Deployment of MagLev Transportation Projects	$0.45
National Corridor Planning/Development of Coordinated Infrastructure Programs	$0.14
Highways for Life Program	$0.45
Highway Use Tax Evasion Projects	$0.12

*All amounts are stated in billions of dollars.

The legislation does recognize the need for massive funding of Interstate highway maintenance, as the system continues to age, with many structural components well past their anticipated service life. It also provides massive funding for the new NHS, which is the successor to the Interstate System in terms of new highways. It also retains the flexibility for local governments to push more funding into public transportation modes.

Although discussions in Congress on a successor act have begun, it is not clear, at this writing, when the next major funding legislation will be passed, or what is will and will not contain.

1.3.3 The National System of Interstate and Defense Highways

The Interstate System has been described as the largest public works project in the history of humankind. In 1919, a young army officer, Dwight Eisenhower, was tasked with moving a complete battalion of troops and military equipment from coast to coast on the nation's highways to determine their utility for such movements in a time of potential war. The trip took months and left the young officer with a keen appreciation for the need to develop a national roadway system. It was no accident that the Interstate System was initiated in the administration of President Dwight Eisenhower, nor that the system now bears his name.

After the end of World War II, the nation entered a period of sustained prosperity. One of the principal signs of that prosperity was the great increase in auto ownership along with the expanding desire of owners to use their cars for daily commuting and for recreational travel. Motorists groups, such as the American Automobile Association (AAA), were formed and began substantial lobbying efforts to expand the nation's highway systems. At the same time, the over-the-road trucking industry was making major inroads against the previous rail monopoly on intercity freight haulage. Truckers also lobbied strongly for improved highway systems. These substantial pressures led to the inauguration of the Interstate System in 1956.

The System Concept

Authorized in 1944 and implemented in 1956, the National System of Interstate and Defense Highways is a 42,500-mile national system of multilane, limited-access facilities. The system was designed to connect all standard metropolitan statistical areas (SMSAs) with 50,000 or greater population with a continuous system of limited-access facilities. The allocation of 90% of the cost of the system to the federal government was justified on the basis of the potential military use of the system in wartime.

System Characteristics

Key characteristics of the Interstate System include the following:

1. All highways have at least two lanes for the exclusive use of traffic in each direction.
2. All highways have full control of access.
3. The system must form a closed loop: All Interstate highways must begin and end at a junction with another Interstate highway.
4. North–south routes have odd two-digit numbers (e.g., I-95).
5. East–west routes have even two-digit numbers (e.g., I-80).
6. Interstate routes serving as bypass loops or acting as a connector to a primary Interstate facility have three-digit route numbers, with the last two digits indicating the primary route.

Figure 1.3 shows a map of the Interstate System.

Status and Costs

By 1994, the system was 99.4% complete. Most of the unfinished sections were not expected to ever be completed for a variety of reasons. The total cost of the system was approximately $125 billion.

Figure 1.3: The Interstate System

The impact of the Interstate System on the nation cannot be understated. The system facilitated and enabled the rapid suburbanization of the United States by providing a means for workers to commute from suburban homes to urban jobs. The economy of urban centers suffered as shoppers moved in droves from traditional CBDs to suburban malls.

The system also had serious negative impacts on some of the environs through which it was built. Following the traditional theory of benefit-cost, urban sections were often built through the low-income parts of communities where land was the cheapest. The massive Interstate highway facilities created physical barriers, partitioning many communities, displacing residents, and separating others from their schools, churches, and local shops. Social unrest resulted in several parts of the country, which eventually resulted in important modifications to the public hearing process and in the ability of local opponents to legally stop many urban highway projects.

Between 1944 and 1956, a national debate was waged over whether the Interstate System should be built into and out of urban areas, or whether all Interstate facilities should terminate in ring roads built around urban areas. Proponents of the ring-road option (including, ironically, Robert Moses who built many highways into and out of urban cities) argued that building these roadways into and out of cities would lead to massive urban congestion. On the other side, the argument was that most of the road users who were paying for the system through their road-user taxes lived in urban areas and should be served. The latter view prevailed, but the predicted rapid growth of urban congestion also became a reality.

1.4 Elements of Traffic Engineering

There are a number of key elements of traffic engineering:

1. Traffic studies and characteristics
2. Performance evaluation
3. Facility design
4. Traffic control
5. Traffic operations
6. Transportation systems management
7. Integration of intelligent transportation system technologies.

Traffic studies and characteristics involve measuring and quantifying various aspect of highway traffic. Studies focus on data collection and analysis that is used to characterize traffic, including (but not limited to) traffic volumes and demands, speed and travel time, delay, accidents, origins and destinations, modal use, and other variables.

Performance evaluation is a means by which traffic engineers can rate the operating characteristics of individual sections of facilities and facilities as a whole in relative terms. Such evaluation relies on measures of performance quality and is often stated in terms of "levels of service." Levels of service are letter grades, from A to F, describing how well a facility is operating using specified performance criteria. Like grades in a course, A is very good, whereas F connotes failure (on some level). As part of performance evaluation, the *capacity* of highway facilities must be determined.

Facility design involves traffic engineers in the functional and geometric design of highways and other traffic facilities. Traffic engineers, per se, are not involved in the structural design of highway facilities but should have some appreciation for structural characteristics of their facilities.

Traffic control is a central function of traffic engineers and involves the establishment of traffic regulations and their communication to the driver through the use of traffic control devices, such as signs, markings, and signals.

Traffic operations involves measures that influence overall operation of traffic facilities, such as one-way street systems, transit operations, curb management, and surveillance and network control systems.

Transportation systems management (TSM) involves virtually all aspects of traffic engineering in a focus on optimizing system capacity and operations. Specific aspects of TSM include high-occupancy vehicle priority systems, car-pooling programs, pricing strategies to manage demand, and similar functions.

Intelligent transportation systems (ITS) refers to the application of modern telecommunications technology to the operation and control of transportation systems. Such systems include automated highways, automated toll-collection systems, vehicle-tracking systems, in-vehicle global positioning systems (GPS) and mapping systems, automated enforcement of traffic lights and speed laws, smart control devices, and others. This is a rapidly emerging family of technologies with the potential to radically alter the way we travel as well as the way in which transportation professionals gather information and control facilities. While the technology continues to expand, society will grapple with the substantial "big brother" issues that such systems invariably create.

This text contains material related to all of these components of the broad and complex profession of traffic engineering.

1.5 Modern Problems for the Traffic Engineer

We live in a complex and rapidly developing world. Consequently, the problems that traffic engineers are involved in evolve rapidly.

Urban congestion has been a major issue for many years. Given the transportation demand cycle, it is not always possible to solve congestion problems through expansion of capacity. Traffic engineers therefore are involved in the development of programs and strategies to manage demand in both time and space and to discourage growth where necessary. A real question is not "how much capacity is needed to handle demand?" but rather, "how many vehicles and/or people can be allowed to enter congested areas within designated time periods?"

Growth management is a major current issue. A number of states have legislation that ties development permits to level-of-service impacts on the highway and transportation system. Where development will cause substantial deterioration in the quality of traffic service, either such development will be disallowed or the developer will be responsible for general highway and traffic improvements that mitigate these negative impacts. Such policies are more easily dealt with in good economic times. When the economy is sluggish, the issue will often be a clash between the desire to reduce congestion and the desire to encourage development as a means of increasing the tax base.

Reconstruction of existing highway facilities also causes unique problems. The entire Interstate System has been aging, and many of its facilities have required major reconstruction efforts. Part of the problem is that reconstruction of Interstate facilities receives the 90% federal subsidy, whereas routine maintenance on the same facility has been, until recently, primarily the responsibility of state and local governments. Deferring routine maintenance on these facilities in favor of major reconstruction efforts has resulted from federal funding policies over the years. Major reconstruction efforts have a substantial major burden not involved in the initial construction of these facilities: maintaining traffic. It is easier to build a new facility in a dedicated right-of-way than to rebuild it while continuing to serve 100,000 or more vehicles per day. Thus issues of long-term and short-term construction detours as well as the diversion of traffic to alternative routes require major planning by traffic engineers.

Since 2001, the issue of security of transportation facilities has come to the fore. The creation of facilities and processes for random and systematic inspection of trucks and other vehicles at critical locations is a major challenge, as is securing major public transportation systems such as railroads, airports, and rapid transit systems.

As the fourth edition of this text is written, we are entering a new era with many unknowns. With the sharp rise in fuel prices through 2008, vehicle use actually began to decline for the first time in decades. Transportation planners and engineers must be careful in determining whether this is a reliable trend with long-term implications or simply a short-term market perturbation. The economic crisis of 2008 and 2009 has caused many additional shifts in the economy, even as the price of fuel came back to more normal levels. Major carmakers in the United States were headed into bankruptcy, with major industry reductions and changes anticipated. Government loans to both banks and industries brought with it more governmental control of private industries. A major shift of U.S. automakers to smaller, more fuel-efficient and "green" vehicles has begun, with no clear appreciation of whether the buying public will sustain the shift. For perhaps the first time in many decades, transportation and traffic demand may be very much dependent on the state of the general economy, not the usual motivators of improved mobility and accessibility. Will people learn new behaviors resulting in fewer and more efficient trips? Will people flock to hybrid or fully electric vehicles to reduce fuel costs? Will public transportation pick up substantial new customers as big-city drivers abandon their cars for the daily commute? It is an unsettling time that will continue to evolve into new challenges for traffic and transportation engineers. With new challenges, however, comes the ability for new and innovative approaches that might not have been feasible only a few years ago.

The point is that traffic engineers cannot expect to practice their profession only in traditional ways on traditional projects. Like any professional, the traffic engineer must be ready to face current problems and to play an important role in any situation that involves transportation and/or traffic systems.

1.6 Standard References for the Traffic Engineer

To remain up to date and aware, the traffic engineer must keep up with modern developments through membership and participation in professional organizations, regular review of key periodicals, and an awareness of the latest standards and criteria for professional practice.

Key professional organizations for the traffic engineer include the Institute of Transportation Engineers (ITE), the Transportation Research Board (TRB), the Transportation Group of the American Society of Civil Engineers (ASCE), ITS America, and others. All of these provide literature and maintain journals, and have local, regional, and national meetings. TRB is a branch of the National Academy of Engineering and is a major source of research papers and reports.

Like many engineering fields, the traffic engineering profession has many manuals and standard references, most of which will be referred to in the chapters of this text. Major references include the following:

- *Traffic Engineering Handbook, 6th edition* [*1*]
- *Uniform Vehicle Code and Model Traffic Ordinance* [2]
- *Manual on Uniform Traffic Control Devices, 2003 (new edition anticipated in 2009–2010)* [*3*]
- *Highway Capacity Manual, 4th edition (5th edition anticipated in 2010)* [*4*]
- *A Policy on Geometric Design of Highways and Streets* (The AASHTO Green Book), *5th edition* [*5*]
- *Traffic Signal Timing Manual, 1st edition* [*6*]
- *Transportation Planning Handbook, 3rd edition* [*7*]
- *Trip Generation, 8th edition* [*8*]
- *Parking Generation, 3rd edition* [*9*]

All of these documents are updated periodically, and the traffic engineering professional should be aware of when updates are published and where they can be accessed.

Other manuals abound and often relate to specific aspects of traffic engineering. These references document the current state of the art in traffic engineering, and those most frequently used should be part of the professional's personal library.

There are also a wide variety of internet sites that are of great value to the traffic engineer. Specific sites are not listed here because they change rapidly. All of the professional organizations, as well as equipment manufacturers, maintain Web sites. The federal Department of Transportation (DOT), Federal Highway Administration (FHWA), National Highway Traffic Safety Administration (NHTSA), and private highway-related organizations maintain Web sites. The entire *Manual on Uniform Traffic Control Devices* is available online through the FHWA Web site, as is the *Manual of Traffic Signal Timing*.

Because traffic engineering is a rapidly changing field, you cannot assume that every standard and analysis process included in this text is current, particularly as the time since publication increases. Although we will continue to produce periodic updates, the traffic engineer must keep abreast of latest developments as a professional responsibility.

1.7 Metric versus U.S. Units

This text is published in English (or standard U.S.) units. Despite several attempts to switch to metric units in the United States, most states now use English units in design and control.

Metric and U.S. standards are not the same. A standard 12-ft lane converts to a standard 3.6-m lane, which is narrower than 12 feet. Standards for a 70-mi/h design speed convert to standards for a 120-km/h design speed, which are not numerically equivalent. This is because even units are used in both systems rather than the awkward fractional values that result from numerically equivalent conversions. That is why a metric set of wrenches for use on a foreign car is different from a standard U.S. wrench set.

Because more states are on the U.S. system than on the metric system (with more moving back to U.S. units) and because the size of the text would be unwieldy if dual units were included, this text continues to be written using standard U.S. units.

1.8 Closing Comments

The profession of traffic engineering is a broad and complex one. Nevertheless, it relies on key concepts and analyses and basic principles that do not change greatly over time. This text emphasizes both the basic principles and current (in 2009) standards and practices. You must keep abreast of changes that influence the latter.

At this writing, drafts of the *Manual on Uniform Traffic Control Devices* (MUTCD), expected to be officially released in late 2009 or 2010, are available online. Also, a great deal of source material that will be in the forthcoming 2010 *Highway Capacity Manual* is available as well. Because of this, they have been incorporated into this text for completeness. It should be remembered, however, that until they are officially released, some of this material is subject to change, even if major changes are not expected. Consult these documents directly to ensure that you are using the official versions of the methodologies and standards included in these important source documents.

References

1. Pline, J., Editor, *Traffic Engineering Handbook*, 6th Edition, Institute of Transportation Engineers, Washington DC, March 2009.
2. *Uniform Vehicle Code and Model Traffic Ordinance*, National Committee on Uniform Traffic Laws and Ordinance, Washington DC, 2002.
3. *Manual on Uniform Traffic Control Devices*, Federal Highway Administration, Washington DC, 2003. (Available on the FHWA Web site, www.fhwa.gov.) Draft of new edition expected in 2009–2010 is also available.
4. *Highway Capacity Manual*, 4th Edition, Transportation Research Board, Washington DC, 2000.
5. *A Policy on Geometric Design of Highways and Streets*, 5th Edition, American Association of State Highway and Traffic Officials, Washington DC, 2004.
6. Koonce, P., et al., *Traffic Signal Timing Manual*, Federal Highway Administration, Washington DC, 2009.
7. Edwards, J.D. Jr., Editor, *Transportation Planning Handbook*, 3rd Edition, Institute of Transportation Engineers, Washington DC, March 2009.
8. *Trip Generation*, 8th Edition, ITE Informational Report, Institute of Transportation Engineers, Washington DC, January 2008.
9. *Parking Generation*, 3rd Edition, ITE Informational Report, Institute of Transportation Engineers, Washington DC, January 2004.

PART 1

Traffic Components and Characteristics

CHAPTER 2

Road User and Vehicle Characteristics

2.1 Overview of Traffic Stream Components

To begin to understand the functional and operational aspects of traffic on streets and highways, it is important to understand how the various elements of a traffic system interact. Further, the characteristics of traffic streams are heavily influenced by the characteristics and limitations of each of these elements. Five critical components interact in a traffic system:

- Road users—drivers, pedestrians, bicyclists, and passengers
- Vehicles—private and commercial
- Streets and highways
- Traffic control devices
- The general environment

This chapter provides an overview of critical road user and vehicle characteristics. Chapter 3 focuses on the characteristics of streets and highways, and Chapter 4 provides an overview of traffic control devices and their use.

The general environment also has an impact on traffic operations, but this is difficult to assess in any given situation. Such factors as weather, lighting, density of development, and local enforcement policies all play a role in affecting traffic operations. These factors are most often considered qualitatively, with occasional supplemental quantitative information available to assist in making judgments.

2.1.1 Dealing with Diversity

Traffic engineering would be a great deal simpler if the various components of the traffic system had uniform characteristics. Traffic controls could be easily designed if all drivers reacted to them in exactly the same way. Safety could be more easily achieved if all vehicles had uniform dimensions, weights, and operating characteristics.

Drivers and other road users, however, have widely varying characteristics. The traffic engineer must deal with elderly drivers as well as 18-year-olds, aggressive drivers and timid drivers, and drivers subject to myriad distractions both inside and outside their vehicles. Simple subjects like reaction time, vision characteristics, and walking speed become complex because no two road users are the same.

Most human characteristics follow the normal distribution (see Chapter 7). The normal distribution is characterized by a strong central tendency (i.e., most people have characteristics falling into a definable range). For example, most pedestrians crossing a street walk at speeds between 3.0 and 5.0 ft/s. However, a few pedestrians walk either much slower or much faster. A normal distribution defines the proportions of the population expected to fall into these ranges. Because of variation, it is not practical to design a system for "average" characteristics. If a signal is timed, for example, to accommodate the average speed of crossing pedestrians, about half of all pedestrians would walk at a slower rate and be exposed to unacceptable risks.

Thus most standards are geared to the "85th percentile" (or "15th percentile") characteristic. In general terms, a percentile is a value in a distribution for which the stated percentage of the population has a characteristic that is less than or equal to the specified value. In terms of walking speed, for example, safety demands that we accommodate slower walkers. The 15th percentile walking speed is used because only 15% of the population walks slower than this. Where driver reaction time is concerned, the 85th percentile value is used because 85% of the population has a reaction time that is numerically equal to or less than this value. This approach leads to design practices and procedures that safely accommodate 85% of the population. What about the remaining 15%? One of the characteristics of normal distributions is that the extreme ends of the distribution (the highest and lowest 15%) extend to plus or minus infinity. In practical terms, the highest and lowest 15% of the distribution represent very extreme values that could not be effectively accommodated into design practices. Qualitatively, the existence of road users who may possess characteristics not within the 85th (or 15th) percentile is considered, but most standard practices and criteria do not directly accommodate them. Where feasible, higher percentile characteristics can be employed.

Just as road-user characteristics vary, the characteristics of vehicles vary widely as well. Highways must be designed to accommodate motorcycles, the full range of automobiles, and a wide range of commercial vehicles, including double- and triple-back tractor-trailer combinations. Thus lane widths, for example, must accommodate the largest vehicles expected to use the facility.

The economic crises of 2008–2009 and the poor condition of the U.S. automobile industry may very well lead to drastic changes in the vehicle fleet. With the emphasis on cleaner and more efficient vehicles, cars may be getting smaller and lighter. Their relative safety within a mixed traffic stream still containing large trucks and buses may become an important issue requiring new planning and design approaches. The traffic professional must be prepared to deal with this and other new issues when they arise.

Some control over the range of road-user and vehicle characteristics is maintained through licensing criteria and federal and state standards on vehicle design and operating characteristics. Although these are important measures, the traffic engineer must still deal with a wide range of road-user and vehicle characteristics.

2.1.2 Addressing Diversity through Uniformity

Although traffic engineers have little control over driver and vehicle characteristics, design of roadway systems and traffic controls is in the core of their professional practice. In both cases, a strong degree of uniformity of approach is desirable. Roadways of a similar type and function should have a familiar "look" to drivers; traffic control devices should be as uniform as possible. Traffic engineers strive to provide information to drivers in uniform ways. Although this does not assure uniform reactions from drivers, it at least narrows the range of behavior as drivers become accustomed to and familiar with the cues traffic engineers design into the system. Chapters 3 and 4 deal with roadways and controls, respectively, and treat the issue of uniformity in greater detail.

2.2 Road Users

Human beings are complex and have a wide range of characteristics that can and do influence the driving task. In a system where the driver is in complete control of vehicle operations, good traffic engineering requires a keen understanding of driver characteristics. Much of the task of traffic engineers is to find ways to provide drivers with information in a clear, effective manner that induces safe and proper responses.

The two driver characteristics of utmost importance are visual acuity factors and the reaction process. The two overlap, in that reaction requires the use of vision for most driving cues. Understanding how information is received and processed is a key element in the design of roadways and controls.

There are other important characteristics as well. Hearing is an important element in the driving task (i.e., horns, emergency vehicle sirens, brakes squealing, etc.). Although noting this is important, however, no traffic element can be designed around audio cues because hearing-impaired and even deaf drivers are licensed. Physical strength may have been important in the past, but the evolution of power-steering and power-braking systems has eliminated this as a major issue, with the possible exception of professional drivers of trucks, buses, and other heavy vehicles.

Of course, one of the most important human factors that influences driving is the personality and psychology of the driver. This, however, is not easily quantified and is difficult to consider in design. It is dealt with primarily through enforcement and licensing procedures that attempt to remove or restrict drivers who periodically display inappropriate tendencies, as indicated by accident and violation experience.

2.2.1 Visual Characteristics of Drivers

When drivers initially apply for, or renew, their licenses, they are asked to take an eye test, administered either by the state motor vehicle agency or by an optometrist or ophthalmologist who fills out an appropriate form for the motor vehicle agency. The test administered is a standard chart-reading exercise that measures *static visual acuity*—that is, the ability to see small stationary details clearly.

Visual Factors in Driving

Although static visual acuity is certainly an important characteristic, it is hardly the only visual factor involved in the driving task. The *Traffic Engineering Handbook* [*1*] provides an excellent summary of visual factors involved in driving, as shown in Table 2.1.

Many of the other factors listed in Table 2.1 reflect the dynamic nature of the driving task and the fact that most objects to be viewed by drivers are in relative motion with respect to the driver's eyes.

Because static visual acuity is the only one of these many visual factors examined as a prerequisite to issuing a driver's license, traffic engineers must expect and deal with significant variation in many of the other visual characteristics of drivers.

Fields of Vision

Figure 2.1 illustrates three distinct fields of vision, each of which is important to the driving task [*2*]:

- *Acute or clear vision cone*—3° to 10° around the line of sight; legend can be read only within this narrow field of vision.

Table 2.1: Visual Factors in the Driving Task

Visual Factor	Definition	Sample Related Driving Task(s)
Accommodation	Change in the shape of the lens to bring images into focus.	Changing focus from dashboard displays to roadway.
Static Visual Acuity	Ability to see small details clearly.	Reading distant traffic signs.
Adaptation	Change in sensitivity to different levels of light.	Adjust to changes in light upon entering a tunnel.
Angular Movement	Seeing objects moving across the field of view.	Judging the speed of cars crossing our paths.
Movement in Depth	Detecting changes in visual image size.	Judging speed of an approaching vehicle.
Color	Discrimination between different colors.	Identifying the color of signals.
Contrast Sensitivity	Seeing objects that are similar in brightness to their background.	Detecting dark-clothed pedestrians at night.
Depth Perception	Judgment of the distance of objects.	Passing on two-lane roads with oncoming traffic.
Dynamic Visual Acuity	Ability to see objects that are in motion relative to the eye.	Reading traffic signs while moving.
Eye Movement	Changing the direction of gaze.	Scanning the road environment for hazards.
Glare Sensitivity	Ability to resist and recover from the effects of glare.	Reduction in visual performance due to headlight glare.
Peripheral Vision	Detection of objects at the side of the visual field.	Seeing a bicycle approaching from the left.
Vergence	Angle between the eyes' line of sight.	Change from looking at the dashboard to the road.

(*Source:* Used with permission of Institute of Transportation Engineers, Dewar, R, "Road Users," *Traffic Engineering Handbook*, 5th Edition, Chapter 2, Table 2-2, p. 8, 1999.)

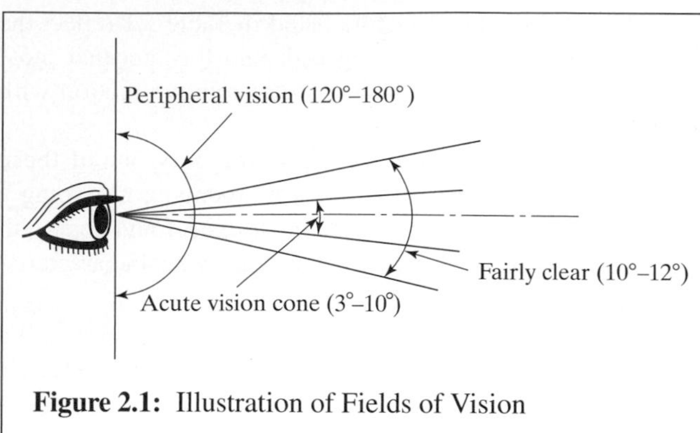

Figure 2.1: Illustration of Fields of Vision

- *Fairly clear vision cone*—10° to 12° around the line of sight; color and shape can be identified in this field.
- *Peripheral vision*—This field may extend up to 90° to the right and left of the centerline of the pupil, and up to 60° above and 70° below the line of sight. Stationary objects are generally not seen in the peripheral vision field, but the movement of objects through this field is detected.

Illustration of Fields of Vision

These fields of vision, however, are defined for a stationary person. In particular, the peripheral vision field narrows, as speed increases, to as little as 100° at 20 mi/h and to 40° at 60 mi/h.

The driver's visual landscape is both complex and rapidly changing. Approaching objects appear to expand in size while other vehicles and stationary objects are in relative motion both to the driver and to each other. The typical driver essentially samples the available visual information and selects appropriate cues to make driving decisions.

The fields of vision affect a number of traffic engineering practices and functions. Traffic signs, for example, are placed so that they can be read within the acute vision field without requiring drivers to change their line of sight. Thus they are generally placed within a 10° range of the driver's expected line of sight, which is assumed to be in line with the highway alignment. This leads to signs that are intended to be read when they are a significant distance from the driver; in turn, this implies how large the sign and its lettering must be to be comprehended at that distance. Objects or other vehicles located in the fairly clear and peripheral vision fields may draw the driver's attention to an important event occurring in that field, such as the approach of a vehicle on an intersection street or driveway or a child running into the street after a ball. Once noticed, the driver may turn his or her head to examine the details of the situation.

Peripheral vision is the single most important factor when drivers estimate their speed. The movement of objects through the peripheral vision field is the driver's single most important indicator of speed. Old studies have demonstrated time and again that drivers deprived of peripheral vision (using blinders in experimental cases) and deprived of a working speedometer have little idea of how fast they are traveling.

2.2.2 Important Visual Deficits

A number of visual problems can affect driver performance and behavior. Unless the condition causes a severe visual disability, drivers affected by various visual deficits often continue to drive. Reference 3 contains an excellent overview and discussion of these.

Some of the more common problems involve cataracts, glaucoma, peripheral vision deficits, ocular muscle imbalance, depth perception deficits, and color blindness. Drivers who have eye surgery to correct a problem may experience temporary or permanent impairments. Other diseases, such as diabetes, can have a significant negative impact on vision if not controlled. Some conditions, like cataracts and glaucoma, if untreated, can lead to blindness.

Although color blindness is not the worst of these conditions, it generally causes some difficulties for the affected driver because color is one of the principal means to impart information. Unfortunately, one of the most common forms of color blindness involves the inability to discern the difference between red and green. In the case of traffic signals, this could have a devastating impact on the safety of such drivers. To ameliorate this difficulty to some degree, some blue pigment has been added to green lights and some yellow pigment has been added to red lights, making them easier to discern by colorblind drivers. Also, the location of colors on signal heads has long been standardized, with red on the top and green on the bottom of vertical signal heads. On horizontal heads, red is on the left and green on the right. Arrow indications are either located on a separate signal head or placed below or to the right of ball indications on a mixed signal head.

2.2.3 Perception–Reaction Time

The second critical driver characteristic is perception–reaction time (PRT). During perception and reaction, the driver must perform four distinct processes [*4*]:

- *Detection.* In this phase, an object or condition of concern enters the driver's field of vision, and the driver becomes consciously aware that something requiring a response is present.

- *Identification.* In this phase, the driver acquires sufficient information concerning the object or condition to allow the consideration of an appropriate response.
- *Decision.* Once identification of the object or condition is sufficiently completed, the driver must analyze the information and make a decision about how to respond.
- *Response.* After a decision has been reached, the response is now physically implemented by the driver.

In some of the literature, the four phases of PRT are referred to as perception, identification, emotion, and volition, leading to the term "PIEV time." This text uses PRT, but you should understand that it is equivalent to PIEV time.

Design Values

Like all human characteristics, perception–reaction times vary widely among drivers, as do a variety of other factors, including the type and complexity of the event perceived and the environmental conditions at the time of the response.

Nevertheless, design values for various applications must be selected. The American Association of State Highway and Transportation Officials (AASHTO) mandates the use of 2.5 seconds for most computations involving braking reactions [*5*], based on a number of research studies [*6–9*]. This value is believed to be approximately a 90th percentile criterion (i.e., 90% of all drivers have a PRT as fast or faster than 2.5 s).

For signal timing purposes, the Institute of Transportation Engineers [*10*] recommends a PRT time of 1.0 s. Because of the simplicity of the response and the preconditioning of drivers to respond to signals, the PRT time is significantly less than that for a braking response on an open highway. Although this is a lower value, it still represents an approximately 85th percentile for the particular situation of responding to a traffic signal.

AASHTO criteria, however, recognize that in certain more complex situations, drivers may need considerably more time to react than 1.0 or 2.5 s. Situations where drivers must detect and react to unexpected events, or a difficult-to-perceive information source in a cluttered highway environment, or a situation in which there is a likelihood of error involving either information reception, decisions, or actions all would result in increased PRT times. Some of the examples cited by AASHTO of locations where such situations might exist include complex interchanges and intersections where unusual movements are encountered and changes in highway cross sections such as toll plazas, lane drops, and areas where the roadway environment is cluttered with visual distractions. Where a collision avoidance maneuver is required, AASHTO criteria call for a PRT of 3.0 s for stops on rural roads and 9.1 s for stops on urban roads. Where collision avoidance requires speed, path, and/or direction changes, AASHTO recommends a PRT of between 10.2 and 11.2 s on rural roads, 12.1 and 12.9 s on suburban roads, and 14.0 and 14.5 s on urban roads.

Expectancy

The concept of expectancy is important to the driving task and has a significant impact on the perception–reaction process and PRT. Simply put, drivers react more quickly to situations they *expect* to encounter as opposed to those that they *do not expect* to encounter. There are three different types of expectancies:

- *Continuity.* Experiences of the immediate past are generally expected to continue. Drivers do not, for example, expect the vehicle they are following to suddenly slow down.
- *Event.* Things that have not happened previously will not happen. If no vehicles have been observed entering the roadway from a small driveway over a reasonable period of time, then the driver will assume that none will enter now.
- *Temporal.* When events are cyclic, such as a traffic signal, the longer a given state is observed, drivers will assume that it is more likely a change will occur.

Figure 2.2 illustrates the impact of expectancy on PRT. This study by Olsen et al. [*11*] in 1984 was a controlled observation of student drivers reacting to a similar hazard when they were unaware it would appear, and again where they were told to look for it. In a third experiment, a red light was added to the dash to initiate the braking reaction. The PRT under the "expected" situation was consistently about 0.5 s faster than under the "unexpected" situation.

Given the obvious importance of expectancy on PRT, traffic engineers must strive to avoid designing "unexpected" events into roadway systems and traffic controls. If there are all right-hand ramps on a given freeway, for example, left-hand ramps should be avoided if at all possible. If absolutely required, guide signs must be very carefully designed to alert drivers to the existence and location of the left-hand ramp, so that when they reach it, it is no longer "unexpected."

Other Factors Affecting PRT

In general, PRTs increase with a number of factors, including (1) age, (2) fatigue, (3) complexity of reaction, and (4) presence of alcohol and/or drugs in the driver's system. Although these trends are well documented, they are generally accounted for in recommended design values, with the

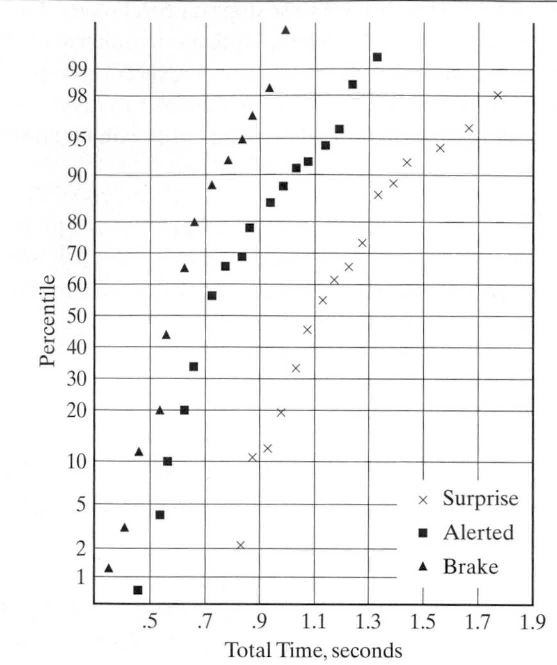

Figure 2.2: Comparison of Perception–Reaction Times between Expected and Unexpected Events

(*Source:* Used with permission of Transportation Research Board, National Research Council, Olson, P., et al., "Parameters Affecting Stopping Sight Distance," *NCHRP Report 270*, Washington DC, 1984.)

exception of the impact of alcohol and drugs. The latter are addressed primarily through enforcement of ever-stricter driving while intoxicated/driving under the influence (DWI/DUI) laws in the various states, with the intent of removing such drivers from the system, especially where repeated violations make them a significant safety risk. Some of the more general affects of alcohol and drugs, as well as aging, on driver characteristics are discussed in a later section.

Reaction Distance

The most critical impact of perception–reaction time is the distance the vehicle travels while the driver goes through the process. In the example of a simple braking reaction, the PRT begins when the driver first becomes aware of an event or object in his or her field of vision and ends when his or her foot is applied to the brake. During this time, the vehicle continues along its original course at its initial speed. Only after the foot is applied to the brake pedal does the vehicle begin to slow down in response to the stimulus.

The reaction distance is simply the PRT multiplied by the initial speed of the vehicle. Because speed is generally in units of mi/h and PRT is in units of seconds, it is convenient to convert speeds to ft/s for use:

$$\frac{1 \text{ mi} * \left(\dfrac{5{,}280 \text{ ft}}{\text{mi}}\right)}{1 \text{ h} * \left(\dfrac{3{,}600 \text{ s}}{\text{h}}\right)} = 1.466666\ldots\frac{\text{ft}}{\text{s}} = 1.47\frac{\text{ft}}{\text{s}}$$

Thus the reaction distance may be computed as:

$$d_r = 1.47\, S\, t \qquad (2\text{-}1)$$

where: d_r = reaction distance, ft
S = initial speed of vehicle, mi/h
t = reaction time, s

The importance of this factor is illustrated in the following sample problem: A driver rounds a curve at a speed of 60 mi/h and sees a truck overturned on the roadway ahead. How far will the driver's vehicle travel before the driver's foot reaches the brake? Applying the AASHTO standard of 2.5 s for braking reactions:

$$d_r = 1.47 * 60 * 2.5 = 220.5 \text{ ft}$$

The vehicle will travel 220.5 ft (approximately 11 to 12 car lengths) before the driver even engages the brake. The implication of this is frightening. If the overturned truck is closer to the vehicle than 220.5 ft when noticed by the driver, not only will the driver hit the truck, he or she will do so at full speed—60 mi/h. Deceleration begins only when the brake is engaged—*after* the perception–reaction process has been completed.

2.2.4 Pedestrian Characteristics

One of the most critical safety problems in any highway and street system involves the interactions of vehicles and pedestrians. A substantial number of traffic accidents and fatalities involve pedestrians. This is not surprising because in any contact between a pedestrian and a vehicle, the pedestrian is at a significant disadvantage.

Virtually all of the interactions between pedestrians and vehicles occur as pedestrians cross the street at intersections and at mid-block locations. At signalized intersections, safe accommodation of pedestrian crossings is as critical as vehicle requirements in establishing an appropriate timing pattern. Pedestrian walking speed in crosswalks is the most important factor in the consideration of pedestrians in signal timing.

At unsignalized crossing locations, gap-acceptance behavior of pedestrians is another important consideration. "Gap acceptance" refers to the clear time intervals between vehicles encroaching on the crossing path and the behavior of pedestrians in "accepting" them to cross through.

Walking Speeds

Table 2.2 shows 50th percentile walking speeds for pedestrians of various ages. Note that these speeds were measured as part of a controlled experiment [*12*] and not specifically at intersection or mid-block crosswalks. Nevertheless, the results are interesting. The standard walking speed used in timing signals is 4.0 ft/s, with 3.5 ft/s recommended where older pedestrians are predominant. Most studies indicate that these standards are reasonable and will accommodate 85% of the pedestrian population. In 2008 and 2009, serious discussion of lowering the general standard walking speed to 3.5 ft/s and 3.0 ft/s, respectively, was underway. These changes are thought to be likely at this writing.

One problem with standard walking speeds involves physically impaired pedestrians. A study of pedestrians with various impairments and assistive devices concluded that average walking speeds for virtually all categories were lower than the standard 4.0 ft/s used in signal timing [*13*]. Table 2.3 includes some of the results of this study. These and similar results of other studies suggest that more consideration needs to be given to the needs of handicapped pedestrians.

Table 2.2: 50th Percentile Walking Speeds for Pedestrians of Various Ages

Age (years)	50th Percentile Walking Speed (ft/s)	
	Males	**Females**
2	2.8	3.4
3	3.5	3.4
4	4.1	4.1
5	4.6	4.5
6	4.8	5.0
7	5.0	5.0
8	5.0	5.3
9	5.1	5.4
10	5.5	5.4
11	5.2	5.2
12	5.8	5.7
13	5.3	5.6
14	5.1	5.3
15	5.6	5.3
16	5.2	5.4
17	5.2	5.4
18	4.9	N/A
20–29	5.7	5.4
30–39	5.4	5.4
40–49	5.1	5.3
50–59	4.9	5.0
60+	4.1	4.1

(*Source:* Compiled from Eubanks, J., and Hill, P., *Pedestrian Accident Reconstruction and Litigation*, 2nd Edition, Lawyers & Judges Publishing Co., Tucson AZ, 1999.)

Gap Acceptance

When a pedestrian crosses at an uncontrolled (either by signals, STOP, or YIELD signs) location, either at an intersection or at a mid-block location, the pedestrian must select an appropriate "gap" in the traffic stream through which to cross. The "gap" in traffic is measured as the time lag between two vehicles in any lane encroaching on the pedestrian's crossing path. As the pedestrian waits to cross, he or she views gaps and decides whether to "accept" or "reject" the gap for a safe crossing. Some studies have used a gap defined as the distance between the pedestrian and the approaching vehicle at the time the pedestrian begins his or her crossing. An early study [*14*] using the latter approach resulted in an 85th percentile gap of approximately 125 ft.

Gap acceptance behavior, however, is quite complex and varies with a number of other factors, including the speed of approaching vehicles, the width of the street, the frequency distribution of gaps in the traffic stream, waiting time, and others. Nevertheless, this is an important characteristic that

Table 2.3: Walking Speeds for Physically Impaired Pedestrians

Impairment/Assistive Device	Average Walking Speed (ft/s)
Cane/Crutch	2.62
Walker	2.07
Wheelchair	3.55
Immobilized Knee	3.50
Below-Knee Amputee	2.46
Above-Knee Amputee	1.97
Hip Arthritis	2.44–3.66
Rheumatoid Arthritis (Knee)	2.46

(*Source:* Compiled from Perry, J., *Gait Analysis*, McGraw-Hill, New York NY, 1992.)

must be considered due to its obvious safety implications. Chapter 18, for example, presents warrants for (conditions justifying) the imposition of traffic signals. One of these is devoted entirely to the safety of pedestrian crossings.

Pedestrian Comprehension of Controls

One of the problems in designing controls for pedestrians is generally poor understanding of and poor adherence to such devices. One questionnaire survey of 4,700 pedestrians [*15*] detailed many problems of misunderstanding. The proper response to a flashing "DON'T WALK" signal, for example, was not understood by 50% of road users, who thought it meant they should return to the curb from which they started. The meaning of this signal is not to start crossing while it is flashing; it is safe to complete a crossing if the pedestrian has already started to do so. Another study [*16*] found that violation rates for the solid "DON'T WALK" signal were higher than 50% in most cities, that the use of the flashing "DON'T WALK" for pedestrian clearance was not well understood, and that pedestrians tend not to use pedestrian-actuated signals. Chapter 22 (on signal timing) discusses some of the problems associated with pedestrian-actuation buttons and their use that compromise both pedestrian comprehension and the efficiency of the signalization. Since this study was completed, the flashing and solid "DON'T WALK" signals have been replaced by the Portland orange "raised hand" symbol.

Thus the task of providing for a safe environment for pedestrians is not an easy one. The management and control of conflicts between vehicles and pedestrians remains difficult.

2.2.5 Impacts of Drugs and Alcohol on Road Users

The effect of drugs and alcohol on drivers has received well-deserved national attention for many years, leading to substantial strengthening of DWI/DUI laws and enforcement. These factors remain, however, a significant contributor to traffic fatalities and accidents. And drivers are not the only road users who contribute to the nation's accident and fatality statistics. Consider that in 1996, 47.3% of fatal pedestrian accidents involved either a driver or a pedestrian with detectable levels of alcohol in their systems. For this group, 12.0% of the drivers and 32.3% of the pedestrians had blood-alcohol levels above 0.10%, the legal definition of "drunk" in many states. More telling is that 7% of the drivers and 6% of the pedestrians had detectable alcohol levels below this limit.

The importance of these isolated statistics is to make the following point: Legal limits for DWI/DUI do not define the point at which alcohol and/or drugs influence the road user. Recognizing this is important for individuals to ensure safe driving it is now causing many states to reduce their legal limits on alcohol to 0.08%, and for some to consider "zero tolerance" criteria (0.01%) for new drivers for the first year or two they are licensed.

Figure 2.3 summarizes some studies on the effects of drugs and alcohol on various driving factors. Note that for many factors, impairment of driver function begins at levels well below the legal limits—for some factors at blood-alcohol levels as low as 0.05%.

What all of these factors add up to is an impaired driver. This combination of impairments leads to longer PRT times, poor judgments, and actions that can and do cause accidents. Because few of these factors can be ameliorated by design or control (although good designs and well-designed controls help both impaired and unimpaired drivers), enforcement and education are critical elements in reducing the incidence of DWI/DUI and the accidents and deaths that result.

The statistics cited in the opening paragraph of this section also highlight the danger caused by pedestrians who are

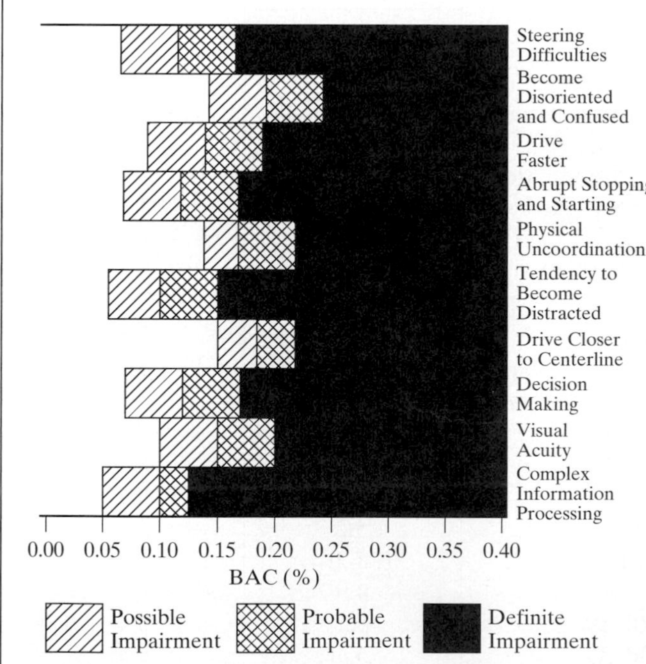

Figure 2.3: Effects of Blood-Alcohol Level on Driving Tasks

(*Source:* Used with permission of Institute of Transportation Engineers, Blaschke, J., Dennis, M., and Creasy, F., "Physical and Psychological Effects of Alcohol and Other Drugs on Drivers," *ITE Journal*, 59, Washington DC, 1987.)

impaired by drugs or alcohol. In the case of impaired pedestrians, the danger is primarily to themselves. Nevertheless, if crossing a street or highway is required, "walking while impaired" is also quite dangerous. Again, enforcement and education are the major weapons in combating the problem because not a great deal can be done through design or control to address the issue.

Both motorists and pedestrians should also be aware of the impact of common prescription and over-the-counter medications on their performance capabilities. Many legitimate medications have effects that are similar to those of alcohol and/or illegal drugs. Users of medications should always be aware of the side effects of what they use (a most frequent effect of many drugs is drowsiness), and to exercise care and good judgment when considering whether or not to drive. Some legitimate drugs can have a direct impact on blood-alcohol levels and can render a motorist legally intoxicated without "drinking."

2.2.6 Impacts of Aging on Road Users

As life expectancy continues to rise, the number of older drivers has risen dramatically over the past several decades. Thus it becomes increasingly important to understand how aging affects driver needs and limitations and how these should impact design and control decisions. Reference 17 is an excellent compilation sponsored by the National Academy of Sciences on a wide range of topics involving aging drivers.

Many visual acuity factors deteriorate with age, including both static and dynamic visual acuity, glare sensitivity and recovery, night vision, and speed of eye movements. Such ailments as cataracts, glaucoma, macular degeneration, and diabetes are also more common as people age, and these conditions have negative impacts on vision.

The increasing prevalence of older drivers presents a number of problems for both traffic engineers and public officials. On one hand, at some point, deterioration of various capabilities must lead to revocation of the right to drive. On the other hand, driving is the principal means of mobility and accessibility in most parts of the nation, and the alternatives for those who can no longer drive are either limited or expensive. The response to the issue of an aging driver population must have many components, including appropriate licensing standards, consideration of some license restrictions on older drivers (e.g., a daytime-only license), provision of efficient and affordable transportation alternatives, and increased consideration of their needs, particularly in the design and implementation of control devices and traffic regulations. Older drivers may be helped, for example, by such measures as larger lettering on signs, better highway lighting, larger and brighter signals, and other measures. Better education can serve to make older drivers more aware of the types of deficits they face and how best to deal with them. More frequent testing of key characteristics such as eyesight may help ensure that prescriptions for glasses and/or contact lenses are frequently updated.

2.2.7 Psychological, Personality, and Related Factors

Over the past decade, traffic engineers and the public in general have become acquainted with the term *road rage*. Commonly applied to drivers who lose control of themselves and react to a wide variety of situations violently, improperly, and almost always dangerously, the problem (which has always existed) is now getting well-deserved attention. *Road rage,* however, is a colloquial term, and is applied to everything from a direct physical assault by one road user on another to a variety of aggressive driving behaviors.

According to the testimony of Dr. John Larsen to the House Surface Transportation Subcommittee on July 17, 1997 (as summarized in Chapter 2 of Reference 1), the following attitudes characterize aggressive drivers:

- The desire to get to one's destination as quickly as possible, leading to the expression of anger at other drivers/pedestrians who impede this desire.
- The need to compete with other fast cars.
- The need to respond competitively to other aggressive drivers.
- Contempt for other drivers who do not drive, look, and act as they do on the road.
- The belief that it is their right to "hit back" at other drivers whose driving behavior threatens them.

Road rage is the extreme expression of a driver's psychological and personal displeasure over the traffic situation he or she has encountered. It does, however, remind traffic engineers that drivers display a wide range of behaviors in accordance with their own personalities and psychological characteristics.

Once again, most of these factors cannot be addressed directly through design or control decisions and are best treated through vigorous enforcement and educational programs.

2.3 Vehicles

In 2007, approximately 240 million registered vehicles were in the United States, a number that represents more than one vehicle per licensed driver. The characteristics of these vehicles vary as widely as those of the motorists who drive them.

In general, motor vehicles are classified by AASHTO [5] into four main categories:

- *Passenger cars*—all passenger cars, SUVs, minivans, vans, and pickup trucks.
- *Buses*—intercity motor coaches, transit buses, school buses, and articulated buses
- *Trucks*—single-unit trucks, tractor-trailer, and tractor-semi-trailer combination vehicles
- *Recreational vehicles*—motor homes, cars with various types of trailers (boat, campers, motorcycles, etc.)

Motorcycles and bicycles also use highway and street facilities but are not isolated as a separate category because their characteristics do not usually limit or define design or control needs.

A number of critical vehicle properties must be accounted for in the design of roadways and traffic controls. These include:

- Braking and deceleration
- Acceleration
- Low-speed turning characteristics
- High-speed turning characteristics

In more general terms, the issues associated with vehicles of vastly differing size, weight, and operating characteristics sharing roadways must also be addressed by traffic engineers.

2.3.1 Concept of the Design Vehicle

Given the immense range of vehicle types using street and highway facilities, it is necessary to adopt standard vehicle characteristics for design and control purposes. For geometric design, AASHTO has defined 20 "design vehicles," each with specified characteristics. The 20 design vehicles are defined as follows:

P	=	passenger car
SU	=	single-unit truck
BUS-40	=	intercity bus with a 40-ft wheelbase
BUS-45	=	intercity bus with a 45-ft wheelbase
CITY-BUS	=	transit bus
S-BUS36	=	conventional school bus for 65 passengers
S-BUS40	=	large school bus for 84 passengers
A-BUS	=	articulated bus
WB-40	=	intermediate semi-trailer (wheelbase = 40 ft)
WB-50	=	intermediate semi-trailer (wheelbase = 50 ft)
WB-62	=	interstate semi-trailer (wheelbase = 62 ft)
WB-65	=	interstate semi-trailer (wheelbase = 65 ft)
WB-67D	=	double trailer combination (wheelbase = 67 ft)
WB-100T	=	triple semi-trailer/trailers (wheelbase = 100 ft)
WB-109D	=	turnpike double semi-trailer/trailer (wheelbase = 109 ft)
MH	=	motor home
P/T	=	passenger car and camper
P/B	=	passenger car and boat trailer
MH/B	=	motor home and boat trailer
TR	=	farm tractor

Wheelbase dimensions are measured from the frontmost axle to the rearmost axle, including both the tractor and all trailers in a combination vehicle.

Design vehicles are primarily employed in the design of turning roadways and intersection curbs, and they are used to help determine appropriate lane widths and such specific design features as lane-widening on curves. Key to such usage, however, is the selection of an appropriate design vehicle for various types of facilities and situations. In general, the design should consider the largest vehicle likely to use the facility with reasonable frequency.

In considering the selection of a design vehicle, it must be remembered that all parts of the street and highway network must be accessible to emergency vehicles, including fire engines, ambulances, emergency evacuation vehicles, and emergency repair vehicles, among others. Therefore the single-unit truck is usually the minimum design vehicle selected for most local street applications. The mobility of hook-and-ladder fire vehicles is enhanced by having rear-axle steering that allows these vehicles to negotiate sharper turns than would normally be possible for combination vehicles, so the use of a single-unit truck as a design vehicle for local streets is not considered to hinder emergency vehicles.

The passenger car is used as a design vehicle only in parking lots, and even there, access to emergency vehicles must be considered. For most other classes or types of highways and intersections, the selection of a design vehicle must consider the expected vehicle mix. In general, the design vehicle selected should easily accommodate 95% or more of the expected vehicle mix.

The physical dimensions of design vehicles are also important considerations. Design vehicle heights range from 4.25 ft for a passenger car to 13.5 ft for the largest trucks. Overhead clearances of overpass and sign structures, electrical wires, and other overhead appurtenances should be sufficient to allow the largest anticipated vehicles to proceed. Because all facilities must accommodate a wide variety of potential emergency vehicles, use of 14.0 ft for minimum clearances is advisable for most facilities.

The width of design vehicles ranges from 7.0 ft for passenger cars to 8.5 ft for the largest trucks (excluding special "wide load" vehicles such as a tractor pulling a prefabricated or motor home). This should influence the design of such features as lane width and shoulders. For most facilities, it is desirable to use the standard 12-ft lane width. Narrower lanes may be considered for some types of facilities when necessary, but given the width of modern vehicles, 10 ft is a reasonable minimum for virtually all applications.

2.3.2 Turning Characteristics of Vehicles

There are two conditions under which vehicles must make turns:

- Low-speed turns ($\leq$10 mi/h)
- High-speed turns ($>$10 mi/h)

Low-speed turns are limited by the characteristics of the vehicle because the minimum radius allowed by the vehicle's steering mechanism can be supported at such speeds. High-speed turns are limited by the dynamics of side friction between the roadway and the tires, and by the superelevation (cross-slope) of the roadway.

Low-Speed Turns

AASHTO specifies minimum design radii for each of the design vehicles, based on the centerline turning radius and minimum inside turning radius of each vehicle. Although the actual turning radius of a vehicle is controlled by the front wheels, rear wheels do not follow the same path. Rear wheels "off-track" as they are dragged through the turning movement.

Reference 5 contains detailed low-speed turning templates for all AASHTO design vehicles. Figure 2.4 shows an example (for a WB-40 combination vehicle). Note that the minimum turning radius is defined by the track of the front outside wheel. The combination vehicle, however, demonstrates considerable "off-tracking" of the rear inside wheel, effectively widening the width of the "lane" occupied by the vehicle as it turns. The path of the inside rear wheel is not circular and has a variable radius.

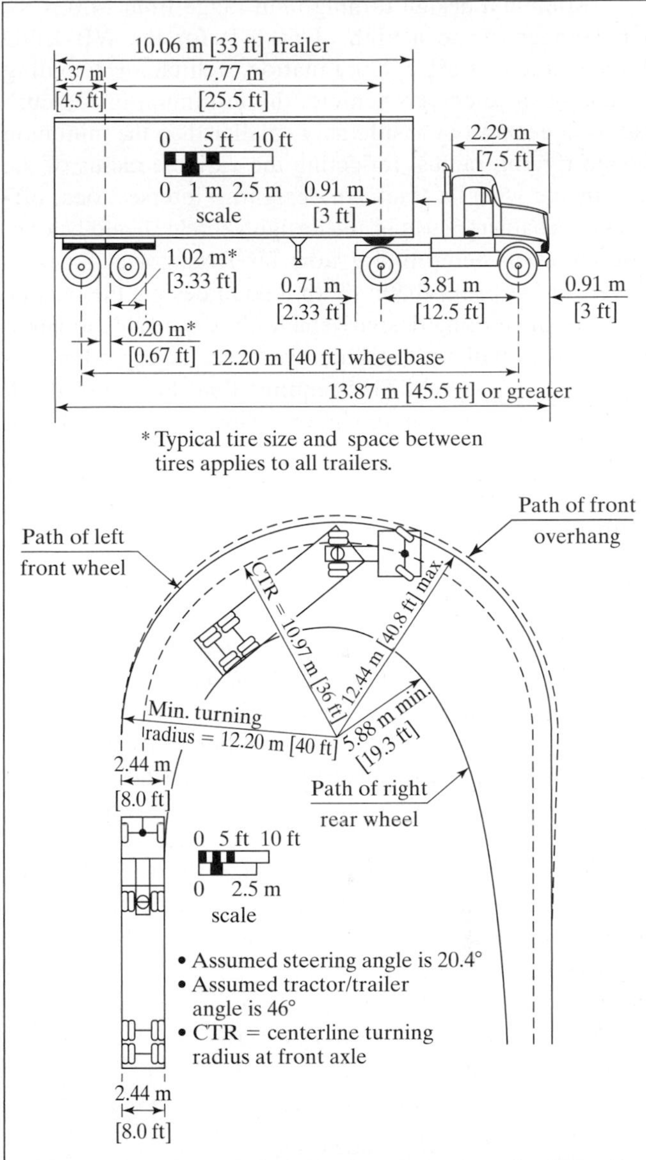

Figure 2.4: Low-Speed Turning Template for WB-40 Combination Vehicles

(*Source:* Used with permission of American Association of State Highway and Transportation Officials, *A Policy on Geometric Design of Highways and Streets,* 5th Edition, 2004, Washington DC, Exhibit 2-13.)

Turning templates provide illustrations of the many different dimensions involved in a low-speed turn. In designing for low-speed turns, the minimum design turning radius is the minimum centerline radius plus half of the width of the front of the vehicle.

Minimum design turning radii range from 24.0 ft for a passenger car to a high of 60.0 ft for the WB-109D double tractor-trailer combination vehicle. Depending on the specific design vehicle, the minimum inside curb radius is generally considerably smaller than the minimum design turning radius, reflecting the variable radius of the rear-inside wheel's track. In designing intersections, off-tracking characteristics of the design vehicle should be considered when determining how far from travel lanes to locate (or cut back) the curb. In a good design, the outside wheel of the turning design vehicle should be able to negotiate its path without "spilling over" into adjacent lanes as the turn is negotiated. This requires that the curb setback must accommodate the maximum off-tracking of the design vehicle.

High-Speed Turns

When involved in a high-speed turn on a highway curve, centripetal forces of momentum are exerted on the vehicle to continue in a straight path. To hold the curve, these forces are opposed by side friction and superelevation.

Superelevation is the cross-slope of the roadway, always with the lower edge in the direction of the curve. The sloped roadway provides an element of horizontal support for the vehicle. Side-friction forces represent the resistance to sliding provided across the plane of the surface between the vehicle's tires and the roadway. From the basic laws of physics, the relationship governing vehicle operation on a curved roadway is:

$$\frac{0.01e + f}{1 - 0.01ef} = \frac{S^2}{gR} \qquad \text{(2-2)}$$

where: e = superelevation rate, %

f = coefficient of side friction

S = speed of the vehicle, ft/s

R = radius of curvature, ft

g = acceleration rate due to gravity, 32.2 ft/s^2

The superelevation rate is the total rise in elevation across the travel lanes of the cross section (ft) divided by the width of the travel lanes (ft), expressed as a percentage (i.e., multiplied by 100). AASHTO [5] expresses superelevation as a percentage in its 2004 criteria, but many other publications still express the superelevation rate as a decimal proportion.

Equation 2-2 is simplified by noting that the term "0.01ef" is extremely small and may be ignored for the normal range of superelevation rates and *side-friction factors*. It is also convenient to express vehicle speed in mi/h. Thus:

$$\frac{0.01e + f}{1} = \frac{(1.47S)^2}{32.2R}$$

$$0.01e + f = \frac{0.067S^2}{R} = \frac{S^2}{15R}$$

This yields the more traditional relationship used to depict vehicle operation on a curve:

$$R = \frac{S^2}{15(0.01e + f)} \qquad \text{(2-3)}$$

where all terms are as previously defined, except that "S" is the speed in mi/h rather than ft/s as in Equation 2-2.

The normal range of superelevation rates is from a minimum of approximately 0.5% to support side drainage to a maximum of 12%. As speed increases, higher superelevation rates are used. Where icing conditions are expected, the maximum superelevation rate is generally limited to 8% to prevent a stalled vehicle from sliding toward the inside of the curve.

Coefficients of side friction for design are based on wet roadway conditions. They vary with speed and are shown in Table 2.4.

Theoretically, a road can be banked to fully oppose centripetal force without using side friction at all. This is, of course, generally not done because vehicles travel at a range of speeds, and the superelevation rate required in many cases would be excessive. High-speed turns on a flat pavement may be fully supported by side friction as well, but this generally limits the radius of curvature or speed at which the curve may be safely traversed.

Chapter 3 treats the design of horizontal curves and the relationships among superelevation, side friction, curve radii, and design speed in greater detail.

Equation 2-3 can be used in a number of ways. In design, a minimum radius of curvature is computed based on maximum values of e and f. For example, if a roadway has a design speed of 65 mi/h, and the maximum values are e = 8% and f = 0.11, the minimum radius is computed as:

$$R = \frac{65^2}{15(0.01 * 8 + 0.11)} = 1{,}482.5 \text{ ft}$$

Table 2.4: Side Friction Factors (f) for Wet Pavements at Various Speeds

Speed (mi/h)	30	40	50	60	70
F	0.16	0.15	0.14	0.12	0.10

It can also be used to solve for a maximum safe speed, given a radius of curvature and maximum values for *e* and *f*. If a highway curve with radius of 800 ft has a superelevation rate of 6%, the maximum safe speed can be estimated. However, doing so requires that the relationship between the *coefficient of side friction*, *f*, and speed, as indicated in Table 2.4, be taken into account. Solving Equation 2-3 for *S* yields:

$$S = \sqrt{15R(0.01e + f)} \tag{2-4}$$

For the example given, the equation is solved for the given values of *e* (6%) and *R* (800 ft) using various values of *f* from Table 2.4. Computations continue until there is closure between the computed speed and the speed associated with the coefficient of side friction selected. Thus:

$$S = \sqrt{15*800*(0.06 + f)}$$

$$S = \sqrt{15*800*(0.06 + 0.10)}$$
$$= 43.8 \text{ mi/h (70 mi/h assumed)}$$

$$S = \sqrt{15*800*(0.06 + 0.12)}$$
$$= 46.5 \text{ mi/h (60 mi/h assumed)}$$

$$S = \sqrt{15*800*(0.06 + 0.14)}$$
$$= 49.0 \text{ mi/h (50 mi/h assumed)}$$

$$S = \sqrt{15*800*(0.06 + 0.15)}$$
$$= 50.2 \text{ mi/h (40 mi/h assumed)}$$

The correct result is obviously between 49.0 and 50.2 mi/h. If straight-line interpolation is used

$$S = 49.0 + (50.2 - 49.0) * \left[\frac{(50.0 - 49.0)}{(50.2 - 49.0) + (50.2 - 40.0)}\right]$$
$$= 49.1 \text{ mi/h}$$

Thus, for the curve as described, 49.1 mi/h is the maximum safe speed at which it should be negotiated.

Note that this is based on the design condition of a wet pavement and that higher speeds would be possible under dry conditions.

2.3.3 Braking Characteristics

Another critical characteristic of vehicles is their ability to stop (or decelerate) once the brakes have been engaged. Again, basic physics relationships are used. The distance traveled during a stop is the average speed during the stop multiplied by the time taken to stop, or:

$$d_b = \left(\frac{S}{2}\right) * \left(\frac{S}{a}\right) = \frac{S^2}{2a} \tag{2-5}$$

where: d_b = braking distance, ft
S = initial speed, ft/s
a = deceleration rate, ft/s^2

It is convenient, however, to express speed in mi/h, yielding:

$$d_b = \frac{(1.47\ S)^2}{2a} = \frac{1.075\ S^2}{a}$$

where S is the speed in mi/h. Note that the 1.075 factor is derived from the more exact conversion factor between mi/h and ft/s (1.4666. . . .). It is often also useful to express this equation in terms of the coefficient of forward rolling or skidding friction, F, where $F = a/g$, and g is the acceleration due to gravity, 32.2 ft/s^2. Then:

$$d_b = \frac{\left(\frac{1.075\ S^2}{32.2}\right)}{\left(\frac{a}{32.2}\right)}$$

where F = coefficient of forward rolling or skidding friction. When the effects of grade are considered, and where a braking cycle leading to a reduced speed other than "0" are considered, the equation becomes:

$$d_b = \frac{S_i^2 - S_f^2}{30(F \pm 0.01G)} \tag{2-6}$$

where: G = grade, %
S_i = initial speed, mi/h
S_f = final speed, mi/h

When there is an upgrade, a "+" is used; a "–" is used for downgrades. This results in shorter braking distances on upgrades, where gravity helps deceleration, and longer braking distances on downgrades, where gravity is causing acceleration.

In previous editions of Reference 5, braking distances were based on coefficients of forward skidding friction that varied with speed. In the latest standards, however, a standard deceleration rate of 11.2 ft/s^2 is adopted as a design rate. This is viewed as a rate that can be developed on wet pavements by most vehicles. It is also expected that 90% of drivers will decelerate at higher rates. This, then, suggests a standard friction factor for braking distance computations of $F = 11.2/32.2 = 0.348$, and Equation 2-6 becomes:

$$d_b = \frac{S_i^2 - S_f^2}{30(0.348 \pm 0.01G)} \tag{2-7}$$

Consider the following case: Once the brakes are engaged, what distance is covered bringing a vehicle traveling at 60 mi/h on a 3% downgrade to a complete stop ($S_f = 0$ mi/h). Applying Equation 2-7:

$$d_b = \frac{60^2 - 0^2}{30(0.348 - 0.01*3)} = 377.4 \text{ ft}$$

The braking distance formula is also a favorite tool of accident investigators. It can be used to estimate the initial speed of a vehicle using measured skid marks and an estimated final speed based on damage assessments. In such cases, actual estimated values of F are used, rather than the standard design value recommended by AASHTO. Thus Equation 2-6 is used.

Consider the following case: An accident investigator estimates that a vehicle hit a bridge abutment at a speed of 20 mi/h, based on his or her assessment of damage. Leading up to the accident location, he or she observes skid marks of 100 ft on the pavement ($F = 0.35$) and 75 ft on the grass shoulder ($F = 0.25$). There is no grade. An estimation of the speed of the vehicle at the beginning of the skid marks is desired.

In this case, Equation 2-6 is used to find the initial speed of the vehicle (S_i) based on a known (or estimated) final speed (S_f). Each skid must be analyzed separately, starting with the grass skid (for which a final speed has been estimated). Then:

$$d_b = 75 = \frac{S^2{}_i - 20^2}{30(0.25)}$$

$$S_i = \sqrt{(75*30*0.25) + 20^2} = \sqrt{962.5}$$

$$= 31.0 \text{ mi/h}$$

This is the estimated speed of the vehicle at the *start* of the grass skid; it is also the speed of the vehicle at the *end* of the pavement skid. Then:

$$d_b = 100 = \frac{S_i^2 - 962.5}{30*0.35}$$

$$S_i = \sqrt{(100*30*0.35) + 962.5} = \sqrt{2012.5}$$

$$= 44.9 \text{ mi/h}$$

It is, therefore, estimated that the speed of the vehicle immediately before the pavement skid was 44.9 mi/h. This, of course, can be compared with the speed limit to determine whether excessive speed was a factor in the accident.

2.3.4 Acceleration Characteristics

The flip side of deceleration is acceleration. Passenger cars are able to accelerate at significantly higher rates than commercial vehicles. Table 2.5 shows typical maximum acceleration rates for a passenger car with a weight-to-horsepower ratio of 30 lbs/hp and a tractor-trailer with a ratio of 200 lbs/hp.

Acceleration is highest at low speeds and decreases with increasing speed. The disparity between passenger cars and trucks is significant. Consider the distance required for a car and a truck to accelerate to 20 mi/h. Converting speed from mi/h to ft/s:

$$d_a = \left(\frac{1.47S}{a}\right) * \left(\frac{1.47S}{2}\right) = 1.075\left(\frac{S^2}{a}\right) \tag{2-8}$$

where: d_a = acceleration distance, ft

S = speed at the end of acceleration (from a stop), mi/h

a = acceleration rate, ft/s^2

Once again, note that the 1.075 factor is derived using the more precise factor for converting mi/h to ft/s (1.46666……). Then:

For a passenger car to accelerate to 20 mi/h at a rate of 7.5 ft/s^2:

$$d_a = 1.075\left(\frac{20^2}{7.5}\right) = 57.3 \text{ ft}$$

Table 2.5: Acceleration Characteristics of a Typical Car vs. a Typical Truck on Level Terrain

Speed Range (mi/h)	Acceleration Rate (ft/s²) for: Typical Car (30 lbs/hp)	Typical Truck (200 lbs/hp)
0–20	7.5	1.6
20–30	6.5	1.3
30–40	5.9	0.7
40–50	5.2	0.7
50–60	4.6	0.3

(*Source:* Compiled from *Traffic Engineering Handbook*, 5th Edition, Institute of Transportation Engineers, Washington DC, 2000, Chapter 3, Tables 3-9 and 3-10.)

For a truck to accelerate to 20 mi/h at a rate of 1.6 ft/s^2:

$$d_a = 1.075\left(\frac{20^2}{1.6}\right) = 268.8 \text{ ft}$$

The disparity is striking. If a car is at a "red" signal behind a truck, the truck will significantly delay the car. If a truck is following a car in a standing queue, a large gap between the two will occur as they accelerate.

Unfortunately, not much can be done about this disparity in terms of design and control. In the analysis of highway capacity, however, the disparity between trucks and cars in terms of acceleration and in terms of their ability to sustain speeds on upgrades leads to the concept of "passenger car equivalency." Depending on the type of facility, severity and length of grade, and other factors, one truck may consume as much roadway capacity as six to seven or more passenger cars. Thus the disparity in key operating characteristics of trucks and passenger cars is taken into account in design by providing additional capacity as needed.

2.4 Total Stopping Distance and Applications

The total distance to bring a vehicle to a full stop, from the time the need to do so is first noted, is the sum of the reaction distance, d_r, and the braking distance, d_b. If Equation 2-1 (for d_r) and Equation 2-7 (for d_b) are combined, the total stopping distance becomes:

$$d = 1.47S_i\, t + \frac{S_i^2 - S_f^2}{30(0.348 \pm 0.01G)} \qquad (2\text{-}9)$$

where: d = total stopping distance, ft

S_i = initial speed, mi/h

S_f = final speed, mi/h

t = reaction time, s

G = grade, %

The concept of total stopping distance is critical to many applications in traffic engineering. Three of the more important applications are discussed in the sections that follow.

2.4.1 Safe Stopping Sight Distance

One of the most fundamental principles of highway design is that the driver must be able to see far enough to avoid a potential hazard or collision. Thus, on all roadway sections, the driver must have a sight distance that is at least equivalent to the total stopping distance required at the design speed.

Essentially, this requirement addresses this critical concern: A driver rounding a horizontal curve and/or negotiating a vertical curve is confronted with a downed tree, an overturned truck, or some other situation that completely blocks the roadway. The only alternative for avoiding a collision is to stop. The design must be such that every point along its length, the driver has a clear line of vision for at least one full stopping distance. By ensuring this, the driver can never be confronted with the need to stop without having sufficient distance to do so.

Consider a section of rural freeway with a design speed of 70 mi/h. On a section of level terrain, what safe stopping distance must be provided? Equation 2-9 is used with a final speed (S_f) of "0" and the AASHTO standard reaction time of 2.5 s. Then:

$$d = 1.47 * 70 * 2.5 + \frac{70^2 - 0^2}{30(0.348)}$$
$$= 257.3 + 469.3 = 726.6 \text{ ft}$$

This means that for the entire length of this roadway section drivers must be able to see at least 726.6 ft ahead. Providing this safe stopping sight distance will limit various elements of horizontal and vertical alignment, as discussed in Chapter 3.

What could happen, for example, if a section of this roadway provided a sight distance of only 500 ft? It would now be possible that a driver would initially notice an obstruction when it is only 500 ft away. If the driver were approaching at the design speed of 70 mi/h, a collision would occur. Again, assuming design values of reaction time and forward skidding friction, Equation 2-9 could be solved for the collision speed (i.e., the final speed of the deceleration cycle), using a known deceleration distance of 500 ft:

$$500 = 1.47 * 70 * 2.5 + \frac{70^2 - S_f^2}{30(0.348)}$$
$$500 - 257.3 = 242.7 = \frac{70^2 - S_f^2}{10.44}$$
$$2{,}533.8 = 4{,}900 - S_f^2$$
$$S_f = \sqrt{4{,}900 - 2{,}533.8} = 48.6 \text{ mi/h}$$

If the assumed conditions hold, a collision at 48.6 mi/h would occur. Of course, if the weather was dry and the driver had faster reactions than the design value (remember, 90% of drivers do), the collision might occur at a lower speed or be avoided altogether. The point is that such a collision *could* occur if the sight distance were restricted to 500 ft.

2.4.2 Decision Sight Distance

Although every point and section of a highway must be designed to provide at least safe stopping sight distance, some sections should provide greater sight distance to allow drivers to react to potentially more complex situations than a simple stop. Previously, reaction times for collision avoidance situations were cited [*5*].

Sight distances based on these collision-avoidance decision reaction times are referred to as *decision sight distances.* AASHTO recommends that decision sight distance be provided at interchanges or intersection locations where unusual or unexpected maneuvers are required; changes in cross section such as lane drops and additions, toll plazas, and intense-demand areas where there is substantial "visual noise" from competing information (e.g., control devices, advertising, roadway elements, etc.).

The decision sight distance is found by using Equation 2-9, replacing the standard 2.5 s reaction time for stopping maneuvers with the appropriate collision avoidance reaction time for the situation.

Consider the decision sight distance required for a freeway section with a 60 mi/h design speed approaching a busy urban interchange with many competing information sources. The approach is on a 3% downgrade. For this case, AASHTO suggests a reaction time up to 14.5 s to allow for complex path and speed changes in response to conditions. The decision sight distance is still based on the assumption that a worst case would require a complete stop. Thus the decision sight distance would be:

$$d = 1.47 * 60 * 14.5 + \frac{60^2 - 0^2}{30(0.348 - 0.01 * 3)}$$

$$= 1{,}278.9 + 377.4 = 1{,}656.3 \text{ ft}$$

AASHTO criteria for decision sight distances do not assume a stop maneuver for the speed/path/direction changes required in the most complex situations. The criteria, which are shown in Table 2.6, replace the braking distance in these cases with maneuver distances consistent with maneuver times between 3.5 and 4.5 s. During the maneuver time, the initial speed is assumed to be in effect. Thus for maneuvers involving speed, path, or direction change on rural, suburban, or urban roads, Equation 2-10 is used to find the decision sight distance.

$$d = 1.47(t_r + t_m)S_i \qquad (2\text{-}10)$$

where: t_r = reaction time for appropriate avoidance maneuver, s

t_m = maneuver time, s

Thus in the sample problem posed previously, AASHTO would not assume a stop is required. At 60 mi/h, a maneuver time of 4.0 s is used with the 14.5 s reaction time, and:

$$d = 1.47 * (14.5 + 4.0) * 60 = 1{,}631.7 \text{ ft}$$

The criteria for decision sight distance shown in Table 2.6 are developed from Equations 2-9 and 2-10 for

Table 2-6: Decision Sight Distances Resulting from Equations 2-9 and 2-10

Design Speed (mi/h)	Assumed Maneuver Time (s)	Decision Sight Distance for Avoidance Maneuver (ft)				
		A (Eq. 2-9)	**B (Eq. 2-9)**	**C (Eq. 2-10)**	**D (Eq. 2-10)**	**E (Eq. 2-10)**
Reaction Time (s)		**3**	**9.1**	**11.2**	**12.9**	**14.5**
30	4.5	219	488	692	767	838
40	4.5	330	688	923	1023	1117
50	4.0	460	908	1117	1242	1360
60	4.0	609	1147	1341	1491	1632
70	3.5	778	1406	1513	1688	1852
80	3.5	966	1683	1729	1929	2117

A: Stop on a rural road.
B: Stop on an urban road.
C: Speed/path/direction change on a rural road.
D: Speed/path/direction change on a suburban road.
E: Speed/path/direction change on an urban road.

the decision reaction times indicated for the five defined avoidance maneuvers.

2.4.3 Other Sight Distance Applications

In addition to safe stopping sight distance and decision sight distance, AASHTO also sets criteria for (1) passing sight distance on two-lane rural highways and (2) intersection sight distances for various control options. These are covered in other chapters of this text. See Chapter 16 for a discussion of passing sight distance on two-lane highways and Chapter 18 for intersection sight distances.

2.4.4 Change (Yellow) and Clearance (All Red) Intervals for a Traffic Signal

The yellow interval for a traffic signal is designed to allow a vehicle that cannot comfortably stop when the green is withdrawn to enter the intersection legally. Consider the situation shown in Figure 2.5.

In Figure 2.5, d is the safe stopping distance. At the time the green is withdrawn, a vehicle at d or less feet from the intersection line will not be able to stop, assuming normal design values hold. A vehicle further away than d would be able to stop without encroaching into the intersection area. The yellow signal is timed to allow a vehicle that cannot stop to traverse distance d at the approach speed (S). A vehicle may legally enter the intersection on yellow.

Having entered the intersection legally, the all-red period must allow the vehicle to cross the intersection width (W) and clear the back end of the vehicle (L) past the far intersection line.

Thus the yellow interval must be timed to allow a vehicle to traverse the safe stopping distance. Consider a case in which the approach speed to a signalized intersection is 40 mi/h. How long should the yellow interval be?

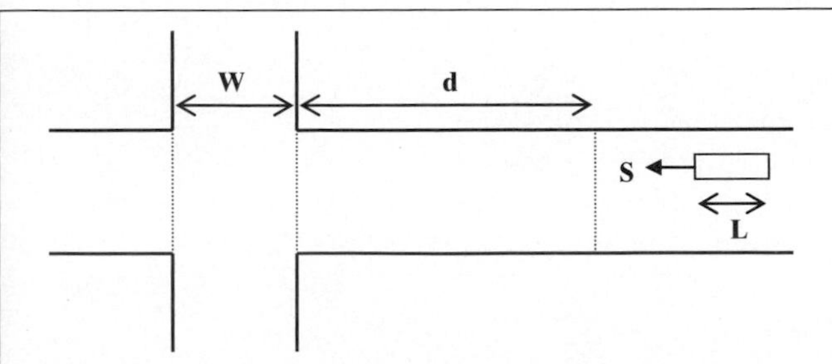

Figure 2.5: Timing Yellow and All-Red Intervals at a Signal

The safe stopping distance is computed using a standard reaction time of 1.0 s for signal timing and level grade:

$$d = 1.47 * 40 * 1.0 + \frac{40^2 - 0^2}{30(0.348)}$$

$$= 58.8 + 153.3 = 212.1 \text{ ft}$$

The length of the yellow signal is the time it takes an approaching vehicle to traverse 212.1 ft at 40 mi/h, or:

$$y = \frac{212.1}{1.47 * 40} = 3.6 \text{ s}$$

In actual practice, the yellow interval is computed using a direct time-based algorithm and a standard deceleration rate. The principle, however, is the same. This example shows how the concept of safe stopping distance is incorporated into signal timing methodologies, which are discussed in detail in Chapter 21.

2.5 Closing Comments

This chapter has summarized some of the key elements of driver, pedestrian, and vehicle characteristics that influence highway design and traffic control. Together with the characteristics of the roadway itself, these elements combine to create traffic streams. As you will see, the characteristics of traffic streams are the result of interactions among and between these elements. The characteristics of human road users and their vehicles have a fundamental impact on traffic streams.

References

1. Dewar, Robert, "Road Users," *Traffic Engineering Handbook*, 5th Edition, Institute of Transportation Engineers, Washington DC, 1999.
2. Ogden, K.W., *Safer Roads: A Guide to Road Safety Engineering*, University Press, Cambridge, England, 1996.
3. Allen, Merrill, et al., *Forensic Aspects of Vision and Highway Safety*, Lawyers and Judges Publishing Co., Inc.,Tucson, AZ, 1996.
4. Olson, Paul, *Forensic Aspects of Driver Perception and Response*, Lawyers and Judges Publishing Co., Inc., Tucson, AZ, 1996.
5. *A Policy on Geometric Design of Highways and Streets*, 5th Edition, American Association of State Highway and Transportation Officials, Washington DC, 2004.

6. Johansson, G. and Rumar, K., "Driver's Brake Reaction Times," *Human Factors*, Vol. 13, No. 1, Human Factors and Ergonomics Society, February 1971.
7. *Report of the Massachusetts Highway Accident Survey*, Massachusetts Institute of Technology, Cambridge, MA, 1935.
8. Normann, O.K., "Braking Distances of Vehicles from High Speeds," *Proceedings of the Highway Research Board*, Vol. 22, Highway Research Board, Washington DC, 1953.
9. Fambro, D.B., et al., "Determination of Safe Stopping Distances," *NCHRP Report 400*, Transportation Research Board, Washington DC, 1997.
10. *Determination of Vehicle Signal Change and Clearance Intervals*, Publication IR-073, Institute of Transportation Engineers, Washington DC, 1994.
11. *Human Factors*, Vol. 28, No. 1, Human Factors and Ergonomics Society, 1986.
12. Eubanks, J.J. and Hill, P.L., *Pedestrian Accident Reconstruction and Litigation*, 2nd Edition, Lawyers and Judges Publishing Co, Inc., Tucson, AZ, 1998.
13. Perry, J., *Gait Analysis*, McGraw-Hill, New York, NY, 1992.
14. Sleight, R.B., "The Pedestrian," *Human Factors in Traffic Safety Research*, John Wiley and Sons, Inc., New York, NY, 1972.
15. Tidwell, J.E. and Doyle, D., *Driver and Pedestrian Comprehension of Pedestrian Laws and Traffic Control Devices*, AAA Foundation for Traffic Safety, Washington DC, 1993.
16. Herms, B.F., "Pedestrian Crosswalk Study: Accidents in Painted and Unpainted Crosswalks," *Pedestrian Protection*, Highway Research Record 406, Transportation Research Board, Washington DC, 1972.
17. "Transportation in an Aging Society," *Special Report 218*, Transportation Research Board, Washington DC, 1988.

Problems

2-1. A driver takes 3.5 s to react to a complex situation while traveling at a speed of 60 mi/h. How far does the vehicle travel before the driver initiates a physical response to the situation (i.e., putting his or her foot on the brake)?

2-2. A driver traveling at 65 mi/h rounds a curve on a level grade to see a truck overturned across the roadway at a distance of 350 ft. If the driver is able to decelerate at a rate of 10 ft/s^2, at what speed will the vehicle hit the truck? Plot the result for reaction times ranging from 0.50 to 5.00 s in increments of 0.5 s. Comment on the results.

2-3. A car hits a tree at an estimated speed of 25 mi/h on a 3% upgrade. If skid marks of 120 ft are observed on dry pavement (F = 0.35) followed by 250 ft (F = 0.25) on a grass-stabilized shoulder, estimate the initial speed of the vehicle just before the pavement skid began.

2-4. Drivers must slow down from 60 mi/h to 40 mi/h to negotiate a severe curve on a rural highway. A warning sign for the curve is clearly visible for a distance of 120 ft. How far in advance of the curve must the sign be located to ensure that vehicles have sufficient distance to decelerate safely? Use the standard reaction time and deceleration rate recommended by AASHTO for basic braking maneuvers.

2-5. How long should the "yellow" signal be for vehicles approaching a traffic signal on a 2% downgrade at a speed of 40 mi/h? Use a standard reaction time of 1.0 s and the standard AASHTO deceleration rate.

2-6. What is the safe stopping distance for a section of rural freeway with a design speed of 80 mi/h on a 3% downgrade?

2-7. What minimum radius of curvature may be designed for safe operation of vehicles at 70 mi/h if the maximum rate of superelevation (e) is 6% and the maximum coefficient of side friction (f) is 0.10?

CHAPTER 3

Roadways and Their Geometric Characteristics

3.1 Highway Functions and Classification

Roadways are a major component of the traffic system, and the specifics of their design have a significant impact on traffic operations. Two primary categories of service are provided by roadways and roadway systems:

- Accessibility
- Mobility

"Accessibility" refers to the direct connection to abutting lands and land uses provided by roadways. This accessibility comes in the form of curb parking, driveway access to off-street parking, bus stops, taxi stands, loading zones, driveway access to loading areas, and similar features. The access function allows a driver or passenger (or goods) to depart the transport vehicle to enter the particular land use in question. "Mobility" refers to the through movement of people, goods, and vehicles from Point A to Point B in the system.

The essential problem for traffic engineers is that the specific design aspects that provide for good access—parking, driveways, loading zones, and so on—tend to retard through movement, or mobility. Thus the two major services provided by a roadway system are often in conflict. This leads to the need to develop roadway systems in a hierarchal manner, with various classes of roadways specifically designed to perform specific functions.

3.1.1 Trip Functions

The American Association of State Highway and Transportation Officials (AASHTO) defines up to six distinct travel movements that may be present in a typical trip:

- Main movement
- Transition
- Distribution
- Collection
- Access
- Termination

The *main movement* is the through portion of trip, making the primary connection between the area of origin and the area of destination. *Transition* occurs when a vehicle transfers from the through portion of the trip to the remaining functions that lead to access and termination. A vehicle

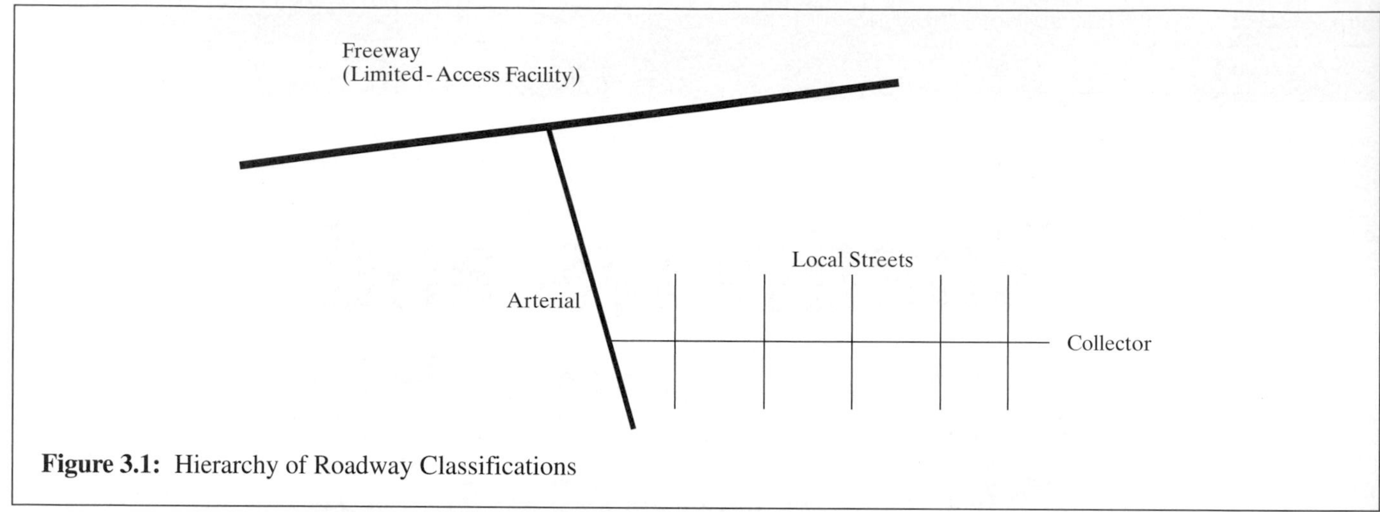

Figure 3.1: Hierarchy of Roadway Classifications

might, for example, use a ramp to transition from a freeway to a surface arterial. The *distribution* function involves providing drivers and vehicles with the ability to leave a major through facility and get to the general area of their destinations. *Collection* brings the driver and vehicle closer to the final destination, and *access* and *termination* result in providing the driver with a place to leave his or her vehicle and enter the land parcel sought. Not all trips involve all of these components.

The hierarchy of trip functions should be matched by the design of the roadways provided to accomplish them. A typical trip has two terminals, one at the origin and one at the destination. At the origin end, the access function provides an opportunity for a trip-maker to enter a vehicle and for the vehicle to enter the roadway system. The driver may go through a series of facilities, usually progressively favoring higher speeds and through movements, until a facility—or set of facilities—is found that will provide the primary through connection. At the destination end of the trip, the reverse occurs, with the driver progressively moving toward facilities favoring access until the specific land parcel desired is reached.

3.1.2 Highway Classification

All highway systems involve a hierarchal classification by the mix of access and mobility functions provided. Four major classes of highways may be identified:

- Limited-access facilities
- Arterials
- Collectors
- Local streets

The *limited-access facility* provides for 100% through movement, or mobility. No direct access to abutting land uses is permitted. *Arterials* are surface facilities designed primarily for through movement but permit some access to abutting lands. *Local streets* are designed to provide access to abutting land uses with through movement only a minor function, if provided at all. The *collector* is an intermediate category between arterials and local streets. Some measure of both mobility and access is provided. The term *collector* comes from a common use of such facilities to collect vehicles from a number of local streets and deliver them to the nearest arterial or limited access facility.

Figure 3.1 illustrates the traditional hierarchy of these categories. The typical trip starts on a local street. The driver seeks the closest collector available, using it to access the nearest arterial. If the trip is long enough, a freeway or limited-access facility is sought. At the destination end of the trip, the process is repeated in reverse order. Depending on the length of the trip and specific characteristics of the area, not all component types of facilities need be included in every trip.

Table 3.1 shows the range of through (or mobility) service provided by the major categories of roadway facility. Many states have their own classification systems that often involve

Table 3.1: Through Service Provided by Various Roadway Categories

Roadway Class	Percent through Service
Freeways (Limited Access Facilities)	100
Arterials	60–80
Collectors	40–60
Local Streets	0–40

Table 3.2: Typical Rural and Urban Roadway Classification Systems

Subcategory	Rural	Urban
	Freeways	
Interstate Freeways	All freeways bearing interstate designation.	All freeways bearing interstate designation.
Other Freeways	All other facilities with full control of access.	All other facilities with full control of access.
Expressways	Facilities with substantial control of access but having some at-grade crossings or entrances.	Facilities with substantial control of access but having some at-grade crossings or entrances.
	Arterials	
Major or Principal Arterials	Serving significant corridor movements, often between areas with populations over 25,000 to 50,000. High-type design and alignment prevail.	Principal service for through movements, with very limited land-access functions that are incidental to the mobility function. High-type design prevails.
Minor Arterials	Provide linkage to significant traffic generators, including towns and cities with populations below the range for principal arterials; serve shorter trip lengths than principal arterials.	Principal service for through movements, with moderate levels of access service also present.
	Collectors	
Major Collectors	Serve generators of intracounty importance not served by arterials; provide connections to arterials and/or freeways.	No subcategories usually used for urban collectors.
Minor Collectors	Link locally important generators with their rural hinterlands; provide connections to major collectors or arterials.	Provide land access and circulation service within residential neighborhoods and/or commercial/industrial areas; collect trips from local generators and channel them to nearby arterials; distribute trips from arterials to their ultimate destination.
	Local Roads	
Residential	No subcategories generally used in rural classification schemes.	Provide land access and circulation within residential neighborhoods.
Commercial	Provide access to adjacent lands of all types; serve travel over relatively short distances.	Provide land access and circulation in areas of commercial development.
Industrial		Provide land access and circulation in areas of industrial development.

subcategories. Table 3.2 provides a general description of frequently used subcategories in highway classification.

It must be emphasized that the descriptions in Table 3.2 are presented as typical. Each highway agency has its own highway classification system, and many have features that are unique to the agency. The traffic engineer should be familiar with highway classification systems, and be able to properly interpret any well-designed system.

3.1.3 Preserving the Function of a Facility

Highway classification systems enable traffic engineers to stratify the highway system by functional purpose. It is important that the intended function of a facility be reinforced through design and traffic controls.

Figure 3.2, for example, illustrates how the design and layout of streets within a suburban residential subdivision can reinforce the intended purpose of each facility. The character of local streets is assured by incorporating sharp curvature into their design and through the use of cul-de-sacs. No local street has direct access to an arterial; collectors within the subdivision provide the only access to arterials. The nature of collectors can be strengthened by not having any residence front on the collector.

The arterials have their function strengthened by limiting the number of points at which vehicles can enter or leave the arterial. Other aspects of an arterial, not obvious here, that could also help reinforce their function include the following:

- Parking prohibitions
- Coordinated signals providing for continuous progressive movement at appropriate speeds
- Median dividers to limit midblock left turns
- Speed limits appropriate to the facility and its environment

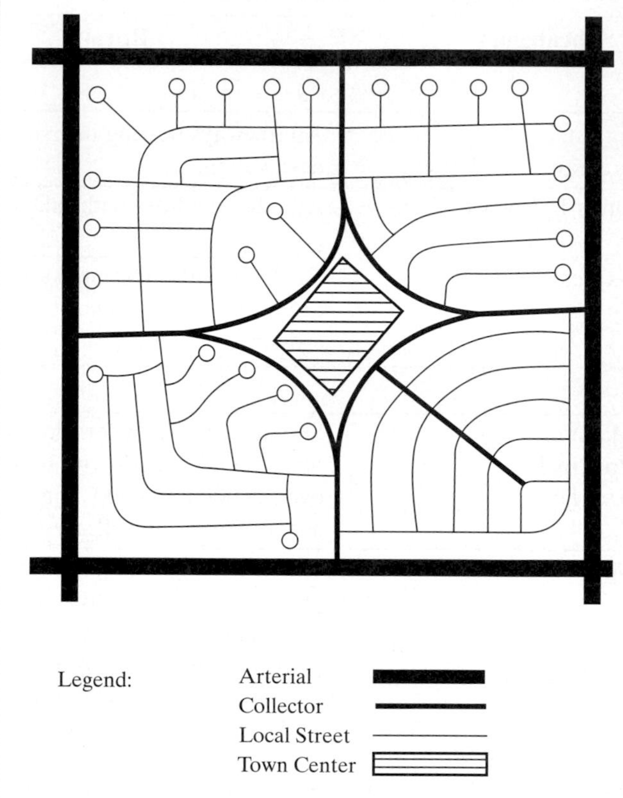

Figure 3.2: Suburban Residential Subdivision Illustrated

In many older cities, it is difficult to separate the functions served by various facilities due to basic design and control problems. The historic development of many older urban areas has led to open-grid street systems. In such systems, local streets, collectors, and surface arterials all form part of the grid. Every street is permitted to intersect every other street, and all facilities provide some land access. Figure 3.3 illustrates this case. The only thing that distinguishes an arterial in such a system is its width and provision of progressive signal timing to encourage through movement.

Such systems often experience difficulties when development intensifies, and all classes of facility, including arterials, are subjected to heavy pedestrian movements, loading and unloading of commercial vehicles, parking, and

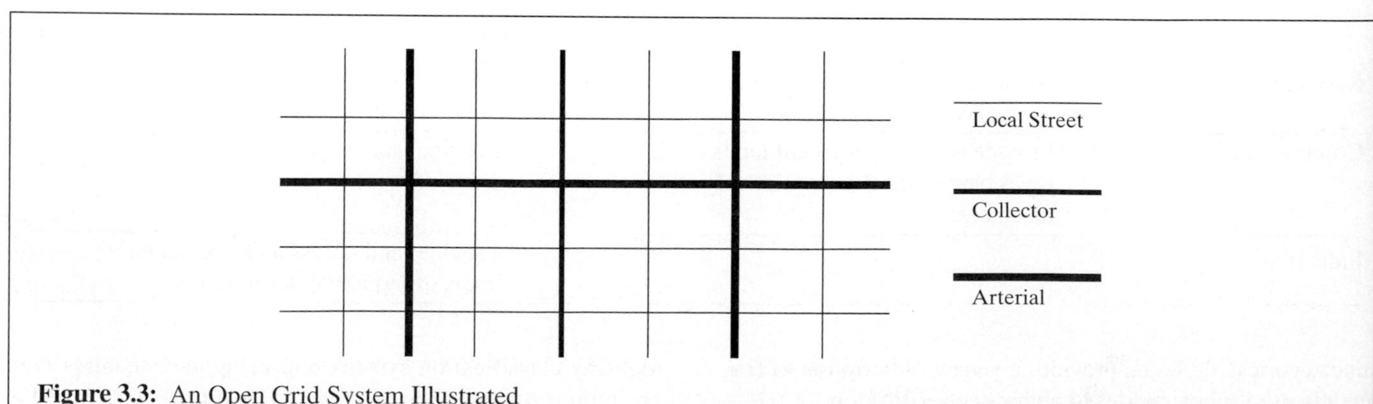

Figure 3.3: An Open Grid System Illustrated

similar functions. Because local streets run parallel to collectors and arterials, drivers experiencing congestion on arterials often reroute themselves to nearby local streets, subjecting them to unwanted and often dangerous heavy through flows.

The importance of providing designs and controls that are appropriate to the intended function of a facility cannot be understated. Chapter 28 provides a more detailed discussion of techniques for doing so.

3.2 Introduction to Highway Design Elements

Highways are complex physical structures involving compacted soil, sub-base layers of aggregate, pavements, drainage structures, bridge structures, and other physical elements.

From an operational viewpoint, it is the geometric characteristics of the roadway that primarily influence traffic flow and operations. Three main elements define the geometry of a highway section:

- Horizontal alignment
- Vertical alignment
- Cross-sectional elements

Virtually all standard practices in geometric highway design are specified by the American Association of State Highway and Transportation Officials in the current version of *Policy on Geometric Design of Highways and Streets* [*1*]. The latest edition of this key reference (at this writing) was published in 2004. Because of severe restrictions on using material directly from Reference 1, this text presents general design practices that are most frequently based on American Association of State Highway and Transportation Officials (AASHTO) standards.

3.2.1 Horizontal Alignment

The horizontal alignment refers to a plan view of the highway. The horizontal alignment includes tangent sections and the horizontal curves and other transition elements that join them.

Highway design is generally initiated by laying out a set of tangents on topographical and development maps of the service area. Selection of an appropriate route and the specific location of these tangent lines involves many considerations and is a complex task. These are some of the more important considerations:

- Forecast demand volumes, with known or projected origin-destination patterns
- Patterns of development
- Topography
- Natural barriers
- Subsurface conditions
- Drainage patterns
- Economic considerations
- Environmental considerations
- Social considerations

The first two items deal with anticipated demand on the facility and the specific origins and destinations that are to be served. The next four are important engineering factors that must be considered. The last three are critically important. Cost is always an important factor, but it must be compared with quantifiable benefits.

Environmental impact statements are required of virtually all highway projects, and much effort is put into providing remedies for unavoidable negative impacts on the environment. Social considerations are also important and cover a wide range of issues. It is particularly critical that highways be built in ways that do not disrupt local communities, either by dividing them, enticing unwanted development, or causing particularly damaging environmental impacts. Although this text does not deal in detail with this complex process of decision making, you should be aware of its existence and of the influence it has on highway programs in the United States.

3.2.2 Vertical Alignment

Vertical alignment refers to the design of the facility in the profile view. Straight grades are connected by vertical curves, which provide for transition between adjacent grades. The *grade* refers to the longitudinal slope of the facility, expressed as "feet of rise or fall" per "longitudinal foot" of roadway length. As a dimensionless value, the grade may be expressed either as a decimal or as a percentage (by multiplying the decimal by 100).

In vertical design, attempts are made to conform to the topography, wherever possible, to reduce the need for costly excavations and landfills as well as to maintain aesthetics. Primary design criteria for vertical curves include:

- Provision of adequate sight distance at all points along the profile
- Provision of adequate drainage

- Maintenance of comfortable operations
- Maintenance of reasonable aesthetics

The specifics of vertical design usually follow from the horizontal route layout and specific horizontal design. The horizontal layout, however, is often modified or established in part to minimize problems in the vertical design.

3.2.3 Cross-Sectional Elements

The third physical dimension, or view, of a highway that must be designed is the cross section. The cross section is a cut across the plane of the highway. Within the cross section, such elements as lane widths, superelevation (cross-slope), medians, shoulders, drainage, embankments (or cut sections), and similar features are established. Because the cross section may vary along the length of a given facility, cross sections are generally designed every 100 ft along the facility length and at any other locations that form a transition or change in the cross-sectional characteristics of the facility.

3.2.4 Surveying and Stationing

In the field, route surveyors define the geometry of a highway by "staking" out the horizontal and vertical position of the route and by similarly marking of the cross section at intervals of 100 ft.

Although this text does not deal with the details of route surveying, it is useful to understand the conventions of "stationing" that are used in the process. Stationing of a new or reconstructed route is generally initiated at the western or northern end of the project. "Stations" are established every 100 ft and are given the notation $xxx + yy$. Values of "xxx" indicate the number of hundreds of feet of the location from the origin point. The "yy" values indicate intermediate distances of less than 100 ft.

Regular stations are established every 100 ft and are numbered 0 + 00, 100 + 00, 200 + 00, and so on. Various elements of the highway are "staked" by surveyors at these stations. If key points of transition occur between full stations, they are also staked and would be given a notation such as 1200 + 52, which signifies a location 1,252 ft from the origin. This notation is used to describe points along a horizontal or vertical alignment in subsequent sections of this chapter. Reference 2 is a text in route surveying, which can be consulted for more detailed information on the subject.

3.3 Horizontal Alignment of Highways

The horizontal alignment of a highway is its path in a plan view of the surrounding terrain. Horizontal curves have critical geometric properties and characteristics that should be well understood. They are also subject to standard design criteria established by AASHTO or by local and state highway agencies.

3.3.1 Quantifying the Severity of Horizontal Curves: Radius and Degree of Curvature

All horizontal curves are circular, that is, formed by an arc with a constant radius. Compound horizontal curves may be formed by consecutive horizontal curves with different radii. On high-speed, high-type facilities, a horizontal curve and a tangent (straight) segment are often joined by a spiral transition curve, which is a curve with a varying radius that starts at "∞" at the connection to the tangent segment and ends at the radius of the circular curve at the connection to the curve.

The severity of a circular horizontal curve is measured by the *radius* or by the *degree of curvature*, which is a related measure. Degree of curvature is most often used because higher values depict sharper, or more severe, curves. Conversely, larger radii depict less severe curves.

Figure 3.4 illustrates two ways of defining degree of curvature. The *chord definition* is illustrated in Figure 3.4 (a). The degree of curvature is defined as the central angle subtending a 100-ft chord on the circular curve. The *arc definition* is illustrated in Figure 3.4 (b), and is the most frequently

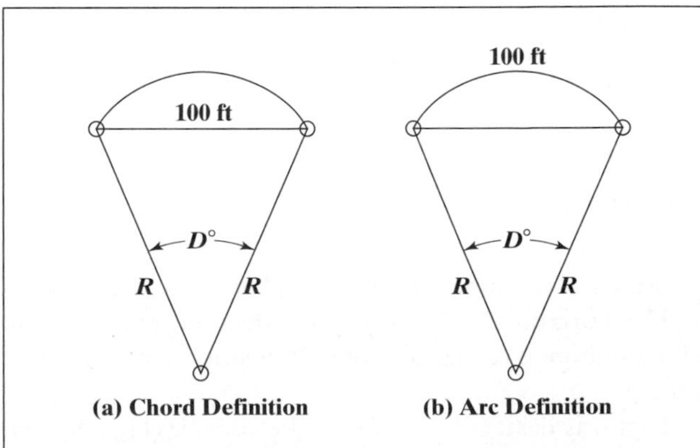

Figure 3.4: Definition of Degree of Curvature

used. In this definition, the degree of curvature is defined as the central angle subtending a 100-ft arc.

Using the arc definition, it is possible to derive the relationship between the radius (R) and the degree of curvature (D). The ratio of the circumference of the circle to 360° is set equal to the ratio of 100 ft to D°. Then:

$$\frac{2\pi R}{360} = \frac{100}{D}$$

$$D = \frac{100(360)}{2\pi R}$$

Noting that $\pi = 3.141592654\ldots$, then:

$$D = \frac{36{,}000}{2(3.1415915)R} = \frac{5{,}729.58}{R} \qquad (3\text{-}1)$$

where: D = degree of curvature, degrees
R = radius of curvature, ft

Thus, for example, a circular curve with a radius of 2,000 ft has a degree of curvature of:

$$D = \frac{5{,}729.58}{2{,}000} = 2.865°$$

Note that for up to 4° curves, there is little difference between the arc and the chord definition of degree of curvature. This text, however, uses only the arc definition illustrated in Figure 3.4 (b).

3.3.2 Review of Trigonometric Functions

The geometry of horizontal curves is described mathematically using trigonometric functions. A brief review of these functions is included as a refresher for those who may not have used trigonometry for some time. Figure 3.5 illustrates a right triangle, from which the definitions of trigonometric functions are drawn.

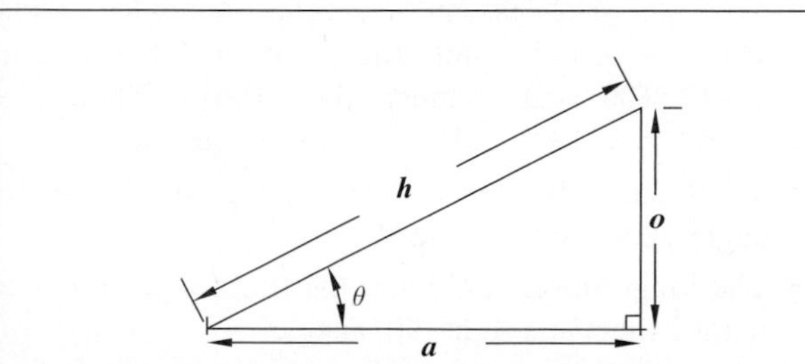

Figure 3.5: Trigonometric Functions Illustrated

In Figure 3.5:

- o = length of the opposite leg of the right triangle
- a = length of the adjacent leg of the right triangle
- h = hypotenuse of the right triangle

Using the legs of the right triangle, the following trigonometric functions are defined:

- Sine $\theta = o/h$
- Cosine $\theta = a/h$
- Tangent $\theta = o/a$

From these primary functions, several derivative functions are also defined:

- Cosecant θ = 1/Sine $\theta = h/o$
- Secant θ = 1/Cosine $\theta = h/a$
- Cotangent θ = 1/Tangent $\theta = a/o$

and:

- Exsecant θ = Secant $\theta - 1$
- Versine $\theta = 1 -$ Cosine θ

Trigonometric functions are tabulated in many mathematics texts and are generally included on most calculators and in virtually all spreadsheet software. When using spreadsheet software or calculators, the user must determine whether angles are entered in *degrees* or *radians*. In a full circle, there are 2π radians and 360°. Thus one radian is equal to 360/2(3.141592654) = 57.3°.

3.3.3 Critical Characteristics of Horizontal Curves

Figure 3.6 depicts a circular horizontal curve connecting two tangent lines. The following points are defined:

- *P.I.* = point of intersection; point at which the two tangent lines meet
- *P.C.* = point of curvature; point at which the circular horizontal curve begins
- *P.T.* = point of tangency; point at which the circular horizontal curve ends
- T = length of tangent, from the *P.C.* to the *P.I.* and from the *P.I.* to the *P.T.*, in feet
- E = external distance, from point 5 to the *P.I.* in Figure 3.6, in feet
- M = middle ordinate distance, from point 5 to point 6 in Figure 3.6, in feet

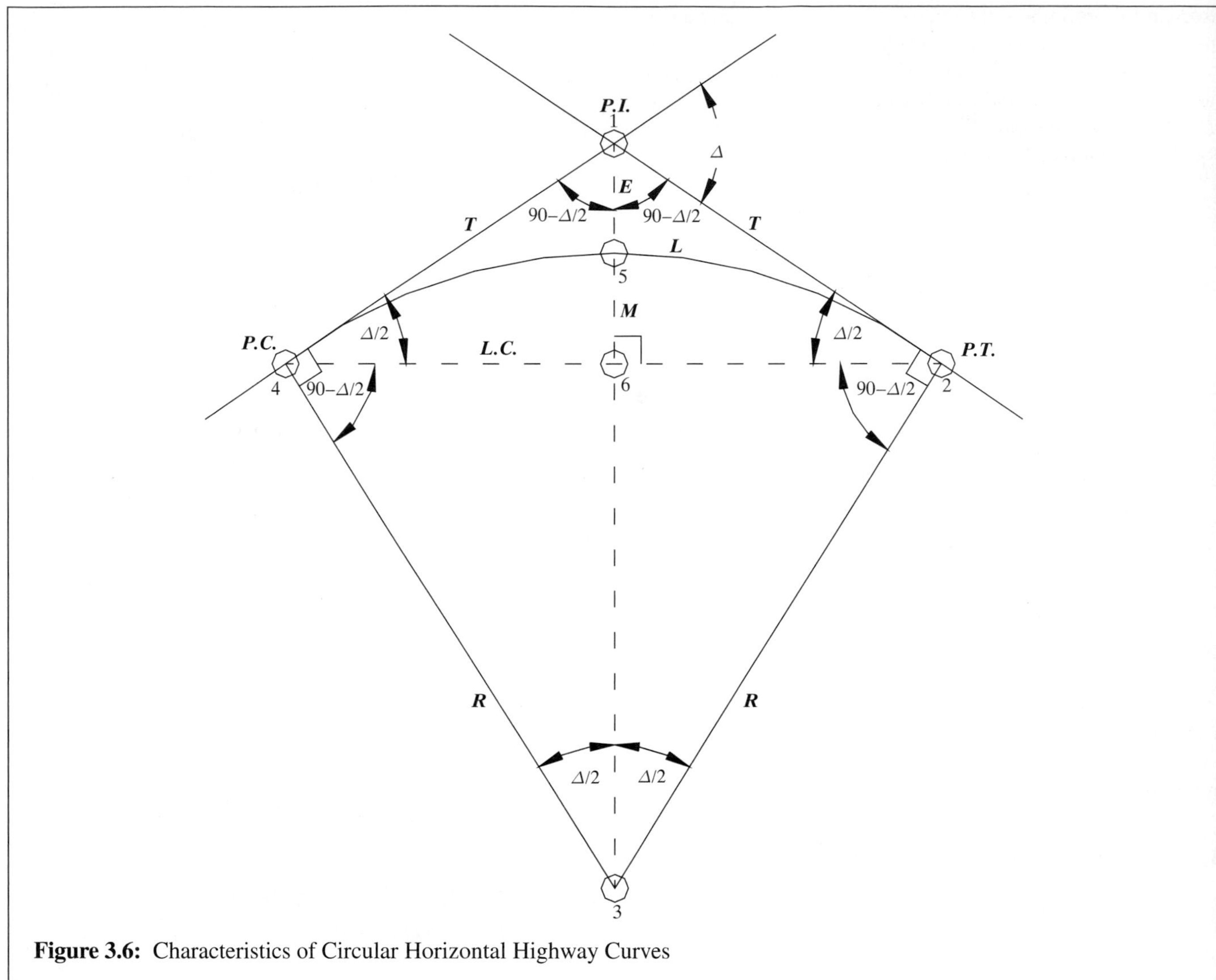

Figure 3.6: Characteristics of Circular Horizontal Highway Curves

- *L.C.* = long chord, from the *P.C.* to the *P.T.*, in feet
- Δ = external angle of the curve, sometimes referred to as the angle of deflection, in degrees
- R = the radius of the circular curve, in feet

A number of geometric characteristics of the circular curve are of interest in deriving important relationships:

- Radii join tangent lines at right (90°) angles at the *P.C.* and *P.T.*
- A line drawn from the *P.I.* to the center of the circular curve bisects $\angle 412$ and $\angle 432$ (numbers refer to Figure 3.6).
- $\angle 412$ equals $180 - \Delta$; thus, $\angle 413$ and $\angle 312$ must be half this, or $90 - \Delta/2$ as shown in Figure 3.6.
- Triangle 412 is an isosceles triangle. Thus, $\angle 142 = \angle 124$, and the sum of these, plus $\angle 412\,(180 - \Delta)$, must be 180°. Therefore, $\angle 142 = \angle 124 = \Delta/2$
- $\angle 346$ and $\angle 326$ must be $90 - \Delta/2$, and the central angle, $\angle 432$, is equal to Δ
- The long chord (*L.C.*) and the line from point 1 to point 3 meet at a right (90°) angle

Given the characteristics shown in Figure 3.6, some of the key relationships for horizontal curves may be derived.

Length of the Tangent (*T*)

Consider the tangent of triangle 143:

$$\text{Tan}(\Delta/2) = \frac{T}{R}$$

Then:

$$T = R\,\text{Tan}(\Delta/2) \qquad (3\text{-}2)$$

Length of the Middle Ordinate (*M*)

The length of the middle ordinate is found by subtracting line segment 3-6 from the radius, which is line segment 3-6-5. Then, considering triangle 362:

$$\text{Cos}(\Delta/2) = \frac{\text{seg } 36}{R}$$

$$\text{seg } 36 = R\,\text{Cos}(\Delta/2)$$

Then:

$$M = R - R\,\text{Cos}(\Delta/2) = R[1 - \text{Cos}(\Delta/2)] \qquad (3\text{-}3)$$

Length of the External Distance (*E*)

Consider triangle 162. Then:

$$\text{Sin}(\Delta/2) = \frac{E + M}{T}$$

$$E = T\,\text{Sin}(\Delta/2) - M$$

Substituting the appropriate equations for *T* and *M*:

$$E = R\,\text{Tan}(\Delta/2)\,\text{Sin}(\Delta/2) - R[1 - \text{Cos}(\Delta/2)]$$

By manipulating the trigonometric functions, this may be rewritten as:

$$E = R\frac{\text{Sin}(\Delta/2)}{\text{Cos}(\Delta/2)}\text{Sin}(\Delta/2) - R + R\,\text{Cos}(\Delta/2)$$

$$E = R\frac{\text{Sin}^2(\Delta/2)}{\text{Cos}(\Delta/2)} - R - R\,\text{Cos}(\Delta/2)$$

$$E = \frac{R\,\text{Sin}^2(\Delta/2) - R\,\text{Cos}(\Delta/2) + R\,\text{Cos}^2(\Delta/2)}{\text{Cos}(\Delta/2)}$$

$$E = \frac{R[1 - \text{Cos}^2(\Delta/2)] - R\,\text{Cos}(\Delta/2) + R\,\text{Cos}^2(\Delta/2)}{\text{Cos}(\Delta/2)}$$

$$E = \frac{R[1 - \text{Cos}(\Delta/2)]}{\text{Cos}(\Delta/2)}$$

and:

$$E = R\left[\left(\frac{1}{\text{Cos}(\Delta/2)}\right) - 1\right] \qquad (3\text{-}4)$$

Length of the Curve (*L*)

The length of the curve derives directly from the arc definition of degree of curvature. A central angle equal to the degree of curvature subtends an arc of 100 ft; the actual central angle (Δ) subtends the length of the curve (L). Thus:

$$\frac{L}{100} = \frac{\Delta}{D}$$

$$L = 100\left(\frac{\Delta}{D}\right) \qquad (3\text{-}5)$$

Length of the Long Chord (*LC*)

Note that the long chord is bisected by line segment 3-6-5. Then, considering triangle 364:

$$\text{Sin}(\Delta/2) = \frac{L.C./2}{R}$$

$$L.C. = 2R\,\text{Sin}(\Delta/2) \qquad (3\text{-}6)$$

An Example Illustrating the Characteristics of a Horizontal Curve

Two tangent lines meet at Station 3,200 + 15. The radius of curvature is 1,200 ft, and the angle of deflection is 14°. Find the length of the curve, the stations for the *P.C.* and *P.T.*, and all other relevant characteristics of the curve (*L.C.*, *M, E*). Figure 3.7 illustrates the case. Using the relationships discussed previously, all of the key measures for the curve of Figure 3.7 may be found:

$$D = \frac{5{,}729.58}{1{,}200} = 4.77°$$

$$L = 100\left(\frac{14}{4.77}\right) = 293.5 \text{ ft}$$

$$L.C. = 2(1{,}200)\,\text{Sin}(14/2) = 292.5 \text{ ft}$$

$$T = 1{,}200\,\text{Tan}(14/2) = 147.3 \text{ ft}$$

$$M = 1{,}200[1 - \text{Cos}(14/2)] = 8.9 \text{ ft}$$

$$E = 1{,}200\left[\frac{1}{\text{Cos}(14/2)} - 1\right] = 9 \text{ ft}$$

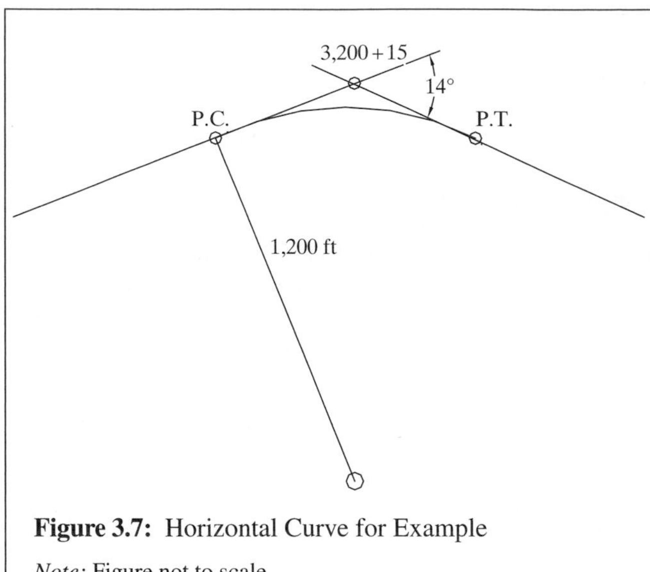

Figure 3.7: Horizontal Curve for Example

Note: Figure not to scale

Obviously, this is a fairly short curve, due to a small angle of deflection (14°). The question also asked for station designations for the *P.C.* and *P.T.* The computed characteristics are used to find these values. The station of the *P.I.* is given as 3,200 + 15, which indicates that it is 3,215 ft from the beginning of the project.

The *P.C.* is found as the *P.I.* − *T*. This is 3,215 − 147.3 = 3,067.7, which is station 3,000 + 67.7. The station of the *P.T.* is found as the *P.C.* + *L*. This is 3,067.7 + 293.5 = 3,361.2, which is station 3,300.0 + 61.2. All station units are in feet.

3.3.4 Superelevation of Horizontal Curves

Most highway curves are "superelevated," or banked, to assist drivers in resisting the effects of centripetal force. Superelevation is quantified as a percentage, computed as follows:

$$e = \left(\frac{\text{total rise in pavement from edge to edge}}{\text{width of pavement}} \right) \tag{3-7}$$

As noted in Chapter 2, the two factors that keep a vehicle on a highway curve are side friction between the tires and the pavement, and the horizontal element of support provided by a banked or "superelevated" pavement. The speed of a vehicle and the radius of curvature are related to the superelevation rate (e) and the coefficient of side friction (f), by the equation:

$$R = \frac{S^2}{15(0.01e + f)} \tag{3-8}$$

where: R = radius of curvature, ft
S = speed of vehicle, mi/h
e = rate of superelevation, %
f = coefficient of side friction

In design, these values become limits: S is the design speed for the facility, e is the maximum rate of superelevation permitted, and f is a design value of the coefficient of side friction representing tires in reasonable condition on a wet pavement. The resulting value of R is the minimum radius of curvature permitted for these conditions.

Maximum Superelevation Rates

AASHTO [*1*] recommends the use of maximum superelevation rates between 4% and 12%. For design purposes, only increments of 2% are used. Maximum rates adopted vary from region to region based on factors such as climate, terrain, development density, and frequency of slow-moving vehicles. Some of the practical considerations involved in setting this range and for selection of an appropriate rate include:

1. Twelve percent (12%) is the maximum superelevation rate in use. Drivers feel uncomfortable on sections with higher rates, and driver effort to maintain lateral position is high when speeds are reduced on such curves. Some jurisdictions use 10% as a maximum practical limit.
2. Where snow and ice are prevalent, a maximum value of 8% is generally used. Many agencies use this as an upper limit regardless, due to the effect of rain or mud on highways.
3. In urban areas, where speeds may be reduced frequently due to congestion, maximum rates of 4% to 6% are often used.
4. On low-speed urban streets or at intersections, superelevation may be eliminated.

It should be noted that on open highway sections, there is generally a minimum superelevation maintained, even on straight sections. This is to provide for cross-drainage of water to the appropriate roadside(s) where sewers or drainage ditches are present for longitudinal drainage. This minimum rate is usually in the range of 1.5% for high-type surfaces and 2.0% to 2.5% for low-type surfaces.

Side-Friction Factors (Coefficients of Side Friction, *f*)

Design values of the side-friction factor vary with design speed. Design values represent wet pavements and tires in reasonable but not top condition. Values also represent frictional forces that can be comfortably achieved; they do not represent, for example, the maximum side friction that is achieved the instant before skidding.

Table 3.3 illustrates commonly used side friction factors (f). Consult Reference 1 directly for a more thorough discussion of side-friction factors. Actual side-friction factors vary with a number of variables, including the superelevation rate.

Determining Design Values of Superelevation

Once a maximum superelevation rate and a design speed are set, the minimum radius of curvature can be found using Equation 3-8. This can be expressed as a maximum degree of curvature using Equation 3-1.

Consider a roadway with a design speed of 60 mi/h, for which a maximum superelevation rate of 6% has been selected. What are the minimum radius of curvature and/or maximum degree of curve that can be included on this facility?

For a design speed of 60 mi/h, Table 3.3 indicates a design value for the coefficient of side friction (f) of 0.120. Then:

$$R_{min} = \frac{S^2}{15(0.01e_{max} + f_{des})} = \frac{60^2}{15(0.01 * 6 + 0.120)}$$

$$= 1{,}333.33 \text{ ft}$$

$$D_{max} = \frac{5{,}729.58}{R_{min}} = \frac{5{,}729.58}{1{,}333.33} = 4.3°$$

Although this limits the degree of curvature to a maximum of 4.3° for the facility, it does not determine the appropriate rate of superelevation for degrees of curvature less than 4.3° (or a radius greater than 1,333.33 ft). The actual rate of superelevation for any curve with less than the maximum degree of curvature (or more than the minimum radius) is found by solving Equation 3-8 for e using the design speed for S and the appropriate design value of f. Then:

$$e = 100\left[\left(\frac{S_{des}^2}{15R}\right) - f_{des}\right] \tag{3-9}$$

For the highway just described, what superelevation rate would be used for a curve with a radius of 1,500 ft? Using Equation 3-9:

$$e = 100\left[\left(\frac{60^2}{15 * 1{,}500}\right) - 0.12\right] = 4.0\%$$

Thus, although the maximum superelevation rate for this facility was set at 6%, a superelevation rate of 4.0% would be used for a curve with a radius of 1,500 ft, which is *larger* than the minimum radius for the design constraints specified for the facility. AASHTO standards [*1*] contain many curves and tables yielding results of such analyses for various specified constraints, for ease of use in design.

Achieving Superelevation

The transition from a tangent section with a normal superelevation for drainage to a superelevated horizontal curve occurs in two stages:

- *Tangent Runoff:* The outside lane of the curve must have a transition from the normal drainage superelevation to a level or flat condition prior to being rotated to the full superelevation for the horizontal curve. The length of this transition is called the *tangent runoff* and is noted as L_t.
- *Superelevation Runoff:* Once a flat cross section is achieved for the outside lane of the curve, it must be

Table 3.3: Side-Friction Factors Used in Design

Design Speed (mi/h)	*f*	Design Speed (mi/h)	*f*	Design Speed (mi/h)	*f*
10	0.38	35	0.17	60	0.12
15	0.32	40	0.16	65	0.11
20	0.26	45	0.15	70	0.10
25	0.23	50	0.14	75	0.09
30	0.20	55	0.13	80	0.08

rotated (with other lanes) to the full superelevation rate of the horizontal curve. The length of this transition is called the *superelevation runoff* and is noted as L_s.

For most undivided highways, rotation is around the centerline of the roadway, although rotation can also be accomplished around the inside or outside edge of the roadway as well. For divided highways, each directional roadway is separately rotated, usually around the inside or outside edge of the roadway.

Figure 3.8 illustrates the rotation of undivided two-lane, four-lane, and six-lane highways around the centerline, although the slopes shown are exaggerated for clarity. The rotation is accomplished in three steps:

1. The outside lane(s) are rotated from their normal cross-slope to a flat condition.
2. The outside lane(s) are rotated from the flat position until they equal the normal cross-slope of the inside lanes.
3. All lanes are rotated from the condition of step 2 to the full superelevation of the horizontal curve.

The tangent runoff is the distance taken to accomplish step 1, whereas the superelevation runoff is the distance taken to accomplish steps 2 and 3. The tangent and superelevation runoffs are, of course, implemented for the transition from tangent to horizontal curve and for the reverse transition from horizontal curve back to tangent.

In effect, the transition from a normal cross-slope to a fully superelevated section is accomplished by creating a grade differential between the rotation axis and the pavement edge lines. To achieve safe and comfortable operations, there are limitations on how much of a differential may be accommodated. The recommended minimum length of superelevation runoff is given as:

$$L_r = \frac{w * n * e_d * b_w}{\Delta} \tag{3-10}$$

where: L_r = minimum length of superelevation runoff, ft
w = width of a lane, ft
n = number of lanes being rotated
e_d = design superelevation rate, %
b_w = adjustment factor for number of lanes rotated
Δ = maximum relative gradient, %

AASHTO-recommended values for the maximum relative gradient, Δ, are shown in Table 3.4. The adjustment factor, b_w, depends on the number of lanes being rotated. A value of 1.00 is used when one lane is being rotated, 0.75 when two lanes are being rotated, and 0.67 when three lanes are being rotated.

Consider the example of a four-lane highway, with a superelevation rate of 4% achieved by rotating two 12-ft lanes around the centerline. The design speed of the highway is 60 mi/h. What is the appropriate minimum length of superelevation runoff? From Table 3.4, the maximum relative gradient

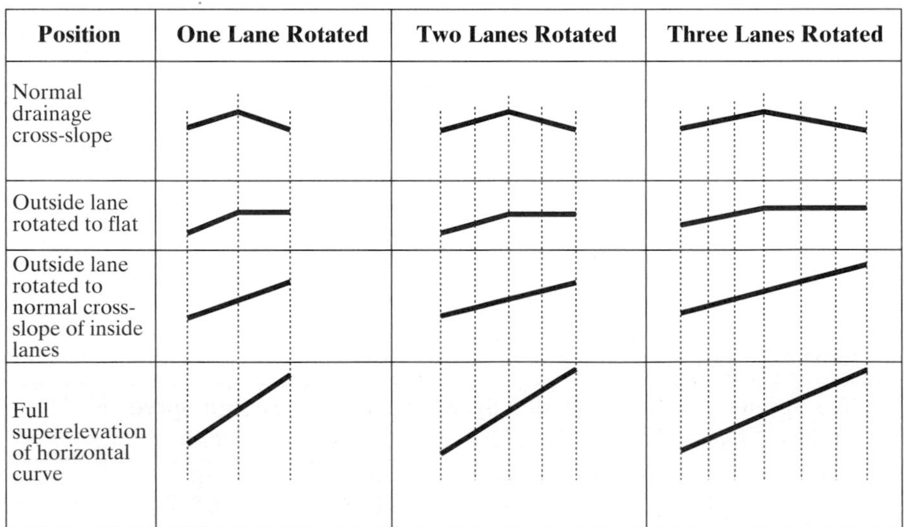

Figure 3-8: Achieving Superelevation by Rotation around a Centerline

Table 3.4: Maximum Relative Gradients (Δ) for Superelevation Runoff

Design Speed (mi/h)	Maximum Relative Gradient (%)	Design Speed (mi/h)	Maximum Relative Gradient (%)
15	0.78	50	0.50
20	0.74	55	0.47
25	0.70	60	0.45
30	0.66	65	0.43
35	0.62	70	0.40
40	0.58	75	0.38
45	0.54	80	0.35

(*Source:* Used with permission of the American Association of State Highway and Transportation Officials, *A Policy on Geometric Design of Highways and Streets,* 5th Edition, condensed from Table 3-30, Pg. 177, Washington DC, 2004.)

for 60 mi/h is 0.45%; the adjustment factor for rotating two lanes is 0.75. Thus:

$$L_r = \frac{w * n * e_d * b_w}{\Delta}$$

$$L_r = \frac{12 * 2 * 4 * 0.75}{0.45} = 160 \text{ ft}$$

Note that although it is a four-lane cross section being rotated, $n = 2$, as rotation is around the centerline. Where separate pavements on a divided highway are rotated around an edge, the full number of lanes on the pavement would be used.

The length of the tangent runoff is related to the length of the superelevation runoff, as follows:

$$L_t = \frac{e_{NC}}{e_d} L_r \tag{3-11}$$

where: L_t = length of tangent runoff, ft
L_r = length of superelevation runoff, ft
e_{NC} = normal cross-slope, %
e_d = design superelevation rate, %

If, in the previous example, the normal drainage cross-slope was 1%, then the length of the tangent runoff would be:

$$L_t = \left(\frac{1}{4}\right)*160 = 40 \text{ ft}$$

The total transition length between the normal cross section to the fully superelevated cross section is the sum of the superelevation and tangent runoffs, or (in this example) $160 + 40 = 200$ ft.

To provide drivers with the most comfortable operation, from 60% to 90% of the total runoff is achieved on the tangent section, with the remaining runoff achieved on the horizontal curve. AASHTO allows states or other jurisdictions to set a constant percentage split anywhere within this range as a matter of policy, but it also gives criteria for optimal splits based on design speed and the number of lanes rotated. For design speeds between 15 and 45 mi/h, 80% (one lane rotated) or 90% (two or more lanes rotated) of the runoff is on the tangent section. For higher design speeds, 70% (one lane rotated), 80% (two lanes rotated), or 85% (three or more lanes rotated) of the runoff is on the tangent section.

Where a spiral transition curve (see next section) is used between the tangent and horizontal curves, the superelevation is achieved entirely on the spiral. If possible, the tangent and superelevation runoff may be accomplished on the spiral.

3.3.5 Spiral Transition Curves

Although not impossible, it is difficult for drivers to travel immediately from a tangent section to a circular curve with a constant radius. A spiral transition curve begins with a tangent (degree of curve, $D = 0$) and gradually and uniformly increases the degree of curvature (decreases the radius) until the intended circular degree of curve is reached.

Use of a spiral transition provides for a number of benefits:

- Provides an easy path for drivers to follow: Centrifugal and centripetal forces are increased gradually
- Provides a desirable arrangement for superelevation runoff
- Provides a desirable arrangement for pavement widening on curves (often done to accommodate off-tracking of commercial vehicles)
- Enhances highway appearance

Figure 3.9: The Visual Impact of a Spiral Transition Curve

(*Source:* Used with permission of Yale University Press, C. Tunnard and B. Pushkarev, *Manmade America*, New Haven CT, 1963.)

The latter is illustrated in Figure 3.9, where the visual impact of a spiral transition curve is obvious. Spiral transition curves are not always used because construction is difficult and construction cost is generally higher than for a simple circular curve. They are recommended for high-volume situations where degree of curvature exceeds 3°. The geometric characteristics of spiral transition curves are complex; they are illustrated in Figure 3.10.

The key variables in Figure 3.10 are defined as follows:

T.S. = transition station from tangent to spiral

S.C. = transition station from spiral to circular curve

C.S. = transition station from circular curve to spiral

S.T. = transition station from spiral to tangent

Δ = angle of deflection (central angle) of original circular curve without spiral

Δ_s = angle of deflection (central angle) of circular portion of curve with spiral

δ = angle of deflection for spiral portion of curve

L_s = length of the spiral, ft

Without going through the very detailed derivations for many of the terms included in Figure 3.10, some of the key relationships are described below.

Length of Spiral, L_s

The length of the spiral can be set in one of two ways: (1) L_s is set equal to the length of the superelevation runoff, as described in the previous section; (2) the length of the spiral can be determined as [*6*]:

$$L_s = \left(\frac{3.15\ S^3}{R * C}\right) \tag{3-12}$$

where: L_s = length of the spiral, ft

S = design speed of the curve, mi/h

R = radius of the circular curve, ft

C = rate of increase of lateral acceleration, ft/s^3

The values of C commonly used in highway design range between 1 and 3 ft/s^3. When a value of 1.97 is used

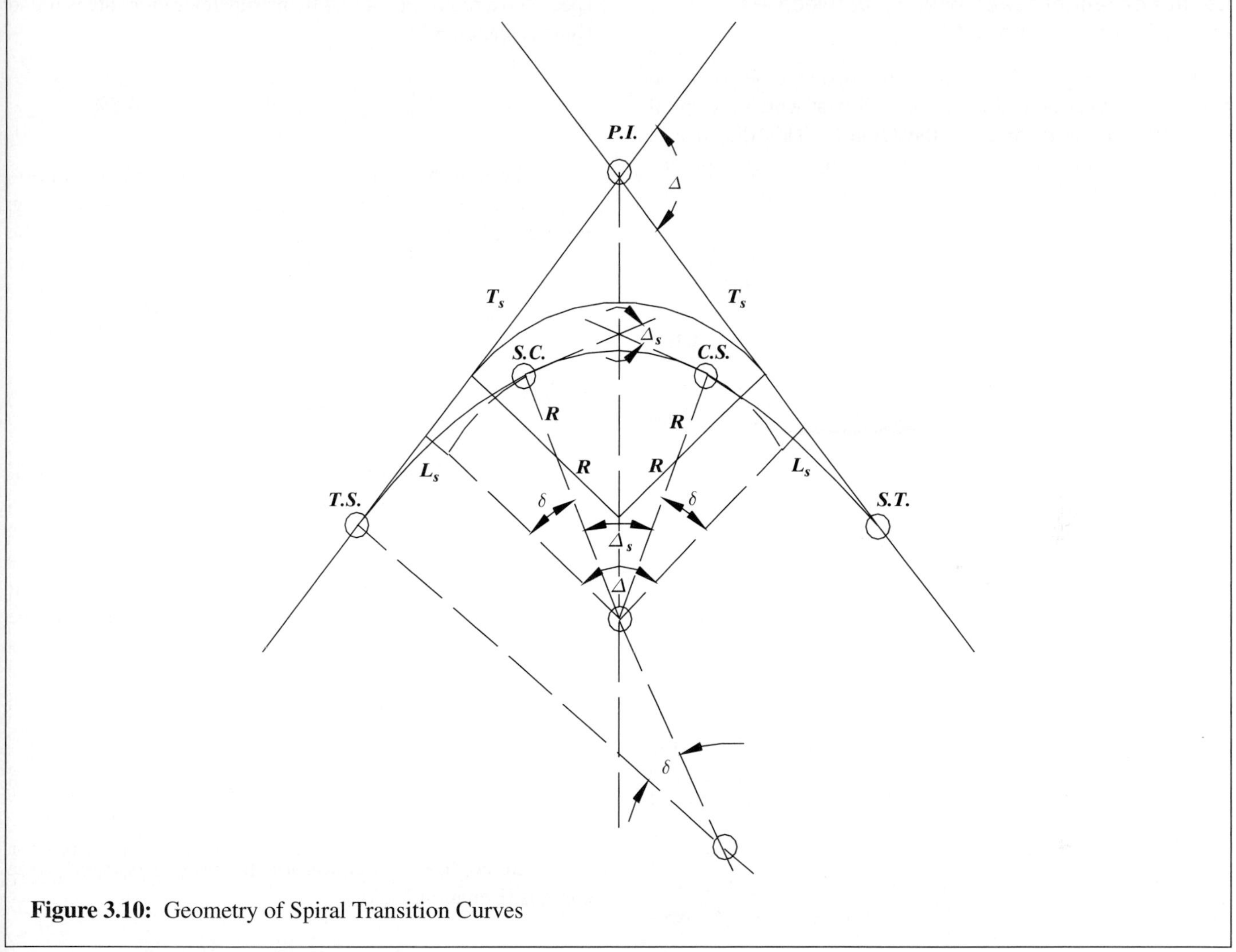

Figure 3.10: Geometry of Spiral Transition Curves

(a common standard value adopted by highway agencies), the equation becomes:

$$L_s = 1.6\frac{S^3}{R} \tag{3-13}$$

Angle of Deflection (Central Angle) for the Spiral, δ

The angle of deflection for the spiral reflects the average degree of curvature along the spiral. As the degree of curvature is uniformly increased from 0 to D, the average degree of curvature for the spiral is $D/2$. Thus the angle of deflection for the spiral is:

$$\delta = \frac{L_s D}{200} \tag{3-14}$$

where: δ = spiral angle of deflection, degrees
L_s = length of the spiral, ft
D = degree of curve for the circular curve, degrees

Angle of Deflection (Central Angle) for Circular Portion of Curve with Spiral Easement, Δ_s

By definition (see Figure 3.10):

$$\Delta_s = \Delta - 2\delta \tag{3-15}$$

where: Δ_s = angle of deflection for circular curve with spiral, degrees or radians
Δ = angle of deflection for circular curve without spiral, degrees or radians
δ = angle of deflection for the spiral, degrees or radians

Length of Tangent Distance, T_s, between *P.I.* and *T.S.* (and *P.I.* and *S.T.*)

As shown on Figure 3.10, this is the distance between the point of intersection (*P.I.*) and the points at which the spiral curve transitions to or from the tangent. This distance is needed to station the curves appropriately. A very complicated derivation, the resulting equation is:

$$T_s = \left[R \operatorname{Tan}\left(\frac{\Delta}{2}\right)\right] + \left[\left(R \operatorname{Cos}(\delta) - R + \frac{L_s^2}{6R}\right) * \operatorname{Tan}\left(\frac{\Delta}{2}\right)\right] + [L_s - R \operatorname{Sin}(\delta)] \quad (3\text{-}16)$$

where: T_s = distance between *P.I.* and *T.S.* (also *P.I.* and *S.T.*), ft
R = radius of circular curve, ft
Δ = angle of deflection for circular curve without spiral, degrees or radians
δ = angle of deflection for spiral, degrees or radians

A Sample Problem

A 4° curve is to be designed on a highway with two 12-ft lanes and a design speed of 60 mi/h. A maximum superelevation rate of 6% has been established, and the appropriate side-friction factor for 60 mi/h is found from Figure 3.8 as 0.120. The normal drainage cross-slope on the tangent is 1%. Spiral transition curves are to be used. Determine the length of the spiral and the appropriate stations for the *T.S.*, *S.C.*, *C.S.*, and *S.T.* The angle of deflection for the original tangents is 38°, and the *P.I.* is at station 1,100 + 62. The segment has a two-lane cross section.

Solution: The radius of curvature for the circular portion of the curve is found from the degree of curvature as (Equation 3-1):

$$R = \frac{5{,}729.58}{D} = \frac{5{,}729.58}{4} = 1{,}432.4 \text{ ft}$$

The length of the spiral may now be computed using Equation 3-13:

$$L_s = 1.6\left(\frac{S^3}{R}\right) = 1.6\left(\frac{60^3}{1{,}432.4}\right) = 241.3 \text{ ft}$$

The minimum length of the spiral can also be determined as the length of the superelevation runoff. For a 60-mi/h design speed and a radius of 1,432.4 ft, the superelevation rate is found using Equation 3-9:

$$e = 100\left[\left(\frac{60^2}{15 * 1{,}432.4}\right) - 0.12\right] = 4.8\%$$

The length of the superelevation and tangent runoffs are computed from Equations 3-10 and 3-11, respectively. For 60 mi/h, the design value of Δ is 0.45 (Table 3.4). The adjustment factor for two lanes being rotated is 0.75. Then:

$$L_r = \frac{w * n * e_d * b_w}{\Delta} = \frac{12 * 2 * 4.8 * 0.75}{0.45} = 192 \text{ ft}$$

$$L_t = \frac{e_{NC}}{e_d} L_r = \left(\frac{1}{4.8}\right)192 = 40 \text{ ft}$$

The spiral must be at least as long as the superelevation runoff, or 192 ft. The result from Equation 3-13 is 241.3 ft, so this value controls. In fact, at 241.3 ft the minimum length of the spiral is sufficient to encompass both the superelevation runoff of 192 ft *and* the tangent runoff of 40 ft. Normally, the length of the spiral would be rounded, perhaps to 250 ft, which will be assumed for this problem.

The angle of deflection for the spiral is computed from Equation 3-14:

$$\delta = \frac{L_s D}{200} = \frac{250 * 4}{200} = 5°$$

The angle of deflection for the circular portion of the curve is (Equation 3-15):

$$\Delta_s = \Delta - 2\delta = 38 - 2(5) = 28°$$

The length of the circular portion of the curve, L_c, is found from Equation 3-6:

$$L_c = 100\left(\frac{\Delta_s}{D}\right) = 100\left(\frac{28}{4}\right) = 700 \text{ ft}$$

From Equation 3-16, the distance between the *P.I.* and the *T.S.* is:

$$T_s = \left[1{,}432.4 \operatorname{Tan}\left(\frac{38}{2}\right)\right] + \left[\left(1.432.4 \operatorname{Cos}(5) - 1{,}432.4 + \frac{250^2}{6 * 1{,}432.4}\right) * \operatorname{Tan}\left(\frac{38}{2}\right)\right] + [250 - (1{,}432.4 \operatorname{Sin}(5))] = 619.0 \text{ ft}$$

From these results, the curve may now be stationed:

$$T.S. = P.I. - T_s = 1{,}162 - 619.0 = 543$$
$$\text{or Sta } 500 + 43$$
$$S.C. = T.S. + L_s = 543 + 250 = 793$$
$$\text{or Sta } 700 + 93$$
$$C.S. = S.C. + L_c = 793 + 700 = 1{,}493$$
$$\text{or Sta } 1400 + 93$$
$$S.T. = C.S. + L_S = 1{,}493 + 250 = 1{,}743$$
$$\text{or Sta } 1700 + 43$$

3.3.6 Sight Distance on Horizontal Curves

One of the most fundamental design criteria for all highway facilities is that a minimum sight distance equal to the safe stopping distance must be provided at every point along the roadway.

On horizontal curves, sight distance is limited by roadside objects (on the inside of the curve) that block drivers' line of sight. Roadside objects such as buildings, trees, and natural barriers disrupt motorists' sight lines. Figure 3.11 illustrates a sight restriction on a horizontal curve.

Figure 3.12 illustrates the effect of horizontal curves on sight distance. Sight distance is measured along the arc of the roadway, using the centerline of the inside travel lane. The middle ordinate, M, is taken as the distance from the centerline of the inside lane to the nearest roadside sight blockage.

Figure 3.11: A Sight Distance Restriction on a Horizontal Curve

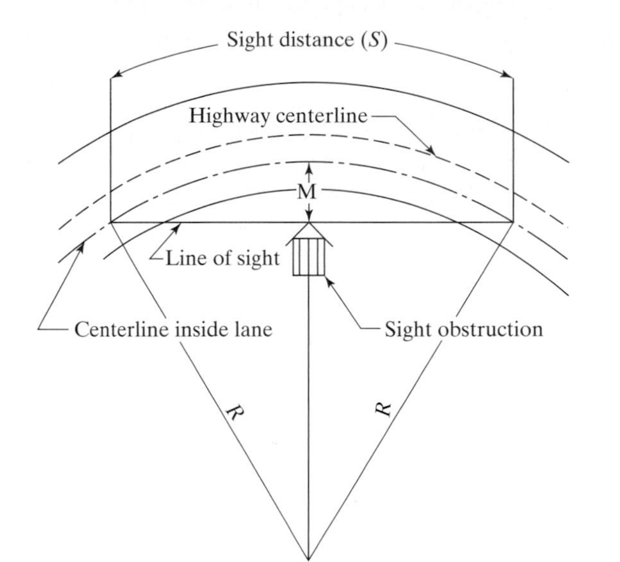

Figure 3.12: Sight Restrictions on Horizontal Curves
(*Source:* Used with permission of American Association of State Highway and Transportation Officials, *A Policy on Geometric Design of Highways and Streets,* 4th Edition, Exhibit 3-58, p. 231, Washington DC, 2001.)

The formula for the middle ordinate was given previously as:

$$M = R\left[1 - \text{Cos}\left(\frac{\Delta}{2}\right)\right]$$

The length of the circular curve has also been defined previously. In this case, however, the length of the curve is set equal to the required stopping sight distance. Then:

$$L = d_s = 100\left(\frac{\Delta}{D}\right)$$

$$\Delta = \frac{d_s D}{100}$$

Substituting in the equation for M:

$$M = R\left[1 - \text{Cos}\left(\frac{d_s D}{200}\right)\right]$$

The equation can be expressed uniformly using either the degree of curvature, D, or the radius of curvature, R:

$$M = \frac{5{,}729.58}{D}\left[1 - \text{Cos}\left(\frac{d_s D}{200}\right)\right]$$

$$M = R\left[1 - \text{Cos}\left(\frac{28.65\, d_s}{R}\right)\right] \qquad (3\text{-}17)$$

Remember (see Chapter 2) that the safe stopping distance used in each of these equations may be computed as:

$$d_s = 1.47\,S\,t + \frac{S^2}{30(0.348 \pm 0.01G)}$$

where: d_s = safe stopping distance, ft
S = design speed, mi/h
t = reaction time, s
G = grade, %

A Sample Problem

A 6° curve (measured at the centerline of the inside lane) is being designed for a highway with a design speed of 70 mi/h. The grade is level, and driver reaction time will be taken as 2.5 s, the AASHTO standard for highway braking reaction. What is the closest any roadside object may be placed to the centerline of the inside lane of the roadway?

Solution: The safe stopping distance, d_s, is computed as:

$$d_s = 1.47(70)(2.5) + \frac{70^2}{30(0.348 + 0.01 * 0)}$$
$$= 257.3 + 469.3 = 726.6 \text{ ft}$$

The minimum clearance at the roadside is given by the middle ordinate for a sight distance of 726.6 ft:

$$M = \frac{5{,}729.58}{6}\left[1 - \text{Cos}\left(\frac{726.6 * 6}{200}\right)\right] = 68.3 \text{ ft}$$

Thus, for this curve, no objects or other sight blockages on the inside roadside may be closer than 68.3 ft to the centerline of the inside lane.

3.3.7 Compound Horizontal Curves

A compound horizontal curve consists of two or more consecutive horizontal curves in a single direction with different radii. Figure 3.13 illustrates such a curve.

Some general criteria for such curves include:

- Use of compound curves should be limited to cases in which physical conditions require it.
- Whenever two consecutive curves are connected on a highway segment, the larger radii should not be more than 1.5 times the smaller. A similar criteria is that the degrees of curvature should not differ by more than 5°.

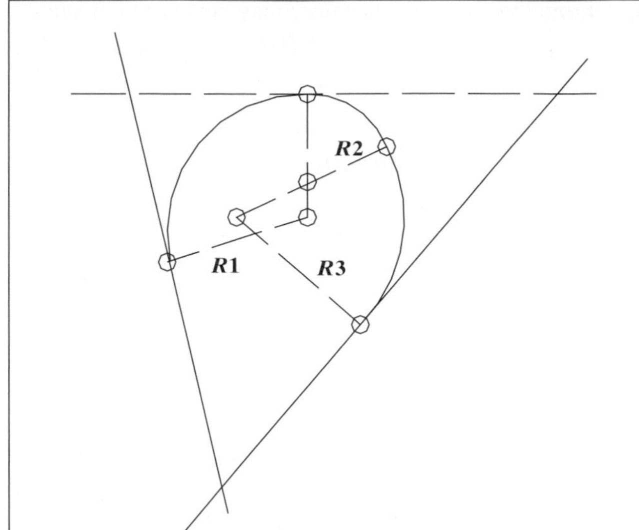

Figure 3.13: Compound Horizontal Curve Illustrated

- Whenever two consecutive curves in the same direction are separated by a short tangent (< 200 ft), they should be combined in a compound curve.
- A compound curve is merely a series of simple horizontal curves subject to the same criteria as isolated horizontal curves.
- AASHTO relaxes some of these criteria for compound curves for ramp design.

3.3.8 Reverse Horizontal Curves

A reverse curve consists of two consecutive horizontal curves in opposite directions. Such a curve is illustrated in Figure 3.14. Two horizontal curves in opposite directions should always be

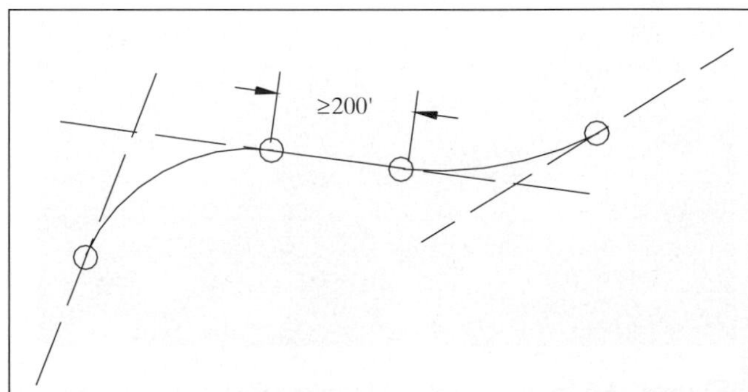

Figure 3.14: A Reverse Horizontal Curve Illustrated

separated by a tangent of at least 200 ft. Use of spiral transition curves is a significant assist to drivers negotiating reverse curves.

3.4 Vertical Alignment of Highways

The vertical alignment of a highway is the *profile* design of the facility in the vertical plane. The vertical alignment is composed of a series of vertical tangents connected by vertical curves. Vertical curves are in the shape of a *parabola*. This provides for a natural transition from a tangent to a curved section as part of the curve characteristics. Therefore, there is no need to investigate or provide transition curves, such as the spiral for horizontal curves.

The longitudinal slope of a highway is called the *grade*. It is generally stated as a percentage.

In vertical design, attempts are made to conform to the topography wherever possible to reduce the need for costly excavations and landfills as well as to maintain aesthetics. Primary design criteria for vertical curves include:

- Provision of adequate sight distance at all points along the profile
- Provision of adequate drainage
- Maintenance of comfortable operations
- Maintenance of reasonable aesthetics

3.4.1 Grades

Vertical tangents are characterized by their longitudinal slope, or grade. When expressed as a percentage, the grade indicates the relative rise (or fall) of the facility in the longitudinal direction as a percentage of the length of the section under study. Thus a 4% grade of 2,000 ft involves a vertical rise of 2,000 * (4/100) = 80 ft. Upgrades have positive slopes and percent grades, whereas downgrades have negative slopes and percent grades.

Maximum recommended grades for use in design depend on the type of facility, the terrain in which it is built, and the design speed. Figure 3.15 presents a general overview of usual practice. These criteria represent a balance between the operating comfort of motorists and passengers and the practical constraints of design and construction in more severe terrains.

The principal operational impact of a grade is that trucks will be forced to slow down as they progress up the grade. This creates gaps in the traffic stream that cannot be effectively filled by simple passing maneuvers. Figure 3.16 illustrates the effect of upgrades on the operation of trucks with a weight-to-horsepower ratio of 200 lbs/hp, which is considered to be operationally typical of heavy trucks on most highways. It depicts deceleration behavior with an assumed entry speed of 70 mi/h.

Because of the operation of trucks on grades, simple maximum grade criteria are not sufficient for design. An example is shown in Figure 3.17. Trucks entering an upgrade with an assumed speed of 70 mi/h begin to slow. The length of the upgrade determines the extent of deceleration. For example, a truck entering a 5% upgrade at 70 mi/h slows to 50 mi/h after 2,000 ft and 33 mi/h after 4,000 ft. Eventually, the truck reaches its *crawl speed*, that constant speed the truck can maintain for any length of grade (of the given steepness). Using the same example, a truck on a 5% upgrade has a crawl speed of 27 mi/h that is reached after approximately 7,800 ft.

Thus the interference of trucks with general highway operations is related not only to the steepness of the grade but to its length as well. For most design purposes, grades should not be longer than the "critical length." For grades entered at 70 mi/h, the *critical length* is generally defined as the length at which the speed of trucks is 15 mi/h less than their speed upon entering the grade. When trucks enter an upgrade from slower speed, a speed reduction of 10 mi/h may be used to define the critical length of grade.

Figure 3.18 shows the relationship between length of grade, percent grade, and speed reduction for 200 lb/hp trucks entering a grade at 70 mi/h. These curves can be used to determine critical length of grade. Note that terrain may make it impossible to limit grades to the critical length or shorter.

A Sample Problem

A rural freeway in rolling terrain has a design speed of 60 mi/h. What is the longest and steepest grade that should be included on the facility?

Solution: From Figure 3.15 (a), for a freeway facility with a design speed of 60 mi/h in rolling terrain, the maximum allowable grade is 4%. Entering Figure 3.18 with 4% on the vertical axis, moving to the "15 mi/h" curve, the critical length of grade is seen to be approximately 1,900 ft.

Again, it must be emphasized that terrain sometimes makes it impossible to follow maximum grade design criteria consistently. This is particularly true for desirable maximum grade lengths. Where the terrain is rising for significant distances, the profile of the roadway must do so as well. It is, however, true that grades longer than the critical length will generally operate poorly, and the addition of a climbing lane may be warranted in such situations.

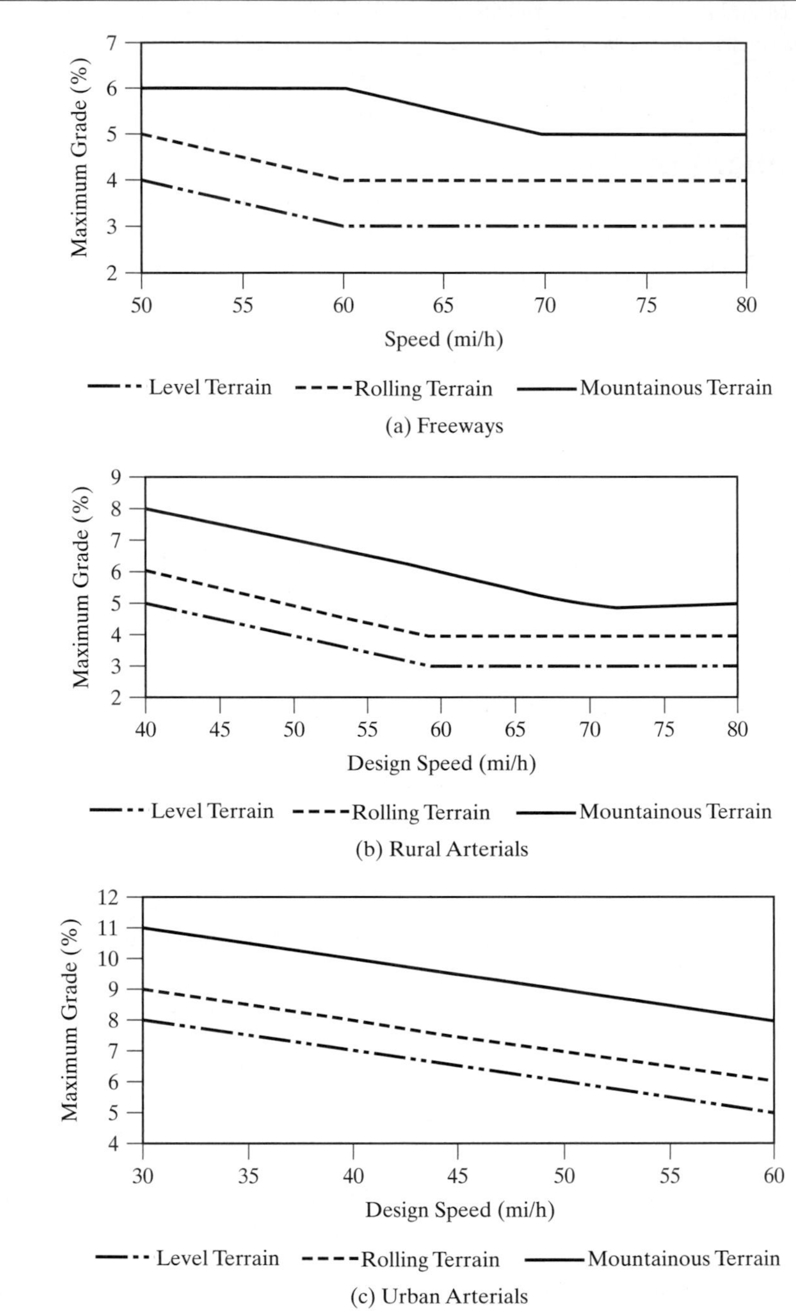

Figure 3.15: General Criteria for Maximum Grades

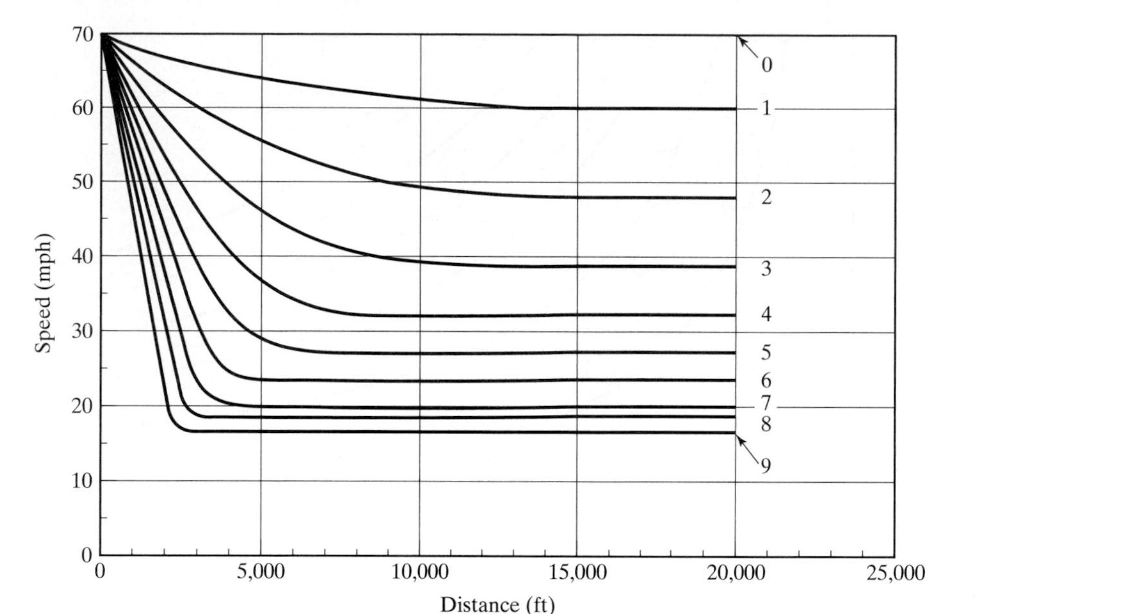

Figure 3.16: Deceleration of Typical Trucks (200 lbs/hp) on Upgrades

(*Source:* Used with permission of American Association of State Highway and Transportation Officials, *A Policy on Geometric Design of Highways and Streets*, 4th Edition, Exhibit 3-59, p. 234, Washington DC, 2004.)

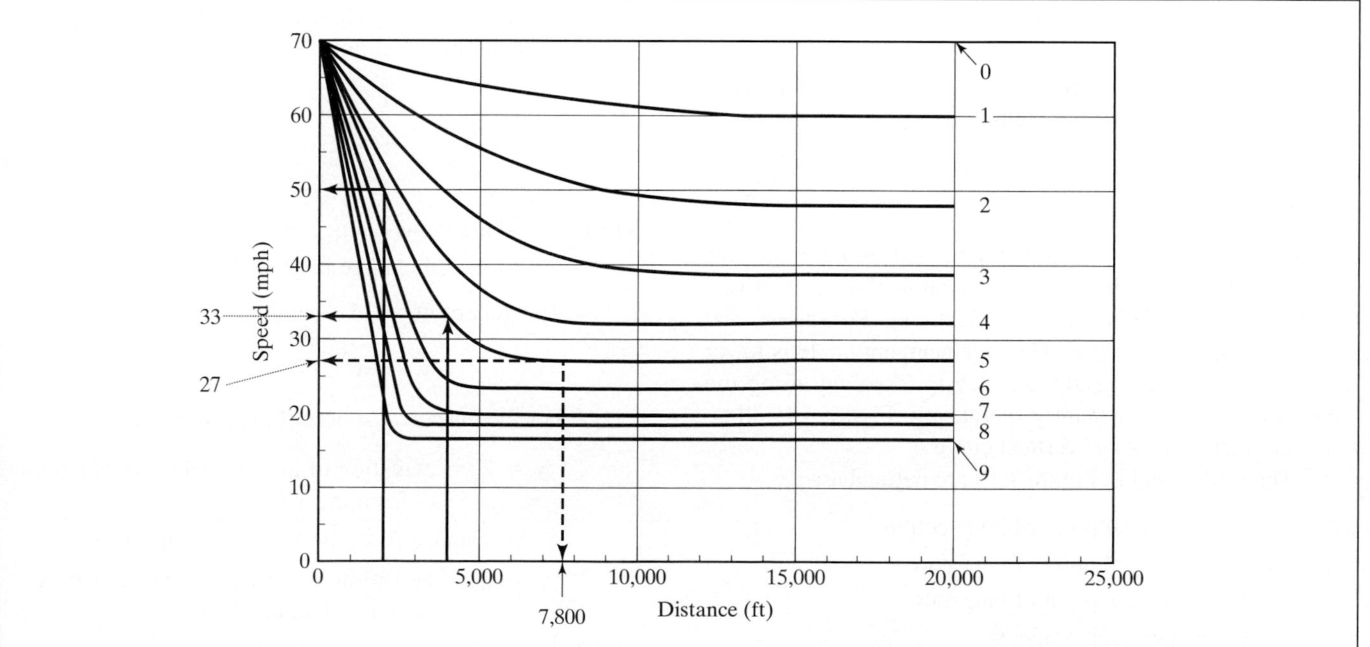

Figure 3.17: An Example of Truck Behavior on an Upgrade

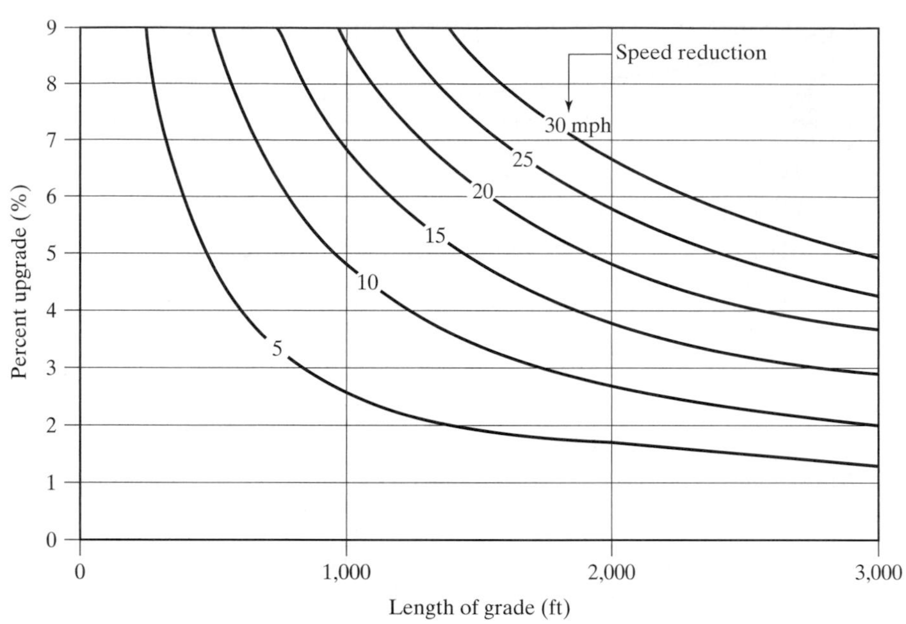

Figure 3.18: Critical Lengths of Grade for a Typical Truck (200 lbs/hp)

(*Source:* Used with permission of American Association of State Highway and Transportation Officials, *A Policy on Geometric Design of Highways and Streets*, 4th Edition, Exhibit 3-63, p. 245, Washington DC, 2001.)

3.4.2 Geometric Characteristics of Vertical Curves

As noted previously, vertical curves are in the shape of a parabola. In general, there are two types of vertical curves:

- Crest vertical curves
- Sag vertical curves

For crest vertical curves, the entry tangent grade is greater than the exit tangent grade. While traveling along a crest vertical curve, the grade is constantly declining. For sag vertical curves, the opposite is true: The entry tangent grade is lower than the exit tangent grade, and while traveling along the curve, the grade is constantly increasing. Figure 3.19 illustrates the various types of vertical curves.

The terms used in Figure 3.19 are defined as:

V.P.I. = vertical point of intersection
V.P.C. = vertical point of curvature
V.P.T. = vertical point of tangency
G_1 = approach grade, %
G_2 = departure grade, %
L = length of vertical curve, in *hundreds* of ft

Length, and all stationing, on a vertical curve is measured in the plan view (i.e., along a level axis). Two other useful variables are defined as follows:

$$A = G_2 - G_1 \tag{3-18}$$

$$r = \frac{G_2 - G_1}{L} \tag{3-19}$$

where: A = algebraic change in grade, percent
r = rate of change in grade per 100 ft

The general form of a parabola is:

$$y = ax^2 + bx + c$$

For the purposes of describing a vertical curve, let:

$y = Y_x$ = elevation of the vertical curve at a point "x" from the V.P.C.
x = distance from the V.P.C in hundreds of ft
$c = Y_0$ = elevation of the *V.P.C.*, which occurs where $x = 0$ hundreds of ft

Then:

$$Y_x = ax^2 + bx + Y_0$$

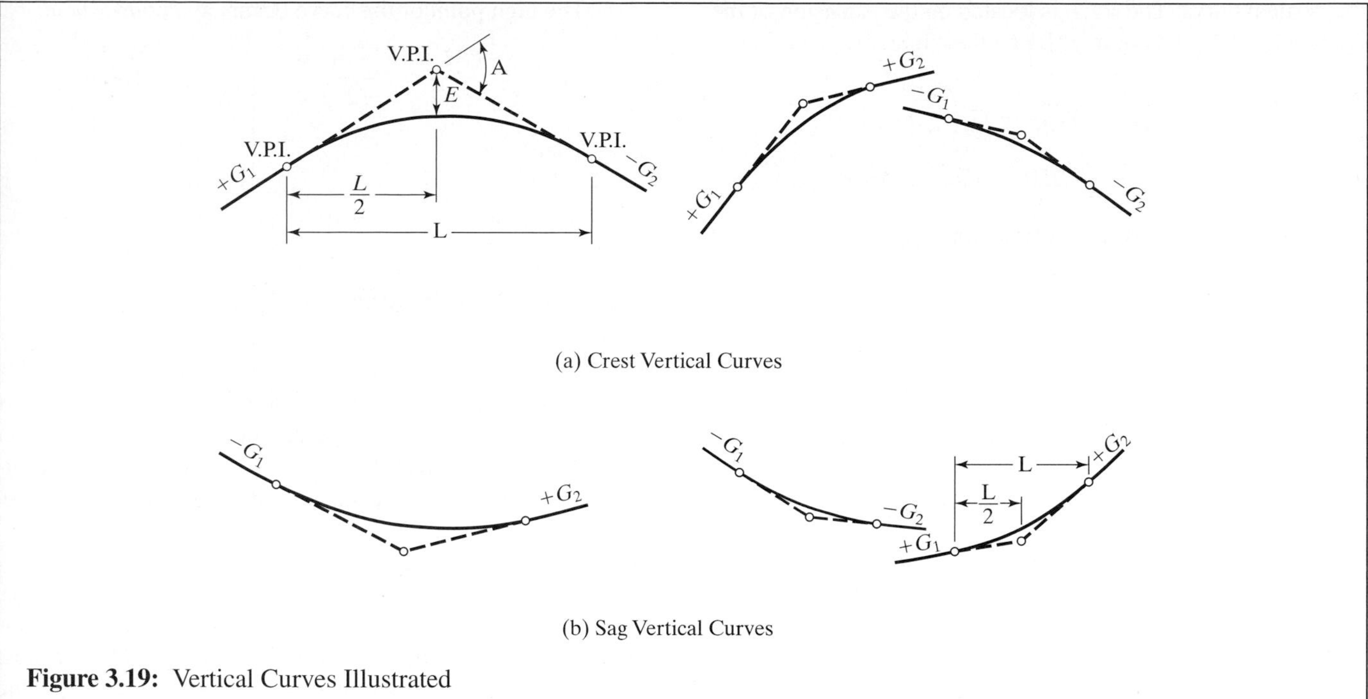

Figure 3.19: Vertical Curves Illustrated

Also, consider that the slope of the curve at any point x is the first derivative of this equation, or:

$$\frac{dY}{dx} = 2ax + b$$

When $x = 0$, the slope is equal to the entry grade, G_1. Thus:

$$\frac{dY}{dx} = G_1 = 2a(0) + b$$

$$b = G_1$$

The second derivative of the equation is equal to the rate of change in slope along the grade, or:

$$\frac{d^2Y}{dx^2} = 2a = r = \frac{G_2 - G_1}{L}$$

$$a = \frac{G_2 - G_1}{2L}$$

Thus, the final form of the equation for a vertical curve is given as:

$$Y_x = \left(\frac{G_2 - G_1}{2L}\right)x^2 + G_1x + Y_0 \qquad \text{(3-20)}$$

The location of the high point (on a crest vertical curve) or the low point (on a sag vertical curve) is at a point where the slope (or first derivative) is equal to "zero." Taking the first derivative of the final curve:

$$\frac{dY_x}{dx} = 0 = \left(\frac{G_2 - G_1}{L}\right)x + G_1$$

$$x = \frac{-G_1L}{G_2 - G_1} \qquad \text{(3-21)}$$

In all of these equations, care must be taken to address the sign of the grade. A negative grade has a minus $(-)$ sign that must be accounted for in the equation. Double negatives become positives in the equation. If both grades are negative (downgrades), the low point on the curve is the V.P.T. If both grades are positive (upgrades), the low point on the curve is the V.P.C.

A Sample Problem

A vertical curve of 600 ft (L = 6) connects a +4% grade to a −2% grade. The elevation of the *V.P.C.* is 1,250 ft. Find the elevation of the *P.V.I.*, the high point on the curve, and the *V.P.T.*

Solution: The elevation of the *V.P.I.* is found from the elevation of the *V.P.C.*, the approach grade, and the length of the vertical curve. The *V.P.I.* is located on the extension of the approach grade at a point $\frac{1}{2}L$ into the curve, or:

$$Y_{VPI} = Y_{VPC} + G_1\left(\frac{L}{2}\right)$$
$$= 1{,}250 + 12 = 1{,}262 \text{ ft}$$

Following the format of Equation 3-20, the equation for this particular vertical curve is:

$$Y_x = \left(\frac{-2-4}{2*6}\right)x^2 + 4x + 1{,}250$$

$$Y_x = -\left(\frac{1}{2}\right)x^2 + 4x + 1{,}250$$

The elevation of the V.P.T. is the elevation of the curve at the end of its length of 600 ft, or where $x = 6$:

$$Y_{VPT} = -\left(\frac{1}{2}\right)6^2 + 4(6) + 1{,}250$$

$$Y_{VPT} = -18 + 24 + 1{,}250 = 1{,}256 \text{ ft}$$

The high point of the curve occurs at a point where:

$$x = \frac{-G_1L}{G_2 - G_1} = \frac{-4(6)}{-2-4} = \frac{-24}{-6} = 4(100 \text{ ft})$$

Then:

$$Y_{HIGH} = -\left(\frac{1}{2}\right)4^2 + 4(4) + 1{,}250 = 1{,}258 \text{ ft}$$

3.4.3 Sight Distance on Vertical Curves

The minimum length of vertical curve is governed by sight-distance considerations. On vertical curves, sight distance is measured from an assumed eye height of 3.5 ft and an object height of 2.0 ft (AASHTO standards). Figure 3.20 shows a situation in which sight distance is limited by vertical curvature.

Crest Vertical Curves

For crest vertical curves, the daylight sight line controls minimum length of vertical curves. The minimum length of a crest vertical curve is given by Equation 3-22 for cases in which the

Figure 3.20: Sight Restriction Due to Vertical Curvature

stopping sight distance is less than the length of the curve ($d_s < L$) or Equation 3-23 for cases in which the stopping sight distance is greater than the length of the vertical curve ($d_s > L$). The two equations yield equal results when $d_s = L$.

$$L = \frac{|G_2 - G_1| * d_s^2}{2{,}158} \quad \text{for } d_s < L \tag{3-22}$$

$$L = 2d_s - \left(\frac{2{,}158}{|G_2 - G_1|}\right) \quad \text{for } d_s > L \tag{3-23}$$

Sag Vertical Curves

For sag vertical curves, the sight distance is limited by the headlamp range during nighttime driving conditions. Again, two equations result. Equation 3-24 is used when $d_s < L$, and Equation 3-25 is used when $d_s > L$. Once again, both equations yield the same results when $d_s = L$.

$$L = \frac{|G_2 - G_1| * d_s^2}{400 + 3.5\, d_s} \quad \text{for } d_s < L \tag{3-24}$$

$$L = 2\, d_s - \left(\frac{400 + 3.5\, d_s}{|G_2 - G_1|}\right) \quad \text{for } d_s > L \tag{3-25}$$

where: L = minimum length of vertical curve, ft
d_s = required stopping sight distance, ft
G_2 = departure grade, %
G_1 = approach grade, %

A Sample Problem

What is the minimum length of vertical curve that must be provided to connect a 5% grade with a 2% grade on a highway with a design speed of 60 mi/h? Driver reaction time is the AASHTO standard of 2.5 s for simple highway stopping reactions.

Solution: This vertical curve is a *crest* vertical curve because the departure grade is less than the approach grade. The safe stopping distance is computed assuming that the vehicle is on a 2% upgrade. This results in a worst-case stopping distance:

$$d_s = 1.47(60)(2.5) + \frac{60^2}{30(0.348 + 0.01 * 2)}$$
$$= 220.5 + 326.1 = 546.6 \text{ ft}$$

Rounding off this number, a stopping sight distance requirement of 547 ft will be used. The first computation is made assuming that the stopping sight distance is less than the resulting length of curve. Using Equation 3-22 for this case:

$$L = \frac{|2 - 5|547^2}{2{,}158} = 416.0 \text{ ft}$$

From this result, it is clear that the initial assumption that $d_s < L$ was not correct. Equation 3-23 is now used:

$$L = 2(547) - \left(\frac{2{,}158}{|2 - 5|}\right) = 1{,}804 - 719.3$$
$$= 347.7 \text{ ft}$$

In this case, $d_s > L$, as assumed, and the 375 ft (rounded) is taken as the result.

3.4.4 Other Minimum Controls on Length of Vertical Curves

There are two other controls on the minimum length of *sag vertical curves only*. For driver comfort, the minimum length of vertical curve is given by:

$$L = \frac{|G_2 - G_1|S^2}{46.5} \tag{3-26}$$

For general appearance, the minimum length of a sag vertical curve is given by:

$$L = 100|G_2 - G_1| \tag{3-27}$$

where: S = design speed, mi/h
all other variables as previously defined

Neither of these controls would enter into the example done previously because it involved a crest vertical curve.

Equations 3-22 through 3-25 for minimum length of vertical curve are based on stopping sight distances only. Consult AASHTO standards [*1*] directly for similar criteria based on passing sight distance (for two-lane roadways) and for sag curves interrupted by overpass structures that block headlamp paths for night vision.

3.4.5 Some Design Guidelines for Vertical Curves

AASHTO gives a number of commonsense guidelines for the design of highway profiles, which are summarized here:

1. A smooth grade line with gradual changes is preferred to a line with numerous breaks and short grades.
2. Profiles should avoid the "roller-coaster" appearance, as well as "hidden dips" in the alignment.

3. Undulating grade lines involving substantial lengths of momentum (down) grades should be carefully evaluated with respect to operation of trucks.
4. Broken-back grade lines (two consecutive vertical curves in the same direction separated by a short tangent section) should be avoided wherever possible.
5. On long grades, it may be preferable to place the steepest grades at the bottom, lightening the grade on the ascent. If this is difficult, short sections of lighter grades should be inserted periodically to aid operations.
6. Where at-grade intersections occur on roadway sections with moderate to steep grades, the grade should be reduced or flattened through the intersection area.
7. Sag vertical curves in cuts should be avoided unless adequate drainage is provided.

3.5 Cross-Section Elements of Highways

The cross section of a highway includes a number of elements critical to the design of the facility. The cross-section view of a highway is a 90° cut across the facility from roadside to roadside. The cross section includes the following features:

- Travel lanes
- Shoulders
- Side slopes
- Curbs
- Medians and median barriers
- Guardrails
- Drainage channels

General design practice is to specify the cross section at each station (i.e., at points 100 ft apart and at intermediate points where a change in the cross-sectional design occurs). The important cross-sectional features are briefly discussed in the sections that follow.

3.5.1 Travel Lanes and Pavement

Paved travel lanes provide the space that moving (and sometimes parked) vehicles occupy during normal operations. The standard width of a travel lane is 12 ft (metric standard is 3.6 m), although narrower lanes are permitted when necessary. The minimum recommended lane width is 9 ft. Lanes wider than 12 ft are sometimes provided on curves to account for the off-tracking of the rear wheels of large trucks. Narrow lanes have a negative impact on the capacity of the roadway and on operations [7]. In general, 9-ft and 10-ft lanes should be avoided wherever possible. Nine-foot (9-ft) lanes are acceptable only on low-volume, low-speed rural or residential roadways, and 10-ft lanes are acceptable only on low-speed facilities.

All pavements have a cross-slope that is provided (1) to provide adequate drainage, and (2) to provide superelevation on curves. For high-type pavements (portland cement concrete, asphaltic concrete), normal drainage cross-slopes range from 1.5% to 2.0%. On low-type pavements (penetration surfaces, compacted earth, etc.), the range of drainage cross-slopes is between 2% and 6%.

How the drainage cross-slope is developed depends on the type of highway and the design of other drainage facilities. A pavement can be drained to *both* sides of the roadway or to one side. Where water is drained to both sides of the pavement, there must be drainage ditches or culverts and pipes on both sides of the pavement. In some cases, water drained to the roadside is simply absorbed into the earth; studies testing whether the soil is adequate to handle maximum expected water loads must be conducted before adopting this approach. Where more than one lane is drained to one side of the roadway, each successive lane should have a cross-slope that is 0.5% steeper than the previous lane. Figure 3.21 illustrates a typical cross-slope for a four-lane pavement.

On superelevated sections, cross-slopes are usually sufficient for drainage purposes, and a slope differential between adjacent lanes is not needed. Superelevated

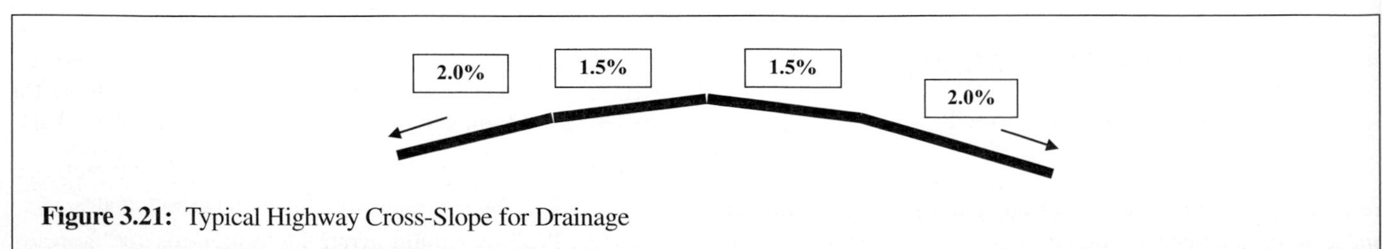

Figure 3.21: Typical Highway Cross-Slope for Drainage

sections, of course, must drain to the inside of the horizontal curve, and the design of drainage facilities must accommodate this.

3.5.2 Shoulders

AASHTO [1]defines shoulders in the following way: "A shoulder is the portion of the roadway contiguous with the traveled way that accommodates stopped vehicles, emergency use, and lateral support of sub-base, base, and surface courses (of the roadway structure)" (p. 316). Shoulders vary widely in both size and physical appearance. For some low-volume rural roads in difficult terrain, no shoulders are provided. Normally, the shoulder width ranges from 2 ft to 12 ft. Most shoulders are "stabilized" (i.e., treated with some kind of material that provides a reasonable surface for vehicles). This can range from a fully-paved shoulder to shoulders stabilized with penetration or stone surfaces or simply grass over compacted earth. For safety, it is critical that the joint between the traveled way and the shoulder be well maintained.

Shoulders are generally considered necessary on rural highways serving a significant mobility function, on all freeways, and on some types of urban highways. In these cases, a minimum width of 10 ft is generally used because this provides for stopped vehicles to be about 2 ft clear of the traveled way. The narrowest 2-ft shoulders should be used only for the lowest classifications of highways. Even in these cases, 6 to 8 ft is considered desirable.

Shoulders serve a variety of functions, including:

- Providing a refuge for stalled or temporarily stopped vehicles
- Providing a buffer for accident recovery
- Contributing to driving ease and driver confidence
- Increasing sight distance on horizontal curves
- Improving capacity and operations on most highways
- Providing space for maintenance operations and equipment
- Providing space for snow removal and storage
- Providing lateral clearance for signs, guardrails, and other roadside objects
- Improving drainage on a traveled way
- Providing structural support for the roadbed

Reference 8 provides an excellent study of the use of roadway shoulders.

Table 3.5: Recommended Cross-Slopes for Shoulders

Type of Surface	Recommended Cross-Slope (%)
Bituminous	2.0–6.0
Gravel or stone	4.0–6.0
Turf	6.0–8.0

Table 3.5 shows recommended cross-slopes for shoulders, based on the type of surface. No shoulder should have a cross-slope of more than 7:1 because the probability of rollover is greatly increased for vehicles entering a more steeply sloped shoulder.

3.5.3 Side-Slopes for Cuts and Embankments

Where roadways are located in cut sections or on embankments, side-slopes must be carefully designed to provide for safe operation. In urban areas, sufficient right-of-way is generally not available to provide for natural side-slopes, and retaining walls are frequently used.

Where natural side-slopes are provided, the following limitations must be considered:

- A 3:1 side-slope is the maximum for safe operation of maintenance and mowing equipment.
- A 4:1 side-slope is the maximum desirable for accident safety. Barriers should be used to prevent vehicles from entering a side-slope area with a steeper slope.
- A 2:1 side-slope is the maximum on which grass can be grown, and only then in good climates.
- A 6:1 side-slope is the maximum that is structurally stable for where sandy soils are predominate.

Table 3.6 shows recommended side-slopes for various terrains and heights of cut and/or fill.

3.5.4 Guardrail

One of the most important features of any cross-section design is the use and placement of guardrail. "Guardrail" is intended to prevent vehicles from entering a dangerous area of the roadside or median during an accident or intended action.

Roadside guardrail is provided to prevent vehicles from entering a cross-slope steeper than 4:1 or from colliding with roadside objects such as trees, culverts, lighting

Table 3.6: Recommended Side-Slopes for Cut and Fill Sections

Height of Cut or Fill (ft)	Terrain		
	Level or Rolling	Moderately Steep	Steep
0–4	6:1	4:1	4:1
4–10	4:1	3:1	2:1
10–15	3:1	2.5:1	1.75:1*
15–20	2:1	2:1	1.5:1*
>20	2:1	1.5:1*	1.5:1*

*Avoid where soils are subject to erosion.

standards, sign posts, and so on. Once a vehicle hits a section of guardrail, the physical design also guides the vehicle into a safer trajectory, usually in the direction of traffic flow.

Median guardrail is primarily provided to prevent vehicles from encroaching into the opposing lane(s) of traffic. It also prevents vehicles from colliding with median objects. The need for median guardrail depends on the design of the median itself. If the median is 20 ft or wider and if there are no dangerous objects in the median, guardrail is usually not provided, and the median is not curbed. Wide medians can effectively serve as accident recovery areas for encroaching drivers.

Narrower medians generally require some type of barrier because the potential for encroaching vehicles to cross the entire median and enter the opposing traffic lanes is significant.

Figure 3.22 illustrates the common types of guardrail in current use. The barriers shown in Figure 3.22 are *all configured for median use* (i.e., they are designed to protect encroachment from either side of the barrier). For roadside use, the same designs are used with only one collision surface.

The major differences in the various designs are the flexibility of guardrail on impact and the strength of the barrier in preventing a vehicle from crossing through the barrier.

The box-beam design, for example, is quite flexible. Upon collision, several posts of the box-beam will give way, allowing the beam to flex as much as 10 to 15 ft. The colliding vehicle is gently straightened and guided back toward the travel lane over a length of the guardrail. Obviously, this type of guardrail is not useful in narrow medians because it could well deflect into the opposing traffic lanes.

The most inflexible design is the concrete median or roadside barrier. These blocks are almost immovable, and it is virtually impossible to crash through them. Thus they are used in narrow roadway medians (particularly on urban freeways) and on roadsides where virtually no deflection would be safe. On collision with such a barrier, the vehicle is straightened out almost immediately, and the friction of the vehicle against the barrier brings it to a stop.

The details of guardrail design are critical. End treatments must be carefully done. A vehicle colliding with a blunt end of a guardrail section is in extreme danger. Thus most W-beam and box-beam guardrails are bent away from the traveled way, with their ends buried in the roadside. Even with this done, vehicles can (with some difficulty) hit the buried end and "ramp up" the guardrail with one or more wheels. Concrete barriers have sloped ends but are usually protected by impact-attenuating devices, such as sand or water barrels or mechanical attenuators.

Connection of guardrail to bridge railings and abutments is also important. Because most guardrails deflect, they cannot be isolated from fixed objects as they could conceivably "guide" a vehicle into a dangerous collision with such an object. Thus where guardrails meet bridge railings or abutments, they are anchored onto the railing or abutment itself to ensure that encroaching vehicles are guided away from the object.

3.6 Closing Comments

This chapter has provided a brief overview of the critical functional and geometric characteristics of highways. Many more details are involved in highway geometry than those illustrated here. Consult the current AASHTO standard, *A Policy on Geometric Design of Highways and Streets,* directly for a more detailed presentation of specific design practices and policies.

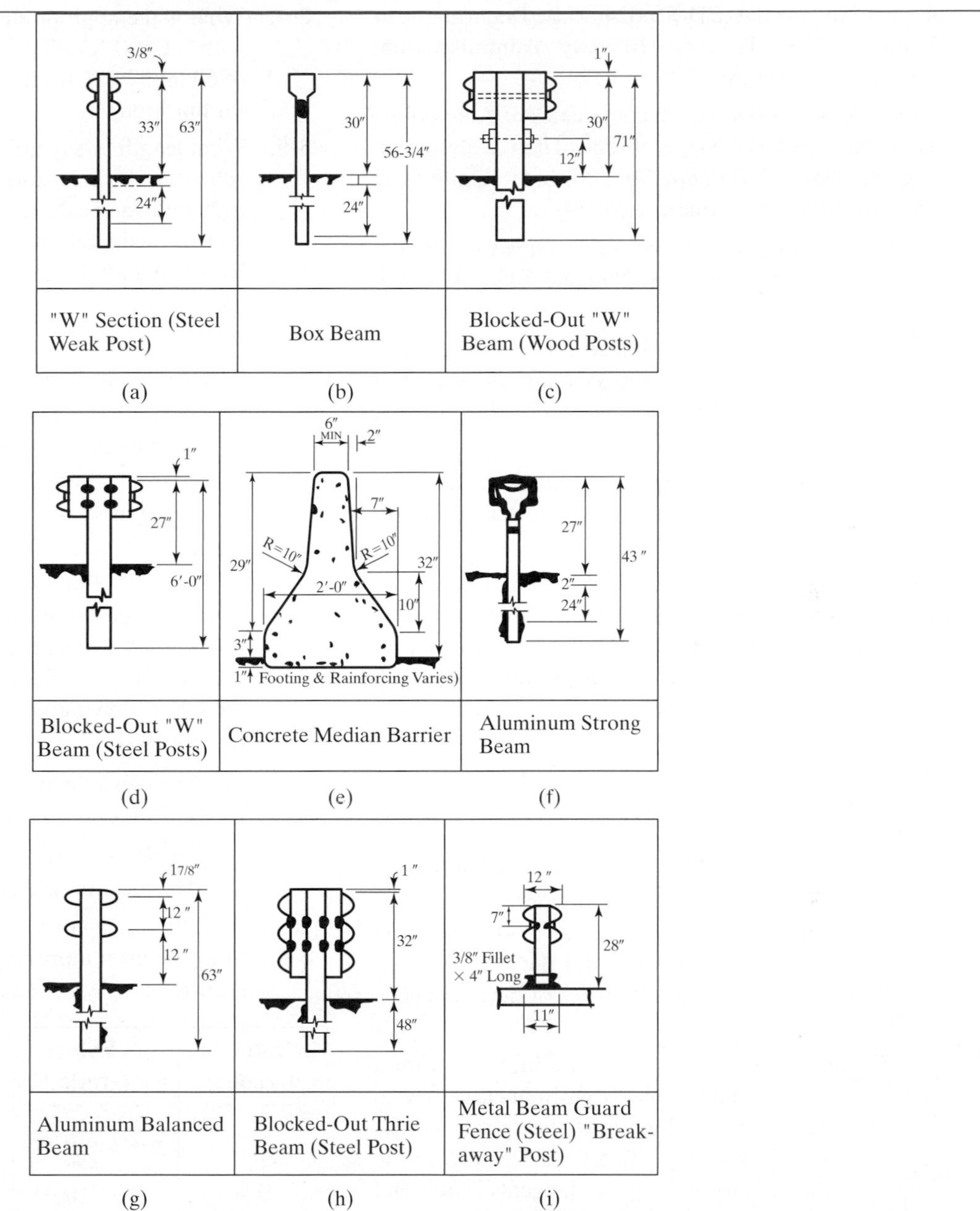

Figure 3.22: Common Types of Median and Roadside Barriers

(*Source:* Used with permission of American Association of State Highway and Transportation Officials, *A Policy on Geometric Design of Highways and Streets,* 2nd Edition, Figure IV-8, p. 398, Washington DC, 1984.)

References

1. *A Policy on Geometric Design of Highways and Streets*, 4th Edition, American Association of State Highway and Transportation Officials, Washington DC, 2004.
2. Kavanagh, B.F., *Surveying with Construction Applications*, 4th Edition, Prentice Hall, Upper Saddle River, NJ, 2001.
3. MacAdam, C.C., Fancher, P.S., and Segal, L., *Side Friction for Superelevation on Horizontal Curves*,

Report No. FHWA-RD-86-024, U.S. Department of Transportation, Federal Highway Administration, McLean, VA, August 1985.

4. Moyer, R.A., "Skidding Characteristics of Automobile Tires on Roadway Surfaces and Their Relation to Highway Safety," *Bulletin No. 120*, Iowa Engineering Experiment Station, Ames, IA, 1934.

5. Hajela, G.P., *Compiler, Résumé of Tests on Passenger Cars on Winter Driving Surfaces*, National Safety Council, Chicago, IL, 1968.

6. Shortt, W.H., "A Practical Method for Improvement of Existing Railroad Curves," *Proceedings: Institution of Civil Engineering*, Vol 76, Institution of Civil Engineering, London ENG, 1909.

7. *Highway Capacity Manual*, 4th Edition, National Research Council, Transportation Research Board, Washington DC, 2000.

8. Zegeer, C.V., Stewart, R., Council, F.M., and Neuman, T.R., "Roadway Widths for Low-Traffic Volume Roads," *National Cooperative Highway Research Report 362,* National Research Council, Transportation Research Board, Washington DC, 1994.

Problems

3-1. The point of intersection (*P.I.*) of two tangent lines is Station 11,500 + 66. The radius of curvature is 1,000 feet, and the angle of deflection is 60°. Find the length of the curve, the stations for the *P.C.* and *P.T.*, and all other relevant characteristics of the curve (*LC*, *M*, *E*).

3-2. A 3.5° curve is to be designed on a highway with a design speed of 60 mi/h. Spiral transition curves are to be used. Determine the length of the spiral and the appropriate stations for the *T.S.*, *S.C.*, *C.S.*, and *S.T.* The angle of deflection for the original tangents is 40°, and the *P.I.* is at station 15,100 + 26. The segment consists of two 12-ft lanes.

3-3. A 5° curve (measured at the centerline of the inside lane) is being designed for a highway with a design speed of 65 mi/h. The curve is on a 2% upgrade, and driver reaction time may be taken as 2.5 seconds. What is the closest any roadside object may be placed to the centerline of the inside lane of the roadway while maintaining adequate stopping sight distance?

3-4. What is the appropriate superelevation rate for a curve with a 1,200-ft radius on highway with a design speed of 60 mi/h? The maximum design superelevation is 6% for this highway.

3-5. What length of superelevation runoff should be used to achieve a superelevation rate of 10%? The design speed is 70 mi/h, and a three-lane cross section (12-ft lanes) is under consideration. Superelevation will be achieved by rotating all three lanes around the inside edge of the pavement.

3-6. Find the maximum allowable grade and critical length of grade for each of the following facilities:

(a) A rural freeway in mountainous terrain with a design speed of 60 mi/h

(b) A rural arterial in rolling terrain with a design speed of 45 mi/h

(c) An urban arterial in level terrain with a design speed of 40 mi/h

3-7. A vertical curve of 1,000 ft is designed to connect a grade of +4% to a grade of −5%. The *V.P.I.* is located at station 1,500 + 55 and has a known elevation of 500 ft. Find the following:

(a) The station of the *V.P.C.* and the *V.P.T.*

(b) The elevation of the *V.P.C.* and the *V.P.T.*

(c) The elevation of points along the vertical curve at 100-ft intervals

(d) The location and elevation of the high point on the curve

3-8. Find the minimum length of curve for the following scenarios:

Entry Grade	Exit Grade	Design Speed	Reaction Time
3%	8%	45 mi/h	2.5 s
−4%	2%	65 mi/h	2.5 s
0%	−3%	70 mi/h	2.5 s

3-9. A vertical curve is to be designed to connect a −4% grade to a +1% grade on a facility with a design speed of 70 mi/h. For economic reasons, a minimum-length curve will be provided. A driver-reaction time of 2.5 seconds may be used in sight distance determinations. The *V.P.I.* of the curve is at station 5,100 + 22 and has an elevation of 1,285 ft. Find the station and elevation of the *V.P.C.* and *V.P.T.*, the high point of the curve, and at 100-ft intervals along the curve.

CHAPTER 4

Introduction to Traffic Control Devices

Traffic control devices are the media through which traffic engineers communicate with drivers. Virtually every traffic law, regulation, or operating instruction must be communicated through the use of devices that fall into three broad categories:

- Traffic markings
- Traffic signs
- Traffic signals

The effective communication between traffic engineer and driver is a critical link if safe and efficient traffic operations are to prevail. Traffic engineers have no direct control over any individual driver or group of drivers. If a motorman violated a RED signal while conducting a subway train, an automated braking system would force the train to stop anyway. If a driver violates a RED signal, only the hazards of conflicting vehicular and/or pedestrian flows would impede the maneuver. Thus it is imperative that traffic engineers design traffic control devices that communicate uncomplicated messages clearly, in a way that encourages proper observance.

This chapter introduces some of the basic principles involved in the design and placement of traffic control devices. Subsequent chapters cover the details of specific applications to freeways, multilane and two-lane highways, intersections, and arterials and streets.

4.1 The Manual on Uniform Traffic Control Devices

The principal standard governing the application, design, and placement of traffic control devices is the current edition of the *Manual on Uniform Traffic Control Devices* (MUTCD) [*1*]. The Federal Highway Administration publishes a national MUTCD that serves as a minimum standard and a model for individual state MUTCDs. Many states simply adopt the federal manual by statute. Others develop their own manuals. In the latter case, the state MUTCD must meet all of the minimum standards of the federal manual, but it may impose additional or more stringent standards. As is the case with most federal mandates in transportation, compliance is enforced through partial withholding of federal-aid highway funds from states deemed in violation of federal MUTCD standards.

4.1.1 History and Background

One of the principal objectives of the MUTCD is to establish *uniformity* in the use, placement, and design of traffic control devices. Communication is greatly enhanced when the same messages are delivered in the same way and in similar circumstances at all times. Consider the potential confusion if each state designed its own STOP sign, with different shapes, colors, and legends.

Varying device design is not a purely theoretical issue. As late as the early 1950s, two-color (red, green) traffic signals had the indications in different positions in different states. Some placed the "red" ball on top; others placed the "green" ball on top. This is a particular problem for drivers with color blindness, the most common form of which is the inability to distinguish "red" from "green." Standardizing the order of signal lenses was a critical safety measure, guaranteeing that even color-blind drivers could interpret the signal by position of the light in the display. More recently, small amounts of blue and yellow pigment have been added to "green" and "red" lenses to enhance their visibility to color-blind drivers.

Early traffic control devices were developed in various locales with little or no coordination on their design, much less their use. The first centerline appeared on a Michigan roadway in 1911. The first electric signal installation is thought to have been in Cleveland, Ohio, in 1914. The first STOP sign was installed in Detroit in 1915, where the first three-color traffic signal was installed in 1920.

The first attempts to create national standards for traffic control devices occurred during the 1920s. Two separate organizations developed two manuals in this period. In 1927, the American Association of State Highway Officials (AASHO, the forerunner of AASHTO), published the *Manual and Specification for the Manufacture, Display, and Erection of U.S. Standard Road Markings and Signs*. It was revised in 1929 and 1931. This manual addressed only rural signing and marking applications. In 1930, the National Conference on Street and Highway Safety (NCSHS) published the *Manual on Street Traffic Signs, Signals, and Markings*, which addressed urban applications.

In 1932, the two groups formed a merged Joint Committee on Uniform Traffic Control Devices and published the first complete MUTCD in 1935, revising it in 1939. This group continued to have responsibility for subsequent editions until 1972 when the Federal Highway Administration formally assumed responsibility for the manual.

The latest official edition of the MUTCD (at this writing) was published in 2003. A new version is expected to receive final approval sometime in 2009 or 2010. This text is based on the draft 2009/2010 manual, which has been available online for review since December 2007. Because this manual is not formally approved at this writing, you should check the online MUTCD on the FHWA Web site (www.fhwa.com) if there is any question about official material.

For an excellent history of the MUTCD and its development, consult a series of articles by Hawkins [*2–5*].

4.1.2 General Principles of the MUTCD

The MUTCD states that the purpose of traffic control devices is "to promote highway safety and efficiency by providing for orderly movement of all road users on streets and highways, bikeways, public facilities, and private property open to public travel throughout the Nation." [Reference 1, p. 41]. It also defines five requirements for a traffic control device to be effective in fulfilling that mission. A traffic control device must:

1. Fulfill a need
2. Command attention
3. Convey a clear, simple message
4. Command respect of road users
5. Give adequate time for a proper response

In addition to the obvious meanings of these requirements, some subtleties should be carefully noted. The first strongly implies that superfluous devices *should not* be used. Each device must have a specific purpose and must be needed for the safe and efficient flow of traffic. The fourth requirement reinforces this. Respect of drivers is commanded only when drivers are conditioned to expect that all devices carry meaningful and important messages. Overuse or misuse of devices encourages drivers to ignore them—it is like "crying wolf" too often. In such an atmosphere, drivers may not pay attention to those devices that are really needed.

Items 2 and 3 affect the design of a device. Commanding attention requires proper visibility and a distinctive design that attracts the driver's attention in what is often an environment filled with visual distractions. Standard use of color and shape coding plays a major role in attracting this attention. Clarity and simplicity of message is critical; the driver is viewing the device for only a few short seconds while traveling at what may be a high speed. Again, color and shape coding is used to deliver as much information as possible. Legend, the hardest element of a device to understand, must be kept short and as simple as possible.

Item 5 affects the placement of devices. A STOP sign, for example, is always placed at the stop line but must be visible for at least one safe stopping distance. Guide signs requiring drivers to make lane changes must be placed well in advance of the diverge area to give drivers sufficient distance to execute the required maneuvers.

4.1.3 Contents of the MUTCD

The MUTCD addresses three critical aspects of traffic control devices. It contains:

1. Detailed standards for the physical design of the device, specifying shape, size, colors, legend types and sizes, and specific legend.
2. Detailed standards and guidelines on where devices should be located with respect to the traveled way.
3. Warrants, or conditions, that justify the use of a particular device.

The most detailed and definitive standards are for the physical design of the device. Little is left to judgment, and virtually every detail of the design is fully specified. Colors are specified by specific pigments and legend by specific fonts. Some variance is permitted with respect to size, with minimum sizes specified and optional larger sizes for use when needed for additional visibility.

Placement guidelines are also relatively definitive but often allow for some variation within prescribed limits. Placement guidelines sometimes lead to obvious problems. One frequent problem involves STOP signs. When placed in the prescribed position, they may wind up behind trees or other obstructions where their effectiveness is severely compromised. Figure 4.1 shows such a case, in which a STOP sign placed at the prescribed height and lateral offset at the stop line winds up virtually hidden by a tree. Common sense must be exercised in such cases if the device is to be effective.

Warrants are given with various levels of specificity and clarity. Signal warrants, for example, are detailed and relatively precise. This is necessary because signal installations represent a significant investment, both in initial investment and in continuing operating and maintenance costs. The warrants for STOP and YIELD signs, however, are far more general and leave substantial latitude for the exercise of professional judgment.

Chapter 18 deals with the selection of an appropriate form of intersection control and covers the warrants for signalization, two-way and multiway STOP signs, and YIELD signs in some detail. Because of the cost of signals, much study has been devoted to the defining of conditions warranting their use. Proper implementation of signal and other warrants in the MUTCD requires appropriate engineering studies to be made to determine the need for a particular device or devices.

Figure 4.1: STOP Sign Hidden by Tree

4.1.4 Legal Aspects of the MUTCD

The MUTCD provides guidance and information in four different categories:

1. *Standard.* A standard is a statement of a required, mandatory, or specifically prohibitive practice regarding a traffic control device. Typically, standards are indicated by the use of the term *shall* or *shall not* in the statement.
2. *Guidance.* Guidance is a statement of recommended, but not mandatory, practice in typical situations. Deviations are allowed if engineering judgment or a study indicates that a deviation is appropriate. Guidance is generally indicated by use of the word *should* or *should not.*
3. *Option.* An option is a statement of practice that is a permissive condition. It carries no implication of requirement or recommendation. Options often contain allowable modifications to a Standard or Guidance. An option is usually stated using the word *may* or *may not.*

4. *Support.* This is a purely information statement provided to supply additional information to the traffic engineer. The words *shall, should,* or *may* do not appear in these statements (nor do their negative counterparts).

The four types of statements given in the MUTCD have legal implications for traffic agencies. Violating a standard leaves the jurisdictional agency exposed to liability for any accident that occurs because of the violation. Thus placing a nonstandard STOP sign would leave the jurisdictional agency exposed to liability for any accident occurring at the location. Guidelines, when violated, also leave some exposure to liability. Guidelines should be modified only after an engineering study has been conducted and documented, justifying the modification(s). Without such documentation, liability for accidents may also exist. Options and Support carry no implications with respect to liability.

It should also be understood that jurisdiction over traffic facilities is established as part of each state's vehicle and traffic law. That law generally indicates what facilities fall under the direct jurisdiction of the state (usually designated state highways and all intersections involving such highways) and specifies the state agency exercising that jurisdiction. It also defines what roadways would fall under control of county, town, and other local governments. Each of those political entities, in turn, would appoint or otherwise specify the local agency exercising jurisdiction.

Many traffic control devices must be supported by a specific law or ordinance enacted by the appropriate level of government. Procedures for implementing such laws and ordinances must also be specified. Many times (such as in the case of speed limits and parking regulations), public hearings and/or public notice must be given before imposition. For example, it would not be legal for an agency to post parking prohibitions during the night and then ticket or tow all parked vehicles without having provided adequate advance public notice, which is most often accomplished using local or regional newspapers.

This chapter presents some of the principles of the MUTCD, and it generally describes the types of devices and their typical applications. Chapter 17 goes into greater detail concerning the use of traffic control devices on freeways, multilane, and two-lane highways. Chapter 19 contains additional detail concerning use of traffic control devices at intersections.

4.1.5 Communicating with the Driver

The driver is accustomed to receiving a certain message in a clear and standard fashion, often with redundancy. A number of mechanisms are used to convey messages. These mechanisms make use of recognized human limitations, particularly with respect to eyesight. Messages are conveyed through the use of these elements:

- *Color.* Color is the most easily visible characteristic of a device. Color is recognizable long before a general shape may be perceived and considerably before a specific legend can be read and understood. The principal colors used in traffic control devices are red, yellow, green, orange, black, blue, and brown. These are used to code certain types of devices and to reinforce specific messages whenever possible.
- *Shape.* After color, the shape of the device is the next element to be discerned by the driver. Particularly in signing, shape is an important element of the message, either identifying a particular type of information that the sign is conveying or conveying a unique message of its own.
- *Pattern.* Pattern is used in the application of traffic markings. In general, double solid, solid, dashed, and broken lines are used. Each conveys a type of meaning with which drivers become familiar. The frequent and consistent use of similar patterns in similar applications contributes greatly to their effectiveness and to the instant recognition of their meaning.
- *Legend.* The last element of a device that the driver comprehends is its specific legend. Signals and markings, for example, convey their entire message through use of color, shape, and pattern. Signs, however, often use a specific legend to transmit the details of the message being transmitted. The legend must be kept simple and short, so that drivers do not divert their attention from the driving task, yet are able to see and understand the specific message being given.

Redundancy of message can be achieved in a number of ways. The STOP sign, for example, has a unique shape (octagon), a unique color (red), and a unique one-word legend (STOP). Any of the three elements alone is sufficient to convey the message. Each provides redundancy for the others.

Redundancy can also be provided through use of different devices, each reinforcing the same message. A left-turn lane may be identified by arrow markings on the pavement, a "This Lane Must Turn Left" sign, and a protected left-turn signal phase indicated by a green arrow. Used together, the message is unmistakable.

The MUTCD provides a set of standards, guidelines, and general advice on how to best communicate various traffic rules and regulations to drivers. The MUTCD, however, is a document that is always developing. The traffic engineer must always consult the latest version of the manual (with all applicable revisions) when considering traffic control options. The most current version of the MUTCD is available online through the Federal Highway Administration Web site.

4.2 Traffic Markings

Traffic markings are the most plentiful traffic devices in use. They serve a variety of purposes and functions and fall into three broad categories:

- Longitudinal markings
- Transverse markings
- Object markers and delineators

Longitudinal and transverse markings are applied to the roadway surface using a variety of materials, the most common of which are paint and thermoplastic. Reflectorization for better night vision is achieved by mixing tiny glass beads in the paint or by applying a thin layer of glass beads over the wet pavement marking as it is placed. The latter provides high initial reflectorization, but the top layer of glass beads is more quickly worn. When glass beads are mixed into the paint before application, some level of reflectorization is preserved as the marking wears. Thermoplastic is a naturally reflective material, and nothing need be added to enhance drivers' ability to see them at night.

In areas where snow and snow plowing is not a problem, paint or thermoplastic markings can be augmented by pavement inserts with reflectors. Such inserts greatly improve the visibility of the markings at night. They are visible in wet weather (often a problem with markings) and resistant to wear. They are generally not used where plowing is common because they can be dislodged or damaged during the process.

Object markers and delineators are small object-mounted reflectors. Delineators are small reflectors mounted on lightweight posts and are used as roadside markers to help drivers in proper positioning during inclement weather when standard markings are not visible.

4.2.1 Colors and Patterns

Five marking colors are in current use: yellow, white, red, blue, and black. In general, they are used as follows:

- *Yellow* markings separate traffic traveling in opposite directions.
- *White* markings separate traffic traveling in the same direction and are used for all transverse markings.
- *Red* markings delineate roadways that will not be entered or used by the viewer of the marking.
- *Blue* markings are used to delineate parking spaces reserved for persons with disabilities.
- *Black* markings are used in conjunction with other markings on light pavements. To emphasize the pattern of the line, gaps between yellow or white markings are filled in with black to provide contrast and easier visibility.

A solid line prohibits or discourages crossing. A double solid line indicates maximum or special restrictions. A broken line indicates that crossing is permissible. A dotted line uses shorter line segments than a broken line. It provides trajectory guidance and often is used as a continuation of another type of line in a conflict area. Normally, line markings are 4 to 6 inches wide. Wide lines, which provide greater emphasis, should be at least twice the width of a normal line. Broken lines normally consist of 10-ft line segments and 30-ft gaps. Similar dimensions with a similar ratio of line segments to gaps may be used as appropriate for prevailing traffic speeds and the need for delineation. Dotted lines usually consist of 2-ft line segments and 4-ft (or longer) gaps. MUTCD suggests a maximum segment-to-gap ratio of 1:3 for dotted lines.

4.2.2 Longitudinal Markings

Longitudinal markings are those markings placed parallel to the direction of travel. The vast majority of longitudinal markings involve centerlines, lane lines, and pavement edge lines.

Longitudinal markings provide guidance for the placement of vehicles on the traveled way cross section and basic trajectory guidance for vehicles traveling along the facility. The best example of the importance of longitudinal markings is the difficulty in traversing a newly paved highway segment on which lane markings have not yet been repainted. Drivers do not automatically form neat lanes without the guidance of longitudinal markings; rather, they tend to place themselves somewhat randomly on the cross section, encountering many difficulties. Longitudinal markings provide for organized flow and optimal use of the pavement width.

Centerlines

The yellow centerline marking is critically important and is used to separate traffic traveling in opposite directions. Use of centerlines on all types of facilities is not mandated by the MUTCD. The applicable standard is:

> Centerline markings shall be placed on all paved urban arterials and collectors that have a traveled way of 20 ft or more and an average daily travel (ADT) of 6,000 veh/day or greater. Centerline markings shall also be placed on all paved, two-way streets or highways that have 3 or more traffic lanes. [MUTCD, p. 227]

Further guidance indicates that placing centerlines is recommended for urban arterials and streets with ADTs of 4,000 or more and a roadway width of 20 ft or more, and on rural highways with a width in excess of 18 ft and ADT more than 3,000 vehicles/day. Caution should be used in placing centerlines on pavements of 16 ft or less, which may increase the incidence of traffic encroaching upon the roadside.

On two-lane, two-way rural highways, centerline markings supplemented by signs are used to regulate passing maneuvers. A double-solid yellow center marking indicates that passing is not permitted in either direction. A solid yellow line with a dashed yellow line indicates that passing is permitted from the dashed side only. Where passing is permissible in both directions, a single dashed yellow centerline is used. Chapter 16 contains additional detail on the use and application of centerlines on two-lane, two-way, rural highways.

There are other specialized uses of yellow markings. Figure 4.2(a) illustrates the use of double dashed-yellow markings to delineate a reversible lane on an arterial. Signing and/or lane-use signals would have to supplement these to denote the directional use of the lane. Figure 4.2(b) shows the markings used for two-way left-turn lanes on an arterial.

Lane Markings

The typical lane marking is a single white dashed line separating lanes of traffic in the same direction. MUTCD standards require the use of lane markings on all freeways and Interstate highways and recommend their use on all highways with two or more adjacent traffic lanes in a single direction. The dashed lane line indicates that lane-changing is permitted. A single solid white lane line is used to indicate that lane-changing is discouraged but not illegal. Where lane-changing is to be prohibited, a double-white solid lane line is used.

Edge Markings

Edge markings are a required standard on freeways, expressways, and rural highways with a traveled way of 20 ft or more in width and an ADT of 6,000 veh/day or greater. They are recommended for rural highways with

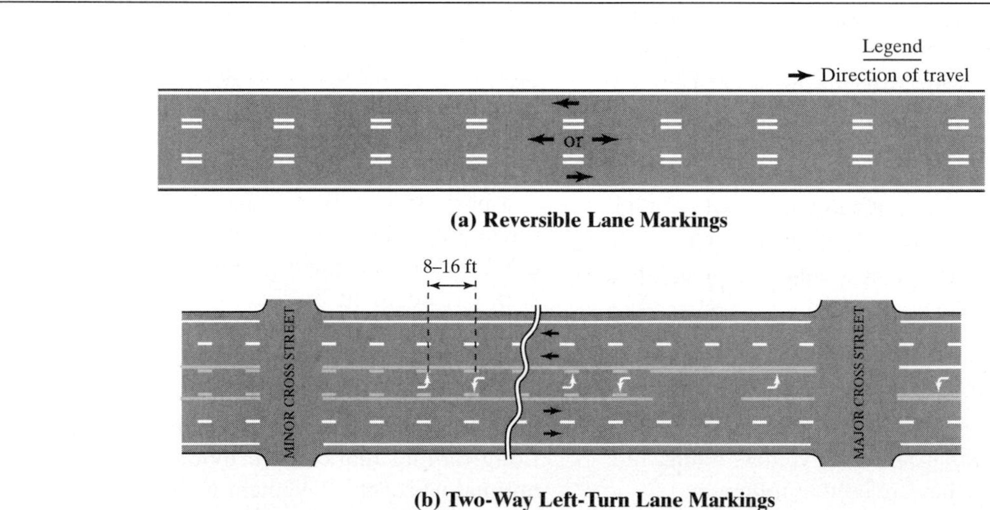

Figure 4.2: Specialized Uses of Yellow Markings

(*Source: Manual of Uniform Traffic Control Devices*, Draft, Federal Highway Administration, Washington DC, December 2007, Figs 3B-6 and 3B-7.)

ADTs over 3,000 veh/day and a 20-ft or wider traveled way. When used, right-edge markings are a single normal solid white line; left-edge markings are a single normal solid yellow line.

Other Longitudinal Markings

The MUTCD provides for many options in the use of longitudinal markings. Consult the manual directly for further detail. The manual also provides standards and guidance for other types of applications, including freeway and nonfreeway merge and diverge areas, lane drops, extended markings through intersections, and other situations.

Chapter 17 contains additional detail on the application of longitudinal markings on freeways, expressways, and rural highways. Chapter 19 includes additional discussion of intersection markings.

4.2.3 Transverse Markings

Transverse markings, as their name implies, include any and all markings with a component that cuts across a portion or all of the traveled way. When used, all transverse markings are white.

STOP and YIELD Lines

STOP lines are generally not mandated by the MUTCD. In practice, STOP lines are almost always used where marked crosswalks exist and in situations where the appropriate location to stop for a STOP sign or traffic signal is not clear. When used, it is recommended that the width of the line be 12 to 24 inches. When used, STOP lines must extend across all approach lanes. STOP lines, however, *shall not* be used in conjunction with YIELD signs; in such cases, the newly introduced YIELD line would be used. The YIELD line is illustrated in Figure 4.3.

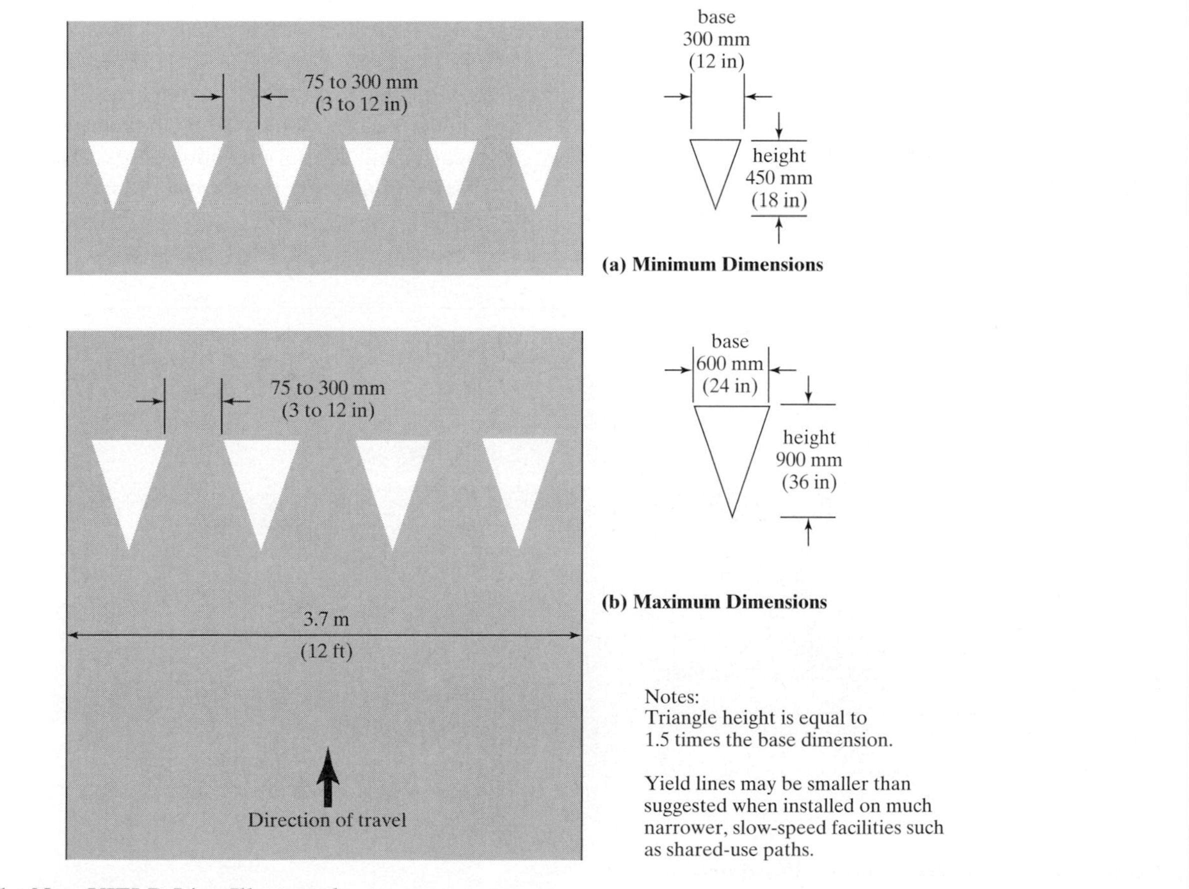

Figure 4.3: The New YIELD Line Illustrated

(*Source: Manual of Uniform Traffic Control Devices*, Draft, Federal Highway Administration, Washington DC, December 2007, Fig 3B-15.)

Crosswalk Markings

Although not mandated by the MUTCD, it is recommended that crosswalks be marked at all intersections at which "substantial" conflict between vehicles and pedestrians exists. They should also be used at points of pedestrian concentration and at locations where pedestrians might not otherwise recognize the proper place and/or path to cross.

A marked crosswalk should be 6 ft or more in width. Figure 4.4 shows the three types of crosswalk markings in general use. The most frequently used is composed of two parallel white lines. Cross-hatching may be added to provide greater focus in areas with heavy pedestrian flows. The use of parallel transverse markings to identify the crosswalk is another option used at locations with heavy pedestrian flows.

The manual also contains a special pedestrian crosswalk marking for signalized intersections where a full pedestrian phase is included. Consult the manual directly for details of this particular marking.

Parking Space Markings

Parking space markings are not purely transverse because they contain both longitudinal and transverse elements. They are officially categorized as transverse markings, however, in the MUTCD. They are always optional and are used to encourage efficient use of parking spaces. Such markings can also help prevent encroachment of parked vehicles into fire-hydrant zones, loading zones, taxi stands and bus stops, and other specific locations at which parking is prohibited. They are also useful on arterials with curb parking because they also clearly demark the parking lane, separating it from travel lanes. Figure 4.5 illustrates typical parking lane markings.

Note that the far end of the last marked parking space should be at least 20 ft away from the nearest crosswalk marking (30 ft on a signalized intersection approach).

Word and Symbol Markings

The MUTCD prescribes a number of word and symbol markings that may be used, often in conjunction with signs and/or signals. These include arrow markings indicating lane-use restrictions. Such arrows (with accompanying signs) are mandatory where a through lane becomes a left- or right-turn-only lane approaching an intersection.

Word markings include "ONLY," used in conjunction with lane use arrows, and "STOP," which can be used only in conjunction with a STOP line and a STOP sign. "SCHOOL" markings are often used in conjunction with signs to demark school and school-crossing zones. The MUTCD contains a listing of all authorized word markings and allows for discretionary use of unique messages where needed.

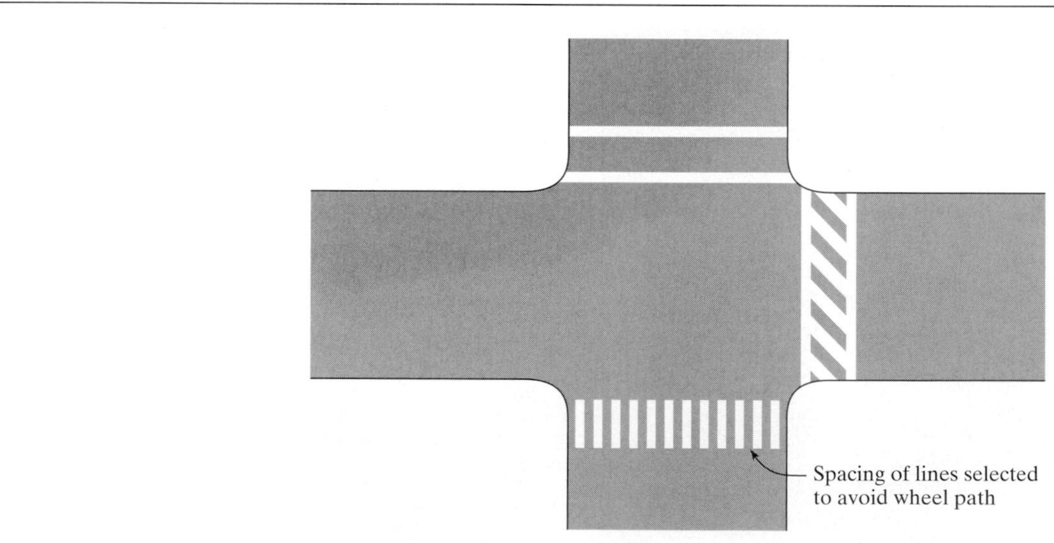

Figure 4.4: Crosswalk Markings Illustrated

(*Source: Manual of Uniform Traffic Control Devices*, Draft, Federal Highway Administration, Washington DC, December 2007, Fig 3B-18.)

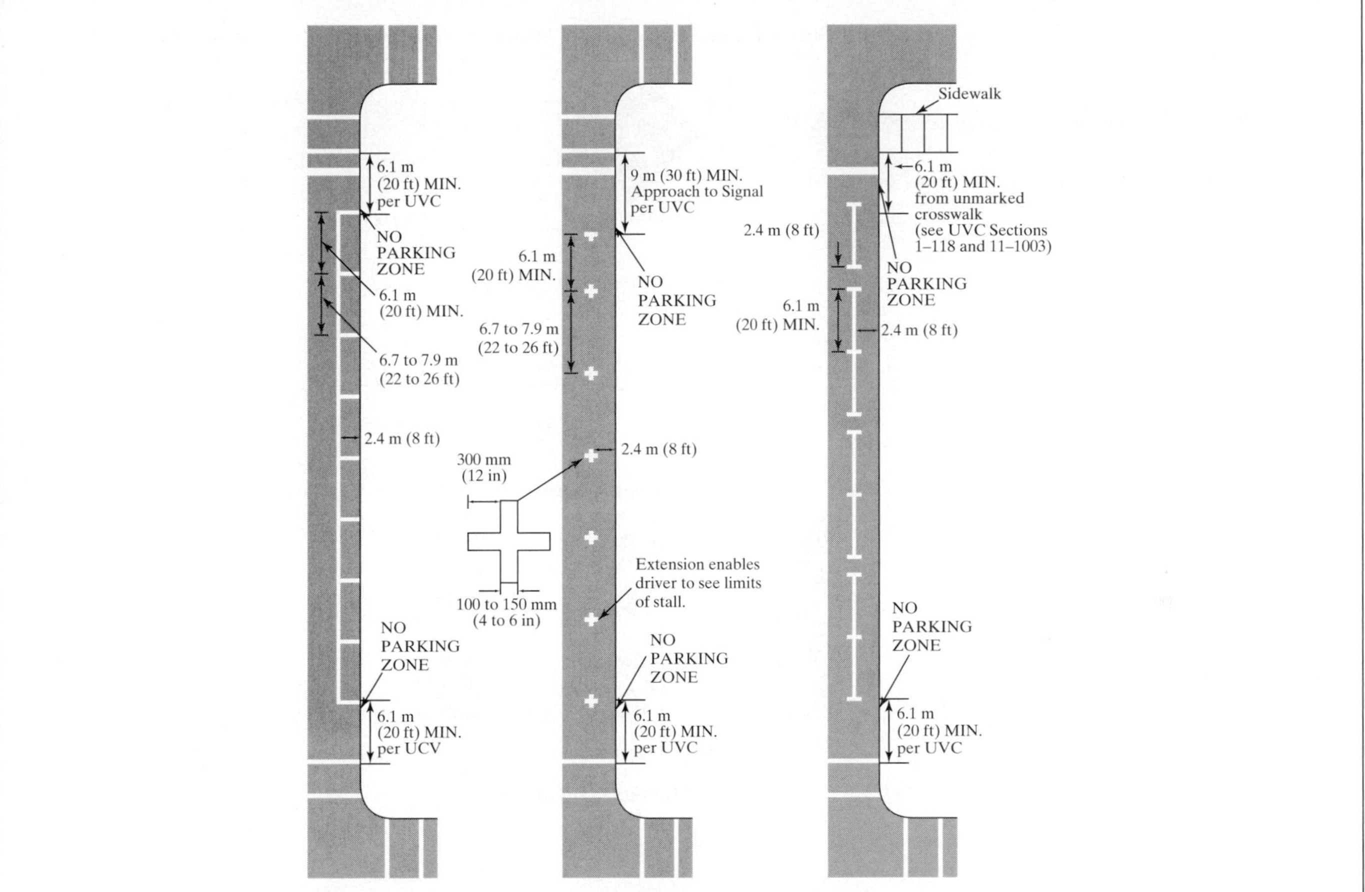

Figure 4.5: Examples of Parking Stall Markings
(*Source: Manual of Uniform Traffic Control Devices*, Draft, Federal Highway Administration, Washington DC, December 2007, Fig 3B-21.)

Other Transverse Markings

Consult the MUTCD directly for examples of other types of markings, including preferential lane markings, curb markings, roundabout and traffic circle markings, and speed-hump markings. Chapter 19 contains a detailed discussion of the use of transverse and other markings at intersections.

4.2.4 Object Markers

Object markers are used to denote obstructions either in or adjacent to the traveled way. Object markers are mounted on the obstruction in accordance with MUTCD standards and guidelines. In general, the lower edge of the marker is mounted a minimum of 4 ft above the surface of the nearest traffic lane (for obstructions 8 ft or less from the pavement edge) or 4 ft above the ground (for obstructions located further away from the pavement edge).

The three types of object markers used are illustrated in Figure 4.6. Obstructions within the roadway *must* be marked using a Type 1 or Type 3 marker. The Type 3 marker, when used, must have the alternating yellow and black stripes sloped downward at a 45° angle toward the side on which traffic is to pass the obstruction. When used to mark a roadside obstruction, the inside edge of the marker must be in line with the inner edge of the obstruction.

4.2.5 Delineators

Delineators are reflective devices mounted at a 4-ft height on the side(s) of a roadway to help denote its alignment. They are

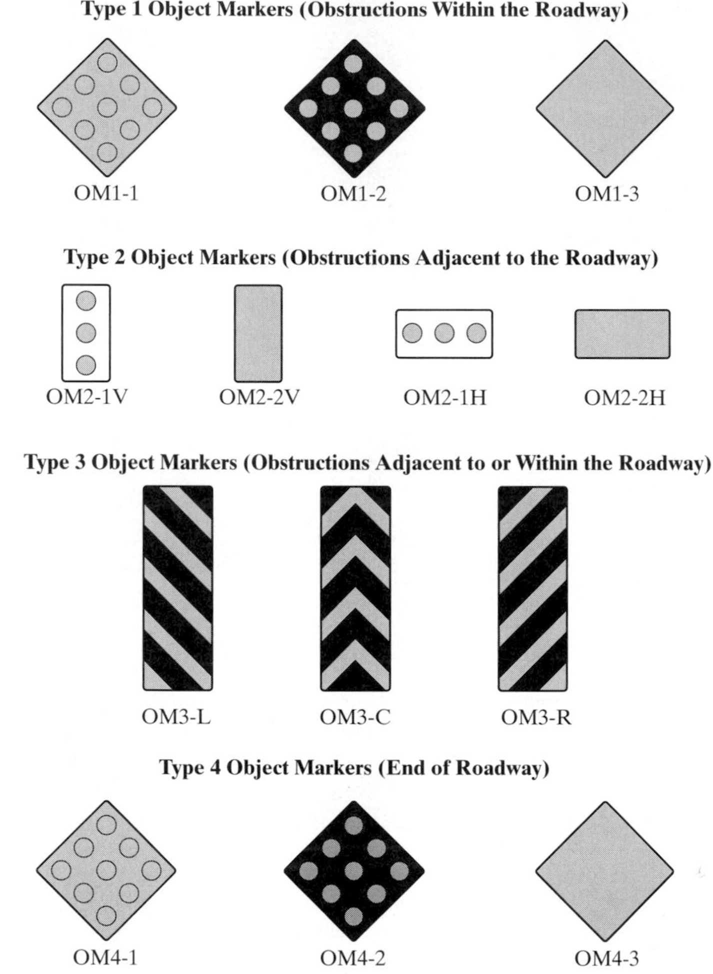

Figure 4.6: Object Markers

(*Source: Manual of Uniform Traffic Control Devices*, Draft, Federal Highway Administration, Washington DC, December 2007, Fig 3B-21.)

particularly useful during inclement weather when pavement edge markings may not be visible. When used on the right side of the roadway, delineators are white; when used on the left side of the roadway, delineators are yellow. The back of delineators may have red reflectors to indicate wrong-way travel on a one-direction roadway.

Delineators are mandated on the right side of freeways and expressways and on at least one side of interchange ramps, with the exception of tangent sections where raised pavement markers are used continuously on all lane lines, where whole routes (or substantial portions thereof) have large tangent sections, or where delineators are used to lead into all curves. They may also be omitted where there is continuous roadway lighting between interchanges. Delineators may be used on an optional basis on other classes of roads.

4.3 Traffic Signs

The MUTCD provides specifications and guidelines for the use of literally hundreds of different signs for myriad purposes. In general, traffic signs fall into one of three major categories:

- *Regulatory signs.* Regulatory signs convey information concerning specific traffic regulations. Regulations may relate to right-of-way, speed limits, lane usage, parking, or a variety of other functions.

- *Warning signs.* Warning signs are used to inform drivers about upcoming hazards that they might not see or otherwise discern in time to react safely.
- *Guide signs.* Guide signs provide information on routes, destinations, and services that drivers may be seeking.

It would be impossible to cover the full range of traffic signs and applications in a single chapter. The sections that follow provide a general overview of the various types of traffic signs and their use.

4.3.1 Regulatory Signs

> Regulatory signs shall be used to inform road users of selected traffic laws or regulations and indicate the applicability of the legal requirements. Regulatory signs shall be installed at or near where the regulations apply. The signs shall clearly indicate the requirements imposed by the regulations and shall be designed and installed to provide adequate visibility and legibility in order to obtain compliance. [Reference 1, p. 69]

Drivers are expected to be aware of many general traffic regulations, such as the basic right-of-way rule at intersections and the state speed limit. Signs, however, should be used in all cases where the driver cannot be expected to know the applicable regulation.

Except for some special signs, such as the STOP and YIELD sign, most regulatory signs are rectangular, with the long dimension vertical. Some regulatory signs are square. These are primarily signs using symbols instead of legend to impart information. The use of symbol signs generally conforms to international practices established at a 1971 United Nations conference on traffic safety. The background color of regulatory signs, with a few exceptions, is white; legend or symbols are black. In symbol signs, a red circle with a bar through it signifies a prohibition of the movement indicated by the symbol. The MUTCD contains many pages of standards for the appropriate size of regulatory signs and should be consulted directly on this issue.

Regulatory Signs Affecting Right-of-Way

The regulatory signs in this category have special designs reflecting the extreme danger that exists when one is ignored. These signs include the STOP and YIELD signs, which assign right-of-way at intersections, and WRONG WAY and ONE WAY signs, indicating directional flow. The STOP and YIELD signs have unique shapes, and they use a red background color to denote danger. The WRONG WAY sign also uses a red background for this purpose. Figure 4.7 illustrates these signs.

Figure 4.7: Regulatory Signs Affecting Right-of-Way

(*Source: Manual of Uniform Traffic Control Devices*, Draft, Federal Highway Administration, Washington DC, December 2007, Figs 2B-1, 2B-14, and 2B-16.)

The "All Way" panel is mounted below a STOP sign where multiway STOP control is in use. Consult Chapter 18 for a detailed presentation and discussion of warrants for use of STOP and YIELD signs at intersections.

Speed Limit Signs

One of the most important issues in providing for safety and efficiency of traffic movement is the setting of appropriate speed limits. To be effective, a speed limit must be communicated to the driver and should be sufficiently enforced to engender general observance. A number of different types of speed limits may be imposed:

- Linear speed limits
- Areawide (statutory) speed limits
- Night speed limits
- Truck speed limits
- Minimum speed limits

Speed limits may be stated in terms of standard U.S. units (mi/h) or in metric units (km/h). Current law allows each

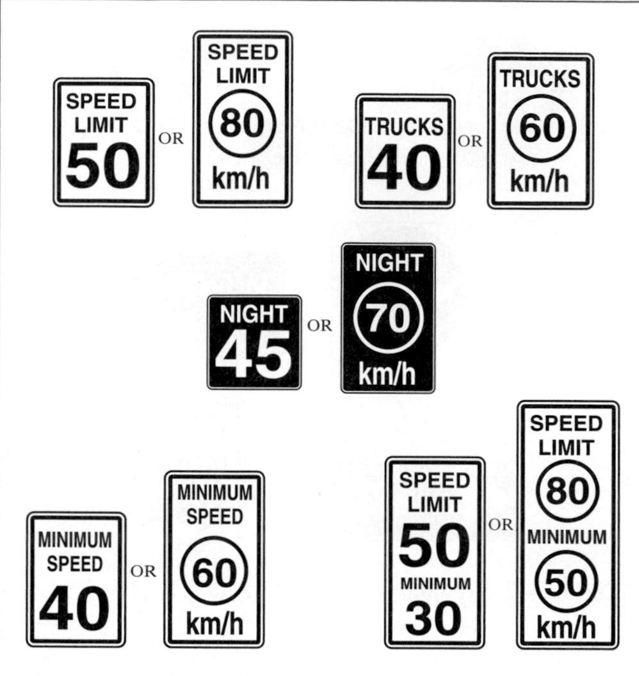

Figure 4.8: Speed Limit Signs

(*Source: Manual of Uniform Traffic Control Devices*, Draft, Federal Highway Administration, Washington DC, December 2007, Figs 2B-1, 2B-3.)

state to determine the system of units to be used. Where the metric system is used, speed limit number is inside a circle to emphasize that metric units are in use.

Figure 4.8 shows a variety of speed signs in common use. Whereas most signs consist of black lettering on a white background, night speed limits are posted using the reverse of this: white lettering on a black background.

Linear speed limits apply to a designated section of roadway. Signs should be posted such that no driver can enter the roadway without seeing a speed limit sign within approximately 1,000 ft. This is not an MUTCD standard but reflects common practice.

Area speed limits apply to all roads within a designated area (unless otherwise posted). A state statutory speed limit is one example of such a regulation. Cities, towns, and other local governments may also enact ordinances establishing a speed limit throughout their jurisdiction. Areawide speed limits should be posted on every facility at the boundary entering the jurisdiction for which the limit is established.

The "reduced speed" or "speed zone ahead" signs (not shown here) should be used wherever engineering judgment indicates a need to warn drivers of a reduced speed limit for compliance. When used, however, the sign must be followed by a speed limit sign posted at the beginning of the section in which the reduced speed limit applies.

Consult Chapter 17 for a discussion of criteria for establishing an appropriate speed limit on a highway or roadway section.

Turn and Movement Prohibition Signs

Where right, left, and/or U-turns, or even through movements, are to be prohibited, one or more of the movement prohibition signs shown in Figure 4.9 are used. In this category, international symbol signs are preferred. The traditional red circle with a bar is placed over an arrow indicating the movement to be banned.

Figure 4.9: Movement Prohibition Signs

(*Source: Manual of Uniform Traffic Control Devices*, Draft, Federal Highway Administration, Washington DC, December 2007, Fig 2B-3.)

Lane-Use Signs

Lane-use control signs are used wherever a given movement or movements are restricted and/or prohibited from designated lanes. Such situations include left-turn- and right-turn-only lanes, two-way left-turn lanes on arterials, and reversible lanes. Lane-use signs, however, may also be used to clarify lane usage even where no regulatory restriction is involved. Where lane usage is complicated, advance lane-use control signs may be used as well. Figure 4.10 illustrates these signs.

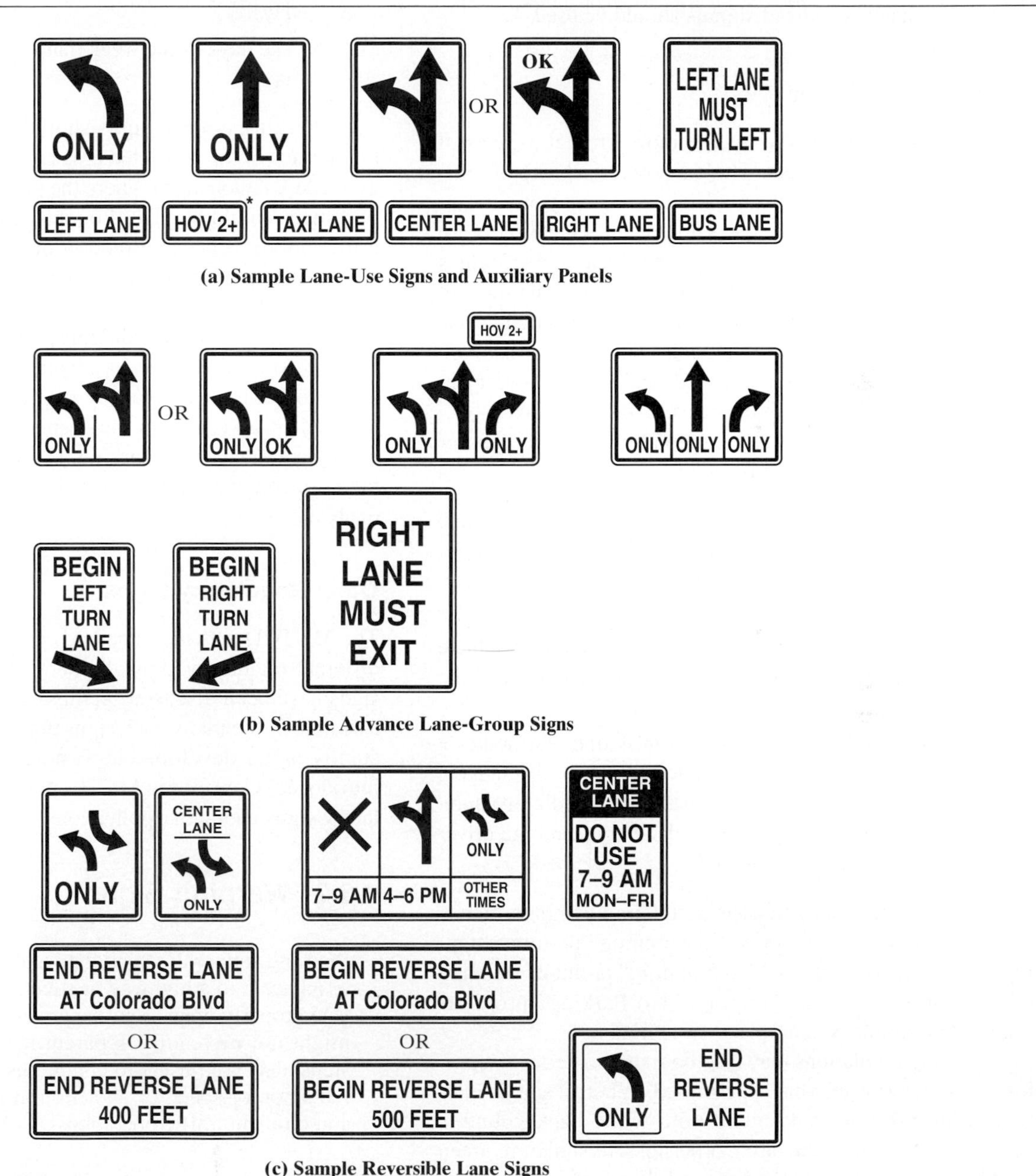

Figure 4.10: Lane-Use Control Signs

(*Source: Manual of Uniform Traffic Control Devices*, Draft, Federal Highway Administration, Washington DC, December 2007, Figs 2B-4, 2B-6.)

Two-way left-turn lane signing must be supplemented by the appropriate markings for such a lane, as illustrated previously. Reversible lane signs must be posted as overhead signs, placed over the lane or lanes that are reversible. Roadside signs may supplement overhead signs. In situations where signing may not be sufficient to ensure safe operation of reversible lanes, overhead signals should be used.

Parking Control Signs

Curb parking control is one of the more critical aspects of urban network management. The economic viability of business areas often depends on an adequate and convenient supply of on-street and off-street parking. At the same time, curb parking often interferes with through traffic and occupies space on the traveled way that might otherwise be used to service moving traffic. Chapter 12 provides a detailed coverage of parking issues and programs. It is imperative that curb parking regulations be clearly signed, and strict enforcement is often necessary to achieve high levels of compliance.

When dealing with parking regulations and their appropriate signing, three terms must be understood:

- *Parking*. A "parked" vehicle is a stationary vehicle located at the curb with the engine not running; whether or not the driver is in the vehicle is not relevant to this definition.
- *Standing*. A "standing" vehicle is a stationary vehicle located at the curb with the engine running and the driver in the car.
- *Stopping*. A "stopping" vehicle is one that makes a momentary stop at the curb to pick up or discharge a passenger; the vehicle moves on immediately upon completion of the pickup or discharge, and the driver does not leave the vehicle.

In legal terms, most jurisdictions maintain a common hierarchal structure of prohibitions. "No Stopping" prohibits stopping, standing, and parking. "No Standing" prohibits standing and parking but permits stopping. "No Parking" prohibits parking, but permits standing and stopping.

Parking regulations may also be stated in terms of a prohibition or in terms of what is permitted. Where a sign is indicating a prohibition, red legend on a white background is used. Where a sign is indicating a permissive situation, green legend on a white background is used. Figure 4.11 illustrates a variety of parking-control signs in common use.

Parking signs must be carefully designed and placed to ensure that the often complex regulations are understood by the majority of drivers. The MUTCD recommends that the following information be provided on parking-control signs, in order from top to bottom of the sign:

- The restriction or prohibition (or condition permitted in the case of a permissive sign).
- The times of the day that it is applicable (if not every day).
- The days of the week that it is applicable (if not every day).

Parking-control signs should always be placed at the boundaries of the restricted area and at intermediate locations as needed. At locations where the parking restriction changes, two signs should be placed on a single support, each with an arrow pointing in the direction of application. Where area-wide restrictions are in effect, the restriction should be signed at all street locations crossing into the restricted area.

In most local jurisdictions, changes in parking regulations must be disclosed in advance using local newspapers and/or other media and/or by placing posters throughout the affected area warning of the change. It is not appropriate, for example, to place new parking restrictions overnight and then ticket or remove vehicles now illegally parked without adequate advance warning.

Other Regulatory Signs

The MUTCD provides standards and guidelines for over 100 different regulatory signs. Some of the most frequently used signs have been discussed in this section, but they are merely a sample of the many such signs that exist. New signs are constantly under development as new types of regulations are introduced. Consult the MUTCD directly for additional regulatory signs and their applications.

4.3.2 Warning Signs

> Warning signs call attention to unexpected conditions on or adjacent to a highway or street, public facility, or private property open to public travel, and to situations that might not be readily apparent to road users. Warning signs alert road users to conditions that might call for a reduction of speed or an action in the interest of safety and efficient traffic operations. [Reference 1, p. 108]

Most warning signs are diamond shaped, with black lettering or symbols on a yellow background. A new lime-green background is being introduced for warning signs dealing with pedestrian and bicycle crossings, and school crossings. A pennant shape is used for the "No Passing Zone" sign,

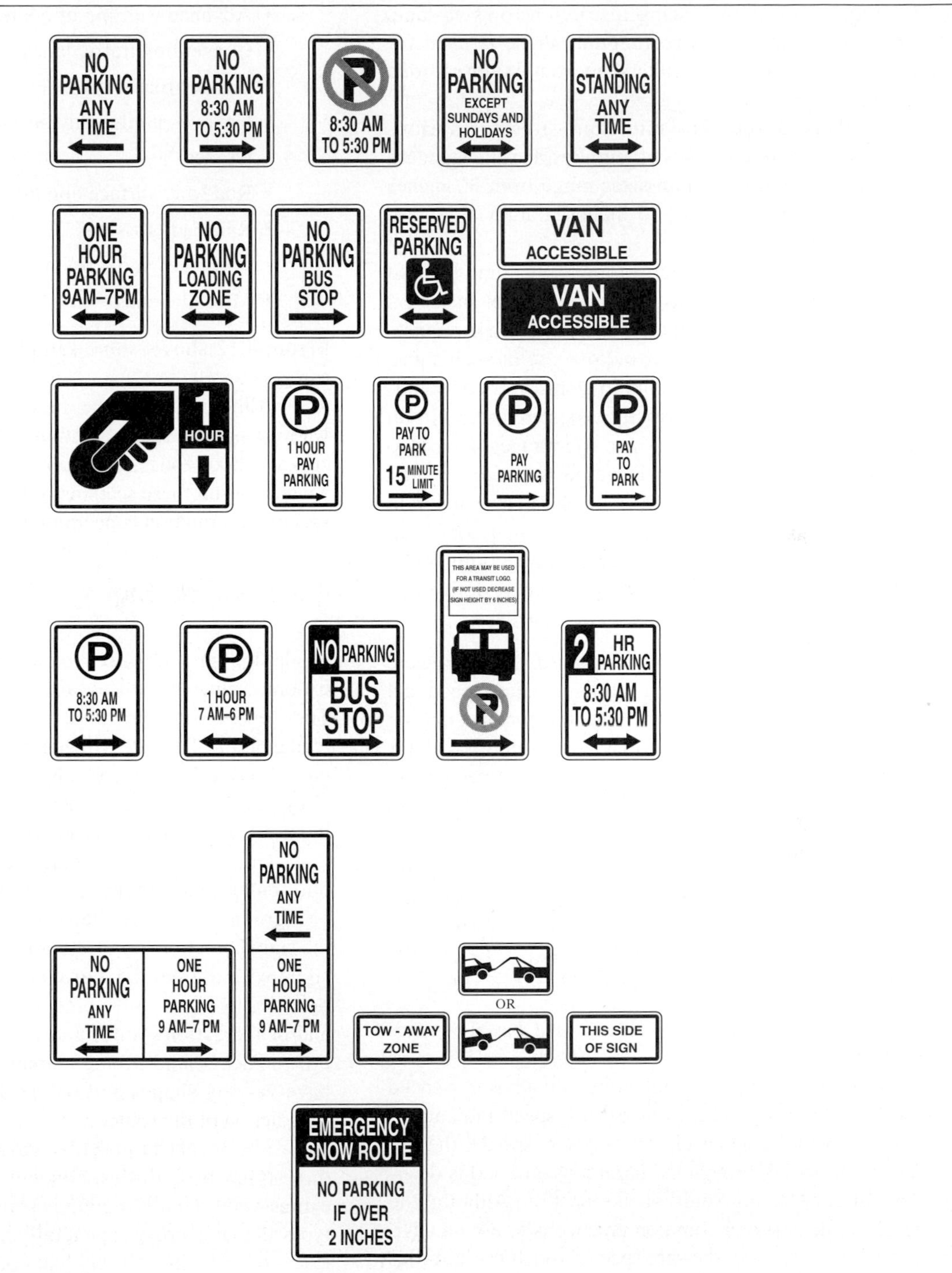

Figure 4.11: Parking-Control Signs

(*Source: Manual of Uniform Traffic Control Devices*, Draft, Federal Highway Administration, Washington DC, December 2007, Fig 2B-7.)

used in conjunction with passing restrictions on two-lane, two-way rural highways. A rectangular shape is used for some arrow indications. A circular shape is used for railroad crossing warnings.

The MUTCD specifies minimum sizes for various warning signs on different types of facilities. For the standard diamond-shaped sign, minimum sizes range from 30 inches by 30 inches to 36 inches by 36 inches. Larger signs are generally permitted.

The MUTCD indicates that warning signs shall be used only in conjunction with an engineering study or based on engineering judgment. Although this is a fairly loose requirement, it emphasizes the need to avoid overuse of such signs. A warning sign should be used only to alert drivers of conditions that they could not be normally expected to discern on their own. Overuse of warning signs encourages drivers to ignore them, which could lead to dangerous situations.

When used, warning signs must be placed far enough in advance of the hazard to allow drivers adequate time to perform the required adjustments. Table 4.1 gives the recommended advance placement distances for two conditions, defined as follows:

- *Condition A: High judgment required.* Applies where the road user must use extra time to adjust speed and change lanes in heavy traffic due to a complex driving situation. Typical applications are warning signs for merging, lane drop, and similar situations. A perception, intellection, emotion, and volition (PIEV) time of 6.7 to 10.0 s is assumed plus 4.5 s for each required maneuver.
- *Condition B: Deceleration to the listed advisory speed for the condition.* Applies in cases where the road user must decelerate to a posted advisory speed to safely maneuver through the hazard. A 1.6-s PIEV time is assumed with a deceleration rate of 10ft/s^2.

In all cases, sign visibility of 175 ft is assumed, based on sign-design standards.

Supplementary panels indicating an advisory speed through the hazard are being replaced by speed indications directly on the warning sign itself. The advisory speed is the recommended safe speed through the hazardous area and is determined by an engineering study of the location. Although no specific guideline is given, common practice is to use an advisory speed panel wherever the safe speed through the hazard is 10 mi/h or more less than the posted or statutory speed limit.

Warning signs are used to inform drivers of a variety of potentially hazardous circumstances, including:

- Changes in horizontal alignment
- Intersections
- Advance warning of control devices
- Converging traffic lanes
- Narrow roadways
- Changes in highway design
- Grades
- Roadway surface conditions
- Railroad crossings
- Entrances and crossings
- Miscellaneous

Figure 4.12 shows some sample warning signs from these categories.

Although not shown here, the MUTCD contains other warning signs in special sections of the manual related to work zones, school zones, and railroad crossings. The practitioner should consult these sections of the MUTCD directly for more specific information concerning these special situations.

4.3.3 Guide Signs

Guide signs provide information to road users concerning destinations, available services, and historical/recreational facilities. They serve a unique purpose in that drivers who are familiar or regular users of a route will generally not need to use them; they provide critical information, however, to unfamiliar road users. They serve a vital safety function: A confused driver approaching a junction or other decision point is a distinct hazard.

Guide signs are rectangular, with the long dimension horizontal, and they have white lettering and borders. The background varies by the type of information contained on the sign. Directional or destination information is provided by signs with a green background; information on services is provided by signs with a blue background; cultural, historical, and/or recreational information is provided by signs with a brown background. Route markers, included in this category, have varying shapes and colors depending on the type and jurisdiction of the route.

The MUTCD provides guide-signing information for three types of facilities: conventional roads, freeways, and expressways. Guide signing is somewhat different from other types in that overuse is generally not a serious issue, unless it leads to confusion. Clarity and consistency of message is the most important aspect of guide signing. Several general principles may be applied:

1. If a route services a number of destinations, the most important of these should be listed. Thus a highway serving Philadelphia as well as several

Table 4.1: Guidelines for Advance Placement of Warning Signs

Posted or 85th-Percentile Speed	Advance Placement Distance[1,2]								
	Condition A: Speed Reduction and Lane Changing in Heavy Traffic[3]	Condition B: Deceleration to the listed advisory speed (mph) for the condition[5]							
		0[4]	10	20	30	40	50	60	70
20 mph	225 ft	100 ft[7]	N/A[6]	—	—	—	—	—	—
25 mph	325 ft	100 ft[7]	N/A[6]	N/A[6]	—	—	—	—	—
30 mph	460 ft	100 ft[7]	N/A[6]	N/A[6]	—	—	—	—	—
35 mph	565 ft	100 ft[7]	N/A[6]	N/A[6]	N/A[6]	—	—	—	—
40 mph	670 ft	125 ft	100 ft[7]	100 ft[7]	N/A[6]	—	—	—	—
45 mph	775 ft	175 ft	125 ft	100 ft[7]	100 ft[7]	N/A[6]	—	—	—
50 mph	885 ft	250 ft	200 ft	175 ft	125 ft	100 ft[7]	—	—	—
55 mph	990 ft	325 ft	275 ft	225 ft	200 ft	125 ft	N/A[6]	—	—
60 mph	1100 ft	400 ft	350 ft	325 ft	275 ft	200 ft	100 ft[7]	—	—
65 mph	1200 ft	475 ft	425 ft	400 ft	350 ft	275 ft	175 ft	100 ft[7]	—
70 mph	1250 ft	550 ft	525 ft	500 ft	450 ft	375 ft	275 ft	150 ft	—
75 mph	1350 ft	650 ft	625 ft	600 ft	550 ft	475 ft	375 ft	250 ft	100 ft

Notes:

[1] For word message warning signs with more than four words or with letter heights of less than 6 inches, the advance placement distance is 100 feet more than the distance shown in this table in order to provide adequate legibility.

[2] The distances are adjusted for a sign legibility distance of 180 feet for Condition A, which is based on a word legend height of 5 inches. The distances for Condition B have been adjusted for a sign legibility distance of 250 feet, which is appropriate for an alignment warning symbol sign.

[3] Typical conditions are locations where the road user must use extra time to adjust speed and change lanes in heavy traffic because of a complex driving situation. Typical signs are Merge and Right Lane Ends. The distances are determined by providing the driver a PRT of 14.0 to 14.5 seconds for vehicle maneuvers (2005 AASHTO Policy, Exhibit 3-3, Decision Sight Distance, Avoidance Maneuver E) minus the legibility distance of 180 feet for the appropriate sign.

[4] Typical condition is the warning of a potential stop situation. Typical signs are Stop Ahead, Yield Ahead, Signal Ahead, and Intersection Warning signs. The distances are based on the 2005 AASHTO Policy, Exhibit 3-1, Stopping Sight Distance, providing a PRT of 2.5 seconds, a deceleration rate of 11.2 feet/second^2, minus the sign legibility distance of 180 feet.

[5] Typical conditions are locations where the road user must decrease speed to maneuver through the warned condition. Typical signs are Turn, Curve, Reverse Turn, or Reverse Curve. The distance is determined by providing a 2.5 second PRT, a vehicle deceleration rate of 10 feet/second^2, minus the sign legibility distance of 250 feet.

[6] No suggested distances are provided for these speeds, as the placement location is dependent on site conditions and other signing. An alignment warning sign may be placed anywhere from the point of curvature up to 100 feet in advance of the curve. However, the alignment warning sign should be installed in advance of the curve and at least 100 feet from any other signs.

[7] The advance placement distance is listed as 100 feet to provide adequate spacing between signs.

(*Source: Manual of Uniform Traffic Control Devices*, Draft, Federal Highway Administration, Washington DC, December 2007, Table 2C-4.)

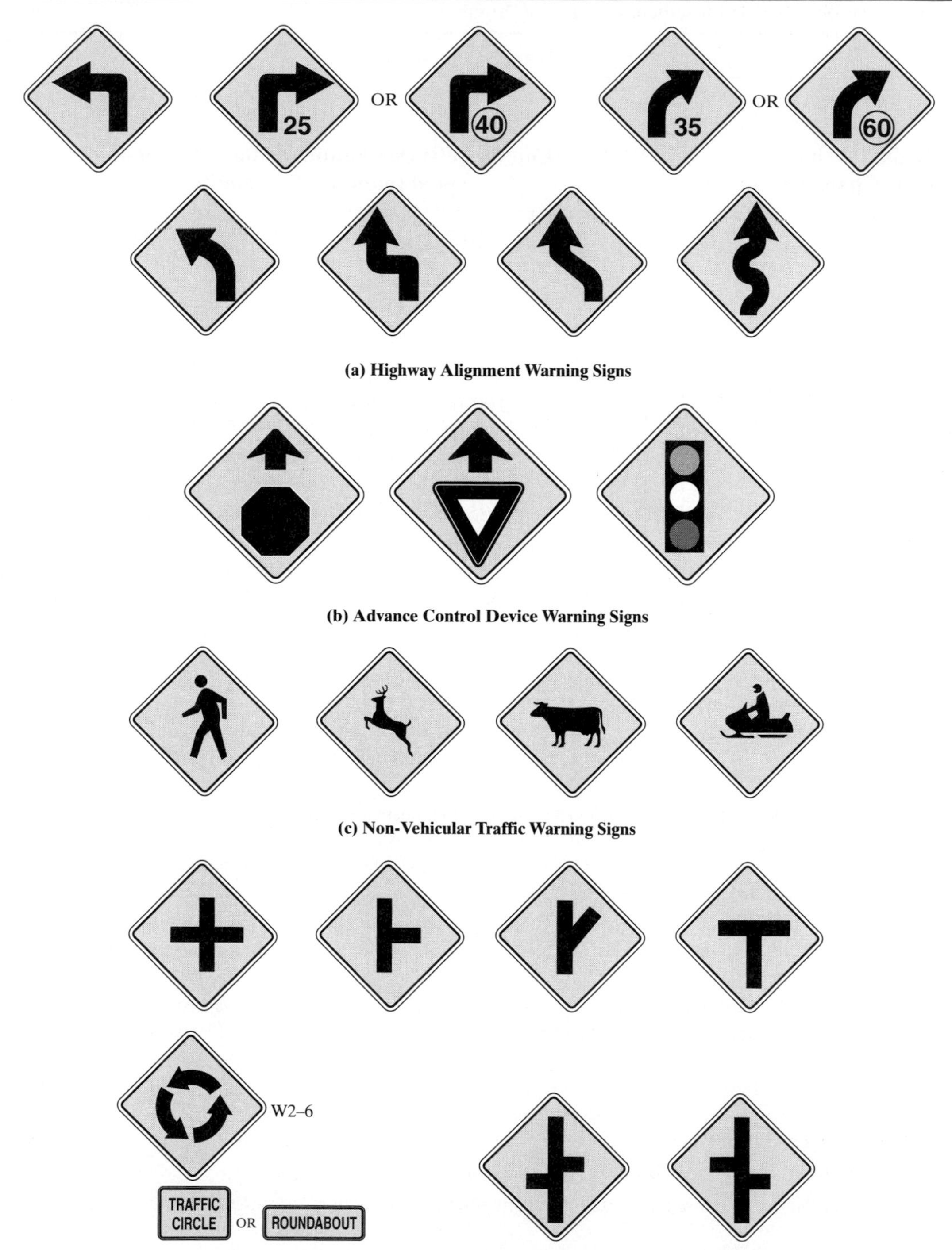

Figure 4.12: Sample Warning Signs

(*Source: Manual of Uniform Traffic Control Devices*, Draft, Federal Highway Administration, Washington DC, December 2007, Figs 2C-1, 2C-6, 2C-10, 2C-12.)

lesser suburbs would consistently list Philadelphia as the primary destination.

2. No guide sign should list more than three (four may be acceptable in some circumstances) destinations on a single sign. This, in conjunction with the first principle, makes the selection of priority destinations a critical part of effective guide signing.
3. Where roadways have both a name and a route number, both should be indicated on the sign if space permits. In cases where only one may be listed, the route number takes precedence. Road maps show route numbers prominently, although not all facility names are included. Unfamiliar drivers, therefore, are more likely to know the route number than the facility name.
4. Wherever possible, advance signing of important junctions should be given. This is more difficult on conventional highways, where junctions may be frequent and closely spaced. On freeways and expressways, this is critical because high approach speeds make advance knowledge of upcoming junctions a significant safety issue.
5. Confusion on the part of the driver must be avoided at all cost. Sign sequencing should be logical and should naturally lead the driver to the desired route selections. Overlapping sequences should be avoided wherever possible. Left-hand exits and other unusual junction features should be signed extremely carefully.

The size, placement, and lettering of guide signs vary considerably, and the manual gives information on numerous options. A number of site-specific conditions affect these design features, and there is more latitude and choice involved than for other types of highway signs. The MUTCD should be consulted directly for this information.

Route Markers

Figure 4.13 illustrates route markers that are used on all numbered routes. The signs have unique designs that signify the type of route involved. Interstate highways have a unique shield shape, with red and blue background and white lettering. The same design is used for designated "business loops." Such loops are generally a major highway that is not part of the interstate system but one that serves the business area of a city from an interchange on the interstate system. U.S. route markers consist of black numerals on a white shield that is placed on a square sign with a black background. State route markers are designed by the individual states and, therefore, vary from state to state. All county route markers, however, follow a standard design, with yellow lettering on a blue background and a unique shape. The name of the county is placed on the route marker. Routes in national parks and/or national forests also have a unique shape and have white lettering on a brown background.

Route markers may be supplemented by a variety of panels indicating cardinal directions or other special purposes. Special-purpose panels include JCT, ALT or ALTERNATE, BY-PASS, BUSINESS, TRUCK, TO, END, and TEMPORARY. Auxiliary panels match the colors of the marker they are supplementing.

Chapter 17 includes a detailed discussion of the use of route markers and various route marker assemblies for freeways, expressways, and conventional roads.

Figure 4.13: Route Markers Illustrated

(*Source: Manual of Uniform Traffic Control Devices*, Draft, Federal Highway Administration, Washington DC, December 2007, Fig 2D-3.)

Destination Signs: Conventional Roads

Destination signs are used on conventional roadways to indicate the distance to critical destinations along the route and to mark key intersections or interchanges. On conventional roads, destination signs use an all-capital white legend on a green background. The distance in miles to the indicated destination may be indicated to the right of the destination.

Destination signs are generally used at intersections of U.S. or state numbered routes with Interstate, U.S., state numbered routes, or junctions forming part of a route to such a numbered route. Distance signs are usually placed on important routes leaving a municipality of a major junction with a numbered route.

Local street name signs are recommended for all suburban and urban junctions as well as for major intersections at rural locations. Local street name signs are categorized as conventional roadway destination signs. Figure 4.14 illustrates a selection of these signs.

Destination Signs: Freeways and Expressways

Destination signs for freeways and expressways are similar, although there are different requirements for size and placement specified in the MUTCD. They differ from conventional road guide signs in a number of ways:

- Destinations are indicated in initial capitals and small letters.
- Numbered routes are indicated by inclusion of the appropriate marker type on the guide sign.
- Exit numbers are included as auxiliary panels located at the upper right or left corner of the guide sign.
- At major junctions, diagrammatic elements may be used on guide signs.

As for conventional roadways, distance signs are frequently used to indicate the mileage to critical destinations along the route. Every interchange and every significant at-grade intersection on an expressway is extensively signed with advance signing as well as with signing at the junction itself.

The distance between interchanges has a major impact on guide signing. Where interchanges are widely separated, advance guide signs can be placed as much as 5 or more miles from the interchange and may be repeated several times as the interchange is approached.

In urban and suburban situations, where interchanges are closely spaced, advanced signing is more difficult to achieve. Advance signing usually gives information only concerning the *next* interchange, to avoid confusion caused by overlapping signing sequences. The only exception to this is a distance sign indicating the distance to the next several interchanges. Thus, in urban and suburban areas with closely spaced interchanges, the advance sign for the next interchange is placed at the last off-ramp of the previous interchange.

A wide variety of sign types are used in freeway and expressway destination signing. A few of these are illustrated in Figure 4.15.

Figure 4.15(a) shows a typical advance exit sign. These are placed at various distances from the interchange in accordance with the overall signing plan. The number and placement of advance exit sign is primarily dependent on interchange spacing. Figure 4.15(b) depicts a "next exit" panel. These may be placed below an exit or advance exit sign. The supplemental multiple exit sign of Figure 4.15(c) is used where separate ramps exist for the two directions of the

Figure 4.14: Destination Signs for Conventional Roads

(*Source: Manual of Uniform Traffic Control Devices*, Draft, Federal Highway Administration, Washington DC, December 2007, Fig 2D-8.)

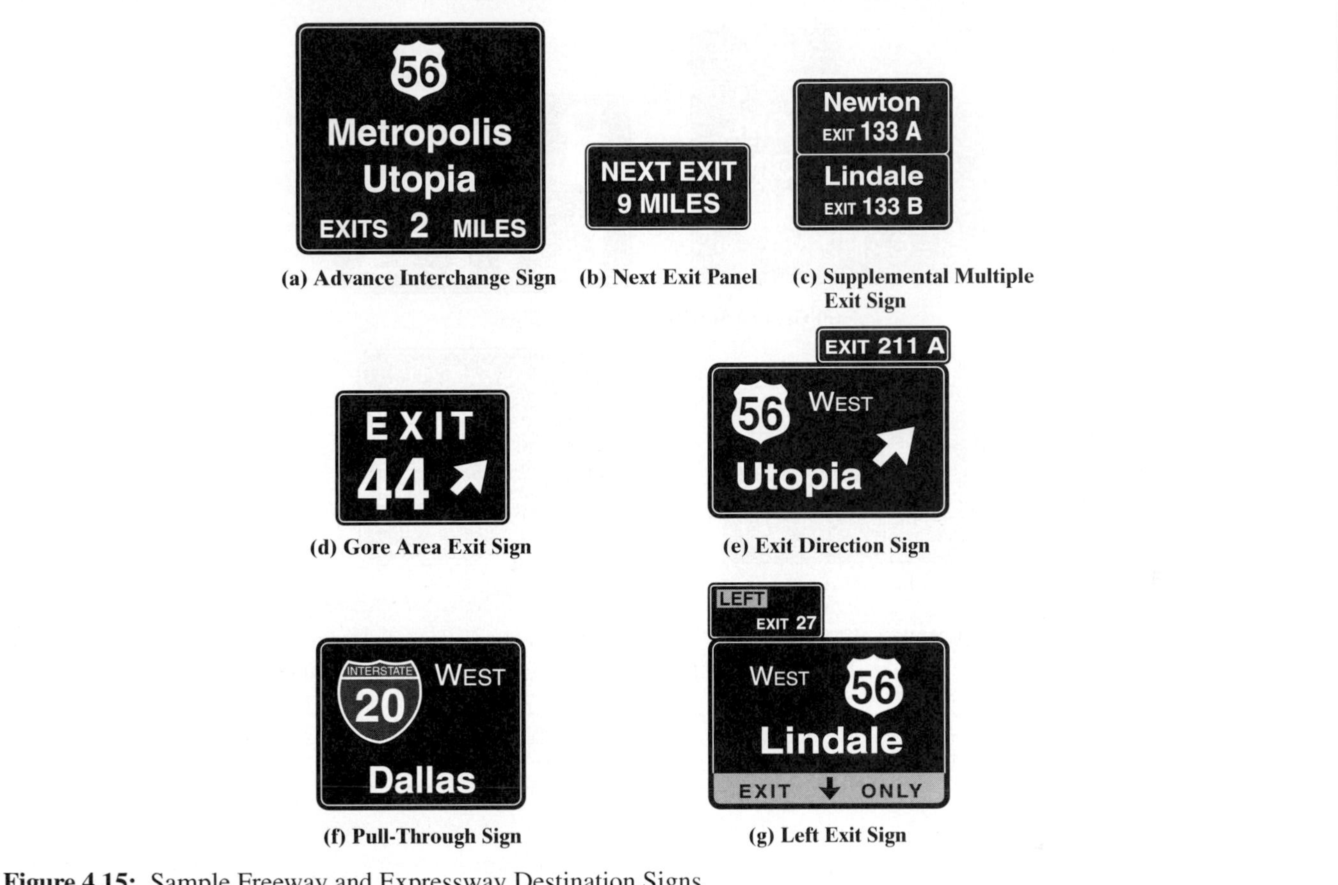

(a) Advance Interchange Sign (b) Next Exit Panel (c) Supplemental Multiple Exit Sign

(d) Gore Area Exit Sign (e) Exit Direction Sign

(f) Pull-Through Sign (g) Left Exit Sign

Figure 4.15: Sample Freeway and Expressway Destination Signs

(*Source: Manual of Uniform Traffic Control Devices*, Draft, Federal Highway Administration, Washington DC, December 2007, Figs 2E-1, 2E-8, 2E-15, 2E-16, 2E-17, 2E-20, 2E-22.)

connecting roadway. It is generally placed after the last advance exit sign and before the interchange itself. The gore area exit sign of Figure 4.15(d) is placed in the gore area and is the last sign associated with a given ramp connection. Such signs are usually mounted on breakaway sign posts to avoid serious damage to vehicles straying into the gore area. Figure 4.15(e) is an exit direction sign, which is posted at the location of the diverge and includes an exit number panel. The "pull-through" sign of Figure 4.15(f) is used primarily in urban or other areas with closely spaced interchanges. It is generally mounted on overhead supports next to the exit direction sign. It reinforces the direction for drivers intending to continue on the freeway. The final illustration, Figure 4.15(g), is a left-side exit sign. The exit number tab is on the left side of the sign and a warning panel provide additional emphasis.

Chapter 17 contains a detailed discussion of guide signing for freeways, expressways, and conventional roadways.

Service Guide Signs

Another important type of information drivers require is directions to a variety of motorists' services. Drivers, particularly those who are unfamiliar with the area, need to be able to easily locate such services as fuel, food, lodging, medical assistance, and similar services. The MUTCD provides for a variety of signs, all using white legend and symbols on a blue background, to convey such information. In many cases, symbols are used to indicate the type of service available. On freeways, large signs using text messages may be used with exit number auxiliary panels. The maximum information is provided by freeway signs that indicate the actual brand names of available services (gas

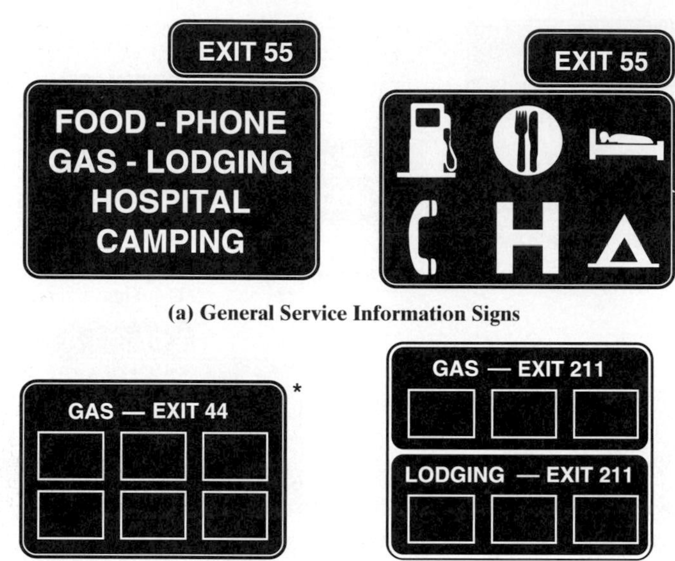

(a) General Service Information Signs

(b) Specific Service Information Signs

Figure 4.16: Service Information Signs

(*Source: Manual of Uniform Traffic Control Devices*, Draft, Federal Highway Administration, Washington DC, December 2007, Fig 2F-3, 2G-1.)

companies, restaurant names, etc.). Figure 4.16 illustrates some of the signs used to provide motorist service information.

There are a number of guidelines for specific service signing. No service should be included that is more than 3 miles from the freeway interchange. No specific services should be indicated where drivers cannot easily reenter the freeway at the interchange.

Specific services listed must also conform to a number of criteria regarding hours of operations and specific functions provided. All listed services must also be in compliance with all federal, state, and local laws and regulations concerning their operation. Consult the MUTCD directly for the details of these requirements.

Service guide signs on conventional highways are similar to those of Figure 4.16 but do not use exit numbers or auxiliary exit number panels.

Recreational and Cultural-Interest Guide Signs

Information on historic, recreational, and/or cultural-interest areas or destinations is given on signs with white legend and/or symbols on a brown background. Symbols are used to depict the type of activity but larger signs with word messages may be used as well. Figure 4.17 shows some examples of these signs. The millennium edition of the MUTCD has introduced many more acceptable symbols and should be consulted directly for illustrations of these.

Mileposts

Mileposts are small 6×9-in vertical white-on-green panels indicating the mileage along the designated route. These are provided to allow drivers to estimate their progress along a route and provide a location system for accident reporting and other emergencies that may occur along the route. Distance numbering is continuous within a state, with "zero" beginning at the south or west state lines or at the southern-most or western-most interchange at which the route begins. Where routes overlap, mileposts are continuous only for *one* of the routes. In such cases, the first milepost beyond the overlap should indicate the total mileage traveled along the route that is *not* continuously numbered and posted. On some freeways, markers are placed every tenth of a mile for a more precise location system.

(a) Directional Signs with Arrow

Yellowstone
National Park
2 MILES

Great Smoky Mts
National Park

(b) Text Legend Cultural-Interest Signs

Figure 4.17: Recreational and Cultural Interest Signs

(*Source: Manual of Uniform Traffic Control Devices*, Draft, Federal Highway Administration, Washington DC, December 2007, Figs 2J-1, 2J-2.)

4.4 Traffic Signals

The MUTCD defines nine types of traffic signals:

- Traffic control signals
- Pedestrian signals
- Emergency vehicle traffic control signals
- Traffic control signals for one-lane, two-way facilities
- Traffic control signals for freeway entrance ramps
- Traffic control signals for movable bridges
- Lane-use control signals
- Flashing beacons
- In-roadway lights

The most common of these is the traffic control signal, used at busy intersections to direct traffic to alternately stop and move.

4.4.1 Traffic Control Signals

Traffic signals are the most complicated form of traffic control devices available to traffic engineers. The MUTCD addresses:

- Physical standards for signal displays, including lens sizes, colors (specific pigments), arrangement of lenses within a single signal head, arrangement and placement of signal heads within an intersection, visibility requirements, and so on.
- Definitions and meaning of the various indications authorized for use.
- Timing and sequence restrictions.
- Maintenance and operations criteria.

There are two important guidelines regarding operation of traffic signals: (1) traffic control signals must be in operation at all times; and (2) STOP signs shall not be used in conjunction with a traffic control signal unless it is operating in the RED-flashing mode at all times.

No traffic signal should ever be "dark," that is, showing no indications. This is particularly confusing to drivers and can result in accidents. Any accidents occurring while a signal is in the dark mode are the legal responsibility of the agency operating the signal in most states. When signals are inoperable, signal heads should be bagged or taken down to avoid such confusion. In power outages, police or other authorized agents should be used to direct traffic at all signalized locations.

The second principle relates to a common past practice—turning off signals at night and using STOP control during these hours. The problem is that during daytime hours, the driver may be confronted with a green signal *and* a STOP sign. This is extremely confusing and is no longer considered appropriate. The use of STOP signs in conjunction with permanent operation of a red flashing light is permissible because the legal interpretation of a flashing red signal is the same as that of a STOP sign.

Signal Warrants

Traffic signals, when properly installed and operated at appropriate locations, provide a number of significant benefits:

- With appropriate physical designs, control measures, and signal timing, the capacity of critical intersection movements is increased.
- The frequency and severity of accidents is reduced for certain types of crashes, including right-angle, turn, and pedestrian accidents.
- When properly coordinated, signals can provide for nearly continuous movement of through traffic along an arterial at a designated speed under favorable traffic conditions.
- They provide for interruptions in heavy traffic streams to permit crossing vehicular and pedestrian traffic to cross safely.

At the same time, misapplied or poorly designed signals can cause excessive delay, signal violations, increased accidents (particularly rear-end accidents), and drivers rerouting their trips to less appropriate routes.

The MUTCD provides very specific warrants for the use of traffic control signals. These warrants are far more detailed than those for other devices, due to their very high cost (relative to other control devices) and the negative impacts of their misapplication. Thus the manual is clear that traffic control signals shall be installed only at locations where an engineering study has indicated that one or more of the specified warrants has been met, and that application of signals will improve safety and/or capacity of the intersection. The manual goes further; if a study indicates that an existing signal is in place at a location that does not meet any of the warrants, it should be removed and replaced with a less severe form of control.

The MUTCD details nine different warrants, any one of which may indicate that installation of a traffic control signal is appropriate. Chapter 18 contains a detailed treatment of these warrants and their application as part of an overall process for determining the appropriate form of intersection control for any given situation.

Signal Indications

The MUTCD defines the meaning of each traffic control signal indication as follows:

- *Green ball.* A steady green circular indication allows vehicular traffic facing the ball to enter the intersection to travel straight through the intersection or to turn right or left, except when prohibited by lane-use controls or physical design. Turning vehicles must yield the right-of-way to opposing through vehicles and to pedestrians legally in a conflicting crosswalk. In the absence of pedestrian signals, pedestrians may proceed to cross the roadway within any legally marked or unmarked crosswalk.
- *Yellow ball.* The steady yellow circular indication is a transition between the Green Ball and the Red Ball indication. It warns drivers that the related green movement is being terminated or that a red indication will immediately follow. In most states, drivers are allowed to legally enter the intersection during a "yellow" display. Some states, however, only allow the driver to enter on "yellow" if they can clear the intersection before the "yellow" terminates. This is very difficult for drivers, however, because they do not know when the "yellow" is timed to end. Where no pedestrian signals are in use, pedestrians may not begin crossing a street during the "yellow" indication.
- *Red ball.* The steady red circular indication requires all traffic (vehicular and pedestrian) facing it to stop at the STOP line, crosswalk line (if no STOP line exists), or at the conflicting pedestrian path (if no crosswalk or STOP line exists). All states allow right-turning traffic to proceed with caution after stopping, unless specifically prohibited by signing or statute. Some states allow left-turners from one one-way street turning into another to proceed with caution after stopping, but this is far from a universal statute.

- *Flashing ball.* A flashing "yellow" allows traffic to proceed with caution through the intersection. A flashing "red" has the same meaning as a STOP sign—the driver may proceed with caution after coming to a complete stop.
- *Arrow indications.* Green, yellow, and red arrow indications have the same meanings as ball indications, except that they apply only to the movement designated by the arrow. A green left-turn arrow is only used to indicate a protected left turn (i.e., a left turn made on a green arrow will not encounter an opposing vehicular through movement). Such vehicles, however, may encounter pedestrians legally in the conflicting crosswalk and must yield to them. A green right-turn arrow is shown only when there are no pedestrians legally in the conflicting crosswalk. Yellow arrows warn drivers that the green arrow is about to terminate. The yellow arrow may be followed by a green ball indication where the protected left- and/or right-turning movement is followed by a permitted movement. A "permitted" left turn is made against an opposing vehicular flow. A "permitted" right turn is made against a conflicting pedestrian flow. It is followed by a red arrow (or red ball) where the movement must stop.

The MUTCD provides additional detailed discussion on how and when to apply various sequences and combinations of indications.

Signal Faces and Visibility Requirements

In general, a signal face should have three to five signal lenses, with some exceptions allowing for a sixth to be shown. Two lens sizes are provided for: 8-inch diameter and 12-inch lenses. The manual now mandates the use of 12-inch lenses for all *new* signal installations, except when used as a supplemental signal for pedestrian use only or when used at very closely spaced intersections where visibility shields cannot be used effectively. Eight-inch lenses at existing installations may be kept in place for their useful service life. If replaced, they must be replaced with 12-inch lenses.

Table 4.2 shows the minimum visibility distances required for signal faces. A minimum of two signal faces must be provided for the major movement on each approach, even if the major movement is a turning movement. This requirement provides some measure of redundancy in case of an unexpected bulb failure.

The arrangement of lenses on a signal face is also limited to approved sequences. In general, the red ball must be at the top of a vertical signal face or at the left of a horizontal signal face, followed by the yellow and green. Where arrow indications are on the same signal face as ball indications, they are located on the bottom of a vertical display or right of a horizontal display. Figure 4.18 shows the most commonly used lens arrangements. The MUTCD contains detailed discussion of the applicability of various signal face designs.

Table 4.2: Minimum Sight Distances for Signal Faces

85th-Percentile Speed (mph)	Minimum Sight Distance (feet)
20	175
25	215
30	270
35	325
40	390
45	460
50	540
55	625
60	715

(*Source: Manual of Uniform Traffic Control Devices*, Draft, Federal Highway Administration, Washington DC, December 2007, Table 4D-1.)

Figure 4.19 illustrates the preferred placement of signal faces. At least one of the two required signal faces for the major movement must be located between 40 and 150 ft of the STOP line, unless the physical design of the intersection prevents it. Horizontal placement should be within 20° of the centerline of the approach, facing straight ahead.

Figure 4.20 illustrates the standard for vertical placement of signal faces that are between 40 and 53 ft from the STOP line. The standard prescribes the maximum height of the top of the signal housing above the pavement.

Operational Restrictions

Continuous operation of traffic control signals is critical for safety. No signal face should ever be "dark" (i.e., with no lens illuminated). In cases where signalization is not deemed necessary at night, signals must be operated in the flashing mode ("yellow" for one street and "red" for the other). Signal operations must also be designed to allow flashing operation to be maintained even when the signal controller is undergoing maintenance or replacement.

When being installed, signal faces should be bagged and turned to make it obvious to drivers that they are not in operation. Signals should be made operational as soon as possible after installation—again, to minimize possible confusion to drivers.

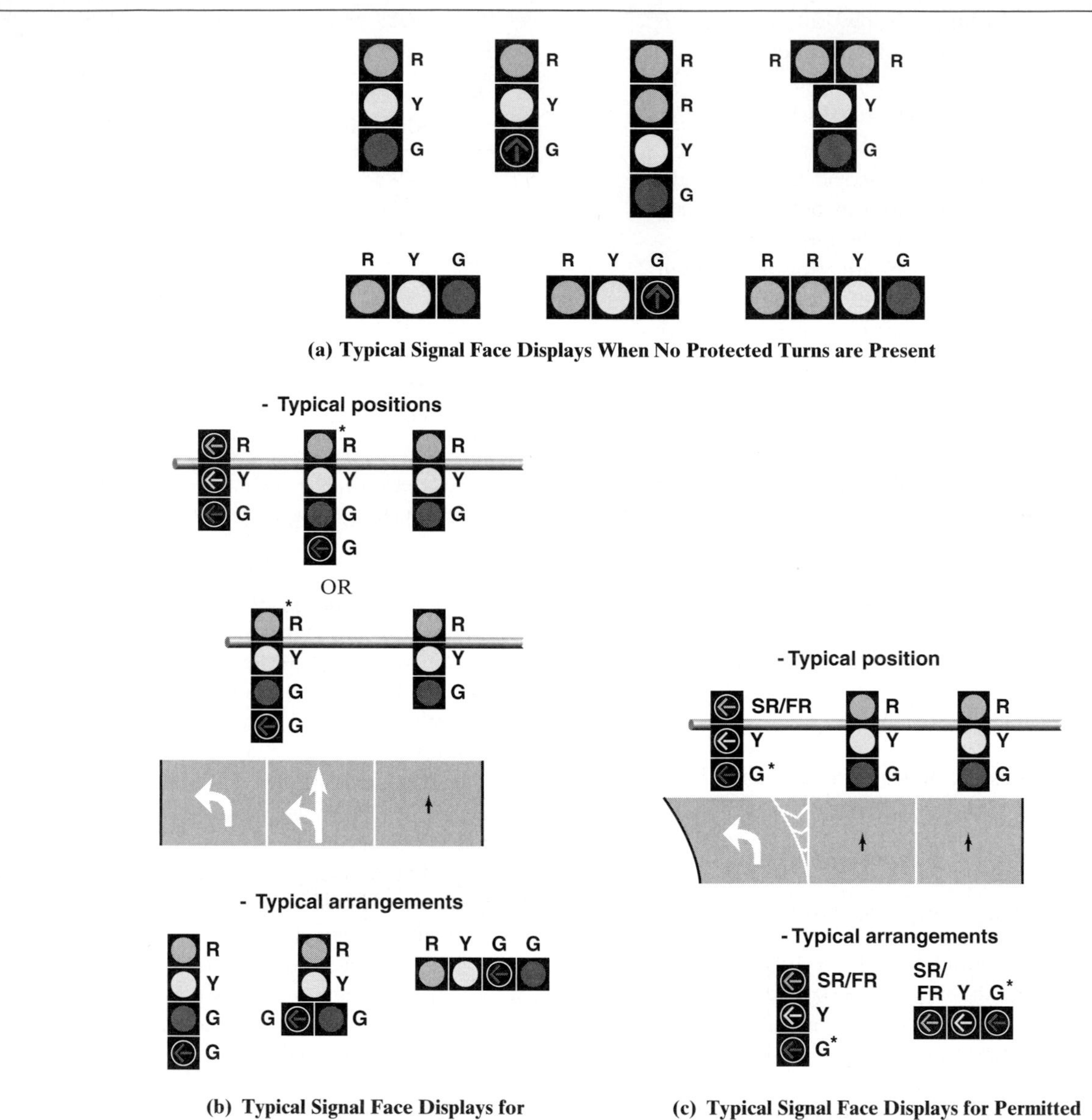

Figure 4.18: Typical Signal Face Arrangements Illustrated

(*Source: Manual of Uniform Traffic Control Devices*, Draft, Federal Highway Administration, Washington DC, December 2007, Figs 4D-2, 4D-8, 4D-9.)

Bulb maintenance is a critical part of safe signal operation because a burned-out bulb can make a signal face appear to be "dark" during certain intervals. A regular bulb-replacement schedule must be maintained. It is common to replace signal bulbs regularly at about 75% to 80% of their expected service life to avoid burnout problems. Other malfunctions can lead to other nonstandard indications appearing, although most controllers are programmed to fall back to the flashing mode in the event of most malfunctions. Most signal agencies maintain a contract with a private maintenance organization that requires rapid response (in the order of 15 to 30 minutes) to any reported malfunction. The agency can also operate its own maintenance group under similar rules. Any accident occurring during a signal malfunction can lead to legal liability for the agency with jurisdiction.

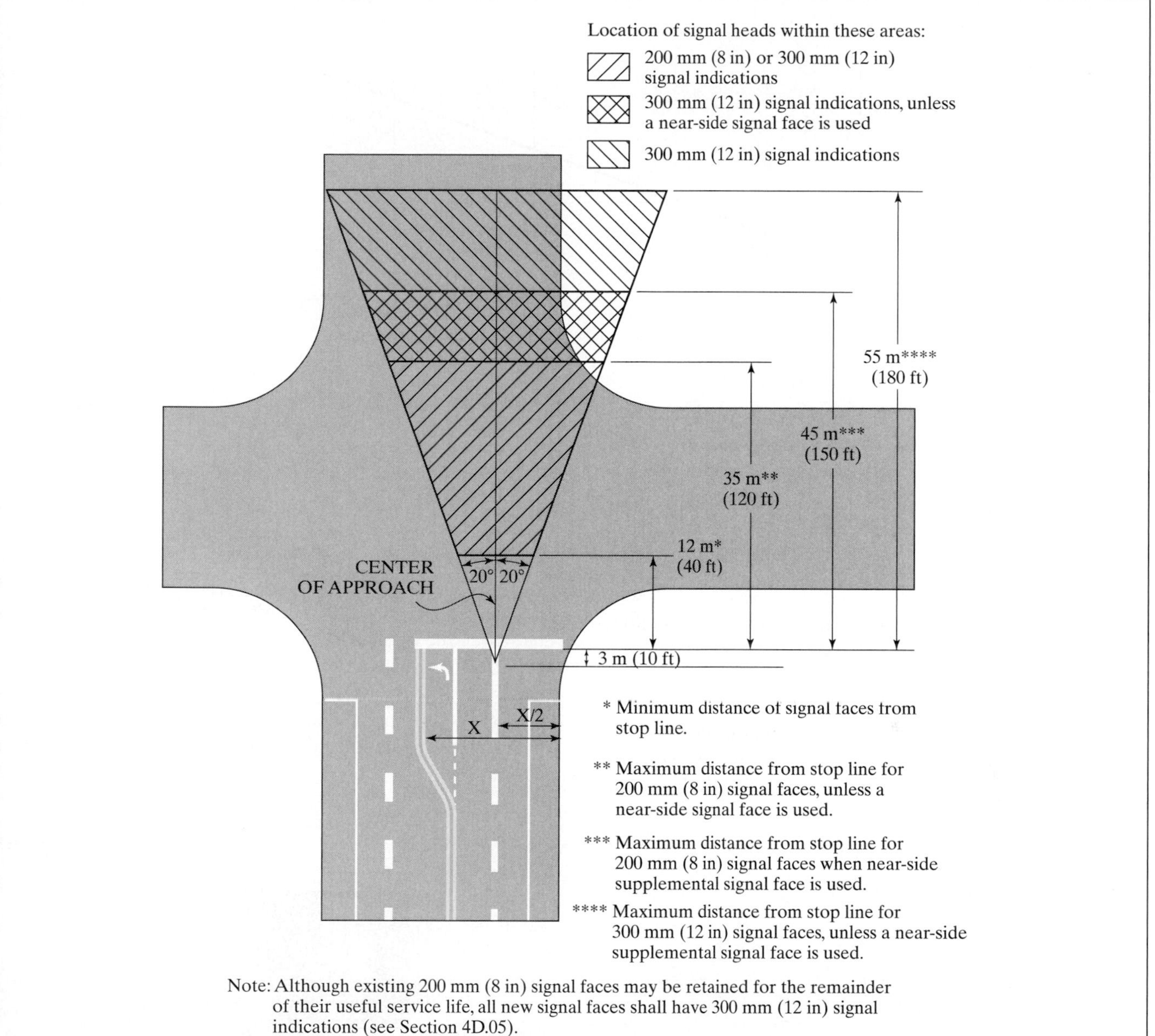

Figure 4.19: Horizontal Location of Signal Faces

(*Source: Manual of Uniform Traffic Control Devices*, Draft, Federal Highway Administration, Washington DC, December 2007, Fig 4D-4.)

4.4.2 Pedestrian Signals

The MUTCD has mandated the use of pedestrian signals that had been introduced as options to replace older "WALK" and "DON'T WALK" designs:

- *Walking man (steady).* The new "WALK" indication is the image of a walking person in the color white. This indicates that it is permissible for a pedestrian to enter the crosswalk to begin crossing the street.
- *Upraised hand (flashing).* The new "DON'T WALK" indication is an upraised hand in the color Portland orange. In the flashing mode, it indicates that no pedestrian may enter the crosswalk to begin crossing the street but that those already crossing may continue safely.

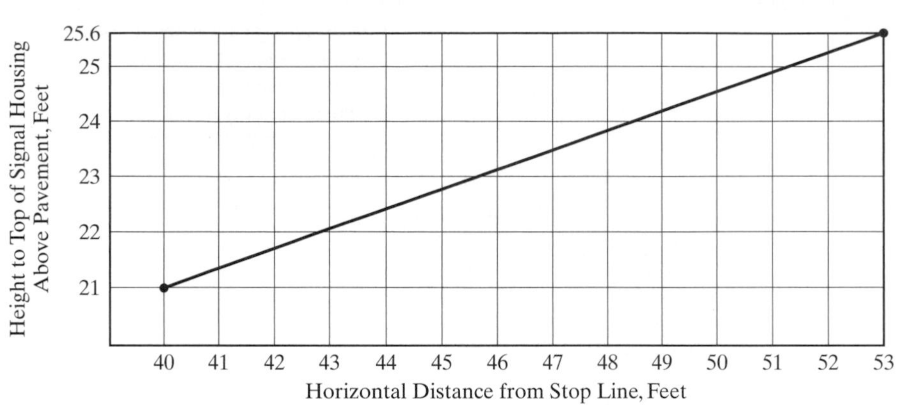

Figure 4.20: Vertical Location of Signal Faces

(*Source: Manual of Uniform Traffic Control Devices*, Draft, Federal Highway Administration, Washington DC, December 2007, Fig 4D-5.)

- *Upraised hand (steady).* In the steady mode, the upraised hand indicates that no pedestrian should begin crossing and that no pedestrian should still be in the crosswalk.

In previous manuals, a flashing "WALK" indication was an option that could be used to indicate that right-turning vehicles may be conflicting with pedestrians legally in the crosswalk. The new manual does not permit a flashing WALKING MAN, effectively discontinuing this practice.

Figure 4.21 shows the new pedestrian signals. Note that both the UPRAISED HAND and WALKING MAN symbols should be shown in the form of a solid image. They may be located side-by-side on a single-section signal or arranged vertically on two-section signal. The UPRAISED HAND is on the left or on top in these displays. When not illuminated, neither symbol should be readily visible to pedestrians at the far end of the crosswalk.

Chapters 21 and 22 discuss the use and application of pedestrian signals in the context of overall intersection control and operation. They include a discussion of when and where pedestrian signals are mandated as part of a signalization design.

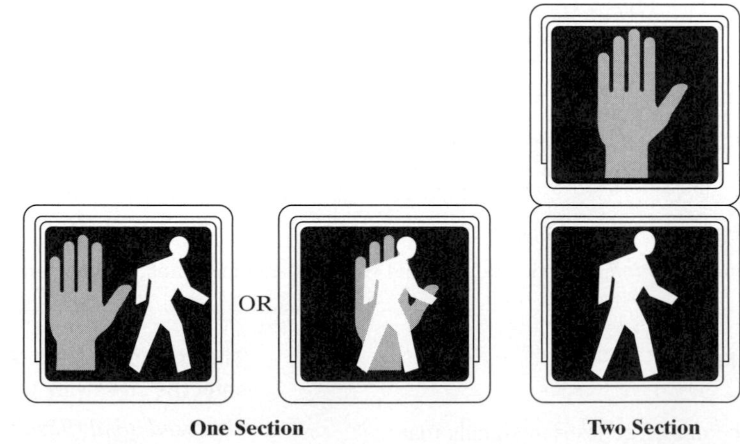

Figure 4.21: Typical Pedestrian Signals

(*Source: Manual of Uniform Traffic Control Devices*, Draft, Federal Highway Administration, Washington DC, December 2007, Fig 4E-1.)

4.4.3 Other Traffic Signals

The MUTCD provides specific criteria for the design, placement, and use of a number of other types of signals, including:

- Beacons
- In-roadway lights
- Lane-use control signals
- Ramp control signals (or ramp meters)

Beacons are generally used to identify a hazard or call attention to a critical control device, such as a speed limit sign, a STOP or YIELD sign, or a DO NOT ENTER sign. Lane-use control signals are used to control reversible lanes on bridges, in tunnels, and on streets and highways.

4.4.4 Traffic Signal Controllers

Modern traffic signal controllers are a complex combination of hardware and software that implements signal timing and ensures that signal indications operate consistently and continuously in accordance with the programmed signal timing. Each signalized intersection has a controller dedicated to implementing the signal-timing plan at that intersection. In addition, master controllers coordinate the operation of many signals, allowing signals along an arterial or in a network to be coordinated to provide progressive movement and/or other arterial or network control policies.

Individual traffic controllers may operate in the *pretimed* or *actuated* mode. In pretimed operation, the sequence and timing of every signal indication is preset and repeated in each signal cycle. In actuated operation, the sequence and timing of some or all of the green indications may change on a cycle-by-cycle basis in response to detected vehicular and pedestrian demand. Chapter 20 discusses the timing and design of pretimed signals. Chapter 21 discusses the timing and design of semiactuated and fully actuated signals.

Traffic controllers implement signal timing designs. They are connected to *display hardware*, which consists of various traffic control and pedestrian signal faces that inform drivers and pedestrians of when they may legally proceed. Such display hardware also includes various types of supporting structures. Where individual intersection control and/or the signal system has demand-responsive elements, controllers must be connected with properly placed *detectors* that provide information on vehicle and/or pedestrian presence that interacts with controllers to determine and implement a demand-responsive signal timing pattern. Chapter 19 contains a more detailed discussion of street hardware, detectors, and their placement as part of an intersection design.

The *Manual of Traffic Control Devices* [*6*] and the *Traffic Detector Handbook* [*7*] are standard traffic engineering references that provide significant detail on all elements of traffic signal hardware.

Standards for Traffic Signal Controllers

The National Electrical Manufacturers Association (NEMA) is the principal trade group for the electronics industry in the United States. Its Traffic Control Systems Section sets manufacturing guidelines and standards for traffic control hardware [*8*]. The group's philosophy is to encourage industry standards that:

- Are based on proven designs
- Are downward compatible with existing equipment
- Reflect state-of-the-art reliability and performance
- Minimize the potential for malfunctions

NEMA standards are not product designs but rather descriptions and performance criteria for various product categories. The standards cover such products as solid-state controllers, load switches, conflict monitors, loop detectors, flashers, and terminals and facilities. Reference 9 provides a good overview of what the NEMA standards mean and imply.

NEMA does not preclude any manufacturer from making and selling a nonconforming product. Many funding agencies, however, require the use of hardware conforming to the latest NEMA standards.

Two other standards also exist. In New York State, standards for the Type 170 controller have been developed and extensively applied. In California, standards for the Type 2070 controller have been similarly developed and implemented. Both have similar features but are implemented using different hardware and software architecture.

4.5 Special Types of Control

Although not covered in this chapter, the MUTCD contains significant material covering special control situations, including:

- School zones
- Railroad crossings
- Construction and maintenance zones
- Pedestrian and bicycle controls

These situations invariably involve a combination of signing, markings, and/or signals for fully effective control. Consult the MUTCD directly for details on these and other applications not covered here.

4.6 Summary and Conclusion

This chapter has provided an introduction and overview to the design, placement, and use of traffic control devices. The MUTCD is not a stagnant document, and updates and revisions are continually being issued. Thus it is imperative that users consult the latest version of the manual and all of its formal revisions. For convenience, the MUTCD can be accessed online at mutcd.fhwa.gov/kno-millennium.htm or through the federal highway administration home Web site—www.fhwa.dot.com. This is a convenient way of using the manual because all updates and revisions are always included. Similarly, virtually every signal manufacturer has a Web site that can be accessed to review detailed specifications and characteristics of controllers and other signal hardware and software. A directory of Web sites related to traffic control devices may be found at www.traffic-signals.com.

The use of the manual is discussed and illustrated in a number of other chapters of this text: Chapter 18 for signal warrants and warrants for STOP and YIELD control, Chapter 17 for application of traffic control devices on freeways and rural highways, and Chapter 19 for application of control devices as part of intersection design.

References

1. *Manual of Uniform Traffic Control Devices*, Draft, Federal Highway Administration, U.S. Department of Transportation, Washington DC, December 2007.
2. Hawkins, H.G., "Evolution of the MUTCD: Early Standards for Traffic Control Devices," *ITE Journal*, Institute of Transportation Engineers, Washington DC, July 1992.
3. Hawkins, H.G., "Evolution of the MUTCD: Early Editions of the MUTCD," *ITE Journal*, Institute of Transportation Engineers, Washington DC, August 1992.
4. Hawkins, H.C., "Evolution of the MUTCD: The MUTCD Since WWII," *ITE Journal*, Institute of Transportation Engineers, Washington DC, November 1992.
5. Hawkins, H.C., "Evolution of the MUTCD Mirrors American Progress Since the 1920's," *Roads and Bridges*, Scranton Gillette, Communications Inc., Des Plaines, IL, July 1995.
6. Kell, J. and Fullerton, I., *Manual of Traffic Signal Design*, 2nd Edition, Institute of Transportation Engineers, Prentice Hall Inc., Englewood Cliffs, NJ, 1991.
7. *Traffic Detector Handbook*, 2nd Edition, JHK & Associates, Institute of Transportation Engineers, Washington DC, n.d.
8. Parris, C., "NEMA and Traffic Control," *ITE Journal*, Institute of Transportation Engineers, Washington DC, August 1986.
9. Parris, C., "Just What Does a NEMA Standard Mean?" *ITE Journal*, Institute of Transportation Engineers, Washington DC, July 1987.

Problems

4-1. Define the following terms with respect to their meaning in the millennium edition of the MUTCD: standard, guideline, option, and support.

4-2. Describe how color, shape, and legend are used to convey and reinforce messages given by traffic control devices.

4-3. Why should overuse of regulatory and warning signs be avoided? Why is this not a problem with guide signs?

4-4. How far from the point of a hazard should the following warning signs be placed?

(a) A "STOP ahead" warning sign on a road with a posted speed limit of 50 mi/h.

(b) A "curve ahead" warning sign with an advisory speed of 30 mi/h on a road with a posted speed limit of 45 mi/h.

(c) A "merge ahead" warning sign on a ramp with an 85th percentile speed of 35 mi/h.

4-5. Select a 1-mile stretch of freeway in your vicinity. Drive one direction of this facility with a friend or colleague. The passenger should count and note the number and type of traffic signs encountered. Are any of them confusing? Suggest improvements as appropriate. Comment on the overall quality of the signing in the test section.

4-6. Select one signalized and one STOP or YIELD controlled intersection in your neighborhood. Note the placement of all devices at each intersection. Do they appear to meet MUTCD standards? Is visibility of all devices adequate? Comment on the effectiveness of traffic controls at each intersection.

CHAPTER 5

Traffic Stream Characteristics

Traffic streams are made up of individual drivers and vehicles interacting with each other and with the physical elements of the roadway and its general environment. Because both driver behavior and vehicle characteristics vary, individual vehicles within the traffic stream do not behave in exactly the same manner. Further, no two traffic streams will behave in exactly the same way, even in similar circumstances, because driver behavior varies with local characteristics and driving habits.

Dealing with traffic, therefore, involves an element of variability. A flow of water through channels and pipes of defined characteristics will behave in an entirely predictable fashion, in accord with the laws of hydraulics and fluid flow. A given flow of traffic through streets and highways of defined characteristics will vary with both time and location. Thus the critical challenge of traffic engineering is to plan and design for a medium that is not predictable in exact terms—one that involves both physical constraints and the complex behavioral characteristics of human beings.

Fortunately, although exact characteristics vary, there is a reasonably consistent range of driver and, therefore, traffic stream behavior. Drivers on a highway designed for a safe speed of 60 mi/h may select speeds in a broad range (perhaps 45 to 65 mi/h); few, however, will travel at 80 mi/h or at 20 mi/h.

In describing traffic streams in quantitative terms, the purpose is both to understand the inherent variability in their characteristics and to define normal ranges of behavior. To do so, key parameters must be defined and measured. Traffic engineers analyze, evaluate, and ultimately plan improvements to traffic facilities based on such parameters and their knowledge of normal ranges of behavior.

This chapter focuses on the definition and description of the parameters most often used for this purpose and on the characteristics normally observed in traffic streams. These parameters are, in effect, the traffic engineer's measure of reality, and they constitute a language with which traffic streams are described and understood.

5.1 Types of Facilities

Traffic facilities are broadly separated into two principal categories:

- Uninterrupted flow
- Interrupted flow

Uninterrupted flow facilities have no external interruptions to the traffic stream. Pure uninterrupted flow exists primarily on freeways, where there are no intersections at grade, traffic signals, STOP or YIELD signs, or other interruptions external to the traffic stream itself. Because such facilities have full control of access, there are no intersections at grade, driveways, or any forms of direct access to abutting lands. Thus the characteristics of the traffic stream are based solely on the interactions among vehicles and with the roadway and the general environment.

Although pure uninterrupted flow exists only on freeways, it can also exist on sections of surface highway, most often in rural areas, where there are long distances between fixed interruptions. Thus uninterrupted flow may exist on some sections of rural two-lane highways and rural and suburban multilane highways. As a very general guideline, it is believed that uninterrupted flow can exist in situations where the distance to the last traffic signal or other significant fixed interruption is more than 2 miles.

Remember that the term *uninterrupted flow* refers to a type of facility, not the quality of operations on that facility. Thus a freeway that experiences breakdowns and long delays during peak hours is still operating under uninterrupted flow. The causes for the breakdowns and delay are not external to the traffic stream but are caused entirely by the internal interactions within the traffic stream.

Interrupted flow facilities are those that incorporate fixed external interruptions into their design and operation. The most frequent and operationally significant external interruption is the traffic signal. The traffic signal alternatively starts and stops a given traffic stream, creating a platoons of vehicles progressing down the facility. Other fixed interruptions include STOP and YIELD signs, unsignalized at-grade intersections, driveways, curb parking maneuvers, and other land-access operations. Virtually all urban surface streets and highways are interrupted flow facilities.

The major difference between uninterrupted and interrupted flow facilities is the impact of time. On uninterrupted facilities, the physical facility is available to drivers and vehicles at all times. On a given interrupted flow facility, movement is periodically barred by "red" signals. The signal timing, therefore, limits access to particular segments of the facility in time. Further, rather than a continuously moving traffic stream, at traffic signals, the traffic stream is periodically stopping and starting again.

Interrupted flow, therefore, is more complex than uninterrupted flow. Although many of the traffic flow parameters described in this chapter apply to both types of facilities, this chapter focuses primarily on the characteristics of uninterrupted flow. Many of these characteristics may also apply within a moving platoon of vehicles on an interrupted flow facility. Specific characteristics of traffic interruptions and their impact on flow are discussed in detail in Chapters 6 and 20.

5.2 Traffic Stream Parameters

Traffic stream parameters fall into two broad categories. *Macroscopic parameters* describe the traffic stream as a whole; *microscopic parameters* describe the behavior of individual vehicles or pairs of vehicles within the traffic stream.

The three principal macroscopic parameters that describe a traffic stream are (1) volume or rate of flow, (2) speed, and (3) density. Microscopic parameters include (1) the speed of individual vehicles, (2) headway, and (3) spacing.

5.2.1 Volume and Rate of Flow

Traffic volume is defined as the number of vehicles passing a point on a highway, or a given lane or direction of a highway, during a specified time interval. The unit of measurement for volume is simply "vehicles," although it is often expressed as "vehicles per unit time." Units of time used most often are "per day" or "per hour."

Daily volumes are used to establish trends over time and for general planning purposes. Detailed design or control decisions require knowledge of hourly volumes for the peak hour(s) of the day.

Rates of flow are generally stated in units of "vehicles per hour" but represent flows that exist for periods of time less than one hour. A volume of 200 vehicles observed over a 15-minute period may be expressed as a rate of $200 \times 4 = 800$ vehicles/hour, even though 800 vehicles would not be observed if the full hour was counted. The 800 vehicles/hour becomes a rate of flow that exists for a 15-minute interval.

Daily Volumes

As noted, daily volumes are used to document annual trends in highway usage. Forecasts based on observed trends can be used to help plan improved or new facilities to accommodate increasing demand.

Four daily volume parameters are widely used in traffic engineering:

- *Average annual daily traffic (AADT)*. The average 24-hour volume at a given location over a full 365-day year; the number of vehicles passing a site in a year divided by 365 days (366 days in a leap year).
- *Average annual weekday traffic (AAWT)*. The average 24-hour volume occurring on weekdays over a full 365-day year; the number of vehicles passing a site on weekdays in a year divided by the number of weekdays (usually 260).

- *Average daily traffic (ADT)*. The average 24-hour volume at a given location over a defined time period less than one year; a common application is to measure an ADT for each month of the year.
- *Average weekday traffic (AWT)*. The average 24-hour weekday volume at a given location over a defined time period less than one year; a common application is to measure an AWT for each month of the year.

All of these volumes are stated in terms of vehicles per day (veh/day). Daily volumes are generally not differentiated by direction or lane but are totals for an entire facility at the designated location.

Table 5.1 illustrates the compilation of these daily volumes based on one year of count data at a sample location.

The data in Table 5.1 is in a form that comes from a permanent count location (i.e., a location where automated detection of volume and transmittal of counts electronically to a central data bank is in place). Average weekday traffic (AWT) for each month is found by dividing the total monthly weekday volume by the number of weekdays in the month (Column 5 ÷ Column 2). The average daily traffic is the total monthly volume divided by the number of days in the month (Column 4 ÷ Column 3). Average annual daily traffic is the total observed volume for the year divided by 365 days/year. Average annual weekday traffic is the total observed volume on weekdays divided by 260 weekdays/year.

The sample data of Table 5.1 gives a capsule description of the character of the facility on which it was measured. Note that ADTs are significantly higher than AWTs in each month. This suggests that the facility is serving a recreational or vacation area, with traffic strongly peaking on weekends. Also, both AWTs and ADTs are highest during the summer months, suggesting that the facility serves a warm-weather recreational/vacation area. Thus, if a detailed study were needed to provide data for an upgrading of this facility, the period to focus on would be weekends during the summer.

Hourly Volumes

Daily volumes, although useful for planning purposes, cannot be used alone for design or operational analysis purposes. Volume varies considerably over the 24 hours of the day, with periods of maximum flow occurring during the morning and evening commuter "rush hours." The single hour of the day that has the highest hourly volume is referred to as the *peak hour*. The traffic volume within this hour is of greatest interest to traffic engineers for design and operational analysis usage.

Table 5.1: Illustration of Daily Volume Parameters

1. Month	2. No. of Weekdays in Month (days)	3. Total Days in Month (days)	4. Total Monthly Volume (vehs)	5. Total Weekday Volume (vehs)	6. AWT 5/2 (veh/day)	7. ADT 4/3 (veh/day)
Jan	22	31	425,000	208,000	9,455	13,710
Feb	20	28	410,000	220,000	11,000	14,643
Mar	22	31	385,000	185,000	8,409	12,419
Apr	22	30	400,000	200,000	9,091	13,333
May	21	31	450,000	215,000	10,238	14,516
Jun	22	30	500,000	230,000	10,455	16,667
Jul	23	31	580,000	260,000	11,304	18,710
Aug	21	31	570,000	260,000	12,381	18,387
Sep	22	30	490,000	205,000	9,318	16,333
Oct	22	31	420,000	190,000	8,636	13,548
Nov	21	30	415,000	200,000	9,524	13,833
Dec	22	31	400,000	210,000	9,545	12,903
Total	**260**	**365**	**5,445,000**	**2,583,000**	—	—

AADT = 5,445,000/365 = 14,918 veh/day

AAWT = 2,583,000/260 = 9,935 veh/day

The peak-hour volume is generally stated as a *directional* volume (i.e., each direction of flow is counted separately).

Highways and controls must be designed to adequately serve the peak-hour traffic volume in the peak direction of flow. Because traffic going one way during the morning peak is going the opposite way during the evening peak, *both* sides of a facility must generally be designed to accommodate the peak directional flow during the peak hour. Where the directional disparity is significant, the concept of reversible lanes is sometimes useful. Washington, DC, for example, makes extensive use of reversible lanes (direction changes by time of day) on its many wide boulevards and some of its freeways.

In design, peak-hour volumes are sometimes estimated from projections of the AADT. Traffic forecasts are most often cast in terms of AADTs based on documented trends and/or forecasting models. Because daily volumes, such as the AADT, are more stable than hourly volumes, projections can be more confidently made using them. AADTs are converted to a peak-hour volume in the peak direction of flow. This is referred to as the "directional design hour volume" (DDHV) and is found using the following relationship:

$$DDHV = AADT * K * D \tag{5-1}$$

where: K = proportion of daily traffic occurring during the peak hour

D = proportion of peak hour traffic traveling in the peak direction of flow.

For design, the K factor often represents the proportion of AADT occurring during the *30th peak hour* of the year. If the 365 peak-hour volumes of the year at a given location are listed in descending order, the 30th peak hour is 30th on the list and represents a volume that is exceeded in only 29 hours of the year. For rural facilities, the 30th peak hour may have a significantly lower volume than the worst hour of the year because critical peaks may occur only infrequently. In such cases, it is not considered economically feasible to invest large amounts of capital in providing additional capacity that will be used in only 29 hours of the year. In urban cases, where traffic is frequently at capacity levels during all daily commuter peaks, the 30th peak hour is often not substantially different from the highest peak hour of the year.

Factors K and D are based on local or regional characteristics at existing locations. Most state highway departments, for example, continually monitor these proportions and publish appropriate values for use in various areas of the state. The K factor decreases with increasing development density in the areas served by the facility. In high-density areas, substantial demand during off-peak periods exists. This effectively lowers the proportion of traffic occurring during the peak hour of the day. The volume generated by high-density development is generally larger than that generated by lower-density areas. Thus it is important to remember that a high proportion of traffic occurring in the peak hour does not suggest that the peak-hour volume itself is large.

The D factor tends to be more variable and is influenced by a number of factors. Again, as development density increases, the D factor tends to decrease. As density increases, it is more likely to have substantial bidirectional demands. Radial routes (i.e., those serving movements into and out of central cities or other areas of activity) will have stronger directional distributions (higher D values) than those that are circumferential (i.e., going around areas of central activity). Table 5.2 indicates general ranges for K and D factors. These are purely illustrative; specific data on these characteristics should be available from state or local highway agencies or should be locally calibrated before application.

Consider the case of a rural highway that has a 20-year forecast of AADT of 30,000 veh/day. Based on the data of Table 5.2, what range of directional design hour volumes might be expected for this situation? Using the values of Table 5.2 for a rural highway, the K factor ranges from 0.15 to 0.25, and the D factor ranges from 0.65 to 0.80. The range of directional design hour volumes, therefore is:

$$DDHV_{LOW} = 30{,}000 * 0.15 * 0.65 = 2{,}925 \text{ veh/h}$$

$$DDHV_{HIGH} = 30{,}000 * 0.25 * 0.80 = 6{,}000 \text{ veh/h}$$

The expected range in DDHV is quite large under these criteria. Thus determining appropriate values of K and D for the facility in question is critical in making such a forecast.

This simple illustration points out the difficulty in projecting future traffic demands accurately. Not only does volume change over time, but the basic characteristics of volume variation may change as well. Accurate projections require the identification of causative relationships that remain stable over time. Such relationships are difficult to discern in

Table 5.2: General Ranges for *K* and *D* Factors

	Normal Range of Values	
Facility Type	***K*-Factor**	***D*-Factor**
Rural	0.15–0.25	0.65–0.80
Suburban	0.12–0.15	0.55–0.65
Urban:		
Radial Route	0.07–0.12	0.55–0.60
Circumferential Route	0.07–0.12	0.50–0.55

the complexity of observed travel behavior. Stability of these relationships over time cannot be guaranteed in any event, making demand forecasting an approximate process at best.

Subhourly Volumes and Rates of Flow

Although hourly traffic volumes form the basis for many forms of traffic design and analysis, the variation of traffic within a given hour is also of considerable interest. The quality of traffic flow is often related to short-term fluctuations in traffic demand. A facility may have sufficient capacity to serve the peak-hour demand, but short-term peaks of flow within the hour may exceed capacity and create a breakdown.

Volumes observed for periods of less than one hour are generally expressed as equivalent hourly rates of flow. For example, 1,000 vehicles counted over a 15-minute interval could be expressed as 1,000 vehs/0.25 h = 4,000 veh/h. The rate of flow of 4,000 veh/h is valid for the 15-minute period in which the volume of 1,000 vehicles was observed. Table 5.3 illustrates the difference between volumes and rates of flow.

The full hourly volume is the sum of the four 15-minute volume observations, or 4,200 veh/h. The rate of flow for each 15-minute interval is the volume observed for that interval divided by the 0.25 hours over which it was observed. In the worst period of time, 5:30 to 5:45 PM, the rate of flow is 4,800 veh/h. This is a *flow rate*, not a volume. The actual volume for the hour is only 4,200 veh/h.

Consider the situation that would exist if the capacity of the location in question were exactly 4,200 vehs/h. Although this is sufficient to handle the full-hour demand indicated in Table 5.3, the demand *rate of flow* during two of the 15-minute periods noted (5:15 to 5:30 PM and 5:30 to 5:45 PM) exceeds the capacity. The problem is that although demand may vary within a given hour, capacity is constant. In each 15-minute period, the capacity is 4,200/4 or 1,050 vehs. Thus, within the peak hour shown, queues will develop in the half-hour period between 5:15 and 5:45 PM, during which the demand exceeds the capacity. Further, although demand is less than capacity in the first 15-minute period (5:00 to 5:15 PM), the unused capacity cannot be used in a later period. Table 5.4 compares the demand and capacity for each of the 15-minute intervals. The queue at the end of each period can be computed as the queue at the beginning of the period plus the arriving vehicles minus the departing vehicles.

Even though the capacity of this segment over the full hour is equal to the peak-hour demand volume (4,200 veh/h), at the end of the hour, there remains a queue of 50 vehicles that has not been served. Although this illustration shows that a queue exists for three of four 15-minute periods within the peak hour, the dynamics of queue clearance may continue to affect traffic negatively for far longer.

Because of these types of impacts, it is often necessary to design facilities and analyze traffic conditions for a period of maximum rate of flow within the peak hour. For most

Table 5.3: Illustration of Volumes and Rates of Flow

Time Interval	Volume for Time Interval (vehs)	Rate of Flow for Time Interval (vehs/h)
5:00–5:15 PM	1,000	1,000/0.25 = 4,000
5:15–5:30 PM	1,100	1,100/0.25 = 4,400
5:30–5:45 PM	1,200	1,200/0.25 = 4,800
5:45–6:00 PM	900	900/0.25 = 3,600
5:00–6:00 PM	Σ = 4,200	

Table 5.4: Queuing Analysis for the Data of Table 5.3

Time Interval	Arriving Vehicles (vehs)	Departing Vehicles (vehs)	Queue Size at End of Period (vehs)
5:00–5:15 PM	1,000	1,050	0
5:15–5:30 PM	1,100	1,050	0 + 1,100 − 1,050 = 50
5:30–5:45 PM	1,200	1,050	50 + 1,200 − 1,050 = 200
5:45–6:00 PM	900	1,050	200 + 900 − 1,050 = 50

practical purposes, 15 minutes is considered to be the minimum period of time over which traffic conditions are statistically stable. Although rates of flow can be computed for any period of time and researchers often use rates for periods of one to five minutes, rates of flow for shorter periods often represent transient conditions that defy consistent mathematical representations. In recent years, however, use of five-minute rates of flow has increased, and there is some thought that these might be sufficiently stable for use in design and analysis. Despite this, most standard design and analysis practices continue to use the 15-minute interval as a base period.

The relationship between the hourly volume and the maximum rate of flow within the hour is defined by the *peak-hour factor*, as follows:

$$PHF = \frac{\text{hourly volume}}{\text{max. rate of flow}}$$

For standard 15-minute analysis period, this becomes:

$$PHF = \frac{V}{4 * V_{m15}} \qquad (5\text{-}2)$$

where: V = hourly volume, vehs
V_{m15} = maximum 15-minute volume within the hour, vehs
PHF = peak-hour factor

For the illustrative data in Tables 5-3 and 5-4:

$$PHF = \frac{4{,}200}{4 * 1{,}200} = 0.875$$

The maximum possible value for the *PHF* is 1.00, which occurs when the volume in each interval is constant. For 15-minute periods, each would have a volume of exactly a quarter of the full hour volume. This indicates a condition in which there is virtually no variation of flow within the hour. The minimum value occurs when the entire hourly volume occurs in a single 15-minute interval. In this case, the *PHF* becomes 0.25 and represents the most extreme case of volume variation within the hour. In practical terms, the *PHF* generally varies between a low of 0.70 for rural and sparsely developed areas to 0.98 in dense urban areas.

The peak-hour factor is descriptive of trip generation patterns and may apply to an area or portion of a street and highway system. When the value is known, it can be used to estimate a maximum flow rate within an hour based on the full-hour volume:

$$v = \frac{V}{PHF} \qquad (5\text{-}3)$$

where: v = maximum rate of flow within the hour, veh/h
V = hourly volume, veh/h
PHF = peak-hour factor.

This conversion is frequently used in the techniques and methodologies covered throughout this text.

5.2.2 Speed and Travel Time

Speed is the second macroscopic parameter describing the state of a traffic stream. Speed is defined as a rate of motion in distance per unit time. Travel time is the time taken to traverse a defined section of roadway. Speed and travel time are inversely related:

$$S = \frac{d}{t} \qquad (5\text{-}4)$$

where: S = speed, mi/h or ft/s
d = distance traversed, mi or ft
t = time to traverse distance d, h or s

In a moving traffic stream, each vehicle travels at a different speed. Thus the traffic stream does not have a single characteristic value but rather a distribution of individual speeds. The traffic stream, taken as a whole, can be characterized using an average or typical speed.

An average speed for a traffic stream can be computed in two ways:

- *Time mean speed* (TMS). The average speed of all vehicles passing a point on a highway or lane over some specified time period.
- *Space mean speed* (SMS). The average speed of all vehicles occupying a given section of highway or lane over some specified time period.

In essence, time mean speed is a point measure, whereas space mean speed describes a length of highway or lane. Figure 5.1 shows an example illustrating the differences between the two average speed measures.

To measure time mean speed (TMS), an observer would stand by the side of the road and record the speed of each vehicle as it passes. Given the speeds and the spacing shown in Figure 5.1, a vehicle will pass the observer in lane A every 176/88 = 2.0 s. Similarly, a vehicle will pass the observer in lane B every 88/44 = 2.0 s. Thus, as long as the traffic stream maintains the conditions shown, for every n vehicles traveling at 88 ft/s, the observer will also observe

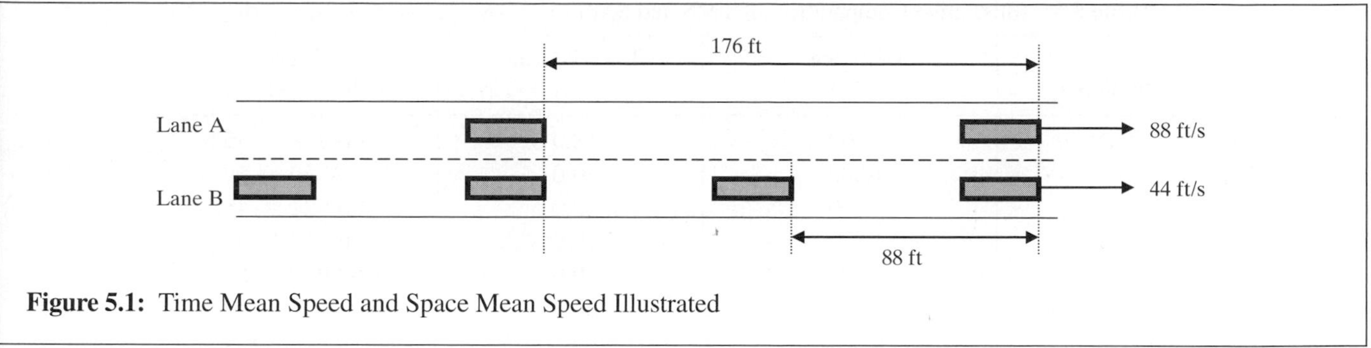

Figure 5.1: Time Mean Speed and Space Mean Speed Illustrated

n vehicles traveling at 44 ft/s. The TMS may then be computed as:

$$TMS = \frac{88.0n + 44.0n}{2n} = 66.0 \text{ ft/s}$$

To measure space mean speed (SMS), an observer would need an elevated location from which the full extent of the section may be viewed. Again, however, as long as the traffic stream remains stable and uniform, as shown, there will be twice as many vehicles in lane B as there are in lane A. Therefore, the SMS is computed as:

$$SMS = \frac{(88.0n) + (44 * 2n)}{3n} = 58.7 \text{ mi/h}$$

In effect, space mean speed accounts for the fact that it takes a vehicle traveling at 44.0 ft/s twice as long to traverse the defined section as it does a vehicle traveling at 88.0 ft/s. The space mean speed weights slower vehicles more heavily, based on the amount of time they occupy a highway section. Thus the space mean speed is usually lower than the corresponding time mean speed, in which each vehicle is weighted equally. The two speed measures may conceivably be equal if all vehicles in the section are traveling at exactly the same speed.

Both the time mean speed and space mean speed may be computed from a series of measured travel times over a specified distance using the following relationships:

$$TMS = \frac{\sum_i (d / t_i)}{n} \tag{5-5}$$

$$SMS = \frac{d}{\left(\sum_i t_i / n\right)} \tag{5-6}$$

where: TMS = time mean speed, ft/s
SMS = space mean speed, ft/s
d = distance traversed, ft
n = number of observed vehicles
t_i = time for vehicle "i" to traverse the section, s

TMS is computed by finding each individual vehicle speed and taking a simple average of the results. SMS is computed by finding the average travel time for a vehicle to traverse the section and using the average travel time to compute a speed. Table 5.5 shows a sample problem in the computation of time mean and space mean speeds.

5.2.3 Density and Occupancy

Density

Density, the third primary measure of traffic stream characteristics, is defined as the number of vehicles occupying a given length of highway or lane, generally expressed as vehicles per mile or vehicles per mile per lane.

Density is difficult to measure directly because an elevated vantage point from which the highway section under study may be observed is required. It is often computed from speed and flow rate measurements, as we discuss later in this chapter.

Density, however, is perhaps the most important of the three primary traffic stream parameters because it is the measure most directly related to traffic demand. Demand does not occur as a rate of flow, even though traffic engineers use this parameter as the principal measure of demand. Traffic is generated from various land uses, injecting a number of vehicles into a confined roadway space. This process creates a density of vehicles. Drivers select speeds that are consistent with how close they are to other vehicles. The speed and density combine to give the observed rate of flow.

Density is also an important measure of the quality of traffic flow because it is a measure of the proximity of other vehicles, a factor that influences freedom to maneuver and the psychological comfort of drivers.

Table 5.5: Illustrative Computation of TMS and SMS

Vehicle No.	Distance d (ft)	Travel Time t (s)	Speed (ft/s)
1	1,000	18.0	1,000/18 = 55.6
2	1,000	20.0	1,000/20 = 50.0
3	1,000	22.0	1,000/22 = 45.5
4	1,000	19.0	1,000/19 = 52.6
5	1,000	20.0	1,000/20 = 50.0
6	1,000	20.0	1,000/20 = 50.0
Total	**6,000**	**119**	**303.7**
Average	**6,000/6 = 1,000**	**119/6 = 19.8**	**303.7/6 = 50.6**

TMS = 50.6 ft/s
SMS = 1,000/19.8 = 50.4 ft/s

Occupancy

Although density is difficult to measure directly, modern detectors can measure *occupancy*, which is a related parameter. Occupancy is defined as the proportion of time that a detector is "occupied," or covered, by a vehicle in a defined time period. Figure 5.2 illustrates.

In Figure 5.2, L_v is the average length of a vehicle (ft), and L_d is the length of the detector (which is normally a magnetic loop detector). If "occupancy" over a given detector is "O," then density may be computed as:

$$D = \frac{5{,}280 * O}{L_v + L_d} \tag{5-7}$$

The lengths of the average vehicle and the detector are added because the detector is generally activated as the front bumper engages the front boundary of the detector and is deactivated when the rear bumper clears the back boundary of the detector.

Consider a case in which a detector records an occupancy of 0.200 for a 15-minute analysis period. If the average length of a vehicle is 28 ft, and the detector is 3 ft long, what is the density?

$$D = \frac{5{,}280 * 0.200}{28 + 3} = 34.1 \text{veh/mi/ln}$$

The occupancy is measured for a specific detector in a specific lane. Thus the density estimated from occupancy is in units of vehicles per mile per lane. If there are adjacent detectors in additional lanes, the density in each lane may be summed to provide a density in veh/mi for a given direction of flow over several lanes.

5.2.4 Spacing and Headway: Microscopic Parameters

Although flow, speed, and density represent macroscopic descriptors for the entire traffic stream, they can be related to microscopic parameters that describe individual vehicles within the traffic stream or specific pairs of vehicles within the traffic stream.

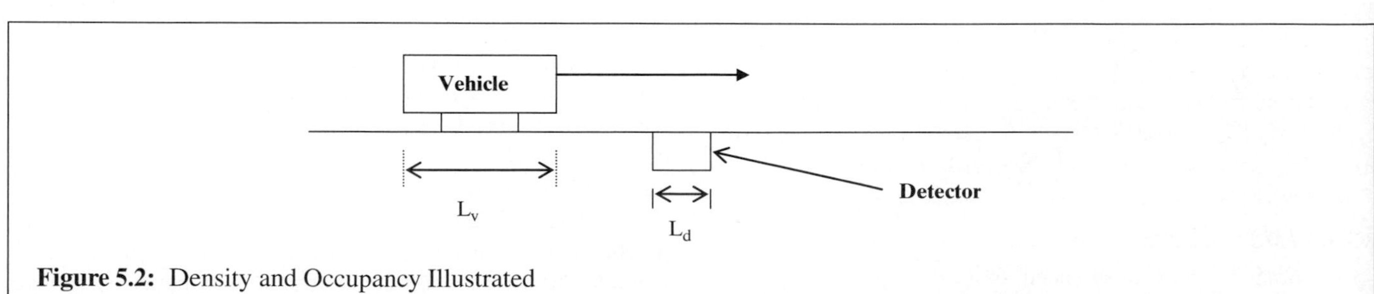

Figure 5.2: Density and Occupancy Illustrated

Spacing

Spacing is defined as the distance between successive vehicles in a traffic lane, measured from some common reference point on the vehicles, such as the front bumper or front wheels. The *average* spacing in a traffic lane can be directly related to the density of the lane:

$$D = \frac{5{,}280}{d_a} \quad (5\text{-}8)$$

where: D = density, veh/mi/ln

da = average spacing between vehicles in the lane, ft

Headway

Headway is defined as the time interval between successive vehicles as they pass a point along the lane, also measured between common reference points on the vehicles. The *average* headway in a lane is directly related to the rate of flow:

$$v = \frac{3{,}600}{h_a} \quad (5\text{-}9)$$

where: v = rate of flow, veh/h/ln

h_a = average headway in the lane, s

Use of Microscopic Measures

Microscopic measures are useful for many traffic analysis purposes. Because a spacing and/or a headway may be obtained for every pair of vehicles, the amount of data that can be collected in a short period of time is relatively large. A traffic stream with a volume of 1,000 vehicles over a 15-minute time period results in a *single* value of rate of flow, space mean speed, and density when observed. There would be, however, 1,000 headway and spacing measurements, assuming that all vehicle pairs were observed.

Use of microscopic measures also allows various vehicle types to be isolated in the traffic stream. Passenger car flows and densities, for example, could be derived from isolating spacing and headway for pairs of passenger cars following each other. Heavy vehicles could be similarly isolated and studied for their specific characteristics. Chapter 14 illustrates such a process for calibrating basic capacity analysis variables.

Average speed can also be computed from headway and spacing measurements as:

$$S = \frac{(d_a / h_a)}{1.47} = 0.68\,(d_a / h_a) \quad (5\text{-}10)$$

where: S = average speed, mi/h

d_a = average spacing, ft

h_a = average headway, s

A Sample Problem

Traffic in a congested multilane highway lane is observed to have an average spacing of 200 ft and an average headway of 3.8 s. Estimate the rate of flow, density, and speed of traffic in this lane.

Solution:

$$v = \frac{3{,}600}{3.8} = 947 \text{ veh/h/ln}$$

$$D = \frac{5{,}280}{200} = 26.4 \text{ veh/mi/ln}$$

$$S = 0.68\,(200/3.8) = 35.8 \text{ mi/h}$$

Note that due to round-off errors, D × S is not exactly equal to v in this case.

5.3 Relationships among Flow Rate, Speed, and Density

The three macroscopic measures of the state of a given traffic stream—flow, speed, and density—are related as follows:

$$v = S * D \quad (5\text{-}11)$$

where: v = rate of flow, veh/h or veh/h/ln

S = space mean speed, mi/h

D = density, veh/mi or veh/mi/ln

Space mean speed and density are measures that refer to a specific *section* of a lane or highway, whereas flow rate is a point measure. Figure 5.3 illustrates the relationship. The space mean speed and density measures must apply to the same defined section of roadway. Under stable flow conditions (i.e., the flow entering and leaving the

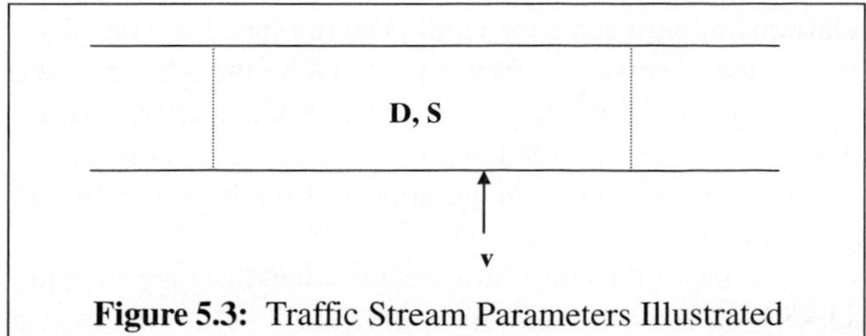

Figure 5.3: Traffic Stream Parameters Illustrated

section are the same; no queues are forming within the section), the rate of flow computed by Equation 5-11 applies to *any* point within the section. Where unstable operations exist (a queue is forming within the section), the computed flow rate represents an average for all points within the section.

If a freeway lane were observed to have a space mean speed of 55 mi/h and a density of 25 veh/mi/ln, the flow rate in the lane could be estimated as:

$$v = 55 * 25 = 1{,}375 \text{ veh/h/ln}$$

As noted previously, this relationship is most often used to estimate density, which is difficult to measure directly, from measured values of flow rate and space mean speed. Consider a freeway lane with a measured space mean speed of 60 mi/h and a flow rate of 1,000 veh/h/ln. The density could be estimated from Equation 5-11 as:

$$D = \frac{v}{S} = \frac{1{,}000}{60} = 16.7 \text{ veh/mi/ln}$$

Equation 5-11 suggests that a given rate of flow (v) could be achieved by an infinite number of speed (S) and density (D) pairs having the same product. Thankfully, this is not what happens because it would make the mathematical interpretation of traffic flow unintelligible. There are additional relationships between pairs of these variables that restrict the number of combinations that can and do occur in the field. Figure 5.4 illustrates the general form of these relationships.

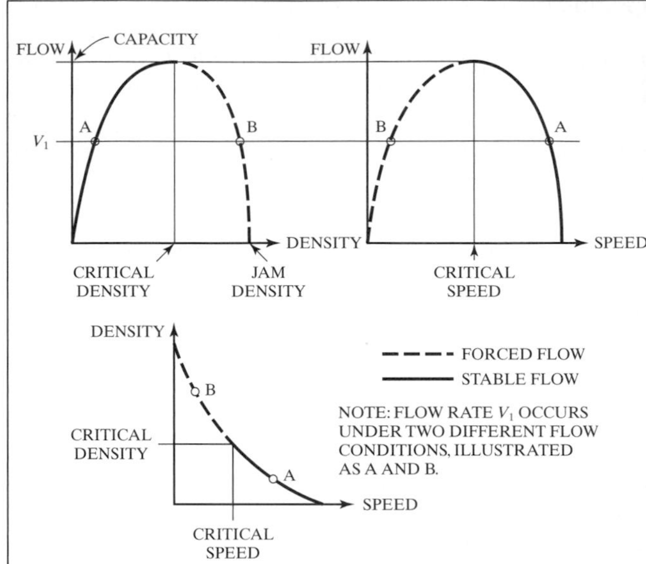

Figure 5.4: Relationships among Speed, Flow, and Density

(*Source:* Used with permission of Transportation Research Board, National Research Council, from *Highway Capacity Manual*, 3rd Edition, *Special Report 209*, pp. 1–7, Washington DC, 1994.)

The exact shape and calibration of these relationships depends on prevailing conditions, which vary from location to location and even over time at the same location.

Note that a flow rate of "0 veh/h" occurs under two very different conditions. When there are no vehicles on the highway, density is "0 veh/mi" and no vehicles can be observed passing a point. Under this condition, speed is unmeasurable and is referred to as "free-flow speed," a theoretical value that exists as a mathematical extension of the relationship between speed and flow (or speed and density). In practical terms, free-flow speed can be thought of as the speed a single vehicle could achieve when there are no other vehicles on the road and the motorist is driving as fast as is practicable given the geometry of the highway and its environmental surroundings.

A flow of "0 veh/h" also occurs when there are so many vehicles on the road that all motion stops. This occurs at a very high density, called the "jam density," and no flow is observed because no vehicle can pass a point to be counted when all vehicles are stopped.

Between these two extreme points on the relationships, there is a peaking characteristic. The peak of the flow-speed and flow-density curves is the maximum rate of flow, or the *capacity* of the roadway. Its value, like everything else about these relationships, depends on the specific prevailing conditions at the time and location of the calibration measurements. Operation at capacity, however, is very unstable. At capacity, with no usable gaps in the traffic stream, the slightest perturbation caused by an entering or lane-changing vehicle, or simply a driver hitting the brakes, causes a chain reaction that cannot be damped. The perturbation propagates upstream and continues until sufficient gaps in the traffic stream allow the event to be effectively dissipated.

The dashed portion of the curves represents *unstable* or *forced* flow. This effectively represents flow within a queue that has formed behind a breakdown location. A breakdown will occur at any point where the arriving flow rate exceeds the downstream capacity of the facility. Common points for such breakdowns include on-ramps on freeways, but accidents and incidents are also common, less predictable causes for the formation of queues. The solid line portion of the curves represents *stable* flow (i.e., moving traffic streams that can be maintained over a period of time).

Except for capacity flow, any flow rate may exist under two conditions:

1. A condition of relatively high speed and low density (on the stable portion of flow relationships)
2. A condition of relatively low speed and high density (on the unstable portion of flow relationships)

Obviously, traffic engineers would prefer to keep all facilities operating on the stable side of the curves.

Because a given volume or flow rate may occur under two very different sets of operating conditions, the volume cannot completely describe flow conditions, nor can it be used as measures of the quality of traffic flow. Values of speed and/or density, however, would define unique points on any of the relationships of Figure 5.4, and both describe aspects of quality that can be perceived by drivers and passengers. Chapter 6, "Traffic Flow Theory," contains additional detail on historic and current studies of the exact characteristics of speed-flow-density relationships [*1–8*].

5.4 Closing Comments

This chapter has introduced the key macroscopic and microscopic parameters used in quantify and describe conditions within an uninterrupted traffic stream and the fundamental relationships that govern them. In Chapter 6, these are discussed in greater detail with an emphasis on specific mathematical models that have been used to describe critical relationships.

In Part III of this text, methodologies used to measure and calibrate these key parameters are presented and illustrated. In Parts IV and V, the use of these parameters across a wide range of traffic engineering applications is discussed in detail.

Like any engineering field, good traffic engineers must understand the medium with which they work. The medium for traffic engineers is traffic streams. Describing them in quantitatively precise terms is critical to the tasks of accommodating and controlling them in such a way as to provide for safe and efficient transportation for people and goods. Thus the foundation of the profession lies in these descriptors.

References

1. Greenshields, B., "A Study of Highway Capacity," *Proceedings of the Highway Research Board*, Vol. 14, Transportation Research Board, National Research Council, Washington DC, 1934.
2. Ellis, R., "Analysis of Linear Relationships in Speed-Density and Speed-Occupancy Curves," *Final Report*, Northwestern University, Evanston, IL, December 1964.
3. Greenberg, H., "An Analysis of Traffic Flows," *Operations Research*, Vol. 7, ORSA, Washington DC, 1959.
4. Underwood, R., "Speed, Volume, and Density Relationships," *Quality and Theory of Traffic Flow*, Yale Bureau of Highway Traffic, Yale University, New Haven, CT, 1961.
5. Edie, L., "Car-Following and Steady-State Theory for Non-Congested Traffic," *Operations Research*, Vol. 9, ORSA, Washington DC, 1961.
6. Duke, J., Schofer, J., and May Jr., A., "A Statistical Analysis of Speed-Density Hypotheses," *Highway Research Record 154*, Transportation Research Board, National Research Council, Washington DC, 1967.
7. *Highway Capacity Manual*, 4th Edition, Transportation Research Board, National Research Council, Washington DC, 2000.
8. Scheon, J., et al., "Speed-Flow Relationships for Basic Freeway Sections," *Final Report*, NCHRP Project 3-45, JHK & Associates, Tucson, AZ, May 1995.

Problems

5-1. A volume of 1,200 veh/h is observed at an intersection approach. Find the peak rate of flow within the hour for the following peak-hour factors: 1.00, 0.90, 0.80, 0.70. Plot and comment on the results.

5-2. A traffic stream displays average vehicle headways of 2.4 s at 55 mi/h. Compute the density and rate of flow for this traffic stream.

5-3. A freeway detector records an occupancy of 0.26 for a 15-minute period. If the detector is 3.5 ft long, and the average vehicle has a length of 18 ft, what is the density implied by this measurement?

5-4. The following traffic count data were taken from a permanent detector location on a major state highway.

1. Month	2. No. of Weekdays in Month (days)	3. Total Days in Month (days)	4. Total Monthly Volume (vehs)	5. Total Weekday Volume (vehs)
Jan	22	31	200,000	170,000
Feb	20	28	210,000	171,000
Mar	22	31	215,000	185,000
Apr	22	30	205,000	180,000
May	21	31	195,000	172,000
Jun	22	30	193,000	168,000
Jul	23	31	180,000	160,000
Aug	21	31	175,000	150,000
Sep	22	30	189,000	175,000
Oct	22	31	198,000	178,000
Nov	21	30	205,000	182,000
Dec	22	31	200,000	176,000

From this data, determine (a) the AADT, (b) the ADT for each month, (c) the AAWT, and (d) the AWT for each month. From this information, what can be discerned about the character of the facility and the demand it serves?

5-5. A lane on a freeway displays the following characteristics: (a) the average headway between vehicles is 2.8 s, and (b) the average spacing between vehicles is 235 ft. What is the rate of flow for the lane? What is the average speed (in mi/h)?

5-6. The following counts were taken on a major arterial during the evening peak period:

Time Period	Volume (vehs)
4:00–4:15 PM	450
4:15–4:30 PM	465
4:30–4:45 PM	490
4:45–5:00 PM	500
5:00–5:15 PM	503
5:15–5:30 PM	506
5:30–5:45 PM	460
5:45–6:00 PM	445

From this data, determine:

(a) The peak hour.

(b) The peak hour volume.

(c) The peak flow rate within the peak hour.

(d) The peak hour factor (PHF).

5-7. A peak-hour volume of 1,200 veh/h is observed on a freeway lane. What is the peak flow rate within this hour if the PHF is 0.87?

5-8. The flow rate on an arterial lane is 1,300 veh/h. If the average speed in the same lane is 35 mi/h, what is the density?

5-9. The AADT for a section of suburban arterial is 50,000 veh/day. Assuming that this is an urban radial facility, what range of directional design hour volumes would be expected?

5-10. The following travel times were measured for vehicles traversing a 1,000-ft segment of an arterial:

Vehicle	Travel Time (s)
1	20.6
2	21.7
3	19.8
4	20.3
5	22.5
6	18.5
7	19.0
8	21.4

Determine the time mean speed (TMS) and space mean speed (SMS) for these vehicles.

CHAPTER 6

Introduction to Traffic Flow Theory

Traffic flow theory is best defined as mathematical models that attempt to relate characteristics of traffic movement to each other and to underlying traffic parameters. The science of traffic flow theory formally began with the work of Bruce Greenshields and the Yale Bureau of Highway Traffic in the 1930s. The field has continued to develop, of course, and plays a major role in traffic engineering.

Virtually every function in traffic engineering, from data collection and analysis, to signal timing, to capacity and level of service analysis, uses analytical models of traffic behavior under a variety of underlying circumstances. These models, and their development and calibration, are the essence of traffic flow theory.

This chapter provides a very brief glimpse into this exciting field and focuses on a few types of models that are applied in other chapters. References 1 to 4 provide excellent sources of comprehensive material on modern traffic flow theory. Reference 1 in particular has chapters that address these topics:

1. Introduction to Traffic Flow Theory
2. Traffic Stream Characteristics
3. Human Factors
4. Car Following
5. Continuum Flow Models
6. Macroscopic Flow Models
7. Traffic Impact Models
8. Unsignalized Intersections
9. Signalized Intersections
10. Traffic Simulation

6.1 Basic Models of Uninterrupted Flow

All of the methodologies related to the analysis of freeway, multilane highway, and two-lane rural highway capacity and level of service analysis are based on the fundamental relationships between the speed, flow, and density of an uninterrupted traffic stream.

6.1.1 Historical Background

The earliest studies of uninterrupted flow characteristics and relationships were conducted by Bruce Greenshields [5]. His and other early studies focused on the relationship between the density and speed of an uninterrupted traffic stream. Greenshields thought that the speed-density relationship was essentially linear.

This focus on the speed-density relationship highlights the fact that this is the relationship that most closely represents driver behavior. Although traffic engineers are most often dealing with volume or flow rate as a quantitative measure of traffic demand, these measures are the *results* of demand. Demand is most immediately represented as a density. Vehicles are injected onto a facility with finite space from a variety of parking facilities serving a multitude of land uses. The injection of vehicles into a defined but limited space establishes a density. Drivers select appropriate speeds based primarily on their proximity to each other and the perceived safety of the situation. Therefore, an average speed results from drivers each selecting a comfortable speed for the density each experiences. Flow rate or volume is the result of the density—speed pairing. Speed-flow and density-speed relationships can be derived from a speed-density relationship knowing that:

$$v = S * D \tag{6-1}$$

where: v = rate of flow, veh/h or veh/h/ln,

D = density, veh/mi or veh/mi/ln

S = average (space mean) speed of traffic stream, mi/h

Later, Ellis [*6*] investigated two- and three-segment linear curves with discontinuities. Greenberg [*7*] hypothesized a logarithmic curve for speed density, whereas Underwood [*8*] used an exponential form. Edie [*9*] combined logarithmic and exponential forms for low- and high-density portions of the curve. Like Ellis, Edie's curves contained discontinuities. May [*10*] suggested using a bell-shaped curve.

Over the years, there have been many suggestions for mathematical descriptions of the relationships between speed, flow, and density on an uninterrupted flow facility. Although there have clearly been changes in driver behavior that influence the shape of these curves, there is no one form that will best fit data from all locations.

6.1.2 Deriving Speed—Flow and Density—Flow Curves from a Speed-Density Curve

Because Equation 6-1 is a fundamental relationship governing speed, flow, and density, once the relationship between speed and density is established, then speed-flow and speed-density curves are also fully determined.

Using Greenshields's simple linear speed-density curve as an example, consider the speed-density relationship shown in Figure 6.1. Two points of interest are the Y- and X-axis intercepts. The Y-intercept is 65.0 mi/h and is called the "free-flow speed," that is, the speed that occurs when density (and therefore flow) is zero. The X-intercept is 110 veh/mi/ln, the density at which all motion stops, making speed zero. This is commonly called the "jam" density.

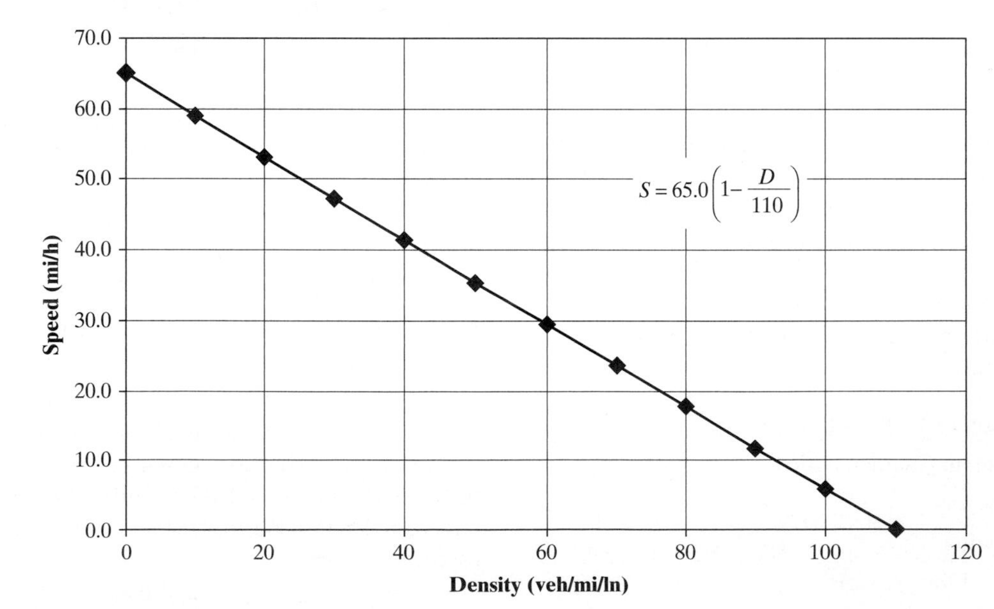

Figure 6.1: Sample Linear Speed-Density Relationship

Given the equation for speed versus density and knowing that $v = S \times D$ (Equation 6.1) is always applicable, the relationship between flow and density is found by substituting $S = v/D$ in the speed-density equation. Then:

$$S = 65.0\left(1 - \frac{D}{110}\right)$$

$$\frac{v}{D} = 65.0\left(1 - \frac{D}{110}\right)$$

$$v = 65D - 0.59091D^2$$

The relationship between flow and speed is found by substituting $D = v/S$ in the speed-density equation. Then:

$$S = 65.0\left(1 - \frac{D}{110}\right)$$

$$S = 65.0\left(1 - \frac{v/S}{110}\right)$$

$$v = 110S - 1.6923S^2$$

As shown in Figure 6.2, both of these curves are parabolic.

Equation 6.1 ($v = S * D$) always applies. Therefore, calibrating any *one* of the relationships between S and D, v and D, or v and S, defines all three relationships. Given one of the three, the other two may be algebraically derived.

Figure 6.3 illustrates a speed-flow curve resulting from a two-segment linear speed-density relationship. It results in *two* parabolas, one for each segment of the discontinuous speed-density curve. It is illustrated because of the discontinuity involved. From the speed-flow curve, it is clear that the discontinuity is near the peak of the curve(s)—that is, in the vicinity of capacity.

The graph is from an old but fascinating study in which various mathematical forms were fit to a set of data from the Eisenhower Expressway in Chicago in the early 1960s. Although the data are not reflective of modern speed-flow behavior on freeways, the study determined that a discontinuous set of curves (not the one shown) best fit the data. Of interest is that Figure 6.3 seems to indicate that there are *two* capacities: one when approached from low speeds (unstable flow) and one when approached from high speeds (stable flow). This characteristic is quite complicated and is discussed further in the sections that follow.

6.1.3 Determining Capacity from Speed-Flow-Density Relationships

Chapter 13 includes a detailed discussion of the concept of capacity and many of the nuances that it contains. One potential understanding of capacity, however, as discussed in Chapter 5, is that capacity is the peak of a speed-flow or flow-density curve. From Figure 6.2, for example, it is clear that the "peak" of either curve occurs at a flow rate slightly less than 1,800 veh/h/ln (hard to read exact number from the scale shown). In Figure 6.3, there are two capacities. The "high" value is also about 1,800 veh/h/ln, which is on the high-speed, or stable portion of the curve. The "low" value is approximately 1,550 veh/h/ln, which is on the low-speed or unstable side of the curve.

The capacity value can also be determined mathematically. Using the curves of Figure 6.2 as an example, it is necessary to determine the speed and density at which capacity occurs. In both cases, this occurs where the slope of the curve (or the first derivative of the curve) is zero.

For the flow-density curve:

$$v = 65.0D - 0.59091D^2$$

$$\frac{dv}{dD} = 0 = 65.0 - 1.18182D$$

$$D = \frac{65.0}{1.18182} = 55.0 \text{ veh/h/ ln}$$

For the speed-flow curve:

$$v = 110S - 1.6923S^2$$

$$\frac{dv}{dS} = 0 = 110 - 3.3846S$$

$$S = \frac{110}{3.3846} = 32.5 \text{ mi/h}$$

The calculus and algebra confirm the obvious: For the linear model of Figures 6.1 and 6.2, capacity occurs when speed is exactly half the free-flow speed and when density is exactly half the jam density. The capacity is then found as the product of the speed and density at which it occurs, or:

$$c = S * D = 32.5 * 55.0 = 1{,}788 \text{ veh/h/ ln}$$

which confirms our observation from Figure 6.2 of "slightly less than 1,800 veh/h/ln."

6.1.4 Modern Uninterrupted Flow Characteristics

"Traffic flow theory" is really a misnomer. Traffic flow does not occur in theory. It occurs on real streets, highways, and freeways all over the world. The mathematical models developed

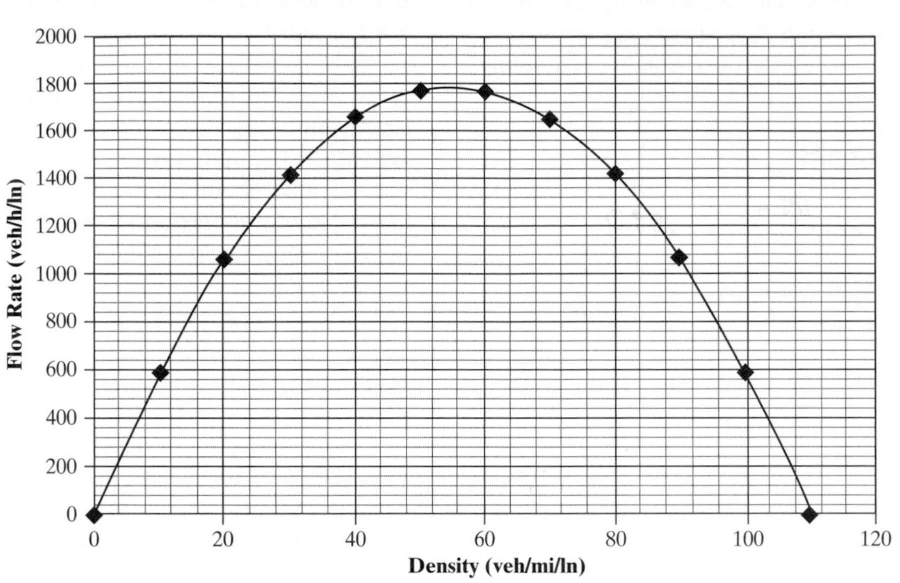

(a) Flow-Density Curve Resulting from Linear Speed-Density Relationship

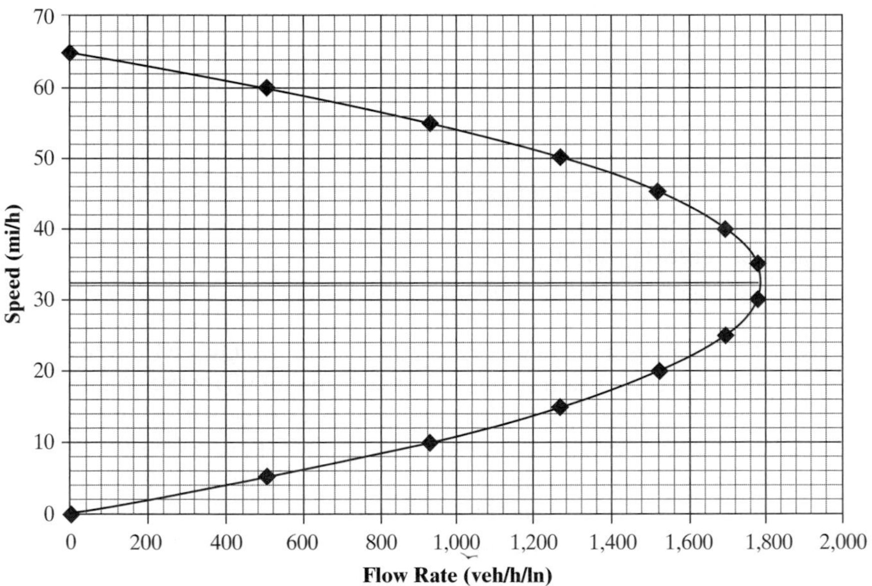

(b) Speed-Flow Curve Resulting from Linear Speed-Density Relationship

Figure 6.2: Flow-Density and Speed-Flow Curves Resulting from a Linear Speed-Density Relationship

by researchers are merely descriptions of driver behavior. Because of this, traffic flow theory is an evolving science. No model is ever static because driver behavior changes over time. Nowhere is this clearer than in speed-flow-density relationships for uninterrupted flow.

The linear model of Greenshields, and most of the other historic models discussed, all have one common characteristic: Speeds decline as flow rates increase. Drivers react to higher densities (which result in higher flows) by slowing down to maintain what they perceive to be safe operations. Modern

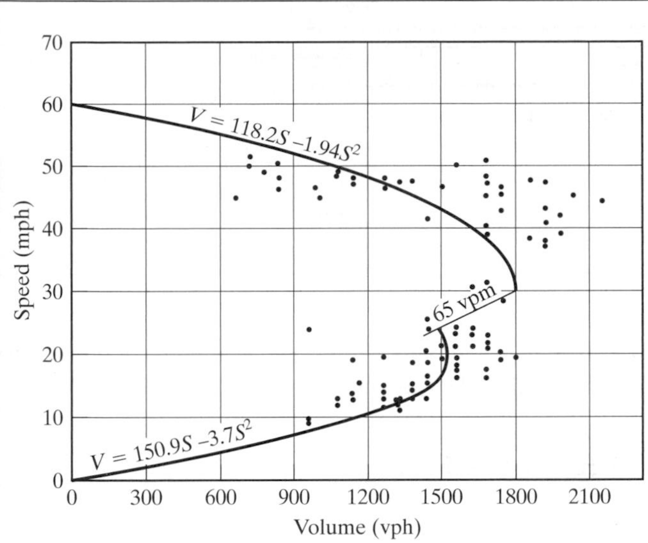

Figure 6.3: A Speed-Flow Curve with Discontinuity in the Vicinity of Capacity

(*Source:* Used with permission of Transportation Research Board, National Research Council, Washington DC, J.S. Drake, J.L. Schofer, and A.D. May Jr., "A Statistical Analysis of Speed-Density Hypotheses," *Transportation Research Record 154,* p. 78, 1967.)

uninterrupted flow, particularly on freeways, does not reflect this characteristic. In fact, drivers maintain high average speeds through a range of flow rates and do not slow down until relatively high flow rates are reached. Figure 6.4 illustrates the general characteristics of uninterrupted flow on a modern freeway.

Figure 6.4 shows three distinct ranges of data: (1) undersaturated (stable) flow, (2) queue discharge flow, and (3) oversaturated (unstable) flow. The speed throughout the undersaturated flow portion of the curve is remarkably stable. If a line was drawn through the center of these points, the speed would range from about 71 mi/h to a low of about 60 mi/h. Further, there seems to be no systematic decline in speeds with flow rate until a flow rate of approximately 1,200 to 1,300 veh/h/ln is reached. The capacity would be the peak of this portion of the curve, or approximately 2,200 veh/h/ln, which is achieved at an astonishingly high speed of approximately 60 mi/h.

Once capacity is reached, and demand in fact exceeds capacity, a queue begins to form. The "queue discharge" portion of the curve reflects vehicles departing from the front of the queue. Such vehicles will begin to accelerate as they move downstream, assuming that no additional downstream congestion exists. The oversaturated portion of the curve is what exists within the queue that forms when demand exceeds capacity at a point.

Although queue discharge rates vary widely at this site, their average is clearly *lower* than the capacity of the undersaturated portion of the curve. It is generally agreed that vehicles cannot depart the head of a queue at the same rate as they pass the same point under stable or undersaturated flow. Is this simply another explanation of the "two capacity" phenomenon of historic curves?

6.1.5 Calibrating a Speed-Flow-Density Relationship

How should data be collected to calibrate a speed-flow-density relationship for a specific uninterrupted flow segment? One of

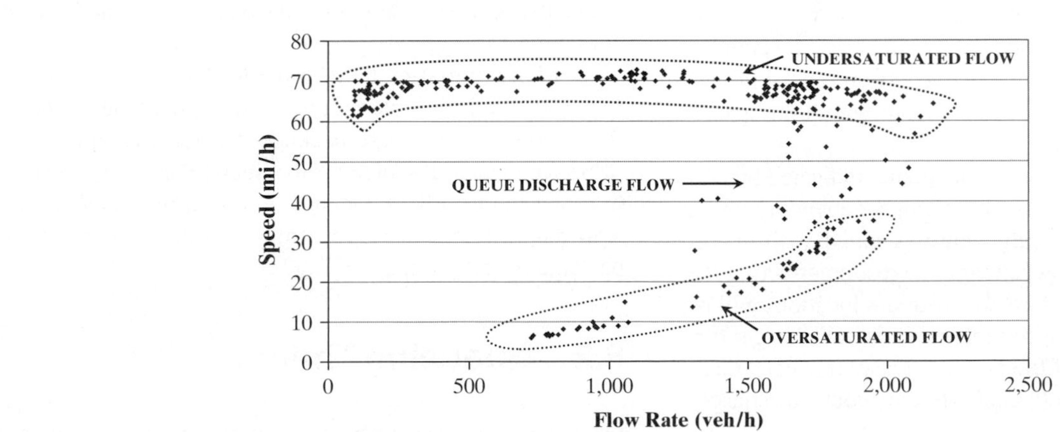

Figure 6.4: Speed-Flow Characteristics for a Modern Freeway

(*Source: Draft Chapter 11, Basic Freeway Segments,* NCHRP Project 3-92, Transportation Research Board, Washington DC, Exhibit 11-1, 2009.)

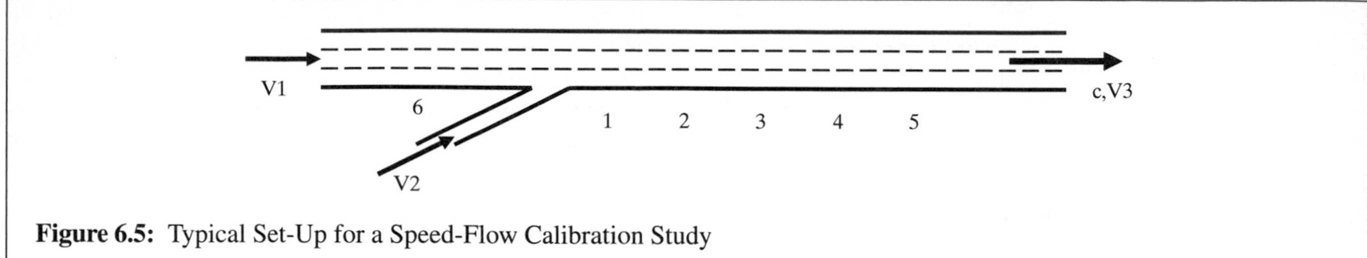

Figure 6.5: Typical Set-Up for a Speed-Flow Calibration Study

the problems involved in interpreting older studies is that it is not clear how or, more importantly, where data were collected.

Although the speed-density relationship is the most descriptive of driver behavior, measuring density in the field is not always a simple task. Speed and flow rate or volume, however, are relatively simple traffic measurements. Most field studies, therefore, focus on calibration of the speed versus flow relationship and derive the others.

If the capacity operation is to be observed, measurements must be taken near a point of frequent congestion. Most of these occur at on-ramps, where the arriving freeway and arriving on-ramp flows may regularly exceed the capacity of the downstream freeway segment. Under these conditions, queues may be expected to form on both the upstream freeway and the ramp roadway. Figure 6.5 illustrates a field setup for taking data to calibrate all regions of the curve.

Under stable flow, flow rates and speeds would be recorded at a point close to, but sufficiently downstream from the merge for ramp vehicles to have accelerated to ambient speed. This is indicated as location 1. The measurements *must* take place downstream of the on-ramp because V2 is part of the downstream demand.

Once queues start forming, stable flow no longer exists. Now observation locations must shift. Because unstable or oversaturated flow exists within the queue forming behind the on-ramp, observations must be made from within the queue, indicated here as location 6.

Downstream of the head of the queue, indicated here as locations 2 through 5 (perhaps including 1 depending on its exact placement), discharging vehicles can be observed. Assuming no additional downstream congestion affecting the study area, the flow rate at these downstream locations will be fairly stable, although the speed increases as vehicles get further away from the head of the queue. Measurements at these locations can be combined to calibrate the "queue discharge" portion of the curve.

These are not simple observations. Care must be taken to avoid the observation of impacts of unseen downstream congestion. Capacity operations are most likely to exist in the last 15-minute intervals before the appearance of queues on the ramp and/or the freeway.

6.1.6 Curve Fitting

Once data have been collected, reduced, and recorded, a mathematical description of the data is sought. A variety of statistical tools are available to accomplish this. Multiple linear and nonlinear regression techniques and software packages are used in the curve-fitting process.

Most of these tools define the "best" fit using an objective function, which the tool seeks to minimize. A common objective function is to minimize the sum of the squared differences between the actual data points and the curve that defines the relationship. The curve, in effect, represents predicted values of the independent variable (in this case speed) that are compared directly to the field-measured values to determine how "good" the fit is.

In some cases, there are not enough data available for formal regression analysis, or the spread of data is so broad as to complicate regression analysis. In such cases, graphic fits using the analyst's best professional judgment is used to define the curve and determine its equation.

Many statistics texts provide detailed treatments of regression and multiple regression analysis. One of the most frequently used software packages for regression analysis is SPSS (Statistical Package for the Social Sciences). Despite its title, it is an excellent package and used by many engineers who must do formal curve fitting. Another commonly used package is Statgraphics.

6.2 Queueing Theory

Queueing theory has many applications in traffic engineering. As noted in the discussion of uninterrupted flow, queues form behind breakdown points at which arriving demand flow exceeds capacity. On interrupted flow facilities, queueing is a systemic process. At every signalized intersection, a queue forms when the signal is "red" and dissipates (ideally) during the "green" that follows. Queueing theory is mathematical modeling and a description of the process by which queues form and dissipate.

6.2.1 One Capacity or Two? An Illustration Using Deterministic Queueing

Deterministic queueing analysis is a very simple approach to the description of queues. It simply treats the length of the queue as the difference between arrivals and departures at a point. Queues are time dependent, and the basic concept is shown in Equation 6-2:

$$Q_{i+1} = Q_i + a_i + d_i \tag{6-2}$$

where: Q_{i+1} = queue size at the beginning of time period $i+1$ (vehs),

Q_i = queue size at the beginning of time period i (vehs),

a_i = arrivals during time period i (vehs), and

d_i = departures during time period i (vehs).

In general, a deterministic queueing analysis would start at a time when there is no queue present at the beginning of time period 1.

Deterministic queueing can be applied to a problem discussed previously: the issue of whether uninterrupted flow relationships produce continuous curves with a single value of capacity or discontinuous curves with two (or more) values of capacity.

An example is used to indicate the practical importance of this seemingly theoretic issue. In practical terms, consider the following sequence of events:

- A driver wakes up at 6:00 AM and hears a radio report that a truck is broken down and is blocking one lane of a three-lane freeway segment on his/her route to work.
- At 6:30 AM, the radio traffic report indicates that the blockage has been cleared.
- The driver leaves for work several hours later but encounters a queue of several miles leading up to the location of the 6:00 AM breakdown.
- On reaching the location (after passing through the queue), traffic rapidly accelerates into an undersaturated downstream freeway.

What exactly happened here? Consider the specifics of this scenario illustrated in Figure 6.6. Two scenarios are

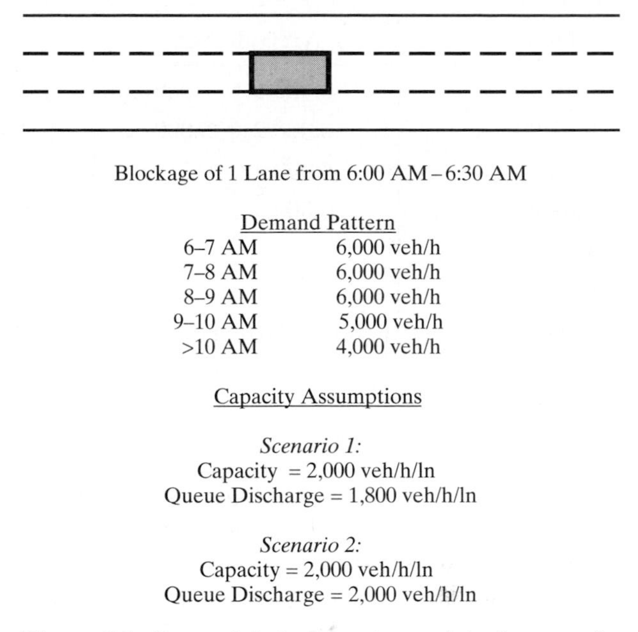

Figure 6.6: Deterministic Queueing and the Impact of a Breakdown

analyzed: (1) There are two capacities, one of 2,000 veh/h/ln on the undersaturated or stable portion of the curve, and one of 1,800 veh/h/ln on the oversaturated or unstable portion of the curve; (2) there is only one capacity of 2,000 veh/h/ln at the peak of the curve.

Tables 6.1 and 6.2 show the results of a deterministic queueing analysis for these two scenarios. Note that the arrival flow rates are 6,000 veh/h and that the normal capacity of the segment (before breakdown) is 3 lanes × 2,000 veh/h/ln = 6,000 veh/h. Once the breakdown occurs, queues begin to form immediately. Once queues form, the capacity of each lane drops to 1,800 veh/h under scenario 1; capacity remains 2,000 veh/h under scenario 2. As seen in Tables 6.1 and 6.2, this seemingly small difference makes a very significant difference in the queues that result.

For each scenario, two questions will be answered: (1) How long does the queue get? (2) How long does it take for the queue to dissipate?

In scenario 1, the instant the breakdown occurs, queues are established, and the capacity is reduced to the queue discharge value of 1,800 veh/h/ln. For the first half hour (when the truck blocks a lane), only 2 lanes are available for movement. After 6:30 AM, all lanes are again available. The problem is that the queue still exists after the blockage is removed. Capacity does not return to its undersaturated value until the entire queue is dissipated.

Table 6.1: Deterministic Queuing Analysis for Scenario 1

Time	Arrivals (veh)	Capacity (veh)	Queue Size (veh)
6:00–6:30 AM	6,000/2 = 3,000	2 × 1,800/2 = 1,800	3,000 – 1,800 = 1,200
6:30–7:00 AM	6,000/2 = 3,000	3 × 1,800/2 = 2,700	1,200 + 3,000 – 2,700 = 1,500
7:00–8:00 AM	6,000	3 × 1800 = 5,400	1,500 + 6,000 – 5,400 = 2,100
8:00–9:00 AM	6,000	5,400	2,100 + 6,000 – 5,400 = 2,700
9:00–10:00 AM	5,000	5,400	2,700 + 5,000 – 5,400 = 2,300
>10:00 AM	4,000 veh/h	5,400 veh/h	Queue decreases by 1,400 veh/h

Queue dissipates 2,300/1,400 = 1.64 h after 10:00 AM, or 11:38 AM.

As shown in Table 6.1, the queue will reach a maximum size of 2,700 vehs at 9:00 AM. Note that the queue continues to grow between 6:30 AM and 9:00 AM, even though the blockage has been removed. This is due to the reduced capacity of queue discharge. After 9:00 AM, demand finally reduces to a value that is less than the queue discharge capacity. At this point, the queue begins to dissipate. It does not disappear, however, until 11:38 AM, at which time, the capacity returns to its normal value of 2,000 veh/h/ln.

The results for scenario 2 are very different. Because the capacity (on a per-lane basis) is never reduced, a queue grows only during the initial half hour of the blockage. It remains stable thereafter until 9:00 AM when the demand becomes less than capacity. The queue never grows to more than 1,000 vehicles and is dissipated by 10:00 AM.

The small difference in queue discharge capacity in the two scenarios makes an enormous difference in the results. The total size of the queue in scenario 1 is more than twice the size of that under scenario 2. Further, it takes 1 hr and 38 minutes longer to clear the queue under scenario 1 than under scenario 2. Obviously, this is more than a theoretical issue.

In both cases, however, drivers arriving hours after the blockage has been cleared still experience the congestion caused by the blockage. When drivers clear the site of the initial blockage, they experience free-flow conditions (assuming no additional downstream congestion exists).

6.2.2 A Problem with Deterministic Queueing

Although conceptually simple and easy to understand, deterministic queueing has a theoretical flaw that can be significant: It assumes that the queue forms at a point, that is, , that queued vehicles are stacked vertically at the point of the breakdown.

This is, of course, not true. In the example of Tables 6.1 and 6.2, queues extend over a significant length of highway. If it is assumed that vehicles in a queue occupy approximately 50 ft of space (including spacing between vehicles), then the maximum lengths of queue are:

$$L_Q(Scenario\ 1) = \left(\frac{2700}{3}\right) * 50 = 45.000 \text{ ft}$$
$$= \frac{45{,}000}{5{,}280} = 8.52 \text{ mi}$$

$$L_Q(Scenario\ 2) = \left(\frac{1000}{3}\right) * 50 = 16{,}667 \text{ ft}$$
$$= \frac{16{,}667}{5{,}280} = 3.16 \text{ mi}$$

Table 6.2: Queuing Analysis for Scenario 2

Time	Arrivals (veh)	Capacity (veh)	Queue Size (veh)
6:00–6:30 AM	6,000/2 = 3,000	2 × 2,000/2 = 2,000	3,000 – 2,000 = 1,000
6:30–7:00 AM	6,000/2 = 3,000	3 × 2,000/2 = 3,000	1,000 + 3,000 – 3,000 = 1,000
7:00–8:00 AM	6,000	3 × 2,000 = 6,000	1,000 + 6,000 – 6,000 = 1,000
8:00–9:00 AM	6,000	6,000	1,000 + 6,000 – 6,000 = 1,000
9:00–10:00 AM	5,000	6,000	1,000 + 5,000 – 6,000 = 0

Queue is dissipated at 10:00 AM.

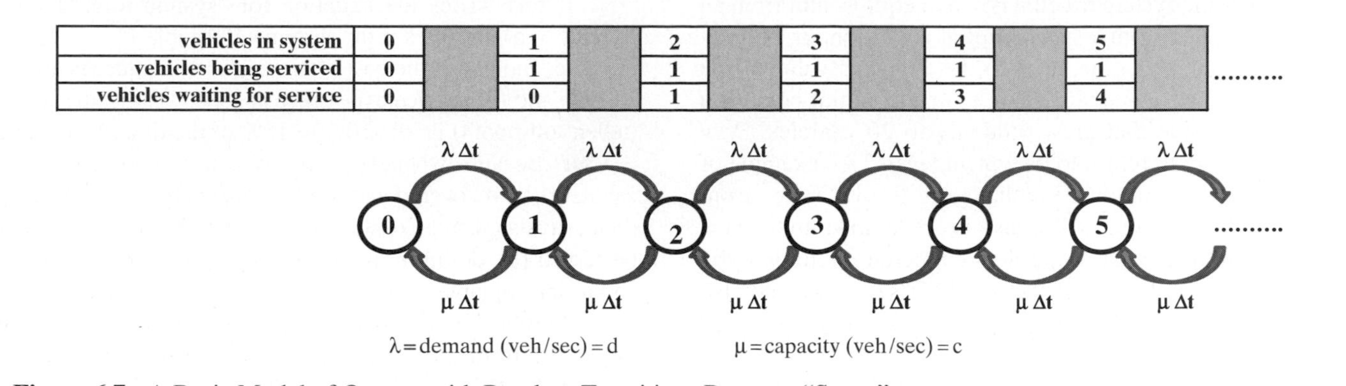

Figure 6.7: A Basic Model of Queues with Random Transitions Between "States"

Because the queue is growing in time, the point at which arriving vehicles join the queue moves upstream from the point of the obstruction or breakdown. Arrival rates, therefore, are increased as related to the upstream propagation speed of the end of the queue.

6.2.3 The Basic Approach to Queueing Analysis: Random Patterns

In the preceding "queueing analyses," the emphasis has been on deterministic computation based on averages, for cases in which demand exceeds capacity ($v/c > 1.00$).

However, in most cases, both the arrival pattern and the service pattern have randomness that leads to the development of queues (i.e., waiting lines). This is a common experience in everyday life, including lines at the bank, department of motor vehicles, and gas stations: Even when demand is less than capacity ($v/c < 1.00$), arrivals are random, service times vary, and queues develop. The most basic case will be addressed now, so that the basic effects and the resultant "rules of thumb" can be understood. See Reference 11 or other complete texts on queueing theory for a full treatment of the subject.

The case considered in this chapter is based on arrivals that are exponentially distributed, with an *average arrival rate* of λ veh/s and an *average* service rate of μ veh/s. The time between arrivals and between serviced vehicles are both exponentially distributed, independent of each other, and independent of prior arrivals/service events, and with a "queue discipline" of first-in-first-out ("FIFO"). Consider the line at a toll booth as an example.

The exponential distribution is popular both because it occurs in nature frequently enough *and* because it leads to some relatively easy mathematical analysis (compared to other arrival patterns and disciplines). It has the interesting mathematical property that the probability of an arrival in the next Δt seconds is $\lambda \Delta t$ *and* that given the passage of time, the distribution of the *remaining* time is itself exponentially distributed. That is, if you have been waiting for an arrival for t_0 seconds, the distribution of the remaining time waiting is still exponential, with the same λ. For this reason, it is referred to as "purely random" events.[1]

Figure 6.7 shows a common model of the various "states" or "conditions" a simple single-line queue can have, namely zero vehicles present (server idle), one vehicle present (server busy, no vehicle waiting), and so forth. The probability of a *transition* in the next Δt from one state/condition to another is determined by the probabilities $\lambda \Delta t$ and $\mu \Delta t$ as shown in Figure 6.7. For instance in State 1 (denoted by the "1" in the circle), one might move to State 2 if there is an arrival in the next Δt, *or* to State 0 if there is a service event in the next Δt. Transitions from State 1 to higher states such as State 3 or beyond would require *two* arrivals, which has a probability of $(\lambda; \Delta t)^2$ and is a negligible number compared to first-order terms (i.e., not raised to a power), particularly as Δt tends to zero.

It is possible to study the *transient* behavior of the system by writing a set of difference equations such as $p_n(t + \Delta t) = \{\lambda \Delta t\}p_{n-1}(t) + \{\mu \Delta t\}p_{n+1}(t) + \{1-(\lambda+\mu) \Delta t\}p_n(t)$ where the last term recognizes the probability of not moving from the existing State "n."

These equations can be used to in conjunction with $\Sigma p_n(t) = 1$ to generate a set of differential equations that can in turn be used to study analytically such questions as the

[1]Over an interval of "T" seconds, the *number* of exponentially distributed events that happen can be shown to follow the Poisson distribution, with an average number of λT events in the interval. This is often applied to traffic counts in five-minute periods, for instance.

time it takes the system to settle down to equilibrium from an abrupt change or from a special initial condition, or to study the set of $p_n(t)$ over time, or even the expected value of the queue over time. An example of a special initial condition might be a queue that grew suddenly to 20 vehicles. How long will it take to return to its typical levels? An example of an abrupt change might be a change in the v/c ratio[2] from 0.60 to 0.90. How long will it take to settle down to the new equilibrium, and what is the new expected queue length? Such tools as *Mathematica* [*12*] or *MATLAB* [*13*] could also be used to calculate the difference or differential equations for studies.

But it is also possible to study the *steady-state* or *equilibrium* conditions more directly by realizing that in equilibrium, the flows between states in the two directions must be equal, so that $\mu\, p_n = \lambda\, p_{n-1}$ or $p_n = \rho p_{n-1}$, where $\rho = \lambda/\mu$, which is the v/c ratio. Combined with $\Sigma p_n = 1$, it is possible to derive that $p_n = \rho^n(1 - \rho)$ and that the expected queue is equal $\rho/(1 - \rho)^2$ because E[# in system] $= \Sigma(n - 1)p_n$. Figure 6.8 shows a plot of expected queue in the system[3] versus the v/c ratio ρ.

Figure 6.8 reveals a number of insights in considering providing service under conditions of randomness:

- The number of vehicles queued (or in the system) is non-zero even when v/c is much less than 1.00 because of the randomness of the arrivals and the service.
- As v/c exceeds 0.85, the curve starts to increase rapidly.
- As v/c passes 0.90, the rate of increase accelerates and the number of vehicles queued (or in the system) grows rapidly, approaching infinity as v/c approaches 1.00.

Therefore, as a useful rule of thumb, it is useful to keep the v/c below 0.85 and certainly below 0.90, so that the queues and such (including customer waiting times) are reasonable. For bank teller lines or for toll plazas, this leads to a rule of thumb to add servers (new teller, new toll booth) whenever the v/c ratio exceeds 0.90 or even 0.85, which can be deduced from the queue size.

[2]Note that the ratio $\rho = \lambda/\mu$ is the ratio of demand to capacity, which is denoted as the v/c ratio in traffic work.

[3]One must be precise because the *expected queue* is the expected number of vehicles *awaiting* service, whereas the *expected number of vehicles in the system* includes the vehicle being served (in any). Further, it is not as simple as "add one" because the system is sometimes entirely idle, with no one being served.

If one writes the equation for "system idle," namely $p_0 = (1 - \rho)$, this means the system will be idle 10% to 15% of the time, and that is the cost of making sure the queues are reasonable and the service experience is good. That is, the worker (teller, toll booth) is idle 10% to 15% of the time. Why is this needed? Because whenever capacity is lost due to idleness, it can *never* be recovered and used productively. Thus when random arrivals mean no customer is present, the customer cannot be served (by definition), the capacity is used up nonetheless and is not available when customers do show up later.

Another insight is that the service experience can be improved if the service time is decreased (i.e., capacity increased). This is done in practice when a higher capacity electronic toll (e.g., EZ Pass) lane is used at a toll plaza.

Many complicated rules can be constructed for service, including (1) multiple servers, customer selects line, (2) multiple service, single waiting line. One can deduce rules of thumb for the desired number of servers and for the effect on queue lengths and their variability (and hence "time in system" and its variability). There are also models that can be constructed of situations that can only accommodate finite queue lengths, and one of the issues is the turn-away rate of potential customers. But that is for a text specializing in queueing theory.

A final note: Figure 6.8 shows the queue exploding to infinity as v/c approaches 1.00, but other treatments in this text (e.g., intersection delay) show curves that have finite values well above v/c of 1.00. How can this be? The answer is straightforward: In those other cases, the analysis is for a burst of time such as 15 minutes during which v/c exceeds 1.00 but is part of a longer period wherein the average v/c is

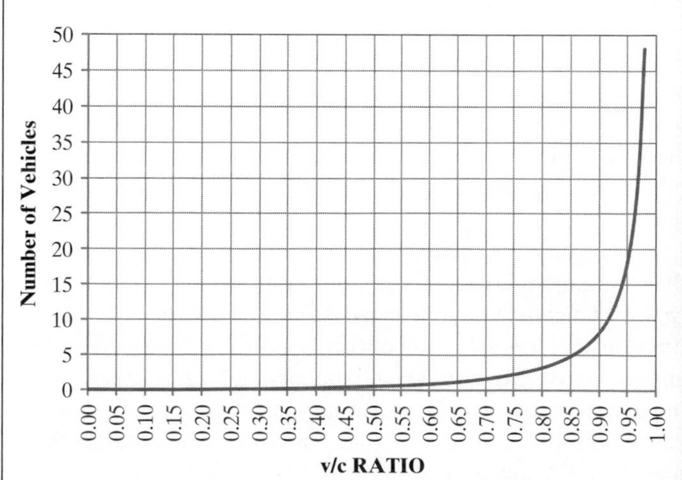

Figure 6.8: Expected Queue for Various v/c Ratios

less than 1.00.[4] The present subsection addresses a long-term equilibrium condition in which v/c cannot exceed 1.00 without a continual accumulation of vehicles, mathematically going to infinity *for this stated analysis condition and load.*

6.3 Shock-Wave Theory and Applications

Within traffic flow theory is another area especially worth mentioning, namely modeling traffic flow as a compressible fluid subject to "shock waves" and disruptions. The classic work on this was Reference 14, and it remains as a landmark paper over 50 years later. In addition to the situations depicted in this section, the authors also covered more detailed cases in which the shock wave forms at a bottleneck and later dissipates as the demand decreases.

6.3.1 Different Flow-Density Curves

A section of roadway may be thought of as defined by its "flow-density" curve, if one appreciates that (1) this curve is related to the speed-density curve, which is more basic, and (2) the analysis to follow is based on the "steady-state" or equilibrium conditions, and transition effects make the depiction more complex. (However, the principal effects are well represented by the analysis to be given.)

Consider the two flow-density relations of Figure 6.9 for sections of a three-lane road (each direction), with Curve I defining Sections 1 and 3, and Curve II defining Section 2. Section 2 clearly has a lower capacity; this might be due to a grade, a surfacing condition, or other factors.

Further, consider that the demand flow rate is Q_1 and is less than the lower capacity value.

- Section 1 has a space mean speed S_1 and a density D_1, as defined by Curve I. The demand is served and flows on into Section 2. At the given demand, Section 2 has a space mean speed and density defined by Curve II, and the density D_2 is *higher* than the prior density D_1. Because the speed is given by the chord of the curve (see the indicated slopes), it follows that the speed S_2 is *lower* than i_n Section 1. Thus the driver experiences lower speed and a greater proximity to other drivers. Nonetheless the demand is served and flows on into Section 3.

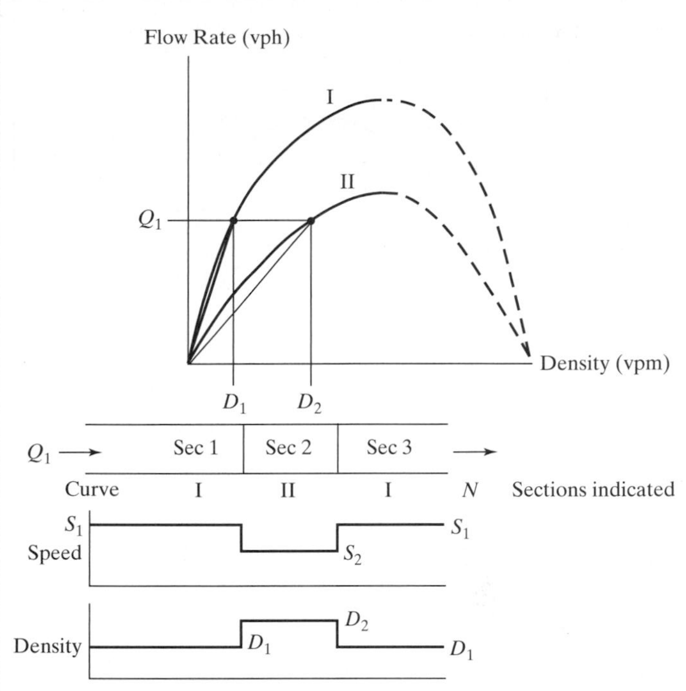

Figure 6.9: Flow-Density Relations for Different Sections of Road

- Section 3 has the same demand and curve as Section 1. There is no reason for the curve to rest on a flow-density pair to the right of the curves (i.e., the dashed parts). Compared to Section 2, the driver experiences better speed and lower density. Indeed, the driver returns to the conditions of Section 1.

The speed and density profiles by section are shown in Figure 6.9 for the given flow. The driver passes through the sections, experiencing this profile.

However, what would have happened if the demand flow rate had been greater than the capacity of the second section? Consider Figure 6.10, which depicts this situation. Note that:

- The demand flow rate Q_1 can be served at some point in Section I, illustrated by the Point 1 in the roadway. In this part of the section, the speed and density are as shown for the corresponding Point 1 and Curve I.
- This demand cannot be passed to Section 2, simply because it exceeds the capacity of Section 2. The most that can be passed is Q_2, corresponding to the capacity of the section.
- Thus, in Section 2, it is the flow Q_2 that exists. This operates at Point 2 on Curve II and is at a lower speed and higher density than the traffic at Point 1.

[4]For example, if the overall v/c ratio in the peak hour is 0.90 but the peak hour factor (PHF) is 0.85, during the busiest 15-minutes, the v/c ratio is 0.90/0.85 = 1.06. The overall effect can still be a finite queue.

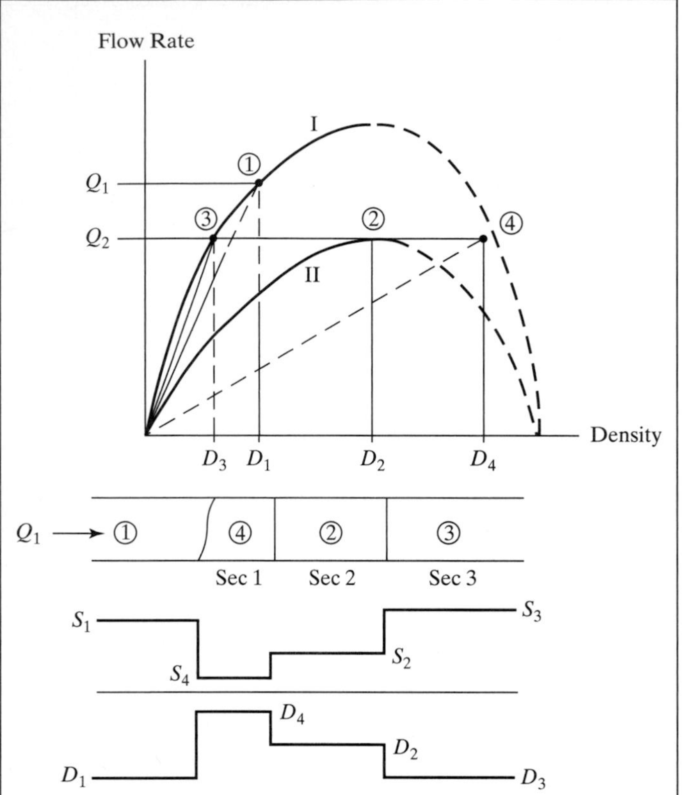

Figure 6.10: The Same Section of Road But with Higher Demand

- Only a flow level Q_2 can be passed to Section 3. Without a mechanism to force operation on the right (dashed) side of the curves, Section 3 operates at Point 3: It has a *higher* speed and *lower* density than even Point 1.

Although this may seem strange, realize that Section 2 serves as the bottleneck, holding back the demand. Section 3 is the same quality road as Section 1 but has less flow—the true demand simply cannot reach it.

This situation is complicated by the fact that there is an accumulation of vehicles somewhere: The difference $(Q_1 - Q_2)$ passes Point 1 but cannot get into Section 2. Thus they are stored upstream of Section 2, actually within Section 1. Thus part of Section 1 stores vehicles *and* experiences an outflow of only Q_2. Clearly, it is not in the same mode of operation as Section 3 and is therefore not operating at Point 3. Rather, it is operating at the same flow rate but on the *right* side of the curve, at Point 4.

The speed and density profiles by section are also shown in Figure 6.10. Note that the *best* speed and lowest density is downstream of the bottleneck; the *worst* speed and highest density is just upstream of the bottleneck. Thus *an assumption that the section with the poorest speed is actually the bottleneck is wrong and would lead to an erroneous identification:* The real bottleneck is the section just *downstream* of the worst section, where the traffic is recovering some performance.

6.3.2 Rate of growth

An interesting problem is the proper identification of the rate of growth of the storage area. This was covered to some extent in Section 6.1.

Figure 6.11 shows the situation at an arbitrary time and at one hour later. The expansion of the storage area now (1) adds the hour's accumulation of $(Q_1 - Q_2)$ vehicles, *and* (2) encompasses an area in which there already were vehicles at density D_1. Thus the growth is defined by a total addition of

$$[(Q_1 - Q_2) + (S)(D_1)]\text{vehicles} \tag{6-3}$$

in the hour, with the speed of growth S not known.

However, it is also logical that the added growth is also defined by its *new* density, D_4. Thus the added growth includes $(S)(D_4)$ vehicles, so that

$$[(Q_1 - Q_2) + (S)(D_1)] = (S)(D_4)$$

or

$$S = (Q_1 - Q_2)/(D_4 - D_1) \tag{6-4}$$

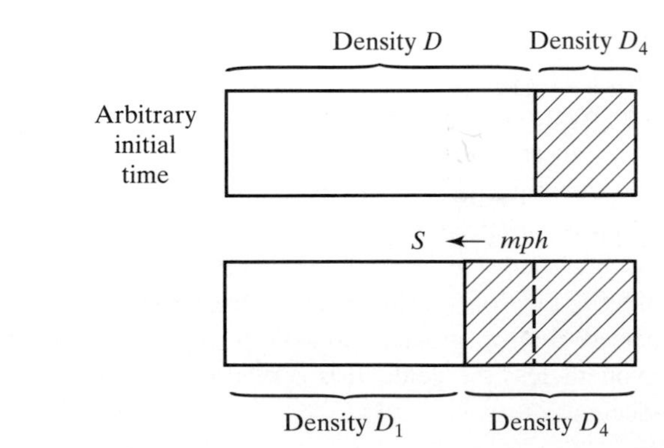

Figure 6.11: Growth of Queue in the Compressible Fluid Model

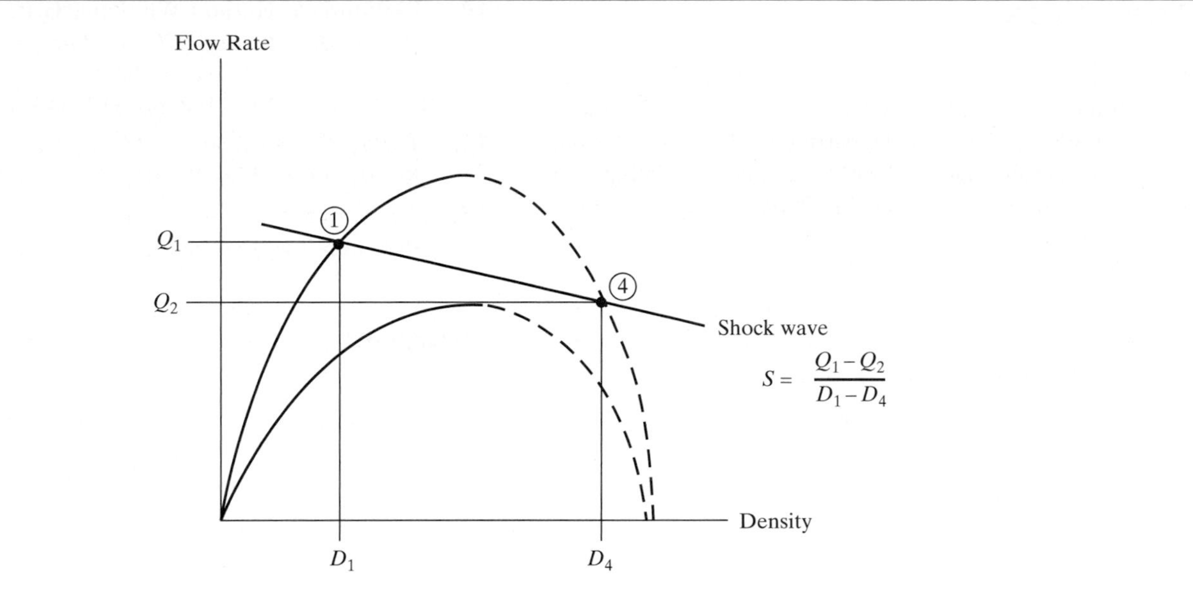

Figure 6.12: Geometric Interpretation of Rate of Growth

which is the speed of the growth of the queued vehicles. This is the "shock wave" traveling *up* the traffic stream from the bottleneck interface, due to the discontinuity.

Equation 6.4 has a fascinating and logical geometric interpretation. As shown in Figure 6.12, it is the chord between Points 1 and 4 and has a *negative* slope, indicating that it travels *against* the direction of the vehicles.

To appreciate the scale, assume that

$$D_1 = 55 \text{ vpm}, \quad D_4 = 100 \text{ vpm}$$

$$Q_1 = 1900 \text{ vph}, Q_2 = 1700 \text{ vph}$$

on a per-lane basis, and note that $S = (1900 - 1700)/(100 - 55) = 4.4$ mph. That is, the queue of stored vehicles grows at 4.4 mph for these numbers.

6.4 Characteristics of Interrupted Flow

The key feature of interrupted flow is the cyclical stopping and restarting of traffic streams at traffic signals and at STOP or YIELD signs.

When traveling along a signalized street or arterial, platoons form, as groups of vehicles proceed in a manner that allows them to move continuously through a number of signals. Within platoons, many of the same characteristics as for uninterrupted traffic streams exist. It is the dynamics of starting and stopping groups of vehicles that adds complexity. Fundamental concepts of interrupted flow are treated in Chapter 20, "Basic Concepts and Principles of Intersection Signalization."

6.5 Closing Comments

The material in this chapter provides only a very elementary introduction to the subject of traffic flow theory. Interested readers should consult the references and other sources of more detailed treatments of this complex and evolving subject. Excellent sources of information is the Web site of the TRB Committee on Traffic Flow Theory and Characteristics at http://www.tft.pdx.edu/index.htm, the 2001 monograph cited as Reference 1,[5] related TRB committees cited on http://www.tft.pdx.edu/index.htm, and the FHWA *Traffic Analysis Toolbox* at http://ops.fhwa.dot.gov/trafficanalysistools/toolbox.htm.

[5]This monograph is the latest in a series of monographs on traffic flow theory. See References 11 and 12, or access them through the Web site cited here.

References

1. *Traffic Flow Theory—A State of the Art Report*, Oak Ridge National Laboratory, Federal Highway Administration, U.S. Department of Transportation, Washington DC, 1999 (available online at www.fhwa.gov.). Reissued in 2001 as "Revised 2001," Project Organized by the TRB Committee on Traffic Flow Theory and Characteristics."
2. "Traffic Flow Theory 2006," *Transportation Research Record 1965*, Transportation Research Board, Washington DC, 2006 (22 papers).
3. "Traffic Flow Theory and Characteristics 2008," *Transportation Research Record 2088*, Transportation Research Board, Washington DC, 2008 (23 papers).
4. "Traffic Flow Theory 2007," *Transportation Research Record 1999*, Transportation Research Board, Washington DC, 2007 (22 papers).
5. Greenshields, B., "A Study of Traffic Capacity," *Proceedings of the Highway Research Board*, Transportation Research Board, Washington DC, 1934.
6. Ellis, R., "Analysis of Linear Relationships in Speed-Density and Speed-Occupancy Curves," *Final Report*, Northwestern University, Evanston IL, December 1964.
7. Greenberg, H., "An Analysis of Traffic Flows," *Operations Research*, Vol. 7, Operations Research Society of America, Washington DC, 1959.
8. Underwood, R., "Speed, Volume, and Density Relationships," *Quality and Theory of Traffic Flow*, Yale Bureau of Highway Traffic, New Haven, CT, 1961.
9. Edie, L., "Car-Following and Steady-State Theory for Non-Congested Traffic," *Operations Research*, Vol. 9, Operations Research Society of America, Washington DC, 1961.
10. Duke, J., Schofer, J., and May, A. Jr., "A Statistical Analysis of Speed-Density Hypotheses," *Highway Research Record 154*, Transportation Research Board, Washington DC, 1967.
11. *Fundamentals of Queueing Theory,* 4th Edition (Wiley Series in Probability and Statistics), Wiley-Interscience, 2008.
12. http://www.wolfram.com/products/mathematica/index.html/.
13. http://www.mathworks.com/.
14. Lighthill, M.H., and Whitham, G.B., *On Kinematic Waves: II. A Theory of Traffic Flow on Long Crowded Roads*. Proceedings of the Royal Society, London Series A229, No. 1178, pp. 317–345, 1957.
15. *Traffic Flow Theory: A Monograph*, Transportation Research Board Special Report 165, 1975.
16. *An Introduction to Traffic Flow Theory*, Highway Research Board Special Report 79, 1964.

Problems

6-1. A freeway with two lanes in one direction has a capacity of 2,100 veh/h/ln under normal operation. On a particular morning, one of these lanes is blocked for 15 minutes, beginning at 7:00 AM. The arrival pattern of vehicles at this location is as follows:

7:00—8:00 AM	4,200 veh/h
8:00—9:00 AM	4,000 veh/h
9:00—10:00 AM	3,800 veh/h
After 10:00 AM	3,000 veh/h

Conduct a deterministic queueing analysis to determine (a) the maximum size and length of the queue, and (b) the time that the queue will fully dissipate. The following two scenarios should be examined:

- Capacity under stable conditions: 2,100 veh/h/ln; queue discharge capacity: 1,950 veh/h/ln
- Capacity under stable conditions: 2,100 veh/h/ln; queue discharge capacity: 2,100 veh/h/ln

6-2. A study of speed-flow-density relationships at a particular site has resulted in the following calibrated relationship:

$$S = 71.2\left(1 - \frac{D}{122}\right)$$

(a) Find the free-flow speed and jam density for this relationship.

(b) Derive equations depicting the relationships between flow and density, and speed and flow. Plot the resulting curves. Show the values of free-flow speed, jam density, and capacity on these curves.

(c) Determine the capacity of the site mathematically.

6-3. Answer all questions as in problem 6-2 for the following speed-density relationship:

$$S = 62e^{-0.015D}$$

PART 2

Traffic Studies and Programs

CHAPTER 7

Statistical Applications in Traffic Engineering

Because traffic engineering involves the collection and analysis of large amounts of data for performing all types of traffic studies, it follows that statistics is also an important element in traffic engineering. Statistics helps us determine how much data will be required, as well as what meaningful inferences can confidently be made based on that data.

Statistics is required whenever it is not possible to directly observe or measure all of the values needed. If a room contained 100 people, the average weight of these people could be measured with 100% certainty by weighing each one and computing the average. In traffic, this is often not possible. If the traffic engineer needs to know the average speed of all vehicles on a particular section of roadway, not all vehicles could be observed. Even if all speeds could be measured over a specified time period (a difficult accomplishment in most cases), speeds of vehicles arriving before or after the study period, or on a different day than the sample day, would be unknown. In effect, no matter how many speeds are measured, there are always more that are not known. For all practical and statistical purposes, the number of vehicles using a particular section of roadway over time is infinite.

Because of this, traffic engineers often observe and measure the characteristics of a finite *sample* of vehicles in a *population* that is effectively infinite. The mathematics of statistics is used to estimate characteristics that cannot be established with absolute certainty, and to assess the degree of certainty that does exist. When this is done, statistical analysis is used to address the following questions:

- How many samples are required (i.e., how many individual measurements must be made)?
- What confidence should I have in this estimate (i.e., how sure can I be that this *sample* measurement has the same characteristics as the *population*)?
- What statistical distribution best describes the observed data mathematically?
- Has a traffic engineering design resulted in a change in characteristics of the population? (For example, has a new speed limit resulted in reduced speeds?)

This chapter explores the statistical techniques used in answering these critical questions and provides some common examples of their use in traffic engineering. This chapter is not, however, intended as a substitute for a course in statistics. References 1 through 4 offer some basic information. For an additional traffic reference addressing statistical tests, see Reference 5. The review that follows assumes that either (1) you have had a previous course in statistics or (2) your instructor will supplement or expand the materials as needed by an individual class.

7.1 Overview of Probability Functions and Statistics

Before exploring some of the more complex statistical applications in traffic engineering, we review some basic principles of probability and statistics that are relevant to these analyses.

7.1.1 Discrete versus Continuous Functions

Discrete functions are made up of discrete variables—that is, they can assume only specific whole values and not any value in between. Continuous functions, made up of continuous variables, in contrast, can assume any value between two given values. For example, Let $N =$ the number of children in a family. N can equal 1, 2, 3, and so on, but not 1.5, 1.6, 2.3. Therefore it is a discrete variable. Let $H =$ the height of an individual. H can equal 5 ft, 5.5 ft, 5.6 ft, and so on, and, therefore, is a continuous variable.

Examples of discrete probability functions are the Bernoulli, binomial, and Poisson distributions, which are discussed in the following sections. Some examples of continuous distributions are the normal, exponential, and chi-square distributions.

7.1.2 Randomness and Distributions Describing Randomness

Some events are very predictable or should be predictable. If you add mass to a spring or a force to a beam, you can expect it to deflect a predictable amount. If you depress the gas pedal a certain amount and you are on level terrain, you expect to be able to predict the speed of the vehicle. However, some events may be totally random. The emission of the next particle from a radioactive sample is said to be completely random.

Some events may have very complex mechanisms and *appear* to be random for all practical purposes. In some cases, the underlying mechanism cannot be perceived, whereas in others we cannot afford the time or money necessary for the investigation.

Consider the question of who turns north and who turns south after crossing a bridge. Most of the time, we simply say there is a probability p that a vehicle will turn north, and we treat the outcome as a random event. However, if we studied who was driving each car and where each driver worked, we might expect to make the estimate a very predictable event, for each and every car. In fact, if we kept a record of their license plates and their past decisions, we could make very predictable estimates. The events—to a large extent—are not random. Obviously, it is not worth that trouble because the random assumption serves us well enough. That, of course, is the crux of engineering: Model the system as simply (or as precisely) as possible (or necessary) *for all practical purposes*. Albert Einstein was once quoted as saying, "Make things as simple as possible, but no simpler."

In fact, a number of things are modeled as random *for all practical purposes*, given the investment we can afford. Most of the time, these judgments are just fine and are very reasonable but, as with every engineering judgment, they can sometimes cause errors.

7.1.3 Organizing Data

When data are collected for use in traffic studies, the raw data can be looked at as individual pieces of data or grouped into classes of data for easier comprehension. Most data fit into a common distribution. Some of the common distributions found in traffic engineering are the normal distribution, the exponential distribution, the chi-square distribution, the Bernoulli distribution, the binomial distribution, and the Poisson distribution.

As part of the process for determining which distribution fits the data, one often summarizes the raw data into classes and creates a frequency distribution table (often the data is collected without even recording the individual data points). This makes the data more easily readable and understood. Consider Table 7.1, which lists the unorganized data of the heights in inches of 100 students in the Engineering 101 class. You could put the data into an array format, which means listing the data points in order from either lowest to highest or highest to lowest. This will give you some feeling for the character of the data, but it is still not very helpful, particularly when you have a large number of data points.

It is more helpful to summarize the data into defined categories and display it as a frequency distribution table, as in Table 7.2.

Even though the details of the individual data points are lost, the grouping of the data adds much clarity to the character of the data. From this frequency distribution table, plots can be made of the frequency histogram, the frequency distribution, the relative frequency distribution, and the cumulative frequency distributions.

Table 7.1: Heights of Students in Engineering 101

62.3	67.5	73	73	63	69.5	70	63.2	74	70.2
72	67	64	67.5	66.1	64.6	67.5	74	66	68
66	61	67	70.5	67	70.1	60.5	67.9	70	61.4
67.5	68	67.5	64.6	69.8	63	66	64.9	68	67.5
73.9	66	66.2	69.2	66.5	67.8	71	69.1	69.4	70.5
64.5	67.4	72.5	61.8	63.7	69	67	68	67.9	64.5
67	73.5	67.4	67.4	69.3	70	66.8	62	64	68
66	64	63	71	66	63.9	70	69	68	67
69.3	69.4	68	64.5	69	66.5	67.2	66.5	63.9	69
71	67	70	67	70.5	71	64.2	70.5	67	67.5

Table 7.2: Frequency Distribution Table: Heights of Students in Engineering 101

Height Group (inches)	Number of Observations
60–62	5
63–65	18
66–68	42
69–71	27
72–74	8

7.1.4 Common Statistical Estimators

In dealing with a distribution, two key characteristics are of interest. These are discussed in the following subsections.

Measures of Central Tendency

Measures of central tendency are measures that describe the center of data in one of several different ways. The *arithmetic mean* is the average of all observed data. The true underlying mean of the population, μ, is an exact number that we do not know, but can estimate as:

$$\bar{x} = \frac{1}{N}\sum_{i=1}^{N} x_i \tag{7-1}$$

where: $\bar{x}$ = arithmetic average or mean of observed values

x_i = ith individual value of statistic

N = sample size

Consider the following example: Estimate the mean from the following sample speeds in mi/h: (53, 41, 63, 52, 41, 39, 55, 34). Using Equation 7-1:

$$\bar{x} = \frac{1}{8}(53 + 41 + 63 + 52 + 41 + 39 + 55 + 34)$$
$$= 47.25$$

Because the original data had only two significant digits, the more correct answer is 47 mi/h.

For grouped data, the average value of all observations in a given group is considered to be the midpoint value of the group. The overall average of the entire sample may then be found as:

$$\bar{x} = \frac{\sum_j f_j m_j}{N} \tag{7-2}$$

where: f_j = number of observations in group j

m_j = middle value of variable in group j

N = total sample size or number of observations

For the height data of Table 7.2:

$$\bar{x} = \frac{61(5) + 64(18) + 67(42) + 70(27) + 73(8)}{100}$$
$$= 67.45 = 67$$

The *median* is the middle value of all data when arranged in an array (ascending or descending order). The median divides a distribution in half: Half of all observed values are higher than the median, and half are lower. For nongrouped data, it is the middle value; for example, for the set of numbers (3, 4, 5, 5, 6, 7, 7, 7, 8), the median is 6. It is the fifth value (in ascending or descending order) in an array of 9 numbers. For grouped data, the easiest way to get the median is to read the 50% percentile point off a cumulative frequency distribution curve (see Chapter 9).

The *mode* is the value that occurs most frequently—that is, the most common single value. For example, in non-grouped data, for the set of numbers (3, 4, 5, 5, 6, 7, 7, 7, 8), the mode is 7. For the set of numbers (3, 3, 4, 5, 5, 5, 6, 7, 8, 8, 8, 9), both 5 and 8 are modes, and the data are said to be bimodal. For grouped data, the mode is estimated as the peak of the frequency distribution curve (see Chapter 9). For a perfectly symmetrical distribution, the mean, median, and mode are the same.

Measures of Dispersion

Measures of dispersion are measures that describe how far the data spread from the center. The *variance and standard deviation* are statistical values that describe the magnitude of variation around the mean, with the variance defined as:

$$s^2 = \frac{\sum (x_i - \bar{x})^2}{N - 1} \tag{7-3}$$

where: s^2 = variance of the data
N = sample size, number of observations

All other variables as previously defined.

The standard deviation is the square root of the variance. It can be seen from the equation that what you are measuring is the distance of each data point from the mean. This equation can also be rewritten (for ease of use) as:

$$s^2 = \frac{1}{N}\sum_{i=1}^{N} x_i^2 - \left(\frac{N}{N-1}\right)\bar{x}^2 \tag{7-4}$$

For grouped data, the standard deviation is found from:

$$s = \sqrt{\frac{\sum fm^2 - N(\bar{x})^2}{N - 1}} \tag{7-5}$$

where all variables are as previously defined. The standard deviation (STD) may also be estimated as:

$$s_{est} = \frac{P_{85} - P_{15}}{2} \tag{7-6}$$

where: P_{85} = 85th percentile value of the distribution (i.e., 85% of all data is at this value or less).
P_{15} = 15th percentile value of the distribution (i.e., 15% of all data is at this value or less).

The *x*th *percentile* is defined as that value below which *x*% of the outcomes fall. P_{85} is the 85th percentile, often used in traffic speed studies; it is the speed that encompasses 85% of vehicles. P_{50} is the 50th percentile speed, or the median.

The *coefficient of variation* is the ratio of the standard deviation to the mean and is an indicator of the spread of outcomes relative to the mean.

The distribution or the underlying shape of the data is of great interest. Is it normal? Exponential? But the engineer is also interested in anomalies in the shape of the distribution (e.g., skewness or bimodality). Skewness is defined as the (mean − mode)/std. If a distribution is negatively skewed, it means that the data are concentrated to the left of the most frequent value (i.e., the mode). When a distribution is positively skewed, the data are concentrated to the right of the mode. The engineer should look for the underlying reasons for skewness in a distribution. For instance, a negatively skewed speed distribution may indicate a problem such as sight distance or pavement condition that is inhibiting drivers from selecting higher travel speeds.

7.2 The Normal Distribution and Its Applications

One of the most common statistical distributions is the *normal distribution*, known by its characteristic bell-shaped curve (Fig. 7.1). The normal distribution is a continuous distribution. Probability is indicated by the area under the probability density function *f*(*x*) between specified values, such as $P(40 < x < 50)$.

The equation for the normal distribution function is:

$$f(x) = \frac{1}{\sigma\sqrt{2\pi}} e^{-\left[\frac{(x-\mu)^2}{2\sigma^2}\right]} \tag{7-7}$$

where: x = normally distributed statistic
μ = true mean of the distribution
σ = true standard deviation of the distribution

The probability of any occurrence between values x_1 and x_2 is given by the area under the distribution function between the two values. The area may be found by integration between the two limits. Likewise, the mean, μ, and the

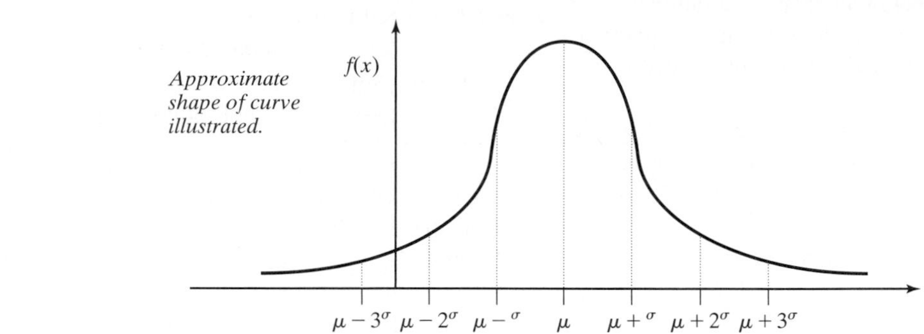

Figure 7.1: The Normal Distribution

variance, σ^2, can be found through integration. The normal distribution is the most common distribution because any process that is the sum of many parts tends to be normally distributed. Speed, travel time, and delay are all commonly described using the normal distribution. The function is completely defined by two parameters: the mean and the variance. All other values in Equation 7-6, including π, are constants. The notation for a normal distribution is x: $N[\mu, \sigma^2]$, which means that the variable x is normally distributed with a mean of μ and a variance of σ^2.

7.2.1 The Standard Normal Distribution

For the normal distribution, the integration cannot be done in closed form due to the complexity of the equation for $f(x)$; thus tables for a "standard normal" distribution, with zero mean ($\mu = 0$) and unit variance ($\sigma^2 = 1$), are constructed. Table 7.3 presents tabulated values of the standard normal distribution. The standard normal is denoted z: $N[0,1]$. Any value of x on any normal distribution, denoted x: $N[\mu, \sigma^2]$, can be converted to an equivalent value of z on the standard normal distribution. This can also be done in reverse when needed. The translation of an arbitrary normal distribution of values of x to equivalent values of z on the standard normal distribution is accomplished as:

$$z = \frac{x - \mu}{\sigma} \tag{7-8}$$

where: z = equivalent statistic on the standard normal distribution, z: $N[0,1]$

x = statistic on any arbitrary normal distribution, x: $N[\mu, \sigma^2]$ other variables as previously defined

Figure 7.2 shows the translation for a distribution of spot speeds that has a mean of 55 mi/h and standard deviation of 7 mi/h to equivalent values of z.

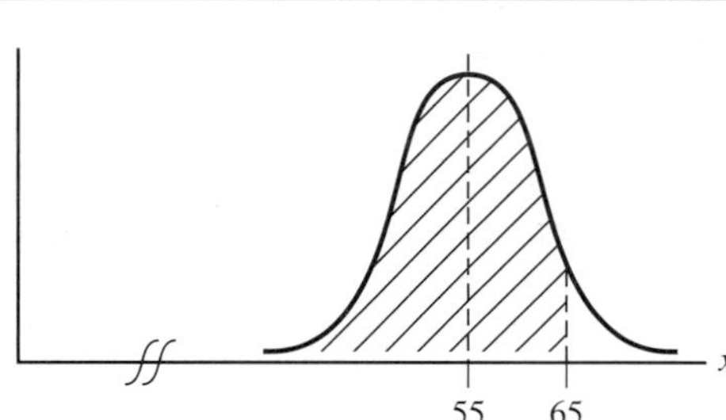

(a) The problem and normal distribution as specified

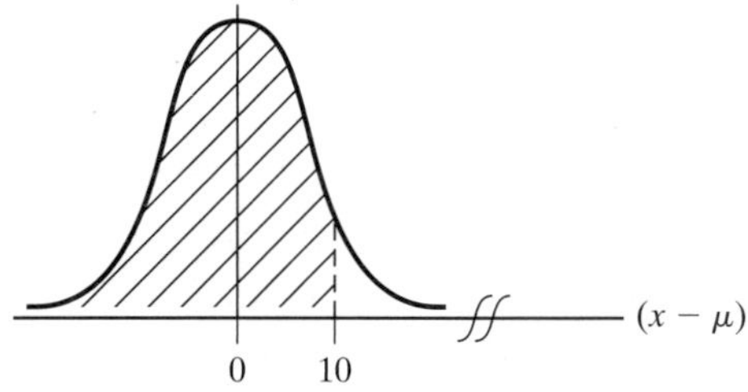

(b) The axis translated to a zero mean

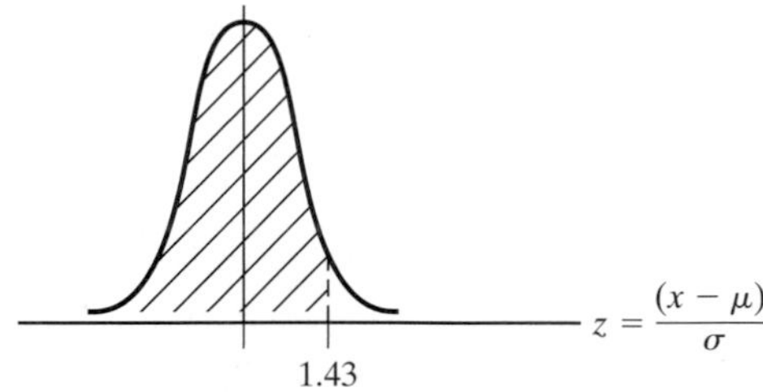

(c) The axis scaled so that the $N(0, 1)$ is used

Figure 7.2: Shifting the Normal Distribution to the Standard Normal Distribution

Consider the following example: For the spot speed distribution of Figure 7.2, x: $N[55,49]$, what is the probability that the next observed speed will be 65 mi/h or less? Translate

Table 7.3: The Standard Normal Distribution

$$F(z) = \int_{-\infty}^{z} \frac{1}{\sqrt{2\pi}} e^{-z^2/2}\, dz$$

z	.00	.01	.02	.03	.04	.05	.06	.07	.08	.09
.0	.5000	.5040	.5080	.5120	.5160	.5199	.5239	.5279	.5319	.5359
.1	.5398	.5438	.5478	.5517	.5557	.5596	.5636	.5675	.5714	.5753
.2	.5793	.5832	.5871	.5910	.5948	.5987	.6026	.6064	.6103	.6141
.3	.6179	.6217	.6255	.6293	.6331	.6368	.6406	.6443	.6480	.6517
.4	.6554	.6591	.6628	.6661	.6700	.6736	.6772	.6808	.6844	.6879
.5	.6913	.6950	.6985	.7019	.7054	.7083	.7123	.7157	.7190	.7224
.6	.7257	.7291	.7324	.7357	.7389	.7422	.7454	.7486	.7517	.7549
.7	.7580	.7611	.7642	.7673	.7704	.7734	.7764	.7794	.7823	.7852
.8	.7881	.7910	.7939	.7967	.7995	.8023	.8051	.8078	.8106	.8133
.9	.8159	.8186	.8212	.8238	.8264	.8289	.8315	.8340	.8365	.8389
1.0	.8413	.8438	.8461	.8485	.8508	.8531	.8554	.8577	.8599	.8621
1.1	.8643	.8665	.8686	.8708	.8729	.8749	.8770	.8790	.8810	.8830
1.2	.8849	.8869	.8888	.8907	.8925	.8944	.8962	.8980	.8997	.9015
1.3	.9032	.9049	.9066	.9082	.9099	.9115	.9131	.9147	.9162	.9177
1.4	.9192	.9207	.9222	.9236	.9251	.9265	.9279	.9292	.9306	.9319
1.5	.9332	.9345	.9357	.9370	.9382	.9394	.9406	.9418	.9429	.9441
1.6	.9432	.9463	.9474	.9484	.9495	.9505	.9515	.9525	.9535	.9545
1.7	.9554	.9564	.9573	.9582	.9591	.9599	.9608	.9616	.9625	.9633
1.8	.9641	.9649	.9658	.9664	.9671	.9678	.9686	.9693	.9699	.9706
1.9	.9713	.9719	.9726	.9732	.9738	.9744	.9750	.9756	.9716	.9767
2.0	.9772	.9778	.9783	.9788	.9793	.9798	.9803	.9808	.9812	.9817
2.1	.9812	.9826	.9830	.9834	.9838	.9842	.9846	.9854	.9854	.9857
2.2	.9861	.9864	.9868	.9871	.9875	.9878	.9881	.9884	.9887	.9890
2.3	.9893	.9896	.9898	.9901	.9904	.9906	.9909	.9911	.9913	.9916
2.4	.9918	.9920	.9922	.9925	.9927	.9929	.9931	.9932	.9934	.9936
2.5	.9938	.9940	.9941	.9943	.9945	.9946	.9948	.9949	.9951	.9952
2.6	.9953	.9955	.9956	.9937	.9959	.9960	.9961	.9962	.9963	.9964
2.7	.9965	.9966	.9967	.9968	.9969	.9970	.9971	.9972	.9973	.9974
2.8	.9974	.9975	.9976	.9977	.9977	.9978	.9979	.9979	.9980	.9981
2.9	.9981	.9982	.9982	.9983	.9984	.9984	.9985	.9985	.9986	.9986
3.0	.9987	.9987	.9987	.9988	.9988	.9989	.9989	.9989	.9990	.9990
3.1	.9990	.9991	.9991	.9991	.9992	.9992	.9992	.9992	.9993	.9993
3.2	.9993	.9993	.9994	.9994	.9994	.9994	.9994	.9995	.9995	.9995
3.3	.9995	.9995	.9995	.9996	.9996	.9996	.9996	.9996	.9996	.9997
3.4	.9997	.9997	.9997	.9997	.9997	.9997	.9997	.9997	.9997	.9998

and scale the x-axis as shown in Figure 7.2. The equivalent question for the standard normal distribution, z: $N[0,1]$, is found using Equation 7-8: Determine the probability that the next value of z will be less than:

$$z = \frac{65 - 55}{7} = 1.43$$

Entering Table 7.3 on the vertical scale at 1.4 and on the horizontal scale at 0.03, the probability of having a value of z less than 1.43 is 0.9236, or 92.36%.[1]

Another type of application frequently occurs: For the case just stated, what is the probability that the speed of the next vehicle is between 55 and 65 mi/h?

The probability that the speed is less than 65 mi/h has already been computed. We can now find the probability that the speed is less than 55 mi/h, which is equivalent to $z = (55 - 55)/7 = 0.00$, so that the probability is 0.50, or 50%, exactly. The probability of being between 55 and 65 mi/h is just the difference of the two probabilities: $(0.9236 - 0.5000) = 0.4236$, or 42.36%.

In similar fashion, using common sense and the fact of symmetry, one can find probabilities less than 0.5000, even though they are not tabulated directly in Table 7.3.

For the case just stated, find the probability that the next vehicle's speed is less than 50 mi/h. Translating to the z-axis, we wish to find the probability of a value being less than $z = (50 - 55)/7 = -0.71$. Negative values of z are not given in Table 7.3, but by symmetry it can be seen that the desired shaded area is the same size as the area *greater* than $+0.71$. Still, we can only find the shaded area less than $+0.71$ (it is 0.7611). However, knowing that the total probability under the curve is 1.00, the remaining area (i.e., the desired quantity) is therefore $(1.0000 - 0.7611) = 0.2389$, or 23.89%.

From these illustrations, three important procedures have been presented: (1) the conversion of values from any arbitrary normal distribution to the standard normal distribution, (2) the use of the standard normal distribution to determine the probability of occurrences, and (3) the use of Table 7.3 to find probabilities less than both positive and negative values of z, and between specified values of z.

7.2.2 Important Characteristics of the Normal Distribution Function

The preceding exercises allow us to compute relevant areas under the normal curve. Some numbers occur frequently in practice, and it is useful to have those in mind. For instance, what is the probability that the next observation will be within one standard deviation of the mean, given that the distribution is normal? That is, what is the probability that x is in the range $(\mu \pm 1.00\sigma)$? By a similar process to those just illustrated, we can find that this probability is 68.3%.

The following ranges have frequent use in statistical analysis involving normal distributions:

- 68.3% of the observations are within $\mu \pm 1.00\sigma$
- 95.0% of the observations are within $\mu \pm 1.96\sigma$
- 95.5% of the observations are within $\mu \pm 2.00\sigma$
- 99.7% of the observations are within $\mu \pm 3.00\sigma$

The total probability under the normal curve is 1.00, and the normal curve is symmetric around the mean. It is also useful to note that the normal distribution is asymptotic to the x-axis and extends to values of $\pm\infty$. These critical characteristics will prove to be useful throughout the text.

7.3 Confidence Bounds

What would happen if we asked everyone in class (70 people) to collect 50 samples of speed data and to compute their own estimate of the mean? How many estimates would there be? What distribution would they have? There would be 70 estimates and the histogram of these 70 means would look normally distributed. Thus the "estimate of the mean" is itself a random variable that is normally distributed.

Usually we compute only one estimate of the mean (or any other quantity), but in this class exercise we are confronted with the reality that there is a range of outcomes. We may, therefore, ask how good our estimate of the mean is. How confident are we that our estimate is correct? Consider that:

- The estimate of the mean quickly tends to be normally distributed.
- The expected value (the true mean) of this distribution is the unknown fixed mean of the original distribution.
- The standard deviation of this new distribution of means is the standard deviation of the original distribution divided by the square root of the number of samples, N. (This assumes independent samples and infinite population.)

The standard deviation of this distribution of the means is called the standard error of the mean (E)

$$E = \sigma/\sqrt{N} \tag{7-9}$$

[1] Probabilities are numbers between zero and one, inclusive, and not percentages. However, many people talk about them as percentages.

where the sample standard deviation, s, is used to estimate σ, and all variables are as previously defined. The same characteristics of any normal distribution apply to this distribution of means as well. In other words, the single value of the estimate of the mean, $\bar{x}_n$, approximates the true mean population, μ, as follows:

$$\mu = \bar{x} \pm \text{E, with 68.3\% confidence}$$

$$\mu = \bar{x} \pm 1.96 \text{ E, with 95\% confidence}$$

$$\mu = \bar{x} \pm 3.00 \text{ E, with 99.7\% confidence}$$

The ± term (E, 1.96E, or 3.00E, depending on the confidence level) in the preceding equation is also called the *tolerance* and is given the symbol e.

Consider the following: 54 speeds are observed, and the mean is computed as 47.8 mi/h, with a standard deviation of 7.80 mi/h. What are the 95% confidence bounds?

$$P[47.8 - 1.96 * (7.80/\sqrt{54})] \leq \mu \leq [47.8 + 1.96 * (7.80/\sqrt{54})] = 0.95 \quad \text{or}$$

$$P(45.7 \leq \mu \leq 49.9) = 0.95$$

Thus it is said there is a 95% probability that the true mean lies between 45.7 and 49.9 mi/h. Further, although not proven here, any random variable consisting of sample means tends to be normally distributed for reasonably large, regardless of the original distribution of individual values.

7.4 Sample Size Computations

We can rewrite the equation for confidence bounds to solve for N, given that we want to achieve a specified tolerance and confidence. Resolving the 95% confidence bound equation for N gives:

$$N \geq \frac{1.96^2 s^2}{e^2} \tag{7-10}$$

where 1.96^2 is used only for 95% confidence. If 99.7% confidence is desired, then the 1.96^2 would be replaced by 3^2.

Consider another example: With 99.7% and 95% confidence, estimate the true mean of the speed on a highway, plus or minus 1 mi/h. We know from previous work that the standard deviation is 7.2 mi/h. How many samples do we need to collect?

$$N = \frac{3^2 * 7.2^2}{1^2} \approx 467 \text{ samples for 99.7\% confidence,}$$

and

$$N = \frac{1.96^2 * 7.2^2}{1^2} \approx 200 \text{ samples for 95\% confidence}$$

Consider further that a spot speed study is needed at a location with unknown speed characteristics. A tolerance of ± 0.2 mph and a confidence of 95% is desired. What sample size is required? Because the speed characteristics are unknown, a standard deviation of 5 mi/h (a most common result in speed studies) is assumed. Then for 95% confidence, $N = (1.96^2 * 5^2)/0.2^2 = 2{,}401$ samples. This number is unreasonably high. It would be too expensive to collect such a large amount of data. Thus the choices are to either reduce the confidence or increase the tolerance. A 95% confidence level is considered the minimum that is acceptable; thus, in this case, the tolerance would be increased. With a tolerance of 0.5 mi/h:

$$N = \frac{1.96^2 * 5^2}{0.5^2} = 384 \text{ vehicles}$$

Thus the increase of just 0.3 mi/h in tolerance resulted in a decrease of 2,017 samples required. Note that the sample size required depends on s, which was assumed at the beginning. After the study is completed and the mean and standard deviation are computed, N should be rechecked. If N is greater (i.e., the actual s is greater than the assumed s), then more samples may need to be taken.

Another example: An arterial is to be studied, and it is desired to estimate the mean travel time to a tolerance of ±5 seconds with 95% confidence. Based on prior knowledge and experience, it is estimated that the standard deviation of the travel times is about 15 seconds. How many samples are required?

Based on an application of Equation 7-10, $N = 1.96^2(15^2)/(5^2) = 34.6$, which is rounded to 35 samples.

As the data is collected, the s computed is 22 seconds, not 15 seconds. If the sample size is kept at $N = 35$, the confidence bounds will be $\pm 1.96(22)/\sqrt{35}$ or about ±7.3 seconds. If the confidence bounds must be kept at ±5 seconds, then the sample size must be increased so that $N \geq 1.96^2(22^2)/(5^2) = 74.4$ or 75 samples. Additional data will have to be collected to meet the desired tolerance and confidence level.

7.5 Addition of Random Variables

One of the most common occurrences in probability and statistics is the summation of random variables, often in the form $Y = a_1X_1 + a_2X_2$ or in the more general form:

$$Y = \sum a_i X_i \tag{7-11}$$

where the summation is over i, usually from 1 to n and a_i is a weighting factor.

It is relatively straightforward to prove that the expected value (or mean) μ_Y of the random variable Y is given by:

$$\mu_Y = \sum a_i \mu_{xi} \tag{7-12}$$

and that if the random variables x_i are independent of each other, the variance $s^2{}_Y$ of the random variable Y is given by:

$$\sigma_Y^2 = \sum a_i^2 \sigma_{xi}^2 \tag{7-13}$$

The fact that the coefficients, a_i, are multiplied has great practical significance for us in all our statistical work.

A Sample Problem: Adding Travel Times

A trip is composed of three parts, each with its own mean and standard deviation as shown here. What are the mean, variance, and standard deviation of the total trip time?

Trip Components	Mean	Standard Deviation
1. Auto	7 min	2 min
2. Commuter Rail	45 min	6 min
3. Bus	15 min	3 min

Solution: The variance may be computed by squaring the standard deviation. The total trip time is the sum of the three components. Equations 7-12 and 7-13 may be applied to yield a mean of 67 minutes, a variance of $(1^2 \times 2^2) + (1^2 \times 6^2) + (1^2 \times 3^2) = 49 \text{ min}^2$, and, therefore, a standard deviation of 7 minutes.

A Sample Problem: Proportion of Female Riders

In a central business district (CBD), there are 10,000 trips per day by subway, 6000 trips by bus, 5,000 trips by light rail, 2,000 walking trips, and 1,000 bicycle trips. The percentile trips that are made by women by mode are as follows:

X	Subway	Bus	Light Rail	Walking	Bicycle
$P(x)$	0.4	0.6	0.6	0.55	0.4

What is the expected number of female riders on public transportation?

Solution: (0.4)(10,000) + (0.6)(6,000) + (0.6)(5,000) + (0.55)(2,000) + (0.4)(1,000) = 12,100

A Sample Problem: Parking Spaces

Based on observations of condos with 100 units, it is believed that the mean number of parking spaces desired is 70, with a standard deviation of 4.6. If we are to build a larger complex with 1,000 units, what can be said about the parking?

Solution: If X = the number of parking spaces for a condo with 100 units, then the mean (μ_x) of this distribution of values has been estimated to be 70 spaces, and the standard deviation (σ_x) has been estimated to be 4.6 spaces. The assumption must be made that the number of parking spaces needed is proportional to the size of the condo development (i.e., that a condo with 1,000 units will need 10 times as many parking spaces as one with 100 units). In Equations 7-12 and 7-13, 10 becomes the value of a. Then, if Y = the number of parking spaces needed for a condo with 1,000 units:

$$\mu_y = 10(70) = 700 \text{ spaces}$$

and:

$$\sigma_y^2 = (10^2)\,(4.6^2) = 2116$$

$$\sigma_y = 46 \text{ spaces}$$

Note that this estimates the *average* number of spaces needed by a condo of size 1,000 units, but it does not address the specific need for parking at any specific development of this size. The estimate is also based on the assumption that parking needs will be proportional to the size of the development, which may not be true. If available, a better approach would have been to collect information on parking at condo developments of varying size to develop a relationship between parking and size of development. The latter would involve regression analysis, a more complex type of statistical analysis.

7.5.1 The Central Limit Theorem

One of the most impressive and useful theorems in probability is that the sum of n similarly distributed random variables tends to the normal distribution, no matter what the initial, underlying distribution is. That is, the random variable $Y = \Sigma X_i$, where the X_i have the same distribution, tends to the normal distribution.

The words "tends to" can be read as "tends to look like" the normal distribution. In mathematical terms, the actual distribution of the random variable Y approaches the normal distribution asymptotically.

Sum of Travel Times

Consider a trip made up of 15 components, all with the same underlying distribution, each with a mean of 10 minutes and standard deviation of 3.5 minutes. The underlying distribution is unknown. What can you say about the total travel time?

Although there might be an odd situation to contradict this, $n = 15$ should be quite sufficient to say that the distribution of total travel times tends to look normal. From Equation 7-12, the mean of the distribution of total travel times is found by adding 15 terms $(a_i \ \mu_i)$ where $a_i = 1$ and $\mu_i = 10$ minutes, or

$$\mu_y = 15 * (1 * 10) = 150 \text{ minutes}$$

The variance of the distribution of total travel times is found from Equation 7-13 by adding 15 terms $(a_i^2 \sigma_i^2)$ where a_i is again 1, and σ_i is 3.5 minutes. Then:

$$\sigma_y^2 = 15 * (1 * 3.5^2) = 183.75 \text{ minutes}^2$$

The standard deviation, σ_y is, therefore, 13.6 minutes.

If the total travel times are taken to be normally distributed, 95% of all observations of total travel time will lie between the mean (150 minutes) ± 1.96 standard deviations (13.6 minutes), or:

$$X_y = 150 \pm 1.96\,(13.6)$$

Thus 95% of all total travel times would be expected to fall within the range of 123 to 177 minutes (values rounded to the nearest minute).

Hourly Volumes

Five-minute counts are taken, and they tend to look rather smoothly distributed but with some skewness (asymmetry). Based on many observations, the mean tends to be 45 vehicles in the five-minute count, with a standard deviation of 7 vehicles. What can be said of the hourly volume?

The hourly volume is the sum of 12 five-minute distributions, which should logically be basically the same if traffic levels are stable. Thus the hourly volume will tend to look normal and will have a mean computed using Equation 7-12, with $a_i = 1$, $\mu_i = 45$ vehicles, and $n = 12$, or $12 * (1 * 45) = 540$ veh/h. The variance is computed using Equation 7-13, with $a_i = 1$, $\sigma_i = 7$, and $n = 12$, or $12 * (1^2 * 7^2) = 588$ (veh/h)^2. The standard deviation is 24.2 veh/h. Based on the assumption of normality, 95% of hourly volumes would be between $540 \pm 1.96\,(24.2) = 540 \pm 47$ veh/h (rounded to the nearest whole vehicle).

Note that the summation has had an interesting effect. The σ/μ ratio for the five-minute count distribution was $7/45 = 0.156$, but for the hourly volumes it was $47/540 = 0.087$. This is due to the summation, which tends to remove extremes by canceling "highs" with "lows" and thereby introduces stability. The mean of the sum grows in proportion to n, but the standard deviation grows in proportion to the square root of n.

Sum of Normal Distributions

Although not proven here, it is also true that the sum of any two normal distributions is itself normally distributed. By extension, if one normal is formed by n_1 summations of one underlying distribution and another normal is formed by n_2 summations of another underlying distribution, the sum of the total also tends to the normal. Thus, in the foregoing travel-time example, not all of the elements had to have exactly the same distribution as long as subgroupings each tended to the normal.

7.6 The Binomial Distribution Related to the Bernoulli and Normal Distributions

7.6.1 Bernoulli and the Binomial Distribution

The *Bernoulli distribution* is the simplest discrete distribution, consisting of only two possible outcomes: yes or no, heads or tails, one or zero, and so on. The first occurs with probability p, and therefore the second occurs with probability $(1 - p = q)$. This is modeled as:

$$P(X = 1) = p$$
$$P(X = 0) = 1 - p = q$$

In traffic, it represents any basic choice—to park or not to park; to take this route or that; to take auto or transit (for one individual). It is obviously more useful to look at more than one individual, however, which leads us to the binomial distribution. The *binomial distribution* can be thought of in two common ways:

1. Observe N outcomes of the Bernoulli distribution, make a record of the number of events that have the outcome "1," and report that number as the outcome X.

2. The binomial distribution is characterized by the following properties:
   - There are N events, each with the same probability p of a positive outcome and $(1-p)$ of a negative outcome.
   - The outcomes are independent of each other.
   - The quantity of interest is the total number X of positive outcomes, which may logically vary between 0 and N.
   - N is a finite number.

The two ways are equivalent, for most purposes.

Consider a situation in which people may choose "transit" or "auto" where each person has the same probability $p = 0.25$ of choosing transit, and each person's decision is independent of that of all other persons. Defining "transit" as the positive choice for the purpose of this example and choosing $N = 8$, note that:

1. Each person is characterized by the Bernoulli distribution, with $p = 0.25$.
2. There are $2^8 = 256$ possible combinations of choices, and some of the combinations not only yield the same value of X but also have the same probability of occurring. For instance, the value of $X = 2$ occurs for both.

TTAAAAAA

and

TATAAAAA

and several other combinations, each with probability of $p^2(1-p)^6$, for a total of 28 such combinations.

Stated without proof is the result that the probability $P(X = x)$ is given by:

$$P(X = x) = \frac{N!}{(N-x)!x!} p^x(1-p)^{N-x} \quad (7\text{-}14)$$

with a mean of Np and a variance of Npq where $q = 1 - p$. The derivation may be found in any standard probability text.

Figure 7.3 shows the plot of the binomial distribution for $p = 0.25$ and $N = 8$. Table 7.4 tabulates the probabilities of each outcome. The mean may be computed as $\mu = Np = 8 \times 0.25 = 2.00$ and the standard deviation as $\sigma = Npq = 8 \times 0.25 \times 0.75 = 1.5$.

There is an important concept that you should master to use statistics effectively throughout this text. Even though on average two out of eight people will choose transit, there is absolutely no guarantee what the next eight randomly selected people will choose, even if they follow the rules (same p, independent decisions, etc.). In fact, the number could range anywhere from $X = 0$ to $X = 8$. And we can expect that the result $X = 1$ will occur 10.0% of the time, $X = 4$ will occur 8.7% of the time, and $X = 2$ will occur only 31.1% of the time.

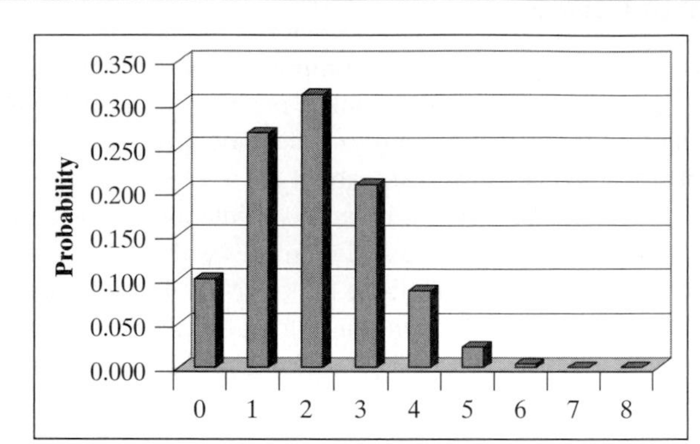

Figure 7.3: Plot of Binomial Results for $N = 8$ and $p = 0.25$

This is the crux of the variability in survey results. If there were 200 people in the senior class and each student surveyed 8 people from the subject population, we would get different results. In general, our results, if plotted and tabulated, would conform to Figure 7.3 and Table 7.4 but would not mimic them perfectly. Likewise, if we average our results, the result would probably be close to 2.00 but would almost surely not be identical to it.

Table 7.4: Tabulated Probabilities for a Binomial Distribution with $N = 8$ and $p = 0.25$

Outcome X People Using Transit	Probability of X
0	0.100
1	0.267
2	0.311
3	0.208
4	0.087
5	0.023
6	0.004
7	0.000
8	0.000

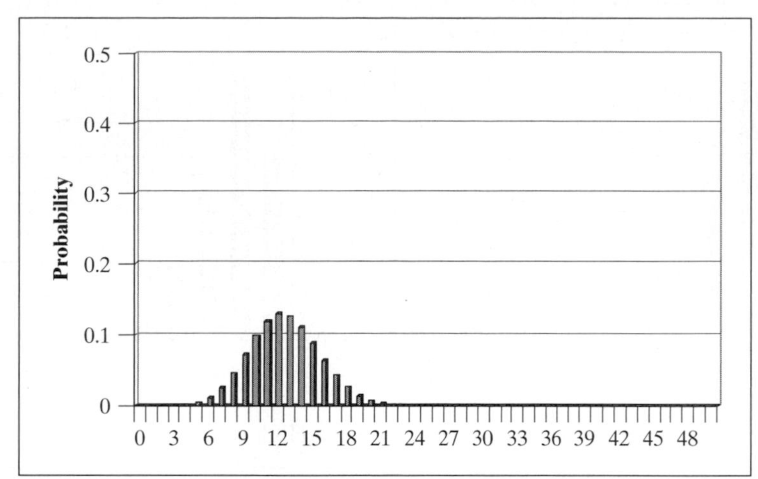

Figure 7.4: Binomial Distribution for $N = 200$ and $p = 0.25$

7.6.2 Asking People Questions: Survey Results

Consider that the commuting population has two choices — $X = 0$ for auto and $X = 1$ for public transit. The probability p is generally unknown, and it is usually of great interest. Assuming the probability is the same for all people (to our ability to discern, at least), then each person is characterized by the Bernoulli distribution.

If we ask $n = 50$ people for their value of X, the resulting distribution of the random variable Y is binomial and may tend to look like the normal. Figure 7.3 shows this exact distribution for $p = 0.25$. Without question, this distribution looks normal. Applying some "quick facts" and noting that the expected value (that is, the mean) is 12.5 (50×0.25), the variance is 9.375 ($50 \times 0.25 \times 0.75$), and the standard deviation is 3.06, we can expect 95% of the results to fall in the range 12.5 ± 6.0 or between 6.5 and 18.5.

If $n = 200$ had been selected, then the mean of Y would have been 50 when $p = 0.25$ and the standard deviation would have been 6.1, so that 95% of the results would have fallen in the range of 38 to 62. Figure 7.4 illustrates the resulting distribution.

7.6.3 The Binomial and the Normal Distributions

The central limit theorem informs us that the sum of Bernoulli distributions (i.e., the binomial distribution) tends to the normal distribution. The only question is How fast? A number of practitioners in different fields use a rule of thumb that says, "For large n and small p the normal approximation can be used without restriction." This is incorrect and can lead to serious errors.

The most notable case in which the error occurs is when rare events are being described, such as auto accidents per million miles traveled or aircraft accidents. Consider Figure 7.5, which is an *exact* rendering of the actual binomial distribution for $p = 0.7(10)^{-6}$ and two values of n—namely $n = 10^6$ and $n = 2(10)^6$, respectively. Certainly p is small and n is large in these cases, and, just as clearly, they do *not* have the characteristic symmetric shape of the normal distribution.

It can be shown that in order for there to be some chance of symmetry—that is, in order for the normal distribution to approximate the binomial distribution—the condition that $np/(1 - p) \geq 9$ is necessary. Clearly, neither of the cases in Figure 7.5 satisfies such a condition.

7.7 The Poisson Distribution

The Poisson distribution is known in traffic engineering as the "counting" distribution. It has the clear physical meaning of a number of events X occurring in a specified counting interval of duration T and is a one-parameter distribution with:

$$P(X = x) = e^{-m} \frac{m^x}{x!} \tag{7-15}$$

with mean $\mu = m$ and variance $\sigma^2 = m$.

The fact that one parameter m specifies both the mean and the variance is a limitation, in that if we encounter field

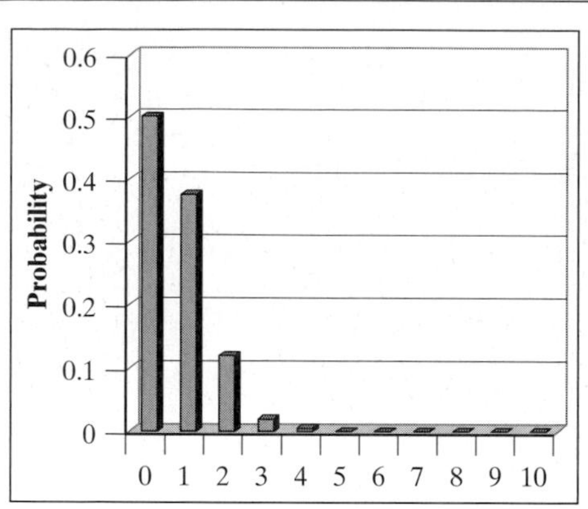

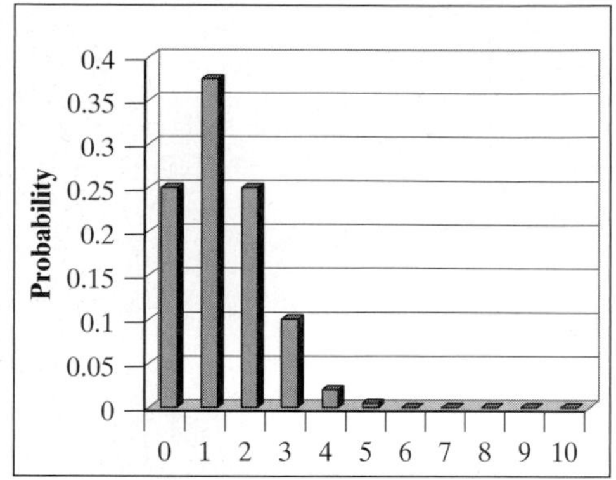

Figure 7.5: Binomial Does Not Always Look Normal for Low p and High N

Table 7.5: A Poisson Distribution with m = 5

Number of Accidents (x)	Probability $P(x)$	Cumulative Probability $P(X \leq x)$
0	.007	.007
1	.034	.041
2	.084	.125
3	.140	.265
4	.177	.442
5	.175	.617
6	.146	.763

data where the variance and mean are clearly different, the Poisson does not apply.

The Poisson distribution often applies to observations per unit of time, such as the arrivals per five-minute period at a toll booth. When headway times are exponentially distributed with mean $\mu = 1/\lambda$, the number of arrivals in an interval of duration T is Poisson distributed with mean $\mu = m = \lambda T$.

Applying the Poisson distribution is done the same way we applied the binomial distribution earlier. For example, say there is an average of five accidents per day on the Florida freeways. Table 7.5 shows the probability of there being 0, 1, 2, and so on, accidents on a given day on the freeways in Florida. The last column shows the cumulative probabilities or the probability $P(X \leq x)$.

7.8 Hypothesis Testing

Very often traffic engineers must make a decision based on sample information. For example, is a traffic control effective or not? To test this, we formulate a hypothesis, H_0, called the null hypothesis and then try to disprove it. The null hypothesis is formulated so there is no difference or no change, and then the opposite hypothesis is called the alternative hypothesis, H_1.

When testing a hypothesis, it is possible to make two types of errors: (1) We could reject a hypothesis that should be accepted (e.g., say an effective control is not effective). This is called a *Type I error*. The probability of making a Type I error is given the variable name, α. (2) We could accept a false hypothesis (e.g., say an ineffective control is effective). This is called a *Type II error*. A Type II error is given the variable name β.

Consider this example: An auto inspection program is going to be applied to 100,000 vehicles, of which 10,000 are "unsafe" and the rest are "safe." Of course, we do not know which cars are safe and which are unsafe.

We have a test procedure but it is not perfect, due to the mechanic and test equipment used. We know that 15% of the unsafe vehicles are determined to be safe, and 5% of the safe vehicles are determined to be unsafe, as seen in Figure 7.6.

We would define H_0: The vehicle being tested is "safe," and H_1: the vehicle being tested is "unsafe." The Type I error, rejecting a true null hypothesis (false negative), is labeling a safe vehicle as "unsafe." The probability of this is called the level of significance, α, and in this case $\alpha = 0.05$. The Type II error, failing to reject a false null hypothesis (false positive), is labeling an unsafe vehicle as "safe." The probability of this, β, is 0.15. In general, for a given test procedure, one can reduce Type I error only by living with a higher Type II error, or vice versa.

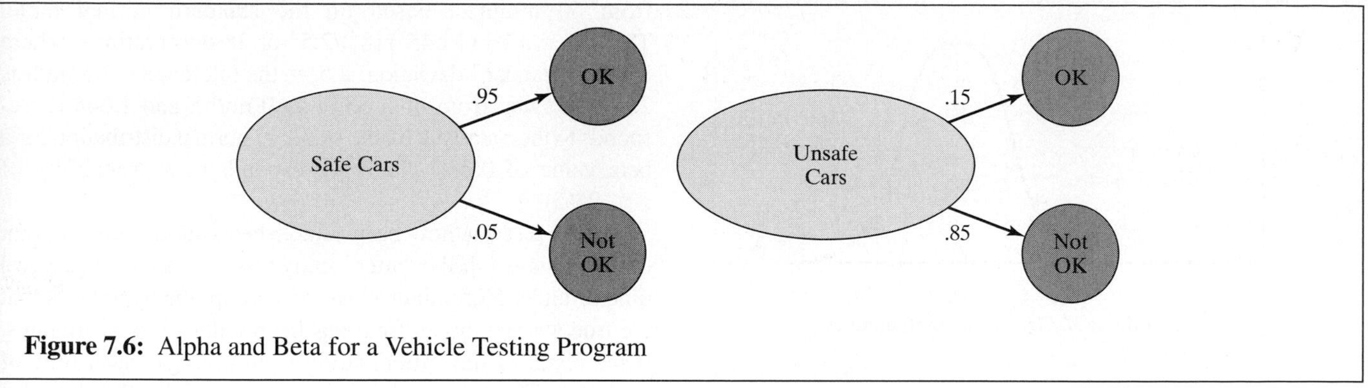

Figure 7.6: Alpha and Beta for a Vehicle Testing Program

7.8.1 Before-and-After Tests with Two Distinct Choices

In a number of situations, there are two clear and distinct choices, and the hypotheses seem almost self-defining:

- Auto inspection (acceptable, not acceptable)
- Disease (have the disease, don't)
- Speed reduction of 5 mi/h (it happened, it didn't)
- Accident reduction of 10% (it happened, it didn't)
- Mode shift by 5 percentage points (it happened, it didn't)

Of course, there is the distinction between *the real truth* (reality, unknown to us) and *the decision* we make, as already discussed and related to Type I and Type II errors. That is, we can decide that some cars in good working order need repairing and we can decide that some unsafe cars do not need repairing.

There is also the distinction that people may not want to reduce the issue to a binary choice or might not be able to do so. For instance, if an engineer expects a 10% decrease in the accident rate, should we test "H_0: no change" against "H_1: 10% decrease" and not allow the possibility of a 5% change? Such cases are addressed in the next section. For the present section, we concentrate on binary choices.

Application: Travel Time Decrease

Consider a situation in which the existing travel time on a given route is known to average 60 minutes, and experience has shown the standard deviation to be about 8 minutes. An "improvement" is recommended that is expected to reduce the true mean travel time to 55 minutes.

This is a rather standard problem, with what is now a fairly standard solution. The logical development of the solution follows.

The first question we might ask ourselves is whether we can consider the mean and standard deviation of the initial distribution to be truly known or whether they must be estimated. Actually, we will avoid this question simply by focusing on whether the after situation has a true mean of 60 minutes or 55 minutes. Note that we do not know the shape of the travel time distribution, but the central limit theorem tells us: a new random variable Y, formed by averaging several travel time observations, will tend to the normal distribution if enough observations are taken. Figure 7.7 shows the shape of Y for two different hypotheses, which we now form:

H_0: The true mean of Y is 60 minutes

H_1: The true mean of Y is 55 minutes

Figure 7.7 also shows a logical decision rule: If the actual observation Y falls to the right of a certain point, Y^*, then accept H_0; if the observation falls to the left of that point, then accept H_1. Finally, Figure 7.7 shows shaded areas that are the probabilities of Type I and Type II errors. Note that:

1. The n travel time observations are all used to produce the one estimate of Y.
2. If the point Y^* is fixed, then the only way the Type I and Type II errors can be changed is to increase n, so that the shapes of the two distributions become narrower because the standard deviation of Y involves the square root of n in its denominator.
3. If the point Y^* is moved, the probabilities of Type I and Type II errors vary, with one increasing while the other decreases.

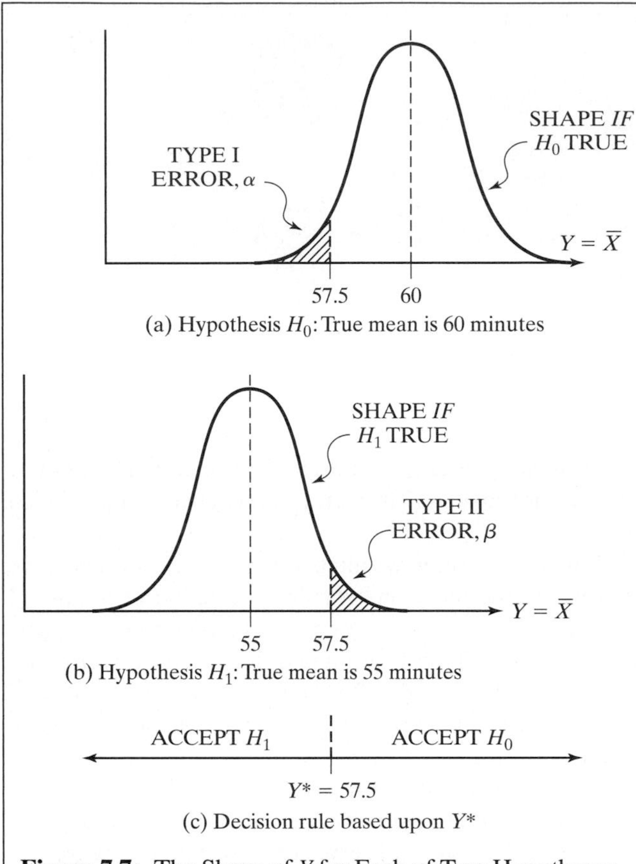

Figure 7.7: The Shape of Y for Each of Two Hypotheses

To complete the definition of the test procedure, the point Y^* must be selected and the Type I and Type II errors determined. It is common to require that the Type I error (also known as the level of significance, α) be set at 0.05, so that there is only a 5% chance of rejecting a true null hypothesis. In the case of two alternative hypotheses, it is common to set both the Type I and Type II errors to 0.05, unless there is very good reason to imbalance them (both represent risks, and the two risks—repairing some cars needlessly versus having unsafe cars on the road, for instance—may not be equal).

Inspecting Figure 7.7, Y^* will be set at 57.5 minutes to equalize the two probabilities. The only way these errors can be equal is if the value of Y^* is set at exactly half the distance between 55 and 60 minutes. The symmetry of the assumed normal distribution requires that the decision point be equally distant from both 55 and 60 minutes, assuming that the standard deviation of both distributions (before and after) remains 8 minutes.

To ensure that both errors are not only equal but have an equal value of 0.05, Y^* must be 1.645 standard deviations away from 60 minutes, based on the standard normal table. Therefore, $n \geq (1.645^2)(8^2)/2.5^2$ or 28 observations, where 8 = the standard deviation, 2.5 = the tolerance (57.5 mi/h is 2.5 mi/h away from both 55 and 60 mi/h), and 1.645 corresponds to the z statistic on the standard normal distribution for a beta value of 0.05 (which corresponds to a probability of $z \leq 95\%$).

The test has now been established with a decision point of 57.5 minutes. If the "after" study results in an average travel time of under 57.5 minutes, we will accept the hypothesis that the true average travel time has been reduced to 55 minutes. If the result of the "after" study is an average travel time of more than 57.5 minutes, the null hypothesis—that the true average travel time has stayed at 60 minutes—is accepted.

Was all this analysis necessary to make the common-sense judgment to set the decision time at 57.5 minutes—halfway between the existing average travel time of 60 minutes and the desired average travel time of 55 minutes? The answer is in two forms: The analysis provides the logical basis for making such a decision. This is useful. The analysis also provided the minimum sample size required for the "after" study to restrict both alpha and beta errors to 0.05. This is the most critical result of the analysis.

Application: Focus on the Travel Time Difference

The preceding illustration assumed that we would focus on whether the underlying true mean of the "after" situation was either 60 minutes or 55 minutes. What are some of the practical objections that people could raise?

Certainly one objection is that we implicitly accepted at face value that the "before" condition truly had an underlying true mean of 60 minutes. Suppose, to overcome that, we focus on the difference between before and after observations.

The n_1 "before" observations can be averaged to yield a random variable Y_1 with a certain mean μ_1 and a variance of σ_1^2/n_1. Likewise, the n_2 "after" observations can be averaged to yield a random variable Y_2 with a (different?) certain mean μ_2 and a variance of σ_2^2/n_2. Another random variable can be formed as $Y = (Y_2 - Y_1)$, which is itself normally distributed and that has an underlying mean of $(\mu_2 - \mu_1)$ and variance $\sigma^2 = \sigma_2^2/n_2 + \sigma_1^2/n_1$. This is often referred to as the normal approximation. Figure 7.8 shows the distribution of Y, assuming two hypotheses.

What is the difference between this example and the previous illustration? The focus is directly on the difference and does not implicitly assume that we know the initial mean. As a result, "before" samples are required. Also, there is more uncertainty, as reflected in the larger variance. A number of practical observations stem from using this result: It is common

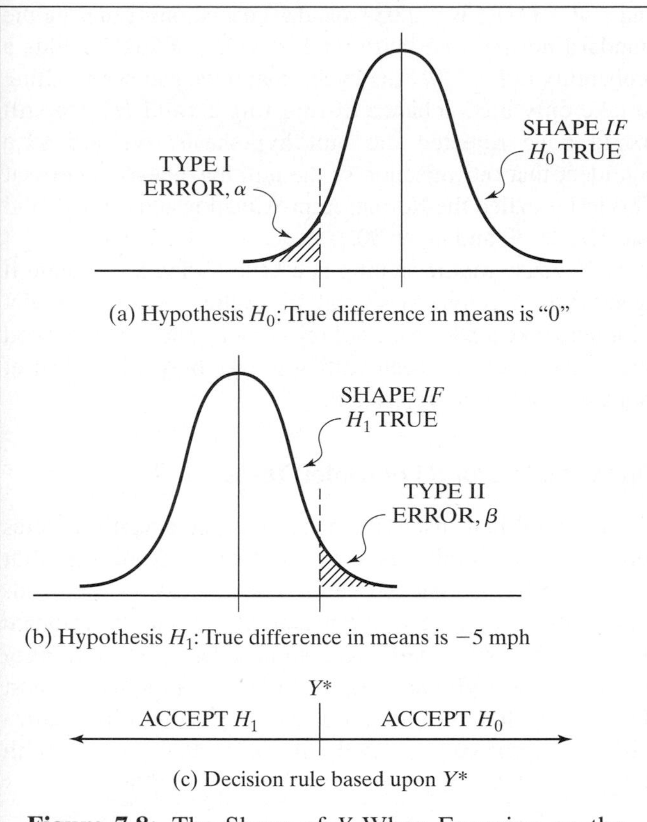

Figure 7.8: The Shape of Y When Focusing on the Differences

that the "before" and "after" variances are equal, in which case the total number of observations can be minimized if $n_1 = n_2$. If the variances are not known, the estimators s_i^2 are used in their place.

If the "before" data was already taken in the past and n_1 is therefore fixed, it may not be possible to reduce the total variance enough (by just using n_2) to achieve a desired level of significance, such as $\alpha = 0.05$. Comparing with the previous problem, note that if both variances are 8^2 and $n_1 = n_2$ is specified, then $n_1 \geq 2*(1.645^2)\ (8^2)/(2.5^2)$ or 55 and $n_2 \geq 55$. The total required is 110 observations. The fourfold increase is a direct result of focusing on the difference of −5 mi/h rather than the two binary choices (60 or 55 minutes).

7.8.2 Before-and-After Tests with Generalized Alternative Hypothesis

It is also common to encounter situations in which the engineer states the situation as "there was a decrease" or "there was a change" versus "there was not," but does not state or claim the magnitude of the change. In these cases, it is standard practice to set up a null hypothesis of "there was no change" ($\mu_1 = \mu_2$) and an alternative hypothesis of "there was a change" ($\mu_1 \neq \mu_2$). In such cases, a level of significance of 0.05 is generally used.

Figure 7.9 shows the null hypotheses for two possible cases, both having a null hypothesis of "no change." The first case implicitly considers that if there were a change, it would be negative—that is, either there was no change or there was a decrease in the mean. The second case does not have any sense (or suspicion) about the direction of the change, if it exists. Note that:

- The first is used when physical reasoning leads one to suspect that if there were a change, it would be a decrease.[2] In such cases, the Type I error probability is concentrated where the error is most likely to occur, in one tail.
- The second is used when physical reasoning leads one to simply assert "there was a change" without any sense of its direction. In such cases, the Type I error probability is spread equally in the two tails.

In using the second case, often we might hope that there was no change and really not want to reject the null hypothesis. That is, not rejecting the null hypothesis in this case is a measure of success. There are, however, other cases in which we wish to prove that there is a difference. The same logic can be used, but in such cases rejecting the null hypothesis is "success."

An Application: Travel Time Differences

Let's assume we have made some improvements and suspect that there is a decrease in the true underlying mean travel time. Figure 7.9 (a) applies. Using information from the previous illustration, let us specify that we wish a level of significance $\alpha = 0.05$. The decision point depends on the variances and the n_i. If the variances are as stated in the prior illustration and $n_1 = n_2 = 55$, then the decision point $Y^* = -2.5$ mph, as before.

Let us now go one step further. The data is collected, and $Y = -3.11$ results. The decision is clear: Reject the null hypothesis of "there is no decrease." But what risk did we take?

[2]The same logic – and illustration – can be used for cases of suspected *increases*, just by a sign change.

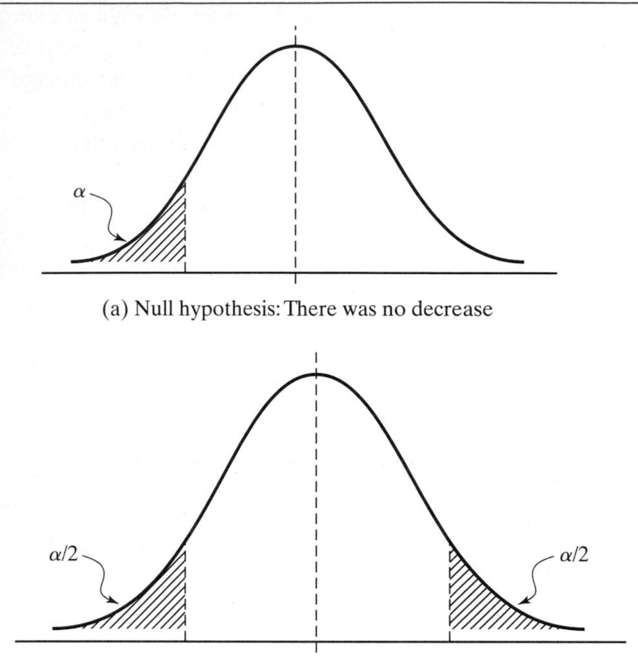

Figure 7.9: Hypothesis Testing for a Generalized Alternative Hypothesis

Consider the following:

- Under the stated terms, had the null hypothesis been valid, we were taking a 5% chance of rejecting the truth. The odds favor (by 19 to 1, in case you are inclined to wager with us) not rejecting the truth in this case.
- At the same time, there is no stated risk of accepting a false hypothesis H_1, for the simple reason that no such hypothesis was stated.
- The null hypothesis was rejected because the value of Y was higher than the decision value of 2.5 mi/h. Because the actual value of -3.11 is considerably higher than the decision value, one could ask about the confidence level associated with the rejection. The point $Y = -3.11$ is 2.033 standard deviations away from the zero point, as can be seen from:

$$\sigma_Y = \sqrt{\frac{\sigma_1^2}{n_1} + \frac{\sigma_2^2}{n_2}}$$

$$= \sqrt{\frac{8^2}{55} + \frac{8^2}{55}} = 1.53$$

and $z = 3.11/1.53 = 2.033$ standard deviations. Entering the standard normal distribution table with $z = 2.033$ yields a probability of 0.9790. This means that if we had been willing to take only a 2% chance of rejecting a valid H_0, we still would have rejected the null hypothesis; we are 98% confident that our rejection of the null hypothesis is correct. This test is called the Normal Approximation and is only valid when $n_1 \geq 30$ and $n_2 \geq 30$.

Because this reasoning is a little tricky, let us state it again: If the null hypothesis had been valid, you were initially willing to take a 5% chance of rejecting it. The data indicated a rejection. Had you been willing to take only a 2% chance, you still would have rejected it.

One-Sided Versus Two-Sided Tests

The material just discussed appears in the statistics literature as "one-sided" tests, for the obvious reason that the probability is concentrated in one tail (we were considering only a decrease in the mean). If there is no rationale for this, a "two-sided" test should be executed, with the probability split between the tails. As a practical matter, this means that one does not use the probability tables with a significance level of 0.05 but rather with $0.05/2 = 0.025$.

7.8.3 Other Useful Statistical Tests

The *t*-Test

For small sample sizes ($N < 30$), the normal approximation is no longer valid. It can be shown that if x_1 and x_2 are from the same population, the statistic t is distributed according to the tabulated t distribution, where:

$$t = \frac{x_1 - x_2}{s_p\sqrt{1/n_1 + 1/n_2}} \tag{7-16}$$

and s_p is a pooled standard deviation, which equals:

$$s_p = \sqrt{\frac{(n_1 - 1)s_1^2 + (n_2 - 1)s_2^2}{n_1 + n_2 - 2}} \tag{7-17}$$

The t distribution depends on the degrees of freedom, f, which refers to the number of independent pieces of data that form the distribution. For the t distribution, the nonindependent pieces of data are the two means, x_1 and x_2. Thus:

$$f = N_1 + N_2 - 2 \tag{7-18}$$

Once the t statistic is determined, the tabulated values of Table 7.6 yield the probability of a t value being greater

Table 7.6: Upper Percentage Points of the *t*-Distribution*

$P(t)$

0 t

$$P(t) = \int_t^{\infty} \frac{(f-1/2)\,!}{(f-2/2)\,!\,\sqrt{\pi f}}\,(1 + t^2/f)^{+(f+1)/2}\, dt$$

Deg of Freedom f	Probability of a Value Equal to or Greater Than t										
	0.40	0.30	0.25	0.20	0.15	0.10	0.05	0.025	0.01	0.005	0.0003
1	0.325	0.727	1.000	1.376	1.936	3.078	6.314	12.706	31.821	63.657	636.619
2	0.289	0.617	0.816	1.061	1.386	1.886	2.290	4.303	6.965	9.925	31.598
3	0.277	0.584	0.765	0.978	1.250	1.638	2.353	3.182	4.541	5.841	12.924
4	0.271	0.569	0.741	0.941	1.190	1.533	2.132	2.776	3.747	4.604	8.610
5	0.267	0.559	0.727	0.920	1.156	1.476	2.015	2.571	3.365	4.032	6.869
6	0.265	0.553	0.718	0.906	1.134	1.440	1.943	2.447	3.143	3.707	5.959
7	0.263	0.549	0.711	0.896	1.119	1.415	1.895	2.365	2.998	3.499	5.608
8	0.262	0.546	0.706	0.889	1.108	1.397	1.860	2.306	2.896	3.355	5.401
9	0.261	0.544	0.703	0.883	1.100	1.383	1.833	2.262	2.821	3.250	4.781
10	0.260	0.542	0.700	0.879	1.093	1.372	1.812	2.228	2.764	3.169	4.587
11	0.260	0.540	0.697	0.876	1.088	1.363	1.796	2.201	2.718	3.106	4.437
12	0.259	0.539	0.695	0.873	1.083	1.356	1.782	2.179	2.681	3.055	4.318
13	0.259	0.538	0.694	0.870	1.079	1.350	1.771	2.160	2.650	3.012	4.221
14	0.258	0.537	0.692	0.866	1.076	1.345	1.761	2.143	2.624	2.977	4.140
15	0.258	0.536	0.691	0.866	1.074	1.341	1.753	2.131	2.602	2.947	4.073
16	0.258	0.535	0.690	0.865	1.071	1.337	1.746	2.120	2.583	2.921	4.015
17	0.257	0.534	0.689	0.865	1.069	1.333	1.740	2.110	2.567	2.898	3.965
18	0.257	0.534	0.688	0.862	1.067	1.330	1.734	2.101	2.552	2.878	3.922
19	0.257	0.533	0.688	0.861	1.066	1.328	1.729	2.093	2.539	2.861	3.883
20	0.257	0.533	0.687	0.860	1.064	1.325	1.725	2.086	2.528	2.845	3.850
21	0.257	0.532	0.686	0.859	1.063	1.323	1.721	2.080	2.518	2.831	3.819
22	0.256	0.532	0.686	0.858	1.061	1.321	1.717	2.074	2.508	2.819	3.792
23	0.256	0.532	0.685	0.858	1.060	1.319	1.714	2.069	2.500	2.807	3.767
24	0.256	0.531	0.685	0.857	1.059	1.318	1.711	2.064	2.492	2.797	3.745
25	0.256	0.531	0.684	0.856	1.058	1.316	1.708	2.060	2.485	2.787	3.735
26	0.256	0.531	0.684	0.856	1.058	1.315	1.706	2.056	2.479	2.779	3.707
27	0.256	0.531	0.684	0.855	1.057	1.314	1.703	2.052	2.473	2.771	3.690
28	0.256	0.530	0.683	0.855	1.056	1.313	1.701	2.048	2.467	2.763	3.674
29	0.256	0.530	0.683	0.854	1.055	1.311	1.699	2.045	2.462	2.756	3.659
30	0.256	0.530	0.683	0.854	1.055	1.310	1.697	2.042	2.457	2.750	3.646
40	0.255	0.529	0.681	0.851	1.050	1.303	1.684	2.021	2.423	2.704	3.551
60	0.254	0.527	0.679	0.848	1.046	1.296	1.671	2.000	2.390	2.660	3.460
120	0.254	0.526	0.677	0.845	1.041	1.289	1.658	1.980	2.358	2.617	3.373
Infinity	0.253	0.524	0.674	0.842	1.036	1.282	1.645	1.960	2.326	2.576	3.291

*Values of "t" are in the body of the table.

(*Source:* Used with permission of U.S. Naval Ordinance Test Station, Crow, Davis, and Maxfield, *Statistics Manual*, Dover NJ, 1960, Table 3.)

than the computed value. To limit the probability of a Type I error to 0.05, the difference in the means will be considered significant only if the probability is less than or equal to 0.05—that is, if the calculated t value falls in the 5% area of the tail, or in other words, if there is less than a 5% probability that such a difference would be found in the same population. If the probability is greater than 5% that such a difference in means would be found in the same population, then the difference is considered insignificant.

Consider the following example: Ten samples of speed data are taken both before and after a change in the speed limit is implemented. The mean and standard deviations found were:

Before		After
35	$\bar{x}$	32
4.0	s	5.0
10	N	10

Then:

$$s_p = \sqrt{\frac{(10 - 1)4^2 + (10 - 1)5^2}{10 + 10 - 2}} = 4.53,$$

$$t = \frac{35 - 32}{4.53\sqrt{1/10 + 1/10}} = 1.48, \text{ and}$$

$$f = 10 + 10 - 2 = 18$$

The probability that $t \geq 1.48$, with 18 degrees of freedom, falls between 0.05 and 0.10. Thus it is greater than 0.05, and we conclude that there is not a significant difference in the means.

The *F*-Test

In using the t-test, and in other areas as well, there is an implicit assumption made that the $\sigma_1 = \sigma_2$. This may be tested with the F distribution, where:

$$F = \frac{s_1^2}{s_2^2} \tag{7-19}$$

(by definition the larger s is always on top)

It can be proven that this F value is distributed according to the F distribution, which is tabulated in Table 7.7. The F distribution is tabulated according to the degrees of freedom in each sample; thus $f_1 = n_1 - 1$ and $f_2 = n_2 - 1$. Because the f distribution in Table 7.7 gives the shaded area in the tail, like the t distribution, the decision rules are as follows:

- If [Prob $F \geq$ F] ≤ 0.05, then the difference is significant.
- If [Prob $F \geq F$] > 0.05, then the difference is not significant.

Consider the following problem: Based on the following data, can we say that the standard deviations come from the same population?

Before		After
30	$\bar{x}$	35
5.0	s	4.0
11	N	21

Thus:

$$F = \frac{5^2}{4^2} = 1.56; \quad f_1 = 11-1 = 10;$$

$$f_2 = 21-1 = 20$$

The f distribution is tabulated for various probabilities, as follows, based on the given degrees of freedom; thus:

when $p = 0.10, \quad F = 1.94$
$p = 0.05, \quad F = 2.35$
$p = 0.025, \quad F = 2.77$

The F values are increasing, and the probability $[F \geq 1.56]$ must be greater than 0.10 given this trend; thus the difference in the standard deviations is not significant. The assumption that the standard deviations are equal, therefore, is valid.

Paired Differences

In some applications, notably simulation, where the environment is controlled, data from the "before" and "after" situations can be paired, and only the differences are important. In this way, the entire statistical analysis is done directly on the differences, and the overall variation can be much lower because of the identity tags.

An example application of paired differences is presented in Table 7.8, which shows the results of two methods for measuring speed. Both methods were applied to the same vehicles, and Table 7.8 considers the data first with no attempt at pairing and then with the data paired.

It is relatively easy to see what appears to be a significant difference in the data. The question is, could it have been statistically detected without the pairing? Using the approach of a one-sided test on the null hypothesis of "no increase," the following computations result:

$$s_1 = 7.74$$
$$s_2 = 7.26$$
$$N_1 = N_2 = 15$$

Table 7.7: Upper Percentage Points of the F Distribution*

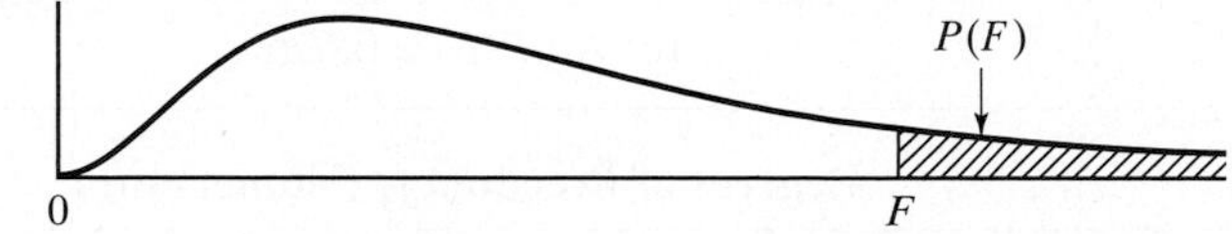

$$P(F) = \int_F^{\infty} \frac{(f_1 + f_2 - 2)/2\,!}{(f_1 - 2)/2\,!\,(f_2 - 2)/2\,!} f^{f_1/2} f^{f_2/2} F^{(f_1 - 2)/2} (f_2 + f_1 F)^{-(f_1 + f_2)/2}\, dF$$

(a) All $P(F) = 0.10$

Degrees of Freedom f_2 Denominator	Degrees of Freedom f_1 (Numerator)								
	1	**5**	**10**	**20**	**30**	**40**	**60**	**120**	**Infinity**
1	39.86	57.24	60.20	61.24	62.26	62.53	62.79	63.06	63.33
5	4.06	3.45	3.30	3.21	3.17	3.16	3.14	3.12	3.10
10	3.28	2.52	2.32	2.20	2.16	2.13	2.11	2.08	2.06
20	2.97	2.16	1.94	1.79	1.74	1.71	1.68	1.64	1.61
30	2.89	2.05	1.82	1.64	1.61	1.57	1.54	1.50	1.46
40	2.84	2.00	1.76	1.61	1.57	1.51	1.47	1.42	1.38
60	2.79	1.95	1.71	1.54	1.51	1.44	1.40	1.35	1.29
120	2.75	1.90	1.65	1.48	1.45	1.37	1.32	1.26	1.19
Infinity	2.71	1.85	1.00	1.42	1.39	1.30	1.24	1.17	1.00

(b) All $P(F) = 0.05$

Degrees of Freedom f_2 Denominator	Degrees of Freedom f_1 (Numerator)								
	1	**5**	**10**	**20**	**30**	**40**	**60**	**120**	**Infinity**
1	161.45	230.16	241.88	248.01	250.09	251.14	252.20	253.25	254.32
5	6.61	5.05	4.74	4.56	4.50	4.46	4.43	4.40	4.36
10	4.96	3.33	2.98	2.77	2.70	2.66	2.62	2.58	2.54
20	4.35	2.71	2.35	2.12	2.04	1.99	1.95	1.90	1.84
30	4.17	2.53	2.16	1.93	1.84	1.79	1.74	1.68	1.62
40	4.08	2.45	2.08	1.84	1.79	1.69	1.64	1.58	1.51
60	4.00	2.37	1.99	1.75	1.65	1.59	1.53	1.47	1.39
120	3.92	2.29	1.91	1.66	1.55	1.50	1.43	1.35	1.23
Infinity	3.84	2.21	1.83	1.57	1.46	1.39	1.32	1.22	1.00

(*Continued*)

Table 7.7: Upper Percentage Points of the *F* Distribution* (*Continued*)

(c) All $P(F) = 0.025$

Degrees of Freedom f_2 Denominator	Degrees of Freedom f_1 (Numerator)								
	1	**5**	**10**	**20**	**30**	**40**	**60**	**120**	**Infinity**
1	647.79	921.85	968.63	993.10	1001.40	1005.60	1009.80	1014.00	1018.30
5	10.01	7.15	6.62	6.33	6.23	6.18	6.12	6.07	6.02
10	6.94	4.24	3.72	3.42	3.31	3.26	3.20	3.14	3.08
20	5.87	3.29	2.77	2.46	2.35	2.29	2.22	2.16	2.09
30	5.57	3.03	2.51	2.20	2.07	2.01	1.94	1.87	1.79
40	5.42	2.90	2.39	2.07	1.94	1.88	1.80	1.72	1.64
60	5.29	2.79	2.27	1.94	1.82	1.74	1.67	1.58	1.48
120	5.15	2.67	2.16	1.82	1.69	1.61	1.53	1.43	1.31
Infinity	5.02	2.57	2.05	1.71	1.57	1.48	1.39	1.27	1.00

*Values of *F* in the body of table.

(*Source:* Condensed from Crow, Davis, and Maxfield, *Statistics Manual*, 7960.)

Table 7.8: Example Showing the Benefits of Paired Testing

Vehicle	Method 1 Speed (mi/h)	Method 2 Speed (mi/h)	Difference by Vehicle *D* (mi/h)	D^2	
1	55	60	−5	25	
2	47	55	−8	64	
3	70	74	−4	16	
4	62	67	−5	25	
5	49	53	−4	16	
6	67	71	−4	16	
7	52	57	−5	25	
8	57	60	−3	9	
9	58	61	−3	9	
10	45	48	−3	9	
11	68	70	−2	4	
12	52	57	−5	25	
13	51	56	−5	25	
14	58	64	−6	36	
15	62	65	−3	9	
			−65.00	313.00	=Sum
MEAN =	56.87	61.20	−4.33		
STD =	7.74	7.26	1.496		

Then:

$$s_p = \sqrt{\frac{(N_1 - 1)s_1^2 + (N_2 - 1)s_2^2}{N_1 + N_2 - 2}}$$

$$s_p = \sqrt{\frac{(15 - 1)7.74^2 + (15 - 1)7.26^2}{15 + 15 - 2}} = 7.51$$

$$t = \frac{\bar{X}_1 - \bar{X}_2}{s_p} = \frac{56.87 - 61.20}{7.51} = -1.5807$$

For a significance level of 0.05 (or 95% confidence), in a one-tailed test, $P(T \le t)$ is 0.063. Given that this probability is greater than 0.05, the null hypothesis of "no increase" *cannot* be rejected, and the test indicates that the two measurement techniques yield *statistically equal* speeds.

However, with the inspection of paired differences, it is seen that *every* measurement resulted in an increase when method 2 is applied. Application of the same statistical approach, but using the N = 15 paired differences yields:

$$s_d = \sqrt{\frac{N\sum(d_i^2) - \left(\sum d_i\right)^2}{N^*(N - 1)}}$$

$$s_d = \sqrt{\frac{15^*313 - (-65)^2}{15^*14}}$$

$$s_d = 1.496$$

$$t = \frac{D}{s_d/\sqrt{N}} = \frac{-4.33}{1.496/\sqrt{15}}$$

$$t = -11.218$$

For a 0.05 level of significance, the critical t for a one-tailed test is 1.761 and the $P(T \le t)$ is 1.1E-08, and thus the null hypothesis of "no increase" is clearly rejected. The two measurement methodologies do yield different results, with method 2 resulting in higher speeds than method 1.

Chi-Square Test: Hypotheses on an Underlying Distribution *f*(*x*)

One of the early problems stated was a desire to "determine" the underlying distribution, such as in a speed study. The most common test to accomplish this is the chi-square (χ^2) goodness-of-fit test.

In actual fact, the underlying distribution will not be determined. Rather, a hypothesis such as "H_0: The underlying distribution is normal" will be made, and we test that it is not rejected, so we may then act as if the distribution were in fact normal.

Table 7.9: Sample Height Data for Chi-Square Test

Height Category	Number of People
5.0 ft–5.2 ft	4
5.2 ft–5.4 ft	6
5.4 ft–5.6 ft	11
5.6 ft–5.8 ft	14
5.8 ft–6.0 ft	21
6.0 ft–6.2 ft	18
6.2 ft–6.4 ft	16
6.4 ft–6.6 ft	5
6.6 ft–6.8 ft	4
6.8 ft–7.0 ft	1
Total	100

The procedure is best illustrated by an example. Consider data on the height of 100 people showed in Table 7.9. To simplify the example, we test the hypothesis that these data are *uniformly* distributed (i.e., there are equal numbers of observations in each group).

To test this hypothesis, the goodness-of-fit test is done by following these steps:

1. Compute the theoretical frequencies, f_i, for each group. Because a uniform distribution is assumed and there are 10 categories with 100 total observations, $f_i = 100/10 = 10$ for all groups.
2. Compute the quantity:

$$\chi^2 = \sum_{i=1}^{N} \frac{(n_i - f_i)^2}{f_i} \tag{7-20}$$

where the summation is done over N data categories or groups. These computations are shown in Table 7.10.

3. As shown in any standard statistical text, the quantity χ^2 is chi-squared distributed, and we expect low values if our hypothesis is correct. (If the observed samples exactly equal the expected, then the quantity is zero.) Therefore, refer to a table of the chi-square distribution (Table 7.11) and look up the number that we would not exceed more than 5% of the time (i.e., $\alpha = 0.05$). To do this, we must also have the number of degrees of freedom, designated df, and defined as $df = N-1-g$, where N is the number of categories and g is the number of things

Table 7.10: Computation of X^2 for the Sample Problem

Height Category (ft)	Observed Frequency n	Theoretical Frequency F	X^2
5.0–5.2	4	10	$(4\text{-}10)^2/10 = 3.60$
5.2–5.4	6	10	$(6\text{-}10)^2/10 = 1.60$
5.4–5.6	11	10	$(11\text{-}10)^2/10 = 0.10$
5.6–5.8	14	10	$(14\text{-}10)^2/10 = 1.60$
5.8–6.0	21	10	$(21\text{-}10)^2/10 = 12.10$
6.0–6.2	18	10	$(17\text{-}10)^2/10 = 6.40$
6.2–6.4	16	10	$(16\text{-}10)^2/10 = 3.60$
6.4–6.6	5	10	$(5\text{-}10)^2/10 = 2.50$
6.6–6.8	4	10	$(4\text{-}10)^2/10 = 1.60$
6.8–7.0	1	10	$(1\text{-}10)^2/10 = 8.10$
Total	100	100	41.20

we estimated from the data in defining the hypothesized distribution. For the uniform distribution, only the sample size was needed to estimate the theoretical frequencies, thus "0" parameters were used in computing $\chi^2 (g = 0)$. Therefore, for this case, $df = 10 - 1 - 0 = 9$.

4. Table 7.11 is now entered with $\alpha = 0.05$ and $df = 9$. A decision value of $\chi^2 = 16.92$ is found. As the value obtained is $43.20 > 16.92$, the hypothesis that the underlying distribution is uniform must be rejected.

In this case, a rough perusal of the data would have led one to believe that the data were certainly *not* uniform, and the analysis has confirmed this. The distribution actually appears to be closer to normal, and this hypothesis can be tested. Determining the theoretical frequencies for a normal distribution is much more complicated, however. An example applying the χ^2 test to a normal hypothesis is discussed in detail in Chapter 9, using spot speed data.

Note that in conducting goodness-of-fit tests, it is possible to show that several different distributions could represent the data. How could this happen? Remember that the test does not prove what the underlying distribution is in fact. At best, it does not oppose the desire to assume a certain distribution. Thus it is possible that different hypothesized distributions for the same set of data may be found to be statistically acceptable.

A final point on this hypothesis testing: The test is not directly on the hypothesized distribution but rather on the expected versus the observed number of samples. Table 7.10 shows this in very plain fashion: The computations involve the expected and observed number of samples; the actual distribution is important only to the extent that it influences the probability of category. That is, the actual test is between two histograms. It is therefore important that the categories be defined in such a way that the "theoretic histogram" truly reflects the essential features and detail of the hypothesized distribution. That is one reason why categories of different size are sometimes used: The categories should each match the fast-changing details of the underlying distribution.

7.9 Summary and Closing Comments

Traffic studies cannot be done without using some basic statistics. In using statistics, the engineer is often faced with the need to act despite the lack of certainty, which is why statistical analysis is employed. This chapter reviewed basic statistics to help you understand and perform everyday traffic studies.

A number of very important statistical techniques were not presented in this chapter that are useful for traffic engineers. They include contingency tables, correlation analysis, analysis of variance (ANOVA), regression analysis, and cluster analysis. See Reference 4 or other basic statistical texts [*1–3, 5*].

Table 7.11: Upper Percentage Points on the Chi-Square Distribution

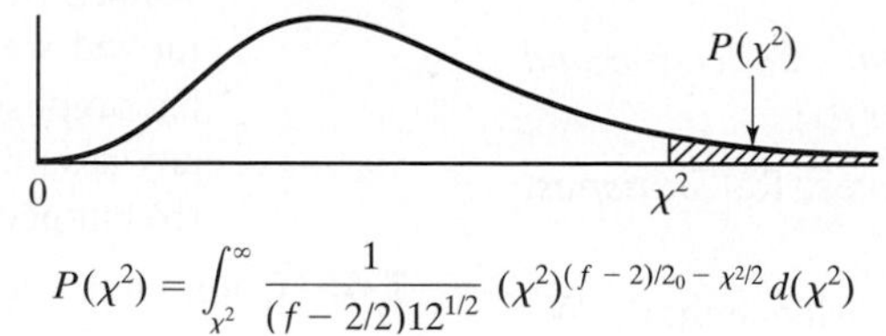

$$P(\chi^2) = \int_{\chi^2}^{\infty} \frac{1}{(f - 2/2)12^{1/2}} (\chi^2)^{(f-2)/2_0 - \chi^{2/2}} d(\chi^2)$$

df	.995	.990	.975	.950	.900	.750	.500	.250	.100	.050	.025	.010	.005
1	3927×10^{-2}	1571×10^{-7}	9821×10^{-7}	3932×10^{-8}	0.01579	0.1015	0.4549	1.323	2.706	3.841	5.024	6.635	7.879
2	0.01003	0.02010	0.05064	0.1026	.2107	.5754	1.386	2.773	4.605	5.991	7.378	9.210	10.60
3	.07172	.1148	.2158	.3518	.5844	1.213	2.366	4.108	6.251	7.815	9.348	11.34	12.34
4	.2070	.2971	.4844	.7107	1.064	1.923	3.357	5.585	7.779	9.488	11.14	13.28	14.86
5	.4117	.5543	.8312	1.145	1.610	2.675	4.351	6.626	9.236	11.07	12.83	15.09	16.75
6	.6757	.8721	1.237	1.635	2.204	3.455	5.348	7.841	10.64	12.59	14.45	16.81	18.55
7	.9893	1.259	1.690	2.167	2.833	4.255	6.346	9.037	12.02	14.07	16.01	18.48	20.28
8	1.344	1.646	2.180	2.733	3.199	5.071	7.344	10.22	13.36	15.51	17.53	20.09	21.98
9	1.735	2.088	2.700	3.325	4.168	5.899	8.343	11.39	14.68	16.92	19.02	21.67	23.59
10	2.150	2.558	3.247	3.940	4.865	6.737	9.342	12.55	15.99	18.31	20.48	23.21	25.19
11	2.603	3.053	3.816	4.575	5.578	7.584	10.34	13.70	17.28	19.68	21.92	24.72	26.76
12	3.074	3.571	4.404	5.226	6.304	8.458	11.34	14.85	18.55	21.03	23.34	26.22	28.30
13	3.565	4.107	5.009	5.892	7.042	9.299	12.34	15.98	19.81	22.36	24.74	27.69	29.82
14	4.075	4.660	5.629	6.571	7.790	10.17	13.34	17.12	21.06	23.68	26.12	29.14	31.32
15	4.601	5.229	6.262	7.261	8.547	11.04	14.34	18.25	22.31	25.00	27.49	30.58	32.80
16	5.142	5.812	6.908	7.962	9.312	11.91	15.34	19.37	23.54	26.30	28.85	32.00	34.27
17	5.697	5.408	7.564	8.672	10.09	12.79	16.34	20.49	24.77	27.59	30.19	33.41	35.72
18	6.265	7.015	8.231	9.390	10.86	13.68	17.34	21.60	25.99	28.87	31.53	34.81	37.16
19	6.844	7.644	8.907	10.12	11.65	14.56	18.34	22.72	27.20	30.14	32.85	36.19	38.58
20	7.434	8.260	9.591	10.85	12.44	15.45	19.34	23.83	28.41	31.41	34.17	37.57	40.00
21	8.034	8.897	10.28	11.59	13.24	16.34	20.34	24.93	29.62	32.67	35.48	38.93	41.40
22	8.643	9.542	10.98	12.34	14.04	17.24	21.34	26.04	30.81	33.92	36.78	40.29	42.60
23	9.260	10.20	11.69	13.09	14.85	18.14	22.34	27.14	32.01	35.17	38.08	41.64	44.18
24	9.886	10.86	12.40	13.85	15.66	19.04	23.24	28.24	33.20	36.42	39.36	42.98	45.58
25	10.52	11.52	13.12	14.61	16.47	19.94	24.34	29.34	34.38	37.65	40.65	44.31	46.93
26	11.16	12.20	13.84	15.38	17.29	20.84	25.34	30.43	35.56	38.89	41.92	45.64	48.29
27	11.81	12.88	14.57	16.15	18.11	21.75	26.34	31.53	36.74	40.11	43.19	46.96	49.64
28	12.46	13.56	15.31	16.93	18.94	22.66	27.34	32.62	37.92	41.34	44.46	48.28	50.99
29	13.12	14.26	16.05	17.71	19.77	23.57	28.34	33.71	39.09	42.58	45.72	49.59	52.34
30	13.79	14.95	16.79	18.49	20.60	24.48	29.34	34.80	43.26	43.77	46.98	50.89	53.67
40	20.71	22.16	24.43	26.51	29.05	33.66	39.34	45.62	51.80	55.76	59.34	63.69	66.77
50	27.99	29.71	32.36	34.76	37.69	42.94	49.33	56.33	63.17	67.50	71.42	76.15	79.49
60	35.53	37.48	40.48	43.19	46.46	52.29	59.33	66.98	79.08	79.08	83.30	88.38	91.95
70	43.28	45.44	48.76	51.74	55.33	61.70	69.33	77.58	85.53	90.53	95.02	100.42	104.22
80	51.17	53.54	57.15	60.39	64.28	71.14	79.33	88.13	96.58	101.88	106.63	112.33	116.32
90	59.20	61.75	65.65	69.13	73.29	80.62	89.33	98.65	107.56	113.14	118.14	124.12	128.30
100	67.33	70.00	74.22	77.93	82.36	90.13	99.33	109.14	118.50	124.34	129.56	135.81	140.17
	−2.576	−2.326	1.960	−1.645	−1.28	−0.6745	0.0000	+0.6745	+1.282	+1.645	+1.960	+2.326	576

(*Source:* E. L. Crow, F. A. Davis, and M. W. Maxwell, *Statistics Manual*, Dover Publications, Mineola, NY, 1960.)

References

1. Vardeman, S.B., and Jobe, J.M., *Data Collection and Analysis*, Duxbury Press, Boston, 2001.
2. Freedman, S., Pisani, R., and Purves, R., *Statistics*, W.W. Norton, 2007.
3. Wackerly, D., Sheaffer, W., and Mendenhall, W., *Mathematical Statistics with Applications*, PWS-KENT, Boston, 2001.
4. Trumbo, Bruce, *Learning Statistics with Real Data*, Duxbury/Thomson Learning, Pacific Grove, CA, 2002.
5. *Statistics with Applications to Highway Traffic Analyses*, 2nd Edition, ENO Foundation for Transportation, Westport, CT, 1978.

Problems

7-1. Experience suggests that spot speed data at a given location is normally distributed with a mean of 57 mi/h and a standard deviation of 7.6 mi/h. What is the speed below which 85% of the vehicles are traveling?

7-2. Travel time data is collected on an arterial, and with 30 runs, an average travel time of 152 seconds is computed over the 2.00-mile length, with a computed standard deviation of 17.3 seconds. Compute the 95% confidence bounds on your estimate of the mean. Was it necessary to make any assumption about the shape of the travel-time distribution?

7-3. Vehicle occupancy data is taken in a high-occupancy vehicle (HOV) lane on a freeway, with the following results:

Vehicle Occupancy	Number of Vehicles Observed
2	120
3	40
4	30
5	10

(a) Compute the estimated mean and standard deviation of the vehicle occupancy in the HOV Lane. Compute the 95% confidence bounds on the estimate of the mean.

(b) When the HOV lane in question has an hourly volume of 900 veh/h, what is your estimate of how many people on average are being carried in the lane? Give a 95% confidence range.

(c) If we observe the lane tomorrow and observe a volume of 900 veh/h, what is the range of persons moved we can expect in that hour? Use a range encompassing 95% of the likely outcomes. List any assumptions or justifications not made in part (b) but necessary in part (c) if any.

7-4. Commuter vans get an average of 18 mi/gal. As the manufacturer of commuter vans, you want to manufacture a more gas-efficient van and be able to say that it gets better mileage than 90% of all other vans. Assuming that gas mileage is normally distributed, what average mile per gallon usage must your new van get?

7-5. What is the probability of a value falling between 45 and 55 on the normal distribution: N[42,16]? On the same distribution, what are the 88% confidence bounds on the mean if you had a sample of 55 to determine the mean and standard deviation?

7-6. Two different procedures are used to measure "delay" on the same intersection approach at the same time. Both are expressed in terms of delay per vehicle. The objective is to see which procedure can be best used in the field. Assume that the following data are available from the field:

(a) A statistician, asked whether the data "are the same," conducts tests on the mean and the variance to determine whether they are the same at a level of significance of 0.05. Use the appropriate tests on the hypothesis that the difference in the means is zero and the variances are equal. Do the computations. Present the conclusions.

Delay (s/veh)	
Procedure 1	**Procedure 2**
8.4	7.2
9.2	8.1
10.9	10.3
13.2	10.3
12.7	11.2
10.8	7.5
15.3	10.7
12.3	10.5
19.7	11.9
8.0	8.7
7.4	5.9
26.7	18.6
12.1	8.2
10.7	8.5
10.1	7.5
12.0	9.5
11.9	8.1
10.0	8.8
22.0	19.8
41.3	36.4

(b) Would it have been more appropriate to use the "paired-t" test in part (a) rather than a simple t-test? Why?

7-7. Based on long-standing observation and consensus, the speeds on a curve are observed to average 57 mi/h with a 6 mi/h standard deviation. This is taken as "given." Some active controls are put in place (signing, flashing lights, etc.), and the engineer is sure that the average will fall to at least 50 mi/h, with the standard deviation probably about the same.

(a) Formulate a null and alternative hypothesis based on taking "after" data only, and determine the required sample size n so that the Type I and Type II errors are each 0.05.

(b) Assume the data have been taken for the required N observations and that the average is 52.2 mi/h and the standard deviation is 6.0 mi/h. What is your decision? What error may have occurred, and what is its probability?

(c) Resolve part (b) with a mean of 52.2 mi/h and a standard deviation of 5.4 mi/h.

CHAPTER 8

Traffic Data Collection and Reduction Methodologies

The starting point for most traffic engineering is a comprehensive description of the current state of the streets and highways that comprise the system, current traffic demands on these facilities, and a projection of future demands.

This requires that information and data that can quantitatively describe the system and its demands be assembled. As the highway system is massive, and demands are both time- and location-sensitive, assembling this information is a massive task. Nevertheless, data must be collected and reduced to some easily interpreted form for analysis.

Collection and reduction of traffic data covers a wide range of techniques and technologies from simple manual techniques (often aided by a variety of handheld or other devices for recording the data) to complex use of the ever-expanding technology of sensors, detectors, transmission, and computer equipment.

This chapter provides a basic overview of data collection and reduction in traffic engineering. The technology applied, as noted, changes rapidly, and traffic engineers need to maintain current knowledge of this field.

8.1 Applications of Traffic Data

The fundamental difficulty in traffic data collection and reduction is the need to observe and quantify the behavior of road users in real time over a large system. Consider the "simple" problem of observing traffic volumes in the midtown Manhattan street network—approximately from 14th Street to 59th Street, and from 1st Avenue to 12th Avenue. Such a network includes over 500 links that ideally would be observed simultaneously for perhaps a full day, or even several. No traffic agency would have access to sufficient equipment and/or personnel to do this successfully. Yet, for any traffic engineering to be done within this network, such information is absolutely necessary.

Traffic engineers collect and reduce data for many reasons and applications:

- *Managing the physical system.* The physical traffic system includes a number of elements that must be monitored, including the roadway itself, traffic control devices, detectors and sensors, and light fixtures. Physical inventories must be maintained simply to know what is "out there." Bulbs in traffic signals and lighting fixtures must not be burned out, traffic markings must be clearly legible, signs must be clean and visible, and so forth. Replacement and maintenance programs must be in place to ensure that all elements are in place, properly deployed, and safe.

- *Establishing time trends.* Traffic engineers need trend data to help identify future transportation needs. Traffic volume trends can identify areas and specific locations that can be expected to congest in the future. Accident data and statistics over time can identify core safety problems and site-specific situations that must be addressed and mitigated. Trend data allows the traffic engineer to anticipate problems and solve them *before* they actually occur.
- *Understanding travel behavior.* A good traffic engineer must understand how and why people (and goods) travel in order to provide an effective transportation system. Studies of how travelers make modal choices, trip time decisions, and destination choices are critical to understanding the nature of traffic demand. Studies of parking and goods delivery characteristics provide information that allows efficient facilities to be provided for these activities.
- *Calibrating basic relationships or parameters.* Fundamental characteristics such as perception-reaction time, discharge headways at signalized intersections, headway and spacing relationships on freeways and other uninterrupted flow facilities, gap acceptance characteristics, and others must be properly quantified and calibrated to existing conditions. Such measures are incorporated into a variety of predictive and assessment models on which a great deal of traffic engineering is based.
- *Assessing the effectiveness of improvements.* When traffic improvements of any kind are implemented, follow-up studies are needed to confirm their effectiveness and to allow for adjustments if all objectives are not met.
- *Assessing potential impacts.* An essential part of traffic engineering is the ability to predict and analyze the traffic impacts of new developments and to provide traffic input into air pollution models.
- *Evaluating facility or system performance.* All traffic facilities and systems must be periodically studied to determine whether they are delivering the intended quantity and quality of access and/or mobility service to the public.

Data and information from traffic studies provide the underpinning for all traffic planning, design, and analysis. If the data are not correct and valid, then any traffic engineering based on the data must be flawed. Some of the tasks involved in data collection and reduction are mundane. Data and information, however, are the foundation of traffic engineering. Without a good foundation, the structure will surely fall.

8.2 Types of Studies

It would be literally impossible to list all of the studies in which traffic engineers get involved. Eleven of the most common types of studies include:

1. *Volume studies.* Traffic counts are the most basic of studies and the primary measure of demand; virtually all aspects of traffic engineering require demand volume as an input, including planning, design, traffic control, traffic operations, detailed signal timing, and others.
2. *Speed studies.* Speed characteristics are strongly related to safety concerns and are needed to assess the viability of existing speed regulations and/or to set new ones on a rational basis.
3. *Travel time studies.* Travel times along sections of roadways constitute a major measure of quality of service to motorists and passengers, and also of relative congestion along the section. Many demand-forecasting models also require good and accurate travel time measures.
4. *Delay studies.* Delay is a term that has many meanings, as we discuss in later chapters. In essence, it is the part or parts of travel time that users find particularly annoying, such as stopping at a traffic signal or because of a midblock obstruction.
5. *Density studies.* Density is rarely directly observed. Some modern detectors can measure "occupancy," which is directly related to density. Density is a major parameter describing quality of operations on uninterrupted flow facilities.
6. *Accident studies.* Because traffic safety is the primary responsibility of the traffic engineer, the focused study of accident characteristics, in terms of systemwide rates, relationships to causal factors, and at specific locations, is a critically important function.

7. *Parking studies.* These involve inventories of parking supply and a variety of counting methodologies to determine accumulations and total parking demand. Interview studies also involve attitudinal factors to determine how and when parking facilities are used.
8. *Goods movement and transit studies.* Inventories of existing truck-loading facilities and transit systems are important descriptors of the transportation system. Because these elements can be significant causes of congestion, proper planning and operational policies are a significant need.
9. *Pedestrian studies.* Pedestrians are an important part of the demand on transportation systems. Their characteristics in using crosswalks at signalized and unsignalized intersections and midblock locations constitute a required input to many analyses. Interview techniques can be used to assess behavioral patterns and to obtain more detailed information.
10. *Calibration studies.* Traffic engineering uses a variety of basic and not-so-basic models and relationships to describe and analyze traffic. Studies are needed to calibrate key values in models to ensure that they are reasonably representative of the conditions they claim to replicate.
11. *Observation studies.* Studies on the effectiveness of various traffic control devices are needed to assess how well controls have been designed and implemented. Observance rates are critical inputs to the evaluation of control measures.

This text includes a detailed treatment of volume studies, speed studies, travel time studies, delay studies, accident studies, and parking studies. Others are noted in passing, but the engineer is encouraged to consult other sources for detailed descriptions of study procedures and methodologies [*1,2*].

8.3 Data Collection Methodologies

In general, data collection takes place at one of three levels:

- Manual studies
- Semiautomated studies
- Fully automated studies

Despite all of the modern technology available to the traffic engineer, some studies are best conducted manually. These studies tend to be for short duration and/or at highly focused locations. The use of automated equipment requires set-up and take-down effort that may not be practical for a one-hour study, for example. Certain types of information are difficult to obtain without manual observation. Turning movements at an intersection are difficult to track without direct manual observation. Vehicle occupancy, frequently of interest, cannot be determined automatically with existing technology. Although the size of vehicles can be used to automate the determination of vehicle class (truck, bus, car), it cannot discern the difference between a private automobile and a taxi. Manual observations can be supported by a variety of handheld devices that assist in the recording of data or can be fully manual using paper forms to record data.

Semiautomated methods rely primarily on the use of pneumatic road tubes and a wide variety of recording devices that can be connected to such tubes. The primary characteristic of a semiautomated study is that the devices used are generally portable—that is, they can be moved from location to location. Other portable devices can also be used.

Fully automated studies rely on a wide variety of permanently installed detectors or sensors, usually in conjunction with connections to a stationary or portable computer station. The growth of sensor technology is ever-accelerating, and the state of the art in this area is a moving target. The same permanent detectors and sensors that can be used to collect traffic data of various sorts are also used to operate actuated and/or adaptive signals. The number of locations that have permanent detectors or sensors is rapidly increasing.

As is the case with satellite data, detection is no longer the restricting element in obtaining traffic data. The principal problem is that the electronic storage and filing capability of traffic agencies is significantly behind the sensor technology. Modern freeways are frequently monitored by large numbers of permanent detectors placed in the pavement or above it. The data at most detector locations, however, are only available in real time or in limited hourly summaries. Capturing and using detailed data often requires that a portable computer system be physically connected to the sensors to download data for storage in real time.

8.3.1 Manual Data Collection Techniques

As we have noted, short time frames, short lead times, and the need for certain types of data lead to manual data collection approaches. The types of studies most often conducted manually are (1) traffic counts at a specific location or small number of such locations, usually when the time frame is less than 24 hours, (2) speed or travel time data at a focused location for short duration, (3) observance studies at specific locations for short durations, and (4) intersection delay studies of short duration.

Traffic Counting Applications

The most likely situation leading to a manual study is the traffic count. A traffic engineer is often faced with a problem requiring some detailed knowledge of existing traffic volumes, which are generally needed quickly. A signal timing problem requires peak hour flow rates for the intersection, perhaps in the AM peak, the PM peak, and midday. Sending a few people out immediately to conduct such a study can produce the required data within a day.

Manual counts can be greatly facilitated through the use of push-button counters, such as those illustrated in Figure 8.1.

Hand counters have between four and six individual registers. On a single counter, an observer can keep track of up to four separate items. This allows an observer to classify volumes by vehicle type, turning movement, and/or vehicle type. When four are placed on a board, a professional traffic observer can keep track of these things for four legs of an intersection or different lanes on a freeway.

The primary difficulty with hand counters is that the data must be manually recorded in the field at periodic intervals. This disrupts thc count. Obviously, while the observer writes down the register counts and resets them for the next period, vehicles are passing by uncounted. To obtain continuous count information on a common basis, short breaks are introduced into the counting procedure. Such breaks must be systematic and uniform for all observers. The system revolves around the *counting period* for the study—the unit of time for which volumes are to be observed and recorded. Common counting periods are 5 minutes, 15 minutes, and 60 minutes, although other times can be, and occasionally must be, used.

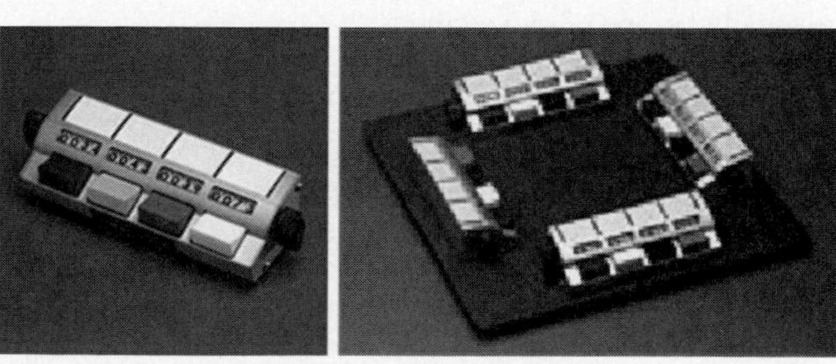

Figure 8.1: Hand Counters Illustrated
(*Source:* Used with permission of the Denominator Company.)

The short breaks are generally arranged in one of two ways:

- A portion of each counting period is set aside for a short break.
- Every other counting period is used as a short break.

In the first option, "x" out of every "y" minutes is counted. Thus, for 5-minute counting periods, an observer might count 4 of every 5 minutes; for a 15-minute counting period, 12 or 13 minutes of 15 minutes might be counted. To provide for a continuous count profile, the volume that is expected to occur during the short breaks must be estimated and added. This is simply done by assuming that the *rate of flow* during the missing minutes is the same as that during the actual count. Then:

$$V_y = V_x(x/y) \tag{8-1}$$

where: V_y = volume for continuous counting period of "y" minutes (vehs)
V_x = volume for discontinuous counts of "x" minutes (veh)
$y-x$ = short break time (minutes)
y = counting period (minutes)
x = counting period minus short break (minutes)

Consider a case where a manual counting survey is conducted by counting vehicles in 4 of every 5-minute period. If a count of 100 vehicles is obtained, what is the estimated count for the full 5-minute counting period? Using Equation 8-1:

$$V_5 = 100(5/4) = 125\text{vehs.}$$

Although not necessary in this case, volumes are always reported as *full vehicles*. No fractions are permitted.

When alternating periods are used as breaks, full period counts are available in alternate counting periods. Missing counts are estimated using straight-line interpolation:

$$V_i = \frac{V_{i-1} + V_{i+1}}{2} \tag{8-2}$$

where: V_i = volume in missing counting period "i" (vehs)
V_{i-1} = volume in counting period "i−1" (vehs)
V_{i+1} = volume in counting period "i+1" (vehs)

Again, any estimated counts are rounded to the nearest whole vehicle.

Table 8.1: Data from an Illustrative Volume Study

		Actual Counts (4 minutes) vehs		**Expanded Counts (x 5/4) vehs**		**Estimated Counts[1] vehs**		**Estimated Flow Rates veh/h**	
Period	**Time (PM)**	**Lane 1**	**Lane 2**	**Lane 1**	**Lane 2**	**Lane 1**	**Lane 2**	**Lane 1**	**Lane 2**
1	5:00	24		30.0		30	*43*	360	516
2	5:05		36		45.0	*33*	35	396	540
3	5:10	28		35.0		35	*37*	420	564
4	5:15		39		48.8	*36*	49	432	588
5	5:20	30		37.5		38	*54*	456	648
6	5:25		47		58.8	*41*	59	492	708
7	5:30	36		45.0		45	*61*	540	732
8	5:35		50		62.5	*44*	63	528	756
9	5:40	34		42.5		43	*61*	516	732
10	5:45		48		60.0	*46*	60	552	720
11	5:50	40		50.0		50	*59*	600	708
12	5:55		46		57.5	*55*	58	660	696
Total		**192**	**266**	**240.0**	**332.6**	**496**	**659**		
% in Lane		41.9%	58.1%	41.9%	58.1%	42.9%	57.1%		

Italics indicate an interpolated or extrapolated value.

In practice, it is often necessary to combine the two approaches. Consider the example shown in Table 8.1. In this case, a single observer is used to count two lanes of traffic on an urban arterial. Using hand counters, the observer alternates between lanes in each counting period. Thus, for each lane, alternating counts are obtained. Because the observer must also take short breaks to record data, only 4 minutes of every 5 minutes are actually counted. Table 8.1 illustrates the three computations that would be involved in this study.

Actual counts are shown, and represent 4-minute observations. These are expanded by a factor of 5/4 (Equation 8-1) to estimate volume in continuous 5-minute counting periods. At this point, each lane has counts for alternating counting periods. Missing counts are then interpolated (Equation 8-2) to estimate the count in each missing period. The first period for lane 2 and the last period for lane 1 cannot be interpolated because they constitute the first and last periods. The counts in these periods must be extrapolated, which is at best an approximate process. All results are rounded to whole vehicles but only in the final step. Thus the "expanded counts" carry a decimal; the rounding is done with the "estimated counts," which include the expansion, interpolation, and extrapolation needed to complete the table. The "estimated counts" constitute 5-minute periods. The flow rate in each 5-minute period is computed by multiplying each count by 12 (twelve 5-minute periods in an hour).

Speed Study Applications

In Chapter 10, the analysis of speed data is fully discussed. Because of the statistical analysis of such data, sample sizes for most speed studies are generally in the range of 100 to 200. Unless there is permanent detection equipment at the desired study location, many speed studies are conducted manually by one of two methods:

- Measurement of elapsed time over a short measured highway segment using a simple stopwatch.
- Direct measurement of speeds using either handheld or fixed-mounted radar meters.

When using a stopwatch to time vehicles as they traverse a short section of highway (often referred to as the "trap"), there are two significant sources of error:

- Parallax (viewing angle)
- Manual operation of the stopwatch

For multiple counts, a real-time communication system is needed that connects all observers to a supervisor. Counts in such situations must all start and stop at exactly the same time; a single supervisor times the study and issues "start" and "stop" orders as appropriate.

For studies longer than 1 or 2 hours, sufficient personnel must be trained to allow for periodic relief of observers. There will always be "no shows" for any large study, so "extra" personnel is also needed to cover these eventualities. Practical limitations must also be recognized: A typical worker not experienced in traffic studies, for example, could be relied on to count and classify one heavy or two light movements, or to measure speeds in one lane.

8.3.2 Portable Traffic Data Equipment/Semiautomated Studies

A large number of traffic studies are conducted using a variety of portable traffic data collection/recording devices. The most common portable device used in traffic studies is the pneumatic road tube. A pneumatic road tube is a closed-end tube in which an air pressure is maintained. When stretched across a roadway, a vehicle (actually an axle) rolling over it creates an air pulse that travels through the tube, which is connected to some form of data capture device. Such tubes are most often used for traffic counting, but they can also be used to measure speed. Figure 8.4 shows a typical field setup using pneumatic road tubes.

Note that Figure 8.4 shows two road tubes attached to the counter. This is to provide backup in case one of the tubes breaks. Pairs of tubes can also be used to measure speed; the recording device measures the elapsed time between the actuation of the first and second tubes. This practice is rarely applied today because radar meters have become the dominant technology to measure speeds. It is, however, the origin of the colloquial term *speed trap*. Police (in the 1940s and 1950s) often used the dual tubes to measure speed; because the distance between them was always referred to as "the trap," the term is now in use for any location at which police enforce speed limits.

Figure 8.4: Road Tubes with Portable Counter Illustrated

Although a variety of traffic counters are available to use with road tubes, the most common types record a total count at preset intervals, so that 5-, 10-, 15-, and 60-minute counts can be automatically recorded.

The primary difficulty with road tubes is that they do not count vehicles—they count axles. A passenger car with two axles crosses the road tube registering a pulse *twice.* A tractor-trailer combination with multiple axles may cause as many as five or six actuations that will be recorded. To obtain an estimate of vehicle counts, manual observations must be made to determine the *average number of axles per vehicle.* Obviously, if this has to be observed for the entire period of the study, there is no point in using road tubes. Representative samples do, however, have to be collected. If a road-tube count is being taken over a week, for example, manual classification counts might be taken for 5 to 10 hours at representative times during the week (AM peak, PM peak, off-peak).

Table 8.2 illustrates a sample classification count. Vehicles are observed and classified by the number of axles on each. As shown, these data can be used to estimate the average number of axles on a vehicle at the study location.

The average number of axles per vehicle at this study location is 1205/515 = 2.34 axles/vehicle. If the recording device shows a count of 7,000 axles for a given day, the estimated vehicle count would be 7,000/2.34 = 2,991 vehicles.

Table 8.2: Sample Classification Count for a Road-Tube Volume Study

Number of Axles	Number of Vehicles Observed	Total Axles for Observed Vehicles
2	400	× 2 = 800
3	75	× 3 = 225
4	25	× 4 = 100
5	10	× 5 = 50
6	5	× 6 = 30
Total	**515**	**1,205**

There are also a number of practical issues with road tubes:

- If not tautly fastened to the pavement (usually accomplished with clamps and epoxy), the road tube can start a "whipping" action when vehicles continually traverse it, causing an eventual breakage. Once broken, the tube cannot relay data. Road tubes are also subject to vandalism. In either event, the tube should be inspected at regular intervals to ensure that it is still functioning.
- If the road tube is stretched across more than one lane, simultaneous actuations are possible, and the recorded counts may low because of this. Obviously, this problem gets worse as the number of lanes and the volume of traffic increases.
- Traffic counters (recording devices) are also subject to vandalism and/or theft and must also be periodically checked to see that they are functioning.

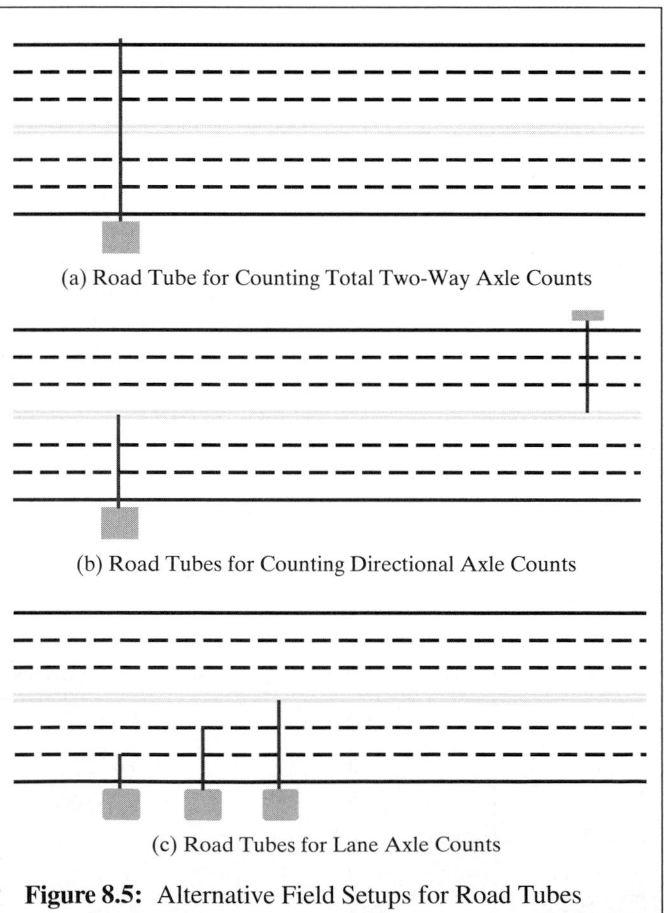

(a) Road Tube for Counting Total Two-Way Axle Counts

(b) Road Tubes for Counting Directional Axle Counts

(c) Road Tubes for Lane Axle Counts

Figure 8.5: Alternative Field Setups for Road Tubes

Figure 8.5 illustrates common setups for road tubes as used in traffic-counting studies. Figure 8.5 (a) shows a single tube across all lanes of a facility. In this configuration, a total two-way count of axles is obtained. Figure 8.5 (b) shows the most common technique—two road tubes set up to record axles in each direction separately. Figure 8.5 (c) shows a less typical setup in which lane counts can be deduced. The tubes must be close enough together that the number of lane changes within the detection range is minimal.

Other types of portable detectors can be used. Portable plate detectors can be fastened to a specific lane. Using wave emissions, such detectors can measure volume, speed, and occupancy and send the data wirelessly to a roadside computer. They can also classify vehicle types by length. They tend to be expensive and can only monitor one lane at a time. Most studies would require several such detectors. An observer generally must be present to monitor the portable computer used to store the data and to ensure that the detectors are not stolen.

There are also devices that operate using electronic tape switches fastened to the pavement. These operate by interpreting current disruptions caused by vehicles crossing over the tape switch. A double wire is generally fastened to the pavement under a tough form of tape and is attached to a low-grade power source. When a vehicle crosses the tape switch, a surge of current occurs that can be monitored by a variety of portable recording devices.

8.3.3 Permanent Detectors

Rapid advances in traffic detector technology are rapidly changing the landscape for traffic studies. Detectors are used for all types of things, from data collection and transmission to the real-time operation of traffic signal systems. As intelligent transportation technology marches forward, there is great interest in real-time monitoring of traffic systems on a massive scale. This requires that the traffic system be instrumented with large numbers of permanent detectors and the capability to observe the data they provide in real time.

In the 1960s, very few permanent detectors were in place, and most were used to operate early versions of actuated signals. Today, thousands of permanent detectors are installed on freeways, at intersections, and other locations. These detectors provide the opportunity to gather large amounts of data on the traffic system. The difficulty is that much of the data is only available in real time—that is, it is not permanently stored in a retrievable form, and the variety of technologies in use makes it difficult to coordinate everything into a seamless system of critical information on the system.

The Traffic Detector Handbook [*3*] provides an excellent overview of current detector and sensor technology. It classifies detectors and sensors into a number of broad categories based on the type of technology used:

- Sound (acoustic)
- Opacity (optical, infrared, video image processing)
- Geomagnetism (magnetic sensors, magnetometers)
- Reflection of transmitted energy (infrared laser radar, ultrasonic, microwave radar)
- Electromagnetic induction (inductive loop detectors)
- Vibration (triboelectric, seismic, inertia switch sensors)

These technologies have advanced dramatically from the earliest detectors. The first widely used detectors were pressure plates. Two steel plates were separated by a rubber spring. The plates were connected to a low-voltage power source. When a vehicle (axle) crossed the pressure plate, the two plates came together, completing the circuit and creating an electrical pulse. Although technologically simple, there were many practical problems with pressure plates. Snow plows and other equipment could dislodge them; they invariably led to deterioration of the pavement around them; or water could get between the plates and freeze, providing a permanent current and rendering the device inoperable. There are still isolated locations where pressure plates are still in use, but none are being installed at this point.

The most prevalent detector in use today is the inductive loop. When a metal object (vehicle) enters the field of the loop detector, inductance properties of the loop are reduced and sensed, thus recording the presence of the vehicle. The loop essentially creates an electromagnetic field that is disturbed when a metal object enters it. Induction loops require that a cut be made in the pavement surface, with one or more wire loops placed in the cut. A pull wire connects the detector to a power source, which is then connected to a controller unit. Figure 8.6 shows a typical installation of an inductive loop.

The standard induction loop detector measures 6 ft × 6 ft and covers one lane. Multiple detectors are needed to monitor multiple lanes. Longer loops are available in sizes ranging from 6 ft × 40 ft to 6 ft × 80 ft. These are often used in conjunction with actuated signals requiring a significant detection area. In some cases, a long detection area is provided by installing a series of 6 ft × 6 ft detectors in each lane.

Inductive loop detectors directly measure the presence and passage of vehicles. Other important measures, such as speed and density, can be deduced using calibrated algorithms, but the accuracy of these is often insufficient for research use.

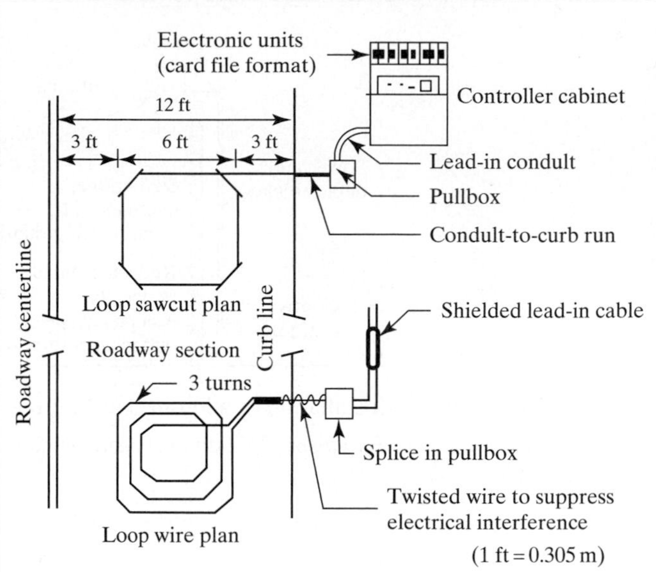

Figure 8.6: Typical Installation of an Induction Loop Detector

(*Source: Traffic Detector Handbook*, 3rd Edition, Federal Highway Administration, Publication No. FHWA-HRT-06-108, Washington DC, 2006, Figure 1-4, pp. 1–12.)

Magnetic detectors come in two forms: the magnetometer and magnetic sensors. Because of the details of their design, the magnetometer can detect both passage and presence, whereas magnetic sensors detect only passage. Both operate on the principle that a moving metal object (vehicle) disturbs the earth's ambient magnetic field and that the disturbance can be monitored and counted. The magnetometer also has a second wire loop that operates on a basis similar to the inductive loop to detect presence. Both types of detectors are placed *under* the roadway on a structure or in the subbase. As a passage-only detector, magnetic sensors have limited applications.

A variety of permanent detectors use the same Doppler principle as handheld radar meters. They all rely on reflected energy from vehicles that can be detected and used to obtain a variety of traffic parameters. These include microwave radar meters, infrared sensors, laser radar meters, and ultrasonic detectors. The difference lies in the wavelength and frequency of energy that is emitted and sensed.

Figure 8.7 illustrates an overhead installation of a microwave radar detector. In this configuration, the detector monitors a single lane or two lanes. Figure 8.8 illustrates an alternative application in which the detector is located at the roadside and covers all lanes of an approach. The illustration

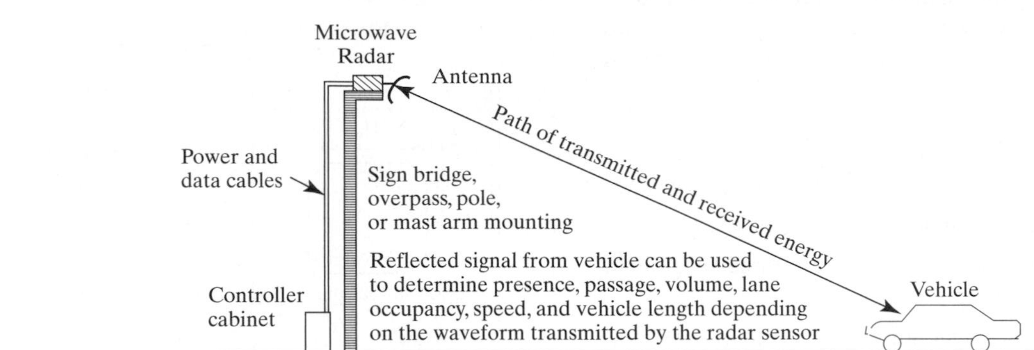

Figure 8.7: Overhead Installation of a Permanent Microwave Radar Detector

(*Source: Traffic Detector Handbook*, 3rd Edition, Federal Highway Administration, Publication No. FHWA-HRT-06-108, Washington DC, 2006, Figure 1-7, p. 1-16.)

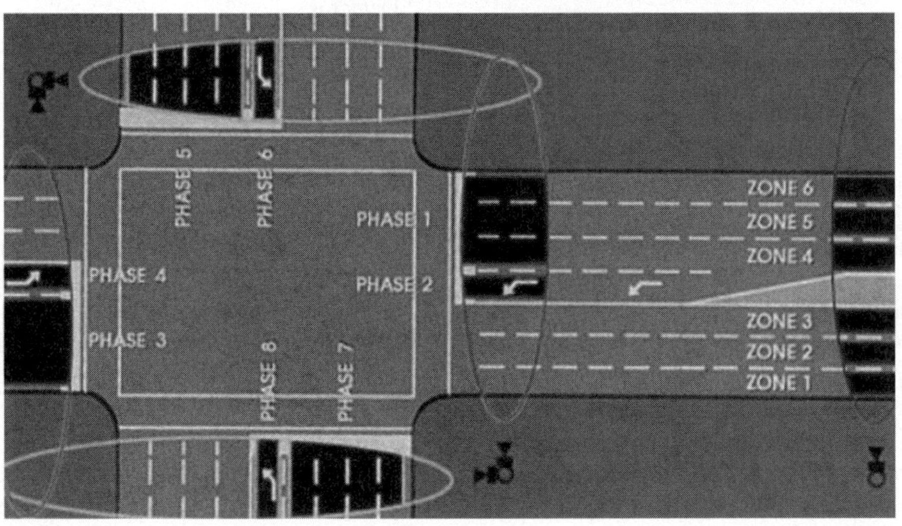

Figure 8.8: Roadside Installation of Permanent Microwave Radar Detectors

(*Source: Traffic Detector Handbook*, 3rd Edition, Federal Highway Administration, Publication No. FHWA-HRT-06-108, Washington DC, 2006, Figure 1-8, p. 1-17.)

of Figure 8.8 is at a signalized intersection, and the detectors are being used to operate an actuated signal.

Figure 8.9 shows an overhead installation of a laser infrared radar detector. Using two beams, it can actually create an image of vehicles within its range, allowing for automated vehicle classifications.

Perhaps the most exciting recent development in real-time traffic detection is the rapid advancement of video image processing (VIP) technologies. Cameras are installed (digital video) typically on a mast arm, often at a signalized intersection location. The camera can be focused on a single lane but can also be used to monitor multiple lanes. The system consists of the camera, a microprocessor to store and interpret images, and software that converts the images to traffic data. In essence, vehicles appear on a video image as a compressed package of pixels moving across the image. The system is calibrated to recognize

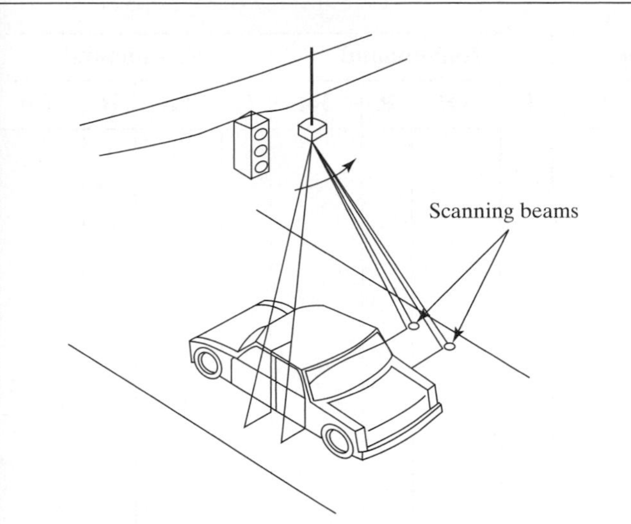

Figure 8.9: An Overhead Laser Infrared Radar Installation

(*Source: Traffic Detector Handbook*, 3rd Edition, Federal Highway Administration, Publication No. FHWA-HRT-06-108, Washington DC, 2006, Figure 1-9, p. 1-18.)

the background image and account for changes due to ambient and artificial light and weather. Available software can discern and classify vehicles by length, count by lane, and provide speeds. VIPs are now being used to operate actuated signals in some locations, and it is almost certain that this technology will advance rapidly over the near and moderate term future.

8.4 Data Reduction

The process of data reduction is critical to the accuracy of any study. The process varies widely depending on the manner in which data is collected, but essentially it involves taking data in whatever raw form it is observed and "reducing" it to some understandable form.

In manual studies, this most often involves taking handwritten forms from various observers at one or more locations and organizing it into a recognizable form. It may involve expansions, interpolations, and extrapolations as discussed previously. It always involves presenting data in a format that allows for efficient and effective analysis.

In the simple case, for example, of a manual count at an intersection, the data would be assembled from the four to six observers involved and entered into tables such that volumes or flow rates of all movements can be discerned for common time intervals. Table 8.3 shows a common format used to depict intersection volumes and/or flow rates.

The tabulated data allows the analyst to see the pattern of volume over time on a movement-by-movement basis. When the data are collected using portable or permanent automated equipment, the data can be placed directly into a spreadsheet for subsequent analysis. Spreadsheet data can also be used to create useful graphs and other visual depictions to further clarify the meaning of the data.

Where data is collected by permanent detectors, it can be sent directly (either through wired or wireless technologies) to centralized computers into a predesigned databank, where standard tabular and graphic displays can be created. As previously discussed, however, data collection technology has advanced far faster than the technology for creating organized databases into which *all* data is entered, and from which *all* data can be conveniently retrieved. Most data that are observed exist only in real time. Portable devices can be connected to a variety of detector inputs, from road tubes to induction loops, to extract data over some period of time for permanent storage.

8.5 Cell Phones

Technology exists that can track the location of cell phones (as long as they are turned on) to reasonably small areas. It is possible, therefore, to use this technology to track vehicles in time and space, yielding data that would be hard to obtain by other means, such as origin destination and routing information. Such use, however, raises a number of issues concerning the rights of privacy and the extent to which people are willing to accept a "big brother" watching our activities. Technology now provides a plethora of ways in which people can be observed without being aware of it. Such technologies are already in use for law enforcement purposes, but they have also been subject to legal challenges. This technology will continue to develop rapidly; its usefulness as a means of collecting useful transportation data will be subject to legal restrictions as determined by the courts and legislatures over time.

8.6 Aerial Photography and Digitizing Technology

As the state of the art moves more toward the need to observe, interpret, and manage system behavior, the approaches to data collection and reduction must change as well. Historically, traffic data collection focused on isolated locations or a limited number of isolated locations. Today, the ability to provide

Table 8.3: Sample Intersection Count Table

Time Period	Eastbound				Westbound				Northbound				Southbound			
	L	TH	R	Tot	L	TH	R	Tot	L	TH	R	Tot	L	TH	R	Tot
1																
2																
3																
4																
5																
6																
7																
8																
9																
10																
.																
.																
.																
.																
.																
N																

individual drivers with real-time traffic and routing information is rapidly advancing. Navigation systems, available on only a few very expensive vehicles a decade ago, can now be purchased for any vehicle at an increasingly reasonable cost. The problems of urban congestion make it increasingly important to understand the state of the overall transportation system and the way in which individuals use it.

The rapid development of detector technology now allows engineers to place large numbers of devices throughout the highway system. The ability to synthesize all of this information into a usable real-time resource is also rapidly developing.

Existing technologies, however, do allow researchers to look at larger parts of the highway system through the use of aerial photography. Some of this technology has been around for a long time. In the late 1960s and early 1970s, time-lapse photography from fixed wing aircraft or helicopters was used to trace vehicles through long segments of highways. The state of the art of film technology at that time limited the photography to one or two frames per second, and it allowed only a few hours of data to be observed during one flight. Modern digital technology allows for continuous photography over longer periods of time. Some current systems can now resolve vehicle locations as many as 16 times each second. Eventually, the use of satellites will become common. Even today, sections of highway can be examined using satellite images available through the Internet. This greatly simplifies the process of examining the geometric characteristics of sites and the selection of individual sites for study. The technology does not, however, provide sufficient time-sequenced images to be used directly for data collection.

Aerial photography is now being used to observe moderately sized highway segments as opposed to individual locations. This technology is useful for studying weaving sections, interchanges, arterial sections, a series of intersections, and similar locations. Because aerial photography can capture an entire segment in the photographic field, it allows individual vehicles to be traced as they traverse the segment. Thus such information as lane changes, queue formation and dissipation, local origin and destination, and similar data can be retrieved, in addition to the more standard counts, vehicle classifications, and speeds.

"Aerial" photography uses three basic platforms: (1) a stationary elevated vantage point, where available (such as a tall building), (2) helicopters, and (3) fixed-wing aircraft. Where available, elevated vantage points are highly desirable because the camera frame remains fixed throughout the study. They are difficult to find, however, and must be within a reasonably short distance from the site to provide a favorable view of the study segment. Helicopters have the ability to hover over a single location but cannot do so for long periods

Figure 8.10: Aerial Photograph of a Freeway Interchange

of time; even when "hovering," the aircraft rotates so that the photographic image does not remain fixed. Fixed-wing aircraft can stay over a given site for longer periods of time, but they do not remain in a fixed location; they fly back and forth in a pattern that keeps a designated highway segment in the image range. The orientation of the photography, however, is constantly changing. Figure 8.10 illustrates an aerial photograph of a freeway interchange taken from a fixed-wing aircraft. The same type of image could have been obtained from a helicopter or a fixed elevated vantage point—although the latter would not be right overhead but at an angle from the fixed vantage point.

A process called "digitizing" captures individual vehicle traces through the photographed segment. This requires that two or more fixed objects with known coordinates be within every frame to be captured. As noted previously, some systems capture a frame up to 16 times per second, although as few a 2 to 4 times per second can also produce reasonable data. Each frame is shown on a large screen. An observer fixes cross hairs over each of the fixed objects with known coordinates; this establishes a grid coordinate system for the frame. The cross hairs are then placed over each vehicle (a fixed position, such as the front left corner of the vehicle is chosen), and the coordinates are automatically entered into a computer. This process is completed for each vehicle in each frame. Even though the process is "automated," it still is labor intensive. When all coordinates for all vehicles in all frames are entered, software analyzes the data and links vehicles in successive frames by assigning probabilities of a match based on logical paths. Where the software does not provide a match from frame to frame, the operator must go back to the frames and manually make the link by identifying the vehicle match from frame to frame. Current technologies can usually match about 50% of all vehicles automatically.

When the linkages are complete, the software produces detailed vehicle traces, showing position of the vehicle in each frame with the appropriate time stamp. Study-specific software is then written to extract the type of data needed for analysis. Position data from frame to frame can provide speed between frames, lane changes between frames, and other microscopic data that allows for very detailed analysis of the operation of the study segment.

Because this type of data collection and reduction is expensive, it is normally reserved for research applications. In a recent study of weaving segment operations [*4*], two hours of such data cost approximately $15,000, using a capture rate of two frames per second. A large quantity of such data is now being collected for the Next Generation Simulation Study (NGSIM) [*5*], using a capture rate of 16 frames per second. It seeks to provide a very detailed database for the development of a new generation of traffic simulation programs.

8.7 Interview Studies

Many types of information cannot be obtained except through direct interviews with travelers. This is particularly true where information on trip purpose and background information on travel choices is required.

8.7.1 Comprehensive Home Interview Studies

The largest scale interview studies involve comprehensive area-wide origin-destination studies, where the typical travel patterns of people within a broad study area are desired. Comprehensive home interviews are conducted in which information on all trips made by all household members the previous day is sought. Sampling is critical to such studies because the sampling rate is often quite low, depending on the size of the area. Setting up the study is complex and involves establishing origin and destination zones that are small enough to provide trip patterns but large enough to allow a reasonable number of interviews to be conducted within each. Appointments must be made for all interviews, and interviewers need formal identification to reassure the interviewees. Law enforcement must be notified of the activity, and any required permits must be acquired. Interviewers need to be well trained and articulate, and a dress code is normally recommended. Interviews are often done in the evening when the maximum number of household members are present. The interviewer extracts detailed information on all trips made the previous day, including time of start and finish, location of start and finish, trip purpose, mode or modes used, transfer locations, and route. Basic demographic information on each household member is also collected, including gender, age, occupation, driver's license status, and so forth. General household income and car ownership is also recorded, as is information on the type of housing.

Comprehensive home interview studies take place over a period of months, perhaps even years, and are very costly. When all data are collected, typical daily origin and destination patterns are created. The origin end is often treated as the "producer" of the trip and is (for about 85% of all trips) a residential location. The destination end is often treated as the "attractor" of the trip and generally involves a location related to the trip purpose (i.e., a place of employment, shopping, recreation, etc.). Models are then calibrated to predict the number of trip productions and attractions in each zone based on known census characteristics of the zone. These are referred to as "trip generation models." Other models are then calibrated to link trip origins to destinations and are referred to as "trip distribution models." "Modal split" models are then developed to predict how travelers will make choices on the transportation mode(s) to be used for each trip. Finally, "traffic assignment" models attempt to predict specific routes that will be used on a typical day. Such studies are normally conducted as part of periodic area-wide transportation planning projects, often in conjunction with a census.

8.7.2 Roadside Interview Studies

Some studies require that motorists be approached during a trip they are making. This is a tricky business because it causes disruption to individual drivers and can cause localized congestion, which annoys motorists. Such interviews need to be short and efficient, and they normally do not consist of more than five simple questions, usually involving such information as trip purpose, trip origin, trip destination, and route.

At locations where drivers will be stopped and interviewed, police security is virtually always required. A typical location might be a roadside location along a nonfreeway highway. Signing and traffic zones must be set up to adequately warn drivers that they may be pulled aside for a brief interview. Generally it is impossible to interview all vehicles passing a location, so police are used to "flag down" those that will be selected for interview. The interview in such cases should take no more than a minute to complete, to avoid undue disruption to individual drivers and to prevent long queues from forming at the interview site.

Drivers might also be stopped on entrance ramps to a freeway. At these locations, an attempt to interview all drivers is made, but each interview should be in the order of 15 seconds. Most often, because the entry point to the freeway is known, information on the destination exit, trip purpose, and trip frequency would be asked.

At signalized intersections, interviewers (multiple) can approach drivers stopped for a RED signal, and ask a series of short questions—which must be completed before the light turns GREEN.

A great deal of useful information can be amassed using roadside interview techniques. The studies must be conducted carefully, however, to avoid causing traffic jams, annoying drivers more than minimally, and to assure security for both motorists and interviewers.

8.7.3 Destination-Based Interview Studies

Another common interview technique is to focus on a known destination. In such cases, the trip destination is known, and interviewees need not be stopped while making their trip. Common locations include shopping centers, sports complexes, airports, beaches and other recreational facilities. Interview stations are established at key locations, and interviewees are asked to participate, as opposed to being obligated to participate by police. Questions on trip origin, route, mode choice (where such choice is available), trip purpose and frequency, and trip times are asked. Again, each interview should take no more than one or two minutes.

8.7.4 Statistical Issues

Chapter 7 addresses a number of statistical issues involved in interview studies, such as determination of sample size and expansion of interview data to reflect a full population.

8.8 Concluding Comments

This chapter attempts to provide a broad overview of the complex subject of data collection and reduction for traffic engineering studies. Subsequent chapters discuss specific types of studies in greater detail and present a more comprehensive picture of how specific types of data are analyzed and appropriate conclusions are drawn. The technology for collecting, storing and retrieval, and reduction of data continues to advance at a rapid pace, and you are encouraged to check the most current literature to get a more up-to-date view of the state of the art.

References

1. *Handbook of Simplified Practice for Traffic Studies*, Center for Transportation Research and Education, Iowa State University, IOWA DOT Project TR-455, November 2002.
2. Lutz, M. (Editor), *Handbook of Transportation Engineering*, McGraw-Hill, New York, 2004.
3. *Traffic Detector Handbook*, 3rd Edition, Federal Highway Administration, Publication No. FHWA-HRT-06-108, Washington DC, 2006.
4. Roess, R., "Analysis of Freeway Weaving Sections." *Final Report*, National Cooperative Highway Research Project 3-75, Polytechnic University, Brooklyn NY, January 2008.
5. *NGSIM Overview*, Federal Highway Administration Publication FHWS-HRT-06-0135, Federal Highway Administration, Washington DC, December 2006.

Problems

1. Traffic volumes on a four-lane freeway (two lanes in each direction) were counted manually from an overhead location, resulting in the data shown here. The desire was to obtain continuous 15-minute counts for each lane of the freeway for a two-hour period surrounding the morning peak hour.

Data for Problem 1

	Eastbound		Westbound	
Time of Count (PM)	**Lane 1**	**Lane 2**	**Lane 1**	**Lane 2**
4:00–4:12		360		310
4:15–4:27	350		285	
4:30–4:42		380		330
4:45–4:57	370		300	
5:00–5:12		370		340
5:15–5:27	345		280	
5:30–5:42		340		310
5:45–5:57	320		260	

From the data shown, determine the following:

(a) Continuous 15-minute volumes for each period and each lane.

(b) The peak hour, peak hour volume, and peak hour factor (PHF) for each direction of flow and for the freeway as a whole.

(c) Directional flow rates during each 15-minute count period.

2. A 24-hour count using a road tube at a rural highway location produces a count of 11,250 actuations. A representative sample count to classify vehicles resulted in the data shown here:

Data for Problem 2

Number of Axles Per Vehicle	Number Observed
2	157
3	55
4	50
5	33
6	8

Based on this sample classification count, how many vehicles were observed during the 24-hour study?

3. A speed study was conducted using manual observation with stopwatches. The trap length was marked for 400 feet. The observer lined up with the entry boundary line, 60 feet away from the vehicles being observed. When viewing the exit boundary, the sight line was at a 60° angle from a perpendicular line from the observer location to the entry boundary. If one of the observed travel times to traverse the trap is 8.3 seconds, at what speed is the vehicle traveling?

4. Briefly describe the types of permanent detectors available for use in traffic studies. Note the benefits and drawbacks of each.

CHAPTER 9

Volume Studies and Characteristics

The most fundamental measure in traffic engineering is volume: how many vehicles are passing defined locations in the roadway system over time, particularly during the peak hour(s) of a typical day. Virtually no decision concerning facility design or traffic control options can be made without knowledge of existing and projected traffic volumes for the location(s) under study.

9.1 Critical Parameters

In Chapter 5, we introduced the concepts of volume and flow rate. Four variables are related to volume:

- Volume
- Rate of flow
- Demand
- Capacity

Sometimes these are used in conjunction with other measures or conditions. The four parameters listed are closely related, and all are expressed in terms of the same or similar units. They are *not,* however, the same.

1. *Volume* is the number of vehicles (or persons) passing a point during a specified time period, which is usually one hour but need not be.
2. *Rate of flow* is the rate at which vehicles (or persons) pass a point during a specified time period less than one hour, expressed as an equivalent hourly rate.
3. *Demand* is the number of vehicles (or persons) that desire to travel past a point during a specified period (also usually one hour). Demand is frequently higher than actual volumes where congestion exists. Some trips divert to alternative routes, and other trips are simply not made.
4. *Capacity* is the maximum rate at which vehicles can traverse a point or short segment during a specified time period. It is a characteristic of the roadway. Actual volume can never be observed at levels higher than the true capacity of the section. However, such results may appear because capacity is most often estimated using standard analysis procedures of the *Highway Capacity Manual* [1]. These estimates may indeed be too low for some locations.

Techniques for collection, reduction, and presentation of volume (and other) traffic data were discussed in Chapter 8. This chapter presents techniques for statistical analysis of volume data and the interpretation and presentation of study results. It also provides an overview of typical volume characteristics found on most highway systems.

9.2 Volume, Demand, and Capacity

We have noted that volume, demand, and capacity are three different measures, even though all are expressed in the same units and may relate to the same location. In practical terms, volume is what *is,* demand is what motorists would like *to be,* and capacity is the physical limit of what *is possible.* In very simple terms, if vehicles were counted at any defined location for one hour:

- Volume would be the number of vehicles counted passing the study location in the hour.
- Demand would be the volume plus the vehicles of motorists wishing to pass the site during the study hour who were prevented from doing so by congestion. The latter would include motorists in queue waiting to reach the study location, motorists using alternative routes to avoid the congestion around the study location, and motorists deciding not to travel at all due to the existing congestion, or choosing to travel to an alternative destination.
- Capacity would be the maximum volume that could be accommodated by the highway at the study location.

Consider the illustration of Figure 9.1. It shows a classic bottleneck location on a freeway, in this case consisting of a major merge area. For each approaching leg, and for the downstream freeway section, the actual volume (v), the demand (d), and the capacity (c) of the segment are given. Capacity is the primary constraint on the facility. As shown in Figure 9.1, the capacity is 2,000 veh/h/ln, so that the capacity of the two-lane approach legs are 4,000 veh/h each, and the capacity of the downstream freeway, which has three lanes, is 6,000 veh/h.

Assuming that the stated capacities are correct, no volume in excess of these capacities can ever be counted. Simply put, you can't carry 6 gallons of water in a 5-gallon bucket. Therefore, it is informative to consider what would be observed for the situation as described. On Approach 1, the true demand is 3,800 veh/h and the capacity is 4,000 veh/h. On Approach 2, the true demand is 3,600 veh/h and the capacity is also 4,000 veh/h. There is no capacity deficiency on either approach. Downstream of the merge, however, the capacity is 6,000 veh/h, but the sum of the approaching demands is 3,800 + 3,600 = 7,400 veh/h. This exceeds the capacity of the segment.

Given this scenario, what can we expect to observe?

- Any volume count downstream of the merge cannot exceed 6,000 veh/h for as long as the illustrated conditions exist. A count of 6,000 veh/h is expected.
- Because of the capacity deficiency downstream of the merge, a queue of vehicles will begin to form and propagate upstream on both approaches.
- If a count of entering vehicles on both approaches is taken upstream of the forming queues, the true demand would be counted on each approach, assuming there has been no diversion of vehicles to alternative routes.
- If a count of approaching vehicles is taken within the forming queues, they would be unstable in both time and space, but their total should not exceed 6,000 veh/h, the capacity of the downstream freeway section.

A final question is also interesting: Given that queues are observed on both approaches, is it reasonable to assume

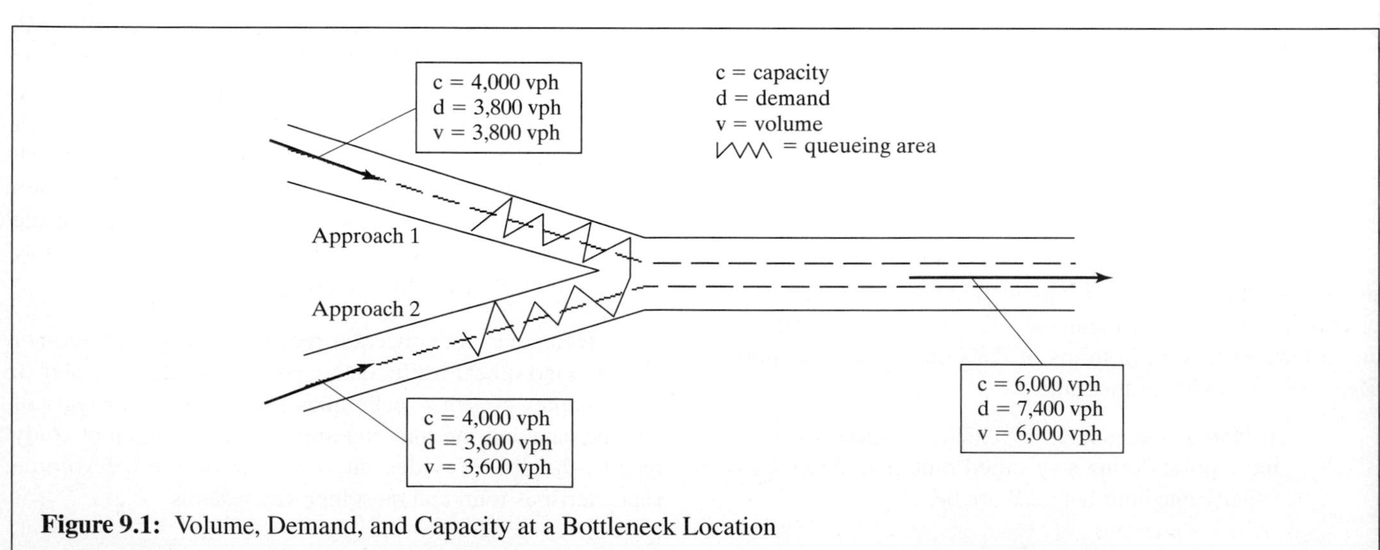

Figure 9.1: Volume, Demand, and Capacity at a Bottleneck Location

that the downstream count of 6,000 veh/h is a *direct measurement* of the capacity of the section? This issue involves a number of subtleties.

The existence of queues on both approaches certainly suggests that the downstream section has experienced capacity flow. Capacity, however, is defined as the maximum flow rate that can be achieved under stable operating conditions (i.e., without breakdown). Thus capacity would most precisely be the flow rate for the period immediately preceding the formation of queues. After the queues have formed, flow is in the "queue discharge" mode, and the flow rates and volumes measured may be equal to, less than, or even more than capacity. In practical terms, however, the queue discharge capacity may be more important than the stable-flow value, which is, in many cases, a transient that cannot be maintained for long periods of time.

As you can see from this illustration, volume (or rate of flow) can be counted anywhere and a result achieved. In a situation where queuing exists, it is reasonable to assume that downstream flows represent either capacity or queue discharge conditions. Demand, however, is much more difficult to address. Although queued vehicles can be added to counts, this is not necessarily a measure of true demand. True demand contains elements that go well beyond queued vehicles at an isolated location. Determining true demand requires an estimation of how many motorists changed their routes to avoid the subject location. It also requires knowledge of motorists who either traveled to alternative destinations or who simply decided to stay home (and not travel) as a result of congestion.

Figure 9.2 illustrates the impact of a capacity constraint on traffic counts. Part (a) shows a plot of demand and capacity. The demand shown would be observable if it were not

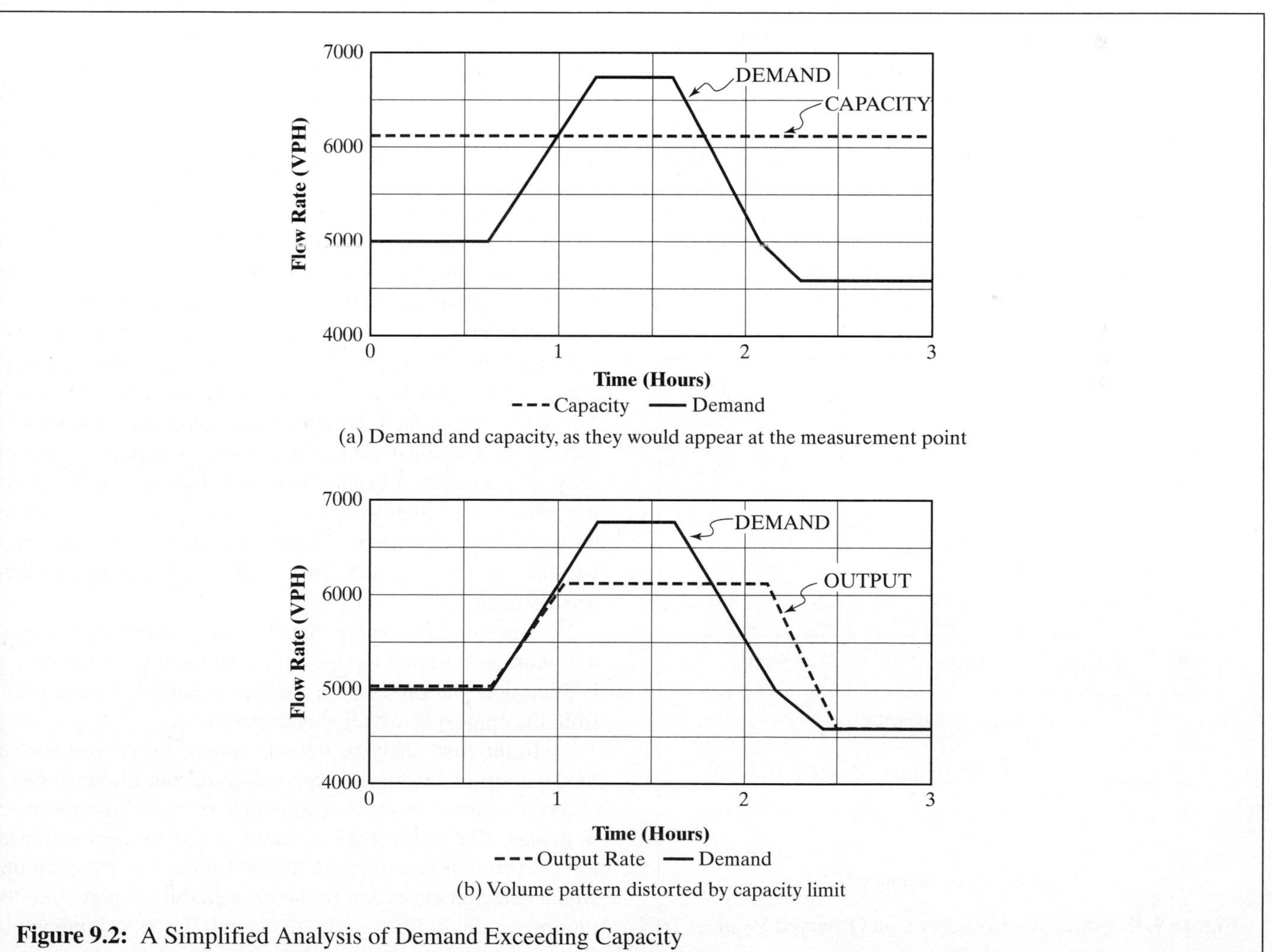

(a) Demand and capacity, as they would appear at the measurement point

(b) Volume pattern distorted by capacity limit

Figure 9.2: A Simplified Analysis of Demand Exceeding Capacity

clear that a capacity constraint is present. Part (b) shows what will actually occur. Volume can never rise to a level higher than capacity. Thus actual counts peak at capacity. Because not all vehicles arriving can be accommodated, the peak period of flow is essentially lengthened until all vehicles can be served. The result is that observed counts will indicate that the peak flow rate is approximately the same as capacity and that it occurs over an extended period of time. The volume distribution looks as if someone took the demand distribution and flattened it out with their hand.

The difference between observed volume counts and true demand can have some interesting consequences when the difference is not recognized and included in planning and design. Figure 9.3 illustrates an interesting case of a freeway section consisting of four ramps.

From Figure 9.3, the arriving demand volume on Segment 3 of 3,700 veh/h exceeds its capacity of 3,400 veh/h. From the point of view of counts taken within each segment, 3,400 veh/h will be observed in Segment 3. Because only 3,400 veh/h are output from Segment 3, the downstream counts in Segments 4 and 5 will be lower than their true demand. In this case, the counts shown reflect a proportional distribution of volume to the various ramps, using the same distribution as reflected in the demand values. The capacities of Segments 4 and 5 are not exceeded by these counts. Upstream counts in Segments 1 and 2 will be unstable and will reflect the transient state of the queue during the count period.

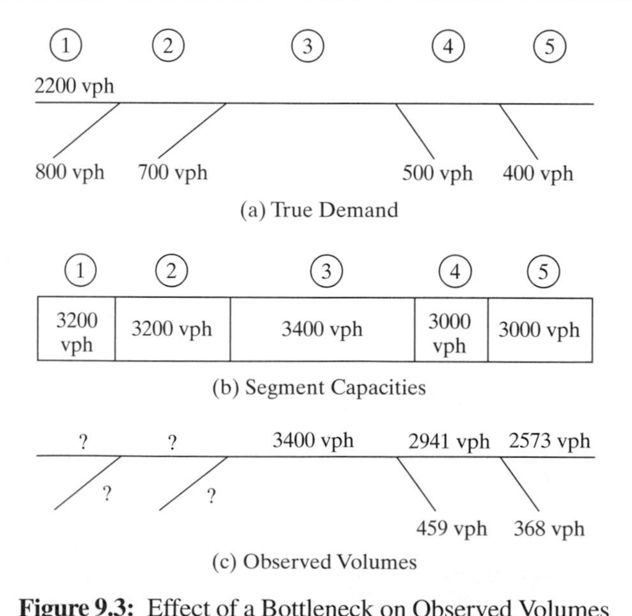

Figure 9.3: Effect of a Bottleneck on Observed Volumes

Assume that as a result of this study, a decision is made to add a lane to Segment 3, essentially increasing its capacity to a value larger than the demand of 3,700 veh/h. Once this is done, the volume now discharged into Segment 4 is 3,200 veh/h, more than the capacity of 3,000 veh/h. This secondary bottleneck, often referred to as a "hidden bottleneck," was not apparent in the volume data originally obtained. It was not obvious because the existing demand was constrained from reaching the segment due to an upstream bottleneck. Such a constraint is often referred to as "demand starvation."

In designing corrective highway improvements, it is critical that all downstream points be evaluated properly to identify such hidden bottlenecks. In the case illustrated, the improvement project would have to address *both* the existing and hidden bottlenecks to achieve a successful result.

The case of a freeway bottleneck is relatively simple to analyze because the number of entry and exit points are limited and the number of available alternative routes is generally small. On arterials, however, the situation is far more complex because every intersection represents a diversion opportunity, and the number of alternative routes is generally quite large. Thus the demand response to an arterial bottleneck is much harder to discern. An arterial may also have a number of overlapping bottlenecks, further complicating the analysis. If several consecutive signalized intersections, for example, are failing, it is difficult to trace the impacts. Is an upstream signal apparently failing because it is inadequate, or because a queue from the downstream signal has blocked the intersection?

On arterial systems, it is often impossible to measure existing demand, except in cases where there are no capacity constraints. Later in this chapter, a method for discerning intersection approach demand from observed volumes in a capacity-constrained case is discussed. It applies, however, only to an isolated breakdown and does not account for the effects of diversion. Congestion in a surface street network severely distorts demand patterns, and observed volumes are more a reflection of capacity constraints than true demand.

You should be aware that the terms *volume* and *demand* are often used imprecisely, even in the technical literature. It is often necessary to discern the true meaning of these terms from the context in which they are used.

In the final analysis, volume counts always result in an observation of "volume." Depending on the circumstances, observed volumes may be equivalent to demand, to capacity, or to neither. The traffic engineer must, however, gain sufficient insight through counting studies and programs to recognize which situation exists and to incorporate this properly into the interpretation of the study data and the development of improvement plans.

9.3 Volume Characteristics

If traffic distributed itself uniformly among the $365 \times 24 = 8{,}760$ hours of the year, there is not a location in the nation that would experience congestion or significant delay. The problem for traffic engineers, of course, is that there are strong peaks during a typical day, caused primarily by commuters going to and from work. Depending on the specific region and location, the peak hour of the day typically contains from 9% to 15% of the 24-hour volume. In remote or rural areas, the percentage can go much higher, but the volumes are much lower in these surroundings.

The traffic engineer, therefore, must deal with the travel preferences of our society in planning, designing, and operating highway systems. In some dense urban areas, policies to induce spreading of the peak have been attempted, including the institution of flex-hours or days and/or variable pricing policies for toll and parking facilities. Nevertheless, the traffic engineer must still face the fundamental problem: Traffic demand varies in time in ways that are quite inefficient. Demand varies by time of day, by day of the week, by month or season of the year, and in response to singular events (both planned and unplanned) such as construction detours, accidents or other incidents, and even severe weather. Modern intelligent transportation system (ITS) technologies increasingly try to manage demand on a real-time basis by providing information on routes, current travel times, and related conditions directly to drivers. This is a rapidly growing technology sector, but its impacts have not yet been well documented.

One of the many reasons for doing volume studies is to document these complex variation patterns and to evaluate the impact of ITS technologies and other measures on traffic demand.

9.3.1 Hourly Traffic Variation Patterns: The Phenomenon of the Peak Hour

When hourly traffic patterns are contemplated, we have been conditioned to think in terms of two "peak hours" of the day: morning and evening. Dominated by commuters going to work in the morning (usually between 7 AM and 10 AM) and returning in the evening (usually between 4 PM and 7 PM), these patterns tend to be repetitive and more predictable than other facets of traffic demand. This so-called typical pattern holds only for weekday travel, and modern evidence may suggest that this pattern is not as typical as we have been inclined to accept.

Figure 9.4 shows a number of hourly variation patterns documented in the *Highway Capacity Manual* [*1*], compiled from References 2 and 3. In part (a) of Figure 9.4, hourly distributions for rural highways are depicted. Only the weekday pattern on a local rural route displays the expected AM and PM peak patterns. Intercity, recreational, and local weekend traffic distributions have only a single, more dispersed peak occurring across the mid- to late afternoon. In part (b), weekday data from four urban sites are shown in a single direction. Sites 1 and 3 are in the opposite direction from Sites 2 and 4, which are only two blocks apart on the same facility. Whereas Sites 2 and 4 show clear AM peaks, traffic after the peak stays relatively high and surprisingly –uniform for most of the day. Sites 1 and 3, in the opposite direction, show evening peaks, with Site 3 also displaying considerable off-peak hour traffic volume. Only Site 1 shows a strong PM peak with significantly less traffic during other portions of the day.

The absence of clear AM and PM peaks in many major urban areas is a spreading phenomenon. On one major facility, the Long Island Expressway (I-495) in New York, a recent study showed that on a typical weekday, only one peak was discernible in traffic volume data—and it lasted for 10 to 12 hours per day. This characteristic is a direct result of system capacity constraints. Everyone who would like to drive during the normal peak hours cannot be accommodated. Because of this, individuals begin to make travel choices that allow them to increasingly travel during the "off-peak" hours. This process continues until off-peak periods are virtually impossible to separate from peak periods.

Figure 9.4 (b) displays another interesting characteristic of note. The outer lines of each plot show the 95% confidence intervals for hourly volumes over the course of one year. Traffic engineers depend on the basic repeatability of peak-hour traffic demands. The variation in these volumes in Figure 9.4 (b), however, is not insignificant. During the course of any given year, there are 365 peak hours at any location, one for each day of the year. This is question for the traffic engineer: Which one should be used for planning, design, and operations?

Figure 9.5 shows plots of peak-hour volumes (as a percentage of annual average daily travel [ADT]) in decreasing order for a variety of facilities in Minnesota. In all cases, there is clearly a "highest" peak hour of the year. The difference between this highest peak and the bulk of the year's peak hours, however, depends on the type of facility. The recreational route has the greatest disparity.

This is not unexpected because traffic on such a route will tend to have enormous peaks during the appropriate season on weekends, with far less traffic on a "normal" day. The main rural route has less of a disparity, as at least some component of traffic consists of regular commuters. Urban roadways show far less of a gap between the highest hour and the bulk of peak hours.

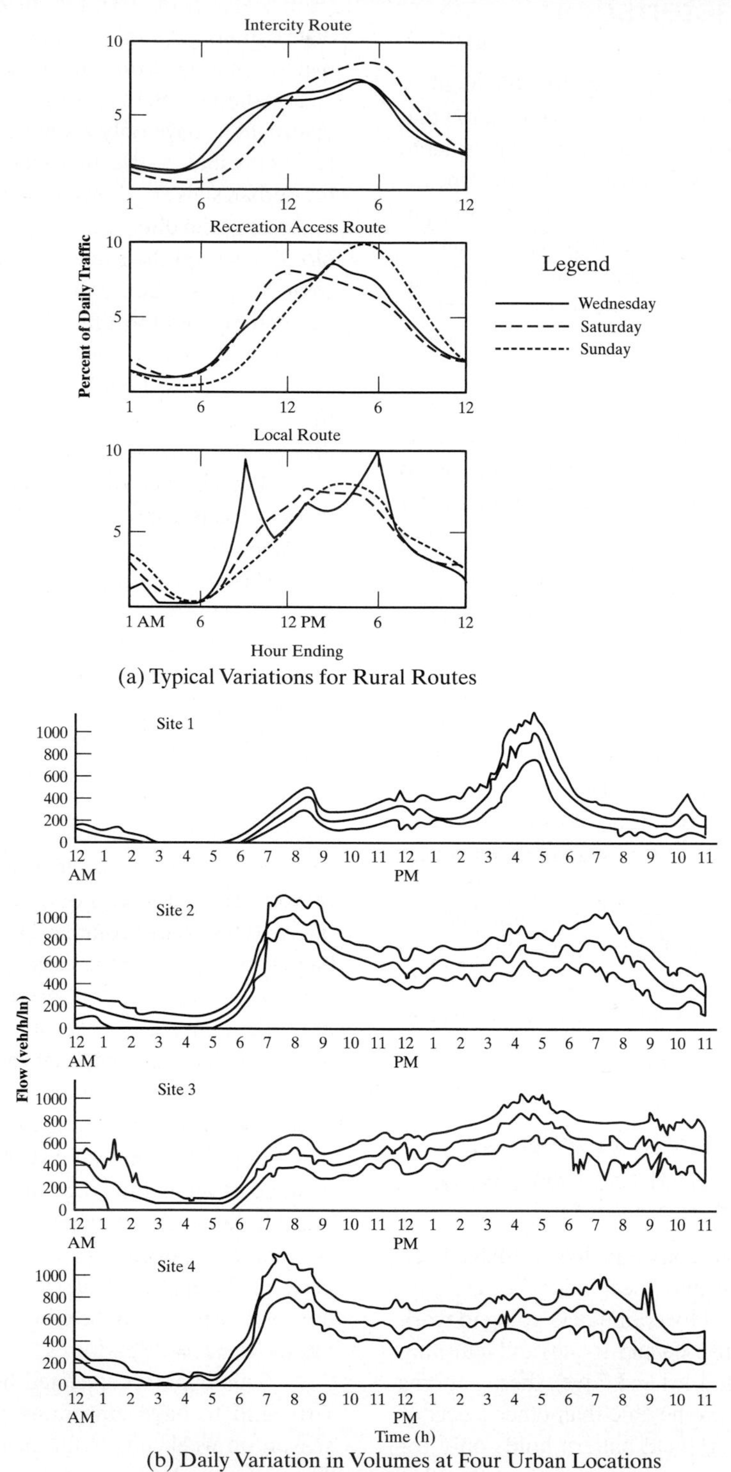

Figure 9.4: Examples of Hourly Volume Variation Patterns

(*Source:* Used with permission of Transportation Research Board, *Highway Capacity Manual,* 4th Edition, Washington DC, 2000, Exhibits 8-6 and 8-7, pp. 8-6 and 8-7, repeated from References 2 and 3.)

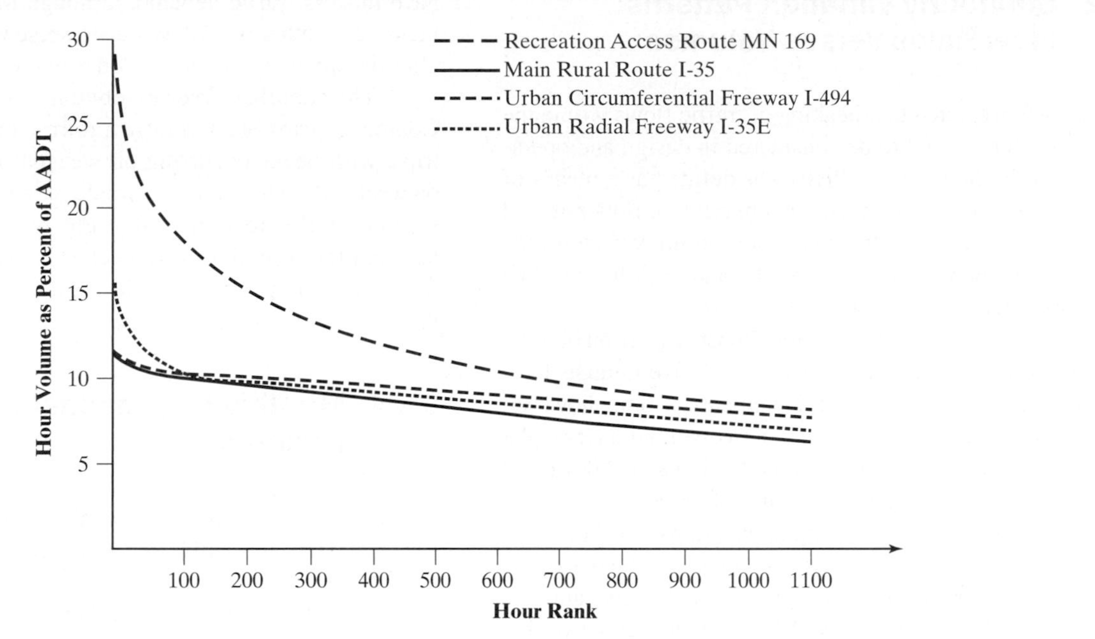

Figure 9.5: Peak Hours as a Percentage of AADT

(*Source:* Used with permission of Transportation Research Board, *Highway Capacity Manual,* 4th Edition, Washington DC, 2000, Exhibit 8-8, p. 8-8.)

It is interesting to examine the various peak hours for the types of facilities illustrated in Figure 9.5, which represents data from various facilities in Minnesota. Table 9.1 tabulates the percentage of AADT occurring within designated peak hours for the facility types represented.

The choice of which peak hour to use as a basis for planning, design, and operations is most critical for the recreational access route. In this case, the highest hour of the year carries twice as much traffic as the 200th peak hour of the year and 1.36 times that of the 30th hour of the year. In the two urban cases, the highest hour of the year is only 1.2 times the 200th highest hour.

Historically, the 30th highest hour has been used in rural planning, design, and operations. There are two primary arguments for such a policy: (1) the target demand would be exceeded only 29 times per year, and (2) the 30th peak hour generally marks a point where subsequent peak hours have similar volumes. The latter defines a point on many relationships where the curve begins to "flatten out," a range of demands where it is deemed economic to invest in additional roadway capacity.

In urban settings, the choice of a design hour is far less clear and has far less impact. Typical design hours selected range from the 30th highest hour to the 100th highest hour. For the facilities of Figure 9.5, this choice represents a range from 10.5% to 10.0% of AADT. With an AADT of 80,000 veh/day, for example, this range is a difference of only 400 veh/h in demand.

Table 9.1: Key Values from Figure 9.5

Type of Facility	Percent of AADT Occurring in the ___ Peak Hour			
	1st	**30th**	**100th**	**200th**
Recreational Access	30.0%	22.0%	18.0%	15.0%
Main Rural	15.0%	13.0%	10.0%	9.0%
Urban Circumferential Freeway	11.5%	10.5%	10.0%	9.5%
Urban Radial Freeway	11.5%	10.5%	10.0%	9.5%

9.3.2 Subhourly Variation Patterns: Flow Rates Versus Volumes

In Chapter 5, we noted that peaking of traffic flows within the peak hour often needed to be considered in design and operations. The peak hour factor (PHF) was defined as a means of quantifying the difference between a maximum flow rate and the hourly volume within the peak hour. Figure 9.6 shows the difference among 5-minute, 15-minute, and peak hourly flow rates from a freeway location in Minnesota.

Flow rates can be measured for almost any period of time. For research purposes, periods from one to five minutes have frequently been used. Very small increments of time, however, become impractical at some point. In a two-second interval, the range of volumes in a given lane would be limited to "0" or "1," and flow rates would be statistically meaningless.

For most traffic engineering applications, 15 minutes is the standard time period used, primarily based on the belief that this is the shortest period of time over which flow rates are "statistically stable." Statistically stable implies that reasonable relationships can be calibrated among flow parameters, such as flow rate, speed, and density. In recent years, there is some thought that five-minute flow rates might qualify as statistically stable, particularly on freeway facilities. Practice, however, continues to use 15 minutes as the standard period for flow rates.

The choice, however, has major implications. In Figure 9.6, the highest 5-minute rate of flow is 2,200 veh/h/ln; the highest 15-minute rate of flow is 2,050 veh/h/ln; the peak hour volume is 1,630 veh/h/ln. Selecting a 15-minute base period for design and analysis means that, in this case, the demand flow rate (assuming no capacity constraints) would be 2,050 veh/h/ln. This value is 7% lower than the peak five-minute flow rate and 20% higher than the peak-hour volume. In real design terms, these differences could translate into a design with one more or fewer lanes or differences in other geometric and control features. The use of 15-minute flow periods also implies that breakdowns of a shorter duration do not cause the kinds of instabilities that accompany breakdowns extending for 15 minutes or more.

9.3.3 Daily Variation Patterns

Traffic volumes also conform to daily variation patterns that are caused by the type of land uses and trip purposes served by the facility. Figure 9.7 illustrates some typical relationships.

The recreational access route displays strong peaks on Fridays and Sundays. This is a typical pattern for such routes because motorists leave the city for recreational areas on Fridays, returning on Sundays. Mondays through Thursdays have far less traffic demand, although Monday is somewhat higher than other weekdays due to some vacationers returning after the weekend rather than on Sunday.

The suburban freeway obviously caters to commuters. Commuter trips are virtually a mirror image of recreational trips, with peaks occurring on weekdays and lower demand on weekends. The main rural route in this exhibit has a pattern similar to the recreational route but with less variation between the weekdays and weekends. The route serves both recreational and commuter trips, and the mix tends to dampen the amount of variation observed.

9.3.4 Monthly or Seasonal Variation Patterns

Figure 9.8 illustrates typical monthly volume variation patterns. Recreational routes will have strong peaks occurring during the appropriate seasons (i.e., summer for beaches, winter for skiing). Commuter routes often show similar patterns with less variability. In Figure 8.8, recreational routes display monthly ADTs that range from 77% to 158% of the AADT. Commuter routes, although showing similar peaking periods, have monthly ADTs ranging from 82% to 119% of AADT.

It might be expected that commuter routes would show a trend opposite to recreational routes (i.e., if recreational routes are peaking in the summer, then commuter routes should have less traffic during those periods). The problem is that few facilities are purely recreational or commuter; there is always some mix present. Further, much recreational travel is done by inhabitants of the region in question; the same motorists may be part of both the recreational and commuter demand during the same months. There are, however, some areas in which commuter traffic does clearly decline during summer recreational months. The distributions shown here are illustrative; different distributions are possible, and they do occur in other regions.

9.3.5 Some Final Thoughts on Volume Variation Patterns

One of the most difficult problems in traffic engineering is that we are continually planning and designing for a demand that represents a peak flow rate within a peak hour on a peak day during a peak season. When we are successful, the resulting facilities are underused most of the time.

It is only through the careful documentation of these variation patterns, however, that the traffic engineer can know the impact of this underutilization. Knowing the volume variation patterns governing a particular area or location is critical to

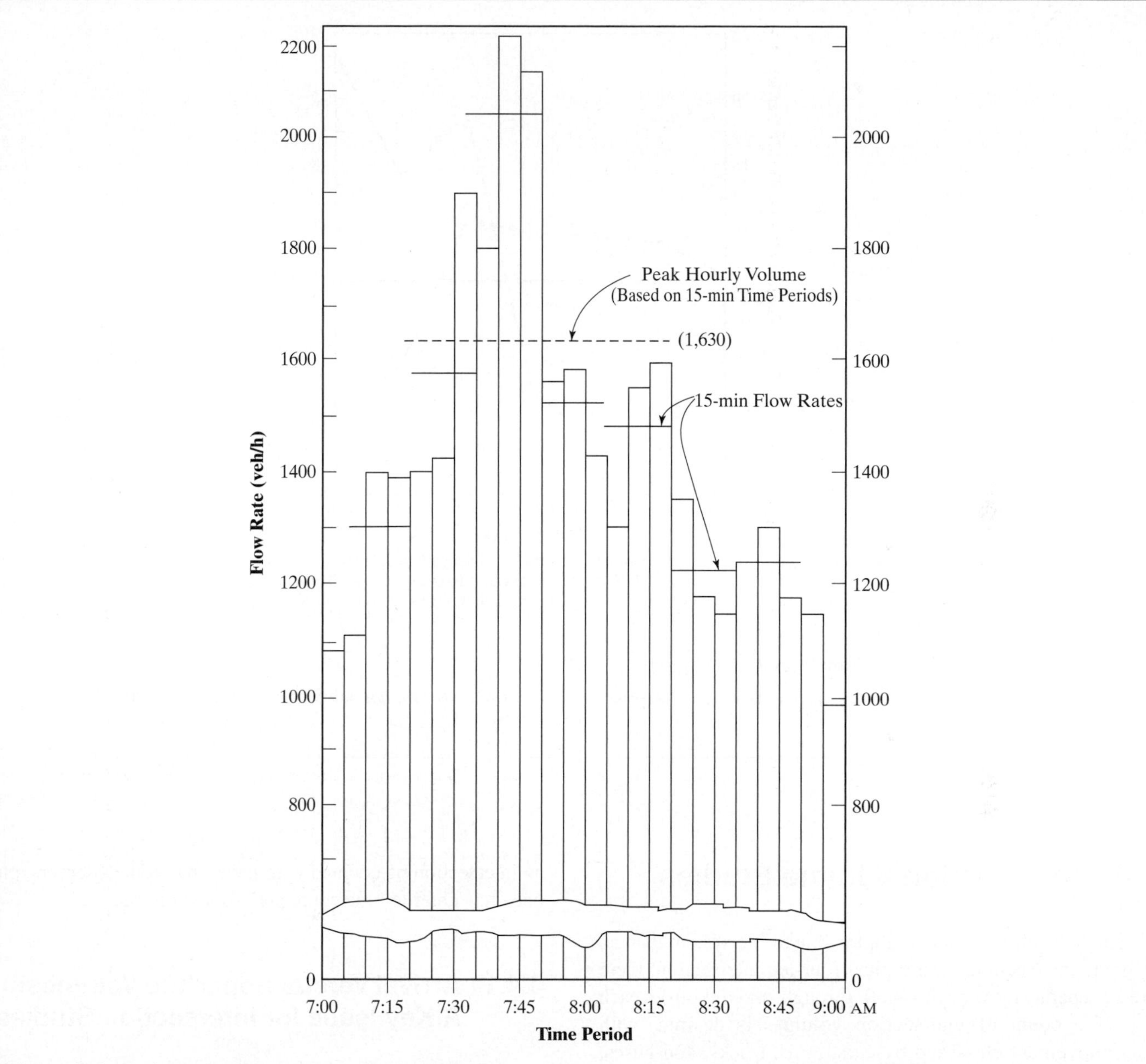

Figure 9.6: Variations of Flow Within the Peak Hour

(*Source:* Used with permission of Transportation Research Board, *Highway Capacity Manual,* 4th Edition, Washington DC, 2000, Fig 8-10, p. 8-10.)

finding appropriate design and control measures to optimize operations. It is also important to document these patterns so that estimates of an AADT can be discerned from data taken for much shorter time periods. It is simply impractical to count every location for a full year to determine AADT and related demand factors. Counts taken over a shorter period of time can, however, be adjusted to reflect a yearly average or a peak occurring during another part of the year, if the variation patterns are known and well documented. These concepts are illustrated and applied in the sections that follow.

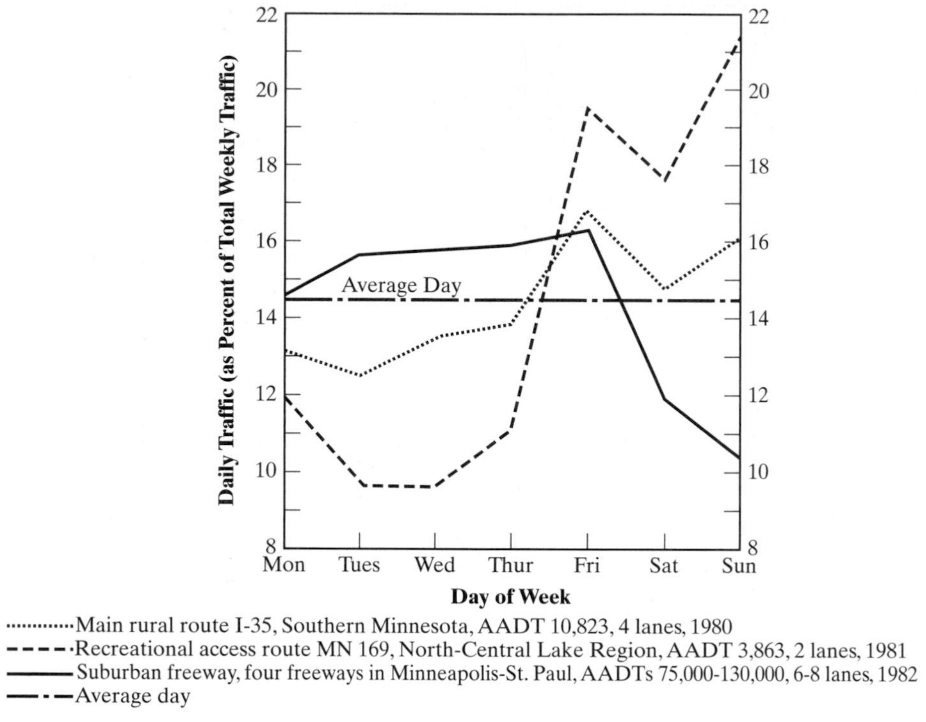

Figure 9.7: Typical Daily Volume Variation Patterns

(*Source:* Used with permission of Transportation Research Board, *Highway Capacity Manual,* 4th Edition, Washington DC, 2000, Exhibit 8-4, p. 8-5.)

9.4 Intersection Volume Studies

No single location is more complex in a traffic system than an at-grade intersection. At a typical four-leg intersection, there are 12 separate movements—left, through, and right from each leg. If a count of intersection volumes is desired, with each movement classified by cars, taxis, trucks, and buses, each count period requires the observation of $12 \times 4 = 48$ separate pieces of data.

When intersections are counted manually (and they often are), observers must be positioned properly to see the movements they are counting. It is doubtful that an inexperienced counter could observe and classify more than one major or two minor movements simultaneously. For heavily used multilane approaches, it may be necessary to use separate observers for different lanes. In manual intersection studies, short-break and alternating-period approaches are almost always combined to reduce the number of observers needed. Rarely, however, can an intersection be counted with fewer than four observers, plus one crew chief to time count periods and breaks.

9.4.1 Arrival Versus Departure Volumes: A Key Issue for Intersection Studies

At most intersections, volumes are counted as they depart the intersection. This is done both for convenience and because turning movements cannot be fully resolved until vehicles exit the intersection. Although this approach is fine where there is no capacity constraint (i.e., an unstable buildup of queues on the approach), it is not acceptable where demand exceeds the capacity of the approach. In such cases, it is necessary to observe *arrival* volumes because these are a more accurate reflection of demand.

At signalized intersections, "unstable queue buildup" is detected when vehicles queued during a red interval are not fully

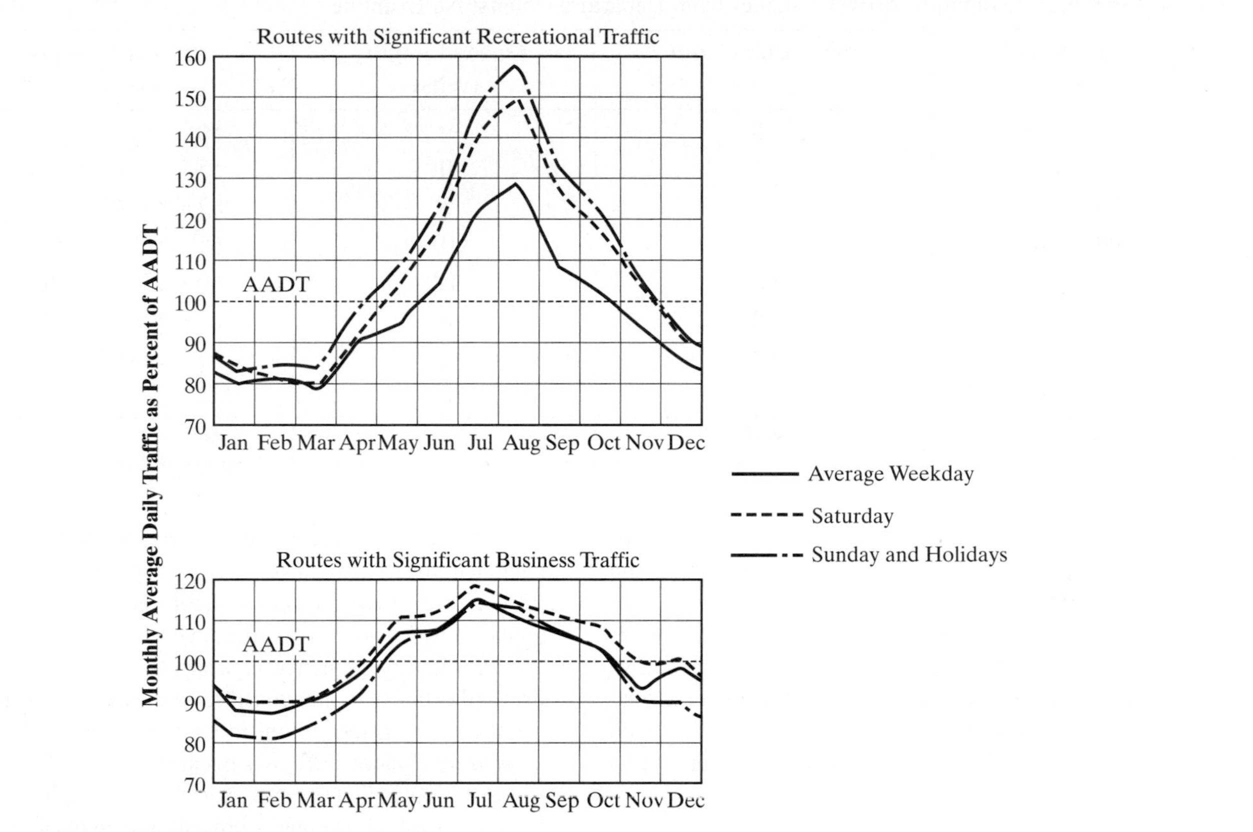

Figure 9.8: Typical Monthly Variation Patterns

(*Source:* Used with permission of Transportation Research Board, *Highway Capacity Manual,* 4th Edition, Washington DC, 2000, Exhibit 8-2, p. 8-3.)

cleared during the next green interval. At unsignalized intersections, "unstable queue buildup" can be identified by queues that become larger during each successive counting period.

Direct observation of arrival volumes at an intersection is difficult because the queue is dynamic. As the queue grows and declines, the point of "arrival" changes. Therefore, the technique used to count arrival volumes is to count departure volumes and the number of queued vehicles at periodic intervals. For signalized approaches, the size of the queue would be recorded *at the beginning of each red phase.* This identifies the "residual queue" of vehicles that arrived during the previous signal cycle but were not serviced. For unsignalized approaches, the queue is counted at the end of each count period. When such an approach is followed, the arrival volume is estimated as follows:

$$V_{ai} = V_{di} + N_{qi} - N_{q(i-1)} \tag{9-1}$$

where: V_{ai} = arrival volume during period i, vehs
V_{di} = departure volume during period i, vehs
N_{qi} = number of queued vehicles at the end of period i, vehs
$N_{q(i-1)}$ = number of queued vehicles at the end of period i-1, vehs

Estimates of arrival volume using this procedure identify only the localized arrival volume. This procedure *does not* identify diverted vehicles or the number of trips that were not made due to general congestion levels. Thus although arrival volumes do represent localized demand, they do not measure diverted or repressed demand. Table 9.2 shows sample study data using this procedure to estimate arrival volumes.

Note that the study is set up so the first and last count periods do not have residual queues. Also, the total departure and arrival counts are the same, but the conversion from

Table 9.2: Estimating Arrival Volumes from Departure Counts: An Example

Time Period (PM)	Departure Count (vehs)	Queue Length (vehs)	Arrival Volume (vehs)
4:00–4:15	50	0	50
4:15–4:30	55	0	55
4:30–4:45	62	5	$62 + 5 = 67$
4:45–5:00	65	10	$65 + 10 - 5 = 70$
5:00–5:15	60	12	$60 + 12 - 10 = 62$
5:15–5:30	60	5	$60 + 5 - 12 = 53$
5:30–5:45	62	0	$62 - 5 = 57$
5:45–6:00	55	0	55
Total	**469**		**469**

departures to arrivals causes a shift in the distribution of volumes by time period. Based on departure counts, the maximum 15-minute volume is 65 vehicles, or a flow rate of 65/0.25 = 260 veh/h. Using arrival counts, the maximum 15-minute volume is 70, or a flow rate of 70/0.25 = 280 veh/h. The difference is important because the higher arrival flow rate (assuming that the study encompasses the peak period) represents a value that would be valid for use in planning, design, or operations.

9.4.2 Special Considerations for Signalized Intersections

At signalized intersections, count procedures are both simplified and more complicated at the same time. For manual observers, the signalized intersection simplifies counting because not all movements are flowing at the same time. An observer who can normally count only one through movement at a time could actually count two such movements in the same count period by selecting, for example, the eastbound and northbound through movements. These two operate during different phases of the signal.

Count periods at signalized intersections, however, must be equal multiples of the cycle length. Further, actual counting times (exclusive of breaks) must also be equal multiples of the cycle length. This is to guarantee that all movements get the same number of green phases within a count period. Thus, for a 60-second signal cycle, a 4 of 5-minute counting procedure may be employed. For a 90-second cycle, however, neither 4 nor 5 minutes are equal multiples of 90 seconds (1.5 minutes). For a 90-second cycle, a counting process of 12 of 15 minutes would be appropriate, as would 4.5 of 6 minutes.

Actuated signals present special problems because both cycle lengths and green splits vary from cycle to cycle. Count periods are generally set to encompass a minimum of five signal cycles, using the maximum cycle length as a guide. The actual counting sequence is arbitrarily chosen to reflect this principle, but it is not possible to assure equal numbers of phases for each movement in each count period. This is not viewed as a major difficulty because the premise of actuated signalization is that green times should be allocated proportionally to vehicle demands present during each cycle.

9.4.3 Presentation of Intersection Volume Data

Intersection volume data may be summarized and presented in a variety of ways. Simple tabular arrays can summarize counts for each count period by movement. Breakdowns by vehicle type are also most easily depicted in tables. More elaborate graphic presentations are most often prepared to depict peak-hour and/or full-day volumes. Figures 9.9 and 9.10 illustrate common forms for display of peak-hour or daily data. The first is a graphic intersection summary diagram that allows simple entry of data on a predesigned graphic form. The second is an intersection flow diagram in which the thickness of flow lines is based on relative volumes.

9.5 Limited Network Volume Studies

Consider the following proposition: A volume study is to be made covering the period from 6 AM to 12 midnight on the

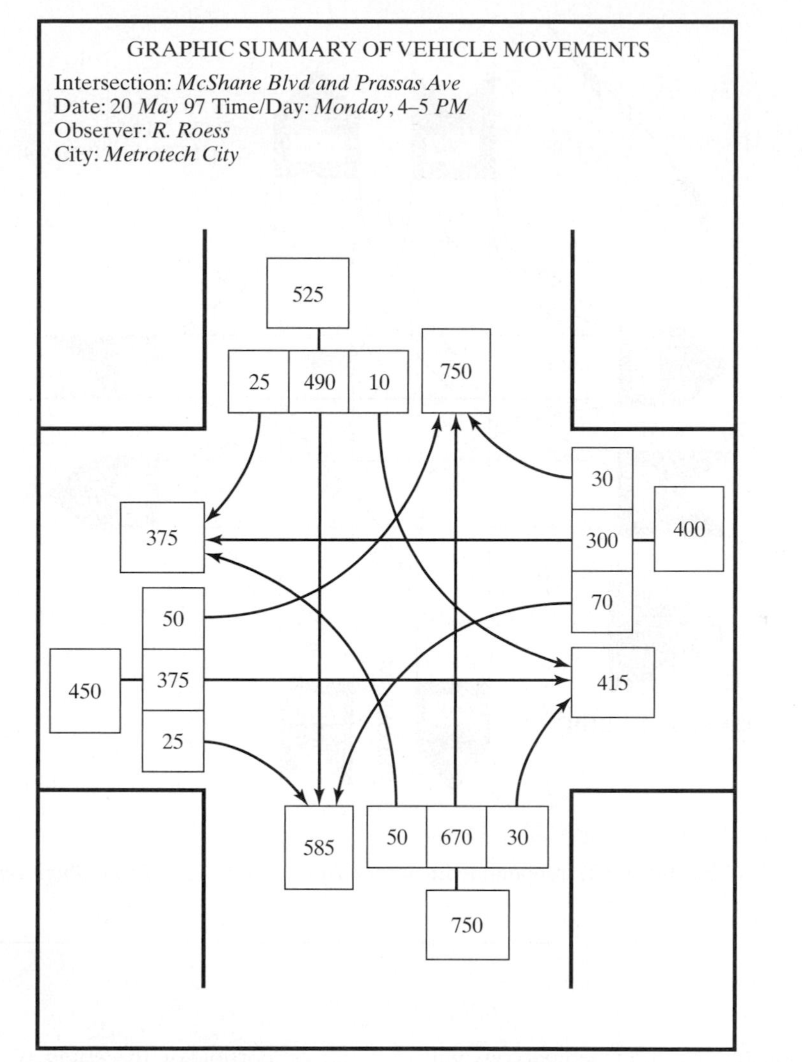

Figure 9.9: Graphic Intersection Summary Diagram

street network comprising midtown Manhattan (i.e., from 14th Street to 59th Street, 1st Avenue to 12th Avenue). Although this is a very big network, including over 500 street links and 500 intersections, it is not the entire city of New York, nor is it a statewide network.

Nevertheless, the size of the network is daunting for a simple reason: It is virtually impossible to acquire and train sufficient personnel to count all of these locations at the same time. Further, it would be impractically expensive to try and acquire sufficient portable counting equipment to do so. To conduct this study, it will be necessary to employ *sampling* techniques (i.e., not all locations within the study area will be counted at the same time or even on the same day). Statistical manipulation based on these samples will be required to produce an hourly volume map of the network for each hour of the intended survey period, or for an average peak period.

Such "limited" networks exist in both small towns and large cities and around other major trip generators, such as airports, sports facilities, shopping malls, and other activity centers. Volume studies on such networks involve individual planning and some knowledge of basic characteristics, such as location of major generators and the nature of traffic on various facilities (local versus through users, for example).

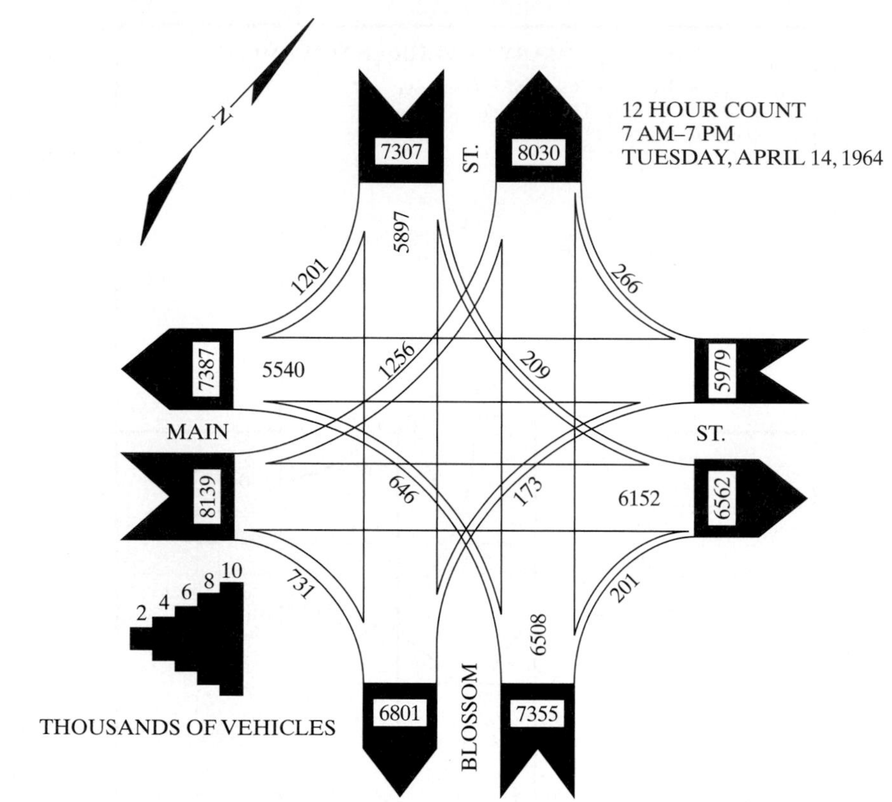

Figure 9.10: An Intersection Flow Diagram

(*Source:* Used with permission of Institute for Transportation Engineers, *Transportation and Traffic Engineering Handbook,* 1st Edition, Washington DC, 1976, p. 410.)

The establishment of a reasonable sampling methodology will require judgment based on such local familiarity.

Sampling procedures rely on the assumption that entire networks, or identifiable subportions of networks, have similar demand patterns in time. If these patterns can be measured at a few locations, the pattern can be superimposed on sample measurements from other locations in the network. To implement such a procedure, two types of counts are conducted:

- *Control counts.* Control counts are taken at selected representative locations to measure and quantify demand variation patterns in time. In general, control counts must be maintained continuously throughout the study period.
- *Coverage counts.* Coverage counts are taken at all locations for which data is needed. They are conducted as samples, with each location being counted for only a portion of the study period, in accordance with a preestablished sampling plan.

These types of counts and their use in volume analysis are discussed in the sections that follow.

9.5.1 Control Counts

Because control counts will be used to expand and adjust the results of coverage counts throughout the network under study, it is critical that representative control-count locations be properly selected. The hourly and daily variation patterns observed at a control count must be representative of a larger portion of the network if the sampling procedure is to be accurate and meaningful. Remember that volume variation patterns are generated by land-use characteristics and by the type of traffic, particularly the percentages of through versus locally generated traffic in the traffic stream. With these principles in mind, some general guidelines can be used in the selection of appropriate control-count locations:

1. There should be one control-count location for every 10 to 20 coverage-count locations to be sampled.
2. Different control-count locations should be established for each class of facility in the network—local streets, collectors, arterials, and so on, because different classes of facilities serve different mixes of through and local traffic.
3. Different control-count locations should be established for portions of the network with markedly different land-use characteristics.

These are only general guidelines. The engineer must exercise judgment and use his or her knowledge of the area under study to identify appropriate control-count locations.

9.5.2 Coverage Counts

All locations at which sample counts will be taken are called *coverage counts.* All coverage counts (and control counts as well) in a network study are taken at midblock locations to avoid the difficulty of separately recording turning movements. Each link of the network is counted at least once during the study period. Intersection turning movements may be approximately inferred from successive link volumes, and, when necessary, supplementary intersection counts can be taken. Counts at midblock locations allow for the use of portable automated counters, although the duration of some coverage counts may be too short to justify their use.

9.5.3 An Illustrative Study

The types of computations involved in expanding and adjusting sample network counts is best described by a simple example. Figure 9.11 shows one segment of a larger network that has been identified as having reasonably uniform traffic patterns in time. The network segment has seven links, one of which has been established as a control-count location. Each of the other six links are coverage-count locations at which sample counts will be conducted. The various proposed study procedures all assume there are only two field crews or automated counters that can be employed simultaneously in this segment of the network. A study procedure is needed to find the volume on each link of the network between 12 noon and 8:00 PM on a typical weekday. Three different approaches are discussed. They are typical and not the only approaches that could be used. However, they illustrate all of the expansion and adjustment computations involved in such studies.

A One-Day Study Plan

It is possible to complete the study in a single day. One of the two available crews or setups would be used to count Control Location A for the entire eight-hour period of the study.

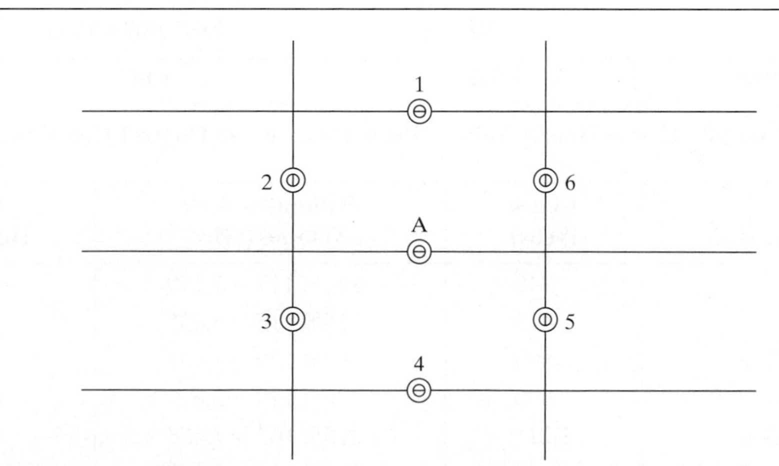

Figure 9.11: A Sample Network Volume Study

The second crew or set-up would be used to count each of Coverage Locations 1 to 6 for one hour. Table 9.3 shows the sample data and analysis resulting from this approach.

Note that full-hour data is shown. This data reflects expansion of actual counts for break periods. If machine counts were conducted, they would also reflect the conversion of axle counts to vehicle counts.

In Table 9.3 (b), the control-count data are used to quantify the hourly variation pattern observed. It is now assumed that this pattern applies to all of coverage locations

Table 9.3: Data and Computations for a One-Day Network Volume Study

Control-Count Data Location A		**Coverage-Count Data**		
Time (PM)	**Count (vehs)**	**Location**	**Time (PM)**	**Count (vehs)**
12–1	825	1	12–1	840
1–2	811	2	1–2	625
2–3	912	3	2–3	600
3–4	975	4	4–5	390
4–5	1,056	5	5–6	1,215
5–6	1,153	6	6–7	1,440
6–7	938			
7–8	397			

(a) Data from a One-Day Study

Time (PM)	**Count (vehs)**	**Proportion of 8-Hour Total**
12–1	825	825/7,067 = 0.117
1–2	811	811/7,067 = 0.115
2–3	912	912/7,067 = 0.129
3–4	975	975/7,067 = 0.138
4–5	1,056	1,056/7,067 = 0.149
5–6	1,153	1,153/7,067 = 0.163
6–7	938	938/7,067 = 0.133
7–8	397	397/7,067 = 0.056
Total	**7,067**	**1.000**

(b) Computation of Hourly Volume Proportions From Control-Count Data

Location	**Time (PM)**	**Count (vehs)**	**Estimated 8-Hr Volume (vehs)**	**Estimated Peak Hour Volume (vehs)**
1	12–1	840	840/0.117 = 7,179	× 0.163 = 1,170
2	1–2	625	625/0.115 = 5,435	× 0.163 = 886
3	2–3	600	600/0.129 = 4,651	× 0.163 = 758
4	4–5	390	390/0.149 = 2,617	× 0.163 = 427
5	5–6	1,215	1,215/0.163 = 7,454	× 0.163 = 1,215
6	6–7	1,440	1,440/0.133 = 10,827	× 0.163 = 1,765

(c) Expansion of Hourly Counts

within the network. Thus a count of 840 vehicles at location 1 would represent 0.117 (or 11.7%) of the eight-hour total at this location. The eight-hour total can then be estimated as 840/0.117 = 7,179 vehicles. Moreover, the peak-hour volume can be estimated as 0.163 × 7,179 = 1,170 vehicles because the hourly distribution shows that the highest volume hour contains 0.163 (or 16.3%) of the eight-hour volume. Note that this expansion of data results in estimates of eight-hour and peak-hour volumes at each of the seven count locations that represent *the day on which the counts were taken.* Daily and seasonal variations have not been eliminated by this study technique. Volumes for the entire network, however, have been estimated for common time periods.

A Multiday Study

In the one-day study approach, each coverage location was counted for one hour. Based on hourly variation patterns documented at the control location, these counts were expanded into eight-hour volume estimates. Hourly variation patterns, however, are not as stable as variations over larger periods of time. For this reason, it could be argued that a better approach would be to count each coverage location for a full eight hours.

Given the limitation to two simultaneous counts due to personnel and/or equipment, such a study would take place over six days. One crew would monitor the control location for the entire period of the study, and the second would count at one coverage location for eight hours on each of six days.

The data and computations associated with a 6-day study are illustrated in Table 9.4. In this case, hourly patterns do not have to be modeled because each coverage location is counted for every hour of the study period. Unfortunately, the counts are spread over six days, over which volume may vary considerably at any given location. In this case, the control data are used to quantify the underlying *daily* variation pattern. These data are used to *adjust* the coverage data.

Daily volume variations are quantified in terms of adjustment factors defined as follows: the volume for a given day multiplied by the factor yields a volume for the average day of the study period. Stated mathematically:

$$V_a = V_i F_{vi} \qquad (9\text{-}2)$$

where V_a = for the average day of the study period, vehs

V_i = for day i

F_{vi} = factor for day i

Using data from the control location, at which the average volume will be known, adjustment factors for each day of the study may be computed as:

$$F_{vi} = V_a / V_i \qquad (9\text{-}3)$$

where all terms are as previously defined. Factors for the sample study are calibrated in Table 9.4 (b). Coverage counts are adjusted using Equation 9-2 in Table 9.4 (c).

The results represent the average eight-hour volumes for all locations for the six-day period of the study. Seasonal variations are not accounted for, nor are weekend days, which were excluded from the study.

A Mixed Approach: A Three-Day Study

The first two approaches can be combined. If a one-day study is not deemed appropriate due to the estimation of eight-hour volumes based on one-hour observations, and the six-day study is too expensive, a three-day study program can be devised in which each coverage location is counted for four hours on one of three days. The control location would have to be counted for the entire three-day study period; results would be used to calibrate the distribution of volume by four-hour period and by day.

In this approach, four-hour coverage counts must be (1) expanded to reflect the full eight-hour study period, and (2) adjusted to reflect the average day of the three-day study period. Table 9.5 illustrates the data and computations for the three-day study approach.

Note that in expanding the four-hour coverage counts to eight hours, the proportional split of volume varied from day to day. The expansions used the proportion appropriate to the day of the count. Because the variation was not great, however, it would have been equally justifiable to use the average hourly split for all three days.

Again, the results obtained represent the particular three-day period over which the counts were conducted. Volume variations involving other days of the week or seasonal factors are not considered.

The three approaches detailed in this section are illustrative. Expansion and adjustment of coverage counts based on control observations can be organized in many different ways, covering any network size and study period. The selection of control locations involves much judgment, and the success of any particular study depends on the quality of the judgment exercised in designing the study. The traffic engineer must design each study to achieve the particular information goals at hand.

Table 9.4: Data and Computations for a Six-Day Study Option

Control-Count Data Location A		Coverage-Count Data		
Day	**8-Hour Count (vehs)**	**Coverage Location**	**Day**	**8-Hour Count (vehs)**
Monday 1	7,000	1	Monday 1	6,500
Tuesday	7,700	2	Tuesday	6,200
Wednesday	7,700	3	Wednesday	6,000
Thursday	8,400	4	Thursday	7,100
Friday	7,000	5	Friday	7,800
Monday 2	6,300	6	Monday 2	5,400

(a) Data for a Six-Day Study

Day	8-Hour Count (vehs)	Adjustment Factor
Monday 1	7,000	7,350/7,000 = 1.05
Tuesday	7,700	7,350/7,700 = 0.95
Wednesday	7,700	7,350/7,700 = 0.95
Thursday	8,400	7,350/8,400 = 0.88
Friday	7,000	7,350/7,000 = 1.05
Monday 2	6,300	7,350/6,300 = 1.17
Total	44,100	
Average	44,100/6 = 7,350	

(b) Computation of Daily Adjustment Factors

Station	Day	8-Hour Count (vehs)	Adjusted 8-Hour Count (vehs)
1	Monday 1	6,500	× 1.05 = 6,825
2	Tuesday	6,200	× 0.95 = 5,890
3	Wednesday	6,000	× 0.95 = 5,700
4	Thursday	7,100	× 0.88 = 6,248
5	Friday	7,800	× 1.05 = 8,190
6	Monday 2	5,400	× 1.17 = 6,318

(c) Adjustment of Coverage Counts

Estimating Vehicle Miles Traveled on a Network

One output of most limited-network volume studies is an estimate of the total vehicle-miles traveled (VMT) on the network during the period of interest. The estimate is done roughly by assuming that a vehicle counted on a link travels the entire length of the link. This is a reasonable assumption because some vehicles traveling only a portion of a link will be counted while others will not, depending on whether they cross the count location. Using the sample network of the previous section, the eight-hour volume results of Table 9.5, and assuming all links are 0.25 miles long, Table 9.6 illustrates the estimation of VMT. In this case, the estimate is the average eight-hour VMT for the three days of the study. It cannot be expanded into an estimate of *annual* VMT without knowing more about daily and seasonal variation patterns throughout the year.

Table 9.5: Data and Computations for a Three-Day Study Option

Time (PM)	Monday		Tuesday		Wednesday		Avg % of 8 Hours
	Count (vehs)	**% of 8 Hours**	**Count (vehs)**	**% of 8 Hours**	**Count (vehs)**	**% of 8 Hours**	
12–4	3,000	42.9%	3,200	42.7%	2,800	43.8%	43.1%
4–8	4,000	57.1%	4,300	57.3%	3,600	56.2%	56.9%
Total	**7,000**	**100.0%**	**7,500**	**100.0%**	**6,400**	**100.0%**	**100.0%**

(a) Control Data and Calibration of Hourly Variation Pattern

Day	8-Hour Control-Count Location A (vehs)	Adjustment Factor
Monday	7,000	6,967/7,000 = 1.00
Tuesday	7,500	6,967/7,500 = 0.93
Wednesday	6,400	6,967/6,400 = 1.09
Total	**20,900**	
Average	20,900/3 = 6,967	

(b) Calibration of Daily Variation Factors

Station	Day	Time (PM)	Count (vehs)	8-Hour Expanded Count (vehs)	8-Hour Adjusted Counts (vehs)
1	Monday	12–4	2,213	2,213/0.429 = 5,159	× 1.00 = 5,159
2	Monday	4–8	3,000	3,000/0.571 = 5,254	× 1.00 = 5,254
3	Tuesday	12–4	2,672	2,672/0.427 = 6,258	× 0.93 = 5,820
4	Tuesday	4–8	2,500	2,500/0.573 = 4,363	× 0.93 = 4,058
5	Wednesday	12–4	3,500	3,500/0.438 = 7,991	× 1.09 = 8,710
6	Wednesday	4–8	3,750	3,750/0.562 = 6,673	× 1.09 = 7,274

(c) Expansion and Adjustment of Coverage Counts

Table 9.6: Estimation of Vehicle-Miles Traveled on a Limited Network: An Example

Station	8-Hour Count (vehs)	Link Length (mi)	Link VMT (veh-miles)
A	6,967	0.25	1,741.75
1	5,159	0.25	1,289.75
2	5,254	0.25	1,313.50
3	5,820	0.25	1,455.00
4	4,058	0.25	1,014.50
5	8,710	0.25	2,177.50
6	7,274	0.25	1,818.50
Network Total			**10,810.50**

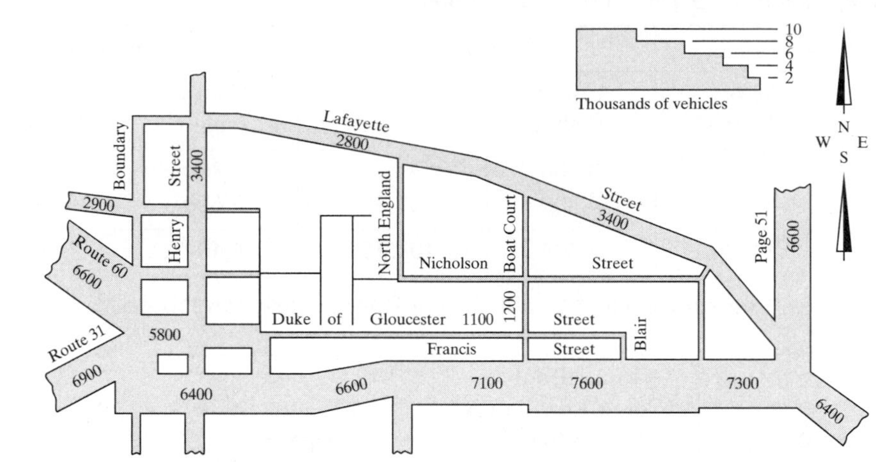

Figure 9.12: A Typical Network Flow Map

(*Source:* Used with permission of Wilbur Smith and Associates, *Traffic, Parking, and Transit – Colonial Williamsburg,* Columbia, South Carolina, 1963.)

Display of Network Volume Results

As was the case with intersection volume studies, most detailed results of a limited network study are presented in tabular form, some of which have been illustrated here. For peak hours or for daily total volumes, it is often convenient to provide a network flow map. This is similar to an intersection flow diagram in that the thickness of flow lines is proportional to the volume. An example of such a map is shown in Figure 9.12.

9.6 Statewide Counting Programs

States generally have a special interest in observing trends in AADT, shifts within the ADT pattern, and vehicle-miles traveled. These trends are used in statewide planning and for the programming of specific highway improvement projects. In recent years, there has been growing interest in person-miles traveled (PMT) and in statistics for other modes of transportation. Similar programs at the local and/or regional level are desirable for non-state highway systems, although the cost is often prohibitive.

Following some general guidelines, as in Reference 4 for example, the state road system is divided into functional classifications. Within each classification, a pattern of control count locations and coverage count locations is established so that trends can be observed. Statewide programs are similar to limited network studies, except that the network involved is the entire state highway system and the time frame of the study is continuous (i.e., 365 days a year, every year).

These are some general principles for statewide programs:

1. The objective of most statewide programs is to conduct a coverage count every year on every 2-mile segment of the state highway system, with the exception of low-volume roadways (AADT < 100 veh/day) Low-volume roadways usually comprise about 50% of state system mileage and are classified as tertiary local roads.
2. The objective of coverage counts is to produce an annual estimate of AADT for each coverage location.
3. One control-count location is generally established for every 20 to 50 coverage-count locations, depending on the characteristics of the region served. Criteria for establishing control locations are similar to those used for limited networks.
4. Control-count locations can be either *permanent counts* or *major or minor control counts,* which use representative samples. In both cases, control-count locations must monitor and calibrate daily variation patterns and monthly or seasonal variation patterns for the full 365-day year.
5. All coverage counts are for a minimum period of 24 to 48 hours, eliminating the need to calibrate hourly variation patterns.

Table 9.7: Calibration of Daily Variation Factors

Day	Yearly Average Volume for Day (vehs/day)	Daily Adjustment Factor (DF)
Monday	1820	1430/1820 = 0.79
Tuesday	1588	1430/1588 = 0.90
Wednesday	1406	1430/1406 = 1.02
Thursday	1300	1430/1300 = 1.10
Friday	1289	1430/1289 = 1.11
Saturday	1275	1430/1275 = 1.12
Sunday	1332	1430/1332 = 1.07
Total	**10,010**	
Estimated AADT	**1,430**	

At permanent count locations, fixed detection equipment with data communications technology is used to provide a continuous flow of volume information. Major and minor control counts are generally made using portable counters and road tubes. Major control counts are generally made for one week during each month of the year. Minor control counts are generally made for one five-day (weekdays only) period in each season.

9.6.1 Calibrating Daily Variation Factors

The illustrative data in Table 9.7 are obtained from a permanent count location. At a permanent count location, data exist for all 52 weeks of the year (i.e., for 52 Sundays, 52 Mondays, 52 Tuesdays, etc.). (Note that in a 365-day year, one day will occur 53 times).

Daily variation factors are calibrated based on the average volumes observed during each day of the week. The base value for factor calibration is the average of the seven daily averages, which is a rough estimate of the AADT (but not exact, due to the 53rd piece of data for one day of the week). The factors can be plotted, as illustrated in Figure 9.13, and display a clear variation pattern that can be applied to coverage count results.

Note that the sum of the seven daily adjustment factors *does not* add up to 7.00 (the actual total is 7.11). This is because of the way in which the factors are defined and computed. The daily averages are in the denominator of the calibration factors. In effect, the average factor is inverse to the average daily volume, so that the totals would not be expected to add to 7.00.

Daily adjustment factors can also be computed from the results of major and/or minor control counts. In a major control count, there would be 12 weeks of data, one week from each month of the year. The daily averages, rather than representing 52 weeks of data, reflect 12 representative weeks of data. The calibration computations, however, are exactly the same.

9.6.2 Calibrating Monthly Variation Factors

Table 9.8 illustrates the calibration of monthly variation (MF) factors from permanent count data. The monthly factors are based on monthly ADTs that have been observed at the permanent count location. Note that the sum of the 12 monthly variation patterns is not 12.00 (the actual sum is 12.29) because the monthly ADTs are in the denominator of the calibration.

Table 9.8 is based on permanent count data, such that the monthly ADTs are directly measured. One seven-day count in each month of the year would produce similar values, except that the ADT for each month would be estimated based on a single week of data, not the entire month. This type of procedure can yield a bias when the week in which the data were collected varies from month to month. In effect, an ADT for a given month is most likely to be observed in the middle of the month (i.e., the 14th to the 16th of any month). This statement is based on the assumption that the volume trend within each month is unidirectional (i.e., volume grows throughout the month or declines throughout the month). Where a peak or low point exists within the month, this statement is not true.

Figure 9.14 illustrates a plot of 12 calibrated monthly variation factors, but one week of data is taken from each

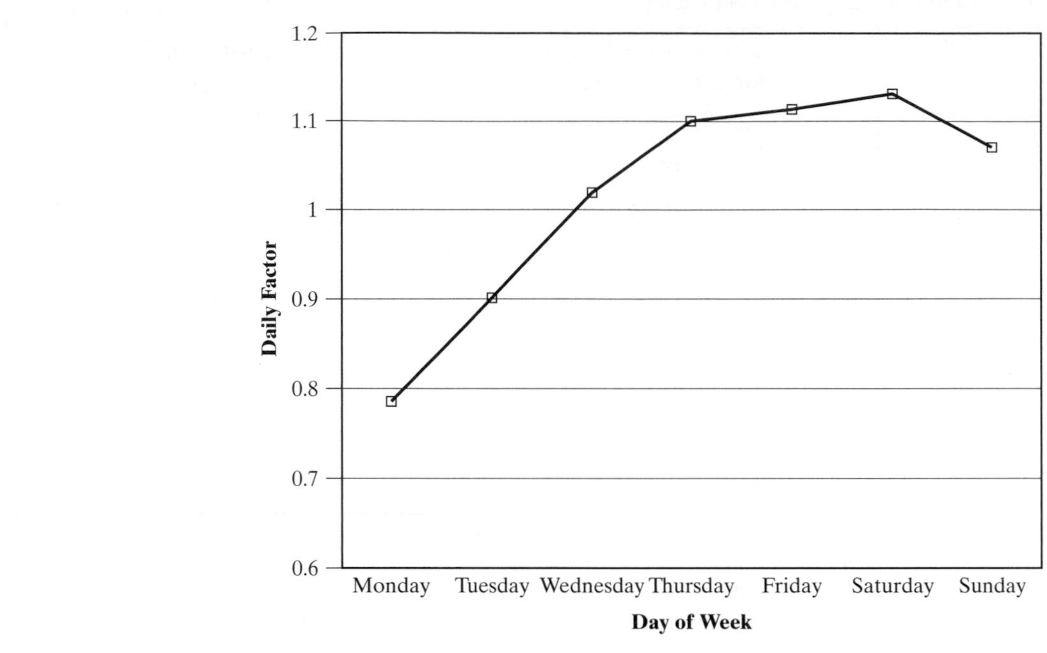

Figure 9.13: Plot of Daily Variation Factors

month. The daily variation factors are plotted against the midpoint of the week in which the data for the month were taken.

This graph may now be entered at the middle of each month (the 15th), and adjusted factors read from the vertical axis. For example, in May the computed factor was 0.93, and the plot indicates that a factor computed for the middle of that month would have resulted in a factor of 0.96. Adjusting the factors in this manner results in a more representative computation based on monthly midpoints.

Table 9.8: Calibration of Monthly Variation Factors

Month	Total Traffic (vehs)	ADT for Month (veh/day)	Monthly Factor (AADT/ADT)
January	19,840	/31 = 640	797/640 = 1.25
February	16,660	/28 = 595	797/595 = 1.34
March	21,235	/31 = 685	797/685 = 1.16
April	24,300	/30 = 810	797/810 = 0.98
May	25,885	/31 = 835	797/835 = 0.95
June	26,280	/30 = 876	797/876 = 0.91
July	27,652	/31 = 892	797/892 = 0.89
August	30,008	/31 = 968	797/968 = 0.82
September	28,620	/30 = 954	797/954 = 0.84
October	26,350	/31 = 850	797/850 = 0.94
November	22,290	/30 = 763	797/763 = 1.07
December	21,731	/31 = 701	797/701 = 1.14
Total	**290,851**	**AADT 290,851/365**	**797 veh/day**

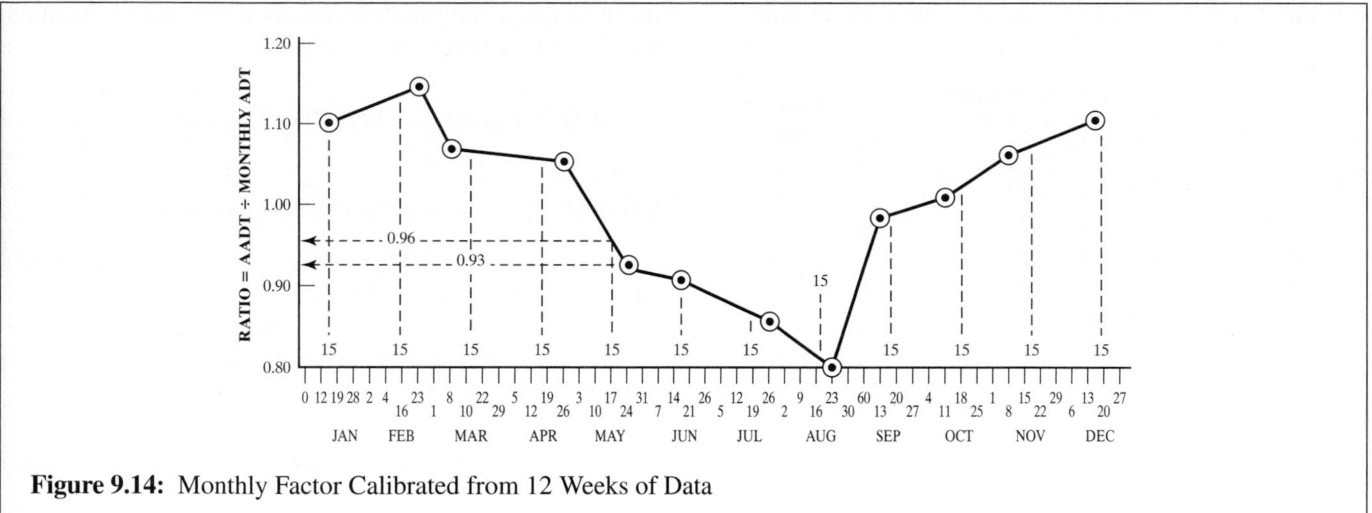

Figure 9.14: Monthly Factor Calibrated from 12 Weeks of Data

9.6.3 Grouping Data from Control Count Locations

On state highway networks and systems, particularly in rural areas, it is possible for a broad region to have similar, if not the same, daily and/or monthly adjustment factors. In such regions, spatially contiguous control stations on the same classification of highway may be combined to form a single control group. The average factors for the group may then be applied over a wide area with similar variation patterns. In general, a statistical standard is applied to such groupings: Contiguous control counts on similar highway types may be grouped if the factors at the individual locations do not differ by more than ± 0.10 from the average for the group.

Consider the example shown in Table 9.9. The daily variation factors for four consecutive control counts on a state highway have been calibrated as shown. It has been hypothesized that the four represent regions with similar daily variation patterns. Average factors have, therefore, been computed for the four grouped stations.

The boldfaced factors indicate cases that violate the statistical rule for grouping (i.e., differences between these factors and the average for the group are more than ± 0.10). This suggests that the proposed grouping is not appropriate. One might be tempted to remove Stations 1 and 4 from the group and combine only Stations 2 and 3. The proper technique, however, is to remove one station from the group at a time because the resulting average factors will change. In this case, a cursory observation indicates that Station 4 does not really display a daily variation pattern similar to the others. This station has its peak traffic (DF < 1.00) occurring during the week, whereas the other stations have their peak traffic on weekends. Thus Station 4 is deleted from the proposed grouping and new averages are computed, as illustrated in Table 9.10.

Now, all factors at individual stations are within ± 0.10 of the average for the group. This would be an appropriate grouping of control stations.

9.6.4 Using the Results

Note that groups for daily factors and groups for monthly factors do not have to be the same. It is convenient if they are, however, and it is not at all unlikely that a set of stations grouped for one type of factor would also be appropriate for the other.

Table 9.9: A Trial Grouping of Four Contiguous Control Stations with Daily Variation Factors

Day	DF for Station Number:				Average DF
	1	2	3	4	
Monday	1.05	1.00	1.06	0.92	1.01
Tuesday	1.10	1.02	1.06	**0.89**	1.02
Wednesday	1.10	1.05	1.11	0.97	1.06
Thursday	1.06	1.06	1.03	1.00	1.04
Friday	1.01	1.03	1.00	0.91	0.98
Saturday	**0.85**	0.94	0.90	**1.21**	0.98
Sunday	0.83	0.90	0.84	1.10	0.92

Note: DF = daily factor.

Table 9.10: A Second Trial Grouping of Control Stations with Daily Variation Factors

Day	DF for Station			Average (DF)
	1	2	3	
Monday	1.05	1.00	1.06	1.04
Tuesday	1.10	1.02	1.06	1.06
Wednesday	1.10	1.05	1.11	1.09
Thursday	1.06	1.06	1.03	1.05
Friday	1.01	1.03	1.00	1.01
Saturday	0.85	0.94	0.90	0.90
Sunday	0.83	0.90	0.84	0.86

Note: DF = daily factor.

The state highway agency will use its counting program to generate basic trend data throughout the state. It will also generate, for contiguous portions of each state highway classification, a set of daily and monthly variation factors that can be applied to any coverage count within the influence area of the subject control grouping. An example of the type of data that would be made available is shown in Table 9.11.

Using these tables, any coverage count for a period of 24 hours or more can be converted to an estimate of the AADT using the following relationship:

$$AADT = V_{24ij} * DF_i * MF_j \tag{9-4}$$

where $AADT$ = average annual daily traffic, vehs/day
V_{24ij} = 24-hour volume for day i in month j, vehs
DF_i = daily adjustment factor for day i
MF_j = monthly adjustment factor for month j

Consider a coverage count taken at a location within the area represented by the factors of Table 9.11. A count of 1,000 vehicles was observed on a Tuesday in July. From Table 9.11, the daily factor (DF) for Tuesdays is 1.121, and the monthly factor (MF) for July is 0.913. Then:

$$AADT = 1{,}000 * 1.121 * 0.913 = 1{,}023 \text{ vehs/day}$$

Estimating Annual Vehicle-Miles Traveled

Given estimates of AADT for every two-mile segment of each category of roadway in the state system (excluding low-volume roads), estimates of annual vehicle-miles traveled can be assembled. For each segment, the annual vehicle-miles traveled is estimated as:

$$VMT_{365} = AADT * L * 365 \tag{9-5}$$

where VMT_{365} = annual vehicle-miles traveled over the segment,
$AADT$ = $AADT$ for the segment, vehs/day, and
L = length of the segment, mi

For any given roadway classification or system, the segment VMTs can be summed to give a regional or statewide total. The question of the precision or accuracy of such estimates is interesting, given that none of the low-volume roads are included and that a real statewide total would need to include inputs for all non–state systems in the state. Regular counting programs at the local level are, in general, far less rigorous than state programs.

There are two other ways commonly used to estimate VMT:

- Use the number of registered vehicles with reported annual mileages, adjusting for out-of-state travel.
- Use fuel tax receipts by category of fuel (which relates to categories of vehicles), and estimate VMT using average fuel consumption ratings for different types of vehicles.

Table 9.11: Typical Daily and Monthly Variation Factors for a Contiguous Area on a State Highway System

Daily Factors (DF)		Monthly Factors (MF)			
Day	Factor	Month	Factor	Month	Factor
Monday	1.072	January	1.215	July	0.913
Tuesday	1.121	February	1.191	August	0.882
Wednesday	1.108	March	1.100	September	0.884
Thursday	1.098	April	0.992	October	0.931
Friday	1.015	May	0.949	November	1.026
Saturday	0.899	June	0.918	December	1.114
Sunday	0.789				

There is interest in improving statewide VMT estimating procedures, and a number of significant research efforts have been sponsored on this topic in recent years. There is also growing interest in nationwide PMT estimates, with appropriate modal categories.

9.7 Specialized Counting Studies

In a number of instances, simple counting of vehicles at a point, or at a series of points, is not sufficient to provide the information needed. Three principal examples of specialized counting techniques are (1) origin and destination counts, (2) cordon counts, and (3) screen-line counts.

9.7.1 Origin and Destination Counts

In many instances, normal point counts of vehicles must be supplemented with knowledge of the origins and destinations of the vehicles counted. In major regional planning applications, origin and destination studies involve massive home-interview efforts to establish regional travel patterns. In traffic applications, the scope of origin and destination counts are often more limited. Common applications include:

- Weaving-area studies
- Freeway studies
- Major activity center studies

Proper analysis of weaving-area operations requires that volume be broken down into two weaving and two nonweaving flows that are present. A total count is insufficient to evaluate performance. In freeway corridors, it is often important to know where vehicles enter and exit the freeway. Alternative routes, for example, cannot be accurately assessed without knowing the underlying pattern of origins and destinations. At major activity centers (sports facilities, airports, regional shopping centers, etc.), traffic planning of access and egress also requires knowledge of where vehicles are coming from when entering the development or going to when leaving the development.

Many ITS technologies hold great promise for providing detailed information on origins and destinations. Automated toll-collection systems can provide data on where vehicles enter and leave toll facilities. Automated license-plate reading technology is used in traffic enforcement and could be used to track vehicle paths through a traffic system. Although these technologies continue to advance rapidly, their use in traditional traffic data collection has been much slower due to the privacy issues that such use raises.

Historically, one of the first origin-destination count techniques was called a *lights-on study.* This method was often applied in weaving areas where vehicles arriving on one leg could be asked to turn on their lights. With the advent of daytime running lights, this methodology is no longer viable.

Conventional traffic origin and destination counts rely primarily on one of three approaches:

- License-plate studies
- Postcard studies
- Interview studies

In a license-plate study, observers (or automated equipment) record the license-plate numbers as they pass designated locations. This is a common method used to track freeway entries and exits at ramps. Postcard studies involve handing out color- or otherwise coded cards as vehicles enter the system under study and collecting them as vehicles leave. In both license-plate and postcard studies, the objective is to match up vehicles at their origin and at their destination. Interview studies involve stopping vehicles (with the approval and assistance of police) and asking a short series of questions concerning their trip, where it began, where it is going, and what route will be followed.

Major activity centers are more easily approached because one end of the trip is known (everyone is at the activity center). Here, interviews are easier to conduct, and license-plate numbers of parked vehicles can be matched to home locations using data from the state Department of Motor Vehicles.

When attempting to match license-plate observations or postcards, sampling becomes a significant issue. If a sample of drivers is recorded at each entry and exit location, then the probability of finding matches is diminished considerably. If 50% of the entering vehicles at Exit 2 are observed, and 40% of the exiting vehicles at Exit 5 are observed, then the statistically expected number of matches of vehicles traveling from Exit 2 to Exit 5 would be $0.50*0.40 = 0.20$ or 20%. When such sampling techniques are used, separate counts of vehicles at all entry and exit points must be maintained to provide a means of expanding the sample data.

Consider the situation illustrated in Figure 9.15. It shows a small local downtown street network with four entry roadways and four exit roadways. Thus there are $4*4 = 16$ possible origin-destination pairs for vehicles accessing or traveling through the area. The data shown reflect both the observed origins and destinations (using license-plate samples) and the full-volume counts observed on each entry and exit leg.

If the columns and rows are totaled, the sums should be equal to the observed total volumes, assuming that a 100% sample of license plates was obtained at each location. This is obviously not the case. Thus the origin-destination volumes must be expanded to reflect the total number of vehicles counted. This can be done in two ways: (1) origin-destination cells can be expanded so that the row totals are correct (i.e., match the measured volume), or (2) origin-destination

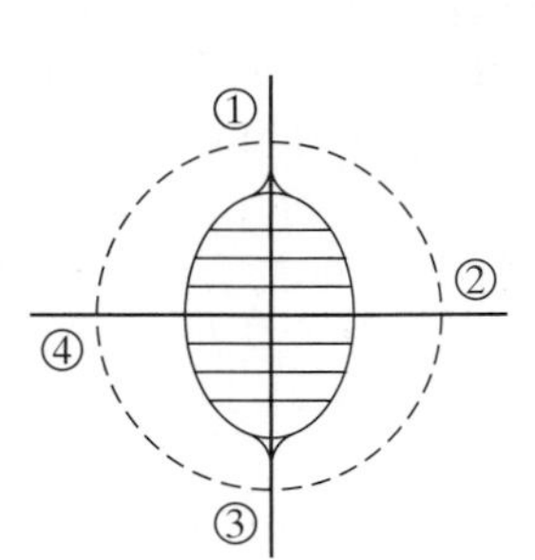

Destination Station	Origin Station				Row Sum T_j	Vol V_j
	1	2	3	4		
1	50	8	20	17	**95**	250
2	10	65	21	10	**106**	310
3	15	12	38	15	**80**	200
4	13	14	18	42	**87**	375
Col Sum T_i	**88**	**99**	**97**	**84**		
Volume V_i	210	200	325	400		**1,135**

Figure 9.15: Data from an Origin-Destination Count Using License-Plate Matching

cells can be expanded so that the column totals are correct. Unfortunately, these two approaches will lead to two different sets of origin-destination volumes.

In practice, the average of the two approaches is adopted. This creates an iterative process because the initial adjustment will still result in column and row totals that are not the same as the measured volumes. Iteration is continued until all row and column totals are within ± 10% of the measured volumes.

The cell volumes, representing matched trips from Station i to Station j, are adjusted using factors based on column closure and row closure:

$$T_{ijN} = T_{ij(N-1)}\left(\frac{F_i + F_j}{2}\right) \tag{9-6}$$

where:

F_i = adjustment factor for origin i = V_i / T_i
F_j = adjustment factor for destination j = V_j / T_j
T_{ijN} = number of trips from station i to station j after the Nth iteration of the data (trips)
$T_{ij}(N–1)$ = number of trips from Station i to Station j after the $(N-1)$th iteration of the data (trips)
T_i = sum of matched trips from Station i (trips)
T_j = sum of matched trips to Station j (trips)
V_i = observed total volume at Station i (vehs)
V_j = observed total volume at Station j (vehs)

The actual data of Figure 9.15 serves as the 0th iteration. Each adjustment cycle results in new values of T_{ij}, T_i, T_j, F_i, and F_j. The observed total volumes, of course, remain constant.

Table 9.12 shows the results of several iterations, with the final O-D counts accepted when all adjustment factors are greater than or equal to 0.90 or less than or equal to 1.10. In this case, the initial expansion of O-D counts was iterated twice to obtain the desired accuracy.

Table 9.12: Sample Expansion of Origin and Destination Data

Destination Station	Origin Station				T_j	V_j	F_j
	1	2	3	4			
1	50	8	20	17	**95**	250	2.63
2	10	65	21	10	**106**	310	2.92
3	15	12	38	15	**80**	200	2.50
4	13	14	18	42	**87**	375	4.31
T_i	**88**	**99**	**97**	**84**	**368**		
V_i	210	200	325	400		**1135**	
F_i	2.39	2.02	3.35	4.76			

(a) Field Data and Factors for Iteration 0

Table 9.12: Sample Expansion of Origin and Destination Data

Destination Station	**Origin Station**						
	1	**2**	**3**	**4**	**T_j**	**V_j**	**F_j**
1	125	19	60	63	**267**	250	0.94
2	27	161	66	38	**292**	310	1.06
3	37	27	111	54	**229**	200	0.87
4	44	44	69	191	**347**	375	1.08
T_i	**232**	**251**	**306**	**346**	**1135**		
V_i	210	200	325	400		**1135**	
F_i	0.90	0.80	1.06	1.16			

(b) Initial Expansion of O-D Matrix (Iteration 0)

Destination Station	**Origin Station**						
	1	**2**	**3**	**4**	**T_j**	**V_j**	**F_j**
1	116	16	60	66	**257**	250	0.97
2	26	150	70	43	**288**	310	1.08
3	33	23	108	55	**218**	200	0.92
4	43	42	74	213	**372**	375	1.01
T_i	**217**	**230**	**311**	**376**	**1135**		
V_i	210	200	325	400		**1135**	
F_i	0.97	0.87	1.04	1.06			

(c) First Iteration of O-D Matrix

Destination Station	**Origin Station**						
	1	**2**	**3**	**4**	**T_j**	**V_j**	**F_j**
1	112	15	60	67	**254**	250	0.98
2	27	145	74	46	**292**	310	1.06
3	31	20	105	55	**211**	200	0.95
4	43	39	76	221	**378**	375	0.99
T_i	**212**	**220**	**316**	**388**	**1135**		
V_i	210	200	325	400		**1135**	
F_i	0.99	0.91	1.03	1.03			

(d) Second Iteration of O-D Matrix

9.7.2 Cordon Counts

A cordon is an imaginary boundary around a study area of interest. It is generally established to define a CBD or other major activity center where the accumulation of vehicles within the area is of great importance. Cordon volume studies require counting volume at all street and highways that cross the cordon, classifying the counts by direction and by 15- to 60-minute times intervals. In establishing the cordon, several principles should be followed:

- The cordoned area must be large enough to define the full area of interest yet small enough so that accumulation estimates will be useful for parking and other traffic planning purposes.
- The cordon is established to cross all streets and highways at *midblock* locations, to avoid the complexity of establishing whether turning vehicles are entering or leaving the cordoned area.
- The cordon should be established to minimize the number of crossing points wherever possible. Natural or manufactured barriers (e.g., rivers, railroads, limited-access highways, and similar features) can be used as part of the cordon.
- Cordoned areas should have relatively uniform land use. Accumulation estimates are used to estimate street capacity and parking needs. Large cordons encompassing different land-use activities will not be focused enough for these purposes.

The accumulation of vehicles within a cordoned area is found by summarizing the total of all counts entering and leaving the area by time period. The cordon counts should begin at a time when the streets are virtually empty. Because this condition is difficult to achieve, the study should start with an estimate of vehicles already within the cordon. This can be done by circulating through the area and counting parked and circulating vehicles encountered. Off-street parking facilities can be surveyed to estimate their overnight population.

Note that an estimate of parking and standing vehicles may *not* reflect true parking demand if supply is inadequate and many circulating vehicles are merely looking for a place to park. Also, demand discouraged from entering the cordoned area due to congestion is not evaluated by this study technique.

When all entry and exit counts are summed, the accumulation of vehicles within the cordoned area during any given period may be estimated as:

$$A_i = A_{i-1} + V_{Ei} - V_{Li} \tag{9-7}$$

where: A_i = accumulation for time period i, vehs
A_{i-1} = accumulation for time period i – 1, vehs
V_{Ei} = total volume entering the cordoned area during time period i, vehs
V_{Li} = total volume leaving the cordoned area during time period i, vehs

Table 9.13: Accumulation Computations for an Illustrative Cordon Study

Time	Vehicles Entering (vehs)	Vehicles Leaving (vehs)	Accumulation (vehs)
4:00–5:00 AM	—	—	250*
5:00–6:00 AM	100	20	250 + 100 – 20 = 330
6:00–7:00 AM	150	40	330 + 150 – 40 = 440
7:00–8:00 AM	200	40	440 + 200 – 40 = 600
8:00–9:00 AM	290	80	600 + 290 – 80 = 810
9:00–10:00 AM	350	120	810 + 350 – 120 = 1,040
10:00–11:00 AM	340	200	1,040 + 340 – 200 = 1,180
11:00–noon	350	350	1,180 + 350 – 350 = 1,180
12:00–1:00 AM	260	300	1,180 + 260 – 300 = 1,140
1:00–2:00 PM	200	380	1,140 + 200 – 380 = 960
2:00–3:00 PM	180	420	960 + 180 – 420 = 720
3:00–4:00 PM	100	350	720 + 100 – 350 = 470
4:00–5:00 PM	120	320	470 + 120 – 320 = 270

*Estimated beginning accumulation.

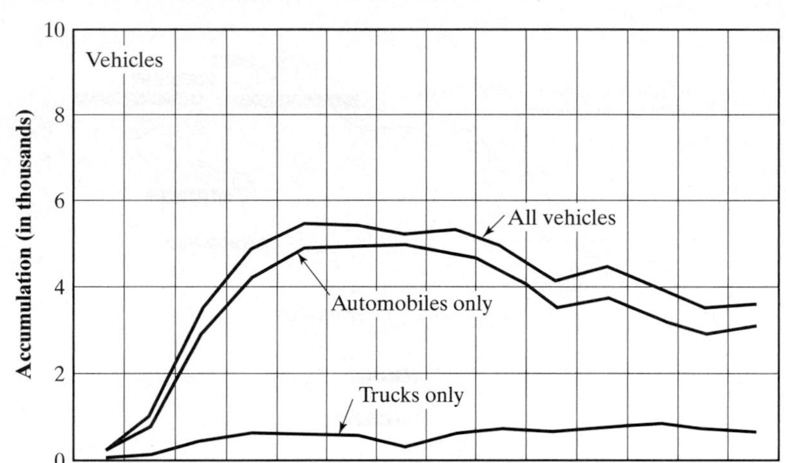

Figure 9.16: Typical Presentation of Accumulation Data
(*Source:* Used with permission of San Diego Area Transportation Study, San Diego CA, 1958.)

LEGEND
Buses
Autos
Trucks
Entering Leaving

2,500 Vehicles
5,000 Vehicles
7,500 Vehicles

N

Vehicles entering and leaving cordon area between 6:00 a.m. – 8:00 p.m. (14 hrs.)

Total vehicles entering – 86,170
Total vehicles leaving – 86,098

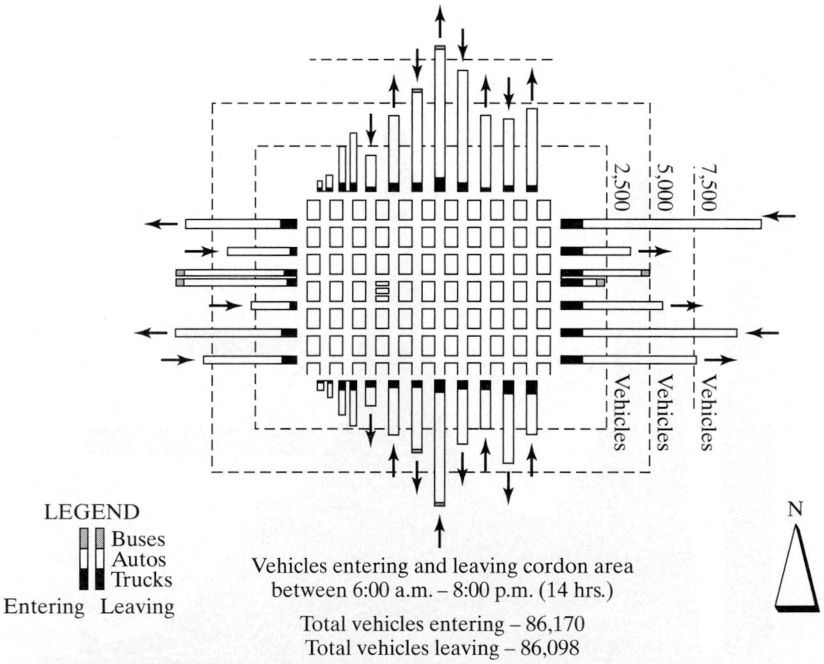

Figure 9.17: Typical Presentation of Daily Cordon Crossings
(*Source:* Used with permission of San Diego Transportation Study, San Diego CA, 1958.)

An example of a cordon volume study and the estimation of accumulation within the cordoned area is shown in Table 9.13. Figure 9.16 illustrates a typical presentation of accumulation data, and Figure 9.17 illustrates an interesting presentation of cordon crossing information.

9.7.3 Screen-Line Counts

Screen-line counts and volume studies are generally conducted as part of a larger regional origin-destination study involving home interviews as the principal methodology. In such regional

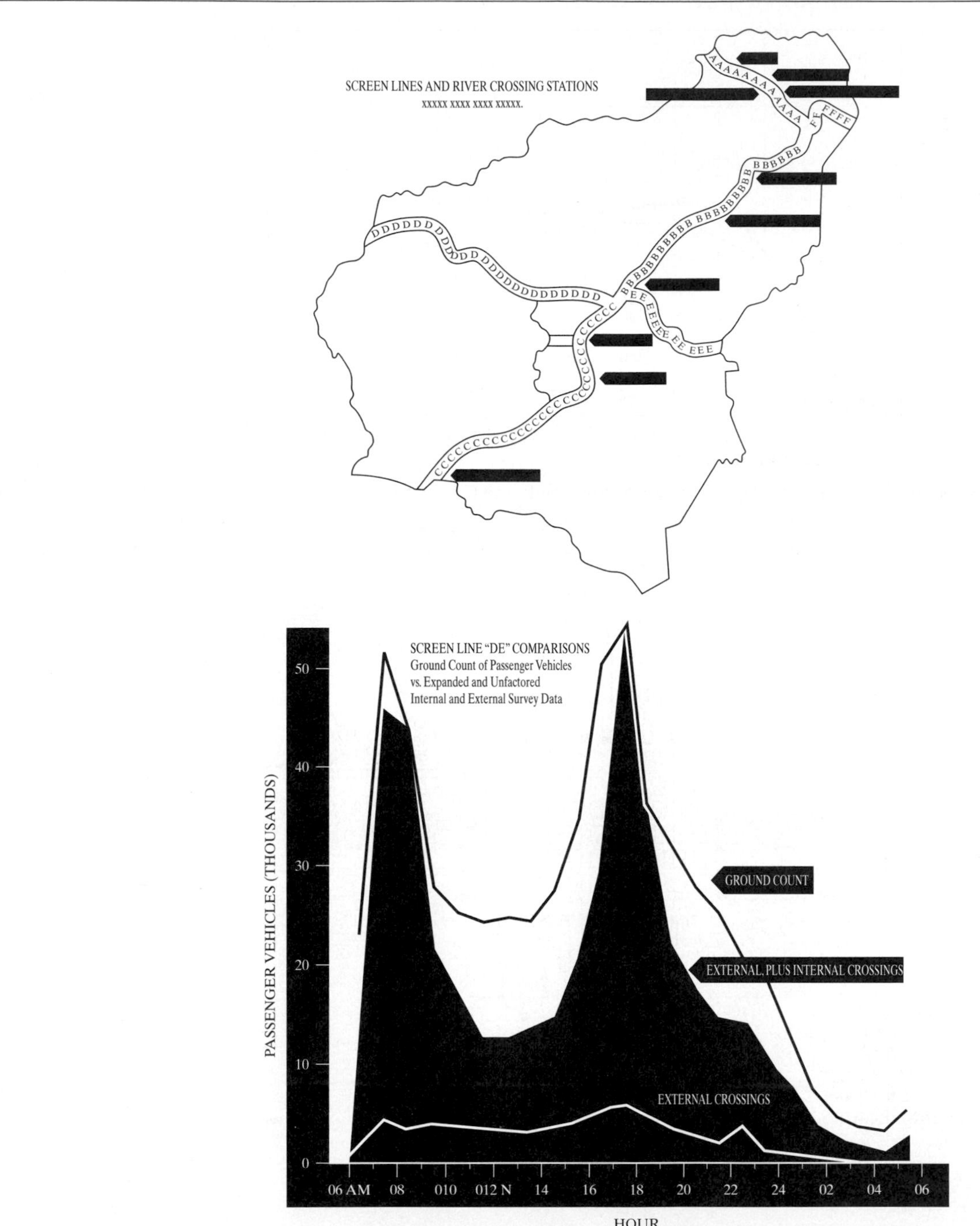

Figure 9.18: Illustration of a Screen-Line Study

(*Source:* Used with permission of Institute of Transportation Engineers, Box, P.C. and Oppenlander, J.C., *Manual of Traffic Engineering Studies*, Washington D.C., 1975, Figure 3-35, pg. 43.)

planning studies, home interview responses constitute a small but detailed sample that is used to estimate the number of trips per day (or some other specified time interval) between defined transportation zones that have been established within the study region.

Because home interview samples are small and additional data are used to estimate trip patterns for those passing through the study area or having only a single trip-end within the study area, it is necessary to use some form of field observations to check on the accuracy of predicted movements.

Screen lines are convenient barriers cutting through the study area with only a limited number of crossing points. Rivers, railroads, limited-access highways, and other features make good screen lines. The zone-to-zone trip estimates of a regional study can be summed in a way that yields the predicted number of trips across the screen line in a defined time period. A screen-line count can then be made to observe the actual number of crossings. The comparison of predicted versus observed crossings provides a means by which predicted zone-to-zone trips can be adjusted.

Figure 9.18 illustrates a study area for which two screen lines have been established. Predicted versus observed crossings are presented in graphic form. The ratio of observed to predicted crossings provides an adjustment factor that can be applied to all zonal trip combinations.

9.8 Closing Comments

The concept is simple: counting vehicles. As reviewed in this chapter, the process is not always simple; nor is the proper use of field results to obtain the desired statistics always straightforward. The field work of volume studies is relatively pedestrian but crucially important. Volume data is one of the primary bases for all traffic engineering analysis, planning, design, and operation.

Volume data must be accurately collected. It must be reduced to understandable forms and properly analyzed to obtain the prescribed objective of the study. It must then be presented clearly and unambiguously for use by traffic engineers and others involved in the planning and engineering process. No geometric or traffic control design can be effective if it is based on incorrect data related to traffic volumes and true demand. The importance, therefore, of performing volume studies properly cannot be understated.

References

1. *Highway Capacity Manual,* 4th Edition, Transportation Research Board, National Research Council, Washington DC, 2000.
2. *Transportation and Traffic Engineering Handbook,* 2nd Edition, Prentice-Hall, Englewood Cliffs, NJ, 1982.
3. McShane, W., and Crowley, K., "Regularity of Some Detector-Observed Arterial Traffic Volume Characteristics," *Transportation Research Record 596,* Transportation Research Board, National Research Council, Washington DC, 1976.
4. *Traffic Monitoring Guide,* Federal Highway Administration, U.S. Department of Transportation, Washington DC, 1985.

Problems

9-1. A limited network counting study was conducted for the network shown here. Because only two sets of road tubes were available, the study was conducted over a period of several days, using Station A as a control location. The network is shown here.

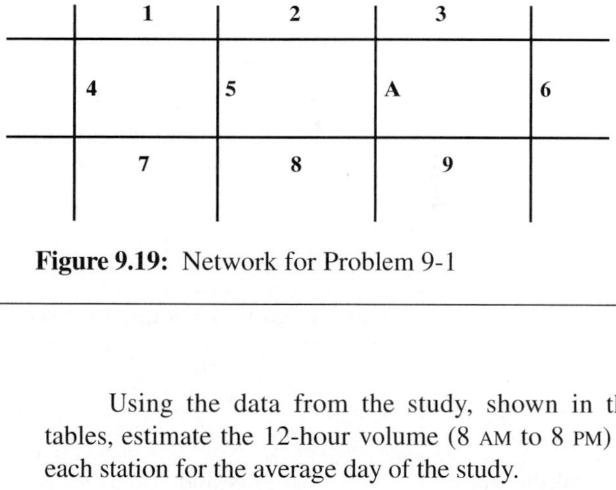

Figure 9.19: Network for Problem 9-1

Using the data from the study, shown in the tables, estimate the 12-hour volume (8 AM to 8 PM) at each station for the average day of the study.

Table 9.14: Axle Counts for Control Station A (Problem 9-1)

	Time Period		
Day	**8:00–11:45**	**12:00–3:45**	**4:00–7:45**
Monday	3,000	2,800	4,100
Tuesday	3,300	3,000	4,400
Wednesday	4,000	3,600	5,000

Table 9.15: Axle-Counts for Coverage Stations (Problem 9-1)

Station	Day	Time	Axle Count
1	Monday	8:00–11:45	1,900
2	Monday	12:00–3:45	2,600
3	Monday	4:00–7:45	1,500
4	Tuesday	8:00–11:45	3,000
5	Tuesday	12:00–3:45	3,600
6	Tuesday	4:00–7:45	4,800
7	Wednesday	8:00–11:45	3,500
8	Wednesday	12:00–3:45	3,200
9	Wednesday	4:00–7:45	4,400

Table 9.16: Sample Vehicle Classification Count (Problem 9-1)

Vehicle Class	Vehicle Count
2-axle	1,100
3-axle	130
4-axle	40
5-axle	6

9-2. The following control counts were made at state-maintained permanent count station. From the information given, calibrate the daily volume variation factors for this station:

Table 9.17: Data for Problem 9-2

Day of Week	Average Annual Volume for Day
Sunday	3,500
Monday	4,400
Tuesday	4,200
Wednesday	4,300
Thursday	3,900
Friday	4,900
Saturday	3,100

9-3. What count period would you select for a volume study at an intersection with a signal cycle length of (a) 60 seconds, (b) 90 seconds, and (c) 120 seconds?

9-4. The following control counts were made at an urban count station to develop daily and monthly variation factors. Calibrate these factors given the data shown here.

Table 9.18: 24-Hour Daily Volumes

First Week in Month of:	Day of Week						
	Mon	Tue	Wed	Thu	Fri	Sat	Sun
January	2,000	2,200	2,250	2,000	1,800	1,500	950
April	1,900	2,080	2,110	1,890	1,750	1,400	890
July	1,700	1,850	1,900	1,710	1,580	1,150	800
October	2,100	2,270	2,300	2,050	1,800	1,550	1,010

Table 9.19: Standard Monthly Volumes

Third Week in Month of:	Average 24-Hour Count (vehs)
January	2,250
February	2,200
March	2,000
April	2,100
May	1,950
June	1,850
July	1,800
August	1,700
September	2,000
October	2,100
November	2,150
December	2,300

9-5. The four control stations shown nearby have been regrouped for the purposes of calibrating daily variation factors. Is the grouping appropriate? If not, what would an appropriate grouping be? What are the combined daily variation factors for the appropriate group(s)? The stations are located sequentially along a state route.

Table 9.20: Daily Variation Factors for Individual Stations

Station	Mon	Tue	Wed	Thu	Fri	Sat	Sun
1	1.04	1.00	0.96	1.08	1.17	0.90	0.80
2	1.12	1.07	0.97	1.06	1.02	0.87	0.82
3	0.97	0.99	0.89	1.01	0.86	1.01	1.06
4	1.01	1.00	1.01	1.09	1.10	0.85	0.85

9-6. Estimate the annual VMT for a section of the state highway system represented by the variation factors of Table 9.11. The coverage counts shown in Table 9.21 are available for the locations within the section.

Table 9.21: Coverage Count Data

Station	Segment Length (Mi)	Coverage Count Date	24-Hour Count (vehs)
1	3.0	Wed in March	9,120
2	2.7	Tue in September	10,255
3	2.5	Fri in August	16,060
4	4.6	Sun in May	21,858
5	1.8	Thu in December	9,508
6	1.6	Fri in January	11,344

9-7. The following origin and destination results were obtained from sample license plate observations at five locations. Expand and adjust the initial trip-table results to reflect the full population of vehicles during the study period.

Table 9.22: Initial Origin and Destination Matches from Sample License-Plate Observations

Destination Station	Origin Station 1	2	3	4	5	Total Destination Count (vehs)
1	50	120	125	210	75	1,200
2	105	80	143	305	100	2,040
3	125	100	128	328	98	1,500
4	82	70	100	125	101	985
5	201	215	180	208	210	2,690
Total Origin Count (vehs)	**1,820**	**1,225**	**1,750**	**2,510**	**1,110**	**8,415**

CHAPTER 10

Speed, Travel Time, and Delay Studies

10.1 Introduction

Speed, travel time, and delay are all related measures commonly used as indicators of performance for traffic facilities. All relate to a factor that is most directly experienced by motorists: How long does it take to get from A to B? Motorists have the obvious desire to complete their trip in the minimum time consistent with safety. The performance of a traffic facility is often described in terms of how well that objective is achieved.

In the *Highway Capacity Manual* [*1*], for example, average travel speed is used as a measure of effectiveness for arterials, for two-lane rural highways, and for more extensive facility evaluations. Control delay is the measure of effectiveness for signalized and STOP-controlled intersections. Whereas freeways use density as a primary measure of effectiveness, speed is an important component of the evaluation of freeway system operation.

Thus traffic engineers must understand how to measure and interpret data on speed, travel time, and delay in ways that yield a basic understanding of the quality of operations on a facility and in ways that directly relate to defined performance criteria. Speed is also an important factor in evaluating high-accident locations as well as in other safety-related investigations.

Speed is inversely related to travel time. The reasons and locations at which speeds or travel times would be measured are, however, quite different. Speed measurements are most often taken at a point (or a short segment) of roadway under conditions of free flow. The intent is to determine the speeds that drivers select, unaffected by the existence of congestion. This information is used to determine general speed trends, to help determine reasonable speed limits, and to assess safety. Such studies are referred to as "spot speed studies" because the focus is on a designated "spot," or location, on a facility.

Travel time must be measured over a distance. Although spot speeds can indeed be measured in terms of travel times over a short measured distance (generally < 1,000 ft), most travel-time measurements are made over a significant length of a facility. Such studies are generally done during times of congestion specifically to measure or quantify the extent and causes of congestion.

In general terms, delay is a portion of total travel time. It is a portion of travel time that is particularly identifiable and unusually annoying to the motorist. Delay along an arterial, for example, might include stopped time due to signals, midblock obstructions, or other causes of congestion.

At signalized and STOP-controlled intersections, delay takes on more importance because travel time is difficult to define for a point location. Unfortunately, delay at intersections, specifically signalized intersections, has many different definitions, and the traffic engineer must be careful to use measurements and criteria that relate to the same delay definition.

Some of the most frequently used forms of intersection delay include the following:

- *Stopped-time delay*—the time a vehicle spends stopped waiting to proceed through a signalized or STOP-controlled intersection.
- *Approach delay*—adds the delay due to deceleration to and acceleration from a stop to stopped time delay.
- *Time-in-queue delay*—the time between a vehicle joining the end of a queue at a signalized or STOP-controlled intersection and the time it crosses the STOP line to proceed through the intersection.
- *Control delay*—the total delay at an intersection caused by a control device (either a signal or a STOP-sign), including both time-in-queue delay plus delays due to acceleration and deceleration.

Control delay was a term introduced in the 1985 *Highway Capacity Manual* [2], and it is used as the measure of effectiveness for signalized and STOP-controlled intersections.

Along routes, another definition of delay may be applied: *Travel-time delay* is the difference between the actual travel time traversing a section of highway and the driver's expected or desired travel time. It is more of a philosophical approach because there are no clearly accurate methodologies for determining the expected travel time of a motorist over a given section of highway. For this reason, it is seldom used for assessing congestion along a highway segment.

Because speeds are generally studied at points under conditions of free flow and travel times and delays are generally studied along sections of roadway under congested conditions, the study techniques for each are quite different, as discussed in Chapter 8. Although sharing many similar elements, the analysis of data and the presentation of results also differ somewhat.

10.2 Spot Speed Studies

Spot speed studies are conducted to document the distribution of vehicle speeds as they pass a point or short segment of the roadway. Because the traffic engineer is interested in conducting spot speed studies under conditions of free flow (i.e., observed speeds are not impeded by volume and density conditions), they are generally not conducted when volumes are in excess of 750 to 1,000 veh/h/ln on freeways or 500 veh/h/ln on other types of uninterrupted flow facilities.

10.2.1 Speed Definitions of Interest

When the speeds of individual vehicles are measured at a given spot or location, the result is a *distribution* of speeds because no two vehicles will be traveling at exactly the same speed. The results of the study, therefore, must describe the observed distribution of speeds as clearly as possible. Several key statistics are used to describe spot speed distributions:

- *Average or time mean speed:* The average speed of all vehicles passing the study location during the period of the study.
- *Standard deviation:* In simplistic terms, the standard deviation of speeds is the average difference between observed speeds and the time mean speed during the period of the study.
- *85th percentile speed:* The speed below which 85% of the vehicles travel.
- *Median:* The speed that equally divides the distribution of spot speeds; 50% of observed speeds are higher than the median; 50% of observed speeds are lower than the median.
- *Pace:* A 10-mi/h increment in speeds that encompasses the highest proportion of observed speeds (as compared with any other 10-mi/h increment).

The desired result of a spot speed study is to determine each of these measures and to determine an adequate mathematical description of the entire observed distribution.

10.2.2 Uses of Spot Speed Data

The results of spot speed studies are used for many different purposes by traffic engineers, including:

- Establishing the effectiveness of new or existing speed limits or enforcement practices.
- Determining appropriate speed limits for application.
- Establishing speed trends at the local, state, and national level to assess the effectiveness of national policy on speed limits and enforcement.
- Specific design applications determining appropriate sight distances, relationships between speed and highway alignment, and speed performance with respect to steepness and length of grades.
- Specific control applications for the timing of "yellow" and "all red" intervals for traffic signals, proper

placement of signs, and development of appropriate signal progressions.

- Investigation of high-accident locations at which speed is suspected to be a contributing cause to the accident experience.

This list is illustrative. It is not intended to be complete because there are myriad situations that may require speed data for a complete analysis. Such studies are of significant importance and are among the tasks most commonly conducted by traffic engineers.

10.2.3 Analysis of Spot Speed Data

The best way to present the analysis of typical spot speed data is by example. The discussions of this section are illustrated using a comprehensive sample application throughout. Figure 10.1 represents a typical set of field data from a spot speed study taken at location of interest on a major arterial. The data are collected as frequencies of observances in predefined speed groups. This method of field data summary is very much related to the statistical analysis that will be applied.

Because the observed speeds form a distribution, they will eventually be described in terms of a continuous distribution function. The mathematical characteristics of a continuous distribution do not allow for the description of the probability of any distinct value occurring—in a continuous function, one discrete speed is one value in a distribution with an infinite number of such values. In more practical terms, a continuous distribution cannot describe the occurrence of a speed of exactly 44.72 mi/h. It can, however, describe the

Route 10 @ *MP* 125.3
LOCATION

July 10, 2003
DATE

1:00 - 4:00 PM
TIME

Good - Clear, Dry
WEATHER CONDITIONS

Asphaltic concrete - good.
ROADWAY SURFACE CONDITIONS

SPEED GROUP		TIME GROUP		PASSENGER CARS	TRUCKS	OTHER	TOTALS			
Lower limit (mph)	Upper limit (mph)	Lower limit (secs)	Upper limit (secs)				PC	Trucks	Other	Total
30	32									
32	34									
34	36			II	II	I	2	2	1	5
36	38			III	II		3	2	0	5
38	40			𝍸	I	I	5	1	1	7
40	42			𝍸 𝍸	III		10	3	0	13
42	44			𝍸 𝍸 𝍸 III	III		18	3	0	21
44	46			𝍸 𝍸 𝍸 𝍸 𝍸 IIII	IIII		29	4	0	33
46	48			𝍸 𝍸 𝍸 𝍸 𝍸 𝍸 𝍸 𝍸 II	II	II	42	2	2	46
48	50			𝍸 𝍸 𝍸 𝍸 𝍸 𝍸 𝍸 𝍸 𝍸 𝍸 𝍸 𝍸	II		60	2	0	62
50	52			𝍸 𝍸 𝍸 𝍸 𝍸 𝍸 𝍸 II			37	0	0	37
52	54			𝍸 𝍸 𝍸 𝍸 III			23	1	0	24
54	56			𝍸 𝍸 III		I	13	0	1	14
56	58			𝍸 II	I	I	7	1	1	9
58	60			𝍸			5	0	0	5
60	62			II			2	0	0	2
62	64									
64	66									
66	68									
68	70									

METHOD OF MEASUREMENT
x Radar
_____ Time over measured course length of _____ ft.
_____ Stop watch/manual
_____ Road tubes w/timer
_____ Electronic contact w/timer

Signature

7/10/03

Figure 10.1: Field Data for an Illustrative Spot Speed Study

occurrence of a speed in the range of 44.7 to 44.8 mi/h. Therefore, the statistical analysis of speed data is based on the number of observed values within a set of defined speed ranges.

The data shown in Figure 10.1 use speed groups that are 2 mi/h in breadth. This is a practical value that is quite typical, although 1 mi/h groups are also used if the sample sizes are large enough. For statistical reasons that are explained later, speed groups of more than 5 mi/h are never used. The number of speed groups defined must relate to the expected range of the data and to the number of speeds that will be observed and recorded. For example, defining 15 speed groups and collecting only 30 speeds would be illogical because there would only be an average of two observations per group. In general, it is customary to collect from 15 to 20 speeds for each defined speed group. This *does not* imply that each group would have 15 to 20 observations; rather, the total number of observations will be sufficient to define the underlying distribution and its characteristics.

Frequency Distribution Table

The first analysis step is to take the data of Figure 10.1 and reformat it into the form of a frequency distribution table, as illustrated in Table 10.1. This tabular array shows the total number of vehicles observed in each speed group. For the convenience of subsequent use, the table includes one speed group at each extreme for which no vehicles were observed. The "middle speed" (S) of the third column is taken as the midpoint value within the speed group. The use of this value is discussed in a later section.

The fourth column of the table shows the number of vehicles observed in each speed group. This value is known as the *frequency* for the speed group. These values are taken directly from the field sheet of Figure 10.1.

In the fifth column, the percentage of total observations in each speed group is computed as:

$$\% = 100 \frac{n_i}{N} \tag{10-1}$$

Table 10.1: Frequency Distribution Table for Illustrative Spot Speed Study

Speed Group							
Lower Limit (mi/h)	Upper Limit (mi/h)	Middle Speed S (mi/h)	Observed Freq. in Group n	% Freq. in Group (%)*	Cum % Freq (%)*	nS**	nS²**
32	34	33	0	0.0%	0.0%	0	0
34	36	35	5	1.8%	1.8%	175	6,125
36	38	37	5	1.8%	3.5%	185	6,845
38	40	39	7	2.5%	6.0%	273	10,647
40	42	41	13	4.6%	10.6%	533	21,853
42	44	43	21	7.4%	18.0%	903	38,829
44	46	45	33	11.7%	29.7%	1,485	66,825
46	48	47	46	16.3%	45.9%	2,162	101,614
48	50	49	62	21.9%	67.8%	3,038	148,862
50	52	51	37	13.1%	80.9%	1,887	96,237
52	54	53	24	8.5%	89.4%	1,272	67,416
54	56	55	14	4.9%	94.3%	770	42,350
56	58	57	9	3.2%	97.5%	513	29,241
58	60	59	5	1.8%	99.3%	295	17,405
60	62	61	2	0.7%	100.0%	122	7,442
62	64	63	0	0.0%	100.0%	0	0
			283	**100.0%**		**13,613**	**661,691**

*All percents computed to two decimal places and rounded to one; this may cause apparent "errors" in cumulative percents due to rounding.
**Computations rounded to the nearest whole number.

where: n_i = number of observations (frequency) in speed group i

N = total number of observations in the sample

For the 40–42 mi/h speed group, there are 13 observations in a total sample of 283 speeds. Thus the percent frequency is 100*(13/283) = 4.6% for this group. The cumulative percent frequency (cum %) is the percentage of vehicles traveling at or below the highest speed in the speed group:

$$cum\% = 100\left(\sum_{1-x} n_i \Big/ N\right) \qquad (10\text{-}2)$$

where: x = consecutive number (starting with the lowest speed group) of the speed group for which the cum % frequency is desired

For the 40–42 mi/h speed group, the sum of the frequencies for all speed groups having a high-speed boundary of 42 mi/h or less is found as 13 + 7 + 5 + 5 + 0 = 30. The cum % frequency is then 100 * (30/283) = 10.6%.

The last two columns of the frequency distribution table are simple multiplications that will be used in subsequent computations.

Frequency and Cumulative Frequency Distribution Curves

The data in Table 10.1 are used to plot two curves that lend a visual impact to the information: (1) a Frequency Distribution Curve and (2) a Cumulative Frequency Distribution Curve. These are illustrated in Figure 10.2 and plotted as follows:

- *Frequency distribution curve.* For each speed group, the % frequency of observations within the group is plotted versus the middle speed of the group (S).
- *Cumulative frequency distribution curve.* For each speed group, the % cumulative frequency of observations is plotted versus the higher boundary of the speed group.

Note that the two frequencies are plotted versus *different* speeds. The middle speed is used for the frequency distribution curve. The cumulative frequency distribution curve, however, results in a very useful plot of speed versus the percent of vehicles traveling at or below the designated speed. For this reason, the upper limit of the speed group is used as the plotting point.

In both cases, the plots are connected by a *smooth* curve that minimizes the total distance of points falling above the line and those falling below the line (on the vertical axis). A smooth curve is defined as one without any breaks in the slope of the curve. The "best fit" is done approximately (by eye), generally a lightly sketched curve in freehand. A French curve may then be used to darken the line. Some statistical packages plot such a line automatically.

It is also convenient to plot the frequency distribution curve directly above the cumulative frequency distribution curve, using the same horizontal scale. This makes it easier to use the curves to extract critical parameters graphically. Figure 10.2 also illustrates the graphic determination of several key variables that help describe the observed distribution. These parameters are defined and their determination explained in the sections that follow.

Common Descriptive Statistics

Common descriptive statistics may be computed from the data in the frequency distribution table or determined graphically from the frequency and cumulative frequency distribution curves. These statistics are used to describe two important characteristics of the distribution:

- *Central tendency:* Measures that describe the approximate middle or center of the distribution.
- *Dispersion:* Measures that describe the extent to which data spreads around the center of the distribution.

Measures of central tendency include the average or mean speed, the median speed, the modal speed, and the pace. Measures of dispersion include the 85th and 15th percentile speeds and the standard deviation.

The Mean Speed: A Measure of Central Tendency The average or mean speed of a distribution is usually easily found as the sum of the observed values divided by the number of observations. In a spot speed study, however, individual values of speed are not recorded; rather, the frequency of observations within defined speed groups is known. Computing the mean speed requires the assumption that *the average speed within a given speed group is the middle speed, S, of the group.* This is the reason that speed groups of more than 5 mi/h are never used. This assumption becomes less valid as the size of the speed groups increases. For 2 mi/h speed groups, as in the illustrative study, the assumption is usually quite good. If this assumption is made, the sum of all speeds in a given speed group may be computed as:

$$n_i S_i$$

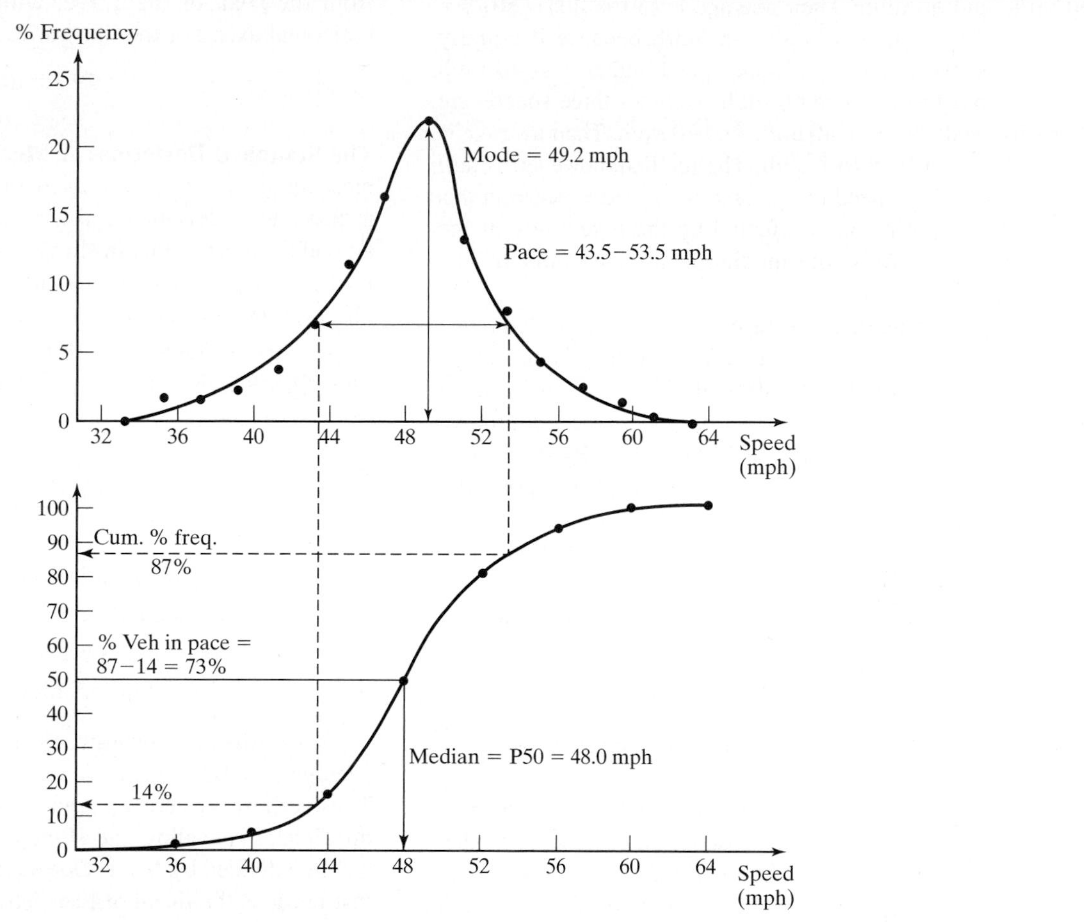

Figure 10.2: Frequency and Cumulative Frequency Distribution Curves for the Illustrative Spot-Speed Distribution

where: n_i = frequency of observations in speed group i
S_i = middle speed of speed group i

The sum of all speeds in the distribution may then be found by adding this product for all speed groups:

$$\sum_i n_i S_i$$

The mean or average speed is then computed as the sum divided by the number of observed speeds:

$$\bar{x} = \frac{\sum_i n_i S_i}{N} \qquad (10\text{-}3)$$

where: $\bar{x}$ = average speed for the sample observations, mi/h
N = total sample size

For the illustrative study data presented in Figure 10.2 and Table 10.1, the average or mean speed is:

$$\bar{x} = \frac{13{,}613}{283} = 48.1 \text{ mi/h}$$

where $\Sigma n_i S_i$ is the sum of the next-to-last column of the frequency distribution table of Table 10.1.

The Median Speed: Another Measure of Central Tendency The median speed is defined as the speed that divides the distribution into equal parts (i.e., there are as many observations of speeds higher than the median as there are lower than the median). It is a positional value and not affected by the absolute value of extreme observations.

The difference between the median and mean is best illustrated by example. Three speeds are observed: 30 mi/h,

40 mi/h, and 50 mi/h. Their average is (30 + 40 + 50)/3 = 40 mi/h. Their median is also 40 mi/h because it equally divides the distribution, with one speed higher than 40 mi/h and one speed lower than 40 mi/h. Another three speeds are then observed: 30 mi/h, 40 mi/h, and 70 mi/h. Their average is (30 + 40 + 70)/3 = 46.7 mi/h. The median, however, is still 40 mi/h, with one speed higher and one speed lower than this observation. The mean is affected by the *magnitude* of the extreme observations; the median is affected only by the *number* of such observations.

Because individual speeds have not been recorded in the illustrative study, however, the "middle value" is not easily determined from the tabular data of Table 10.1. It is easier to estimate the median graphically using the cumulative frequency distribution curve of Figure 10.2. By definition, the median equally divides the distribution. Therefore, 50% of all observed speeds should be less than the median. This is exactly what the cumulative frequency distribution curve plots. If the curve is entered at 50% on the vertical axis, the median speed is found, as illustrated in Figure 10.2. For the illustrative study:

$$P_{50} = 48.0 \text{ mi/h}$$

where P_{50} is the median or 50th percentile speed.

The Pace: Another Measure of Central Tendency The pace is a traffic engineering measure not commonly used for other statistical analyses. It is defined as *the 10-mi/h increment in speed in which the highest percentage of drivers is observed.* It is also found graphically using the frequency distribution curve of Figure 10.2. The solution recognizes that the area under the frequency distribution curve between any two speeds approximates the percentage of vehicles traveling between those two speeds, where the total area under the curve is 100%.

The pace is found as follows: A 10-mi/h template is scaled from the horizontal axis. Keeping this template horizontal, place an end on the lower left side of the curve and move slowly along the curve. When the right side of the template intersects the right side of the curve, the pace has been located. This procedure identifies the 10-mi/h increment that intersects the peak of the curve; this contains the most area and, therefore, the highest percentage of vehicles. The pace is shown in Figure 10.2 as:

$$43.5 - 53.5 \text{ mi/h}$$

The Modal Speed: Another Measure of Central Tendency The mode is defined as the single value of speed that is most likely to occur. Because no discrete values were recorded, the modal speed is also determined graphically from the frequency distribution curve. A vertical line is dropped from the peak of the curve, with the result found on the horizontal axis. For the illustrative study, the modal speed is:

$$49.2 \text{ mi/h}$$

The Standard Deviation: A Measure of Dispersion The most common statistical measure of dispersion in a distribution is the standard deviation. It is a measure of how far data spreads around the mean value. In simple terms, the standard deviation is the average value of the difference between individual observations and the average value of those observations. Where discrete values of a variable are available, the equation for computing the standard deviation is:

$$s = \sqrt{\frac{\sum_i (x_i - \bar{x})^2}{N - 1}} \tag{10-4}$$

where: s = the standard deviation

x_i = observation i

$\bar{x}$ = average of all observations

N = number of observations

The difference between a given data point and the average is a direct measure of the magnitude of dispersion. These differences are squared to avoid positive and negative differences canceling and summed for all data points. They are then divided by $N - 1$. One statistical *degree of freedom* is lost because the mean of the distribution is known and used to compute the differences. If there are three numbers and it is known that the differences between the values and the mean for the first two are "3" and "2," then the third or last difference must be "–5," because the sum of all differences must be zero. Only the first "$N - 1$" observations of differences are statistically random. A square root is taken of the results because the values of the differences were squared to begin the computation.

Because discrete values of speed are not recorded, Equation 10-3 is modified to reflect group frequencies:

$$s = \sqrt{\frac{\sum n_i (S_i - \bar{x})^2}{N - 1}}$$

which may be manipulated into a more convenient form, as follows:

$$s = \sqrt{\frac{\sum n_i S_i^2 - N\bar{x}^2}{N - 1}} \tag{10-5}$$

where all terms are as previously defined. This form is most convenient because the first term is the sum of the last column of the frequency distribution table of Table 10.1. For the illustrative study, the standard deviation is:

$$s = \sqrt{\frac{661{,}691 - 283^{*}48.1^2}{183 - 1}} = 4.96\,mi/h$$

Most observed speed distributions have standard deviations that are close to 5 mi/h because this represents most driver behavior patterns reasonably well. Unlike averages and other central speeds, which vary widely from location to location, most speed studies yield similar standard deviations.

The 85th and 15th Percentile Speeds The 85th and 15th percentile speeds give a general description of the high and low speeds observed by most reasonable drivers. It is generally thought that the upper and lower 15% of the distribution represents speeds that are either too fast or too slow for existing conditions. These values are found graphically from the cumulative frequency distribution curve of Figure 10.2. The curve is entered on the vertical axis at values of 85% and 15%. The respective speeds are found on the horizontal axis, as shown in Figure 10.2. For the illustrative study, these speeds are:

$$P_{85} = 52.7 \text{ mi/h}$$
$$P_{15} = 43.7 \text{ mi/h}$$

The 85th and 15th percentile speeds can be used to roughly estimate the standard deviation of the distribution, although this is not recommended when the data are available for a precise determination:

$$s_{est} = \frac{P_{85} - P_{15}}{2} \tag{10-6}$$

where all terms are as previously defined. For the illustrative spot speed study:

$$s_{est} = \frac{52.7 - 43.7}{2} = 4.5 \text{ mi/h}$$

In this case, the estimated value is relatively close to the actual computed value of 4.96 mi/h.

The 85th and 15th percentile speeds give insight to both the central tendency and dispersion of the distribution. As these values get closer to the mean, less dispersion exists and the stronger the central tendency of the distribution becomes.

Percentage Vehicles Within the Pace The pace itself is a measure of the center of the distribution. The percentage of vehicles traveling within the pace speeds is a measure of both central tendency and dispersion. The smaller the percentage of vehicles traveling within the pace, the greater the degree of dispersion in the distribution.

The percentage of vehicles within the pace is found graphically using both the frequency distribution and cumulative frequency distribution curves of Figure 10.2. The pace speeds were determined previously from the frequency distribution curves. Lines from these speeds are dropped vertically to the cumulative frequency distribution curve. The percentage of vehicles traveling at or below each of these speeds can then be determined from the vertical axis of the cumulative frequency distribution curve, as shown. Then:

% Vehicles under 53.5 mi/h = 87.0%
% Vehicles under 43.5 mi/h = 14.0%
% Veh between 43.5 and 53.5 mi/h = 73.0%

Even though speeds between 34 and 62 mi/h were observed in this study, over 70% of the vehicles traveled at speeds between 43.5 and 53.5 mi/h. This represents expected traffic behavior with a standard deviation of approximately 5 mi/h.

Using the Normal Distribution in the Analysis of Spot Speed Data

Most speed distributions tend to be statistically normal (i.e., they can be reasonably represented by a normal distribution). Chapter 7 contains a detailed description of the normal distribution and its properties that should be reviewed in conjunction with this section.

If observed speeds are assumed to be normally distributed, then several additional analyses of the data may be conducted. Recall that the standard notation $x{:}N[40,25]$ signifies that the variable "x" is normally distributed with a mean of "40" and a variance of "25." The standard deviation is the square root of the variance, or "5" in this case. Recall also that a value of "x" on any normal distribution can be converted to an equivalent value of "z" on the standard normal distribution, where $z{:}\ N[0,1]$:

$$z_i = \frac{x_i - \mu}{\sigma} \tag{10-7}$$

where: x_i = a value on any normal distribution $x{:}N[\mu, \sigma^2]$
μ = the true mean of the distribution of values x_i
σ = the true standard deviation of the distribution of values x_i
z_i = equivalent value on the standard normal distribution $z{:}N[0,1]$

In practical terms, the true values, μ and σ are unknown. What results from a spot speed study are estimates of the true mean and standard deviation of the distribution based on a measured sample, $\bar{x}$ and s. A table of values of the standard normal distribution is included in Chapter 7 and used in the analysis of the illustrative data.

Precision and Confidence Intervals When a spot speed study is conducted, a single value of the mean speed is computed. For the illustrative study of this chapter, the mean is 48.1 mi/h, based on a sample of 283 observations. In effect, this value, based on a finite number of measured speeds, is being used to estimate the true mean of the underlying distribution of all vehicles traversing the site under uncongested conditions. The number of such vehicles, for all practical and statistical purposes, is infinite. The measured value of $\bar{x}$ is being used as an estimate for μ. The first statistical question that must be answered is: How good is this estimate?

In Chapter 7, the standard error of the mean, E, was introduced and defined. If a variable x is normally distributed:

$$x{:}N[\mu, \sigma^2]$$

it can be shown that the distribution of sample means (of a set of means with a constant sample size, n) is also normally distributed, as follows:

$$\bar{x}_n{:}N\left[\mu, \left(\sigma^2 / n\right)\right]$$

Assume that 100 speed observations had an average value of 50 mi/h. The speeds are then arranged in 10 groups of 10 speeds, and 10 separate averages are computed (one for each group). The average of the 10 group averages would still be 50 mi/h because the mean of the distribution of sample means is the same as the mean of the original distribution. The standard deviations, however, would be different because the grouping and averaging process significantly reduces the occurrence of extreme values. For example, in a distribution with an average speed of 50 mi/h, it is conceivable that some observations of 70 mi/h or more would be obtained. However, at the same site, it is highly unlikely that the average of any 10 observed speeds would be 70 mi/h or higher.

The standard error of the mean, E, is simply the standard deviation of a distribution of sample means with a constant group size of n:

$$E = s/\sqrt{n} \qquad (10\text{-}8)$$

where: E = standard error of the mean

s = standard deviation of the original distribution of individual values

n = number of samples in each group of observations

The characteristics of the normal distribution are also discussed in Chapter 7. These characteristics, together with the standard error of the mean, can be used to quantify the quality of the sample estimate of the true mean of the underlying distribution. In effect, the entire illustrative spot speed study (with its sample size of 283 values) is considered to be a single point on a distribution of sample means, all with a group size of 283. Assuming a normal distribution, it is known that 95% of all values lie between the mean $\pm$ 1.96 standard deviations; 99.7% of all values lie between the mean $\pm$3.00 standard deviations. Thus it is 95% certain that the sample mean (48.1 mi/h) is within the range of the true mean $\pm$1.96 standard deviations. The standard deviation is, in this case, the standard error of the mean. Then:

$$\bar{x} = \mu \pm 1.96E \Rightarrow \mu = \bar{x} \pm 1.96E \qquad (10\text{-}9)$$

95% of the time. The percentage is referred to as the confidence interval, whereas the precision of the measurement is given by the term 1.96 E. For the illustrative spot speed study:

$$E = \frac{4.96}{\sqrt{283}} = 0.295 \text{ mi/h}$$
$$\mu = 48.1 \pm 1.96(0.295) = 48.1 \pm 0.578$$
$$\mu = 47.522 - 48.678 \text{ mi/h}$$

Rounding off the values, it can be stated that we are *95% confident* that the true mean of the underlying speed distribution lies between 47.5 and 48.7 mi/h. For a 99.7% confidence level:

$$\bar{x} = \mu \pm 3.00E \Rightarrow \mu = \bar{x} \pm 3.00E$$
$$\mu = 48.1 \pm 3.00(0.295)$$
$$\mu = 48.1 \pm 0.885$$
$$\mu = 47.215 - 48.985 \text{ mi/h} \qquad (10\text{-}10)$$

Again rounding off these values, it can be stated that we are *99.7%* confident that the true mean of the underlying speed distribution lies between 47.2 and 49.0 mi/h.

These statements provide a quantitative description of the precision of the measurement and the confidence with which the estimate is given. Note that as the confidence level increases, the precision of the estimate decreases (i.e., the range of the estimate increases). Given that speeds are normally distributed, we can be 100% confident that the true mean speed lies between 48.10 mi/h $\pm \infty$.

Such a statement is useless in engineering terms. Because spot speed studies represent a sample of measurements selected from a virtually infinite population, the average can never be measured with complete precision and 100% confidence. The most common approach uses the 95% confidence interval to compute the precision and confidence of the sample mean as an estimator of the true mean of the underlying distribution.

Estimating the Required Sample Size Although it is useful to know the confidence level and precision of a measured sample mean after the fact, it is more useful to determine what sample size is required to obtain a measurement that satisfies a predetermined precision and confidence level. Given that the *precision* or *tolerance* (e) of the estimate is the $\pm$ range around the mean:

$$95\% : e = 1.96E = 1.96(s/\sqrt{n})$$

$$99.7\% : e = 3.00E = 3.00(s/\sqrt{n})$$

These equations can now be solved for the sample size, n. To obtain a desired precision with 95% confidence:

$$n = \frac{3.84s^2}{e^2} \qquad (10\text{-}11)$$

To obtain a desired precision with 99.7% confidence:

$$n = \frac{9.0s^2}{e^2} \qquad (10\text{-}12)$$

where all variables are as previously defined.

Consider the following problem: How many speeds must be collected to determine the true mean speed of the underlying distribution to within $\pm$1.0 mi/h with 95% confidence? How do the results change if the tolerance is changed to $\pm$0.5 mi/h and the confidence level to 99.7%?

The first problem is that the standard deviation of the distribution, s, is not known because the study has not yet been conducted. Here, practical use is made of the knowledge that most speed distributions have standard deviations of approximately 5.0 mi/h. This value is assumed, and the results are shown in Table 10.2.

A sample size of 96 speeds is required to achieve a tolerance of $\pm$1.0 mi/h with 95% confidence. To achieve a tolerance of $\pm$0.5 mi/h with 99.7% confidence, the required sample size must be almost 10 times greater. For most traffic engineering studies, a tolerance of $\pm$1.0 mi/h and a confidence level of 95% are quite sufficient.

Before and After Spot Speed Studies

In many situations, existing speeds at a given location should be reduced. This occurs in situations where a high accident and/or accident severity rate is found to be related to excessive speed. It also arises where existing speed limits are being exceeded by an inordinate number of drivers.

Many traffic engineering actions can help reduce speeds, including lowered speed limits, stricter enforcement measures, warning signs, installation of rumble strips, and others. The major study issue, however, is to demonstrate that speeds have indeed been successfully reduced.

This is not an easy issue. Consider the following scenario: Assume that a new speed limit has been installed at a given location in an attempt to reduce the average speed by 5 mi/h. A speed study is conducted before implementing the reduced speed limit, and another is conducted several months after the new speed limit is in effect. Note that the "after" study is normally conducted after the new traffic engineering measures have been in effect for some time. This is done so that stable driver behavior is observed, rather than a transient response to something new. It is observed that the average speed of the "after" study is 3.5 mi/h less than the average speed of the "before" study. Statistically, these two questions must be answered:

- Is the observed reduction in average speeds real?
- Is the observed reduction in average speeds the intended 5 mi/h?

Although both questions appear to have obvious answers, they in fact do not. There are two reasons that a reduction in

Table 10.2: Sample Size Computations Illustrated

Tolerance e (mi/h)	Confidence Level 95%	Confidence Level 99.7%
1.0	$n = \frac{3.84(5)^2}{(1.0)^2} = 96$	$n = \frac{9.0(5)^2}{(1.0)^2} = 225$
0.5	$n = \frac{3.84(5)^2}{(0.5)^2} = 384$	$n = \frac{9.0(5)^2}{(0.5)^2} = 900$

average speeds could have occurred: (1) the observed 3.5-mi/h reduction could occur because the new speed limit caused the true mean speed of the underlying distribution to be reduced; (2) the observed 3.5-mi/h reduction could also occur because two different samples were selected from an underlying distribution that did not change. In statistical terms, the first is referred to as a *significant* reduction in speeds, and the latter is statistically *not significant.*

The second question is equally tricky. Assuming that the observed 3.5-mi/h reduction in speeds is found to be statistically significant, it is necessary to determine whether the true mean speed of the underlying distribution has likely been reduced by 5 mi/h. Statistical testing will be required to answer both questions. Further, it will not be possible to answer either question with 100% certainty or confidence.

Chapter 7 introduced the concepts and methodologies for before-and-after testing for the significance of observed differences in sample means. The concept of truth tables was also discussed. The statistical tests for the significance of observed differences have four possible results: (1) the actual difference is significant, and the statistical test determines that it is significant; (2) the actual difference is not significant, and the statistical test determines that it is not significant; (3) the actual difference is significant and the statistical test determines that it is not significant; and (4) the actual difference is not significant and the statistical test determines that it is significant. The first two outcomes result in an accurate assessment of the situation; the last two represent erroneous results. In statistical terms, outcome (4) is referred to as a Type I, or α error; outcome (3) is referred to as a Type II, or β error.

In practical terms, the traffic engineer must avoid making a Type I error. In this case, it will appear that the problem (excessive speed) has been solved, when in fact it has not been solved. This may result in additional accidents, injuries, and/or deaths before the "truth" becomes apparent. If a Type II error is made, additional effort will be expended to entice lower speeds. Although this might involve additional expense, it is unlikely to lead to any negative results.

The statistical test applied to assess the significance of an observed reduction in mean speeds is the normal approximation. As discussed in Chapter 7, this test is applicable as long as the "before" and "after" sample sizes are more than or equal to 30, which will always be the case in properly conducted speed studies. To certify that an observed reduction is significant, we wish to be 95% confident that this is so. In other words, we wish to ensure that the chance of making a Type I error is less than 5%.

The normal approximation is applied by converting the observed reduction in mean speeds to a value of z on the standard normal distribution:

$$z_d = \frac{(\bar{x}_1 - \bar{x}_2) - 0}{s_y}$$

$$s_y = \sqrt{\frac{s_1^2}{N_1} + \frac{s_2^2}{N_2}} \tag{10-13}$$

where: z_d = standard normal distribution equivalent to the observed difference in sample speeds

$\bar{x}_1$ = mean speed of the "before" sample, mi/h

$\bar{x}_2$ = mean speed of the "after" sample, mi/h

s_y = pooled standard deviation of the distribution of sample mean differences

s_1 = standard deviation of the "before" sample, mi/h

s_2 = standard deviation of the "after" sample, mi/h

N_1 = sample size of the "before" study (must be $\geq$ 30)

N_2 = sample size of the "after" study (must be $\geq$ 30)

The standard normal distribution table of Table 7.3 (Chapter 7) is used to find the probability that a value equal to or less than z_d occurs when both sample means are from the same underlying distribution. Then:

- If Prob $(z \leq z_d) \geq 0.95$, the observed reduction in speeds is *statistically significant.*
- If Prob $(z \leq z_d) < 0.95$, the observed reduction in speeds is *not statistically significant.*

In the first case, it means that the observed difference in sample means would be exceeded less than 5% of the time, assuming that the two samples came from the same underlying distribution. Given that such a value was observed, this may be interpreted as being less than 5% probable that the observed difference came from the same underlying distribution and more than 95% probable that it resulted from a change in the underlying distribution.

Note that a *one-sided* test is conducted (i.e., we are testing the significance of an observed *reduction* in sample means, *not* an observed *difference* in sample means). If the observations revealed an increase in sample means, no statistical test is conducted because it is obvious that the desired result was not achieved.

If the observed reduction is found to be statistically significant, the second question can be entertained (i.e., was the target speed reduction achieved?). This is done using only the results of the "after" distribution. Note that from the normal distribution characteristics, it is 95% probable that the true mean of the distribution is:

$$\mu = \bar{x} \pm 1.96E$$

If the target speed lies within this range, it can be stated that it was successfully achieved.

Consider the following results of a before-and-after spot speed study conducted to evaluate the effectiveness of a new speed limit intended to reduce the average speed at the location to 60 mi/h:

Before Results		After Results
65.3 mi/h	$\bar{x}$	63.0 mi/h
5.0 mi/h	s	6.0 mi/h
50	N	60

A normal approximation test is conducted to determine whether the observed reduction in sample means is statistically significant:

Step 1: Compute the pooled standard deviation.

$$s_y = \sqrt{\frac{5.0^2}{50} + \frac{6.0^2}{60}} = 1.05\text{mi/h}$$

Step 2: Compute z_d.

$$z_d = \frac{(65.3 - 63.0) - 0}{1.05} = 2.19$$

Step 3: Determine the prob ($z \leq 2.19$) from Table 7.3.

$$\text{Prob } (z \leq 2.19) = 0.9857$$

Step 4: Compare results with the 95% criteria.

As 98.57% > 95% the results indicate that the observed reduction in sample means was statistically significant.

Given these results, it is now possible to investigate whether or not the target speed of 60 mi/h was successfully achieved in the "after" sample. The 95% confidence interval for the "after" estimate of the true mean of the underlying distribution is:

$$E = 6/\sqrt{60} = 0.7746$$

$$\mu = 63.0 \pm 1.96(0.7746)$$

$$\mu = 63.0 \pm 1.52$$

$$\mu = 61.48 - 64.52\text{mi/h}$$

Because the target speed of 60 mi/h does not lie in this range, it cannot be stated that it was successfully achieved.

In this case, although a significant reduction of speeds was achieved, it was not sufficient to achieve the target value of 60 mi/h. Additional study of the site would be undertaken and additional measures enacted to achieve additional speed reduction.

The 95% confidence criteria for certifying a significant reduction in observed speeds should be well understood. If a before-and-after study results in a confidence level of 94.5%, it would not be certified as statistically significant. This decision limits the probability of making a Type I error to less than 5%. When we state that the observed difference in mean speeds is not statistically significant in this case, however, it is 94.5% probable that we are making a Type II error. Before expending large amounts of funds on additional speed-reduction measures, a larger "after" speed sample should be taken to see whether or not 95% confidence can be achieved with an expanded database.

Testing for Normalcy: The Chi-Square Goodness-of-Fit Test

Virtually all of the statistical analyses of this section start with the basic assumption that the speed distribution can be mathematically represented as normal. For completeness, it is therefore necessary to conduct a statistical test to confirm that this assumption is correct. As described in Chapter 7, the chi-square test is used to determine whether the difference between an observed distribution and its assumed mathematical form is significant. For grouped data, the chi-squared statistic is computed as:

$$\chi^2 = \sum_{N_G} \frac{(n_i - f_i)^2}{f_i} \qquad (10\text{-}14)$$

where: χ^2 = chi-squared statistic

n_i = frequency of observations in speed group i

f_i = theoretical frequency in speed group i, assuming that the assumed distribution exists

N_G = number of speed groups in the distribution

Table 10.3 shows these computations for the illustrative spot speed study. Speed groups are already specified, and the observed frequencies are taken directly from the field sheet of Figure 10.1.

For convenience, the speed groups are listed from highest to lowest. This is to coordinate with the standard normal distribution table of Chapter 7, which gives probabilities of $z \leq z_d$. The upper limit of the highest group is adjusted to "infinity" because the theoretical normal distribution extends to both positive and negative infinity. The remaining columns of Table 10.3 focus on determining the theoretical frequencies, f_i and on determining the final value of χ^2.

Table 10.3: Chi-Square Test for Normalcy on Illustrative Spot Speed Data

Average Speed = 48.10 mi/h Standard Deviation = 4.96 mi/h Sample Size = 283

Speed Group									
Upper Limit (mi/h)	**Lower Limit (mi/h)**	**Observed Frequency** n	**Upper Limit (Std. Normal)** z_d	**Prob.** $z \leq z_d$ **Table 7.3**	**Prob. of Occurrence in Group**	**Theoretical Frequency** f	**Combined Groups** n	**Combined Groups** f	χ^2 **Group**
∞	60	2	∞	1.0000	0.0082	2.3206			
60	58	5	2.40	0.9918	0.0146	4.1318	7	6.4524	0.0465
58	56	9	2.00	0.9772	0.0331	9.3673	9	9.3673	0.0144
56	54	14	1.59	0.9441	0.0611	17.2913	14	17.2913	0.6265
54	52	24	1.19	0.8830	0.0978	27.6774	24	27.6774	0.4886
52	50	37	0.79	0.7852	0.1372	38.8276	37	38.8276	0.0860
50	48	62	0.38	0.6480	0.1560	44.1480	62	44.1480	7.2188
48	46	46	−0.02	0.4920	0.1548	43.8084	46	43.8084	0.1096
46	44	33	−0.42	0.3372	0.1339	37.8937	33	37.8937	0.6320
44	42	21	−0.83	0.2033	0.0940	26.6020	21	26.6020	1.1797
42	40	13	−1.23	0.1093	0.0577	16.3291	13	16.3291	0.6787
40	38	7	−1.63	0.0516	0.0309	8.7447	7	8.7447	0.3481
38	36	5	−2.04	0.0207	0.0134	3.7922	10	5.8581	2.9285
36	34	5	−2.44	0.0073	0.0073	2.0659			
Total					**1.0000**	**283**	**283**	**283**	**14.3574**

$\chi^2 = 14.3574$
Degrees of Freedom = 12 − 3 = 9

The theoretical frequencies are the numbers of observations that would have occurred in the various speed groups *if the distribution were perfectly normal.* To find these values, the probability of an occurrence within each speed group must be determined from the standard normal table of Chapter 7. This is done in columns 4 through 7 of Table 10.3, as follows:

1. The upper limit of each speed group (in mi/h) is converted to an equivalent value of z on the standard normal distribution, using Equation 10-8. This computation is illustrated for the speed group with an upper limit of 60 mi/h:

$$z_{60} = \frac{60.00 - 48.10}{4.96} = 2.40$$

Note that the mean speed and standard deviation of the illustrative spot speed study are used in this computation.

2. Each computed value of z is now looked up on the standard normal table of Chapter 7. From this, the probability of $z \leq z_d$ is found and entered into column 5 of Table 10.3.

3. Consider the 48 to 50 mi/h speed group in Table 10.3. From column 5, 0.6480 is the probability of speed ≤ 50 mi/h occurring on a normal distribution; 0.4920 is the probability of a speed ≤ 48 mi/h occurring. Thus the probability of an occurrence between 48 and 50 mi/h is 0.6480 − 0.4920 = 0.1560. The probabilities of column 6 are computed via sequential subtractions as shown here. The result is the probability of a speed being in any speed group, assuming a normal distribution.

4. The theoretical frequencies of column 7 are found by multiplying the sample size by the probability of an occurrence in that speed group. Fractional results are permitted for theoretical frequencies.

5. The chi-square test is valid only when all values of the theoretical frequency are 5 or more. To achieve this, the first two and last two speed groups must be combined. The observed frequencies are similarly combined.

6. The value of chi-square for each speed group is computed as shown. The computation for the 40 to 42 mi/h speed group is illustrated here:

$$\chi_i^2 = \frac{(n_i - f_i)^2}{f_i} = \frac{(13 - 16.3291)^2}{16.3291} = 0.6787$$

These values are summed to yield the final value of x^2 for the distribution, which is 14.3574.

To assess this result, the chi-square table of Chapter 7 is used. Probability values are shown on the horizontal axis of the table. The vertical axis shows *degrees of freedom.* For a chi-square distribution, the number of degrees of freedom is the number of data groups (after they are combined to yield theoretical frequencies of 5 or more), minus 3. Three degrees of freedom are lost because the computation of χ^2 requires that three characteristics of the measured distribution be known: the mean, the standard deviation, and the sample size. Thus, for the illustrative spot speed study, the number of degrees of freedom is $12 - 3 = 9$.

The values of χ^2 are shown in the body of chi-square table. For the illustrative data, the value of χ^2 lies between the tabulated values of 11.39 (Prob = 0.25) and 14.68 (Prob = 0.10). Note also that the probabilities shown in table represent the probability of a value being *greater* than or equal to χ^2. Interpolation is used to determine the precise probability level associated with a value of 14.3574 on a chi-square distribution with 9 degrees of freedom:

Value	Probability
11.3900	0.25
14.3574	p
14.6800	0.10

$$p = \text{Prob}(\chi^2 \geq 14.3574)$$

$$= 0.10 + (0.15)\left[\frac{14.6800 - 14.3574}{14.6800 - 11.3900}\right] = 0.1147$$

From this determination, it is 11.47% probable that a value of 14.3574 or higher would exist if the distribution were statistically normal. The decision criteria are the same as for other statistical tests (i.e., to say that the data and the assumed mathematical description are *significantly different*, we must be 95% confident that this is true). For tables that yield a probability of a value *less than or equal to* the computed statistic, the probability must be 95% or more to certify a significant difference. This was the case in the normal approximation test. The corresponding decision point using a table with probabilities greater than or equal to the computed statistic is that the probability must be *5% or less* to certify a significant statistical difference. In the case of the illustrative data, the probability of a value of 14.3574 or greater is 11.47%. This is more than 5%. Thus the data and the assumed mathematical description are *not significantly different*, and its normalcy is successfully demonstrated.

A chi-square test is rarely actually conducted on spot speed results because they are virtually always normal. If the data are seriously skewed or take a shape obviously different from the normal distribution, this will be relatively obvious, and the test can be conducted. It is also possible to compare the data with other types of distributions. A number of distributions have the same general shape as the normal distribution but have skews to the low or high end of the distribution. It is also possible that a given set of data can be reasonably described using a number of different distributions. This does not negate the validity of a normal description when it occurs. As long as speed data can be described as normal, all of the manipulations described here are valid.

If a speed distribution is found to be not normal, then other distributions can be used to describe it, and other statistical tests can be performed. These are not covered in this text, and we refer you to standard statistics textbooks.

10.3 Travel-Time Studies

Travel-time studies involve significant lengths of a facility or group of facilities forming a route. Information on the travel time between key points within the study area is sought and is used to identify those segments in need of improvements. Travel-time studies are often coordinated with delay observations at points of congestion along the study route.

Travel-time information is used for many purposes, including the following:

- To identify problem locations on facilities by virtue of high travel times and/or delay.
- To measure arterial level of service, based on average travel speeds and travel times.
- To provide necessary input to traffic assignment models, which focus on link travel time as a key determinant of route selection.
- To provide travel-time data for economic evaluation of transportation improvements.
- To develop time contour maps and other depictions of traffic congestion in an area or region.

10.3.1 Field Study Techniques

Because significant lengths of roadway are involved, it is difficult to remotely observe vehicles as they progress through the study section. The most common techniques for conducting travel time studies involve driving test cars through the study section, while an observer records elapsed times through the section, and at key intermediate points within the section. The observer is equipped with a field sheet pre-defining the intermediate points for which travel times are desired. The observer uses a stop-watch that is started when the test vehicle enters the study section, and records the elapsed time at each intermediate point, and when the end of the study section is reached. A second stop-watch is used to measure the length of midblock and intersection stops. Their location is noted, and if the cause can be identified, it is also noted.

To maintain some consistency of results, test-car drivers are instructed to use one of three driving strategies:

1. *Floating Car Technique:* In this technique, the test-car driver is asked to pass as many vehicles as pass the test car. In this way, the vehicle's relative position in the traffic stream remains unchanged, and the test car approximates the behavior of an average vehicle in the traffic stream.
2. *Maximum Car Technique:* In this procedure, the driver is asked to drive as fast as is safely practical in the traffic stream, without ever exceeding the design speed of the facility.
3. *Average Car Technique:* The driver is instructed to drive at the approximate average speed of the traffic stream.

The floating car and average car techniques result in estimates of the average travel time through the section. The floating car technique is generally applied only on two –lane highways, where passing is rare, and the number of passings can be counted and balanced relatively easily. On a multilane freeway, such a driving technique would be difficult at best, and might cause dangerous situations to arise, as a test vehicle attempts to "keep up" with the number of vehicles that have passed it. The average car technique yields similar results with less stress applied to the driver of the test vehicle.

The maximum car technique does not result in measurement of average conditions in the traffic stream. Rather, the measured travel times represent the lower range of the distribution of travel times. Travel times are more indicative of a 15th percentile speed than an average. Speeds computed from these travel times are approximately indicative of the 85th percentile speed.

It is important, therefore, that all test car runs in a given study follow the same driving strategy. Comparisons of travel times measured using different driving techniques will not yield valid results.

Issues related to sample size are handled similarly to spot speed studies. When specific driving strategies are followed, the standard deviation of the results is somewhat constrained, and fewer samples are needed. This is important. As a practical issue, too many test cars released into the traffic stream over a short period of time will affect its operation, in effect altering the observed results. For most common applications, the number of test car runs that will yield travel time measurements with reasonable confidence and precision ranges from a low of 6-10 to a high of 50, depending upon the type of facility and the amount of traffic. The latter is difficult to achieve without affecting traffic, and may require that runs be taken over an extended time period, such as during the evening peak hour over several days.

Another technique may be used to collect travel times. Roadside observers can record license plate numbers as vehicles pass designated points along the route. The time of passage is noted along with the license plate number. The detail of delay information at intermediate points is lost with this technique. Sampling is quite difficult, as it is virtually impossible to record every license plate and time. Assume that a sample of 50% of all license plates is recorded at every study location. The probability that a license plate match occurs at two locations is 0.50×0.50, or 0.25 (25%). The probability that a license plate match occurs across three locations is $0.50 \times 0.50 \times 0.50 = 12.5\%$. Also, as there is no consistent driving strategy among drivers in the traffic stream, many more license plate matches are required than test-car runs to obtain similar precision and confidence in the results.

In some cases, elevated vantage points may be available to allow an entire study section to be viewed. The progress of individual vehicles in the traffic stream can be directly observed. This type of study generally involves videotaping the study section, so that many, or even all, vehicle travel times can be observed and recorded.

An alternative to the use of direct observation is to equip the test vehicle with one of several devices that plots speed vs. distance as the vehicle travels through the test section. Data can be extracted from the plot to yield checkpoint travel times, and the locations and time of stopped delays can be determined.

The sample data sheet of Table 10.4 is for a seven-mile section of Lincoln Highway, which is a major suburban multilane highway of six lanes. Checkpoints are defined in terms of mileposts. As an alternative, intersections or other known geographic markers can be used as identifiers. The elapsed stopwatch time to each checkpoint is noted. Section data refer to the distance between the previous checkpoint and the checkpoint noted. Thus, for the section labeled MP 16, the section data refers to the section between mileposts 16 and 17. The total stopped delay experienced in each section is noted, along with the number of stops. The "special notes" column contains the observer's determination of the cause(s) of the delays noted. Section travel times are computed as the difference between cumulative times at successive checkpoints.

In this study, the segments ending in mileposts 18 and 19 display the highest delays and therefore the highest travel times. If this is consistently shown in *all* or *most* of the test runs, these sections would be subjected to more detailed study. Because the delays are indicated as caused primarily by traffic control signals, their timing and coordination would be examined carefully to see if they can be improved. Double parking is also noted as a cause in one segment. Parking regulations would be reviewed, along with available legal parking supply, as would enforcement practices.

10.3.2 Travel Time Along an Arterial: An Example of the Statistics of Travel Times

Given the cost and logistics of travel-time studies (test cars, drivers, multiple runs, multiple days of study, etc.), there is a natural tendency to keep the number of observations, N, as small as possible. This case considers a hypothetical arterial on which the true mean running time is 196 seconds over a three-mile section. The standard deviation of the running time is 15 seconds. The distribution of running

Table 10.4: A Sample Travel-Time Field Sheet

Site: Lincoln Highway			*Run No. 3*		*Start Location: Milepost 15.0*	
Recorder: William McShane			*Date: Aug 10, 2002*		*Start Time: 5:00 PM*	
			Per Section			
Checkpoint	**Cum. Dist. Along Route (mi)**	**Cum. Trav. Time (min:sec)**	**Stopped Delay (s)**	**No. of Stops**	**Section Travel Time (min:sec)**	**Special Notes**
MP 16	1.0	1:35	0.0	0	1:35	
MP 17	2.0	3:05	0.0	0	1:30	
MP 18	3.0	5:50	42.6	3	2:45	Stops due to signals at: MP17.2 MP17.5 MP18.0
MP 19	4.0	7:50	46.0	4	2:00	Stops due to signal MP18.5 and double parked cars.
MP 20	5.0	9:03	0.0	0	1:13	
MP 21	6.0	10:45	6.0	1	1:42	Stop due to School bus.
MP 22	7.0	12:00	0.0	0	1:15	
Section Totals	**7.0**		**88.6**	**8**	**12:00**	

times is normal. Note that the discussion is, at this point, limited to *running times*. These do not include stopped delays encountered along the route and are not equivalent to *travel times*.

Given the normal distribution of running times, the mean running time for the section is 196 seconds, and 95% of all running times would fall within 1.96(15) = 29.4 seconds of this value. Thus the 95% interval for travel times would be between 196 – 29.4 = 166.6 seconds and 196 + 29.4 = 225.4 seconds. The speeds corresponding to these running times (including the average) are:

$$S_1 = \frac{3\text{mi}}{225.4\text{s}} * \frac{3600\text{s}}{\text{h}} = 47.9\,\text{mi/h}$$

$$S_{av} = \frac{3\text{mi}}{196\text{s}} * \frac{3600\text{s}}{\text{h}} = 55.1\,\text{mi/h}$$

$$S_2 = \frac{3\text{mi}}{1666.6\text{s}} * \frac{3600\text{s}}{\text{h}} = 64.8\,\text{mi/h}$$

Note that the average of the two 95% confidence interval limits is (47.9 + 64.8)/2 = 56.4 mi/h, *not* 55.1 mi/h. This discrepancy is due to the fact that the *running times* are normally distributed and are therefore symmetric. The resulting running speed distribution is skewed. The distribution of speeds, which are inverse to running times, cannot be normal if the running times are normal. The 55.1 mi/h value is the appropriate average speed, based on the observed average running time over the three-mile study section.

So far, this discussion considers only the *running times* of test vehicles through the section. The actual *travel-time* results of 20 test-car runs are illustrated in Figure 10.3.

This distribution does not look normal. In fact, it is not normal at all because the total travel time represents the *sum* of running time (which is normally distributed) and stop time delay that follows another distribution entirely. Specifically, it is postulated that:

No. of Signal Stops	Probability of Occurrence	Duration of Stops
0	0.569	0s
1	0.300	40s
2	0.131	80s

The observations of Table 10.4 result from the combination of random driver selection of running speeds and signal delay effects that follow the relationship just specified.

The actual mean travel time of the observations in Figure 10.3 is 218.5 seconds, with a standard deviation of 38.3 seconds. The 95% confidence limits on the average are:

$$218.5 \pm 1.96(38.3/\sqrt{20}) = 218.5 \pm 16.79$$

$$201.71 - 235.29\text{s}$$

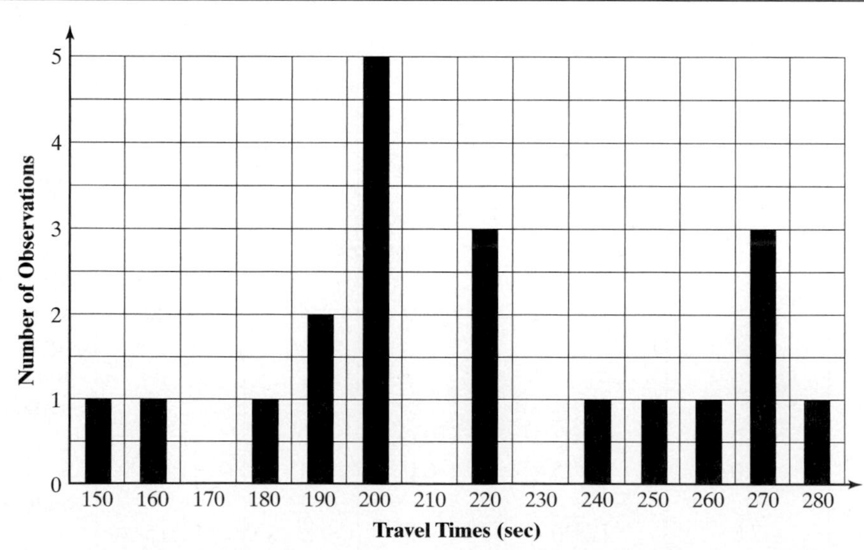

Figure 10.3: Histogram of Hypothetical Travel-Time Data for 20 Runs over a 3-Mile Arterial Segment

The speeds associated with these average and limiting travel times are:

$$S_1 = \frac{3\text{mi}}{235.29\text{s}} * \frac{3600s}{h} = 45.9\,\text{mi/h}$$

$$S_{av} = \frac{3\text{mi}}{218.5\text{s}} * \frac{3600s}{h} = 49.4\,\text{mi/h}$$

$$S_2 = \frac{3\text{mi}}{201.71\text{s}} * \frac{3600s}{h} = 53.5\,\text{mi/h}$$

Another way of addressing the average travel time is to add the average running time (196 seconds) to the average delay time, which is computed from the probabilities just noted as:

$$d_{av} = (0.569*0) + (0.300*40) + (0.131*80) = 22.5s$$

The average travel time is then expected to be 196.0 + 22.5 = 218.5s, which is the same average obtained from the histogram of measurements.

10.3.3 Overriding Default Values: Another Example of Statistical Analysis of Travel-Time Data

Figure 10.4 shows a default curve calibrated by a local highway jurisdiction for average travel speed along four-lane arterials within the jurisdiction. As with all "standard" values, the use of another value is always permissible as long as there are specific field measurements to justify replacing the standard value.

Assume that a case exists in which the default value of travel speed for a given volume, V_1 is 40 mi/h. Based on three travel-time runs over a two-mile section, the measured average travel speed is 43 mi/h. The analysts would like to replace the standard value with the measured value. Is this appropriate?

The statistical issue is whether or not the observed 3 mi/h difference between the standard value and the measured value is *statistically significant.* As a practical matter (in this hypothetical case), practitioners generally believe that the standard values of Figure 10.4 are too low and that higher values are routinely observed. This suggests that a one-sided hypothesis test should be used.

Figure 10.5 shows a probable distribution of the random variable $Y = \Sigma t_i/N$, the estimator of the average travel time through the section. Based on the standard and measured average travel speeds, the corresponding travel times over a two-mile section of the roadway are (2/40) * 3,600 = 180.0s and (2/43) * 3,600 = 167.4 s. These two values are formulated, respectively, as the null and alternative hypotheses, as illustrated in Figure 10.5. The following points relate to Figure 10.5:

- Type I and Type II errors are equalized and set at 5% (0.05).
- From the standard normal table of Chapter 7, the value of z_d corresponding to Prob. $(z \leq z_d) = 0.95$ (corresponding to a one-sided test with Type I and II errors set at 5%) is 1.645.
- The difference between the null and alternate hypotheses is a travel time of 180.0 − 167.4 = 12.6, noted as Δ.
- The standard deviation of travel times is known to be 28.0s.

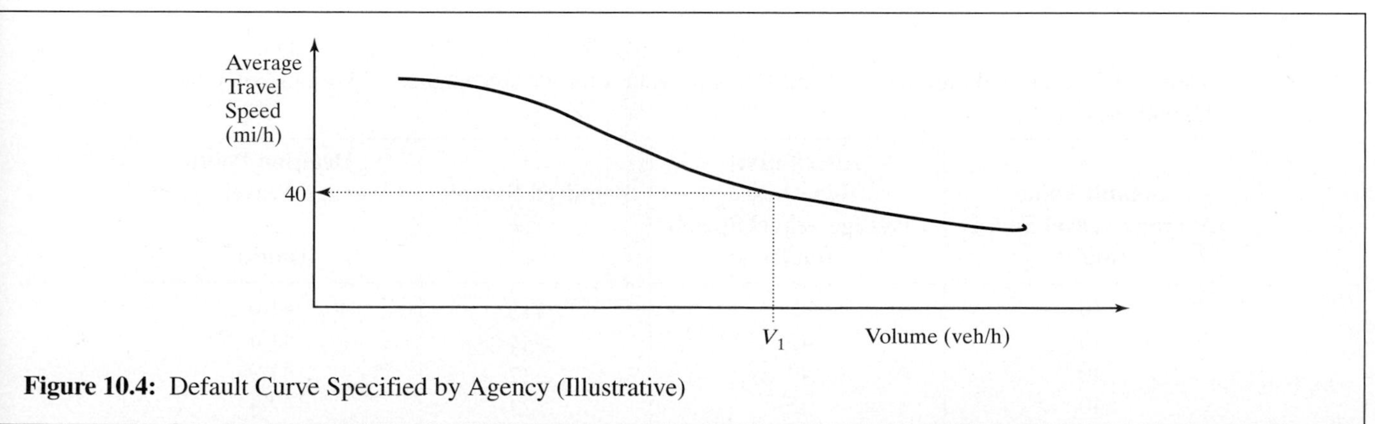

Figure 10.4: Default Curve Specified by Agency (Illustrative)

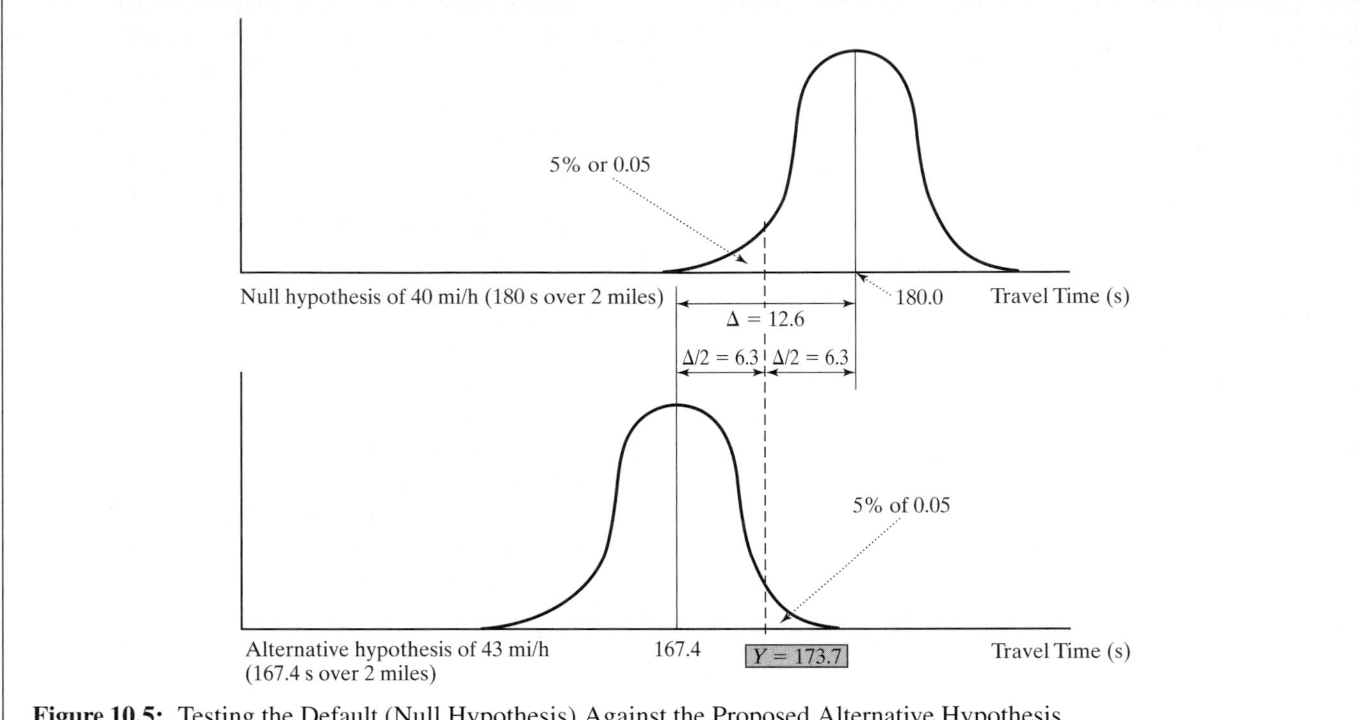

Figure 10.5: Testing the Default (Null Hypothesis) Against the Proposed Alternative Hypothesis

From Figure 10.5, for the difference between the default and alternative hypotheses to be statistically significant, the value of Δ/2 must be equal to or larger than 1.645 times the standard error for travel times, or:

$$\Delta/2 \geq 1.645\left(s \Big/ \sqrt{N}\right)$$

$$6.3 \geq 1.645\left(28 \Big/ \sqrt{3}\right) = 26.6$$

Obviously, in this case the difference is not significant, and the measured value of 43 mi/h cannot be accepted in place of the default value. This relationship can, of course, be solved for N:

$$N \geq \frac{8{,}486}{\Delta^2}$$

using the known value of the standard deviation (28). Remember that Δ is stated in terms of the difference in *travel times* over the two-mile test course, not the difference in average travel speeds. Table 10.5 shows the sample size

Table 10.5: Required Sample Sizes and Decision Values for the Acceptance of Various Alternative Hypotheses

Default Value (Average Travel Speed) (mi/h)	**Alternative Hypothesis (Average Travel Speed) (mi/h)**	**Required Sample Size *N***	**Decision Point (Average Travel Speed) *Y* (mi/h)**
40	42	≥115	41.0
40	43	≥54	41.4
40	44	≥32	41.9
40	45	≥22	42.4

requirements for accepting various alternative average travel speeds in place of the default value. For the alternative hypothesis of 43 mi/h to be accepted, a sample size of $8{,}486/(12.6)^2 = 54$ would have been required. However, as illustrated in Figure 10.5, had 54 samples been collected, the alternative hypothesis of 43 mi/h would have been accepted as long as the average travel time was less than 173.7 s (i.e., the average travel speed was greater than $(2/173.7) * 3{,}600 =$ 41.5 mi/h. Table 10.5 shows a number of different alternative hypotheses, along with the required sample sizes and decision points for each to be accepted.

Although this problem illustrates some of the statistical analyses that can be applied to travel-time data, you should examine whether the study, as formulated, is appropriate. Should the Type II error be equalized with the Type I error? Does the existence of a default value imply that it should not? Should an alternative value higher than any measured value ever be accepted? (For example, should the alternative hypothesis of 43 mi/h be accepted if the average travel speed from a sample of 54 or more measurements is 41.6 mi/h, which is greater than the decision value of 41.5 mi/h?)

Given the practical range of sample sizes for most travel-time studies, it is very difficult to justify overriding default values for individual cases. However, a compendium of such cases—each with individually small sample sizes—can and should motivate an agency to review the default values and curves in use.

10.3.3 Travel-Time Displays

Travel-time data can be displayed in many interesting and informative ways. One method that is used for overall traffic planning in a region is the development of a travel-time contour map, of the type shown in Figure 10.6. Travel times along all major routes entering or leaving a central area are measured. Time contours are then plotted, usually in increments of 15 minutes. The shape of the contours gives an immediate visual assessment of corridor travel times in various directions. The closer together contour lines plot, the longer the travel time to progress any set distance. Such plots can be used for overall planning purposes and for identifying corridors and segments of the system that require improvement.

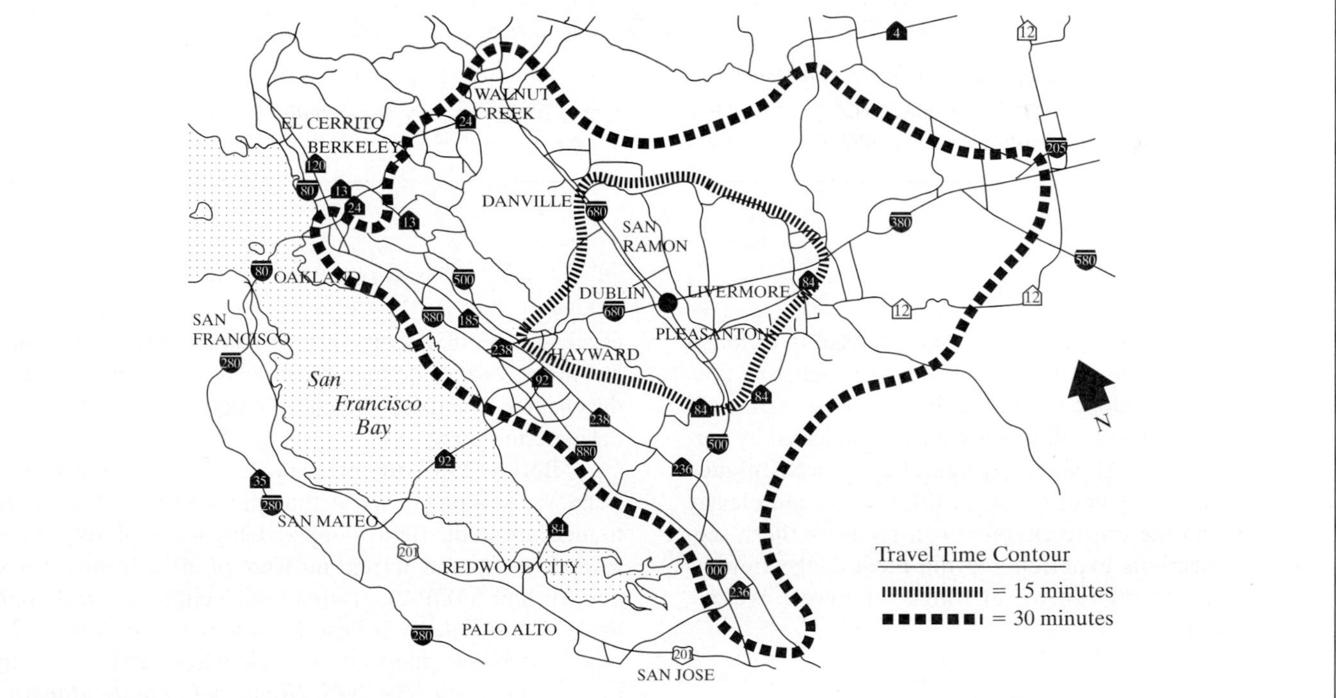

Figure 10.6: A Travel-Time Contour Map

(*Source:* Used with permission of Prentice-Hall Inc, from Pline, J., Editor, *Traffic Engineering Handbook*, 4th Edition, Institute of Transportation Engineers, Washington DC, 1992, p. 69.)

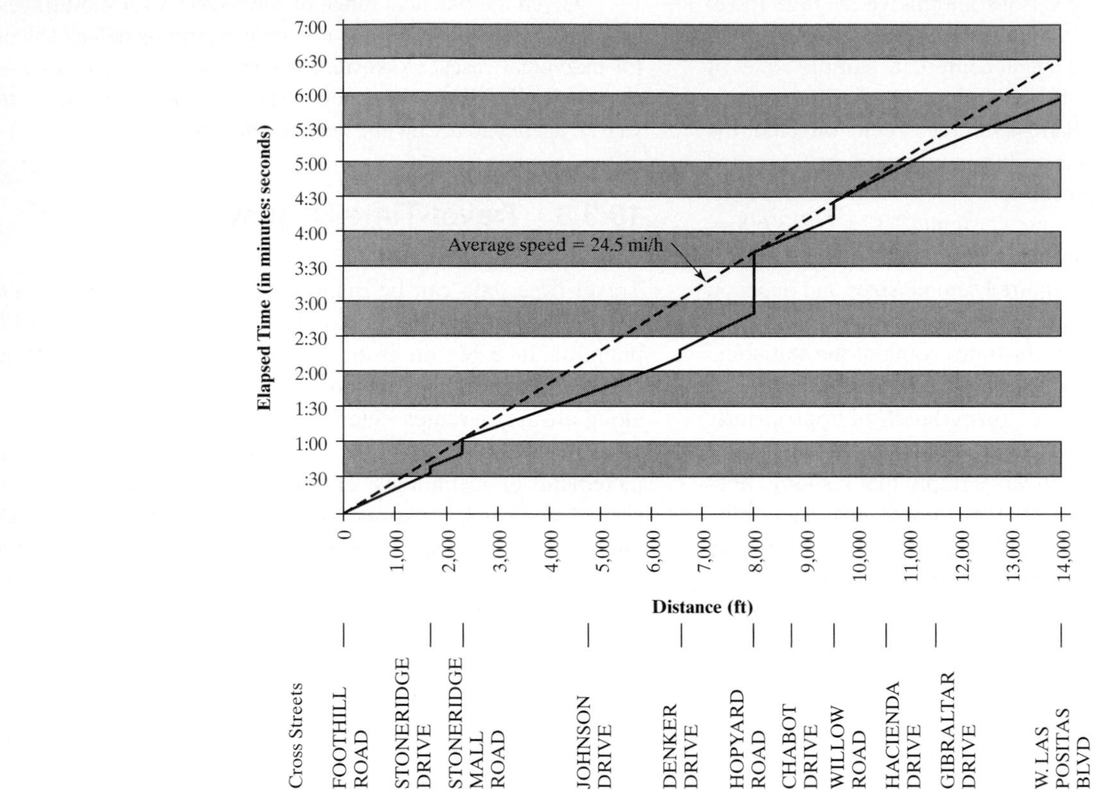

Figure 10.7: A Plot of Elapsed Time Versus Distance

(*Source:* Used with permission of Prentice-Hall Inc, from Pline, J., Editor, *Traffic Engineering Handbook*, 4th Edition, Institute of Transportation Engineers, Washington DC, 1992.)

Travel time along a route can be depicted in different ways as well. Figure 10.7 shows a plot of cumulative time along a route. The slope of the line in any given segment is speed (ft/s), and stopped delays are clearly indicated by vertical lines. Figure 10.8 shows average travel speeds plotted against distance. In both cases, problem areas are clearly indicated, and the traffic engineer can focus on those sections and locations experiencing the most congestion, as indicated by the highest travel times (or lowest average travel speeds).

10.4 Intersection Delay Studies

Some types of delay are measured as part of a travel-time study by noting the location and duration of stopped periods during a test run. A complicating feature for all delay studies lies in the various definitions of delay, as reviewed earlier in the chapter. The measurement technique must conform to the delay definition.

Before 1997, the primary delay measure at intersections was stopped delay. Although no form of delay is easy to measure in the field, stopped delay was certainly the easiest. However, the current measure of effectiveness for signalized and STOP-controlled intersections is *total control delay.* Control delay is best defined as time-in-queue delay plus time losses due to deceleration from and acceleration to ambient speed. The 2000 *Highway Capacity Manual* [*1*] defines a field measurement technique for control delay, using the field sheet shown in Figure 10.9.

The study methodology recommended in the *Highway Capacity Manual* is based on direct observation of

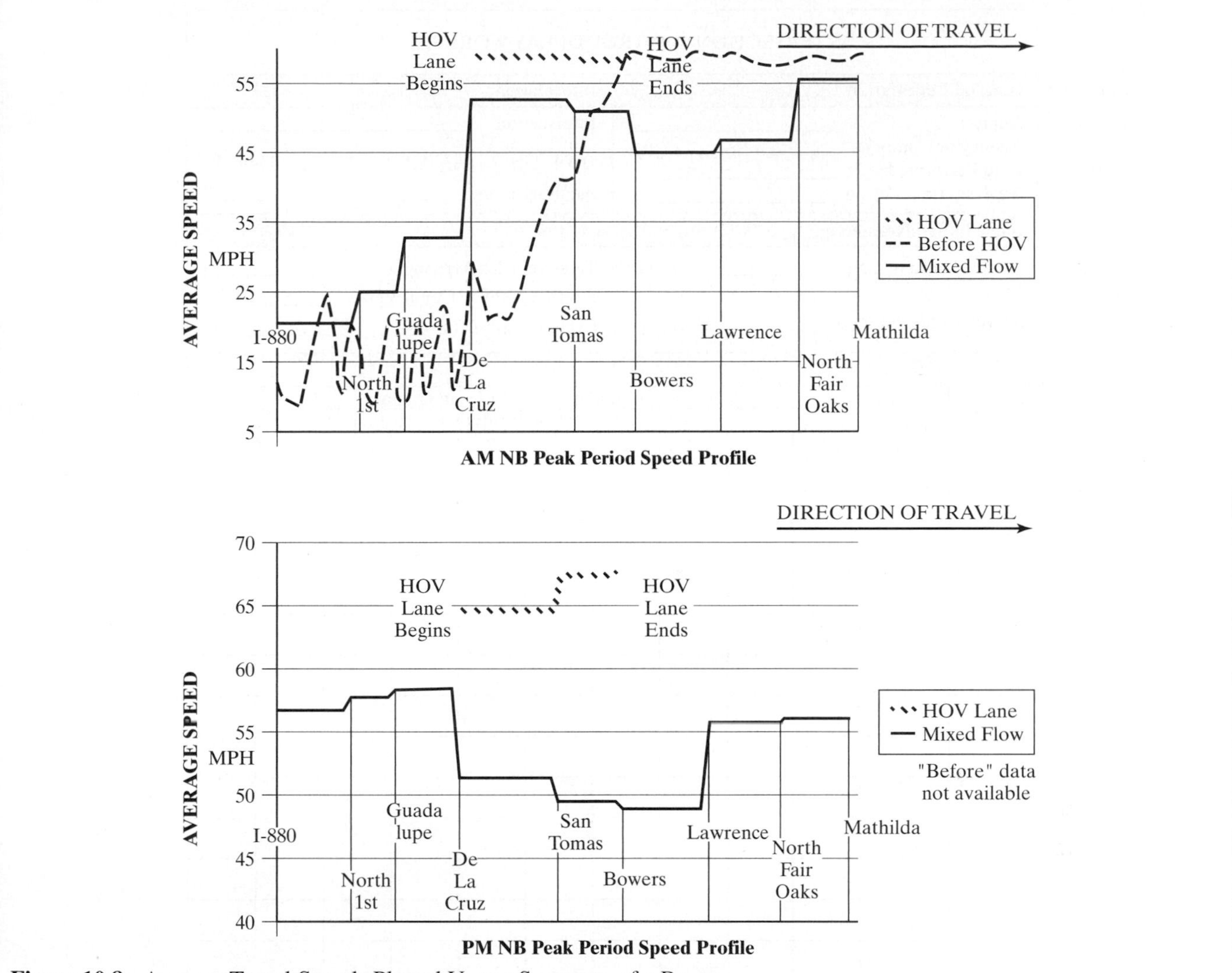

Figure 10.8: Average Travel Speeds Plotted Versus Segments of a Route

(*Source:* Used with permission of Prentice-Hall Inc, from Pline, J., Editor, *Traffic Engineering Handbook*, 4th Edition, Institute of Transportation Engineers, Washington DC, 1992.)

vehicles-in-queue at frequent intervals and requires a minimum of two observers. The following should be noted:

1. The method is intended for undersaturated flow conditions, and for cases where the maximum queue is about 20 to 25 vehicles.
2. The method does not directly measure acceleration-deceleration delay but uses an adjustment factor to estimate this component.
3. The method also uses an adjustment to correct for errors that are likely to occur in the sampling process.
4. Observers must make an estimate of free-flow speed before beginning a detailed survey. This is done by driving a vehicle through the intersection during periods when the light is green and there are no queues and/or by measuring approach speeds at a position where they are unaffected by the signal.

Actual measurements start at the beginning of the red phase of the subject lane group. There should be no overflow queue from the previous green phase when

INTERSECTION CONTROL DELAY WORKSHEET											
General Information						**Information**					
Analyst ______						Intersection ______					
Agency or Company ______						Area Type ☐ CBD ☐ Other					
Date Performed ______											
Analysis Time Period ______						Analysis Year ______					
Input Initial Parameters											
Number of Lanes, N ______						Total Vehicles Arriving V_T ______					
						Stopped Vehicle Count V_{STOP} ______					
Survey Count Interval I_s ______						Cycle Length D (s) ______					
Input Field Data											
Clock Time	Cycle Number	Number of Vehicles in Queue									
		Count Interval									
		1	2	3	4	5	6	7	8	9	10
Total											

Figure 10.9: Field Sheet for Signalized Intersection Delay Studies

(*Source:* Used with permission of Transportation Research Board, *Highway Capacity Manual*, 4th Edition, Washington DC, p. 16-173.)

measurements start. The following tasks are performed by the two observers:

Observer 1

- Keeps track of the end of standing queues for each cycle by observing the last vehicle in each lane that stops due to the signal. This count includes vehicles that arrive on green but stop or approach within one car length of queued vehicles that have not yet started to move.
- At intervals between 10 seconds and 20 seconds, the number of vehicles in queue are recorded on the field sheet. The regular intervals for these observations should be an integral divisor of the cycle length. Vehicles in queue are those that are included in the queue of stopping vehicles (as just defined) and have not yet exited the

intersection. For through vehicles, "exiting the intersection" occurs when the rear wheels cross the STOP line; for turning vehicles, "exiting" occurs when the vehicle clears the opposing vehicular or pedestrian flow to which it must yield and begins to accelerate.

- At the end of the survey period, vehicle-in-queue counts continue until all vehicles that entered the queue during the survey period have exited the intersection.

Observer 2

- During the entire study period, separate counts are maintained of vehicles arriving during the survey period and of vehicles that stop one or more times during the survey period. Stopping vehicles are counted only once, regardless of how many times they stop.

For convenience, the survey period is defined as an integer number of cycles, although an arbitrary length of time (e.g., 15 minutes) could also be used and would be necessary where an actuated signal is involved.

Each column of the vehicle-in-queue counts is summed; the column sums are then added to yield the total vehicle-in-queue count for the study period. It is than assumed that the average time-in-queue for a counted vehicle is the time interval between counts. Then:

$$T_Q = \left(I_s * \frac{\sum V_{iq}}{V_T} \right) * 0.90 \qquad (10\text{-}15)$$

where: T_Q = average time-in-queue, s/veh
I_s = time interval between time-in-queue counts, s
$\sum V_{iq}$ = sum of all vehicle-in-queue counts, vehs
V_T = total number of vehicles *arriving* during the study period, vehs
0.9 = empirical adjustment factor

The adjustment factor (0.9) adjusts for errors that generally occur when this type of sampling technique is used. Such errors usually result in an overestimate of delay.

A further adjustment for acceleration/deceleration delay requires that two values be computed: (1) the average number of vehicles stopping per lane, per cycle, and (2) the proportion of vehicles arriving that actually stop. These are computed as:

$$V_{SLC} = \frac{V_{STOP}}{N_c * N_L} \qquad (10\text{-}16)$$

where: V_{SLC} = number of vehicles stopping per lane, per cycle (veh/ln/cycle)
V_{STOP} = total count of stopping vehicles, vehs
N_c = number of cycles included in the survey
N_L = number of lanes in the survey lane group

$$FVS = \frac{V_{STOP}}{V_T} \qquad (10\text{-}17)$$

where FVS = fraction of vehicles stopping
other variables as previously defined

Using the number of stopping vehicles per lane, per cycle, and the measured free-flow speed for the approach in question, an adjustment factor is found in Table 10.6.

The final estimate of control delay is then computed as:

$$d = T_Q + (FVS*CF) \qquad (10\text{-}18)$$

where: d = total control delay, s/veh
CF = correction factor from Table 10.6
other variables as previously defined

Table 10.7 shows a facsimile of a field sheet, summarizing the data for a survey on a signalized intersection approach. The approach has two lanes, and the signal cycle length is 60 seconds. Ten cycles were surveyed, and the vehicle-in-queue count interval is 20 seconds.

The average time-in-queue is computed using Equation 10-15:

$$T_Q = \left(20 * \frac{132}{120} \right) * 0.90 = 19.8 s/veh$$

To find the appropriate correction factor from Table 10.6, the number of vehicles stopping per lane per cycle is computed using Equation 9-16:

$$V_{SLC} = \frac{75}{10 * 2} = 3.75 \text{vehs}$$

Using this and the measured free-flow speed of 35 mi/h, the correction factor is +5 seconds. The control delay is now estimated using Equations 10-17 and 10-18:

$$\text{FVS} = \frac{75}{120} = 0.625$$

$$d = 19.8 + (0.625 * 5) = 22.9 \text{s/veh}$$

A similar technique and field sheet can be used to measure stopped time delay as well. In this case, the interval counts

Table 10.6: Adjustment Factor for Acceleration/Deceleration Delay

Free-Flow Speed (mi/h)	Vehicles Stopping Per Lane, Per Cycle (V_{SLC})		
	≤7 vehs	8–19 vehs	20–30 vehs
≤37	+5	+2	−1
>37 − 45	+7	+4	+2
>45	+9	+7	+5

(*Source:* Used with permission of Transportation Research Board, *Highway Capacity Manual*, 4th Edition, Washington DC, 2000, Exhibit A16-2, p. 16-91.)

Table 10.7: Sample Data for a Signalized Intersection Delay Study

Clock Time	Cycle Number	Number of Vehicles in Queue		
		+0 s	+20 s	+40 s
5:00 PM	1	4	7	5
5:01 PM	2	6	6	5
5:02 PM	3	3	5	5
5:03 PM	4	2	6	4
5:04 PM	5	5	3	3
5:05 PM	6	5	4	5
5:06 PM	7	6	8	4
5:07 PM	8	3	4	3
5:08 PM	9	2	4	3
5:09 PM	10	4	3	5
	Total	**40**	**50**	**42**

$\Sigma V_{qi} = 132$ vehs $V_T = 120$ vehs $V_{STOP} = 75$ FFS = 35 mi/h

include only vehicles stopped within the intersection queue area, not those moving within it. No adjustment for acceleration/deceleration delay would be added.

10.5 Closing Comments

Time is one of the key commodities that motorists and other travelers invest in getting from here to there. Travelers most often wish to minimize this investment by making their trips as short as possible. Travel-time and delay studies provide the traffic engineer with data concerning congestion, section travel times, and point delays. Through careful examination, the causes of congestion, excessive travel times, and delays can be determined and traffic engineering measures developed to ameliorate problems.

Speed is the inverse of travel time. Although travelers wish to maximize the speed of their trip, they also wish to do so consistent with safety. Speed data provide insight into many factors, including safety, and are used to help time traffic signals, set speed limits, locate signs, and in a variety of other important traffic engineering activities.

References

1. *Highway Capacity Manual*, 4th Edition, Transportation Research Board, National Science Foundation, Washington DC, 2000.

2. Highway Capacity Manual, 3rd Edition, *Special Report 209*, Transportation Research Board, National Science Foundation, Washington DC, 1985.

Problems

10-1. Consider the spot speed data here, collected at a rural highway site under conditions of uncongested flow:

Speed Group (mi/h)	Number of Vehicles Observed (N)
15–20	0
20–25	4
25–30	9
30–35	18
35–40	35
40–45	42
45–50	32
50–55	20
55–60	9
60–65	0

(a) Plot the frequency and cumulative frequency curves for this data.

(b) Determine the median speed, the modal speed, the pace, and the percentage of vehicles in the pact from the curves, and show how each was found.

(c) Compute the mean and standard deviation of the speed distribution.

(d) What are the confidence bounds of the estimate of the true mean speed of the distribution with 95% confidence? With 99.7% confidence?

(e) Based on the results of this study, a second is to be conducted to achieve a tolerance of ± 0.8 mi/h with 95% confidence. What sample size is required?

(f) Can this data be adequately described as "normal?"

10-2. A before-and after speed study was conducted to determine the effectiveness of a series of rumble strips installed approaching a toll plaza to reduce approach speeds to 40 mi/h.

Item	Before Study	After Study
Average Speed	43.5 mi/h	40.8 mi/h
Standard Deviation	4.8 mi/h	5.3 mi/h
Sample Size	120	108

(a) Were the rumble strips effective in reducing average speeds at this location?

(b) Were the rumble strips effective in reducing average speeds to 40 mi/h?

10-3. The following data were collected during a delay study on a signalized intersection approach. The cycle length of the signal is 60 seconds.

(a) Estimate the time spent in queue for the average vehicle.

(b) Estimate the average control delay per vehicle on this approach.

Clock Time	Cycle Number	Number of Vehicles in Queue			
		+ 0 s	+ 15 s	+ 30 s	+ 45 s
9:00 AM	1	3	4	2	4
9:01 AM	2	1	2	3	3
9:02 AM	3	4	3	3	4
9:03 AM	4	2	3	3	4
9:04 AM	5	0	1	2	3
9:05 AM	6	2	1	1	2
9:06 AM	7	4	3	4	3
9:07 AM	8	5	5	6	4
9:08 AM	9	2	3	4	3
9:09 AM	10	0	3	2	2
9:10 AM	11	1	2	3	1
9:11 AM	12	1	0	1	0
9:12 AM	13	2	2	1	2
9:13 AM	14	2	3	2	2
9:14 AM	15	4	3	3	3

Note: $V_T = 435$ vehs; $V_{STOP} = 305$ vehs; FFS = 35 mi/h.223.

10-4. A series of travel time runs are to be made along an arterial section. Tabulate the number of runs required to estimate the overall average travel time with 95% confidence to within ±2 min, ±5 min, ±10 min, for standard deviations of 5, 10, and 15 minutes. Note that a 3 × 3 table of values is desired.

10-5. The results of a travel time study are summarized in the table that follows. For this data:

(a) Tabulate and graphically present the results of the travel time and delay runs. Show the average travel speed and average running speed for each section.

(b) Note that the number of runs suggested in this Problem (5) is not necessarily consistent with the results of Problem 10-3. Assuming that *each vehicle* makes five runs, how many test vehicles would be needed to achieve a tolerance of ± 3 with 95% confidence?

Erin Blvd Recorder: XYZ Summary of 5 Runs

Checkpoint Number	Cumulative Section Length (mi)	Cumulative Travel Time (min:sec)	Per Section	
			Delay (s)	No. of Stops
1	—	—	—	—
2	1.00	2:05	10	1
3	2.25	4:50	30	1
4	3.50	7:30	25	1
5	4.00	9:10	42	2
6	4.25	10:27	47	1
7	5.00	11:54	14	1

CHAPTER 11

Highway Traffic Safety: Studies, Statistics, and Programs

11.1 Introduction

In 2007, 41,059 people were killed in accidents on U.S. highways in a total of 6.024 million police-reported accidents. Because police-reported accidents are generally believed to make up only 50% of all accidents occurring, this implies a staggering total of almost 12 million accidents for the year. A more complete set of statistics for the year 2007 is shown in Table 11.1 [*1*].

To fully appreciate these statistics, some context is needed: More people have been killed in highway accidents in the United States than in all of the wars in which the nation has been involved, from the Revolutionary War through the current wars in Iraq and Afghanistan. The fatalities for 2007 are equal to an average of one vehicular accident death occurred every 13 minutes in the United States [*2*].

A great deal of effort on every level of the profession is focused on the reduction of these numbers. In fact, U.S. highways have become increasingly safe for a variety of reasons. The number of annual fatalities peaked in 1972 with 54,589 deaths in highway-related accidents. Although the total number of fatalities has declined by over 23% since 1972, the decline in the fatality rate per 100 million vehicle-miles traveled (VMT) is even more spectacular. In 1966, the rate stood at 5.5 fatalities per 100 million VMT; in 2007, the rate was 1.37 fatalities per 100 million VMT, a decline of 75%.

There are many reasons for this decline. Highway design has incorporated many safety improvements that provide for a more "forgiving" environment for drivers. Better alignments, vast improvements in roadside and guardrail design, use of breakaway supports for signs and lighting, impact-attenuating devices, and other features have made the highway environment safer. Vehicle design has also improved. Federal requirements now dictate such common features as padded dashboards, seat belts and shoulder harnesses, airbags, and energy-absorbing crumple zones. Drivers are more familiar with driving in congested urban environments and on limited-access facilities. Despite vastly improved accident and fatality *rates,* the number of fatalities remains high. This is because Americans continue to drive more vehicle-miles each year. In 2007, U.S. motorists drove more than 3 *trillion* vehicle-miles. In 2008, preliminary statistics indicate that traffic fatalities were reduced to 37,261. This is at least partially due to a reduction in VMT, which fell below 3 trillion due to the rapid increase in fuel costs and general economic difficulties.

Table 11.1: National Highway Accident Statistics for 2008

Accident Type	Number of Deaths/Injuries	Number of Accidents*	Number of Vehicles Involved
Fatal	37,261	34,017	48,904
Injury	2,346,000	1,630,000	2,875,000
Property Damage Only (PDO)	NA	4,146,000	7,115,000

*Includes only accidents reported by police.

(*Source:* Compiled from National Traffic Safety Administration, U.S. Department of Transportation, *Traffic Safety Facts 2008.*)

Figure 11.1 shows the trend of total highway fatalities for 1966 through 2007. Figure 11.2 shows the trend of fatality rates over the same period.

Although the fatality rate trend has been continuously declining, the fatality trend has been less consistent, despite an overall downward trend. Significant decreases in both fatalities and fatality rates in the mid-1970s coincided with the imposition of the national 55-mi/h speed limit. Originally enacted as a measure to reduce fuel consumption, the dramatic decrease in fatalities and fatality rates was a critical, somewhat unanticipated benefit. This is why the 55-mi/h speed limit survived long after the fuel crises of the 1970s had passed.

The statistical connection between the reduction in the national speed limit and improved safety on U.S. highways has been difficult to make. Although the 55-mi/h speed limit was a single change that occurred at a definable point in time, other changes in highway design were taking place at the same time. Further, speed limit enforcement practices changed substantially, with more emphasis shifting from lower-speed surface facilities to high-speed facilities such as freeways.

Between 1996 and 1998, the federal government gradually, then completely, eliminated the 55-mi/h speed limit. The national speed limit had been loosely enforced at best. States had to file formal reports on speed behavior each year and were threatened with partial loss of federal-aid highway funds if too high a percentage of drivers were found to be exceeding 55 mi/h. By the mid-1990s, many states were in technical violation of these requirements. Either the restriction had to be changed, or the federal statutes would have to be enforced.

Despite dire predictions, the lifting of the 55-mi/h national speed limit did not lead to dramatic increases in fatalities or fatality rates. The rates, as shown, continued to decline, and the number of fatalities has been relatively stable over the past five years. States, now free to enact their own statutory speed limits, began systematically to increase speed limits on freeways to 65 mi/h or more. Because there has been no clear statistical increase in fatality rates, the clamor against these speed limit increases has declined over the last few years. Speeding, however, is a common contributing factor in fatal

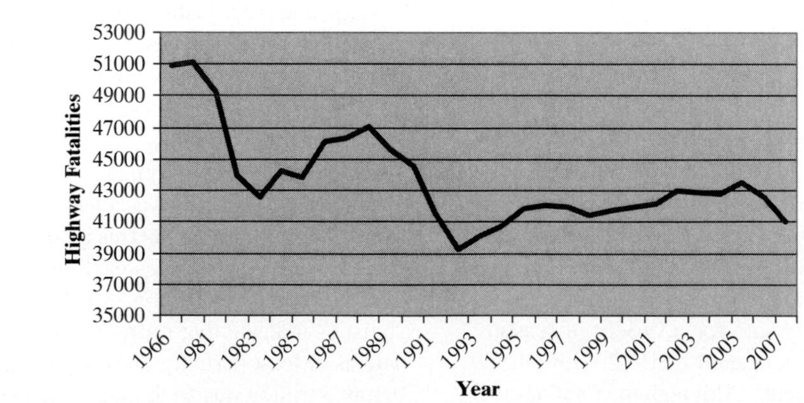

Figure 11.1: Highway Fatality Trend, 1966–2008

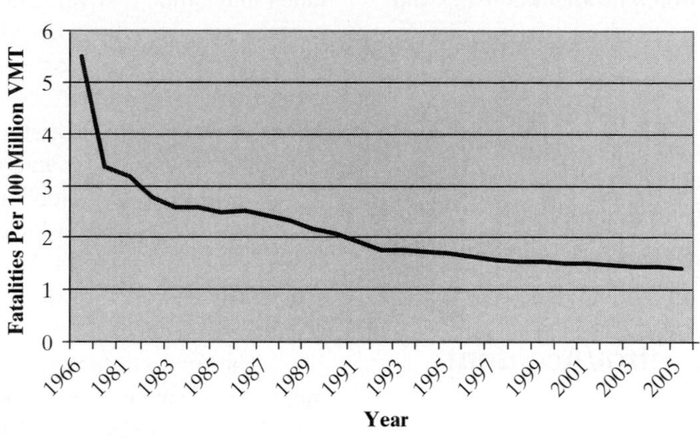

Figure 11.2: Fatality Rate Trend, 1966–2006

crashes. In 2007, 31% of all fatal crashes had speeding as a contributing factor [2]. Additionally, in 2008, which saw huge increases in fuel prices, we also saw an unexpectedly large decrease in fatalities. Forty-two of the 50 states reported a significant decrease in the number of fatalities from 2007 to 2008 [3]. Drivers were driving less, choosing the trips that were important enough to make, and driving slower, all to save gasoline. Because of this, fatalities have decreased to a level not seen since the 1960s, and for the first time in over 50 years, the number of fatalities dropped below 40,000 for 2008.

11.2 Approaches to Highway Safety

Improving highway safety involves consideration of three elements influencing traffic operations: the driver, the vehicle, and the roadway. Unfortunately, the traffic engineer has effective control over only one of these elements: the roadway. Traffic engineers can also play the role of informed advocates for improved driver education and licensing procedures, and for the required incorporation of safety features in vehicle design. The latter factors, however, are subject to the political and legislative process and are not under the direct control of the engineer.

One of the basic references in highway safety is the Institute of Transportation Engineers (ITE) *Traffic Safety Toolbox* [4]. It lists three basic strategies that may be employed to improve traffic safety:

1. Exposure control
2. Accident risk control
3. Injury control

Another basic reference, *Safer Roads: A Guide to Road Safety Engineering* [5], cites similar strategies, separated into five categories:

1. Exposure control
2. Accident prevention
3. Behavior modification
4. Injury control
5. Postinjury management

11.2.1 Exposure Control

Exposure control is common to both lists and involves strategies that reduce the number of vehicle-miles of travel by motorists. This, of course, has proven to be a very difficult strategy to implement in the United States, given that the automobile is the overwhelming choice of travelers. The march toward ever-increasing national vehicle-miles traveled has continued unabated for over 35 years until the recent increase (2008) in gasoline prices.

Efforts to reduce auto use and travel cover a wide range of policy, planning, and design issues. Policies and practices that attempt to reduce auto use include:

- Diversion of travel to public transportation modes
- Substitution of telecommunications for travel
- Implementation of policies, taxes, and fees to discourage auto ownership and use
- Reorganization of land uses to minimize travel distances for various trip purposes

- Driver and vehicle restrictions through licensing and registration restrictions

Most of these strategies must take place over long time periods, and many require systemic physical changes to the urban infrastructure and behavioral changes in the traveling public. Some require massive investments (such as providing good public transportation alternatives and changing the urban land-use structure); others have not yet demonstrated the potential to affect large changes in travel behavior.

11.2.2 Accident Risk Control/Accident Prevention

Accident risk control and *accident prevention* are similar terms with a number of common features. They are not, however, the same. Accident prevention implies actions that reduce the number of accidents that occur for a given demand level. Accident risk control incorporates this but also includes measures that reduce the *severity* of an accident when it occurs. Reduction of accident severity overlaps accident risk control and injury control strategies.

Accident prevention involves a number of policy measures, including driver and pedestrian training, removal of drivers with "bad" driving records (through the suspension or revocation of licenses), and provision of better highway designs and control devices that encourage good driving practices and minimize the occurrence of driver error.

Risk control, or reduction of severity, often involves the design and protection of roadside and median environments. Proper guardrail and/or impact-attenuating devices reduce the impact energy transferred to the vehicle in an accident and can direct the path of a vehicle away from objects or areas that would result in a more serious collision.

11.2.3 Behavior Modification

This category, separately listed in Reference 5, is an important component of strategies for accident prevention and exposure reduction. Affecting mode choice is a major behavior modification action that is hard to achieve successfully. Often, this requires providing very high-class and convenient public transportation alternatives and implementing policies that make public transportation a much more attractive alternative than driving for commuter and other types of trips. This is an expensive process, often involving massive subsidies to keep the cost of public transportation reasonable, coupled with high parking and other fees associated with driving. Use of high-occupancy vehicle lanes and other restricted-use lanes to speed public transportation, providing a visual travel-time differential between public transportation and private automobiles, is another useful strategy.

If drivers and motorists cannot be successfully diverted to alternative modes, driver and pedestrian training programs are a common strategy for behavior modification. Many states offer insurance discounts if a basic driving safety course is completed every three years. However, little statistical evidence indicates that driver training has any measurable effect on accident prevention.

The final strategy in behavior modification is enforcement. This can be very effective, but it is also expensive. Speed limits will be more closely obeyed if enforcement is strict, and the fines for violations are expensive. In recent years, the use of automated systems for ticketing drivers who violate red lights have become quite popular. Automated speed enforcement is also possible with current technologies. The issues involved in automated enforcement are more legal than technical at present. Although the license plate of a vehicle running a red light can be automatically recorded, it does not prove who is driving the vehicle. In most states, automated ticketing results in a fine but does not include "points" on the owner's license because it cannot be proved that the owner was the driver at the time of the violation.

11.2.4 Injury Control

Injury control focuses on crash survivability of occupants in a vehicular accident. This is primarily affected by better vehicle design that is generally "encouraged by an act of Congress." Vehicle design features that have been implemented with improved crash survivability in mind include:

- Seat belts and shoulder harnesses, and laws requiring their use
- Child-restraint seats and systems, and laws requiring their use
- Anti-burst door locks
- Padded instrument panels
- Energy-absorbing steering posts and crumple zones
- Side door beams
- Air bags
- Head rests and restraints
- Shatterproof glass
- Forgiving interior fittings

11.2.5 Postinjury Management

Although included as part of injury control in Reference 4, this is treated as a separate category in Reference 5. Traffic fatalities tend to occur during three critical time periods [*6*]:

- During the accident occurrence, or within minutes of it. Death is usually related to head or heart trauma or extreme loss of blood.
- Within one to two hours of the accident occurrence. In this period, death is usually due to the same causes just noted: head or heart trauma and/or loss of blood.
- Within 30 days of admission to the hospital. Death usually results from cessation of brain activity, organ failure, or infection.

About 50% of traffic fatalities occur in the first category, 35% in the second, and 15% in the third. Little can be done for deaths occurring during the accident or immediately thereafter, and the latter category is difficult to improve in developed countries with high-quality medical care systems. The biggest opportunity for improvement is in the second category.

Deaths within one to two hours of an accident can be reduced by systems that ensure speedy emergency medical responses along with high-quality emergency care at the site and during transport to a hospital facility. Such systems involve speedy notification of emergency services, fast dispatch of appropriate equipment to the site, well-trained emergency medical technicians attending to immediate medical needs of victims, and well-staffed and equipped trauma centers at hospitals. Because survival often depends on quickly stabilizing a victim at the crash site and speedy transport to a trauma center, communications and dispatch systems must be in place to respond to a variety of needs. A simple decision on whether to dispatch an ambulance or a med-evac helicopter is often a life-or-death decision.

11.2.6 Planning Actions to Implement Policy Strategies

Table 11.2 provides a shopping list of planning strategies and actions that can be effective in implementing the various safety approaches discussed.

Table 11.2: Traffic Planning and Operations Measures Related to Highway Safety Strategies

Strategy	Actions to Implement Strategy
Exposure Control	
Reducing transport demand and the amount of road traffic.	1. Urban and transport policies, pricing, and regulation.
	2. Urban renewal (increased density, short distances).
	3. Telecommunications (tele-working, tele-shopping).
	4. Informatics for pre-trip, on-board information.
	5. TDM, mobility management (car pools, ride-sharing).
	6. Logistics (rail, efficient use of transport fleets).
Promoting safe, comfortable walking and biking.	7. Areawide pedestrian and bike networks.
	8. Land use integrated with public transport.
Providing and promoting public transportation.	9. Efficient service (bus lanes, fare systems, etc.)
Accident Risk Control	
Through homogenization of the traffic flow.	1. Standards for geometric design.
	2. Classification of links with regard to function.
	3. Traffic management, pedestrian zones, auto restrictions.
	4. Traffic calming; speed management.
Through separation between traffic streams.	5. Grade separation (multilevel interchanges).
	6. At-grade separation (traffic signals, roundabouts).
	7. Channelization (medians, road markings).
Through traffic control and road management.	8. Travel time distribution (staggered hours and holidays).
	9. Traffic control (information, warning, flexible signs).
	10. Road maintenance and inspection.

(*Continued*)

Table 11.2: Traffic Planning and Operations Measures Related to Highway Safety Strategies (*Continued*)

Injury Control	
Reducing consequences, preventative measures.	1. Emergency zones without obstacles; breakaway posts.
	2. Installation of median and lateral barriers.
Reducing consequences, efficient rescue service.	3. Establishment of rescue service.
	4. Emergency operation (traffic regulation, rerouting).
Reinstalling the traffic apparatus.	5. Road reparation and inspection.

(*Source:* Used with permission of Institute of Transportation Engineers, *Traffic Safety Toolbox: A Primer on Highway Safety,* Washington DC, 1999, Table 2-2, p. 19.)

Note that the list of implementing actions includes a range of measures that traffic engineers normally work into facility design and control. This is not unexpected because the principal objective of traffic engineering is the provision of safe and efficient traffic operations. The list includes design functions such as interchange design and layout, horizontal and vertical alignments, roadside design and protection, and other measures. It includes the full range of control device implementation: markings, signs (warning, regulatory, guide), and traffic signals. It includes making use of modern Intelligent Transportation Systems (ITS), such as motorist information services and rapid dispatch of emergency vehicles.

It also includes broader planning areas that are not the exclusive domain of the traffic engineer. These include public transportation planning and policies to promote its use, as well as major changes in land-use structures and policies. Although this text does not deal in detail with these broader issues, the traffic engineer must be aware of them and must actively participate in planning efforts to create appropriate policies and implementation strategies.

The engineering aspects of these implementing actions are treated throughout this text in the appropriate chapters.

11.2.7 National Policy Initiatives

Some aspects of traffic accident and fatality abatement can be addressed through imposition of broad policy initiatives and programs. Such programs generally are initiated in federal or state legislation. At the federal level, compliance is encouraged by tying implementation to receipt of federal-aid highway funds. Some examples of such policies include:

- State vehicle-inspection programs and requirements
- National speed limit (eliminated in 1996)
- National 21-year-old drinking age
- Reduction in driving while intoxicated (DWI) requirements to 0.08 blood content (from 0.10 and 0.12)
- State driving under the influence (DUI)/DWI programs
- Federal vehicle design standards

Traffic engineers should be involved in creating and implementing these programs and in providing guidance and input to policy makers through professional and community organizations. As long as traffic accidents and fatalities occur, all levels of government will attempt to deal with the problem programmatically. These programs should be well founded in research and must concentrate on specific ways to improve safety.

Stricter DUI/DWI policies have been developed in most states over the past decade. Public groups, such as Mothers Against Drunk Driving (MADD) and Students Against Drunk Driving (SADD) have played a major role in forcing federal and state governments to focus on this problem. They provide a compelling case study of the impact of community involvement in a major public policy area.

11.3 Accident Data Collection and Record Systems

Two national accident data systems are maintained by the National Highway Traffic Safety Administration (NHTSA). The *Fatality Analysis Reporting System* (FARS) is basically the repository of data on all fatal highway accidents from all 50 states. The primary source of information is police accident reports, which are virtually always filed in the case of a fatal accident. Descriptions of accidents are provided, and as many as 90 coded variables are used in the description. Data can be isolated by geographic location. The *General Estimates System* (GES) is a more general system including information on all types of highway accidents, fatal, injury, and property damage only (PDO) accidents. It uses a sample of police accident reports to estimate national statistics. The sample used is approximately 45,000 police accident reports each year.

The study of traffic accidents is fundamentally different from methods employed to observe other traffic parameters. Because accidents occur infrequently and at unpredictable times and locations, they cannot be directly observed and studied in the field. All accident data come from secondary sources—primarily police and motorist accident reports. All basic information and data originates in these reports, and a system for collecting, storing, and retrieving this information in a usable and efficient form, is an absolute necessity. The information is needed for a wide variety of purposes, including:

1. Identification of locations at which unusually high numbers of accidents occur.
2. Detailed functional evaluations of high-accident locations to determine contributing causes of accidents.
3. Development of general statistical measures of various accident-related factors to give insight into general trends, common causal factors, driver profiles, and other factors.
4. Development of procedures that allow the identification of hazards *before* large numbers of accidents occur.

11.3.1 Accident Reporting

The ultimate basis for all accident information is the individual accident report. These reports come in two types:

- *Motorists' accident reports*—filed by each involved motorist in a traffic accident; required by state law for all accidents with total property damage exceeding a prescribed limit, and for all accidents involving injuries and fatalities.
- *Police accident reports*—filed by an attendant police officer for all accidents at which an officer is present. These would generally include all fatal accidents, most accidents involving a serious injury requiring emergency and/or hospital treatment, and PDO accidents involving major damage. It is estimated that police accident reports are filed for approximately 50% of all traffic accidents that occur.

These reports take a variety of forms, but they have a number of common features that always appear. The centerpiece is a schematic diagram illustrating the accident. Although poorly done in many cases, these documents are a principal source of information for traffic engineers. Information on all drivers and vehicles involved is requested, as are notations of probable causal elements. The names of people injured or killed in the accident must be given, along with a general assessment of their condition at the time of the accident. Of course, information on the location, time, and prevailing environmental conditions of the accident are also included.

Police accident reports are the most reliable source of information because the officer is trained and is a disinterested party. Motorist accident reports reflect the bias of the motorist. Also, there is one police accident report for an accident, but each involved motorist files an accident report, creating several, often conflicting, descriptions of the accident. As noted, the national data systems (FARS, GES) are based entirely on police accident reports.

11.3.2 Manual Filing Systems

Although statewide and national computer-based record systems are extremely valuable in developing statistical analyses and a general understanding of the problem, the examination of a particular site requires the details that exist only in the written accident report. Thus it is still customary to maintain manual files where written police accident reports for a given location can be retrieved and reviewed.

Police accident reports are generally sent and stored in three different locations:

- A copy of each form goes to the state motor vehicle bureau for entry into the state's accident data systems.
- A copy of the form is sent to the central filing location for the municipality or district in which the accident occurred.
- A copy of the form is retained by the officer in his or her precinct as a reference for possible court testimony.

The central file of written accident reports is generally the traffic engineer's most useful source of information for the detailed examination of high-accident locations, and the state's (or municipality's) computer record systems are most useful for the generation and analysis of statistical information.

Location Files

For any individual accident report to be useful, it must be easily retrievable for some time after the occurrence of the accident. Because the traffic engineer must deal with the observed safety deficiencies at a particular location, it is critical that the filing system be organized *by location.* In such a system, the traffic engineer is able to retrieve all of the accident reports for a given location (for some period of time) easily.

There are many ways of coding accident reports for location. For urban systems, a primary file is established for each street or facility. Subfiles are then created for each intersection along the street and for midblock sections, which are normally identified by a range of street addresses. For example, the following file structure might be set up to store accident reports for all accidents occurring on Main Street:

Main Street

First Avenue
Second Avenue
Third Avenue
Foster Blvd.
Fourth Avenue
Lincoln Road
Fifth Avenue
100–199 Main Street
200–299 Main Street
300–399 Main Street
Etc.

Primary files must select a primary direction. For example, if east-west is the primary direction, all primary files for east-west facilities would include intersection and midblock subfiles. Primary files for north-south facilities are also needed but would contain subfiles only for midblock locations. All intersection accident files would use the east-west artery as the primary locator.

In rural situations, and on freeways and some expressways, the distances between intersections and street addresses are generally too large to provide a useful accident-locating system. Tenth-mile markers and mileposts are used where these exist. Where these are not in place, landmarks and recognizable natural features are used in addition to addresses and crossroads to define accident locations.

Central accident files are generally kept current for one to three years. It is preferable to use a rotating system. At the beginning of each month, all records from the previous (or most distant) year for that month are removed. There is generally a "dead file" maintained for three to five years, after which most records are discarded or removed to a warehouse location. Use of microfiche can preserve files and allow them to be filed for longer periods of time.

Accident Summary Sheets

A common approach to maintaining records for a longer period of time is to prepare summary sheets of each year's accident records. These may be kept indefinitely, whereas the individual accident forms are discarded in three to five years.

Figure 11.3 illustrates such a summary sheet, on which all of the year's accidents at one location can be reduced to a single coded sheet. The form retains the basic type of accident, the number and types of vehicles involved, their cardinal direction, weather and roadway conditions, and numbers of injuries and/or fatalities. Of course, such summary information can also be maintained in a computer file.

11.3.3 Computer Record Systems

Two national computer record systems have been noted previously: FARS and GES. Every state—as well as many large municipalities and/or counties—maintains a computerized accident record system. Computer record systems have the advantage of being able to maintain a large number of accident records, keyed to locations. Computer systems are also able to correlate accident records with other data and information. The detail of the individual accident diagram with its accompanying descriptions is often lost, and information is limited to material that can be easily expressed as a series of alphanumeric codes.

Computer record systems serve two vital functions: (1) They produce regular statistical reports at prescribed intervals (usually annually), sorting accident data in ways that provide overall insight into accident trends and problems, and (2) they can produce (on request) large amounts of data on accidents at a specific location or set of locations. Moreover, most computer record systems tie accident files to a number of other statewide information systems, including a highway system network code, traffic volume files from regular counting programs, and project improvement files. These can be correlated to compute statistics and to perform a variety of statistical analyses.

Common types of statistical reports that are available from state computer accident systems include:

- Numbers of accidents by location, type of accident, type of vehicle, driver characteristics, time of day, weather conditions, and other stratifications
- Accident rates by highway location and/or segment, driver characteristics, highway classification, and other variables
- Correlation of types of accidents versus contributing factors
- Correlation of improvement projects with accident experience

TRAFFIC ACCIDENT RECORD

LOCATION MARCUS AVE. & NEW HYDE PARK RD. TSL

VILLAGE N. NEW HYDE PARK PRECINCT

Number	Date	Time	Private Car	Private Car	Private Car	Private Car	Commercial	Omnibus	Bicycle	Pedestrian	Other	Accident Type	Weather	Road Conditions	Road Surface	Injured	Fatal	
267	1.23.78	1645	N				N					E	C	W				
282	1.24.78	1300	S				S					E	C	SLUSH				
463	2.11.78	1805	S	S	S	S						B	C	W		1		
855	3.11.78	1205	N	N								B	C	D				
1462	5.6.78	1612	W	S								A	C	D		1		
1513	5.11.78	1315	N/W	S								A	C	D				
1528	5.12.78	1645	S	S	S							B	C	D				
1569	5.15.78	1525	S	W								A	C	D		2		
1675	5.24.78	1130	W	S								A	R	W		1		
1801	6.3.78	2039	W	S ←	←	Bicycle						A	R	W		1		
1831	6.6.78	1510	N				N/E					A	C	D				
2216	7.8.78	1201	N	W								A	C	D				
2219	7.8.78	0720	S	W								A	C	D		2		
2248	7.10.78	1700	E	E								B	C	D				
2375	7.21.78	1610	S/E								MC/N	A	C	D		1		
2759	8.19.78	2015	N	E								A	C	D				
2787	8.22.78	0850	S	W								A	C	D				
2791	8.22.78	1715	S					School S				B	C	D				
3310	10.1.78	1920	S				W					A	R	W				
3432	10.11.78	1235	S	N/W								A	C	D				
3531	10.19.78	0950	W	W								E	C	D				
4217	12.7.78	0925	S				S					B	C	D				
4234	12.8.78	1030	S	E/S								A	R	W		2		
4281	12.10.78	1900	S	N/W								A	C	D				

Codes: A, right angle collision; B, rear-end collision; C, head-on collision; D, mv sideswiped (opposite direction); E, mv sideswiped (same direction); F, mv leaving curb; G, mv collided with parked mv; H, mv collided with fixed object; I, mv executing U-turn; K, mv executing improper left turn; L, mv executing improper right turn; M, left-turn, head-on collision; N, right turn, head-on collision; O, pedestrian struck by mv; P, unknown; Q, hole in roadway; R, mv backing against traffic; S, operator or occupant fell out of mv; T, person injured while hitchhiking ride on mv; U, mv collided with separated part or object of another mv; V, mv and train collision; W, mv struck by thrown or fallen object; X, parked mv (unattended) rolled into another mv or object; Y, towed mv or trailer broke free of towing vehicle; Z, mv and bicycle in collision.

Figure 11.3: Illustrative Accident Summary Sheet

(*Source:* Courtesy of Nassau County Traffic Safety Board, Mineola, NY.)

Most states also honor requests from traffic professionals for special reports and statistical correlations, although these take somewhat longer to obtain because such inquiries must be specifically programmed. The regular statistical reports of most states, however, provide a broad range of useful information for the engineer, who should always be aware of the capabilities of the specific state and local systems available for use.

11.4 Accident Statistics

Table 11.1 gave some basic information on traffic accidents in the United States during 2007. Accident statistics are measures (or estimates) of the number and severity of accidents. They should be presented in a way that is intended to provide insight into the general state of highway safety and into systematic contributing causes of accidents. These insights can help develop policies, programs, and specific site improvements intended to reduce the number and severity of accidents. Care must be taken, however, in interpreting such statistics because incomplete or partial statistics can be misleading. Further, it is important to understand what each statistic cited means and (even more important) what it *does not* mean.

Consider a simple statistical statement: In the State of X, male drivers are involved in three times as many accidents as female drivers. This is a simple statement of fact. Its implications, however, are not as obvious as they might appear. Are female drivers safer than male drivers in the State of X? It is really not possible to say, given this fact alone. We would need to know how many licensed male and female drivers exist in the State of X. We would need to know how many miles per year the typical male or female driver in the State of X drives. If male drivers compile three times as many vehicle-miles as do female drivers, then one might expect them to have three times as many accidents. However, if males and females each account for half the vehicle-miles in the State of X, the higher number of accidents for male drivers would be quite relevant. It is important, therefore, to understand what various statistics mean and how they are numerically constructed.

11.4.1 Types of Statistics

Accident statistics generally address and describe one of three principal informational elements:

- Accident occurrence
- Accident involvements
- Accident severity

Accident occurrence relates to the numbers and types of accidents that occur, which are often described in terms of rates based on population or vehicle-miles traveled. *Accident involvement* concerns the numbers and types of vehicles and drivers involved in accidents, with population-based rates a very popular method of expression. *Accident severity* is generally dealt with by proxy: The numbers of fatalities and fatality rates are often used as a measure of the seriousness of accidents.

Statistics in these three categories can be stratified and analyzed in an almost infinite number of ways, depending on the factors of interest to the analyst. Some common types of analyses include:

- Trends over time
- Stratification by highway type or geometric element
- Stratification by driver characteristics (gender, age)
- Stratification by contributing cause
- Stratification by accident type
- Stratification by environmental conditions

Such analyses allow the correlation of accident types with highway types and specific geometric elements, the identification of high-risk driver populations, quantifying the extent of DUI/DWI influence on accidents and fatalities, and other important determinations. Many of these factors can be addressed through policy or programmatic approaches. Changes in the design of guardrails have resulted from the correlation of accident and fatality rates with specific types of installations. Changes in the legal drinking age and in the legal definition of DUI/DWI have resulted partially from statistics showing the very high rate of involvement of this factor in fatal accidents. Improved federal requirements on vehicle safety features (air bags, seat belts and harnesses, energy-absorbing steering columns, padded dashboards) have occurred partially as a result of statistics linking these features to accident severity. All of these changes also involved heavy lobbying of interested groups and special studies demonstrating the impact of specific vehicle and/or highway design changes. These types of statistics, however, direct policy makers to key areas requiring attention and research.

11.4.2 Accident Rates

Simple statistics citing total numbers of accidents, involvements, injuries, and/or deaths can be quite misleading because they ignore the base from which they arise. An increase in the number of highway fatalities in a specific jurisdiction from one year to the next must be matched against population and vehicle-usage patterns to make any sense. For this reason, many accident statistics are presented in the form of rates.

Population-Based Accident Rates

Accident rates generally fall into one of two broad categories: *population-based* rates, and *exposure-based* rates. Some common bases for population-based rates include:

- Area population
- Number of registered vehicles
- Number of licensed drivers
- Highway mileage

These values are relatively static (they do not change radically over short periods of time) and do not depend on vehicle usage or the total amount of travel. They are useful in quantifying overall risk to individuals on a comparative basis. Numbers of registered vehicles and licensed drivers may also partially reflect usage.

Exposure-Based Accident Rates

Exposure-based rates attempt to measure the amount of travel as a surrogate for the individual's exposure to potential accident situations. The two most common bases for exposure-based rates are:

- Vehicle-miles traveled
- Vehicle-hours traveled

The two can vary widely depending on the speed of travel, and comparisons based on mileage can yield different insights from those based on hours of exposure. For point locations, such as intersections, vehicle-miles or vehicle-hours have very little significance. Exposure rates for such cases are "event-based" using total volume passing through the point to define "events."

True "exposure" to risk involves a great deal more than just time or mileage. Exposure to vehicular or other conflicts that are susceptible to accident occurrence varies with many factors, including volume levels, roadside activity, intersection frequency, degree of access control, alignment, and many others. Data requirements make it difficult to quantify all of these factors in defining exposure. The traffic engineer should be cognizant of these and other factors when interpreting exposure-based accident rates.

Common Bases for Accident and Fatality Rates

In computing accident rates, numbers should be scaled to produce meaningful values. A fatality rate per mile of vehicle-travel would yield numbers with many decimal places before the first significant digit and would be difficult to conceptualize. The following list indicates commonly used forms for stating accident and fatality rates:

Population-based rates are generally stated according to:

- Fatalities, accidents, or involvements per 100,000 area population
- Fatalities, accidents, or involvements per 10,000 registered vehicles
- Fatalities, accidents, or involvements per 10,000 licensed drivers
- Fatalities, accidents, or involvements per 1,000 miles of highway

Exposure-based rates are generally stated according to:

- Fatalities, accidents, or involvements per 100,000,000 vehicle-miles traveled
- Fatalities, accidents, or involvements per 10,000,000 vehicle-hours traveled
- Fatalities, accidents, or involvements per 1,000,000 entering vehicles (for intersections only)

An Example in Computing Accident and Fatality Rates

The following are sample gross accident statistics for a relatively small urban jurisdiction in 2008:

Fatalities:	75
Fatal Accidents:	60
Injury Accidents:	300
PDO Accidents:	2,000
Total Involvements:	4,100
Vehicle-Miles Traveled:	1,500,000,000
Registered Vehicles:	100,000
Licensed Drivers:	150,000
Area Population:	300,000

In general terms, all rates are computed as:

$$Rate = Total^{*}\left(Scale \middle/ Base \right) \tag{11-1}$$

where: *Total* = total number of accidents, involvements, or fatalities

Scale = scale of the base statistic, as "X" vehicle-miles traveled

Base = total base statistic for the period of the rate

Using this formula, the following fatality rates can be computed using the sample data:

$$Rate\ 1 = 75*\left(100{,}000 \middle/ 300{,}000\right)$$
$$= 25 \text{ deaths per } 100{,}000 \text{ population}$$
$$Rate\ 2 = 75*\left(10{,}000 \middle/ 100{,}000\right)$$
$$= 7.5 \text{ deaths per } 10{,}000 \text{ registered vehicles}$$
$$Rate\ 3 = 75*\left(10{,}000 \middle/ 150{,}000\right)$$
$$= 5.0 \text{ deaths per } 10{,}000 \text{ licensed drivers}$$
$$Rate\ 4 = 75*\left(100{,}000{,}000 \middle/ 1{,}500{,}000{,}000\right)$$
$$= 5.0 \text{ deaths per } 100{,}000{,}000 \text{ veh-mile}$$

Similar rates may also be computed for accidents and involvements but are not shown here.

Accident and fatality rates for a given county, city, or other jurisdiction should be compared against past years, as well as against state and national norms for the analysis year. Such rates may also be subdivided by highway type, driver age and sex groupings, time of day, and other useful breakdowns for analysis.

Severity Index

A widely used statistic for the description of relative accident severity is the severity index (SI), defined as the number of fatalities per accident. For the data of the previous example, there were 75 fatalities in a total of 2,360 accidents. This yields a severity index of:

$$SI = \frac{75}{2360} = 0.0318 \text{ deaths per accident}$$

The severity index is another statistic that should be compared with previous years and state and national norms, so that conclusions may be drawn with respect to the general severity of accidents in the subject jurisdiction.

11.4.3 Statistical Displays and Their Use

Graphic and tabular displays of accident statistics can be most useful in transmitting information in a clear and understandable manner. If a picture is worth 1,000 words, then a skillfully prepared graph or table is at least as useful in forcefully depicting facts. Figure 11.4 shows just one example, a graphic depiction of fatality statistics from the present to the early years of motor vehicles and the availability of traffic fatality records.

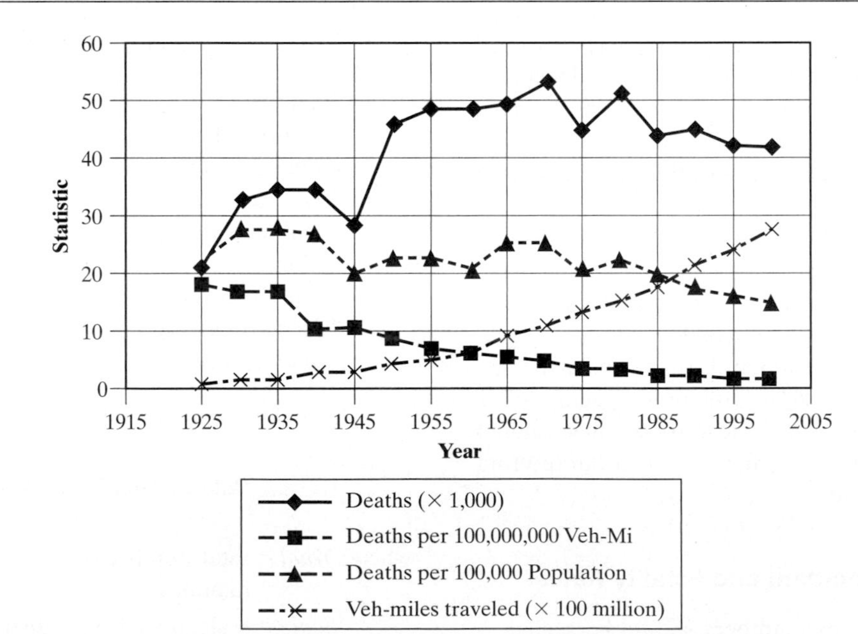

Figure 11.4: Fatality Statistics: 1925–2000

Death rates have steadily declined since the 1930s, even though number of deaths peaked in the 1960s and 1970s. Declining death rates since the 1970s are due to a number of factors, many of which have already been noted: the imposition of a national 55 mi/h speed limit, improved highway and roadside design (including better guardrail, breakaway posts, and impact-attenuating devices), and improved vehicle design. Increased use of seat belts and harnesses, encouraged by many state laws requiring their use, has also helped cut down on fatalities and fatality rates.

The combined presentation of fatalities, fatality rates, and vehicle-miles traveled on a single graph points out that the most significant underlying problem in reducing fatalities is the ever-rising trend in vehicle-miles traveled. Although engineers have been able to provide a safer environment for motorists, increased highway travel keeps the total number of fatalities high.

The statistics shown also mirror other factors in U.S. history. During World War II, the increase in vehicle-miles traveled leveled off, and total highway fatalities and population-based fatality rates dipped. Between 1945 and 1950, however, traffic fatalities increased alarmingly as service men and women returned home and onto the nation's highways.

Careful displays of accident statistics can tell a compelling story, identify critical trends, and spotlight specific problem areas. Care should be taken in the preparation of such displays to avoid misleading the reviewer; when the engineer reviews such information, he or she must analyze what the data says, and (more importantly), what it *does not* say.

11.4.4 Identifying High-Accident Locations

A primary function of an accident record system is to regularly identify locations with an unusually high rate of accidents and/or fatalities. Accident spot maps are a tool that can be used to assist in this task.

Figure 11.5 shows a sample accident spot map. Coded pins or markers are placed on a map. Color or shape codes are used to indicate the category and/or severity of the accident. Modern computer technology allows such maps to be electronically generated. To allow this, the system must contain a location code system sufficient to identify specific accident locations.

Computer record systems can also produce lists of accident locations ranked by either total number of accidents occurring or by defined accident or fatality rate. It is useful to examine both types of rankings because they may yield significantly different results. Some locations with high accident numbers reflect high volumes and have a relatively low accident rate. Conversely, a small number of accidents

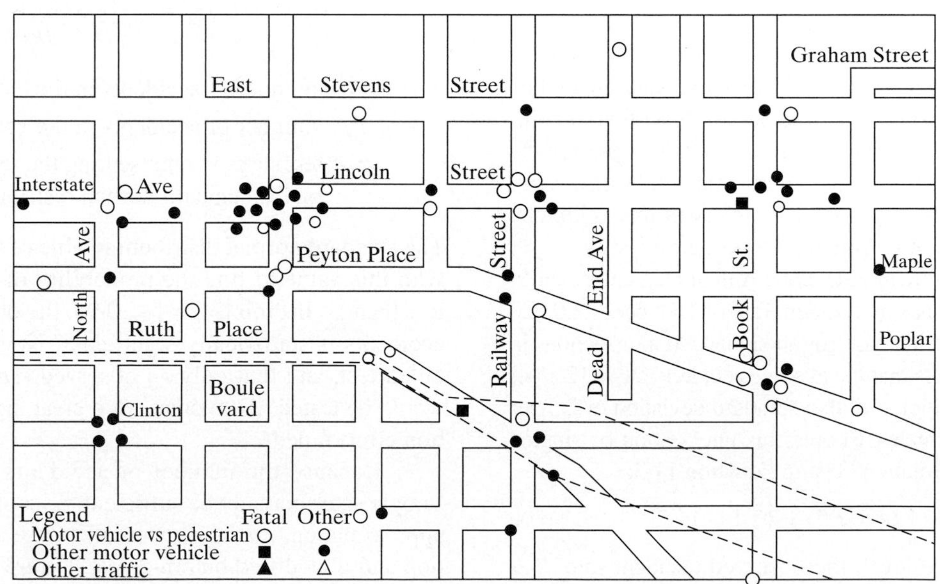

Figure 11.5: Typical Accident Spot Map

(*Source:* Used with permission of Prentice-Hall, Inc, *Manual of Traffic Engineering Studies,* Institute of Transportation Engineers, Washington DC, 1994, p. 400.)

occurring at a remote location with very little demand can produce a very high accident rate. Although statistical rankings give the engineer a starting point, judgment must still be applied in the identification and selection of sites most in need of improvement during any given budget year.

One common approach to determining which locations require immediate attention is to identify those with accident rates that are significantly higher than the average for the jurisdiction under study. To say that the accident rate at a specific location is "significantly" higher than the average, only those locations with accident rates in the highest 5% of the (normal) distribution would be selected. In a one-tailed test, the value of z (on the standard normal distribution) for Prob(z) < 0.95 is 1.645. The actual value of z for a given accident location is computed as:

$$z = \frac{x_1 - \bar{x}}{s} \tag{11-2}$$

where: x_1 = accident rate at the location under consideration

$\bar{x}$ = average accident rate for locations within the jurisdiction under study

s = standard deviation of accident rates for locations within the jurisdiction under study

If the value of z must be at least 1.645 for 95% confidence, the minimum accident rate that would be considered to be significantly higher than the average may be taken to be:

$$x_1 \geq 1.645s + \bar{x} \tag{11-3}$$

Locations with a higher accident rate than this value would be selected for specific study and remediation. Note that in comparing average accident rates, similar locations should be grouped (i.e., accident rates for signalized intersections are compared to those for other signalized intersections; midblock rates are compared to other midblock rates, etc.).

Consider the following example: A major signalized intersection in a small city has an accident rate of 15.8 per 1,000,000 entering vehicles. The database for all signalized intersections in the jurisdiction indicates that the average accident rate is 12.1 per 1,000,000 entering vehicles, with a standard deviation of 2.5 per 1,000,000 entering vehicles. Should this intersection be singled out for study and remediation? Using Equation 11-3:

$$15.8 \geq (1.645^{*}2.5) + 12.1 = 16.2$$

For a 95% confidence level, the observed accident rate *does not* meet the criteria for designation as a significantly higher accident rate.

An important factor that tempers statistical identification of high-accident locations is the budget that can be applied to remediation projects in any given year. Ranking systems are important because they can help set priorities. Priorities are necessary whenever funding is insufficient to address all locations identified as needing study and remediation. A jurisdiction may have 15 locations that are identified as having significantly higher accident rates than the average. However, if funding is available to address only 8 of them in a given budget year, priorities must be established to select projects for implementation.

11.4.5 Before-and-After Accident Analysis

When an accident problem has been identified, and an improvement implemented, the engineer must evaluate whether or not the remediation has been effective in reducing the number of accidents and/or fatalities. A before-and-after analysis must be conducted. The length of time considered before and after the improvement must be long enough to observe changes in accident occurrence. For most locations, periods ranging from three months to one year are used. The length of the "before" period and the "after" period must be the same.

The normal approximation test is often used to make this determination. This test is more fully discussed in Chapters 7 and 10. The statistic z is computed as:

$$z_1 = \frac{f_B - f_A}{\sqrt{f_A + f_B}} \tag{11-4}$$

where: f_A = number of accidents in the "after" period

f_B = number of accidents in the "before" period

z_1 = test statistic representing the reduction in accidents on the standard normal distribution

The standard normal distribution table of Chapter 7 is entered with this value to find the probability of z being equal to or less than z_1. If Prob $[z \leq z_1] \geq 0.95$, the observed reduction in accidents is statistically significant. Note that this is a one-tailed test, and that only an observed *reduction* in accidents would be tested. An increase is a clear sign that the remediation effort failed.

Because the number of accidents may be small, the sample sizes may not be sufficient to justify use of the normal approximation. It is more accurate to use the Poisson distribution and a modified binomial test. Figure 11.6 shows graphic criteria for rejecting the null hypothesis (i.e., that there has been no change in accident occurrence). The curve is entered with the number of accidents in the "before" period, and the percentage decrease in accidents in the "after" period. If the point plots above the appropriate decision line (one line is for

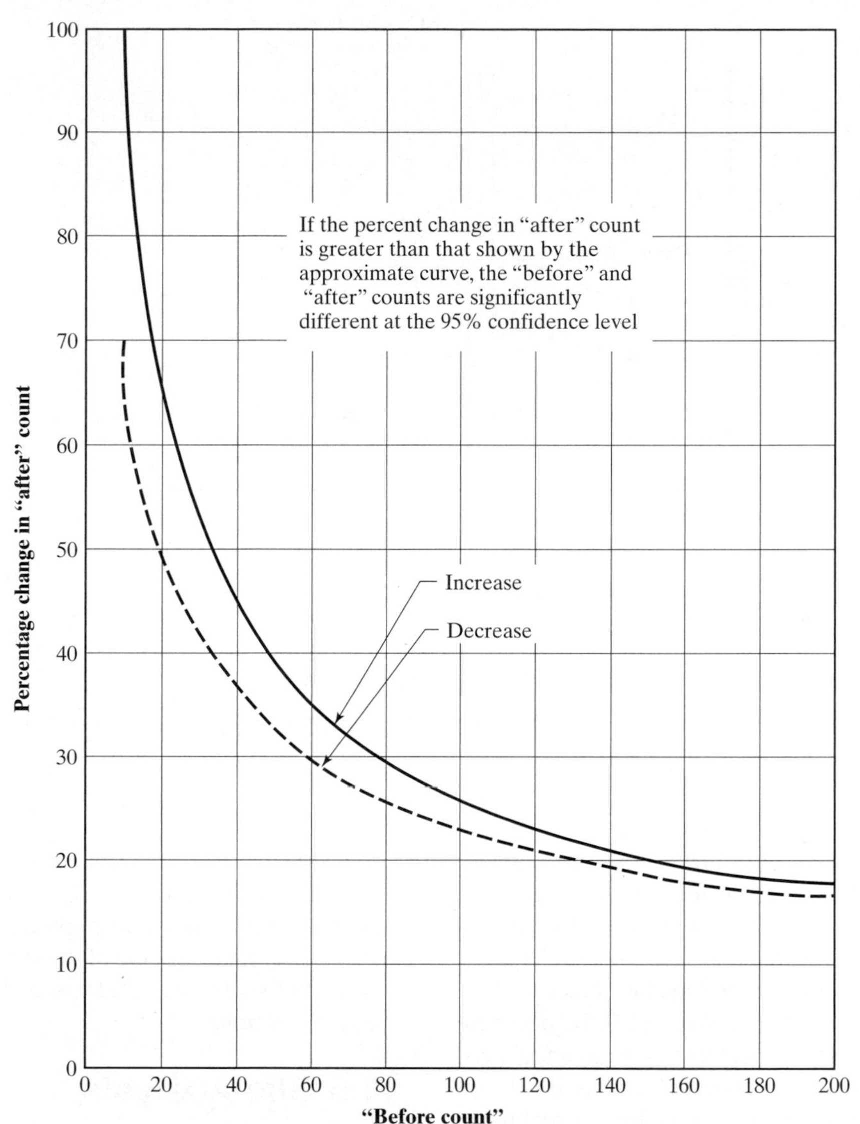

Figure 11.6: Rejection Criteria for Poisson-Distributed Data Based on the Binomial Test

(*Source:* Used with permission of Transportation Research Board, Weed, R., "Research Decision Criteria for Before-and-After Analyses," *Transportation Research Record 1069,* Washington DC, 1986, p. 11.)

accident reductions, the other for accident increases), the observed change is statistically significant.

Consider the following problem: A signal is installed at a high-accident location to reduce the number of right-angle accidents that are occurring. In the six-month period prior to installing the signal, 10 such accidents occurred. In the six-month period following the installation of the signal, six such accidents occurred. Was this reduction statistically significant?

Using the normal approximation test, the statistic z is computed as:

$$z = \frac{10 - 6}{\sqrt{10 + 6}} = \frac{4}{4} = 1.00$$

From the standard normal distribution table of Chapter 7, Prob $[z \leq 1.00] = 0.8413 < 0.95$. The reduction in accidents observed is *not* statistically significant.

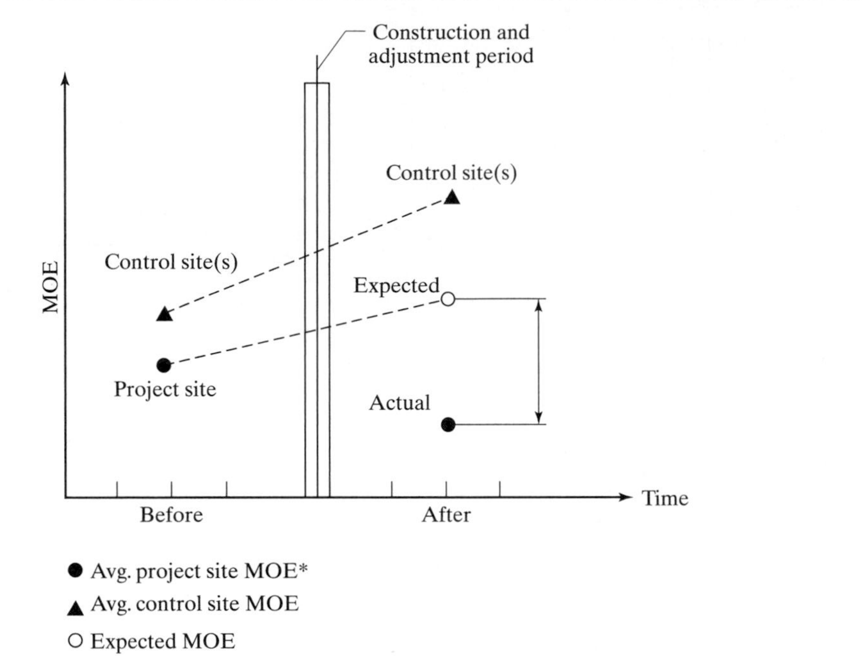

Figure 11.7: Before and After Experiment with Control Conditions

(*Source:* Used with permission of Prentice-Hall Inc., *Manual of Traffic Engineering Studies,* Institute of Transportation Engineers, Washington DC, 1994, p. 368.)

Given the small sample size involved, it may be better to use the modified binomial test. The percent reduction in accidents for the "after" period is 4/10, or 40%. A point is plotted on Figure 11.6 at (10, 40). This point is clearly below the decision line for accident reductions. Therefore, the observed accident reduction is *not* statistically significant. If the decision curve is entered with 10 "before" accidents, the minimum percentage reduction needed for a "significant" result is 70%.

Technically, there is a serious flaw in the way that most before-and-after accident analyses are conducted. There is generally a base assumption that any observed change in accident occurrence (or severity) is due to the corrective actions implemented. Because the time span involved in most studies is long, however, this may not be correct in any given case.

If possible, a control experiment or experiments should be established. These control experiments involve locations with similar accident experience that have not been treated with corrective measures. The controls establish the expected change in accident experience due to general environmental causes not influenced by corrective measures. For the subject location, the null hypothesis is that the change in accident experience is not significantly different from the change at observed control locations. Figure 11.7 illustrates this technique.

Although desirable from a statistical point of view, the establishment of control conditions is often a practical problem, requiring that some high-accident locations be left untreated during the period of the study. For this reason, many before-after studies of accidents are conducted without such control conditions.

11.5 Site Analysis

One of the most important tasks in traffic safety is the study and analysis of site-specific accident information to identify contributing causes and to develop site remediation measures that will lead to improved safety.

Once a location has been statistically identified as a "high-accident" location, detailed information is required in two principal areas:

1. Occurrence of accidents at the location in question
2. Environmental and physical conditions existing at the location

The analysis of this information must identify the environmental and physical conditions that potentially or actually contribute to the observed occurrence of accidents. Armed with such analyses, engineers may then develop countermeasures to alleviate the problem(s).

The best information on the occurrence of accidents is compiled by reviewing all accident reports for a given location over a specified study period. This can be done using computer accident records, but the most detailed data will be available from the actual police accident reports on file. Environmental and physical conditions are established by a thorough site investigation conducted by appropriate field personnel. Two primary graphical outputs are then prepared:

1. Collision diagram
2. Condition diagram

11.5.1 Collision Diagrams

A collision diagram is a schematic representation of all accidents occurring at a given location over a specified period. Depending on the accident frequency, the "specified period" usually ranges from one to three years.

Each collision is represented by a set of arrows, one for each vehicle involved, which schematically represents the type of accident and directions of all vehicles. Arrows are generally labeled with codes indicating vehicle types, date and time of accident, and weather conditions.

The arrows are placed on a schematic (not-to-scale) drawing of the intersection with no interior details shown. One set of arrows represents one accident. Note that arrows are not

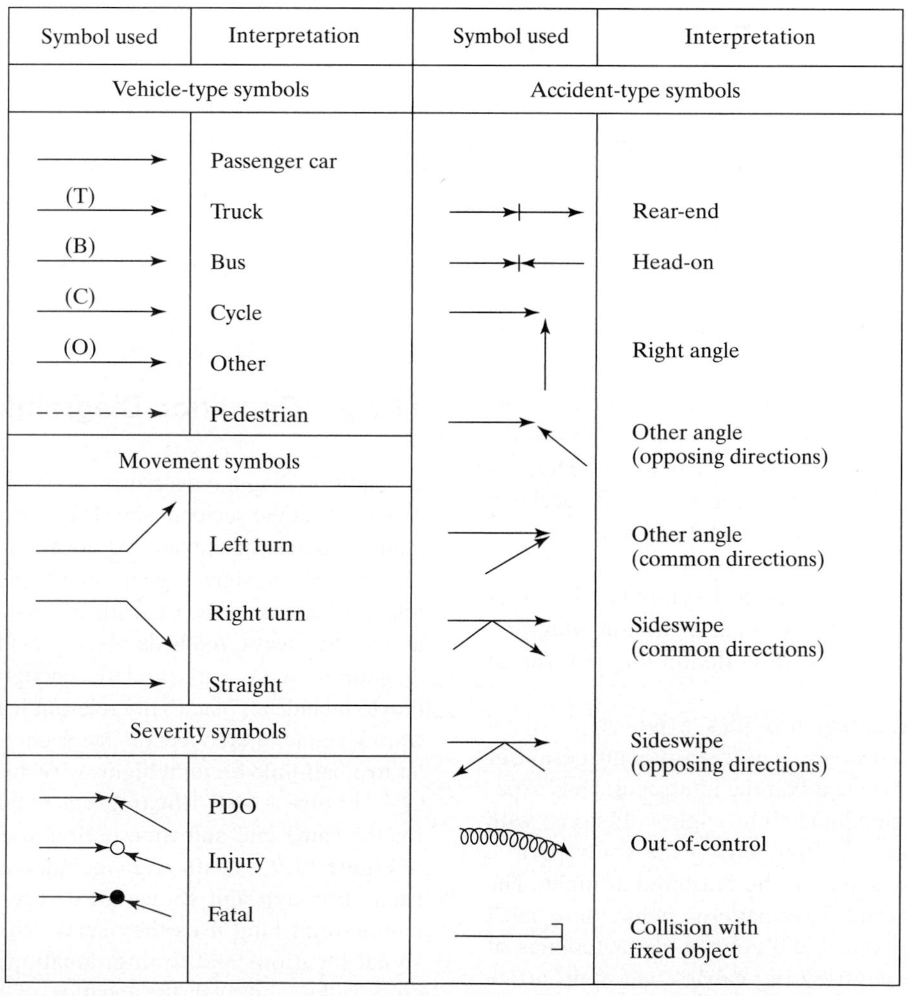

Figure 11.8: Symbols Used in Collision Diagrams

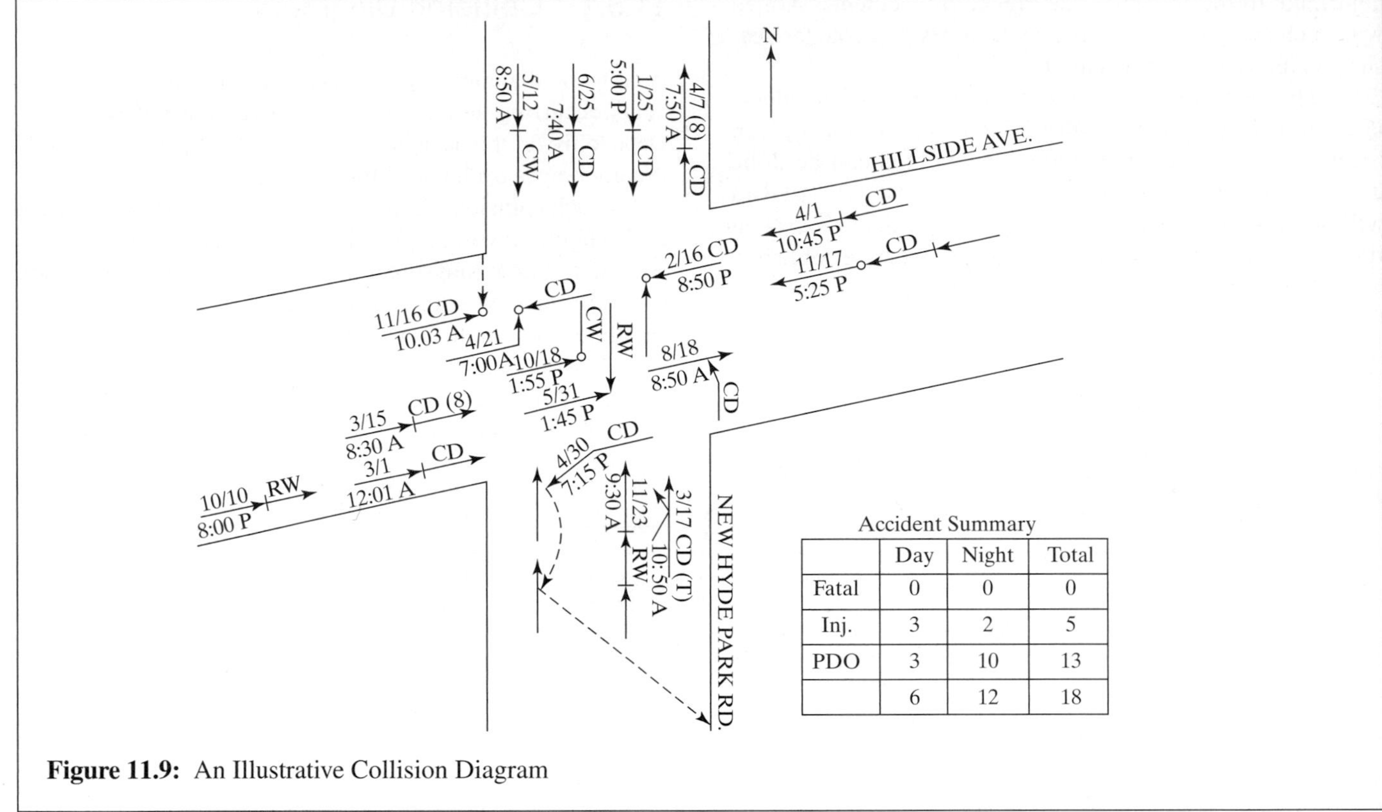

	Day	Night	Total
Fatal	0	0	0
Inj.	3	2	5
PDO	3	10	13
	6	12	18

Figure 11.9: An Illustrative Collision Diagram

necessarily placed at the exact spot of the accident on the drawing. There could be several accidents that occurred at the same spot, but separate sets of arrows would be needed to depict them. Arrows illustrate the occurrence of the accident and are placed as close to the actual spot of the accident as possible.

Figure 11.8 shows the standard symbols and codes used in the preparation of a typical collision diagram. Figure 11.9 shows an illustrative collision diagram for an intersection.

The collision diagram provides a powerful visual record of accident occurrence over a significant period of time. In Figure 11.9, it is clear that the intersection has experienced primarily rear-end and right-angle collisions, with several injuries but no fatalities during the study period. Many of the accidents appear to be clustered at night. The diagram clearly points out these patterns, which now must be correlated to the physical and control characteristics of the site to determine contributing causes and appropriate corrective measures.

11.5.2 Condition Diagrams

A condition diagram describes all physical and environmental conditions at the accident site. The diagram must show all geometric features of the site, the location and description of all control devices (signs, signals, markings, lighting, etc.), and all relevant features of the roadside environment, such as the location of driveways, roadside objects, land uses, and so on. The diagram must encompass a large enough area around the location to include all potentially relevant features. This may range from several hundred feet on intersection approaches to a quarter to a half mile on rural highway sections.

Figure 11.10 illustrates a condition diagram. It is for the same site and time period as the collision diagram of Figure 11.9. The diagram includes several hundred feet of each approach and shows all driveway locations and the commercial land uses they serve. Control details include signal locations and timing, location of all stop lines and crosswalks, and even the location of roadside trees, which could conceivably affect visibility of the signals.

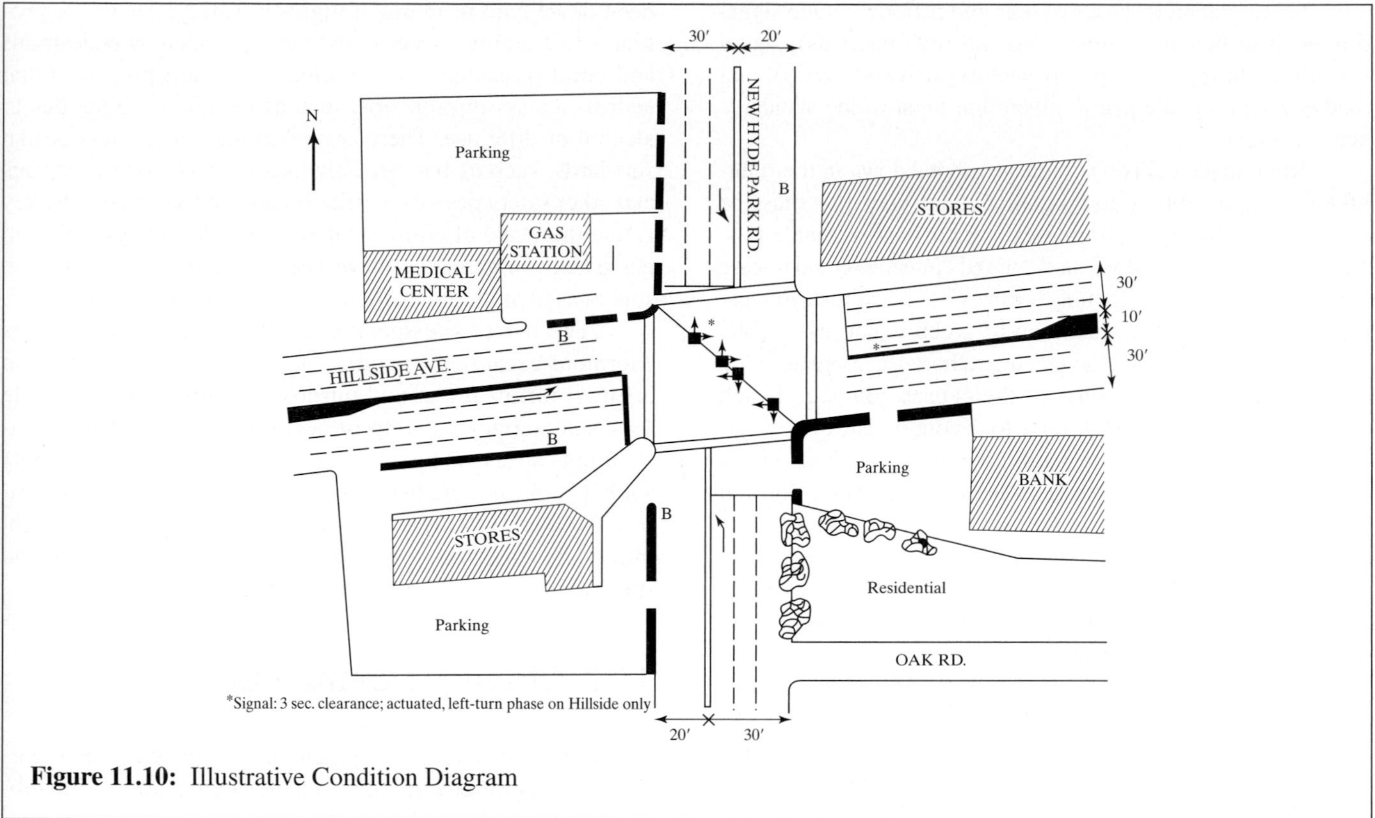

Figure 11.10: Illustrative Condition Diagram

11.5.3 Interpretation of Condition and Collision Diagrams

This brief overview chapter cannot fully discuss and present all types of accident site analyses. The objective in analyzing collision and condition diagrams is straightforward: Find contributing causes to the observed accidents shown in the collision diagram among the design, control, operational, and environmental features summarized on the condition diagram. Doing so involves virtually all of the traffic engineer's knowledge, experience, and insight, and the application of professional judgment.

Accidents are generally grouped by type. Predominant types of accidents shown in Figure 11.9 are rear-end and right-angle collisions. For each type of accident, three questions should be asked:

1. What driver actions led to the occurrence of such accidents?
2. What existing conditions at the site could contribute to drivers taking such actions?
3. What changes can be made to reduce the chances of such actions taking place?

Rear-end accidents occur when the lead vehicle stops suddenly or unexpectedly and/or when the trailing driver follows too closely for the prevailing speeds and environmental conditions. Although "tailgating" by a following driver cannot be easily corrected by design or control measures, a number of factors evident in Figures 11.9 and 11.10 may contribute to vehicles stopping suddenly or unexpectedly.

The condition diagram shows a number of driveways allowing access to and egress from the street at or near the intersection itself. Unexpected movements into or out of these driveways could cause mainline vehicles to stop suddenly. Because of these driveways, STOP lines are located well back from the sidewalk line, particularly in the northbound direction. Vehicles, therefore, are stopping at positions not normally expected, and following drivers may be surprised and unable to respond in time to avoid a collision. Potential corrective actions include closing some or all of these driveways and moving STOP lines closer to their normal positions.

Other potential causes of rear-end actions include signal timing (insufficient "yellow" and "all red" intervals), signal visibility (do trees block approaching drivers' views), and roadway lighting adequacy (given that most of the accidents occur at night).

Right-angle collisions indicate a breakdown in the right-of-way assignment by the signal. Signal visibility must be checked and the signal timing examined for reasonableness. Again, insufficient "yellow" and "all red" intervals could release vehicles before the competing vehicles have had time to clear the intersection. If the allocation of green is not reasonable, some drivers will "jump" the green or otherwise disregard it.

At this location, some of the causes compound each other. The setback of STOP lines to accommodate driveways, for example, lengthens the requirements for "all red" clearance intervals and, therefore, amplifies the effect of a shortfall in this factor.

This analysis is illustrative. The number of factors that can affect accident occurrence and/or severity at any given location is large indeed. A systematic approach, however, is needed if all relevant factors are to be identified and dealt with in an effective way. Traffic safety is not an isolated subject for study by traffic engineers. Rather, everything traffic engineers do is linked to a principal objective of safety. The importance of building safety into all traffic designs, control measures, and operational plans is emphasized throughout this text.

11.6 Development of Countermeasures

The ultimate goal of any general or site-specific safety analysis is the development of programmatic or site-specific improvements to mitigate the circumstances leading to those accidents. Each case, however, has its own unique characteristics that must be studied and analyzed in detail. Program development must consider national, regional, and local statistics; site remediation requires detailed collision and condition information.

Programmatic countermeasures are used to attack systemic safety problems that prevail throughout the highway system. These measures generally involve education and/or control of drivers, vehicles, or highway design features. Table 11.3 provides a sampling of systemic problems and the types of programmatic approaches that are used to address them.

Table 11.3 presents only a small sampling of the thousands of legislative and programmatic measures that have been developed to improve highway safety. There are programs that address specific user groups, such as pedestrians and child passengers in vehicles. There are programs that address a class of problems, such as impaired driving due to alcohol or drug use. There are programs that address design standards, such as bumper heights and provision of air bags and other safety devices in vehicles and on roadways. The key is that this scale of effort is intended to address systemic and persistent problems that have been studied, and where such study has identified potential ways of improving the situation.

The list of site-specific accident countermeasures is enormous because the number of accident situations that can arise is, for all intent and purposes, almost infinite. Table 11.4 summarizes a listing of accident types, probable contributing causes, and potential countermeasures from a 1981 study [7]. Again, the listing is intended to be purely illustrative and it is certainly not exhaustive. It does provide insight into the linkages between observed accident occurrence and specific measures designed to combat it.

11.7 Closing Comments

This chapter provides a very general overview of the important subject of highway safety and accident studies. The subject is complex and covers a vast range of material. Everything the traffic engineer does, from field studies, to planning and design, to control and operations, is related to the provision of a safe system for vehicular travel. The traffic engineer is not alone in the focus on highway safety. Many other professionals, from urban planners to lawyers to public officials, also have an abiding interest in safe travel.

References

1. *Traffic Safety Facts 2007,* National Highway Traffic Safety Administration, U.S. Department of Transportation, Washington DC, 2008.
2. *Traffic Safety Facts, 2007 Data,* Overview, National Highway Traffic Safety Administration, Publication Number: DOT HS 810 993.
3. "Deaths Down on America's Roads," *U.S. News,* February 2009, Number: DOT HS 810 993. Available at: http://www.usatoday.com/news/nation/2009-02-04-traffic-deaths_N.htm
4. *The Traffic Safety Toolbox: A Primer on Traffic Safety,* Institute of Transportation Engineers, Washington DC, 1999.

Table 11.3: Illustrative Programmatic Safety Approaches

Target Group	Problem Area	Sample Strategies or Programs
Drivers	Ensuring driver competency	1. Implement graduated licensing system. 2. Develop and implement improved competency training and assessment procedures for entry-level drivers. 3. Increase effectiveness of license suspension and revocation procedures. 4. Develop and provide technical aids such as simulators and electronic media for private self-assessment and improvement of driver skills.
Drivers	Reducing impaired driving	1. Implement stronger legislation to reduce drinking and driving. 2. Develop and implement sobriety checkpoints and saturation enforcement blitzes. 3. Develop and implement a comprehensive public awareness campaign.
Drivers	Keeping drivers alert	1. Retrofit rural and other fatigue-prone facilities with shoulder rumble strips. 2. Provide 24-hour "coffee stops" along fatigue-prone facilities.
Vehicles	Increasing safety enhancements in vehicles	1. Implement educational programs on the use of antilock brake systems and other vehicle safety features. 2. Strengthen regulations requiring incorporation of safety features on vehicles and their use by motorists and passengers.
Highways	Keeping vehicles on the roadway	1. Improve driver guidance through installation of better pavement markings and delineation. 2. Implement a targeted rumble-strip program. 3. Improve highway maintenance. 4. Develop better guidance to control variance in speed through combinations of geometric, control, and enforcement techniques.
Highways	Reducing the consequences of leaving the roadway	1. Provide improved practices for the selection, installation, and maintenance of upgraded roadside safety hardware (guardrail, impact-attenuating devices, etc.) 2. Implement a program to remove hazardous trees or other natural roadside hazards. 3. Install breakaway sign and lighting posts.
Intersections	Reducing intersection accidents	1. Install and use automated methods to monitor and enforce intersection traffic control. 2. Implement more effective access control strategies with a safety perspective.

(*Source:* Excerpted from Institute of Transportation Engineers, *The Traffic Safety Toolbox: A Primer on Traffic Safety,* Washington DC, pp. 8–11.)

Table 11.4: Illustrative Site-Specific Accident Countermeasures

Accident Pattern	Probable Cause[a]	Possible Countermeasures[b]
Left turn, head-on	A	1–11
	B	3, 6, 12–15
	C	16, 17
	D	3
	E	15
Rear-end at unsignalized intersection	A	4, 13, 18
	E	15
	F	14
	G	15, 19–22
	H	23
	I	10, 24
	J	25
Rear-end at signalized intersection	A	3, 4, 13, 18
	G	15, 19–22
	H	23
	J	25, 26
	K	12, 14, 15, 27–32
	L	16, 17, 33
	M	34
Right angle at signalized intersection	B	6, 12, 14, 15, 35, 36
	E	15, 16, 37
	H	23
	K	14, 27–32, 38
	L	11, 16, 17, 33, 39, 40
	N	14
	O	2, 11
Right angle at unsignalized intersection	B	6, 10, 12, 14, 15, 24, 35, 36, 41, 42
	E	15, 16, 37
	H	23
	N	14
	O	10, 43
	P	44, 45
Pedestrian-vehicle	B	12, 25, 35, 46
	E	14, 15, 45, 47
	H	23
	I	10, 25, 26
	L	11
	P	26
	Q	47, 48
	R	49
	S	14, 15, 47, 50
	T	51–53
Wet pavement	G	15, 19–22, 62
	T	53
Ran off roadway	E	15
	G	15, 19–22
	H	23
	K	54
	U	55–58
	V	14, 53, 59
	W	60
	X	6
	Y	61
Fixed object	E	15
	G	20, 22, 55, 62
	H	23
	T	53
	U	14, 63
	Z	58, 64–67
	AA	68
Parked or parking vehicle	F	15
	T	69
	BB	35
	CC	70
	DD	45, 50, 71
	EE	1, 43
Sideswipe or head-on	E	15, 72, 73
	T	53
	U	1, 55
	W	60
	X	6, 13, 74
	Y	61
	FF	38, 75
Driveway-related	A	13, 18, 35, 55, 72, 76
	B	12, 15, 32, 35
	E	15
	H	23
	GG	77–81
	HH	43, 79, 82
	II	6, 10, 74
Train-vehicle	B	12, 14, 24, 83–85
	E	15
	G	62
	K	23, 54
	T	36, 42, 53
	JJ	11
	KK	86
	LL	87
	MM	88
Night	K	14, 23, 59
	V	14, 59, 89
	X	14, 53, 59, 89
	FF	44, 90

(Continued)

Table 11.4: Illustrative Site-Specific Accident Countermeasures *(Continued)*

[a]Key to probable causes:

A	Large turn volume	T	Inadequate or improper pavement markings
B	Restricted sight distance	U	Inadequate roadway design for traffic conditions
C	Amber phase too short	V	Inadequate delineation
D	Absence of left-turn phase	W	Inadequate shoulder
E	Excessive speed	X	Inadequate channelization
F	Driver unaware of intersection	Y	Inadequate pavement maintenance
G	Slippery surface	Z	Fixed object in or too close to roadway
H	Inadequate roadway lighting	AA	Inadequate TCDs and guardrail
I	Lack of adequate gaps	BB	Inadequate parking clearance at driveway
J	Crossing pedestrians	CC	Angle parking
K	Poor traffic control device (TCD) visibility	DD	Illegal parking
L	Inadequate signal timing	EE	Large parking turnover
M	Unwarranted signal	FF	Inadequate signing
N	Inadequate advance intersection warning signs	GG	Improperly located driveway
O	Large total intersection volume	HH	Large through traffic volume
P	Inadequate TCDs	II	Large driveway traffic volume
Q	Inadequate pedestrian protection	JJ	Improper traffic signal preemption timing
R	School crossing area	KK	Improper signal or gate warning time
S	Drivers have inadequate warning of frequent midblock crossings	LL	Rough crossing surface
		MM	Sharp crossing angle

[b]Key to possible countermeasures:

1	Create one-way street	30	Install signal back plates
2	Add lane	31	Relocate signal
3	Provide left-turn signal phase	32	Add signal heads
4	Prohibit turn	33	Provide progression through a set of signalized intersections
5	Reroute left-turn traffic	34	Remove signal
6	Provide adequate channelization	35	Restrict parking near corner/crosswalk/driveway
7	Install stop sign	36	Provide markings to supplement signs
8	Revise signal-phase sequence	37	Install rumble strips
9	Provide turning guidelines for multiple left-turn lanes	38	Install illuminated street name sign
10	Provide traffic signal	39	Install multidial signal controller
11	Retime signal	40	Install signal actuation
12	Remove signal	41	Install yield sign
13	Provide lurn lane	42	Install limit lines
14	Install or improve warning sign	43	Reroute through traffic
15	Reduce speed limit	44	Upgrade TCDs
16	Adjust amber phase	45	Increase enforcement
17	Provide all-red phase	46	Reroute pedestrian path
18	Increase curb radii	47	Install pedestrian barrier
19	Overlay pavement	48	Install pedestrian refuge island
20	Provide adequate drainage	49	Use crossing guard at school crossing area
21	Groove pavement	50	Prohibit parking
22	Provide "slippery when wet" sign	51	Install thermoplastic markings
23	Improve roadway lighting	52	Provide signs to supplement markings
24	Provide stop sign	53	Improve or install pavement markings
25	Install or improve pedestrian crosswalk TCDs	54	Increase sign size
26	Provide pedestrian signal	55	Widen lane
27	Install overhead signal	56	Relocate island
28	Install 12-inch signal lenses	57	Close curb lane
29	Install signal visors	58	Install guardrail

(*Continued*)

Table 11.4: Illustrative Site-Specific Accident Countermeasures *(Continued)*

59	Improve or install delineation	75	Install advance guide sign
60	Upgrade roadway shoulder	76	Increase driveway width
61	Repair road surface	77	Regulate minimum driveway spacing
62	Improve skid resistance	78	Regulate minimum corner clearance
63	Provide proper superelevation	79	Move driveway to side street
64	Remove fixed object	80	Install curb to define driveway location
65	Install barrier curb	81	Consolidate adjacent driveways
66	Install breakaway posts	82	Construct a local service road
67	Install crash cushioning device	83	Reduce grade
68	Paint or install reflectors on obstruction	84	Install train-actuated signal
69	Mark parking stall limits	85	Install automatic flashers or flashers with gates
70	Convert angle to parallel parking	86	Retime automatic flashers or flashers with gates
71	Create off-street parking	87	Improve crossing surface
72	Install median barrier	88	Rebuild crossing with proper angle
73	Remove constriction such as parked vehicle	89	Provide raised markings
74	Install acceleration or deceleration lane	90	Provide illuminated sign

(*Source:* Reprinted with permission of Prentice-Hall Inc., from Robertson, Hummer, and Nelson (Editors), *Manual of Traffic Engineering Studies,* Institute of Transportation Engineers, Washington DC, 1994, pp. 214, 215.)

5. Ogden, K.W., *Safer Roads: A Guide to Road Safety Engineering,* Avebury Technical, Brookfield, VT, 1996.

6. Trinca, G., et al., *Reducing Traffic Injury—A Global Challenge,* Royal Australian College of Surgeons, Melbourne, Australia, 1988.

7. Robertson, H.D., Hummer, J.E., and Nelson, D.C. (Editors), *Manual of Transportation Engineering Studies,* Institute of Transportation Engineers, Prentice Hall, Upper Saddle River, NJ, 1994.

Problems

11-1. Consider the following data for 2008 in a small suburban community:

- Number of accidents 360
  - Fatal 10
  - Injury 36
  - PDO 314
- Number of fatalities 15
- Area population 50,000s
- Registered vehicles 35,000
- Annual VMT 12,000,000
- Average speed 30 mi/h

Compute all relevant exposure- and population-based accident and fatality rates for this data. Compare these to national norms for the current year. (Hint: Use the Internet to locate current national norms.)

11-2. A before-and-after accident study results in 25 accidents during the year before a major improvement to an intersection, and 15 the year after. Is this reduction in accidents statistically significant? What statistical test is appropriate for this comparison? Why?

11-3. Consider the collision and condition diagrams illustrated here. Discuss probable causes of the accidents observed. Recommend improvements, and illustrate them on a revised condition diagram.

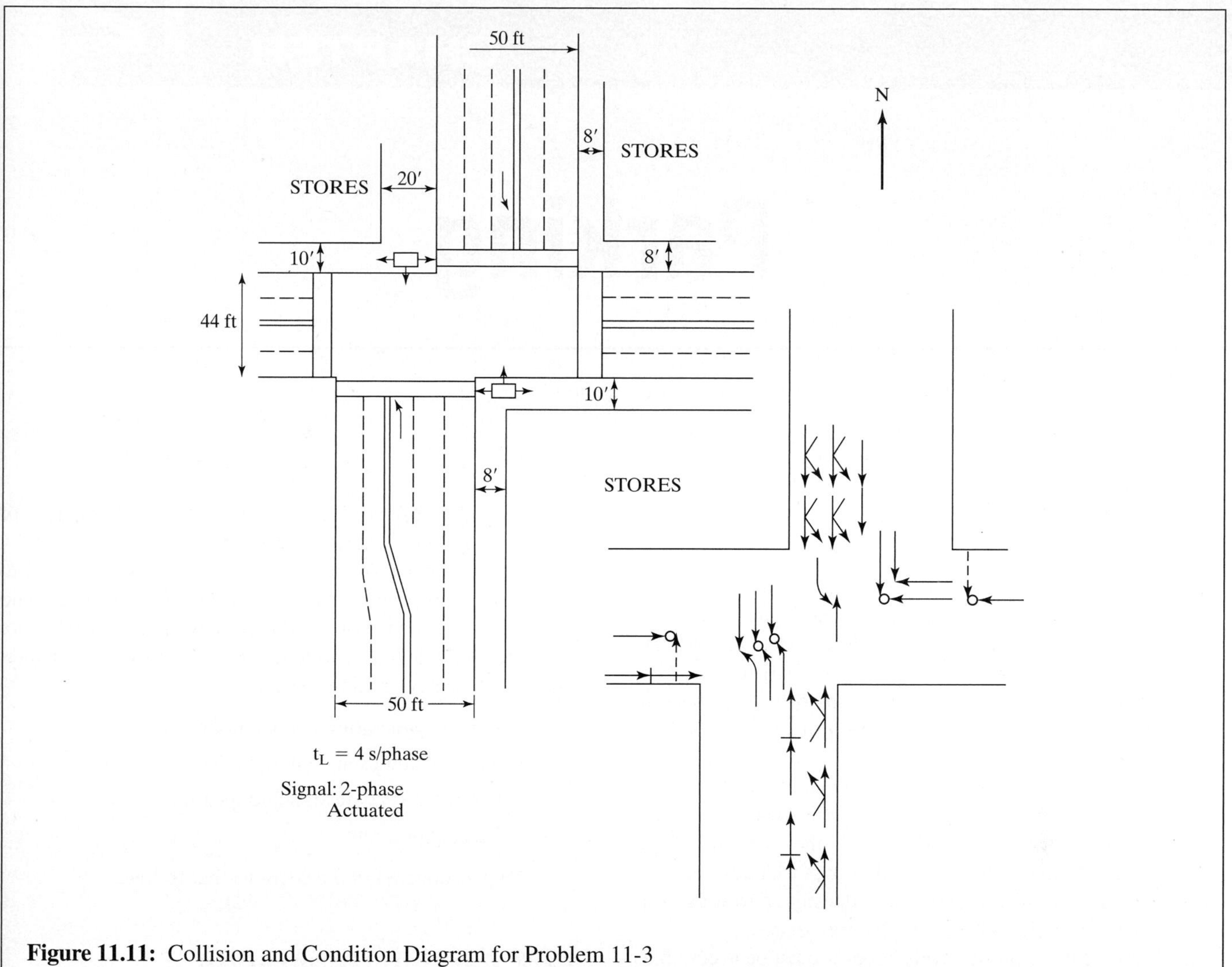

Figure 11.11: Collision and Condition Diagram for Problem 11-3

CHAPTER 12

Parking

12.1 Introduction

Every person starts and ends each trip as a pedestrian. With the exception of drive-through facilities now present at such varied destinations as banks and fast-food restaurants, travelers generally leave from their origins as pedestrians and enter their destinations as pedestrians. In terms of trips using private automobiles, the pedestrian portion of the trip starts or ends at a parking space.

At the residential trip end, private vehicles are accessed in private driveways and garages, in on-street parking spaces, or in nearby off-street lots or garages. At the other end of the trip, the location and nature of parking opportunities depends heavily on the land-use function and density as well as on a wide variety of public policy and planning issues.

For land to be productively used, it must be accessible. Although public transportation can be a major part of providing accessibility in dense urban areas, for the most part, accessibility depends on the supply, convenience, and cost of parking facilities. Major activity centers, from regional shopping malls to sports facilities to airports, rely on significant parking supply to provide site accessibility. Without such supply, these facilities could not operate profitably over a substantial period of time.

The economic survival of most activity centers, therefore, is directly related to parking and other forms of access. Parking supply must be balanced with other forms of access (public transportation), the traffic conditions created by such access, and the general environment of the activity center. Although economic viability is most directly related to the availability of parking, the environmental impacts of generated traffic may have negative effects as well.

This chapter provides an overview of issues related to parking. The coverage is not intended to be exhaustive, and you are encouraged to consult the available literature for more complete and detailed treatments of the subject. This chapter addresses four key parking issues:

- Parking generation and supply needs
- Parking studies and characteristics
- Parking facility design and location
- Parking programs

Each is covered in the sections that follow.

12.2 Parking Generation and Supply Needs

The key issue in parking is a determination of how many spaces are required for a particular development and where they should be located. These requirements lead to locally based zoning regulations on the minimum numbers of spaces that need to be provided when a development is built.

The need for parking spaces depends on many factors, some of which are difficult to assess. The type and size of land use(s) in a development is a major factor, but so is the general density of the development environment and the amount and quality of public transportation access available.

12.2.1 Parking Generation

The most comprehensive source of parking generation information is the Institute of Transportation Engineers' *Parking Generation* [*1*]. The third edition of this basic reference was published in 2004, but updates are provided periodically, and you are encouraged to consult the latest edition directly for up-to-date criteria. Material in this text is based on the third edition.

Parking generation relates the maximum observed number of occupied parking spaces to *one* underlying variable that is used as a surrogate for the size or activity level of the land use involved. Early studies reported in *Parking Principles* [*2*] established preferred and alternative variables for establishing parking generation rates. These variables are listed in Table 12.1.

A summary of parking generation rates and relationships, compiled from Reference 1, is shown in Table 12.2.

Table 12.2 shows only a sample of the parking generation data from *Parking Generation.* Data for other uses are included in *Parking Generation*, but many categories are backed up with only small sample sizes. Even for those land uses included, the number of sites used to calibrate the values is not always impressive, and the R^2 values often connote significant variability in parking characteristics.

For this reason, it is always preferable to base projections of parking needs on locally calibrated values, using similar types of land uses and facilities as a basis.

Consider the case of a small office building, consisting of 25,000 square feet of office space. What is the peak parking load expected to be at this facility? From Table 12.2 for

Table 12.1: Typical Parking Generation Specification Units

Type of Land Use	Parking-Related Unit	
	Preferred	**Alternate**
Single-Family Residential	Per Dwelling Unit	Per Dwelling Unit with range by number of bedrooms
Apartment Residential	Per Dwelling Unit with range by number of bedrooms	Per Dwelling Unit
Shopping Center	Per 1,000 sq ft GLA*	N/A
Other Retail	Per 1,000 sq ft GFA**	N/A
Office	Per Employee	Per 1,000 sq ft GFA**
Industrial	Per Employee	Per 1,000 sq ft GFA**
Hospital	Per Employee	Per Bed
Medical/Dental	Per Doctor	Per Office
Nursing Home	Per Employee	Per Bed
Hotel/Motel	Per Unit	N/A
Restaurant	Per Seat	Per 1,000 sq ft GFA**
Bank	Per 1,000 sq ft GFA**	N/A
Public Assembly	Per Seat	N/A
Bowling Alley	Per Lane	Per 1,000 sq ft GFA**
Library	Per 1,000 sq ft GFA**	N/A

*GLA = gross leaseable area.
**GLA = gross floor area.
(*Source:* Used with permission of Transportation Research Board, "Parking Principles," *Special Report 125*, Washington DC, 1971, Table 3-1, p. 34.)

Table 12.2: Typical Parking Generation Rates

Land Use*	Avg Rate	Per	Equation†	R^2	No. of Studies
Residential—Low/Mid-Rise Apartment (Wkdy)	1.20	Dwelling Unit	$P = 1.43\,X - 46.0$	0.93	19
Residential—High-Rise Apartment (Wkdy)	1.37	Dwelling Unit	$P = 1.04\,X + 130.0$	0.85	7
Residential—Condominium/Townhouse (Wkdy)	1.46	Dwelling Unit	$P = 96.8 \ln X - 272$	0.90	32
Hotel (Wkdy)	0.91	Room	$P = 1.13\,X - 60$	0.75	14
Motel (Wkdy)	0.90	Room	$P = 1.03\,X - 24$	0.76	5
Resort Hotel (Wkdy)	1.42	Room	N/A	N/A	3
Industrial—Light (Wkdy)	0.75	1,000 sq ft GFA	$P = 0.61\,X + 6$	0.81	7
Industry—Industrial Park (Wkdy)	1.27	1,000 sq ft GFA	$P = 0.76\,X + 26$	0.66	8
Industry—Warehousing (Wkdy)	0.41	1,000 sq ft GFA	$P = 0.41\,X - 5$	0.87	13
Medical—Urban Hospital (Wkdy)	1.47	Bed	N/A	N/A	23
Medical—Clinic (Wkdy)	4.33	1,000 sq ft GFA	$P = 4.24\,X + 1$	0.99	6
Office—Office Building (Wkdy)	2.84	1,000 sq ft GFA	$P = 2.51\,X + 27$	0.91	173
Shopping—Shopping Center (Sat-December)	4.74	1,000 sq ft GLA	$P = 4.59\,X + 140$	0.84	82
Restaurant—Quality Restaurant (Sat)	17.20	1,000 sq ft GFA	N/A	N/A	7
Restaurant—Urban Family Restaurant	10.1	1,000 sq ft GFA	N/A	N/A	21
Recreation—Movie Theater (Sat)	0.26	Seat	$P = 0.60\,X - 542$	0.65	6
Recreation—Health/Fitness Club (Wkdy)	5.19	1,000 sq ft GFA	$P = 3.62\,X + 27$	0.61	20
Religion—Church or Synagogue (Sat/Sun)	7.81	1,000 sq ft GFA	N/A	N/A	11

*Parking generation shown for peak day of the week.

†P = peak number of parking spaces occupied; X = appropriate underlying variable shown in the Per column.

(*Source:* Compiled from "Parking Generation," 3rd Edition, Institute of Transportation Engineers, Washington DC, 2004.)

office buildings, the average peak parking occupancy is 2.84 per thousand square feet of building area, or in this case, 2.84*25 = 71 parking spaces. A more precise estimate might be obtained using the equation related to facility size:

$$P = 2.51X + 27 = (2.51*25) + 27 = 90 \text{ spaces}$$

This presents a significant range to the engineer—from 71 to 90 parking spaces needed. Thus, although these general guidelines can provide some insight into parking needs, it is important to do localized studies of parking generation to augment national norms.

A 1998 study provides additional data on parking generation [*3*]. Over 400 shopping centers were surveyed, resulting in the establishment of recommended "parking ratios," the number of spaces provided per 1,000 square feet of gross leaseable area (GLA). Centers were categorized by total size (in GLA), and by the percentage of total center GLA occupied by movie houses, restaurants, and other entertainment uses. The results are summarized in Table 12.3.

The guidelines were established such that the 20th peak parking hour of the year is accommodated (i.e., there are only 19 hours of the year when parking demand would exceed the recommended values). Parking demands accommodate both patrons and employees.

Table 12.3: Recommended Parking Ratios from a 1998 Study

Center Size (Total GLA)	**% Usage by Movie Houses, Restaurants, and Other Entertainment**				
	0%	**5%**	**10%**	**15%**	**20%**
0–399,999	4.00	4.00	4.00	4.15	4.30
400,000–419,999	4.00	4.00	4.00	4.15	4.30
420,000–439,999	4.06	4.06	4.06	4.21	4.36
440,000–459,999	4.11	4.11	4.11	4.26	4.41
460,000–479,999	4.17	4.17	4.17	4.32	4.47
480,000–499,999	4.22	4.22	4.22	4.37	4.52
500,000–519,999	4.28	4.28	4.28	4.43	4.58
520,000–539,999	4.33	4.33	4.33	4.48	4.63
540,000–559,999	4.39	4.39	4.39	4.54	4.69
560,000–579,000	4.44	4.44	4.44	4.59	4.74
580,000–599,999	4.50	4.50	4.50	4.65	4.80
600,000–2,500,000	4.50	4.50	4.50	4.65	4.80

(*Source:* Used with permission of Urban Land Institute, *Parking Requirements for Shopping Centers*, 2nd Edition, Washington DC, 1999, compiled from Appendix A, Recommended Parking Ratios.)

Where movie theaters, restaurants, and other entertainment facilities occupy more than 20% of the GLA, a "shared parking" approach is recommended. Parking requirements would be predicted for shopping facilities, and for movies, restaurants, and entertainment facilities separately. Local studies would be used to establish the amount of overlapping usage that might occur (e.g., spaces used by shoppers in the afternoon would be used by movie patrons in the evening).

Consider the following case: a new regional shopping center with 1,000,000 square feet of GLA is to be built. It is anticipated that about 15% of the GLA will be occupied by movie theaters, restaurants, or other entertainment facilities. How many parking spaces should be provided? From Table 12.3, the center as described would require a parking ratio of 4.65 spaces per 1,000 square feet GLA, or 4.65*1,000 = 4,650 parking spaces.

Reference 4 presents a more detailed model for predicting peak parking needs. Because the model is more detailed, additional input information is needed to apply it. Peak parking demand may be estimated as:

$$D = \frac{NKRP*pr}{O} \tag{12-1}$$

where: D = parking demand, spaces
N = size of activity measured in appropriate units (floor area, employment, dwelling units, or other appropriate land-use parameters)
K = portion of destinations that occur at any one time
R = person-destinations per day (or other time period) per unit of activity
P = proportion of people arriving by car
O = average auto occupancy
pr = proportion of persons with primary destination at the designated study location

Consider the case of a 400,000-sq-ft retail shopping center in the heart of a central business district (CBD). The following estimates have been made:

- Approximately 40% of all shoppers are in the CBD for other reasons ($pr = 0.60$).
- Approximately 70% of shoppers travel to the retail center by automobile ($P = 0.70$).
- Approximate total activity at the center is estimated to be 45 person-destinations per 1,000 square feet of gross leasable area, of which 20% occur during the peak parking accumulation period ($R = 45$; $K = 0.20$).
- The average auto occupancy of travelers to the shopping center is 1.5 persons per car ($0 = 1.5$).

Because the unit of size is 1,000 square feet of gross leasable area, $N = 400$ for this illustration. The peak parking demand may now be estimated using Equation 12-1 as:

$$D = \frac{400*45*0.20*0.70*0.60}{1.5}$$
$$= 1{,}008 \text{ parking spaces}$$

This is equivalent to 1,008/400 = 2.52 spaces per 1,000 ft of GLA.

Although this technique is analytically interesting, it requires that a number of estimates be made concerning parking activity. For the most part, these would be based on data from similar developments in the localized area or region or on nationwide activity information if no local information is available.

12.2.2 Zoning Regulations

Control of parking supply for significant developments is generally maintained through zoning requirements. Local zoning regulations generally specify the minimum number of parking spaces that must be provided for developments of specified type and size. Zoning regulations also often specify needs for handicapped parking and set minimum standards for loading zones.

Reference 4 contains a substantial list of recommended zoning requirements for various types of development in "suburban" settings. "Suburban" settings have little transit access, no significant ride sharing, and little captive walk-in traffic to reduce parking demands. The recommendations are based on satisfying the 85th percentile demand (i.e., a level of demand that would be exceeded only 15% of the time) and are summarized in Table 12.4. Zoning requirements are generally set 5% to 10% higher than the 85th percentile demand expectation.

The recommended zoning requirements of Table 12.4 would be significantly reduced in urban areas with good transit access, captive walk-in patrons (people working or living in the immediate vicinity of the development), or organized car-pooling programs. In such areas, the modal split characteristics of users must be determined, and parking spaces may be reduced accordingly. Such a modal split estimate must consider local conditions because this can vary widely. In a typical small urban community, transit may provide 10% to 15% of total access; in Manhattan (New York City), less than 5% of major midtown and downtown access is by private automobile.

In any parking facility, handicapped spaces must be provided as required by federal and local laws and ordinances. Such standards affect both the number of spaces that must be required and their location. The Institute of Transportation Engineers [4] recommends the following minimum standards for provision of handicapped spaces:

- *Office*—0.02 spaces per 1,000 sq ft GFA
- *Bank*—1 to 2 spaces per bank
- *Restaurant*—0.30 spaces per 1,000 sq ft GFA
- *Retail* (< 500,000 sq ft GFA)—0.075 spaces per 1,000 sq ft GFA
- *Retail* (≥ 500,000 sq ft GFA)—0.060 spaces per 1,000 sq ft GFA

In all cases, there is an effective minimum of one handicapped space.

12.3 Parking Studies and Characteristics

A number of characteristics of parkers and parking have a significant influence on planning. Critical to parking supply needs are the duration, accumulation, and proximity requirements of parkers. Duration and accumulation are related characteristics. If parking capacity is thought of in terms of "space-hours," then vehicles parked for a longer duration consume more of that capacity than vehicles parked for only a short period. In any area, or at any specific facility, the goal is to provide enough parking spaces to accommodate the maximum accumulation on a typical day.

12.3.1 Proximity: How Far Will Parkers Walk?

Maximum walking distances that parkers will tolerate vary with trip purpose and urban area size. In general, tolerable walking distances are longer for work trips than for any other type of trip, perhaps because of the relatively long duration involved. Longer walking distances are tolerated for off-street parking spaces as opposed to on-street (or curb) parking spaces. As the urban area population increases, longer walking distances are experienced.

The willingness of parkers to walk certain distances to (or from) their destination to their car must be well understood because it will have a significant influence over where parking capacity must be provided. Under any conditions, drivers tend to seek parking spaces as close as possible to their destination. Even in cities of large population (1 to 2 million), 75% of drivers park within a quarter mile of their final destination.

Table 12.5 shows the distribution of walking distances between parking places and final destinations in urban areas.

Table 12.4: Recommended Parking Space Zoning Requirements in Suburban Settings

Land Use	Unit	Parking Spaces Per Unit	
		Peak Parking Demand[a]	**Recommended Zoning Requirement**
Residential			
Single Family	Dwelling Unit	2.0	2.0
Multiple Family	Dwelling Unit	2.0	2.0
Efficiency/Studio	Dwelling Unit	1.0	1.0
1 Bedroom Apt	Dwelling Unit	1.5	1.5
2 or more Bedroom Apt	Dwelling Unit	2.0	2.0
Elderly Housing	Dwelling Unit	0.7	0.5 + 1 space/day shift employee
Accessory D.U.	Dwelling Unit	1.0	1.0
Commercial Lodging			
Hotel/Motel	Bedroom	1.2	1.0 + spaces for restaurant, lounge, meeting rooms + 0.25 per day shift employee
Sleeping Rooms	Bedroom	1.0	1.0 + 2.0 for resident manager
Medical Treatment			
Hospitals	Bed	2.5	**Higher of:** 2.7 *-OR -* 0.33 + 0.4/employee + 0.2/outpatient + 0.25/staff physician
Medical Center	Bed	5.5	**Higher of:** 6.0 *–OR –* 0.5 + 0.4/employee + 0.2/outpatient + 0.25/staff physician + 0.33/student
Business Offices			
General Office:	1,000 sq ft GFA	3.0	
≤ 30,000 sq ft			4.0
>30,000 sq ft			3.3
Banks	1,000 sq ft GFA	3.3	3.6
Branch Drive-In	1,000 sq ft GFA	3.5	5.6
Bank w/walk-in window services			
Retail Services			
General Retail	1,000 sq ft GFA	2.2	2.4
Personal Care	1,000 sq ft GFA	3.5	**Higher of:** 4.0 *–OR-* 2/treatment station
Coin Operated Laundries	Wash/Dry Clean Machine	0.5	0.5

(Continued)

Table 12.4: Recommended Parking Space Zoning Requirements in Suburban Settings (*Continued*)

Land Use	Unit	Parking Spaces Per Unit	
		Peak Parking Demand[a]	**Recommended Zoning Requirement**
Retail Goods			
General Retail	1,000 sq ft GFA	3.0	3.3
Convenience Store	1,000 sq ft GFA	4.0	4.4
Hard Goods Store	1,000 sq ft GFA	3.0	2.5 + 1.5/1,000 sq ft interior storage and exterior display/storage
Shopping Centers:	1,000 sq ft GFA		
<400,000 sq ft GLA		4.5	4.7
400,001–600,000		5.0	5.2
>600,000 sq ft GLA		5.5	5.8
Food and Beverage			
Quality Restaurant	1,000 sq ft GLA	20.0	22.0 + banquet/meeting room needs
Family Restaurant	1,000 sq ft GLA	11.2	12.3 + banquet/meeting room needs
Fast-Food Restaurant	1,000 sq ft GLA	15.4	16.9 (kitchen, serving counter, waiting areas) + 0.5/seat
Educational			
Elem/Secondary	Classroom	1.5	1.5 (include classrooms and other rooms used by students and faculty) + 0.25/student of driving age
College/University	Population	N/A	1.0/daytime staff and faculty + 0.5/resident and commuting student
Day-Care Center	Employee	N/A	1.0 + 0.1/licensed enrollment capacity
Cultural, Entertainment, and Recreational			
General Public Assembly	Max. Occupancy	N/A	0.25
General Recreation	Max. Occupancy	N/A	0.33
Auditorium, Theater or Stadium	Seat	0.35	0.38 0.50
Church	Seat	N/A	
Industrial			
General	Employee	0.60–1.00	1.0 + 1.0/1,000 sq ft GFA
Storage, Wholesale or Utility			
General	1,000 sq ft GFA	N/A	0.50 + required spaces for office or sales areas

[a]Typically, the 85th percentile demand, based on analysis of comparative studies.

(*Source:* Used with permission of Eno Foundation for Transportation, Weant, R., and Levinson, H., *Parking*, Westport CT, 1990; reformatted from Table 3-2, pp. 42 and 43.)

Table 12.5: CBD Walking Distances to Parking Spaced

Distance		% Walking This Distance or Further	
Feet	**Miles**	**Mean**	**Range**
0	0	100	
250	0.05	70	60–80
500	0.10	50	40–60
750	0.14	35	25–45
1,000	0.19	27	17–37
1,500	0.28	16	8–24
2,000	0.38	10	5–15
3,000	0.57	4	0–8
4,000	0.76	3	0–6
5,000+	0.95+	1	0–2

(*Source:* Used with permission of Eno Foundation for Transportation, Weant, R., and Levinson, H., *Parking*, Westport CT, 1990, Table 6-3, p. 98.)

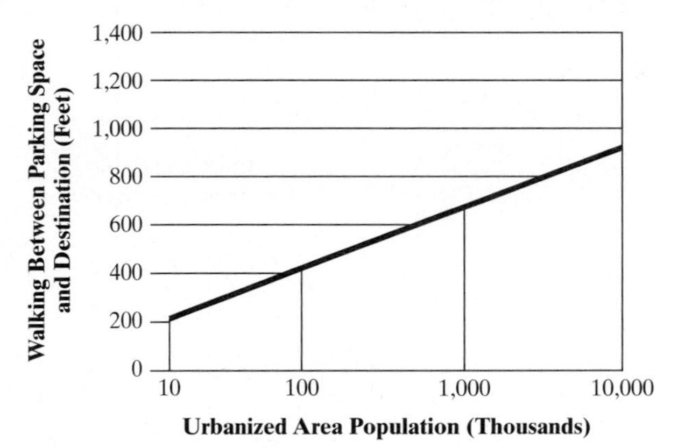

Figure 12.1: Average Walking Distance by Urbanized Area Population

(*Source:* Used with permission of Eno Foundation for Transportation, Weant, R., and Levinson, H., *Parking*, Westport CT, 1990, Fig. 6.5, p. 98.)

The distribution is based on studies in five different cities (Atlanta, Pittsburgh, Dallas, Denver, and Seattle), as reported in Reference 4.

As indicated in this table, parkers like to be close to their destination. Half (50%) of all drivers park within 500 feet of their destination. Figure 12.1 shows average walking distances to and from parking spaces versus the total urban area population.

Again, these data emphasize the need to place parking capacity in close proximity to the destination(s) served. Even in an urban region of over 10 million population, the average walking distance to a parking place is approximately 900 feet.

Trip purpose and trip duration also affect the walking distances drivers are willing to accommodate. For shopping or other trips where things must be carried, shorter walking distances are sought. For short-term parking, such as to get a newspaper or a takeout order of food, short walking distances are also sought. Drivers will not walk 10 minutes if they are going to be parked for only 5 minutes. In locating parking capacity, general knowledge of parkers' characteristics is important, but local studies would provide a more accurate picture. In many cases, however, application of common sense and professional judgment is also an important component.

12.3.2 Parking Inventories

One of the most important studies to be conducted in any overall assessment of parking needs is an inventory of existing parking supply. Such inventories include observations of the number of parking spaces and their location, time restrictions on use of parking spaces, and the type of parking facility (e.g., on-street, off-street lot, off-street garage). Most parking inventory data is collected manually, with observers canvassing an area on foot, counting and noting curb spaces and applicable time restrictions, as well as recording the location, type, and capacity of off-street parking facilities. Use of intelligent transportation system technologies have begun to enhance the quantity of information available and the ease of accessing it. Some parking facilities have begun to use electronic tags (such as EZ Pass) to assess fees. Such a process, however, can also keep track of parking durations and accumulations on a real-time basis. Smart parking meters can provide the same types of information for curb parking spaces.

To facilitate the recording of parking locations, the study area is usually mapped and precoded in a systematic fashion. Figure 12.2 illustrates a simple coding system for blocks and block faces. Figure 12.3 illustrates the field sheets that would be used by observers.

Curb parking places are subdivided by parking restrictions and meter duration limits. Where several lines of a field sheet are needed for a given block face, a subtotal is prepared and shown. Where curb spaces are not clearly marked, curb lengths are used to estimate the number of available spaces, using the following guidelines:

- Parallel parking: 23 ft/stall
- Angle parking: 12.0 ft/stall
- 90-degree parking: 9.5 ft/stall

Although the parking inventory basically counts the number of spaces available during some period of

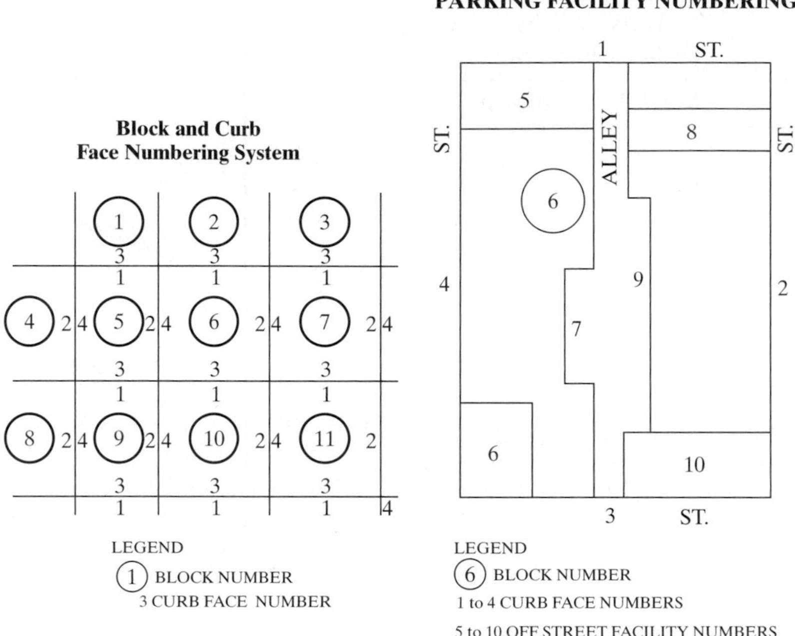

Figure 12.2: Illustrative System for Parking Location Coding

(*Source:* Used with permission of Institute of Transportation Engineers, Box, P., and Oppenlander, J., *Manual of Traffic Engineering Studies*, 4th Edition, Washington DC, 1976, Figs. 10-1 and 10-2, p. 131.)

interest—often the 8- to 11-hour business day—parking supply evaluations must take into account regulatory and time restrictions on those spaces and the average parking duration for the area. Total parking supply can be measured in terms of how many vehicles can be parked during the period of interest within the study area:

$$P = \left(\frac{\sum_{n} NT}{D} \right) * F \qquad (12\text{-}2)$$

where: P = parking supply, vehs

N = no. of spaces of a given type and time restriction

T = time that N spaces of a given type and time restriction are available during the study period, hrs

D = average parking duration during the study period, hrs/veh

F = insufficiency factor to account for turnover—values range from 0.85 to 0.95 and increase as average duration increases

Consider an example in which a 11-hour study of an area revealed that there were 450 spaces available for the full 12 hours, 280 spaces available for 6 hours, 150 spaces available for 7 hours, and 100 spaces available for 5 hours. The average parking duration in the area was 1.4 hours. Parking supply in this study area is computed as:

$$P = \left\{ \frac{[(450*12) + (280*6) + (150*7) + (100*5)]}{1.4} \right\} *0.90$$

$$= 5{,}548 \text{ vehs}$$

where an insufficiency factor of 0.90 is used.

This result means that 5,548 vehicles could be parked in the study area over the 11-hour period of the study. It *does not* mean that all 5,548 vehicles could be parked at the same time. This analysis, however, requires that the average parking duration be known. Determining this important factor is discussed in the next section.

Inventory data can be displayed in tabular form, usually similar to that illustrated in Figure 12.3, or it can be

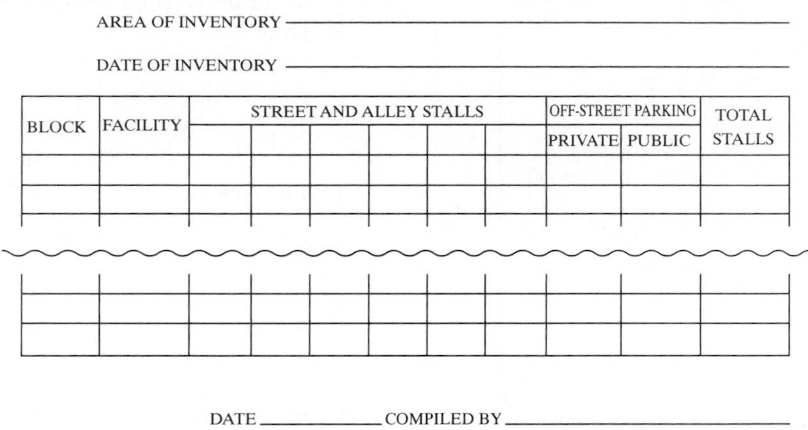

AREA OF INVENTORY ______

DATE OF INVENTORY ______

BLOCK	FACILITY	STREET AND ALLEY STALLS						OFF-STREET PARKING		TOTAL STALLS
								PRIVATE	PUBLIC	

DATE ______ COMPILED BY ______

Figure 12.3: A Parking Inventory Field Sheet

(*Source:* Used with permission of Institute of Transportation Engineers, Box, P., and Oppenlander, J., *Manual of Traffic Engineering Studies,* 4th Edition, Washington DC, 1976, Fig. 10-3, p. 133.)

graphically displayed on coded maps. Maps provide a good overview but cannot contain the detailed information provided in tabular summaries. Therefore, maps and other graphic displays are virtually always accompanied by tables.

12.3.3 Accumulation and Duration

Parking accumulation is defined as the total number of vehicles parked at any given time. Many parking studies seek to establish the distribution of parking accumulation over time to determine the peak accumulation and when it occurs. Of course, observed parking accumulations are constrained by parking supply; thus parking demand constrained by lack of supply must be estimated using other means.

Nationwide studies have shown that parking accumulation in most cities has increased over time. Total accumulation in an urban area, however, is strongly related to the urbanized area population, as illustrated in Figure 12.4.

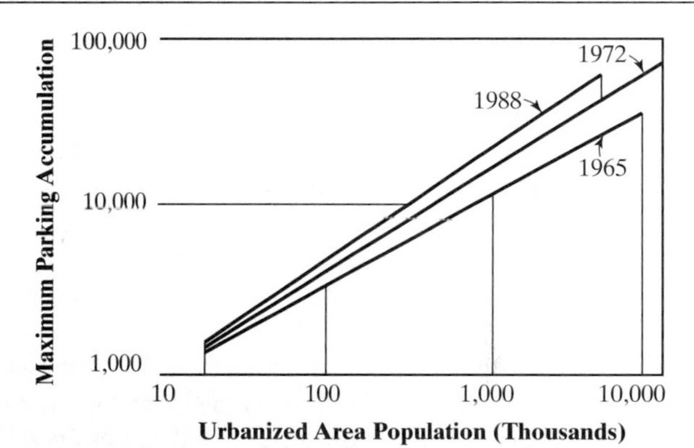

Figure 12.4: Parking Accumulation in Urbanized Areas by Population

(*Source:* Used with permission of Eno Foundation for Transportation, Weant, R., and Levinson, H., *Parking*, Westport CT, 1990, Fig. 6.8, p. 100.)

Parking duration is the length of time that individual vehicles remain parked. This characteristic, therefore, is a distribution of individual values, and both the distribution and the average value are of great interest.

Like parking accumulation, average parking durations are related to the size of the urban area, with average duration increasing with urban area population, as shown in Figure 12.5. Average duration also varies considerably with trip purpose, as indicated in Table 12.6, which summarizes the results of studies in Boston in 1972 and Charlotte in 1987.

From Table 12.6, it is obvious that durations vary widely from location to location. The Charlotte results are quite different from those obtained in Boston. Thus local studies of both parking duration and parking accumulation are important elements of an overall approach to the planning and operation of parking facilities.

The most commonly used technique for observing duration and accumulation characteristics of curb parking and surface parking lots is the recording of license plate numbers of parked vehicles. At regular intervals ranging from 10 to 30 minutes, an observer walks a particular route (usually up

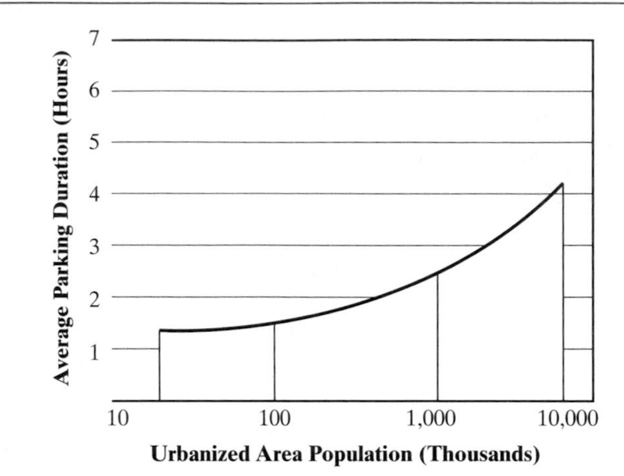

Figure 12.5: Parking Duration Versus Urbanized Area Population

(*Source:* Used with permission of Eno Foundation for Transportation, Weant, R., and Levinson, H., *Parking*, Westport CT, 1990, Fig. 6.4, p. 97.)

Table 12.6: Average Urban Parking Durations by Trip Purpose

	Average Duration (hours, minutes)	
Trip Purpose	**Boston (1972)**	**Charlotte (1987)**
Work		
Manager	5h, 30m	8h, 8m
Employee	5h, 59m	
All		
Personal Business	2h, 6m	1h, 5m
Sales/Employment Business	2h, 14m	3h, 32m
Service	2h, 9m	
Recreational	2h, 18m	1h, 29m
Shopping	1h, 57m	4h, 17m
Other	3h, 12m	
All Purposes (Average)	4h, 20m	1h, 41m

(*Source:* Used with permission of Eno Foundation for Transportation, Weant, R., and Levinson, H., *Parking*, Westport CT, 1990, Table 6-2, p. 97.)

one block face and down the opposite block face) and records the license plate numbers of vehicles occupying each parking space. A typical field sheet is shown in Figure 12.6.

Each defined parking space is listed on the field sheet prepared for the specific study, along with any time restrictions associated with it. A variety of special notations can be used to indicate a variety of circumstances, such as "T" for truck, "TK" for illegally parked and ticketed vehicle, and so on. One observer can be expected to observe up to 60 spaces every 15 minutes. Study areas, therefore, must be carefully mapped to allow planning of routes for complete data coverage.

Analysis of the data involves several summaries and computations that can be made using the field sheet information:

- *Accumulation totals.* Each column of each field sheet is summed to provide the total accumulation of parked vehicles within each time period on each observer's route.
- *Duration distribution.* By observing the license plate records of each space, vehicles can be classified as having been parked for one interval, two intervals, three intervals, and so on. By examining each line of each field sheet, a duration distribution is created.
- *Violations.* The number of vehicles illegally parked, either because they occupy an illegal space or have exceeded the legal time restriction of a space, should be noted.

The average parking duration is computed as:

$$D = \frac{\sum_x (N_x * X * I)}{N_T} \qquad (12\text{-}3)$$

where: D = average parking duration, h/veh
N_x = number of vehicles parked for x intervals
X = number of intervals parked
I = length of the observation interval, h
N_T = total number of parked vehicles observed

Another useful statistic is the parking turnover rate, *TR*. This rate indicates the number of parkers that, on average, use a parking stall over a period of one hour. It is computed as:

$$TR = \frac{N_T}{P_S * T_S} \qquad (12\text{-}4)$$

LICENSE PLATE CHECK FIELD DATA SHEET

City________ Date 10 MAY 1978 Recorded by JONES Side of Street W

Street WRIGHT between 5th and 6th

Codes: 000 last three digits of license number. ✓ for repeat number from prior circuit __ for empty space

Space and Regulation	Time circuit begins											
	07	07^{30}	08	08^{30}	09	09^{30}	10	10^{30}	11	11^{30}	12^{00}	
5th												
X-WALK	–	–	–									
NPHC	–	–	–									
I HR M	–	713	✓	✓TK								
″ M	631	✓	(✓)	971								
″ M	512	34✓	✓	019								
DRIVEWAY	–	–	–	–								
″	–	–	–	613								
I HR M	–	–	418	✓								
″ M	117	220	✓	989								
″ M	–	148	096	✓								
FIRE HYD	–	–	–	–								
I HR M	042	–	216	✓								
N PHC	–	–	–	774								
X-WALK	–	–	–	–								
6th												

Figure 12.6: A License-Plate Parking Survey Sheet

(*Source:* Used with permission of Institute of Transportation Engineers, Box, P., and Oppenlander, J., *Manual of Traffic Engineering Studies,* 4th Edition, Washington DC, 1976, Fig. 10-6, p. 140.)

where: TR = parking turnover rate, veh/stall/h

N_T = total number of parked vehicles observed

P_S = total number of legal parking stalls

T_S = duration of the study period, h

The average duration and turnover rate may be computed for each field sheet, for sectors of the study area, and/or for the study area as a whole. Table 12.7 shows a typical field sheet resulting from one observer's route. Table 12.8 shows how data from individual field sheets can be summarized to obtain areawide totals.

$\Sigma = 2{,}118$ total parkers observed.

Note that the survey includes only the study period. Thus vehicles parked at 3:00 PM will have a duration that ends at that time, even though they may remain parked for an additional time period outside the study limits. For convenience, only the last three numbers of the license plates are recorded; in most states, the initial two or three letters/numbers represent a code indicating where the plate registration was issued. Thus these letters/numbers are often repetitive on many plates.

The average duration for the study area, based on the summary of Table 12.8 (b) is:

$$D = \frac{(875*1*0.5) + (490*2*0.5) + (308*3*0.5) + (275*4*0.5) + (143*5*0.5) + (28*6*0.5)}{2119}$$

$$D = 1.12 \text{ h/veh}$$

The turnover rate is:

$$TR = \frac{2119}{1500*7} = 0.20 \text{ veh/stall/h}$$

The maximum observed accumulation occurs at 11:00 AM from Table 12.8 (a), and is 1,410 vehicles, which represents use of $(1{,}410/1{,}500)*100 = 94\%$ of available spaces.

Table 12.7: Summary and Computations from a Typical Parking Survey Field Sheet

Pkg*	Time														
Space	**8:00**	**8:30**	**9:00**	**9:30**	**10:00**	**10:30**	**11:00**	**11:30**	**12:00**	**12:30**	**1:00**	**1:30**	**2:00**	**2:30**	**3:00**
1	–	–	861	√	√	–	136	–	140	√	–	–	201	√	√
2	470	√	380	–	–	412	307	–	900	√	√	√	√	–	070
3	–	211	√	√	√	400	√	√	–	–	666	–	855	999	–
4	175	√	√	500	√	222	–	–	616	√	√	√	√	√	–
5	333	–	–	380	√	√	420	√	707	–	–	–	–	–	–
hydrant	–	–	–	–	–	–	–	242TK	–	–	–	–	–	–	–
1-hr	–	–	484	√	909	–	811	√	√	158	√	√	685	√	–
1-hr	301	–	–	525	√	√	696	√	422	–	299	√	√	–	892
1-hr	–	675	895	√	√	703	√	819	–	401	√	√	288	–	412
1-hr	406	–	442	781	882	√	√	√	444	–	903	√	–	–	–
1-hr	–	–	115	√	618	√	818	√	√	906	√	–	–	893	√
2-hr	–	509	√	√	–	705	√	√	√	688	√	696	–	–	807
2-hr	–	–	214	√	√	√	209	–	248	√	797	√	√	√	√
2-hr	101	√	√	√	–	531	√	–	940	√	√	√	628	√	√
2-hr	–	392	√	√	√	251	√	772	–	835	√	√	√	–	–
Accum.	**6**	**7**	**12**	**13**	**11**	**12**	**13**	**10**	**11**	**10**	**12**	**11**	**11**	**8**	**9**

*All data for Block Face 61; timed spaces indicate parking meter limits; √ = same vehicle parked in space.

For off-street facilities, the study procedure is somewhat altered, with counts of the number of entering and departing vehicles recorded by 15-minute intervals. Accumulation estimates are based on a starting count of occupancy in the facility and the difference between entering and departing vehicles. A duration distribution for off-street facilities can also be obtained if the license-plate numbers of entering and departing vehicles are also recorded.

As noted earlier, accumulation and duration observations cannot reflect repressed demand due to inadequacies in the parking supply. Several findings, however, would serve to indicate that deficiencies exist:

- Large numbers of illegally parked vehicles
- Large numbers of vehicles parked unusually long distances from primary generators
- Maximum accumulations that occur for long periods of the day and/or where the maximum accumulation is virtually equal to the number of spaces legally available

Even these indications do not reflect trips either not made at all or those diverted to other locations because of parking constraints. A cordon-count study (see Chapter 9) may be used to estimate the total number of vehicles both parked and circulating within a study area, but trips not made are still not reflected in the results.

12.3.4 Other Types of Parking Studies

A number of other techniques can be used to gain information concerning parked vehicles and parkers. Origins of parked vehicles can be obtained by recording the license-plate numbers of parked vehicles and petitioning the state motor vehicle agency for home addresses (which are assumed to be the origins). This technique, which requires special permission from state authorities, is frequently used at shopping centers, stadiums, and other large trip attractors.

Interviews of parkers are also useful and are most easily conducted at large trip attraction locations. Basic information on trip purpose, duration, distance walked, and so on, can be obtained. In addition, however, attitudinal and background parker characteristic information can also be obtained to gain greater insight into how parking conditions affect users.

Table 12.8: Summary Data for an Entire Study Area Parking Survey

Block	**Accumulation for Interval (1,500 Total Stalls)**														
No.	**8:00**	**8:30**	**9:00**	**9:30**	**10:00**	**10:30**	**11:00**	**11:30**	**12:00**	**12:30**	**1:00**	**1:30**	**2:00**	**2:30**	**3:00**
61	6	7	12	13	11	12	13	10	11	10	12	11	11	8	9
62	5	10	15	14	16	18	17	15	15	10	9	9	7	7	8
.	.	.	.	.	.	.	.	.	.	.	.	.	.	.	.
.	.	.	.	.	.	.	.	.	.	.	.	.	.	.	.
.	.	.	.	.	.	.	.	.	.	.	.	.	.	.	.
180	7	8	13	13	18	14	15	15	11	14	16	10	9	9	6
181	7	5	18	16	12	14	13	11	11	10	10	10	6	6	5
Total	**806**	**900**	**1106**	**1285**	**1311**	**1300**	**1410**	**1309**	**1183**	**1002**	**920**	**935**	**970**	**726**	**694**

(a) Summarizing Field Sheets for Accumulation Totals

Block	**Number of Intervals Parked**					
Face No.	**1**	**2**	**3**	**4**	**5**	**6**
61	28	17	14	9	2	1
62	32	19	20	7	1	3
.	.	.	.	.	.	.
.	.	.	.	.	.	.
.	.	.	.	.	.	.
180	24	15	12	10	3	0
181	35	17	11	9	4	2
Total	**875**	**490**	**308**	**275**	**143**	**28**

(b) Summarizing Field Sheets for Duration Distribution

12.4 Design Aspects of Parking Facilities

Off-street parking facilities are provided as surface lots or parking garages. The latter may be above ground, below ground, or a combination of both. The construction costs of both surface lots and garages vary significantly depending on location and specific site conditions. In general, surface lots are considerably cheaper than garages. Typically, surface lots cost between \$1,000 and \$3,000 per space provided. Garages are more complex, and below-ground garages are far more costly than above-ground structures. Typical costs for above-ground garages range from \$8,000 to \$15,000 per space; below-ground garages may cost between \$20,000 and \$35,000 per space (*1*). The decision of how to provide off-street parking involves many considerations, including the availability of land, the amount of parking needed, and the cost to provide it.

Reference 4 lists three key objectives in the design of a parking facility:

- A parking facility must be convenient and safe for the intended users.
- A parking facility should be space efficient and economical to operate.
- A parking facility should be compatible with its environs.

Convenience and safety involve many issues, including proximity to major destinations, adequate access and egress facilities (including reservoir space), a simple and efficient internal circulation system, adequate stall dimensions, and basic security. The latter refers to security against theft of vehicles and security from muggings and other personal crimes. Space efficiency implies that although appropriate circulation, stall, and reservoir space must be provided, parking facilities should be designed to maximize parking capacity and minimize wasted space. The third objective involves issues of architectural beauty and ensuring that the facility and the vehicle-trips it generates do not present a visual or auditory disruption to the environment in its immediate area.

12.4.1 Some Basic Parking Dimensions

Basic parking dimensions are based on one of two "design vehicles." Some parking facilities make use of separated parking areas for "small cars" to maximize total parking capacity. Figure 12.7 illustrates the basic criteria for the two design vehicles used in parking facility design:

- Large cars
- Small cars

Parking Stall Width

Parking stalls must be wide enough to encompass the vehicle and allow for door-opening clearance. The minimum door-opening clearance is 22 inches, but this may be increased to 26 inches where turnover rates are high. Only one door-opening clearance is provided per stall because the parked vehicle and its adjacent neighbor can use *the same clearance space.* For large cars, the parking stall width should range between 77 + 22 = 99 (8.25 ft) and 77 + 26 = 103 (8.58 feet).

Reference 5 recommends the use of four parking classes, depending on turnover rates and typical users. Recommended design guidelines for large-car stall widths are shown in Table 12.9.

For small cars, these guidelines suggest a parking stall width between 66 + 22 = 88 inches (7.3 ft) and 66 + 26 = 92 inches (7.7 ft). A 7.6-foot design standard is often applied to small-car stall widths. Reference 5 suggests a design width of 8.0 ft for parking classes A and B (Table 12.9), and 7.5 ft for parking classes C and D.

Parking Stall Length and Depth

Parking stall length is measured parallel to the parking angle. It is generally taken as the length of the design vehicle plus 6 inches for bumper clearance. This implies a length of 215 + 6 = 221 inches (18.4 feet) for large cars and 175 + 6 = 181 inches (15.1 feet) for small cars.

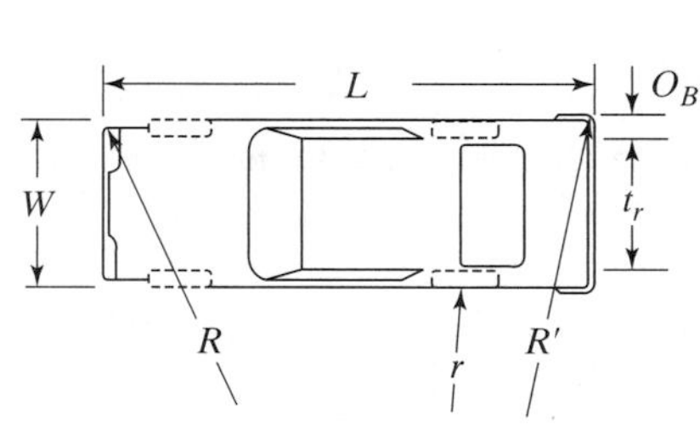

W = overall width, inches
L = overall length, inches
O_R = rear overhang, ft
O_B = body overhang from center of rear tire, ft
t_r = width from center of rear tires, ft

Minimum Turning Radius
r = inside rear wheel, ft
R = outside point, front bumper, ft
R' = outside point, rear bumper, ft

Dimension	Design Vehicle	
	Large Car	Small Car
Width, W (inches)	77	66
Length, L (inches)	215	175
Outside Front Bumper Radius, R (ft)	20.5	18.0
Inside Rear Wheel Radius, r (ft)	12.0	9.6
Rear Width, t_r (ft)	5.1	4.6
Body Overhang, Rear Tire, O_B (ft)	0.63	0.46
Rear Radius, R' (ft)	17.4	15.0

Figure 12.7: Design Vehicles for Parking Design

(*Source:* Used with permission of Eno Foundation for Transportation, Weant, R., and Levinson, H., *Parking*, Westport CT, 1990, reformatted from Table 8-1, p. 157.)

Table 12.9: Stall Width Design Criteria for Various Parking Classifications

Parking Class	Stall Width (ft)	Typical Turnover			Typical Uses
		Low	Medium	High	
A	9:00			X	Retail customers, banks, fast foods, other very high turnover facilities.
B	8.75		X	X	Retail customers, visitors.
C	8.50	X	X		Visitors, office employees, residential, airport, hospitals.
D	8.25	X			Industrial, commuter, universities.

(*Source:* Used with permission of Institute of Transportation Engineers, *Guidelines for Parking Facility Location and Design: A Recommended Practice of the ITE*, Washington DC, 1994, Table 1, p. 7.)

The depth of a parking stall is the 90° projection of the design vehicle length and 6-inch bumper clearance. For a 90° parking stall, the length and depth of the stall are equivalent. For other-angle parking, the depth of the stall is smaller than the length.

Aisle Width

Aisles in parking lots must be sufficiently wide to allow drivers to enter and leave parking stalls safely and conveniently in a minimum number of maneuvers, usually one on entry and two on departure. As stalls become narrower, the aisles need to be a bit wider to achieve this. Aisles also carry circulating traffic and accommodate pedestrians walking to or from their vehicles. Aisle width depends on the angle of parking and on whether the aisle serves one-way or two-way traffic.

12.4.2 Parking Modules

A "parking module" refers to the basic layout of one aisle with a set of parking stalls on both sides of the aisle. There are many potential ways to lay out a parking module. For 90° stalls, two-way aisles are virtually always used because vehicles may enter parking stalls conveniently from either approach direction. Where angle parking is used, vehicle may enter a stall in only one direction of travel and must depart in the same direction. In most cases, angle parking is arranged using one-way aisles, and stalls on both sides of the aisle are arranged to permit entries and exits from and to the same direction of travel. Angle stalls can also be arranged such that stalls on one side of the aisle are approached from the opposite direction as those on the other side of the aisle. In such cases, two-way aisles must be provided. Figure 12.8 defines the basic dimensions of a parking module.

Note that Figure 12.8 shows four different ways of laying out a module. One module width applies if both sets of stalls butt up against walls or other horizontal physical barriers. Another applies if both sets of stalls are "interlocked" (i.e., stalls interlock with those of the next adjacent parking module). A third applies if one set of stalls is against a wall while the other is interlocked. Yet another module reflects only a single set of stalls against a wall. Table 12.10 summarizes critical dimensions for various types of parking modules.

Providing separate parking areas for large and small vehicles presents a number of operating problems. Areas must be clearly signed, and circulation must allow drivers to get to both types of parking space easily. Because the mix of small and large cars is a variable, there may be times when large cars try to force themselves into small-car stalls and times when small cars occupy large-car stalls. Large cars will have difficulty not only in fitting into small-car stalls but in maneuvering in aisles designed for small cars.

Some designers advocate using a single size for all parking stalls. Ideally, large-car dimensions would be provided for all spaces. A policy colloquially referred to as "one size fits all" (OSFA) is based on uniform stall and modular dimensions that are taken to be the weighted average of small- and large-car criteria. The weighting is on the basis of expected proportions of users in each category. Some traffic engineers have advocated this policy as a means of providing more efficient use of scarce off-street parking space. All of the problems, however, associated with large cars attempting to use small-car spaces and aisles would exist throughout such a facility. Thus the engineer must carefully weight the negative operating impacts of OSFA against the increased use of space OSFA provides.

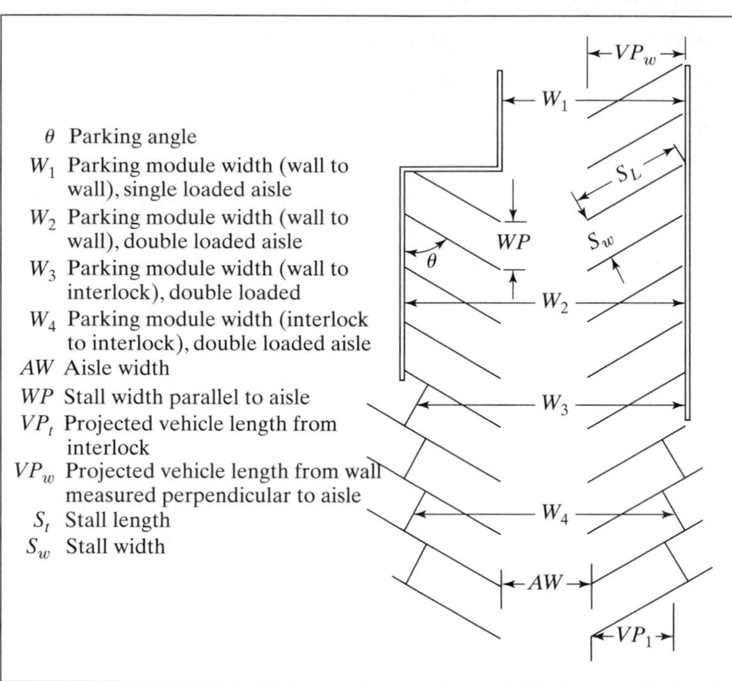

Figure 12.8: Dimensional Elements of Parking Modules

(*Source:* Used with permission of Institute of Transportation Engineers, *Guidelines for Parking Facility Location and Design: A Recommended Practice of the ITE*, Washington DC, 1994, Fig. 3, p. 6.)

12.4.3 Separating Small and Large Vehicle Areas

Reference 5 recommends a number of different techniques for separating (or integrating) small and large vehicle stalls where both are provided. Figure 12.9 illustrates various patterns for integrating small and large vehicle stalls.

- *Complete separation of small and large size spaces.* Maximum parking layout efficiency can be obtained by completely separating small and large vehicle parking areas. In this case, all of the reduced dimensions of the small-car requirements can be fully utilized. The downside is that careful signing must be used to direct vehicles to the appropriate areas, and one type of vehicle user is virtually guaranteed to be disadvantaged by having spaces further away from primary generators.
- *Mixing small and large size spaces in alternating rows.* In this pattern, large and small car modules are alternated, as shown at the top of Figure 12.9, producing about a 50-50 split in the number of each type of stall. The advantage is that both types of stall are about equally convenient, and drivers do not have the opportunity to enter a completely "wrong" area for their vehicles. Problems arise if a 50-50 split is inappropriate for the prevailing mix of vehicles. This pattern can be modified by using a double-alternating layout of two large-car aisles with two small-car aisles.
- *Small and large size spaces in the same row (or module).* This pattern is illustrated in the center of Figure 12.9. A portion of each row is allotted to each type of vehicle. The advantages of this layout are that any mix of small versus large vehicle spaces can be implemented, and no driver will ever wind up in the "wrong" row. One disadvantage is that all aisles must conform to large-car criteria.
- *Cross-aisle separation.* This layout is illustrated in the bottom of Figure 12.9. Small-car stalls are provided on one side of the aisle, and large-car stalls are provided on the other. In this case, small-car stalls are always placed at a 90° angle; large-car spaces are at a shallower angle.

Table 12.10: Parking Module Layout Dimension Guidelines

Basic Layout	Parking Class	S_w Stall Width (ft)	WP Stall Width (ft)	VP_w Stall Depth to Wall (ft)	VP_i Stall Depth to Interlock (ft)	AW Aisle Width (ft)	Modules: W_2 Wall to Wall (ft)	Modules: W_4 Interlock to Interlock (ft)
Large Cars								
2-Way Aisle—90°	A	9.00	9.00	17.5	17.5	26.0	61.0	61.0
	B	8.75	8.75	17.5	17.5	26.0	61.0	61.0
	C	8.50	8.50	17.5	17.5	26.0	61.0	61.0
	D	8.25	8.25	17.5	17.5	26.0	61.0	61.0
2-Way Aisle—60°	A	9.00	10.4	18.0	16.5	26.0	62.0	59.0
	B	8.75	10.1	18.0	16.5	26.0	62.0	59.0
	C	8.50	9.8	18.0	16.5	26.0	62.0	59.0
	D	8.25	9.5	18.0	16.5	26.0	62.0	59.0
1-Way Aisle—75°	A	9.00	9.3	18.5	17.5	22.0	59.0	57.0
	B	8.75	9.0	18.5	17.5	22.0	59.0	57.0
	C	8.50	8.8	18.5	17.5	22.0	59.0	57.0
	D	8.25	8.5	18.5	17.5	22.0	59.0	57.0
1-Way Aisle—60°	A	9.00	10.4	18.0	16.5	18.0	54.0	51.0
	B	8.75	10.1	18.0	16.5	18.0	54.0	51.0
	C	8.50	9.8	18.0	16.5	18.0	54.0	51.0
	D	8.25	9.5	18.0	16.5	18.0	54.0	51.0
1-Way Aisle—45°	A	9.00	12.7	16.5	14.5	15.0	48.0	44.0
	B	8.75	12.4	16.5	14.5	15.0	48.0	44.0
	C	8.50	12.0	16.5	14.5	15.0	48.0	44.0
	D	8.25	11.7	16.5	14.5	15.0	48.0	44.0
Small Cars*								
2-Way Aisle—90°	A/B	8.0	8.0	15.0	15.0	21.0	51.0	51.0
	C/D	7.5	7.5	15.0	15.0	21.0	51.0	51.0
2-Way Aisle—60°	A/B	8.0	9.3	15.4	14.0	21.0	52.0	50.0
	C/D	7.5	8.7	15.4	14.0	21.0	52.0	50.0
1-Way Aisle—75°	A/B	8.0	8.3	16.0	15.1	17.0	49.0	47.0
	C/D	7.5	7.8	16.0	15.1	17.0	49.0	47.0
1-Way Aisle—60°	A/B	8.0	9.3	15.4	14.0	15.0	46.0	43.0
	C/D	7.5	8.7	15.4	14.0	15.0	46.0	43.0
1-Way Aisle—45°	A/B	8.0	11.3	14.2	12.3	13.0	42.0	38.0
	B/C	7.5	10.6	14.2	12.3	13.0	42.0	38.0

*Although various angles are presented, the vast majority of small car layouts are for 90° parking.
(*Source:* Used with permission of Institute of Transportation Engineers, *Guidelines for Parking Facility Location and Design: A Recommended Practice of the ITE*, Washington DC, 1994, Tables 2 and 3, pp. 8 and 9.)

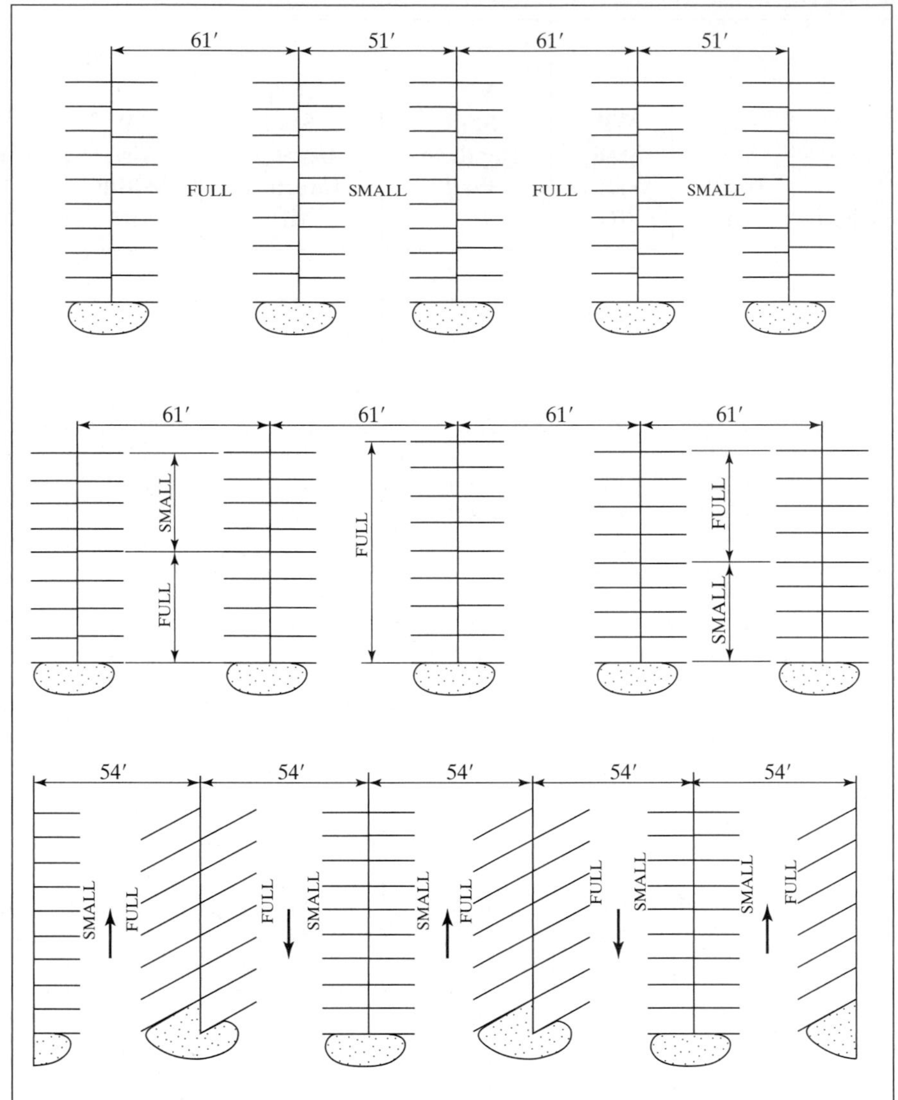

Figure 12.9: Various Layouts for Integrating Small-Car and Large-Car Stalls in Parking Facilities

(*Source:* Used with permission of Institute of Transportation Engineers, *Guidelines for Parking Facility Location and Design: A Recommended Practice of the ITE*, Washington DC, 1994, Fig. 5, p. 10.)

Use of spaces is somewhat self-enforced because drivers of large cars find it difficult to maneuver into small-car spaces. If this pattern is used throughout the parking lot, however, the balance of spaces is more in favor of small cars (more than 50%), which may not be appropriate.

Although it is preferable to use one layout throughout a parking facility (so as not to confuse drivers), it is always possible to carefully implement more than a single pattern. This may be necessary to obtain maximum use of space, but it must be accompanied by careful signing.

12.4.4 Parking Garages

Parking garages are subject to the same stall and module requirements as surface parking lots and have the same requirements for reservoir areas and circulation. The structure of a parking garage, however, presents additional constraints, such as building dimensions and the location of structural columns and other features. Ideal module and stall dimensions must sometimes be compromised to work around these structural features.

Parking garages have the additional burden of providing vertical as well as horizontal circulation for vehicles. This involves a general design and layout that includes a ramp system, at least where self-parking is involved. Some smaller attendant-parking garages use elevators to move vertically, but this is a slow and often inefficient process.

Ramping systems fall into two general categories:

- *Clearway systems.* Ramps for interfloor circulation are completely separated from ramps providing entry and exit to and from the parking garage.
- *Adjacent parking systems.* Part or all of the ramp travel is performed on aisles that provide direct access to adjacent parking spaces.

The former provides for easier and safer movement with minimum delays. Such systems, however, preempt a relatively large amount of potential parking space and are therefore usually used only in very large facilities.

Figure 12.10, on the next few pages, illustrates a number of alternative ramp layouts that may be used in parking garages.

In some attendant-park garages and surface lots, mechanical stacking systems are used to increase the parking capacity of the facility. Mechanical systems are generally slow,

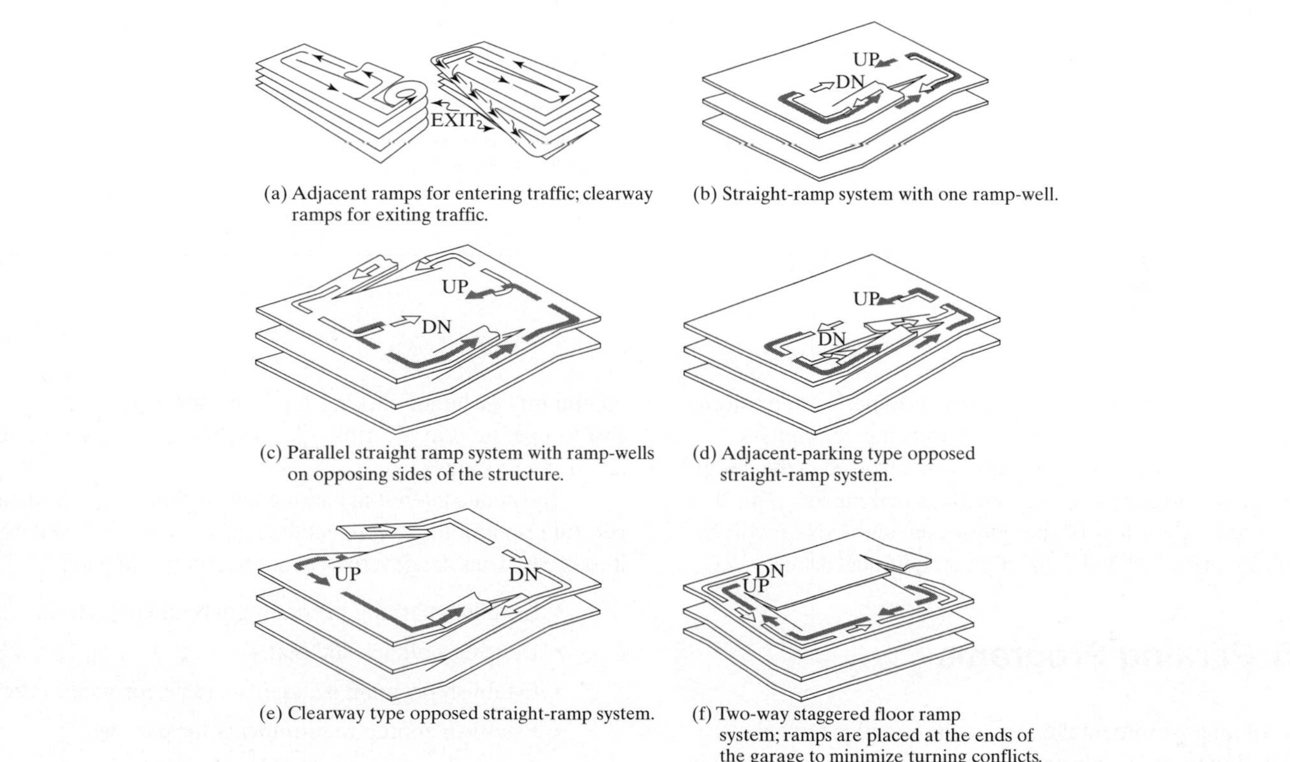

(a) Adjacent ramps for entering traffic; clearway ramps for exiting traffic.

(b) Straight-ramp system with one ramp-well.

(c) Parallel straight ramp system with ramp-wells on opposing sides of the structure.

(d) Adjacent-parking type opposed straight-ramp system.

(e) Clearway type opposed straight-ramp system.

(f) Two-way staggered floor ramp system; ramps are placed at the ends of the garage to minimize turning conflicts.

Figure 12.10: Parking Garage Circulation Systems *(Continued)*

(*Source:* Used with permission of Eno Foundation for Transportation, Weant, R., and Levinson, H., *Parking*, Westport CT, 1990, Figs. 9.5 – 9.16, pp. 188–192.)

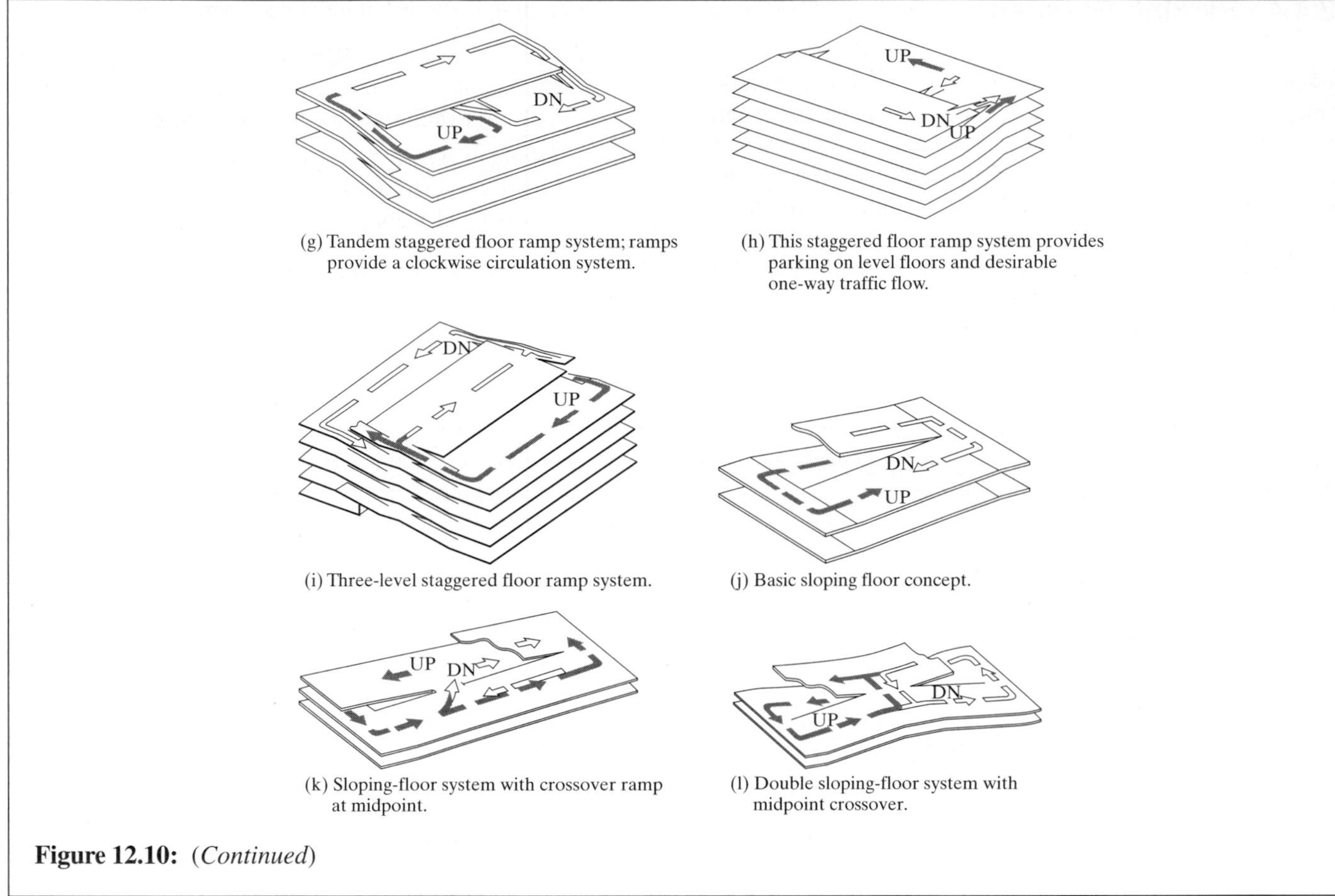

(g) Tandem staggered floor ramp system; ramps provide a clockwise circulation system.

(h) This staggered floor ramp system provides parking on level floors and desirable one-way traffic flow.

(i) Three-level staggered floor ramp system.

(j) Basic sloping floor concept.

(k) Sloping-floor system with crossover ramp at midpoint.

(l) Double sloping-floor system with midpoint crossover.

Figure 12.10: (*Continued*)

however, and are most suited to longer-term parking durations, such as the full-day parking needs of working commuters.

Of course, many intricate details are involved in the design and layout of parking garages and surface parking lots. This text has covered only a few of the major considerations involved. Consult References 4 and 5 directly for additional detail.

12.5 Parking Programs

Every urban governmental unit must have a plan to deal effectively with parking needs and associated problems. Parking is often a controversial issue because it is of vital concern to the business community in general and to particular businesses that are especially sensitive to parking. Further, parking has enormous financial aspects as well. In addition to the impact of parking on accessibility and the financial health of the community at large, parking facilities are expensive to build and to operate. On the flip side, revenues from parking fees are also enormous.

The public interest in parking falls within the government's general responsibility to protect the health, safety, and welfare of its citizens. Thus the government has a responsibility to [*4*]:

- Establish parking program goals and objectives
- Develop policies and plans
- Establish program standards and performance criteria
- Establish zoning requirements for parking
- Regulate commercial parking
- Provide parking for specific public uses
- Manage and regulate on-street parking and loading
- Enforce laws, regulations, and codes concerning parking, and adjudicate offenses

There are a number of organizational approaches to implementing the public role effectively. Parking can be placed under the authority of an existing department of the government. In small communities, where there is no professional traffic engineer or traffic department, a department of public works might be tasked with parking. In some cases, police departments have been given this responsibility (as an adjunct to their enforcement responsibilities), but this is not considered an optimal solution given that it will be subservient to the primary role of police departments. Where traffic departments exist, responsibility for parking can logically be placed there. In larger municipalities, separate departments can be established for parking. Parking boards may be created with appointed and/or elected members supervising the process. Because of the revenues and costs involved in parking, separate public parking authorities may also be established.

Parking facilities may be operated directly by governmental units or can be franchised to private operators. This is often a critical part of the process and may have a substantial impact on the net revenues from parking that find their way into the public coffers.

Parking policy varies widely depending on local circumstances. In some major cities, parking supply is deliberately limited, and costs are deliberately kept high as a discouragement to driving. Such a policy works only where there is significant public transportation supply to maintain access to the city's businesses. Where parking is a major part of access, the planning, development, and operation of off-street parking facilities becomes a major issue. Private franchisees are often chosen to build, operate, and manage parking facilities. Although this generally provides a measure of expertise and relieves the government of the immediate need to finance and operate such facilities directly, the city must negotiate and assign a significant portion of parking revenues to the franchisee. Of course, parking lots and garages can be fully private, although such facilities are generally regulated.

Revenues are also earned from parking meter proceeds and from parking violations. Metering programs are implemented for two primary reasons: to regulate turnover rates and to earn revenue. The former is accomplished through time limits. These limits are established in conjunction with localized needs. Meters at a commuter rail station would have long-term time limits, for example, because most people would be parking for a full working day. Parking spots near local businesses such as convenience stores, barber shops, fast-food restaurants, florists, and similar uses would have relatively short-term time limits to encourage turnover and multiple users. Fees are set based on revenue needs and are influenced by general policy on encouragement or discouragement of parking.

No matter how the effort is organized and managed, parking programs must deal with the following elements:

1. *Planning and policy.* Overall objectives must be established and plans drafted to achieve them; general policy on parking must be set as part of the planning effort.
2. *Curb management.* Curb space must be allocated to curb parking, transit stops, taxi stands, loading areas, and other relevant uses; amounts and locations to be allocated must be set and the appropriate regulations implemented and signed.
3. *Construction, maintenance, and operation of off-street parking facilities.* Whether through private or governmental means, the construction of needed parking facilities must be encouraged and regulated; the financing of such facilities must be carefully planned so as to guarantee feasible operation while providing a revenue stream for the local government.
4. *Enforcement.* Parking and other curb-use regulations must be strictly enforced if they are to be effective; this task may be assigned to local police, or a separate parking violations bureau may be established; adjudication may also be accomplished through a separate traffic court system or through the regular local court system of the community.

To be most effective, parking policies should be integrated into an overall accessibility plan for central areas. Provision and/or improvement of public transportation services may mitigate some portion of parking demands while maintaining the fiscal viability of the city centers.

As was the case with other parking topics, this text can only scratch the surface of the complex issues involved in effective parking programs. We urge you to consult the literature, particularly Reference 4, for a more complete and detailed coverage of the subject.

12.6 Closing Comments

Without a place to park at both ends of a trip, the automobile would be a very ineffective transportation medium. Because our society relies so heavily on the private automobile for mobility and access, the subject of parking needs and the provision of adequate parking facilities is a critical element of the transportation system.

References

1. *Parking Generation*, 3rd Edition, Institute of Transportation Engineers, Washington DC, 2004.
2. "Parking Principles," *Special Report 125*, Transportation Research Board, Washington DC, 1971.
3. *Parking Requirements for Shopping Centers*, 2nd Edition, Urban Land Institute, Washington DC, 1999.
4. Weant, R., and Levinson, H., *Parking*, Eno Foundation for Transportation, Westport, CT, 1990.
5. *Guidelines for Parking Facility Design and Location: A Recommended Practice*, Institute of Transportation Engineers, Washington DC, April 1994.

Problems

12-1. A high-rise apartment complex with 600 dwelling units is to be built. What is the expected peak parking demand for such a facility, assuming it is in an area without significant transit access?

12-2. A shopping center with 600,000 square feet of gross leasable floor area is planned. It is expected that 10% of the floor area will be devoted to movie theaters and restaurants. What peak parking demand would be expected for such a development?

12-3. Based on typical zoning regulations, what number of parking spaces should the developers of problems 1 and 2 be asked to provide?

12-4. A new office complex will house 2,000 back-office workers for the securities industries. Few external visitors are expected at this site. Each worker will account for 1.0 person-destinations per day. Of these, 35% are expected to occur during the peak hour. Only 7% of the workers will arrive by public transportation. Average car occupancy is 1.3. What peak parking demand can be expected at this facility?

12-5. A parking study has found that the average parking duration in the city center is 35 minutes, and that the following spaces are available within the 14-hour study period (6 AM to 8 PM) with a 90% efficiency factor. How many vehicles may be parked in the study area in one 14-hour day?

Number of Spaces	Time Available
100	6:00 AM–8:00 PM
150	12:00 noon–8:00 PM
200	6:00 AM–12:00 noon
300	8:00 AM–6:00 PM

12-6. Consider the license-plate data in the table that follows for a study period from 7:00 AM to 2:00 PM. For this data:

(a) Find the duration distribution and plot it as a bar chart.

(b) Plot the accumulation pattern.

(c) Compute the average parking duration.

(d) Summarize the overtime and parking violation rates.

(e) Compute the parking turnover rate.

Is there a surplus or deficiency of parking supply on this block? How do you know this?

Parking Space	**7:00**	**7:30**	**8:00**	**8:30**	**9:00**	**9:30**	**10:00**	**10:30**	**11:00**	**11:30**	**12:00**	**12:30**	**1:00**	**1:30**	**2:00**
1 hr meter	100	√	–	150	√	√	246	385	–	691	√	√	–	810	√
1 hr	–	468	√	630	√	485	–	711	888	927	√	√	108	√	–
1 hr	848	911	√	√	221	747	922	√	–	787	√	452	√	–	289
1 hr	–	–	206	√	242	√	√	–	899	√	205	603	812	√	√
1 hr	–	–	566	665	√	333	848	√	999	–	720	–	802	√	–
1 hr	–	690	–	551	√	√	347	√	265	835	486	√	–	721	855
Hydrant	–	–	–	–	–	–	–	777	–	–	–	–	–	–	–
2 hr meter	–	–	940	√	√	505	608	√	√	√	121	123	√	–	880
2 hr	636	√	√	√	√	–	582	√	√	811	919	√	711	√	√
2 hr	–	399	√	√	401	904	√	√	789	√	556	√	√	√	232
2 hr	–	416	√	√	√	√	√	–	658	√	292	844	493	√	√
2 hr	188	√	√	–	665	558	√	√	√	213	√	–	779	√	√
2 hr	–	–	–	277	√	336	409	√	√	884	√	√	713	895	431
2 hr	–	–	837	√	√	418	575	√	952	√	√	√	√	–	762
2 hr	–	506	√	√	–	786	√	√	√	527	606	√	385	√	√
Hydrant	–	–	–	–	–	518	–	–	–	758	–	–	–	–	–
3 hr	–	079	√	√	√	√	√	√	√	–	441	√	611	√	√
3 hr	256	√	√	√	√	–	295	√	√	338	√	–	499	√	√
3 hr	–	–	848	√	√	√	√	√	–	933	√	√	√	√	√
Bus stop	–	–	–	–	–	740	142	–	–	–	–	–	–	–	–
Bus stop	–	–	–	–	–	915	–	–	–	–	–	–	–	–	–
Bus stop	–	–	–	–	–	–	–	–	–	–	–	–	–	–	–
Bus stop	–	–	–	–	–	–	–	–	–	–	–	–	–	–	818
Bus stop	–	–	–	888	–	175	755	–	–	–	–	–	–	–	397

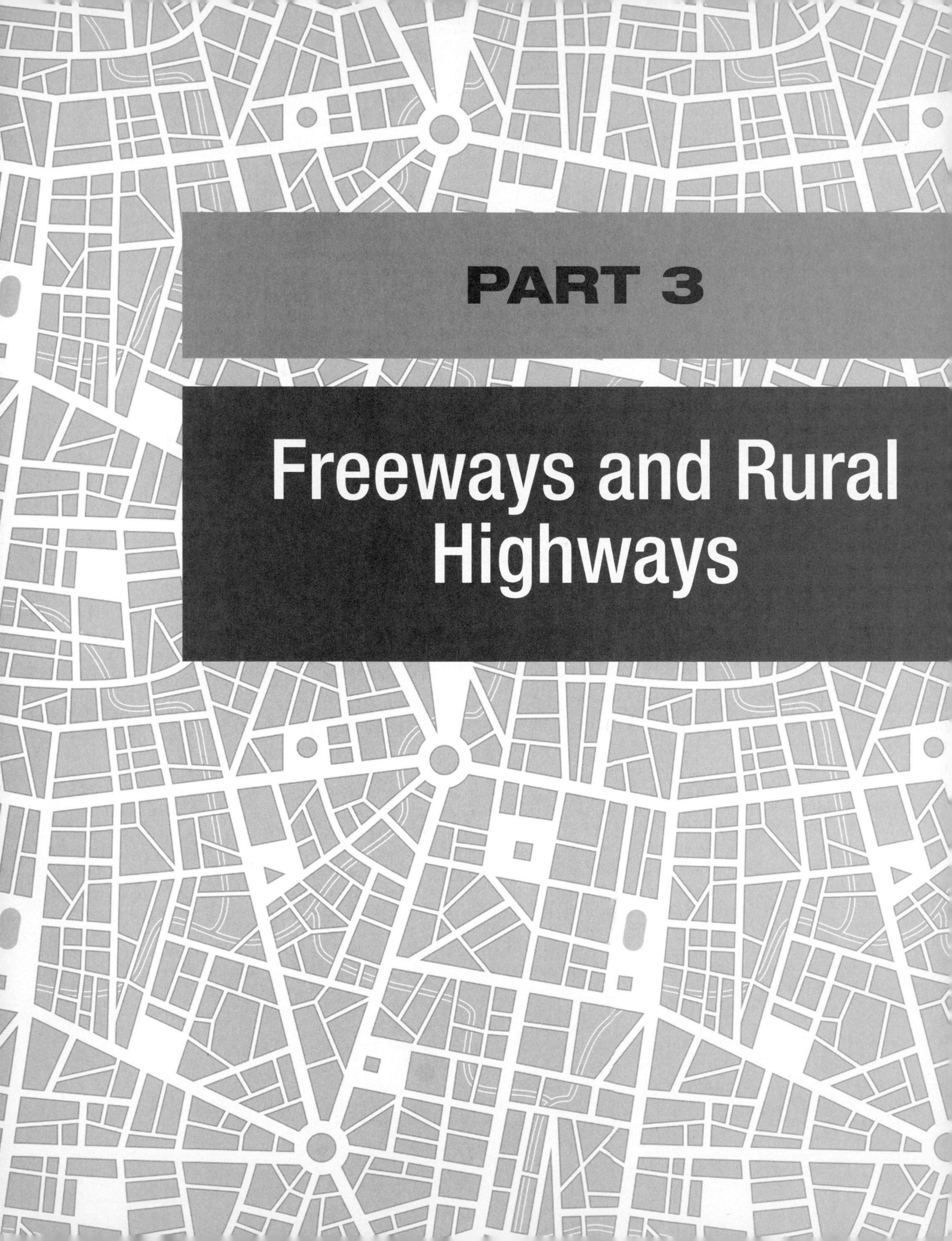

PART 3

Freeways and Rural Highways

CHAPTER 13

Fundamental Concepts for Uninterrupted Flow Facilities

Uninterrupted flow exists in its purest form on freeways. On these facilities, all entering and exiting vehicles do so by high-speed ramps designed to minimize the disruption to the through vehicle traffic stream. There is no direct access to any land parcel, and there are no fixed causes of interruption to traffic external to the traffic stream (i.e., no traffic signals, STOP signs, at-grade intersections, etc.).

This is not to suggest that a driver can expect a stable, enjoyable, uncongested trip on any freeway at all times. It simply means that all interruptions are due to internal causes (i.e., the interaction of vehicles within the traffic stream). Many freeways, particularly in urban areas, experience congested conditions during peak hours.

Because uninterrupted flow facilities, by definition, do not involve signalization or other complex forms of external control (aside from speed limits and some lane-use restrictions), traffic engineers do not have many tools to affect operational quality beyond design and redesign. Modern techniques, such as ramp metering and high-occupancy vehicle (HOV) lanes can be helpful, but they are directed more toward control of basic demand rather than direct operational measures.

It is critical, therefore, that engineers have a clear understanding of *how much* traffic a given uninterrupted flow facility can accommodate and *under what operating conditions.* It is only during design and redesign that engineers have the ability to affect operations significantly. Such opportunities occur at infrequent intervals. The traffic engineer must be ready to seize such opportunities to optimize results. The tools used to do this fall under the heading of *highway capacity and level of service analysis.* This part of the text focuses on capacity and level of service analysis tools for uninterrupted flow facilities and their application in design and redesign.

13.1 Types of Uninterrupted Flow Facilities

Although pure uninterrupted flow occurs only on freeways, it can also exist on some surface facilities in rural and/or suburban areas with long distances between points of fixed interruption. In general terms, on a surface facility, flow that is more than 2 miles from the nearest point of fixed interruption (e.g., signal, STOP sign, YIELD sign) can be essentially uninterrupted. In specific terms, this means that flow is virtually random (as it is on a freeway), with no characteristics of platoons formed by traffic signals. There is nothing magic, however, about the 2-mile distance from the nearest signal. In some cases, uninterrupted flow might exist a shorter distances from a fixed operation. The number of driveways and other uncontrolled access points and the general environment has much to do with it. It is generally agreed, however, that at 2 miles or more from points of fixed interruption, most facilities are operating under uninterrupted flow.

Because of this, three primary types of facilities operate under uninterrupted flow:

- *Freeways:* As noted, these facilities offer pure uninterrupted flow.
- *Multilane highways:* Sections of surface multilane highway (four or six lanes) that are more than 2 miles from the nearest point of fixed operation.
- *Rural two-lane highways:* Sections of two-lane highway (one lane in each direction) that are more than 2 miles from the nearest point of fixed operation.

Freeways and uninterrupted flow sections of multilane highway are quite similar in many ways. The existence of two or more lanes for exclusive use of each direction provides for passing opportunities limited only by traffic conditions. Although similar to freeways, multilane highways have adjacent development with direct access and a more complex general environment that will affect traffic flow. Nevertheless, the methodologies used for the analysis of freeways and multilane highways share many common elements and even values.

Uninterrupted flow sections of two-lane highway have one critical characteristic that makes them unique: Passing maneuvers must take place in the opposing lane of flow. Traffic in one direction limits passing opportunities in the other. On freeways and multilane highways, the two directions of flow operate independently; on two-lane highways, they interact and have effects on each other.

Analysis methodologies for freeways and multilane highways are presented in Chapter 14, and procedures for two-lane highways are covered in Chapter 16.

13.2 The Highway Capacity Manual

The U.S. standard for capacity and level of service analyses is the *Highway Capacity Manual* (HCM), a publication of the Transportation Research Board (TRB) of the National Academy of Engineering. Its content is controlled by the Committee on Highway Capacity and Quality of Service (HCQSC) of TRB. The committee consists of 30 regular members, with several hundred others participating through a number of subcommittees focused on particular methodologies in the HCM. Members are generally rotated off the committee after nine years of service.

The development of material for the manual is supported by a number of federal agencies through funding of basic and applied research. These agencies include the National Cooperative Highway Research Program (NCHRP), which is directly funded by state highway and transportation departments, and the Federal Highway Administration (FHWA).

The first edition of the HCM was published in 1950 [*1*]. At that time, the HCQSC consisted of full-time employees of the then Bureau of Public Roads (later FHWA) and several volunteers. Its original objective was to provide a measure of consistency in design practice for the nation's rapidly expanding postwar highway construction program.

The second edition was published in 1965 [*2*], the first formally published by TRB, which had adopted the committee and its members by that time. It introduced significant new material on limited access facilities and introduced the level of service concept for the first time.

The third edition appeared in 1985 [*3*] and included refinements to the level of service concept and added material on transit and pedestrian facilities. It was also the first edition published in loose-leaf form, which allowed two significant interim updates that appeared in 1994 [*4*] and 1997 [*5*]. It was also the first edition for which implementing software was developed, albeit several years after the manual itself appeared. The principal implementing software for that and subsequent editions is the *Highway Capacity Software,* maintained at the McTrans Center of University of Florida at Gainesville. As the methodologies of the manual have become increasingly complex, the software has become necessary for efficient use and application.

The fourth edition of the HCM was published in December 2000 [*6*] and is generally referred to as HCM2000. It added significant new material on corridors and networks, in addition to significant updates to other methodologies. It was originally prepared in metric units because standing U.S. law at the time required conversion of all highway and other agencies to the metric system by 2000. Late in preparation, the legislation was modified, making conversion to metric optional. Because this left a significant number of states using different systems, a version of the manual in U.S. standard units was also prepared.

As this text is being drafted, efforts are underway for a fifth edition of the HCM, which is expected to be published late in 2010. It will be in U.S. standard units only because most states have now returned to this system. It will contain significant upgrades to existing methodologies in many areas, and it will add material on freeway interchanges, and on multimodal analysis of arterials.

This text includes material from research source materials that is expected to be included in the forthcoming fifth edition of the HCM. In some cases, these methodologies have been reviewed and approved by the Highway Capacity and Quality of Service Committee of the Transportation Research Board. In other cases, the material is still under review. In any case, when the fifth edition of the HCM is officially released, you should consult it directly for final approved analysis procedures.

In a few cases, where material intended for the fifth edition is not available, this text uses the fourth edition as a source. Sources of material on highway capacity and level of service analysis are detailed in each chapter for clarity. A paper by Kittelson [7] provides an excellent history and discussion of the development of the HCM and its key concepts.

13.3 The Capacity Concept

13.3.1 The Current Definition

The HCM2000 defines capacity as follows:

> *The capacity of a facility is the maximum hourly rate at which persons or vehicles can be reasonably expected to traverse a point or a uniform section of a lane or roadway during a given time period under prevailing roadway, traffic, and control conditions.* (HCM2000, p. 2-2)

This definition contains a number of significant concepts that must be understood when applying capacity analysis procedures:

- Capacity is defined as a *maximum hourly rate.* For most cases, the rate used is for the peak 15 minutes of the peak hour, although HCM2000 allows for some discretion in selecting the length of the analysis period. In any analysis, care must be taken to express both the demand and the capacity in the same terms.
- Capacity may be defined in terms of *persons* or *vehicles.* This reflects the increasing importance of transit and pedestrians, HOV lanes, and multimodal facilities, where person-capacity may have more importance than vehicle-capacity.
- Capacity is defined for *prevailing traffic, roadway, and control conditions.* Roadway conditions refer to geometric characteristics such as number of lanes, lane width, shoulder width, and free-flow speed. Traffic conditions refer to the composition of the traffic stream in terms of cars, trucks, buses, and recreational vehicles (RVs). Control conditions refer mainly to interrupted flow facilities, where signal timing can have significant impacts on capacity. The important concept is that a change in any of the prevailing conditions changes the capacity.
- Capacity is defined for a *point* or *uniform segment* of a facility. This correlates to the prevailing conditions discussed earlier. The segment becomes nonuniform when any of these change. A uniform segment must have consistent prevailing conditions.
- Capacity refers to the vague concept of *reasonable expectancy.* Because capacity, like any traffic characteristic, includes some elements of stochastic variation, and may vary with time and/or location, the meaning of this term is important. In general, it implies that stated capacity values should be achievable on the vast majority of facilities with similar characteristics, regardless of time and location. Thus capacity is *not* the single highest traffic flow ever observed on any freeway or even at any given location. It is basically a value that is regularly observed over a variety of times and locations (of similar prevailing characteristics). The vagueness of this term has led to consideration of a more precise statistical definition of "capacity," but none has emerged at this writing.

13.3.2 Historical Background

Historically, when introduced in 1950, capacity was stated in terms of an hourly volume. Shorter time periods were not addressed. In 1965, use of the peak hour factor was introduced but only at levels of service C and D (see next section). Capacity was still defined in terms of a full-hour volume. The definition was changed to a flow rate for a standard 15-minute analysis period in 1985 and has been retained through subsequent editions of the manual.

As the 2010 edition approaches, a significant change in the concept is not expected, although some discussion of hourly volumes versus flow rates has taken place. The issue of defining capacity for more complex traffic systems, such as a multimodal facility, a corridor, or a network has not been addressed, and it will not be in the next edition.

Table 13.1: Capacity Under Ideal Conditions for Uninterrupted Flow Facilities

Type of Facility	Free-Flow Speed (mi/h)	Capacity
Freeways	≥70	2,400 pc/h/ln
	65	2,350 pc/h/ln
	60	2,300 pc/h/ln
	55	2,250 pc/h/ln
Multilane Highways	≥60	2,200 pc/h/ln
	55	2,100 pc/h/ln
	50	2,000 pc/h/ln
	50	1,900 pc/h/ln
Two-Lane Highways	All	3,200 pc/h (total, both dir) 1,700 pc/h (max. one dir)

13.3.3 Current Values of Capacity for Uninterrupted Flow Facilities

Table 13.1 shows the current basic capacities for uninterrupted flow facilities. These values are stated in pc/h/ln under equivalent ideal conditions, and would, in practice, be converted to prevailing conditions for application. These will not change in HCM 2010.

For freeways and multilane highways, capacity depends on the free-flow speed of the facility—the theoretical value of speed when flow is "0." In practice, the free-flow speed is relatively constant for low flow rates (<1,000 veh/h/ln) and is measured at such values. Lower free-flow speeds lead to lower capacities.

Free-flow speed decreases with a number of characteristics, including narrow lane widths, restricted lateral clearances, median type (multilane highways only), increasing ramp density (freeways), and increasing roadside access point density (multilane highways).

For two-lane highways, capacity does not depend on free-flow speed and is often stated as a total in both directions because passing maneuvers cause the two-directional flows to interact.

13.4 The Level of Service Concept

Although capacity is an important concept, operating conditions at capacity are generally quite poor, and it is difficult (but not impossible) to maintain operation at capacity without breakdowns for long periods of time. At capacity, there are virtually no usable gaps in the traffic stream. Any vehicle entering the traffic stream, or even changing lanes within the traffic stream, trailing vehicles to drop back to make room for it. This sets up a chain reaction that propagates upstream until sufficient space in the traffic stream is available to absorb the impact.

The level of service (LOS) concept was introduced in the second edition of the HCM in 1965 as a convenient way to describe the general quality of operations on a facility with defined traffic, roadway, and control conditions. Using a simple letter-grade scale from A to F, a terminology for operational quality was created that has become an important tool in communicating complex issues to decision makers and the general public.

13.4.1 Historical Development of the Level of Service Concept

The First Edition of the HCM (1950)

The first edition of the HCM [*1*] did not mention the words "level of service" or define a concept that was similar. It did, however, define three different "capacities," one of which can be considered a predecessor to the LOS concept:

- *Basic Capacity:* "The maximum number of passenger cars that can pass a given point on a lane or roadway during one hour under the most nearly ideal roadway and traffic conditions which can possibly be attained." (Reference 1, p. 6)
- *Possible Capacity:* "The maximum number of vehicles that can pass a given point on a lane or roadway during one hour under prevailing roadway and traffic conditions." (Reference 1, p. 6)
- *Practical Capacity:* "The maximum number of vehicles that can pass a given point on a roadway or in a designated lane during one hour without the traffic density being to great as to cause unreasonable delay, hazard, or restriction to the drivers' freedom to maneuver under the prevailing roadway and traffic conditions." (Reference 1, p. 7)

In terms of current terminology, *basic capacity* is similar to "capacity under ideal conditions," *possible capacity* is similar to the current "capacity," and *practical capacity* is most similar to a service volume for LOS C. A key difference is that the first edition of the HCM dealt only with hourly volumes, not flow rates for 15-minute periods.

The Second Edition of the HCM (1965)

The formal introduction of the term *level of service* came with the second edition of the HCM [*2*] in 1965. The concept was defined as follows, with italics added for emphasis:

> Level of service is a term which, broadly interpreted, denotes any one of an infinite number of different combinations of operating conditions that may occur on a given lane or roadway when it is accommodating various traffic volumes. *Level of service is a qualitative measure of the effect of a number of factors, which include speed and travel time, traffic interruptions, freedom to maneuver, safety, driving comfort and convenience, and operating cost.* (Reference 2, p. 7)

It follows the definition with the following explanation—which introduces the concept of driver perceptions into the equation:

> From the viewpoint of the driver, low flow rates or volumes on a given lane or roadway provide higher levels of service than greater flow rates or volumes on the same roadway. Thus, the level of service for any particular lane or roadway varies inversely as some function of the flow or volume, or of the density. (Reference 2, pp. 7, 8)

The issue of how to treat driver or other road user perceptions as part of the level of service concept has attracted a great deal of recent interest. Note that in the definition of LOS, of all of the parameters mentioned as possible quantifiers, all of them can be perceived by a driver within a traffic stream. Missing from the list, however, is volume or rate of flow. Volume or flow rate cannot be perceived by a driver within the traffic stream because it requires observation of vehicles passing a fixed point on the roadway. By definition, the observer must be *outside* the traffic stream.

The second edition presumes that the driver associates lower flow rates, volumes, or densities with better levels of service. This was a somewhat odd presumption, given that volume or flow rate cannot be observed by a driver within the traffic stream. Although safety, operating cost, and comfort and convenience *are* included in the definition, no serious effort has been expended in actually using these as measures of effectiveness. In fact, the second edition used flow rates or volumes almost exclusively in defining LOS boundaries:

> After careful consideration, the Committee has selected travel speed as the major factor for use in identifying the level of service. The Committee also uses a second factor—either the ratio of demand volume to capacity, or the ratio of service volume to capacity, depending upon the particular problem situation—in making this determination. (Reference 2, pp. 7, 8)

The methodologies of the second edition, however, actually reverse this statement, using v/c ratio as the primary determinant, with speed used as a secondary measure in some cases and completely ignored in others.

The Third Edition of the HCM (1985)

The third edition of the HCM [*3*] made some subtle changes in the definition of the LOS concept, even though it is quite similar to the definition in the second edition.

> The concept of level of service is defined as a qualitative measure describing operational conditions within a traffic stream and their perception by motorists and/or passengers. A level-of-service definition generally prescribes these conditions in terms of such factors as speed and travel time, freedom to maneuver, traffic interruptions, comfort and convenience, and safety. (Reference 3, pp. 1–3)

This is the first direct inclusion of road users' perceptions in the concept definition. Levels of service are to be qualitative measures of operational conditions within a traffic stream, which clearly excludes volumes and flow rates explicitly. The inclusion of operating costs is removed, but safety remains—primarily as a goal because it is not included in any of the analysis methodologies of the third edition.

In a related, but extremely important change, capacity is defined in terms of a flow rate over a 15-minute period, *not* an hourly volume. The third edition of the HCM actually went through two partial revisions in 1994 [*4*] and 1997 [*5*]. By 1997, all uses of flow rates and/or volumes as measures of effectiveness had been eliminated.

13.4.2 The Fourth Edition of the HCM (2000): The Current Definition

If the third-edition HCM was a radical departure from past editions, the fourth edition catapulted the manual into significant areas of uncharted territory. The LOS concept was driven by an important research effort specifically focused on defining and interpreting the LOS concept. Sponsored by the National Cooperative Highway Research Program, Project 3-55 [*8*] broke new ground with respect to system and network analysis, and it presented some interesting approaches to dealing with failure: LOS F.

Although it would be impossible to summarize all of the results of this project here, it forced the HCQSC to consider and resolve some very thorny issues. Some of the critical policy determinations were as follows:

1. Analysis inputs must focus on *demand* rather than existing volumes or flow rates. The traditional v/c ratio should be viewed as a demand-to-capacity ratio (d/c).
2. System analysis should be based on a common approach to disaggregation of the traffic system. Four levels of disaggregation were recommended: points, segments, facilities, and subsystems. The final aggregation would consist of the entire transportation system, taken as a whole.
3. System performance measures should focus on trip time and trip delay.
4. Five levels of service (A–E) should continue to describe undersaturated operations.
5. In defining boundaries between LOS E and F, individual chapter methodologies could adopt one of two policies: (a) LOS F occurs when the demand-to-capacity ratio exceeds 1.00, or (b) LOS F occurs when a prescribed measure of effectiveness limit is exceeded.
6. Division of LOS F into sublevels is recommended, as is the use of at least one measure of effectiveness describing quality of operations within LOS F. (This was a recommendation that was *not* adopted for the fourth edition).
7. A deterministic queuing approach is recommended to analyze the spatial and time impacts of a breakdown.

Item 5 was a key compromise. By 1997, all interrupted flow methodologies used definition (a), and most interrupted flow methodologies used definition (b). This continued in the fourth edition. It should be noted that option (b) contradicts recommendation 4 because it allows undersaturated flow to be described by LOS F. In another recommendation, the study endorses the use of multiple measures of effectiveness to define levels of service for given facility types.

The fourth edition of the HCM [*6*] implemented many of the recommendations of NCHRP Project 3-55(*4*) and dove into the issues of corridors and networks, and it became the first edition to deal explicitly with other tools—primarily simulation. It was almost three times as long as the third edition and was produced in two versions: standard U.S. units and metric units.

In terms of the LOS concept, the definition provided in the fourth edition is not very different than its predecessors:

> Level of service (LOS) is a quality measure describing operational conditions within a traffic stream, generally in terms of such service measures as speed and travel time, freedom to maneuver, traffic interruptions, and comfort and convenience. (Reference 6, p. 2-2)

Safety is eliminated from the list of potential measures of effectiveness. There is no mention of road user perceptions, but a later statement makes it clear that road user perceptions are still to be considered: "Each level of service represents a range of operating conditions and the driver's perception of those conditions" (Reference 6, p. 2-3).

Table 13.2 shows the measures of effectiveness used to define LOS in the fourth edition. For the first time, the HCQSC specifically declined to define levels of service for a methodology. Freeway facilities and all corridor and network applications do not have defined levels of service. At the facility level of analysis, this presents an inconsistency: Freeway facilities have no levels of service; arterials and streets do have levels of service.

13.4.3 Incorporating Road User Perceptions into Levels of Service

In the late 1990s and early 2000s, a variety of researchers began seriously to study the role of user perceptions in establishing level of service criteria. A paper by Flannery, McLeod, and Pederson provided an excellent overview of the field [*9*]. The paper was based on studies conducted by the Florida and Maryland DOTs and early results from an NCHRP project [*10*] that focused on developing multimodal level of service criteria for urban streets.

The paper highlighted the evidence that user perceptions were significantly affected by a variety of nonoperational factors, and it argued for inclusion of some of these in a level of service structure. These included environmental and aesthetic factors such as landscaping. In some cases, basic fixed geometric characteristics also affected the perception of service quality. Traffic factors such as heavy vehicles presence, quality of traveler information, and speed differentials were also important factors.

The paper also concludes that in Florida, drivers on rural freeways view LOS very differently from those on urban freeways, and it presents an argument for having different threshold values for LOS based on whether the freeway is in an urban area or a rural area.

Table 13.2: Measures of Effectiveness in the HCM2000

Element	HCM Chapter	Measure of Effectiveness
Vehicular, Interrupted Flow		
Urban streets	15	Average Travel Speed
Signalized intersections	16	Control Delay
Two-way STOP-controlled intersections	17	Control Delay
All-way STOP-controlled intersections	17	Control Delay
Roundabouts	17	*None*[1]
Interchange ramp terminals	26	Control Delay
Vehicular, Uninterrupted Flow		
Two-lane highways	20	%Time Spent Following Average Travel Speed
Multilane highways	21	Density
Basic freeway segments	23	Density
Freeway weaving areas	24	Density
Freeway ramp junctions	25	Density
Freeway facilities	22	*None*[2]
Other Road Users		
Transit	27	[3]
Pedestrians	18	Space, delay
Bicycle	19	Event, delay

1. HCM does not predict any performance measures for roundabouts.
2. By recommendation of the Uninterrupted Flow Group, a decision was made by the HCQSC *not* to define levels of service for freeway facilities.
3. Chapter 27 uses several measures to define levels of service for transit.

The entire discussion of how to incorporate the results of research on user perceptions into LOS definitions in the forthcoming 2010 HCM was a very important and contentious issue for the HCQSC. In general terms, the new traveler-based perception measures are to be used to define levels of service for bicycle and pedestrian facilities, whereas more traditional operational measures will be used for the automobile mode.

Among the most difficult issues under discussion are two key conclusions from the NCHRP study [*10*]. The first is that users do not perceive six different levels of service. At most, they perceive three distinct levels; in some cases, only two. The second is that environmental and aesthetic variables do influence user perceptions. For HCM 2010, it has been decided that environmental factors will not be used to define LOS and the current system of six levels will be retained.

Obviously, LOS is not a fixed concept. It has evolved over time and will doubtless continue to evolve. The next few years, however, may shape the approaches taken for decades into the future. We encourage you to consult the latest literature on level of service to keep up to date with this rapidly developing field.

13.5 Service Flow Rates and Service Volumes

Closely associated with the concept of LOS is the concept of service flow rates and service volumes. A service flow (SF) rate is defined as the maximum rate of flow that can be reasonably expected on a lane or roadway under prevailing roadway, traffic, and control conditions *while maintaining a particular level of service.* It is essentially a "capacity" for a specified level of service (e.g., the most traffic that can be accommodated at LOS "X"). Figure 13.1 illustrates the concept.

The most important thing illustrated in Figure 13.1 is that a given level of service covers a *range* of operating conditions and, therefore, flow rates. Two freeway segments

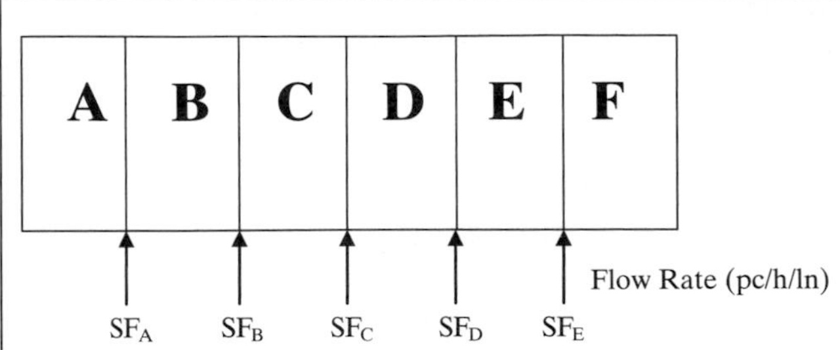

Figure 13.1: Illustration of Levels of Service and Service Flow Rates

operating at levels of service C and D may, in fact, be quite similar, if both are operating near the boundary between the two levels. A greater difference can exist between two segments operating *at the same level of service,* if they are at opposite ends of the range represented by the level. Levels of service represent a *step function* that is applied to a defining measure of effectiveness that is continuous. Thus, although freeways have six discrete levels of service (A to F), those levels are defined by density, which is a continuous variable. Because of this, care must be taken in using the "language" of LOS; in the absence of specific values of the quantifying variable (density), LOS designations alone can be misleading.

The SF rate is defined as the maximum flow rate that can be sustained without exceeding the maximum density defined for the LOS. There are only *five* SF rates, not six. LOS F represents unstable flow, which exists within a queue formed behind a point or segment where arrival flow exceeds capacity. Because it is unstable, no fixed SF rate can be assigned. Thus SF rates exist only for levels of service A through E.

For all present methodologies on uninterrupted flow facilities, the SF rate for level of service E (SF_E) is synonymous with "capacity" (i.e., the maximum flow rate for LOS E is capacity). This relationship *does not* hold for many interrupted flow facilities, as you will see in later chapters.

The term *service volume* (SV) is a vestige of early editions of the HCM in which capacity and level of service described conditions that existed over a full hour as opposed to the standard 15-minute periods used in current standards. The relationship between SF rates and SVs is the same as the relationship between an actual flow rate and an actual hourly volume. They are related by the peak hour factor (PHF), and SVs can be specified from SF rates as follows:

$$SV_i = SF_i * PHF \tag{13-1}$$

where: SV_i = service volume for LOS i, (veh/h)
SF_i = service flow rate for LOS i, (veh/h)
PHF = peak hour factor

13.6 The v/c Ratio and Its Use in Capacity Analysis

One of the most important measures resulting from a capacity and/or a level of service analysis is the v/c ratio: the ratio of the current or projected demand flow to the capacity of the facility. This ratio is used as a measure of the sufficiency of the existing or proposed capacity of the facility.

It is, of course, desirable that all facilities be designed to provide sufficient capacity to handle present and projected demands (i.e., that the v/c ratio be maintained at a value less than 1.00).

When estimating a v/c ratio, care must be taken to understand the origin of demand (v) and capacity (c) values. In existing cases, *true demand* consists of the actual arrival flow rate plus traffic that has diverted to other facilities, other times, or other destinations due to congestion limitations. If existing flow rate observations consist of *departing vehicles* from the study point, there is no guarantee that this represents true demand. However, a count of departing vehicles *cannot* exceed the actual capacity of the segment (you cannot put 5 gallons in a 4-gallon jug!). Therefore, if a departing flow rate is compared to the capacity of a section, and a v/c ratio more than 1.00 results, the conclusion must be that either (a) the counts were incorrect (too high), or (b) that the capacity was underestimated. The latter is the usual culprit. Capacities are estimated using methodologies presented in the HCM. They are based on national averages and on the principle of "reasonable expectation." Actual capacity can be larger than the estimate produced by these methodologies at any specific location and time.

When dealing with future situations and forecast demand flow rates, both the demand flow *and* the capacity are estimates. A v/c ratio in excess of 1.00 implies that the forecast demand flow will exceed the estimated capacity of the facility.

When the true ratio of demand flow to capacity exceeds 1.00 (either in the present or the future), the implication is that queuing will occur and propagate upstream from the segment in question. The extent of queues and the time required to clear them depends on many conditions, including the time period over which v/c is more than 1.00 and by how much it exceeds 1.00. It depends on the demand profile over time because queues can start to dissipate only when demand flow decreases to levels less than the capacity of the segment. Further, drivers presented with queuing situations tend to seek alternative routes to avoid congestion. Thus the occurrence of v/c more than 1.00 often causes a dynamic shift in demand patterns that could significantly impact operations in and around the subject segment.

In any event, the comparison of true demand flows to capacity is a principal objective of capacity and level of service analysis. Thus, in addition to LOS criteria, the v/c ratio is a major output of such an analysis.

13.7 Problems in Use of Level of Service

The difficulty in interpreting LOS results due to its step-function nature has been discussed. The simple six-letter grade scale was introduced in 1965 for two primary purposes: (1) to provide a descriptor of service quality where no prediction of operational parameters was possible, and (2) to provide a simple language that could be used to explain complex situations to decision makers and the public, who are most often not engineers.

The first reason no longer exists: Since 1997, every LOS methodology includes the prediction of at least one numerical measure of effectiveness. The second remains a significant need. Periodically, discussions arise as to whether or not it is time to abandon LOS completely and rely entirely on numeric measures. As the methodologies themselves become more complex and deal with more complex factors, a double-edged sword emerges:

- Because the methodologies are so complex, it is more important than ever to have a simple, well-understood language to use in communication with nonprofessionals.
- Because the methodologies reflect increasingly complex operating conditions, a simplistic six-letter grade scale is an oversimplification, allowing decision makers, and even professionals, to ignore many important factors.

Both arguments have merit. For the HCM 2010, LOS will be retained. The importance of quantitative measures beyond the LOS, however, will be emphasized, perhaps setting the stage for eliminating the concept in a sixth edition in the future. For professionals, however, it is critical that LOS not be used as a single, all-encompassing result. Many methodologies produce multiple operational measures (such as both speed and density for freeways and multilane highways), and produce estimates of capacity and v/c ratios. Engineers must make use of all available descriptors to fully understand any given operation and to make the best possible judgments on the best course of action to eliminate or reduce existing problems.

LOS as a descriptor also raises the problem of comparability. There is a fundamental question: Does LOS C mean the same thing in New York City as it does in Peoria? A freeway density of 40 veh/h/ln may be a common occurrence in New York City but a rare and unusual event in Peoria. New Yorkers may be much more accommodating to a high density (or high delay at an intersection) than someone from a small rural community. AASHTO design standards, for example, suggest that a good target level of service in D in an urban area but C or even B in a rural area. Unfortunately, because the letter grades we use have general connotations from other uses (think school grades), few are willing to say that LOS D or LOS E is "acceptable" under any circumstances. User perceptions, however, are different at different times and in different environments. The LOS system does not prevent engineers from dealing with these differences, but it makes it harder, and it complicates the issue of interpreting LOS for nontechnical groups.

If the relativity issue was not difficult enough, many highway agencies have incorporated LOS criteria in their development standards and mitigation requirements. Some states have incorporated specific LOS references in law. This can and does create problems when a new edition of the HCM is released. In virtually every one, LOS criteria are subject to small, and sometimes, major changes. In effect, release of a new HCM changes development policy and law—certainly *not* the job of the HCQSC, and certainly an unintended consequence of LOS becoming such a commonly accepted standard.

13.8 Closing Comments

Capacity and LOS are two of the most important concepts used in traffic and transportation engineering. They have been introduced here, with a specific emphasis on uninterrupted flow facilities. Chapter 14 applies the concepts to basic freeway sections and multilane highways, Chapter 15 to weaving areas and ramp junctions on freeway and/or multilane highways, and Chapter 16 to two-lane rural highways.

References

1. *Highway Capacity Manual,* Bureau of Public Roads, Washington DC, 1950.
2. *Highway Capacity Manual,* "Special Report 87," Transportation Research Board, Washington DC, 1965.
3. *Highway Capacity Manual,* "Special Report 209," Transportation Research Board, Washington DC, 1985.
4. *Highway Capacity Manual,* "Special Report 209," (as revised in 1994), Transportation Research Board, Washington DC, 1994.

5. *Highway Capacity Manual,* "Special Report 209," (as revised in 1997), Transportation Research Board, Washington DC, 1997.

6. *Highway Capacity Manual,* Transportation Research Board, Washington DC, 2000 (Metric and Standard U.S. versions).

7. Kittelson, W., "Historical Overview of the Committee on Highway Capacity and Quality of Service," *Proceedings of the Fourth International Symposium on Highway Capacity,* Transportation Research Circular E-C0-18, Maui, Hawaii, June 27–July 1, 2000.

8. May, A., Jr., "Performance Measures and Levels of Service in the Year 2000 Highway Capacity Manual," *Final Report,* National Cooperative Highway Research Program Project 3-55 (4), Transportation Research Board, Washington DC, 2000.

9. Flannery, A., McLeod, D., and Pederson, N., "Customer-Based Measures of Level of Service," *ITE Journal,* Institute of Transportation Engineers, Washington DC, May 2006.

10. Dowling, R., et al., "Multimodal Level of Service Analysis for Urban Streets," *Draft Final Report,* National Cooperative Highway Research Program Project 3-70, Transportation Research Board, Washington DC, September 2007.

Problems

13-1. Explain the difference between "capacity" and a "service flow rate" for an uninterrupted flow facility.

13-2. Explain the difference between the following terms: (a) capacity under ideal conditions, (b) capacity, and (c) service flow rate for LOS C.

13-3. The following service flow rates have been determined for a six-lane freeway in one direction:

LOS A: 3,250 veh/h; LOS B: 3,900 veh/h; LOS C: 4,680 veh/h; LOS D: 5,810 veh/h; LOS E: 7,000 veh/h.

(a) What is the capacity of this facility?

(b) Determine the service volumes for this freeway, which has a peak hour factor (PHF) of 0.95.

(c) Determine service flow rates under ideal conditions for this freeway given that the following adjustment factors apply: $f_{HV} = 0.935$; $f_p = 1.00$.

13-4. Explain the differences between freeways, multilane highways, and two-lane rural highways. What are the key differentiating characteristics of each?

CHAPTER 14

Basic Freeway Segments and Multilane Highways

The procedures described in this chapter cover the analysis of multilane uninterrupted flow facilities. These include basic freeway sections and sections of surface multilane highways with sufficient distances between fixed interruptions (primarily traffic signals) to allow uninterrupted (random, or nonplatoon) flow between points of interruption.

14.1 Facility Types

Freeways are the only types of facilities providing pure uninterrupted flow. All entries and exits from freeways are made using ramps designed to allow such movements to occur without interruption to the freeway traffic stream. There are no at-grade intersections (either signalized or unsignalized), no driveway access, and no parking permitted within the right-of-way. Full control of access is provided. Freeways are generally classified by the total number of lanes provided in both directions (e.g., a six-lane freeway has three lanes in each direction). Common categories are four-, six-, and eight-lane freeways, although some freeway sections in major urban areas may have 10 or more lanes in specific segments.

Multilane surface facilities should be classified and analyzed as urban streets (arterials) if signal spacing is less than one mile. Uninterrupted flow can exist on multilane facilities where the signal spacing is more than two miles. Where signal spacing is between one and two miles, the existence of uninterrupted flow depends on prevailing conditions. There are, unfortunately, no specific criteria to guide traffic engineers in making this determination, which could easily vary over time. In the majority of cases, signal spacings between one and two miles do not result in the complete breakdown of platoon movements unless the signals are not coordinated. Thus most of these cases are best analyzed as arterials.

Multilane highway segments are classified by the number of lanes and the type of median treatment provided. Surface multilane facilities generally consist of four- or six-lane alignments. They can be *undivided* (i.e., having no median but with a double-solid-yellow marking separating the two directions of flow), or *divided,* with a physical median separating the two directions of flow. In suburban areas, a third median treatment is also used: the two-way left-turn lane. This treatment requires an alignment with an odd number of lanes—most commonly three, five, or seven. The center lane is used as a continuous left-turn lane for both directions of flow.

The median treatment of a surface multilane highway can have a significant impact on operations. A physical median prevents midblock left turns across the median except at locations where a break in the median barrier is provided. Midblock left turns can be made at any point on an undivided alignment. Where a two-way left-turn lane is provided, midblock left turns are permitted without restriction, but vehicles waiting to turn do so in the special lane and do not unduly restrict through vehicles.

In terms of capacity analysis procedures, both basic freeway sections and multilane highways are categorized by the *free-flow speed.* By definition, the free-flow speed is the speed intercept when flow is "zero" on a calibrated speed-flow curve. In practical terms, it is the average speed of the traffic stream when flow rates are less than approximately 1,000 veh/h/ln. Refer to Chapter 5 for a more complete discussion of speed-flow relationships and characteristics. Figure 14.1 illustrates some common freeway and multilane alignments.

14.2 Basic Freeway and Multilane Highway Characteristics

14.2.1 Speed-Flow Characteristics

The basic characteristics of uninterrupted flow were presented in detail in Chapter 5. Capacity analysis procedures for freeways and multilane highways are based on calibrated speed-flow curves for sections with various free-flow speeds operating under *base conditions.* Base conditions for freeways and multilane highways include:

- No heavy vehicles in the traffic stream
- A driver population dominated by regular or familiar users of the facility

Figures 14.2 and 14.3 show the standard curves calibrated for use in the capacity analysis of basic freeway sections and multilane highways. These exhibits also show the density lines that define levels of service for uninterrupted flow facilities. Table 14.1 provides equations for the freeway curves of Figure 14.2, which are comprised of three linear segments. Complex equations for multilane highways are shown in Figure 14.3, but the curves were originally graphic fits, and direct graphic use of the curves is recommended.

Modern drivers maintain high average speeds at relatively high rates of flow on freeways and multilane highways. This is clearly indicated in Figures 14.2 and 14.3. For freeways, the free-flow speed is maintained until flows reach 1,000 to 1,800 pc/h/ln, depending upon the free-flow speed. Multilane highway characteristics are similar, with free-flow speeds maintained for flow rates up to 1,400 pc/h/ln for all free-flow speeds. Thus, on most uninterrupted flow facilities, the transition from stable to unstable flow occurs very quickly and with relatively small increments in flow.

14.2.2 Levels of Service

For freeways and multilane highways, the measure of effectiveness used to define levels of service is *density.* The use of density, rather than speed, is based primarily on the shape of the speed-flow relationship depicted in Figures 14.2 and 14.3. Because average speed remains constant through most of the range of flows and because the total difference between free-flow speed and the speed at capacity is relatively small, defining five level-of-service boundaries based on this parameter would be very difficult.

If flow rates vary while speeds remain relatively stable, then density must be varying throughout the range of flows, given the basic relationships that $v = S \times D$. Further, density describes the proximity of vehicles to each other, which is the principal influence on freedom to maneuver. Thus, it is an appropriate descriptor of service quality.

For uninterrupted flow facilities, the density boundary between levels of service E and F is defined as the density at which capacity occurs. The speed-flow curves determine this critical boundary. For freeways, the curves indicate a constant density of 45 pc/mi/ln. For multilane highways, capacity occurs at densities ranging from 40 to 45 pc/mi/ln, depending on the free-flow speed of the segment.

Other level-of-service boundaries are set judgmentally by the Highway Capacity and Quality of Service Committee (HCQSC) to provide reasonable ranges of both density and service flow rates. Table 14.2 shows the defined level-of-service criteria for basic freeway sections and multilane highways.

The general operating conditions for these levels of service can be described as follows:

- *Level of service A* is intended to describe free-flow operations. At these low densities, the operation of each vehicle is not greatly influenced by the presence of others. Speeds are not affected by flow in this level of service, and operation is at the free-flow speed. Lane changing, merging, and diverging maneuvers are easily accomplished because many

(a) A Typical 8-Lane Freeway

(b) A Divided Multilane Rural Highway

(c) A Divided Multilane Suburban Highway

(d) An Undivided Multilane Suburban Highway

(e) A Multilane Highway w/TWLTL

(f) An Undivided Multilane Rural Highway

Figure 14.1: Typical Freeway and Multilane Highway Alignments

(*Sources:* Photo (a) courtesy of J. Ulerio; (b),(c),(d),(f) Used with permission of Transportation Research Board, National Research Council, "Highway Capacity Manual," *Special Report 209*, 1994, Illustrations 7-1 through 7-4, p. 7-3; (e) Used with permission of Transportation Research Board, National Research Council, *Highway Capacity Manual*, December 2000, Illustration 12-8, p. 12-6.)

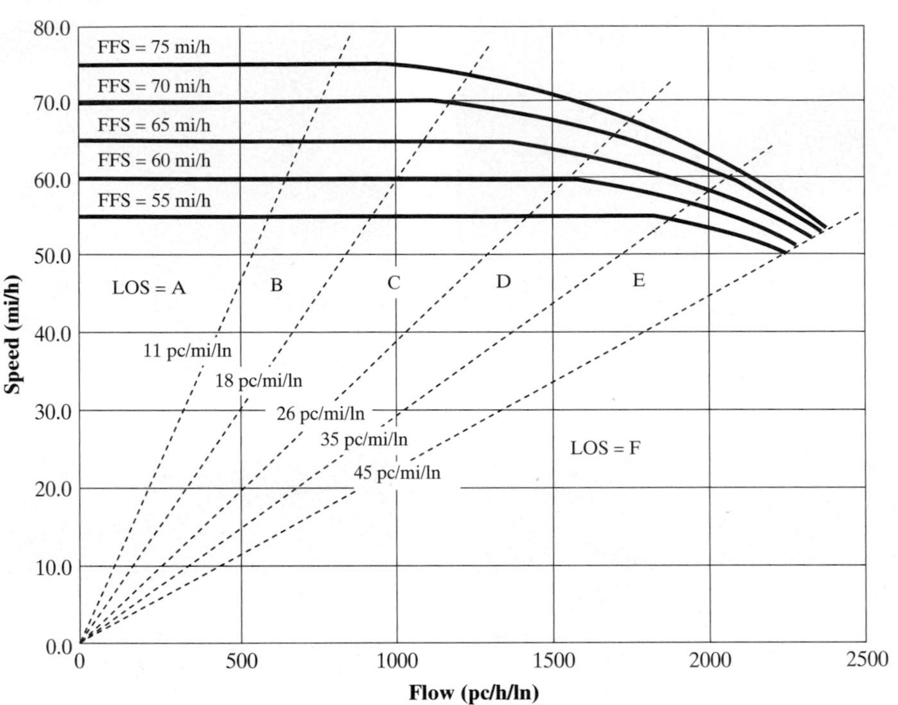

Figure 14.2: Base Speed-Flow Curves for Freeways

(*Source: Basic Freeway Segments*, Draft Chapter 11, NCHRP Project 3-92, Production of the 2010 *Highway Capacity Manual*, Kittelson and Associates, Portland OR, 2009, Exhibit 11-6, p. 11-8.)

Table 14.1: Equations for Curves in Figure 14.1

FFS (mi/h)	Break-Point (pc/h/ln)	Flow Rate Range ≥ 0 ≤ Break-Point	> Break-Point ≤ Capacity
75	1,000	75	$75 - 0.00001107\ (v_p - 1{,}000)^2$
70	1,200	70	$70 - 0.00001160\ (v_p - 1{,}200)^2$
65	1,400	65	$65 - 0.00001418\ (v_p - 1{,}400)^2$
60	1,600	60	$60 - 0.00001816\ (v_p - 1{,}600)^2$
55	1,800	55	$55 - 0.00002469\ (v_p - 1{,}800)^2$

Notes:
1. FFS = free-flow speed.
2. Maximum flow rate for the equations is capacity: 2,400 pc/h/ln for 70- and 75-mph FFS; 2,350 pc/h/ln for 65-mph FFS; 2,300 pc/h/ln for 60-mph FFS; and 2,250 pc/h/ln for 55-mph FFS.

(*Source: Basic Freeway Segments*, Draft Chapter 11, NCHRP Project 3-92, Production of the 2010 *Highway Capacity Manual*, Kittelson and Associates, Portland OR, 2009, Exhibit 11-3, p. 11-4.)

large gaps in lane flow exist. Short-duration lane blockages may cause the level of service to deteriorate somewhat but do not cause significant disruption to flow. Average spacing between vehicles is a minimum of 480 ft, or approximately 24 car lengths at this level of service.

- At *level of service B,* drivers begin to respond to the existence of other vehicles in the traffic stream,

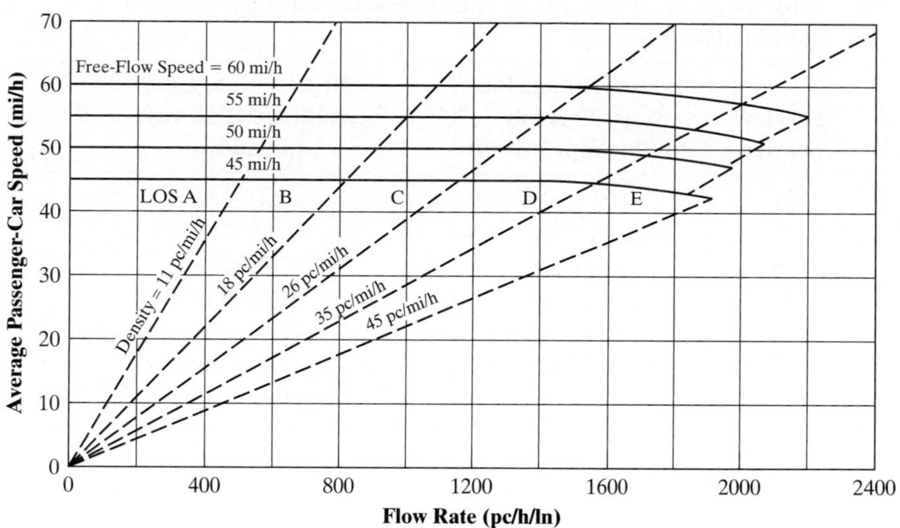

Equations for Curves

FFS (mi/h)	For v ≤ 1,400 pc/h/ln S (mi/h)	For v > 1,400 pc/h/ln S (mi/h)
60	S = 60	$S = 60 - \left[5.00\left(\frac{v_p - 1{,}400}{800}\right)^{1.31}\right]$
55	S = 55	$S = 55 - \left[3.78\left(\frac{v_p - 1{,}400}{700}\right)^{1.31}\right]$
50	S = 50	$S = 50 - \left[3.49\left(\frac{v_p - 1{,}400}{600}\right)^{1.31}\right]$
45	S = 45	$S = 45 - \left[2.78\left(\frac{v_p - 1{,}400}{500}\right)^{1.31}\right]$

Figure 14.3: Base Speed-Flow Curves for Multilane Highways

(*Source:* Used with permission of Transportation Research Board, National Research Council, from *Highway Capacity Manual*, December 2000, Exhibit 21-3, p. 21–4.)

although operation is still at the free-flow speed. Maneuvering within the traffic stream is still relatively easy, but drivers must be more vigilant in searching for gaps in lane flows. The traffic stream still has sufficient gaps to dampen the impact of most minor lane disruptions. Average spacing is a minimum of 293 feet, or approximately 15 car lengths.

- At *level of service C,* the presence of other vehicles begins to restrict maneuverability within the traffic stream. Operations remain at the free-flow speed, but drivers now need to adjust their course to find gaps they can use to pass or merge. A significant increase in driver vigilance is required at this level. Although there are still sufficient gaps in the traffic stream to dampen the impact of minor lane blockages, any significant blockage could lead to breakdown and queuing. Average spacing is a minimum of 203 feet, or approximately 10 car lengths.

Table 14.2: Level of Service Criteria for Basic Freeway Segments and Multilane Highways

Level of Service	Density Range for Basic Freeway Sections (pc/mi/ln)	Density Range for Multilane Highways (pc/mi/ln)
A	$\geq 0 \leq 11$	$\geq 0 \leq 11$
B	$> 11 \leq 18$	$> 11 \leq 18$
C	$> 18 \leq 26$	$> 18 \leq 26$
D	$> 26 \leq 35$	$> 26 \leq 35$
E	$> 35 \leq 45$	$> 35 \leq$ (40–45) depending on FFS
F	Demand Exceeds Capacity > 45	Demand Exceeds Capacity $>$ (40–45) depending on FFS

- *Level of service D* is the range in which average speeds begin to decline with increasing flows. Density deteriorates more quickly with flow in this range. At level of service D, breakdowns can occur quickly in response to small increases in flow. Maneuvering within the traffic stream is now quite difficult, and drivers often have to search for gaps for some time before successfully passing or merging. The ability of the traffic stream to dampen the impact of even minor lane disruptions is severely restricted, and most such blockages result in queue formation unless removed very quickly. Average spacing is a minimum of 151 feet, or approximately seven car lengths.
- *Level of service E* represents operation in the vicinity of capacity. The maximum density limit of level of service E is capacity operation. For such an operation there are few or no usable gaps in the traffic stream, and any perturbation caused by lane-changing or merging maneuvers will create a shock wave in the traffic stream. Even the smallest lane disruptions may cause extensive queuing. Maneuvering within the traffic stream is now very difficult because other vehicles must give way to accommodate a lane-changing or merging vehicle. The average spacing is a minimum of 117 feet, or approximately six car lengths.
- *Level of service F* describes operation within the queue that forms upstream of a breakdown point. Such breakdowns may be caused by accidents or incidents, or they may occur at locations where arrival demand exceeds the capacity of the section on a regular basis. Actual operating conditions vary widely and are subject to short-term perturbations. As vehicles "shuffle" through the queue, there are times when they are standing still and times when they move briskly for short distances. Level of service *F* is also used to describe the point of the breakdown, where demand flow (v) exceeds capacity (c). In reality, operation at the point of the breakdown is usually good because vehicles discharge from the queue. Nevertheless, it is insufficient capacity at the point of breakdown that causes the queue, and level of service F provides an appropriate descriptor for this condition.

Note that in Table 14.2, LOS F is identified when "demand exceeds capacity," which is equivalent to a v/c ratio more than 1.00. That is because no density can be predicted for such cases. Thus, in analysis, LOS F exists when demand exceeds capacity, and the density will be higher than 45 pc/h/ln (for freeways) or the designated values for multilane highways.

14.2.3 Service Flow Rates and Capacity

Maximum service flow rates (MSF) for the various levels of service for freeways and multilane highways are shown in Tables 14.3 and 14.4.

The values in these tables are taken directly from the curves of Figures 14.2 and 14.3. Maximum service flow rates are stated in terms of pc/mi/ln and reflect the base conditions defined previously. As in the HCM, all values are rounded to the nearest 10 pc/h/ln.

Table 14.3: Maximum Service Flow Rates for Basic Freeway Sections

FFS (mi/h)	Level of Service				
	A	B	C	D	E
75	820	1,310	1,750	2,110	2,400
70	770	1,250	1,690	2,080	2,400
65	710	1,170	1,630	2,030	2,350
60	660	1,080	1,560	2,010	2,300
55	600	990	1,430	1,900	2,250

Note: All values rounded to the nearest 10 pc/h/ln.

(*Source: Draft Chapter 11: Basic Freeway Segments,* National Cooperative Highway Research Program Project 3–92, Transportation Research Board, Washington DC, Exhibit 11-18, p. 11-24.)

Table 14.4: Maximum Service Flow Rates for Multilane Highways

FFS (mi/h)	Level of Service				
	A	B	C	D	E
60	660	1,080	1,550	1,980	2,200
55	600	990	1,430	1,850	2,100
50	550	900	1,300	1,710	2,000
45	490	810	1,170	1,550	1,900

Note: All values rounded to the nearest 10 pc/h/ln.

(*Source:* Used with permission of Transportation Research Board, National Research Council, from *Highway Capacity Manual,* Dec 2000, Exhibit 21-2, p. 21-3, Modified.)

14.3 Analysis Methodologies for Basic Freeway Sections and Multilane Highways

The characteristics and criteria described for freeways and multilane highways in the previous section apply to facilities with base traffic and roadway conditions. In most cases, base conditions do not exist, and a methodology is required to address the impact of prevailing conditions on these characteristics and criteria.

Analysis methodologies are provided that account for the impact of a variety of prevailing conditions, including:

- Lane widths
- Lateral clearances
- Type of median (multilane highways)
- Frequency of ramps (freeways) or access points (multilane highways)
- Presence of heavy vehicles in the traffic stream
- Driver populations dominated by occasional or unfamiliar users of a facility

Some of these factors affect the free-flow speed of the facility; others affect the equivalent demand flow rate on the facility.

14.3.1 Types of Analysis

Three types of analysis can be conducted for basic freeway sections and multilane highways:

- Operational analysis
- Service flow rate and service volume analysis
- Design analysis

In addition, the HCM defines "planning analysis." This, however, consists of beginning the analysis with an AADT as a demand input, rather than a peak hour volume. Planning analysis begins with a conversion of an AADT to a directional design hour volume (DDHV) using the traditional procedure as described in Chapter 5.

All forms of analysis require the determination of the free-flow speed of the facility in question. Field measurement and estimation techniques for making this determination are discussed in a later section.

Operational Analysis

The most common form of analysis is *operational analysis.* In this form of analysis, all traffic, roadway, and control conditions are defined for an existing or projected highway section, and the expected level of service and operating parameters are determined.

The basic approach is to convert the existing or forecast demand volumes to an equivalent flow rate under ideal conditions:

$$v_p = \frac{V}{PHF^*N^*f_{HV}^*f_p} \quad (14\text{-}1)$$

where: v_p = demand flow rate per lane under equivalent ideal conditions, pc/h/ln

PHF = peak-hour factor

N = number of lanes (in one direction) on the facility

f_{HV} = adjustment factor for presence of heavy vehicles

f_p = adjustment factor for presence of occasional or non-familiar users of a facility

This result is used to enter either the standard speed-flow curves of Figure 14.2 (freeways) or 14.3 (multilane highways). Using the appropriate free-flow speed, the curves may be

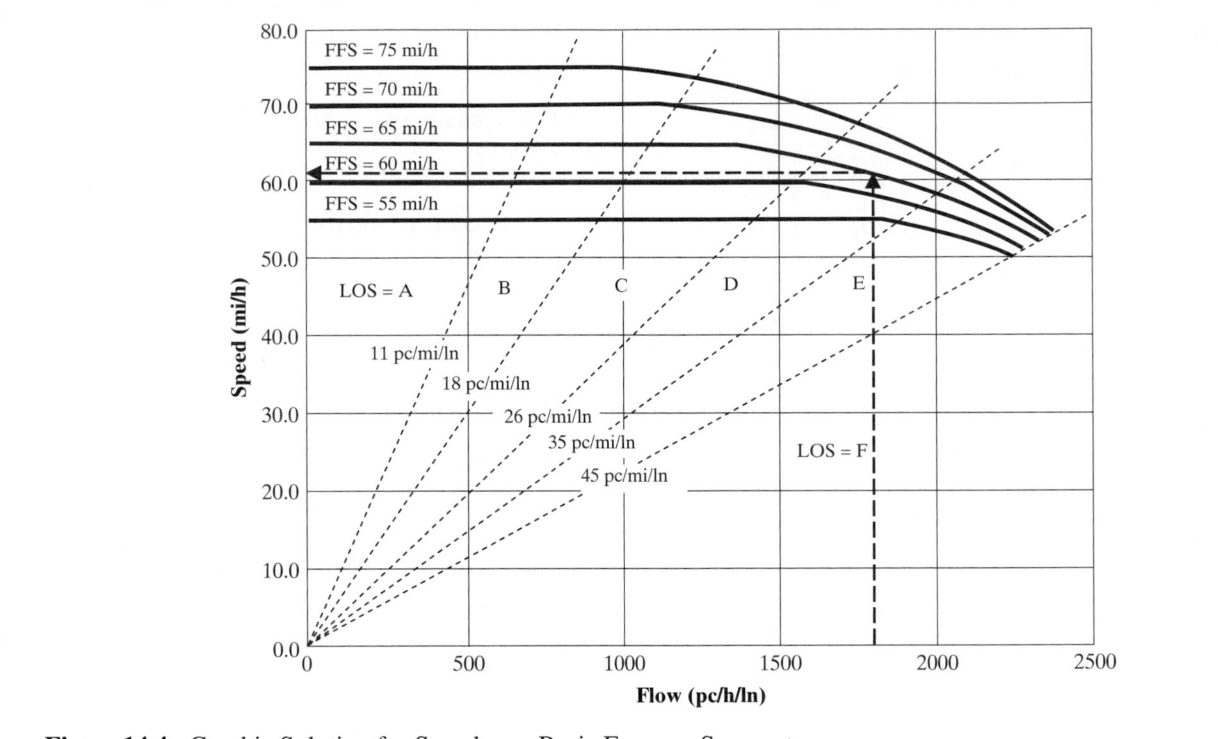

Figure 14.4: Graphic Solution for Speed on a Basic Freeway Segment

entered on the x-axis with the demand flow rate, v_p, to determine the level of service and the expected average speed. This technique is illustrated in Figure 14.4.

In the example shown, an adjusted demand flow (v_p) of 1,800 pc/h/ln is traveling on a freeway with a free-flow speed of 65 mi/h. For this condition, the expected speed is determined as 62 mi/h. Density may then be estimated as the flow rate divided by speed, in this case, 1,800/62 = 29.0 pc/mi/ln. Level of service may be determined on the basis of the computed density, or by examination of the curves: the intersection of 1,800 pc/h/ln with a 65 mi/h free-flow speed obviously falls within the range of level of service D.

Because this is a freeway example, the equations of Table 14.1 could be used to obtain a more precise result. Using the equation cited for a 65-mi/h freeway, with a demand flow rate above 1,400 pc/h/ln:

$$S = 0.00001418(1800 - 1400)^2 = 62.7 \text{ mi/h}$$

In this case, the graphic solution produced a result close to the computed value.

Methods for determining the free-flow speed and adjustment factors for heavy vehicles and driver population are presented in later sections.

Service Flow Rate and Service Volume Analysis

It is often useful to determine the service flow rates and service volumes for the various levels of service under prevailing conditions. Various demand levels may then be compared to these estimates for a speedy determination of expected level of service. The service flow rate for a given level of service is computed as:

$$SF_i = MSF_i{*}N{*}f_{HV}{*}f_p \qquad (14\text{-}2)$$

where: SF_i = service flow rate for level of service "i," veh/h

MSF_i = maximum service flow rate for level of service "i," pc/h/ln

N, f_{HV}, f_p as previously defined

The maximum service flow rates for each level of service, MSF_i are taken from Table 14.3 (for freeways) and Table 14.4 (for multilane highways). The tables are entered with the appropriate free-flow speed.

Service flow rates are stated in terms of peak flows within the peak hour, usually for a 15-minute analysis period. It is often convenient to convert service flow rates to service volumes over the full peak hour. This is done using the peak-hour factor:

$$SV_i = SF_i{*}PHF \qquad (14\text{-}3)$$

where: SV_i = service volume over a full peak hour for level of service "i"

SF_i, PHF as previously defined

Design Analysis

In design analysis, an existing or forecast demand volume is used to determine the number of lanes needed to provide for a specified level of service. The number of lanes may be computed as:

$$N_i = \frac{DDHV}{PHF{*}MSF_i{*}f_{HV}{*}f_p} \quad (14\text{-}4)$$

where: N_i = number of lanes (in one direction) required to provide level of service "i"

$DDHV$ = directional design hour volume, veh/h

MSF_i, f_{HV}, f_p as previously defined

Equation 14-4, however, will almost always result in a fractional answer. If the equation indicates that 3.1 lanes are needed to provide LOS i, then 4 lanes will have to be provided. Because of this, it is often more convenient to compute the service flow rate and service volume for the desired level of service for a range of reasonable values of N (usually 2, 3, 4, and possibly 5 lanes). Then the demand volume or flow rate can be compared to the results for a simpler determination of the required number of lanes.

14.3.2 Determining the Free-Flow Speed

The free-flow speed of a facility is best determined by field measurement. Given the shape of speed-flow relationships for freeways and multilane highways, an average speed measured when flow is less than or equal to 1,000 veh/h/ln may be taken to represent the free-flow speed.

It is not always possible, however, to measure the free-flow speed. When new facilities or redesigned facilities are under consideration, it is not possible to measure free-flow speeds. Even for existing facilities, the time and cost of conducting field studies may not be warranted. For such cases, models have been developed that allow the analyst to estimate the free-flow speed based on characteristics of the segment under study.

Freeways

The free-flow speed of a freeway can be estimated as:

$$FFS = 75.4 - f_{LW} - f_{LC} - 3.22\ TRD^{0.84} \quad (14\text{-}5)$$

where: FFS = free-flow speed of the freeway, mi/h

f_{LW} = adjustment for lane width, mi/h

f_{LC} = adjustment for right-side lateral clearance, mi/h

TRD = total ramp density, ramps/mi

Lane Width Adjustment The base condition for lane width is an average width of 12 feet or greater. For narrower lanes, the base free-flow speed is reduced by the factors shown in Table 14.5.

Table 14.5: Adjustment to Free-Flow Speed for Lane Width on a Freeway

Lane Width (ft)	Reduction in Free-Flow Speed, f_{LW} (mi/h)
≥12	0.0
11	1.9
10	6.6

(*Source:* Used with permission of Transportation Research Board, National Research Council, *Highway Capacity Manual*, December 2000, Exhibit 23-4, p. 23-6.)

Lateral Clearance Adjustment Base lateral clearance is 6 feet or greater on the right side and 2 feet or greater on the median, or left, side of the basic freeway section. Adjustments for right-side lateral clearances less than 6 feet are given in Table 14.6. There are no adjustments provided for median clearances less than 2 feet because such conditions are considered rare.

Table 14.6: Adjustment to Free-Flow Speed for Lateral Clearance on a Freeway

Right Shoulder Lateral Clearance (ft)	Reduction in Free-Flow Speed, f_{LC} (mi/h)			
	Lanes in One Direction			
	2	3	4	≥5
≥6	0.0	0.0	0.0	0.0
5	0.6	0.4	0.2	0.1
4	1.2	0.8	0.4	0.2
3	1.8	1.2	0.6	0.3
2	2.4	1.6	0.8	0.4
1	2.0	2.0	1.0	0.5
0	3.6	2.4	1.2	0.6

(*Source:* Used with permission of Transportation Research Board, National Research Council, *Highway Capacity Manual*, December 2000, Exhibit 23-5, p. 23-6.)

Care should be taken in assessing whether an "obstruction" exists on the right side of the freeway. Obstructions may be continuous, such as a guardrail or retaining wall, or they may be periodic, such as light supports and bridge abutments. In some cases, drivers may become accustomed to some obstructions, and the impact of these on free-flow speeds may be minimal.

Right-side obstructions primarily influence driver behavior in the right lane. Drivers "shy away" from such obstructions, moving further to the left in the lane. Drivers in adjacent lanes may also shift somewhat to the left in response to vehicle placements in the right lane. The overall affect is to cause vehicles to travel closer to each other laterally than would normally be the case, thus making flow less efficient. This is the same affect as for narrow lanes. Because the primary impact is on the right lane, the total impact on free-flow speed declines as the number of lanes increases.

Total Ramp Density (TRD) Total ramp density is the total number of on-ramps and off-ramps within ±3 miles of the midpoint of the study segment, divided by 6 miles. Ramp density is a surrogate measure that relates to the intensity of land use activity in the vicinity of the study segment. In practical terms, drivers drive at lower speeds where there are frequent on- and off-ramps creating turbulence in the traffic stream.

Multilane Highways

The free-flow speed for a multilane highway may be estimated as:

$$FFS = BFFS - f_{LW} - f_{LC} - f_M - f_A \tag{14-6}$$

where: FFS = free-flow speed of the multilane highway, mi/h
$BFFS$ = base free-flow speed (as discussed below)
f_{LW} = adjustment for lane width, mi/h
f_{LC} = adjustment for lateral clearance, mi/h
f_M = adjustment for type of median, mi/h
f_A = adjustment for access points, mi/h

A base free-flow speed of 60 mi/h may be used for rural and suburban multilane highways, if no field data are available. It may also be estimated using the posted speed limit. The base free-flow speed is approximately 7 mi/h higher than the posted speed limit, for speed limits of 40 and 45 mi/h. For speed limits of 50 and 55 mi/h, the base free-flow speed is approximately 5 mi/h higher than the limit.

Lane Width Adjustment The base lane width for multilane highways is 12 feet, as was the case for freeways. For narrower lanes, the free-flow speed is reduced by the values shown in Table 14.5. This adjustment is the same for multilane highways as for freeways.

Lateral Clearance Adjustment For multilane highways, this adjustment is based on the *total lateral clearance,* which is the sum of the lateral clearances on the right side of the roadway and on the left (median) side of the roadway. Although this seems like a simple concept, there are some details that must be observed:

- A lateral clearance of 6 feet is the base condition. Thus, no right- or left-side lateral clearance is ever taken to be greater than 6 feet, even if greater clearance physically exists. Thus the base total lateral clearance is 12 feet (6 feet for the right side, 6 feet for the left or median side).
- For an undivided multilane highway, there is no left- or median-side lateral clearance. However, there is a separate adjustment taken for type of median, including the undivided case. To avoid double-counting the impact of an undivided highway, the left or median lateral clearance on an undivided highway is assumed to be 6 feet.
- For multilane highways with two-way left-turn lanes, the left or median lateral clearance is also taken as 6 feet.
- For a divided multilane highway, the left- or median-side lateral clearance may be based on the location of a median barrier, periodic objects (light standards, abutments, etc.) in the median, or the distance to the opposing traffic lane. As noted previously, the maximum value is 6 feet.

The adjustments to free-flow speed for total lateral clearance on a multilane highway are shown in Table 14.7.

Median-Type Adjustment The median-type adjustment is shown in Table 14.8. A reduction of 1.6 mi/h is made for undivided configurations, whereas divided multilane highways, or multilane highways with two-way left-turn lanes, represent base conditions.

Access-Point Density Adjustment A critical adjustment to base free-flow speed is related to access-point density. Access-point density is the average number of unsignalized driveways or roadways per mile that provide access to the multilane highway on *the right* side of the roadway (for the subject direction of traffic).

Driveways or other entrances with little traffic, or that, for other reasons, do not affect driver behavior, should not be included in the access-point density. Adjustments are shown in Table 14.9.

Table 14.7: Adjustment to Free-Flow Speed for Total Lateral Clearance on a Multilane Highway

4-Lane Multilane Highways		**6-Lane Multilane Highways**	
Total Lateral Clearance (ft)	**Reduction in Free-Flow Speed, f_{LC} (mi/h)**	**Total Lateral Clearance (ft)**	**Reduction in Free-Flow Speed, f_{LC} (mi/h)**
≥12	0.0	≥12	0.0
10	0.4	10	0.4
8	0.9	8	0.9
6	1.3	6	1.3
4	1.8	4	1.7
2	3.6	2	2.8
0	5.4	0	3.9

(*Source:* Used with permission of Transportation Research Board, National Research Council, *Highway Capacity Manual*, December 2000, Exhibit 21-5, p. 21-6.)

Table 14.8: Adjustment to Free-Flow Speed for Median Type on Multilane Highways

Median Type	**Reduction in Free-Flow Speed, f_M (mi/h)**
Undivided	1.6
TWLTLs	0.0
Divided	0.0

(*Source:* Used with permission of Transportation Research Board, National Research Council, *Highway Capacity Manual*, December 2000, Exhibit 21-6, p. 21-6.)

Table 14.9: Adjustment to Free-Flow Speed for Access-Point Density on a Multilane Highway

Access Density (access Points/mi)	**Reduction in Free-Flow Speed, f_A (mi/h)**
0	0.0
10	2.5
20	5.0
30	7.5
≥40	10.0

(*Source:* Used with permission of Transportation Research Board, National Research Council, *Highway Capacity Manual*, December 2000, Exhibit 21-7, p. 21-7.)

Sample Problems in Free-Flow Speed Estimation

Example 14-1: An Urban Freeway

An old 6-lane urban freeway has the following characteristics: 11-foot lanes; frequent roadside obstructions located 2 feet from the right pavement edge; and a total ramp density of 3 ramps/mile. What is the free-flow speed of this freeway?

Solution: The free-flow speed of a freeway may be estimated using Equation 12-5:

$$FFS = 75.4 - f_{LW} - f_{LC} - 3.22 TRD^{0.84}$$

The following values are used in this computation:

f_{LW} = 1.9 mi/h (Table 14.5, 11-ft lanes)

f_{LC} = 1.6 mi/h (Table 14.6, 2-ft lateral clearance, 3 lanes)

TRD = 3 ramps/mi

Then:

$$FFS = 75.4 - 1.9 - 1.6 - 3.22\,(3^{0.84})$$
$$= 75.4 - 11.6 = 63.8 \text{ mi/h}$$

Example 14-2: A Four-Lane Suburban Multilane Highway

A four-lane undivided multilane highway in a suburban area has the following characteristics: posted speed limit = 50 mi/h; 11-foot lanes; frequent obstructions located 4 feet from the right pavement edge; 30 access points/mi on the right side of the facility. What is the free-flow speed for the direction described?

Solution: The free-flow speed for a multilane highway is computed using Equation 14-6:

$$FFS = BFFS - f_{LW} - f_{LC} - f_M - f_A$$

The base free-flow speed for a multilane highway may be taken as 60 mi/h as a default or may be related to the posted speed limit. In the latter case, for a posted speed limit of 50 mi/h, the base free-flow speed may be taken to be 5 mi/h more than the limit, or 50 + 5 = 55 mi/h. This is the value that will be used.

Adjustments to the base free-flow speed are as follows:

f_{LW} = 1.9 mi/h (Table 14.5, 11-ft lanes)

f_{LC} = 0.4 mi/h (Table 14.7, total lateral clearance = 10 ft, 4-lane highway)

f_M = 1.6 mi/h (Table 14.8, undivided highway)

f_A = 7.5 mi/h (Table 14.9, 30 access points/mi)

Then:

$$FFS = 55.0 - 1.9 - 0.4 - 1.6 - 7.5 = 43.6 \text{ mi/h}$$

Note that in selecting the adjustment for lateral clearance, the total lateral clearance is 4 feet (for the right side) plus an assumed value of 6.0 feet (for the left or median side) of an undivided highway.

Choosing a Free-Flow Speed Curve

Measured or estimated free-flow speeds can yield results to varying degrees of accuracy. It is generally recommended that free-flow speeds be measured or predicted to the nearest 0.1 mi/h. Once the result has been obtained, the appropriate curve in Figure 14.2 (freeways) or 14.3 (multilane highways) must be selected. It is *not* recommended that analysts interpolate a speed-flow curve between those that have been calibrated as part of the methodology. Table 14.10 provides guidelines for selecting an appropriate curve based on the measured or estimated value of *FFS*.

14.3.3 Determining the Heavy-Vehicle Factor

The principal adjustment to demand volume is the heavy-vehicle factor, which adjusts for the presence of heavy vehicles in the traffic stream. A *heavy vehicle* is defined as any vehicle with more than four tires touching the pavement during normal operation. Two categories of heavy vehicle are used:

- Trucks and buses
- Recreational vehicles (RVs)

Table 14.10: Selecting a Speed-Flow Curve in Figures 14.2 and 14.3

Free-Flow Speed is: (mi/h)	Use Speed-Flow Curve for a *FFS* of: (mi/h)
≥72.5 < 77.5	75
≥67.5 < 72.5	70
≥62.5 < 67.5	65
≥57.5 < 62.5	60
≥52.5 < 57.5	55
≥47.5 < 52.5	50
≥42.5 < 47.5	45

Trucks and buses have similar characteristics and are placed in the same category for capacity analysis purposes. These are primarily commercial vehicles, with the exception of some privately owned small trucks. Trucks vary widely in size and characteristics and range from small panel and single-unit trucks to double-back tractor-trailer combination vehicles. Factors in the HCM 2000 are based on a typical mix of trucks with an average weight-to-horsepower ratio of approximately 150:1 (150 lbs/hp).

Recreational vehicles also vary in size and characteristics. Unlike trucks and buses, which are primarily commercial vehicles operated by professional drivers, RVs are mostly privately owned and operated by drivers not specifically trained in their use and who often make only occasional trips in them. RVs include self-contained motor homes and a variety of trailer types hauled by a passenger car, SUV, or small truck. RVs generally have better operating conditions than trucks or buses, and they have typical weight-to-horsepower ratios in the range of 75 to 100 lbs/hp.

The effect of heavy vehicles on uninterrupted multilane flow is the same for both freeways and multilane highways. Thus the procedures described in this section apply to both types of facility.

The Concept of Passenger-Car Equivalents and Their Relationship to the Heavy-Vehicle Adjustment Factor

The heavy-vehicle adjustment factor is based on the concept of passenger-car equivalents. A *passenger-car equivalent* is the number of passenger cars displaced by one truck, bus, or RV in a given traffic stream under prevailing conditions. Given that two categories of heavy vehicle are used, two passenger-car equivalent values are defined:

E_T = passenger-car equivalent for trucks and buses in the traffic stream under prevailing conditions

E_R = passenger-car equivalent for RV's in the traffic stream under prevailing conditions

The relationship between these equivalents and the heavy-vehicle adjustment factor is best illustrated by example: Consider a traffic stream of 1,000 veh/h, containing 10% trucks and 2% RVs. Field studies indicate that for this particular traffic stream, each truck displaces 2.5 passenger cars (E_T) from the traffic stream, and each RV displaces 2.0 passenger cars (E_R) from the traffic stream. What is the total number of equivalent passenger cars/h in the traffic stream?

Note that from the passenger-car equivalent values, it is known that:

1 truck = 2.5 passenger cars
1 RV = 2.0 passenger cars

The number of equivalent passenger cars in the traffic stream is found by multiplying the number of each class of vehicle by its passenger-car equivalent, noting that the passenger-car equivalent of a passenger car is 1.0 by definition. Passenger-car equivalents are computed for each class of vehicle:

Trucks:	1,000*0.10*2.5 =	250 pce/h
RVs:	1,000*0.02*2.0 =	40 pce/h
Cars:	1,000*0.88*1.0 =	880 pce/h
TOTAL:		1,170 pce/h

Thus the prevailing traffic stream of 1,000 veh/h operates as if it contained 1,170 passenger cars per hour.

By definition, the heavy-vehicle adjustment factor, f_{HV} converts veh/h to pc/h when divided into the flow rate in veh/h. Thus:

$$V_{pce} = \frac{V_{vph}}{f_{HV}} \quad (14\text{-}7)$$

where: V_{pce} = flow rate, pce/h
V_{vph} = flow rate, veh/h

In the case of the illustrative computation:

$$1{,}170 = \frac{1{,}000}{f_{HV}}$$

$$f_{HV} = \frac{V_{vph}}{V_{pce}} = \frac{1{,}000}{1{,}170} = 0.8547$$

In the example, the number of equivalent passenger cars per hour for each vehicle type was computed by multiplying the total volume by the proportion of the vehicle type in the traffic stream and by the passenger-car equivalent for the appropriate vehicle type. The number of passenger-car equivalents in the traffic stream may be expressed as:

$$V_{pce} = (V_{vph}{*}P_T{*}E_T) + (V_{vph}{*}P_R{*}E_R) + (V_{vph}{*}(1 - P_T - P_R)) \quad (14\text{-}8)$$

where: P_T = proportion of trucks and buses in the traffic stream
P_R = proportion of RVs in the traffic stream
E_T = passenger-car equivalent for trucks and buses
E_R = passenger-car equivalent for RVs

The heavy-vehicle factor may now be stated as:

$$f_{HV} = \frac{V_{vph}}{V_{pce}}$$

$$= \frac{V_{vph}}{(V_{vph}{*}P_T{*}E_T) + (V_{vph}{*}P_R{*}E_R) + (V_{vph}{*}(1 - P_T - P_R))}$$

that may be simplified as:

$$f_{HV} = \frac{1}{1 + P_T\,(E_T - 1) + P_R\,(E_R - 1)} \quad (14\text{-}9)$$

For the illustrative computation:

$$f_{HV} = \frac{1}{1 + 0.10(2.5 - 1) + 0.02(2.0 - 1)}$$

$$= \frac{1}{1.170} = 0.8547$$

This, as expected, agrees with the original computation.

Passenger-Car Equivalents for Extended Freeway and Multilane Highway Sections

HCM 2000 specifies passenger-car equivalents for trucks and buses and RVs for extended sections of roadway in general terrain categories and for specific grade sections of significant impact.

A long section of roadway may be considered as a single extended section if no one grade of 3% or greater is longer than 0.25 miles, and if no grade of less than 3% is longer than 0.5 miles. Such general terrain sections are designated in one of three general terrain categories:

- *Level terrain.* Level terrain consists of short grades, generally less than 2% in severity. The combination of horizontal and vertical alignment permits trucks and other heavy vehicles to maintain the same speed as passenger cars in the traffic stream.
- *Rolling terrain.* Rolling terrain is any combination of horizontal and vertical alignment that causes trucks and other heavy vehicles to reduce their speeds substantially below those of passenger cars but does not require heavy vehicles to operate at crawl speed for extended distances. Crawl speed is defined as the minimum speed that a heavy vehicle can sustain on a given segment of highway.
- *Mountainous terrain.* Mountainous terrain is severe enough to cause heavy vehicles to operate at crawl speed either frequently or for extended distances.

Note that, in practical terms, mountainous terrain is a rare occurrence. It is difficult to have an extended section of highway that forces heavy vehicles to crawl speed frequently and/or for long distances without violating the limits for extended section analysis. Such situations usually involve longer and steeper grades that would require analysis as a specific grade.

Table 14.11 shows passenger-car equivalents for freeways and multilane highways on extended sections of general terrain.

In analyzing extended general sections, it is the alignment of the roadway itself that determines the type of terrain, not the topography of the surrounding landscape. Thus, for

Table 14.11: Passenger-Car Equivalents for Trucks, Buses, and RVs on Extended General Terrain Sections of Freeways or Multilane Highways

	Type of Terrain		
Factor	**Level**	**Rolling**	**Mountainous**
E_T	1.5	2.5	4.5
E_R	1.2	2.0	4.0

(*Source:* Used with permission of Transportation Research Board, National Research Council, *Highway Capacity Manual*, December 2000, Exhibit 23-8, p. 23-9.)

example, many urban freeways or multilane highways in a relatively level topography have a rolling terrain based on underpasses and overpasses at major cross streets. Further, because the definitions for each category depend on the operation of heavy vehicles, the classification may depend somewhat on the mix of heavy vehicles present in any given case.

Passenger-Car Equivalents for Specific Grades on Freeways and Multilane Highways

Any grade less than 3% that is longer than 0.50 miles and any grade of 3% or steeper that is longer than 0.25 miles must be considered a specific grade. This is because a long grade may have a significant impact on both heavy-vehicle operation and the characteristics of the entire traffic stream.

HCM 2000 (these will not change in HCM 2010) specifies passenger-car equivalents for:

- Trucks and buses on specific upgrades (Table 14.12)
- RVs on specific upgrades (Table 14.13)
- Trucks and buses on specific downgrades (Table 14.14)

The passenger-car equivalent for RVs on downgrade sections is taken to be the same as that for level terrain sections, or 1.2.

Over time, the operation of heavy vehicles has improved relative to passenger cars. Trucks in particular, now have considerably more power than in the past, primarily due to turbo-charged engines. Thus the maximum passenger-car equivalent shown in Tables 14.12 and 14.13 is 7.0. In the 1965 HCM, these values were as high as 17.0.

Tables 14.12 through 14.14 indicate the impact of heavy vehicles on the traffic stream. In the worst case, a single truck can displace as many as 7.0 to 7.5 passenger cars from the traffic stream. This displacement accounts for the both the size of heavy vehicles and the fact that they cannot maintain the same speed as passenger cars in many situations. The latter is a serious impact that often creates large gaps between heavy vehicles and passenger cars that cannot be continuously filled by passing maneuvers.

Note that in these tables, there are some consistent trends. Obviously, as the grades get steeper and/or longer (either upgrade or downgrade), the passenger-car equivalents increase, indicating a harsher impact on the operation of the mixed traffic stream.

A less obvious trend is that the passenger-car equivalent in any given situation decreases as the proportion of trucks, buses, and RVs increases. Remember that the values given in Tables 14.12 through 14.14 are passenger-car equivalents (i.e., the number of passenger cars displaced by *one* truck, bus, or RV). The maximum impact of a single heavy vehicle is

Table 14.12: Passenger-Car Equivalents for Trucks and Buses on Upgrades

Upgrade (%)	Length (mi)	E_T								
		Percentage of Trucks and Buses (%)								
		2	**4**	**5**	**6**	**8**	**10**	**15**	**20**	**≥25**
< 2	All	1.5	1.5	1.5	1.5	1.5	1.5	1.5	1.5	1.5
>2–3	0.00–0.25	1.5	1.5	1.5	1.5	1.5	1.5	1.5	1.5	1.5
	>0.25–0.50	1.5	1.5	1.5	1.5	1.5	1.5	1.5	1.5	1.5
	>0.50–0.75	1.5	1.5	1.5	1.5	1.5	1.5	1.5	1.5	1.5
	>0.75–1.00	2.0	2.0	2.0	2.0	1.5	1.5	1.5	1.5	1.5
	>1.00–1.50	2.5	2.5	2.5	2.5	2.0	2.0	2.0	2.0	2.0
	>1.50	3.0	3.0	2.5	2.5	2.0	2.0	2.0	2.0	2.0
>3–4	0.00–0.25	1.5	1.5	1.5	1.5	1.5	1.5	1.5	1.5	1.5
	>0.25–0.50	2.0	2.0	2.0	2.0	2.0	2.0	1.5	1.5	1.5
	>0.50–0.75	2.5	2.5	2.0	2.0	2.0	2.0	2.0	2.0	2.0
	>0.75–1.00	3.0	3.0	2.5	2.5	2.5	2.5	2.0	2.0	2.0
	>1.00–1.50	3.5	3.5	3.0	3.0	3.0	3.0	2.5	2.5	2.5
	>1.50	4.0	3.5	3.0	3.0	3.0	3.0	2.5	2.5	2.5
>4–5	0.00–0.25	1.5	1.5	1.5	1.5	1.5	1.5	1.5	1.5	1.5
	>0.25–0.50	3.0	2.5	2.5	2.5	2.0	2.0	2.0	2.0	2.0
	>050–0.75	3.5	3.0	3.0	3.0	2.5	2.5	2.5	2.5	2.5
	>0.75–1.00	4.0	3.5	3.5	3.5	3.0	3.0	3.0	3.0	3.0
	>1.00	5.0	4.0	4.0	4.0	3.5	2.5	3.0	3.0	3.0
>5–6	0.00–0.25	2.0	2.0	1.5	1.5	1.5	1.5	1.5	1.5	1.5
	>0.25–0.30	4.0	3.0	2.5	2.5	2.0	2.0	2.0	2.0	2.0
	>0.30–0.50	4.5	4.0	3.5	3.0	2.5	2.5	2.5	2.5	2.5
	>0.50–0.75	5.0	4.5	4.0	3.5	3.0	3.0	3.0	3.0	3.0
	>0.75–1.00	5.5	5.0	4.5	4.0	3.0	3.0	3.0	3.0	3.0
	>1.00	6.0	5.0	5.0	4.5	3.5	3.5	3.5	3.5	3.5
>6	0.00–0.25	4.0	3.0	2.5	2.5	2.5	2.5	2.0	2.0	2.0
	>0.25–0.30	4.5	4.0	3.5	3.5	3.5	3.0	2.5	2.5	2.5
	>0.30–0.50	5.0	4.5	4.0	4.0	3.5	3.0	2.5	2.5	2.5
	>0.50–0.75	5.5	5.0	4.5	4.5	4.0	3.5	3.0	3.0	3.0
	>0.75–1.00	6.0	5.5	5.0	5.0	4.5	4.0	3.5	3.5	3.5
	>1.00	7.0	6.0	5.5	5.5	5.0	4.5	4.0	4.0	4.0

(*Source:* Used with permission of Transportation Research Board, National Research Council, *Highway Capacity Manual*, December 2000, Exhibit 29-8, p. 23-10.)

when it is relatively isolated in the traffic stream; as the flow of heavy vehicles increases, they begin to form their own platoons, within which they can operate more efficiently. The cumulative impact, however, of more heavy vehicles is a reduction in operating quality.

In some cases, the downgrade impact of a heavy vehicle is worse than the same heavy-vehicle situation on a similar upgrade. Downgrade impacts depend on whether or not trucks and other heavy vehicles must shift to low gear to avoid losing control of the vehicle. This is a particular problem for trucks on downgrades steeper than 4%; for lesser downgrades, values of E_T for level terrain are used.

Composite Grades

The passenger-car equivalents given in Tables 14.12 through 14.14 are based on a constant grade of known length. In most situations, however, highway alignment leads to composite

Table 14.13: Passenger-Car Equivalents for RVs on Upgrades

Grade (%)	Length (mi)	E_R								
		Percentage of RVs (%)								
		2	**4**	**5**	**6**	**8**	**10**	**15**	**20**	**≥25**
≤2	All	1.2	1.2	1.2	1.2	1.2	1.2	1.2	1.2	1.2
>2–3	0.00–0.50	1.2	1.2	1.2	1.2	1.2	1.2	1.2	1.2	1.2
	>0.50	3.0	1.5	1.5	1.5	1.5	1.5	1.2	1.2	1.2
>3–4	0.00–0.25	1.2	1.2	1.2	1.2	1.2	1.2	1.2	1.2	1.2
	>0.25–0.50	2.5	2.5	2.0	2.0	2.0	2.0	1.5	1.5	1.5
	>0.50	3.0	2.5	2.5	2.5	2.0	2.0	2.0	1.5	1.5
>4–5	0.00–0.25	2.5	2.0	2.0	2.0	1.5	1.5	1.5	1.5	1.5
	>0.25–0.50	4.0	3.0	3.0	3.0	2.5	2.5	2.0	2.0	2.0
	>0.50	4.5	3.5	3.0	3.0	3.0	2.5	2.5	2.0	2.0
>5	0.00–0.25	4.0	3.0	2.5	2.5	2.5	2.5	2.0	2.0	1.5
	>0.25–50	6.0	4.0	4.0	4.0	3.5	3.0	2.5	2.5	2.0
	>0.50	6.0	4.5	4.0	4.0	4.0	3.5	3.0	2.5	2.0

(*Source:* Used with permission of Transportation Research Board, National Research Council, *Highway Capacity Manual*, December 2000, Exhibit 23-10, p. 23-10.)

Table 14.14: Passenger-Car Equivalents for Trucks and Buses on Downgrades

Downgrade (%)	Length (mi)	E_T			
		Percentage Trucks and Buses (%)			
		5	**10**	**15**	**≥20**
< 4	All	1.5	1.5	1.5	1.5
≥4–5	≤4	1.5	1.5	1.5	1.5
	>4	2.0	2.0	2.0	1.5
>5–6	≤4	1.5	1.5	1.5	1.5
	>4	5.5	4.0	4.0	3.0
>6	≤4	1.5	1.5	1.5	1.5
	>4	7.5	6.0	5.5	4.5

(*Source:* Used with permission of Transportation Research Board, National Research Council, *Highway Capacity Manual*, December 2000, Exhibit 23-11, p. 23-11.)

grades (i.e., a series of upgrades and/or downgrades of varying steepness). In such cases, an equivalent uniform grade must be used to determine the appropriate passenger-car equivalent values.

Consider the following composite grade profile: 3,000 feet of 3% upgrade followed by 5,000 feet of 5% upgrade. What heavy vehicle equivalents should be used in this case?

The general approach is to find a uniform upgrade, 8,000 feet in length, which has the same impact on the traffic stream as the composite described.

Average Grade Technique One approach to this problem is to find the average grade over the 8,000-foot length of the composite grade. This involves finding the total rise in the composite profile, as follows:

Rise on 3% Grade: 3,000*0.03 = 90 ft

Rise on 5% Grade: 5,000*0.05 = 250 ft

TOTAL RISE ON COMPOSITE: 340 ft

The average grade is then computed as the total rise divided by the total length of the grade, or:

$$G_{AV} = \frac{340}{8{,}000} = 0.0425\, or 4.25\%$$

The appropriate values would now be entered to find passenger-car equivalents using a 4.25% grade, 8,000 feet (1.52 miles) in length.

The average grade technique is a good approximation when all subsections of the grade are less than 4%, or when the total length of the composite grade is less than 4,000 feet. Note that for the example given, this is not the case. One portion of the grade is more than 4%, and the total length of the grade more than 4,000 feet.

Composite Grade Technique For more severe grades, a more exact technique is used. In this procedure, a percent grade of 8,000 feet is found that results in the same final operating speed of trucks as the composite described. This is essentially a graphic technique and requires a set of grade performance curves for a "typical" truck, with a weight-to-horsepower ratio of 200. Figure 14.5 shows these performance curves, and Figure 14.6 illustrates their use to find the equivalent grade for the example previously solved using the average-grade technique.

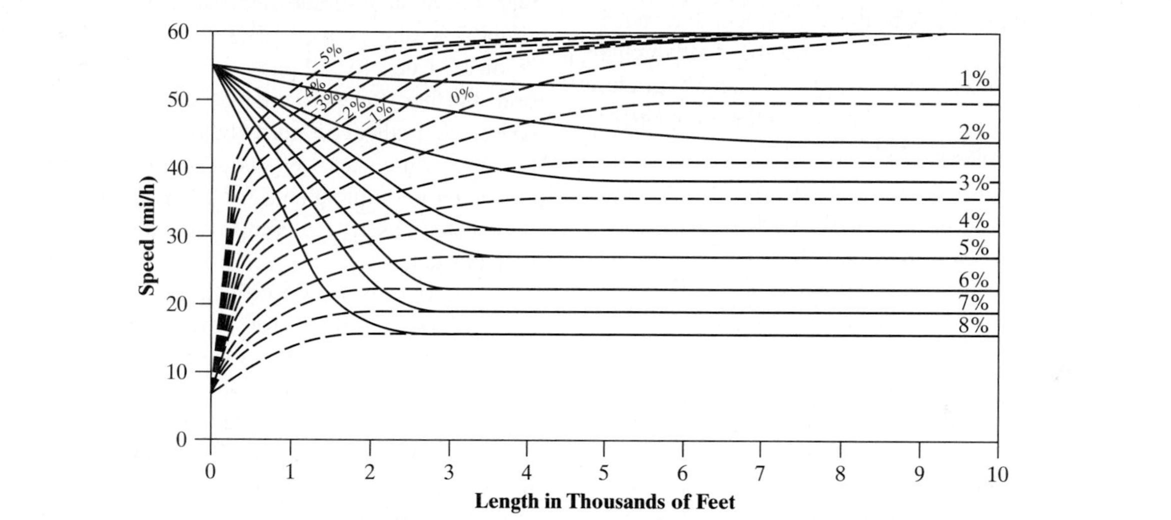

Figure 14.5: Performance of a Typical Truck on Grades

(*Source:* Used with permission of Transportation Research Board, National Research Council, *Highway Capacity Manual*, December 2000, Exhibit A23-2, p. 23-30.)

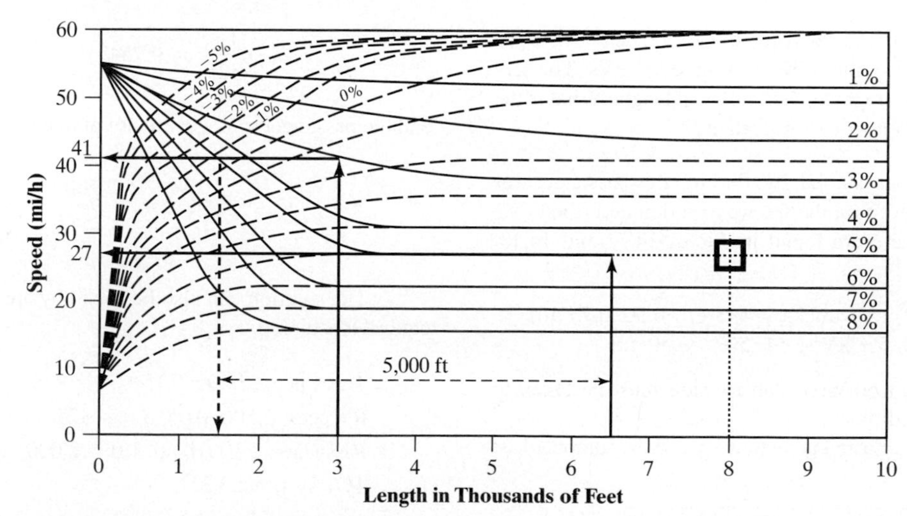

Figure 14.6: Composite Grade Solution for Example

The following steps are followed in Figure 14.6 to find the composite grade equivalent:

- The curves are entered on the x-axis at 3,000 feet, the length of the first portion of the grade. The intersection of a vertical line constructed at this point with the 3% grade curve is found.
- The intersection of 3,000 feet with a 3% grade is projected horizontally back to the y-axis, where a speed of 41 mi/h is determined. This is the speed at which a typical truck is traveling after 3,000 ft on a 3% grade. It is also, however, the speed at which the truck *enters* the 5% segment of the grade.
- The intersection of 41 mi/h with the 5% grade curve is found and projected vertically back to the x-axis. The "length" found is approximately 1,350 ft. Thus, when the truck enters the 5% grade after 3,000 feet of 3% grade, it is as if the truck were on the 5% grade for 1,350 feet.
- The truck will now travel another 5,000 ft on the 5% grade—to an equivalent length of 1350 + 5000 = 6350 ft. A vertical line is constructed at this point, and the intersection with the 5% curve is found. When projected horizontally to the y-axis, it is seen that a typical truck will be traveling at 27 mi/h at the end of the composite grade as described.
- To find the equivalent grade, the intersection of 27 mi/h and the length of the composite grade (8,000 feet) is found. This is the solution point, indicating that the equivalent grade is a 5% grade of 8,000 feet.

Although this methodology is considered "more" exact than the average-grade approach, it embodies a number of simplifications as well. The selection of 200 lbs/hp as the "typical" truck for all cases is certainly one such simplification. Further, the performance curves assume that the truck enters the grade at 55 mi/h and never accelerates to more than 60 mi/h, which are very conservative assumptions. Finally, passenger-car equivalents for all types of heavy vehicles will be selected based on a composite grade equivalent based on a typical truck.

Despite these simplifications, this is viewed as a more appropriate technique for severe grade profiles than the average-grade technique. A composite grade equivalent can be found using this technique for any number of subsections and may include upgrade and downgrade segments.

It should also be noted that for analysis purposes, the impact of a grade is worst at the end of its steepest (uphill) section. Thus, if 1,000 feet of 4% grade were followed by 1,000 feet of 3% grade, passenger-car equivalents would be found for a 1,000 feet, 4% grade.

Example 14-3: Determination and Use of the Heavy-Vehicle Adjustment Factor

Consider the following situation: A volume of 2,500 veh/h traverses a section of freeway and contains 15% trucks and 5% RVs. The section in question is on a 5% upgrade, 0.75 miles in length. What is the equivalent volume in passenger-car equivalents?

Solution: The solution is started by finding the passenger car equivalent of trucks and RVs on the freeway section described (5% upgrade, 0.75 miles). These are found in Tables 14.12 and 14.13, respectively:

$E_T = 2.5$ (Table 14.12, 15% trucks, >4–5%, >0.50–0.75 mi)
$E_R = 3.0$ (Table 14.13, 5% RV's, >4–5%, >0.50 mi)

In entering values from these tables, care must be taken to observe the boundary conditions.

The heavy-vehicle adjustment factor may now be computed as:

$$f_{HV} = \frac{1}{1 + 0.15(2.5 - 1) + 0.05(3.0 - 1)}$$

$$= \frac{1}{1.325} = 0.7547$$

and the passenger-car equivalent volume may be estimated as:

$$V_{pce} = \frac{V_{vph}}{f_{HV}} = \frac{2{,}500}{0.7547} = 3{,}313 \text{ pc/h}$$

The solution can also be found by applying the passenger-car equivalents directly:

Truck pces: 2,500*0.15*2.5 = 938
RV pces: 2,500*0.05*3.0 = 375
Pass Cars: 2,500*0.80*1.0 = 2,000
TOTAL pces: 3,313

14.3.4 Determining the Driver Population Factor

The base procedures for freeways and multilane highways assume a driver population of commuters or drivers familiar with the roadway and its characteristics. On some recreational routes, the majority of drivers may not be familiar with the route. This can have a significant impact on operations.

This adjustment factor is not well defined and depends on local conditions. In general, the factor ranges between a value of 1.00 (for commuter traffic streams) to 0.85 as a lower limit for other driver populations. Unless specific evidence for a lower value is available, a value of 1.00 is generally used in analysis. Where unfamiliar users dominate a route, field studies comparing their characteristics to those of commuters are suggested to obtain a better estimate of this factor. Where a future situation is being analyzed, and recreational users dominate the driver population, a value of 0.85 is suggested because it represents a "worst-case" scenario.

14.4 Sample Problems

Example 14-4: Analysis of an Older Urban Freeway

Figure 14.7 shows a section of an old freeway in New York City. It is a four-lane freeway (additional service roads are shown in the picture) with the following characteristics:

- Ten-foot travel lanes
- Lateral obstructions at 0 feet at the roadside
- Total ramp density is 4.5 ramps/mile
- Rolling terrain

The roadway has a current peak demand volume of 3,500 veh/h. The peak-hour factor is 0.95, and there are no trucks, buses, or RVs in the traffic stream because the roadway is classified as a parkway and such vehicles are prohibited. At what level of service will the freeway operate during its peak period of demand?

Solution:

Step 1: *Determine the Free-Flow Speed of the Freeway* The free-flow speed of the freeway is estimated using Equation 14-5 as:

$$FFS = 75.4 - f_{LW} - f_{LC} - 3.22\ TRD^{0.84}$$

where: f_{LW} = 6.6 mi/h (Table 14.5, 10-ft lanes)

f_{LC} = 3.6 mi/h (Table 14.6, 2 lanes, 0-ft obstructions)

TRD = 4.5 ramps/mi

Thus:

$$FFS = 75.4 - 6.6 - 3.6 - 3.22(4.5^{0.84}) = 53.8 \text{ mi/h}$$

Because this free-flow speed lies between 52.5 and 57.5 mi/h (Table 14.10), the 55-mi/h speed-flow curve is used to represent base conditions.

Step 2: *Determine the Demand Flow Rate in Equivalent pce Under Base Conditions* The demand volume may be converted to an equivalent flow rate under base conditions using Equation 14-1:

$$v_p = \frac{V}{PHF^*N^*f_{HV}^*f_p}$$

where: V = 3,500 veh/h (given)

PHF = 0.95 (given)

N = 2 lanes (given)

f_{HV} = 1.00 (no trucks, buses, or RVs in the traffic stream)

f_p = 1.00 (assumed commuter driver population)

Then:

$$v_p = \frac{3500}{0.95^*2^*1.00^*1.00} = 1{,}842 \text{ pc/h/ln}$$

Step 3: *Find the Level of Service and the Speed and Density of the Traffic Stream* The demand flow of 1,842 veh/h is used to enter the 55-mi/h FFS curve of Figure 14.2 to find level of service and speed.

From Figure 14.8, a demand flow of 1,842 pc/h/ln on a freeway with a 55 - mi/h free-flow speed will result in an approximate operating speed of 54.9 mi/h. Even at the relatively high demand flow, the free-flow speed will be maintained. It can also be seen that the level of service for this freeway is D, just barely missing LOS E.

Figure 14.7: Freeway Segment for Sample Problem 1

(*Source:* Used with permission of Transportation Research Board, National Research Council, *Highway Capacity Manual*, December 2000, Illustration 12-1, p. 12-6.)

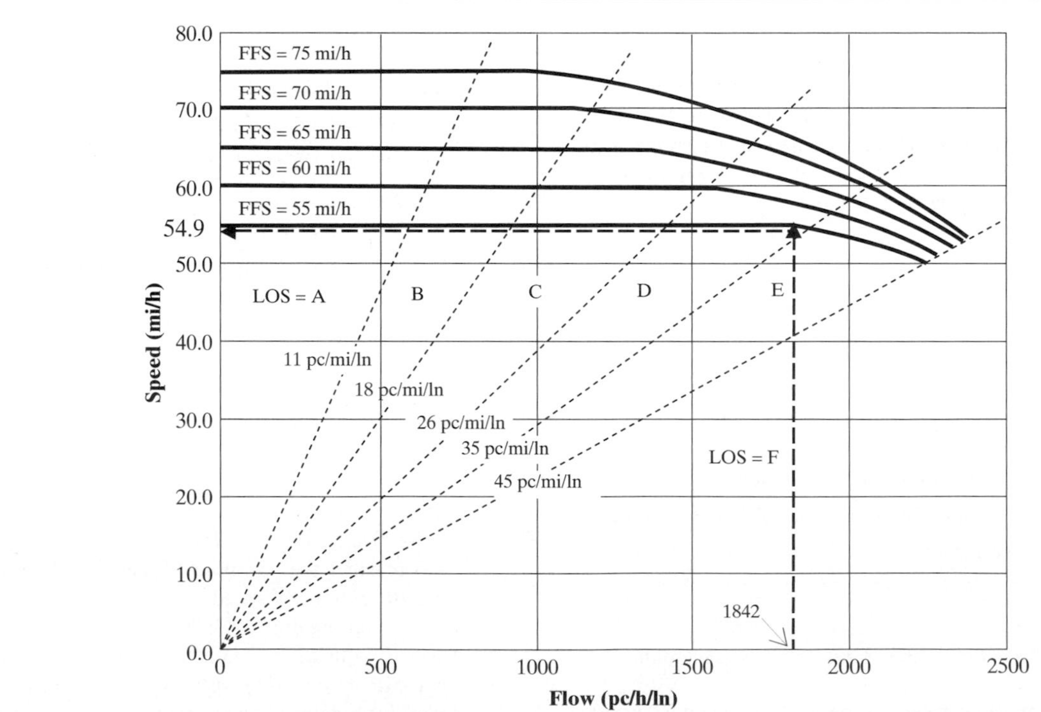

Figure 14.8: Speed Determination for Sample Problem 14.4

The density on the freeway may be estimated as the demand flow rate divided by the speed, or:

$$D = \frac{1842}{54.9} = 33.6 \text{ pc/mi/ln}$$

This value can be used to enter Table 14.2 to confirm the level of service, which is D, falling within the defined boundaries of 26–35 pc/mi/ln for LOS D.

Example 14-5: Analysis of a Multilane Highway Section

A four-lane multilane highway section with a full median carries a peak-hour volume of 2,600 veh/h in the heaviest direction. There are 12% trucks and 2% RVs in the traffic stream. Motorists are primarily regular users of the facility. The section under study is on a 3% sustained grade, 1 mile in length. The *PHF* is 0.88.

Field studies have been conducted to determine that free-flow speed of the facility is 55.0 mi/h.

At what level of service will this facility operate during the peak hour?

Solution: As the free-flow speed has been found from field data, it is not necessary to estimate it using Equation 14-6. The analysis section is a sustained grade. Because the peak volume would be expected to travel upgrade during one peak and downgrade during the other, it will be necessary to examine the downgrade as well as the upgrade under peak demand conditions.

Step 1: *Determine the Upgrade Demand Flow Rate in Equivalent pces Under Base Conditions* Equation 14-1 is used to convert the peak hour demand volume to an equivalent flow rate in pces under base conditions:

$$v_p = \frac{V}{PHF{*}N{*}f_{HV}{*}f_p}$$

where: V = 2,600 veh/h (given)

PHF = 0.88 (given)

N = 2 lanes (given)

f_p = 1.00 (regular users)

The heavy-vehicle factor, f_{HV} is computed using Equation 14-9:

$$f_{HV} = \frac{1}{1 + P_T(E_T - 1) + P_R(E_R - 1)}$$

where: P_T = 0.12 (given)

P_R = 0.02 (given)

E_T = 1.5 (Table 14.12, ≥2–3%, >0.75–1.00 mi, 12% trucks)

E_R = 3.0 (Table 14.13, >2–3%, >0.50 mi, 2% RVs)

Then:

$$f_{HV} = \frac{1}{1 + 0.12(1.5 - 1) + 0.02(3.0 - 1)} = \frac{1}{1.10} = 0.909$$

and:

$$v_p = \frac{2{,}600}{0.88{*}2{*}0.909{*}1.00} = 1{,}625 \text{ pc/h/ln}$$

Step 2: *Determine the Downgrade Demand Flow Rate in Equivalent pces Under Base Conditions* The downgrade computation follows the same procedure as the upgrade computation, except that the passenger-car equivalents for trucks and RVs are selected for the downgrade condition. These are found as follows:

E_T = 1.5 (Table 14.14, <4%, all lengths, 12% trucks)

E_R = 1.2 (Table 14.11, level terrain)

Note that passenger-car equivalents for downgrade RVs are found assuming level terrain. Then:

$$f_{HV} = \frac{1}{1 + 0.12(1.5 - 1) + 0.02(1.2 - 1)} = \frac{1}{1.064} = 0.940$$

$$v_p = \frac{2{,}600}{0.88{*}2{*}0.940{*}1.00} = 1{,}572 \text{ pc/h/ln}$$

Step 3: *Find the Level of Service and the Speed and Density of the Traffic Stream* Level of service and speed determinations are made using Figure 14.3 for multilane highways, as shown in Figure 14.9. Remember that the free-flow speed was field-measured as 55 mi/h.

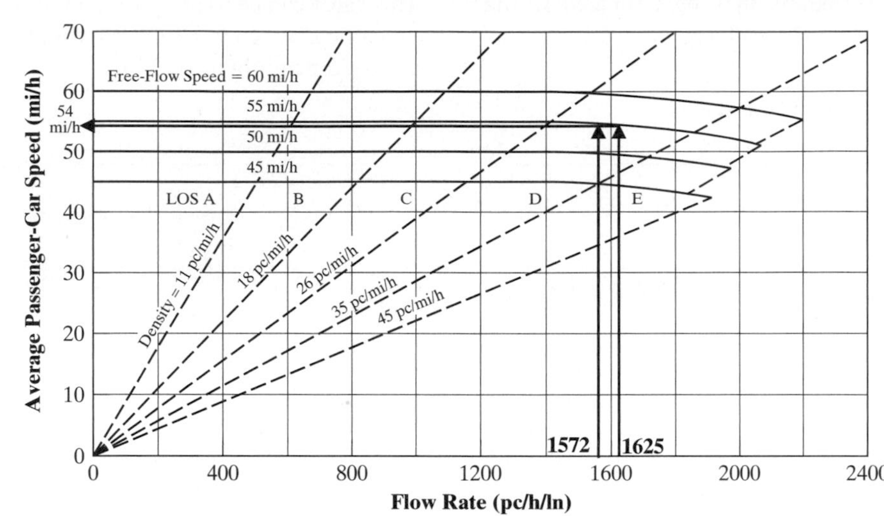

Figure 14.9: Estimating Speeds for Sample Problem 2

From Figure 14.9, it can be seen that the expected speeds for both the upgrade and downgrade sections are approximately 54 mi/h, although it might be argued that the upgrade speed is a fraction less than the downgrade speed. The scale of Figure 14.3 makes it difficult to estimate this small difference. The level of service for both upgrade and downgrade sections is expected to be D during periods of peak demand flow.

The density for upgrade and downgrade sections is estimated as the flow rate divided by the expected speed, or:

$$D_{up} = \frac{1625}{54} = 30.1 \text{ pc/mi/ln}$$

$$D_{dn} = \frac{1572}{54} = 29.1 \text{ pc/mi/ln}$$

both of which are between the limits defined for LOS D–26–35 pc/mi/ln.

Example 14-6: Finding Service Flow Rates and Service Volumes for a Freeway Segment

A six-lane urban freeway has the following characteristics: 12-foot lanes, 6-feet clearances on the right side of the roadway, rolling terrain, a ramp density of 2.8 ramps per mile, and a PHF of 0.92. The traffic consists of 8% trucks and no RVs, and all drivers are regular users of the facility.

The peak-hour volume on the facility is currently 3,600 veh/h, which is expected to grow at a rate of 6% a year for the next 20 years.

What is the current level of service on the facility, and what levels of service can be expected in 5 years? In 10 years? In 15 years? In 20 years?

Solution: These questions could be answered by conducting five separate operational analyses, determining the expected LOS for the various demand flows both now and expected in the future target years. It is often easier to solve a problem involving multiple demand levels by simply computing the service flow rates (SF) and service volumes (SV) for the section for each level of service. Then, demand volumes can be easily compared to the results to determine the LOS for each target demand level.

Determine the Free-Flow Speed of the Freeway:

Step 1: *The free-flow speed of the facility is found using Equation 14-5:*

$$FFS = 75.4 - f_{LW} - f_{LC} - 3.22\,TRD^{0.84}$$

where: f_{LW} = 0.0 mi/h (Table 14.5, 12-ft lanes)

f_{LC} = 0.0 mi/h (Table 16.6, 6-ft lateral clearance)

Then:

$$FFS = 75.4 - 0.0 - 0.0 - 3.22\,(2.8^{.84}) = 67.8 \text{ mi/h}$$

Because this value is between 67.5 and 72.5 mi/h, the speed-flow curve for 70 mi/h is used in this solution.

Table 14.15: Spreadsheet Computation of Service Flow Rates and Service Volumes

Level of Service	MSF (pc/h/ln)	N	f_{HV}	f_p	SF (veh/h)	PHF	SV (veh/h)
A	770	3	0.863	1.000	**1,994**	0.92	**1,834**
B	1,250	3	0.863	1.000	**3,236**	0.92	**2,977**
C	1,690	3	0.863	1.000	**4,375**	0.92	**4,301**
D	2,080	3	0.863	1.000	**5,385**	0.92	**4,954**
E	2,400	3	0.863	1.000	**6,214**	0.92	**5,617**

Step 2: *Determine the Maximum Service Flow Rates for Each Level of Service* Maximum service flow (MSF) rates for each level of service are drawn from Table 14.3 for a freeway with a 70-mi/h free-flow speed. These values are A: 770 pc/h/ln, B: 1,250 pc/h/ln, C: 1,690 pc/h/ln, D: 2,080 pc/h/ln, and E: 2,400 pc/h/ln.

Step 3: *Determine the Heavy-Vehicle Factor* The heavy-vehicle factor is computed as:

$$f_{HV} = \frac{1}{1 + P_T(E_T - 1) + P_R(E_R - 1)}$$

where: $P_T = 0.08$ (given)

$P_R = 0.00$ (given)

$E_T = 2.5$ (Exhibit 14-11, rolling terrain)

Then:

$$f_{HV} = \frac{1}{1 + 0.08(2.5 - 1)} = \frac{1}{1.12} = 0.893$$

Step 4: *Determine the Service Flow Rates and Service Volumes for Each Level of Service* Service flow rates and service volumes are computed using Equations 14-2 and 14-3:

$$SF_i = MSF_i * N * f_{HV} * f_p$$
$$SV_i = SF_i * PHF$$

where: MSF_i = as determined in Step 2

$N = 3$ (given)

$f_{HV} = 0.893$ (as computed in Step 3)

$f_p = 1.00$ (regular users)

$PHF = 0.92$ (given)

These computations are shown in Table 14.15.

The service flow rates (SF) refer to the peak 15-minute interval; service volumes apply to peak-hour volumes.

Step 5: *Determine Target-Year Peak-Demand Volumes* The problem statement indicates that present demand is 3,600 veh/h and that this volume will increase by 6% per year for the foreseeable future. Future demand volumes may be computed as:

$$V_j = V_o(1.06^n)$$

where: V_j = peak-hour demand volume in target year j

V_o = peak-hour demand volume in year 0, 3,600 veh/h

N = number of years to target year

Then:

$$V_o = 3{,}600 \text{ veh/h}$$
$$V_5 = 3{,}600(1.06^5) = 4{,}818 \text{ veh/h}$$
$$V_{10} = 3{,}600(1.06^{10}) = 6{,}447 \text{ veh/h}$$
$$V_{15} = 3{,}600(1.06^{15}) = 8{,}628 \text{ veh/h}$$
$$V_{20} = 3{,}600(1.06^{20}) = 11{,}546 \text{ veh/h}$$

Step 6: *Determine Target Year Levels of Service* The target year demand volumes are stated as full peak-hour volumes. They are, therefore, compared to the *service volumes* computed in Table 14.15 to determine LOS. The results are shown in Table 14.16.

As indicated in Table 14.16, level of service F prevails in target years 10, 15, and 20. In each of these years, demand exceeds capacity. Clearly, the point at which capacity is reached occurs between years 5 and 10. Capacity, stated in terms of a full peak hour, is 5,617 veh/h (Table 14.15). The exact year that demand reaches capacity may be found as follows:

$$5617 = 3600(1.06^n)$$
$$n = 7.63$$

Table 14.16: Levels of Service for Sample Problem

Target Year	Demand Volume (veh/h)	Level of Service
0	3,600	C
5	4,818	D
10	6,447	F
15	8,628	F
20	11,546	F

Analysis The results of this analysis indicate that demand will reach the capacity of the freeway in 7.63 years. If no action is taken, users can expect regular breakdowns during the peak hour in this freeway section. To avoid this situation, action must be taken to either reduce demand and/or increase the capacity of the section.

Increasing the capacity of the section suggests adding a lane. Computations would be redone using a four-lane, one-direction cross section to see whether sufficient capacity was added to handle the 20-year demand forecast. Reduction in demand is more difficult and would involve intensive study of the nature of demand on the freeway section in question. Reduction would require diversion of users to alternative routes or alternative modes, encouraging users to travel at different times or to different destinations, encouraging car-pooling and other actions to increase auto occupancy. Given the constraints of capacity on the current cross section, it is also unlikely that demand would grow to the levels indicated in later years because queuing and congestion would reach intolerable levels. In Year 20, the projected demand of 11,546 veh/h is more than twice the capacity of the current cross section.

As is the case in many uses of the HCM, this analysis identifies and gives insight into a problem. It does not definitively provide a solution—unless engineers are prepared to more than double the current capacity of the facility or modify alternative routes to provide the additional capacity needed. Even these options involve judgments. Capacity and level-of-service analyses of the various alternatives would provide additional information on which to base those judgments, but would not, taken alone, dictate any particular course of action. Economic, social, and environmental issues would obviously also have to be considered as part of the overall process of finding a remedy to the forecasted problem.

Example 14-7: A Design Application

A new freeway is being designed through a rural area. The directional design hour volume (DDHV) has been forecast to be 2,700 veh/h during the peak hour, with a PHF of 0.85 and 15% trucks in the traffic stream. The total ramp density is 0.50 ramps/mi. A long section of the facility will have level terrain characteristics, but one 2-mile section involves a sustained grade of 4%. If the objective is to provide level of service C, with a minimum acceptable level of D, how many lanes must be provided?

Solution: The problem calls for the determination of the number of required lanes for three distinct sections of freeway: (1) a level terrain section, (2) a 2-mile, 4% sustained upgrade, and (3) a 2-mile, 4% downgrade.

Step 1: *Determine the Free-Flow Speed of the Freeway* This is a design situation. Unless additional information concerning the terrain suggested otherwise, it would be assumed that lane widths (12 ft) and lateral clearances (≥6 ft) conform to modern standards and meet base conditions. The free-flow speed is estimated using Equation 14-5:

$$FFS = 75.4 - f_{LW} - f_{LW} - 3.22\ TRD^{0.84}$$

$$FFS = 75.4 - 0.0 - 0.0 - 3.22(0.50^{0.84}) = 74.8\text{ mi/h} \rightarrow 75\text{mi/h}$$

Step 2: *Determine the Maximum Service Flow Rate (MSF) for Levels of Service C and D* Because the target level of service is C, with a minimum acceptable level of D, it is necessary to determine the maximum service flow rates that would be permitted if these levels of service are to be maintained. These are found from Table 14.3 for a 75 mi/h free-flow speed:

$$MSF_C = 1{,}750\text{ pc/h/ln}$$
$$MSF_D = 2{,}110\text{ pc/h/ln}$$

Step 3: *Determine the Number of Lanes Required for the Level, Upgrade, and Downgrade Freeway Sections* The required number of lanes is found using Equation 14-4:

$$N_i = \frac{DDHV}{PHF{*}MSF_i{*}f_{HV}{*}f_p}$$

where: $DDHV = 2{,}700$ veh/h (given)

$PHF = 0.85$ (given)

$MSF_C = 1{,}795$pc/h/ln (determined in Step 2)

$f_p = 1.00$ (regular users assumed)

Three different heavy-vehicle factors (f_{HV}) must be considered: one for level terrain, one for the upgrade, and one for the downgrade. With no RVs in the traffic stream, the heavy-vehicle factor is computed as:

$$f_{HV} = \frac{1}{1 + P_T\,(E_T - 1)}$$

where: $P_T = 0.15$ (given)
E_T (level) = 1.5 (Table 14.11, level terrain)
E_T (up) = 2.5 (Table 14.12, >3–4%, > 1.5 mi, 15% trucks)
E_T (down) = 1.5 (Table 14.14, >4–5%, >4 mi, 15% trucks)

Then:

$$f_{HV}\text{ (level, down)} = \frac{1}{1 + 0.15(1.5 - 1)} = \frac{1}{1.075} = 0.930$$

$$f_{HV}\text{ (up)} = \frac{1}{1 + 0.15(2.5 - 1)} = \frac{1}{1.225} = 0.816$$

and:

$$N\text{ (level, down)} = \frac{2700}{0.85*1795*0.93*1.00} = 1.9\text{ lanes}$$

$$N\text{(up)} = \frac{2700}{0.85*1795*0.816*1.00} = 2.2\text{ lanes}$$

This suggests that the level and downgrade sections require two lanes in each direction but that the upgrade requires three lanes. The computed values are minimal to provide the target level of service. This suggests that the facility should be constructed as a four-lane freeway with a truck-climbing lane on the sustained upgrade.

Analysis The results of such an analysis most often result in fractional lanes. An operational analysis could be performed using the DDHV and a four-lane freeway cross section to determine the resulting LOS. It is at least possible that the level of service will be better than the target.

Because level of service of D would have been minimally acceptable, it is useful to examine what LOS would have resulted on the upgrade if only two lanes were provided. Then:

$$v_p = \frac{V}{PHF*N*f_{HV}*f_p} = \frac{2{,}700}{0.85*2*0.816*1.00}$$

$$= 1{,}949\text{ pc/h/ ln}$$

Because this is less than the MSF_D of 2,110 pc/h/ln determined in Step 2, provision of two lanes on the upgrade would provide LOS D.

Thus a choice must be made of providing the target LOS C (or better) on the upgrade by building a three-lane cross section with a truck-climbing lane, or accepting the minimal LOS D with a two-lane cross section. A compromise solution might be to build two lanes but to acquire sufficient right-of-way and build all structures so that a climbing lane could be added later.

Also note that the analysis of a three-lane upgrade cross-section is approximate. If one of the lanes is a truck-climbing lane, then there will be substantial segregation of heavy vehicles from passenger cars in the traffic stream. The HCM freeway methodology assumes a mix of vehicles across all lanes.

Again, it must be emphasized that although the results of the analysis provide the engineer with a great deal of information to assist in making a final design decision on the upgrade section, it does not dictate such a decision. Economic, environmental, and social factors would also have to be considered.

14.5 Calibration Speed-Flow-Density Curves

The analysis methodologies of the HCM for basic freeway sections and multilane highways rely on defined speed-flow curves for base conditions, and on a variety of adjustment factors applied to determine free-flow speed and the demand flow rate in equivalent pce.

It is important to understand some of the issues involved in calibrating these basic relationships, both as background knowledge and because the HCM allows traffic engineers to substitute locally calibrated relationships and values where they are available. Chapter 6, "Traffic Flow Theory," provides a more detailed discussion of basic speed-flow-density curves and how they are calibrated.

14.6 Calibrating Passenger-Car Equivalents

The heavy-vehicle adjustment factor is the most significant in converting a demand volume under prevailing conditions to a flow rate in equivalent passenger-car equivalents under base conditions. As noted, the adjustment is based on calibrated passenger-car equivalents for trucks and buses and for RVs under various conditions of terrain. The calibration of these equivalents, therefore, is of interest. To simplify the presentation, assume that only one category of heavy vehicle and one passenger-car equivalent value (E_H) is being used. Recalling several previous relationships:

$$f_{HV} = \frac{v_{vph}}{v_{pce}}$$

$$f_{HV} = \frac{1}{1 + P_H(E_H - 1)}$$

If these equations are manipulated to solve for E_H, the following is obtained:

$$E_H = \left[\frac{\left(\frac{v_{pce}}{v_{vph}} \right) - 1}{P_H} \right] \quad (14\text{-}10)$$

The values of v_{pce} and v_{vph} can be related to headway measurements in any traffic stream. If, for example, headways in a traffic stream are categorized by both leading and trailing vehicles in each pair and if only two types of vehicle are used, the following classifications, defined previously, result: *P-P, P-H, H-P,* and *H-H*. Using the general relationship of flow rates to headways:

$$v_{pce} = \frac{3{,}600}{h_{aPP}} \quad (14\text{-}11)$$

and:

$$v_{vph} = \frac{3600}{h_a} \quad (14\text{-}12)$$

where h_a is the average of all vehicle headways, which may be expressed as:

$$h_a = P_H^2 h_{aHH} + P_H(1 - P_H)h_{aHP} + (1 - P_H)P_H h_{aPH} + (1 - P_H)^2 h_{aPP} \quad (14\text{-}13)$$

where: P_H = proportion of heavy vehicles in the traffic stream
h_{aHH} = average headway for heavy vehicles following heavy vehicles, s
h_{aHP} = average headway for heavy vehicles following passenger cars, s
h_{aPH} = average headway for passenger cars following heavy vehicles, s
h_{aPP} = average headway for passenger cars following passenger cars, s

Inserting all of this into Equation 14-10 results in:

$$E_H = \frac{(1 - p_H)*(h_{aPH} + h_{aHP} - h_{aPP}) + P_H h_{aHH}}{h_{aPP}} \quad (14\text{-}14)$$

where all terms are as previously defined.

Where headways are classified only by the type of following or trailing vehicles, the following assumptions are implicit:

$$h_{aPP} = h_{aPH} = h_{aP}$$
$$h_{aHH} = h_{aHP} = h_{aH}$$

When these assumptions are included in Equation 14-14, the relationship simplifies to:

$$E_H = \frac{h_{aH}}{h_{aP}} \quad (14\text{-}15)$$

This is a convenient form where a number of different categories of heavy vehicles exist.

The entire basis for these computations, however, is the identification of a flow of mixed vehicles that is "equivalent" to a flow of passenger cars only. Once such an equivalence is identified, headways can be used to compute or calibrate values of E_H. The major issue, then, is how to define "equivalent" conditions for use in calibration. The following sections discuss several different approaches.

14.6.1 Driver-Determined Equivalence

Krammas and Crowley [*1*] used a very straightforward means to establish equivalence. A given traffic stream represents an equilibrium condition in which individual drivers have adjusted their vehicles' operation consistent with their subjective perceptions of optimality. Krammas and Crowley suggest that individual headways within a given traffic stream represent the drivers' view of "equivalent" operational quality or level of service. This is an important concept. Because the service flow rate is defined for a given level of service, it is rational to use the concept of level of service as a basis for equivalence.

Using this approach, during each 15-minute period of observations, headways would be classified by type, and equivalents would be computed using either Equation 14-14 or Equation 14-15.

Consider the following example: The data shown in Table 14.17 were obtained during a single 15-minute interval on a freeway. The traffic stream during this interval consisted of 10% trucks (49) and 90% (440) passenger cars.

Table 14.17: Data for Sample Problem in Driver-Determined Equivalence

Type of Headway	Number Observed	Average Headway (s)
P-P	400	3.0
P-T	40	3.4
T-P	40	4.2
T-T	9	4.6

Note that the distribution of headway types is consistent with the distribution of trucks (10%) and passenger cars (90%) in the traffic stream. Using trailing vehicle types, there are 440 passenger cars and 49 trucks in the traffic stream. If lead vehicles are counted, the same distribution arises. With headways classified by both leading and trailing vehicle types, Equation 14-14 is used to determine the value of E_T:

$$E_T = \frac{(1 - 0.10)*(3.4 + 4.2 - 3.0) + (0.10*4.6)}{3.0}$$

$$= 1.53$$

Based on the Krammas and Crowley theory of equivalence, each truck consumes as much space or capacity as 1.53 passenger cars.

It would also be interesting to see how the results would be affected by using the simpler Equation 14-15, where headways are classified only by the type of trailing vehicle. To do this, weighted average headways for trailing passenger cars and trailing trucks would have to be computed from the data in Table 14.17. The weighted average headways are:

$$h_{aT} = \frac{(40*4.2) + (9*4.6)}{(40 + 9)} = 4.27 \text{ s}$$

$$h_{aP} = \frac{(400*3.0) + (40*3.4)}{(400 + 40)} = 3.04 \text{ s}$$

Then, using Equation 14-15:

$$E_T = \frac{4.27}{3.04} = 1.40$$

This equation, however, is based on the assumption that headways depend only on the type of trailing vehicle. From the data in Table 14.17, this is obviously not true in this case—$h_{aPP} \neq h_{aPT}$ and $h_{aTT} \neq h_{aTP}$. Thus the second computation is more approximate than the first. The results in the two cases, however, differ by only 0.13, indicating that use of Equation 14-15 with headways classified only by the trailing vehicle may be a reasonable approach, particularly where multiple vehicle types exist.

This example, however, illustrates the calibration of *one* value of E_T. E_T, however, varies with type of terrain, the proportion of trucks in the traffic stream, and the length and severity of a sustained grade. Some studies have suggested that these values vary by flow levels and/or level of service as well.

Calibrating a complete set of E_T and/or E_R values would, therefore, require a large database covering a wide range of underlying conditions. This is both expensive and time consuming. For practical reasons, most studies of heavy-vehicle equivalence have relied at least partially on simulations to produce the desired data set. Field studies are then conducted to validate a selection of cells in the multivariable space. Validation results may then be used to adjust equivalents to reflect discrepancies between field data and simulation values.

14.6.2 Equivalence Based on Constant Spacing

Another approach to equivalence is to select headways of various vehicle types such that their *spacing* is constant. This is done by plotting headways of various types of vehicles at a given location (using only the trailing vehicle to classify) vs. their spacing. This is an interesting approach, illustrated in Figure 14.10. Spacing is related to density ($D = 5{,}280/S_a$), and density is the measure that defines level of service for freeways and multilane highways. Thus this process results in passenger-car equivalents that define traffic streams of equal density and, therefore, of equal level of service.

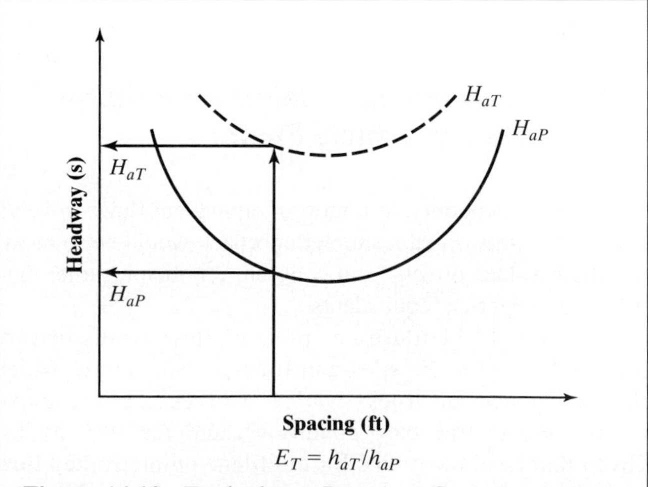

Figure 14.10: Equivalence Based on Constant Spacing

The plot of headways versus spacing can be made from data covering any time period sufficient to define the relationships. The plots would, however, be valid only for the site in question and would relate to the size of the facility (number of lanes), terrain, grade, and length of grade present. Passenger-car equivalents might be shown to vary with spacing (density) as the result of any given set of plots. This approach, however, would not reveal any variation in E_T based on the proportion of trucks in the traffic stream, P_T. Implicit in this approach is the assumption that there is no such variation.

Note also that the equivalent headways selected using this procedure do not have to occur within the same 15-minute period because the plots are calibrated using a set of points for many 15-minute periods. This is fundamentally different from the Krammas and Crowley approach. It defines "equivalent" in terms of traffic streams having equal average spacings.

14.6.3 Equivalence Based on Constant Speed

It is also possible to prepare a plot of average headways versus space mean speed for each 15-minute interval. Similar to the approach of Figure 14.10, equivalents may then be based on traffic streams having equal space mean speeds. This is an attractive alternative if speed is the measure defining level of service. Thus, although this method was used to calibrate passenger-car equivalents for two-lane rural highways in the 1965 HCM, it is not particularly well suited to such calibrations for freeways or multilane highways. In the 1985 HCM, speed is a secondary service measure for two-lane highways [2].

14.6.4 Macroscopic Calibration of the Heavy-Vehicle Factor

Because f_{HV} is defined as a ratio of equivalent flows in veh/h and pc/h, it seems that a simpler approach would be to measure these values directly and calibrate the factor, rather than using passenger-car equivalents.

Figure 14.11 illustrates plots of flow versus density for similar geometric sites and time periods during which the percentage of trucks varies. Curves are illustrated for 0% trucks (the base condition), and for 10% trucks. Given that level of service for multilane uninterrupted flow facilities is defined by density, equivalent flow levels can be found by holding density constant, as shown in the figure. The ratio of these flows now yields a value of f_{HV} directly.

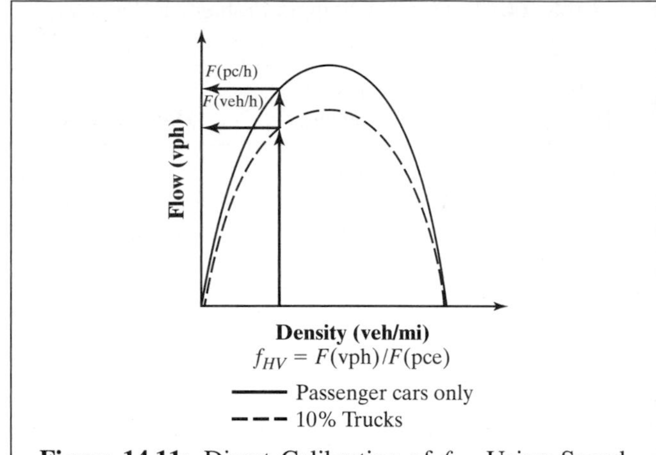

Figure 14.11: Direct Calibration of f_{HV} Using Speed-Density Curves

The data collection for such a calibration is difficult. A family of curves representing different percentages of trucks would be needed as would sites with varying conditions of terrain and/or grade. Finding sites sufficient to produce a curve for truck percentages ranging from 0% to as high as 25% to 30% is a significant challenge that makes this approach less practical than others discussed previously.

A similar approach can be taken using speed-flow curves, but this would not be as philosophically consistent because speed is not the measure of service quality used in the HCM.

14.6.5 Additional References on Heavy Vehicle Factors

The literature contains numerous theoretical and practical treatments of passenger-car equivalents and the heavy-vehicle factor [3–5]. No common concept of equivalence has emerged, nor do all researchers agree on specific calibration approaches. Reference 6 explains the approach to multilane uninterrupted-flow passenger-car equivalents used in HCM 2000.

14.7 Calibrating the Driver Population Factor

The adjustment factor for driver population, f_p virtually always requires local calibration, if a value other than 1.00 (for a commuting traffic stream, or familiar users of a facility) is to be used.

This is because the potential "other" driver populations and their impacts on operations are widely variable, depending on the specific situation in question. Recreational routes frequently involve nonregular users who may not be familiar with local characteristics. However, familiarity, or lack thereof, is not a "yes" or "no" proposition but a matter of degrees. Further, recreational and other routes may have a range of driver characteristics, including some familiar or regular users. Geometry also matters. An unfamiliar user of a roadway with severe geometry in a mountainous area will be far more cautious than an unfamiliar user on a roadway that is basically straight and level.

The same facility may have markedly different driver populations on weekdays and weekends if it serves both work-related and recreational destinations. Calibration is often done by observing capacity operations on a single facility on weekdays and on weekends, or by comparing observed capacities of similar routes with different driver populations. Such calibration is always approximate because the extension of results from one facility (or a group of facilities) to others cannot ensure accuracy. Nevertheless, this is the best approach available at present and is a reasonable approach to getting a local estimate of a very volatile adjustment factor.

14.8 Adjustment Factors to Free-Flow Speed

The HCM 2000 applies some adjustments to the prediction of free-flow speed. These adjustments (for lane width, lateral clearance, number of lanes or median type, and interchange or access-point density) are different from the driver population and heavy-vehicle factors, in that they are subtractive rather than multiplicative. The data on which these factors are based were quite sparse. Lane width and lateral clearance adjustments, for example, were calibrated using some field data and by projecting former adjustments to flow rates for these factors onto the speed axis using standard speed-flow curves.

Free-flow speed for any facility is easily measured as the average speed (of passenger cars uninfluenced by heavy vehicles) when flow rates are low. For freeways and multilane highways, measurements taken when flow is less than 800 to 1,000 veh/h/ln will provide reasonable estimates. Calibration requires controlled experiments in which one variable (i.e., lane width) is varied while others are held constant. Sites fulfilling this objective will be difficult to find. Another approach is to collect data at a number of sites with varying conditions and use regression analysis to establish a relationship for predicting free-flow speed. The relationship might then be used to examine the impact of individual features on the result.

14.9 Software

One major software package claims to replicate the methodologies of the HCM 2000 and will be updated to reflect changes and new methodologies of the HCM 2010. The *Highway Capacity Software* package (HCS) continues to be maintained and available through McTrans Center at the University of Florida, Gainesville.

Note that the Highway Capacity and Quality of Service Committee of the Transportation Research Board *does not* examine, certify, or endorse any software product. The burden of demonstrating that a software package faithfully replicates the current HCM is entirely that of the software producers.

14.10 Source Documents

The methodologies for basic freeway segments and multilane highways are based on two National Cooperative Highway Program projects (*7,8*). Reference 9 was used to further update material related to basic freeway segments. Reference 10 is a good overview of the history of the development of HCM procedures.

References

1. Krammas, R., and Crowley, K., "Passenger Car Equivalents for Trucks on Level Freeway Segments," *Transportation Research Record 1194,* Transportation Research Board, Washington DC, 1988.
2. Craus, J., Polus, A., and Grinberg, A. "A Revised Method for the Determination of Passenger Car Equivalents," *Transportation Research,* Vol. 14A, No. 4, Pergamon Press, London, UK, 1980.
3. Linzer, E., Roess, R., and McShane, W., "Effect of Trucks, Buses, and Recreational Vehicles on Freeway Capacity and Service Volume," *Transportation Research Record 699,* Transportation Research Board, Washington DC, 1979.

4. Cunagin, W., and Messer, C., "Passenger Car Equivalents for Rural Highways," *Transportation Research Record 905,* Transportation Research Board, Washington DC, 1983.
5. Roess, R., and Messer, C., "Passenger Car Equivalents for Uninterrupted Flow: Revision of the Circular 212 Values," *Transportation Research Record 971,* Transportation Research Board, Washington DC, 1984.
6. Webster, L., and Elefteriadou, A., "A Simulation Study of Truck Passenger Car Equivalents (PCE) on Basic Freeway Sections," *Transportation Research B,* Vol. 33, No. 5, Pergamon Press, London, UK, 1999.
7. Reilly, W., Harwood, D., and Schoen, J., "Capacity and Quality of Flow of Multilane Highways," *Final Report,* JHK & Associates, Tucson AZ, 1988.
8. Schoen, J., May, A. Jr., Reilly, W., and Urbanik, T., "Speed-Flow Relationships for Basic Freeway Sections," *Final Report,* JHK & Associates, Tucson AZ, and Texas Transportation Institute, Texas A&M University, College Station TX, December 1994.
9. Roess, R., *TASK 6: Re-Calibration of the 75-mi/h Speed-Flow Curve and the FFS Prediction Algorithm for HCM 2010 (Draft 2)*, Technical Memorandum, NCHRP Project 3-92, Polytechnic Institute of NYU, Brooklyn NY, January 2009.
10. Kittelson, W., "Historical Overview of the Committee on Highway Capacity and Quality of Service," *Proceedings of the Fourth International Symposium on Highway Capacity,* Transportation Research Circular E-C018, Maui, Hawaii, June 27–July 1, 2000.

Problems

14-1. Estimate the free-flow speed of a four-lane undivided multilane highway having the following characteristics:

(a) Base free-flow speed = 60 mi/h

(b) Average lane width = 11 ft.

(c) Lateral clearance = 3 ft at both roadsides

(d) Access-point density = 15/mi on each side of the roadway.

14-2. Estimate the free-flow speed of a six-lane suburban freeway with 12-ft lanes, right-side lateral clearance of 2 ft, and a ramp density of 3.5/mi.

14-3. Find the appropriate composite grade for each of the following grade sequences:

(a) 2,000 ft of 3% grade followed by 1,000 ft of 2% grade, followed by 900 ft of 4% grade.

(b) 2,000 ft of 4% grade, followed by 5,000 ft of 3% grade, followed by 3,000 ft of 5% grade.

(c) 4,000 ft of 5% grade, followed by 3,000 ft of 3% grade.

14-4. A freeway operating in generally rolling terrain has a traffic composition of 12% trucks and 3% RVs. If the observed peak hour volume is 3,200 veh/h, what is the equivalent volume in pc/h?

14-5. Find the upgrade and downgrade service flow rates and service volumes for an eight-lane urban freeway with the following characteristics:

(a) 11-ft lanes

(b) 2-ft right-side lateral clearance

(c) 4.2 ramps/mi

(d) 3% trucks, no recreational vehicles

(e) Driver population consisting of regular facility users

The section in question is on a 4% sustained grade of 1.5 mile. The PHF is 0.92.

14-6. An existing six-lane divided multilane highway with a field-measured free-flow speed of 45 mi/h serves a peak-hour volume of 4,000 veh/h, with 12% trucks and no RVs. The PHF is 0.88. The highway has rolling terrain. What is the likely level of service for this section?

14-7. A long section of suburban freeway is to be designed on level terrain. A level section of 5 miles is, however, followed by a 5% grade, 2.0 mi in length. If the DDHV is 2,500 veh/h with 10% trucks and 3% RV's, how many lanes will be needed on the

(a) Upgrade

(b) Downgrade

(c) Level terrain section

to provide for a minimum of level of service C? Assume that base conditions of lane width and lateral clearance exist and that ramp density is 0.50/mi. The PHF = 0.92.

14-8. An old urban four-lane freeway has the following characteristics:

(a) 11-ft lanes

(b) No lateral clearances (0 ft)

(c) A ramp density of 4.5/mi

(d) 7% trucks, no RVs

(e) PHF = 0.90,

(f) Rolling terrain

The present peak-hour demand on the facility is 2,100 veh/h, and anticipated growth is expected to be 3% per year. What is the present level of service expected? What is the expected level of service in 5 years? 10 years? 20 years? To avoid breakdown (LOS F), when will substantial improvements be needed to this facility or alternative routes?

14-9. The following headways are observed during a 15-minute period on an urban freeway:

Type of Headway	Number Observed	Average Value (s)
P-P	128	3.1
P-T	32	3.8
T-P	32	4.3
T-T	8	4.9

Compute the effective passenger-car equivalent (pce) for trucks in this case, assuming that all headway types are different. Recompute the pce for trucks, assuming that headway values depend only on the type of following vehicle. Are the results different? Why? Use the "driver selected equivalent" approach in solving this problem.

CHAPTER 15

Weaving, Merging, and Diverging Movements on Freeways and Multilane Highways

15.1 Turbulence Areas on Freeways and Multilane Highways

In Chapter 14, we presented and illustrated capacity and level-of-service (LOS) analysis approaches for basic freeway and multilane highway sections. Segments of such facilities that accommodate weaving, merging, and/or diverging maneuvers, however, experience additional turbulence as a result of these movements. This additional turbulence in the traffic stream results in operations that cannot be simply analyzed using basic segment techniques.

Although there are no generally accepted measures of "turbulence" in the traffic stream, the basic distinguishing characteristic of weaving, merging, and diverging segments is the additional lane-changing these maneuvers cause. Other elements of turbulence include the need for greater vigilance on the part of drivers, more frequent changes in speed, and average speeds that may be somewhat lower than on similar basic sections.

Figure 15.1 illustrates the basic maneuvers involved in weaving, merging, and diverging segments. *Weaving* occurs when one movement must cross the path of another along a length of facility without the aid of signals or other control devices, with the exception of guide and/or warning signs. Such situations are created when a merge area is closely followed by a diverge area. The flow entering on the left leg of the merge and leaving on the right leg of the diverge must cross the path of the flow entering on the right leg of the merge and leaving on the left leg of the diverge. Depending on the specific geometry of the segment, these maneuvers may require lane changes to be successfully completed. Further, other vehicles in the segment (i.e., those that do not weave from one side of the roadway to the other) may make additional lane changes to avoid concentrated areas of turbulence within the segment.

Merging occurs when two separate traffic streams join to form a single stream. Merging can occur at an on-ramp to a freeway or multilane highway or when two significant facilities join to form one. Merging vehicles often make lane changes to align

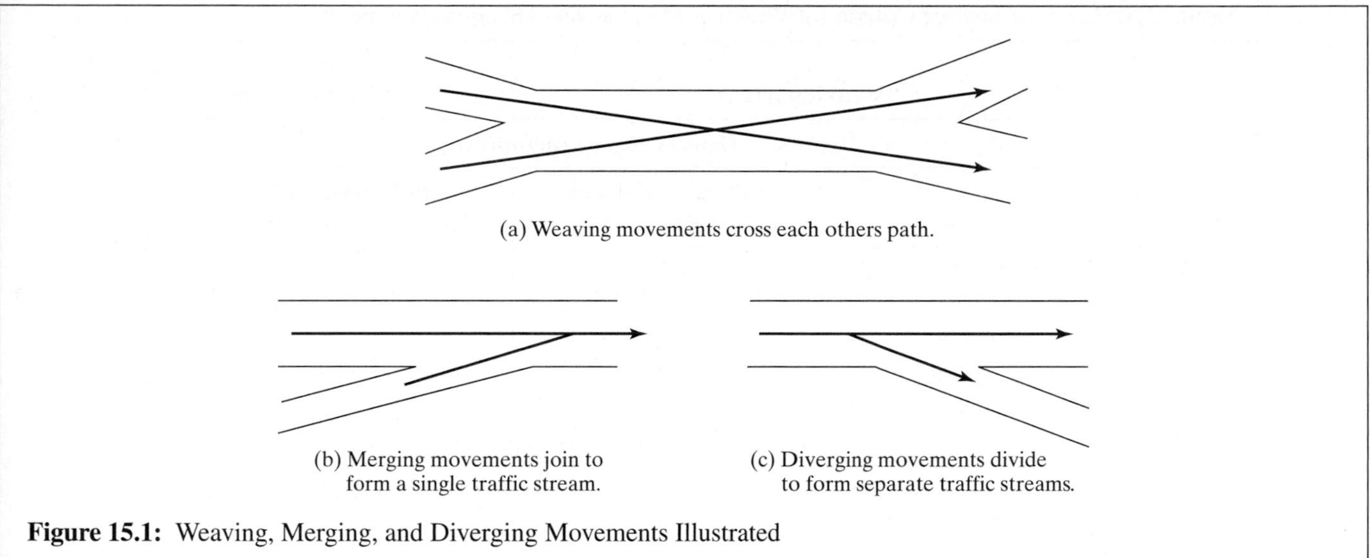

Figure 15.1: Weaving, Merging, and Diverging Movements Illustrated

themselves in lanes appropriate to their desired movement. Nonmerging vehicles also make lane changes to avoid the turbulence caused by merging maneuvers in the segment.

Diverging occurs when one traffic stream separates to form two separate traffic streams. This occurs at off-ramps from freeways and multilane highways, but it can also occur when a major facility splits to form two separate facilities. Again, diverging vehicles must properly align themselves in appropriate lanes, thus inducing lane-changing; nondiverging vehicles also make lane changes to avoid the turbulence created by diverge maneuvers.

The difference between weaving and separate merging and diverging movements is unclear at best. Weaving occurs when a merge segment is "closely followed" by a diverge segment. The exact meaning of "closely followed" is not well defined. The HCM 2000 [*1*] employed a uniform 2,500-foot length as the maximum for weaving operations; the latest research [*15*] indicates that this length is variable. At some point, however, the merge and diverge end of the weaving segment are far enough apart to operate independently. Even where the distance between a merge and diverge is less than the maximum, the classification of the movement depends on the details of the configuration. For example, a one-lane, right-hand, on-ramp followed by a one-lane, right-hand, off-ramp is considered a weaving section only if the two are connected by a continuous auxiliary lane. If the on-ramp and off-ramp have separate, discontinuous acceleration and deceleration lanes, they are treated as isolated merge and diverge areas, respectively, independent of the distance between them. The 1965 HCM [*2*] recognized weaving movements over distances up to 8,000 feet, but this was based on a small number of data points, and lengths greater than 2,500 feet were subsequently removed from consideration as weaving areas.

Even though the nature of lane-changing and other turbulence factors is similar in weaving, merging, and diverging segments, the methodologies for analysis of weaving segments are different from those for merging and diverging segments. This is primarily an accident of research history because conceptually, similar procedures would seem to be more appropriate. Research efforts on these subjects have been done at different times using different databases, as mandated by sponsoring agencies. In writing material for the HCM 2000 and for the forthcoming 2010 edition, however, considerable effort was expended to make the two analytic approaches more consistent, particularly in terms of level-of-service measures and criteria.

HCM methodologies for weaving and for merging/diverging segments are calibrated for freeways. These methods have some limited application to multilane highways with uncontrolled weaving or merging/diverging operations, but they must generally be considered more approximate in these cases.

15.2 Level-of-Service Criteria

The measure of effectiveness for weaving, merging, and diverging segments is density. This is consistent with freeway and multilane highway methodologies. Level-of-service criteria are shown in Table 15.1.

Table 15.1: Level-of-Service Criteria for Weaving, Merging, and Diverging Segments

	Weaving Areas		Merge or Diverge Areas
	Density Range (pc/mi/ln)		
Level of Service	**On Freeways**	**On Multilane Highways or C-D Roadways**	**On Freeways, Multilane Highways, or C-D Roadways**
A	0–10	0–12	0–10
B	>10–20	>12–24	>10–20
C	>20–28	>24–32	>20–28
D	>28–35	>32–36	>28–35
E	>35	>36	>35
F	Demand Exceeds Capacity		

(*Source:* Used with permission of Transportation Research Board, National Research Council, *Highway Capacity Manual*, 2000. Compiled from Exhibit 24-2, p. 24-3, and Exhibit 25-4, p. 25-5.)

For weaving areas, separate criteria are specified for segments on freeways and on multilane highways. Boundary conditions for multilane highways are set at somewhat higher densities than for freeways, reflecting users' lower expectations on multilane highways. This is somewhat inconsistent with the criteria for basic sections, which are the same for freeways and multilane highways, except for the LOS E/F boundary. For merge and diverge junctions, only one set of criteria are specified, which is applied to both freeways and multilane highways.

For weaving segments, merge segments, and diverge segments, LOS F occurs when demand exceeds capacity of the segment (i.e., when v/c is more than 1.00). The limit of LOS E is defined as the capacity of the segment.

For basic freeway segments and multilane highways, a maximum density is used to define the boundary between levels of service E and F. Basic speed-flow curves for both basic freeway and multilane highways show capacity occurring at fixed values of density. Therefore, for any given segment, there is a one-to-one correlation between capacity and the density at which it occurs. In terms of application, LOS F for both basic freeway segments and multilane highways is identified when demand exceeds capacity (i.e., v/c is more than 1.00). This is because the analytic procedures do not allow the estimation of a specific density value. Thus although it is known that density will be greater than the boundary value, an exact value cannot be computed. The criteria of Table 15.1 are compiled from the HCM 2000, and they will not change in the forthcoming fifth edition in 2010.

The spatial definitions of influence areas for weaving segments are different, however, than those for merging and diverging segments. For weaving areas, the density reflects an average for all vehicles across all lanes of the segment between the entry and exit points of the segment plus 500 feet upstream and downstream of the segment. For merge and diverge areas, densities reflect the "merge/diverge influence area," which consists of lanes 1 and 2 (right and next-to-right lanes of the freeway) and the acceleration or deceleration lane for a distance 1,500 feet upstream of a diverge or 1,500 feet downstream of a merge. These influence areas are illustrated in Figure 15.2.

In some cases, these definitions cause overlaps. For example if an on-ramp is followed by an off-ramp less than 3,000 feet away, the two 1,500-feet influence areas will at least partially overlap. In such cases, the worst density or LOS is applied to the overlap area. Other overlaps between ramp and weaving segments and/or basic segments are similarly treated.

The most significant inconsistency between the analysis methodologies for weaving segments and merge and diverge segments is the manner in which capacity is determined and in the results obtained. For merge and diverge segments, capacity is most often controlled by the upstream (diverge) or downstream (merge) freeway or multilane highway segment. Thus ramp junctions rarely limit the capacity of the basic facility. For weaving sections, capacity values are generally less than a basic segment capacity for the same number of lanes. This inconsistency is not, however, unreasonable. In weaving segments, some lanes, particularly auxiliary lanes, cannot be fully used due to the distribution of weaving and nonweaving demands in the segment.

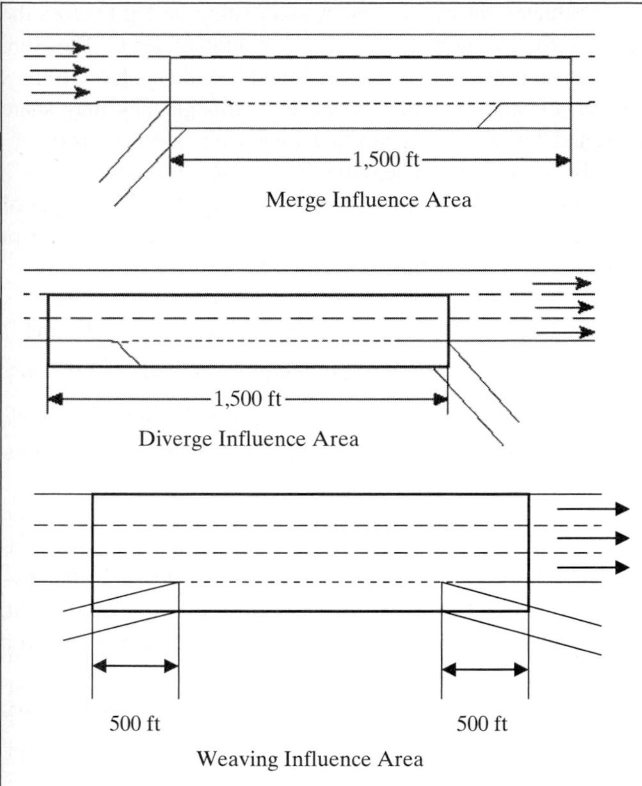

Figure 15.2: Influence Areas for Merge, Diverge, and Weaving Segments

(Source: Used with permission of Transportation Research Board, National Research Council, modified from *Highway Capacity Manual*, 2000, Exhibit 13-13, p. 13-21.)

15.3 A Common Point: Converting Demand Volumes

Both procedures for weaving areas and for merge/diverge areas rely on algorithms calibrated in terms of demand flow rates in passenger car units for base or ideal conditions. Thus a point common to both is that all component demand volumes must be converted before proceeding to use either methodology.

$$v_i = \frac{V_i}{PHF^* f_{HV}^* f_p} \quad (15\text{-}1)$$

where: v_i = demand flow rate, pc/h, under equivalent base conditions

V_i = demand volume, veh/h, under prevailing conditions

PHF = peak hour factor

f_{HV} = heavy-vehicle adjustment factor

f_p = driver-population adjustment factor

The heavy-vehicle and driver-population factors are the same ones used for basic freeway and multilane highway segments. They are found using the methods and values presented in Chapter 14.

15.4 Weaving Segments: Basic Characteristics and Variables

Weaving areas have been the subject of a great deal of research since the late 1960s, yet many features of current procedures continue to rely, at least partially, on judgment. This is primarily due to the great difficulty and cost of collecting comprehensive data on weaving operations. Weaving areas cover significant lengths and generally require videotaping from elevated vantage points or time-linked separate observation of entry and exit terminals and visual matching of vehicles. Further, a large number of variables affect weaving operations, and therefore a large number of sites reflecting these variables would be needed to provide a statistically desirable database.

The first research study leading up to the third edition 1985 HCM focused on weaving areas [*3*]. This was unfortunate because basic section models would be revised later, causing judgmental modification in weaving models for consistency. It relied on 48 sets of data collected by the then Bureau of Public Roads in the late 1960s and an additional 12 sets collected specifically for the study. The methodology that resulted was complex and iterative. It was later modified as part of a study of all freeway-related methodologies [*4*] in the late 1970s. In 1980, a set of interim analysis procedures was published by TRB [*5*], which included the modified weaving analysis procedure. It also contained an independently developed methodology that often produced substantially different results. The latter methodology was documented in a subsequent study [*6*]. To resolve the differences between these two methodologies, another study was conducted in the early 1980s, using a new database consisting of 10 sites [*7*]. This study produced yet a third methodology, substantially different from the first two. As the publication date of the 1985 HCM approached, the three methodologies were judgmentally merged, using the 10 sites from the 1980s for general validation purposes [*8*]. A number of studies throughout the 1980s and 1990s continued to examine the various weaving approaches, with no common consensus emerging [*9–12*].

It was, therefore, no surprise that a new study, relying on some new data but primarily on simulation, was commissioned as part of the research for the HCM 2000 [*13*]. Unfortunately, the simulation approach was not particularly successful, and it yielded a number of trends that were judged (by the Highway Capacity and Quality of Service Committee of the Transportation Research Board) to be counterintuitive. The method of the HCM 2000 resulted from a further judgmental modification of earlier procedures [*14*].

The weaving segment analysis methodology presented in this text is drawn from the National Cooperative Highway Research Program Project 3-75, Analysis of Freeway Weaving Sections [*15*]. This procedure was developed for inclusion in the 2010 HCM and was formally approved by the HCQSC at its 2009 summer meeting. Because the HCM 2010 is not yet (at this writing) in production, it is possible that some minor revisions will be made before inclusion in the HCM.

15.4.1 Flows in a Weaving Area

In a typical weaving area, four component flows may exist. By definition, the two that cross each other's path are called *weaving flows;* those that do not are called *nonweaving, or outer, flows.* Figure 15.3 illustrates.

Vehicles entering on leg A and exiting on leg D cross the path of vehicles entering on leg B and exiting on leg C. These are the weaving flows. Movements A-C and B-D do not have to cross the path of any other movement, even though they may share lanes, and they are referred to as nonweaving, or outer, flows.

By convention, weaving flows use the subscript "w"; outer or nonweaving flows use the subscript "o." The larger of the two outer or weaving flows is given the second subscript "1"; the smaller uses the subscript "2." Thus:

v_{o1} = larger outer flow, pc/h, equivalent base conditions

v_{o2} = smaller outer flow, pc/h, equivalent base conditions

v_{w1} = larger weaving flow, pc/h, equivalent base conditions

v_{w2} = smaller weaving flow, pc/h, equivalent base conditions

The schematic line drawing of Figure 15.3 is called the *weaving diagram.* In block form, it shows the weaving and nonweaving flows and their relative positions on the roadway. By convention, it is always drawn with traffic moving from left to right. It is a convenient form to illustrate the component flows in a consistent way for analysis.

Other critical variables, used in analysis algorithms, may be computed from these:

$$v_w = \text{total weaving flow, pc/h} = v_{w1} + v_{w2}$$

$$v_{nw} = \text{total non-weaving or outer flow, pc/h} = v_{o1} + v_{o2}$$

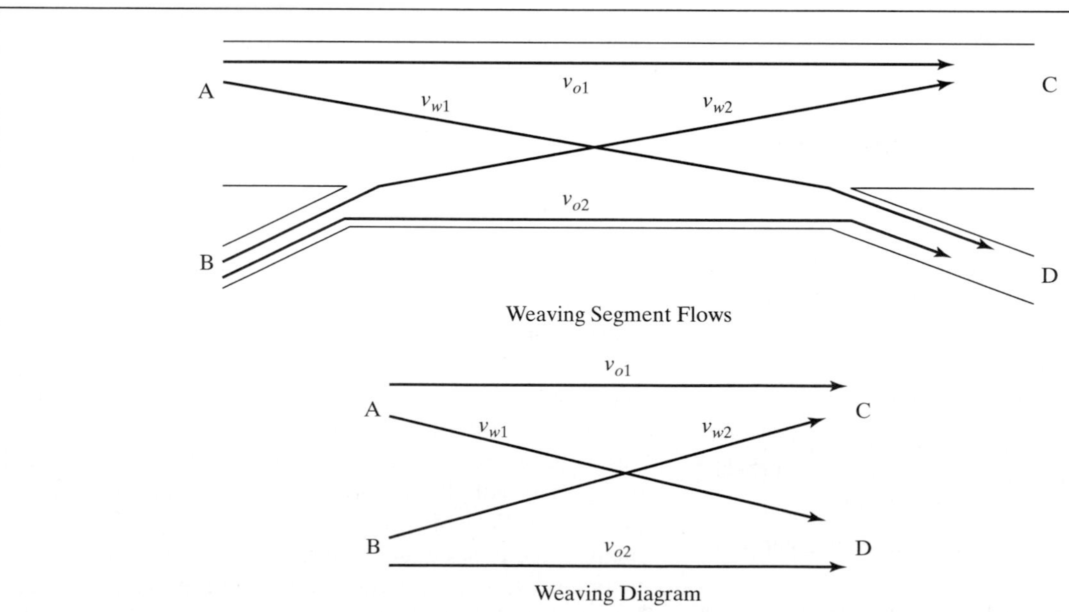

Figure 15.3: Flows in a Weaving Segment and the Weaving Diagram

v = total demand flow, pc/h = $v_w + v_{nw}$

VR = volume ratio = v_w/v

R = weaving ratio = v_{w2}/v_w

15.4.2 Critical Geometric Variables

Three geometric variables have a significant effect on the quality of weaving operations:

- Lane configuration
- Length of the weaving area (in feet)
- Width (number of lanes) in the weaving area

Each of these has an impact on the amount of lane-changing that must or may occur and its intensity.

Lane Configuration

Lane configuration refers to the manner in which entry and exit legs "connect" with each other. This is a critical characteristic because it ultimately determines how many lane changes *must* be made by weaving vehicles to successfully complete their weaving maneuver. These are mandatory lane changes because they must be made for the weaving vehicle to get from its entry leg to its desired exit leg. By definition, these lane changes must be made within the weaving section.

Lane Configuration Classifications Many lane configurations may exist based on the number and location of entry and exit lanes, and the number of lanes within the weaving segment. Weaving segments are categorized in two ways:

- One-sided versus two-sided weaving segments
- Ramp-weave versus major weaving segments

In a one-sided weaving segment, weaving movements are substantially restricted to lanes on one side of the facility, usually (but not always) the right side. In two-sided weaving sections, at least one of the weaving movements must use lanes on both sides of the facility. Weaving turbulence in one-sided segments is more localized, whereas in two-sided segments, it may spread across most or all lanes of the facility. In more specific terms, the following definitions apply:

- A *one-sided weaving segment* is one in which no weaving maneuver requires more than two lane changes.
- A *two-sided weaving segment* is one in which one weaving maneuver requires three or more lane changes, or one in which a one-lane on-ramp on one side of the facility is closely followed by a one-lane off-ramp on the other side of the facility.

Weaving segments may also be classified as *ramp weaves* or *major weaves*. The ramp-weave segment is very common and has a standard characteristic: A one-lane on-ramp is followed by a one-lane off-ramp (on the *same* side of the facility) and are connected by a continuous auxiliary lane. In major weaving segments, at least three of the entry and exit legs have more than one lane. In ramp-weaves, ramp roadways generally have design speeds that are lower, sometimes significantly, than that of the main facility. Because of this, on-ramp and off-ramp vehicles are most often accelerating or decelerating as they traverse the weaving segment. In major weave segments, entry and exit legs are often designed to standards that are closer to those of the main facility. Consequently, there is less acceleration and deceleration within the segment than for ramp-weaves. Figure 15.4 illustrates some of these characteristics.

Figure 15.4 (a) shows a one-sided ramp-weave segment. Created by an on-ramp followed by an off-ramp connected by a continuous auxiliary lane, every weaving vehicle must make at least one lane change: on-ramp vehicles from the auxiliary lane to the right lane of the facility, and off-ramp vehicles from the right lane of the facility to the auxiliary lane. Because both ramps are on the right side of the freeway, these lane changes are somewhat restricted to one side of the facility. Figure 15.4 (b) is a major weave segment because three of the four entry and exit legs have two lanes. Once again, however, the focus of weaving lane changes is on one side (the right) of the facility. Figure 15.4 (c) is the most common two-sided configuration. In this case, a left-side on-ramp is closely followed by a right-side off-ramp; the reverse arrangement produces a similar configuration Ramp-to-ramp vehicles must cross the entire facility and will occupy every lane within the segment for some period of time. Figure 15.4 (d) is a major weave, again because three entry and exit lanes have two or more lanes. It is clearly also a two-sided configuration because ramp-to-ramp vehicles again must cross most of the lanes of the facility, making at least three lane changes.

Numerical Characteristics of One-Sided Weaving Configurations Three numerical descriptors have been defined that quantify the key element of configuration. It is noted that these definitions apply only to one-sided weaving segments, in which the ramp-to-facility and facility-to-ramp movements are the weaving movements:

LC_{RF} = minimum number of lanes changes that a ramp-to-facility weaving vehicle must make to successfully complete the ramp-to-facility movement.

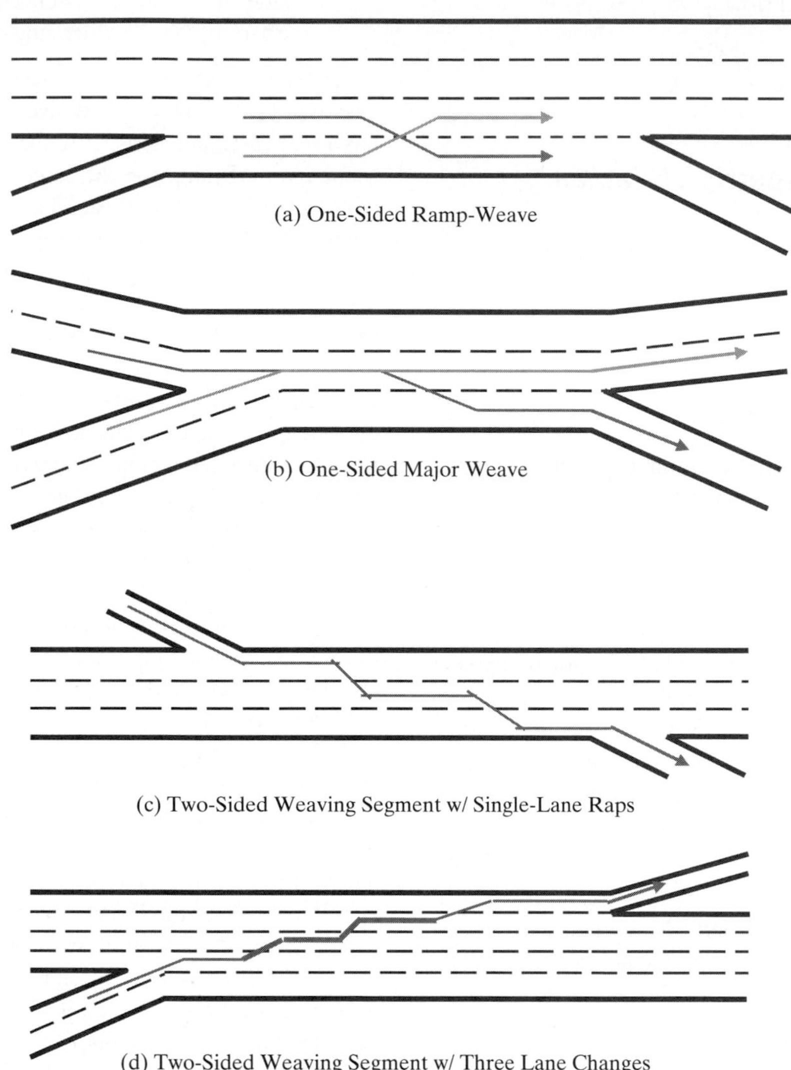

Figure 15.4: Weaving Configurations Illustrated

(*Source:* Roess, R., et al., *Analysis of Freeway Weaving Sections*, Final Report, Draft Chapter for the HCM, National Cooperative Highway Research Program Project 3-75, Polytechnic University and Kittelson and Associates, Brooklyn, NY, September 2007, Exhibits 24-3 and 24-4, p. 5.)

LC_{FR} = minimum number of lane changes that a facility-to-ramp weaving vehicle must make to successfully complete the facility-to-ramp movement.

N_{WV} = number of lanes *from which* a weaving maneuver may be completed with one lane change, or no lane change.

Figure 15.5 illustrates these critical parameters. The values of LC_{RF} and LC_{FR} are determined by assuming that every weaving vehicle enters the segment in the lane closest to its desired exit and leaves the segment in the lane closest to its entry. In Figure 15.5 (a), all ramp-to-facility vehicles enter in the auxiliary lane and leave in the right-most lane of the facility. Facility-to-ramp vehicles enter in the right-most lane

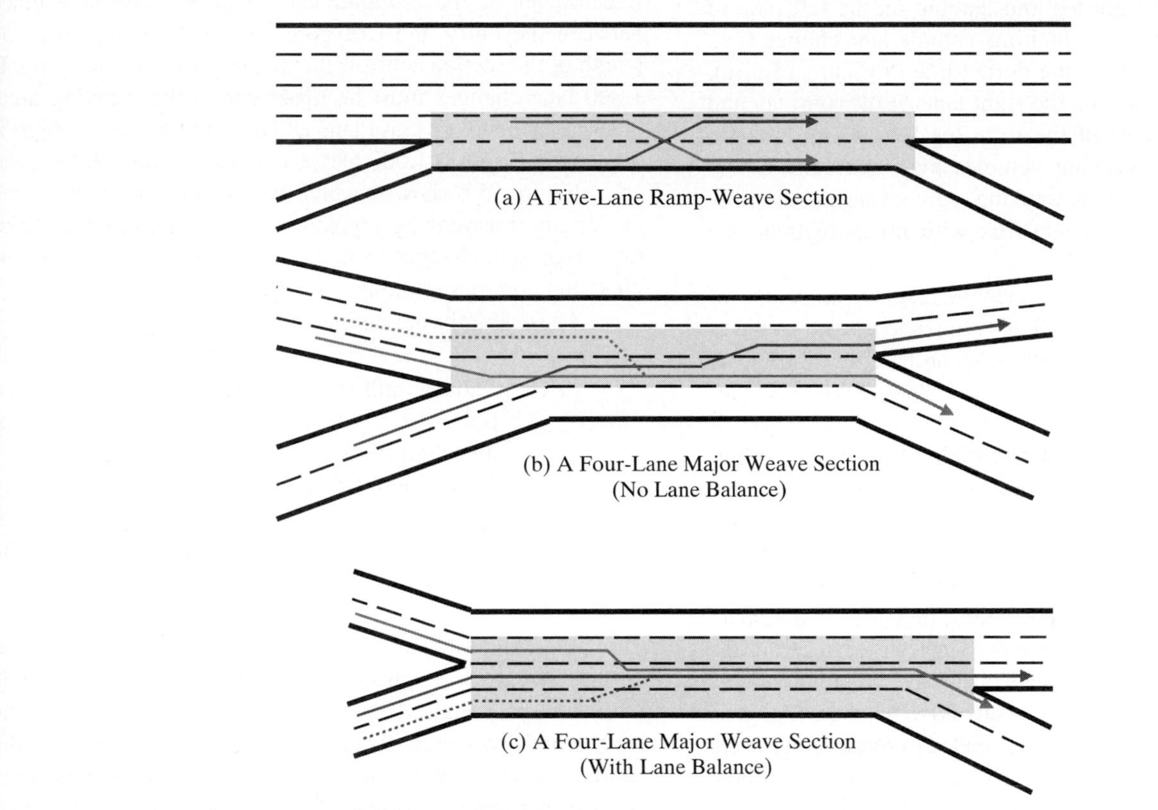

Figure 15.5: Configuration Parameters Illustrated

(*Source:* Roess, R., et al., *Analysis of Freeway Weaving Sections*, Final Report, Draft Chapter for the HCM, National Cooperative Highway Research Program Project 3-75, Polytechnic University and Kittelson and Associates, Brooklyn, NY, September 2007, Exhibit 24-4, p. 7.)

of the facility and leave in the auxiliary lane. Each vehicle in both flows must make one lane change to successfully complete its desired maneuver. For this case, $LC_{RF} = LC_{FR} = 1$. Any weaving vehicle entering or leaving on a facility lane that is *not* the right-most lane would have to make two or more lane changes. Thus the only lanes in which weaving may be accomplished with a single lane change are the auxiliary lane and the right lane of the facility (i.e., $N_{WV} = 2$).

Figure 15.1 (b) is a major weaving segment. Vehicles weaving from right to left are assumed to enter on the left lane of the on-ramp and leave on the right lane of the left exit leg. The configuration requires that one lane change be made to do this (i.e. $LC_{RF} = 1$). Weaving vehicles moving from the left leg to the right leg have a simpler task. A vehicle entering on the right lane of the left entry leg and leaving on the left lane of the right entry leg can do so without making any lane changes. This occurs because two entry leg lanes merge into a single lane. In this case, $LC_{FR} = 0$. As shown by the dotted line in Figure 15.5 (b), a left-to-right weaving vehicle may also enter in the second lane of the left leg and leave in the left lane of the right leg by making a single lane change. Because of this, weaving vehicles may enter the segment on either of the two middle lanes and weave with no more than one lane change (i.e., $N_{WV} = 2$).

Figure 15.5 (c) is also a major weave section. Its most distinctive characteristic occurs at the exit gore area: lane balance. *Lane balance* exists at an exit gore when the number of lanes leaving the exit gore is *one more* than the number of lanes entering it. In this case, four lanes approach the exit gore, but five lanes depart it. One approaching lane splits to two at the exit. This provides for great flexibility in use of that lane. Vehicles approaching in that lane may access either exit leg without making a lane change. Vehicles entering on the left lane of the right entry leg and exiting on the right lane of the left exit leg can do so without making a lane change (i.e., $LC_{RF} = 0$). Vehicles entering on the left lane of the right leg and leaving on the left lane of the right leg may do so by making a single lane change (i.e., $LC_{FR} = 1$). As shown by the dotted line in Figure 15.5 (c), vehicles may also enter on the right lane of the right leg and leave on the right lane of the right leg by making a single lane change. Thus weaving vehicles may enter any of the three right-most lanes of the weaving segment and successfully complete their desired maneuvers with no more than one lane change (i.e., $N_{WV} = 3$).

In terms of one-sided weaving segments, values of LC_{FR} and LC_{RF} are normally 0 or 1. In some cases, a value of 2 is also possible. The value of N_{WV} can be either 2 or 3; no other values are possible.

Numerical Characteristics of Two-Sided Weaving Configurations In two-sided configurations, ramp-to-facility and facility-to-ramp movements are *not* the weaving flows. In such configurations, the ramp-to-ramp vehicles weave across facility-to-facility vehicles. Although the through vehicles on the facility actually weave in such sections, they are the dominant movement and do not have to make any lane changes in the segment. Therefore, in a two-sided weaving configuration, only the ramp-to-ramp vehicles are considered to be "weaving." The minimum number of lane changes needed to successfully move from ramp to ramp is the key characteristic – LC_{RR}. In both Figures 15.4 (b) and (c), this value is 3. By definition, in all two-sided weaving segments, N_{WV} is set to "0."

Length of the Weaving Area

Although configuration has a tremendous impact on the number of lane changes that must be made within the confines of the weaving area, the length of the section is a critical determinant of the *intensity* of lane-changing within the section. Because all of the *required* lane changes must take place between the entry and exit gores of the weaving area, the length of the section controls the intensity of lane-changing. If 1,000 lane changes must be made within the weaving area, then the intensity of those lane changes will be half as high if the section length is 1,000 feet as compared with 500 feet.

Figure 15.6 shows several two potential ways in which the length of a weaving segment could be measured. Both of these represent changes from the definition of length in HCM 2000 and previous editions.

These lengths are defined as:

L_S = Short length (ft); the distance between the end points of any barrier markings that prohibit or discourage lane changing.

L_B = Base length (ft); the distance between points in the respective gore areas where the left edge of the ramp travel lanes and the right edge of the facility travel lanes meet.

Although logic would indicate that the base length would be the best measure, all of the algorithms calibrated for this methodology produced significantly better results when the short length was used. Therefore, the methodology used the short length as the input parameter in all elements. This is not to suggest that there is no lane-changing over a barrier line in a weaving segment. Lane changes can generally be observed over barrier lines, and indeed, even painted gore areas. Such barrier markings do, however, act as a partial deterrent, and the majority of lane changes do take place over the dashed line.

In some cases, barrier markings are not used, and the two lengths are the same. If an analysis of a future situation is conducted, the appropriate length should be based on local or agency policy regarding the marking of weaving segments.

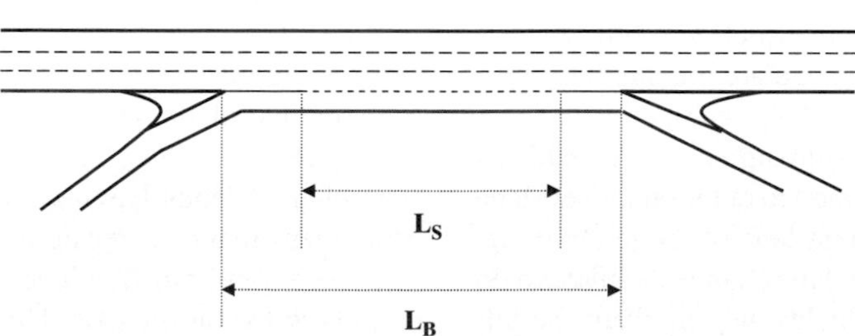

Figure 15.6: Measuring the Length of a Weaving Segment

(*Source:* Roess, R., et al., *Analysis of Freeway Weaving Sections*, Final Report, Draft Chapter for the HCM, National Cooperative Highway Research Program Project 3-75, Polytechnic University and Kittelson and Associates, Brooklyn, NY, September 2007, Exhibit 24-2, p. 2.)

Where even that is not available, a default value (based on the database used in developing the methodology) may be used in which $L_S = 0.77*L_B$.

Width of a Weaving Area

The total width of the weaving area is measured as the total number of lanes available for all flows, *N*. The width of the section has an impact on the total number of lane changes that drivers can choose to make.

15.5 Computational Procedures for Weaving Area Analysis

The computational procedures for weaving areas are most easily used in the operational analysis mode (i.e., all geometric and traffic conditions are specified), and the analysis results in a determination of LOS and weaving segment capacity. The steps in the procedure are illustrated by the flow chart in Figure 15.7.

As with most analysis methodologies, the first step is always to specify the segment under study and its demand flows. For an existing case, these will be based on measured characteristics. For future cases, the geometry would be based on a proposed plan or design, and the demand flows (and characteristics) would be based on forecasts. Where not all information is available, default values may be used; these can be based on regional or agency policies, or on national recommendations. Such recommendations are included in the 2010 edition of the HCM.

The second step has already been discussed. All demand volumes must be converted to flow rates in pc/h for equivalent ideal conditions. This is done using Equation 15-1 and adjustment factors from Chapter 14 for basic freeway and multilane highway segments.

The remainder of the methodology is based on four types of models:

- Algorithms to predict the total rate of lane-changing taking place in the weaving segment. This includes both required and optional lane changes made by weaving vehicles and lane changes made by non-weaving vehicles. The total rate of lane-changing is a measure of turbulence, and it reflects both demand flow rates and configuration characteristics.
- Algorithms to predict the average speed of weaving and nonweaving vehicles within the weaving segment, given stable operations (i.e., *not* LOS F).
- Algorithms to predict the capacity of the weaving segment under both ideal and prevailing conditions.
- An algorithm to estimate the maximum length at which weaving operations exist. Longer segments, even if an apparent weaving configuration exists, operate as if the merge and diverge operations were separate. In such cases, the entry and exit gore areas are separately analyzed using the merge and diverge methodologies presented later in this chapter.

15.5.1 Parameters Used in Weaving Segment Analysis

A very large number of variables are used as input to, output from, or intermediate values in the overall methodology. It is convenient to define them in one place, rather than spread them across the chapter. Figure 15.8 illustrates and defines variables used in the analysis of one-sided weaving segments. Figure 15.9 does so for two-sided weaving segments. As discussed, the basic definition of weaving and nonweaving flows is different in one-sided and two-sided segments, and this influences several portions of the methodology.

15.5.2 Volume Adjustment (Step 2)

Equation 15-1, presented previously, is used to convert all component demand volumes to demand flow rates under equivalent base (or ideal) conditions.

15.5.3 Determining Configuration Characteristics (Step 3)

Two parameters quantify the impact of configuration on lane-changing. One of these is the number of lanes *from which* weaving maneuvers can be completed with no more than one lane change, N_{WV}, which has been previously discussed and defined. The second is LC_{MIN}. This is defined as the minimum rate at which weaving vehicles must change lanes to successfully complete *all* weaving maneuvers in lane-changes per hour (lc/h). It is easily determined from the values of LC_{FR}, LC_{RF}, and LC_{RR}, which have also been defined previously:

For *one-sided* weaving segments:

$$LC_{MIN} = (LC_{FR}*v_{FR}) + (LC_{RF}*v_{RF}) \tag{15-2}$$

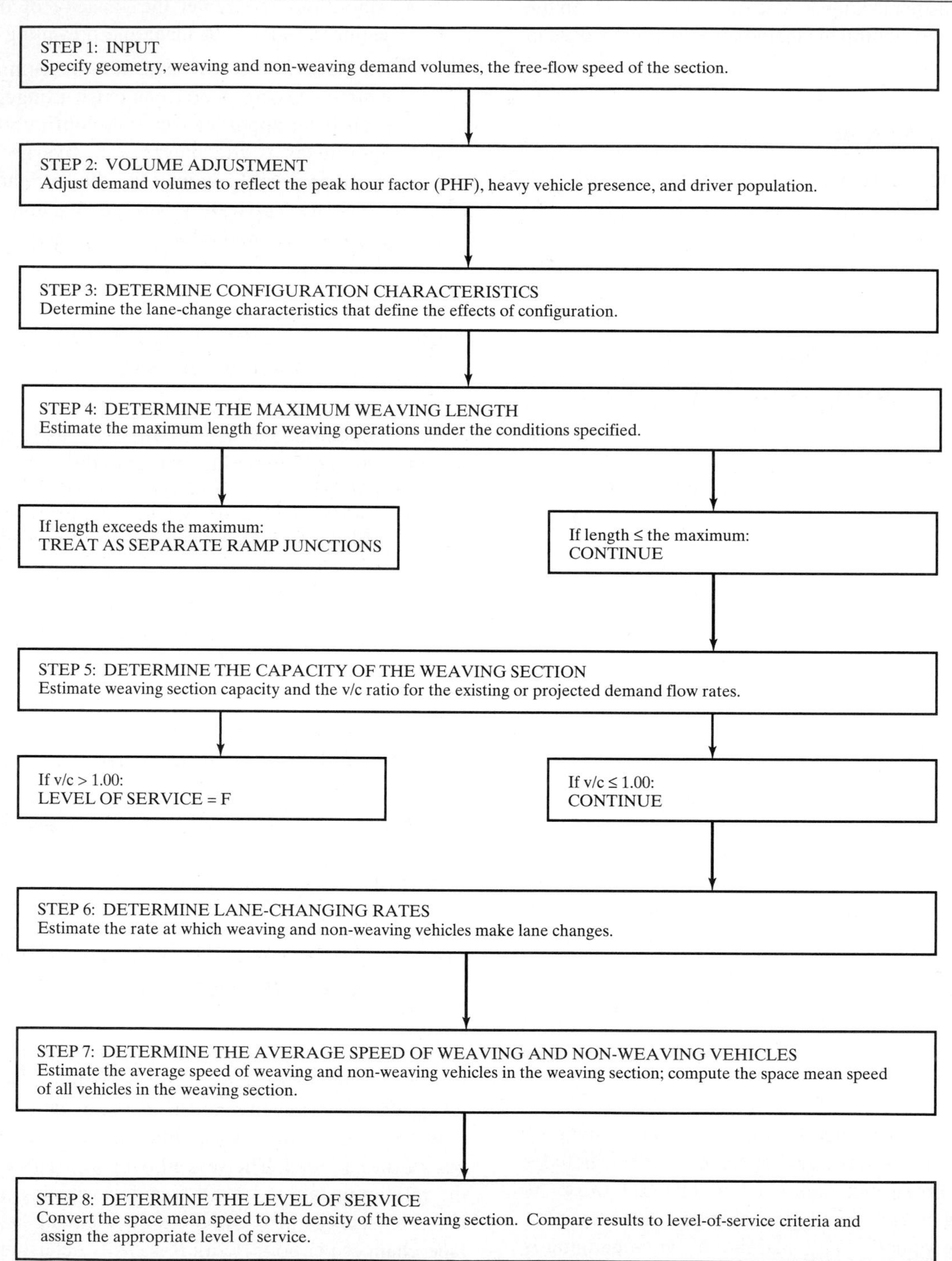

Figure 15.7: Flow Chart for the Weaving Segment Analysis Methodology

(*Source:* Roess, R., et al., *Analysis of Freeway Weaving Sections*, Final Report, Draft Chapter for the HCM, National Cooperative Highway Research Program Project 3-75, Polytechnic University and Kittelson and Associates, Brooklyn, NY, September 2007, Exhibit 24-6, p. 11.)

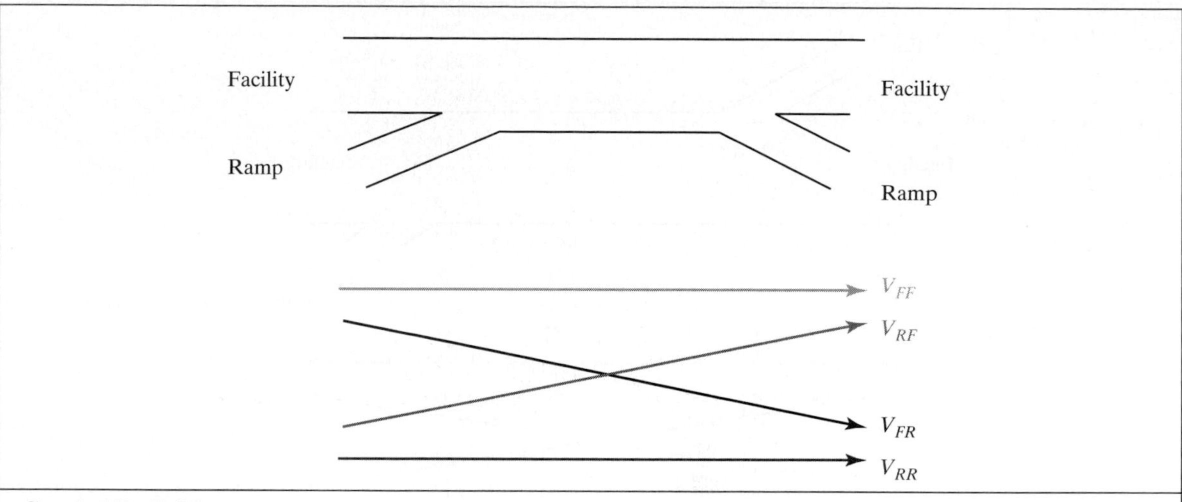

Symbol Definition

Symbol	Definition
v_{FF}	freeway-to-freeway demand flow rate in the weaving section (pc/h)
v_{RF}	ramp-to-freeway demand flow rate in the weaving section (pc/h)
v_{FR}	freeway-to-ramp demand flow rate in the weaving section (pc/h)
v_{RR}	ramp-to-ramp demand flow rate in the weaving section (pc/h)
v_W	weaving demand flow rate in the weaving section (pc/h): $v_W = v_{RF} + v_{FR}$
v_{NW}	non-weaving demand flow rate in the weaving section (pc/h); $v_{NW} = v_{FF} + v_{RR}$
v	total demand flow rate in the weaving section (pc/h), $v = v_W + v_{NW}$
VR	volume ratio: $VR = v_W/v$
N	number of lanes within the weaving section
N_W	number of lanes *from which* a weaving maneuver may be made with one or no lane changes.
S_W	average speed of weaving vehicles within the weaving section (mi/h)
S_{NW}	average speed of non-weaving vehicles within the weaving section (mi/h)
S	average speed of all vehicles within the weaving section (mi/h)
FFS	free-flow speed of the weaving section (mi/h)
D	average density of all vehicles within the weaving section (pc/mi/ln)
W	weaving intensity factor
L_S	length of the weaving section (ft), based on short length definition.
LC_{RF}	minimum number of lane changes that must be made by a single weaving vehicle moving from the on-ramp to the facility.
LC_{FR}	minimum number of lane changes that must be made by a single weaving vehicle moving from the facility to the ramp.
LC_{MIN}	minimum rate of lane changing that must exist for *all* weaving vehicles to successfully complete their weaving maneuvers (lc/h) $LC_{MIN} = (LC_{RF} \times v_{RF}) + (LC_{FR} \times v_{FR})$
LC_W	total rate of lane changing by weaving vehicles within the weaving section (lc/h)
LC_{NW}	total rate of lane changing by non-weaving vehicles within the weaving section (lc/h)
LC_{ALL}	total lane-changing rate of all vehicles within the weaving section (lc/h) $LC_{ALL} = LC_W + LC_{NW}$

Figure 15.8: Weaving Variables Defined for One-Sided Weaving Segments

(*Source:* Roess, R., et al., *Analysis of Freeway Weaving Sections*, Final Report, Draft Chapter for the HCM, National Cooperative Highway Research Program Project 3-75, Polytechnic University and Kittelson and Associates, Brooklyn, NY, September 2007, Exhibit 24-7, p. 12.)

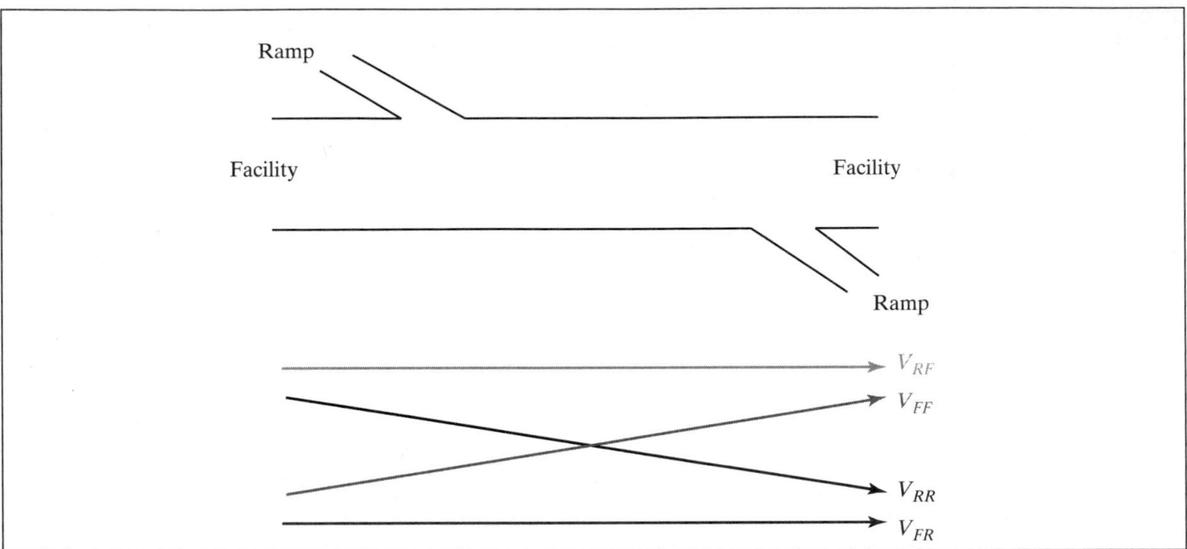

All variables are defined as in Figure 15.8, except for the following variables related to flow designations and lane-changing variables.

Symbol	Definition
v_W	total weaving demand flow rate within the weaving section (pc/h) $v_W = v_{RR}$
v_{NW}	total non-weaving demand flow rate within the weaving section (pc/h) $v_{NW} = v_{FR} + v_{RF} + v_{FF}$
LC_{RR}	minimum number of lane changes that must be made by *one* ramp-to-ramp vehicle to complete a weaving maneuver.
LC_{MIN}	minimum rate of lane changing that must exist for *all* weaving vehicles to successfully complete their weaving maneuvers (lc/h) $LC_{MIN} = (LC_{RR} \times v_{RR})$

Figure 15.9: Weaving Variables Defined for Two-Sided Weaving Segments

(*Source:* Roess, R., et al., *Analysis of Freeway Weaving Sections*, Final Report, Draft Chapter for the HCM, National Cooperative Highway Research Program Project 3-75, Polytechnic University and Kittelson and Associates, Brooklyn, NY, September 2007, Exhibit 24-8, p. 13.)

For *two-sided* weaving segments:

$$LC_{MIN} = LC_{RR} * v_{RR} \tag{15-3}$$

where all variables are as defined in Figures 15.8 and 15.9.

LC_{MIN} effectively defines the hourly rate of lane changes that *must* be made by all weaving vehicles to successfully reach their desired destinations. It is *not* the total lane-changing rate in the segment, which is determined later in the methodology. Total lane-changing includes optional lane changes made by weaving vehicles and all lane changes made by nonweaving vehicles. The importance of LC_{MIN} is that it is primarily a function of the configuration, which forces all of these lane changes to be made *within the confines of the weaving segment.* Optional lane changes, whether made by weaving or nonweaving vehicles, can be made within the weaving segment but could just as easily be made upstream or downstream of the weaving section.

15.5.4 Determining the Maximum Weaving Length (Step 4)

"Weaving" implies that vehicles involved in such maneuvers are using the length of the segment to complete their maneuvers. When the length of the segment is long enough, however, merging at the entry gore and diverging at the exit gore are physically separate, and weaving does not exist. Analytically, such cases are treated as separate merge and diverge segments, with the potential for some length of basic facility between them.

Defining the "maximum length," however, can be accomplished using two different interpretations. In general terms, it is the length at which weaving turbulence no longer has an impact on operations in or capacity of the segment. Unfortunately, basing the maximum length on operational equivalence to basic facilities results in far longer distances compared to basing it on capacity equivalence. If the operational definition were used, however, the resulting capacities of the weaving segments could be significantly higher than the capacities of similar basic facility segments. Therefore, the methodology bases the determination of maximum weaving length on capacity equivalence. The following regression equation has been calibrated:

$$L_{MAX} = [5{,}728\,(1 + VR)^{1.6}] - [1{,}566\,N_{WV}] \qquad (15\text{-}4)$$

where all variables are as defined in Figures 15.8 and 15.9.

The model indicates that the maximum weaving length increases as VR, the volume ratio, increases. This is quite logical because when more of the total traffic is weaving, the impact of weaving is expected to extend over a longer distance. The maximum weaving length decreases with N_{WV}. This variable can only be 2 or 3, and it represents the number of lanes from which a weaving maneuver can be completed with one or fewer lane changes. Given the same flow and split of weaving vehicles, there will be fewer lane changes in a segment in which N_{WV} is 3 than in one in which the value is 2.

Once estimated, the actual weaving length of the segment under study (short length definition) must be compared to the maximum:

- If $L_{MAX} \geq L_S$, continue the analysis using the weaving methodology.
- If $L_{MAX} < L_S$, use merge and diverge analysis methodologies presented later in this chapter.

Noted that Equation 15-4 was calibrated for freeways. Its application to weaving segments on multilane highways and C-D roadways is highly approximate.

15.5.5 Determine the Capacity of the Weaving Segment (Step 5)

The methodology calls for determining capacity before investigating operating parameters and LOS. This is because the models used in estimating densities and speeds within the weaving segment are only valid for cases in which flow is stable, that is, LOS is *not* F. LOS F exists when the demand flow exceeds the capacity of the segment. Logically, then, capacity must be known to determine if stable flow exists; only then can valid estimates of density and speed be made.

There are two situations in which breakdown is expected in a weaving segment:

- Breakdown of a weaving section is expected when the total demand flow exceeds the total capacity of the segment. In practice, this breakdown occurs when an approximate density of 43 pc/mi/ln is reached in the weaving segment.
- Breakdown of a weaving section is expected when the total weaving flow rate exceeds the capacity of the segment to handle weaving flows. The following criteria define the maximum weaving flow rates (total, both weaving flows) that can be accommodated in a weaving segment:

  2,400 pc/h when $N_{WV} = 2$ lanes.
  3,500 pc/h when $N_{WV} = 3$ lanes.

Capacity of a Weaving Segment Based on Breakdown Density

The breakdown density of 43 pc/mi/ln is a logical extension of the calibrated breakdown density on basic freeway segments: 45 pc/h/ln. Given the additional turbulence present in weaving segments, it is logical to assume that breakdown would occur at a lower density. Further, the research behind this methodology [*15*] found no stable operations at higher densities. Fortunately, the methodology does not require trial-and-error computations until the breakdown density is reached. A relatively straightforward regression relationship was calibrated that estimates the capacity at which the density occurs.

Because of turbulence in the weaving segment, and the fact that some weaving segment lanes cannot be used to full advantage due to the existing split between component flows, the capacity controlled by a density of 43 pc/h/ln must be less than the capacity of a lane on a basic facility segment with the same free-flow speed as the weaving segment. Therefore, the algorithm for estimating this capacity is essentially a deduction from the basic freeway segment capacity:

$$c_{IWL} = c_{IFL} - [438.2\,(1 + VR)^{1.6}] + [0.0765\,L_S] + [119.8\,N_{WV}] \qquad (15\text{-}5)$$

where: c_{IWL} = capacity per lane of the weaving section under ideal conditions (pc/h/ln)

c_{IFL} = capacity per lane of a basic facility segment with the same free-flow speed as the weaving segment (pc/h/ln).

All other variables are as previously defined.

Values of basic facility capacity under ideal conditions are taken from Chapter 14 but are repeated in Table 15.2 for convenience.

The weaving segment capacity per lane under ideal conditions must now be converted to a total capacity for the weaving segment under prevailing conditions:

$$c_{WI} = c_{IWL} N f_{HV} f_p \qquad (15\text{-}6)$$

where: c_{WI} = capacity of the weaving section based on breakdown density, veh/h.

All other variables are as previously defined.

Capacity of a Weaving Segment Based on Maximum Weaving Flow Rates

It is possible for the split among component flows to be such that the number of weaving vehicles reaches its capacity *before* the density of the entire weaving segment reaches 43 pc/h/ln. In these cases, the effective controls on the capacity of the segment are the limiting values of weaving flow rate noted earlier. Because the proportion of weaving vehicles is a traffic characteristic of the demand (i.e., fixed for any given analysis), weaving turbulence can cause a breakdown while there is still "capacity" available for nonweaving vehicles. In this type of breakdown, on-ramp vehicles queue on the ramp, whereas off-ramp vehicles queue on the approaching facility segment. Freer flow may exist in the most distant outside lane(s). Capacity of the weaving segment based on maximum weaving flow rates is found as:

$$c_{IW} = \frac{2{,}400}{VR} \quad for \quad N_{WV} = 2$$

$$c_{IW} = \frac{3{,}500}{VR} \quad for \quad N_{WV} = 3 \qquad (15\text{-}7)$$

where: c_{IW} = capacity of the weaving segment under ideal conditions (pc/h)

All other variables are as previously defined.

Note that unlike Equation 15-5, which defined weaving capacity under ideal conditions *per lane*, Equation 15-7 defines the total capacity of the weaving segment under ideal conditions. This, of course, must be converted to prevailing conditions:

$$c_{W2} = c_{IW} f_{HV} f_p \qquad (15\text{-}8)$$

where: c_{W2} = capacity of a weaving segment based on maximum weaving flow rate, veh/h

All other variables are as previously defined.

Final Capacity of the Weaving Segment and the v/c Ratio

Because there are two controls on capacity of a weaving segment, the actual capacity is based on the smallest of the two values computed in Equations 15-6 and 15-8:

$$c_W = min\,(c_{WI}, c_{W2}) \qquad (15\text{-}9)$$

The effective demand-to-capacity ratio is simply the ratio of the total demand flow to the estimated capacity. At this point in the methodology, the demand flow rate, v, is expressed in pc/h under equivalent ideal conditions, whereas capacity, c_W, is expressed in veh/h under prevailing conditions. Thus, to find the appropriate ratio, one must be converted so that both are stated in the same terms:

$$v/c = \frac{v f_{HV} f_p}{c_w} \qquad (15\text{-}10)$$

where all terms have been previously defined.

Table 15.2: Basic Facility Capacity Values (c_{IFL}) for Use in Equation 15-5

Freeways		Multilane Highways and C–D Roadways	
FFS (mi/h)	**Capacity (pc/h/ln)**	**FFS (mi/h)**	**Capacity (pc/h/ln)**
≥70	2,400	≥60	2,200
65	2,350	55	2,100
60	2,300	50	2,000
55	2,250	45	1,900

Final Assessment of Capacity

If the v/c ratio exceeds 1.00, LOS F is automatically assigned, and all computations cease. If the v/c ratio is less than or equal to 1.00, computations continue to find speed and density within the weaving segment.

15.5.6 Determining Total Lane-Changing Rates Within the Weaving Segment (Step 6)

Three types of lane-changing maneuvers exist within a weaving segment:

- **Required** lane changes made by **weaving vehicles:** These lane changes must be made to successfully complete a weaving maneuver. They represent the *absolute minimum lane-changing rate* that can exist in the weaving section for the defined demands. By definition, these lane changes must be made within the confines of the weaving segment. This has been discussed previously, and the rate for such lane changes is defined as LC_{MIN}, and was determined in Step 3 of the methodology.
- **Optional** lane changes made by **weaving vehicles:** These involve lane changes by weaving vehicles that choose to enter segment on a lane *that is not the closest to their desired destination*, or leave the segment on a lane *that is not the closest to their entry leg.* Such entries and exits require additional lane changes to be made within the weaving segment, and act to increase turbulence.
- **Optional** lane changes made by **nonweaving vehicles:** Nonweaving vehicles are never required to make lane changes within a weaving segment. They may, however, *choose* to make lane changes to avoid perceived turbulence.

Although LC_{MIN} is known based on the segment configuration and component demand flow rates, the last two categories of optional lane changing are estimated based on regression equations developed in Reference 15. Total lane-changing rates are separately determined for weaving vehicles and nonweaving vehicles.

Total Lane-Changing Rate for Weaving Vehicles

The total lane-changing rate for weaving vehicles in a weaving segment is estimated as:

$$LC_W = LC_{MIN} + 0.39\left[(L_S - 300)^{0.5} N^2 (1 + ID)^{0.8}\right] \quad (15\text{-}11)$$

where: LC_W = total lane-changing rate for weaving vehicles within a weaving segment, lc/h.
N = number of lanes in the weaving segment.
ID = interchange density, interchanges/mi.

Other variables are as previously defined.

The term L_S–300 is interesting. It suggests that for a segment shorter than 300 feet, weaving vehicles *do not make any optional lane changes.* Because the second term of the equation cannot be negative (LC_W can never be less than LC_{MIN}), for all weaving lengths less than 300 feet (a hopefully very rare event), L_S must be set at 300 ft.

The equation is logical in its form. As length increases, weaving vehicles have more distance and time to make optional lane changes. As the number of lanes, N, increases, more possible lane changes can be made.

This is the first place that interchange density shows up in capacity and LOS analysis. Whereas interchange density was used in HCM 2000 to predict the FFS of a basic freeway segment, in HCM 2010, interchange density has been replaced by total ramp density. A higher interchange density yields more lane-changing as weaving vehicles align themselves as a result of upstream or downstream turbulence.

For weaving sections, interchange density considers a facility segment 3 miles upstream and downstream of the middle of the weaving segment. The weaving segment itself counts as *one* interchange within this 6-mile range.

When applying Equation 15-11 to a weaving segment on a multilane highway, interchange density is replaced by the density of roadside access points in the analysis direction. *Only* significant unsignalized access points should be considered in this density. Application of Equation 15-11 to multilane highway weaving segments is highly approximate.

Total Lane-Changing Rate for Nonweaving Vehicles

Because no nonweaving vehicle *must* make a lane change within the weaving segment, all such lane changes are optional. This makes them far more difficult to predict than weaving vehicle lane changes, which are tied to the configuration of the weaving segment and the demand flow rates. The methodology has *two* basic equations that are used to estimate nonweaving lane-changing rates:

$$LC_{NW1} = (0.206\, v_{NW}) + (0.542\, L_S) - (192.6\, N)$$

$$LC_{NW2} = 2{,}135 + 0.223(v_{NW} - 2{,}000) \quad (15\text{-}12)$$

where: LC_{NW1} = first estimate, nonweaving vehicle lane-changing rate, lc/h.

LC_{NW2} = second estimate, nonweaving vehicle lane-changing rate, lc/h.

All other variables are as previously defined.

The first equation covers the majority of situations. It presents a logical set of trends. As nonweaving flow increases, nonweaving lane-changing also increases. As the length of the segment increases, nonweaving lane-changing increases because such vehicles have more distance and time to make such movements. Nonweaving lane-changing *decreases* as the number of lanes in the weaving segment increases. This is less obvious. As the width of the weaving segment increases, nonweaving vehicles have a better opportunity to segregate from weaving vehicles in outer lanes. This would tend to decrease their desire to make lane changes out of these lanes. The first equation has an arbitrary minimum of "0."

The two equations, unfortunately, are *very* discontinuous. Therefore, it is critical to have a methodology that provides for smooth transitions from one equation to the other without distorting the results. This is done using a lane-changing index, I_{NW}:

$$I_{NW} = \frac{L_S ID v_{NW}}{10{,}000} \tag{15-13}$$

where all variables are as previously defined. The origin of this index is to explain when the second Equation 15-12 is used. It applies to cases in which long lengths, high interchange densities, and/or high nonweaving flows conspire to create far more lane-changing among such vehicles than normally expected. In calibrating these algorithms [*15*], the first equation applies to cases in which $I_{NW} \leq 1{,}300$. The second applies to cases in which $I_{NW} \geq 1{,}950$. For values in between, a straight-line interpolation of the two equations is used. Thus:

If $I_{NW} \leq 1{,}300$:

$$LC_{NW} = LC_{NW1} \tag{15-14}$$

If $I_{NW} \geq 1{,}950$:

$$LC_{NW} = LC_{NW2} \tag{15-15}$$

If $1{,}300 < I_{NW} < 1{,}950$:

$$LC_{NW} = LC_{NW1} + (LC_{NW2} - LC_{NW1})\left(\frac{I_{NW} - 1{,}300}{650}\right) \tag{15-16}$$

Total Lane-Changing in a Weaving Segment

The total lane-changing rate in any weaving segment is simply the sum of the lane-changing rate for weaving vehicles and the lane-changing rate for nonweaving vehicles:

$$LC_{ALL} = LC_W + LC_{NW} \tag{15-17}$$

where: LC_{ALL} = total lane-changing rate in a weaving segment, lc/h.

All other variables are as previously defined.

15.5.7 Determining the Average Speed of Vehicles Within a Weaving Segment (Step 7)

The heart of the methodology for weaving segments is the estimation of average speeds within the weaving segment. The average speeds of weaving and nonweaving vehicles are estimated separately because they are affected by different factors, and can be quite different in some cases. Estimated speeds, together with known demand flow rates, will yield a density estimate, which is used to determine the LOS. Thus, although speed is a secondary performance measure for weaving segments, it must be computed to obtain an estimate of density, the primary measure of effectiveness used for weaving segments.

Average Speed of Weaving Vehicles

The general algorithm for prediction of the average speed of weaving vehicles in a weaving segment is basically the same as that in HCM 2000:

$$S_W = S_{MIN} + \left(\frac{S_{MAX} - S_{MIN}}{1 + W}\right) \tag{15-18}$$

where: S_W = average speed of weaving vehicles, mi/h.

S_{MIN} = minimum average speed of weaving vehicles expected in a weaving segment, mi/h.

S_{MAX} = maximum average speed of weaving vehicles expected in a weaving segment, mi/h.

W = weaving intensity factor.

Although this basic algorithm is the same as that in the HCM 2000, the new methodology changes some of the values used in it. The maximum speed of weaving vehicles is the free-flow speed of the facility. In HCM 2000, the maximum was set as the FFS + 5, to correct for a tendency in the algorithm to underpredict high speeds. The minimum average speed remains 15 mi/h. The most significant change, however, is a change in the algorithm specifying the weaving intensity factor, W:

$$W = 0.226\left(\frac{LC_{ALL}}{L_S}\right)^{0.789} \quad (15\text{-}19)$$

where all variables are as previously defined. The term LC_{ALL}/L_S is essentially a measure of lane-changing intensity over length—total lane changes per foot of weaving segment length. Thus lane-changing behavior becomes the primary measure of weaving intensity. Then:

$$S_W = 15 + \left(\frac{FFS - 15}{1 + W}\right) \quad (15\text{-}20)$$

where all terms are as previously defined.

The term (1+*W*) is used instead of *W* because *W* can be less than or greater than 1.00. Dividing by a number that can be less than or more than 1.00 creates inconsistent arithmetic results. The *(1+W)* assures that all denominators are more than 1.00 and that as *W* increases, speed decreases.

Average Speed of Nonweaving Vehicles

The average speed of nonweaving vehicles is treated as a reduction from the free-flow speed according to the following algorithm:

$$S_{NW} = FFS - (0.0072 LC_{MIN}) - (0.0048\ {}^{v}/_{N}) \quad (15\text{-}21)$$

where all terms are as previously defined.

Nonweaving speed obviously decreases as *v/N* increases. More surprising is the appearance of LC_{MIN} in the equation. Because this is a regression equation, its appearance is as a measure of weaving turbulence. For nonweaving speeds, it was a stronger statistical predictor than other measures, such as *W* or LC_{ALL}.

Average Speed of All Vehicles

Given estimates of both average speed of weaving vehicles and average speed of nonweaving vehicles, a space mean speed for all vehicles may be computed as:

$$S = \frac{v_W + v_{NW}}{\left(\frac{v_W}{S_W}\right) + \left(\frac{v_{NW}}{S_{NW}}\right)} \quad (15\text{-}22)$$

where all variables are as previously defined.

15.5.8 Determining Density and Level of Service in a Weaving Segment (Step 8)

The final computation in the analysis of weaving segments is the conversion of average speed and demand flow rate into an estimate of density, from which LOS is determined using Table 15.1.

$$D = \frac{({}^{v}/_{N})}{S} \quad (15\text{-}23)$$

where D is the density in pc/mi/ln and all other variables are as previously defined.

The methodology results in estimating both and average speed and an average density of all vehicles within the weaving segment, and a determination of the prevailing LOS given the geometric characteristics of the segment and the demand characteristics. The capacity of the weaving segment is also determined for the prevailing conditions specified. This information provides for significant insight into the expected operational characteristics of the segment, as well as insight into existing or potential problems

15.6 Basic Characteristics of Merge and Diverge Segment Analysis

As illustrated in Figure 15.2, analysis procedures for merge and diverge areas focus on the merge or diverge influence area that encompasses lanes 1 and 2 (shoulder and adjacent) freeway lanes and the acceleration or deceleration lane for a distance of 1,500 feet upstream of a diverge or 1,500 feet downstream of a merge area.

Analysis procedures provide algorithms for estimating the density in these influence areas. Estimated densities are compared to the criteria of Table 15.1 to establish the LOS.

Because the analysis of merge and diverge areas focuses on influence areas including only the two right-most lanes of the freeway, a critical step in the methodology is the estimation of the lane distribution of traffic immediately upstream of the merge or diverge. Specifically, a determination of the approaching demand flow remaining in lanes 1 and 2 immediately upstream of the merge or diverge is required. Figure 15.10 shows the key variables involved in the analysis.

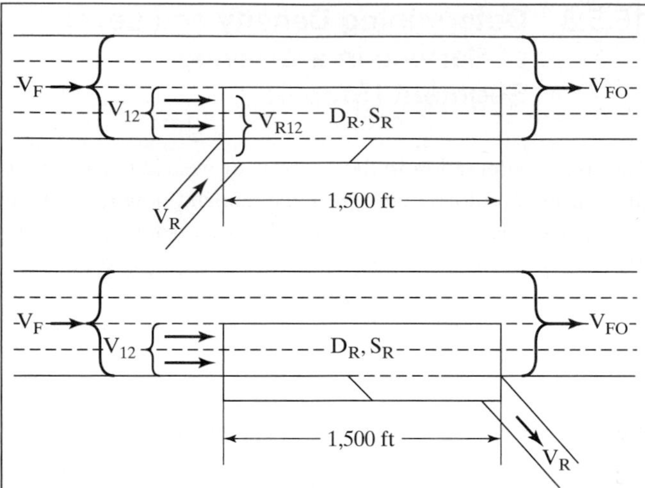

Figure 15.10: Critical Variables in Merge and Diverge Analysis

(*Source:* Used with permission of Transportation Research Board, National Research Council, *Highway Capacity Manual*, 2000, Exhibit 25-2, p. 25-2.)

The variables included in Figure 15.10 are defined as follows:

v_F = freeway demand flow rate immediately upstream of merge or diverge junction, in pc/h under equivalent base conditions

v_{12} = freeway demand flow rate in lanes 1 and 2 of the freeway immediately upstream of the merge or diverge junctions, in pc/h under equivalent base conditions

v_R = ramp demand flow rate, in pc/h under equivalent base conditions

v_{R12} = total demand flow rate entering a merge influence area, $v_R + v_{12}$, in pc/h under equivalent base conditions

v_{FO} = total outbound demand flow continuing downstream on the freeway, pc/h under equivalent base conditions

D_R = average density in the ramp influence area, pc/mi/ln

S_R = space mean speed of all vehicles in the ramp influence area, mi/h

Other than the standard geometric characteristics of the facility that are used to determine its free-flow speed and adjustments to convert demand volumes in veh/h to pc/h under equivalent base conditions (Equation 15-1), there are two specific geometric variables of importance in merge and diverge analysis:

$L_{a \text{ or } d}$ = length of the acceleration or deceleration lane, ft

$RFFS$ = free-flow speed of the ramp, mi/h

The length of the acceleration or deceleration lane is measured from the point at which the ramp lane and lane 1 of the main facility touch to the point at which the acceleration or deceleration lane begins or ends. This definition includes the taper portion of the acceleration or deceleration lane and is the same for both parallel and tapered lanes. Figure 15.11 illustrates the measurement of length of acceleration and deceleration lanes.

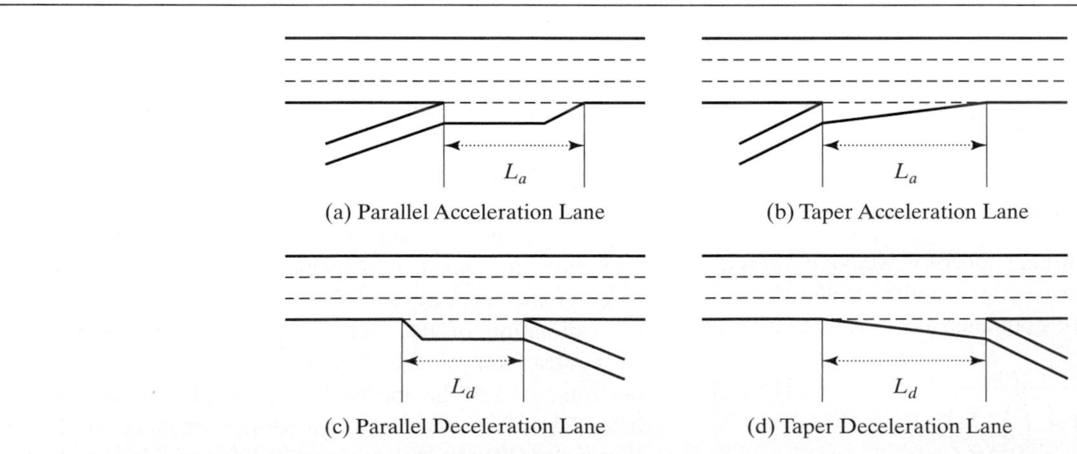

Figure 15.11: Measuring the Length of Acceleration and Deceleration Lanes

The free-flow speed of the ramp is best observed in the field but may be estimated as the design speed of the most restrictive element of the ramp. Many ramps include compound horizontal curves or a number of separate horizontal or vertical curves. The free-flow speed is generally controlled by the design speed (or maximum safe operating speed) of the most severe of these.

15.7 Computational Procedures for Merge and Diverge Segments

15.7.1 Overview

Figure 15.12 is a flow chart of the analysis methodology for merge or diverge junctions. It illustrates the following five fundamental steps:

1. Specify all traffic and roadway data for the junction to be analyzed: peak-hour demands, peak-hour factor (PHF), traffic composition, driver population, and geometric details of the site, including the free-flow speed for the facility and for the ramp. Convert all demand volumes to flow rates in pc/h under equivalent base conditions using Equation 15-1.
2. Determine the demand flow in lanes 1 and 2 of the facility immediately upstream of the merge or diverge junction using the appropriate algorithm as specified.
3. Determine whether the demand flow exceeds the capacity of any critical element of the junction. Where demand exceeds capacity, LOS F is assigned and the analysis is complete.
4. If operation is determined to be stable, determine the density of all vehicles within the ramp influence area. Table 15.1 is then used to determine LOS based on the density in the ramp influence area.
5. If the operation is determined to be stable, determine the speed of all vehicles within the ramp influence area and across all facility lanes as secondary measures of performance.

Once all input characteristics of the merge or diverge junction are specified and all demand volumes have been converted to flow rates in pc/h under equivalent base conditions, remaining parts of the methodology may be completed.

Note that the base methodology for merge and diverge segments is based on single-lane, right-hand on- and off-ramps. There are a many other types of configuration, including multi-lane on- and off-ramp junctions, left-hand ramps, major merge and diverge segments, and ramps on five-lane (one direction) facility segments. These are handled as "Special Cases," which are included in the appendix to this chapter. These cases involve logical modifications to the base methodology for each case. In few cases, substantial databases were available to calibrate these modifications. In most cases, the modifications are based on theoretical models and informed judgment of the HCQSC.

15.7.2 Estimating Demand Flow Rates in Lanes 1 and 2 (Step 2)

The starting point for analysis is the determination of demand flow rates in lanes 1 and 2 (the two right-most lanes) of the facility immediately upstream of the merge or diverge junction. This is done using a series of regression-based algorithms developed as part of a nationwide study of ramp-freeway junctions [*16*]. Different algorithms are used for merge and diverge areas.

On-Ramps (Merge Segments)

For merge areas, the flow rate remaining in lanes 1 and 2 immediately upstream of the junction is computed simply as a proportion of the total approaching facility flow:

$$v_{12} = v_F * P_{FM} \tag{15-24}$$

where: P_{FM} = proportion of approaching vehicles remaining in lanes 1 and 2 immediately upstream of the merge junction, in decimal form

All other variables are as previously defined.

The value of P_{FM} varies with the number of lanes on the facility, demand flow levels, the proximity of adjacent ramps (in some cases), the length of the acceleration lane (in some cases), and the free-flow speed on the ramp (in some cases.) Table 15.3 summarizes algorithms for estimating P_{FM}, and contains a matrix for determining which algorithm is appropriate for various analysis scenarios.

For four-lane facilities (two lanes in each direction), the value is trivial because the entire flow is in lanes 1 and 2. For such cases, P_{FM} is 1.00. For six- and eight-lane facilities,

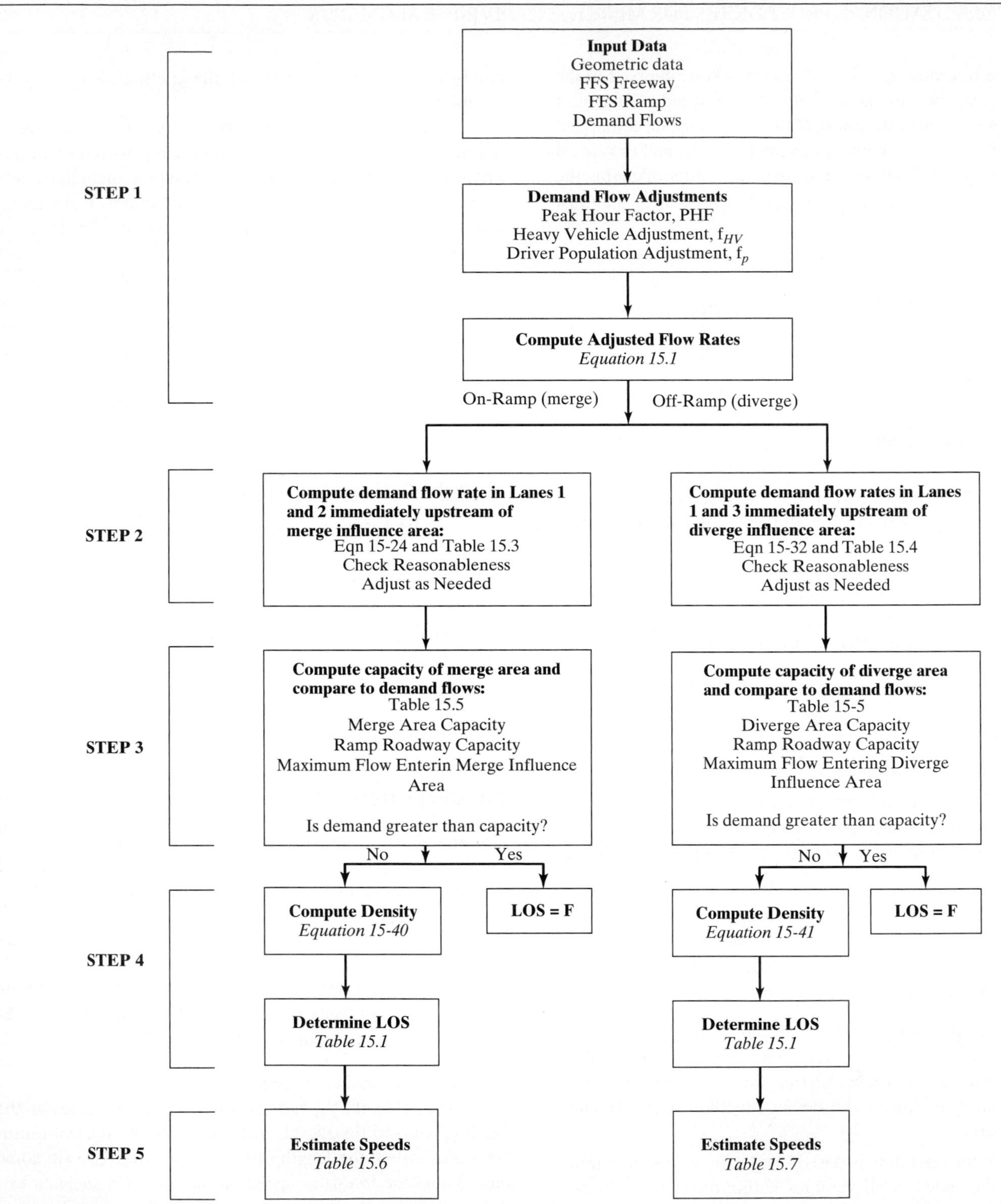

Figure 15.12: Flow Chart for Analysis of Ramp-Facility Junctions

(*Source: Draft Chapter 13, HCM2010,* National Cooperative Highway Research Program Project 3-29, Kittelson and Associates, Portland OR, Exhibit 13-4, p. 13-8, 2008.)

Table 15.3: Estimating P_{FM} at On-Ramps (Merge Segments)

No. of Freeway Lanes[a]	Model for Determining P_{FM}	
4	$P_{FM} = 1.000$	
6	$P_{FM} = 0.5775 + 0.000028L_A$	Equation 15-25
	$P_F = 0.7289 - 0.0000135(V_F + v_R) - 0.003296S_{FR} + 0.000063L_{UP}$	Equation 15-26
	$P_{FM} = 0.5487 + 0.2628(v_D/L_{DOWN})$	Equation 15-27
8	For v_F/RFFS $\leq 72: P_{FM} = 0.2178 - 0.000125v_R + 0.0115(L_A/S_{FR})$	Equation 15-28
	For v_F/RFFS > 72: $P_{FM} = 0.2178 - 0.000125\ v_R$	Equation 15-29

Selecting Equations for P_{FM} for 6-Lane Freeways

Adjacent Up stream Ramp	Subject Ramp	Adjacent Downstream Ramp	Equation(s) Used
None	On	None	Equation 15-25
None	On	On	Equation 15-25
None	On	Off	Equation 15-27 or 15-25
On	On	None	Equation 15-25
Off	On	None	Equation 15-26 or 15-25
On	On	On	Equation 15-25
On	On	Off	Equation 15-27 or 15-25
Off	On	On	Equation 15-27 or 15-25
Off	On	Off	Equation 15-25, 15-26, or 15-27.

Note: [a]4 lanes = 2 lanes in each direction.

(*Source:* Modified from *Draft Chapter 13, HCM2010*, National Cooperative Highway Research Program Project 3-92, Kittelson and Associates, Portland OR, Exhibit 13-6, p. 13-12, 2008.)

the values are established using the appropriate algorithm shown in Table 15.3. All variables are as previously defined. L_{up} is the distance to the adjacent upstream ramp in feet, and L_{dn} is the distance to the adjacent downstream ramp.

For six- and eight-lane facilities, it is believed that the flow remaining in lanes 1 and 2 depends on the distance to and flow rate on adjacent upstream and downstream ramps. A driver entering the facility on a nearby upstream on-ramp is more likely to remain in lanes 1 and 2 if the distance between ramps is insufficient to allow the driver to make two lane-changes to reach outer lanes. Likewise, a driver knowing he or she has to exit at a nearby downstream off-ramp is more likely to move into lanes 1 and 2 than a driver proceeding downstream on the main facility. Although these are logical expectations, the database on ramp junctions was sufficient to establish these relationships for only a few scenarios on six-lane freeways. Thus Equation 15-26 considers the impact of adjacent upstream off-ramps, whereas Equation 15-27 considers the impact of adjacent downstream off-ramps on lane distribution at merge areas on six-lane facilities. There are no relationships considering the impact of adjacent upstream or downstream on-ramps on six-lane facilities or for any adjacent ramps on eight-lane facilities. Equation 15-25 is used for isolated merge areas on six-lane freeways and is used as the default algorithm for all merge areas on six-lane freeways that cannot be addressed using Equations 15-26 or 15-27.

As shown in the selection matrix of Table 15.3, there are situations in which more than one algorithm may be appropriate. In such cases, it must be determined whether or not the subject ramp is far enough away from the adjacent ramp to be considered "isolated."

Equation 15-26 or 15-25 may be used to analyze merge areas on six-lane facilities with an adjacent upstream off-ramp. The selection of the appropriate equation for use is based on the distance at which the two equations yield the same result, or:

$$L_{EQ} = 0.214(v_F + v_R) + 0.444\ L_a + 53.32\ RFFS - 2{,}403 \qquad (15\text{-}30)$$

where: L_{EQ} = equivalence distance, ft

All other variables are as previously defined.

When the actual distance to the upstream off-ramp is greater than or equal to the equivalence distance ($L_{up} \geq L_{EQ}$), the subject ramp may be considered to be "isolated." Equation 15-25 is used. Where the reverse is true ($L_{up} < L_{EQ}$), the effect of the adjacent upstream off-ramp must be considered. Equation 15-26 is used.

Similarly, Equation 15-27 or 15-25 may be used to analyze merge areas on six-lane facilities with an adjacent downstream off-ramp. Again, the equivalence distance is established:

$$L_{EQ} = \frac{v_d}{0.1096 + 0.000107L_a} \quad (15\text{-}31)$$

where: L_{EQ} = equivalence distance, ft
v_d = demand flow rate on the adjacent down-stream off-ramp, pc/h under equivalent base conditions

All other variables are as previously defined.

As previously, if the actual distance to the adjacent downstream off-ramp (L_{dn}) is greater than or equal to the equivalence distance ($L_{dn} \geq L_{EQ}$), the subject merge may be considered to be "isolated." Equation 15-25 would be used. If the reverse is true ($L_{dn} < L_{EQ}$), the effect of the adjacent downstream off-ramp must be considered, and Equation 15-27 is used.

Equations 15-28 and 15-29 both apply to isolated on-ramp junctions on eight-lane freeways (four lanes in each direction). Two equations are used to avoid a potential solution in which lengthening the acceleration lane leads to an increase in density.

It is also possible that two equations will legitimately apply to a given situation. Where a merge area has both an upstream and a downstream adjacent off-ramp, it would be considered first in combination with the adjacent downstream ramp, then with the adjacent upstream ramp. Two different answers may result. In such situations, the approach yielding the worst prediction of operating conditions is used.

Diverge Areas

The general approach to estimating the demand flow rate in lanes 1 and 2 immediately upstream of a diverge is somewhat different from the one used for merge areas. This is because all of the off-ramp traffic is assumed to be in lanes 1 and 2 at this point. Thus the flow in lanes 1 and 2 is taken as the off-ramp flow plus a proportion of the through traffic on the facility.

$$v_{12} = v_R + (v_F - v_R)P_{FD} \quad (15\text{-}32)$$

where: P_{FD} = proportion of approaching vehicles remaining in lanes 1 and 2 immediately upstream of the diverge junction, in decimal form

All other variables are as previously defined.

Table 15.4 shows the algorithms used to estimate P_{FD} as well as a matrix that may be consulted when selecting the appropriate algorithm. L_{up} is the distance to the adjacent upstream ramp, and L_{dn} is the distance to the adjacent downstream ramp.

Again, the database for ramp junctions was sufficient to establish relationships for the impact of adjacent ramp activity for only two cases, restricted to six-lane facilities: Equation 15-34 considers the impact of an adjacent upstream on-ramp and Equation 15-35 considers the impact of an adjacent downstream off-ramp. There are no algorithms for the effects of adjacent upstream off-ramps or adjacent downstream on-ramps on six-lane facilities. For eight-lane facilities, the value of P_{FD} is taken to be a constant (0.436) primarily due to a sparse database of off-ramps on an eight-lane facility.

When an adjacent upstream on-ramp on a 6-lane freeway exists, an equilibrium distance must be established beyond which the subject off-ramp is considered to be isolated:

$$L_{EQ} = \frac{v_u}{0.071 + 0.000023v_F - 0.000076v_R} \quad (15\text{-}36)$$

where: L_{EQ} = equivalence distance, ft
v_u = demand flow rate on the adjacent up-stream on-ramp pc/h under equivalent base conditions

All other variables are as previously defined.

If $L_{up} \geq L_{EQ}$, the subject ramp may be considered to be isolated, and Equation 15-33 would be used. If $L_{up} \leq L_{EQ}$, the effect of the adjacent upstream on-ramp is taken into account by using Equation 15-34.

When an adjacent downstream off-ramp on a 6-lane freeway exists, an equivalence distance must also be computed:

$$L_{EQ} = \frac{v_d}{1.15 - 0.000032v_F - 0.000369v_R} \quad (15\text{-}37)$$

where all variables have been previously defined. If $L_{dn} \geq L_{EQ}$, the subject off-ramp may be considered to be isolated, and Equation 15-33 is applied. If $L_{dn} < L_{EQ}$, the effect of the downstream off-ramp is considered by using Equation 15-35.

Table 15.4: Estimating P_{FD} at Off-Ramps (Diverge Segments)

No. of Freeway Lanes[a]	Model for Determining P_{FD}	
4	$P_{FD} = 1.000$	
6	$P_{FD} = 0.760 - 0.000025\, v_F - 0.000046\, v_R$	Equation 15-33
	$P_{FD} = 0.717 - 0.000039\, v_F + 0.604\, (v_U / L_{UP})$	Equation 15-34
	$P_{FD} = 0.616 - 0.000021\, v_F + 0.124\, (v_D / L_{DOWN})$	Equation 15-35
8	$P_{FD} = 0.436$	

Selecting Equations for P_{FD} for 6-Lane Freeways

Adjacent Upstream Ramp	Subject Ramp	Adjacent Downstream Ramp	Equation(s) Used
None	Off	None	Equation 15-33
None	Off	On	Equation 15-33
None	Off	Off	Equation 15-35 or 15-33
On	Off	None	Equation 15-34 or 15-33
Off	Off	None	Equation 15-33
On	Off	On	Equation 15-34 or 15-33
On	Off	Off	Equation 15-33, 15-34, or 15-35

Note: [a]4 lanes = 2 lanes in each direction.
(*Source:* Modified from *Draft Chapter 13, HCM2010*, National Cooperative Highway Research Program Project 3-92, Kittelson and Associates, Portland OR, Exhibit 13-6, p. 13-12, 2008.)

Checking the Reasonableness of Lane Distribution Predictions

Once the flow rate for lanes 1 and 2 have been predicted, it is necessary to subject the results to a "reasonableness" check. Because the algorithms used are regression based, when used outside the boundaries of the database used to calibrate them, results can lead to illogical lane distributions. The estimated lane distribution must meet these two conditions:

1. Average flow rate in the outer lanes may not exceed 2,700 pc/h/ln.
2. Average flow rate in the outer lanes may not be more than 1.5 times the average flow rate in lanes 1 and 2.

Obviously, the size of the freeway determines the number of outer lanes. For four-lane freeways (two lanes in each direction), there are no outer lanes, and all vehicles approach in lanes 1 and 2. For six-lane freeways (three lanes in each direction), there is one outer lane (lane 3). For eight-lane freeways (four lanes in each direction), there are two outer lanes (lanes 3 and 4).

If either or both of these criteria are violated by the predicted lane distribution, the flow rate in lanes 1 and 2 must be adjusted to accommodate these limits. If the average flow rate in outer lanes exceeds 2,700 pc/h/ln, it is set at 2,700 pc/h/ln, and the flow rate in lanes 1 and 2 is recomputed as:

$$V_{12} = V_F - 2700 N_O \tag{15-38}$$

where N_O is the number of outer lanes. If the average flow rate in outer lanes exceeds 1.5 times the average flow rate in lanes 1 and 2, the outer lane flow is set at 1.5 times the average flow in lanes 1 and 2, and the flow rate in lanes 1 and 2 is recomputed as:

$$For\ N_0 = 1:\quad V_{12} = {}^{V_F}/_{1.75}$$

$$For\ N_0 = 2:\quad V_{12} = {}^{V_F}/_{2.50}$$

$$For\ N_0 > 2\quad V_{12} = 2 v_F/(1.5 N_0 + 2) \tag{15-39}$$

In cases where both limitations are violated, the revision that meets *both* criteria is used.

15.7.3 Capacity Considerations

The analysis procedure for merge and diverge areas determines whether the segment in question has failed (LOS = F) based on a comparison of demand flow rates to critical capacity values. In general, the basic capacity of the facility is not affected by merging or diverging activities. Because of this, the basic facility capacity must be checked immediately upstream and/or downstream of the merge or diverge. Ramp roadway capacities must also be examined for adequacy. When demand flows exceed any of these capacities, a failure is expected, and the LOS is determined to be F.

The total flow entering the ramp influence area is also checked. Although a maximum desirable value is set for this flow, exceeding it does not imply LOS F if no other capacity value is exceeded. In cases where only this maximum is violated, expectations are that service quality will be less than that predicted by the methodology. Capacity values are given in Table 15.5. The freeway capacity values shown are the same as those for basic freeway sections used in Chapter 14. When applying these procedures to merging or diverging segments on multilane highways or collector-distributor roadways, use the values indicated in Chapter 14 for multilane highway sections directly. Other values shown in Table 15.5 may be approximately applied to merging or diverging multilane highway segments.

The specific checkpoints that should be compared to the capacity criteria of Table 15.5 may be summarized as follows:

- For merge areas, the maximum facility flow occurs downstream of the merge. Thus the facility capacity is compared with the downstream facility flow ($v_{FO} = v_F + v_R$).
- For diverge areas, the maximum facility flow occurs upstream of the diverge. Thus the facility capacity is compared to the approaching upstream facility flow, v_F.
- Where lanes are added or dropped at a merge or diverge, both the upstream (v_F) and downstream (v_{FO}) facility flows must be compared to capacity criteria.
- For merge areas, the flow entering the ramp influence area is $V_{R12} = V_{12} + V_R$. This sum is compared to the maximum desirable flow indicated in Table 15.5.
- For diverge areas, the flow entering the ramp influence area is v_{12}, as the off-ramp flow is already

Table 15.5: Capacity Values for Ramp Checkpoints

Freeway FFS (mi/h)	Maximum Freeway Flow Upstream/Downstream of Merge or Diverge (pc/h) — Number of Lanes in One Direction: 2	3	4	≥5	Maximum Desirable Flow Entering Merge Influence Area (pc/h)	Maximum Desirable Flow Entering Diverge Influence Area (pc/h)
≥70	4,800	7,200	9,600	2,400/ln	4,600	4,400
65	4,700	7,050	9,400	2,350/ln	4,600	4,400
60	4,600	6,900	9,200	2,300/ln	4,600	4,400
55	4,500	6,750	9,000	2,250/ln	4,600	4,400

Ramp Free-Flow Speed RFFS (mi/h)	Capacity of Ramp Roadway(pc/h): Single-Lane Ramps	Two-Lane Ramps
>50	2,200	4,400
>40–50	2,100	4,100
>30–40	2,000	3,800
≥20–30	1,900	3,500
<20	1,800	3,200

(*Source:* Used with permission of Transportation Research Board, National Research Council, *Highway Capacity Manual*, 2000, compiled from Exhibits 25-3, p. 25-4, 25-7, p. 25-9, and 25-14, p. 25-14.)

included. It is compared directly with the maximum desirable flow indicated in Table 15.5.

- All ramp flows, v_R must be checked against the ramp capacities given in Table 15.5.

The ramp capacity check is most important for diverge areas. Diverge segments rarely fail unless the capacity of one of the diverging legs is exceeded by the demand flow. This is most likely to happen on the off-ramp. Also note that the capacities shown in Table 15.5 for two-lane ramps may be quite misleading. They refer to the ramp roadway itself, not to the junction with the main facility. There is no evidence, for example, that a two-lane on-ramp junction can accommodate any greater flow than a one-lane junction. It is unlikely that a two-lane on-ramp can handle more than 2,250 to 2,400 pc/h through the merge area. For higher on-ramp demands, a two-lane on-ramp would have to be combined with a lane addition at the facility junction.

15.7.4 Determining Density and Level of Service in the Ramp Influence Area

If all facility and ramp capacity checks indicate that stable flow prevails in the merge or diverge area, the density in the ramp influence area may be estimated using Equation 15-40 for merge areas and Equation 15-41 for diverge areas:

$$D_R = 5.475 + 0.00734v_R + 0.0078v_{12} - 0.00627L_a \qquad (15\text{-}40)$$

$$D_R = 4.252 + 0.0086v_{12} - 0.009L_d \qquad (15\text{-}41)$$

where all variables have been previously defined. In both cases, the density in the ramp influence area depends on the flow entering it (v_R and v_{12} for merge areas, and v_{12} for diverge areas) and the length of the acceleration or deceleration lane. The density computed by Equation 15-40 or 15-41 is directly compared to the criteria of Table 15.1 to determine the expected LOS.

15.7.5 Determining Expected Speed Measures

Although it is not a measure of effectiveness, and the determination of an expected speed is not required to estimate density (as was the case for weaving areas), it is often convenient to have an average speed as an additional measure or as an input to system analyses. Because speed behavior in the vicinity of ramps (vicinity = 1,500-ft segment encompassing the ramp influence area) is different from basic sections, three algorithms are provided for merge areas and three for diverge areas as follows:

- Estimation algorithm for average speed within the ramp influence area.
- Estimation algorithm for average speed in outer lanes (where they exist) within the 1,500-foot boundaries of the ramp influence area.
- Algorithm for combining the above into an average space mean speed across all lanes within the 1,500-foot boundaries of the ramp influence area.

Table 15.6 summarizes these algorithms for merge areas, and Table 15.7 summarizes them for diverge areas.

Table 15.6: Estimating Average Speeds in Merge Areas

Avg Spd In ________	Estimation Algorithm
Ramp Influence Area	$S_R = FFS - (FFS - 42)\, M_S$ $M_S = 0.321 + 0.0039_e{}^{(v_{R12}/1000)} - 0.002(La * RFFS/1{,}000)$
Outer Lanes	$S_o = FFS \quad v_{oa} < 500$ pc/h $S_o = FFS - 0.0036(v_{oa} - 500) \quad v_{oa} = 500 - 2{,}300$ pc/h $S_o = FFS - 6.53 - 0.006(v_{oa} - 2300) \quad v_{oa} > 2{,}300$ pc/h
All Lanes	$S = \dfrac{V_{R12} + V_{OA}N_o}{\left(\dfrac{V_{R12}}{S_R}\right) + \left(\dfrac{V_{OA}N_O}{S_O}\right)}$

Table 15.7: Estimating Average Speeds in Diverging Areas

Avg Spd In ________	**Estimation Algorithm**
Ramp Influence Area	$S_R = FFS - (FFS - 42)D_S$ $D_S = 0.883 + 0.00009v_{12} - 0.013RFFS$
Outer Lanes	$S_o = 1.097FFS \quad v_{oa} < 1{,}000$ pc/h $S_o = 1.097FFS - 0.0039(v_{oa} - 1{,}000) \quad v_{oa} \geq 1000$ pc/h
All Lanes	$S = \dfrac{V_{12} + V_{OA}N_o}{\left(\dfrac{V_{12}}{S_R}\right) + \left(\dfrac{V_{OA}N_O}{S_O}\right)}$

The variables in Tables 15.6 and 15.7 are defined as follows:

S_R = space mean speed of vehicles within the ramp influence area; v_{R12} for merge areas; v_{12} for diverge areas; mi/h

S_o = space mean speed of vehicles traveling in outer lanes (lanes 3 and 4 where they exist) within the 1,500 ft length range of the ramp influence area, mi/h

S = space mean speed of vehicles in all lanes within the 1,500-ft range of the ramp influence area, mi/h

M_s = speed proportioning factor for merge areas

D_s = speed proportioning factor for diverge areas

v_{oa} = average demand flow rate in outer lanes, computed as $(v_F - v_{12})/N_o$, pc/h/ln

N_o = number of outer lanes (one for three-lane segments, two for four-lane segments)

All other variables are as previously defined.

15.7.6 Special Cases

As we noted previously, the methodology for merge and diverge analysis presented here applies directly to one-lane, right-hand on-ramps and off-ramps. A number of special modifications can be used to apply them to a variety of other merge and diverge situations. These special cases are described in the appendix to this chapter.

15.8 Sample Problems in Weaving, Merging, and Diverging Analysis

Example 15.1: Analysis of a Ramp-Weave Area

Figure 15.13 illustrates a typical ramp-weave section on a six-lane freeway (three lanes in each direction). The analysis is to determine the expected LOS and capacity for the prevailing conditions shown.

Solution:

Steps 1 and 2: *Convert All Demand Volumes to Flow Rates in pc/h Under Equivalent Base Conditions* Each of the component demand volumes is converted to a demand flow rate in pc/h under equivalent base conditions using Equation 15-1:

$$v_p = \frac{V}{PHF^* f_{HV}{}^* f_p}$$

where: $PHF = 0.9$ (given)

$f_p = 1.00$ (assume drivers are familiar with the site)

The heavy-vehicle factor, f_{HV}, is computed using Equation 14-9 and a value of E_T selected from Table 14.13 for trucks on level terrain ($E_T = 1.5$). Then:

$$f_{HV} = \frac{1}{1 + P_T(E_T - 1)} = \frac{1}{1 + 0.10\,(1.5 - 1)}$$

$$= 0.952$$

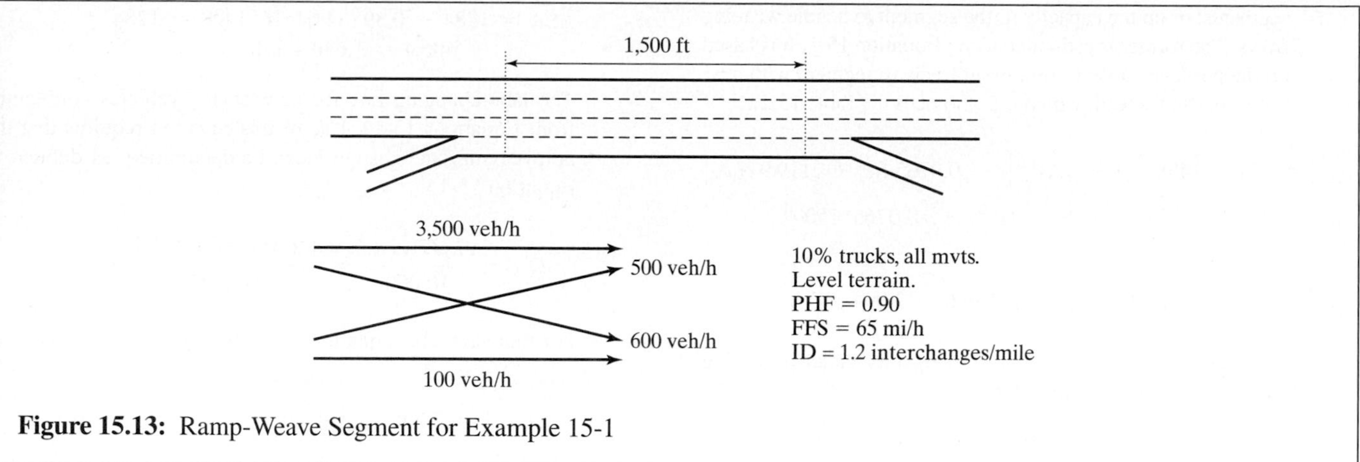

Figure 15.13: Ramp-Weave Segment for Example 15-1

and:

$$v_{o1} = \frac{3,500}{0.90*0.952*1.00} = 4,085 \text{ pc/h}$$

$$v_{o2} = \frac{100}{0.90*0.952*1.00} = 117 \text{ pc/h}$$

$$v_{w1} = \frac{600}{0.90*0.952*1.00} = 700 \text{ pc/h}$$

$$v_{w2} = \frac{500}{0.90*0.952*1.00} = 584 \text{ pc/h}$$

Other critical variables used in the analysis may now be computed and/or summarized:

$$v_w = 700 + 584 = 1,284 \text{ pc/h}$$

$$v_{nw} = 4,085 + 117 = 4,202 \text{ pc/h}$$

$$v = 1,284 + 4,202 = 5,486 \text{ pc/h}$$

$$v/N = 5,486/4 = 1,372 \text{ pc/h/ln}$$

$$VR = 1,284/5,485 = 0.23$$

$$L = 1,500 \text{ ft}$$

Step 3: *Determine Configuration Characteristics* The two critical numeric variables that define configuration are N_{WV}, the number of lanes from which a weaving movement can be successfully executed with no more than one lane change, and LC_{MIN}, the minimum number of lane changes that must be made by all weaving vehicles to complete their maneuvers successfully.

The number of weaving lanes, N_{WV}, is determined by perusing the site drawing (Figure 15.13) and comparing it to the illustration of Figure 15.5. As a ramp-weave, the value of N_{WV} is 2. The value of LC_{MIN} is found from Equation 15-2:

$$LC_{MIN} = (LC_{FR}*v_{FR}) + (LC_{RF}*v_{RF})$$

where: LC_{FR} = minimum number of lane changes for a freeway-to-ramp vehicle needed to execute a weaving maneuver successfully; from Figure 15.13, this value is 1.

v_{FR} = freeway-to-ramp demand flow rate, pc/h = v_{w1} = 700 pc/h.

LC_{RF} = minimum number of lane changes for a ramp-to-freeway vehicle needed to execute a weaving maneuver successfully; from Figure 15.13, this valuc is 1.

v_{RF} = ramp-to-freeway demand flow rate, pc/h = v_{w2} = 584 pc/h.

Then:

$$LC_{MIN} = (1*700) + (1*584) = 1,284 \text{ lc/h}$$

Step 4: *Determine the Maximum Weaving Length* The maximum length for which this segment may be considered to be a "weaving segment" is estimated using Equation 14-4:

$$L_{MAX} = \left[5,728\,(1 + VR)^{1.6}\right] - \left[1,566\, N_{WV}\right]$$

$$L_{MAX} = \left[5,728\,(1 + 0.23)^{1.6}\right] - \left[1,566*2\right]$$

$$L_{MAX} = 7,977 - 3,132 = 4,845 \text{ ft}$$

Because the actual length of the segment, 1,500 feet, is far less than this maximum, the segment is operating as a weaving segment, and the analysis may continue.

Step 5: *Determine the Capacity of the Weaving Segment* The capacity of the weaving segment can be determined by overall operation at a density of 43 pc/h/ln, the density at which it is believed breakdown occurs in weaving segments, or on the capacity of the segment to handle weaving flows. The former is estimated using Equation 15-5. It is based on the per lane capacity of a basic freeway segment with a 65 mi/h free-flow speed, which is 2,350 pc/h/ln (Table 15.2):

$$c_{IWL} = c_{IFL} - \left[438.2\,(1 + VR)^{1.6}\right] + \left[0.0765\,L_S\right] + \left[119.9\,N_{WV}\right]$$

$$c_{IWL} = 2{,}350 - \left[438.2\,(1 + 0.23)^{1.6}\right] + \left[0.0765{*}1500\right] + \left[119.9{*}2\right]$$

$$c_{IWL} = 2{,}350 - 610.3 + 114.8 + 239.6 = 2{,}094 \text{ pc/h/ln}$$

This value must be converted to a capacity under prevailing conditions using Equation 15-6:

$$c_{w1} = c_{IWL}\,N f_{HV} f_p$$

$$c_{w1} = 2094{*}4{*}0.952{*}1 = 7{,}974 \text{ veh/h}$$

The capacity based on maximum weaving demand flow rate, based on $N_{WV} = 2$, is estimated using Equation 15-7:

$$C_{IW} = \frac{2400}{VR} = \frac{2400}{0.23} = 10{,}235 \text{ pc/h}$$

This value must also be converted to prevailing conditions using Equation 15-8:

$$c_{w2} = c_{IW} f_{HV} f_p = 10{,}235{*}0.952{*}1 = 9{,}744 \text{ veh/h}$$

The limiting capacity is obviously based on the density condition (i.e., 7,974 veh/h). As with any capacity, this is defined in terms of a maximum demand flow rate that the segment can accommodate without breakdown. This must be compared with the demand flow rate, also under prevailing conditions. The total demand volume, V, is given as 4,700 veh/h (Figure 15.13). This must be converted to a flow rate:

$$v = {}^{V}\!/_{PHF} = {}^{4700}\!/_{0.90} = 5{,}222 \text{ veh/h}$$

Because the demand flow rate is less than the capacity of the segment ($v/c = 5222/7974 = 0.655$), operations will be stable, and LOS F *does not* exist in the segment. The analysis may move forward to estimate density, LOS, and speed within the segment.

Step 6: *Determine Lane-Changing Rates* To estimate speed and density in the weaving segment, total lane-changing rates within the segment must be estimated. Lane-changing rates for weaving and nonweaving vehicles are separately estimated. The lane-changing rate for weaving vehicles is computed using Equation 15-11:

$$LC_W = LC_{MIN} + 0.39\left[(L_S - 300)^{0.5}\,N^2\,(1 + ID)^{0.8}\right]$$

$$LC_W = 1284 + 0.39\left[(1500 - 300)^{0.5}\,4^2\,(1 + 1.2)^{0.8}\right]$$

$$LC_W = 1284 + 0.39\,(34.64{*}16{*}1.88) = 1284 + 406.4 = 1{,}690.4 \text{ lc/h}$$

The lane-changing rate for nonweaving vehicles is obtained from Equations 15-12. Use of this equation requires that the nonweaving lane-change index be determined, as defined in Equation 15-13:

$$I_{NW} = \frac{L_S I D v_{NW}}{10{,}000} = \frac{1500{*}1.2{*}4202}{10{,}000} = 756.4$$

For this value, the equation 15-12 for $I_{NW} \leq 1{,}300$ is used:

$$LC_{NW} = (0.206\,v_{NW}) + (0.542\,L_S) - (192.6\,N)$$

$$LC_{NW} = (0.206{*}4202) + (0.542{*}1500) - (192.6{*}4)$$

$$LC_{NW} = 856.6 + 813.0 - 770.4 = 899.2 \text{ lc/h}$$

The total lane-changing rate in the segment is the sum of the weaving vehicle rate and the nonweaving vehicle rate, or:

$$LC_{ALL} = LC_W + LC_{NW} = 1690.4 + 899.2 = 2{,}589.6 \text{ lc/h}$$

Step 7: *Determine the Average Speed of Weaving and Nonweaving Vehicles* The average speed of weaving vehicles in the weaving segment is estimated using Equations 15-19 and 15-20. Equation 15-19 is used to find the weaving intensity factor, W:

$$W = 0.226\left(\frac{LC_{ALL}}{L_S}\right)^{0.789} = 0.226\left(\frac{2{,}589.6}{1{,}500}\right)^{0.789} = 0.348$$

Then:

$$S_W = 15 + \left(\frac{FFS - 15}{1 + W}\right) = 15 + \left(\frac{65 - 15}{1 + 0.348}\right) = 52.1 \text{ mi/h}$$

The average speed of nonweaving vehicles in the weaving segment is estimated using Equation 15-21:

$$S_{NW} = FFS - (0.0072 LC_{MIN}) - (0.0048\ {}^{v}\!/_{N})$$

$$S_{NW} = 65 - (0.0072{*}1284) - \left(0.0048{*}\frac{5486}{4}\right)$$

$$S_{NW} = 65 - 9.2 - 6.6 = 49.2 \text{ mi/h}$$

These results indicate that weaving vehicles are actually traveling somewhat faster than nonweaving vehicles within the weaving segment. Although unusual for ramp-weaving segments, this is entirely possible given the dominance of the through freeway flow in the segment. Nonweaving vehicles may be crowding into the two outer freeway lanes to avoid the weaving turbulence, and they may therefore experience slightly lower speeds (and higher densities) than weaving vehicles.

The average speed of all vehicles in the segment is computed from Equation 15-22:

$$S = \frac{v_W + v_{NW}}{\left(\frac{v_W}{S_W}\right) + \left(\frac{v_{NW}}{S_{NW}}\right)} = \frac{1284 + 4202}{\left(\frac{1284}{52.1}\right) + \left(\frac{4202}{49.2}\right)}$$

$$= \frac{5486}{24.64 + 85.41} = 49.9\,\text{mi/h}$$

Step 8: *Determine Density and Level of Service in the Weaving Segment* The average density in the weaving segment is computed from Equation 15-23:

$$D = \frac{\left({}^{v}/_{N}\right)}{S} = \frac{\left({}^{5486}/_{4}\right)}{49.9} = 27.5\,\text{pc/mi/ln}$$

From Table 5.1, this is LOS C, but it is very close to the LOS D boundary of 28 pc/h/ln.

This ramp-weave segment is operating acceptably at LOS C. Because the density is very close to the LOS D boundary, virtually any growth in demand will cause the segment to enter LOS D operations. The capacity of the segment is 2,752 vehs/h (as a flow rate) larger than the current demand (7,974 pc/h versus 5,222 pc/h), so that demand can grow by 52.7% (2752*100/5222) before capacity is reached.

Example 15.2: Analysis of a Major Weaving Area

The freeway weaving area shown in Figure 15.14 is to be analyzed to determine the expected LOS for the conditions shown and the capacity of the weaving area. For convenience, all demand volumes have already been converted to flow rates in pc/h under equivalent base conditions. For information purposes, the following values were used to make these conversions: $PHF = 0.95$, $f_{HV} = 0.93, f_p = 1.00$.

Solution:

Steps 1 and 2: *Convert All Demand Volumes to Flow Rates in pc/h Under Equivalent Base Conditions* Because all demands are specified as flow rates in pc/h under equivalent base conditions, no further conversion of these is necessary. Key analysis variables are summarized here:

$$v_w = 800 + 1{,}700 = 2{,}500 \text{ pc/h}$$

$$v_{nw} = 1{,}700 + 1{,}500 = 3{,}200 \text{ pc/h}$$

$$v = 2{,}500 + 3{,}200 = 5{,}700 \text{ pc/h}$$

$$v/N = 5{,}700/3 = 1{,}900 \text{ pc/h/ln}$$

$$VR = 2{,}500/5{,}700 = 0.439$$

$$L = 2{,}000 \text{ ft}$$

Note that this is a major weaving configuration. The weave from left to right can be made with no lane changes ($LC_{FR} = 0$), whereas the weave from right to left requires one lane change ($LC_{RF} = 1$). Successful weaving maneuvers can be made from

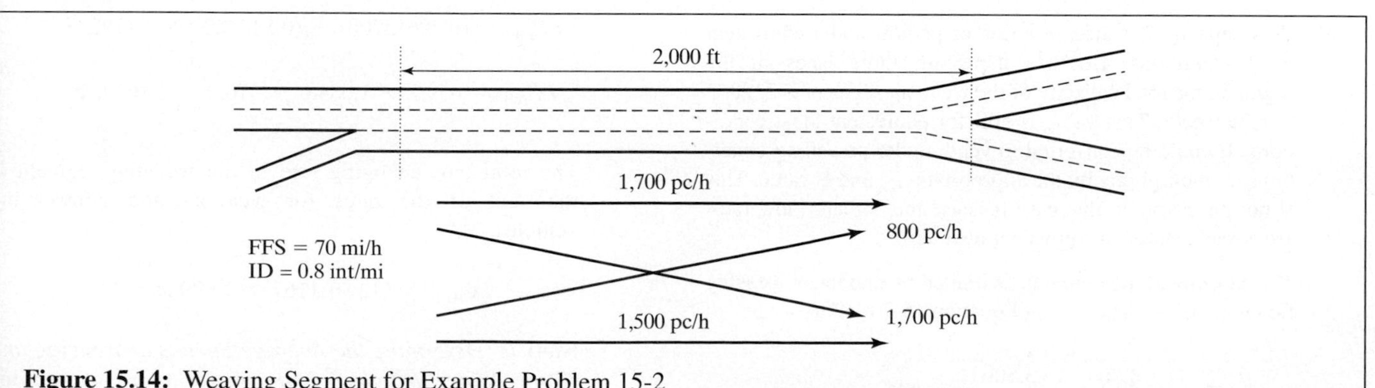

Figure 15.14: Weaving Segment for Example Problem 15-2

any of the three lanes in the segment with no more than one lane change (i.e., $N_{WV} = 3$).

Step 3: *Determine Configuration Characteristics* One of the configuration characteristics, N_{WV}, was determined to be 3. The second configuration characteristic needed is LC_{MIN}, as determined by Equation 15-2:

$$LC_{MIN} = (LC_{FR}\, v_{FR}) + (LC_{RF}\, v_{RF})$$

$$LC_{MIN} = (0*1{,}700) + (1*800) = 800 \text{ lc/h}$$

Step 4: *Determine the Maximum Weaving Length* The maximum weaving length is estimated using Equation 15-4:

$$L_{MAX} = [5{,}728\,(1 + VR)^{1.6}] - [1{,}566\, N_{WV}]$$

$$L_{MAX} = \left[5{,}728\,(1 + 0.439)^{1.6}\right] - \left[1{,}544^{*}3\right] = 5{,}556 \text{ ft}$$

Because the actual length of the weaving segment is only 2,000 feet, it falls within this limit, and the analysis of the segment as a weaving segment may continue.

Step 5: *Determine the Capacity of the Weaving Segment* To determine whether stable operations prevail, the capacity of the weaving segment must be determined. Capacity may be determined in two ways. It may be limited by a breakdown density of 43 pc/h/ln or by a maximum weaving flow rate the segment can accommodate. Capacity, as determined by a breakdown density of 43 pc/h/ln, is estimated using Equation 15-5:

$$c_{IWL} = c_{IFL} - \left[438.2\,(1+VR)^{1.6}\right] + \left[0.0765 L_s\right] + \left[119.8\, N_{WV}\right]$$

$$c_{IWL} = 2400 - \left[438.2\,(1+0.439)^{1.6}\right] + \left[0.0765*2000\right] + \left[119.8*3\right]$$

$$c_{IWL} = 2400 - 784.5 + 153 + 359.4 = 2{,}128 \text{ pc/h/ln}$$

This capacity is stated in terms of pc/h/ln under equivalent ideal conditions. Because there are three lanes in the segment, the total capacity of the weaving segment is 2128*3 = 6,384 pc/h. This value is still for equivalent ideal conditions. It could be converted to veh/h under prevailing conditions by multiplying by the appropriate f_{HV} and f_P value. This is not necessary in this case because the demand flow rates are already stated in equivalent ideal terms.

The capacity of the segment, as limited by maximum weaving flow rate, is estimated using Equation 15-7 for $N_{WV} = 3$:

$$C_{IW} = \frac{3{,}500}{VR} = \frac{3{,}500}{0.439} = 7{,}973 \text{ pc/h}$$

Because this value is larger than capacity limited by density, the smaller value is used. The capacity of the weaving segment is 6,384 pc/h under equivalent ideal conditions.

Because the total demand flow rate is 5,700 pc/h, capacity is sufficient. The v/c ratio is 5,700/6,384 = 0.893, which means that demand flows are quite near capacity. A 10.7% increase in demand would create a LOS F situation.

Step 6: *Determine Lane-Changing Rates* To estimate speeds in the weaving segments, and then density and LOS, the total lane-changing rate within the weaving segment must be determined using Equations 15-11, 15-12, and 15-13. Lane-changing rates are separately estimated for weaving and nonweaving vehicles. Equation 15-11 is used to estimate the lane-changing rate for weaving vehicles:

$$LC_W = LC_{MIN} + 0.39\left[(L_s - 300)^{0.5}\, N^2\, (1+ID)^{0.8}\right]$$

$$LC_W = 800 + 0.39\left[(2000 - 300)^{0.5}\, 3^2\, (1+0.8)^{0.8}\right]$$

$$LC_W = 800 + 0.30*\left[41.23*9*1.6\right] = 1{,}032 \text{ lc/h}$$

Equation 15-12 is used to estimate the rate of lane-changing among nonweaving vehicles. There are two equations 15-12, and the lane-changing index must be computed to interpret the results from theses equations. The index is computed using Equation 15-13:

$$I_{NW} = \frac{L_s ID V_{NW}}{10{,}000} = \frac{2000*0.8*3200}{10{,}000} = 512 < 1{,}300$$

Because the index is less than 1,300, the *first* equation 15-12 is used:

$$LC_{NW} = (0.206\, V_{NW}) + (0.542\, L_s) - (192\, N)$$

$$LC_{NW} = (0.206*3200) + (0.542*2000) - (192*3)$$

$$LC_{NW} = 659.2 + 1{,}084.0 - 576.0 = 1{,}167 \text{ lc/h}$$

The total lane-changing rate in the weaving segment is the sum of the rates for weaving and nonweaving vehicles:

$$LC_{ALL} = 512 + 1167 = 1{,}679 \text{ lc/h}$$

Step 7: *Determine the Average Speeds of Weaving and Nonweaving Vehicles* The average speed of weaving vehicles within the weaving segment is estimated using equations 15-19 and 15-20. Equation 15-19 determines the weaving intensity factor, W:

$$W = 0.226\left(\frac{LC_{ALL}}{L_S}\right)^{0.789} = 0.226\left(\frac{1679}{2000}\right)^{0.789} = 0.197$$

Then:

$$S_W = 15 + \left(\frac{FFS - 15}{1 + W}\right) = 15 + \left(\frac{70 - 15}{1 + 0.197}\right) = 60.9\,\text{mi/h}$$

The average speed of nonweaving vehicles in the segment is estimated using Equation 15-21:

$$S_{NW} = FFS - (0.0072{*}LC_{MIN}) - (0.0048{*}\,{}^{v}/_{N})$$

$$S_{NW} = 70 - (0.0072{*}800) - (0.0048{*}1900) = 55.1\,\text{mi/h}$$

In this case, nonweaving vehicles will be traveling over 5 mi/h slower than weaving vehicles. This is not unexpected. Weaving vehicles dominate this segment (3200 pc/h versus 2500 pc/h), and the configuration favors weaving vehicles.

The average speed of all vehicles is computed Using equation 15-22:

$$S = \frac{v_W + v_{NW}}{\left(\frac{v_W}{S_W}\right) + \left(\frac{v_{NW}}{S_{NW}}\right)} = \frac{2500 + 3200}{\left(\frac{2500}{55.1}\right) + \left(\frac{3200}{60.9}\right)}$$

$$= \frac{5700}{45.37 + 52.55} = 58.2\,\text{mi/h}$$

Step 8: *Determine the Density and Level-of-Service* Density in the weaving segment is computed using Equation 15-23:

$$D = \frac{({}^{v}/_{N})}{S} = \frac{1900}{58.2} = 32.6\ \text{pc/mi/h}$$

From Table 15.2, this is LOS D.

The weaving segment is currently operating stably in LOS D but not far from the LOS E boundary. Although speeds appear to be acceptable, the demand is almost 90% of the capacity, and there is little room for growth in demand at this location. The bottom line is that with virtually any traffic growth, this segment will reach capacity. Operations will deteriorate rapidly with demand growth. Even if only ambient growth is expected (as opposed to growth caused by new development), the segment should be looked at now for potential improvements.

Example 15-3: Analysis of an Isolated On-Ramp

An on-ramp to a busy eight-lane urban freeway is illustrated in Figure 15.15. An analysis of this merge area is to determine the likely LOS under the prevailing conditions shown.

Solution:

Step 1: *Convert All Demand Volumes to Flow Rates in pc/h Under Equivalent Ideal Conditions* The freeway and ramp flows approaching the merge area must be converted to flow rates in pc/h under equivalent base conditions using Equation 15-1. In this case, note that the truck percentages and PHF are different for the two. From Chapter 14, the passenger car equivalent for trucks (E_T) is 2.5. It Is assumed that drivers are familiar users, and that f_p = 1.00. Then for the ramp demand flow:

$$f_{HV} = \frac{1}{1 + 0.10\,(2.5 - 1)} = 0.870$$

$$v_R = \frac{900}{0.89{*}0.870{*}1.00} = 1{,}162\ \text{pc/h}$$

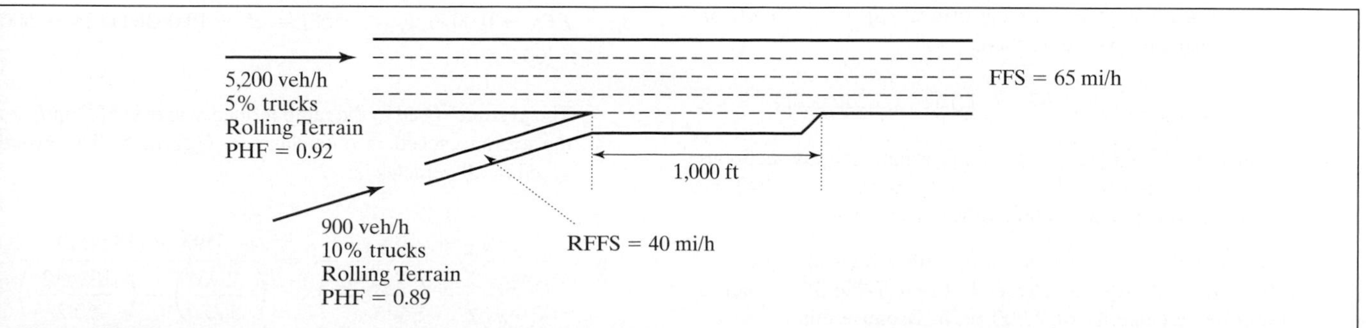

Figure 15.15: On-Ramp Segment for Example Problem 15-3

For the freeway demand flow:

$$f_{HV} = \frac{1}{1 + 0.05\,(2.5 - 1)} = 0.930$$

$$v_F = \frac{5200}{0.92*0.930*1.00} = 6{,}078 \text{ pc/h}$$

Step 2: *Determine the Demand Flow Remaining in Lanes 1 and 2 Immediately Upstream of the Merge* Table 15.3 gives values of P_{FM}, the proportion of freeway vehicles remaining in lanes 1 and 2 immediately upstream of a merge. For an eight-lane freeway (four lanes in each direction), Table 15.3 contains two equations. Selection depends on the value of v_F/RFFS = 6078/40 = 151.9. Because this is more than 72, Equation 15-29 is used. Then, Equation 15-32 is used to find the flow rate in lanes 1 and 2:

$$P_{FM} = 0.2178 - 0.000125\,v_R = 0.2178 - (0.000125*1162) = 0.0726$$

$$v_{12} = v_F*P_{FM} = 6078*0.0726 = 441 \text{ pc/h}$$

This prediction must be checked for "reasonableness." The average flow rate in lanes 1 and 2 is 441/2 = 221 pc/h is very low by any judgment. This leaves 6,078 − 441 = 5,637 pc/h in the two outer lanes (lanes 3 and 4), or 5637/2 = 2,819 pc/h/ln. This violates the maximum reasonable limit of 2,700 pc/h/ln. It also violates the 1.5 rule: 2819 > 1.5*221= 332 pc/h/ln. In this case, the 1.5 rule is violated by a great deal. The expected flow rate in lanes 1 and 2, therefore, must be revised in accordance with Equation 15-39:

$$v_{12} = {}^{v_F}/_{2.50} = {}^{6078}/_{2.50} = 2{,}431 \text{ pc/h}$$

With this value for v_{12}, the outer lanes would carry 6,078 − 2,431 = 3,647 pc/h, or 3647/2 = 1,824 pc/h/ln, which now satisfies both "reasonableness" criteria. The example will move forward using this value.

Step 3: *Check Capacity of Merge Area and Compare to Demand Flows* To determine whether the section will fail (LOS F), the capacity values of Table 15.5 must be consulted. For a merge section, the critical capacity check is on the downstream freeway section.

$$v_{FO} = v_F + v_R = 6{,}078 + 1{,}162 = 7{,}240 \text{ pc/h}$$

From Table 15.3, the capacity of a four-lane freeway section is 9,400 pc/h when the FFS is 65 mi/h. As 9,400 > 7,240, no failure is expected due to total downstream flow.

The capacity of a one-lane ramp with a free-flow speed of 40 mi/h must also be checked. From Table 15.5, such a ramp has a capacity of 2,000 pc/h. Because this is greater than the ramp demand flow of 1,162 pc/h, this element will not fail either.

Total flow entering the merge influence area is:

$$v_{R12} = v_R + v_{12} = 1{,}162 + 2{,}431 = 3{,}593 \text{ pc/h}$$

Because the maximum desirable entering flow for single-lane merge area is 4,600 pc/h, this element is also acceptable.

Step 4: *Estimate Density and Level of Service in the Ramp Influence Area* Because stable operations are expected, Equation 15-40 may be used to estimate the density in the ramp influence area:

$$D_R = 5.475 + 0.00734\,v_R + 0.0078\,v_{12} - 0.00627\,L_a$$

$$D_R = 5.475 + (0.00734*1162) + (0.0078*2431) - (0.00627*1000)$$

$$D_R = 5.475 + 8.529 + 18.962 - 6.270 = 26.7 \text{ pc/mi/ln}$$

From the criteria in Table 15.1, this is LOS C, but close to the LOS D boundary of 28 pc/mi/ln.

Step 5: *Estimate Speed Parameters for Information* Although not used to determine LOS, the algorithms of Table 15.6 may be used to estimate speed parameters of interest:

$$M_S = 0.321 + 0.0039\,e^{(v_{R12}/1000)} - 0.002\,(L_a*RFFS\,/\,1000)$$

$$M_S = 0.321 + \left[0.0039\,e^{(3593\,/\,1000)}\right] - \left[0.002\,(1000*40\,/\,1000)\right]$$

$$M_S = 0.321 + 0.142 - 0.080 = 0.383$$

$$S_R = FFS - (FFS - 42)\,M_S = 65 - (65 - 42)*0.383 = 56.2 \text{ mi/h}$$

$$S_O = FFS - 0.0036\,(v_{OA} - 500) = 65 - [0.0036\,(1824 - 500)] = 60.2 \text{ mi/h}$$

The average speed in the ramp influence area is 56.2 mi/h, and the average speed in outer lanes is 60.2 mi/h. The average speed of all vehicles is:

$$S = \frac{v_{R12} + v_{OA}N_O}{\left(\frac{v_{R12}}{S_R}\right) + \left(\frac{v_{OA}N_O}{S_O}\right)} = \frac{3593 + (1824*2)}{\left(\frac{3593}{56.2}\right) + \left(\frac{1824*2}{60.2}\right)}$$

$$= \frac{7241}{63.93 + 60.60} = 57.2 \text{ mi/h}$$

Several additional items may be of interest. The lane distribution of the incoming freeway flow (v_F) was checked for reasonableness, and altered as a result.

It is also useful to check the LOS on the downstream basic freeway section. It carries a total of 7,240 pc/h in four lanes, or 1,810 pc/h/ln. Using the standard speed-flow curve for FFS = 65, this is LOS D.

What does this mean, considering that the LOS for the ramp influence area is determined to be C? It means that the total freeway flow is the determining element in overall LOS. This is as it should be because it is always undesirable to have minor movements (in this case, the on-ramp), controlling the overall operation of the facility.

Example 15-4: Analysis of a Sequence of Freeway Ramps

Figure 15.16 shows a series of three ramps on a six-lane freeway (three lanes in each direction). All three ramps are to be analyzed to determine the LOS expected under the prevailing conditions shown.

Solution:

Step 1: *Convert All Demand Volumes to Flow Rates in pc/h Under Equivalent Ideal Conditions* Before applying any of the models for ramp analysis, all demand volumes must be converted to flow rates in pc/h under equivalent base conditions. This is done using Equation 15-1. Peak-hour factors for each movement are given, as are truck percentages. The heavy-vehicle factor is computed using E_T values from Chapter 14. For level terrain, $E_T = 1.5$ for all movements. It is assumed that the driver population consists primarily of familiar users and that f_p, therefore, is 1.00. The conversion computations are shown here:

$$f_{HV}(\text{freeway}) = \frac{1}{1 + 0.10\,(1.5 - 1)} = 0.952$$

$$v_F = \frac{4{,}000}{0.90*0.952*1.00} = 4{,}669 \text{ pc/h}$$

$$f_{HV}(\text{Ramp1}) = \frac{1}{1 + 0.15\,(1.5 - 1)} = 0.930$$

$$v_{R1} = \frac{500}{0.95*0.930*1.00} = 566 \text{ pc/h}$$

$$f_{HV}(\text{Ramp 2}) = \frac{1}{1 + 0.05\,(1.5 - 1)} = 0.976$$

$$v_{R2} = \frac{600}{0.92*0.976*1.00} = 668 \text{ pc/h}$$

$$f_{HV}(\text{Ramp 3}) = \frac{1}{1 + 0.12\,(1.5 - 1)} = 0.943$$

$$v_{R3} = \frac{400}{0.91*0.943*1.00} = 466 \text{ pc/h}$$

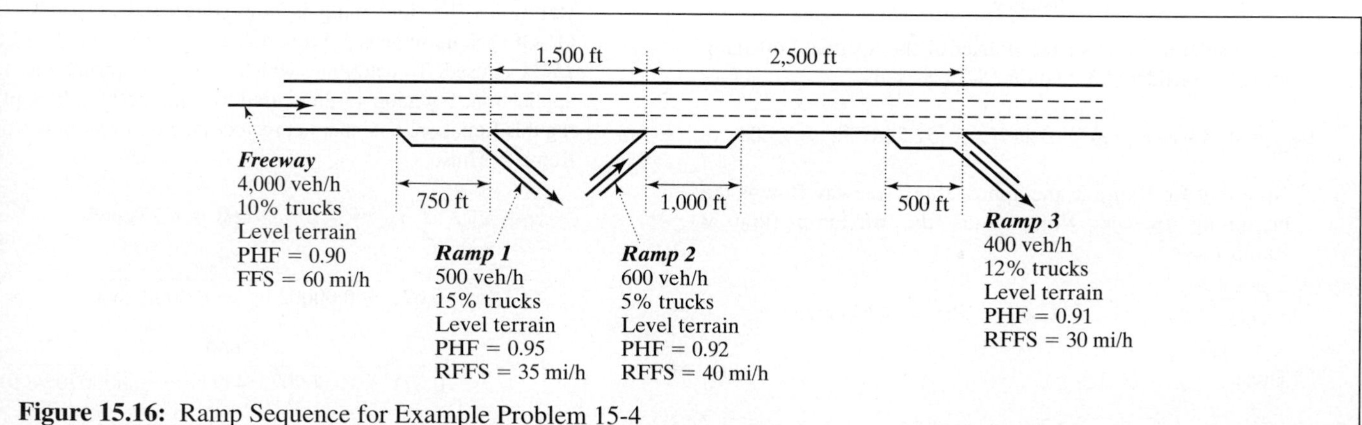

Figure 15.16: Ramp Sequence for Example Problem 15-4

Step 2: *Determine the Flow in Lanes 1 and 2 Immediately Upstream of Each Ramp in the Sequence* Each of the ramps in the analysis section must now be considered for the potential impact of the adjacent ramp(s) on lane distribution.

Ramp 1: The first ramp is part of a three-ramp sequence that can be described as None-OFF-On (no upstream adjacent ramp; an adjacent downstream on-ramp). Using Table 15.4, for a six-lane freeway and the sequence indicated, Equation 15-33 should be used to determine v_{12}.

$$v_{12(1)} = v_{R1} + (v_F - v_{R1})P_{FD}$$

$$P_{FD} = 0.760 - 0.000025v_F - 0.000046v_{R1}$$

$$P_{FD} = 0.760 - (0.000025*4669) - (0.000046*566) = 0.617$$

$$v_{12(1)} = 566 + (4{,}669 - 566)^{*}0.617 = 3{,}098 \text{ pc/h}$$

The resulting lane distribution must be checked for reasonableness. A six-lane freeway has only one outer lane, which would carry 4669 − 3098 = 1,571 pc/h. Because this is less than 2,700 pc/h, and less than 1.5*(3098/2) = 2,324 pc/h, the predicted lane distribution is reasonable and will be used.

Ramp 2: The second ramp is an on-ramp that can be described as part of an Off-ON-Off sequence. From Table 15.3, there are three potential equations that might apply: Equation 15-26, which considers the effect of the upstream off-ramp; Equation 15-27, which considers the effect of the downstream off-ramp; or Equation 15-25, which treats the ramp as if it were isolated. It is even possible that two of these apply, in which case the equation yielding the larger v_{12} estimate is used. To determine which of these apply, however, requires the use of Equations 15-30 and 15-31 to determine equivalence distances.

In considering whether the impact of the upstream off-ramp must be considered, Equation 15-30 is used:

$$L_{EQ} = 0.214(v_R + v_F) + 0.444L_a + 52.32RFFS - 2{,}403$$

Note that for Ramp 2, the approaching freeway flow is the beginning freeway flow minus the off-ramp flow at Ramp 1:

$$v_{F2} = v_F - v_{R1} = 4{,}669 - 566 = 4{,}103 \text{ pc/h}$$

Thus:

$$L_{EQ} = 0.214\,(v_F + v_R) + 0.444\,L_a + 53.32\,RFFS - 2{,}403$$

$$L_{EQ} = 0.214\,(4103 + 668) + (0.444*1000) + (53.32*40) - 2{,}403$$

$$L_{EQ} = 1{,}021.0 + 444.0 + 2{,}132.8 - 2{,}403.0 = 1{,}195 \text{ ft}$$

Because the actual distance to the upstream ramp is 1,500 ft > 1,195 ft, the impact of the upstream off-ramp need not be considered, and Equation 15-25 is used.

To determine whether or not the effect of the downstream off-ramp must be considered, Equation 15-31 is used to compute L_{EQ}:

$$L_{EQ} = \frac{v_d}{0.1096 + 0.000107L_a} = \frac{466}{0.1096 + (0.000107*1{,}000)} = 2{,}151 \text{ ft}$$

The actual distance to the downstream off-ramp is 2,500 ft (>2,151 ft). Thus the impact of the downstream off-ramp is also not considered, and Equation 15-25 is used. Through the determination of these equivalence distances, it is seen that Ramp 2 may be considered an isolated ramp. Only one—Equation 15-25—applies to the estimation of $v_{12(2)}$.

$$v_{12(2)} = v_{F2}{*}P_{FM}$$

$$P_{FM} = 0.5775 + 0.000028L_a = 0.5775 + (0.000028*1{,}000) = 0.6055$$

$$v_{12(2)} = 4{,}103*0.6055 = 2{,}484 \text{ pc/h}$$

This distribution must also be tested for reasonableness. The outer lane carries 4103 − 2482 = 1,621 pc/h < 2,700 pc/h. It also carries less than 1.5*(2484/2) = 1,863 pc/h. Therefore, the predicted lane distribution is reasonable and will be used.

Ramp 3: The third ramp is now considered as part of an On-OFF-None sequence. From Table 15.4, Equation 15-33 or 15-34 is used. To determine which is the appropriate one for application, Equation 15-36 is used to compute L_{EQ}. In applying this Equation, note that v_{F3} includes the on-ramp flow from Ramp 2. Thus:

$$v_{F3} = v_{F2} + v_{R2} = 4{,}103 + 668 = 4{,}771 \text{ pc/h}$$

$$L_{EQ} = \frac{v_u}{0.071 + 0.000023v_F - 0.000076v_R} = \frac{668}{0.071 + (0.000023*4771) - (0.000076*466)}$$

$$L_{EQ} = 4{,}597 \text{ ft}$$

Because the actual distance to the upstream on-ramp is only 2,500 ft (<4,597 ft), Equation 15-34 is used to consider the impact of Ramp 2 on lane distribution at Ramp 3:

$$v_{12(3)} = v_{R3} + (v_{F3} - v_{R3})P_{FD}$$

$$P_{FD} = 0.717 - 0.000039\, v_{F3} + 0.604(\, ^{v_U}/_{L_{UP}}\,)$$

$$P_{FD} = 0.717 - (0.000039*4771) + (0.604*\, ^{668}/_{2500}\,)$$

$$P_{FD} = 0.717 - 0.186 + 0.161 = 0.692$$

$$v_{12(3)} = 466 + (4771 - 466)*0.692 = 3{,}445\,\text{pc/h}$$

Again, the predicted lane distribution should be checked for reasonableness. The outer lane flow is 4771 − 3445 = 1,326 pc/h/ln < 2,700 pc/h/ln. It is also less than 1.5*(3445/2) = 2,582 pc/h. Therefore the distribution is reasonable and will be used.

Summarizing the results for v_{12} immediately upstream of each of the three ramps:

$$v_{12(1)} = 3{,}098\ \text{pc/h}$$

$$v_{12(2)} = 2{,}484\ \text{pc/h}$$

$$v_{12(3)} = 3{,}445\ \text{pc/h}$$

Step 3: *Check Capacities* The capacities and limiting values of Table 15.5 must now be checked to see whether operations are stable or whether LOS F exists. The freeway flow check is made between Ramps 2 and 3 because this is the point where total freeway flow is greatest (v_{F3}). These checks are performed in Table 15.8. Remember that the freeway FFS is 60 mi/h.

None of the demand flows exceed the capacities or limiting values of Table 15.5. Thus stable operation is expected throughout the section.

Step 4: *Determine Densities and Levels of Service in Each Ramp Influence Area* The density in the ramp influence area is estimated using Equations 15-40 for on-ramps and 15-41 for off-ramps:

$$D_{R1} = 4.252 + 0.0086 v_{12(1)} - 0.009 L_d$$

$$D_{R1} = 4.252 + (0.0086*3098) - (0.009*750) = 24.1\ \text{pc/mi/ln}$$

Table 15.8: Capacity Checks for Example Problem 15.4

Item	Demand Flow	Capacity (Table 15.5)
v_{F3}	4,771 pc/h	6,900 pc/h
$v_{12(1)}$	3,098 pc/h	4,400 pc/h
$v_{R12(2)}$	2,484 + 668 = 3,152 pc/h	4,600 pc/h
$v_{12(3)}$	3,446 pc/h	4,400 pc/h
v_{R1}	566 pc/h	2,000 pc/h (RFFS = 35 mi/h)
v_{R2}	668 pc/h	2,000 pc/h (RFFS = 40 mi/h)
v_{R3}	466 pc/h	1,900 pc/h (RFFS = 30 mi/h)

$$D_{R2} = 5.475 + 0.00734 v_R + 0.0078 v_{12(2)} - 0.00627 L_a$$

$$D_{R2} = 5.475 + (0.00734*668) + (0.0078*2{,}484) - (0.00627*1{,}000) = 23.5\ \text{pc/mi/ln}$$

$$D_{R3} = 4.252 + (0.0086*3446) - (0.009*500) = 29.4\ \text{pc/mi/ln}$$

From Table 15.1, Ramp 1 operates at LOS C, Ramp 2 at LOS C, and Ramp 3 at LOS D.

Step 5: *Determine Speeds for Each Ramp* As was done in example problem 15.3, the algorithms of Tables 15.6 and 15.7 may be used to estimate space mean speeds within each ramp influence area and across all freeway lanes within the 1,500-foot range of each ramp influence area. Because of the length of these computations, they are not shown here. Each would follow the sequence illustrated in Example 15-3. Note that the ramp influence areas of Ramps 2 and 3 overlap for a distance of 500 ft (1,500 + 1,500 − 2,500). For this overlapping segment, the influence area having the highest density and lowest LOS would be used. In this case, Ramp 3, with LOS D, controls this overlap area.

Again, it is interesting to check the basic freeway LOS associated with the controlling (or largest) total freeway flow, which occurs between Ramps 2 and 3. The demand flow per lane for this segment is 4,771/3 = 1,591 pc/h/ln. From Chapter 14 and a FFS of 60 mi/h, the LOS is found to be almost exactly on the border between levels of service C and D. This is consistent with the Ramp 3 LOS, which is D but is only barely above the LOS C/D boundary. Thus the operation of the freeway as a whole and ramp sequence are somewhat in balance, a desirable condition.

15.9 Analysis of Freeway Facilities

The HCM 2000 contains a methodology for the analysis of long stretches of freeway facilities, containing many basic, weaving, merge, and/or diverge sections, which will be updated and expanded in HCM 2010. The methodology is reasonably straightforward for cases in which no segments fail (i.e., LOS F) but is extremely complex in cases that encompass segment failures. The models involved cannot be easily implemented by hand, so the overall procedure is outlined here without detail. The methodology is best implemented through the use of FREEVAL2010, a computational engine that will be made available through HCM 2010.

15.9.1 Segmenting the Freeway

The analysis of a freeway facility must begin by breaking the facility into component sections. Sections are fairly easily established using the definitions of basic, weaving, merge, and diverge areas as defined in Chapter 14. All weaving, merge, and diverge areas are isolated as segments; all other sections, by definition, are basic freeway segments.

For analysis, however, segments must be further divided into subsegments. In addition to boundaries between segments, subsegment boundaries must be established at all points where a change in geometric or traffic conditions occurs. Because the influence areas of ramp junctions extend 1,500 feet downstream of an on-ramp and 1,500 feet upstream of an off-ramp, longer acceleration or deceleration lanes will be treated as basic freeway segments outside the influence area; a separate basic freeway segment would then have to be established at the point where the acceleration lane ends (or the deceleration lane begins). A basic freeway segment might have to be divided into several subsegments if there are changes in geometry, such as a change in terrain or specific grades. The general process of segmenting the facility for analysis is illustrated in Figure 15.17.

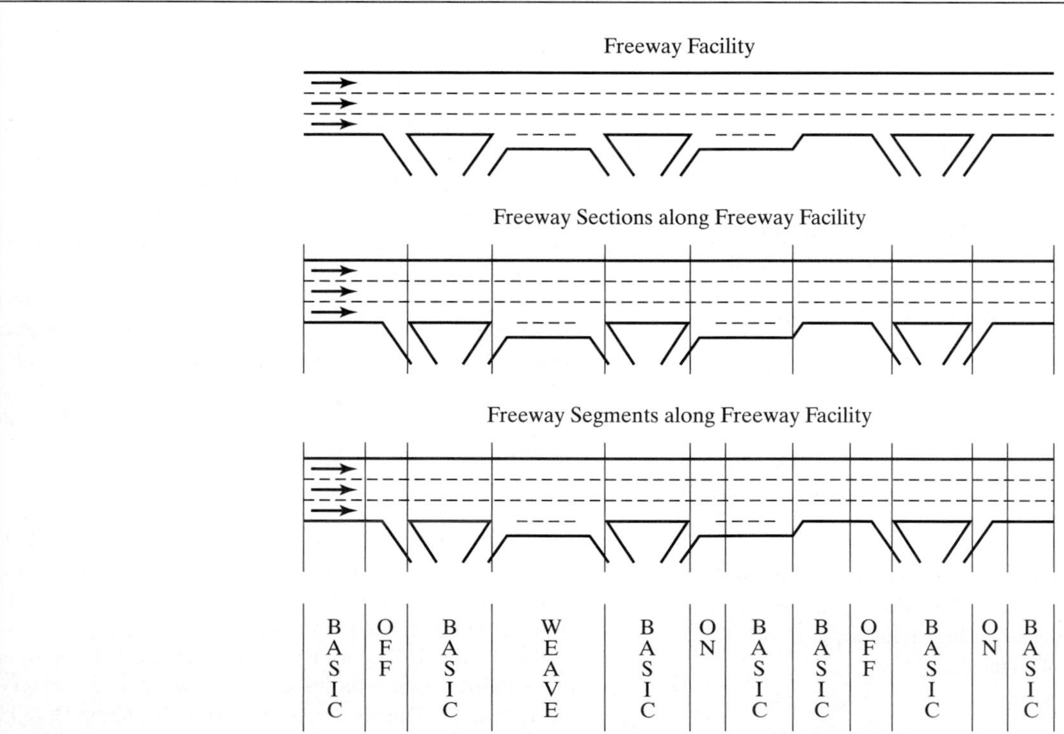

Figure 15.17: Segmenting a Freeway Facility for Analysis

(*Source:* Used with permission of Transportation Research Board, *Highway Capacity Manual*, 4th Edition, Washington DC, 2000, Exhibit 22-3, p. 22-5.)

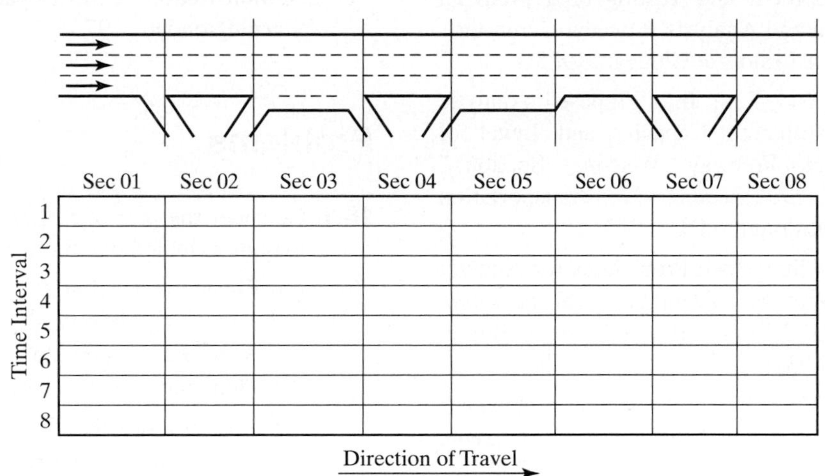

Figure 15.18: The Time-Space Domain for Analysis of a Freeway Facility

(*Source:* Used with permission of Transportation Research Board, *Highway Capacity Manual*, 4th Edition, Washington DC, 2000, Exhibit 22-2, p. 22-4.)

15.9.2 Analysis Models

The analysis procedure treats the freeway facility as a time-space domain, as illustrated in Figure 15.18. Input demands are established for each section during each time interval. Each time interval is usually 15 minutes, but longer periods may be used. Each time-section cell is then analyzed, establishing appropriate subsegments within each and obtaining LOS and speed estimates. If any cell breaks down within the analysis period, additional models track the spread of queuing to upstream segments, and shift the demand in both space and time that results from such queuing. The time-space matrix should be set up so that the beginning and ending time periods have no segment failures or residual queues.

Using this procedure, the models produce LOS and speed predictions for all segments and time periods and show both the buildup and dissipation of queues due to segment failures.

References

1. *Highway Capacity Manual*, Transportation Research Board, Washington DC, 2000.
2. *Highway Capacity Manual*, "Special Report 87," Transportation Research Board, Washington DC, 1965.
3. Pignataro, L., et al., "Weaving Areas—Design and Analysis," *NCHRP Report 159*, Transportation Research Board, Washington DC, 1975.
4. Roess, R., et al., "Freeway Capacity Analysis Procedures," *Final Report*, Project No. DOT-FH-11-9336, Polytechnic University, Brooklyn, NY, 1978.
5. "Interim Procedures on Highway Capacity," *Circular 212*, Transportation Research Board, Washington DC, 1980.
6. Leisch, J., "Completion of Procedures for Analysis and Design of Traffic Weaving Areas," *Final Report*, Vols. 1 and 2, U.S. Department of Transportation, Federal Highway Administration, Washington DC, 1983.
7. Reilly W., et al., "Weaving Analysis Procedures for the New Highway Capacity Manual," *Technical Report*, JHK & Associates, Tucson, AZ, 1983.
8. Roess, R., "Development of Weaving Area Analysis Procedures for the 1985 Highway Capacity Manual," *Transportation Research Record 1112*, Transportation Research Board, Washington DC, 1987.

9. Fazio, J., "Development and Testing of a Weaving Operational Design and Analysis Procedure," master's thesis, University of Illinois at Chicago, 1985.
10. Cassidy, M., and May, A., Jr., "Proposed Analytic Technique for Estimating Capacity and Level of Service of Major Freeway Weaving Sections," *Transportation Research Record 1320*, Transportation Research Board, Washington DC, 1992.
11. Ostram, B., et al., "Suggested Procedures for Analyzing Freeway Weaving Sections, *Transportation Research Record 1398*, Transportation Research Board, Washington DC, 1993.
12. Windover, J., and May, A., Jr., "Revisions to Level D Methodology of Analyzing Freeway Ramp-Weaving Sections, *Transportation Research Record 1457*, Transportation Research Board, Washington DC, 1995.
13. "Weaving Zones," *Draft Report*, NCHRP Project 3-55(5), Viggen Corporation, Sterling, VA, 1998.
14. Roess, R., and Ulerio, J., "Weaving Area Analysis in the HCM 2000," *Transportation Research Record*, Transportation Research Board, Washington DC, 2000.
15. Roess, R., et al., "Analysis of Freeway Weaving Sections," *Final Report*, National Cooperative Highway Research Program Project 3-75, Polytechnic University, Brooklyn NY, 11201, January 2008.
16. Roess, R., and Ulerio, J., "Capacity of Ramp-Freeway Junctions," *Final Report*, Polytechnic University, Brooklyn, NY, 1993.
17. Leisch, J., *Capacity Analysis Techniques for Design and Operation of Freeway Facilities*, Federal Highway Administration, U.S. Department of Transportation, Washington DC, 1974.

Problems

15-1. Consider the pair of ramps shown in Figure 15.19. It may be assumed that there is no ramp-to-ramp flow.

(a) Given the existing demand volumes and other prevailing conditions, at what LOS is this section expected to operate? If problems exist, which elements appear to be causing the difficulty?

(b) It is proposed that the acceleration and deceleration lanes be joined to form a continuous auxiliary lane. How will this affect the operation? What LOS would be expected? Would you recommend this change?

(c) What is the capacity of the section under the two scenarios described in parts (a) and (b)?

15-2. Consider the weaving area in Figure 15.20. All demands are shown as flow rates in pc/h under equivalent base conditions.

(a) Describe the critical characteristics of the segment.

(b) What is the expected LOS for these conditions?

(c) What is the capacity of the weaving section under equivalent ideal conditions?

(d) If all demands include 10% trucks in rolling terrain, and all drivers are assumed to be familiar with the facility, and the PHF = 0.92, what is the capacity of the segment under prevailing conditions?

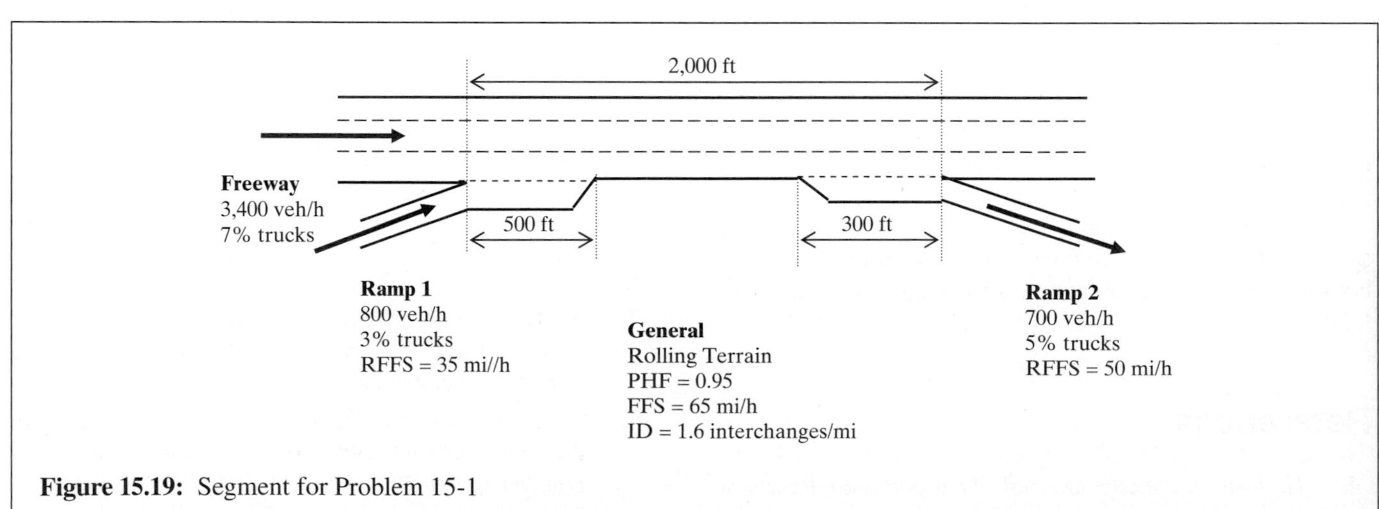

Figure 15.19: Segment for Problem 15-1

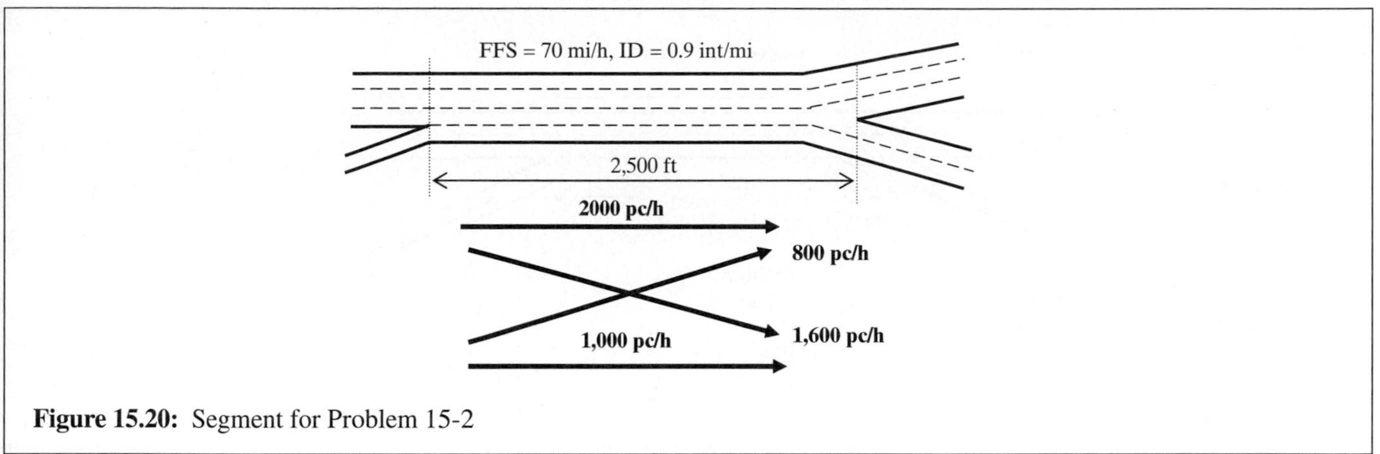

Figure 15.20: Segment for Problem 15-2

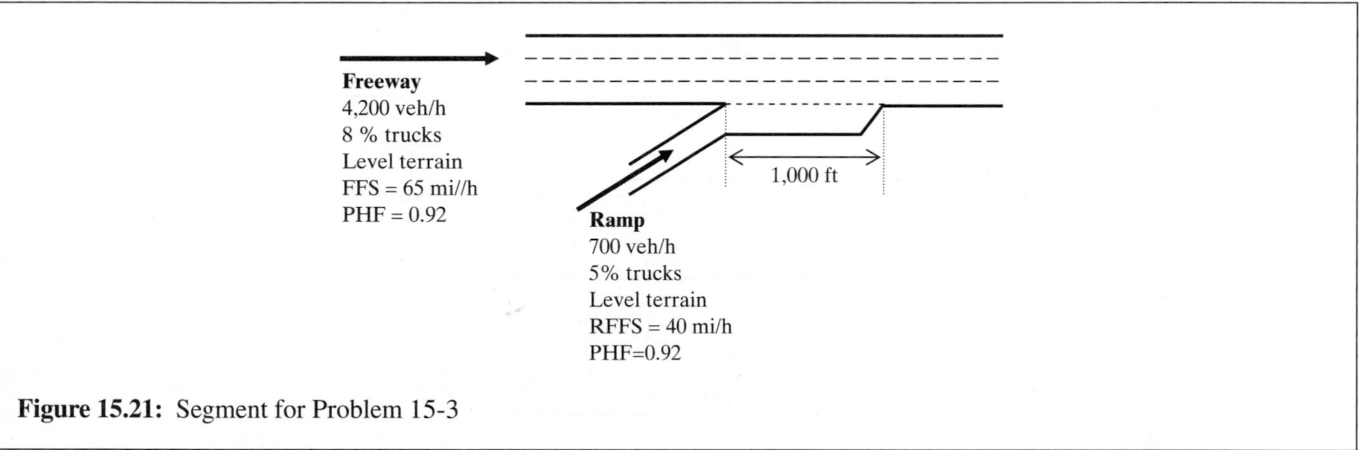

Figure 15.21: Segment for Problem 15-3

15-3. Consider the on-ramp shown in Figure 15.21.

(a) At what LOS would the merge area be expected to operate?

(b) A new development nearby opens and increases the on-ramp volume to 1,000 veh/h. How does this affect the LOS?

15-4. Figure 15.22 illustrates two consecutive ramps on an older freeway. It may be assumed that there is a ramp-to-ramp flow of 150 veh/h.

(a) What is the expected LOS for the conditions shown?

(b) Several improvement plans are under consideration:

i. *Connect the two ramps with a continuous auxiliary lane.*

ii. *Add a third lane to the freeway and extend the length of acceleration and deceleration lanes to 300 feet.*

iii. *Provide a lane addition at the on-ramp that continues past the off-ramp on the downstream freeway section. The off-ramp deceleration lane remains 200 feet long.*

Which of these three improvements would you recommend? Why? Justify your answer.

FFS = 60 mi/h; ID = 2 int/mi

15-5. What is the expected LOS and capacity for the weaving area shown in Figure 15.23? All demands are stated in terms of flow rates in pc/h under equivalent base conditions.

Comment on the results. Are they acceptable? If not, suggest any solutions that might mitigate some of the problems. You do not have to analyze the suggested improvements but should provide a verbal description of why they would improve on current operations.

ID = 1.3 interchanges/mi

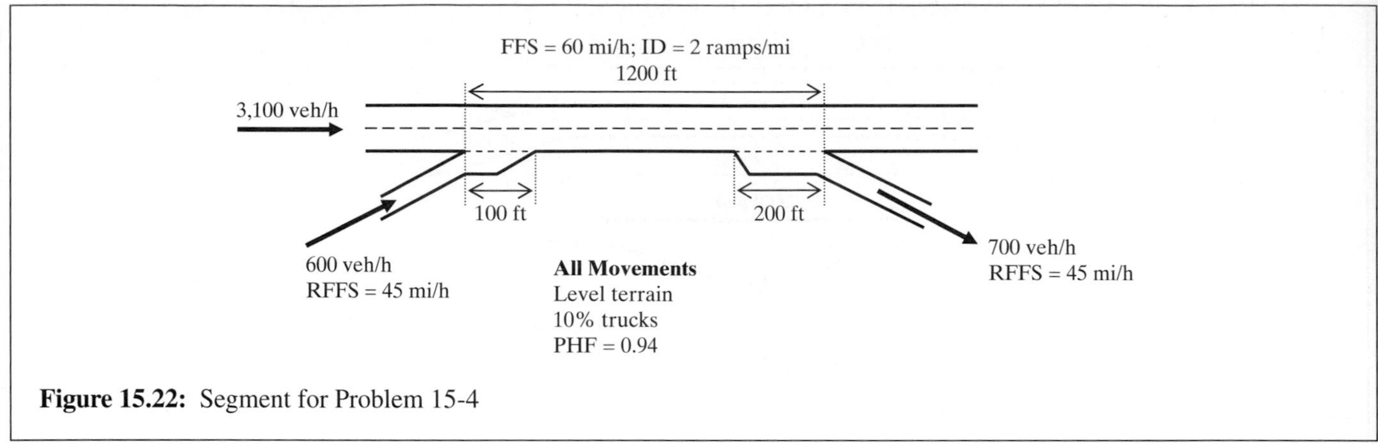

Figure 15.22: Segment for Problem 15-4

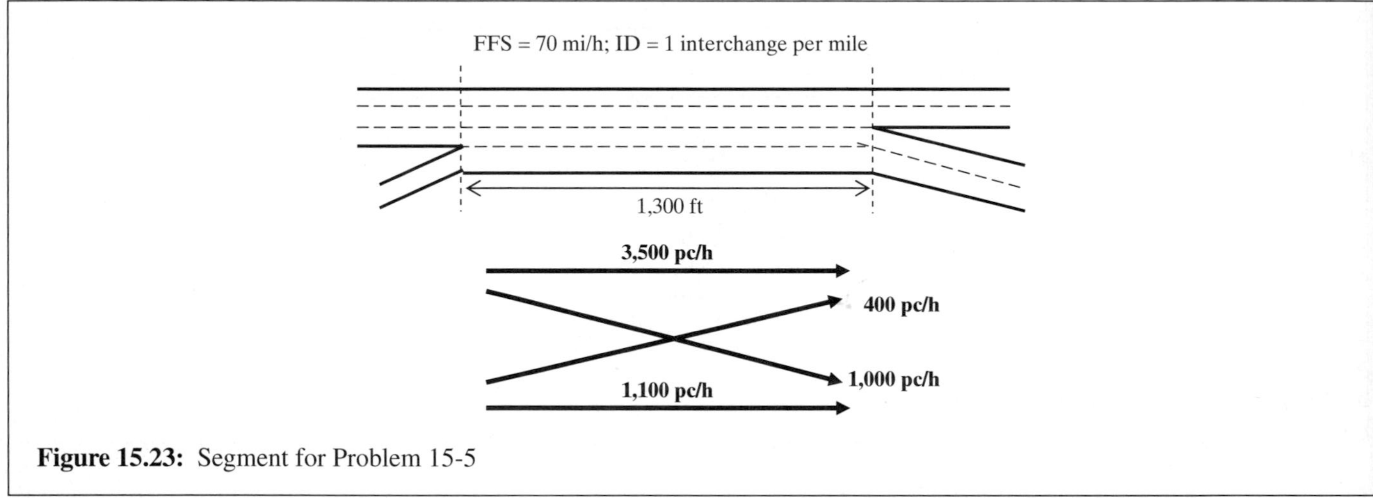

Figure 15.23: Segment for Problem 15-5

Appendix

Special Cases in Merge and Diverge Analysis

As noted in the body of the chapter, merge and diverge analysis procedures were calibrated primarily for single-lane, right-hand on- and off-ramps. Modifications have been developed so that a broad range of merge and diverge geometries can be analyzed using these procedures. These "special cases" include:

- Two-lane, right-hand on- and off-ramps
- On- and off-ramps on five-lane (one direction) freeway sections
- One-lane, left-hand on- and off-ramps
- Major merge and diverge areas
- Lane drops and lane additions

Each of these special cases is addressed in the sections that follow.

Two-Lane on-Ramps

Figure 15.A1 illustrates the typical geometry of a two-lane on-ramp. Two lanes join the freeway at the merge point. There are, in effect, two acceleration lanes. First, the right ramp lane merges into the left ramp lane; subsequently, the left ramp lane merges into the right freeway lane. The lengths of these two acceleration lanes are as shown in the figure.

The general procedure for on-ramps is modified in two ways. When estimating the demand flow in lanes 1 and 2 immediately upstream of the on-ramp (v_{12}), the standard equation is used:

$$v_{12} = v_F * P_{FM}$$

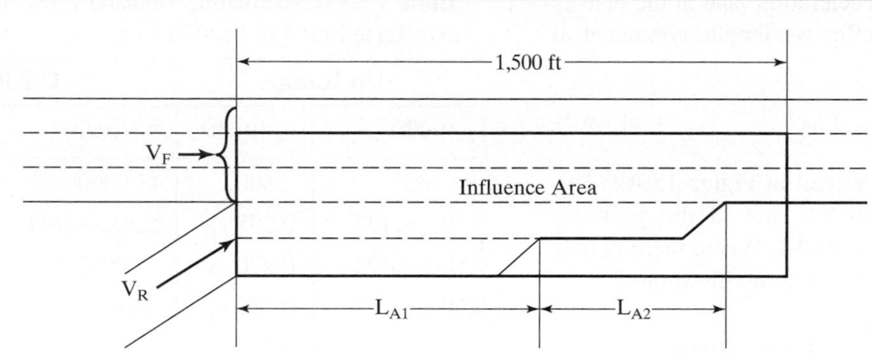

Figure 15.A1: Typical Two-Lane On-Ramp

(*Source:* Used with permission of Transportation Research Board, National Research Council, *Highway Capacity Manual*, 2000, Exhibit 25-8, p. 25-9.)

Instead of using the standard equations to find P_{FM} the following values are used:

$$P_{FM} = 1.000 \text{ four-lane freeways}$$

$$P_{FM} = 0.555 \text{ six-lane freeways}$$

$$P_{FM} = 0.209 \text{ eight-lane freeways}$$

In addition, in the density equation, the length of the acceleration lane is replaced by an effective length that considers both lanes of the two-lane merge area:

$$L_{aEFF} = 2L_{A1} + L_{A2} \tag{15-A1}$$

where L_{A1} and L_{A2} are defined in Figure 15.A1. Occasionally, a two-lane on-ramp is used at a location where one or two lanes are being added to the downstream freeway section. Depending on the details of such merge areas, they could be treated as lane additions or as major merge areas.

Two-Lane Off-Ramps

Figure 15.A2 illustrates two common geometries used with two-lane off-ramps. The first is a mirror image of a typical two-lane on-ramp junction, with two deceleration lanes provided. The second provides a single deceleration lane, with the left-hand ramp lane originating at the diverge point without a separate deceleration lane.

As was the case with two-lane on-ramps, the standard procedures are applied to the analysis of two-lane off-ramps with two modifications. In the standard equation,

$$v_{12} = v_R + (v_F - v_R)P_{FD}$$

the following values are used for P_{FD}:

$$P_{FD} = 1.000 \text{ four-lane freeways}$$

$$P_{FD} = 0.450 \text{ six-lane freeways}$$

$$P_{FD} = 0.260 \text{ eight-lane freeways}$$

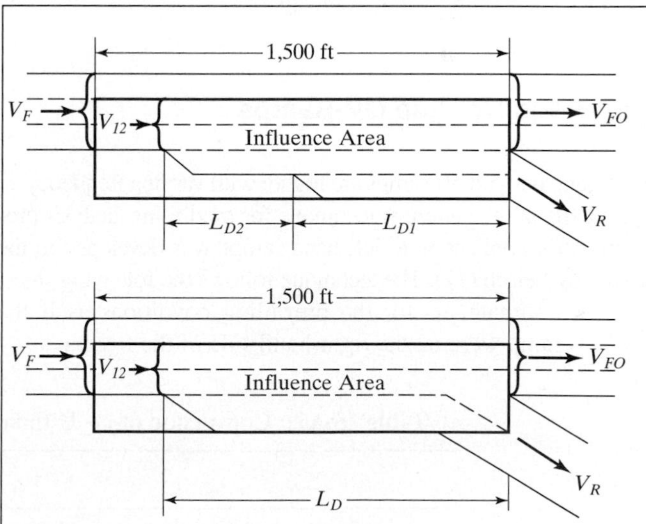

Figure 15.A2: Typical Geometries for Two-Lane Off-Ramps

(*Source:* Used with permission of Transportation Research Board, National Research Council, *Highway Capacity Manual*, 2000, Exhibit 25-15, p. 25-15.)

Also, the length of the acceleration lane in the density equation is replaced with an effective length, computed as follows:

$$L_{dEFF} = 2L_{D1} + L_{D2} \quad (15\text{-A2})$$

where L_{D1} and L_{D2} are defined in Figure 15.A2. This modification is applied only in the case of the geometry shown in the first part of Figure 15.A2. Where there is only one deceleration lane, it is used without modification.

On- and Off-Ramps on Five-Lane Freeway Sections

In some areas of the country, freeway sections with five lanes in a single direction are not uncommon. The procedure for analyzing right-hand ramps on such sections is relatively simple: An estimate of the demand flow in lane 5 (the left-most lane) of the section is made. This is deducted from the total approaching freeway flow; the remaining flow is in the right four lanes of the section. Once this deduction is made, the section can be analyzed as if it were a ramp on an eight-lane freeway (four lanes in one direction). Table 15.A1 gives simple algorithms for determining the flow in lane 5 (v_5). Then:

$$v_{4EFF} = v_F - v_5 \quad (15\text{-A3})$$

and the remainder of the problem is analyzed using v_{4EFF} as the approaching freeway flow on a four-lane (one direction) freeway section.

Left-Hand On- and Off-Ramps

Left-hand on- and off-ramps are found, with varying frequency, in most parts of the nation. A technique for modifying analysis procedures for application to left-hand ramps was developed in the 1970s by Leisch (*17*). The technique follows the following steps:

- Estimate v_{12} for the prevailing conditions as if the ramp were on the right-hand side of the freeway.
- To estimate the traffic remaining in the two left-most lanes of the freeway (v_{12} for a four-lane freeway, v_{23} for a six-lane freeway, v_{34} for an eight-lane freeway), multiply the result by the appropriate factor selected from Table 15.A2.
- Using the demand flow in the two left-most freeway lanes instead of v_{12}, check capacities and estimate density in the ramp influence area without further modification to the methodology.
- Speed algorithms should be viewed as only very rough estimates for left-hand ramps. Speed predictions for "outer lanes" may not be applied.

Table 15.A1: Estimating Demand Flow in Lane 5 of a Five-Lane Freeway Section

On-Ramps		Off-Ramps	
v_F (pc/h)	v_5 (pc/h)	v_F (pc/h)	v_5 (pc/h)
≥8,500	2,500	≥7,000	0.200 v_F
7,500–8,499	0.295 v_F	5,500–6,999	0.150 v_F
6,500–7,499	0.270 v_F	4,000–5,499	0.100 v_F
5,500–6,499	0.240 v_F	<4,000	0
<5,500	0.220 v_F		

(*Source:* Used with permission of Transportation Research Board, National Research Council, *Highway Capacity Manual,* 2000, compiled from Exhibits 25-11 and 15-18, pp. 25-11 and 25-17.)

Lane Additions and Lane Drops

Many merge and diverge junctions involve the addition of a lane (at a merge area) or the deletion of a lane (at a diverge area). In general, these areas are relatively straightforward to analyze, applying the following general principles:

- Where a single-lane ramp adds a lane (at a merge) or deletes a lane (at a diverge), the capacity of the

Table 15.A2: Conversion of v_{12} Estimates for Left-Hand Ramps

$v_{xy} = v_{12} {}^* f_{LH}$		
	Adjustment Factor	
To Estimate:	**For On-Ramps**	**For Off-Ramps**
v_{12} on four-lane freeways (2 lanes ea. dir.)	1.00	1.00
v_{23} on six-lane freeways (3 lanes ea. dir.)	1.12	1.05
v_{34} on eight-lane freeways (4 lanes ea. dir.)	1.20	1.10

ramp is determined by its free-flow speed, and it is analyzed as a ramp roadway using the criteria of Table 15.5. LOS criteria for basic freeway sections are applied to upstream and downstream freeway segments, which will have a different number of lanes.

- Where a two-lane ramp results in a lane addition or a lane deletion, it is treated as a major merge or diverge area. The techniques described in the next section are applied.

Major Merge and Diverge Areas

A major merge area is formed when two multilane roadways join to form a single freeway or multilane highway segment. A major diverge area occurs when a freeway or multilane highway segment splits into two multilane downstream roadways. These multilane merge and diverge situations may be part of major freeway interchanges or may involve significant multilane ramp connections to surface streets. The typical characteristic of these roadways is that they are often designed to accommodate relatively high speeds, which somewhat changes the dynamics of merge and diverge operations.

At a major merge area, a lane may be dropped, or the number of lanes in the downstream section may be the same as the total approaching the merge. Similarly, at a diverge area, a lane may be added, or the total lanes leaving the diverge area may be equal to the number on the approaching facility segment. Figure 15.A3 illustrates these configurations.

The analysis of major merge and diverge areas is generally limited to an examination of the demand-capacity balance of approaching and departing facility segments. No LOS criteria are applied.

For major diverge areas, an algorithm has been developed to roughly estimate the density across all approaching freeway lanes for a segment 1,500 feet upstream of the diverge:

$$D = 0.0109\left({}^{v_F}/_{N} \right) \qquad \text{(15-A4)}$$

where: D = density across all freeway lanes, from diverge to a point 1,500 ft upstream of the diverge, pc/mi/ln

v_F = approaching freeway demand flow, pc/h

N = number of freeway lanes approaching the diverge

This is an approximation at best and not used to assign a LOS to the diverge area.

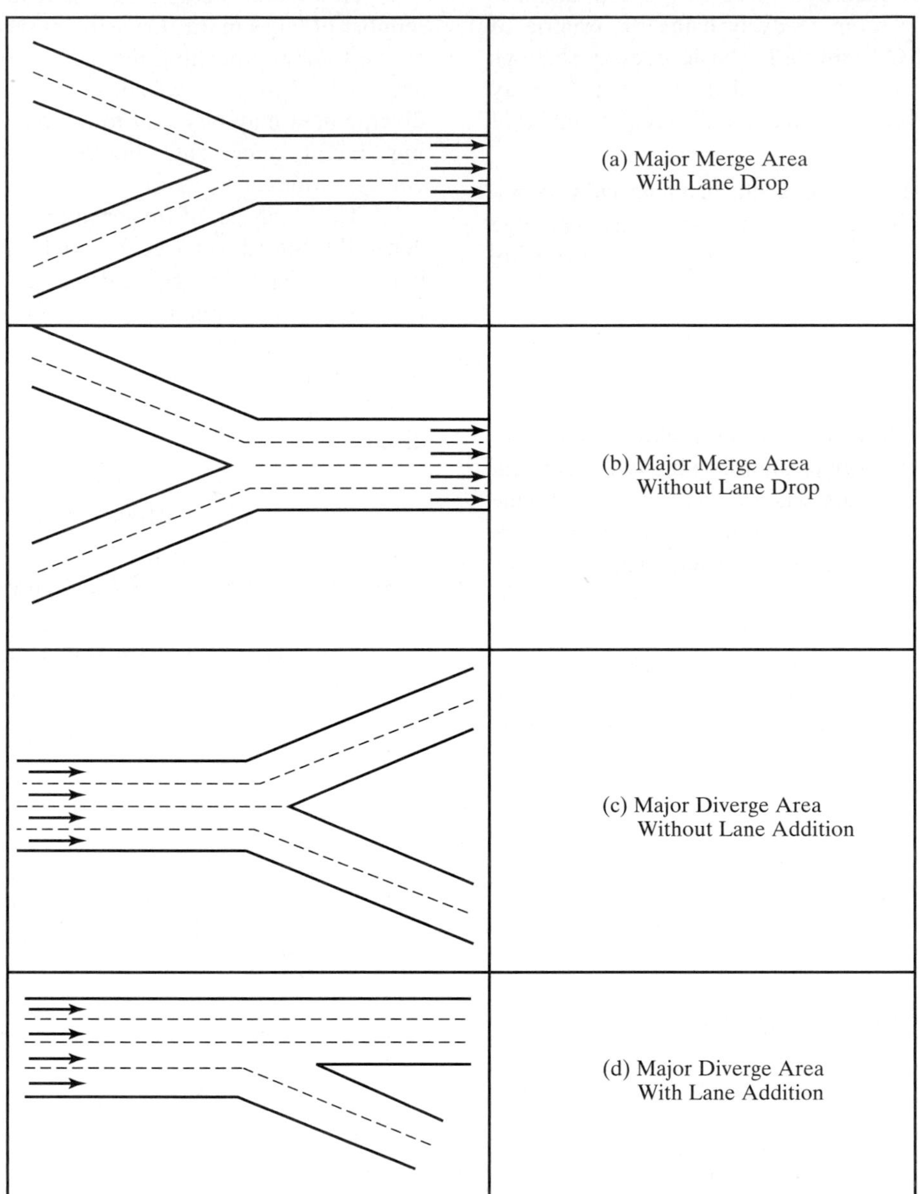

Figure 15.A3: Major Merge and Diverge Areas

(*Source:* Used with permission of Transportation Research Board, National Research Council, *Highway Capacity Manual*, 2000, Exhibits 25-9 and 25-10, p. 25-10, and Exhibits 15-16 and 15-17, p. 25-16.)

CHAPTER 16

Two-Lane Highways

16.1 Introduction

A significant portion of the nation's almost 4 million miles of paved highway are considered to be "rural." Of these, most are two-lane highways (i.e., one lane for traffic in each direction). On such highways, passing takes place in the opposing lane when the sight distance and opposing traffic conditions permit. Because of these passing operations, the two-lane, two-way rural highway is the only type of highway link on which traffic in one direction has a distinct operational impact on traffic in the other direction.

These roadways range from heavily traveled intercity routes to sparsely traveled links to isolated areas. They provide a vast network connecting the fringes of urban areas, agricultural regions, resource development areas, and remote outposts. Because of the varied functions they serve, two-lane rural highways are built to widely varying geometric standards and display a wide range of operating characteristics.

Rural two-lane highways serve two primary functions in the nation's highway network:

- Mobility
- Accessibility

As part of state and county primary highway systems, they serve a critical mobility function. Large numbers of road users rely on these highways for regular trips of significant length. Design standards for this type of two-lane highway generally reflect their use in serving higher demand flows. Higher design speeds reflect the primary mobility service provided.

Many two-lane rural highways, however, serve low demands, sometimes under 100 veh/day. The primary function of such highways is to provide for basic all-weather access to remote or sparsely-developed areas. Because such highways are not used by large numbers of people or vehicles, their design speeds and related geometric features are often not a major concern.

Because of the broad diversity of use on these highways, the 2000 edition of the *Highway Capacity Manual* (HCM 2000) [*1*] created two distinct classes of rural two-lane, two-way highways:

- *Class I.* These are highways on which motorists expect to travel at relatively high speeds, including major intercity routes, primary arterials, and daily commuter routes.
- *Class II.* These are highways on which motorists do not necessarily expect to travel at high speeds, including access routes, scenic and recreational routes that are not primary arterials, and routes through rugged terrain.

Class I two-lane highways serve primarily mobility needs, whereas Class II two-lane highways serve primarily access needs.

Even this categorization does not completely describe the diversity in the "look and feel" of such highways. Routes through rugged terrain are classified as Class II, primarily because the terrain limits the geometry of the roadway, forcing low-speed operation and providing few or no passing opportunities. Nevertheless, some of these roads must serve mobility needs where demand is sufficient.

The American Association of State Highway and Transportation Officials (AASHTO) classifies two-lane rural highways as "rural local roads," "rural collectors," or "rural arterials."

Unfortunately, these categories overlap the HCM classifications. Virtually all rural arterials would be Class I facilities. Rural collectors, however, could fall into either HCM class depending on the specifics of terrain and geometry. Virtually all rural local roads would fall into Class II. Some judgment, therefore, is needed to properly classify rural highways.

The Florida Department of Transportation (FDOT) has also defined a Class III two-lane highway. Class III two-lane highways may serve more developed areas. Often, rural highways become the "main street" of small towns and communities in an otherwise rural environment. Such highways often have reduced speed limits, tight restrictions on passing, and more roadside driveways and unsignalized junctions. Class III highways may also include recreational routes where scenic beauty is a primary characteristic. Such highways may also have reduced speed limits and more limited passing opportunities. Because of this, the operation of Class III highways is somewhat different from Class I and Class II two-lane highways. As you will see, level-of-service (LOS) criteria for two-lane highways differ for each of the three defined classes of two-lane highways [*2,3*].

Figure 16.1 contains illustrations of two-lane highways, together with their likely classifications according to the HCM and the AASHTO.

(a) A Class I Rural Arterial

(b) A Class I Collector

(c) A Class II Rural Collector

(d) A Class II Rural Local Road

(e) Class III Arterial through a Developed Area

Figure 16.1: Rural Two-Lane Highways Illustrated

Class I highways generally feature gentle geometries allowing for higher speed operation, as illustrated in Figures 16.1 (a) and (b). Usable paved shoulders and/or stabilized roadside recovery areas, full pavement markings, and signage are generally present. Figure 16.1 (c) illustrates a Class II rural collector. The rural local road depicted in Figure 16.1 (d) is clearly a Class II facility. It lacks pavement markings and usable shoulders, and a low-type penetration pavement is evident. Figure 16.1 (e) shows a Class III facility running through a developed area in an otherwise rural environment.

Because of the wide diversity in the function and physical characteristics of rural roadways, both design standards and LOS criteria must be flexible and must address the full range of situations in which two-lane, two-way highways exist.

This chapter provides an overview of the following subjects related to two-lane, two-way highways:

- Design standards
- Passing sight distance requirements and the impact of "No Passing" zones
- Capacity and LOS analysis of two-lane highways

Each of these is treated in the major sections that follow.

16.2 Design Standards

Design standards for urban, suburban, and rural highways are set by AASHTO in the current version of *A Policy on Geometric Design of Highways and Streets* [*4*], often referred to as the "Green Book" because of the color of its cover in recent editions. The latest version of the Policy (at this writing) is the fifth edition, published in 2004.

The single most important design factor controlling the specifics of geometry is the design speed. Every element of a highway (horizontal alignment, vertical alignment, cross section) must be designed to allow safe operation at the "design speed." Thus a design speed of 60 mi/h, for example, requires that every element and segment of the facility must allow for safe operation at 60 mi/h. In Chapter 3, the impacts of speed on horizontal and vertical alignment are discussed in detail.

Table 16.1 shows recommended design speeds for rural two-lane highways based on function, classification, terrain, and ADT demand volumes. AASHTO does not have a similar classification to Class III highways as defined by FDOT.

Figure 16.2 shows reasonable recommended design criteria for maximum grades on two-lane, two-way rural highways. Maximum grade relates to the function of the facility, its design speed, and the terrain in which it is located. The design speed, once selected, limits horizontal and vertical alignment. Consult Chapter 3 for specific relationships between design speed and horizontal and vertical curvature.

Note that the criteria of Table 16.1 and Figure 16.2 represent *minimum* recommended design speeds and *maximum* grades. When conditions permit, higher design speeds and less severe grades should be used.

Table 16.1: Recommended Minimum Design Speeds for Rural Two-Lane Highways (mi/h)

Type of Facility	ADT (veh/day)	*Minimum Design Speed in*		
		Level Terrain	**Rolling Terrain**	**Mountainous Terrain**
Rural Local Roads	**<50**	30	20	20
	50–249	30	30	20
	250–399	40	30	20
	400–1499	50	40	30
	1500–1999	50	40	30
	≥ 2000	50	40	30
Rural Collectors	**<400**	40	30	20
	400–2000	50	40	30
	>2000	60	50	40
Rural Arterials	**All**	60	50	40

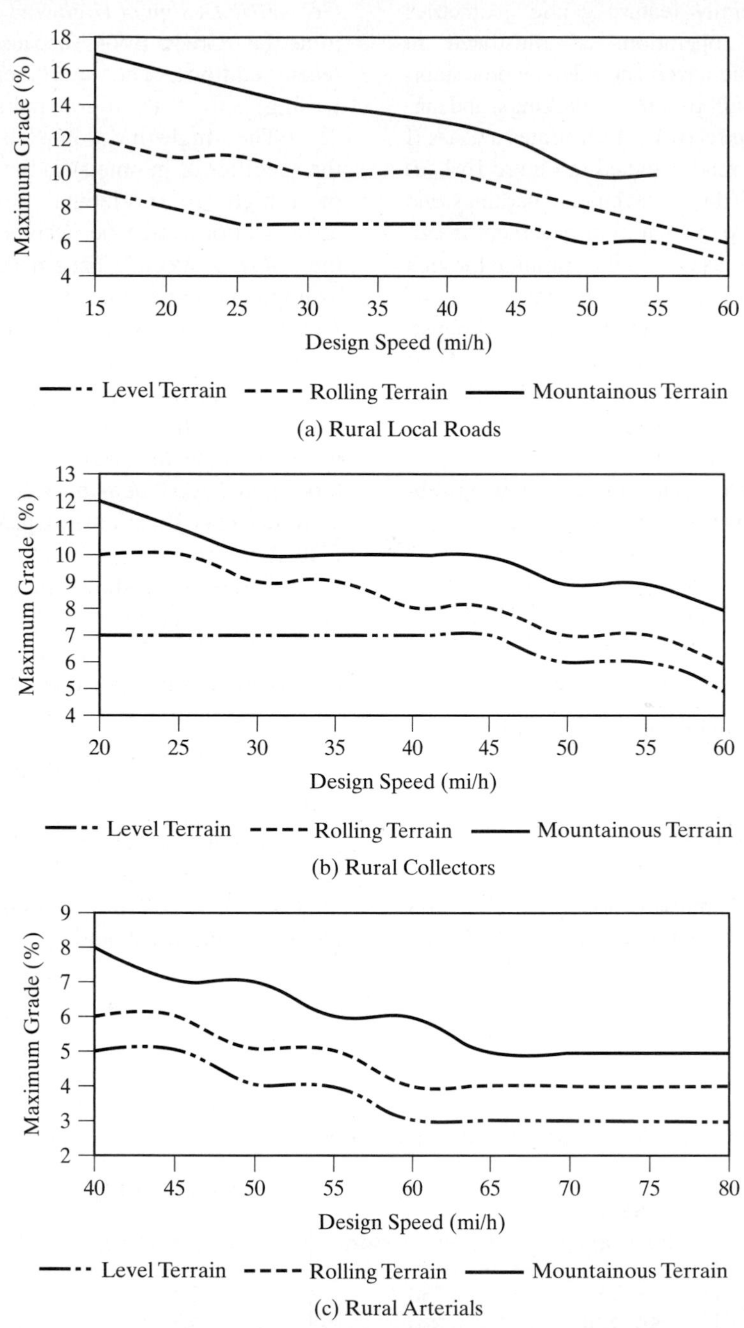

Figure 16.2: Design Criteria for Maximum Grades on Two-Lane Rural Highways

16.3 Passing Sight Distance on Two-Lane Highways

In Chapters 2 and 3, the importance of safe stopping sight distance was discussed and illustrated. The safe stopping sight distance *must* be provided at every point on every roadway for the selected design speed. No driver should ever be confronted with a sudden obstacle on the roadway and insufficient distance (and time) to stop before colliding with it.

On two-lane highways, another critical safety feature is the passing sight distance. Because vehicles pass by using the opposing traffic lane, passing maneuvers on two-lane rural highways are particularly dangerous. Passing sight distance is the minimum sight distance required to safely begin and complete a passing maneuver under the assumed conditions for the highway.

Passing sight distance need not be provided at every point along a two-lane rural highway. Passing should not, however, be permitted where passing sight distance is not available. Roadway markings and "No Passing" zone signs are used to prohibit passing where sight distance is insufficient to do so safely. Note that on unmarked rural two-lane highways, drivers *may not* assume that passing sight distance is available. On unmarked facilities, driver judgment is the only control on passing maneuvers.

AASHTO criteria for minimum passing sight distances are based on a number of assumptions concerning driver behavior. These assumptions accommodate most drivers and are not based on averages.

The minimum passing sight distance is determined as the sum of four component distances, as illustrated in Figure 16.3. The component distances are defined as:

d_1 = distance traversed during perception and reaction time and during the initial acceleration to the point of encroachment on the left lane.

d_2 = distance traveled while the passing vehicle occupies the left lane.

d_3 = distance between the passing vehicle at the end of its maneuver and the opposing vehicle.

d_4 = distance traversed by the opposing vehicle for two thirds of the time the passing vehicle occupies the left lane, or two thirds of d_2 above.

Criteria for the speeds and distances used in determining passing sight distances are based on extensive field studies [5] and validations [6,7]. The component distances are computed as follows:

$$d_1 = 1.47 t_1 \left(S - m + \frac{a t_1}{2} \right) \tag{16-1}$$

where: t_1 = reaction time, s

S = speed of passing vehicle, mi/h

m = difference between speed of passing vehicle and passed vehicle, mi/h

a = average acceleration of passing vehicle, mi/h/s

$$d_2 = 1.47\, S t_2 \tag{16-2}$$

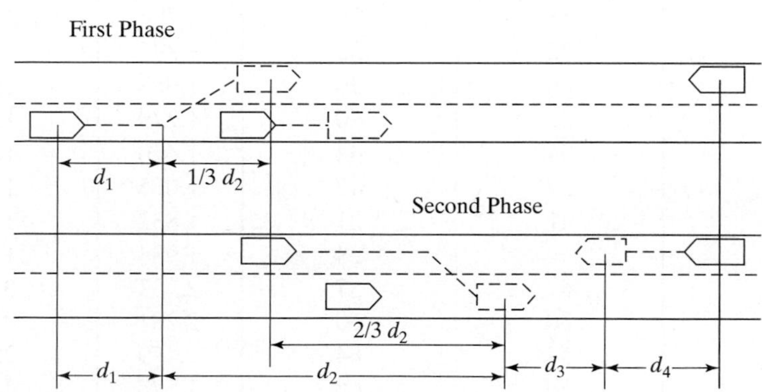

Figure 16.3: Elements of Passing Sight Distance

(*Source:* Used with Permission of the American Association of State Highway and Transportation Officials, *A Policy on Geometric Design of Highways and Streets,* 5th Edition, Washington DC, 2004, Exhibit 3-4, p. 119.)

where: t_2 = time passing vehicle occupies the left lane, s
S = speed of the passing vehicle, mi/h

$$d_3 = 100 - 300 \text{ ft} \qquad (16\text{-}3)$$

$$d_4 = \left(\frac{2}{3}\right)d_2 \qquad (16\text{-}4)$$

In using these equations, AASHTO assumes that the speed of the passing vehicle is 10 mi/h (m) greater than the speed of the passed vehicle. Acceleration rates between 1.4 and 1.5 mi/h/s are used. Reaction times range between 3.6 and 4.5 s (t_1). The time that the left lane is occupied is based on speed parameters and ranges between 9.3 and 11.3 s (t_2). The clearance distance range is as shown, with the lower end of the scale used for slower speeds.

To be useful in design, safe passing sight distances must be related to the design speed of the facility. Table 16.2 shows AASHTO passing sight distance criteria based on design speed. The assumed passed and passing vehicle speeds are such that they accommodate the vast majority of potential passing maneuvers occurring on a two-lane rural highway.

It is always desirable to have passing sections provided as frequently as possible. The effects of "No Passing" zones on operations are made clear by the capacity and level-of-service (LOS) models presented later in this chapter. Where passing is permitted, it is also desirable to provide as much sight distance as is practical, using the values of Table 16.2 as minimums.

The design criteria of Table 16.2 are not the same as the warrants for signing and marking "No Passing" zones presented in the MUTCD [*8*]. The MUTCD requirements, shown in Table 16.3, are substantially lower than the design criteria. The assumptions used in establishing the warrants are different from those described for design standards. When a centerline is used, and where sight distances are less than the criteria of Table 16.3, "No Passing" zone markings and signs must be installed. A more conservative, but safer, practice would be to post "No Passing" signs and markings wherever the design criteria for passing sight distance are not met.

16.4 Capacity and Level-of-Service Analysis of Two-Lane Rural Highways

The HCM 2000 methodology for the analysis of two-lane, two-way rural highways was a new procedure based on an extensive research study conducted at the Midwest Research

Table 16.2: Design Values for Passing Sight Distance on Two-Lane Highways

Design Speed (mi/h)	Assumed Speeds (mi/h)		Passing Sight Distance (ft)	
	Passed Veh	Passing Veh	Exact	Rounded for Design
20	18	28	706	710
25	22	32	897	900
30	26	36	1,088	1,090
35	30	40	1,279	1,280
40	34	44	1,470	1,470
45	37	47	1,625	1,625
50	41	51	1,832	1,835
55	44	54	1,984	1,985
60	47	57	2,133	2,135
65	50	60	2,281	2,285
70	54	64	2,479	2,480
75	56	66	2,578	2,580
80	58	68	2,677	2,680

(*Source:* Used with Permission of the American Association of State Highway and Transportation Officials, *A Policy on Geometric Design of Highways and Streets,* 5th Edition, Washington DC, 2004, Exhibit 3-7, p. 124.)

Table 16.3: Minimum Passing Sight Distance Criteria for Placement of "No Passing" Markings and Signs

85th Percentile Speed or Statutory Speed Limit (mi/h)	Minimum Passing Sight Distance (ft)
25	450
30	500
35	550
40	600
45	700
50	800
55	900
60	1,000
65	1,100
70	1,200

(*Source:* Used with Permission of Federal Highway Administration, U.S. Department of Transportation, *Manual of Uniform Traffic Control Devices,* Washington DC, 2003, Table 3B-1, p. 3B-9.)

Institute [9]. It builds conceptually on the methodology in use since 1985 but introduces a number of new elements. In the forthcoming HCM 2010, the procedures have once again been updated but less dramatically. This chapter is based on source materials for the HCM 2010. Although the procedures included here have been approved by the Highway Capacity and Quality of Service Committee of the Transportation Research Board, consult the HCM 2010, when it becomes available, to determine whether additional changes have been incorporated.

One of the principal difficulties in studying the behavior of two-lane highways is the fact that few operate under conditions that are even near capacity. Thus research into two-lane highway operations relies heavily on simulation modeling in addition to limited field observations.

Among the unique features of this methodology are the definition of three classes of two-lane highways (described previously) and the use of three measures of effectiveness to define levels of service.

16.4.1 Capacity

The capacity of a two-lane highway has been significantly increased from previous values to accommodate frequent, higher observations throughout North America. The capacity of a two-lane highway under base conditions has been established as 3,200 pc/h in both directions, with a maximum of 1,700 pc/h in one direction. These values were established in HCM2000 and are maintained in HCM 2010. The base conditions for which this capacity is defined include:

- 12-foot (or greater) lanes
- 6-foot (or greater) usable shoulders
- Level terrain
- No heavy vehicles
- 100% passing sight distance available (no "No Passing" zones)
- 50-50 directional split of traffic
- No traffic interruptions

As with all capacity values, these standards reflect "reasonable expectancy" (i.e., most two-lane highway segments operating under base conditions should be able to achieve such capacities most of the time). Isolated observations of higher volumes do not negate the standard.

In the HCM 2010, LOS F for two-lane highways continues to exist only when demand exceeds the capacity of a segment.

16.4.2 Level of Service

LOS for two-lane rural highways is defined in terms of three measures of effectiveness:

- Average travel speed (ATS)
- Percent time spent following (PTSF)
- Percent free-flow speed (PFFS)

Average travel speed is the average speed of all vehicles traversing the defined analysis segment for the specified time period, which is usually the peak 15 minutes of a peak hour. *Percent time spent following* is similar to "percent time delay," which was used in the 1985 through 1997 *Highway Capacity Manuals.* It is the aggregate percentage of time that all drivers spend in queues, unable to pass, with the speed restricted by the queue leader. A surrogate measure for PTSF is the percentage of vehicles following others at headways of 3.0 seconds or less. *Percent free-flow speed* is based on the comparison of the prevailing speed to the free-flow speed, expressed as a percentage.

LOS criteria for two-lane highways are shown in Table 16.4. The criteria vary for Class I, II, and III highways. On Class I highways, users expect both high speeds and the ability to pass. Both ATS and PTSF criteria are used. The LOS is based on the measure that indicates the poorest service quality. On Class II highways, speed is rarely an issue because many such highways have restricted design speeds and low speed limits. The ability to pass to get out of

Table 16.4: Level of Service Criteria for Two-Lane Rural Highways

Level of Service	Class I Highways		Class II Highways	Class III Highways
	ATS (mi/h)	PTSF (%)	PTSF (%)	PFFS (%)
A	>55	≤35	≤40	>91.7
B	>50–55	>35–50	>40–55	>83.3–91.7
C	>45–50	>50–65	>55–70	>75.0–83.3
D	>40–45	>65–80	>70–85	>66.7–75.0
E	≤40	>80	>85	≤66.7

(*Source: Two-Lane Highways,* Draft Chapter 15, National Cooperative Highway Research Program Project 3-92, Exhibit 15-3.)

slow-moving platoons is critical. For Class II highways, therefore, only PTSF is used to determine the LOS. On Class III highways, speed limits are often low due to the surrounding development, and passing is generally very much restricted by design. For Class III highways, therefore, the LOS is based on the PFFS that can be maintained under prevailing conditions.

Figure 16.4 illustrates the relationships between PTSF and two-way flow rate on a two-lane highway with base conditions. This figure is used only for illustrative purposes. Although the HCM2000 originally allowed for analysis of the two directions of a two-lane highway simultaneously, this often produced results that were inconsistent with two separate single-direction analyses. The illustration, however, is important. It shows that high PTSF levels occur when flow rates are quite low. In the illustration, 70% PTSF is reached with a total flow rate of only 1,500 pc/h, less than half the nominal capacity of a two-lane highway.

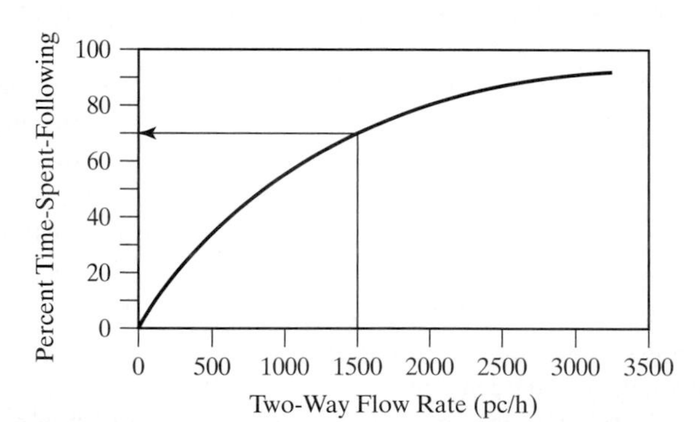

Figure 16.4: Two-Way Flow vs. Percent Time Spent Following

(*Source:* Used with permission of Transportation Research Board, National Research Council, *Highway Capacity Manual,* 4th Edition, Washington DC, 2000, Exhibit 12-6(b), p. 12-14.)

This characteristic explains why few two-lane highway segments are observed operating at flows near capacity. Long before demands approach capacity, operational quality has seriously deteriorated. On two-lane highways, this can lead to safety problems because drivers attempt unsafe passing maneuvers to avoid these delays. Most two-lane highways, therefore, are reconfigured or reconstructed because of such problems, which occur long before demand levels approach the capacity of the facility.

As you will see as a result of the analysis methodology, poor levels of service on two-lane highways can exist at extremely low demand flow rates. No other type of traffic facility exhibits this characteristic.

16.4.3 Types of Analysis

Previous editions of the HCM required that analysis of two-lane rural highways operations be limited to a composite analysis of both directions. This requirement reflected the operational interaction between directional flows on a two-lane highway. Despite this, there are many occasions where operational characteristics in the two directions of a two-lane highway varied considerably, particularly where significant grades were involved. Because of this, the HCM 2010 mandates single-direction analysis. There are currently two distinct methodologies for analysis of two-lane highways:

- Single-directional analysis of general extended sections (≥2.0 mi) in level or rolling terrain
- Single-direction analysis of specific grades

For specific grades, single-direction analysis of the upgrade and downgrade is particularly important because these tend to differ significantly. In what is usually referred to as "mountainous" terrain, all analysis is on the basis of specific grades comprising that terrain. Any grade of 3% or more and at least 0.6 miles long must be addressed using specific grade procedures.

16.4.4 Free-Flow Speed

As was the case for multilane highways and freeways, the free-flow speed of a two-lane highway is a significant variable used in estimating expected operating conditions. The HCM recommends that free-flow speeds be measured in the field where practical, but it also offers a methodology for their estimation where measurement is not practical.

Field Measurement of Free-Flow Speed

The free-flow speed of a two-lane highway may be measured directly in the field. The speed study should be conducted at a representative site within the study section. Free-flow speeds may be directly measured as follows:

- A representative speed sample of 100 or more vehicles should be obtained.
- Total two-way traffic flow should be 200 pc/h or less.
- All vehicle speeds should be observed during the study period, or a systematic sampling (such as 1 vehicle of every 10) should be applied.
- The speed sample should be selected only from the direction under study.

If field measurements must be made at total flow levels higher than 200 pc/h, the free-flow speed may be estimated as:

$$FFS = S_m + 0.00776\left(\frac{v_f}{f_{HV}}\right) \qquad (16\text{-}5)$$

where: FFS = free-flow speed for the facility, mi/h

S_m = mean speed of the measured sample (where total flow >200 pc/h), mi/h

v_f = observed flow rate for the period of the speed sample, veh/h

f_{HV} = heavy vehicle adjustment factor

Chapter 10 of this text details procedures for measuring speeds in the field.

Estimating Free-Flow Speeds

If field observation of free-flow speed is not practical, free-flow speed on a two-way rural highway may be estimated as follows:

$$FFS = BFFS - f_{LS} - f_A \qquad (16\text{-}6)$$

where: FFS = free-flow speed for the facility, mi/h

$BFFS$ = base free-flow speed for the facility, mi/h

f_{LS} = adjustment for lane and shoulder width, mi/h

f_A = adjustment for access point density, mi/h

Unfortunately, the HCM does not provide any detailed criteria for the base free-flow speed, *BFFS*. It is limited to a range of 45 to 65 mi/h, with Class I highways usually in the 55 to 65 mi/h range, Class II highways usually in the 45 to 50 mi/h range, and Class III highways in the 40 to 50 mi/h range. Design speed and statutory speed limits may be used as inputs to establishing an appropriate value for *BFFS*. The design speed, which represents the maximum safe speed for the horizontal and vertical alignment of the highway, is a reasonable surrogate for the *BFFS*. Adjustment factors account for the impact of lane and shoulder width and access points, none of which are accounted for in the basic design speed of the facility. Speed limits are not as good a guide because various jurisdictions apply different philosophies and policies in setting such limits but may be used as a last resort. In general, the *BFFS* should be from 5 to 7 mi/h higher than the posted speed limit.

Adjustment factors for lane and shoulder width are shown in Table 16.5; adjustment factors for access point density are shown in Table 16.6. Access point density is computed by dividing the total number of driveways and intersections on both sides of the highway by the total length of the segment in miles.

A Sample Problem

Find the free-flow speed of a two-lane rural highway segment in rolling terrain. The lane width is 11.0 ft with 2.0-ft shoulders. Access point density is 30 per mile, and the base free-flow speed may be taken to be 60 mi/h.

The estimation technique is used. The free-flow speed is determined as follows:

$$FFS = BFFS - f_{LS} - f_A$$

where: $BFFS$ = 60 mi/h (given)

f_{LS} = 3.0 mi/h (Table 16.5)

f_A = 7.5 mi/h (Table 16.6)

$$FFS = 60.0 - 3.0 - 7.5 = 49.5 \text{ mi/h}$$

Table 16.5: Free-Flow Speed Adjustments for Lane and Shoulder Width

Lane Width (ft)	***Reduction in FFS (mi/h), f_{LS}***			
	Shoulder Width (ft)			
	≥0 < 2	**≥2 < 4**	**≥4 < 6**	**≥6**
≥9 < 10	6.4	4.8	3.5	2.2
≥10 < 11	5.3	3.7	2.4	1.1
≥11 < 12	4.7	3.0	1.7	0.4
≥12	4.2	2.6	1.3	0.0

(*Source:* Used by Permission of Transportation Research Board, National Science Council, *Highway Capacity Manual,* 4th Edition, Washington DC, 2000, Exhibit 20-5, p. 20-6.)

Table 16.6: Free-Flow Speed Adjustments for Access point Density

Access Points Per Mi	**Reduction in FFS (mi/h) f_A**
0	0.0
10	2.5
20	5.0
30	7.5
40	10.0

(*Source:* Used by Permission of Transportation Research Board, National Science Council, *Highway Capacity Manual,* 4th Edition, Washington DC, 2000, Exhibit 20-6, p. 20-6.)

Note that in selecting the lane and shoulder width adjustment, f_{LS}, care must be taken to observe the boundary conditions in Table 16.5. The lane width of 11.0 feet falls within the "≥11 <12" category; the 2.0 feet shoulders fall within the "≥2 <4" category.

16.4.5 Estimating Demand Flow Rate

As for most HCM methodologies, a critical computational step is the determination of a demand flow rate reflecting the base conditions for the facility type being analyzed. This requires that an hourly volume reflecting prevailing conditions be adjusted to reflect peak flow rates within the hour and base conditions. For two-lane highways, this adjustment is made as follows:

$$v = \frac{V}{PHF * f_{HV} * f_G} \tag{16-7}$$

where: v = demand flow rate, pc/h

V = hourly demand volume under prevailing conditions, veh/h

PHF = peak hour factor

f_{HV} = adjustment for heavy vehicle presence

f_G = adjustment for grades

The methodology, however, becomes more complex. Adjustment factors depend on whether the performance measure to be estimated is *ATS* or *PTSF.* The *PFFS* for Class III highways is based on *ATS,* so it does not involve a third flow rate determination. Thus for single-direction analysis, there are four different determinations of demand flow rate: two demand flow rates in *each direction,* one for *ATS* determination, and one for *PTSF* determination.

Determining Grade Adjustment Factors

For every computation, two grade adjustment factors are required: one for the ATS determination and one for the PTSF determination. Selection of appropriate adjustment factors also depends on the type of analysis being conducted. Grade adjustment factors are found as follows:

- One-direction analysis of general terrain segments for both ATS and PTSF determinations: Table 16.7.
- One-direction analysis of specific upgrades for ATS determination: Table 16.8.
- One-direction analysis of specific upgrades for PTSF determination: Table 16.9.
- One-direction analysis of specific downgrades for both ATS and PTSF determination: Table 16.7.

Table 16.7: Grade Adjustment Factor (f_G) for General Terrain Segments and Specific Downgrades (ATS and PTSF Determinations)

One Direction Demand Flow Rate, $v = V/PHF$ (veh/h)	Level Terrain and Specific Downgrades, ATS and PTSF	Rolling Terrain ATS	Rolling Terrain PTSF
≤100	1.00	0.67	0.73
200	1.00	0.75	0.80
300	1.00	0.83	0.85
400	1.00	0.90	0.90
500	1.00	0.95	0.96
600	1.00	0.97	0.97
700	1.00	0.98	0.99
800	1.00	0.99	1.00
≥900	1.00	1.00	1.00

Note: Interpolation to the nearest 0.01 is recommended.

(*Source: Two-Lane Highways,* Draft Chapter 15, National Cooperative Highway Research Program Project 3-92, Exhibits15-9 and 15-16.)

Determining the Heavy-Vehicle Adjustment Factor

The heavy-vehicle adjustment factors for ATS and PTSF determinations are found from passenger-car equivalents as follows:

$$f_{HV} = \frac{1}{1 + P_T(E_T - 1) + P_R(E_R - 1)} \tag{16-8}$$

where: f_{HV} = heavy-vehicle adjustment factor
P_T = proportion of trucks and buses in the traffic stream
P_R = proportion of recreational vehicles in the traffic stream
E_T = passenger-car equivalent for trucks and buses
E_R = passenger-car equivalent for recreational vehicles

As in multilane methodologies, the *passenger-car equivalent* is the number of passenger cars displaced by one truck (or RV) under the prevailing conditions on the analysis segment.

Passenger-car equivalents depend on which measure of effectiveness is being predicted (ATS or PTSF) and the terrain. Passenger-car equivalents are found from the following tables:

- One-direction analysis of general terrain segments for both ATS and PTSF determination: Table 16.10.
- One-direction analysis of specific upgrades for ATS determination: *Trucks:* Table 16.11; *RV's:* Table 16.12.
- One-direction analysis of specific upgrades for PTSF determination: Table 16.13.
- One-direction analysis of specific downgrades: Table 16.10.

Some specific downgrades are steep enough to require some trucks to shift into low gear and travel at crawl speeds to avoid loss of control. In such situations, the effect of trucks traveling at crawl speed may be taken into account by replacing Equation 16-8 with the following when computing the heavy vehicle adjustment factor, f_{HV}, for ATS determination only:

$$f_{HV} = \frac{1}{1 + P_{TC} * P_T(E_{TC} - 1) + (1 - P_{TC}) * P_T(E_T - 1) + P_R(E_R - 1)} \tag{16-9}$$

where: P_{TC} = proportion of heavy vehicles forced to travel at crawl speeds
E_{TC} = passenger car equivalents for trucks at crawl speed (Table 16.14)

All other variables are as previously defined.

Table 16.8: Grade Adjustment Factor (f_G) for Specific Upgrades: ATS Determinations

Grade (%)	Grade Length (mi)	Directional Demand Flow Rate, v_{vph}(veh/h)								
		≤100	200	300	400	500	600	700	800	≥900
≥3 <3.5	0.25	0.78	0.84	0.87	0.91	1.00	1.00	1.00	1.00	1.00
	0.50	0.75	0.83	0.86	0.90	1.00	1.00	1.00	1.00	1.00
	0.75	0.73	0.81	0.85	0.89	1.00	1.00	1.00	1.00	1.00
	1.00	0.73	0.79	0.83	0.88	1.00	1.00	1.00	1.00	1.00
	1.50	0.73	0.79	0.83	0.87	0.99	0.99	1.00	1.00	1.00
	2.00	0.73	0.79	0.82	0.86	0.98	0.98	0.99	1.00	1.00
	3.00	0.73	0.78	0.82	0.85	0.95	0.96	0.96	0.97	0.98
	≥4.00	0.73	0.78	0.81	0.85	0.94	0.94	0.95	0.95	0.96
≥3.5 <4.5	0.25	0.75	0.83	0.86	0.90	1.00	1.00	1.00	1.00	1.00
	0.50	0.72	0.80	0.84	0.88	1.00	1.00	1.00	1.00	1.00
	0.75	0.67	0.77	0.81	0.86	1.00	1.00	1.00	1.00	1.00
	1.00	0.65	0.73	0.77	0.81	0.94	0.95	0.97	1.00	1.00
	1.50	0.63	0.72	0.76	0.80	0.93	0.95	0.96	1.00	1.00
	2.00	0.62	0.70	0.74	0.79	0.93	0.94	0.96	1.00	1.00
	3.00	0.61	0.69	0.74	0.78	0.92	0.93	0.94	0.98	1.00
	≥4.00	0.61	0.69	0.73	0.78	0.91	0.91	0.92	0.96	1.00
≥4.5 <5.5	0.25	0.71	0.79	0.83	0.88	1.00	1.00	1.00	1.00	1.00
	0.50	0.60	0.70	0.74	0.79	0.94	0.95	0.97	1.00	1.00
	0.75	0.55	0.65	0.70	0.75	0.91	0.93	0.95	1.00	1.00
	1.00	0.54	0.64	0.69	0.74	0.91	0.93	0.95	1.00	1.00
	1.50	0.52	0.62	0.67	0.72	0.88	0.90	0.93	1.00	1.00
	2.00	0.51	0.61	0.66	0.71	0.87	0.89	0.92	0.99	1.00
	3.00	0.51	0.61	0.65	0.70	0.86	0.88	0.91	0.98	0.99
	≥4.00	0.51	0.60	0.65	0.69	0.84	0.86	0.88	0.95	0.97
≥5.5 <6.5	0.25	0.57	0.68	0.72	0.77	0.93	0.94	0.96	1.00	1.00
	0.50	0.52	0.62	0.66	0.71	0.87	0.90	0.92	1.00	1.00
	0.75	0.49	0.57	0.62	0.68	0.85	0.88	0.90	1.00	1.00
	1.00	0.46	0.56	0.60	0.65	0.82	0.85	0.88	1.00	1.00
	1.50	0.44	0.54	0.59	0.64	0.81	0.84	0.87	0.98	1.00
	2.00	0.43	0.53	0.58	0.63	0.81	0.83	0.86	0.97	0.99
	3.00	0.41	0.51	0.56	0.61	0.79	0.82	0.85	0.97	0.99
	≥4.00	0.40	0.50	0.55	0.61	0.79	0.82	0.85	0.97	0.99
≥6.5	0.25	0.54	0.64	0.68	0.73	0.88	0.90	0.92	1.00	1.00
	0.50	0.43	0.53	0.57	0.62	0.79	0.82	0.85	0.98	1.00
	0.75	0.39	0.49	0.54	0.59	0.77	0.80	0.83	0.96	1.00
	1.00	0.37	0.45	0.50	0.54	0.74	0.77	0.81	0.96	1.00
	1.50	0.35	0.45	0.49	0.54	0.71	0.75	0.79	0.96	1.00
	2.00	0.34	0.44	0.48	0.53	0.71	0.74	0.78	0.94	0.99
	3.00	0.34	0.44	0.48	0.53	0.70	0.73	0.77	0.93	0.98
	≥4.00	0.33	0.43	0.47	0.52	0.70	0.73	0.77	0.91	0.95

Note: Straight-line interpolation of $f_{g,ATS}$ for length of grade and demand flow permitted to the nearest 0.01.

(*Source: Two-Lane Highways,* Draft Chapter 15, National Cooperative Highway Research Program Project 3-92, Exhibit 15-10.)

Table 16.9: Grade Adjustment Factor (f_G) for Specific Upgrades: PTSF Determinations

Grade (%)	Grade Length (mi)	Directional Demand Flow Rate, $v = V/PHF$ (veh/h)								
		≤100	200	300	400	500	600	700	800	≥900
≥3 <3.5	0.25	1.00	0.99	0.97	0.96	0.92	0.92	0.92	0.92	0.92
	0.50	1.00	0.99	0.98	0.97	0.93	0.93	0.93	0.93	0.93
	0.75	1.00	0.99	0.98	0.97	0.93	0.93	0.93	0.93	0.93
	1.00	1.00	0.99	0.98	0.97	0.93	0.93	0.93	0.93	0.93
	1.50	1.00	0.99	0.98	0.97	0.94	0.94	0.94	0.94	0.94
	2.00	1.00	0.99	0.98	0.98	0.95	0.95	0.95	0.95	0.95
	3.00	1.00	1.00	0.99	0.99	0.97	0.97	0.97	0.96	0.96
	≥4.00	1.00	1.00	1.00	1.00	1.00	0.99	0.99	0.97	0.97
≥3.5 <4.5	0.25	1.00	0.99	0.98	0.97	0.94	0.93	0.93	0.92	0.92
	0.50	1.00	1.00	0.99	0.99	0.97	0.97	0.97	0.96	0.95
	0.75	1.00	1.00	0.99	0.99	0.97	0.97	0.97	0.96	0.96
	1.00	1.00	1.00	0.99	0.99	0.97	0.97	0.97	0.97	0.97
	1.50	1.00	1.00	0.99	0.99	0.97	0.97	0.97	0.97	0.97
	2.00	1.00	1.00	0.99	0.99	0.98	0.98	0.98	0.98	0.98
	3.00	1.00	1.00	1.00	1.00	1.00	1.00	1.00	1.00	1.00
	≥4.00	1.00	1.00	1.00	1.00	1.00	1.00	1.00	1.00	1.00
≥4.5 <5.5	0.25	1.00	1.00	1.00	1.00	1.00	0.99	0.99	0.97	0.97
	≥0.50	1.00	1.00	1.00	1.00	1.00	1.00	1.00	1.00	1.00
≥5.5	All	1.00	1.00	1.00	1.00	1.00	1.00	1.00	1.00	1.00

Note: Interpolation for length of grade and demand flow rate to the nearest 0.01 is recommended.

(*Source: Two-Lane Highways,* Draft Chapter 15, National Cooperative Highway Research Program Project 3-92, Exhibit 15-17.)

Table 16.10: Passenger Car Equivalents for General Terrain Segments: ATS and PTSF Determinations

Vehicle Type	Range of One-Way Flows, v (veh/h) $v = V/PHF$	For ATS Determination		For PTSF Determination	
		Level Terrain Specific Downgrades	Rolling Terrain	Level Terrain Specific Downgrades	Rolling Terrain
Trucks and Buses E_T	≤100	1.9	2.7	1.1	1.9
	200	1.5	2.3	1.1	1.8
	300	1.4	2.1	1.1	1.7
	400	1.3	2.0	1.1	1.6
	500	1.2	1.8	1.0	1.4
	600	1.1	1.7	1.0	1.2
	700	1.1	1.6	1.0	1.0
	800	1.1	1.4	1.0	1.0
	≥900	1.0	1.3	1.0	1.0
Recreational Vehicles E_R	All	1.0	1.1	1.0	1.0

(*Source: Two-Lane Highways,* Draft Chapter 15, National Cooperative Highway Research Program Project 3-92, Exhibits 15-10 and 15-18.)

Table 16.11: Passenger Car Equivalents for Trucks (E_T) on Specific Upgrades: ATS Determination

Grade (%)	Grade Length (mi)	Directional Demand Flow Rate, $v = V/PHF$ (veh/h)								
		≤100	200	300	400	500	600	700	800	≥900
≥3 <3.5	0.25	2.6	2.4	2.3	2.2	1.8	1.8	1.7	1.3	1.1
	0.50	3.7	3.4	3.3	3.2	2.7	2.6	2.6	2.3	2.0
	0.75	4.6	4.4	4.3	4.2	3.7	3.6	3.4	2.4	1.9
	1.00	5.2	5.0	4.9	4.9	4.4	4.2	4.1	3.0	1.6
	1.50	6.2	6.0	5.9	5.8	5.3	5.0	4.8	3.6	2.9
	2.00	7.3	6.9	6.7	6.5	5.7	5.5	5.3	4.1	3.5
	3.00	8.4	8.0	7.7	7.5	6.5	6.2	6.0	4.6	3.9
	≥4.00	9.4	8.8	8.6	8.3	7.2	6.9	6.6	4.8	3.7
≥3.5 <4.5	0.25	3.8	3.4	3.2	3.0	2.3	2.2	2.2	1.7	1.5
	0.50	5.5	5.3	5.1	5.0	4.4	4.2	4.0	2.8	2.2
	0.75	6.5	6.4	6.5	6.5	6.3	5.9	5.6	3.6	2.6
	1.00	7.9	7.6	7.4	7.3	6.7	6.6	6.4	5.3	4.7
	1.50	9.6	9.2	9.0	8.9	8.1	7.9	7.7	6.5	5.9
	2.00	10.3	10.1	10.0	9.9	9.4	9.1	8.9	7.4	6.7
	3.00	11.4	11.3	11.2	11.2	10.7	10.3	10.0	8.0	7.0
	≥4.00	12.4	12.2	12.2	12.1	11.5	11.2	10.8	8.6	7.5
≥4.5 <5.5	0.25	4.4	4.0	3.7	3.5	2.7	2.7	2.7	2.6	2.5
	0.50	6.0	6.0	6.0	6.0	5.9	5.7	5.6	4.6	4.2
	0.75	7.5	7.5	7.5	7.5	7.5	7.5	7.5	7.5	7.5
	1.00	9.2	9.2	9.1	9.1	9.0	9.0	9.0	8.9	8.8
	1.50	10.6	10.6	10.6	10.6	10.5	10.4	10.4	10.2	10.1
	2.00	11.8	11.8	11.8	11.8	11.6	11.6	11.5	11.1	10.9
	3.00	13.7	13.7	13.6	13.6	13.3	13.1	13.0	11.9	11.3
	≥4.00	15.3	15.3	15.2	15.2	14.6	14.2	13.8	11.3	10.0
≥5.5 <6.5	0.25	4.8	4.6	4.5	4.4	4.0	3.9	3.8	3.2	2.9
	0.50	7.2	7.2	7.2	7.2	7.2	7.2	7.2	7.2	7.2
	0.75	9.1	9.1	9.1	9.1	9.1	9.1	9.1	9.1	9.1
	1.00	10.3	10.3	10.3	10.3	10.3	10.3	10.3	10.2	10.1
	1.50	11.9	11.9	11.9	11.9	11.8	11.8	11.8	11.7	11.6
	2.00	12.8	12.8	12.8	12.8	12.7	12.7	12.7	12.6	12.5
	3.00	14.4	14.4	14.4	14.4	14.3	14.3	14.3	14.2	14.1
	≥4.00	15.4	15.4	15.3	15.3	15.2	15.1	15.1	14.9	14.8
>6.5	0.25	5.1	5.1	5.0	5.0	4.8	4.7	4.7	4.5	4.4
	0.50	7.8	7.8	7.8	7.8	7.8	7.8	7.8	7.8	7.8
	0.75	9.8	9.8	9.8	9.8	9.8	9.8	9.8	9.8	9.8
	1.00	10.4	10.4	10.4	10.4	10.4	10.4	10.4	10.3	10.2
	1.50	12.0	12.0	12.0	12.0	11.9	11.9	11.9	11.8	11.7
	2.00	12.9	12.9	12.9	12.9	12.8	12.8	12.8	12.7	12.6
	3.00	14.5	14.5	14.5	14.5	14.4	14.4	14.4	14.3	14.2
	≥4.00	15.4	15.4	15.4	15.4	15.3	15.3	15.3	15.2	15.1

Note: Straight line interpolation of E_T for length of grade permitted to nearest 0.1.

(*Source: Two-Lane Highways,* Draft Chapter 15, National Cooperative Highway Research Program Project 3-92, Exhibit 15-12.)

Table 16.12: Passenger Car Equivalents of RV's (E_R) for Specific Upgrades: ATS Determination

Grade (%)	Grade Length (mi)	**Directional Demand Flow Rate, $v = V/PHF$ (veh/h)**								
		≤100	**200**	**300**	**400**	**500**	**600**	**700**	**800**	**≥900**
≥3 <3.5	≤0.25	1.1	1.1	1.1	1.0	1.0	1.0	1.0	1.0	1.0
	>0.25 ≤0.75	1.2	1.2	1.1	1.1	1.0	1.0	1.0	1.0	1.0
	>0.75 ≤1.25	1.3	1.2	1.2	1.1	1.0	1.0	1.0	1.0	1.0
	>1.25 ≤2.25	1.4	1.3	1.2	1.1	1.0	1.0	1.0	1.0	1.0
	>2.25	1.5	1.4	1.3	1.2	1.0	1.0	1.0	1.0	1.0
3.5 <4.5	≤0.75	1.3	1.2	1.2	1.1	1.0	1.0	1.0	1.0	1.0
	>0.75 ≤3.50	1.4	1.3	1.2	1.1	1.0	1.0	1.0	1.0	1.0
	≥3.50	1.5	1.4	1.3	1.2	1.0	1.0	1.0	1.0	1.0
≥4.5 <5.5	≤2.50	1.5	1.4	1.3	1.2	1.0	1.0	1.0	1.0	1.0
	>2.50	1.6	1.5	1.4	1.2	1.0	1.0	1.0	1.0	1.0
≥5.5 <6.5	≤0.75	1.5	1.4	1.3	1.1	1.0	1.0	1.0	1.0	1.0
	>0.75 ≤2.50	1.6	1.5	1.4	1.2	1.0	1.0	1.0	1.0	1.0
	>2.50 ≤3.50	1.6	1.5	1.4	1.3	1.2	1.1	1.0	1.0	1.0
	>3.50	1.6	1.6	1.6	1.5	1.5	1.4	1.3	1.2	1.1
≥6.5	≤2.50	1.6	1.5	1.4	1.2	1.0	1.0	1.0	1.0	1.0
	>2.50 ≤3.50	1.6	1.5	1.4	1.2	1.3	1.3	1.3	1.3	1.3
	>3.50	1.6	1.6	1.6	1.5	1.5	1.5	1.4	1.4	1.4

Note: Interpolation in this table is not recommended.

(*Source: Two-Lane Highways,* Draft Chapter 15, National Cooperative Highway Research Program Project 3-92, Exhibit 15-13.)

Table 16.13: Passenger Car Equivalents of Trucks (E_T) and RVs (E_R) on Specific Upgrades: PTSF Determination

Grade (%)	Grade Length (mi)	**Directional Demand Flow Rate, v_{vph} (veh/h)**								
		≤100	**200**	**300**	**400**	**500**	**600**	**700**	**800**	**≥900**
		Passenger Car Equivalents for Trucks (E_T)								
≥3 <3.5	≤2.00	1.0	1.0	1.0	1.0	1.0	1.0	1.0	1.0	1.0
	3.00	1.5	1.3	1.3	1.2	1.0	1.0	1.0	1.0	1.0
	≤4.00	1.6	1.4	1.3	1.3	1.0	1.0	1.0	1.0	1.0
≥3.5 <4.5	≤1.00	1.0	1.0	1.0	1.0	1.0	1.0	1.0	1.0	1.0
	1.50	1.1	1.1	1.0	1.0	1.0	1.0	1.0	1.0	1.0
	2.00	1.6	1.3	1.0	1.0	1.0	1.0	1.0	1.0	1.0
	3.00	1.8	1.4	1.1	1.2	1.2	1.2	1.2	1.2	1.2
	≥4.00	2.1	1.9	1.8	1.7	1.4	1.4	1.4	1.4	1.4
≥4.5 <5.5	≤1.00	1.0	1.0	1.0	1.0	1.0	1.0	1.0	1.0	1.0
	1.50	1.1	1.1	1.1	1.2	1.2	1.2	1.2	1.2	1.2
	2.00	1.7	1.6	1.6	1.6	1.5	1.4	1.4	1.3	1.3
	3.00	2.4	2.2	2.2	2.1	1.9	1.8	1.8	1.7	1.7
	≥4.00	3.5	3.1	2.9	2.7	2.1	2.0	2.0	1.8	1.8
≥5.5 <6.5	≤0.75	1.0	1.0	1.0	1.0	1.0	1.0	1.0	1.0	1.0
	1.00	1.0	1.0	1.1	1.1	1.2	1.2	1.2	1.2	1.2

(*Continued*)

Table 16.13: Passenger Car Equivalents of Trucks (E_T) and RVs (E_R) on Specific Upgrades: PTSF Determination (*Continued*)

Grade (%)	Grade Length (mi)	Directional Demand Flow Rate, v_{vph} (veh/h)								
		≤100	200	300	400	500	600	700	800	≥900
		Passenger Car Equivalents for Trucks (E_T)								
	1.50	1.5	1.5	1.5	1.6	1.6	1.6	1.6	1.6	1.6
	2.00	1.9	1.9	1.9	1.9	1.9	1.9	1.9	1.8	1.8
	3.00	3.4	3.2	3.0	2.9	2.4	2.3	2.3	1.9	1.9
	≥4.00	4.5	4.1	3.9	3.7	2.9	2.7	2.6	2.0	2.0
≥6.5	≤0.50	1.0	1.0	1.0	1.0	1.0	1.0	1.0	1.0	1.0
	0.75	1.0	1.0	1.0	1.0	1.1	1.1	1.1	1.0	1.0
	1.00	1.3	1.3	1.3	1.4	1.4	1.5	1.5	1.4	1.4
	1.50	2.1	2.1	2.1	2.1	2.0	2.0	2.0	2.0	2.0
	2.00	2.9	2.8	2.7	2.7	2.4	2.4	2.3	2.3	2.3
	3.00	4.2	3.9	3.7	3.6	3.0	2.8	2.7	2.2	2.2
	≥4.00	5.0	4.6	4.4	4.2	3.3	3.1	2.9	2.7	2.5
		Passenger Car Equivalents for RVs (E_R)								
All	All	1.0	1.0	1.0	1.0	1.0	1.0	1.0	1.0	1.0

Note: Interpolation for both length of grade and demand flow rate to the nearest 0.1 is recommended.

(*Source: Two-Lane Highways,* Draft Chapter 15, National Cooperative Highway Research Program Project 3-92, Exhibit 15-19.)

Table 16.14: Passenger Car Equivalents for Trucks Operating at Crawl Speeds on Specific Downgrades - ATS Determination

Difference Between FFS and Truck Crawl Speed (mi/h)	Directional Demand Flow Rate, $v = V/PHF$ (veh/h)								
	≤100	200	300	400	500	600	700	800	≥900
≤15	4.7	4.1	3.6	3.1	2.6	2.1	1.6	1.0	1.0
20	9.9	8.9	7.8	6.7	5.8	4.9	4.0	2.7	1.0
25	15.1	13.5	12.0	10.4	9.0	7.7	6.4	5.1	3.8
30	22.0	19.8	17.5	15.6	13.1	11.6	9.2	6.1	4.1
35	29.0	26.0	23.1	20.1	17.3	14.6	11.9	2.2	6.5
≥40	35.9	32.3	28.6	24.9	21.4	18.1	14.7	11.3	7.9

Note: interpolation of E_{TC} for speed difference and demand flow rate is permitted to nearest 0.1.

(*Source: Two-Lane Highways,* Draft Chapter 15, National Cooperative Highway Research Program Project 3-92, Exhibit 15-14.)

In applying Equation 16-9, note that P_{TC} is stated as a proportion of the *truck* population, not of the entire traffic stream. Thus a P_{TC} of 0.50, means that 50% of the *trucks* are operating down the grade at crawl speeds. Note also that for two-lane highways, all composite grades are treated using the *average grade* of the analysis section. The average grade for any segment is the total change in elevation (feet) divided by the length of the segment (feet).

16.4.6 Estimating Average Travel Speed

Once the appropriate demand flow rate(s) are computed, the average travel speed in the section is estimated using Equation 16-10:

$$ATS_d = FFS - 0.00776(v_d + v_o) - f_{npA} \qquad (16\text{-}10)$$

where: ATS_d = average travel speed in the direction of analysis, mi/h

FFS_d = free-flow speed in the direction of analysis, mi/h

v_d = demand flow rate in the direction of analysis, pc/h

v_o = demand flow rate in the opposing direction, pc/h

f_{npA} = adjustment to ATS for the existence of "No Passing" zones in the study segment

Values of the adjustment factor, f_{npA}, are given in Table 16.15. The adjustment is based on flow rates, the percentage of the analyses segment for which passing is prohibited, and the free-flow speed of the facility.

16.4.7 Determining Percent Time Spent Following

Percent time spent following (PTSF) is determined using Equation 16-11:

$$PTSF_d = BPTSF_d + f_{npP}\left(\frac{V_d}{V_d + V_o}\right)$$

$$BPTSF_d = 100\left[1 - \exp\left(av_d^b\right)\right] \qquad (16\text{-}11)$$

where: $PTSF_d$ = percent time spent following, single direction, %

$BPTSF_d$ = percent time spent following, single direction, %

v_d = demand flow rate in analysis direction, pc/h

v_o = demand flow rate in the opposite direction, pc/h

f_{npP} = adjustment to PTSF for the effect of percent "No Passing" zones in the study segment, %

a, b = calibration constants based on opposing flow rate in single-direction analysis

The adjustment factor f_{npP} is found in Table 16.16 and calibration constants "a" and "b" are found in Table 16.17.

16.5 Sample Problems in Analysis of Rural Two-Lane Highways

16.5.1 Analysis of a Class I Rural Two-Lane Highway in Rolling Terrain

A Class I two-lane highway in rolling terrain has a peak demand volume of 500 veh/h, with 15% trucks and 5% RVs. The highway serves as a main link to a popular recreation area. The directional split of traffic is 60-40 during peak periods, and the peak-hour factor is 0.88. The 10-mile section under study has 40% "No Passing" zones. The base free-flow speed of the facility may be taken to be 60 mi/h. Lane widths are 12 feet, and shoulder widths are 2 feet. There are 10 access points per mile along this 10-mile section.

Step 1: *Estimate the Free-Flow Speed.* The free-flow speed (*FFS*) is estimated from the base free-flow speed (*BFFS*) and applicable adjustment factors f_{LS} (Table 16.5) and f_A (Table 16.6). Then:

$$FFS = BFFS - f_{LS} - f_A$$

$$FFS = 60.0 - 2.6 - 2.5 = 54.9 \text{ mi/h}$$

where f_{LS} = 2.6 (for 12-foot lanes and 2-foot shoulders) and f_A = 2.5 (for 10 access points per mile)

Step 2: *Compute the Directional Demand Flow Rates for ATS and PTSF Determinations* Because each direction of the two-lane highway must be separately analyzed, it is necessary that the demand of 500 veh/h be separated by direction. Note that the two directional analyses may be done concurrently because the directional demand in one case is the opposing demand in the other. Given the specified 60-40 split:

$$V_1 = 500 * 0.60 = 300 \text{ veh/h}$$

$$V_2 = 500 * 0.40 = 200 \text{ veh/h}$$

Both of these values have to be converted to base passenger-car flow rates.

Four demand flows will be computed. Both the directional and opposing volumes must be separately converted for ATS determination and for PTSF determination. The initial selection of adjustment factors would be based on flow rates, which are found using the demand volume

Table 16.15: Adjustment to *ATS* for %NPZ, Opposing Demand Flow, and *FFS*

Opposing Demand Flow Rate, v_o (pc/h)	Percent No Passing Zones				
	≤20	40	60	80	100
	FFS = ≥ 65 mi/h				
≤100	1.1	2.2	2.8	3.0	3.1
200	2.2	3.3	3.9	4.0	4.2
400	1.6	2.3	2.7	2.8	2.9
600	1.4	1.5	1.7	1.9	2.0
800	0.7	1.0	1.2	1.4	1.5
1,000	0.6	0.8	1.1	1.1	1.2
1,200	0.6	0.8	0.9	1.0	1.1
1,400	0.6	0.7	0.9	0.9	0.9
≥1,600	0.6	0.7	0.7	0.7	0.8
	FFS = 60 mi/h				
≤100	0.7	1.7	2.5	2.8	2.9
200	1.9	2.9	3.7	4.0	4.2
400	1.4	2.0	2.5	2.7	3.9
600	1.1	1.3	1.6	1.9	2.0
800	0.6	0.9	1.1	1.3	1.4
1,000	0.6	0.7	0.9	1.1	1.2
1,200	0.5	0.7	0.9	0.9	1.1
1,400	0.5	0.6	0.8	0.8	0.9
≥1,600	0.5	0.6	0.7	0.7	0.7
	FFS = 55 mi/h				
≤100	0.5	1.2	2.2	2.6	2.7
200	1.5	2.4	3.5	3.9	4.1
400	1.3	1.9	2.4	2.7	2.8
600	0.9	1.1	1.6	1.8	1.9
800	0.5	0.7	1.1	1.2	1.4
1,000	0.5	0.6	0.8	0.9	1.1
1,200	0.5	0.6	0.7	0.9	1.0
1,400	0.5	0.6	0.7	0.7	0.9
≥1,600	0.5	0.6	0.6	0.6	0.7
	FFS = 50 mi/h				
≤100	0.2	0.7	1.9	2.4	2.5
200	1.2	2.0	3.3	3.9	4.0
400	1.1	1.6	2.2	2.6	2.7
600	0.6	0.9	1.4	1.7	1.9
800	0.4	0.6	0.9	1.2	1.3

(*Continued*)

Table 16.15: Adjustment to *ATS* for %NPZ, Opposing Demand Flow, and *FFS* (*Continued*)

Opposing Demand Flow Rate, v_o (pc/h)	Percent No Passing Zones				
	≤20	40	60	80	100
FFS = 50 mi/h					
1,000	0.4	0.4	0.7	0.9	1.1
1,200	0.4	0.4	0.7	0.8	1.0
1,400	0.4	0.4	0.6	0.7	0.8
≥1,600	0.4	0.4	0.5	0.5	0.5
FFS ≤ 45 mi/h					
≤100	0.1	0.4	1.7	2.2	2.4
200	0.9	1.6	3.1	3.8	4.0
400	0.9	0.5	2.0	2.5	2.7
600	0.4	0.3	1.3	1.7	1.8
800	0.3	0.3	0.8	1.1	1.2
1,000	0.3	0.3	0.6	0.8	1.1
1,200	0.3	0.3	0.6	0.7	1.0
1,400	0.3	0.3	0.6	0.6	0.7
≥1,600	0.3	0.3	0.4	0.4	0.6

Note: Interpolation of $f_{np,ATS}$ for percent no-passing zones, demand flow rate, and FFS to the nearest 0.1 is recommended.

(*Source: Two-Lane Highways,* Draft Chapter 15, National Cooper ative Highway Research Program Project 3-92, Exhibit 15-15.)

Table 16.16: Adjustment (f_{npP}) for the Effect of "No Passing" Zones on PTSF

Total Two-Way Flow Rate, $v = v_d + v_o$ (pc/h)	Percent No-Passing Zones					
	0	20	40	60	80	100
Directional Split = 50/50						
≤200	9.0	29.2	43.4	49.4	51.0	52.6
400	16.2	41.0	54.2	61.6	63.8	65.8
600	15.8	38.2	47.8	53.2	55.2	56.8
800	15.8	33.8	40.4	44.0	44.8	46.6
1,400	12.8	20.0	23.8	26.2	27.4	28.6
2,000	10.0	13.6	15.8	17.4	18.2	18.8
2,600	5.5	7.7	8.7	9.5	10.1	10.3
3,200	3.3	4.7	5.1	5.5	5.7	6.1

(*Continued*)

Table 16.16: Adjustment (f_{npP}) for the Effect of "No Passing" Zones on PTSF (*Continued*)

Total Two-Way Flow Rate, $v = v_d + v_o$ (pc/h)	**Percent No-Passing Zones**					
	0	**20**	**40**	**60**	**80**	**100**
Directional Split = 60-40						
≤200	11.0	30.6	41.0	51.2	52.3	53.5
400	14.6	36.1	44.8	53.4	55.0	56.3
600	14.8	36.9	44.0	51.1	52.8	54.6
800	13.6	28.2	33.4	38.6	39.9	41.3
1,400	11.8	18.9	22.1	25.4	26.4	27.3
2,000	9.1	13.5	15.6	16.0	16.8	17.3
2,600	5.9	7.7	8.6	9.6	10.0	10.2
Directional Split = 70/30						
≤200	9.9	28.1	38.0	47.8	48.5	49.0
400	10.6	30.3	38.6	46.7	47.7	48.8
600	10.9	30.9	37.5	43.9	45.4	47.0
800	10.3	23.6	28.4	33.3	34.5	35.5
1,400	8.0	14.6	17.7	20.8	21.6	22.3
2,000	7.3	9.7	12.7	13.3	14.0	14.5
Directional Split = 80/20						
≤200	8.9	27.1	37.1	47.0	47.4	47.9
400	6.6	26.1	34.5	42.7	43.5	44.1
600	4.0	24.5	31.3	38.1	39.1	40.0
800	3.8	18.5	23.5	28.4	29.1	29.9
1,400	3.5	10.3	13.3	16.3	16.9	32.2
2,000	3.5	7.0	8.5	10.1	10.4	10.7
Directional Split = 90/10						
≤200	4.6	24.1	33.6	43.1	43.4	43.6
400	0.0	20.2	28.3	36.3	36.7	37.0
600	−3.1	16.8	23.5	30.1	30.6	31.1
800	−2.8	10.5	15.2	19.9	20.3	20.8
1,400	−1.2	5.5	8.3	11.0	11.5	11.9

Note: Straight-line interpolation of f_{npP} for % "No Passing" zones, demand flow rate, and directional split is permitted to the nearest 0.1.

(*Source: Two-Lane Highways,* Draft Chapter 15, National Cooperative Highway Research Program Project 3-92, Exhibit 15-21.)

and the peak hour factor: $v_1 = 300/0.88 = 341$ veh/h and $v_2 = 200/0.88 = 227$ veh/h. Interpolation for the grade factor to the nearest 0.01 and passenger car equivalents to the nearest 0.1 is required. Table 16.18 summarizes the results of these determinations.

Using these values, the following heavy vehicle factors are computed:

$$f_{HV} = \frac{1}{1 + P_T(E_T - 1) + P_R(E_R - 1)}$$

Table 16.17: Coefficients "a" and "b" for Use in Equation 16-11

Opposing Demand Flow Rate, v_o (pc/h)	Coefficient a	Coefficient b
≤200	−0.0014	0.973
400	−0.0022	0.923
600	−0.0033	0.870
800	−0.0045	0.833
1,000	−0.0049	0.829
1,200	−0.0054	0.825
1,400	−0.0058	0.821
≥1,600	−0.0062	0.817

Note: Straight-line interpolation of "a" and "b" for opposing demand flow is permitted to the nearest 0.001.

(*Source: Two-Lane Highways,* Draft Chapter 15, National Cooperative Highway Research Program Project 3-92, Exhibit 15-20.)

Table 16.18: Values Used in Converting Volumes to Flow Rates Under Equivalent Base Conditions

Factor	Source Table	Directional Demand Volume	
		V_1 = 341 veh/h	V_2 = 227 veh/h
$f_G\,(ATS)$	Table 16.7 (Rolling Terrain)	0.86	0.77
$E_T\,(ATS)$	Table 16.10 (Rolling Terrain)	2.1	2.3
$E_R\,(ATS)$	Table 16.10 (Rolling Terrain)	1.1	1.1
$f_G\,(PTSF)$	Table 16.7 (Rolling Terrain)	0.87	0.81
$E_T\,(PTSF)$	Table 16.10 (Rolling Terrain)	1.7	1.8
$E_R\,(PTSF)$	Table 16.10 (Rolling Terrain)	1.0	1.0

$$f_{HV1}\,(ATS) = \frac{1}{1 + 0.15(2.1 - 1) + 0.05(1.1 - 1)} = 0.85$$

$$f_{HV1}\,(PTSF) = \frac{1}{1 + 0.15(1.7 - 1) + 0.05(1.0 - 1)} = 0.90$$

$$f_{HV2}\,(ATS) = \frac{1}{1 + 0.15(2.3 - 1) + 0.05(1.1 - 1)} = 0.83$$

$$f_{HV2}\,(PTSF) = \frac{1}{1 + 0.15(1.8 - 1) + 0.05(1.0 - 1)} = 0.89$$

Then:

$$v = \frac{V}{PHF * f_G * f_{HV}}$$

$$v_1\,(ATS) = \frac{341}{0.88 * 0.86 * 0.85} = 530 \text{ pc/h}$$

$$v_1\,(PTSF) = \frac{341}{0.88 * 0.87 * 0.90} = 495 \text{ pc/h}$$

$$v_2\,(ATS) = \frac{227}{0.88 * 0.77 * 0.83} = 404 \text{ pc/h}$$

$$v_2\,(PTSF) = \frac{277}{0.88 * 0.81 * 0.89} = 358 \text{ pc/h}$$

Step 3: *Determine ATS Values for Each Direction*

These values are estimated as:

$$ATS_d = FFS - 0.00776\,(v_d + v_o) - f_{npA}$$

where: FFS_d = 54.9 mph (directions 1 and 2, previously computed)
v_1 = 530 pc/h (previously computed)
v_2 = 404 pc/h (previously computed)
f_{npA} *(dir 1)* = 1.9 mi/h (Table 16.15, 55 mi/h, 40% No Passing Zones, v_o = 404 pc/h, interpolated)
f_{npA} *(dir 2)* = 1.4 mi/h (Table 16.15, 55 mi/h, 40% No Passing Zones, v_o = 530 pc/h, interpolated)

Then:

$$ATS_{d1} = 54.9 - 0.00776(530 + 404) - 1.9 = 45.8 \text{ mi/h}$$

$$ATS_{d2} = 54.9 - 0.00776(404 + 530) - 1.4 = 46.3 \text{ mi/h}$$

Step 4: *Determine PTSF Values for Each Direction*

These values are determined as:

$$PTSF_d = BPTSF_d + f_{npP}\left(\frac{V_d}{V_d + V_o}\right)$$

$$BPTSF_d = 100\left[1 - \exp\left(av_d^b\right)\right]$$

where: v_1 = 495 pc/h (computed previously)
v_2 = 358 pc/h (computed previously)
a_1 = −0.0020 (Table 16.17, v_o = 358 pc/h, interpolated)
a_2 = −0.0027 (Table 16.17, v_o = 495 pc/h, interpolated)
b_1 = 0.934 (Table 16.17, v_o = 358 pc/h, interpolated)
b_2 = 0.920 (Table 16.17, v_o = 495 pc/h, interpolated)

f_{npP} *(dir 1)* = f_{npP} *(dir 2)* = 32.4%

(Table 16.16, FFS = 55 mi/h, 40% no passing zones, v = 495 + 358 = 853 pc/h). Note from Table 16.16 that the adjustment is the same in *both* directions and is based on the total two-way flow rate.
Then:

$$BPTSF_{d1} = 100\left[1 - \exp\left(-0.0020 * 495^{0.934}\right)\right] = 48.2\%$$

$$PTSF_{d1} = 48.2 + 32.4\left(\frac{495}{853}\right) = 67.0\%$$

$$BPTSF_{d2} = 100\left[1 - \exp\left(-0.0027 * 358^{0.920}\right)\right] = 45.3\%$$

$$PTSF_{d2} = 45.3 + 32.4\left(\frac{358}{853}\right) = 58.9\%$$

From Table 16.4, Direction 1 has an LOS of C based on an ATS of 45.8 mi/h and an LOS of C based on a PTSF of 67.0%. The overall LOS is C. Direction 2 has an LOS of C based on an ATS of 46.3 mi/h and an LOS of C based on a PTSF of 58.9%. The overall LOS is C.

Even though the two-lane highway has a 60-40 directional split, the LOS is the same for both directions: C. In the 40% direction, there is less demand, but passing is more difficult due to the heavier opposing flow.

Estimating Capacity

Although the LOS of the segment has been determined in two directions, it may also be useful to estimate the capacity of the segment in *both* directions and in *each* direction. The capacity of a two-lane highway under base conditions is 1,700 pc/h in one direction, with a limit of 3,200 pc/h in both directions. These values are stated for equivalent base conditions and should be converted to prevailing conditions, as follows:

$$c = 1{,}700 * f_g * f_{HV}$$

There will be two values, however, depending on whether the adjustment factors for ATS or PTSF are used. The factors leading to the *lowest* value would be used. Moreover, the adjustment factors for a flow rate at or near capacity must be used, not those for the existing demand flow rates given for this problem. In all cases, adjustment factors for a demand flow rate ≥ 900 veh/h would be used.

From Table 16.7, the grade adjustment factor is 1.0 for both ATS and PTSF. From Table 16.10, passenger car equivalents are higher for the ATS determination and are therefore used. E_T = 1.3 and E_R = 1.1. Then:

$$f_{HV} = \frac{1}{1 + 0.10(1.3 - 1) + 0.05(1.1 - 1)} = 0.966$$

and:

$$c = 1{,}700 * 1.00 * 0.966 = 1{,}642 \text{ veh/h}$$

As all capacities, this is stated as a flow rate for the peak 15-minute period of the analysis hour. This is the single-direction capacity in the peak direction. Given a 60-40 directional split, the implied two-way capacity is 1,642/0.60 = 2,732 veh/h, which is less than the nominal value of 3,200 * 1.00 * 0.966 = 3,091 veh/h.

This difference is seen more clearly if the capacity of the 40% direction is considered. Because the two directions have the same characteristics, the capacity of direction 2 is the same as direction 1: 1,642 veh/h. This, however, implies a two-way capacity of 1,642/0.40 = 4,105 veh/h, well in excess of the nominal value of 3,091 veh/h. The problem is that the directional split is one of the prevailing traffic conditions. When the peak direction flow rate hits a capacity of 1,642 veh/h, the maximum flow rate in the other direction is limited to 2,737 − 1,642 = 1,095 veh/h.

This illustration is included to emphasize the difficulty in assessing capacity on a two-way rural highway. The two-way nominal capacity is an almost completely meaningless concept. When a capacity is determined for the peak direction, the "capacity" in the other direction is limited by the existing directional split of traffic.

16.5.2 Single-Direction Analysis of a Specific Grade

A Class II two-lane highway serving a rural logging area has a 2-mile grade of 4%. Peak demand on the grade is 250 veh/h, with a 70-30 directional split and a PHF of 0.82. Because of active logging operations in the area, demand contains 20% trucks. There are also 5% RVs using the facility. The grade has 100% "No Passing" zones. The free-flow speed of the facility has been measured to be 45 mi/h. Twenty percent of all trucks operate at a crawl speed that is 20 mi/h lower than other vehicles on the downgrade. At what LOS does the facility operate, assuming that 70% of the traffic is on the upgrade?

Step 1: *Compute Demand Flow Rates for PTSF Determination* Both the upgrade and downgrade must be analyzed in this case. Because this is a Class II facility, however, PTSF is the only parameter that need be established. It is necessary to divide the demand volume into the upgrade volume, V_u (250 * 0.70 = 175 veh/h), and downgrade volume, V_d (250 * 0.30 = 75 veh/h). As the upgrade and downgrade are analyzed, these will serve as directional and opposing volumes.

Base passenger-car equivalent demand flows are computed as:

$$v = \frac{V}{PHF * f_G * f_{HV}}$$

Selecting values of f_G, E_T and E_R from the appropriate tables for PTSF determination results in the values shown in Table 16.19. All tables are entered with the flow rate: v_{up} = 175/0.82 = 213 veh/h; v_{dn} = 75/0.82 = 91 veh/h. Note that the fact that 20% of trucks travel at crawl speeds on the downgrade is not relevant to the predication of *PTSF*, and so this information is not used here.

Then:

$$f_{HV}\ (\text{up}) = \frac{1}{1 + 0.20(1.3 - 1) + 0.05(1 - 1)} = 0.943$$

$$f_{HV}\ (\text{dn}) = \frac{1}{1 + 0.20(1.1 - 1) + 0.05(1 - 1)} = 0.980$$

And:

$$v_{up} = \frac{175}{0.82 * 1.00 * 0.943} = 226 \text{ pc/h}$$

$$v_{down} = \frac{75}{0.82 * 1.00 * 0.980} = 93 \text{ pc/h}$$

Step 2: *Estimate the PTSF for the Specific Upgrade and Specific Downgrade* The percent time spent following is estimated as:

$$PTSF_d = BPTSF_d + f_{npP}\left(\frac{V_d}{V_d + V_o}\right)$$

$$BPTSF_d = 100\left[1 - \exp\left(av_d^b\right)\right]$$

where:

f_{npP} (up and down) = 48.9% (Table 16.16, 100% "No Passing" zones, 70/30 split, and v = 226 + 93 = 319 pc/h, interpolated)

Table 16.19: Critical Values for Demand Flow Rate Computation

Direction	f_G	E_T	E_R
Upgrade	1.00 (Table 16.9)	1.3 (Table 16.13)	1.0 (Table 16.13)
Downgrade	1.00 (Table 16.7)	1.1 (Table 16.10)	1.0 (Table 16.10)

a (up) = –0.0014 (Table 16.17, $v_o < 200$ pc/h)
b (up) = 0.973 (Table 16.17, $v_o < 200$ pc/h)
a (down) = –0.0015 (Table 16.17, $v_o = 226$ pc/h, interpolated)
b (down) = 0.967 (Table 16.17, $v_o = 226$ pc/h, interpolated)

Then:

$$BPTSF_{up} = 100\left[1 - \exp\left(-0.0014 * 226^{0.973}\right)\right] = 23.9\%$$

$$PTSF_{up} = 23.9 + 48.9\left(\frac{226}{226 + 93}\right) = 58.5\%$$

$$BPTSF_{down} = 100\left[1 - \exp\left(-0.0015 * 93^{0.967}\right)\right] = 11.3\%$$

$$PTSF_{down} = 11.3 + 48.9\left(\frac{93}{93 + 226}\right) = 21.9\%$$

Step 3: *Determine Level of Service* From Table 16.4, this is LOS C for the upgrade and A for the downgrade. Because this was a Class II two-lane highway, only the PTSF criterion was used. Speeds, however, will be quite slow, given the FFS of 45 mi/h and the existence of many trucks traveling at a crawl speed of 20 mi/h downgrade. The speed criterion, however, does not apply to Class II two-lane highways when determining the LOS. It is interesting to note that the upgrade has deteriorated to LOS C at a volume of only 175 vehs/h.

16.6 The Impact of Passing and Truck Climbing Lanes

The single operational aspect that makes two-lane highways unique is the need to pass in the opposing lane of traffic. Both PTSF and ATS are improved when passing lanes are periodically provided or where truck climbing lanes exist on significant upgrades. These components allow platoons (in the direction of the passing or truck climbing lane) to break up through unrestricted passing for the length of the lane. If such lanes are provided periodically, the formation of long platoons behind a single slow-moving vehicle can be ameliorated to a degree.

16.6.1 Evaluating the Impact of Passing Lanes

The HCM 2000 provides a methodology for estimating the impact of a passing lane on a single-direction segment of a two-lane highway in level or rolling terrain. The procedure is intended for use on a directional segment containing only *one* passing lane of appropriate length, based on the criteria of Table 16.20.

Table 16.20: Optimal Lengths of Passing Lanes on Two-Lane Highways

Directional Flow Rate (pc/h)	Optimal Length of Passing Lane (mi)
100	≤0.50
200	>0.50–0.75
400	>0.75–1.00
≥ 700	>1.00–2.00

(*Source:* Used with permission of Transportation Research Board, National Research Council, *Highway Capacity Manual,* 4th Edition, Washington DC, 2000, Exhibit 12-12, p. 12-18).

The first step in the procedure is to complete a single-direction analysis for the cross section as if the passing lane did not exist. This will result in an estimation of ATS_d and $PTSF_d$ for the section without a passing lane. The analysis segment is now broken into four subsegments, as follows:

- The subsection upstream of the passing lane, length L_u (mi)
- The passing lane, including tapers, length L_{pl} (mi)
- The effective downstream length of the passing lane, length L_{de} (mi)
- The subsection downstream of the effective length of the passing lane, length L_d (mi)

The sum of the four lengths must be equal to the total length of the directional segment.

The "effective downstream length of the passing lane" reflects observations indicating that the passing lane improves both ATS and PTSF for a distance downstream of the actual passing lane itself. These "effective" distances differ depending on which measure of effectiveness is involved as well as by the demand flow rate. The "effective" distance *does not* include the passing lane itself, and it is measured from the end of the passing lane taper. Effective lengths, L_{de}, are shown in Table 16.21.

Lengths L_u and L_{pl} are known from existing conditions or plans. Length L_{de} is obtained from Table 16.21. Then:

$$L_d = L - (L_u + L_{pl} + L_{de}) \qquad (16\text{-}12)$$

where: L = length of the single-direction analysis segment, mi

All other variables are as previously defined. Note that both L_{de} and L_d will differ for ATS and PTSF determinations.

Table 16.21: Effective Downstream Length of Passing Lanes

Directional Flow Rate (pc/h)	Downstream Length of Roadway Affected, L_{de} (mi)	
	ATS	PTSF
≤200	1.7	13.0
400	1.7	8.1
700	1.7	5.7
≤1,000	1.7	3.6

Note: Straight-line interpolation for L_{de} for directional flow rate is permitted to the nearest 0.1 mile.

(*Source:* Used with permission of Transportation Research Board, National Research Council, *Highway Capacity Manual,* 4th Edition, Washington DC, 2000, Exhibit 20-23, p. 20-24).

Effect of Passing Lanes on PTSF

It is assumed that PTSF for the upstream subsection, L_u, and downstream subsection, L_d, are the same as if the passing lane did not exist (i.e., $PTSF_d$). Within the passing lane segment, L_{pl}, PTSF is usually between 58% and 62% of the upstream value. With the effective length segment, L_{de}, PTSF is assumed to rise linearly from its passing lane value to the downstream segment value. Thus the adjusted value of PTSF, accounting for the impact of the passing lane on the single-direction segment found as:

$$PTSF_{pl} = \frac{PTSF_d\left[L_u + L_d + f_{pl}L_{pl} + \left(\frac{1+f_{pl}}{2}\right)L_{de}\right]}{L} \quad (16\text{-}13)$$

where: $PTSF_{pl}$ = percent time spent following, adjusted to account for the impact of a passing lane in a directional segment (level or rolling terrain only), %

$PTSF_d$ = percent time spent following, assuming no passing lane in the directional segment, %

f_{pl} = adjustment factor for the effect of the passing lane on PTSF (Table 16.22)

All lengths are as previously defined.

Effect of Passing Lane on ATS

Again, it is assumed that the ATS upstream (L_u) and downstream (L_d) of the passing lane and its effective length is the same as if no passing lane existed, (i.e., ATS_d). Within the passing lane, speeds are generally 8% to 11% higher than the upstream value. Within the effective length, L_{de}, ATS is assumed to move linearly from the passing lane value to the downstream value. Thus the adjusted value of ATS is found as follows:

$$ATS_{pl} = \frac{ATS_d * L}{L_u + L_d + \left(\frac{L_{pl}}{f_{pl}}\right) + \left(\frac{2L_{de}}{1 + f_{pl}}\right)} \quad (16\text{-}14)$$

where: ATS_{pl} = average travel speed, adjusted to account for the effect of a passing lane, mi/h

ATS_d = average travel speed for the directional segment without a passing lane, mi/h

f_{pl} = adjustment factor for the effect of the passing lane on ATS (Table 16.21)

A Sample Problem

Consider the 15.0-mile segment of two-lane rural highway described in sample problem 16.5.1. Demand flow rates were found to be 530 pc/h for ATS determination and 495 pc/h for PTSF determination. In the heaviest direction of travel, measures of effectiveness were determined to be: ATS = 45.8 mi/h, PTSF = 67.0%.

Table 16.22: Adjustment Factors for Passing Lanes in Single-Direction Segments of Two-Lane Highways

Directional Demand Flow Rate (pc/h)	Adjustment Factor, f_{pl} For ATS	Adjustment Factor, f_{pl} for PTSF
0–300	1.08	0.58
>300–600	1.10	0.61
>600	1.11	0.62

(*Source:* Used with permission of Transportation Research Board, National Research Council, *Highway Capacity Manual,* 4th Edition, Washington DC, 2000, Exhibit 20-24, p. 20-26.)

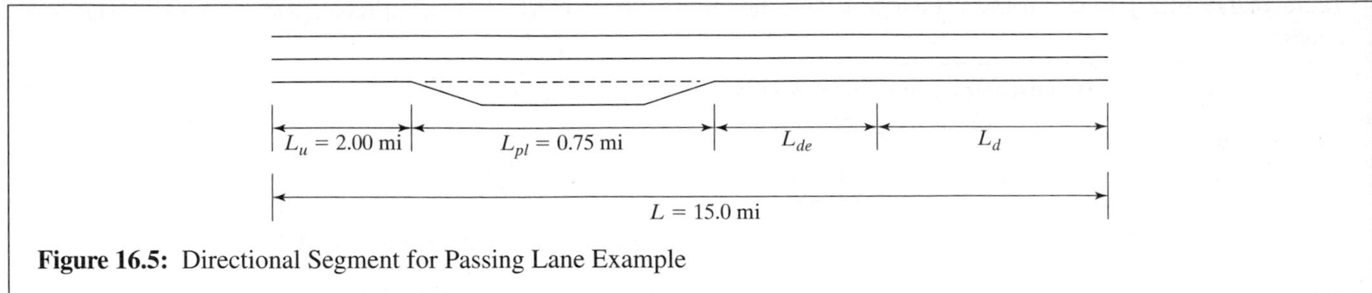

Figure 16.5: Directional Segment for Passing Lane Example

A passing lane of 0.75 miles is to be added to this directional segment as illustrated in Figure 16.5. Determine the impact of the passing lane on ATS and PTSF for the section.

From Table 16.21, L_{de} is 1.7 miles for ATS and 7.3 miles for PTSF. The latter is interpolated based on a demand flow rate of 495 pc/h (obtained from Example 16.5.1, previous section). From Table 16.21, f_{pl} is 1.10 for ATS and 0.61 for PTSF. Then:

$$L_{dATS} = 15.0 - (2.00 + 0.75 + 1.7) = 10.55 \text{ mi}$$

$$L_{dPTSF} = 15.0 - (2.00 + 0.75 + 7.3) = 4.95 \text{ mi}$$

and:

$$ATS_{pl} = \frac{45.8 * 15.0}{2.00 + 10.55 + \left(\frac{0.75}{1.10}\right) + \left(\frac{2 * 1.7}{1 + 1.10}\right)}$$

$$= 46.2 \text{ mi/h}$$

$$PTSF_{pl} = \frac{67.0\begin{bmatrix} 2.00 + 3.95 + (0.61 * 0.75) \\ + \left(\frac{1 + 0.61}{2}\right)7.3 \end{bmatrix}}{15.0}$$

$$= 54.9\%$$

Thus the placement of a single 0.75-mile passing lane in this 15-mile directional segment increases the ATS slightly (from 45.8 mi/h to 46.2 mi/h) and decreases the PTSF from 67.0% to 54.9%.

Although the single passing lane as described has a small impact on operations over the 15-mile analysis segment, increasing the length of the passing lane would have a larger impact. A greater impact would also be achieved by having passing lanes at closer spacings, say one in every 5-mile segment. This would provide for three unopposed passing opportunities in the 15-mile segment analyzed. The analysis, however, would have to be broken into three segments, each having a single passing lane.

The HCM 2000 provides a modified procedure for segments in which L_{de} would extend beyond the boundaries of the analysis segment. Consult the manual in Reference 1 directly for this detail.

16.6.2 Evaluating the Impact of Climbing Lanes

On specific upgrades, climbing lanes are often added to avoid development of long queues behind slow-moving vehicles, particularly heavy vehicles unable to maintain speed on upgrades. According to AASHTO criteria [2], climbing lanes are warranted on two-lane rural highways when:

- The directional flow rate on the upgrade exceeds 200 veh/h.
- The directional flow rate for trucks on the upgrade exceeds 20 veh/h.
- Any of the following conditions apply: (1) a speed reduction of 10 mi/h for a typical heavy truck, (2) LOS E or F exists on the upgrade, or (3) the LOS on the upgrade is two or more levels below that existing on the approach to the upgrade.

Values of ATS and PTSF for the upgrade may be modified to take into account the existence of a climbing lane using Equations 16-13 and 16-14 from the previous section, except that:

- L_u, L_{de}, and L_d are generally equal to 0.0 mile
- Adjustment factors f_{pl} are selected from Table 16.23 instead of Table 16.22.

Table 16.23: Adjustment Factors for Climbing Lanes in Single-Direction Segments of Two-Lane Highways

Directional Demand Flow Rate (pc/h)	Adjustment Factor, f_{pl} for ATS	Adjustment Factor, f_{pl} for PTSF
0–300	1.02	0.20
>300–600	1.07	0.21
>600	1.14	0.23

(*Source:* Used with permission of Transportation Research Board, National Research Council, *Highway Capacity Manual,* 4th Edition, Washington DC, 2000, Exhibit 20-27, p. 20-29.)

16.7 Summary

Two-lane rural highways continue to form a vital part of the nation's intercity and rural roadway networks. Their unique operating characteristics require careful consideration because poor operations are often a harbinger of safety problems, particularly on Class I roadways. Passing and climbing lanes are often used to improve the operation of two-lane highways, but serious accident or operational problems generally require more comprehensive solutions. Such solutions may involve creating a continuous three-lane alignment with alternating passing zones for each direction of flow or reconstruction as a multilane or limited-access facility.

Operational quality is not as serious a concern on Class II highways, where local service or all-weather access to remote areas is the primary function served. Class III highways, because they serve more built-up areas (or scenic recreational routes) must consider operational quality but on a different scale (*PFFS*) than Class I two-lane highways.

References

1. *Highway Capacity Manual,* 4th Edition, Transportation Research Board, National Research Council, Washington DC, 2000.
2. *A Policy on Geometric Design of Highways and Streets,* 5th Edition, American Association of State Highway and Transportation Officials, Washington DC, 2004.
3. Washburn, S.S., McLeod, D.S., and Courage, K.G., "Adaptation of the HCM2000 Planning Level Analysis of Two-Lane and Multilane Highways in Florida," *Transportation Research Record 1802,* Transportation Research Board, Washington DC, 2002, pp. 62–68.
4. *Level of Service Handbook,* Florida Department of Transportation, Tallahassee FL, 2005.
5. Prisk, C.W., "Passing Practices on Two-Lane Highways," *Proceedings of the Highway Research Board,* Transportation Research Board, National Research Council, Washington DC, 1941.
6. Weaver, G.D., and Glennon, J.C., *Passing Performance Measurements Related to Sight Distance Design,* Report No. 134-6, Texas Transportation Institute, Texas A&M University, College Station TX, July 1971.
7. Weaver, G.D., and Woods, D.L., *Passing and No-Passing Signs, Markings, and Warrants,* Report No. FHWA-RD-79-5, Federal Highway Administration, U.S. Department of Transportation, Washington DC, September 1978.
8. *Manual of Uniform Traffic Control Devices,* Millennium Edition, Federal Highway Administration, U.S. Department of Transportation, Washington DC, 2000.
9. Harwood, D.W., et al., *Capacity and Quality of Service of Two-Lane Highways,* Final Report, NCHRP Project 3-55(3), Midwest Research Institute, Kansas City MO, 2000.

Problems

16-1. A Class I rural two-lane highway segment of 20 miles in rolling terrain has a base free-flow speed (BFFS) of 60 mi/h. It consists of 12-foot lanes, with 4-foot clear shoulders on each side of the roadway. There is an average of eight access points per mile along the segment in question, and 60% of the segment consists of "no-passing" zones. Current traffic on the facility is 400 veh/h (total, both ways), including 15% trucks and 2% RVs. The directional distribution of traffic is 60-40, and the PHF is 0.84.

(a) For the current conditions described, at what LOS would each direction of the facility be expected to operate?

(b) If traffic is growing at a rate of 8% per year, how many years will it be before the capacity of this facility is reached?

16-2. A Class I rural highway segment of 10 miles in level terrain has a base free-flow speed (BFFS) of 65 mi/h. The cross section has 12-foot lanes and 6-foot clear shoulders on both sides of the road. There is an average of 15 access points per mile along the segment, and 25% of the segment consists of "no-passing" zones. Current traffic on the facility is 800 veh/h, with 8% trucks and no RVs. The directional distribution of traffic is 70-30, and the PHF is 0.83. What is the LOS expected in each direction on this segment?

16-3. A Class I rural two-lane highway segment contains a significant grade of 5%, 3 miles in length. The base free-flow speed (BFFS) of the facility is 55 mi/h, and the cross section is 11-foot lanes with 2-foot clear shoulders on both sides. There are five access points per mile along the segment, and 80% of the segment consists of "no-passing" zones. Current traffic on the facility is 250 veh/h, including 20% trucks and 5% RVs. Directional distribution is 70-30 (70% traveling upgrade), and the PHF is 0.85. What LOS is expected on the upgrade and downgrade of this segment?

16-4. A Class III two-lane highway in rolling terrain services a small built-up area. It has a base free-flow speed (BFFS) of 45 mi/h. The highway has 11-foot lanes and 4-foot clear shoulders on both sides. There are 20 access points/mile, and 100% of the segment is in "no-passing" zones. Current traffic is 500 veh/h, with 10% trucks, 3% RVs, and a PHF of 0.81. The directional distribution is 60-40. What LOS is expected in each direction on this facility?

16-5. Reconsider problem 16-1. How would the resulting ATS and PTSF be affected by adding a single 3-mile passing lane beginning 3.0 miles from the beginning of the segment?

16-6. What would be the impact of adding a truck climbing lane to the upgrade described in problem 16-3?

CHAPTER 17

Signing and Marking for Freeways and Rural Highways

The principal forms of traffic control implemented on freeways and rural highways involve road markings and signing. At-grade intersections occur on most rural highways (multilane and two-lane) and in some cases signalized intersections. The conventions and norms for signalization are the same as for urban and suburban signalization, with the exception that higher approach speeds are generally present in such cases. This, however, is taken into account as part of the normal design and analysis procedures for signalized intersections.

This chapter presents some of the special conventions for application of traffic signs and markings on freeways and rural highways.

17.1 Traffic Markings on Freeways and Rural Highways

Traffic markings on freeways and rural highways involve lane lines, edge markings, gore area, and other specialized markings at and in the vicinity of ramps or interchanges. On two-lane rural highways, centerline markings are used in conjunction with signs to designate passing and no-passing zones.

17.1.1 Freeway Markings

Figure 17.1 illustrates typical mainline markings on a freeway. Standard lane lines are provided to delineate lanes for travel. Right-edge markings consist of a solid white line, and left-edge markings consist of a solid yellow line. On freeways, lane markings and edge markings are mandated by the MUTCD [*1*].

17.1.2 Rural Highway Markings

General highway segment marking conventions for rural highways vary according to the specific configuration that is present. MUTCD criteria for placement of centerline and edge markings on rural highways are as follows:

- Centerline markings *shall* (standard) be placed on all paved two-way streets or highways that have three or more traffic lanes.
- Centerline markings *should* (guidance) be placed on all rural arterials and collectors that have a travelled width of more than 18 ft *and* an ADT of 3,000 veh/day or greater.

Figure 17.1: Typical Markings on a Freeway

- Edge lane markings *shall* (standard) be placed on paved streets or highways with the following characteristics: freeways, expressways, or rural highways with a traveled way of 20 ft or more in width and an ADT of 6,000 veh/day or greater.
- Edge lines *should* (guidance) be placed on rural highways with a traveled way of 20 ft or more in width and an ADT of 3,000 veh/day.

From these standards and guidelines, two-lane rural highways with an ADT of less than 3,000 veh/day are the only types of highways for which centerline and edge-line markings need not be used. In many of these cases, it would be wise to provide centerlines, if only to clarify whether or not passing is permitted. On an unmarked, two-lane rural highway, it is the driver's responsibility to recognize situations in which a passing maneuver would be unsafe.

Typical two-lane, two-way rural highway markings are illustrated in Figure 17.2. Both edge markings are white, and the centerline markings are yellow. Standard centerline markings for two-lane highways are as follows:

- A single dashed yellow line signifies that passing is permitted from either lane.
- A double solid yellow line signifies that passing is prohibited from either lane.
- A double yellow line, one solid, one dashed, signifies that passing is permitted from the lane adjacent to the dashed yellow line and that passing is prohibited from the lane adjacent to the solid yellow line.

The "No Passing Zone" sign is classified as a warning sign, although it has a unique pennant shape. It is placed on the left side of the road at the beginning of a marked "No Passing" zone.

Most rural highways of four or six lanes are marked similarly to the freeway example shown in Figure 17.1. Right- and left-edge markings are provided, as well as lane lines and centerlines. Whereas centerline markings are mandated for four- and six-lane highways, edge markings are required only where average daily travels (ADTs) exceed 6,000 veh/day. There is no requirement for use of lane lines on surface rural four- or six-lane highways. Nevertheless, usual practice calls for use of both centerlines and lane lines as a minimum on four-lane and six-lane rural highways. In addition, such highways would not normally be provided for low ADTs, and edge markings are therefore often installed as well.

One rural highway alignment calls for special markings. In the 1940s and 1950s, three-lane rural highway alignments were provided in areas where two-lane highways were becoming congested or where additional capacity beyond what a two-lane highway could provide was considered necessary. They were initially viewed as a low-cost alternative to improve two-lane highway operations without creating a full multilane highway of four to six lanes.

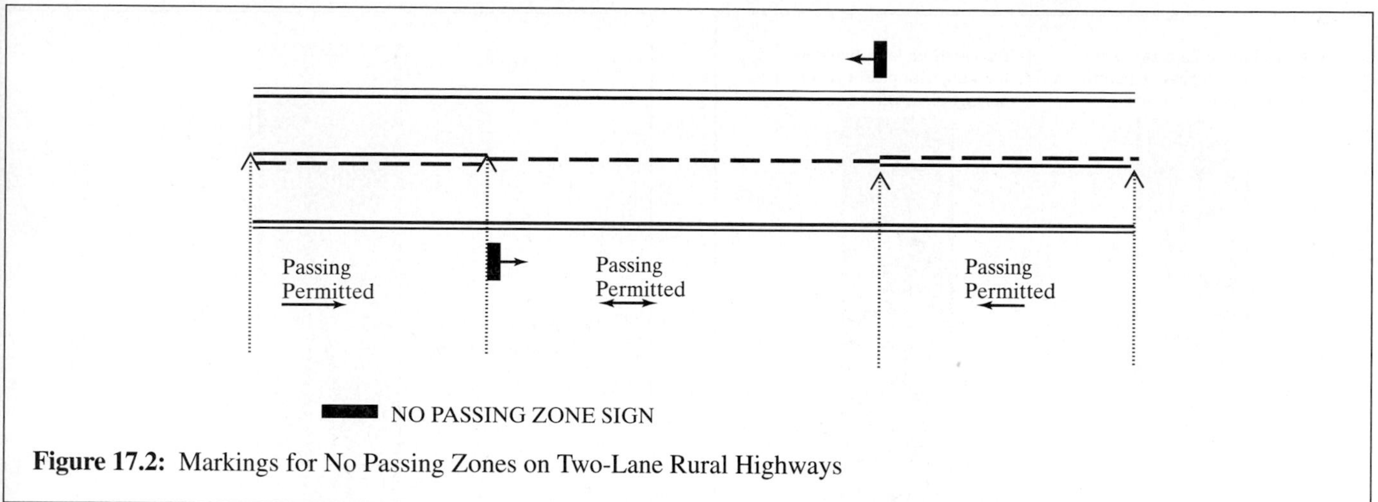

Figure 17.2: Markings for No Passing Zones on Two-Lane Rural Highways

Initially, they were marked to provide for one lane in each direction and a center passing lane shared by both directions of flow. Because there were no restrictions to the use of the passing lane, these roadways eventually produced high accident and fatality rates. Modern practice is to assign two lanes to one direction and one lane to the other, alternating the assignment at appropriate intervals and locations. The marking conventions for this type of operation are shown in Figure 17.3.

Where passing from the single-lane direction is permitted, it is necessary that passing sight distance be available to drivers in that direction. These sight distances are the same as those defined for two-lane highways, as discussed in Chapter 16. Where the additional lane assignment is alternated at frequent intervals (less than every two miles or so), passing from the single-lane direction is usually not permitted.

As the direction of the center lane in a three-lane alignment must be periodically alternated, special transition markings must be provided at these locations, as illustrated in Figure 17.4.

The advance warning distance, d, is found from Chapter 4, where advance warning sign distances are prescribed. A warning sign indicating the merging of two lanes into one would be posted at this point, and the lane line would be discontinued as shown.

Distance L is the length of the taper markings. These are related to the posted or statutory speed limit as indicated in Equations 17-1:

For speed limits of 70 mi/h or more:

$$L = WS \tag{17-1a}$$

For speed limits less than 70 mi/h:

$$L = \frac{WS^2}{60} \tag{17-1b}$$

where: L = length of taper, ft
W = width of the center lane, ft
S = 85th percentile speed, or speed limit, mi/h

Distance B is the buffer distance between the two transition markings. It should be a minimum of 50 ft long. If there is a no passing restriction in both directions at the location of the transition zone, the buffer zone is the distance between the beginning of the no-passing zones in each direction. This is an issue only in cases marked to allow passing from the single-lane direction.

17.1.3 Ramp Junction Markings

Ramp junctions occur at all freeway interchanges and on all types of rural highways at locations where grade separated interchanges are provided. Figure 17.5 illustrates typical markings for off-ramps.

Two basic design types are used: parallel deceleration lanes and tapered deceleration lanes. The former are the more common type. Typical edge and lane markings are carried through the off-ramp junction; the right-edge marking, however, follows the right edge of the deceleration lane and ramp. In the case of a parallel deceleration lane, the lane line between the deceleration lane and lane 1 (right lane) of the freeway or highway is carried half the distance between

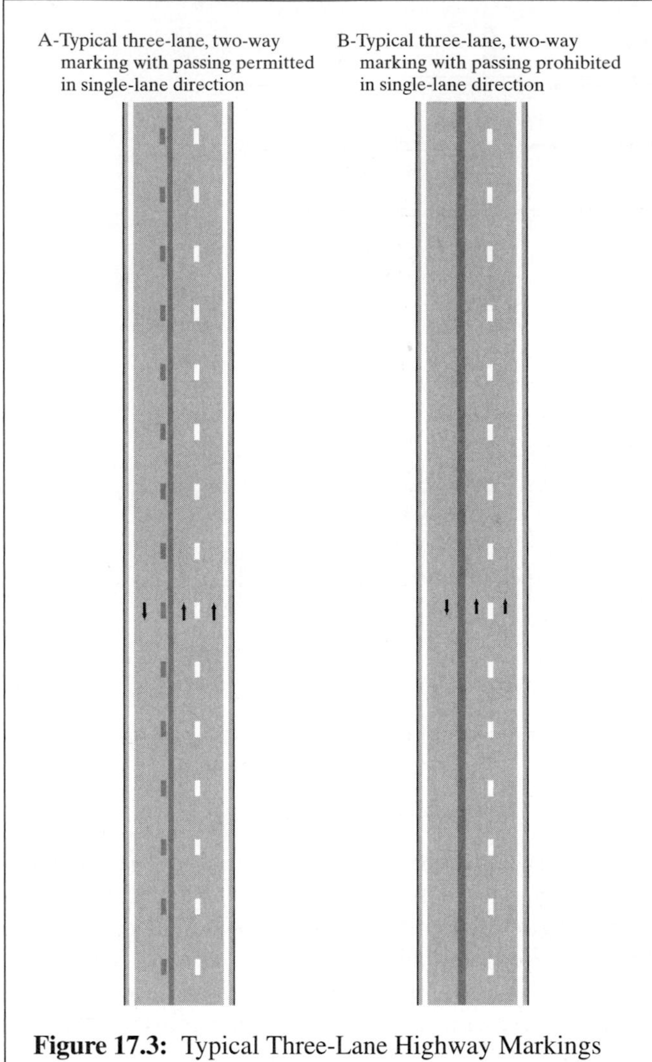

Figure 17.3: Typical Three-Lane Highway Markings

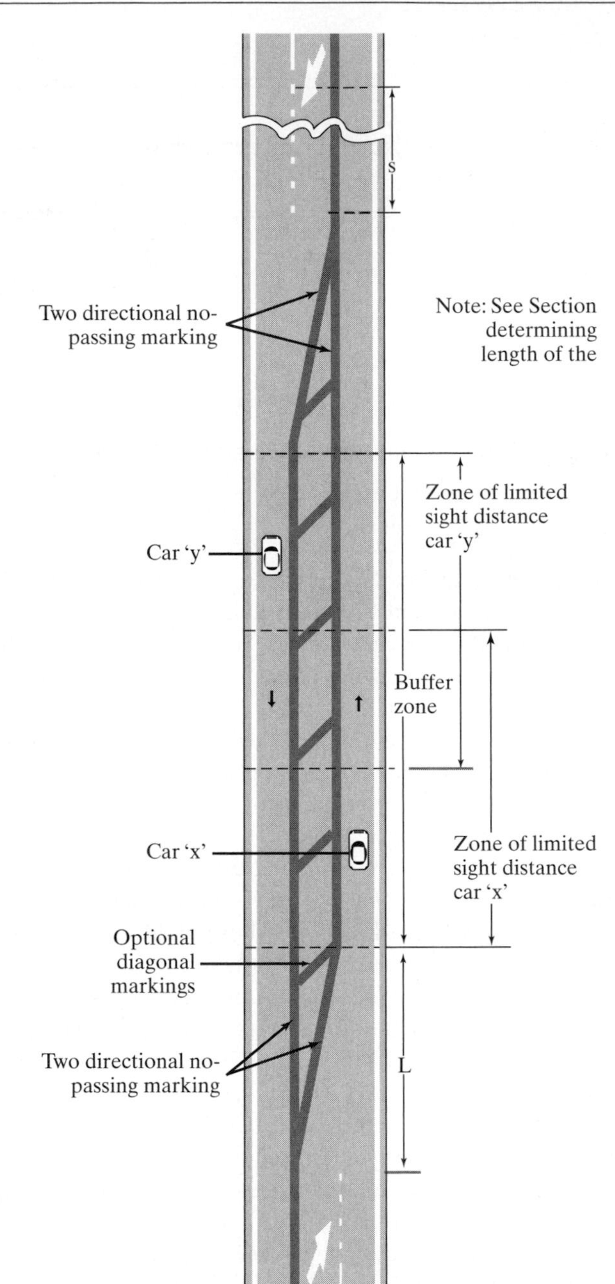

Figure 17.4: Three-Lane Highway Transition Markings

(*Source: Manual of Uniform Traffic Control Devices* and *Manual of Uniform Traffic Control Devices,* Draft, Federal Highway Administration, Washington Draft, Federal Highway Administration, Washington DC, December 2007, Figures 3B-2 and 3B-4.)

the beginning of the deceleration lane (taper included) and the gore area. A dotted line extension of this lane line is optional and often included. In the case of a tapered deceleration lane, a dotted line is used between the deceleration lane and the right lane of the freeway or highway.

The gore area itself is delineated with channelizing lines. The theoretical gore point is the beginning of the channelizing line of the gore area. The interior of the gore area is often marked with chevron markings at off-ramps, as shown. Note the orientation of the chevron markings. They are positioned to visually guide a driver who encroaches onto the gore area back into the appropriate lane (either the ramp or right lane of the freeway or highway).

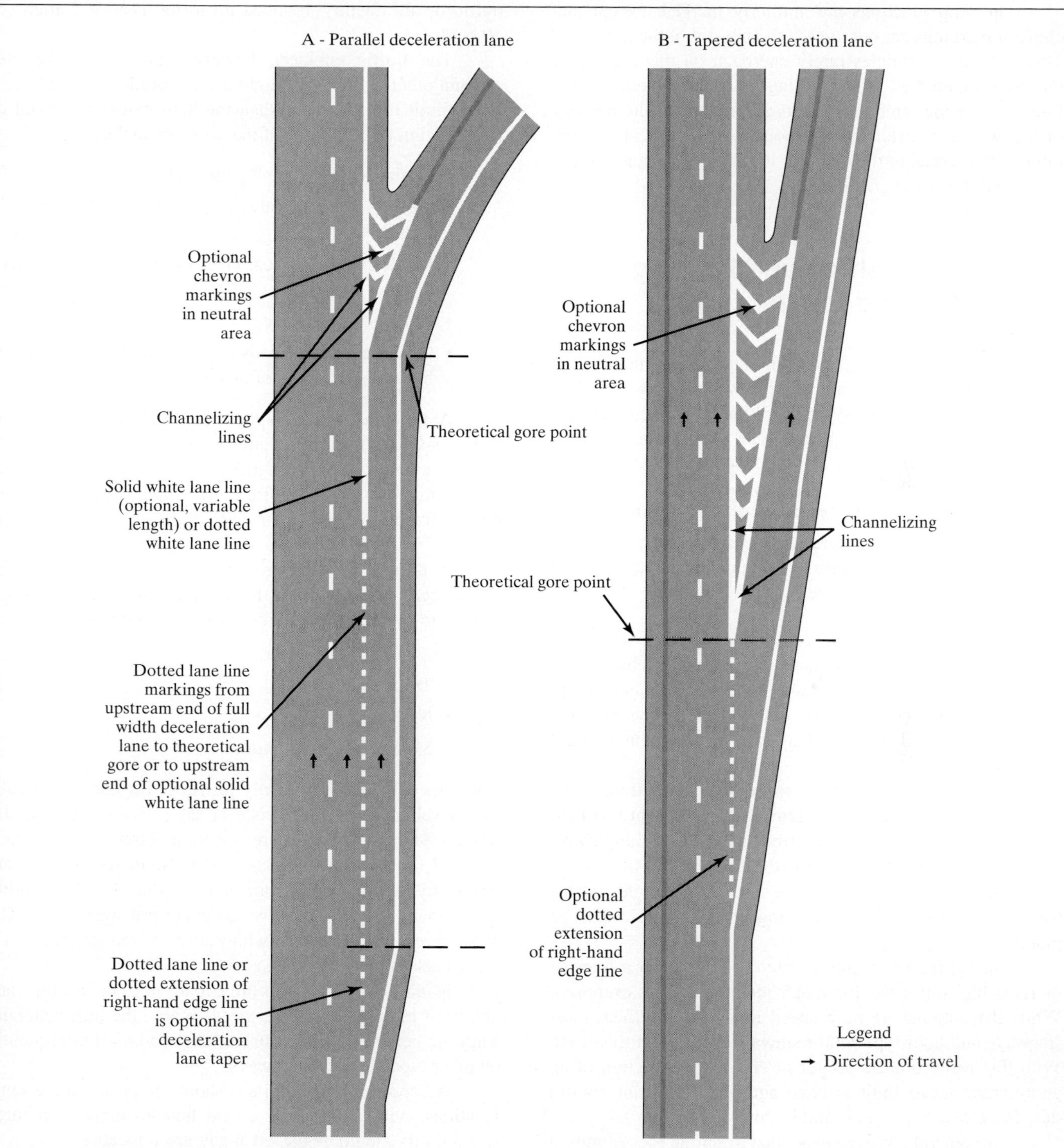

Figure 17.5: Typical Off-Ramp Markings
(*Source: Manual of Uniform Traffic Control Devices*, Draft, Federal Highway Administration, Washington DC, December 2007, Figure 3B-8.)

On-ramp junctions are similarly marked, except that chevron markings are not used in the interior of the gore area. This is because vehicles rarely encroach on this area at an on-ramp. Tapered on-ramp markings involve an extended lane line between the ramp lane and the right lane of the freeway or highway. It is extended to a point where the width of the ramp lane is equal to the width of the right-most lane. This is illustrated in Figure 17.6.

17.2 Establishing and Posting of Speed Limits

A brief discussion of speed limit signs was included in Chapter 4. On freeways and rural highways, speed limits are generally of the linear type (i.e., applying to a specified linear section of a designated highway). The MUTCD requires the posting of speed limit signs at:

- Points of change from one speed limit to another
- Beyond major intersections and at other locations where it is necessary to remind drivers of the speed limit that is applicable.

In practical terms, this is generally interpreted to mean that speed-limit signs should be located at points of change and within 1,000 feet of major entry locations. "Major" entry locations would include all on-ramps on freeways or other rural highways and significant at-grade intersections on rural highways.

Where the state statutory speed limit is in effect, signs should be periodically posted reminding drivers of this fact. Placement of such signs along freeways and rural highways follows the same pattern as for the posting of other speed limits. At borders between states, signs indicating the statutory speed limit of the state being entered should also be placed.

The setting of an appropriate speed limit for a freeway or rural highway calls for much judgment to be exercised. While the national 55 mi/h speed limit was in effect, most freeways and high-type rural highways were set at this level. With this restriction no longer in effect, the selection of an appropriate speed limit is once again a significant control decision for freeways and rural highways.

The general philosophy applied to setting speed limits is that the majority of drivers are not suicidal. They will, with no controls imposed, tend to select a range of speeds that is safe for the conditions that exist. Using this approach, speed limits should be set at the 85th percentile speed of free-flowing traffic on the facility, rounded up to the nearest 5 mi/h (or 10 km/h).

The traffic engineer, however, must also take into account other factors that might make it prudent to establish a speed limit that is slower than the 85th percentile speed of free-flowing traffic. Some of these factors include:

- Design speed of the facility section
- Details of the roadway geometry, including sight distances
- Roadside development intensity and roadside environment
- Accident experience
- Observed pace speeds (10 mi/h increment with the highest percentage of drivers)

A reduction of the speed limit below the 85th percentile speed is usually required in traffic environments that contain elements that are difficult for drivers to perceive. Such limits, however, require vigilant enforcement to deter drivers from following their own judgments on appropriate speeds.

A number of different types of speed limits may be established. In addition to the primary speed limit, which applies to all vehicles, additional speed limits may be set as follows:

- Truck speed limits
- Night speed limits
- Minimum speed limits

Truck speed limits only apply to trucks (as defined in each state's vehicle and traffic code or law). They are generally introduced in situations where operation of trucks at the general speed limits involves safety issues. Night speed limits are frequently used in harsh terrain, where reduced night visibility would make it unsafe to drive at the general speed limit. The night speed limit sign has white reflectorized lettering on a black background.

Minimum speed limits are employed to reduce the variability of individual vehicle speeds within the traffic stream. They are generally applied to freeways and used infrequently on other types of rural highways.

All applicable speed limits should be posted at the same locations, with the general caveat that no more than three speed limits should be posted at any given location.

Additional speed signs can be used to inform drivers of forthcoming changes in the speed limit. These include signs indicating "Reduced Speed Ahead" or "Speed Zone Ahead."

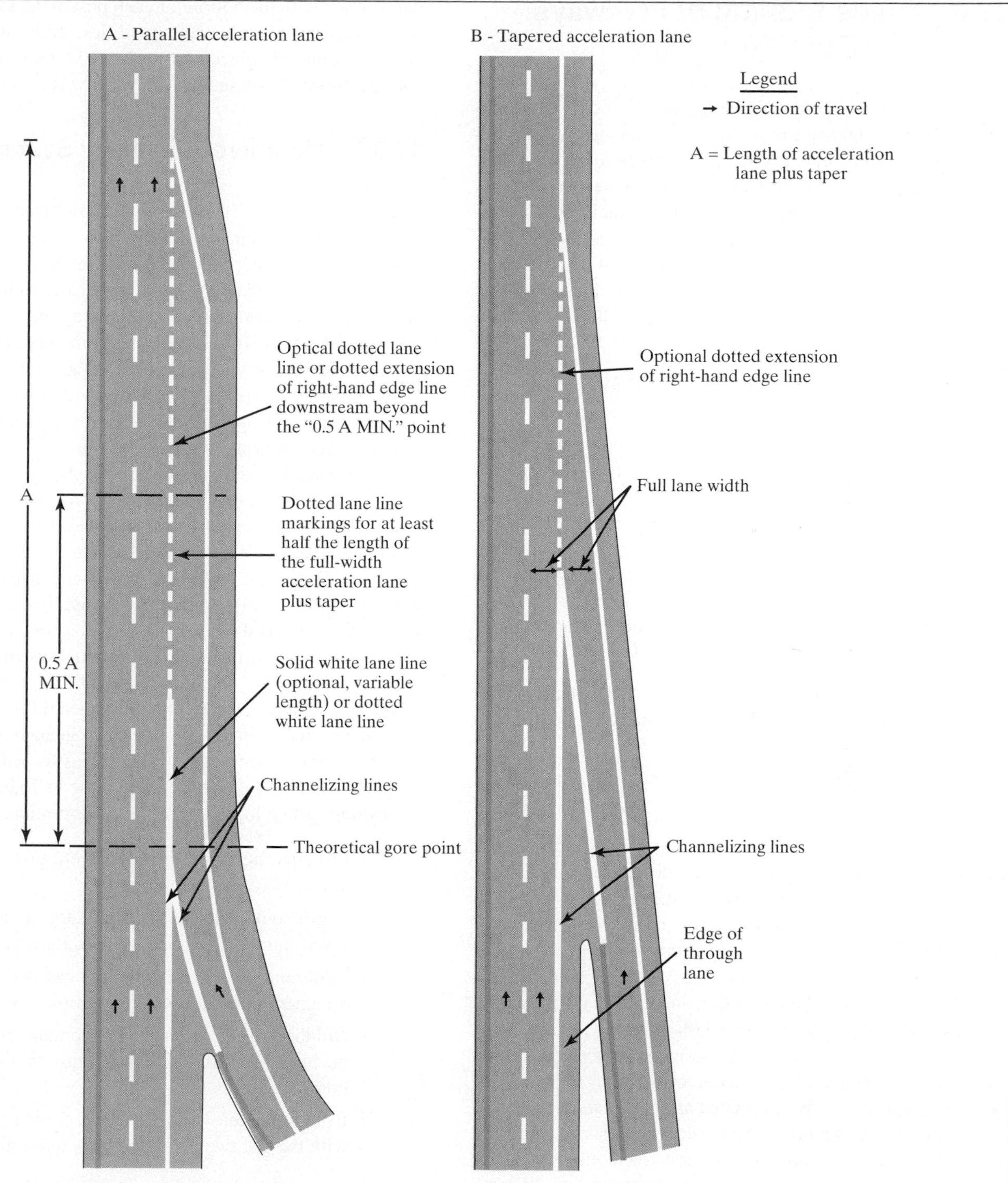

Figure 17.6: Typical On-Ramp Markings

(*Source: Manual of Uniform Traffic Control Devices,* Draft, Federal Highway Administration, Washington DC, December 2007, Figure 3B-9.)

17.3 Guide Signing of Freeways and Rural Highways

One of the most important elements of traffic control on freeways and rural highways is proper guide signing. These types of highways often serve significant numbers of drivers who are not regular users of the facility. These drivers are the primary target of guide signing. A confused driver is a dangerous driver. Thus it is important that unfamiliar users be properly informed in terms of routing to key destinations that they may be seeking. There are many elements involved in doing this effectively, as discussed in the sections that follow.

17.3.1 Reference Location Posts

Reference location posts provide a location system along highways on which they are installed. Formerly (and still often) referred to as "mileposts," reference posts indicate the number of miles (or kilometers in states using a metric system) along a highway from a designated terminus. They provide a location system for accidents and emergencies and may be used as the basis for exit numbering.

The numbering system must be continuous within a state. By convention, mile (or kilometer) "0" is located:

- At the southern state boundary or southernmost point on a route beginning within a state, on north-south routes.
- At the western state boundary or westernmost point on a route beginning within a state, on east-west routes.

Cardinal directions used in highway designations recognize only two axes: north-south or east-west. Each highway is designated based on the general direction of the route within the state. Thus a north-south route may have individual sections that are aligned east-west and vice versa. In general, if a straight line between the beginning and end of a route within a given state makes an angle of less than 45° *or* more than 135° with the horizontal, it is designated as an east-west route. If such a line makes an angle of between 45° and 135° with the horizontal, it is designated as a north-south route.

The MUTCD requires that reference posts be placed on all freeway facilities and on expressway facilities that are "located on a route where there is reference post continuity." Other rural highways may also use reference posts, but their use is optional.

When used, reference posts are placed every mile (or kilometer) along the route. They are located in line with delineators, with the bottom of each post at the same height as the delineator. On interstate facilities, additional reference posts are generally placed for each 1/10th mile. Typical reference posts are shown in Figure 17.7.

17.3.2 Numbered Highway Systems

Four types of highway systems are numbered in the United States. The two national systems are the Eisenhower National System of Interstate and Defense Highways (the Interstate System), and the U.S. system. These are numbered by the American Association of State Highway and Transportation Officials (AASHTO) based on recommendations from individual state highway departments in accordance with published policies [*2,3*]. State and county roadway systems are numbered by the appropriate agencies in accordance with standards and criteria established by each state.

The oldest system of numbered highways is the U.S. system, developed in an often-heated series of meetings of the then American Association of State Highway Officials (AASHO) and representatives from state highway agencies between 1923 and 1927 [*4*]. A loose system of named national routes existed at the time, with most of these "named" routes (such as the Lincoln Highway) sponsored by private organizations and motorists' clubs. This was to be replaced by a national system of numbered highways. Highways that afforded significant travel over more than one state were eligible to be considered for inclusion. The U.S. system was initially envisioned to include approximately 50,000 miles of highway, but when it was formally established on November 11, 1926, it included close to 75,000 miles. The numbering system loosely followed a convention:

- Principal north-south routes were given numbers (of one or two digits) ending in "1."
- North-south routes of secondary importance were given numbers (of one or two digits) ending in "5."
- Transcontinental and principal east-west routes were assigned route numbers in multiples of 10.
- Numbers of principal and secondary routes were to be in numerical order from east to west and from north to south.
- Branch routes were assigned three-digit numbers, with the last two being the principal route.

Many of these conventions were adapted for the Interstate System route designations:

- All primary east-west routes have two-digit even numbers.

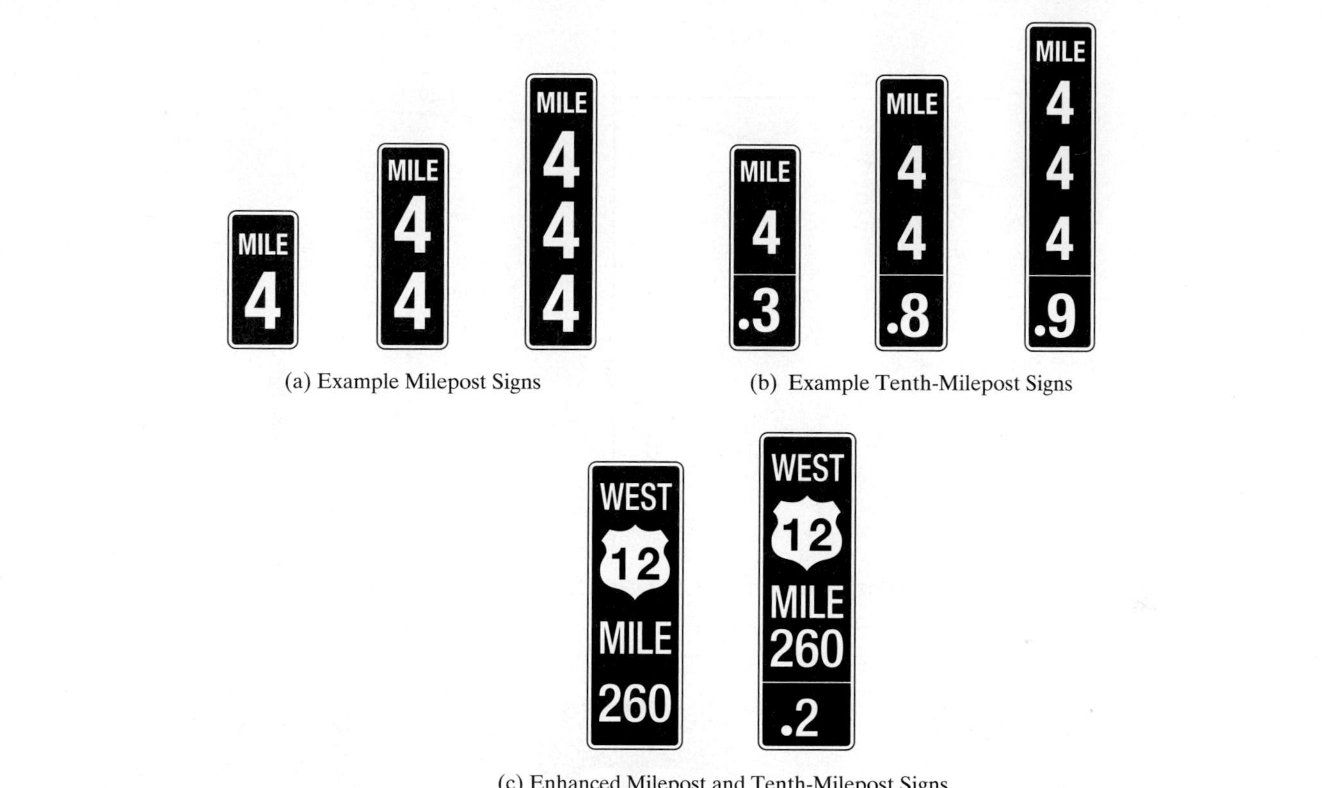

(a) Example Milepost Signs (b) Example Tenth-Milepost Signs

(c) Enhanced Milepost and Tenth-Milepost Signs

Figure 17.7: Reference Location Posts for Freeways and Rural Highways

(*Source: Manual of Uniform Traffic Control Devices,* Draft, Federal Highway Administration, Washington DC, December 2007, Figures 2I-1, 2I-2, 2I-3.)

- All primary north-south routes have two-digit odd numbers.
- All branch routes have three-digit numbers, with the last two representing the primary route.

Figure 17.8 illustrates this system, depicting the primary interstate routes emanating from New York City. The principal north-south route serving the entire East Coast is I-95. The principal east-west routes serving New York are I-80 and I-78. Branch routes connecting with I-95, such as the Baltimore and Washington Beltways, have three-digit numbers ending in "95." Note that three-digit route numbers need not be unique. There are several Interstate routes 495, 695, and 895 along the East Coast, all providing major connections to I-95. Principal Interstate route numbers are unique, however.

Numbered routes are identified by the appropriate shield bearing the route number and an auxiliary panel indicating the cardinal direction of the route. Standard shield designs are illustrated in Figure 17.9. The U.S. and Interstate shields each have a standard design. The shape of each state shield is designated by the state transportation agency. County shields are uniform throughout the nation; the name of the county appears as part of the shield design.

When numbered routes converge, both route numbers are signed using the appropriate shields. All route shields are posted at common locations. Because cardinal directions indicate the general direction of the route, as described previously, it is possible to have a given section of highway with multiple route numbers signed with different cardinal directions. For example, a section of the New York State Thruway (a north-south route) that is convergent with the Cross-Westchester Expressway (an east-west route) is signed as "I-87 North" and "I-287 West." The reverse direction bears "south" and "east" cardinal directions.

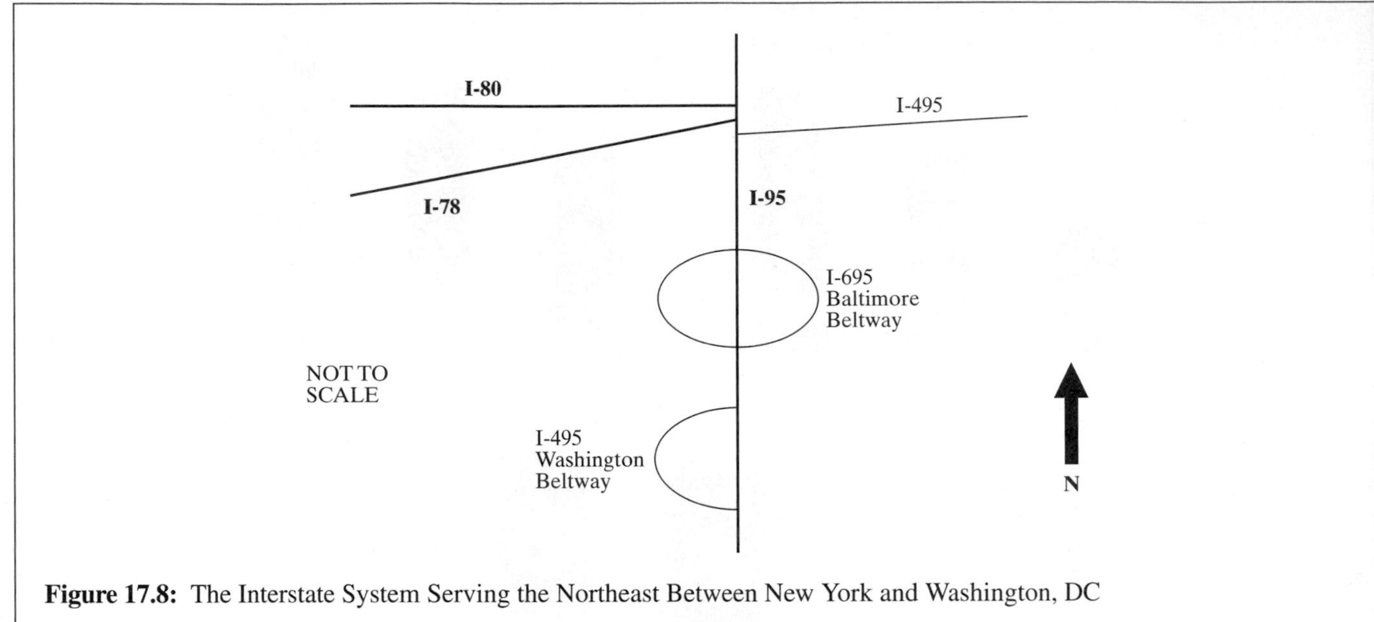

Figure 17.8: The Interstate System Serving the Northeast Between New York and Washington, DC

Figure 17.9: Route Marker Signs

(*Source: Manual of Uniform Traffic Control Devices,* Draft, Federal Highway Administration, Washington DC, December 2007, Figure 2D-3.)

17.3.3 Exit Numbering Systems

On freeways and some expressways, interchanges are numbered using one of two systems:

- *Milepost numbering.* The exit number is the milepost number closest to the interchange.
- *Sequential numbering.* Exits are sequentially numbered; Exit 1 begins at the westernmost or southernmost interchange within the state.

Milepost numbering is now the preferred system according to the MUTCD, and many states have been converting from sequential numbering to this system. Although milepost numbers are not sequential, drivers can use the exit number in conjunction with mileposts to determine the distance to the desired interchange. Another advantage is that interchanges may be added without disrupting the numbering sequence. In sequential systems, addition of an interchange forces the use of "A" and "B" supplemental designations, which are also used for interchanges with separate ramps for each direction of the intersecting route. This dual use of supplemental designations can cause confusion for unfamiliar motorists.

The most significant argument in favor of sequential numbering is that it is a system with which drivers have become familiar and comfortable. Historically, all freeways were initially signed using sequential numbering, with milepost numbering introduced primarily over the past 25 years.

Where routes converge, mileposts and interchange numbers are continuous for only one route. In terms of hierarchy, Interstate routes take precedence over all other systems, followed (in order) by U.S. routes, state routes, and county routes. Where two routes of similar precedence converge, primary routes take precedence over branch routes; where two routes have exactly the same precedence, the state transportation agencies determines which one is continuously numbered.

Mileposts, numbered routes, and interchange numbers are important elements in route guidance for both freeways and conventional rural highways. All can be used in guide signing to present information in a clear and consistent fashion that will minimize confusion to motorists.

17.3.4 Route Sign Assemblies

A route sign assembly is any posting of single or multiple route number signs. Where numbered routes converge or divide or intersect with other numbered routes, the proper design of route sign assemblies is a critical element in providing directional guidance to motorists. The MUTCD defines a number of different route sign assemblies, as follows:

- *Junction assembly.* Used to indicate an upcoming intersection with another numbered route(s).
- *Advance route turn assembly.* Used to indicate that a turn must be made at an upcoming intersection to remain on the indicated route.
- *Directional assembly.* Used to indicate required turn movements for route continuity at an intersection of numbered routes as well as the beginning and end of numbered routes.
- *Confirming or reassurance assemblies.* Used after motorists have passed through an intersection of numbered routes. Within a short distance, such an assembly assures motorists they are on the intended route.
- *Trailblazer assemblies.* Used on non-numbered routes that lead to a numbered route; "To" auxiliary panel is used in conjunction with the route shield of the numbered route.

The MUTCD gives relatively precise guidelines on the exact placement and arrangement of these assemblies. A junction assembly, for example, must be placed in the block preceding the intersection in urban areas or at least 400 feet in advance of the intersection in rural areas. Further, it must be at least 200 feet in advance of the directional guide sign for the intersection and 200 feet in advance of the advance route turn assembly.

The directional assembly is an arrangement of route shields and supplementary directional arrows for all intersection routes. Routes approaching the intersection from the left are posted at the top and/or left of the assembly. Routes approaching the intersection from the right are posted at the right and/or bottom of the assembly. Routes passing straight through the intersection are posted in the center of a horizontal or vertical display.

Figure 17.10 shows two typical examples of the use of route marker assemblies. Both show signing in only one direction; each approach to the intersection would have similar signing.

In both illustrations, signing is for a driver approaching from the south. The first assembly in both cases is the junction assembly. Note that the two illustrations show two different styles of sign that can be used for this purpose. The junction assembly is followed by a directional guide sign (form for conventional highways). Neither example includes an advance turn assembly, as drivers approaching from the south do not have to execute a turn to stay on the same route. The next placement is the directional assembly. The standard location for this is on the far side of the intersection. In one of the illustrations, a duplicate is provided on the near side of the intersection. Once the intersection is crossed, a confirming assembly is posted within 200 feet of the intersection, so that drivers may be assured they are indeed on their desired route. These are followed by optional destination distance guide signs.

Route sign assemblies provide critical information to motorists on numbered routes. Figure 17.11 illustrates the typical situation that such assemblies must clarify.

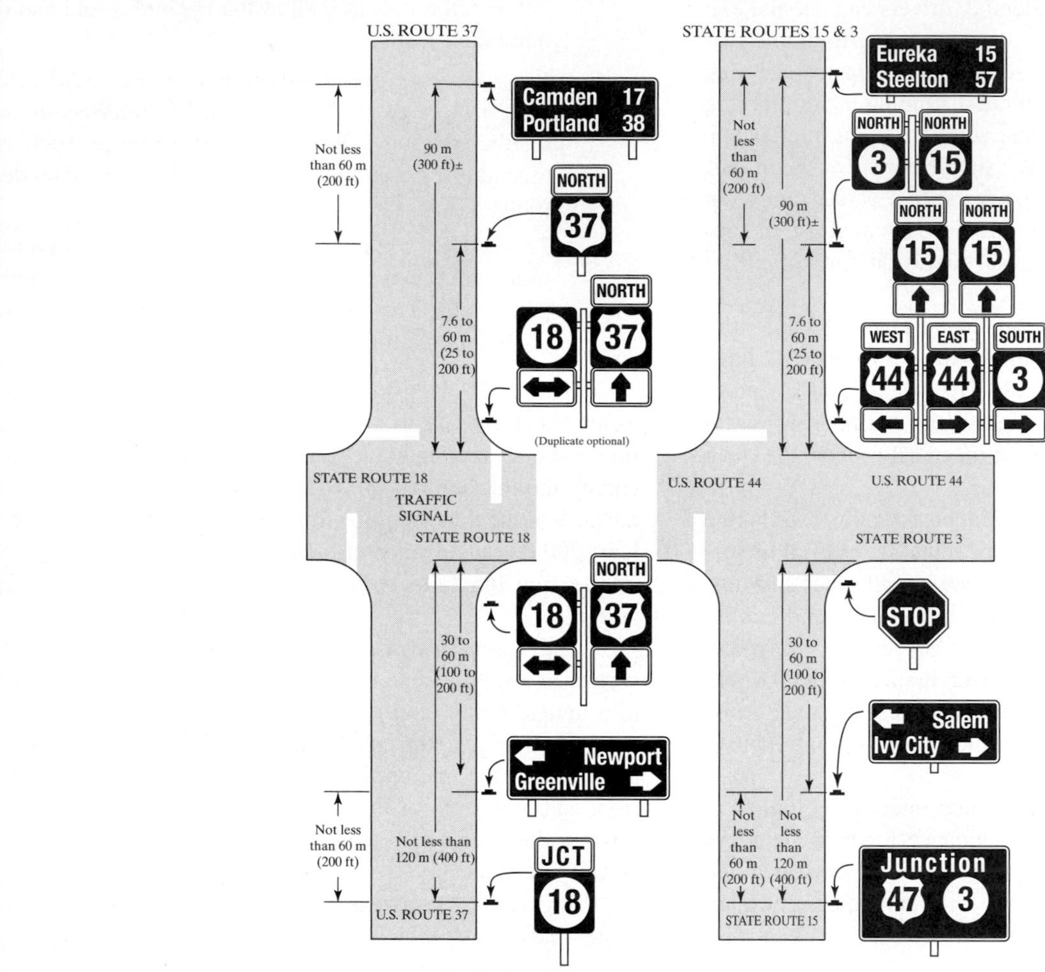

Figure 17.10: Typical Use of Route Sign Assemblies

(*Source: Manual of Uniform Traffic Control Devices,* Draft, Federal Highway Administration, Washington DC, December 2007, Figure 2D-7.)

Before the age of the Interstate System, numbered U.S. and state routes served most interstate or long-distance intercity travel by vehicle. Unlike freeways, however, most surface routes go through every significant town and city along the way. Thus Figure 17.11 illustrates a typical condition: Many numbered routes converging on a town or city, diverging again on the other side of town. These routes, however, are generally not continuous, and they often converge and diverge several times within the town or city. The only way for an unfamiliar driver to negotiate the proper path through the town to either stay on a particular route or turn onto another is to follow appropriate route sign assemblies.

17.3.5 Freeway and Expressway Guide Signing

Freeways and most expressways have numbered exits and reference posts, and guide signing is keyed to these features. As noted in Chapter 4, guide signing provides route or directional guidance, information on services, and information on historical and natural locations. Signs are rectangular, with the long dimension horizontal, and are color coded: white on green for directional guidance, white on blue for service information, and white on brown for information on historical destinations.

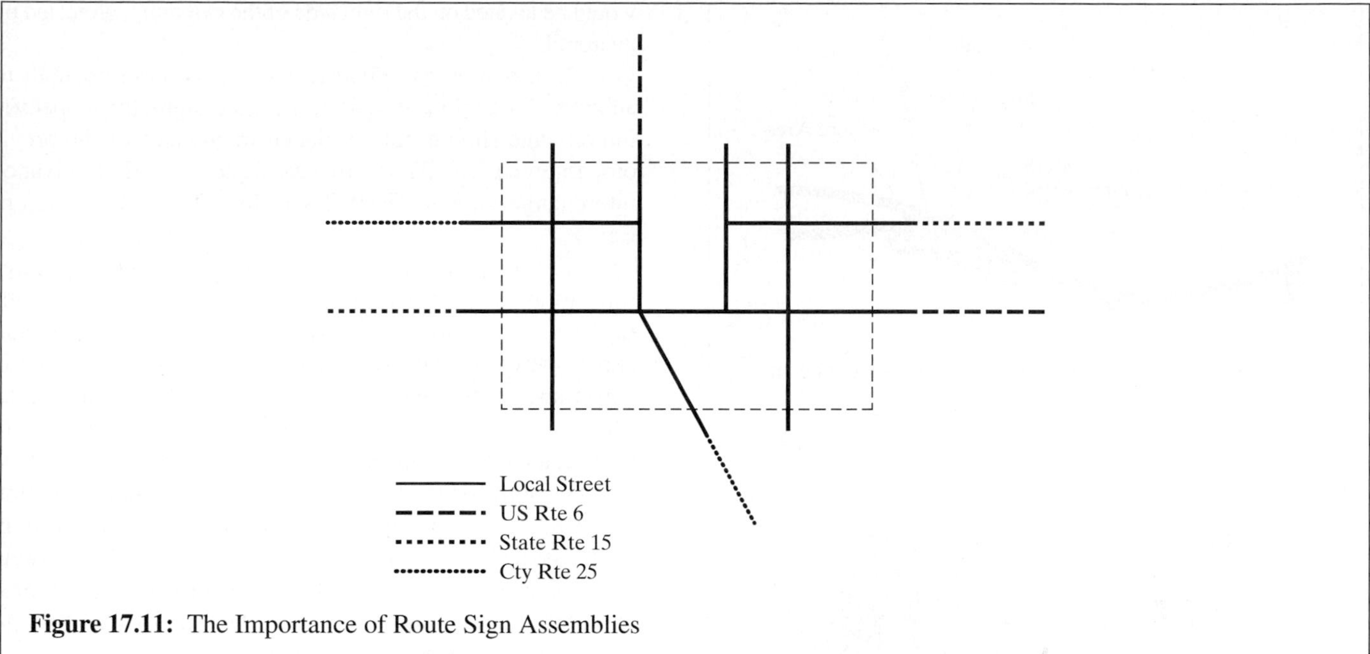

Figure 17.11: The Importance of Route Sign Assemblies

Directional guide signs, however, make up the majority of these signs and are the most important in assuring safe and unconfused operation of vehicles. A typical freeway guide sign is illustrated in Figure 17.12. In this case, it is a guide sign located just at the exit. It contains a number of informative features:

- The ramp leads to U.S. Route 56, West (U.S. shield used).
- The primary destination reached by selecting this ramp is the city or town of Utopia.
- The exit number is 211A. Assuming this is a milepost-numbered sign, the exit is located at milepost 211. Exit 211A refers to the ramp leading to Route 56 WB. Exit 211B would lead to Route 56 EB.
- The exit number tab is located on the right side of the sign, indicating this is a right-hand exit. For a left-hand exit, the tab would be located on the left side of the sign.

In general, drivers should be given as much advance warning of interchanges and destinations as possible. This leads to very different signing approaches in urban and rural areas. In rural areas, advance signing is much easier to accomplish because there are long distances between interchanges. Figure 17.13 shows a typical signing sequence for a rural area.

The first advance directional guide sign can be as far as 10 miles away, assuming there are no other exits between the sign and the subject interchange. If a 10-mile sign is placed, then the usual sequence would be to repeat advance signs at 5 miles, 2 miles, 1 mile, and ½ mile from the interchange. Where distance between interchanges permits, the first advance sign should be *at least* 2 miles from the exit. The ½-mile sign is optional but often helpful. At the point of the exit, a large directional sign of the type shown in Figure 17.12 is placed. Another small exit number sign, of the type illustrated in Figure 17.14, is located in the gore area, as shown in Figure 17.13.

Figure 17.12: A Typical Freeway Directional Guide Sign

(*Source: Manual of Uniform Traffic Control Devices,* Draft, Federal Highway Administration, Washington DC, December 2007, Figure 2E-20.)

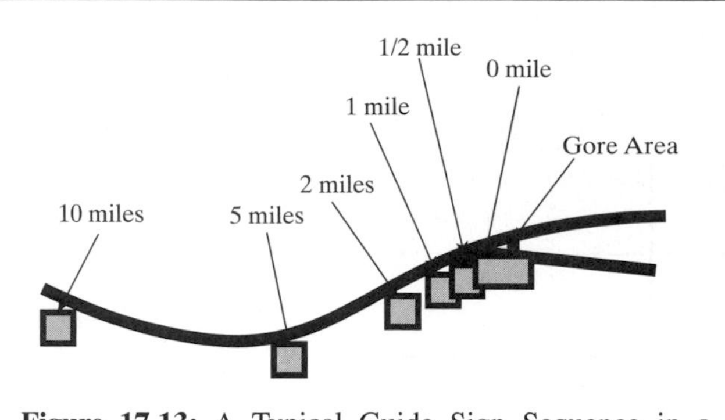

Figure 17.13: A Typical Guide Sign Sequence in a Rural Area

Figure 17.14: Gore Area Exit Sign with Speed Warning Panel

(*Source: Manual of Uniform Traffic Control Devices,* Draft, Federal Highway Administration, Washington DC, December 2007, Figure 2E-22.)

The location of a final directional guide sign in the gore area of the off-ramp is desirable from an information point of view, but it is a problem from the point of view of safety. The gore area is an area that unsure or confused drivers often encroach on. Thus signs located in this area are subject to being hit in accidents. Such signs should be mounted on breakaway sign supports to avoid injury to motorists and extensive damage to vehicles that enter the gore area. Where warranted, the gore-area exit panel can be replaced by the final destination sign mounted on an overhead cantilevered sign support such that the sign is situated over a portion of the gore area. The sign support would be located on the right side of the exit ramp, protected by guardrail.

In urban areas, advance signing is more difficult to achieve. To avoid confusion, advance signs for a specific interchange should not be placed in advance of the previous interchange. Thus, for example, the first advance interchange sign for Exit 2 should not be placed before Exit 1.

In urban situations, signing must be done very carefully to avoid confusing unfamiliar drivers. "Sign spreading" is a technique employed on urban freeways and expressways to avoid confusing sign sequences and the appearance of too many signs at one location. Figure 17.15 illustrates the concept. The sign for Exit 7 is placed on an overhead cantilevered support over the gore area. The first advance sign for Exit 8 is *not* placed at the same location but rather is on another overhead support (in this case an overpass) a short distance beyond Exit 7. This is preferable to an older practice in which the advance sign for Exit 8 was placed at the same location as the Exit 7 gore area sign on a single overhead sign structure.

The signing of freeway and expressway interchanges is complex, and virtually every situation contains unique features that must be carefully addressed. Several examples from the MUTCD are included and discussed for illustration. Additional examples exist in the MUTCD and should be consulted directly.

Figure 17.16 illustrates guide signing for a series of closely spaced interchanges on an urban freeway. Distances are such that only one advance sign is provided for each exit, using the sign-spreading technique previously discussed. Note also that sequential exit numbers are used, with Exits 22A and 22B representing separate interchanges, doubtless due to the addition of one after the original exit numbering had been assigned. The unique feature here is the use of "interchange sequence signs" to provide additional advance information on upcoming destinations. The signs list three upcoming interchanges each but do not list exit numbers. This avoids overlapping advance exit number sequences. Street names are given for each interchange with the mileage to each one. This effectively supplements the single advance interchange sign for each exit without presenting a confusing numbering sequence.

Figure 17.17 illustrates a complex interchange between two interstate highways, I-42 and I-17. Signing is shown for only two of the four approach directions. In the eastbound (EB) direction, the critical feature is that a single off-ramp serves both directions of I-17. The single ramp, however, splits shortly after leaving I-42. Thus two

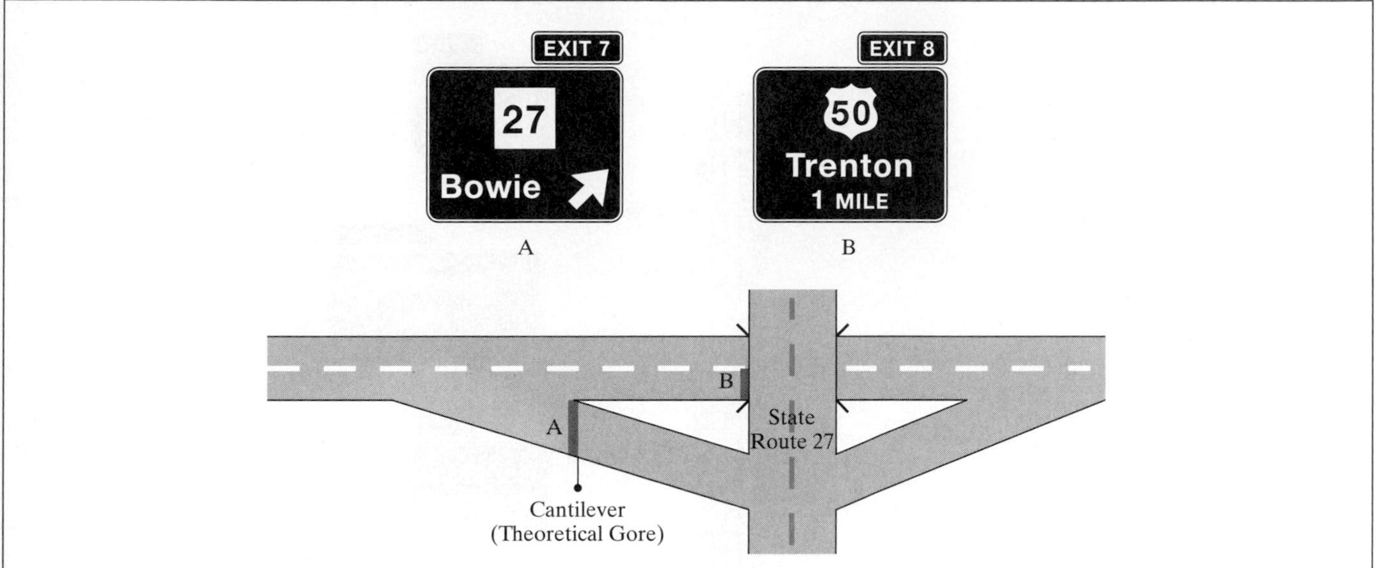

Figure 17.15: Illustration of Sign Spreading at an Urban Interchange

(*Source: Manual of Uniform Traffic Control Devices,* Draft, Federal Highway Administration, Washington DC, December 2007, Figure 2E-1.)

closely spaced sequential diverges must be negotiated by motorists. Advance exit signs are placed at 2 miles, 1 mile, and ½ mile from the exit. Arrows indicate that the exit has two lanes, and the exit tab indicates that it is a right-side ramp. Pull-through signs for Springfield are located to the left of exit signs at the 1-mile, ½ mile, and exit location. All of the advance exit signs indicate both Miami and Portland as destinations. Because of the distances involved, a single set of overhead signs indicates I-17 North and I-17 South at the location where the ramp divides. The signing of the EB approach to the interchange involves nine signs and five sign support structures.

In the northbound (NB) direction, separate ramps are provided for the EB and westbound (WB) directions on I-42. "A" and "B" exits are used to differentiate between the two, but advance signs for both are placed at the same locations, except for the first advance sign at two miles, which lists both ramps on the same sign. Pull-through signs for Miami are used at each advance location, except at two miles. For the NB approach, there are 12 signs and 5 sign-support structures.

Proper signing of a major interchange can be a considerable expense. In the case shown in Figure 17.17, 21 signs and 10 support structures are needed for two of the four approach directions. The expense is necessary, however, to provide drivers with the required information to navigate the interchange safely.

Diagrammatic guide signs may be of great utility when approaching major diverge junctions on freeways and expressways. MUTCD, however, *does not* permit their use in signing for cloverleaf interchanges because studies have shown these to be more confusing than helpful.

Figure 17.18 illustrates the signing of a major diverge of two interstate routes, I-50 West and I-79 South. Advance signs at 2 miles, 1 mile, and ½ mile are diagrammatic. The diagram shows how the three approach lanes split at the diverge and allow drivers to move into a lane appropriate to their desired destinations well in advance of the gore area. The last sign at the diverge uses arrows to indicate which two lanes can be used for each route.

The guiding principle for destination guide signing is to keep it as simple as possible. Consult Chapter 4 for guidelines on how to keep the content of each sign understandable. Chapter 4 also contains guidelines and illustrations of the application of service information and historical/cultural/recreational information signing.

17.3.6 Guide Signing for Conventional Roads

Guide signing for conventional roadways primarily consists of route sign assemblies, as previously discussed,

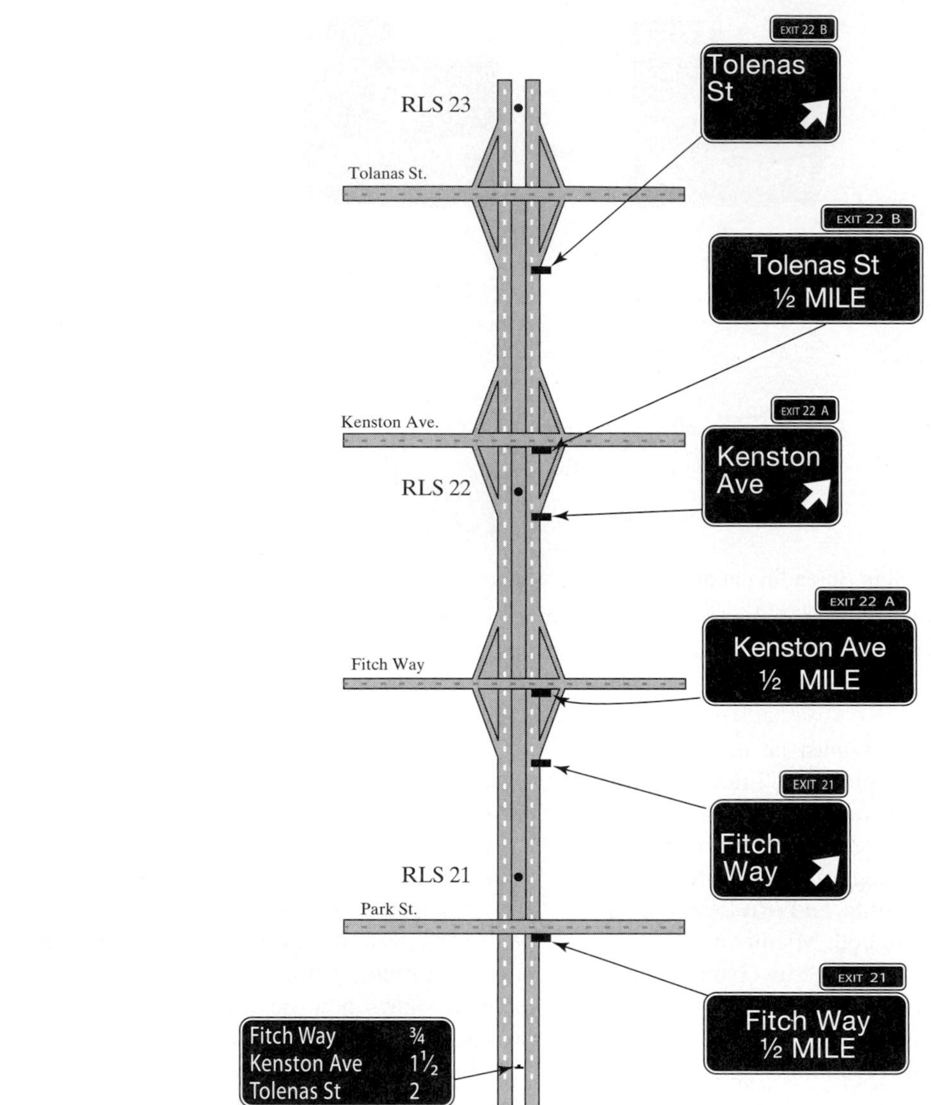

Figure 17.16: Guide Signs for a Series of Closely Spaced Urban Interchanges

(*Source: Manual of Uniform Traffic Control Devices,* Draft, Federal Highway Administration, Washington DC, December 2007, Figure 2E-24.)

and destination signing. Because numbered routes are not involved in most cases, destination names become the primary means of conveying information. Advance destination signs are generally placed at least 200 feet from an intersection, with confirming destination signs located after passing through an intersection with route marker assemblies at the intersection. A sample of guide signs for conventional roads is shown in Figure 17.19.

17.4 Other Signs on Freeways and Rural Highways

Other than regulatory and guide signs, freeways and expressways do not have extensive additional signing. Warning signs are relatively rare on freeways and expressways but would be used to provide advance warning of

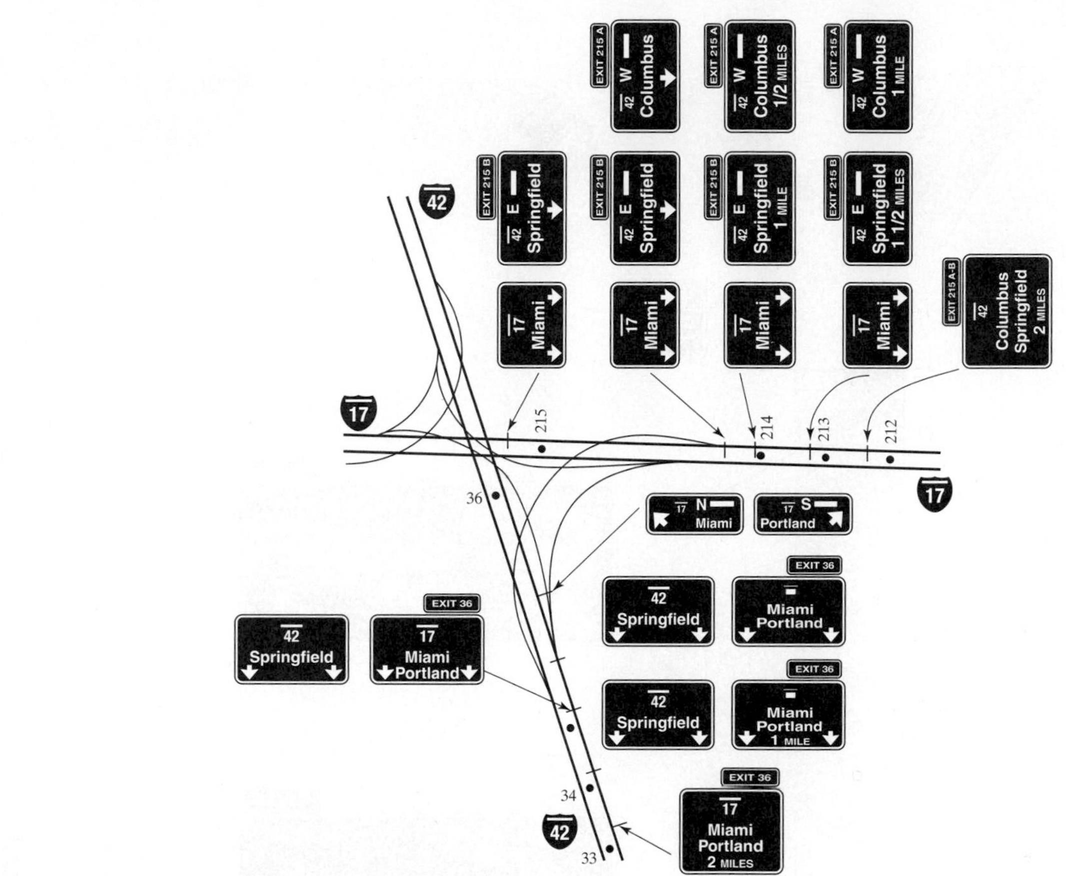

Figure 17.17: A Complex Interchange Between Two Interstate Facilities

(*Source:* Used with permission of Federal Highway Administration, *Manual on Uniform Traffic Control Devices,* Millennium Edition, Washington DC, Figure 2E-25, p. 2E-58.)

at-grade intersections on expressways to warn of extended downgrades (primarily for trucks). Rarely are other elements warranting warning signs present in these types of facilities, except for animal-crossing warnings, particularly in deer country.

On conventional rural highways, warning signs are frequently warranted to warn of various hazards, including unexpected restricted geometric elements, blind driveways and intersections, crossings of various types, and advance warning of control devices, such as STOP signs and signals. Consult Chapter 4 for a detailed discussion of the application of warning signs on all types of highways.

References

1. *Manual of Uniform Traffic Control Devices,* Draft, Federal Highway Administration, U.S. Department of Transportation, Washington DC, December 2007.
2. "Purpose and Policy in the Establishment and Development of United States Numbered Highways," American Association of State Highway and Transportation Officials, Washington DC, revised September 15, 1970.
3. "In the Establishment of a Marking System of Routes Comprising the National System of Interstate and

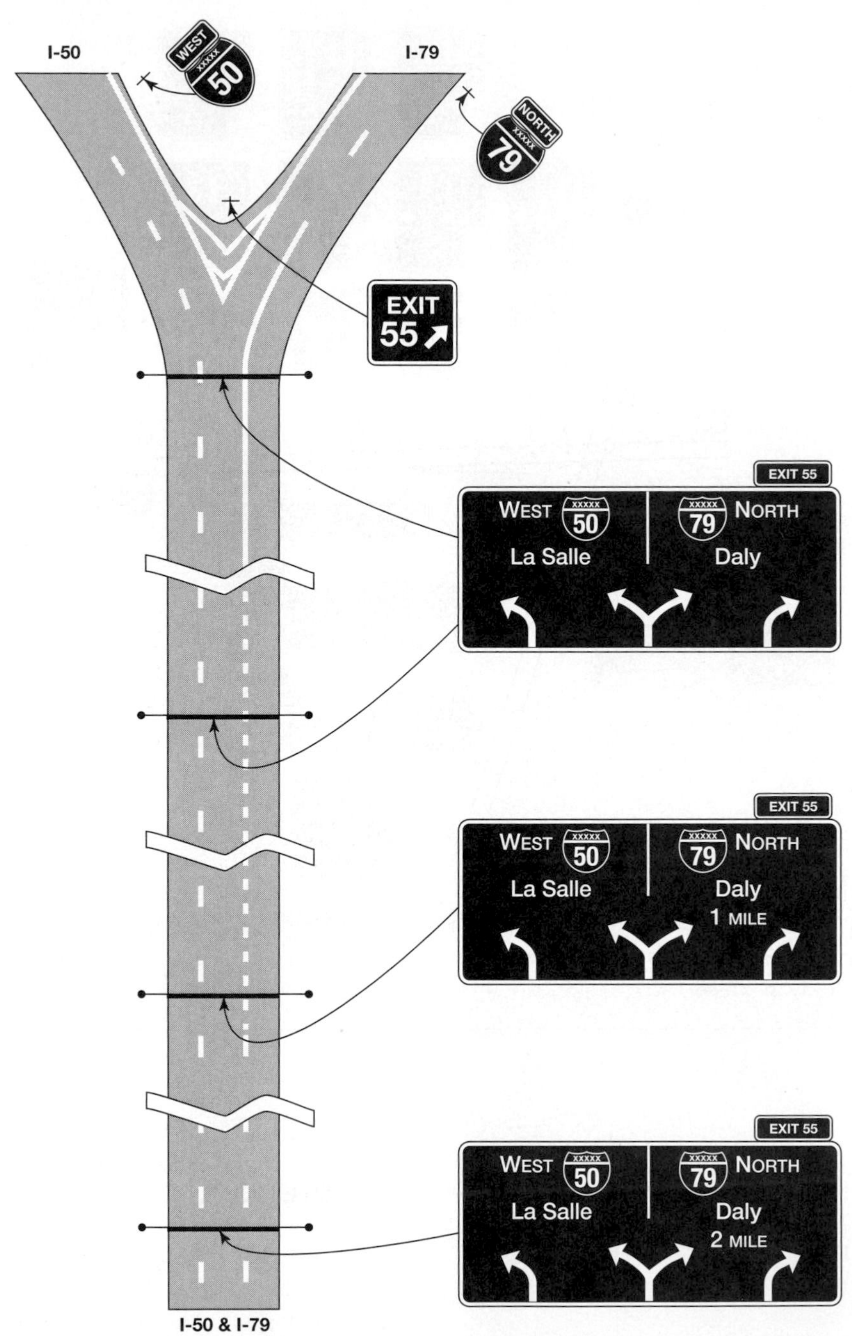

Figure 17.18: Diagrammatic Signing for a Major Diverge of Interstate Routes

(*Source: Manual of Uniform Traffic Control Devices,* Draft, Federal Highway Administration, Washington DC, December 2007, Figure 2E-6.)

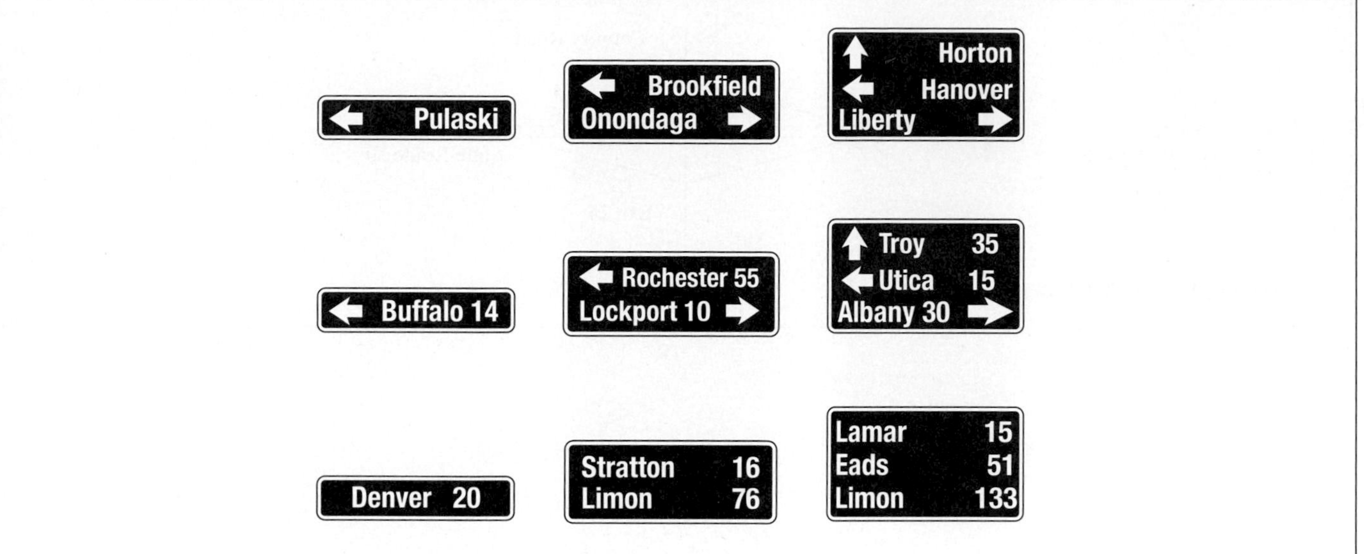

Figure 17.19: Illustration of Conventional Road Guide Signs

(*Source: Manual of Uniform Traffic Control Devices,* Draft, Federal Highway Administration, Washington DC, December 2007, Figure 2D-8.)

Defense Highways," American Association of State Highway and Transportation Officials, Washington DC, adopted August 14, 1954, revised August 10, 1973.

4. Weingroff, R.F., "From Names to Numbers: The Origins of the U.S. Numbered Highway System," *AASHTO Quarterly,* American Association of State Highway and Transportation Officials, Washington DC, Spring 1997.

Problems

17-1. A three-lane rural highway has 12-foot lanes and a speed limit of 55 mi/h. There is no passing permitted in the direction with a single lane. What is the minimum length of the transition and buffer markings at a location where the center-lane direction is to be switched?

17-2. An expressway in a suburban area has a design speed of 65 mi/h and an 85th percentile speed of 72 mi/h. It is experiencing a high accident rate compared to similar highways in the same jurisdiction. What speed limit would you recommend? What additional information would you like to have before making such a recommendation?

17-3. This class project should be assigned to groups with a minimum of two persons in each group. Select a 5-mile section of freeway or rural highway in your area. Survey the section in both directions, making note of all traffic signs and markings that exist. Evaluate the effectiveness of these signs and markings, and suggest improvements that might result in better communication with motorists. Write a report on your findings, including photographs where appropriate to illustrate your comments.

17-4. Figure 17.20 illustrates a diamond interchange between a state-numbered freeway and a county road. The diamond–county road intersections are STOP controlled. Indicate what guide signs and route signs you would place, specifying their location. Prepare a rough sketch of each sign to indicate its precise content. Note that there are no other exits on State Route 50 within 25 miles of this location.

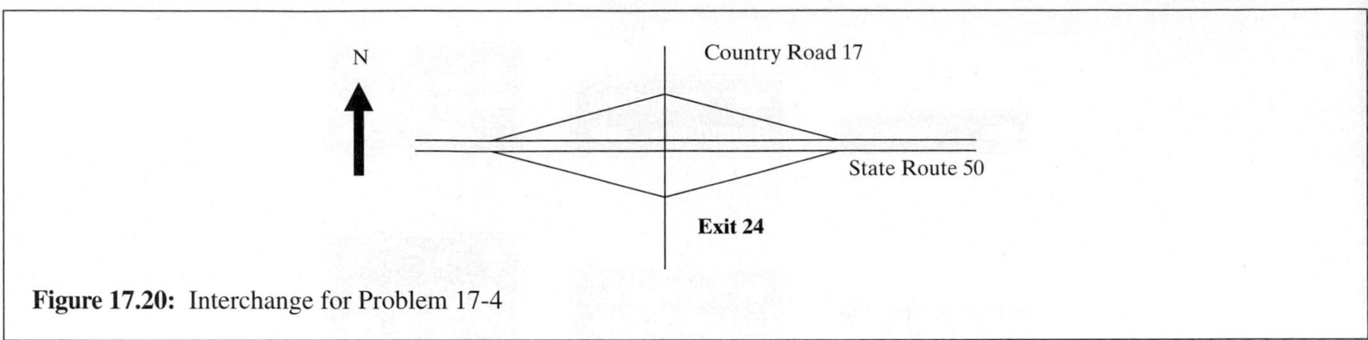

Figure 17.20: Interchange for Problem 17-4

PART 4

The Intersection

CHAPTER 18

The Hierarchy of Intersection Control

The most complex individual locations within any street and highway system are at-grade intersections. At a typical intersection of two two-way streets, there are 12 legal vehicular movements (left turn, through, and right turn from four approaches) and four legal pedestrian crossing movements. As indicated in Figure 18.1, these movements create many potential conflicts where vehicles and/or pedestrian paths may try to occupy the same physical space at the same time.

As illustrated, there are a total of 16 potential vehicular crossing conflicts: four between through movements from the two streets, four between left-turning movements from the two streets, and eight between left-turning movements and through movements from the two streets. In addition, there are eight vehicular merge conflicts as right- and left-turning vehicles merge into a through flow at the completion of their desired maneuver. Pedestrians add additional potential conflicts to the mix.

The critical task of the traffic engineer is to control and manage these conflicts in a manner that ensures safety and provides for efficient movement through the intersection for both motorists and pedestrians.

Three basic levels of control can be implemented at an intersection:

- *Level I*—Basic rules of the road
- *Level II*—Direct assignment of right-of-way using YIELD or STOP signs
- *Level III*—Traffic signalization

There are variations within each level of control as well. The selection of an appropriate level of control involves a determination of which (and how many) conflicts a driver should be able to perceive and avoid through the exercise of judgment. Where it is not reasonable to expect a driver to perceive and avoid a particular conflict, traffic controls must be imposed to assist.

Two factors affect a driver's ability to avoid conflicts: (1) a driver must be able to see a potentially conflicting vehicle or pedestrian in time to implement an avoidance maneuver, and (2) the volume levels that exist must present reasonable opportunities for a safe maneuver to take place. The first involves considerations of sight distance and avoidance maneuvers, and the second involves an assessment of demand intensity, the complexity of potential conflicts that exist at a given intersection, and finally, the gaps available in major movements.

A rural intersection of two farm roads contains all of the potential conflicts illustrated in Figure 18.1. However, pedestrians are rare, and vehicular flows may be extremely low. There is a low probability of any two vehicles and/or pedestrians attempting to use a common physical point simultaneously. At the junction between two major urban arterials, the probability of vehicles or pedestrians on conflicting paths arriving simultaneously is quite high. The sections that follow discuss how a determination of an appropriate form of intersection control can be made, highlighting the important factors to consider in making such critical decisions.

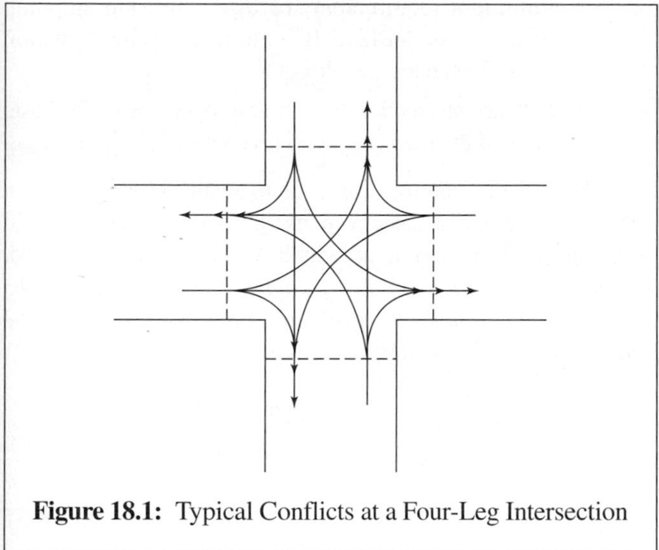

Figure 18.1: Typical Conflicts at a Four-Leg Intersection

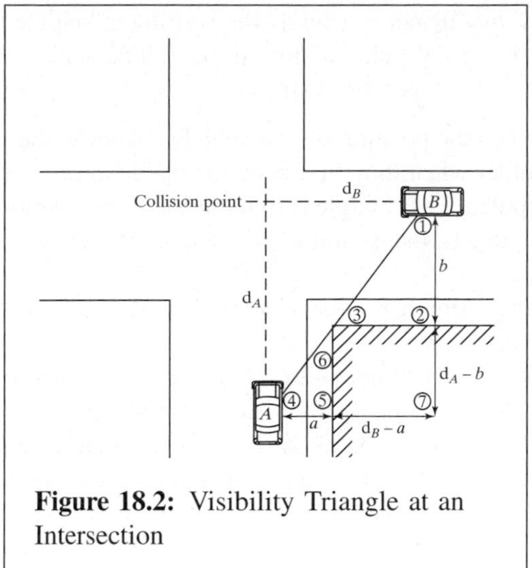

Figure 18.2: Visibility Triangle at an Intersection

18.1 Level I Control: Basic Rules of the Road

Basic rules of the road apply at any intersection where right-of-way is not explicitly assigned through the use of traffic signals, STOP, or YIELD signs. These rules are spelled out in each state's vehicle and traffic law, and drivers are expected to know them. At intersections, all states follow a similar format. In the absence of control devices, the driver on the left must yield to the driver on the right when the vehicle on the right is approaching in a manner that may create an impending hazard. In essence, the responsibility for avoiding a potential conflict is assigned to the vehicle on the left. Most state codes also specify that through vehicles have the right-of-way over turning vehicles at uncontrolled intersections.

Operating under basic rules of the road does not imply that no control devices are in place at or in advance of the intersection, although that could be the case. Use of street-name signs, other guide signs, or advance intersection warning signs do not change the application of the basic rules. They may, however, be able to contribute to the safety of the operation by calling the driver's attention to the existence and location of the intersection.

To safely operate under basic rules of the road, drivers on conflicting approaches must be able to see each other in time to assess whether an "impending hazard" is imposed and to take appropriate action to avoid an accident. Figure 18.2 illustrates a visibility triangle at a typical intersection. Sight distances must be analyzed to ensure that they are sufficient for drivers to judge and avoid conflicts.

At intersections, sight distances are normally limited by buildings or other sight-line obstructions located on or near the corners. There are, of course, four sight triangles at every intersection with four approaches.

All must provide adequate visibility for basic rules of the road to be considered. At the point where the drivers of both approaching vehicles first see each other, Vehicle A is located a distance of d_A from the collision or conflict point, and Vehicle B is located a distance d_B from the collision point. The sight triangle must be sufficiently large to ensure that at no time could two vehicles be on conflicting paths at distances and speeds that might lead to an accident, without sufficient time and distance being available for either driver to take evasive action.

Note that the sight line forms three similar triangles with sides of the sight obstruction: Δ123, Δ147, and Δ645. From the similarity of the triangles, a relationship between the critical distances in Figure 18.2 can be established:

$$\frac{b}{d_B - a} = \frac{d_A - b}{a}$$

$$d_B = \frac{a\,d_A}{d_A - b} \qquad (18\text{-}1)$$

Where: d_A = distance from Vehicle A to the collision point, ft

d_B = distance from Vehicle B to the collision point, ft

a = distance from driver position in Vehicle A to the sight obstruction, measured parallel to the path of Vehicle B, ft

b = distance from driver position in Vehicle B to the sight obstruction, measured parallel to the path of Vehicle A, ft

Thus, when the position of one vehicle is known, the position of the other when they first become visible to each other can be computed. The triangle is dynamic, and the position of one vehicle affects the position of the other when visibility is achieved.

The American Association of State Highway and Transportation Officials (AASHTO) suggests that to ensure safe operation with no control, both drivers should be able to stop before reaching the collision point when they first see each other. In other words, both d_A and d_B should be equal to or greater than the safe stopping distance at the points where visibility is established. AASHTO standards [*1*] suggest that a driver reaction time of 2.5 seconds be used in estimating safe stopping distance and that the 85th percentile speed of immediately approaching vehicles be used. AASHTO does suggest, however, that drivers slow from their midblock speeds when approaching uncontrolled intersections, and it recommends use of an immediate approach speed that is assumed to be lower than the design speed of the facility. From Chapter 2, the safe stopping distance is given by:

$$d_s = 1.47\, S_i t + \frac{S_i^2}{30(0.348 \pm 0.01G)} \qquad (18\text{-}2)$$

where: d_s = safe stopping distance, ft
S_i = initial speed of vehicle, mi/h
G = grade, %
t = reaction time, s
0.348 = standard friction factor for stopping maneuvers

Using this equation, the following analysis steps may be used to test whether an intersection sight triangle meets these sight distance requirements:

1. Assume that Vehicle A is located one safe stopping distance from the collision point (i.e., $d_A = d_s$), using Equation 18-2. By convention, Vehicle A is generally selected as the vehicle on the *minor* street.
2. Using Equation 18-1, determine the location of Vehicle B when the drivers first see each other. This becomes the actual position of Vehicle B when visibility is established, d_{Bact}.
3. Because the avoidance rule requires that both vehicles have one safe stopping distance available, the minimum requirement for d_B is the safe stopping distance for Vehicle B, computed using Equation 18-2. This becomes d_{Bmin}.
4. For the intersection to be safely operated under basic rules of the road (i.e., with no control), $d_{Bact} \geq d_{Bmin}$.

Historically, another approach to ensuring safe operation with no control has also been used. In this case, to avoid collision from the point at which visibility is established, *Vehicle A must travel 18 feet past the collision point in the same time that Vehicle B travels to a point 12 feet before the collision point.* This can be expressed as:

$$\frac{d_A + 18}{1.47\, S_A} = \frac{d_B - 12}{1.47\, S_B}$$

$$d_B = (d_A + 18)\frac{S_B}{S_A} + 12 \qquad (18\text{-}3)$$

where all variables are as previously defined. This, in effect, provides another means of estimating the minimum required distance, d_{Bmin}. In conjunction with the four-step analysis process outlined previously, it can also be used as a criterion to ensure safe operation.

At any intersection, all of the sight triangles must be checked and must be safe to implement basic rules of the road. If, for any of the sight triangles, $d_{Bact} < d_{Bmin}$, then operation with no control cannot be permitted. When this is the case, there are three potential remedies:

- Implement intersection control, using STOP- or YIELD-control, or traffic signals.
- Lower the speed limit on the major street to a point where sight distances are adequate.
- Remove or reduce sight obstructions to provide adequate sight distances.

The first is the most common result. The exact form of control implemented would require consideration of warrants and other conditions, as discussed in subsequent portions of this chapter. The second approach is viable where sight distances at series of uncontrolled intersections can be remedied by a reduced but still reasonable speed limit. The latter depends on the type of obstruction and ownership rights.

Consider the intersection illustrated in Figure 18.3. It shows an intersection of a one-way minor street and a two-way major street. In this case, two sight triangles must be analyzed. The 85th percentile immediate approach speeds are shown.

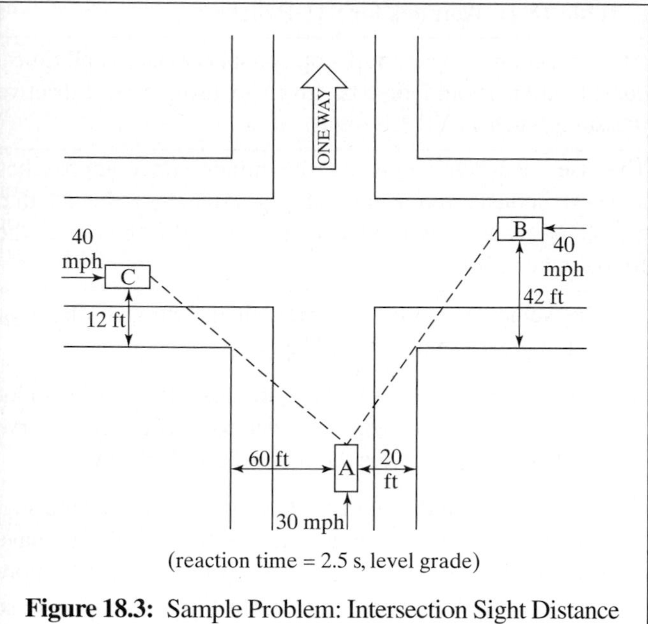

Figure 18.3: Sample Problem: Intersection Sight Distance

First, it is assumed that Vehicle A is one safe stopping distance from the collision point:

$$d_A = 1.47 * 30 * 2.5 + \frac{30^2}{30(0.348 + 0)}$$
$$= 110.3 + 86.2 = 196.5 \text{ ft}$$

where 2.5 s is the standard driver reaction time used in safe stopping sight distance computations. Using Equation 18-1, the actual position of Vehicle B when it is first visible to the driver of Vehicle A is found:

$$d_{Bact} = \frac{a\, d_A}{d_A - b} = \frac{20 * 196.5}{196.5 - 42} = \frac{3{,}930}{154.5} = 25.4 \text{ ft}$$

This must be compared with the minimum requirement for d_B, estimated as either one safe stopping distance (Equation 18-2), or using Equation 18-3:

$$d_{Bmin} = 1.47 * 40 * 2.5 + \frac{40^2}{30(0.348 + 0)}$$
$$= 147.0 + 153.3 = 300.3 \text{ ft}$$

or:

$$d_{Bmin} = (196.5 + 18)\frac{40}{30} + 12 = 298.0 \text{ ft}$$

In this case, both of the minimum requirements are similar, and both are far larger than the actual distance of 25.4 ft. Thus the sight triangle between Vehicles A and B fails to meet the criteria for safe operation under basic rules of the road.

Consider the actual meaning of this result. Clearly, if Vehicle A is 196.5 feet away from the collision point when Vehicle B is only 25.4 feet away from it, they will not collide. Why, then, is this condition termed "unsafe?" It is unsafe because there could be a Vehicle B, further away than 25.4 feet, on a collision path with Vehicle A and the drivers would not be able to see each other.

Because the sight triangle between Vehicles A and B did not meet the sight-distance criteria, it is not necessary to check the sight triangle between vehicles A and C. Basic rules of the road may not be permitted at this intersection without reducing major street speeds or removing sight obstructions. This implies that, in many cases, YIELD or STOP control should be imposed on the minor street as a minimum form of control.

Even if the intersection met the sight-distance criteria, this does not mean that basic rules of the road should be applied to the intersection. Adequate sight distance is a *necessary*, but *not sufficient*, condition for adopting a "no-control" option. Traffic volumes or other conditions may make a higher level of control desirable or necessary.

18.2 Level II Control: YIELD and STOP Control

If a check of the intersection sight triangle indicates it would not be safe to apply the basic rules of the road, then as a minimum, some form of level II control is often imposed. Even if sight distances are safe for operating under no control, there may be other reasons to implement a higher level of control as well. Usually, these would involve the intensity of traffic demand and the general complexity of the intersection environment.

The *Manual of Uniform Traffic Control Devices* (MUTCD) [2] gives some guidance as to conditions for which imposition of STOP or YIELD control is justified. It is not very specific, and it requires the exercise of engineering judgment. The warrants shown are taken from the draft of the 2010 MUTCD, which has been available for review on line since December 2007. At this writing, final approval to the contents of this edition has not been obtained. Final approval and official publication is expected in late 2009 or early 2010.

The MUTCD gives some very general "guidance" for the imposition of *either* STOP signs *or* YIELD signs. These shown in Table 18.1.

Table 18.1: Warrants for Using 2-Way STOP or YIELD Control at an Intersection

STOP or YIELD signs should be used at an intersection if one or more of the following conditions exist:

A. An intersection of a less important road with a main road where application of the normal right-of-way rule would not be expected to provide reasonable compliance with the law;

B. A street entering a designated through highway; and/or

C. An unsignalized intersection within a signalized area.

(*Source: Manual on Uniform Traffic Control Devices,* Draft, Federal Highway Administration, Washington DC, December 2007, p. 70, available at www.fhwa.com.)

These are very general. The first condition simply addresses a situation in which the sight triangle is insufficient to provide for safety. STOP or YIELD signs can be used to help establish a major or through road. If all unsignalized approaches to a major road are controlled by STOP or YIELD signs, through drivers have a clear right-of-way. The last condition addresses a situation in which virtually *all* intersections in an area or along an arterial are signalized. If a few isolated locations do not need to be signalized, then they should *at least* have STOP or YIELD signs.

Table 18.2: Warrants for STOP Signs

At intersections where a full stop is not necessary at all times, consideration should first be given to using less restrictive measures, such as YIELD signs.

The use of STOP signs on the minor street approaches should be considered if engineering judgment indicates that a stop is always required because of one or more of the following conditions:

A. The vehicular traffic volumes on the through street or highway exceed 6,000 veh/day;

B. A restricted view exists that requires road users on the minor street approach to stop in order to adequately observe conflicting traffic on the through street or highway.

C. Crash records indicate that 3 or more crashes that are susceptible to correction by installation of a STOP sign have been reported within a 12-month period, or that 5 or more such crashes have been reported within a 2-year period. Such crashes include right-angle collisions involving road users on the minor street approach failing to yield the right-of-way to traffic on the through street or highway.

(*Source: Manual of Uniform Traffic Control Devices,* Draft, Federal Highway Administration, Washington DC, December 2007, p. 71, available at www.fhwa.com.)

18.2.1 Two-Way Stop Control

The most common form of Level II control is the two-way STOP sign. In fact, such control may involve one or two STOP signs, depending on the number of intersection approaches. It is not all-way STOP control, which is discussed later in this chapter.

Under the heading of "guidance," the MUTCD suggests several conditions under which the use of STOP signs would be justified. Table 18.2 shows these warrants.

Warrant A establishes a reasonable level of major street traffic that would require use of a STOP sign to allow minor-street drivers to select an appropriate gap in a busy traffic stream. Warrant B merely restates the need for STOP (or YIELD) control where a sight triangle at the intersection is found to be inadequate. Warrant C establishes criteria for using a STOP sign to correct a perceived accident problem.

The MUTCD is somewhat more explicit in dealing with inappropriate uses of the STOP sign. Under the heading of a "standard" (i.e., a mandatory condition), STOP (or YIELD) signs *shall not* be installed at intersections where traffic control signals are installed and operating, except where signal operation is a flashing red at all times, or where a channelized right turn exists. This disallows a past practice in which some jurisdictions turned signals off at night, leaving STOP signs in place for the evening hours. During the day, however, an unfamiliar driver approaching a green signal with a STOP sign could become significantly confused. The manual also disallows the use of portable or part-time STOP signs except for emergency and temporary traffic control.

Under the heading of "guidance," STOP signs *should not* be used for speed control, although this is frequently done on local streets designed in a straight grid pattern. In modern designs, street layout and geometric design would be used to discourage excessive speeds on local streets.

In general, STOP signs should be installed in a manner that minimizes the number of vehicles affected, which generally means installing them on the minor street.

AASHTO [*1*] also provides sight distance criteria for STOP-controlled intersections. A methodology based on observed gap acceptance behavior of drivers at STOP-controlled intersections is used. A standard stop location is assumed for the minor street vehicle (Vehicle A in Figure 18.2). The distance to the collision point (d_A) has three components:

- Distance from the driver's eye to the front of the vehicle (assumed to be 8 feet)
- Distance from the front of the vehicle to the curb line (assumed to be 10 feet)
- Distance from the curb line to the center of the right-most travel lane approaching from the left, or from the curb line to the left-most travel lane approaching from the right

Thus:

$$d_{A\text{-}STOP} = 18 + d_{cl} \qquad (18\text{-}4)$$

where: $d_{A\text{-}STOP}$ = distance of Vehicle A on a STOP-controlled approach from the collision point, ft

d_{cl} = distance from the curb line to the center of the closest travel lane from the direction under consideration, ft

The required sight distances for Vehicle B, on the major street for STOP-controlled intersections is found as follows:

$$d_{Bmin} = 1.47 * S_{maj} * t_g \qquad (18\text{-}5)$$

where: d_{Bmin} = minimum sight distance for Vehicle B approaching on major (uncontrolled) street, ft

S_{maj} = design speed of major street, mi/h

t_g = average gap accepted by minor street driver to enter the major road, s

Average gaps accepted are best observed in the field for the situation under study. In general, they range from 6.5 seconds to 12.5 seconds depending on the minor street movement and vehicle type, as well as some of the specific geometric conditions that exist.

For most STOP-controlled intersections, the design vehicle is the passenger car, and the criteria for left-turns are used because they are the most restrictive. Trucks or combination vehicles are considered only when they make up a substantial proportion of the total traffic on the approach. Values for right-turn and through movements are used when no left-turn movement is present. For these typical conditions, AASHTO recommends the use of $t_g = 7.5$ s.

Consider the case of a STOP-controlled approach at an intersection with a two-lane arterial with a design speed of 40 mi/h, as shown in Figure 18.4.

Using Equation 18-4, the position of the stopped vehicle on the minor approach can be determined.

$$d_{A\text{-}STOP}(\text{from left}) = 18.0 + 6.0 = 24.0 \text{ ft}$$
$$d_{A\text{-}STOP}(\text{from right}) = 18.0 + 18.0 = 36.0 \text{ ft}$$

The minimum sight distance requirement for Vehicle B is determined from Equation 18-5, using a time gap (t_g) of 7.5 seconds for typical conditions.

$$d_{Bmin} = 1.47 * 40 * 7.5 = 441 \text{ ft}$$

Now the actual distance of Vehicle B from the collision point when visibility is established is determined using Equation 18-1:

$$d_{Bact}(\text{from left}) = \frac{36 * 24}{24 - 20} = 216 \text{ ft} < 441 \text{ ft}$$

$$d_{Bact}(\text{from right}) = \frac{16 * 34}{36 - 35} = 576 \text{ ft} > 441 \text{ ft}$$

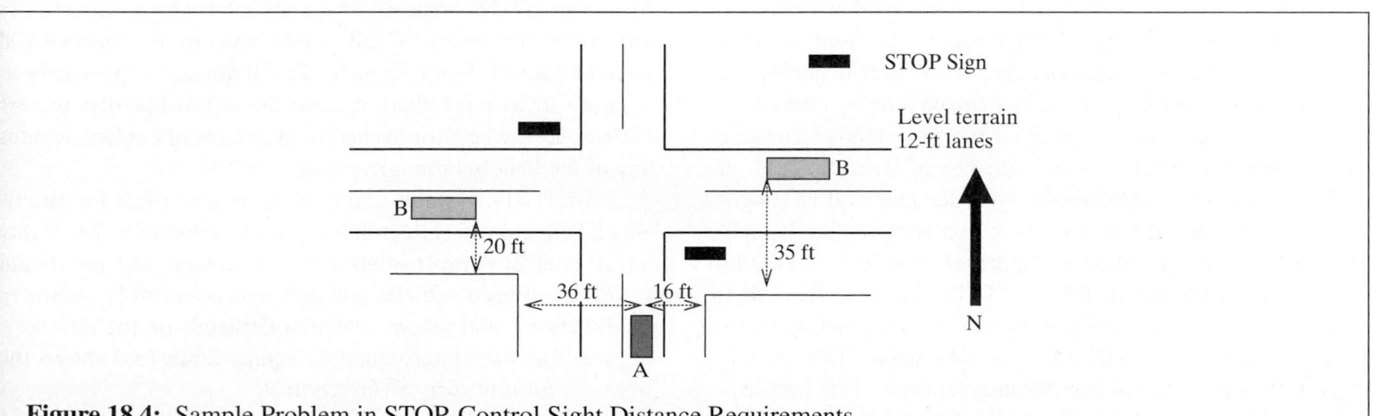

Figure 18.4: Sample Problem in STOP-Control Sight Distance Requirements

In the case of a major street Vehicle B approaching from the left, there is not sufficient sight distance to meet the criteria. The sight distance for Vehicle B approaching from the right meets the criteria. Note that it is possible for d_{Bact} to be negative. This would indicate there was no sight obstruction from the direction analyzed.

Where the STOP-sign sight-distance criterion is not met, it is recommended that speed limits be reduced (with signs posted) to a level that would allow appropriate sight distance to the minor street. Removal or cutting back of sight obstructions is also a potential solution, but this is often impossible in developed areas, where buildings are the principal obstructions.

18.2.2 Yield Control

A YIELD sign assigns right-of-way to the major uncontrolled street. It requires vehicles on the minor approach(es) to slow and yield the right-of-way to any major street vehicle approaching at a distance and speed that would present an impending hazard to the minor street vehicle if it entered the major street. Most state laws require that drivers on YIELD-controlled approaches slow to 8 to 10 mi/h before entering the major street.

Warrants for YIELD control in the MUTCD are hardly definitive, and they are given only under the heading of "options," except for one relatively new mandatory usage. The warrants are summarized in Table 18.3.

The principal uses of the YIELD sign emanate from their mandatory use at roundabouts and Warrants B, C, and E. Warrant B is a common application where medians exist and are wide enough to store at least one crossing vehicle. In such cases, a vehicle crosses the first set of lanes, and may stop again in the median to seek another gap to cross the second set of lanes. Warrant C allows use of the YIELD sign to control channelized right turns at signalized and unsignalized intersections, and Warrant E allows their use at on-ramp or other merge situations. The latter is a frequent use in which adequate sight distance or geometry (i.e., inadequate length of the acceleration lane) make an uncontrolled merge potentially unsafe.

There has been some controversy over the use of YIELD signs at normal crossings. Because YIELD signs require drivers to slow down, the sight triangle may be analyzed using the legal reduced approach speed. In 2000, the Millennium Edition of the MUTCD required that sight distance sufficient for safety at the normal approach speed be present whenever a YIELD sign was used. This greatly discouraged their use at regular intersections. This prescription is expected to be removed in the forthcoming 2010 edition of the manual.

Table 18.3: Warrants for YIELD Signs

A YIELD sign *shall* be used to assign right-of-way at the entrance to a roundabout. YIELD signs at roundabouts *shall* be used to control the approach roadways and *shall not* be used to control the circulatory roadway.

YIELD signs may be installed:

A. On approaches to a through street or highway where conditions are such that a stop is not always required.

B. At the second crossroad of a divided highway, where the median width at the intersection is 30 ft or greater. In this case, a STOP or YIELD sign may be installed at the entrance to the first roadway, and a YIELD sign may be installed at the entrance to the second roadway.

C. On a channelized turn lane that is separated from the adjacent travel lane by an island, even if the adjacent lanes at the intersection are controlled by a highway traffic control signal or by a STOP sign.

D. At an intersection where a special problem exists and where engineering judgment indicates the problem to be susceptible to correction by the use of YIELD signs.

E. Facing the entering roadway for a merge-type movement if engineering judgment indicates that the control is needed because acceleration geometry and/or sign distance is not sufficient for merging traffic operation.

(*Source: Manual of Uniform Traffic Control Devices,* Draft, Federal Highway Administration, Washington DC, p. 73, available at www.fhwa.com.)

18.2.3 Multiway Stop Control

Multiway STOP control, where all intersection approaches are controlled using STOP signs, remains a controversial form of control. Some agencies find it attractive, primarily as a safety measure. Others believe the confusion that drivers often exhibit when confronted by this form of control negates any of the benefits it might provide.

MUTCD warrants and provisions with regard to multiway STOP control reflect this ongoing controversy. Multiway STOP control is most often used where there are significant conflicts between vehicles and pedestrians and/or bicyclists in all directions, and where vehicular demands on the intersecting roadways are approximately equal. Table 18.4 shows the warrants for multiway STOP control.

Note that such control is generally implemented as a safety measure because operations at such locations are often

Table 18.4: Warrants for Multiway STOP Signs

The following criteria should be considered in the engineering study for a multiway STOP sign:
A. Where traffic control signals are justified, the multiway STOP is an interim measure that can be installed quickly to control traffic while arrangements are being made for the installation of the traffic control signal.
B. Five or more reported crashes in a 12-month period that are susceptible to correction by a multiway STOP installation. Such crashes include right- and left-turn collisions as well as right-angle collisions.
C. Minimum volumes:
1. The vehicular volume entering the intersection from the major street approaches (total of both approaches) averages at least 300 veh/h for any 8 hours of an average day, and
2. The combined vehicular, pedestrian, and bicycle volume entering the intersection from the minor street approaches (total of both approaches) averages at least 200 units/h for the same 8 hours, with an average delay to minor-street vehicular traffic of at least 30 s/veh during the highest hour, but
3. If the 85th percentile approach speed of the major highway exceeds 40 mi/h, the minimum vehicular volume warrants are 70% of the above values.
D. Where no single criterion is satisfied, but where criteria B., C1. and C2 are all satisfied to 80% of the minimum values. Criterion C3 is excluded from this condition.

(*Source: Manual on Uniform Traffic Control Devices,* Draft, Federal Highway Administration, Washington DC, December 2007, p. 72, available at www.fhwa.com.)

not very efficient. The fourth edition of the *Highway Capacity Manual,* [*3*] includes a methodology for analysis of the capacity and level of service provided by multiway STOP control.

18.3 Level III Control: Traffic Control Signals

The ultimate form of intersection control is the traffic signal. Because it alternately assigns right-of-way to specific movements, it can substantially reduce the number and nature of intersection conflicts as no other form of control can.

If drivers obey the signal, then driver judgment is not needed to avoid some of the most critical intersection conflicts. Imposition of traffic signal control does not, however, remove all conflicts from the realm of driver judgment. At two-phase signals, where all left-turns are made against an opposing vehicular flow, drivers must still evaluate and select gaps in opposing traffic through which to safely turn. At virtually all signals, some pedestrian-vehicle and bicycle-vehicle conflicts remain between legal movements, and driver vigilance and judgment are still required to avoid accidents. Nevertheless, drivers at signalized intersections do not have to negotiate the critical conflicts between crossing vehicle streams, and where exclusive left-turn phases are provided, critical conflicts between left turns and opposing through vehicles are also eliminated through signal control. This chapter deals with the issue of whether or not signal control is warranted or needed. Given that it is needed, Chapter 20 deals with the design of a specific phasing plan and the timing of the signal.

Although warrants and other criteria for STOP and YIELD signs are somewhat general in the MUTCD, warrants for signals are quite detailed. The cost involved in installation of traffic signals (e.g., power supply, signal controller, detectors, signal heads, and support structures, and other items) is considerably higher than for STOP or YIELD signs and can run into the hundreds of thousands of dollars for complex intersections. Because of this, and because traffic signals introduce a fixed source of delay into the system, it is important that they not be overused; they should be installed only where no other solution or form of control would be effective in assuring safety and efficiency at the intersection.

18.3.1 Advantages of Traffic Signal Control

The Millennium Edition of the MUTCD lists the following advantages of traffic control signals that are "properly designed, located, operated, and maintained" [MUTCD, Millennium Edition, p. 4b-2]. These advantages include:

1. They provide for the orderly movement of traffic.
2. They increase the traffic-handling capacity of the intersection if proper physical layouts and control measures are used and if the signal timing is reviewed and updated on a regular basis (every two years) to ensure that it satisfies the current traffic demands.
3. They reduce the frequency and severity of certain types of crashes, especially right-angle collisions.

4. They are coordinated to provide for continuous or nearly continuous movement at a definite speed along a given route under favorable conditions.
5. They are used to interrupt heavy traffic at intervals to permit other traffic, vehicular or pedestrian, to cross.

These specific advantages address the primary reasons why a traffic signal would be installed: to increase capacity (thereby improving level of service), to improve safety, and to provide for orderly movement through a complex situation. Coordination of signals provides other benefits, but not all signals are necessarily coordinated.

18.3.2 Disadvantages of Traffic Signal Control

The description of the second advantage in the earlier list indicates that capacity is increased by a well-designed signal at a well-designed intersection. Poor design of either the signalization or the geometry of the intersection can significantly reduce the benefits achieved or negate them entirely. Improperly designed traffic signals, or the placement of a signal where it is not justified, can lead to some of the following disadvantages [MUTCD, Millennium Edition, p. 4B-3]:

1. Excessive delay
2. Excessive disobedience of the signal indications
3. Increased use of less adequate routes as road users attempt to avoid the traffic control signal
4. Significant increases in the frequency of collisions (especially rear-end collisions)

Item 4 is of some interest. Even when they are properly installed and well designed, traffic signal controls can lead to increases in rear-end accidents because of the cyclical stopping of traffic.

Where safety is concerned, signals can reduce the number of right-angle, turning, and pedestrian/bicycle accidents; they might cause an increase in rear-end collisions (which tend to be less severe); they will have almost no impact on head-on or sideswipe accidents, or on single-vehicle accidents involving fixed objects.

Excessive delay can result from an improperly installed signal, but it can also occur if the signal timing is inappropriate. In general, excessive delay results from cycle lengths that are either too long or too short for the existing demands at the intersection. Further, drivers tend to assume that a signal is broken if they experience an excessive wait, particularly when there is little or no demand occurring on the cross street.

18.3.3 Warrants for Traffic Signals

The forthcoming 2009/2010 MUTCD specifies nine different warrants that justify the installation of a traffic signal. The ninth is just being added in this edition. It covers the installation of a signal in coordination with a railroad crossing. Satisfying one or more of the warrants for signalization does *not* require or justify the installation of a signal. The manual *requires,* however, that a comprehensive engineering study be conducted to determine whether or not installation of a signal is justified. The study *must* include applicable factors reflected in the specified warrants but could extend to other factors as well. However, traffic signal control *should not* be implemented if none of the warrants are met. The warrants, therefore, still require the exercise of engineering judgment. In the final analysis, if engineering studies and/or judgment indicate that signal installation *will not* improve the overall safety or operational efficiency at a candidate location, it should not be installed.

Although offered only under the heading of an option, the MUTCD suggests that the following data be included in an engineering study of the need for a traffic signal [2009/2010 MUTCD, Draft, p. 268]:

1. The number of vehicles entering the intersection from each approach during 12 hours of an average day. It is desirable that the hours selected contain the greatest percentage of the 24-hour traffic volume.
2. Vehicular volumes for each traffic movement, from each approach, classified by vehicle type (heavy trucks, passenger cars and light trucks, public-transit vehicles, and in some locations, bicycles), during each 15-minute period of the 2 hours in the morning and 2 hours in the afternoon during which total traffic entering the intersection is greatest.
3. Pedestrian volume counts on each crosswalk during the same periods as the vehicular counts in Item 2 above and during hours of highest pedestrian volume. Where young, elderly, and/or persons with physical or visual disabilities need special consideration, the pedestrians and their crossing times may be classified by general observation.
4. Information about nearby facilities and activity centers that serve the young, elderly, and/or persons with disabilities, including requests from persons with disabilities for accessible parking improvements at the location under study. These persons might not be adequately reflected in the pedestrian volume count if the absence of a signal restrains their mobility.

5. The posted or statutory speed limit or the 85th percentile speed on the uncontrolled approaches to the location.
6. A condition diagram showing details of the physical layout, including such features as intersection geometrics, channelization, grades, sight distance restrictions, transit stops and routes, parking conditions, pavement markings, roadway lighting, driveways, nearby railroad crossings, distance to nearest traffic control signals, utility poles and fixtures, and adjacent land use.
7. A collision diagram showing crash experience by type, location, direction of movement, severity, weather, time of day, date, and day of week for at least one year.

MUTCD also recommends collection of stopped-time delay data and queuing information at some locations where these are thought to be problems.

This data will allow the engineer to fully evaluate whether or not the intersection satisfies the requirements of one or more of the following warrants:

- *Warrant 1:* Eight-Hour Vehicular Volume
- *Warrant 2:* Four-Hour Vehicular Volume
- *Warrant 3:* Peak Hour
- *Warrant 4:* Pedestrian Volume
- *Warrant 5:* School Crossing
- *Warrant 6:* Coordinated Signal System
- *Warrant 7:* Crash Experience
- *Warrant 8:* Roadway Network
- *Warrant 9:* Intersection Near a Highway-Rail Crossing

It also provides a sufficient base for the exercise of engineering judgment in determining whether a traffic signal should be installed at the study location. Each of these warrants is presented and discussed in the sections that follow.

In most cases, an engineering study includes data from an existing location. In some cases, however, consideration of signalization relates to a future situation or design. In such cases, forecast demand volumes may be used to compare with the criteria in the warrants.

Warrant 1: Eight-Hour Vehicular Volume

The eight-hour vehicular volume warrant represents a merging of three different warrants in the pre-2000 MUTCD (old Warrants 1, 2, and 8). It addresses the need for signalization for conditions that exist over extended periods of the day (a minimum of eight hours). Two of the most fundamental reasons for signalization are addressed:

- Heavy volumes on conflicting cross-movements that make it impractical for drivers to select gaps in an uninterrupted traffic stream through which to safely pass. This requirement is often referred to as the "minimum vehicular volume" condition (Condition A).
- Vehicular volumes on the major street are so heavy that no minor-street vehicle can safely pass through the major-street traffic stream without the aid of signals. This requirement is often referred to as the "interruption of continuous traffic" condition (Condition B).

Details of this warrant are shown in Table 18.5. The warrant is met when:

- Either Condition A or Condition B is met to the 100% level.
- Either Condition A or Condition B is met to the 70% level, where the intersection is located in an isolated community of population 10,000 or less, or where the major-street approach speed is 40 mi/h or higher.
- Both Conditions A and B are met to the 80% level.

Note that in applying these warrants, the major-street volume criteria are related to the total volume in both directions, whereas the minor-street volume criteria are applied to the highest volume in one direction. The volume criteria in Table 18.5 must be met for a minimum of eight hours on a typical day. The eight hours do not have to be consecutive, and they often involve four hours around the morning peak and four hours around the evening peak. Major- and minor-street volumes must be for the same eight hours, however.

Either of the intersecting streets may be treated as the "major" approach, but the designation must be consistent for a given application. If the designation of the "major" street is not obvious, a warrant analysis can be conducted considering each as the "major" street in turn. Although the designation of the major street may not be changed within any one analysis, the direction of peak one-way volume for the minor street need not be consistent.

The 70% reduction allowed for rural communities of population 10,000 or less reflects the fact that drivers in small communities have little experience in driving under congested situations. They will require the guidance of traffic signal control at volume levels lower than those for drivers more used to driving in congested situations. The same reduction applies where the major-street speed limit is 40 mi/h or greater. Because gap selection is more difficult through a higher-speed major-street flow, signals are justified at lower volumes.

Table 18.5: Warrant 1: Eight-Hour Vehicular Volume

Condition A: Minimum Vehicular Volume

Number of lines for moving traffic on each approach		Vehicles per hour on major street (total, both approaches)			Vehicles per hour on higher-volume minor street approach (one direction only)		
Major Street	Minor Street	100%	80%	70%	100%	80%	70%
1	1	500	400	350	150	120	105
2 or more	1	600	480	420	150	120	105
2 or more	2 or more	600	480	420	200	160	140
1	2 or more	500	400	350	200	160	140

Condition B: Interruption of Continuous Traffic

Number of lanes for moving traffic on each approach		Vehicles per hour on major street (total, both approaches)			Vehicles per hour on higher-volume minor street approach (one direction only)		
Major Street	Minor Street	100%	80%	70%	100%	80%	70%
1	1	750	600	525	75	60	53
2 or more	1	900	720	630	75	60	53
2 or more	2 or more	900	720	630	100	80	70
1	2 or more	750	600	525	100	80	70

(*Source:* Used with permission of Federal Highway Administration, US Department of Transportation, *Manual on Uniform Traffic Control Devices,* Millennium Edition, Table 4C-1, p. 4C-5, Washington DC, 2000.)

The various elements of the eight-hour vehicular volume warrant are historically the oldest of the warrants, having been initially formulated and disseminated in the 1930s.

Warrant 2: Four-Hour Vehicular Volume

The four-hour vehicular volume warrant was introduced in the 1970s to assist in the evaluation of situations where volume levels requiring signal control might exist for periods shorter than eight hours. Prior to the MUTCD Millennium Edition, this was old Warrant 9. Figure 18.5 shows the warrant, which is in the form of a continuous graph. Because this warrant is expressed as a continuous relationship between major and minor street volumes, it addresses a wide variety of conditions. Indeed, Conditions A and B of the eight-hour warrant represent two points in such a continuum for each configuration, but the older eight-hour warrant did not investigate or create criteria for the full range of potential conditions.

Figure 18.5 (a) is the warrant for normal conditions, and Figure 18.5 (b) reflects the 70% reduction applied to isolated small communities (with population less than 10,000) or where the major-street speed limit is above 40 mi/h. Because the four-hour warrant represents a continuous set of conditions, there is no need to include an 80% reduction for two discrete conditions within the relationship.

To test the warrant, the two-way major-street volume is plotted against the highest one-way volume on the minor street for each hour of the study period. To meet the warrant, at least four hours must plot *above* the appropriate decision curve. The three curves represent intersections of (1) two streets with one lane in each direction, (2) one street with one lane in each direction with another having two or more lanes in each direction, and (3) two streets with more than one lane in each direction. In Case (2), the distinction between which intersecting street has one lane in each direction (major or minor) is no longer relevant, except for the footnotes.

Warrant 3: Peak Hour

Warrant 3 addresses two critical situations that might exist for only one hour of a typical day. The first is a volume condition, similar in form to Warrant 2, and shown in Figure 18.6 (old Warrant 11). The second is a delay warrant (old Warrant 10). If either condition is satisfied, the peak-hour warrant is met.

The volume portion of the warrant is implemented in the same manner as the four-hour warrant. For each hour of

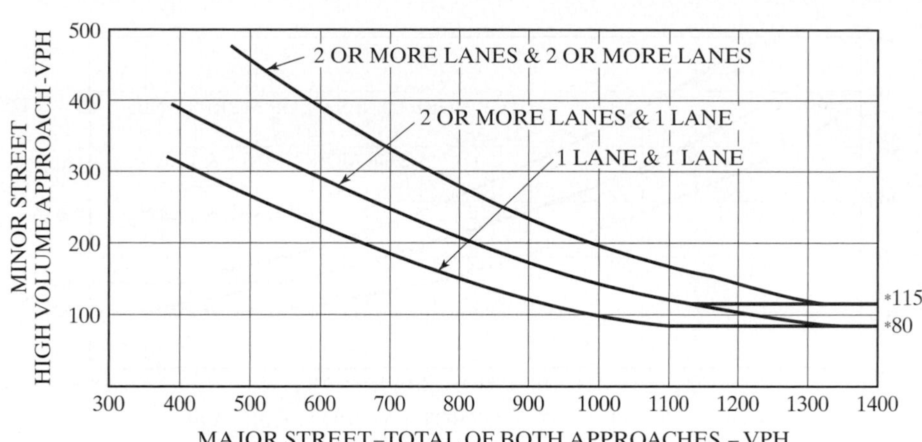

*Note: 115 vph applies as the lower threshold volume for a minor street approach with two or more lanes and 80 vph applies as the lower threshold volume for a minor street approach with one lane.

(a) Normal Conditions

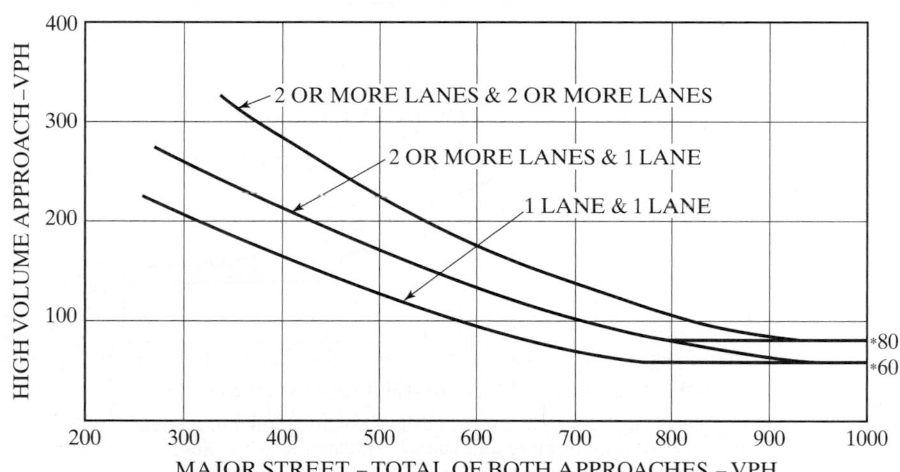

*Note: 80 vph applies as the lower threshold volume for a minor street approach with two or more lanes and 60 vph applies as the lower threshold volume for a minor street approach with one lane.

(b) Criteria for Small Communities (pop <10,000) or High Major Street Approach Speed (≥40 mi/h)

Figure 18.5: Warrant 2: Four-Hour Vehicular Volume

(*Source:* Used with permission of Federal Highway Administration, U.S. Department of Transportation, *Manual on Uniform Traffic Control Devices,* Millennium Edition, Figures 4C-1, 4C-2, p. 4C-7, Washington DC, 2000.)

the study, the two-way major street volume is plotted against the high single-direction volume on the minor street. For the Peak-Hour Volume Warrant, however, only one hour must plot above the appropriate decision line to meet the criteria. Criteria are given for normal conditions in Figure 18.6 (a), and the 70% criteria for small isolated communities and high major-street speeds are shown in Figure 18.6 (b). The Peak-Hour Delay Warrant is summarized in Table 18.6.

It is important to recognize that the delay portion of Warrant 3 applies only to cases in which STOP control is already in effect for the minor street. Thus delay during the peak hour is not a criterion that allows going from no control or YIELD control to signalization directly.

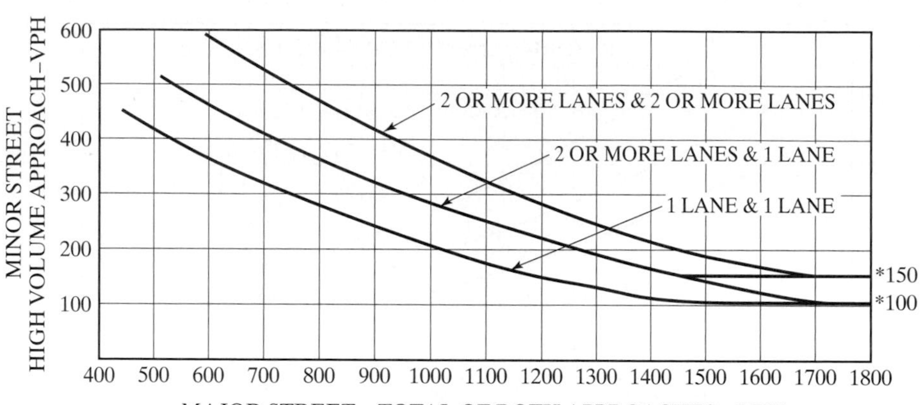

*Note: 150 vph applies as the lower threshold volume for a minor street approach with two or more lanes and 100 vph applies as the lower threshold volume for a minor street approach with one lane.

(a) Normal Conditions

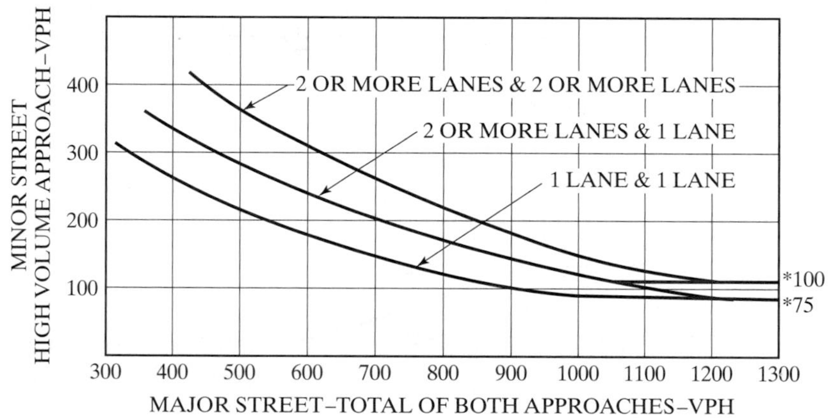

*Note: 100 vph applies as the lower threshold volume for a minor street approach with two or more lanes and 75 vph applies as the lower threshold volume for a minor street approach with one lane.

(b) Criteria for Small Communities (Pop <10,000) or High Major Street Approach Speed (≥40 mi/h)

Figure 18.6: Warrant 3A: Peak Hour Volume

(*Source:* Used with permission of Federal Highway Administration, U.S. Department of Transportation, *Manual on Uniform Traffic Control Devices,* Millennium Edition, Figures 4C-3, 4C-4, p. 4C-9, Washington DC, 2000.)

The MUTCD also emphasizes that the Peak-Hour Warrant should be applied only in special cases, such as office complexes, manufacturing plants, industrial complexes, or high-occupancy vehicle facilities that attract or discharge large numbers of vehicles over a short time.

Warrant 4: Pedestrians

The Pedestrian Warrant addresses situations in which the need for signalization is the frequency of vehicle-pedestrian conflicts and the inability of pedestrians to avoid such conflicts due to the volume of traffic present. Signals may be placed under this warrant at midblock locations, as well as at intersections.

This warrant is met when any four hourly plots of total pedestrians crossing the major street and the total major street vehicular traffic falls over the line in Figure 18.7 (a), or when any one similar hourly plot falls above the line in Figure 18.8 (a). If the location is in a built-up area of a small community (population less than 10,000) or where the posted

Table 18.6: Warrant 3B: Peak-Hour Delay

The need for a traffic control signal shall be considered if an engineering study finds that . . . all three of the following conditions exist for the same 1 hour (any four consecutive 15-minute periods) of an average day:

1. The total stopped-time delay experienced by traffic on one minor street approach (one direction only) controlled by a STOP sign equals or exceeds: 4 veh-hours for a one-lane approach; or 5 veh-hours for a two lane approach, and
2. The volume on the same minor street approach (one direction only) equals or exceeds 100 veh/h for one moving lane of traffic; or 150 veh/h for two moving lanes, and
3. The total entering volume serviced during the hour equals or exceeds 650 veh/h for intersections with three approaches, or 800 veh/h for intersections with four or more approaches.

(*Source: Manual of Uniform Traffic Control Devices,* Draft, Federal Highway Administration, Washington DC, 2007, p. 270.)

or statutory speed limit, or the 85th percentile approach speed exceeds 35 mi/h, Figures 18.7 (b) and 18.8 (b) may be used.

The figures address cases in which a steadier pedestrian flow over four hours requires signal control and the case in which a single peak hour has pedestrian-vehicle conflicts that must be signal controlled. The (b) figures apply the same 70% reduction in criteria that is used in conjunction with vehicular volume criteria in Warrants 1, 2, and 3.

If the traffic signal is justified at an intersection by this warrant only, it will usually be at least a semi-actuated signal (a full actuated signal is also a possibility at an isolated intersection) with pedestrian pushbuttons and signal heads for pedestrians crossing the major street. If it is within a coordinated signal system, it would also be coordinated into the system. If such a signal is located in midblock, it will always be pedestrian actuated, and parking and other sight restrictions should be eliminated within 20 feet of both sides of the crosswalk. Standard reinforcing markings and signs should also be provided.

If the intersection meets this warrant but also meets other vehicular warrants, any type of signal could be installed as appropriate to other conditions. Pedestrian signal heads would be required for major-street crossings. Pedestrian pushbuttons would be installed unless the vehicular signal timing safely accommodates pedestrians in every signal cycle.

A signal would not normally be implemented under this warrant if there is another signal within 300 feet of the location. Placement of a signal so close to another would only be permitted if did not disrupt progressive flow on the major street.

Pedestrian volume criteria may be reduced by as much as 50% if the 15th-percentile crossing speed is less than 3.5 mi/h, as might be the case where elderly, very young, or disabled pedestrians are present in significant percentages.

Warrant 5: School Crossing

This warrant is similar to the pedestrian warrant but is limited to application at designated school crossing locations, either at intersections or at midblock locations. The warrant requires the study of available gaps to see whether they are "acceptable" for children to cross through. An acceptable gap would include the crossing time, buffer time, and an allowance for groups of children to start crossing the street. The frequency of acceptable gaps should be no less than one for each minute during which school children are crossing. The minimum number of children crossing the major street is 20 during the highest crossing hour.

Traffic signals are rarely implemented under this warrant. Children do not usually observe and obey signals regularly, particularly if they are very young. Thus traffic signals would have to be augmented by crossing guards in most cases. Except in unusual circumstances involving a very heavily traveled major street, the crossing guard, perhaps augmented with STOP signs, would suffice under most circumstances without signalization. Where extremely high volumes of school children cross a very wide and heavily traveled major street, overpasses or underpasses should be provided with barriers preventing entry onto the street.

Warrant 6: Coordinated Signal System

Chapters 25 and 26 of this text addresses signal coordination and progression systems for arterials and networks. Critical to such systems is the maintenance of platoons of vehicles moving together through a "green wave" as they progress along an arterial. If the distance between two adjacent coordinated signals is too large, platoons begin to dissipate and the positive impact of the progression is sharply reduced. In such cases, the traffic engineer may place a signal at an intermediate intersection where it would not otherwise be warranted to reinforce the coordination scheme and to help maintain platoon coherence. The application of this warrant, shown in Table 18.7, should not result in signal spacing of

*Note: 107 pph applies as the lower threshold volume.

(a) Normal Criteria.

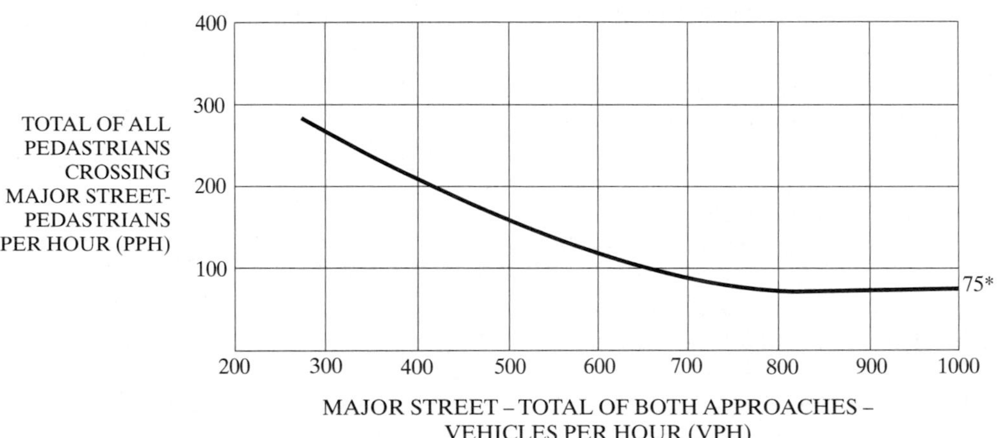

*Note: 75 pph applies as the lower threshold volume.

(b) Criteria for Small Communities (Pop <10,000) or High Major Street Approach Speed (> 35 mi/h)

Figure 18.7: Four-Hour Pedestrian Warrant

(*Source: Manual of Uniform Traffic Control Devices,* Draft, Federal Highway Administration, Washington DC, December 2007, Figures 4C-5 and 4C-7.)

less than 1,000 feet. Such signals, when placed, are often referred to as "spacer signals."

The two criteria are similar but not exactly the same. Inserting a signal in a one-way progression is always possible without damaging the progression. On a two-way street, it is not always possible to place a signal that will maintain the progression in both directions acceptably. This issue is discussed in greater detail in Chapter 25.

Warrant 7: Crash Experience

The Crash Experience Warrant addresses cases in which a traffic control signal would be installed to alleviate an observed high-accident occurrence at the intersection. The criteria are summarized in Table 18.8.

The requirement for an adequate trial of alternative methods means that either YIELD or STOP control is already

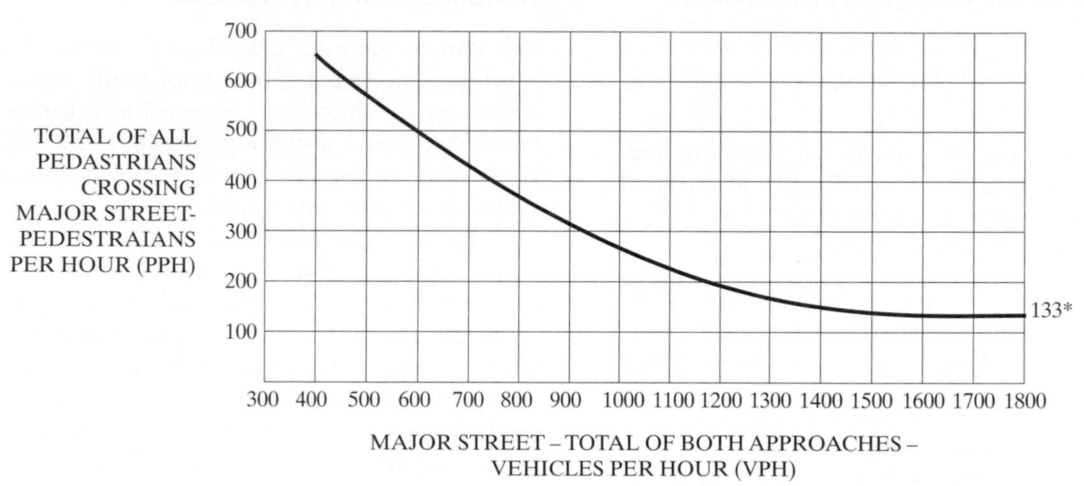

*Note: 133 pph applies as the lower threshold volume.

(a) Normal Criteria.

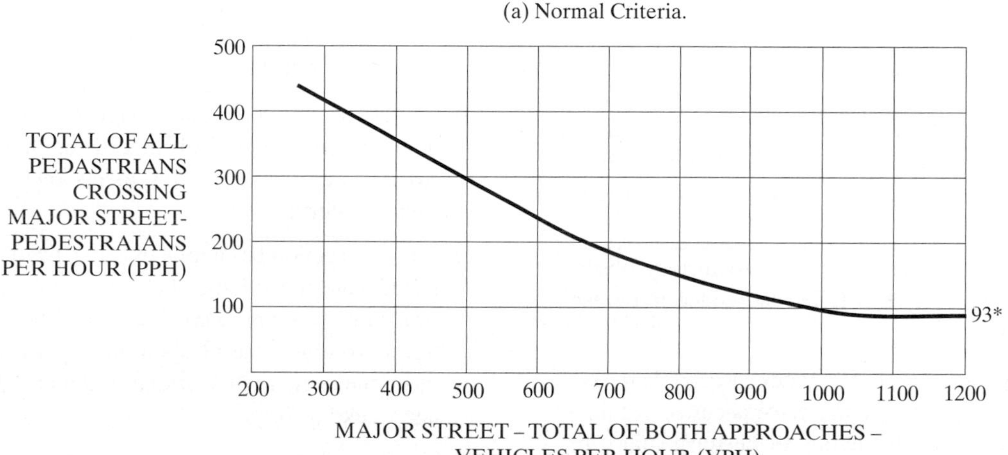

(b) Criteria for Small Communities (Pop <10,000) or High Major Street Approach Speed (> 35 mi/h)

Figure 18.8: Peak-Hour Pedestrian Warrant

(*Source: Manual of Uniform Traffic Control Devices,* Draft, Federal Highway Administration, Washington DC, Draft 2007, Figures 4C-6 and 4C-8.)

in place and properly enforced. These types of control can also address many of the same accident problems as signalization. Thus a signal is justified only when these lesser measures have failed to address the situation adequately.

Accidents that are susceptible to correction by signalization include right-angle accidents, accidents involving turning vehicles from the two streets, and accidents between vehicles and pedestrians crossing the street on which the vehicle is traveling. Rear-end accidents are often increased with imposition of traffic signals (or STOP/YIELD signs) because some drivers may be induced to stop quickly or suddenly. Head-on and sideswipe collisions are not addressed by signalization; accidents between vehicles and fixed objects at corners are also not correctable through signalization.

Table 18.7: Warrant 6: Coordinated Signal System

The need for a traffic control signal shall be considered if an engineering study finds that one of the following criteria is met:

1. On a one-way street or a street that has traffic predominantly in one direction, the adjacent traffic control signals are so far apart that they do not provide the necessary degree of vehicular platooning.
2. On a two-way street, adjacent traffic control signals do not provide the necessary degree of platooning and the proposed and adjacent traffic control signals will collectively provide a progressive operation.

(*Source:* Used by permission of Federal Highway Administration, US Dept. of Transportation, *Manual on Uniform Traffic Control Devices,* Millennium Edition, Washington DC, 2001, p. 4C-12.)

Table 18.8: Warrant 7: Crash Experience

The need for a traffic control signal shall be considered if an engineering study finds that all of the following criteria are met:

1. Adequate trial of alternatives with satisfactory observance and enforcement has failed to reduce the crash frequency, and
2. Five or more reported crashes of types susceptible to correction by a traffic control signal have occurred within a 12-month period, each involving an personal injury or property damage apparently exceeding the applicable requirements for a reportable crash, and
3. For each of any 8 hours of the day, vehicles per hour (vph) given in both of the 80% columns of Condition A (in Warrant 1) or the vph in both of the 80% columns of Condition B (in Warrant 1) exists on the major-street and the higher-volume minor-street approach, respectively, to the intersection, or the volume of pedestrian traffic is not less than 80% of the requirements specified in the Pedestrian Volume warrant. These major-street and minor-street volumes shall be for the same 8 hours. On the minor street, the higher volume shall not be required to be on the same approach during each of the 8 hours.

(*Source: Manual of Uniform Traffic Control Devices,* Draft, Federal Highway Administration, Washington DC, December 2007, p. 273.)

Warrant 8: Roadway Network

This warrant addresses a developing situation (i.e., a case in which present volumes would not justify signalization but where new development is expected to generate substantial traffic that would justify signalization). The MUTCD also allows other warrants to be applied based on properly forecast vehicular and pedestrian volumes.

Large traffic generators, such as regional shopping centers, sports stadiums and arenas, and similar facilities, are often built in areas that are sparsely populated and where existing roadways have light traffic. Such projects often require substantial roadway improvements that change the physical layout of the roadway network and create new or substantially enlarged intersections that will require signalization. Generally, the "existing" situation is irrelevant to the situation being assessed. The warrant is described in Table 18.9.

Table 18.9: Warrant 8: Roadway Network

The need for a traffic control signal shall be considered if an engineering study finds that the common intersection of two or more major routes meets one or both of the following criteria:

1. The intersection has a total existing, or immediately projected, entering volume of at least 1,000 veh/h during the peak hour of a typical weekday, and has 5-year projected traffic volumes, based upon an engineering study, that meet one or more of Warrants 1, 2 and 3 during an average weekday, or
2. The intersection has a total existing of immediately projected entering volume of at least 1,000 veh/h for each of any 5 hours of a non-normal business day (Saturday or Sunday).

A major route as used in this warrant shall have one or more of the following characteristics:

1. It is part of the street or highway system that serves as the principal roadway network for through traffic flow, or
2. It includes rural or suburban highways outside, entering, or traversing a city, or
3. It appears as a major route on an official plan, such as a major street plan in an urban area traffic and transportation study.

(*Source:* Used by permission of Federal Highway Administration, US Dept. of Transportation, *Manual on Uniform Traffic Control Devices,* Millennium Edition, Washington DC, 2000, pp. 4C-13, 4C-14.)

"Immediately projected" generally refers to the traffic expected on day one of the opening of new facilities and/or traffic generators that create the need for signalization.

Warrant 9: Intersection Near a Highway-Rail Grade Crossing

This is a new warrant being added to the forthcoming 2010 MUTCD. It addresses a unique situation: an intersection that does not meet any other warrant for signalization but that is close enough to a highway-railroad crossing to present a hazard. Table 18.10 shows the detailed criteria for the warrant.

Figure 18.9 applies when there is only one lane approaching the intersection at the track-crossing location, and Figure 18.10 applied where there are two or more lanes approaching the track-crossing location.

The minor-street volume used in entering either Figure 18.9 or 18.10 may be multiplied by up to three adjustment factors: (1) an adjustment for train volume (Table 18.11), (2) an adjustment for presence of high-occupancy buses (Table 18.12), and (3) an adjustment for truck presence (Table 18.13). The base conditions for Figures 18.9 and 18.10 include four trains per day, no buses, and 10% trucks.

Table 18.10: Warrant 9: Intersection Near a Highway-Rail Grade Crossing

The need for a traffic control signal shall be considered if an engineering study finds that both of the following criteria are met:

1. A highway-rail grade crossing exists on a approach controlled by a STOP or YIELD sign and the center of the track nearest to the intersection is within 140 ft of the stop line on the approach, and
2. During the highest traffic volume hour during which trains use the crossing, the plotted point representing the vehicles per hour on the major street (total of both approaches) and the corresponding vehicles per hour on the minor-street approach that crosses the track (one direction only) falls above the applicable curve in Figure 18.9 or 18.10 for the existing combination of approach lanes over the track and distance D, which is the clear storage distance (between the grade crossing stop line and the near curb line of the major street).

(*Source: Manual of Uniform Traffic Control Devices,* Draft, Federal Highway Administration, Washington DC, December 2007, pp. 273–274.)

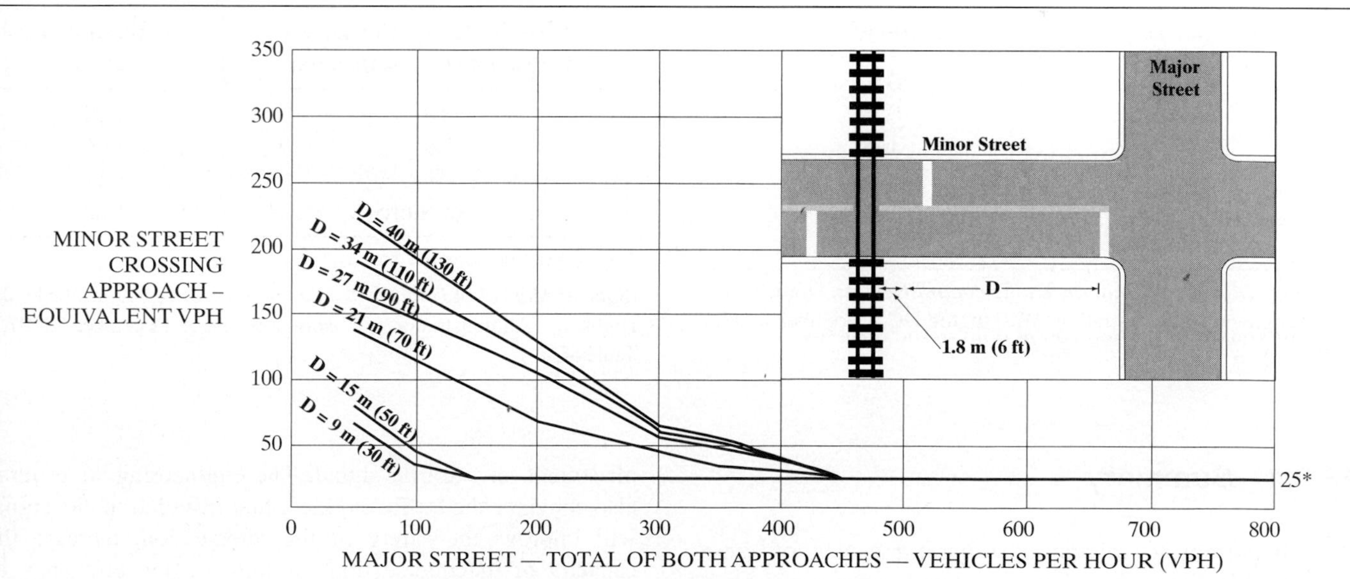

Figure 18.9: Warrant 9: Railroad Crossings for One-Lane Approaches

(*Source: Manual of Uniform Traffic Control Devices,* Draft, Federal Highway Administration, Washington DC, December 2007, Figure 4C-9.)

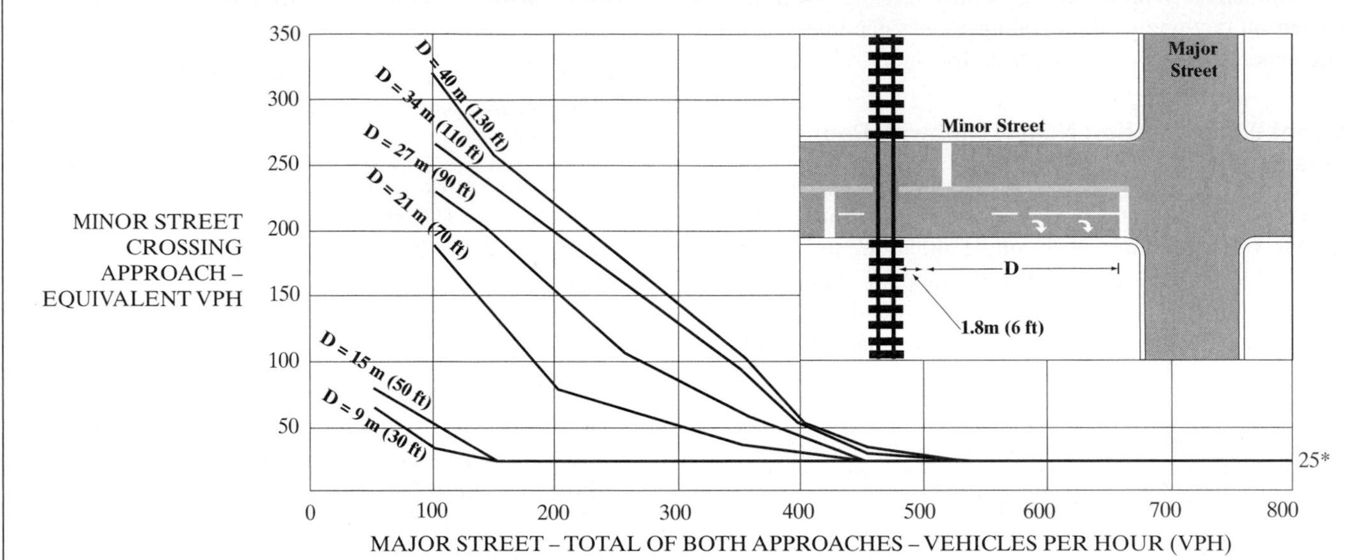

Figure 18.10: Warrant 9: Railroad Crossings for Two or More-Lane Approaches

(*Source: Manual of Uniform Traffic Control Devices,* Draft, Federal Highway Administration, Washington DC, December 2007, Figure 4C-9.)

Table 18.11: Adjustment Factor for Train Frequency

Trains per Day	Adjustment Factor
1	0.67
2	0.91
3–5	1.00
6–8	1.18
9–11	1.25
12 or more	1.33

(*Source: Manual of Uniform Traffic control Devices,* Draft, Federal Highway Administration, Washington DC, December 2007, Table 4C-2.)

Table 18.12: Adjustment Factor for High-Occupancy Buses

% of High-Occupancy Buses* on Minor-Street Approach	Adjustment Factor
0%	1.00
2%	1.09
4%	1.19
6% or more	1.32

*20 or more persons per bus.

(*Source: Manual of Uniform Traffic control Devices,* Draft, Federal Highway Administration, Washington DC, December 2007, Table 4C-3.)

18.3.4 Summary

It is important to reiterate the basic meaning of these warrants. No signal should be placed without an engineering study showing that the criteria of at least one of the warrants are met. However, meeting one or more of these warrants *does not* necessitate signalization. Note that every warrant uses the language "The need for a traffic control signal *shall be considered . . .*" (emphasis added). Although the "shall" is a mandatory standard, it calls only for consideration, not placement, of a traffic signal. The engineering study must also convince the traffic engineer that installation of a signal will improve the safety of the intersection, increase the capacity of the intersection, or improve the efficiency of operation at the intersection before the signal is installed. That is why the recommended information to be collected during an "engineering study" exceeds that needed to simply apply the nine warrants of the MUTCD. In the end, engineering judgment is called for, as is appropriate in any professional practice.

Table 18.13: Adjustment Factor for Tractor-Trailer Trucks

% of Tractor-Trailer Trucks on Minor-Street Approach	Adjustment Factor	
	D Less Than 70 ft	**D of 70 ft or More**
0%–2.5%	0.50	0.50
2.6%–7.5%	0.75	0.75
7.6%–12.5%	1.00	1.00
12.6%–17.5%	2.30	1.15
17.6%–22.5%	2.70	1.35
22.6%–27.5%	3.28	1.64
More than 27.5%	4.18	2.09

(*Source: Manual of Uniform Traffic control Devices,* Draft, Federal Highway Administration, Washington DC, December 2007, Table 4C-4.)

18.3.5 A Sample Problem in Application of Signal Warrants

Consider the intersection and related data shown in Figure 18.11.

Note that the data are formatted in a way that is conducive to comparing with warrant criteria. Thus a column adding the traffic in each direction on the major street is included, and a column listing the "high volume" in one direction on the minor street is also included. Pedestrian volumes are summarized for those crossing the major street because this is the criterion used in the pedestrian warrant. As you will see, not every warrant applies to every intersection, and data for some warrants are not provided. The following analysis is applied:

- *Warrant 1:* There is no indication that the 70% reduction factor applies, so it is assumed that either

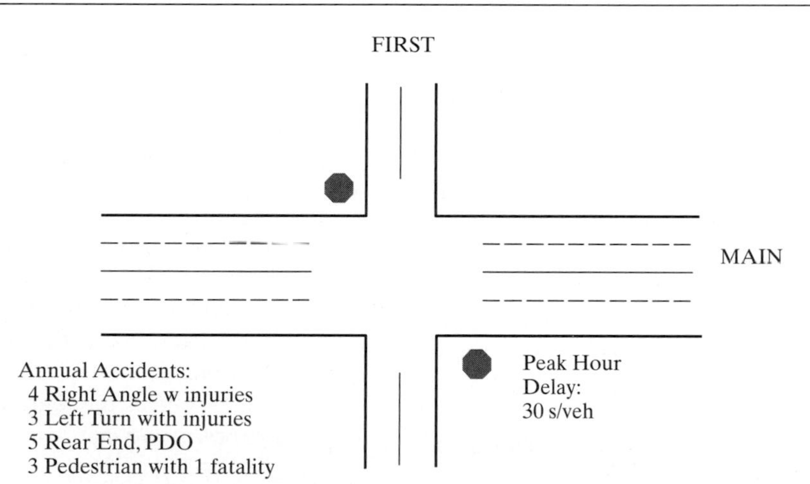

Time	Main Street Volume (veh/h)			First Ave Volume (veh/h)			Ped Volume (ped/h)
	EB	**WB**	**TOT**	**NB**	**SB**	**High Vol**	**Xing Main**
11 AM–12	400	425	**825**	75	80	**80**	115
12–1 PM	450	465	**915**	85	85	**85**	120
1–2 PM	485	500	**985**	90	100	**100**	125
2–3 PM	525	525	**1,050**	110	115	**115**	130
3–4 PM	515	525	**1,040**	100	95	**100**	135
4–5 PM	540	550	**1,090**	90	100	**100**	140
5–6 PM	550	580	**1,130**	110	125	**125**	120
6–7 PM	545	525	**1,070**	96	103	**103**	108
7–8 PM	505	506	**1,011**	90	95	**95**	100
8–9 PM	485	490	**975**	85	75	**85**	90
9–10 PM	475	475	**950**	75	60	**75**	50
10–11 PM	400	410	**810**	50	55	**55**	25

Figure 18.11: Intersection and Data for Sample Problem in Signal Warrant

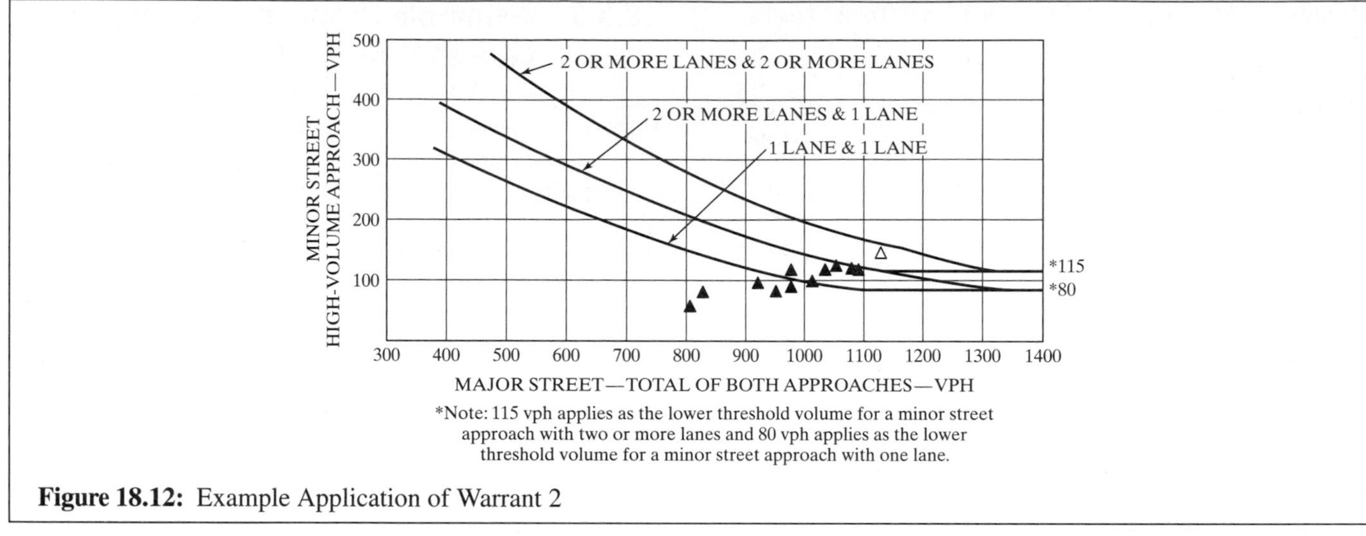

Figure 18.12: Example Application of Warrant 2

Condition A or Condition B must be met at 100%, or both must be met at 80%. Condition A requires 600 veh/h in both directions on the multilane major street and 150 veh/h in the high-volume direction on the one-lane minor street. Although all 12 hours on the major street shown in Figure 18.11 have more than 600 veh/h (total, both directions), none have a one-way volume equal to or higher than 150 veh/h on the minor street. Condition A is not met. Condition B requires 900 veh/h on the major street (both directions) and 75 veh/h on the minor street (one direction). The 10 hours between 12:00 noon and 10:00 PM meet the major-street criterion. The same 10 hours meet the minor-street criterion as well. Therefore, Condition B is met. Because one condition is met at 100%, the consideration of whether both conditions are met at 80% is not necessary. *Warrant 1 is satisfied.*

- *Warrant 2:* Figure 18.12 shows the hourly volume data plotted against the four-hour warrant graph. The center decision curve (one street with multilane approaches, one with one-lane approaches) is used. Only one of the 12 hours of data is above the criterion. To meet the warrant, four are required. *The warrant is not met.*

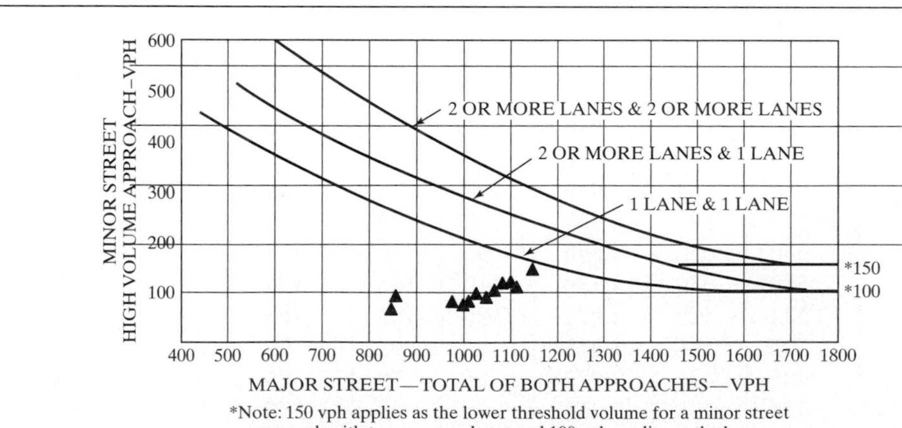

Figure 18.13: Example Application of Warrant 3

- *Warrant 3:* Figure 18.13 shows the hourly volume data plotted against the peak-hour volume warrant graph. Again, the center decision curve is used. None of the 12 hours of data is above the criterion. *The volume portion of this warrant is not met.*

  The delay portion of the peak-hour warrant requires 4 vehicle-hours of delay in the high-volume direction on a STOP-controlled approach. The intersection data indicate that each vehicle experiences 30 seconds of delay. The peak one-direction volume is 125 veh/h, resulting in 125 * 30 = 3,750 veh-secs of aggregate delay, or 3,750/3,600 = 1.04 veh-hrs of delay. This is less than that required by the warrant. *The delay portion of this warrant is not met.*

- *Warrant 4:* This warrant includes both a four-hour criterion and a peak-hour criterion, only one of which must be met to satisfy it. Figure 18.14 illustrates the solution.

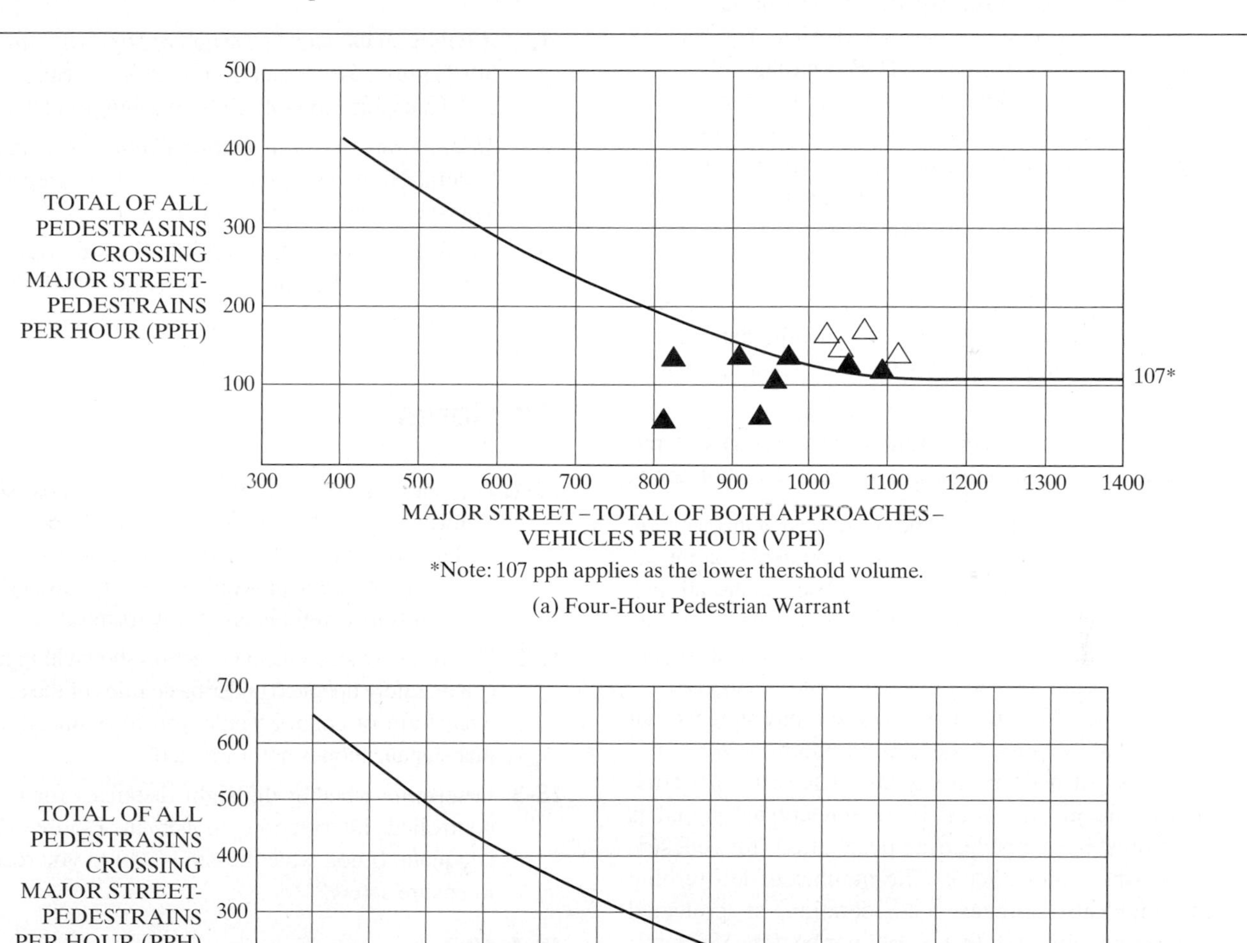

(a) Four-Hour Pedestrian Warrant

(b) Peak-Hour Pedestrian Warrant

Figure 18.14: Example Application of Warrant 4

The four-hour pedestrian warrant is met, and the peak-hour pedestrian warrant is not met. Because only one condition must be satisfied, *the pedestrian warrant is met.*

- *Warrant 5:* The school-crossing warrant does not apply. This is not a school crossing.
- *Warrant 6:* No information on signal progression is given, so this warrant cannot be applied.
- *Warrant 7:* The crash experience warrant has several criteria: Have lesser measures been tried? Yes, because the minor street is already STOP-controlled. Have five accidents susceptible to correction by signalization occurred in a 12-month period? Yes—four right-angle, three left-turn, and three pedestrian. Are the criteria for Warrants 1A or 1B met to the extent of 80%? Yes, Warrant 1B is met at 100%. Therefore, *the crash experience warrant is met.*
- *Warrant 8:* There is no information given concerning the roadway network, and the data reflect an existing situation. This warrant is not applicable in this case.
- *Warrant 9:* Because this situation is not a highway-rail grade crossing location, this warrant does not apply.

In summary, a signal should be considered at this location because the criteria for Warrants 1B (Interruption of Continuous Traffic), 4 (Pedestrians), and 7 (Crash Experience) are all met. Unless unusual circumstances are present, it would be reasonable to expect that the accident experience will improve with signalization, and it is, therefore, likely that one would be placed.

The fact that Warrant 1B is satisfied may suggest that a semiactuated signal be considered. In addition, Warrant 4 requires the use of pedestrian signals, at least for pedestrians crossing the major street. If a semiactuated signal is installed, it must have a pedestrian pushbutton (for pedestrians crossing the major street). The number of left-turning accidents may also suggest consideration of protected left-turn phasing, although this would not be done if a semiactuated signal is used.

18.4 Closing Comments

In selecting an appropriate type of control for an intersection, the traffic engineer has many factors to consider, including sight distances and warrants. In most cases, the objective is to provide the minimum level of control that will assure safety and efficient operations. In general, providing unneeded or excessive control leads to additional delay to drivers and passengers. With all of the analysis procedures and guidelines, however, engineering judgment is still required to make intelligent decisions. It is always useful to view the operation of existing intersections in the field in addition to reviewing study results before making recommendations on the best form of control.

References

1. *A Policy on Geometric Design of Highways and Streets,* 5th Edition, American Association of State Highway and Transportation Officials, Washington DC, 2004.
2. *Manual on Uniform Traffic Control Devices,* Draft, Federal Highway Administration, U.S. Department of Transportation, Washington DC, 2007.
3. *Highway Capacity Manual,* 4th Edition, Transportation Research Board, National Research Council, Washington DC, 2000.

Problems

18-1. For the intersection of two rural roads shown in Figure 18.15, determine whether or not operation under basic rules of the road would be safe. If not, what type of control would you recommend, assuming that traffic signals are not warranted?

18-2. Determine whether the intersection shown in Figure 18.16 can be safely operated under basic rules of the road. If not, what form of control would you recommend, assuming that signalization is not warranted?

18-3. Determine whether the sight distances for the STOP-controlled intersection shown in Figure 18.17 are adequate. If not, what measures would you recommend to ensure safety?

18-4–18-7. For each of the intersections shown in the following figures, determine whether the data support each of the nine signal warrants. For each problem, and each warrant, indicate whether the warrant is:

(a) met

(b) not met

(c) not applicable

(d) insufficient information given to assess.

For each problem, indicate (a) whether a signal is warranted, (b) the type of signalization that should be

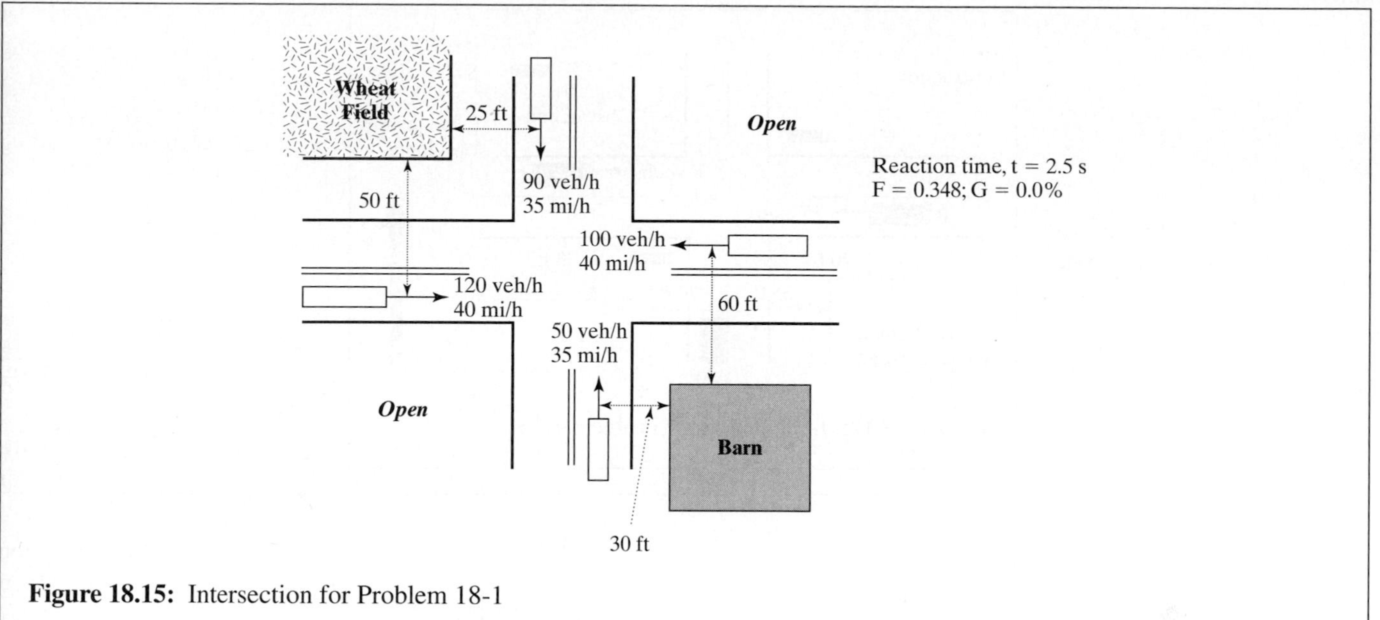

Figure 18.15: Intersection for Problem 18-1

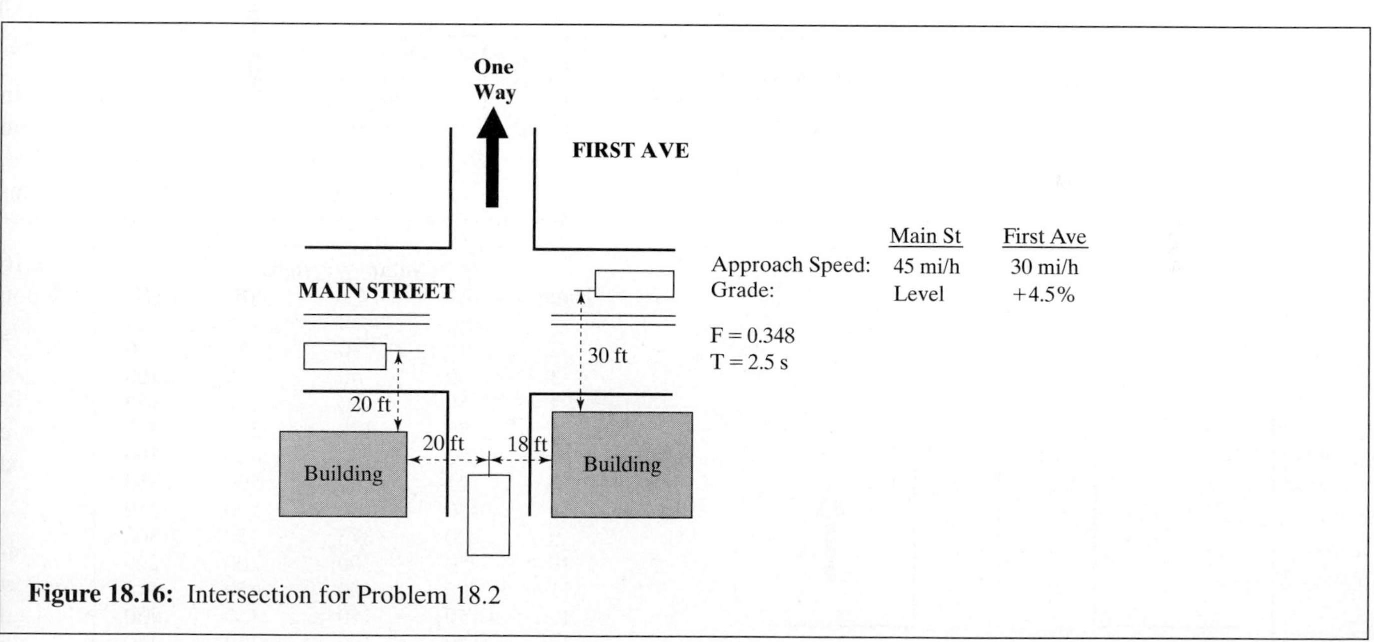

Figure 18.16: Intersection for Problem 18.2

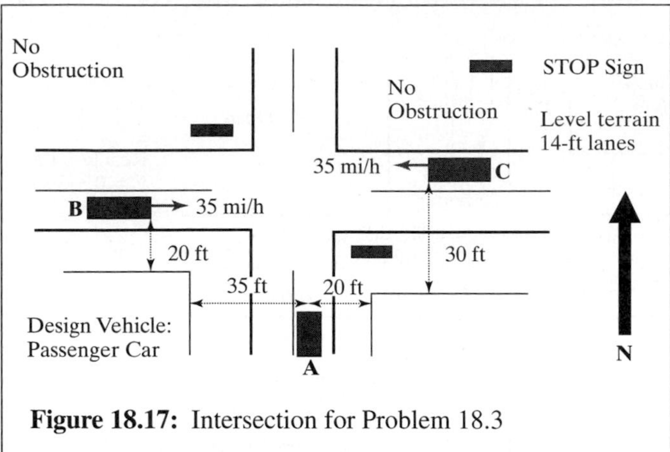

Figure 18.17: Intersection for Problem 18.3

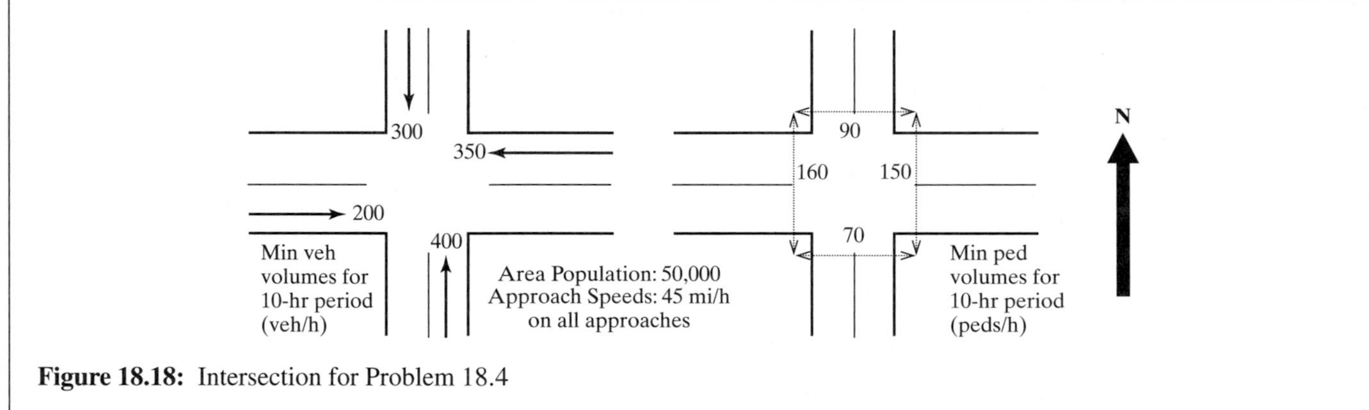

Figure 18.18: Intersection for Problem 18.4

N

Area Population: 75,000
Approach Speeds: 35 mi/h
4-Way STOP Control in place.

Hour	EB	WB	NB	SB
	Volumes (veh/h)			
1	30	30	25	25
2	30	30	50	50
3	50	50	75	100
4	50	50	150	150
5	75	100	250	200
6	100	250	400	300
7	125	400	500	350
8	150	450	500	350
9	200	375	450	300
10	250	300	200	200
11	200	300	150	150
12	150	150	150	150
13	100	100	150	150
14	100	100	150	200
15	100	75	150	200
16	250	100	200	250
17	325	125	350	250
18	375	150	400	300
19	400	150	350	450
20	425	150	350	450
21	325	100	200	200
22	150	75	100	100
23	100	50	50	50
24	50	25	50	50

Figure 18.19: Intersection and Data for Problem 18.5

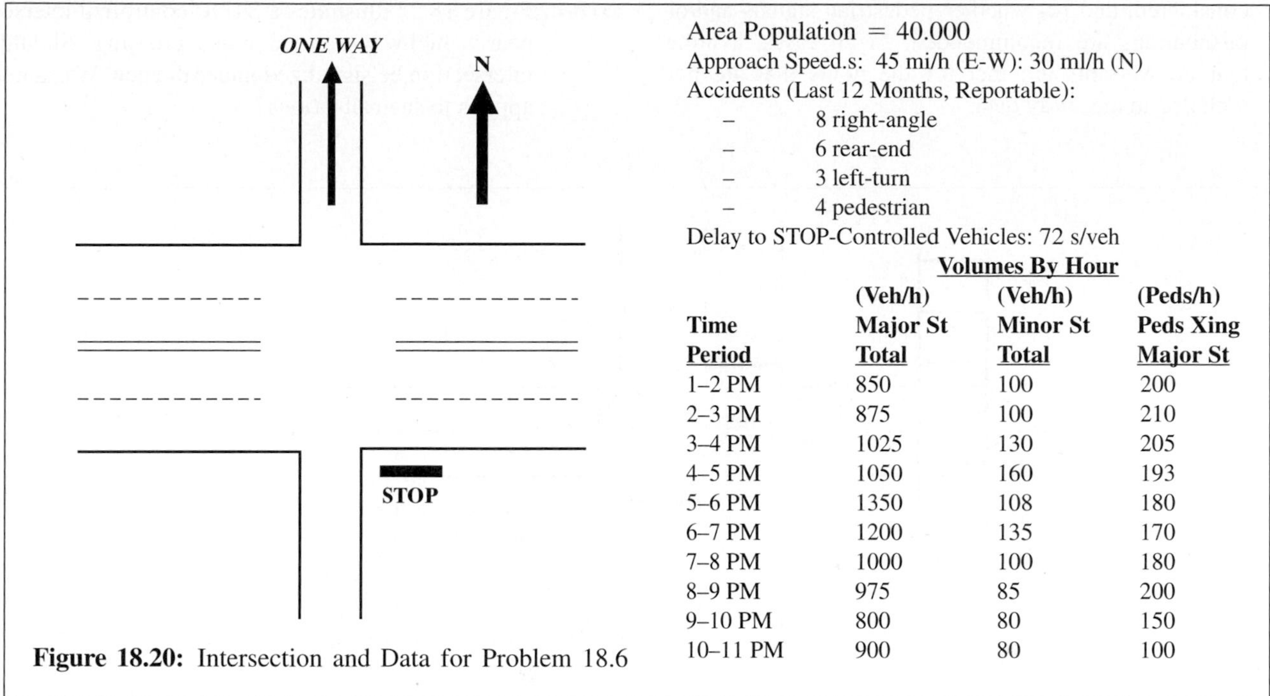

Area Population = 40.000
Approach Speed.s: 45 mi/h (E-W): 30 ml/h (N)
Accidents (Last 12 Months, Reportable):

- 8 right-angle
- 6 rear-end
- 3 left-turn
- 4 pedestrian

Delay to STOP-Controlled Vehicles: 72 s/veh

	Volumes By Hour		
Time Period	(Veh/h) Major St Total	(Veh/h) Minor St Total	(Peds/h) Peds Xing Major St
1–2 PM	850	100	200
2–3 PM	875	100	210
3–4 PM	1025	130	205
4–5 PM	1050	160	193
5–6 PM	1350	108	180
6–7 PM	1200	135	170
7–8 PM	1000	100	180
8–9 PM	975	85	200
9–10 PM	800	80	150
10–11 PM	900	80	100

Figure 18.20: Intersection and Data for Problem 18.6

PARK PLACE

Approach Speeds:
45 mi/h on Broad St.
30 mi/h on Park Place

Annual Accidents:

3 Right-Angle with injuries
5 Left-Turn with injuries
4 Rear-End, PDO
6 Pedestrian with 2 fatalities

BROAD STREET

Peak-Hour Delay:
45 s/veh

Time	Broad Street Vol (veh/h)			Park Place Volume (veh/h)			Ped Vol (ped/h) Xing Broad
	EB	WB	TOT	NB	SB	High Vol	
11 AM – 12	300	400	**700**	100	90	**100**	120
12 – 1 PM	325	450	**775**	110	125	**125**	115
1 – 2 PM	400	475	**875**	140	150	**150**	109
2 – 3 PM	450	480	**930**	150	165	**165**	122
3 – 4 PM	455	475	**930**	155	170	**170**	135
4 – 5 PM	445	425	**870**	160	160	**160**	140
5 – 6 PM	400	380	**780**	152	155	**155**	125
6 – 7 PM	385	350	**735**	140	150	**150**	121
7 – 8 PM	350	350	**700**	145	152	**152**	120
8 – 9 PM	350	375	**725**	130	156	**156**	105
9 – 10 PM	325	325	**650**	122	120	**122**	85
10 – 11 PM	300	300	**600**	120	95	**120**	85

Figure 18.21: Intersection and Data for Problem 18.7

considered, and (c) whether pedestrian signals and/or pushbuttons are recommended. In all cases, assume that no warrants are met for the hours that are not included in the study data.

18-8. Figure 18.22 illustrates a STOP-controlled intersection near a highway-railroad grade crossing. Should this intersection be signalized under the new Warrant 9 that applies to such situations?

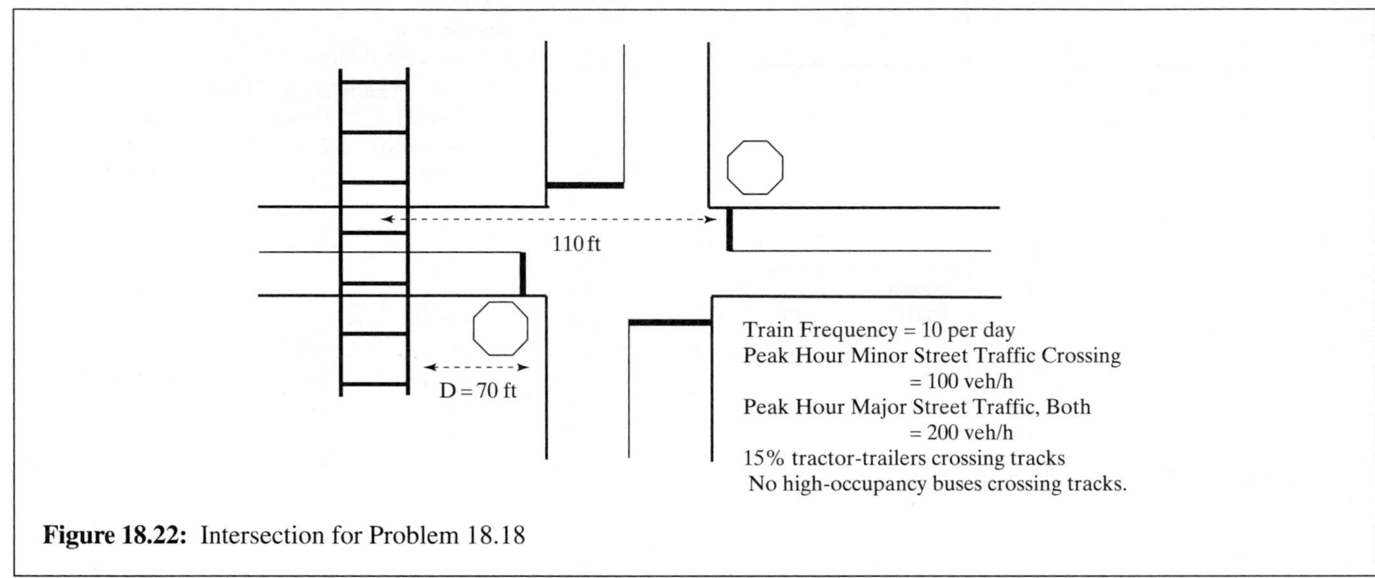

Figure 18.22: Intersection for Problem 18.18

CHAPTER 19

Elements of Intersection Design and Layout

In Chapter 18, the selection of appropriate control measures for intersections was addressed. Whether signalized or unsignalized, the control measures implemented at an intersection must be synergistic with the design and layout of the intersection. In this chapter, an overview of several important intersection design features is provided. We emphasize that this is only an overview because the details of intersection design could be the subject of a textbook on its own.

The elements treated here include techniques for determining the appropriate number and the use of lanes at an intersection approach, channelization, right- and left-turn treatments, special safety issues at intersections, and location of intersection signs and signal displays. There are a number of standard references for more detail on these and related subject areas, including the AASHTO *Policy on Geometric Design of Highways and Streets* [*1*], the *Manual on Uniform Traffic Control Devices* [*2*], the *Manual of Traffic Signal Design* [*3*], the *Traffic Detector Handbook* [*4*], and the *Highway Capacity Manual* [*5*].

19.1 Intersection Design Objectives and Considerations

As in all aspects of traffic engineering, intersection design has two primary objectives: (1) to ensure safety for all users, including drivers, passengers, pedestrians, bicyclists, and others, and (2) to promote efficient movement of all users (motorists, pedestrians, bicyclists, etc.) through the intersection. Achievement of both is not an easy task because safety and efficiency are often competing rather than mutually reinforcing goals.

In developing an intersection design, AASHTO [*1*] recommends that the following elements be considered:

- Human factors
- Traffic considerations
- Physical elements
- Economic factors
- Functional intersection area

Human factors must be taken into account. Thus intersection designs should accommodate reasonable approach speeds, user expectancy, decision and reaction times, and other user characteristics. Design should, for example, reinforce natural movement paths and trajectories, unless doing so presents a particular hazard.

Traffic considerations include provision of appropriate capacity for all user demands; the distribution of vehicle types and turning movements; approach speeds; and special requirements for transit vehicles, pedestrians, and bicyclists.

Physical elements include the nature of abutting properties, particularly traffic movements generated by these properties (parking, pedestrians, driveway movements, etc.). They also include the intersection angle, existence and location of traffic control devices, sight distances, and specific geometric characteristics, such as curb radii.

Economic factors include the cost of improvements (construction, operation, maintenance), the effects of improvements on the value of abutting properties (whether used by the expanded right-of-way or not), and the effect of improvements on energy consumption.

Finally, intersection design must encompass the full functional intersection area. The operational intersection area includes approach areas that fully encompass deceleration and acceleration zones as well as queuing areas. The latter are particularly critical at signalized intersections.

19.2 A Basic Starting Point: Sizing the Intersection

One of the most critical aspects of intersection design is the determination of the number of lanes needed on each approach. This is not an exact science because the result is affected by the type of control at the intersection, parking conditions and needs, availability of right-of-way, and a number of other factors that are not always directly under the control of the traffic engineer. Further, considerations of capacity, safety, and efficiency all influence the desirable number of lanes. As is the case in most design exercises, there is no one correct answer, and many alternatives may be available that provide for acceptable safety and operation.

19.2.1 Unsignalized Intersections

Unsignalized intersections may be operated under basic rules of the road (no control devices other than warning and guide signs), or under STOP or YIELD control.

When totally uncontrolled, intersection traffic volumes are generally light, and there is rarely a clear "major" street with significant volumes involved. In such cases, intersection areas do not often require more lanes than on the approaching roadway. Additional turning lanes are rarely provided. Where high speeds and/or visibility problems exist, channelization may be used in conjunction with warning signs to improve safety.

The conditions under which two-way (or one-way at a T-intersection or intersection of one-way roadways) STOP or YIELD control are appropriate are treated in Chapter 18. The existence of STOP- or YIELD-controlled approach(es), however, adds some new considerations into the design process:

- Should left-turn lanes be provided on the major street?
- Should right-turn lanes be provided on the major street?
- Should a right-turn lane be provided on minor approaches?
- How many basic lanes does each minor approach require?

Most of these issues involve capacity considerations. For convenience, however, some general guidelines are presented here.

When left turns are made from a mixed lane on the major street, there is the potential for unnecessary delay to through vehicles that must wait while left-turners find a gap in the opposing major-street traffic. The impact of major-street left turns on delay to all major-street approach traffic becomes noticeable when left turns exceed 150 veh/h. This may be used as a general guideline indicating the probable need for a major-street left-turn lane, although a value as low as 100 veh/h could be justified.

Right-turning vehicles from the major street do not have a major impact on the operation of STOP- or YIELD-controlled intersections. Although they do not technically conflict with minor-street movements when they are made from shared lanes, they may impede some minor-street movements when drivers do not clearly signal that they are turning or approach the intersection at high speed. When major-street right turns are made from an exclusive lane, their intent to turn is more obvious to minor-street drivers. Right-turn lanes for major-street vehicles can be easily provided where on-street parking is permitted. In such situations, parking may be prohibited for 100 to 200 feet from the STOP line, thus creating a short right-turn lane.

Most STOP-controlled approaches have a single lane shared by all minor-street movements. Occasionally, two lanes are provided. Any approach with sufficient demand to

Table 19.1: Guidelines for Number of Lanes at STOP-Controlled Approaches[1]

Total Volume on Minor Approach (veh/h)	Total Volume on Major Street (veh/h)			
	500	**1,000**	**1,500**	**2,000**
100	1 lane	1 lane	1 lane	2 lanes
200	1 lane	1 lane	2 lanes	NA
300	1 lane	2 lanes	2 lanes	NA
400	1 lane	2 lanes	NA	NA
500	2 lanes	NA	NA	NA
600	2 lanes	NA	NA	NA
700	2 lanes	NA	NA	NA
800	2 lanes	NA	NA	NA

[1]Not including multiway STOP-controlled intersections.
NA = STOP control probably not appropriate for these volumes.

require three lanes is probably inappropriate for STOP control. Approximate guidelines for the number of lanes required may be developed from the unsignalized intersection analysis methodology of the *Highway Capacity Manual.* Table 19.1 shows various combinations of minor-approach demand versus total crossing traffic on the major street, along with guidelines as to whether one or two lanes would be needed. They are based on assumptions that (1) all major-street traffic is through traffic, (2) all minor- approach traffic is through traffic, and (3) various impedances and other nonideal characteristics reduce the capacity of a lane to about 80% of its original value.

The other issue for consideration on minor STOP-controlled approaches is whether or not a right-turning lane should be provided. Because the right-turn movement at a STOP-controlled approach is much more efficient than crossing and left-turn movements, better operation can usually be accomplished by providing a right-turn lane. This is often as simple as banning parking within 200 feet of the STOP line, and it prevents right-turning drivers from being stuck in a queue when they could easily be executing their movements. Where a significant proportion of the minor-approach traffic is turning right (>20%), provision of a right-turning lane should always be considered.

Note that the lane criteria of Table 19.1 are approximate. Any finalized design should be subjected to detailed analysis using the appropriate procedures of the HCM 2000 (or the forthcoming HCM 2010).

Consider the following example: two-lane major roadway carries a volume of 800 veh/h, of which 10% turn left and 5% turn right at a local street. Both approaches on the local street are STOP-controlled and carry 150 veh/h, with 50 turning left and 50 turning right. Suggest an appropriate design for the intersection.

Given the relatively low volume of left turns (80/h) and right turns (40/h) on the major street, neither left- nor right-turn lanes would be required, although they could be provided if space is available. From Table 19.1, it appears that one lane would be sufficient for each of the minor-street approaches. The relatively heavy percentage of right turns (33%), however, suggests that a right-turn lane on each minor approach would be useful.

19.2.2 Signalized Intersections

Approximating the required size and layout of a signalized intersection involves many factors, including the demands on each lane group, the number of signal phases, and the signal cycle length.

Determining the appropriate number of lanes for each approach and lane group is not a simple design task. Like so many design tasks, there is no absolutely unique result, and many different combinations of physical design and signal timing can provide for a safe and efficient intersection.

The primary control on number of lanes is the *maximum sum of critical-lane volumes* that the intersection can support. This concept is more thoroughly discussed and illustrated in Chapter 20. The concept involves finding the single lane during a signal cycle that carries the most intense traffic, which means it would be the one that consumes the most green time of all movements to process its demand. Each signal phase has a critical-lane volume, and the cycle length of the signal is set to accommodate the sum of these critical volumes for each phase in the signal plan. This is the equation governing the maximum sum of critical-lane volumes:

$$V_c = \frac{1}{h}\left[3{,}600 - Nt_L\left(\frac{3{,}600}{C}\right)\right] \qquad (19\text{-}1)$$

where: V_c = maximum sum of critical-lane volumes, veh/h
h = average headway for prevailing conditions on the lane group or approach, s/veh
N = number of phases in the cycle
t_L = lost time per phase, s/phase
C = cycle length, s

Table 19.2 gives approximate maximum sums of critical-lane volumes for typical prevailing conditions. An average headway of 2.6 s/veh is used, along with a typical lost time per phase of 4.0 s (t_L). Maximum sums are tabulated for a number of combinations of N and C.

Consider the case of an intersection between two major arterials. Arterial 1 has a peak directional volume of 900 veh/h; Arterial 2 has a peak directional volume of 1,100 veh/h. Turning volumes are light, and a two-phase signal is anticipated. As a preliminary estimate, what number of lanes is needed to accommodate these volumes, and what range of cycle lengths might be appropriate?

From Table 19.2, the range of maximum sums of critical-lane volumes is between 1,015 veh/h for a 30-second cycle length and 1,292 veh/h for a 120-second cycle length. The two critical volumes are given as 900 veh/h and 1,100 veh/h. If only one lane is provided for each, then the sum of critical-lane volumes is 900 + 1,100 = 2,000 veh/h, well outside the range of maximum values for reasonable cycle lengths. Table 19.3 shows a number of reasonable scenarios for the number of lanes on each critical approach along with the resulting sum of critical-lane volumes.

With one lane on Arterial 1 and 3 lanes on Arterial 2, the sum of critical-lane volumes is 1,267 veh/h. From Table 19.2, this would be a workable solution with a cycle length over 100 seconds. With two lanes on each arterial, the sum of critical-lane volumes is 1,000 veh/h. This situation would be workable at any cycle length between 30 and 120 seconds. All other potentially workable scenarios in Table 19.3 could accommodate any cycle length between 30 and 120 seconds as well.

This type of analysis does not yield a final design or cycle length because it is approximate. But it does give the traffic engineer a basic idea of where to start. In this case, providing two lanes on each arterial in the peak direction appears to be a reasonable solution. Because peaks tend to be reciprocal (what goes one way in the morning comes back the opposite way in the evening), two lanes would also be provided for the off-peak directions on each arterial as well.

The signal timing should then be developed using the methodology of Chapter 21. The final design and timing should then be subjected to analysis using the *Highway*

Table 19.2: Maximum Sums of Critical-Lane Volumes for a Typical Signalized Intersection

Cycle Length (s)	No. of Phases		
	2	**3**	**4**
30	1,015	831	646
40	1,108	969	831
50	1,163	1,052	942
60	1,200	1,108	1,015
70	1,226	1,147	1,068
80	1,246	1,177	1,108
90	1,262	1,200	1,138
100	1,274	1,218	1,163
110	1,284	1,234	1,183
120	1,292	1,246	1,200

Table 19.3: Sum of Critical-Lane Volumes (veh/h) for Various Scenarios: Sample Problem

No. of Lanes on Arterial 2	Critical-Lane Volume for Arterial (veh/h)	No. of Lanes on Arterial[1]		
		1	**2**	**3**
		900/1 = 900	900/2 = 450	900/3 = 300
1	1,100/1 = 1,100	*2,000*	*1,550*	*1,400*
2	1,100/2 = 550	*1,450*	*1,000*[1]	*850*[1]
3	1,100/3 = 367	*1,267*[1]	*817*[1]	*667*[1]

[1]Acceptable lane plan with V_c acceptable at some cycle length.

Capacity Manual (see Chapter 24) or some other appropriate analysis technique.

The number of anticipated phases is, of course, critical to a general analysis of this type. Suggested criteria for determining when protected left-turn phases are needed are given in Chapter 21. Because there is a critical-lane volume for *each* signal phase; a four-phase signal involves four critical-lane volumes, for example.

Exclusive left-turn lanes must be provided whenever a fully protected left-turn phase is used and is highly desirable when compound left-turn phasing (protected + permitted or vice versa) is used.

19.3 Intersection Channelization

19.3.1 General Principles

Channelization can be provided through the use of painted markings or by installation of raised channelizing islands. The AASHTO *Policy on Geometric Design of Highways and Streets* [*1*] gives a number of reasons for considering channelization at an intersection:

- Vehicle paths may be confined so that no more than two paths cross at any one point.
- The angles at which merging, diverging, or weaving movements occur may be controlled.
- Pavement area may be reduced, decreasing the tendency to wander and narrowing the area of conflict between vehicle paths.
- Clearer indications of proper vehicle paths may be provided.
- Predominant movements may be given priority.
- Areas for pedestrian refuge may be provided.
- Separate storage lanes may be provided to permit turning vehicles to wait clear of through-traffic lanes.
- Space may be provided for the mounting of traffic control devices in more visible locations.
- Prohibited turns may be physically controlled.
- Vehicle speeds may be somewhat reduced.

The decision to channelize an intersection depends on a number of factors, including the existence of sufficient right-of-way to accommodate an effective design. Factors such as terrain, visibility, demand, and cost also enter into the decision. Channelization supplements other control measures but can sometimes be used to simplify other elements of control.

19.3.2 Some Examples

It is difficult to discuss channelization in the abstract. A selection of examples illustrates the implementation of the principles noted previously.

Figure 19.1 shows the intersection of a major street (E–W) with a minor crossroad (N–S). A median island is provided on the major street. Partial channelization is provided for the southbound (SB) right turn, and a left-turn lane is provided for the eastbound (EB) left turn. The two channelized turns are reciprocal, and the design reflects a situation in which these two turning movements are significant. The design illustrated minimizes the conflict between SB right turns and other movements and provides a storage lane for EB left turns, removing the conflict with EB through movements. The lack of any channelization for other turning movements suggests they have light demand. The design does not provide for a great deal of pedestrian refuge, except for the wide median on the east leg of the intersection. This suggests that pedestrian volumes are relatively low at this location; if this is so, the crosswalk markings are optional. The channelization at this intersection is appropriate for both an unsignalized and a signalized intersection.

Figure 19.2 shows a four-leg intersection with similar turning movements as in Figure 19.1. In this case, however, the SB-EB and EB-SB movements are far heavier and require a more dramatic treatment. Here channelization is used to create two additional intersections to handle these dominant turns. Conflicts between the various turning movements are minimized in this design.

Figure 19.3 is a similar four-leg intersection with far greater use of channelization. All right turns are channelized,

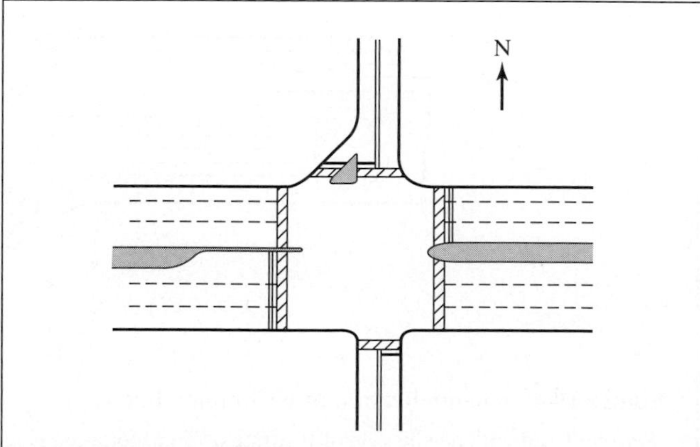

Figure 19.1: A Four-Leg Intersection with Partial Channelization for SB-EB and EB-SB Movements

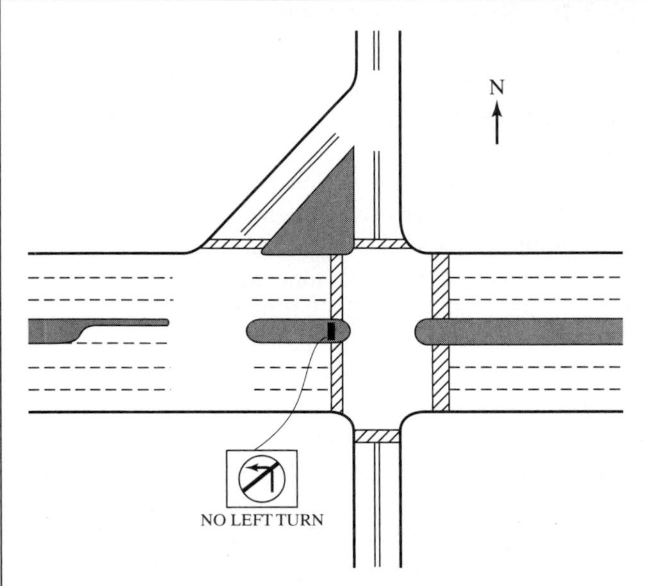

Figure 19.2: A Four-Leg Intersection Channelization for Major SB-EB and EB-SB Movements

and both major street left-turning movements have an exclusive left-turn lane. This design addresses a situation in which turning movements are more dominant. Pedestrian refuge is provided only on the right-turn channelizing islands, which may be limited by the physical size of the islands. Again, the

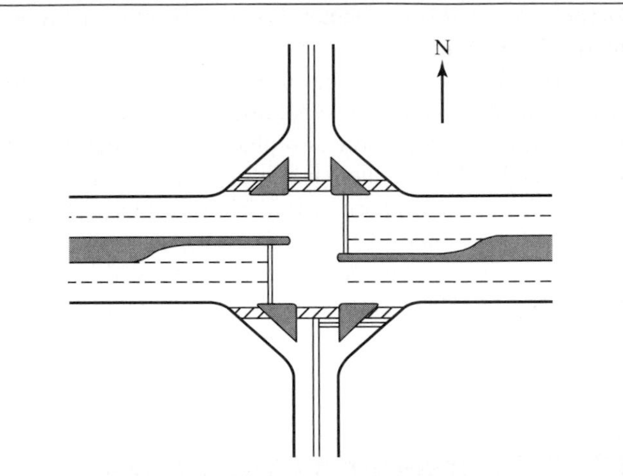

Figure 19.3: A Four-Leg Intersection with Full Channelization of Right Turns

channelization scheme is appropriate for either signalized or unsignalized control.

Channelization can also be used at locations with significant traffic volumes to simplify and reduce the number of conflicts and to make traffic control simpler and more effective. Figure 19.4 illustrates such a case.

In this case, a major arterial is fed by two major generators, perhaps two large shopping centers, on opposite sides of the roadway. Through movements across the arterial are

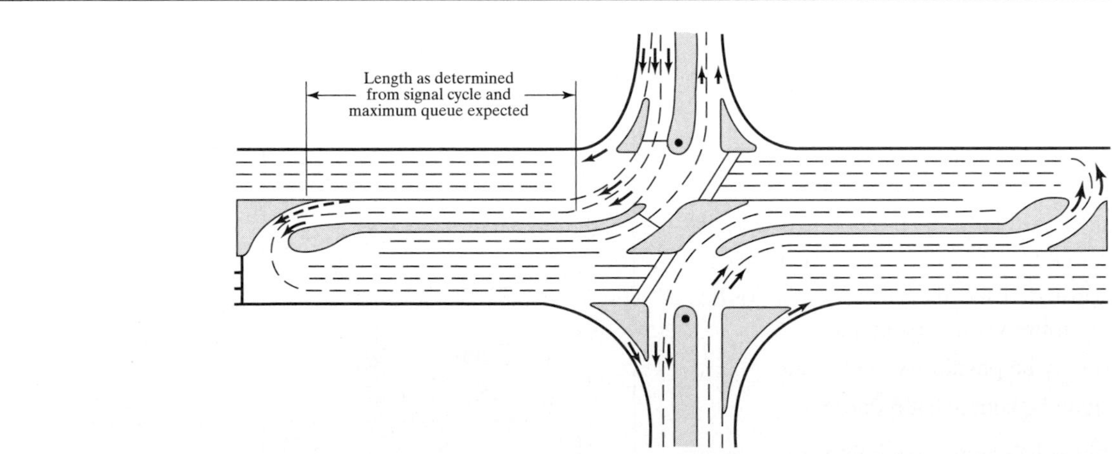

Figure 19.4: Channelization of a Complex Intersection

(*Source:* Used with permission of Institute of Transportation Engineers, R.P. Kramer, "New Combination of Old Techniques to Rejuvenate Jammed Suburban Arterials," *Strategies to Alleviate Traffic Congestion,* Washington DC, 1988.)

prevented by the channelization scheme as are left turns from either generator onto the arterial. The channelization allows only the following movements to take place:

- Through movements on the arterial
- Right-turn movements into either generator
- Left-turn movements into either generator
- Right-turn movements onto the arterial

Double-left-turn lanes on the arterial are provided for storage and processing of left turns entering either generator. A wide median is used to nest a double U-turn lane next to the left-turn lanes. These U-turn lanes allow vehicles to exit either generator and accomplish either a left-turning movement onto the arterial or a through movement into the opposite generator. In this case, it is highly likely the main intersection and the U-turn locations would be signalized. However, all movements at this complex location could be handled with two-phase signalization because the channelization design limits the signal to the control of two conflicting movements at each of the three locations. The distance between the main intersection and the U-turn locations must consider the queuing characteristics in the segments between intersections to avoid spill back and related demand starvation issues. From these examples, you can see that channelization of intersections can be a powerful tool to improve both the safety and the efficiency of intersection operation.

19.3.3 Channelizing Right Turns

When space is available, it is virtually always desirable to provide a channelized path for right-turning vehicles. This is especially true at signalized intersections where such channelization accomplishes two major benefits:

- Where "right-turn on red" regulations are in effect, channelized right turns minimize the probability of a right-turning vehicle or vehicles being stuck behind a through vehicle in a shared lane.
- Where channelized, right turns can effectively be removed from the signalization design because they would, in most cases, be controlled by a YIELD sign and would be permitted to move continuously.

The accomplishment of these benefits, however, depends on some of the details of the channelization design.

Figure 19.5 shows three different schemes for providing channelized right turns at an intersection. In Figure 19.5 (a), a simple channelizing triangle is provided. This design has limited benefits for two reasons: (1) through vehicles in the right lane may queue during the "red" signal phase, blocking access to the channelized right-turn lane, and (2) high right-turn volumes may limit the usefulness of the right-hand lane to through vehicles during "green" phases.

In the second design, shown in Figure 19.5 (b), acceleration and deceleration lanes are added for the channelized right turn. If the lengths of the acceleration and deceleration lanes are sufficient, this design can avoid the problem of queues blocking access to the channelized right turn.

In the third design, Figure 19.5 (c), a very heavy right-turn movement can run continuously. A lane drop on the approach leg and a lane addition to the departure leg provide a continuous lane and an unopposed path for right-turning vehicles. This design requires unique situations in which the lane drop and lane addition are appropriate for the arterials involved. To be effective, the lane addition on the departure leg cannot be removed too close to the intersection. It should be carried for at least several thousand feet before it is dropped, if necessary.

Right-turn channelization can simplify intersection operations, particularly where the movement is significant. It can also make signalization more efficient because channelized right turns, controlled by a YIELD sign, do not require green time to be served.

19.4 Special Situations at Intersections

This section deals with four unique intersection situations that require attention: (1) intersections with junction angles less than 60° or more than 120°, (2) T-intersections, (3) offset intersections, and (4) special treatments for heavy left-turn movements.

19.4.1 Intersections at Skewed Angles

Intersections, both signalized and unsignalized, work best when the angle of the intersection is 90°. Sight distances are easier to define, and drivers tend to expect intersections at right angles. Nevertheless, in many situations the intersection angle is not 90°. Such angles may present special challenges to the traffic engineer, particularly when they are less than 60° or more than 120°. These occur relatively infrequently. Drivers are generally less familiar with their special characteristics, particularly vis-à-vis sight lines and distances.

Skewed-angled intersections are particularly hazardous when uncontrolled and combined with high intersection-approach speeds. Such cases generally occur in rural areas

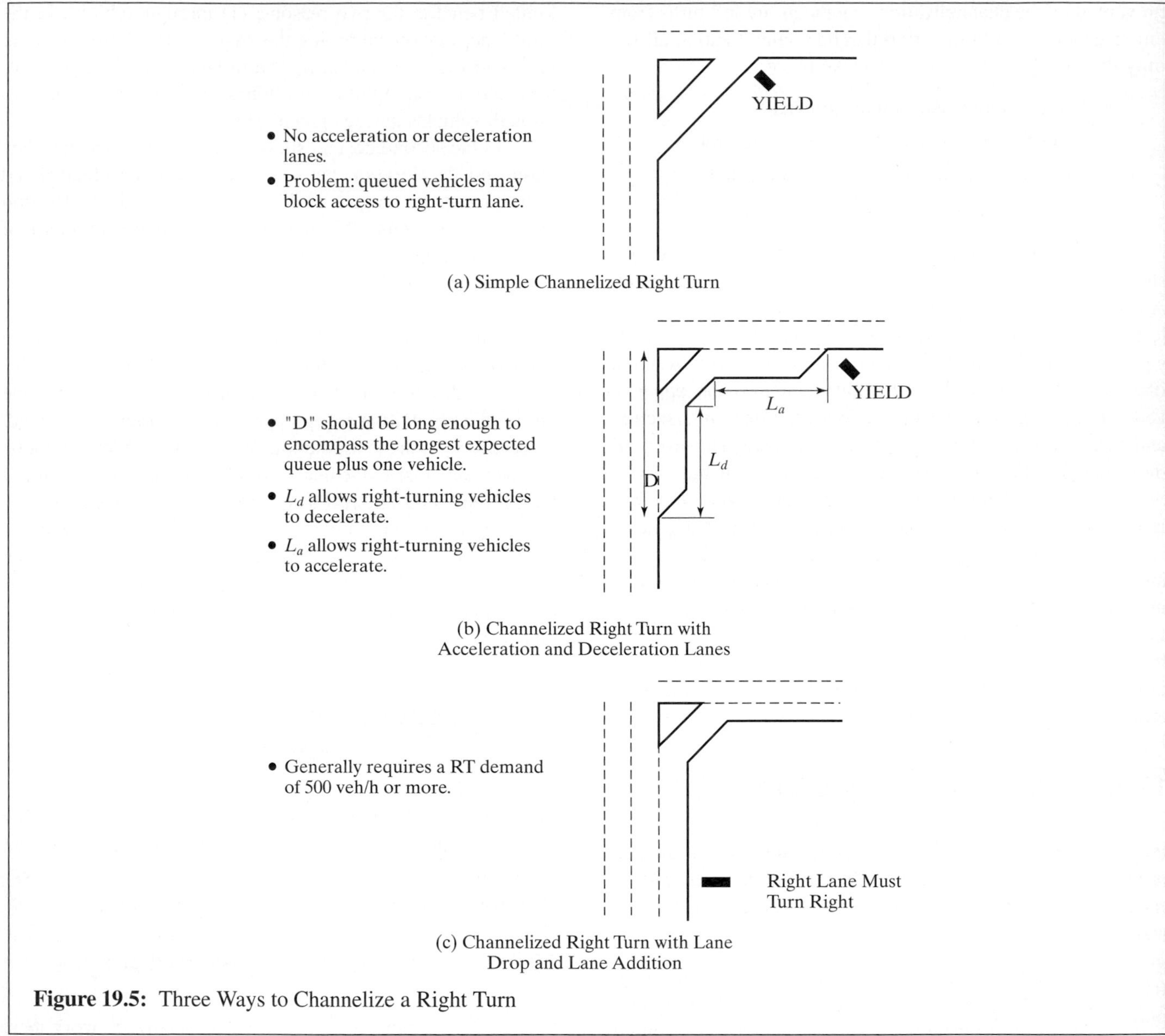

Figure 19.5: Three Ways to Channelize a Right Turn

and involve primary state and/or county routes. The situation illustrated in Figure 19.6 provides an example.

The example is a rural junction of two-lane, high-speed arterials, Routes 160 and 190. Given relatively gentle terrain, low volumes, and the rural setting, speed limits of 50 mi/h are in effect on both facilities. Figure 19.6 also illustrates the two movements representing a hazard. The conflict between the WB movement on Route 160 and the EB movement on Route 190 is a significant safety hazard. At the junction shown, both roadways have similar designs. Thus there is no visual cue to the driver indicating which route has precedence or right-of-way. Given that signalization is rarely justifiable in low-volume rural settings, other means must be considered to improve the safety of operations at the intersection.

The most direct means of improving the situation is to change the alignment of the intersection, making it clear which of the routes has the right-of-way. Figure 19.7 illustrates the two possible realignments. In the first case, Route 190 is given clear preference; vehicles arriving or departing on the east leg

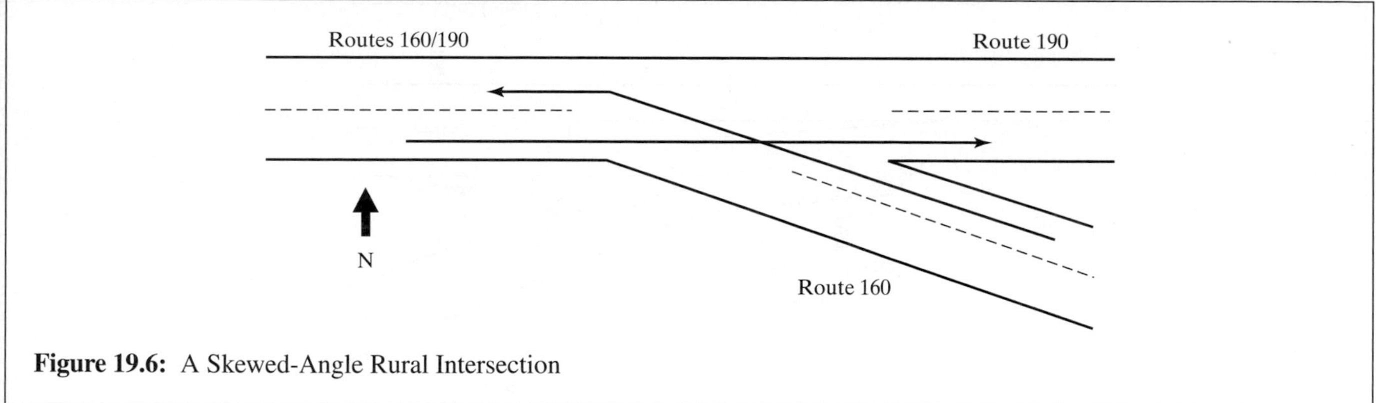

Figure 19.6: A Skewed-Angle Rural Intersection

of Route 160 must go through a 90° intersection to complete their maneuver. In the second case, Route 160 is dominant, and those arriving or departing on the east leg of Route 190 go through the 90° intersection. In either case, the 90° intersection would be controlled using a STOP sign to clearly designate right-of-way.

Although basic realignment is the best solution for high-speed odd-angle intersections, it requires that right-of-way be available to implement the change. Even in a rural setting, sufficient right-of-way to realign the intersection may not always be available. Other solutions can also be considered. Channelization can be used to better define the intersection movements, and control devices can be used to designate right-of-way. Figure 19.8 shows another potential design that requires less right-of-way than full realignment.

In this case, only the WB movement on Route 106 was realigned. Although this would still require some right-of-way, the amount needed is substantially less than for full realignment. Additional channelization is provided to separate EB movements on Routes 106 and 109. In addition to the regulatory signs indicated in Figure 19.8, warning and directional guide signs would be placed on all approaches to the intersection. In this solution, the WB left turn from Route 109 must be prohibited; an alternative route would have to be provided and appropriate guide signs designed and placed.

The junction illustrated is, in essence, a three-leg intersection. Skewed-angle four-leg intersections also occur in rural, suburban, and urban settings and present similar problems. Again, total realignment of such intersections is the most desirable solution. Figure 19.9 shows an intersection and the potential realignments that would eliminate the odd-angle junction. Where a four-leg intersection is involved, however, the realignment solution creates two separate intersections. Depending on volumes and the general traffic environment of the intersection, the realignments proposed in Figure 19.9 could result in signalized or unsignalized intersections.

In urban and suburban settings, where right-of-way is a significant impediment to realigning intersection, signalization of the odd-angle intersection can be combined with channelization to achieve safe and efficient operations. Channelized right turns would be provided for acute-angle turns, and left-turn lanes (and signalization) would be provided as needed.

In extreme cases, where volumes and approach speeds present hazards that cannot be ameliorated through normal traffic engineering measures, consideration may be given to providing a full or partial interchange with the two main roadways grade-separated. Providing grade separation would also involve some expansion of the traveled way, and overpasses in some suburban and urban surroundings may involve visual pollution and/or other negative environmental impacts.

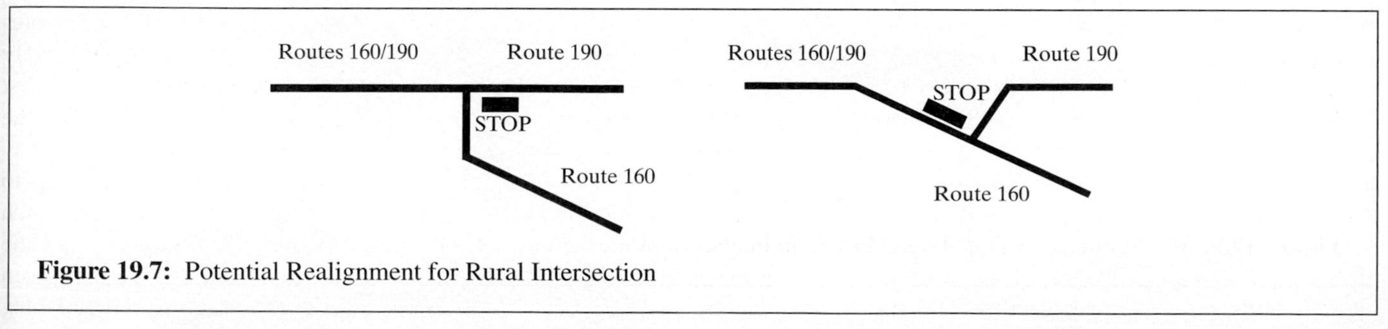

Figure 19.7: Potential Realignment for Rural Intersection

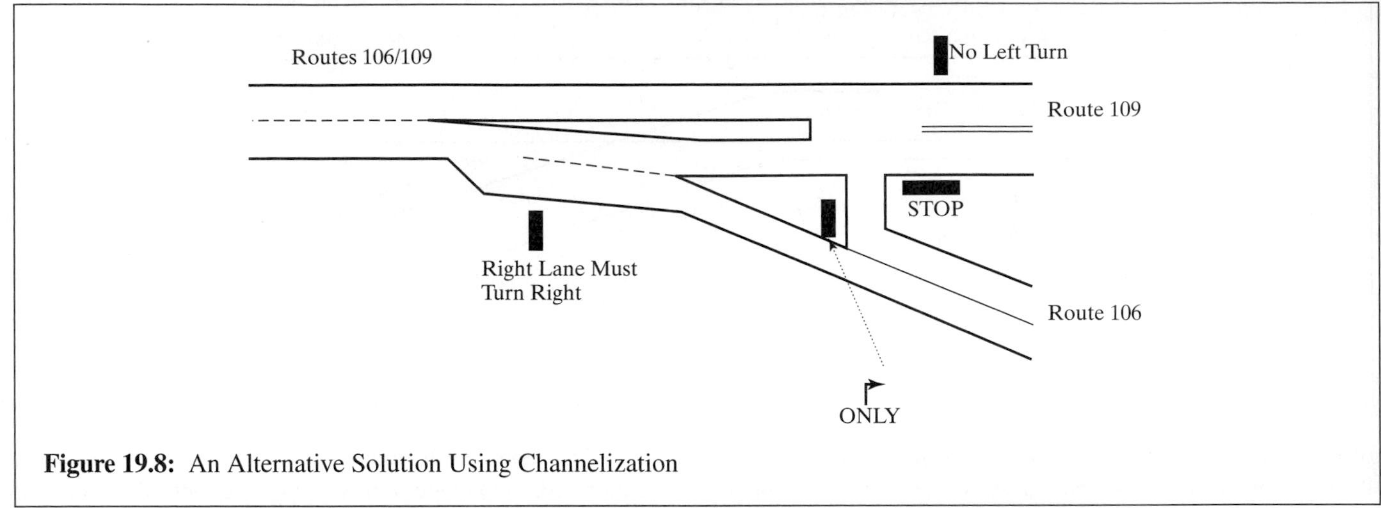

Figure 19.8: An Alternative Solution Using Channelization

19.4.2 T-Intersections: Opportunities for Creativity

In many ways, T-intersections are far simpler than traditional four-leg intersections. The typical four-leg intersection contains 12 vehicular movements and 4 crossing pedestrian movements. At a T-intersection, only six vehicular movements exist and there are only three crossing pedestrian movements. These are illustrated in Figure 19.10.

Note that in the set of T-intersection vehicular movements, there is only one opposed left turn—the WB left-turn movement in this case. Because of this, conflicts are easier to manage, and signalization, when necessary, is easier to address.

Control options include all generally applicable alternatives for intersection control:

- Uncontrolled (warning and guide signs only)
- STOP or YIELD control
- Signal control

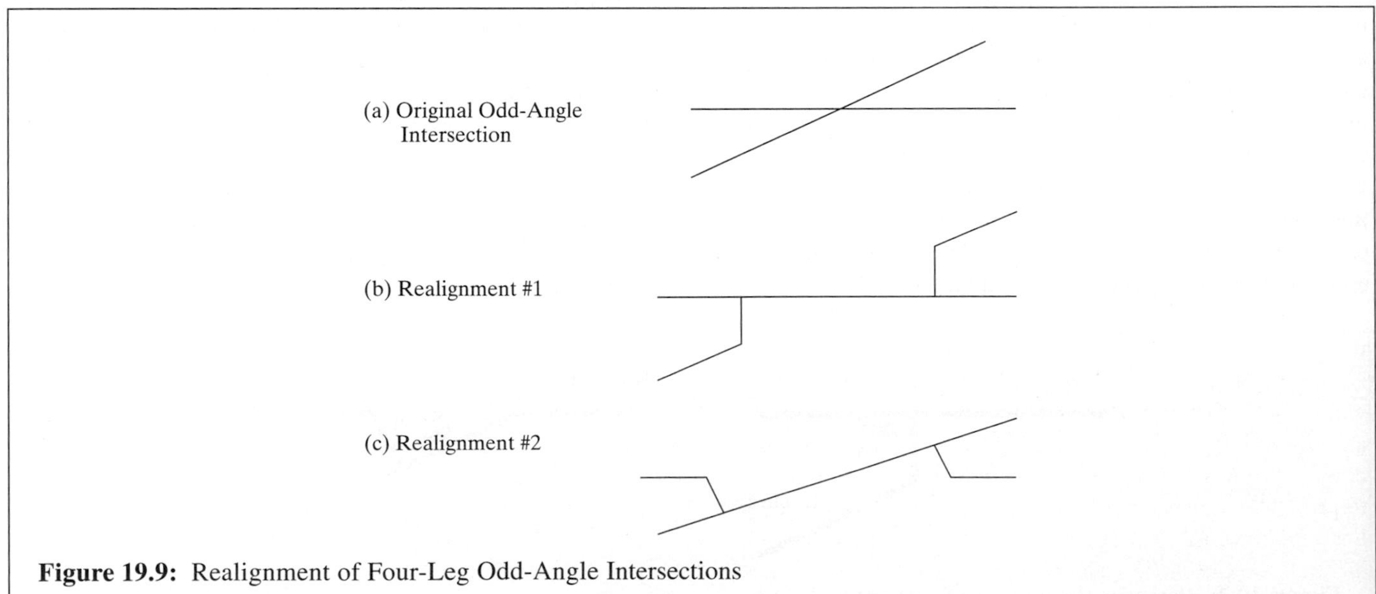

Figure 19.9: Realignment of Four-Leg Odd-Angle Intersections

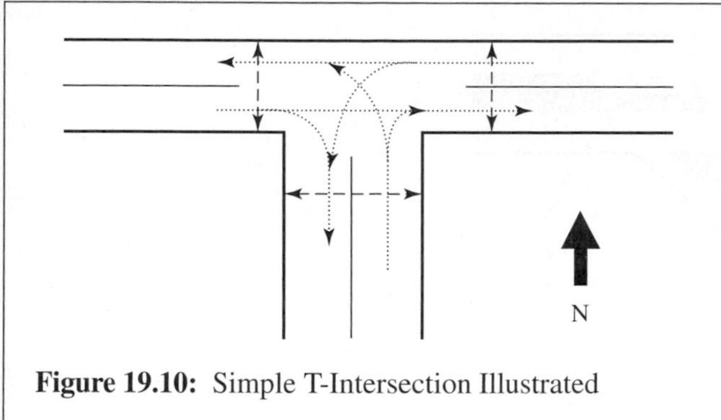

Figure 19.10: Simple T-Intersection Illustrated

The intersection shown in Figure 19.10 has one lane for each approach. There are no channelized movements or left-turn lanes. If visibility is not appropriate for uncontrolled operation under basic rules of the road, then the options of STOP/YIELD control or signalization must be considered. The normal warrants would apply.

The T-intersection form, however, presents some relatively unique characteristics that influence how control is applied. STOP-control is usually applied to the stem of the T-intersection, although it is possible to apply two-way STOP control to the cross street if movements into and out of the stem dominate.

If needed, the form of signalization applied to the intersection of Figure 19.10 depends entirely on the need to protect the (WB) opposed left turn. A protected phase is normally suggested if the left-turn volume exceeds 200 veh/h or the cross-product of the left-turn volume and the opposing volume per lane exceeds 50,000. If left-turn protection is not needed, a simple two-phase signal plan is used. If the opposed left-turn must be protected and there is no left-turn lane available (as in Figure 19.10), a three-phase plan must be used. Figure 19.11 illustrates the possible signal plans for the T-intersection of Figure 19.10. The three-phase plan is relatively inefficient because a separate phase is needed for each of the three approaches.

Where a protected left-turn phase is desirable, the addition of an exclusive left-turn lane would simplify the signalization. Channelization and some additional right-of-way would be required to do this. Channelization can also be applied in other ways to simplify the overall operation and control of the intersection. Channelizing islands can be used to create separated right-turn paths for vehicles entering and leaving the stem via right turns. Such movements would be YIELD-controlled, regardless of the primary form of intersection control.

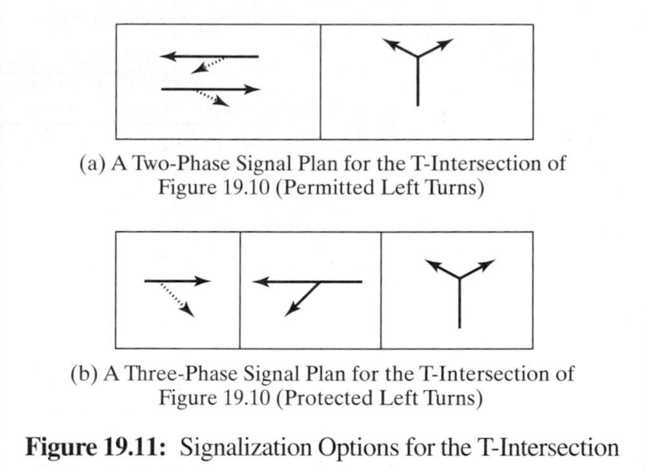

(a) A Two-Phase Signal Plan for the T-Intersection of Figure 19.10 (Permitted Left Turns)

(b) A Three-Phase Signal Plan for the T-Intersection of Figure 19.10 (Protected Left Turns)

Figure 19.11: Signalization Options for the T-Intersection of Figure 19.10

Figure 19.12 shows a T-intersection in which a left-turn lane is provided for the opposed left turn. Right turns are also channelized. Assuming that a signal with a protected left turn is needed at this location, the signal plan shown could be implemented. This plan is far more efficient than that of Figure 19.11 because EB and WB through flows can move simultaneously. Right turns move more or less continuously through the YIELD-controlled channelized turning roadways. The potential for queues to block access to the right-turn roadways, however, should be considered in timing the signal.

Right turns can be completely eliminated from the signal plan if volumes are sufficient to allow lane drops or additions for the right-turning movements, as illustrated in Figure 19.13. Right turns into and out of the stem of the T-intersection become continuous movements.

19.4.3 Offset Intersections

One of the traffic engineer's most difficult problems is the safe operation of high-volume offset intersections. Figure 19.14 illustrates such an intersection with a modest right offset. In the case illustrated, the driver needs more sight distance (when compared with a perfectly aligned 90° intersection) to observe vehicles approaching from the right. The obstruction caused by the building becomes a more serious problem because of this. In addition to sight-distance problems, the offset intersection distorts the normal trajectory of all movements, creating accident risks that do not exist at aligned intersections.

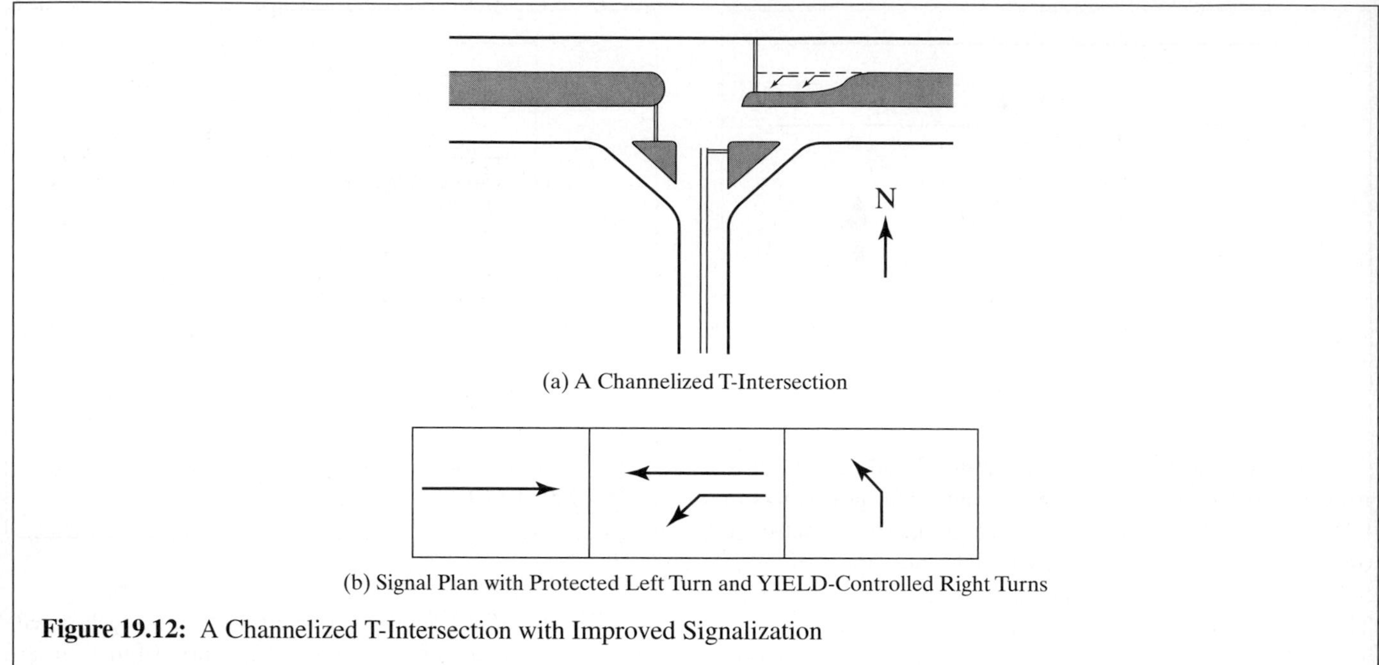

(a) A Channelized T-Intersection

(b) Signal Plan with Protected Left Turn and YIELD-Controlled Right Turns

Figure 19.12: A Channelized T-Intersection with Improved Signalization

Offset intersections are rarely consciously designed. They are necessitated by a variety of situations, generally involving long-standing historic development patterns. Figure 19.15 illustrates a relatively common situation in which offset intersections occur.

In many older urban or suburban developments, zoning and other regulations were (and in some cases, still are) not particularly stringent. Additional development was considered to be an economic benefit because it added to the property tax base of the community involved. Firm control over the specific design of subdivision developments, therefore, is not always exercised by zoning boards and authorities.

The situation depicted in Figure 19.15 occurs when Developer A obtains the land to the south of a major arterial and lays out a circulation system that will maximize the number of building lots that can be accommodated on the parcel. At a later time, Developer B obtains the rights to

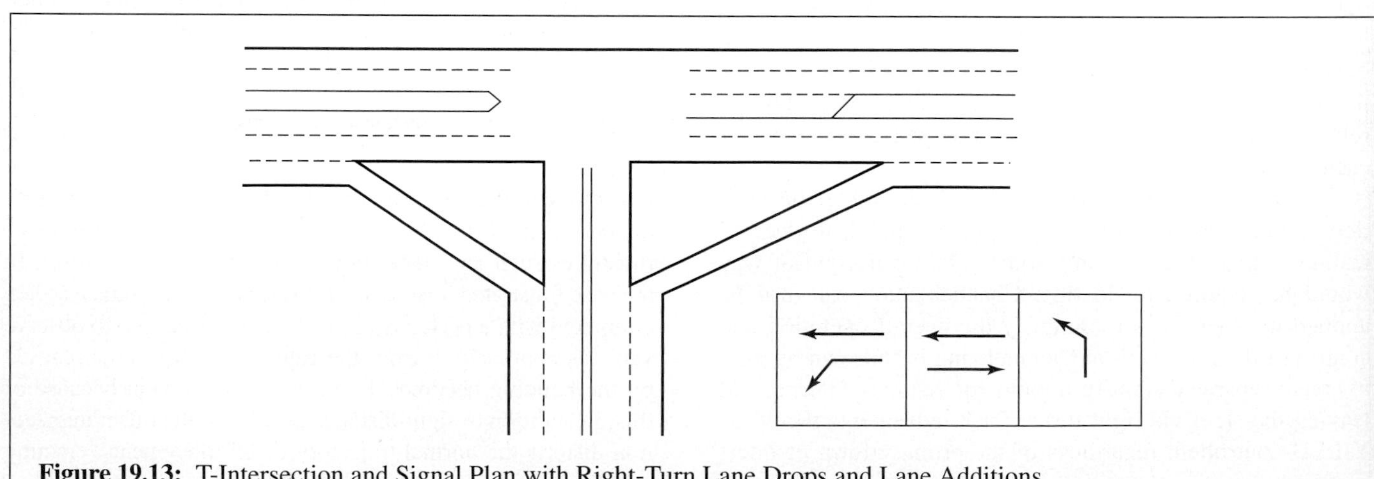

Figure 19.13: T-Intersection and Signal Plan with Right-Turn Lane Drops and Lane Additions

Figure 19.14: Offset Intersection with Sight Distance and Trajectory Problems

land north of the same arterial. Again, an internal layout that provides the maximum number of development parcels is selected. Without a strong planning board or other oversight group requiring it, there is no guarantee that opposing local streets will "line up." Offsets can and do occur frequently in such circumstances. In urban and suburban environments, it is rarely possible to acquire sufficient right-of-way to realign the intersections; therefore, other approaches to control and operation of such intersections must be considered.

Two major operational problems are posed by a right-offset intersection, as illustrated in Figure 19.16. In Figure 19.16 (a), the left-turn trajectories from the offset legs involve a high level of hazard. Unlike the situation with an aligned intersection, a vehicle turning left from either offset leg is in conflict with the opposing through vehicle almost immediately after crossing the STOP line. To avoid this conflict, left-turning vehicles must bear right as if they were going to go through to the opposite leg, beginning their left turns only when they are approximately halfway through the intersection. This, of course, is not a natural movement, and a high incidence of left-turn accidents often result at such intersections.

In Figure 19.16 (b), the hazard to pedestrians crossing the aligned roadway is highlighted. Two paths are possible, and both are reasonably intuitive for pedestrians: They can cross from corner to corner, following an angled crossing path, or they can cross perpendicularly. The latter places one end of

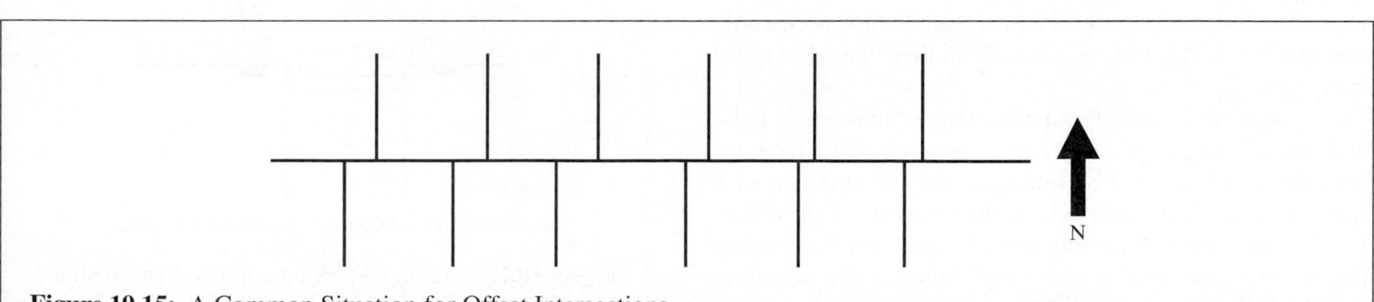

Figure 19.15: A Common Situation for Offset Intersections

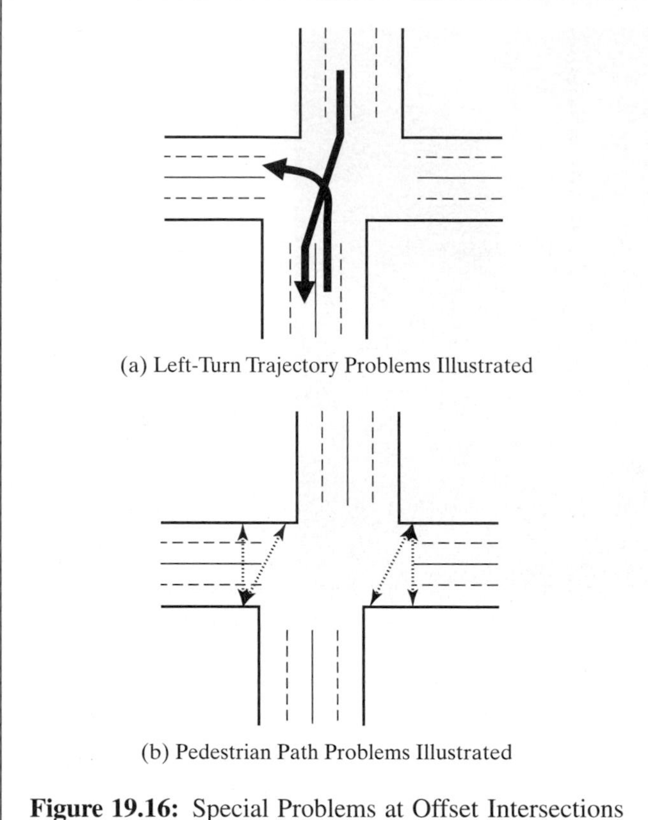
(a) Left-Turn Trajectory Problems Illustrated

(b) Pedestrian Path Problems Illustrated

Figure 19.16: Special Problems at Offset Intersections

their crossing away from the street corner. Perpendicular crossings, however, minimize the crossing time and distance. However, right-turning vehicles encounter the pedestrian conflict at an unexpected location, after they have virtually completed their right turn. Diagonal crossings increase the exposure of pedestrians, but conflicts with right-turning vehicles are closer to the normal location.

Yet another special hazard at offset intersections, not clearly illustrated by Figure 19.16, is the heightened risk of sideswipe accidents as vehicles cross between the offset legs. Because the required angular path is not necessarily obvious, more vehicles will stray from their lane during the crossing.

There are, however, remedies that will minimize these additional hazards. Where the intersection is signalized, the left-turn conflict can be eliminated through the use of a fully protected left-turn phase in the direction of the offset. In this case, the left-turning vehicles will not be entering the intersection area at the same time as the opposing through vehicles. This requires, however, that one of the existing lanes be designated an exclusive turning lane or that a left-turn lane can be added to each offset leg. If this is not possible, a more extreme remedy is to provide each of the offset legs with an exclusive signal phase. Although this separates the left-turning vehicles from the opposing flows, it is an inefficient signal plan and can lead to four-phase signalization if left-turn phases are needed on the aligned arterial.

For pedestrian safety, it is absolutely necessary that the traffic engineer clearly designate the intended path they are to take. This is done through proper use of markings, signs, and pedestrian signals, as shown in Figure 19.17.

Crosswalk locations influence the location of STOP-lines and the position of pedestrian signals, which must be located in the line of sight (which is the walking path) of pedestrians. Vehicular signal timing is also influenced by the crossing paths implemented. Where perpendicular crossings are used, the distance between STOP-lines on the aligned street can be considerably longer than for diagonal crossings.

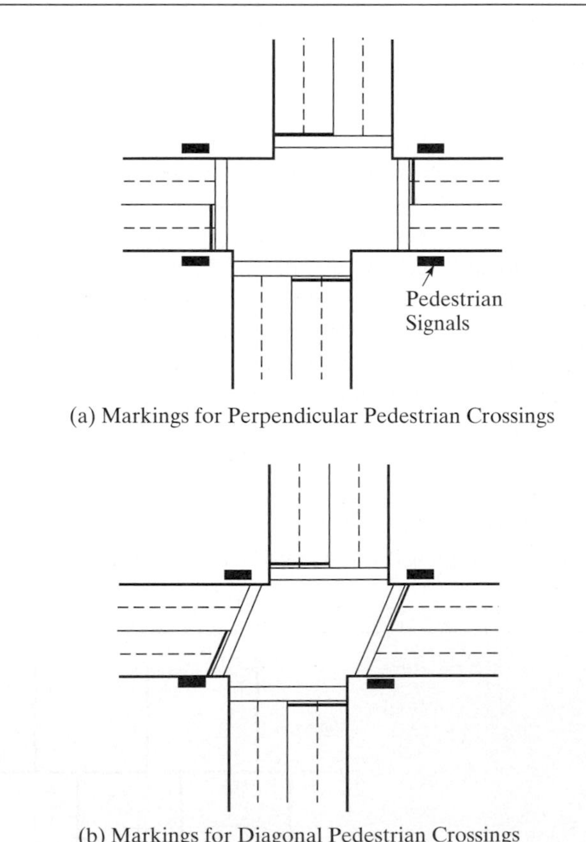

(a) Markings for Perpendicular Pedestrian Crossings

(b) Markings for Diagonal Pedestrian Crossings

Figure 19.17: Signing, Markings, and Pedestrian Signals for a Right-Offset Intersection

This increases the length of the all-red interval for the aligned street and adds lost time to the signal cycle.

In extreme cases, where enforcement of perpendicular crossings becomes difficult, barriers can be placed at normal street corner locations, preventing pedestrians from entering the street at an inappropriate or unintended location.

To help vehicles follow appropriate paths through the offset intersection, dashed lane and centerline markings through the intersection may be added, as illustrated in Figure 19.18. The extended centerline marking would be yellow, and the lane lines would be white.

Left-offset intersections share some of the same problems as right-offset intersections. The left-turn interaction with the opposing through flow is not as critical, however. The pedestrian–right-turn interaction is different but potentially just as serious. Figure 19.19 illustrates.

The left-turn trajectory through the offset intersection is still quite different from an aligned intersection, but the left-turn movement does not thrust the vehicle immediately into the path of the oncoming through movement, as in a right-offset intersection. Sideswipe accidents are still a risk, and extended lane markings would be used to minimize this risk.

At a left-offset intersection, the diagonal pedestrian path is more difficult because it brings the pedestrian into immediate conflict with right-turning vehicles more quickly than at an aligned intersection. For this reason, diagonal crossings are generally not recommended at left-offset intersections. The signing, marking, and signalization of perpendicular pedestrian crossings is similar to that used at a right-offset intersection.

When at all possible, offset intersections should be avoided. If sufficient right-of-way is available, basic realignment should be seriously considered. When confronted with such a situation, however, the traffic engineering approaches discussed here can ameliorate some of the fundamental concerns associated with offset alignments. The traffic engineer should recognize that many of these measures will negatively affect capacity of the approaches due to the additional signal phases and longer lost times often involved. This is, however, a necessary price paid to optimize safety of intersection operation.

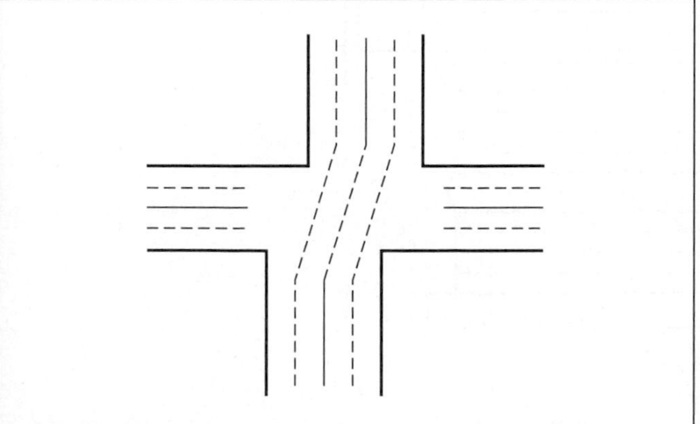

Figure 19.18: Dashed Lane and Centerline Through an Offset Intersection

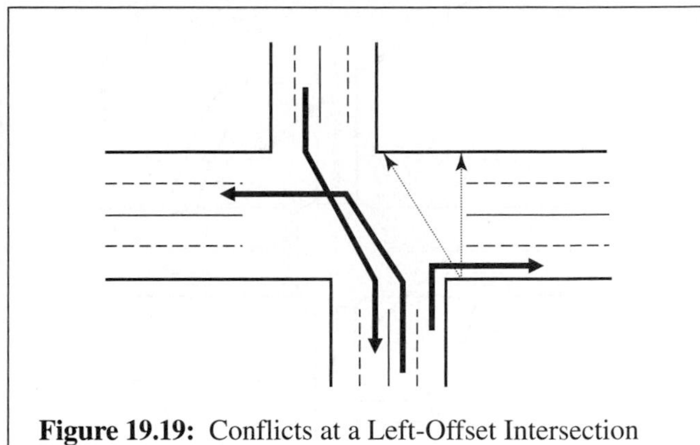

Figure 19.19: Conflicts at a Left-Offset Intersection

19.4.4 Special Treatments for Heavy Left-Turn Movements

Some of the most difficult intersection problems to solve involve heavy left-turn movements on major arterials. Accommodating such turns usually requires the addition of protected left-turn phasing, which often reduces the effective capacity to handle through movements. In some cases, adding an exclusive left-turn phase or phases is not practical, given the associated losses in through capacity.

Alternative treatments must be sought to handle such left-turn movements, with the objective of maintaining two-phase signalization at the intersection. Several design and control treatments are possible, including:

- Prohibition of left turns
- Provision of jug-handles
- Provision of at-grade loops and diamond ramps
- Provision of a continuous-flow intersection
- Provision of U-turn treatments

Prohibition of left turns is rarely a practical option for a heavy left-turn demand. Alternative paths would be needed to

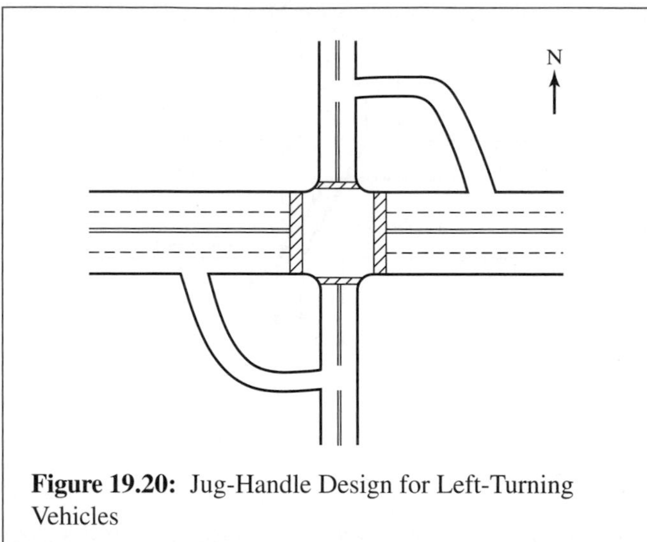

Figure 19.20: Jug-Handle Design for Left-Turning Vehicles

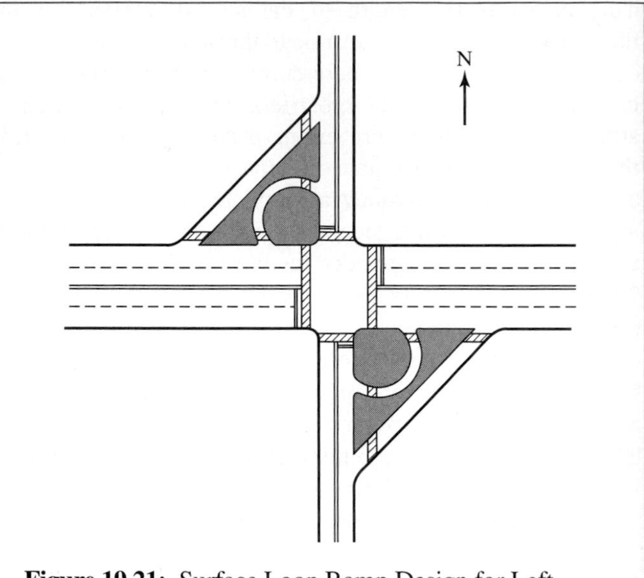

Figure 19.21: Surface Loop Ramp Design for Left Turns

accommodate the demand for this movement, and diversion of a heavy flow onto an "around-the-block" or similar path often creates problems elsewhere.

Figure 19.20 illustrates the use of jug-handles for handling left turns. In effect, left-turners enter a surface ramp on the right, executing a left turn onto the cross street. The jug-handle may also handle right-turn movements. The design creates two new intersections. Depending on volumes, these may require signalization or could be controlled with STOP signs. In either case, queuing between the main intersection and the two new intersections is a critical issue. Queues should not block egress from either of the jug-handle lanes. The provision of jug-handles also requires sufficient right-of-way available to accommodate the solution. In some extreme cases, existing local streets may be used to form a jug-handle pattern.

Figure 19.21 illustrates the use of surface loop ramps to handle heavy left-turning movements at an arterial intersection. These are generally combined with surface diamond ramps to

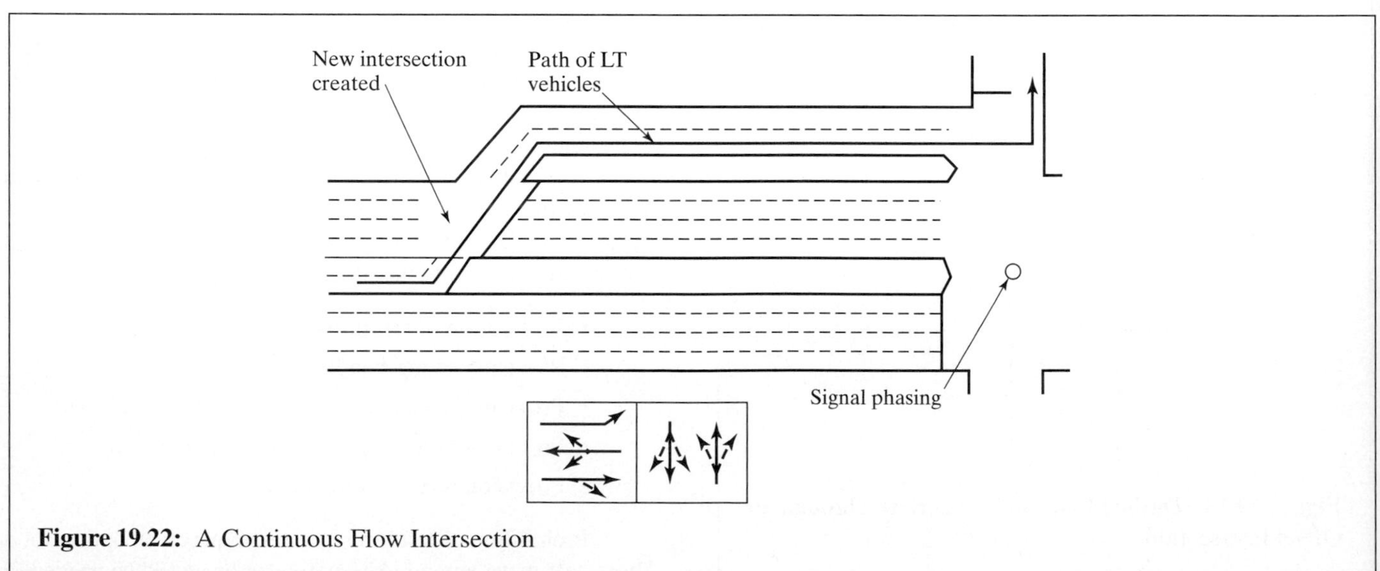

Figure 19.22: A Continuous Flow Intersection

handle right turns from the cross street, thus avoiding the conflict between normal right turns and the loop ramp movements on the arterial. Once again, queuing could become a problem if left-turning vehicles back up along the loop ramp far enough to affect the flow of vehicles that can enter the loop ramp. This option also consumes considerable right-of-way and may be difficult to implement in high-density environments.

Figure 19.22 illustrates a continuous-flow intersection, a relatively novel design approach developed during the late 1980s and early 1990s. The continuous-flow intersection [*6*] takes a single intersection with complex multiphase signalization and separates it into two intersections, each of which can be operated with a two-phase signal and coordinated. At the new intersection, located upstream of the left-turn location, left-turning vehicles are essentially transferred to a separate roadway on the left side of the arterial. At the main intersection, the left turns can then be made without a protected phase, regardless of the demand level. The design requires sufficient right-of-way on one side of the arterial to create the new left-turn roadway and a median that is wide enough to provide one or two left-turning lanes at the new intersection. Queuing from the main intersection can become a problem if left-turning vehicles are blocked from entering the left-turn lane(s) at the new upstream intersection.

Although a few continuous-flow intersections have been built, they have not seen the widespread use that was originally anticipated. In most cases, right-of-way restrictions make this solution somewhat impractical.

As a last resort, left turns may be handled in a variety of ways as U-turn movements. Figure 19.23 illustrates four

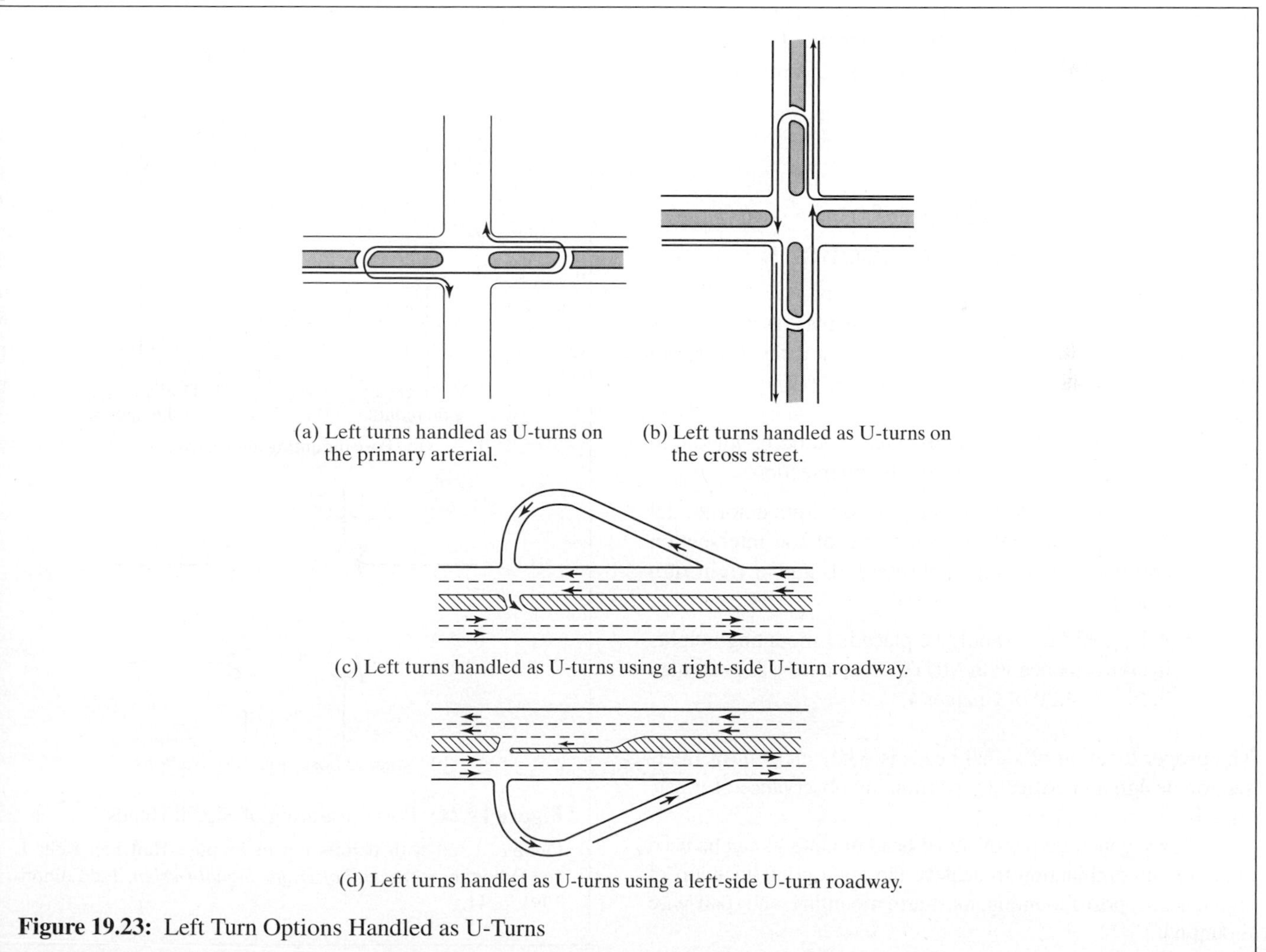

(a) Left turns handled as U-turns on the primary arterial.

(b) Left turns handled as U-turns on the cross street.

(c) Left turns handled as U-turns using a right-side U-turn roadway.

(d) Left turns handled as U-turns using a left-side U-turn roadway.

Figure 19.23: Left Turn Options Handled as U-Turns

potential designs for doing this. In Figure 19.23 (a), left-turning vehicles go through the intersection and make a U-turn through a wide median downstream. The distance between the U-turn location and the main intersection must be sufficient to avoid blockage by queued vehicles and must provide sufficient distance for drivers to execute the required number of lane changes to get from the median lane to right lane. In Figure 19.23 (b), left-turning vehicles turn right at the main intersection, then execute a U-turn on the cross street. Queuing and lane-changing requirements are similar to those described for Figure 19.23 (a). Where medians are narrow, the U-turn paths of (a) and (b) cannot be provided. Figures 19.23 (c) and (d) use U-turn roadways built to the right and left sides of the arterial (respectively) to accommodate left-turning movements. These options require additional right-of-way.

The safe and efficient accommodation of heavy left-turn movements on arterials often requires creative approaches that combine both design and control elements. The examples shown here are intended to be illustrations, not a complete review of all possible alternatives.

19.5 Street Hardware for Signalized Intersections

In Chapter 4, the basic requirements for display of signal faces at a signalized intersection were discussed in detail. These are the key specifications:

- A minimum of two signal faces should be visible to each primary movement in the intersection.
- All signal faces should be placed within a horizontal 20° angle around the centerline of the intersection approach (including exclusive left- and/or right-turn lanes).
- All signal faces should be placed at mounting heights in conformance with MUTCD standards, as presented in Figure 4.20 of Chapter 4.

The proper location of signal heads is a key element of intersection design and critical to maximizing observance of traffic signals.

Three general types of signal-head mountings can be used alone or in combination to achieve the appropriate location of signal heads: post-mounting, mast-arm mounting, and span-wire mounting.

Figure 19.24 illustrates post-mounting. The signal head can be oriented either vertically or horizontally, as shown. Post-mounted signals are located on each street corner. A post-mounted signal head generally has two faces, oriented such that a driver sees two faces located on each of the far intersection corners. Because they are located on street

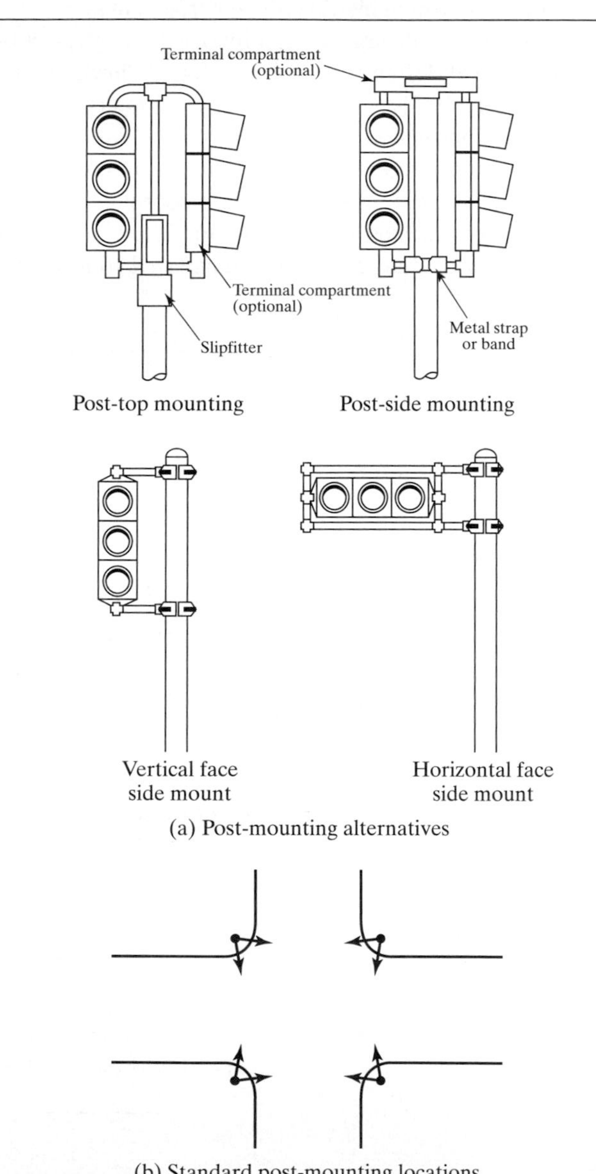

Figure 19.24: Post-Mounting of Signal Heads

(*Source:* Used with permission of Prentice-Hall Inc, Kell, J. and Fullerton, I., *Manual of Traffic Signal Design,* 2nd Edition, 1991, p. 44.)

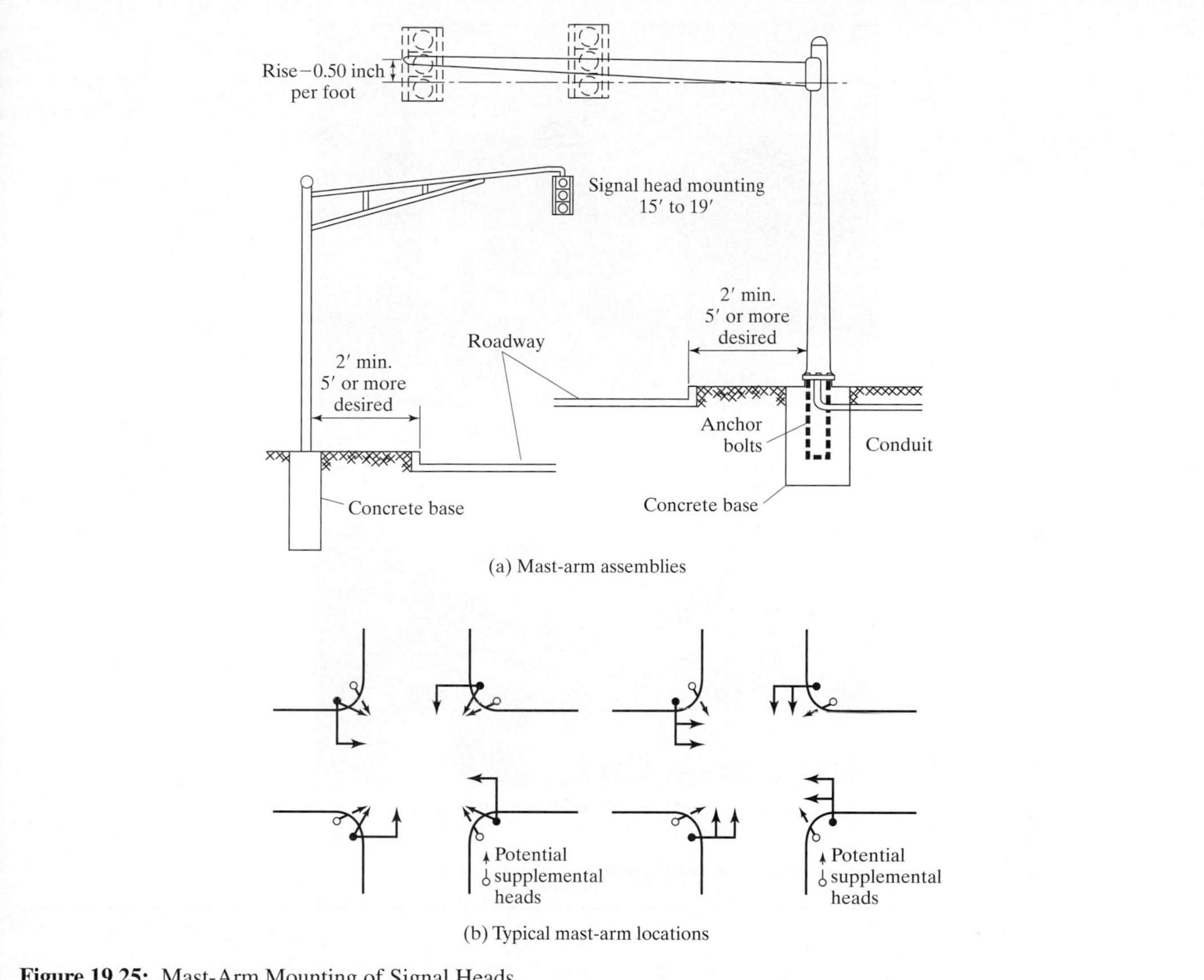

Figure 19.25: Mast-Arm Mounting of Signal Heads

(*Source:* Used with permission of Prentice-Hall Inc, Kell, J. and Fullerton, I., *Manual of Traffic Signal Design,* 2nd Edition, 1991, p. 57.)

corners, care must be taken to ensure that post-mounted signals fall within the required 20° angle of the approach centerline. Post-mounted signals are often inappropriate for use at intersections with narrow streets because street corners in such circumstance lie outside of the visibility requirement.

Figure 19.25 illustrates mast-arm mounting of signal heads. Typically, the mast arm is perpendicular to the intersection approach. They are located so that drivers are looking at a signal face or faces on the far side of the intersection. Mast arms can be long enough to accommodate two signal heads, but they are rarely used for more than two signal heads.

Figure 19.26 shows two typical mast-arm signal installations. The first (a) shows mast-arm signals at a four-leg intersection, with the mast-arm oriented perpendicular to the direction of traffic. Note that the mast-arm signal heads are supplemented by a post-mounted signal in the gore of the four-leg intersection. The second (b) represents a very efficient scheme for mounting signal heads at a simple intersection of two two-lane streets. Two mast

(a) Mast-Arm Mounted Signals at a Y-Intersection

(b) Mast-Arm Mounted Signals at the Intersection of Two 2-Lane Streets

Figure 19.26: Two Examples of Mast-Arm Mounted Signals

arms are used, each extending diagonally across the intersection. Only two signal heads are used, each with a full four faces. In this way, using only two signal heads, all movements have two signal faces displaying the same signal interval.

In the case of both post-mounted and mast-arm–mounted signal heads, power lines are carried to the signal head within the hollow structure of the post or mast arm.

The most common method for mounting signal heads is span wire because it is the most flexible and can be used in a variety of configurations. Figure 19.27 shows four basic configurations in which span wires can be used. The first is a single diagonal span wire between two intersection corners. The span wire allows the installation of a number of signal heads, each having between one and three faces, depending on the exact location. Such installations are generally supplemented by post-mounted signals on the two other intersection corners. The second installation illustrated is a "box" design. Four span wires are installed across each intersection leg. Signal heads are oriented much in the same way as with mast arms. Most signal heads have a single face and are visible from the far side of the intersection. The third example is a "modified box," in which the box is suspended over the middle of the intersection. This is done to accomplish signal-face locations that are more visible and more clearly aligned with specific lanes of each intersection approach. The final example of Figure 19.27 is a "lazy Z" pattern in which the primary span wire is anchored on

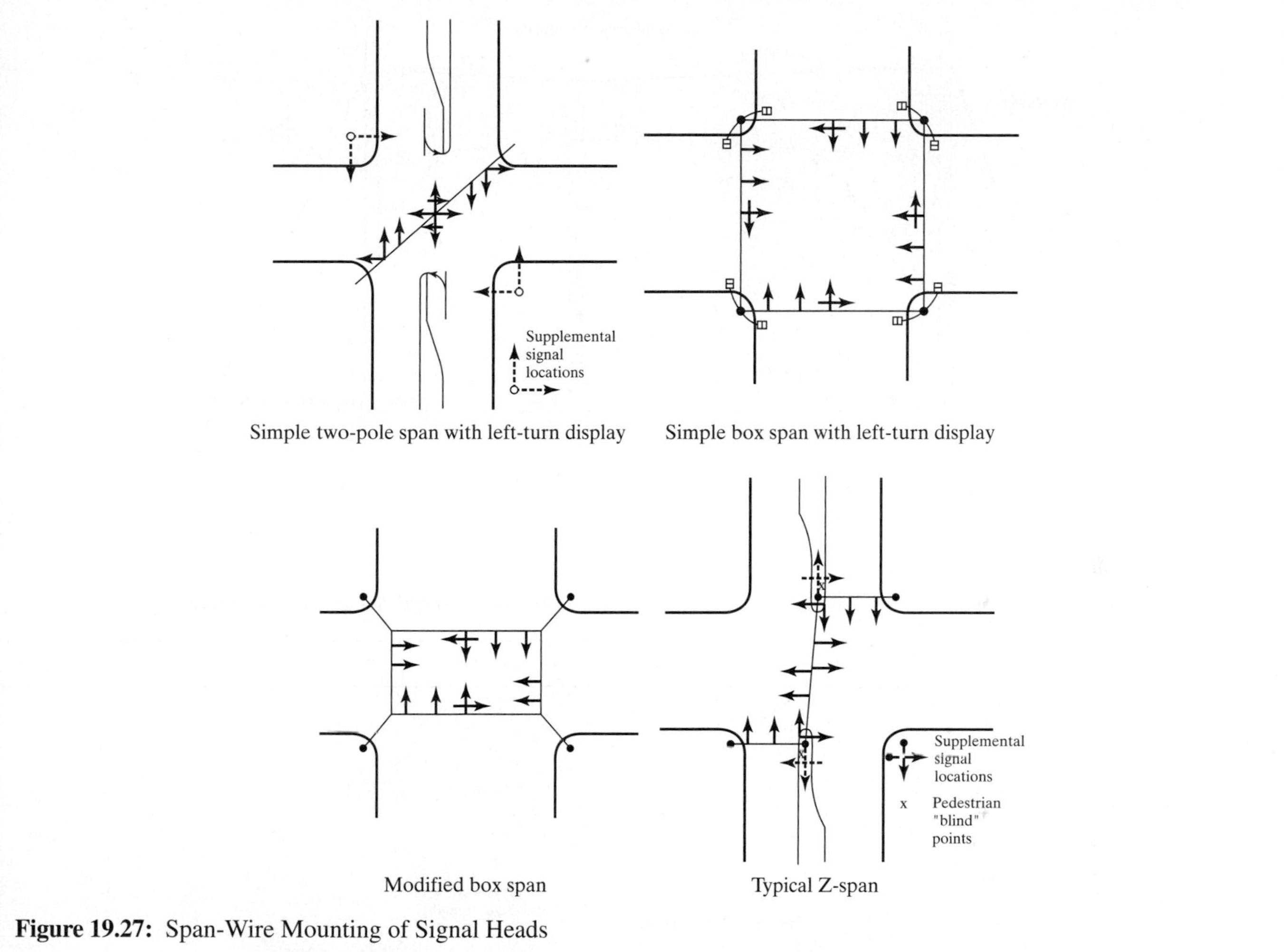

Figure 19.27: Span-Wire Mounting of Signal Heads

(*Source:* Used with permission of Prentice-Hall Inc, Kell, J. and Fullerton, I., *Manual of Traffic Signal Design,* 2nd Edition, 1991, pp. 51–53.)

opposing medians. This latter design is possible only where opposite medians exist.

Span wire allows the traffic engineer to place signal faces in almost any desired position and is often used at complex intersections where a signal face for each entering lane is desired.

Figure 19.28 illustrates how signal heads are anchored on span wires. In general, the main cable supports each signal head from above. Signal heads so mounted can and do sway in the wind. Where wind is excessive or where the exact orientation of the signal face is important, a tether wire may be attached to the bottom of the signal head for restraint. This is most important where Polaroid signal lenses are used. These lenses are visible only when viewed from a designated angle. They are often used at closely spaced signalized intersections, where the traffic engineer uses them to prevent drivers from reacting to the next downstream signal.

Figure 19.29 illustrates how power is supplied to a span-wire mounted signal head. A shielded power cable is wrapped around the primary support wire and connected to each signal head.

Figure 19.30 shows a typical field installation of span-wire mounted signals. In this case, a single span wire supports six signal heads that are sufficient to control all movements, including a left-turn phase on the major street.

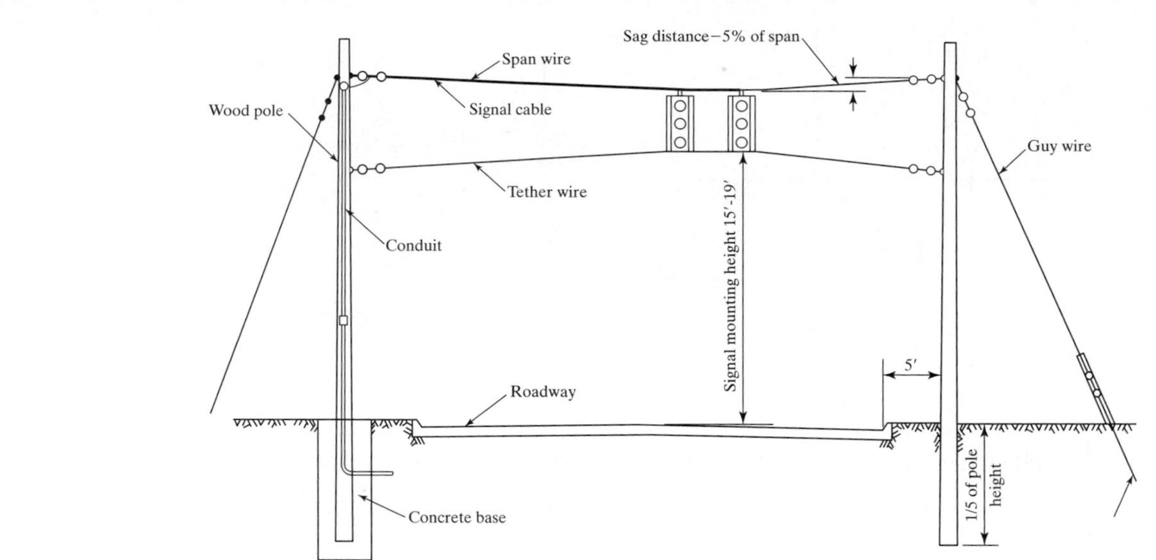

Figure 19.28: Use of Span Wire and Tether Wire Illustrated

(*Source:* Used with permission of Prentice-Hall Inc, Kell, J. and Fullerton, I., *Manual of Traffic Signal Design,* 2nd Edition, 1991.)

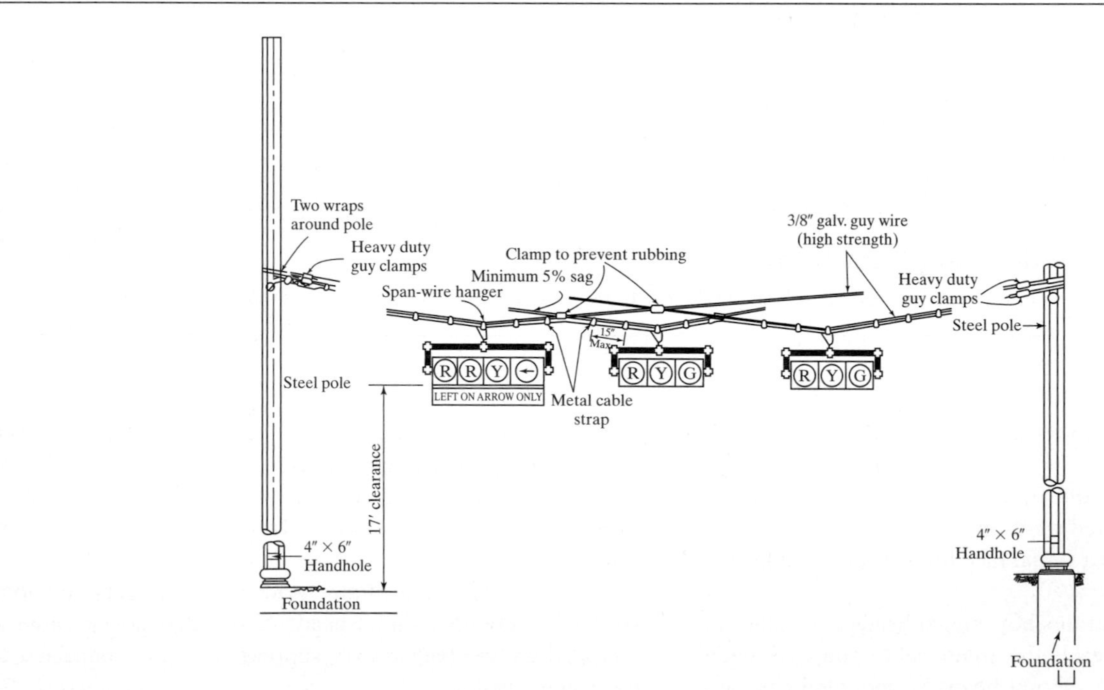

Figure 19.29: Providing Power to Span-Mounted Signals

(*Source:* Used with permission of Prentice-Hall Inc, Kell, J. and Fullerton, I., *Manual of Traffic Signal Design,* 2nd Edition, 1991.)

Figure 19.30: A Typical Span-Wire Signal Installation

Using the three signal mounting options (post mounted, mast-arm mounted, span-wire mounted), either alone or in combination, the traffic engineer can satisfy all of the posting requirements of the MUTCD and present drivers with clear and unambiguous operating instructions. Achieving this goal is critical to ensuring safe and efficient operations at signalized intersections.

19.6 Closing Comments

This chapter has provided an overview of several important elements of intersections design. It is not intended to be exhaustive, and we encourage you to consult standard references for additional relevant topics and detail.

References

1. *A Policy on Geometric Design of Highways and Streets,* 5th Edition, American Association of State Highway and Transportation Officials, Washington DC, 2004.
2. *Manual of Uniform Traffic Control Devices,* Federal Highway Administration, U.S. Department of Transportation, 2009.
3. Kell, J., and Fullerton, I., *Manual of Traffic Signal Design,* 2nd Edition, Institute of Transportation Engineers, Washington DC, 1991.
4. *Traffic Detector Handbook,* JHK & Associates, Institute of Transportation Engineers, Washington DC, n.d.
5. *Highway Capacity Manual,* 4th Edition, Transportation Research Board, Washington DC, 2000.
6. Hutchinson, T., "The Continuous Flow Intersection: The Greatest New Development Since the Traffic Signal?" *Traffic Engineering and Control,* 36, no. 3, Printhall Ltd., London, UK, 1995.

Problems

19-1–19-2. Each of the sets of demands shown in Figures 19.31 and 19.32 represent the forecast flows (already adjusted for peak hour factor) expected at new intersections that are created as a result of large new developments. Assume that each intersection will be signalized. In each case, propose a design for the intersection, including a detailing of where and how signal heads would be located.

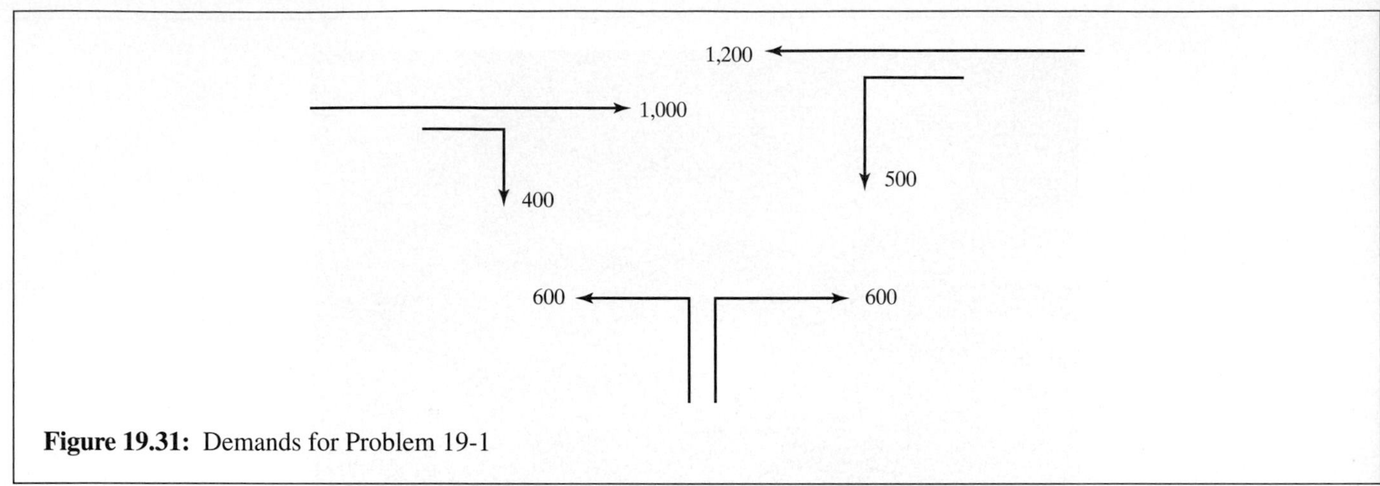

Figure 19.31: Demands for Problem 19-1

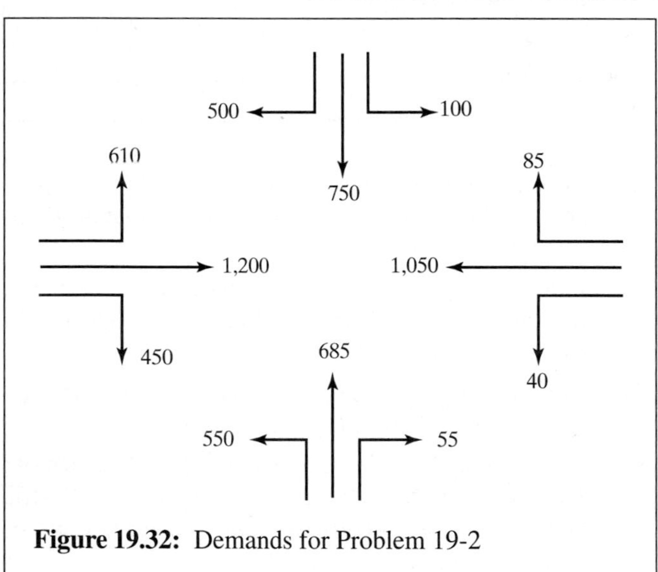

Figure 19.32: Demands for Problem 19-2

CHAPTER 20

Basic Principles of Intersection Signalization

In Chapter 18, various options for intersection control were presented and discussed. Warrants for implementation of traffic control signals at an intersection, presented in the *Manual on Uniform Traffic Control Devices* [1], provide general and specific criteria for selection of an appropriate form of intersection control. At many intersections, the combination of traffic volumes, potential conflicts, overall safety of operation, efficiency of operation, and driver convenience lead to a decision to install traffic control signals.

The operation of signalized intersections is often complex, involving competing vehicular and pedestrian movements. Appropriate methodologies for design and timing of signals and for the operational analysis of signalized intersections require the behavior of drivers and pedestrians at a signalized intersection to be modeled in a form that can be easily manipulated and optimized. This chapter discusses some of the fundamental operational characteristics at a signalized intersection and the ways in which they may be effectively modeled.

In Chapter 21, these principles are applied to a signalized intersection design and timing process for pretimed signals. In Chapters 23 and 24, they are augmented and combined into overall models of signalized intersection operations. The particular model presented in Chapter 24 is that of the *Highway Capacity Manual* [2].

This chapter focuses on four critical aspects of signalized intersection operation:

1. Discharge headways, saturation flow rates, and lost times
2. Allocation of time and the critical-lane concept
3. The concept of left-turn equivalency
4. Delay as a measure of service quality

Other aspects of signalized intersection operation are also important, and the *Highway Capacity Manual* analysis model addresses many of them. These four, however, are central to understanding traffic behavior at signalized intersections and are highlighted here.

20.1 Terms and Definitions

Traffic signals are complex devices that can operate in a variety of different modes. A number of key terms and definitions should be understood before pursuing a more substantive discussion.

20.1.1 Components of a Signal Cycle

The following terms describe portions and subportions of a signal cycle. The most fundamental unit in signal design and timing is the *cycle,* as defined here.

1. *Cycle.* A signal cycle is one complete rotation through all of the indications provided. In general, every legal vehicular movement receives a "green" indication during each cycle, although there are some exceptions to this rule.
2. *Cycle length.* The cycle length is the time (in seconds) that it takes to complete one full cycle of indications. It is given the symbol "C."
3. *Interval.* The interval is a period of time during which no signal indication changes. It is the smallest unit of time described within a signal cycle. There are several types of intervals within a signal cycle:
   a. *Change interval.* The change interval is the "yellow" indication for a given movement. It is part of the transition from "green" to "red," in which movements about to lose "green" are given a "yellow" signal while all other movements have a "red" signal. It is timed to allow a vehicle that cannot safely stop when the "green" is withdrawn to enter the intersection legally. The change interval is given the symbol "y_i" for movement(s) i.
   b. *Clearance interval.* The clearance interval is also part of the transition from "green" to "red" for a given set of movements. During the clearance interval, all movements have a "red" signal. It is timed to allow a vehicle that legally enters the intersection on "yellow" to safely cross the intersection before conflicting flows are released. The clearance interval is given the symbol "ar_i" (for "all red") for movement(s) i.
   c. *Green interval.* Each movement has one green interval during the signal cycle. During a green interval, the movements permitted have a "green" light while all other movements have a "red" light. The green interval is given the symbol "G_i" for movement(s) i.
   d. *Red interval.* Each movement has a red interval during the signal cycle. All movements not permitted have a "red" light while those permitted to move have a "green" light. In general, the red interval overlaps the green, yellow, and all red intervals for all other movements in the intersection. The red interval is given the symbol "R_i" for movement(s) i.

   Note that for a given movement or set of movements, the "red" signal is present during both the *clearance* (all red) and *red* intervals.
4. *Phase.* A signal phase consists of a green interval, plus the change and clearance intervals that follow it. It is a set of intervals that allows a designated movement or set of movements to flow and to be safely halted before release of a conflicting set of movements.

20.1.2 Types of Signal Operation

The traffic signals at an individual intersection can operate on a pretimed basis or may be partially or fully actuated by arriving vehicles or pedestrians sensed by detectors.

1. *Pretimed operation.* In pretimed operation, the cycle length, phase sequence, and timing of each interval are constant. Each cycle of the signal follows the same predetermined plan. Modern signal controllers allow different pretimed settings to be established. An internal clock is used to activate the appropriate timing for each defined time period. In such cases, it is typical to have at least an AM peak, a PM peak, and an off-peak signal timing.
2. *Semi-actuated operation.* In semi-actuated operation, detectors are placed on the minor approach(es) to the intersection; there are no detectors on the major street. The light is green for the major street at all times except when a "call" or actuation is noted on one of the minor approaches. Then, subject to limitations such as a minimum major-street green, the green is transferred to the minor street. The green returns to the major street when the maximum minor-street green is reached or when the detector senses there is no further demand on the minor street. Semi-actuated operation is often used where the primary reason for signalization is "interruption of continuous traffic," as discussed in Chapter 18.
3. *Full actuated operation.* In full actuated operation, every lane of every approach must be monitored by a detector. Green time is allocated in accordance with information from detectors and programmed "rules" established in the controller for capturing and retaining the green. In full actuated operation, the cycle length, sequence of phases, and green time split may vary from cycle to cycle. Chapter 22 presents more detailed descriptions of actuated signal operation, along with a methodology for timing such signals.

In most urban and suburban settings, signalized intersections along arterials and in arterial networks are close enough to have a significant impact on adjacent signalized intersection operations. In such cases, it is common to coordinate signals into a signal system. When coordinated, such systems attempt to keep vehicles moving through sequences of individual signalized intersections without stopping for as long as possible. This is done by controlling the "offsets" between adjacent green signals; that is, the green at a downstream signal initiates "x" seconds after its immediate upstream neighbor. Coordinated signal systems must operate on a common cycle length because offsets cannot be maintained from cycle to cycle if cycle lengths vary at each intersection. Coordination is provided using a variety of technologies:

1. *Master controllers.* A "master controller" provides a linkage between a limited set of signals. Most such controllers can connect from 20 to 30 signals along an arterial or in a network. The master controller provides fixed settings for each offset between connected signals. Settings can be changed for defined periods of the day.
2. *Computer control.* In a computer-controlled system, the computer acts as a "supersized" master controller, coordinating the timings of a large number (hundreds) of signals. The computer selects or calculates an optimal coordination plan based on input from detectors placed throughout the system. In general, such selections are made only once in advance of an AM or PM peak period. The nature of a system transition from one timing plan to another is sufficiently disruptive to be avoided during peak-demand periods in a traditional system. Individual signals in a computer-controlled system generally operate in the pretimed mode.
3. *Adaptive traffic control systems (ATCS).* Since the early 1990s, there has been rapid development and implementation of "adaptive" traffic control systems. In such systems, both individual intersection signal timings and offsets are continually modified in real time based on advanced detection system inputs. In many cases, such systems use actuated controllers at individual intersections. Even though the system still requires a fixed cycle length (which can be changed periodically based on detector input), the allocation of green within a fixed cycle length has been found to be useful in reducing delay and travel times. A critical part of adaptive traffic control systems is the underlying logic of software used to monitor the system and continually update timing patterns. A number of software systems are in use, and the list of products is increasing each year. Some of the more popular systems (in 2009) include SCOOT (Split Cycle Offset Optimization Technique), SCATS (Sydney Coordinated Adaptive Traffic System), RHODES (Real-Time Hierarchical Optimized Distributed Effective System), OPAC (Optimization Policies for Adaptive Control), and ACS-Lite (Adaptive Control System–Lite). In addition to the standard features of signal coordination, such systems usually also incorporate other features, such as bus priority, emergency vehicle priority, traffic gating, and incident detection.

Table 20.1 summarizes the various types of individual signal controllers with key characteristics and guidelines on their most common uses.

Dramatic changes have occurred in the use of traffic signal control technology over the past two decades. Before 1990, all coordinated traffic signal systems on arterials and in networks used pretimed signal controllers exclusively. Today, actuated controllers are regularly coordinated, although, as shown in Table 20.1, they lose one of their principal variable features: cycle length. To coordinate signals, cycle lengths must be common during any given time period, so that the offset between the initiation of green at an upstream intersection and the adjacent downstream intersection is constant for every cycle. Pretimed signals, because they are the cheapest to implement and maintain, are still a popular choice where demands are relatively constant throughout major periods of the day. Where demand levels (and relative demands for various movements) vary significantly during all times of the day, actuated signals are the most likely choice for use. Even when coordinated and using a constant cycle length, the allocation of green times among the defined phases can significantly reduce delay.

20.1.3 Treatment of Left Turns and Right Turns

The modeling of signalized intersection operation would be straightforward if left turns did not exist. Left turns at a signalized intersection can be handled in one of three ways:

1. *Permitted left turns.* A "permitted" left turn movement is one that is made across an opposing flow of vehicles. The driver is permitted to cross through the opposing flow but must select an appropriate gap in the opposing traffic stream through which to turn. This is the most common form of left-turn phasing at

Table 20.1: Signal Controllers and Types of Intersection Control

Type of Operation	Pretimed		Actuated		
	Isolated	**Coordinated**	**Semi-Actuated**	**Fully Actuated**	**Coordinated**
Fixed Cycle Length?	Yes	Yes	No	No	Yes
Conditions Where Applicable	Where detection is not available.	Where traffic is consistent, closely spaced intersections, and where cross street is consistent.	Where defaulting to one movement is desirable, major road is posted <40 mi/h and cross road carries light traffic demand.	Where detection is provided on all approaches, isolated locations where posted speed is >40 mi/h.	Arterial where traffic is heavy and adjacent intersections are nearby.
Example Application	Work zones.	Central business districts, interchanges.	Highway operations.	Locations without nearby signals; rural high-speed locations; intersections of two arterials.	Suburban arterial.
Key Benefit	Temporary application keeps signals operational.	Predictable operations. Lowest cost of equipment and maintenance.	Lower cost for highway maintenance.	Responsive to changing traffic patterns, efficient allocation of green time, reduced delay, and improved safety.	Lower arterial delay, potential reduction in delay for the system, depending upon the settings.

(*Source:* Koonce, P., et al., *Traffic Signal Timing Manual,* Final Report, FHWA Contract No. DTFH61-98-C-00075, Kittelson and Associates Inc, Portland, OR, June 2008, Table 5-1, p. 5-3.)

signalized intersections, used where left-turn volumes are reasonable and where gaps in the opposing flow are adequate to accommodate left turns safely.

2. *Protected left turns.* A "protected" left turn movement is made without an opposing vehicular flow. The signal plan protects left-turning vehicles by stopping the opposing through movement. This requires that the left turns and the opposing through flow be accommodated in separate signal phases and leads to multiphase (more than two) signalization. In some cases, left turns are "protected" by geometry or regulation. Left turns from the stem of a T-intersection, for example, face no opposing flow because there is no opposing approach to the intersection. Left turns from a one-way street similarly do not face an opposing flow.
3. *Compound left turns.* More complicated signal timing can be designed in which left turns are protected for a portion of the signal cycle and are permitted in another portion of the cycle. Protected and permitted portions of the cycle can be provided in any order. Such phasing is also referred to as *protected plus permitted* or *permitted plus protected,* depending on the order of the sequence.

The permitted left turn movement is very complex. It involves the conflict between a left turn and an opposing through movement. The operation is affected by the left-turn flow rate and the opposing flow rate, the number of opposing lanes, whether left turns flow from an exclusive left-turn lane or from a shared lane, and the details of the signal timing. Modeling the interaction among these elements is a complicated process, one that often involves iterative elements.

The terms *protected* and *permitted* may also be applied to right turns. In this case, however, the conflict is between the right-turn vehicular movement and the pedestrian movement in the conflicting crosswalk. The vast majority of right turns at signalized intersections are handled on a permitted basis. Protected right turns generally occur at locations where there are overpasses or underpasses provided for pedestrians. At these locations, pedestrians are prohibited from making surface crossings; barriers are often required to enforce such a prohibition.

20.2 Discharge Headways, Saturation Flow, Lost Times, and Capacity

The fundamental element of a signalized intersection is the periodic stopping and restarting of the traffic stream. Figure 20.1 illustrates this process. When the light turns GREEN, there is a queue of stored vehicles that were stopped during the preceding RED interval, waiting to be discharged. As the queue of vehicles moves, headway measurements are taken as follows:

- The first headway is the time lapse between the initiation of the GREEN signal and the time that the front wheels of the first vehicle cross the stop line.
- The second headway is the time lapse between the time that the first vehicle's front wheels cross the stop line and the time that the second vehicle's front wheels cross the stop line.
- Subsequent headways are similarly measured.
- Only headways through the last vehicle in queue (at the initiation of the GREEN light) are considered to be operating under "saturated" conditions.

If many queues of vehicles are observed at a given location and the average headway is plotted versus the queue position of the vehicle, a trend similar to that shown in Figure 20.1 (b) emerges.

The first headway is relatively long. The first driver must go through the full perception-reaction sequence, move his or her foot from the brake to the accelerator, and accelerate through the intersection. The second headway is shorter because the second driver can overlap the perception-reaction and acceleration process of the first driver. Each successive headway is a little bit smaller than the last. Eventually, the headways tend to level out. This generally occurs when

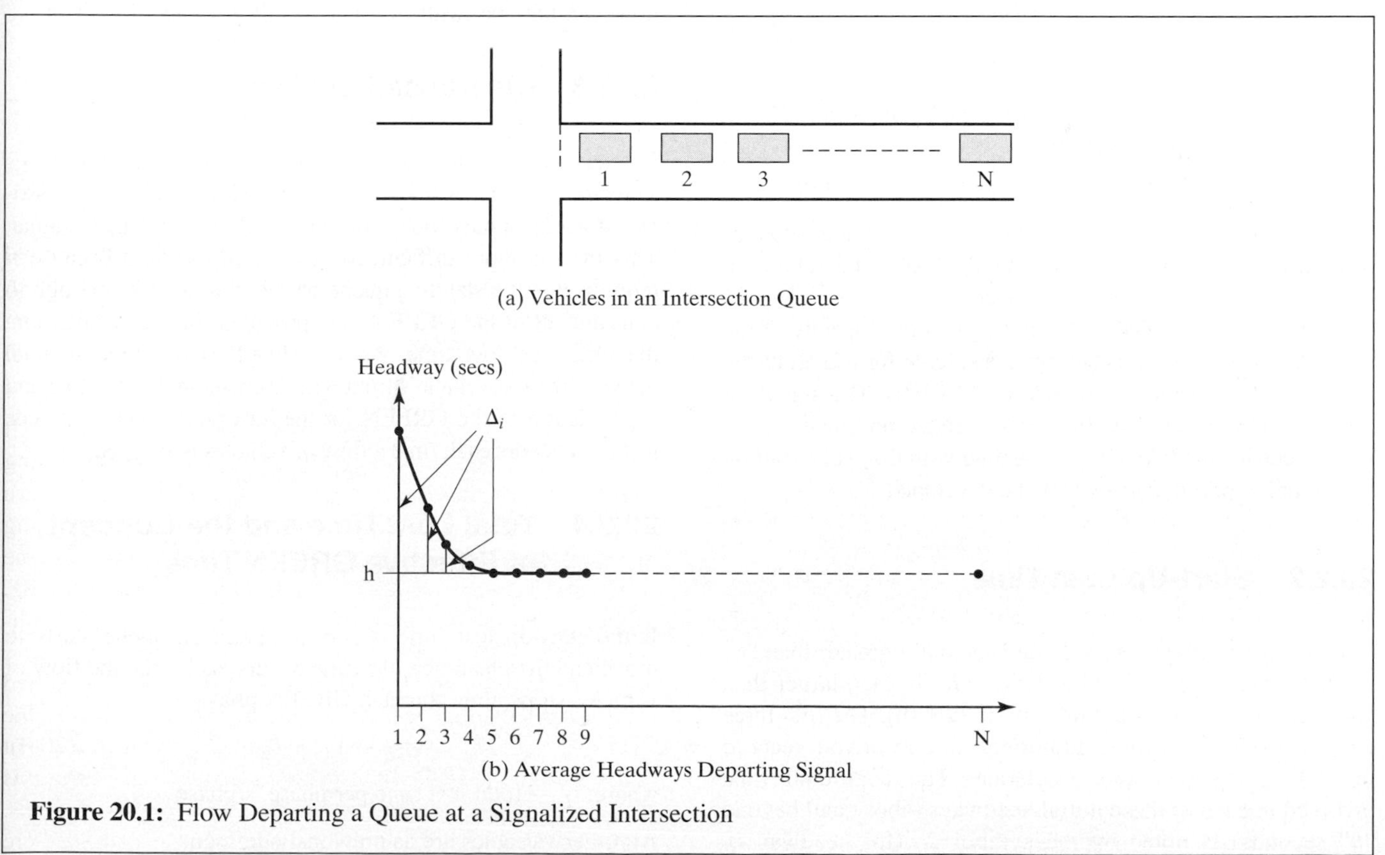

Figure 20.1: Flow Departing a Queue at a Signalized Intersection

queued vehicles have fully accelerated by the time they cross the stop line. At this point, a stable moving queue has been established.

20.2.1 Saturation Headway and Saturation Flow Rate

As noted, average headways tend toward a constant value. In general, this occurs from the fourth or fifth headway position. The constant headway achieved is referred to as the *saturation headway* because it is the average headway that can be achieved by a saturated, stable moving queue of vehicles passing through the signal. It is given the symbol "*h*," in units of s/veh.

It is convenient to model behavior at a signalized intersection by assuming that every vehicle (in a given lane) consumes an average of "*h*" seconds of green time to enter the intersection. If every vehicle consumes "*h*" seconds of green time and if the signal were *always* green, then "*s*" vehicles per hour could enter the intersection. This is referred to as the *saturation flow rate:*

$$s = \frac{3{,}600}{h} \tag{20-1}$$

where: s = saturation flow rate, vehicles per hour of green per lane (veh/hg/ln)
h = saturation headway, seconds/vehicle (s/veh)

Saturation flow rate can be multiplied by the number of lanes provided for a given set of movements to obtain a saturation flow rate for a lane group or approach.

The saturation flow rate, in effect, is the capacity of the approach lane or lanes if they were available for use all of the time (i.e., if the signal were always GREEN). The signal, of course, is not always GREEN for any given movement. Thus, some mechanism (or model) for dealing with the cyclic starting and stopping of movements must be developed.

20.2.2 Start-Up Lost Time

The average headway per vehicle is actually greater than "*h*" seconds. The first several headways are, in fact, larger than "*h*" seconds, as illustrated in Exhibit 20.1 (b). The first three or four headways involve additional time as drivers react to the GREEN signal and accelerate. The additional time involved in each of these initial headways (above and beyond "*h*" seconds) is noted by the symbol Δ_i (for headway *i*). These additional times are added and referred to as the *start-up lost time:*

$$\ell_1 = \sum_i \Delta_i \tag{20-2}$$

where: ℓ_1 = start-up lost time, s/phase
Δ_i = incremental headway (above "*h*" seconds) for vehicle *i*, s

Thus it is possible to model the amount of GREEN time required to discharge a queue of "*n*" vehicles as:

$$T_n = \ell_1 + nh \tag{20-3}$$

where: T_n = GREEN time required to move queue of "*n*" vehicles through a signalized intersection, s
ℓ_1 = start-up lost time, s/phase
n = number of vehicles in queue
h = saturation headway, s/veh

Although this particular model is not of great use, it does illustrate the basic concepts of saturation headway and start-up lost times. The start-up lost time is thought of as a period of time that is "lost" to vehicle use. Remaining GREEN time, however, may be assumed to be usable at a rate of *h* s/veh.

20.2.3 Clearance Lost Time

The start-up lost time occurs every time a queue of vehicles starts moving on a GREEN signal. There is also a lost time associated with stopping the queue at the end of the GREEN signal. This time is more difficult to observe in the field because it requires that the standing queue of vehicles be large enough to consume all of the GREEN time provided. In such a situation, the clearance lost time, ℓ_2, is defined as the time interval between the last vehicle's front wheels crossing the stop line and the initiation of the GREEN for the *next* phase. The clearance lost time occurs each time a flow of vehicles is stopped.

20.2.4 Total Lost Time and the Concept of Effective GREEN Time

If the start-up lost time occurs each time a queue starts to move and the clearance lost time occurs each time the flow of vehicles stops, then for each GREEN phase:

$$t_L = \ell_1 + \ell_2 \tag{20-4}$$

where: t_L = total lost time per phase, s/phase

All other variables are as previously defined.

The concept of lost times leads to the concept of *effective green time.* The actual signal goes through a sequence of intervals for each signal phase:

- Green
- Yellow
- All-red
- Red

The "yellow" and "all-red" intervals are a transition between GREEN and RED. This must be provided because vehicles cannot stop instantaneously when the light changes. The "all-red" is a period of time during which all lights in all directions are red. During the RED interval for one set of movements, another set of movements goes through the green, yellow, and all-red intervals. These intervals are defined more precisely in Chapter 21.

In terms of modeling, there are really only two time periods of interest: *effective green time* and *effective red time.* For any given set of movements, *effective green time* is the amount of time that vehicles can move (at a rate of one vehicle every h seconds). The *effective red time* is the amount of time that they cannot move (at a rate of one vehicle every h seconds). Effective green time is related to actual green time as follows:

$$g_i = G_i + Y_i - t_{Li} \qquad (20\text{-}5)$$

where: g_i = effective green time for movement(s) i, s
G_i = actual green time for movement(s) i, s
Y_i = sum of yellow and all red intervals for movement(s) i, ($Y_i = y_i + ar_i$)
y_i = yellow interval for movement(s) i, s
ar_i = all-red interval for movement(s) i, s
t_{Li} = total lost time for movement(s) i, s

This model results in an effective green time that may be fully used by vehicles at the saturation flow rate (i.e., at an average headway of h s/veh).

20.2.5 Capacity of an Intersection Lane or Lane Group

The saturation flow rate(s) represents the capacity of an intersection lane or lane group assuming that the light is always GREEN. The portion of real time that is effective green is defined by the "green ratio," the ratio of the effective green time to the cycle length of the signal (g/C). The capacity of an intersection lane or lane group may then be computed as:

$$c_i = s_i\left(\frac{g_i}{C}\right) \qquad (20\text{-}6)$$

where: c_i = capacity of lane or lane group i, veh/h
s_i = saturation flow rate for lane or lane group i, veh/hg
g_i = effective green time for lane or lane group i, s
C = signal cycle length, s

A Sample Problem

These concepts are best illustrated using a sample problem. Consider a given movement at a signalized intersection with the following known characteristics:

- Cycle length, $C = 60$ s
- Green time, $G = 27$ s
- Yellow plus all-red time, $Y = 4$ s.
- Saturation headway, $h = 2.4$ s/veh
- Start-up lost time, $\ell_1 = 2.0$ s
- Clearance lost time, $\ell_2 = 2.0$ s

For these characteristics, what is the capacity (per lane) for this movement?

The problem will be approached in two different ways. In the first, a ledger of time within the hour is created. Once the amount of time per hour used by vehicles at the saturation flow rate is established, capacity can be found by assuming that this time is used at a rate of one vehicle every h seconds. Because the characteristics stated are given on a *per phase* basis, these would have to be converted to a *per hour* basis. This is easily done knowing the number of signal cycles that occur within an hour. For a 60-second cycle, there are 3,600/60 = 60 cycles within the hour. The subject movements will have one GREEN phase in each of these cycles. Then:

- Time in hour: 3,600 s
- RED time in hour: $(60 - 27 - 4) \times 60 = 1{,}740$ s
- Lost time in hour: $(2.0 + 2.0) \times 60 = 240$ s
- Remaining time in hour: $3{,}600 - 1{,}740 - 240 = 1{,}620$ s

The 1,620 remaining seconds of time in the hour represent the amount of time that can be used at a rate of one vehicle every h seconds, where $h = 2.4$ s/veh in this case. This number was calculated by deducting the periods during which no vehicles (in the subject movements) are effectively moving. These periods include the RED time as well as the start-up and clearance lost times in each signal cycle. The capacity of this movement may then be computed as:

$$c = \frac{1620}{2.4} = 675 \text{ veh/h/ln}$$

A second approach to this problem uses Equation 20-6, with the following values:

$$s = \frac{3{,}600}{2.4} = 1.500 \text{ veh/h/ln}$$

$$g = 27 + 4 - 4 = 27 \text{ s}$$

$$c = 1{,}500\left(^{27}/_{60}\right) = 675 \text{ veh/h/ln}$$

The two results are, as expected, the same. Capacity is found by isolating the effective green time available to the subject movements and by assuming that this time is used at the saturation flow rate (or headway).

20.2.6 Notable Studies on Saturation Headways, Flow Rates, and Lost Times

For purposes of illustrating basic concepts, subsequent sections of this chapter assume that the value of saturation flow rate (or headway) is known. In reality, the saturation flow rate varies widely with a variety of prevailing conditions, including lane widths, heavy-vehicle presence, approach grades, parking conditions near the intersection, transit bus presence, vehicular and pedestrian flow rates, and other conditions.

The first significant studies of saturation flow were conducted by Bruce Greenshields in the 1940s [*3*]. His studies resulted in an average saturation flow rate of 1,714 veh/hg/ln and a start-up lost time of 3.7 seconds. The study, however, covered a variety of intersections with varying underlying characteristics. A later study in 1978 [*4*] reexamined the Greenshields hypothesis; it resulted in the same saturation flow rate (1,714 veh/hg/ln) but a lower start-up lost time of 1.1 seconds. The latter study had data from 175 intersections, covering a wide range of underlying characteristics.

A comprehensive study of saturation flow rates at intersections in five cities was conducted in 1987–1988 [*5*] to determine the effect of opposed left turns. It also produced a good deal of data on saturation flow rates in general. Some of the results are summarized in Table 20.2.

These results show generally lower saturation flow rates (and higher saturation headways) than previous studies. The data, however, reflect the impact of opposed left turns, truck presence, and a number of other "nonstandard" conditions, all of which have a significant impeding effect. The most remarkable result of this study, however, was the wide variation in measured saturation flow rates, both over time at the same site and from location to location. Even when underlying conditions remained fairly constant, the variation in observed saturation flow rates at a given location was as large as 20% to 25%. In a doctoral dissertation using the same data, Prassas demonstrated that saturation headways and flow rates have a significant stochastic component, making calibration of stable values difficult [*6*].

The study also isolated saturation flow rates for "ideal" conditions, which include all passenger cars, no turns, level grade, and 12-foot lanes. Even under these conditions, saturation flow rates varied from 1,240 pc/hg/ln to 2,092 pc/hg/ln for single-lane approaches and from 1,668 pc/hg/ln to 2,361 pc/hg/ln for multilane approaches. The difference between observed saturation flow rates at single and multilane approaches is also interesting. Single-lane approaches have a number of unique characteristics that are addressed in the *Highway Capacity Manual* model for analysis of signalized intersections (see Chapter 24).

Current standards in the *Highway Capacity Manual* [*1*] use an ideal saturation flow rate of 1,900 pc/hg/ln for both

Table 20.2: Saturation Flow Rates from a Nationwide Survey

Item	Single-Lane Approaches	Two-Lane Approaches
Number of Approaches	14	26
Number of 15-Minute Periods	101	156
Saturation Flow Rates		
Average	1,280 veh/hg/ln	1,337 veh/hg/ln
Minimum	636 veh/hg/ln	748 veh/hg/ln
Maximum	1,705 veh/hg/ln	1,969 veh/hg/ln
Saturation Headways		
Average	2.81 s/veh	2.69 s/veh
Minimum	2.11 s/veh	1.83 s/veh
Maximum	5.66 s/veh	4.81 s/veh

single and multilane approaches. This ideal rate is then adjusted for a variety of prevailing conditions. The manual also provides default values for lost times. The default value for start-up lost time (ℓ_1) is 2.0 seconds. For the clearance lost time (ℓ_2), the default value varies with the "yellow" and "all-red" timings of the signal:

$$\ell_2 = y + ar - e \tag{20-7}$$

where: ℓ_2 = clearance lost time, s
y = length of yellow interval, s
ar = length of all-red interval, s
e = encroachment of vehicles into yellow and all-red, s

A default value of 2.0 s is used for e.

20.3 The Critical-Lane and Time-Budget Concepts

In signal analysis and design, the "critical-lane" and "time budget" concepts are closely related. The time budget, in its simplest form, is the allocation of time to various vehicular and pedestrian movements at an intersection through signal control. Time is a constant: There are always 3,600 seconds in an hour, and all of them must be allocated. In any given hour, time is "budgeted" to legal vehicular and pedestrian movements and to lost times.

The "critical-lane" concept involves the identification of specific lane movements that will control the timing of a given signal phase. Consider the situation illustrated in Figure 20.2. A simple two-phase signal controls the intersection. Thus all E-W movements are permitted during one phase, and all N-S movements are permitted in another phase. During each of these phases, there are four lanes of traffic (two in each direction) moving simultaneously. Demand is not evenly distributed among them; one of these lanes will have the most intense traffic demand. The signal must be timed to accommodate traffic in this lane—the "critical lane" for the phase.

In the illustration of Figure 20.2, the signal timing and design must accommodate the total demand flows in lanes 1 and 2. Because these lanes have the most intense demand, if the signal accommodates them, all other lanes will be accommodated as well. Note that the critical lane is identified as the lane with the most *intense traffic demand,* not the lane with the highest volume. This is because many variables are affecting traffic flow. A lane with many left-turning vehicles, for example, may require more time than an adjacent lane with no turning vehicles, but a higher volume. Determining the intensity of traffic demand in a lane involves accounting for prevailing conditions that may affect flow in that particular lane.

In establishing a time budget for the intersection of Figure 20.2, time would have to be allocated to four elements:

- Movement of vehicles in critical lane 1
- Movement of vehicles in critical lane 2
- Start-up and clearance lost times for vehicles in critical lane 1
- Start-up and clearance lost times for vehicles in critical lane 2

This can be thought of in the following way: Lost times are not used by any vehicle. When deducted from total time, remaining time is effective green time and is allocated to critical-lane

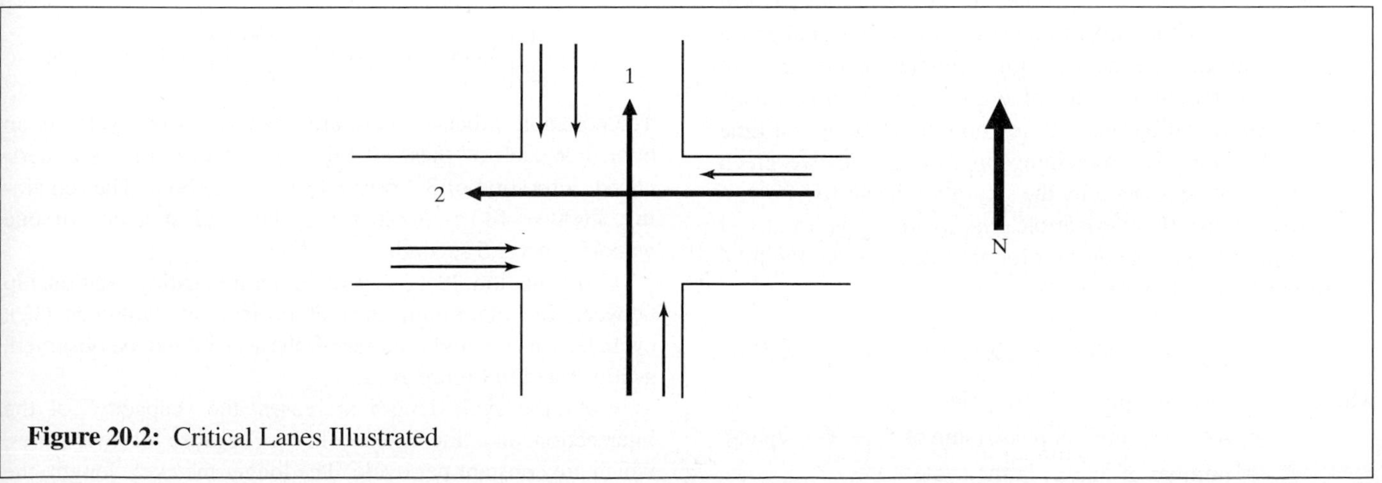

Figure 20.2: Critical Lanes Illustrated

demands—in this case, in lanes 1 and 2. The total amount of effective green time, therefore, must be sufficient to accommodate the total demand in lanes 1 and 2 (the critical lanes). These critical demands must be accommodated one vehicle at a time because they cannot move simultaneously.

The example of Figure 20.2 is a relatively simple case. In general, the following rules apply to the identification of critical lanes:

a. There is a critical lane and a critical-lane flow for each discrete signal phase provided.
b. Except for lost times, when no vehicles move, there must be *one* and *only one* critical lane moving during every second of effective green time in the signal cycle.
c. Where there are overlapping phases, the potential combination of lane flows yielding the highest sum of critical lane flows while preserving the requirement of item (b) identifies critical lanes.

Chapter 21 contains a detailed discussion of how to identify critical lanes for any signal timing and design.

20.3.1 The Maximum Sum of Critical-Lane Volumes: One View of Signalized Intersection Capacity

It is possible to consider the maximum possible sum of critical-lane volumes to be a general measure of the "capacity" of the intersection. This is not the same as the traditional view of capacity presented in the *Highway Capacity Manual,* but it is a useful concept to pursue.

By definition, each signal phase has one and only one critical lane. Except for lost times in the cycle, one critical lane is always moving. Lost times occur for each signal phase and represent time during which *no* vehicles in any lane are moving. The maximum sum of critical-lane volumes may, therefore, be found by determining how much total lost time exists in the hour. The remaining time (total effective green time) may then be divided by the saturation headway.

To simplify this derivation, it is assumed the total lost time per phase (t_L) is a constant for all phases. Then, the total lost time per signal cycle is:

$$L = N * t_L \tag{20-8}$$

where: L = lost time per cycle, s/cycle
t_L = total lost time per phase (sum of $\ell_1 + \ell_2$), s/phase
N = number of phases in the cycle

The total lost time in an hour depends on the number of cycles occurring in the hour:

$$L_H = L\left(\frac{3{,}600}{C}\right) \tag{20-9}$$

where: L_H = lost time per hour, s/hr
L = lost time per cycle, s/cycle
C = cycle length, s

The remaining time within the hour is devoted to effective green time for critical-lane movements:

$$T_G = 3{,}600 - L_H \tag{20-10}$$

where: T_G = total effective green time in the hour, s

This time may be used at a rate of one vehicle every h seconds, where h is the saturation headway:

$$V_c = \frac{T_G}{h} \tag{20-11}$$

where: V_c = maximum sum of critical-lane volumes, veh/h
h = saturation headway, s/veh

Merging Equations 20-8 through 20-11, the following relationship emerges:

$$V_c = \frac{1}{h}\left[3{,}600 - Nt_L\left(\frac{3{,}600}{C}\right)\right] \tag{20-12}$$

All variables are as previously defined.

Consider the example of Figure 20.2 again. If the signal at this location has two phases, a cycle length of 60 seconds, total lost times of 4 s/phase, and a saturation headway of 2.5 s/veh, the maximum sum of critical-lane flows (the sum of flows in lanes 1 and 2) is:

$$V_c = \frac{1}{2.5}\left[3{,}600 - 2 * 4 * \left(\frac{3{,}600}{60}\right)\right] = 1{,}248 \text{ veh/h}$$

The equation indicates there are 3,600/60 = 60 cycles in an hour. For each of these, 2 * 4 = 8 s of lost time is experienced, for a total of 8 * 60 = 480 s in the hour. The remaining 3,600 − 480 = 3,120 s may be used at a rate of one vehicle every 2.5 seconds.

If Equation 20-12 is plotted, an interesting relationship between the maximum sum of critical-lane volumes (V_c), cycle length (C), and number of phases (N) may be observed, as illustrated in Figure 20.3.

As the cycle length increases, the "capacity" of the intersection also increases. This is because of lost times, which are constant per cycle. The longer the cycle length, the

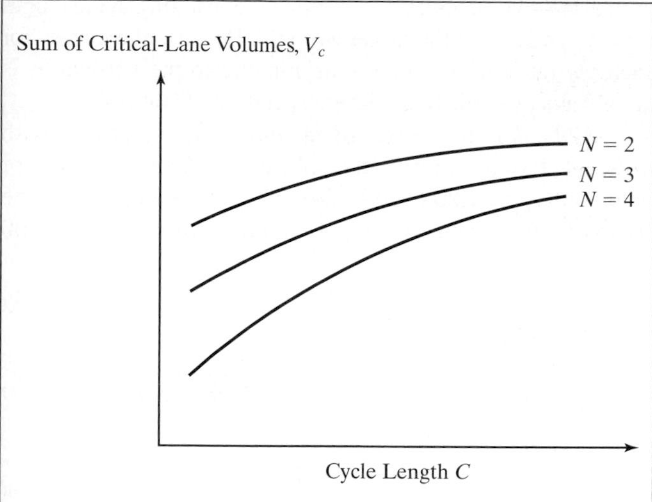

Figure 20.3: Maximum Sum of Critical-Lane Volumes Plotted

fewer cycles there are in an hour. This leads to less lost time in the hour, more effective green time in the hour, and a higher sum of critical-lane volumes. Note, however, that the relationship gets flatter as cycle length increases. As a general rule, increasing the cycle length may result in small increases in capacity. However, capacity can rarely be increased significantly by only increasing the cycle length. Other measures, such as adding lanes, are often also necessary.

Capacity also decreases as the number of phases increases. This is because for each phase, there is one full set of lost times in the cycle. Thus a two-phase signal has only two sets of lost times in the cycle, and a three-phase signal has three.

These trends provide insight but also raise an interesting question: Given these trends, it appears that all signals should have two phases and that the maximum practical cycle length should be used in all cases. After all, this combination would, apparently, yield the highest "capacity" for the intersection.

Using the maximum cycle length is not practical unless truly needed. Having a cycle length that is considerably longer than what is needed causes increases in delay to drivers and passengers. The increase in delay is because there will be times when vehicles on one approach are waiting for the green while there is no demand on conflicting approaches. Shorter cycle lengths yield less delay. Further, there is no incentive to maximize the cycle length. There will always be 3,600 seconds in the hour, and increasing the cycle length to accommodate increasing demand over time is quite simple, requiring only a resetting of the local signal controller. The shortest cycle length consistent with a v/c ratio in the range of 0.80 to 0.95 is generally used to produce optimal delays. Thus the view of signal capacity is quite different from that of pavement capacity. When deciding on the number of lanes on a freeway (or on an intersection approach), it is desirable to build excess capacity (i.e., achieve a low v/c ratio). This is because once built, it is unlikely that engineers will get an opportunity to expand the facility for 20 or more years, and adjacent land development may make such expansion impossible. The 3,600 seconds in an hour, however, are immutable, and retiming the signal to allocate more of them to effective green time is a simple task requiring no field construction.

20.3.2 Finding an Appropriate Cycle Length

If it is assumed that the demands on an intersection are known and the critical lanes can be identified, then Equation 20-12 could be solved using a known value of V_c to find a minimum acceptable cycle length:

$$C_{\min} = \frac{Nt_L}{1 - \left(\dfrac{V_c}{3{,}600/h}\right)} \qquad (20\text{-}13)$$

Thus, if in the example of Figure 20.2, the actual sum of critical-lane volumes was determined to be 1,000 veh/h, the minimum feasible cycle length would be:

$$C_{\min} = \frac{2 * 4}{1 - \left(\dfrac{1{,}000}{3{,}600/2.5}\right)} = 26.2 \text{ s}$$

The cycle length could be reduced, in this case, from the given 60 seconds to 30 seconds (the effective minimum cycle length used). This computation, however, assumes that the demand (V_c) is uniformly distributed throughout the hour and that every second of effective green time will be used. Neither of these assumptions is very practical. In general, signals would be timed for the flow rates occurring in the peak 15 minutes of the hour. Equation 20-13 could be modified by dividing V_c by a known peak-hour factor (PHF) to estimate the flow rate in the worst 15-minute period of the hour. Similarly, most signals would be timed to have somewhere between 80% and 95% of the available capacity actually used. Due to the normal stochastic variations in demand on a cycle-by-cycle and daily basis, some excess capacity must be provided to avoid failure of individual cycles or peak periods on a specific day. If demand, V_c, is also divided by the expected utilization of capacity (expressed in decimal form), then this is also accommodated. Introducing these changes transforms Equation 20-13 to:

$$C_{des} = \frac{Nt_L}{1 - \left[\frac{1{,}000}{(3{,}600/h) * PHF * (v/c)}\right]} \quad (20\text{-}14)$$

where: C_{des} = desirable cycle length, s
PHF = peak hour factor
v/c = desired volume to capacity ratio

All other variables are as previously defined.

Returning to the example, if the PHF is 0.95 and it is desired to use no more than 90% of available capacity during the peak 15-minute period of the hour, then:

$$C_{des} = \frac{2 * 4}{1 - \left[\frac{1{,}000}{(3{,}600/25) * 0.95 * 0.90}\right]} = \frac{8}{0.188} = 42.6 \text{ s}$$

In practical terms, this would lead to the use of a 45-second cycle length.

The relationship between a desirable cycle length, the sum of critical-lane volumes, and the target v/c ratio is quite interesting and is illustrated in Figure 20.4.

Figure 20.4 illustrates a typical relationship for a specified number of phases, saturation headway, lost times, and PHF. If a vertical is drawn at any specified value of V_c (sum of critical-lane volumes), it is clear that the resulting cycle length is very sensitive to the target v/c ratio. Because the curves for each v/c ratio are eventually asymptotic to the vertical, it is not always possible to achieve a specified v/c ratio.

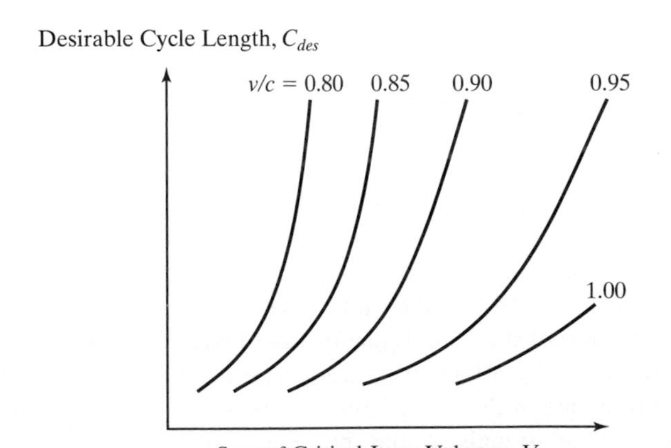

Figure 20.4: Desirable Cycle Length Versus Sum of Critical-Lane Volumes

Consider the case of a three-phase signal, with t_L = 4 s/phase, a saturation headway of 2.2 s/veh, a *PHF* of 0.90, and V_c = 1,200 veh/h. Desirable cycle lengths will be computed for a range of target v/c ratios varying from 1.00 to 0.80.

$$C_{des} = \frac{3 * 4}{1 - \left[\frac{1{,}200}{(3{,}600/2.2) * 0.90 * 1.00}\right]} = \frac{12}{0.1852} = 64.8 \Rightarrow 65\,s$$

$$C_{des} = \frac{3 * 4}{1 - \left[\frac{1{,}200}{(3{,}600/2.2) * 0.90 * 0.95}\right]} = \frac{12}{0.1423} = 84.3 \Rightarrow 85\,s$$

$$C_{des} = \frac{3 * 4}{1 - \left[\frac{1{,}200}{(3{,}600/2.2) * 0.90 * 0.90}\right]} = \frac{12}{0.0947} = 126.7 \Rightarrow 130\,s$$

$$C_{des} = \frac{3 * 4}{1 - \left[\frac{1{,}200}{(3{,}600/2.2) * 0.90 * 0.85}\right]} = \frac{12}{0.0414} = 289.9 \Rightarrow 290\,s$$

$$C_{des} = \frac{3 * 4}{1 - \left[\frac{1{,}200}{(3{,}600/2.2) * 0.90 * 0.80}\right]} = \frac{12}{-0.0185} = -648.6\,s$$

For this case, reasonable cycle lengths can provide target v/c ratios of 1.00 or 0.95. Achieving v/c ratios of 0.90 or 0.85 would require long cycle lengths beyond the practical limit of 120 seconds for pretimed signals. The 130-second cycle needed to achieve a v/c ratio of 0.90 might be acceptable for an actuated signal location, or in some extreme cases warranting a longer pretimed signal cycle. However, a v/c ratio of 0.80

cannot be achieved under any circumstances. The negative cycle length that results signifies there is not enough time within the hour to accommodate the demand with the required green time plus the 12 seconds of lost time per cycle. In effect, more than 3,600 seconds would have to be available in the hour to accomplish this.

A Sample Problem

Consider the intersection shown in Figure 20.5. The critical directional demands for this two-phase signal are shown with other key variables. Using the time-budget and critical-lane concepts, determine the number of lanes required for each of the critical movements and the minimum desirable cycle length that could be used. Note that an initial cycle length is specified but will be modified as part of the analysis.

Assuming that the initial specification of a 60-second cycle is correct and given the other specified conditions, the maximum sum of critical lanes that can be accommodated is computed using Equation 20-12:

$$V_c = \frac{1}{2.3}\left[3{,}600 - 2 * 4 * \left(\frac{3{,}600}{60}\right)\right] = 1{,}357 \text{ veh/h}$$

The critical SB volume is 1,200 veh/h, and the critical EB volume is 1,800 veh/h. The number of lanes each must be divided into is now to be determined. Whatever combination is used, the sum of the critical-lane volumes for these two approaches must be below 1,357 veh/h. Figure 20.6 shows a number of possible lane combinations and the resulting sum of critical-lane volumes. As you can see from the scenarios of Figure 20.6, to have a sum of critical-lane volumes less than 1,357 veh/h, the SB approach must have at least two lanes, and the EB approach must have three lanes. Realizing that these demands probably reverse in the other peak hour (AM or PM), the N-S artery would probably require four lanes, and the E-W artery six lanes.

This is a very basic analysis, and it would have to be modified based on more specific information regarding individual movements, pedestrians, parking needs, and other factors.

If the final scenario is provided, V_c is only 1,200 veh/h. It is possible that the original cycle length of 60 seconds could be reduced. A desirable cycle length may be computed from Equation 20-14:

$$C_{des} = \frac{2 * 4}{1 - \left[\dfrac{1{,}200}{(3{,}600/2.3) * 0.95 * 0.90}\right]} = 77.7 \Rightarrow 80\,\text{s}$$

The resulting cycle length is *larger* than the original 60 seconds because the equation takes both the *PHF* and target v/c ratios into account. Equation 20-12 for computing the maximum value of V_c does not; it assumed full use of capacity ($v/c = 1.00$) and no peaking within the hour. In essence, the 2×3 lane design proposal should be combined with an 80-second cycle length to achieve the desired results.

This problem illustrates the critical relationship between number of lanes and cycle lengths. Clearly, other scenarios would produce desirable results. Additional lanes could be provided in either direction, which would allow the use of a shorter cycle length. Unfortunately, for many cases, signal timing is considered with a fixed design already in place. Only where right-of-way is available or a new intersection is being constructed can major changes in the number of lanes be considered. Allocation of lanes to various movements is also a consideration. Optimal solutions are generally

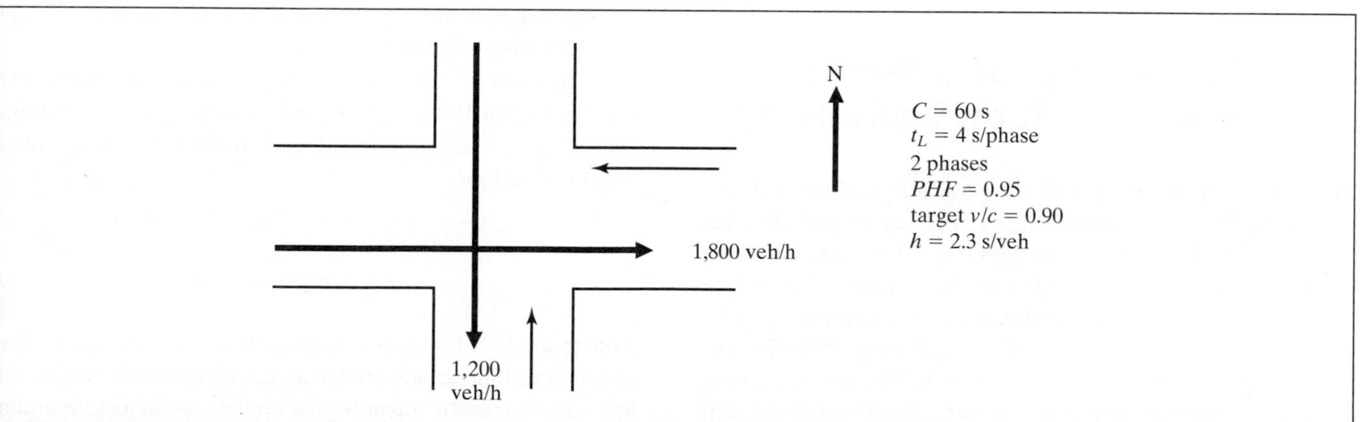

Figure 20.5: Sample Problem Using the Time-Budget and Critical-Lane Concepts

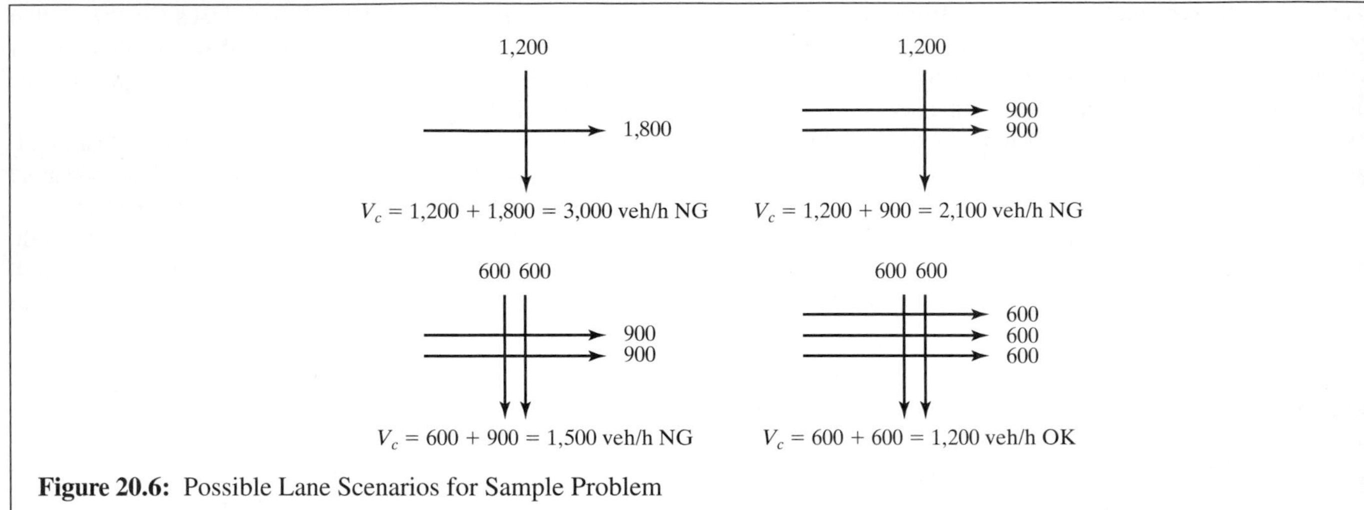

Figure 20.6: Possible Lane Scenarios for Sample Problem

found more easily when the physical design and signalization can be treated in tandem.

If, in the problem of Figure 20.5, space limited both the EB and SB approaches to two lanes, the resulting V_c would be 1,500 veh/h. Would it be possible to accommodate this demand by lengthening the cycle length? Again, Equation 20-14 is used:

$$C_{des} = \frac{2 * 4}{1 - \left[\dfrac{1{,}500}{(3{,}600/2.3) * 0.95 * 0.90}\right]}$$

$$= \frac{8}{-0.121} = -66.1 \text{ s} \quad NG$$

The negative result indicates no cycle length can accommodate a V_c of 1,500 veh/h at this location.

20.4 The Concept of Left-Turn (and Right-Turn) Equivalency

The most difficult process to model at a signalized intersection is the left turn. Left turns are made in several different modes using different design elements. Left turns may be made from a lane shared with through vehicles (shared-lane operation) or from a lane dedicated to left-turning vehicles (exclusive-lane operation). Traffic signals may allow for permitted or protected left turns, or some combination of the two.

Whatever the case, however, a left-turning vehicle will consume more effective green time traversing the intersection than will a similar through vehicle. The most complex case is that of a permitted left turn made across an opposing vehicular flow from a shared lane. A left-turning vehicle in the shared lane must wait for an acceptable gap in the opposing flow. While waiting, the vehicle blocks the shared lane, and other vehicles (including through vehicles) in the lane are delayed behind it. Some vehicles will change lanes to avoid the delay while others are unable to and must wait until the left-turner successfully completes the turn.

Many models of the signalized intersection account for this in terms of "through vehicle equivalents" (i.e., how many through vehicles would consume the same amount of effective green time traversing the stop-line as *one* left-turning vehicle?). Consider the situation depicted in Figure 20.7. If both the left lane and the right lane were observed, an equivalence similar to the following statement could be determined: *In the same amount of time, the left lane discharges five through vehicles and two left-turning vehicles while the right lane discharges eleven through vehicles.*

In terms of effective green time consumed, this observation means that 11 through vehicles are equivalent to 5 through vehicles plus 2 left turning vehicles. If the left-turn equivalent is defined as E_{LT}:

$$11 = 5 + 2E_{LT}$$

$$E_{LT} = \frac{11 - 5}{2} = 3.0$$

Note that this computation holds only for the prevailing characteristics of the approach during the observation period. The left-turn equivalent depends on a number of factors, including how left turns are made (protected, permitted, compound), the opposing traffic flow, and the number of opposing lanes.

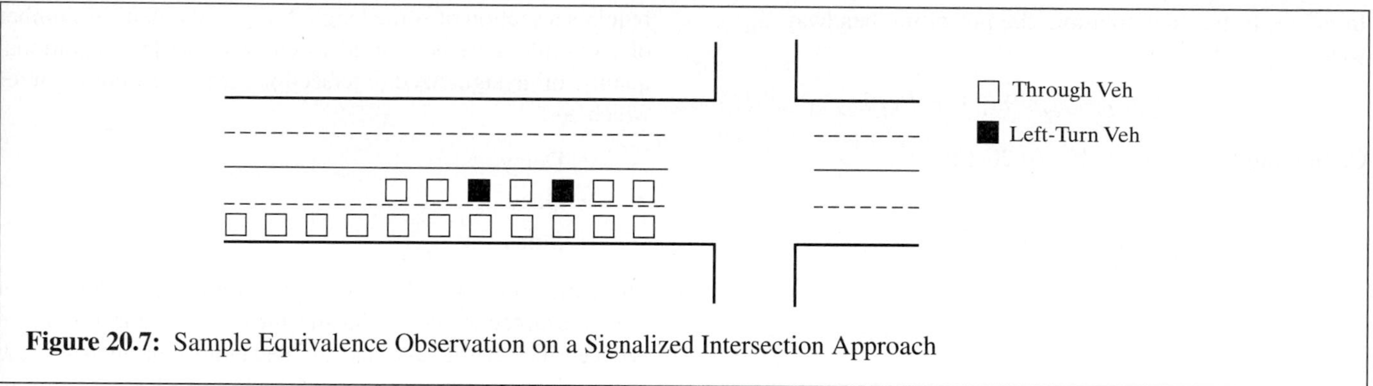

Figure 20.7: Sample Equivalence Observation on a Signalized Intersection Approach

Figure 20.8 illustrates the general form of the relationship for through vehicle equivalents of *permitted* left turns.

The left-turn equivalent, E_{LT}, increases as the opposing flow increases. For any given opposing flow, however, the equivalent decreases as the number of opposing lanes is increased from one to three. This latter relationship is not linear because the task of selecting a gap through multilane opposing traffic is more difficult than selecting a gap through single-lane opposing traffic. Further, in a multilane traffic stream, vehicles do not pace each other side by side, and the gap distribution does not improve as much as the per-lane opposing flow decreases. To illustrate the use of left-turn equivalents in modeling, consider the following problem:

> An approach to a signalized intersection has two lanes, permitted left-turn phasing, 10% left-turning vehicles, and a left-turn equivalent of 5.0. The saturation headway for through vehicles is 2.0 s/veh. Determine the equivalent saturation flow rate and headway for all vehicles on this approach.

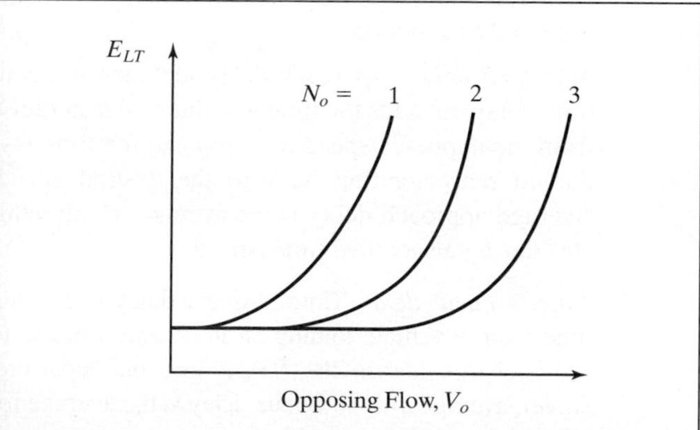

Figure 20.8: Relationship Among Left-Turn Equivalents, Opposing Flow, and Number of Opposing Lanes

The first way to interpret the left-turn equivalent is that each left-turning vehicle consumes 5.0 times the effective green time as a through vehicle. Thus, for the situation described, 10% of the traffic stream has a saturation headway of 2.0 × 5.0 = 10.0 s/veh, and the remainder (90%) has a saturation headway of 2.0 s/veh. The average saturation headway for all vehicles, therefore, is:

$$h = (0.10 * 10.0) + (0.90 * 2.0) = 2.80 \text{ s/veh}$$

This corresponds to a saturation flow rate of:

$$s = \frac{3{,}600}{2.80} = 1{,}286 \text{ veh/hg/ln}$$

A number of models, including the *Highway Capacity Manual* approach, calibrate a multiplicative adjustment factor that converts an ideal (or through) saturation flow rate to a saturation flow rate for prevailing conditions:

$$s_{prev} = s_{ideal} * f_{LT}$$

$$f_{LT} = \frac{s_{prev}}{s_{ideal}} = \frac{(3{,}600/h_{prev})}{(3{,}600/h_{ideal})} = \frac{h_{ideal}}{h_{prev}} \tag{20-15}$$

where: s_{prev} = saturation flow rate under prevailing conditions, veh/hg/ln

s_{ideal} = saturation flow rate under ideal conditions, veh/hg/ln

f_{LT} = left-turn adjustment factor

h_{ideal} = saturation headway under ideal conditions, s/veh

h_{prev} = saturation headway under prevailing conditions, s/veh

In effect, in the first solution, the prevailing headway, h_{prev}, was computed as follows:

$$h_{prev} = (P_{LT}E_{LT}h_{ideal}) + [(1 - P_{LT})h_{ideal}] \quad (20\text{-}16)$$

Combining Equations 20-15 and 20-16:

$$f_{LT} = \frac{h_{ideal}}{(P_{LT}E_{LT}h_{ideal}) + [(1 - P_{LT})h_{ideal}]}$$

$$f_{LT} = \frac{1}{P_{LT}E_{LT} + (1 - P_{LT})} = \frac{1}{1 + P_{LT}(E_{LT} - 1)} \quad (20\text{-}17)$$

The problem posed may now be solved using a left-turn adjustment factor. Note that the saturation headway under ideal conditions is 3,600/2.0 = 1,800 veh/hg/ln.

Then:

$$f_{LT} = \frac{1}{1 + 0.10\,(5 - 1)} = 0.714$$

$$s_{prev} = 1800 * 0.714 = 1{,}286 \text{ veh/hg/ln}$$

This, of course, is the same result.

It is important that the concept of left-turn equivalence be understood. Its use in multiplicative adjustment factors often obscures its intent and meaning. The fundamental concept, however, is unchanged—the equivalence is based on the fact that the effective green time consumed by a left-turning vehicle is E_{LT} times the effective green time consumed by a similar through vehicle.

A similar case can be made for describing the effects of right turns. Right turns are typically made through a conflicting pedestrian flow in the crosswalk to the immediate right of the approach. Like left turns, this interaction causes right turns to consume more effective green time than through movements. An equivalent, E_{RT}, is used to quantify these effects and is used in the same manner as described for left-turn equivalents.

Signalized intersection and other traffic models use other types of equivalents as well. Heavy-vehicle and local-bus equivalents have similar meanings and result in similar equations. Some of these have been discussed in previous chapters, and others will be discussed in subsequent chapters.

20.5 Delay as a Measure of Effectiveness

Signalized intersections represent point locations within a surface street network. As point locations, the measures of operational quality or effectiveness used for highway sections are not relevant. Speed has no meaning at a point, and density requires a section of some length for measurement. A number of measures have been used to characterize the operational quality of a signalized intersection, the most common of which are:

- Delay
- Queuing
- Stops

These measures are all related. Delay refers to the amount of time consumed in traversing the intersection—the difference between the arrival time and the departure time, where these may be defined in a number of different ways. Queuing refers to the number of vehicles forced to queue behind the stop line during a RED signal phase; common measures include the average queue length or a percentile queue length. Stops refer to the percentage or number of vehicles that must stop at the signal.

20.5.1 Types of Delay

The most common measure used to describe operational quality at a signalized intersection is delay, with queuing and/or stops often used as a secondary measure. Although it is possible to measure delay in the field, it is a difficult process, and different observers may make judgments that could yield different results. For many purposes, it is, therefore, convenient to have a predictive model for the estimate of delay. Delay, however, can be quantified in many different ways. The most frequently used forms of delay are defined as follows:

1. *Stopped-time delay.* Stopped-time delay is defined as the time a vehicle is stopped in queue while waiting to pass through the intersection; average stopped-time delay is the average for all vehicles during a specified time period.
2. *Approach delay.* Approach delay includes stopped-time delay but adds the time loss due to deceleration from the approach speed to a stop and the time loss due to reacceleration back to the desired speed. Average approach delay is the average for all vehicles during a specified time period.
3. *Time-in-queue delay.* Time-in-queue delay is the total time from a vehicle joining an intersection queue to its discharge across the STOP line on departure. Again, average time-in-queue delay is the average for all vehicles during a specified time period.
4. *Travel time delay.* This is a more conceptual value. It is the difference between the driver's expected travel

time through the intersection (or any roadway segment) and the actual time taken. Given the difficulty in establishing a "desired" travel time to traverse an intersection, this value is rarely used, other than as a philosophical concept.

5. *Control delay.* The concept of control delay was developed in the 1994 *Highway Capacity Manual* and is included in the current HCM. It is the delay caused by a control device, either a traffic signal or a STOP sign. It is approximately equal to time-in-queue delay plus the acceleration-deceleration delay component.

Figure 20.9 illustrates three of these delay types for a single vehicle approaching a RED signal.

Stopped-time delay for this vehicle includes only the time spent stopped at the signal. It begins when the vehicle is fully stopped and ends when the vehicle begins to accelerate. Approach delay includes additional time losses due to deceleration and acceleration. It is found by extending the velocity slope of the approaching vehicle as if no signal existed; the approach delay is the horizontal (time) difference between the hypothetical extension of the approaching velocity slope and the departure slope after full speed is achieved. Travel time delay is the difference in time between a hypothetical desired velocity line and the actual vehicle path. Time-in-queue delay cannot be effectively shown using one vehicle because it involves joining and departing a queue of several vehicles.

Delay measures can be stated for a single vehicle, as an average for all vehicles over a specified time period, or as an aggregate total value for all vehicles over a specified time period. Aggregate delay is measured in total *vehicle-seconds, vehicle-minutes,* or *vehicle-hours* for all vehicles in the specified time interval. Average individual delay is generally stated in terms of s/veh for a specified time interval.

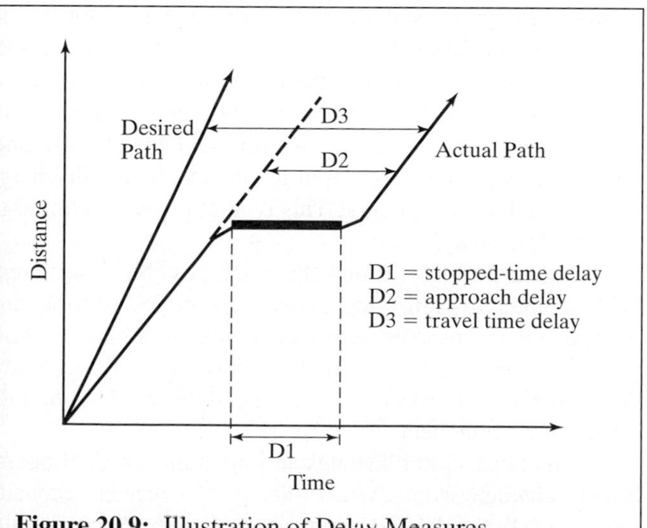

Figure 20.9: Illustration of Delay Measures

20.5.2 Basic Theoretical Models of Delay

Virtually all analytic models of delay begin with a plot of cumulative vehicles arriving and departing versus time at a given signal location. The time axis is divided into periods of effective green and effective red as illustrated in Figure 20.10.

Vehicles are assumed to arrive at a uniform rate of flow of v vehicles per unit time, seconds in this case. This is shown by the constant slope of the arrival curve. Uniform arrivals assume that the inter-vehicle arrival time between vehicles is a constant. Thus, if the arrival flow rate, v is 1,800 vehs/h, then one vehicle arrives every 3,600/1,800 = 2.0 s.

Assuming no preexisting queue, vehicles arriving when the light is GREEN continue through the intersection (i.e., the departure curve is the same as the arrival curve). When the light turns RED, however, vehicles continue to arrive, but none depart. Thus the departure curve is parallel to the x-axis during the RED interval. When the next effective GREEN begins, vehicles queued during the RED interval depart from the intersection, now at the saturation flow rate, s, in veh/s. For stable operations, depicted here, the departure curve "catches up" with the arrival curve before the next RED interval begins (i.e., there is no residual or unserved queue left at the end of the effective GREEN).

This simple depiction of arrivals and departures at a signal allows the estimation of three critical parameters:

- The total time that any vehicle i spends waiting in the queue, $W(i)$, is given by the horizontal time-scale difference between the time of arrival and the time of departure.
- The total number of vehicles queued at any time t, $Q(t)$, is the vertical vehicle-scale difference between the number of vehicles that have arrived and the number of vehicles that have departed.
- The aggregate delay for all vehicles passing through the signal is the area between the arrival and departure curves (vehicles × time).

Note that because the plot illustrates vehicles arriving in queue and departing from queue, this model most closely represents what has been defined as *time-in-queue delay.* There

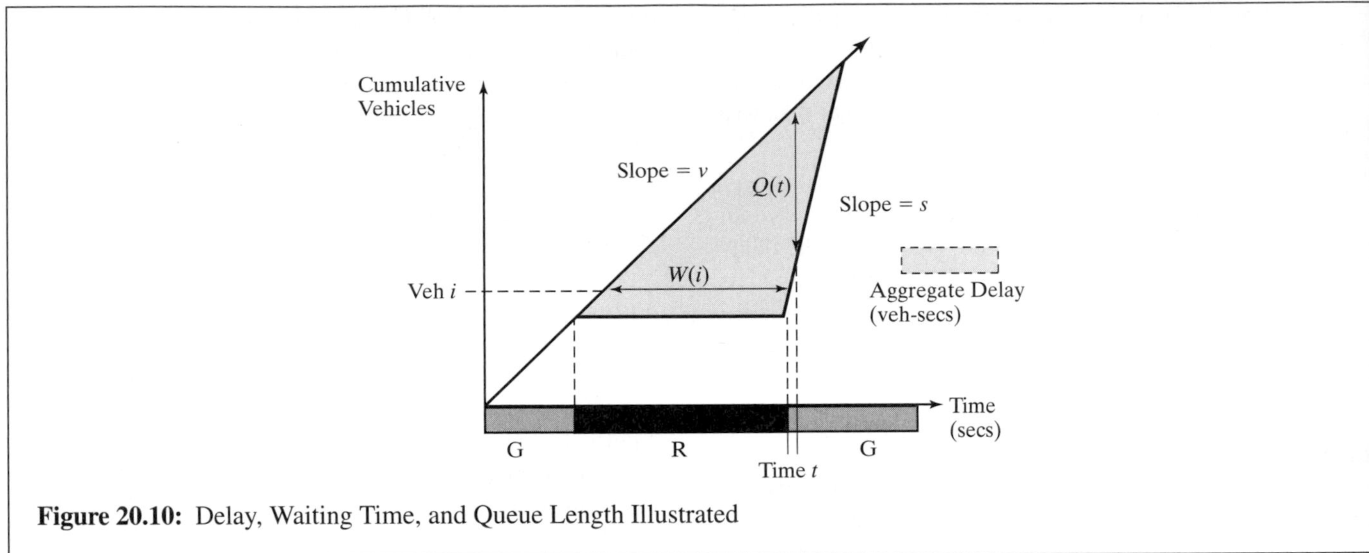

Figure 20.10: Delay, Waiting Time, and Queue Length Illustrated

are many simplifications that have been assumed, however, in constructing this simple depiction of delay. It is important to understand the two major simplifications:

- The assumption of a uniform arrival rate is a simplification. Even at a completely isolated location, actual arrivals would be *random* (i.e., would have an average rate over time), but inter-vehicle arrival times would vary around an average rather than being constant. Within coordinated signal systems, however, vehicle arrivals are usually in platoons.
- It is assumed that the queue is building at a point location (as if vehicles were stacked on top of one another). In reality, as the queue grows, the rate at which vehicles arrive at its end is the arrival rate of vehicles (at a point), plus a component representing the backward growth of the queue in space.

Both of these can have a significant effect on actual results. Modern models account for the former in ways that we discuss subsequently. The assumption of a "point queue," however, is imbedded in many modern applications.

Figure 20.11 expands the range of Figure 17.10 to show a series of GREEN phases and depicts three different types of operation. It also allows for an arrival function, $a(t)$, that varies while maintaining the departure function, $d(t)$, described previously.

Figure 20.11 (a) shows stable flow throughout the period depicted. No signal cycle "fails" (i.e., ends with some vehicles queued during the preceding RED unserved). During every GREEN phase, the departure function "catches up" with the arrival function. Total aggregate delay during this period is the total of all the triangular areas between the arrival and departure curves. This type of delay is often referred to as "uniform delay."

In Figure 20.11 (b), some of the signal phases "fail." At the end of the second and third GREEN intervals, some vehicles are not served (i.e., they must wait for a second GREEN interval to depart the intersection). By the time the entire period ends, however, the departure function has "caught up" with the arrival function and there is no residual queue left unserved. This case represents a situation in which the overall period of analysis is stable (i.e., total demand does not exceed total capacity). Individual cycle failures within the period, however, have occurred. For these periods, there is a second component of delay in addition to uniform delay. It consists of the area between the arrival function and the dashed line, which represents the capacity of the intersection to discharge vehicles and has the slope c. This type of delay is referred to as "overflow delay."

Figure 20.11 (c) shows the worst possible case: Every GREEN interval "fails" for a significant period of time, and the residual, or unserved, queue of vehicles continues to grow throughout the analysis period. In this case, the overflow delay component grows over time, quickly dwarfing the uniform delay component.

The latter case illustrates an important practical operational characteristic. When demand exceeds capacity ($v/c > 1.00$), the delay depends on *the length of time* that the condition exists. In Figure 20.11 (b), the condition exists for

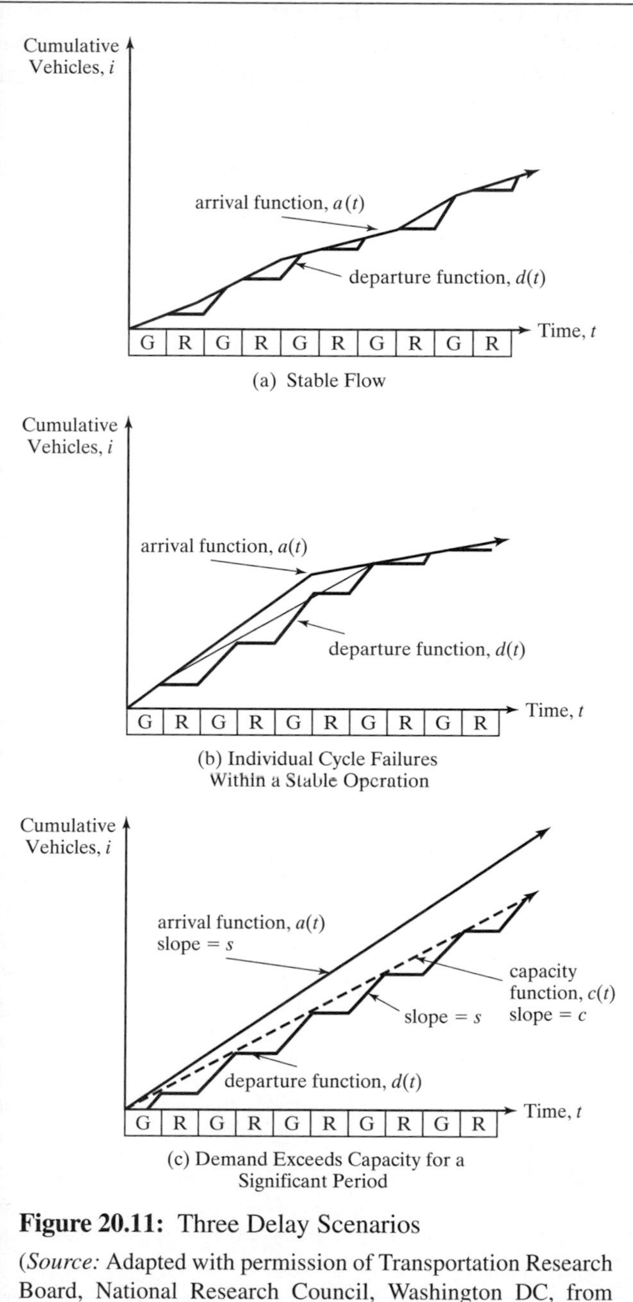

Figure 20.11: Three Delay Scenarios

(*Source:* Adapted with permission of Transportation Research Board, National Research Council, Washington DC, from V.F. Hurdle, "Signalized Intersection Delay Model: A Primer for the Uninitiated," *Transportation Research Record 971,* pp. 97, 98, 1984.)

only two phases. Thus the queue and the resulting overflow delay are limited. In Figure 20.11 (c), the condition exists for a long time, and the delay continues to grow throughout the oversaturated period.

Components of Delay

In analytic models for predicting delay, three distinct components of delay may be identified:

- *Uniform delay* is the delay based on an assumption of uniform arrivals and stable flow with no individual cycle failures.
- *Random delay* is the additional delay, above and beyond uniform delay, because flow is randomly distributed rather than uniform at isolated intersections.
- *Overflow delay* is the additional delay that occurs when the capacity of an individual phase or series of phases is less than the demand or arrival flow rate.

In addition, the delay impacts of platoon flow (rather than uniform or random) have been historically treated as an adjustment to uniform delay. Many modern models combine the random and overflow delays into a single function, which is referred to as "overflow delay," even though it contains both components.

The differences between uniform, random, and platooned arrivals are illustrated in Figure 20.12. As noted, the analytic basis for most delay models is the assumption of uniform arrivals, which are depicted in Figure 20.12 (a). Even at isolated intersections, however, arrivals would be random, as shown in Figure 20.12 (b). With random arrivals, the underlying rate of arrivals is a constant, but the inter-arrival times are exponentially distributed around an average. In most urban and suburban cases, where a signalized intersection is likely to be part of a coordinated signal system, arrivals will be in organized platoons that move down the arterial in a cohesive group, as shown in Figure 20.12 (c). The exact time that a platoon arrives at a downstream signal has an enormous

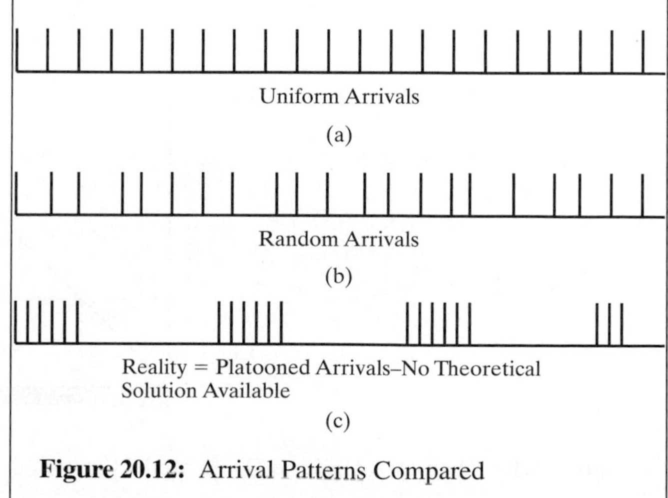

Figure 20.12: Arrival Patterns Compared

potential effect on delay. A platoon of vehicles arriving at the beginning of the RED forces most vehicles to stop for the entire length of the RED phase. The same platoon of vehicles arriving at the beginning of the GREEN phase may flow through the intersection without any vehicles stopping. In both cases, the arrival flow, v, and the capacity of the intersection, c, are the same. The resulting delay, however, would vary significantly. The existence of platoon arrivals, therefore, necessitates a significant adjustment to models based on theoretically uniform or random flow.

Webster's Uniform Delay Model

Virtually every model of delay starts with Webster's model of uniform delay. Initially published in 1958 [7], this model begins with the simple illustration of delay depicted in Figure 20.13, with its assumptions of stable flow and a simple uniform arrival function. As noted previously, aggregate delay can be estimated as the area between the arrival and departure curves in the figure. Thus Webster's model for uniform delay is the area of the triangle formed by the arrival and departure functions. For clarity, this triangle is shown again in Figure 20.13.

The area of the aggregate delay triangle is simply half the base times the height, or:

$$UD_a = \frac{1}{2} RV$$

where: UD_a = aggregate uniform delay, veh-sec
R = length of the RED phase, s
V = total vehicles in queue, veh

By convention, traffic models are not developed in terms of RED time. Rather, they focus on GREEN time. Thus Webster substitutes the following equivalence for the length of the RED phase:

$$R = C\left[1 - \left(g/C\right)\right]$$

where: C = cycle length, s
g = effective green time, s

In words, the RED time is the portion of the cycle length that is not effectively green.

The height of the triangle, V, is the total number of vehicles in the queue. In effect, it includes vehicles arriving during the RED phase, R, plus those that join the end of the queue while it is moving out of the intersection (i.e., during time t_c in Figure 20.13). Thus determining the time it takes for the queue to clear, t_c, is an important part of the model. This is done by setting the number of vehicles arriving during the period $R + t_c$ equal to the number of vehicles departing during the period t_c, or:

$$v(R + t_c) = st_c$$

$$R + t_c = \left(\frac{s}{v}\right)t_c$$

$$R = t_c\left[\left(\frac{s}{v}\right) - 1\right]$$

$$t_c = \frac{R}{\left[\left(\frac{s}{v}\right) - 1\right]}$$

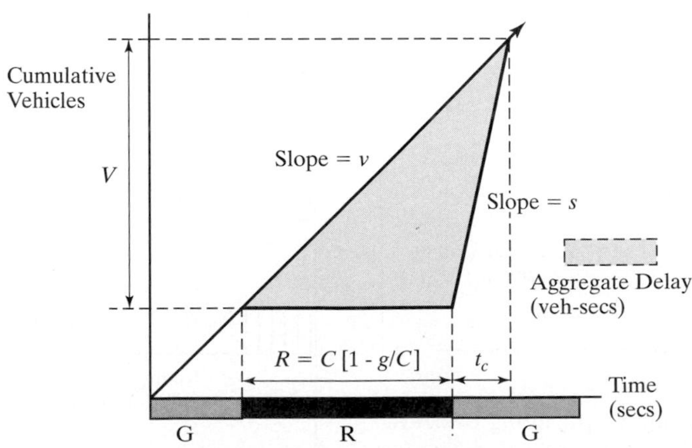

Figure 20.13: Webster's Uniform Delay Model Illustrated

Then, substituting for t_c:

$$V = v(R + t_c) = v\left[R + \frac{R}{\left(\frac{s}{v} - 1\right)}\right] = R\left(\frac{vs}{s - v}\right)$$

and for R:

$$V = C\left[1 - \left(g/C\right)\right]\left[\frac{vs}{s - v}\right]$$

Then, aggregate delay can be stated as:

$$UD_a = \frac{1}{2}RV = \frac{1}{2}C^2\left[1 - \left(g/C\right)\right]\left[\frac{vs}{s - v}\right] \quad (20\text{-}18)$$

where all variables are as previously defined.

Equation 20-18 estimates aggregate uniform delay in vehicle-seconds for one signal cycle. To get an estimate of average uniform delay per vehicle, the aggregate is divided by the number of vehicles arriving during the cycle, vC. Then:

$$UD = \frac{1}{2}C\frac{\left[1 - g/C\right]^2}{\left[1 - v/s\right]} \quad (20\text{-}19)$$

Another form of the equation uses the capacity, c, rather than the saturation flow rate, s. Noting that $s = c/(g/C)$, the following form emerges:

$$UD = \frac{1}{2}C\frac{\left[1 - \left(g/C\right)\right]^2}{\left[1 - \left(g/C\right)\left(v/c\right)\right]} = \frac{0.50\,C\left[1 - \left(g/C\right)\right]^2}{1 - \left(g/C\right)X} \quad (20\text{-}20)$$

where: UD = average uniform delay per vehicles, s/veh
C = cycle length, s
g = effective green time, s
v = arrival flow rate, veh/h
c = capacity of intersection approach, veh/h
X = v/c ratio, or degree of saturation

This average includes the vehicles that arrive and depart on green, accruing no delay. This is appropriate. One of the objectives in signalizations is to minimize the number or proportion of vehicles that must stop. Any meaningful quality measure would have to include the positive impact of vehicles that are not delayed.

In Equation 20-20, note that the maximum value of X (the v/c ratio) is 1.00. As the uniform delay model assumes no overflow, the v/c ratio cannot be more than 1.00.

Modeling Random Delay

The uniform delay model assumes that arrivals are uniform and that no signal phases fail (i.e., that arrival flow is less than capacity during every signal cycle of the analysis period).

At isolated intersections, vehicle arrivals are more likely to be random. A number of stochastic models have been developed for this case, including those by Newall [*8*], Miller [*9,10*], and Webster [*7*]. Such models assume that inter-vehicle arrival times are distributed according to the Poisson distribution, with an underlying average arrival rate of v vehicles/unit time. The models account for both the underlying randomness of arrivals and the fact that some individual cycles within a demand period with $v/c < 1.00$ could fail due to this randomness. This additional delay is sometimes referred to as "overflow delay," but it does not address situations in which $v/c > 1.00$ for the entire analysis period. This text refers to additional delay due to randomness as "random delay," *RD*, to distinguish it from true overflow delay when $v/c > 1.00$. The most frequently used model for random delay is Webster's formulation:

$$RD = \frac{X^2}{2v(1 - X)} \quad (20\text{-}21)$$

where: RD = average random delay per vehicle, s/veh
X = v/c ratio

This formulation was found to somewhat overestimate delay, and Webster proposed that total delay (the sum of uniform and random delay) be estimated as:

$$D = 0.90(UD + RD) \quad (20\text{-}22)$$

where: D = sum of uniform and random delay

Modeling Overflow Delay

"Oversaturation" is used to describe extended time periods during which arriving vehicles exceed the capacity of the intersection approach to discharge vehicles. In such cases, queues grow, and overflow delay, in addition to uniform delay, accrues. Because overflow delay accounts for the failure of an extended series of phases, it encompasses a portion of random delay as well.

Figure 20.14 illustrates a time period for which $v/c > 1.00$. Again, as in the uniform delay model, it is assumed the arrival function is uniform. During the period of oversaturation, delay consists of both uniform delay (in the triangles between the capacity and departure curves) and overflow delay (in the growing triangle between the arrival and capacity curves). The formula for the uniform delay component may be simplified in this case because the

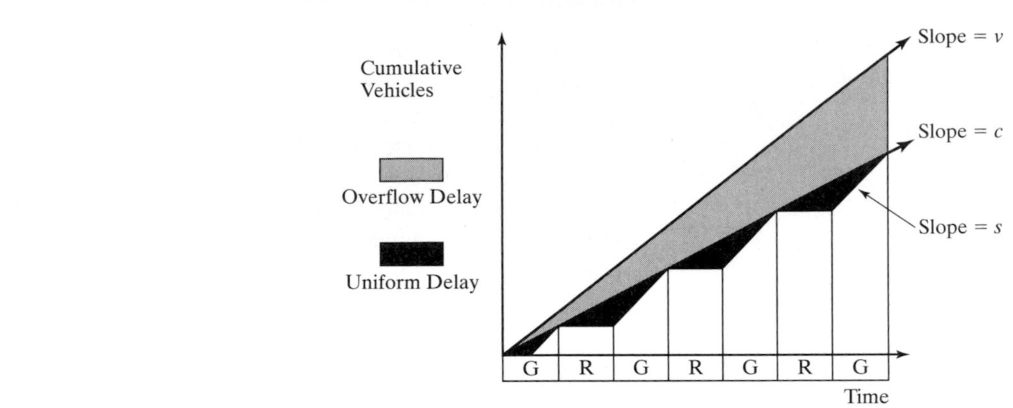

Figure 20.14: An Oversaturated Period Illustrated

v/c ratio (X) is the maximum value of 1.00 for the uniform delay component. Then:

$$UD_o = \frac{0.50C[1 - (g/C)]^2}{1 - (g/C)X} = \frac{0.50C[1 - (g/C)]^2}{1 - (g/C)1.00}$$
$$= 0.50C[1 - (g/C)] \qquad (20\text{-}23)$$

To this, the overflow delay must be added. Figure 20.15 illustrates how the overflow delay is estimated. The aggregate and average overflow delay can be estimated as:

$$OD_a = \frac{1}{2}T(vT - cT) = \frac{T^2}{2}(v - c)$$
$$OD = \frac{T}{2}[X - 1] \qquad (20\text{-}24)$$

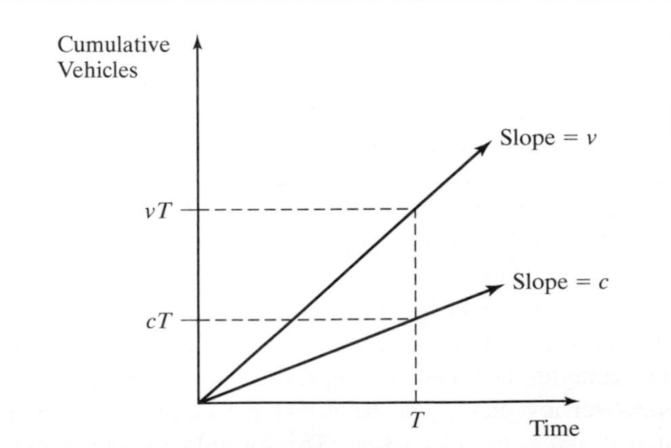

Figure 20.15: Derivation of the Overflow Delay Formula

where: OD_a = aggregate overflow delay, veh-sec
OD = average overflow delay per vehicle, s/veh

Other parameters are as previously defined.

In Equations 20-24, the average overflow delay is obtained by dividing the aggregate delay by *the number of vehicles discharged within time T, cT.* Unlike the formulation for uniform delay, where the number of vehicles arriving and the number of vehicles discharged during a cycle were the same, the overflow delay triangle includes vehicles that arrive within time T but are not discharged within time T. The delay triangle, therefore, includes only the delay accrued by vehicles through time T and excludes additional delay that vehicles still "stuck" in the queue will experience after time T.

Equations 20-24 may use any unit of time for "T." The resulting overflow delay, *OD,* will have the same units as specified for T, on a per-vehicle basis.

Equations 20-24 are time dependent (i.e., the longer the period of oversaturation exists, the larger delay becomes). The predicted delay per vehicle is averaged over the entire period of oversaturation, T. This masks, however, a significant issue: Vehicles arriving early during time T experience far less delay than vehicles arriving later during time T. A model for average overflow delay during a time period T_1 through T_2 may be developed, as illustrated in Figure 20.16. Note that the delay area formed is a trapezoid, not a triangle.

The resulting model for average delay per vehicle during the time period T_1 through T_2 is:

$$OD = \frac{T_1 + T_2}{2}(X - 1) \qquad (20\text{-}25)$$

where all terms are as previously defined. Note that the trapezoidal shape of the delay area results in the $T_1 + T_2$

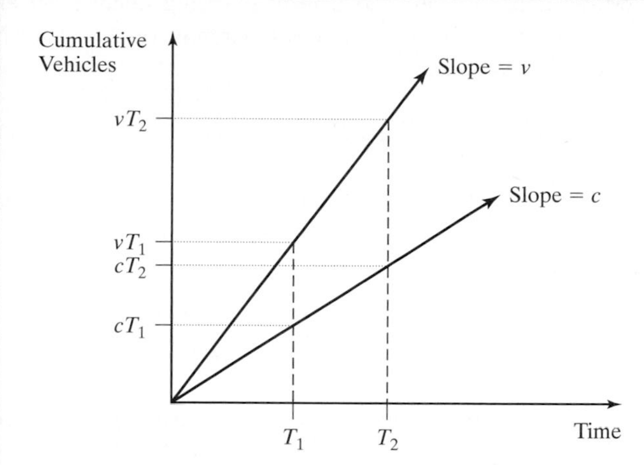

Figure 20.16: A Model for Overflow Delay Between Times T_1 and T_2

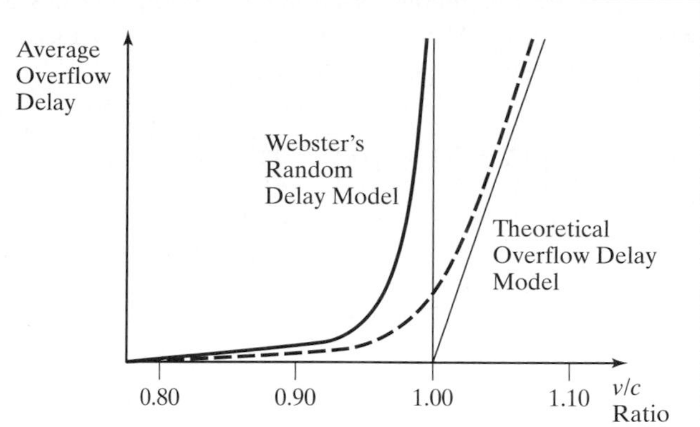

Figure 20.17: Random and Overflow Delay Models Compared

(*Source:* Adapted with permission of Transportation Research Board, National Research Council, Washington DC, from Hurdle, V.F. "Signalized Intersection Delay Model: A Primer for the Uninitiated, *Transportation Research Record 971,* p. 101, 1984.)

formulation, emphasizing the growth of delay as the oversaturated condition continues over time. Also, this formulation predicts the average delay per vehicle that occurs during the specified interval, T_1 through T_2 Thus delays to vehicles arriving before time T_1 but discharging after T_1 are included only to the extent of their delay within the specified times, not any delay they may have experienced in queue before T_1 Similarly, vehicles discharging after T_2 do have a delay component after T_2 that is not included in the formulation.

The three varieties of delay—uniform, random, and overflow delay—can be modeled in relatively simple terms as long as simplifying assumptions are made in terms of arrival and discharge flows, and in the nature of the queuing that occurs, particularly during periods of oversaturation. The next section begins to consider some of the complications that arise from the direct use of these simplified models.

20.5.3 Inconsistencies in Random and Overflow Delay

Figure 20.17 illustrates a basic inconsistency in the random and overflow delay models previously discussed. The inconsistency occurs when the v/c ratio (X) is in the vicinity of 1.00. When the v/c ratio is below 1.00, a random delay model is used because there is no "overflow" delay in this case. Webster's random delay model (Equation 20-22), however, contains the term (1-X) in the denominator. Thus as X approaches a value of 1.00, random delay increases asymptotically to an infinite value. When the v/c ratio (X) is greater than 1.00, an overflow delay model is applied. The overflow delay model of Equation 17-24, however, has an overflow delay of 0 when $X = 1.00$, and increases uniformly with increasing values of X thereafter.

Neither model is accurate in the immediate vicinity of $v/c = 1.00$. Delay does not become infinite at $v/c = 1.00$. There is no true "overflow" at $v/c = 1.00$, although individual cycle failures due to random arrivals do occur. Similarly, the overflow model, with overflow delay $= 0.0$ s/veh at $v/c = 1.00$, is also unrealistic. The additional delay of individual cycle failures due to the randomness of arrivals is not reflected in this model.

In practical terms, most studies confirm that the uniform delay model is a sufficient predictive tool (except for the issue of platooned arrivals) when the v/c ratio is 0.85 or less. In this range, the true value of random delay is minuscule, and there is no overflow delay. Similarly, the simple theoretical overflow delay model (when added to uniform delay) is a reasonable predictor when $v/c \geq 1.15$ or so. The problem is that the most interesting cases fall in the intermediate range ($0.85 < v/c < 1.15$), for which neither model is adequate. Much of the more recent work in delay modeling involves attempts to bridge this gap, creating a model that closely follows the uniform delay model at low v/c ratios and approaches the theoretical overflow delay model at high v/c ratios ($\geq$1.15), producing "reasonable" delay estimates in between. Figure 20.17 illustrates this as the dashed line.

The most commonly used model for bridging this gap was developed by Akcelik for the Australian Road Research Board's signalized intersection analysis procedure [*11,12*]:

$$OD = \frac{cT}{4}\left[(X-1) + \sqrt{(X-1)^2 + \left(\frac{12(X-X_0)}{cT}\right)}\right]$$

$$X_0 = 0.67 + \left(\frac{sg}{600}\right) \qquad (20\text{-}26)$$

where: T = analysis period, h
X = v/c ratio
c = capacity, veh/h
s = saturation flow rate, veh/sg (veh/s of green)
g = effective green time, s

The only relatively recent study resulting in large amounts of delay measurements in the field was conducted by Reilly et al. [*13*] in the early 1980s to calibrate a model for use in the 1985 edition of the *Highway Capacity Manual.* The study concluded that Equation 17-26 substantially overestimated field-measured values of delay and recommended that a factor of 0.50 be included in the model to adjust for this. The version of the delay equation that was included in the 1985 *Highway Capacity Manual* ultimately did not follow this recommendation and included other empirical adjustments to the theoretical equation.

20.5.4 Delay Models in the HCM

The delay model incorporated into the HCM 2000 [2] includes the uniform delay model, a version of Akcelik's overflow delay model, and a term covering delay from an existing or residual queue at the beginning of the analysis period. The model is:

$$d = d_1 PF + d_2 + d_3 \qquad (20\text{-}27)$$

where: d = control delay, s/veh
d_1 = uniform delay component, s/veh
PF = progression adjustment factor
d_2 = overflow delay component, s/veh
d_3 = delay due to preexisting queue, s/veh

The progression factor was an empirically calibrated adjustment to uniform delay that accounts for the effect of platooned arrival patterns. This adjustment is discussed in greater detail in Chapter 24. The delay due to preexisting queues, d_3, is found using a relatively complex model (see Chapter 24).

A significant revision has been included in the forthcoming HCM 2010. Traditional delay models have been replaced by Incremental Queue Analysis (IQA). Chapter 24 contains a more detailed discussion and presentation of this approach.

In the final analysis, all delay modeling is based on the determination of the area between an arrival curve and a departure curve on a plot of cumulative vehicles versus time. As the arrival and departure functions are permitted to become more complex and as rates are permitted to vary for various subparts of the signal cycle, the models become more complex as well.

20.5.5 Examples in Delay Estimation

Example 20-1:

Consider the following situation: An intersection approach has an approach flow rate of 1,000 veh/h, a saturation flow rate of 2,800 veh/hg, a cycle length of 90 seconds, and a *g/C* ratio of 0.55. What average delay per vehicle is expected under these conditions?

Solution:
To begin, the capacity and *v/c* ratio for the intersection approach must be computed. This will determine what model(s) are most appropriate for application in this case:

$$C = s(g/C) = 2{,}800 * 0.55 = 1.540 \text{ veh/h}$$

$$v/c = X = \frac{1{,}000}{1{,}540} = 0.649$$

Because this is a relatively low value, the uniform delay equation (Equation 20-19) may be applied directly. There is little random delay at such a *v/c* ratio and no overflow delay to consider. Thus:

$$d = \left(\frac{C}{2}\right)\frac{[1-(g/C)]^2}{1-(v/s)} = \left(\frac{90}{2}\right)\frac{(1-0.55)^2}{1-\left(\frac{1{,}000}{2{,}800}\right)} = 142 \text{ s/veh}$$

Note that this solution assumes that arrivals at the subject intersection approach are random. Platooning effects are not taken into account.

Example 20-2:

How would the preceding result change if the demand flow rate increased to 1,600 veh/h for a one-hour period?

Solution:

In this case, the *v/c* ratio now changes to 1,600/1,540 = 1.039. This is in the difficult range of 0.85 to 1.15 for which neither the simple random flow model nor the simple overflow delay model is accurate. The Akcelik model of Equation 20-26 will be used. Total delay, however, includes both uniform delay and overflow delay. The uniform delay component when $v/c > 1.00$ is given by Equation 20-23:

$$UD = 0.50C\left[1 - \left[{}^{g}\!/\!_{C}\right]\right] = 0.50 * 90 * (1 - 0.55) = 20.3 \text{ s/veh}$$

Use of Akcelik's overflow delay model requires that the analysis period be selected or arbitrarily set. Using a one-hour time period, as specified, then:

$$OD = \frac{cT}{4}\left[(X - 1) + \sqrt{(X - 1)^2 + \left(\frac{12(X - X_0)}{cT}\right)}\right]$$

$$X_0 = 0.67 + \left(\frac{sg}{600}\right) = 0.67 + \left(\frac{0.778 * 49.5}{600}\right) = 0.734$$

$$OD = \frac{1{,}540 * 1}{4}\left[(1.039 - 1) + \sqrt{(1.039 - 1)^2 + \left(\frac{12 * (1.039 - 0.734)}{1540 * 1}\right)}\right]$$

$$= 39.1 \text{ s/veh}$$

where: $g = 0.55 * 90 = 49.5$ s

$s = 2{,}800/3{,}600 = 0.778$ veh/sg

In this case, even with the "overflow" quite small (approximately 4% of the demand flow), the additional average delay due to this overflow is considerable. The total expected delay in this situation is the sum of the uniform and overflow delay terms, or:

$$d = 20.3 + 39.1 = 59.4 \text{ s/veh}$$

Note that this computation, as in Example 20-1, assumes random arrivals on this intersection approach.

Example 20-3:

How would the result change if the demand flow rate increased to 1,900 veh/h over a two-hour period?

Solution:

The *v/c* ratio in this case is now 1,900/1,540 = 1.23. In this range, the simple theoretical overflow model is an adequate predictor. As in Example 20-2, the Uniform Delay component must also be included; this computation is the same as in Example 20-2: 20.3 s/veg.

The overflow delay component may be estimated using the simple theoretical Equation 20-24:

$$OD = \frac{T}{2}(X - 1) = \frac{7{,}200}{2}(1.23 - 1) = 828.0 \text{ s/veh}$$

Because the period of oversaturation is given as two hours, and a result in seconds is desired, T is entered as $2 * 3{,}600 = 7{,}200$ s. The total delay experienced by the average motorist is the sum of uniform and overflow delay, or:

$$d = 20.3 + 828.0 = 848.3 \text{ s/veh}$$

This is a very large value but represents an average over the full two-hour period of oversaturation. Equation 20-25 may be used to examine the average delay to vehicles arriving in the first 15 minutes of oversaturation to those arriving in the last 15 minutes of oversaturation:

$$OD_{first\,15} = \frac{T_1 + T_2}{2}(X - 1)$$

$$= \frac{0 + 900}{2}(1.23 - 1) = 103.5 \text{ s/veh}$$

$$OD_{last\,15} = \frac{6{,}300 + 7{,}200}{2}(1.23 - 1) = 1{,}552.5 \text{ s/veh}$$

As previously noted, the delay experienced during periods of oversaturation is very much influenced by the length of time that oversaturated operations have prevailed. Total delay for each case would also include the 20.3 s/veh of uniform delay. As in Examples 20-1 and 20-2, random arrivals are assumed.

20.6 Overview

This chapter has reviewed four key concepts necessary to understand the operation of signalized intersections:

1. Saturation flow rate and lost times
2. The time budget and critical lanes
3. Left-turn equivalency
4. Delay as a measure of effectiveness

These fundamental concepts are also the critical components of models of signalized intersection analysis. In Chapter 21, some of these concepts are implemented in a simple methodology for signal timing. In Chapter 24, all are used as parts of the HCM analysis procedure for signalized intersections.

References

1. *Manual of Uniform Traffic Control Devices,* Millennium Edition, Federal Highway Administration, U.S. Department of Transportation, Washington DC, 2003.
2. *Highway Capacity Manual,* 4th Edition, Transportation Research Board, National Research Council, Washington DC, 2000.
3. Greenshields, B., "Traffic Performance at Intersections," *Yale Bureau Technical Report No. 1,* Yale University, New Haven CT, 1947.
4. Kunzman, W., "Another Look at Signalized Intersection Capacity," *ITE Journal,* Institute of Transportation Engineers, Washington DC, August 1978.
5. Roess, R., et al., "Level of Service in Shared, Permissive Left-Turn Lane Groups," *Final Report,* FHWA Contract DTFH-87-C-0012, Transportation Training and Research Center, Polytechnic University, Brooklyn NY, September 1989.
6. Prassas, E., "Modeling the Effects of Permissive Left Turns on Intersection Capacity," doctoral dissertation, Polytechnic University, Brooklyn NY, December 1994.
7. Webster, F., "Traffic Signal Settings," *Road Research Paper No. 39,* Road Research Laboratory, Her Majesty's Stationery Office, London, UK, 1958.
8. Newall, G., "Approximation Methods for Queues with Application to the Fixed-Cycle Traffic Light," *SIAM Review,* vol. 7, 1965.
9. Miller, A., "Settings for Fixed-Cycle Traffic Signals," *ARRB Bulletin 3,* Australian Road Research Board, Victoria, Australia, 1968.
10. Miller, A., "The Capacity of Signalized Intersections in Australia," *ARRB Bulletin 3,* Australian Road Research Board, Victoria, Australia, 1968.
11. Akcelik, R., "Time-Dependent Expressions for Delay, Stop Rate, and Queue Lengths at Traffic Signals," *Report No. AIR 367-1,* Australian Road Research Board, Victoria, Australia, 1980.
12. Akcelik, R., "Traffic Signals: Capacity and Timing Analysis," *ARRB Report 123,* Australian Road Research Board, Victoria, Australia, March 1981.
13. Reilly, W., and Gardner, C., "Technique for Measuring Delay at Intersections," *Transportation Research Record 644,* Transportation Research Board, National Research Council, Washington DC, 1977.

Problems

20-1. Consider the headway data shown in the table here. Data were taken from the center lane of a three-lane intersection approach for a total of 10 signal cycles. For the purposes of this analysis, the data may be considered to have been collected under ideal conditions.

a. Plot the headways versus position in queue for the data shown. Sketch an approximate best-fit curve through the data.

b. Using the approximate best-fit curve constructed in (a), determine the saturation headway and the start-up lost time for the data.

c. What is the saturation flow rate for this data?

Data for Problem 20-1

Q Pos	Headways (s) for Cycle No.									
	1	2	3	4	5	6	7	8	9	10
1	3.6	3.4	3.2	3.5	3.5	3.3	3.6	3.5	3.4	3.5
2	2.8	2.7	2.6	2.7	2.5	2.6	2.9	2.6	2.7	2.8
3	2.2	2.4	2.3	2.1	2.5	2.4	2.4	2.4	2.6	2.4
4	2.0	2.2	2.1	2.1	2.3	2.1	2.0	2.2	2.2	2.2
5	2.1	1.9	2.0	2.2	2.1	2.0	2.1	1.8	1.9	1.8
6	1.9	2.0	2.1	2.0	1.8	2.1	2.0	1.8	2.0	1.7
7	1.9	2.0	1.8	2.1	1.9	1.9	2.1	1.9	2.0	2.0
8	x	2.1	1.8	1.9	2.0	2.0	2.0	1.8	x	1.9
9	x	1.8	x	2.0	x	2.0	1.9	x	x	1.8
10	x	1.9	x	1.8	x	x	2.0	x	x	1.8

20-2. A signalized intersection approach has three lanes with no exclusive left- or right-turning lanes. The approach has a 40-second green out of a 75-second cycle. The yellow plus all-red intervals for the phase total 4.0 seconds. If the start-up lost time is 2.3 s/phase, the clearance lost time is 1.1 s/phase, and the saturation headway is 2.48 s/veh under prevailing conditions, what is the capacity of the intersection approach?

20-3. An equation has been calibrated for the amount of time required to clear N vehicles through a given signal phase:

$$T = 2.04 + 2.35N$$

a. What start-up lost time does this equation suggest exists?

b. What saturation headway and saturation flow rate is implied by the equation?

20-4. What is the maximum sum of critical-lane volumes that may be served by an intersection having three phases, a cycle length of 100 s, a saturation headway of 2.35 s/veh, and a total lost time per phase of 4.3 seconds?

20-5–20-6. For the two intersections illustrated here, find the appropriate number of lanes for each lane group needed. Assume that all volumes shown have been converted to compatible "through-car equivalent" values for the conditions shown. Assume that critical volumes reverse in the other daily peak hour.

20-7. For the intersection of Problem 20-5, consider a case in which the E-W arterial has two lanes in each direction and the N-S arterial has only one lane in each direction. For this case:

a. What is the absolute minimum cycle length that could be used?

b. What cycle length would be required to provide for a ratio of 0.90 during the worst 15 minutes of the hour if the PHF is 0.92?

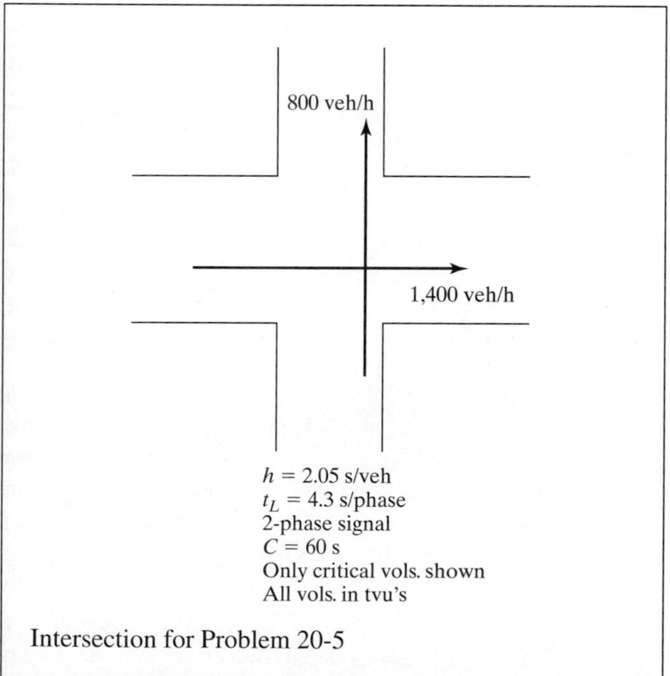

Intersection for Problem 20-5

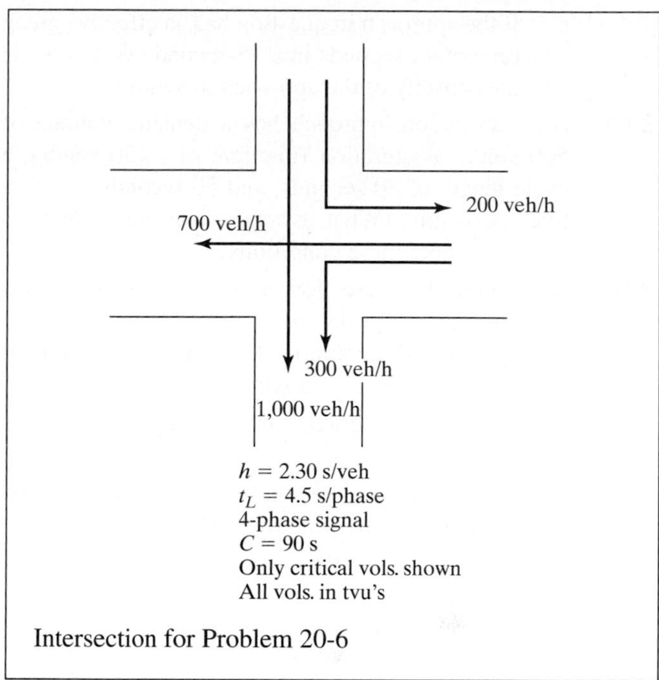

Intersection for Problem 20-6

20-8. At a signalized intersection, one lane is observed to discharge 20 through vehicles in the same time as the left lane discharges 10 through vehicles and 5 left-turning vehicles. For this case:

a. What is the through-vehicle equivalent, E_{LT}, for left-turning vehicles?

b. What is the left-turn adjustment factor, f_{LT}, for the case described?

c. What variables can be expected to affect the observed value of E_T?

20-9. An intersection approach volume is 1,350 veh/h and includes 8% left turns with a through-vehicle equivalent of 2.7 tvu/left turn. What is the total equivalent through volume on the approach?

20-10. An intersection approach has three lanes, permitted left turns, and 6% left-turning volume with a through vehicle equivalent of 5.0 tvu/left turn. The saturation flow rate for *through vehicles* under prevailing conditions is 1,700 veh/hg/ln.

a. What is the left-turn adjustment factor for the case described?

b. Determine the saturation flow rate and saturation headway for the approach, including the impact of left-turning vehicles.

c. If the approach in question has an effective green time of 45 seconds in a 75-second cycle, what is the capacity of the approach in veh/h?

20-11. An intersection approach has a demand volume of 500 veh/h, a saturation flow rate of 1,450 veh/hg, a cycle length of 80 seconds, and 50 seconds of effective green time. What average delay per vehicle is expected under these conditions?

20-12. A signalized intersection approach operates at an effective ratio of 1.05 for a peak 30-minute period each evening. If the approach has a *g/C* ratio of 0.60 and the cycle length is 75 seconds:

a. What is the average control delay for the entire 30-minute period?

b. What is the average control delay during the first 5 minutes of the peak period?

c. What is the average control delay per vehicle during the last 5 minutes of the peak period? Why is this period significant?

20-13. A signalized intersection approach experiences chronic oversaturation for a 1-hour period each day. During this time, vehicles arrive at a rate of 2,000 veh/h. The saturation flow rate for the approach is 3,250 veh/hg, with a 100-second cycle length, and 55 seconds of effective green.

a. What is the average control delay per vehicle for the full hour?

b. What is the average control delay per vehicle for the first 15 minutes of the peak period?

c. What is the average control delay per vehicle for the last 15 minutes of the peak hour?

CHAPTER 21

Fundamentals of Signal Timing and Design: Pretimed Signals

Signal timing and design involve several important components, including the physical design and layout of the intersection itself. Physical design is treated in some detail in Chapter 19. This chapter focuses on the design and timing of traffic control signals.

These are the key steps involved in signal design and timing:

1. Development of a safe and effective phase plan and sequence
2. Determination of vehicular signal needs:
    a. Timing of "yellow" (change) and "all-red" (clearance) intervals for each signal phase
    b. Determination of the sum of critical-lane volumes (V_c)
    c. Determination of lost times per phase (t_L) and per cycle (L)
    d. Determination of an appropriate cycle length (C)
    e. Allocation of effective green time to the various phases defined in the phase plan—often referred to as "splitting" the green
3. Determination of pedestrian signal needs:
    a. Determine minimum pedestrian "green" times
    b. Check to see if vehicular greens meet minimum pedestrian needs
    c. If pedestrian needs are unmet by the vehicular signal timing, adjust timing and/or add pedestrian actuators to ensure pedestrian safety

Although most signal timings are developed for vehicles and checked for pedestrian needs, it is critical that signal timing designs provide safety and relative efficiency for both. Approaches vary with relative vehicular and pedestrian flows, but every signal timing must consider and provide for the requirements of both groups.

Many aspects of signal timing are tied to the principles discussed in Chapter 20 and elsewhere in this text. The process, however, is not exact, nor is there often a single "right" design and timing for a traffic control signal. Thus signal timing does involve judgmental elements and represents true engineering design in a most fundamental way.

All of the key elements of signal timing are discussed in some detail in this chapter, and various illustrations are offered. Note, however, that it is virtually impossible to develop a complete and final signal timing that will not be subject to subsequent fine tuning when the proposed design is analyzed using the HCM 2000 (and forthcoming HCM 2010) analysis model or some other analysis model or simulation. This is because no straightforward signal design and timing process can hope to include and fully address all of the potential complexities that may exist in any given situation. Thus initial design and timing is often a starting point for analysis using a more complex model.

Chapters 23 and 24 of this text discuss two analysis models: one based on critical lane analysis, which is essentially signal timing in reverse, and the second the HCM analysis model. The latter includes some of the updates currently under development for the HCM 2010. Because this text will be published *before* the HCM 2010 is finalized and released, there may be additional changes that are not incorporated into Chapter 24. Although a number of other models are available, the HCM is widely used and accepted in the United States and a number of other countries.

21.1 Development of Signal Phase Plans

The most critical aspect of signal design and timing is the development of an appropriate phase plan. Once this is done, many other aspects of the signal timing can be analytically treated in a deterministic fashion. The phase plan and sequence involves the application of engineering judgment while applying a number of commonly used guidelines. In any given situation, there may be a number of feasible approaches that will work effectively.

21.1.1 Treatment of Left Turns

The single most important feature that drives the development of a phase plan is the treatment of left turns. As we discussed in Chapter 20, left turns may be handled as permitted movements (with an opposing through flow), as protected movements (with the opposing vehicular through movement stopped), or as a combination of the two (compound phasing). The simplest signal phase plan has two phases, one for each of the crossing streets. In this plan, all left turns are permitted. Additional phases may be added to provide protection for some or all left turns, but additional phases add lost time to the cycle. Thus the consideration of protection for left turns must weigh the inefficiency of adding phases and lost time to the cycle against the improved efficiency in operation of left-turning and other vehicles gained from that protection.

Two general guidelines provide initial insight into whether or not a particular left-turn movement requires a protected or a partially protected phase. Such phasing should be considered whenever there is an opposed left turn that satisfies one of the two following criteria:

$$v_{LT} \geq 200 \text{ veh/h} \tag{21-1a}$$

$$xprod = v_{LT} * \left({}^{v_o}\!/_{N_o} \right) \geq 50{,}000 \tag{21-1b}$$

where: v_{LT} = left-turn flow rate, veh/h

v_o = opposing through movement flow rate, veh/h

N_o = number of lanes for opposing through movement

Equation 21-1b is often referred to as the "cross-product" rule. Various agencies may use different forms of this particular guideline. These criteria, however, are not absolute. They provide a starting point for considering whether or not left-turn protection is needed for a particular left-turn movement.

There are other considerations. Left-turn protection, for example, is rarely provided when left-turn flows are less than two vehicles per cycle. It is generally assumed that in the worst case, where opposing flows are so high that *no* left turns may filter through it, an average of two vehicles each cycle will wait in the intersection until the opposing flow is stopped and then complete their turns. Such vehicles are usually referred to as "sneakers." Where a protected phase is needed for one left-turning movement, it is often convenient to provide one for the opposing left turn, even if it does not meet any of the normal guidelines. Sometimes, a left turn that does not meet any of the guidelines will present a particular problem that is revealed during the signal timing or in later analysis, and protection will be added.

The *Traffic Engineering Handbook* [*1*] provides some additional criteria for consideration of protected or partially protected left-turn movements, based on a research study [2]. Permitted phasing should be provided when the following conditions exist:

1. The left-turn demand flow within the peak hour, as plotted on Figure 21.1 against the speed limit for opposing traffic, falls within the "permitted" portion of the exhibit.
2. The sight distance for left-turning vehicles is not restricted.
3. Fewer than eight left-turn accidents have occurred within the last three years at any one approach with permitted-only phasing.

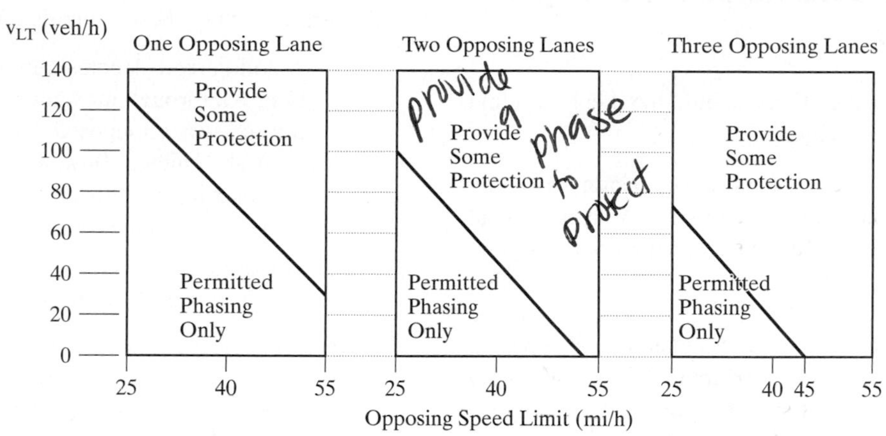

Figure 21.1: Recommended Selection Criteria for Left-Turn Protection

(*Source:* Used with permission of Transportation Research Board, National Research Council, Asante, S., Ardekani, S., and Williams, J., "Selection Criteria for Left-Turn Phasing and Indication Sequence," *Transportation Research Record 1421*, Washington DC, 1993, p. 11.)

The third criterion may be applied only to intersections that are already signalized using permitted left-turn phasing.

Guidance is also provided concerning the choice between fully protected and compound phasing. Fully protected phasing is recommended when any *two* of the following criteria are met:

1. Left-turn flow rate is greater than 320 veh/h.
2. Opposing flow rate is greater than 1,100 veh/h.
3. Opposing speed limit is greater than or equal to 45 mi/h.
4. There are two or more left-turn lanes.

Fully protected phasing is also recommended when any *one* of the following conditions exists:

1. There are three opposing traffic lanes, and the opposing speed is 45 mi/h or greater.
2. Left-turn flow rate is greater than 320 veh/h, and the percentage of heavy vehicles exceeds 2.5%.
3. The opposing flow rate exceeds 1,100 veh/h, and the percentage of left turns exceeds 2.5%.
4. Seven or more left-turn accidents have occurred within three years under compound phasing.
5. The average stopped delay to left-turning traffic is acceptable for fully protected phasing, and the engineer judges that additional left-turn accidents would occur under the compound phasing option.

A complex criterion based on observed left-turn conflict rates is also provided. The conflict criteria as well as criteria 4 and 5 apply only where an existing intersection is being operated with compound phasing.

These criteria essentially indicate when compound phasing *should not* be implemented. They suggest that compound phasing may be considered when left-turn protection is needed but none of these criteria are met.

Note, however, that use of compound phasing is also subject to many local agency policies that may supersede or supplement the more general guidelines presented here. Use of compound phasing varies widely from jurisdiction to jurisdiction. Some agencies use it only as a last resort for the reasons stated previously. Others provide compound phasing wherever a left-turn phase is needed because of the delay reductions usually achieved by compound versus fully protected left-turn phasing.

In extreme cases, it may be necessary to ban left turns entirely. This must be done, however, with the utmost care. It is essential that alternative paths for vehicles wishing to turn left are available and that they do not unduly inconvenience the affected motorists. Further, the additional demands on alternative routes should not cause worse problems at nearby intersections. Special design treatments for left turns are also discussed in Chapter 19.

21.1.2 General Considerations in Signal Phasing

Several important considerations should be kept in mind when establishing a phase plan:

1. Phasing can be used to minimize accident risks by separating competing movements. A traffic signal always eliminates the basic crossing conflicts present at intersections. Addition of left-turn protection can also eliminate some or all of the conflicts between left-turning movements and their opposing through movements. Additional phases generally lead to additional delay, which must be weighed against the safety and improved efficiency of protected left turns.
2. Although increasing the number of phases increases the total lost time in the cycle, the offsetting benefit is an increase in affected left-turn saturation flow rates.
3. All phase plans must be implemented in accordance with the standards and criteria of the MUTCD [*3*], and they must be accompanied by the necessary signs, markings, and signal hardware needed to identify appropriate lane usage.
4. The phase plan must be consistent with the intersection geometry, lane-use assignments, volumes and speeds, and pedestrian crossing requirements.

For example, it is not practical to provide a fully protected left-turn phase where there is no exclusive left-turn lane. If such phasing were implemented with a shared lane, the first vehicle in queue may be a through vehicle. When the protected left-turn green is initiated, the through vehicle blocks all left-turning vehicles from using the phase. Thus protected left-turn phases *require* exclusive left-turn lanes for effective operation.

21.1.3 Phase and Ring Diagrams

A number of typical and a few not-so-typical phase plans are presented and discussed here. Signal phase plans are generally illustrated using *phase diagrams* and *ring diagrams*. In both cases, movements allowed during each phase are shown using arrows. Here, only those movements allowed in each phase are shown; in some of the literature, movements not allowed are also shown with a straight line at the head of the arrow, indicating that the movement is stopped during the subject phase. Figure 21.2 illustrates some of the basic conventions used in these diagrams.

A more complete definition and discussion of the use and interpretation of these symbols follows:

1. A solid arrow denotes a movement without opposition. All through movements are unopposed by definition. An unopposed left turn has no opposing through vehicular flow. An unopposed right turn has no opposing pedestrian movement in the crosswalk through which the right turn is made.
2. Opposed left- and or right-turn movements are shown as a dashed line.
3. Turning movements made from a shared lane(s) are shown as arrows connected to the through movement that shares the lane(s).
4. Turning movements from an exclusive lane(s) are shown as separate arrows, not connected to any through movement.

Although not shown in Figure 21.2, pedestrian paths may also be shown on phase or ring diagrams. They are generally shown as dotted lines with a double arrowhead, denoting movement in both directions in the crosswalk.

A *phase diagram* shows all movements being made in a given phase within a single block of the diagram. A *ring diagram* shows which movements are controlled by which "ring" on a signal controller. A "ring" of a controller generally controls one set of signal faces. Thus, although a phase involving two opposing through movements would be shown in one block of a phase diagram, each movement would be separately shown in a ring diagram. As you will see in the next section, the ring diagram is more informative, particularly where overlapping phase sequences are involved. Chapter 19 describes signal hardware and the operation of signal controllers in more detail.

21.1.4 Common Phase Plans and Their Use

Simple two-phase signalization is the most common plan in use. If guidelines or professional judgment indicate the need to fully or partially protect one or more left-turn movements, a variety of options are available for doing so. The following sections illustrate and discuss the most common phase plans in general use.

Basic Two-Phase Signalization

Figure 21.3 illustrates basic two-phase signalization. Each street receives one signal phase, and all left and right turns are

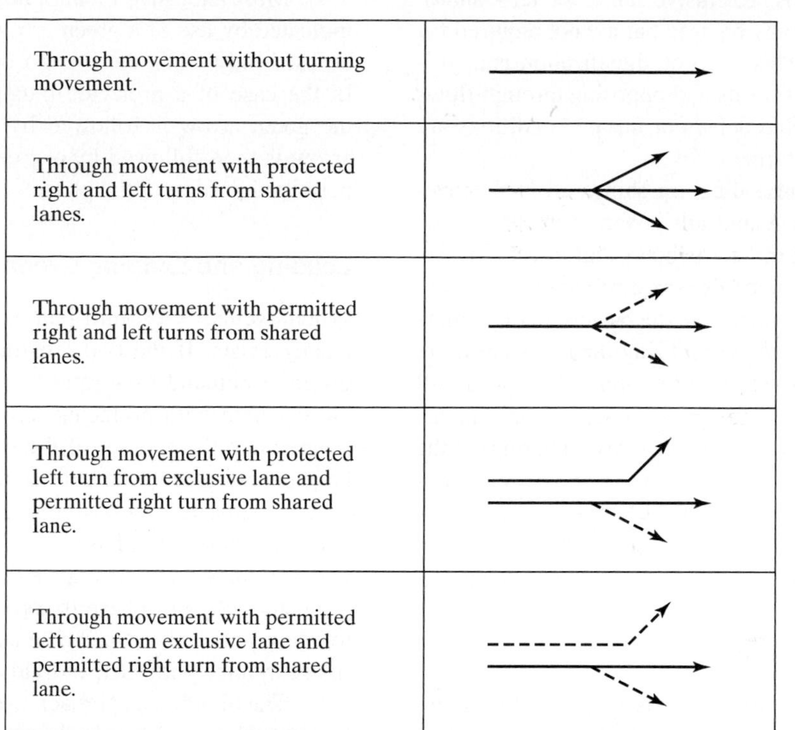

Through movement without turning movement.	
Through movement with protected right and left turns from shared lanes.	
Through movement with permitted right and left turns from shared lanes.	
Through movement with protected left turn from exclusive lane and permitted right turn from shared lane.	
Through movement with permitted left turn from exclusive lane and permitted right turn from shared lane.	

Figure 21.2: Selected Signal Phase Arrows Illustrated

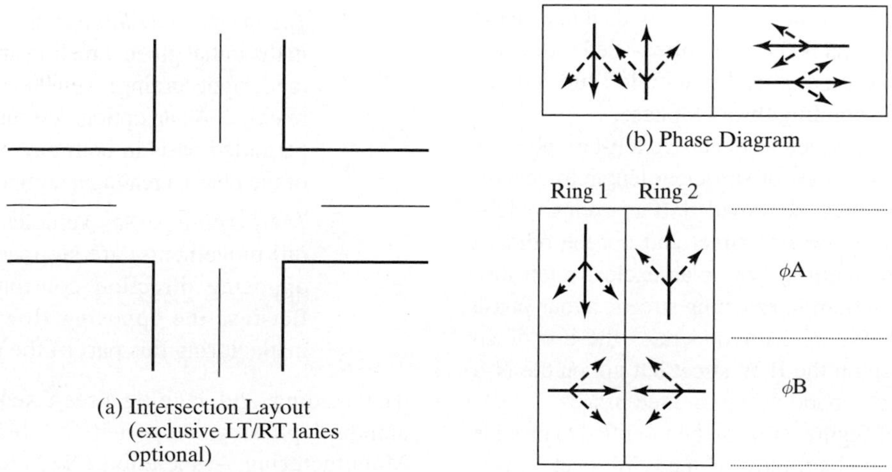

(a) Intersection Layout (exclusive LT/RT lanes optional)

(b) Phase Diagram

Ring 1	Ring 2	
		ϕA
		ϕB

(c) Ring Diagram

Figure 21.3: Illustration of a Two-Phase Signal

made on a permitted basis. Exclusive lanes for left- and/or right-turning movements may be used but are not required for two-phase signalization. This form of signalization is appropriate where the mix of left turns and opposing through flows is such that no unreasonable delays or unsafe conditions are created by and/or for left-turners.

In this case, the phase diagram shows all N-S movements occurring in Phase A and all E-W movements occurring in Phase B. The ring diagram shows that in each phase, each set of directional movements is controlled by a separate ring of the signal controller. Because the basic signalization is relatively simple, both the phase and ring diagrams are quite similar, and both are relatively easy to interpret. As you will see, this is not the case for more complex signal phase plans.

Note that all phase boundaries cut across both rings of the controllers, meaning that all transitions occur at the same times in both rings. Also, it would make little difference which movements appear in which rings. The combination shown could be easily reversed without affecting the operation of the signal.

Exclusive Left-Turn Phasing

When a need for left-turn protection is indicated by guidelines or professional judgment, the simplest way to provide it is through the use of an exclusive left-turn phase(s). Two opposing left-turn movements are provided with a simultaneous and exclusive left-turn green, during which the two through movements on the subject street are stopped. An exclusive left-turn phase may be provided either before or after the through/right-turn phase for the subject street, although the most common practice is to provide it *before* the through phase. Because this is the most often-used sequence, drivers have become more comfortable with left-turn phases placed before the corresponding through phase.

As noted previously, when an exclusive left-turn phase is used, an exclusive left-turn lane of sufficient length to accommodate expected queues must be provided. If an exclusive left-turn phase is implemented on one street and not the other, a three-phase signal plan emerges. Where an exclusive left-turn phase is implemented on both intersecting streets, a four-phase signal plan is formed. Figure 21.4 illustrates the use of an exclusive left-turn phase on the E-W street but not on the N-S street, where left turns are made on a permitted basis.

The phase plan of Figure 21.4 can be modified to provide for protected plus permitted left turns on the E-W street. This is done by adding a permitted left-turn movement to Phase B. In general, such compound phasing is used where the combination of left turns and opposing flows is so heavy that provision of fully protected phasing leads to undesirably long or unfeasible cycle lengths. Compound phasing is more difficult for drivers to comprehend and is more difficult to display.

Most exclusive or unopposed left-turn movements are indicated by use of a green arrow. The arrow indication may be used only when there is no opposing through movement. In the case of a protected-plus-permitted compound phase, the green arrow is followed by a yellow arrow; the yellow arrow is then followed by a green ball indication during the permitted portion of the phase.

Leading and Lagging Green Phases

When exclusive left-turn phases are used, a potential inefficiency exists. If the two left-turning movements have very different demand flow rates (on a per-lane basis), then providing them with protected left-turn phases of equal length assures that the smaller of the two left-turn movements will have excess green time that cannot be used. Where this inefficiency leads to excessive or unfeasible cycle lengths and/or excessive delays, a phase plan in which opposing protected left-turn phases are separated should be considered. If a NB protected left-turn phase is separated from the SB protected left-turn phase, the two can be assigned different green times in accordance with their individual demand flow rates.

The traditional approach to accomplishing this is referred to as "leading and lagging" green phases. A leading and lagging green sequence for a given street has three components:

1. *The leading green.* Vehicles in one direction get the green while vehicles in the opposing direction are stopped. Thus the left-turning movement in the direction of the "green" is protected.
2. *The overlapping through green.* Left-turning vehicles in the initial green direction are stopped while through (and right-turning) vehicles in both directions are released. As an option, left turns may be allowed on a permitted basis in both directions during this portion of the phase, creating a compound phase plan.
3. *The lagging green.* Vehicles in the initial direction (all movements) are stopped while vehicles in the opposing direction continue to have the green. Because the opposing flow is stopped, left turns made during this part of the phase are protected.

The leading and lagging green sequence is no longer a standard phasing supported by the National Electronics Manufacturing Association (NEMA), which creates standards for signal controllers and other electronic devices. Such controllers, however, are still available, and this sequence is still used in some jurisdictions.

Figure 21.5 illustrates a leading and lagging green sequence in the E-W direction. A similar sequence can be used in the N-S direction as well. Again, an exclusive left-turn

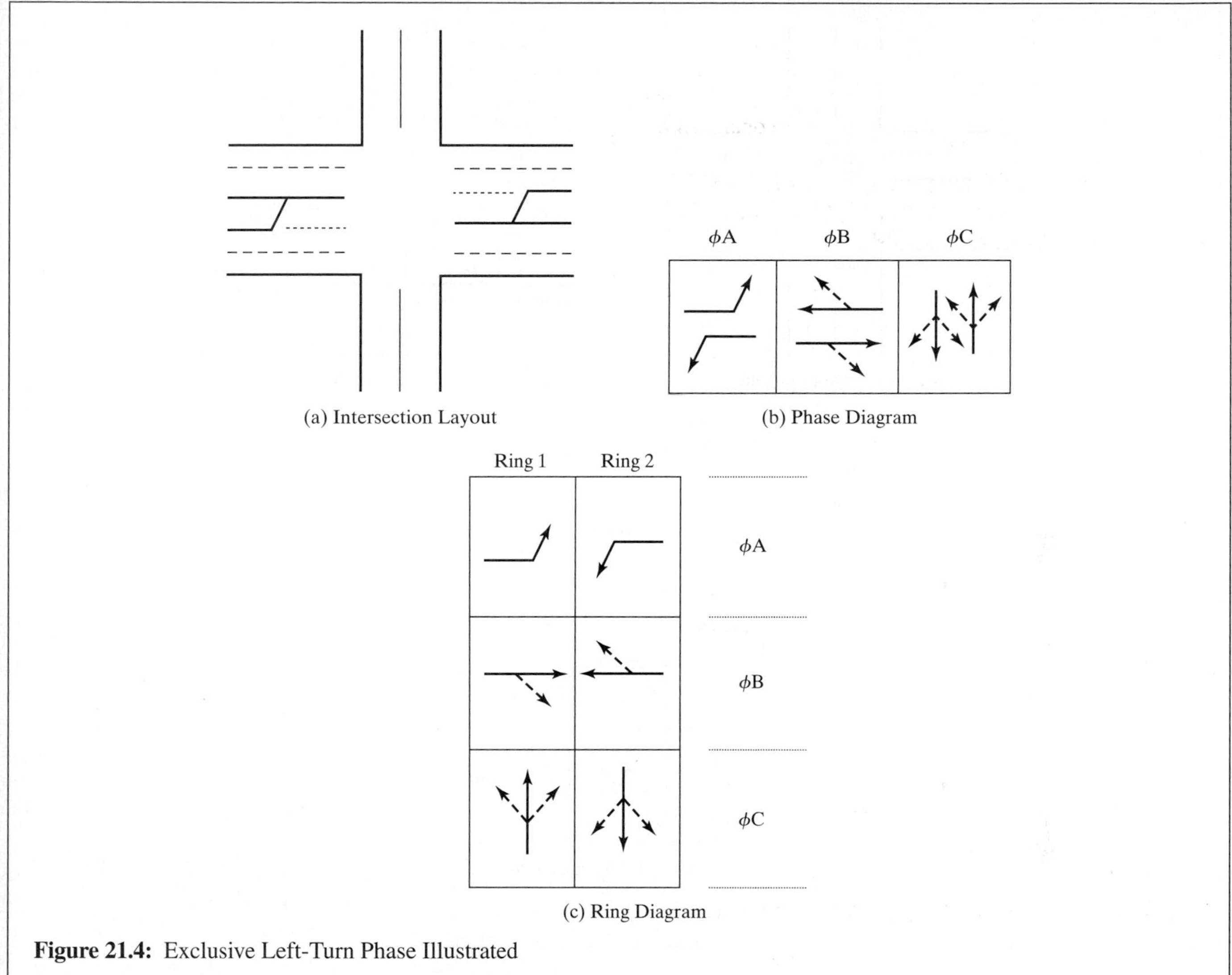

Figure 21.4: Exclusive Left-Turn Phase Illustrated

lane must be provided when a leading and lagging green is implemented.

The leading and lagging green phase plan involves "overlapping" phases. The EB through is moving in Phases A1 and A2 while the WB through is moving in Phases A2 and A3. One critical question arises in this case: How many phases are there in this plan? It might be argued that there are *four* distinct phases: A1, A2, A3, and B. It might also be argued that Phases A1, A2, and A3 form a single overlapping phase and that the plan therefore involves only *two* phases. In fact, both analyses are incorrect.

The ring diagram is critical in the analysis of overlapping phase plans. At the end of Phase A1, only Ring 1 goes though a transition, transferring the green from the EB left turn to the WB through and right-turn movements. At the end of Phase A2, only Ring 2 goes through a transition, transferring the green from the EB through and right-turn movements to the WB left turn. Each ring, therefore, goes through *three* transitions in a cycle. In effect, this is a *three*-phase signal plan. The ring diagram makes the difference between partial and full phase boundaries clear, whereas the phase diagram can easily mask this important feature.

This distinction is critical to subsequent signal-timing computations. For each phase transition, a set of lost times (start-up plus clearance) is experienced. If the sum of the lost times per phase (t_L) were 4.0 seconds per phase, then a two-phase signal would have 8.0 seconds of lost time per cycle (L), a three-phase signal would have 12.0 seconds of lost time per cycle, and a four-phase signal would have 16.0 seconds of lost time per cycle. As you will see, the lost time per cycle has a dramatic affect on the required cycle length.

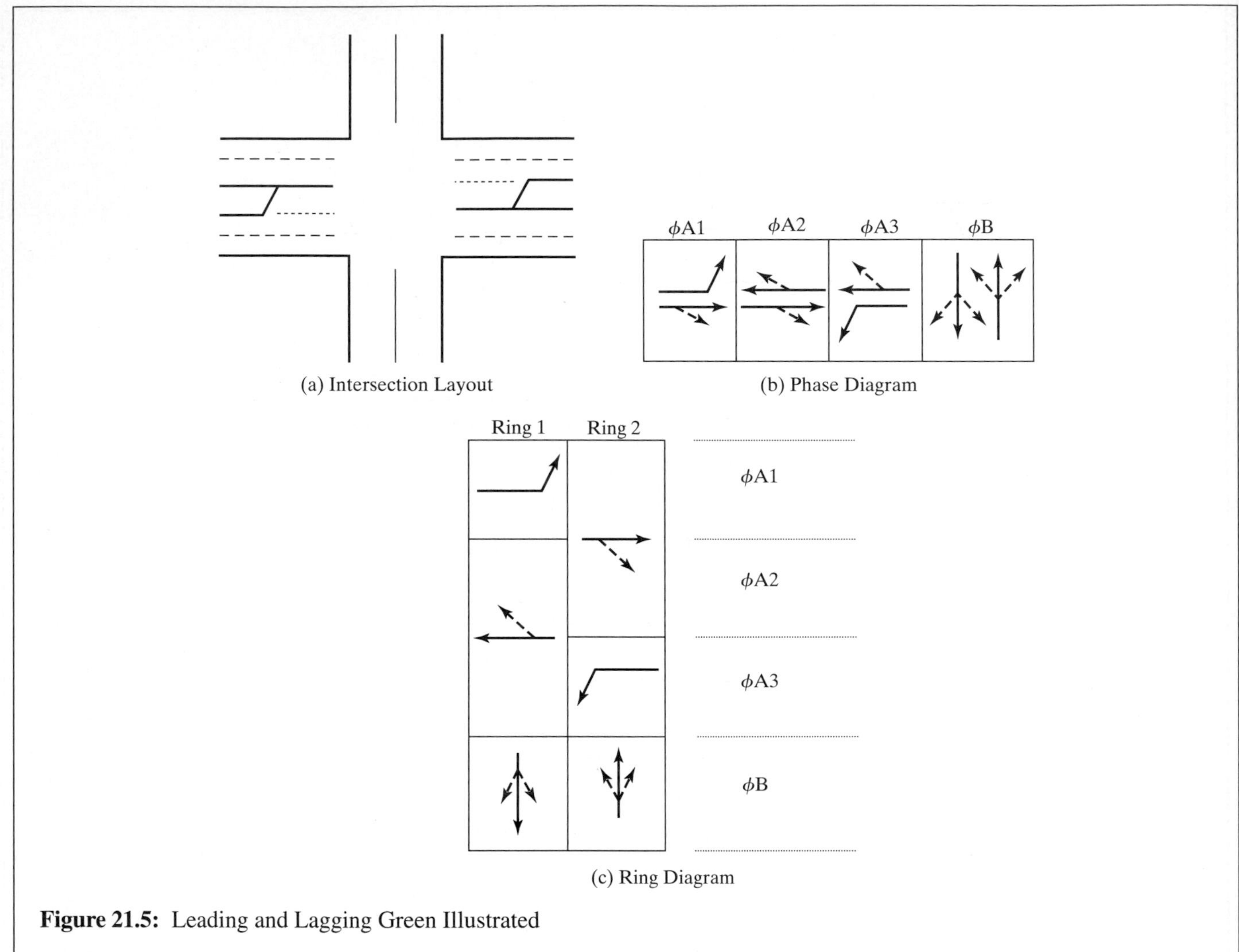

Figure 21.5: Leading and Lagging Green Illustrated

A number of interesting options to the leading and lagging phase plan could be implemented:

1. A leading green may be used *without* a lagging green or vice versa. This is usually done where a one-way street or T-intersection creates a case in which there is only one left turn from a two-way major street.
2. A compound phasing can be created by allowing permitted left turns in Phase A2. This would create a protected-plus-permitted phase for the EB left turn and a permitted-plus-protected phase for the WB left turn.
3. A leading and/or lagging green may be added to the N-S street, assuming that an exclusive left-turn lane could be provided within the right-of-way.

Exclusive Left-Turn Phase with Leading Green

It was previously noted that NEMA does not have a set of controller specifications to implement a leading and lagging green phase plan. The NEMA standard phase sequence for providing unequal protected left-turn phases employs an exclusive left-turn phase followed by a leading green phase in the direction of the heaviest left-turn demand flow. In effect, this sequence provides the same benefits as the leading and lagging green, but it allows all protected left-turn movements to be made before the opposing through movements are released. Figure 21.6 illustrates such a phase plan for the E-W street.

In the case illustrated, the EB left-turn movement receives the leading green as the heavier of the two left-turn demand flows. If the WB left turn required the leading green,

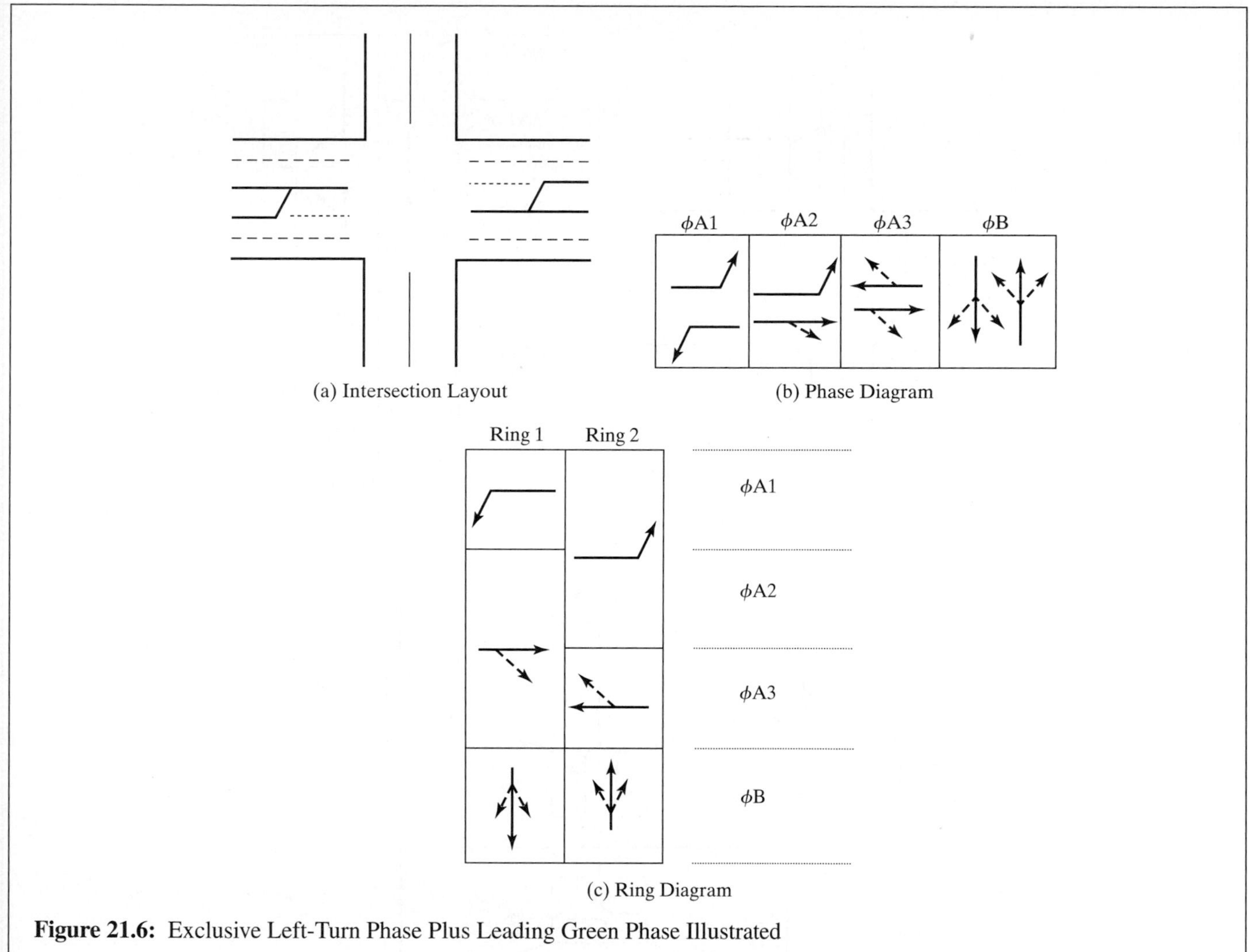

(a) Intersection Layout

(b) Phase Diagram

(c) Ring Diagram

Figure 21.6: Exclusive Left-Turn Phase Plus Leading Green Phase Illustrated

this is easily accomplished by reversing the positions of the partial boundaries between Phases A1 and A2 and between A2 and A3.

Note there is a similarity between the leading and lagging green phase plan and the exclusive left-turn plus leading green phase plan. In both cases, the partial transition in each ring is between a protected left turn and the opposing through and right-turn movements, or vice versa. Virtually all overlapping phase sequences involve such transfers.

A compound phase can be implemented by allowing EB and WB permitted left-turn movements in Phase A2. In both cases, this creates a protected-plus-permitted phase sequence. This phasing can also be implemented for the N-S street if needed, as long as an exclusive left-turn lane is provided.

The issue of number of phases is also critical in this phase plan. The phase plan of Figure 21.6 involves *three* discrete phases and *three* phase transitions on each ring.

Eight-Phase Actuated Control

Any of the previous phase plans may be implemented using a pretimed or an actuated controller with detectors. However, actuated controllers offer the additional flexibility of skipping phases when no demand is detected. This is most often done for left-turn movements. Protected left-turn phases may be skipped in any cycle where detectors indicate no left-turn demand. The most flexible controller follows the phase sequence of an exclusive left-turn phase plus a leading green. Figure 21.7 shows the actuated phase plan for such a controller.

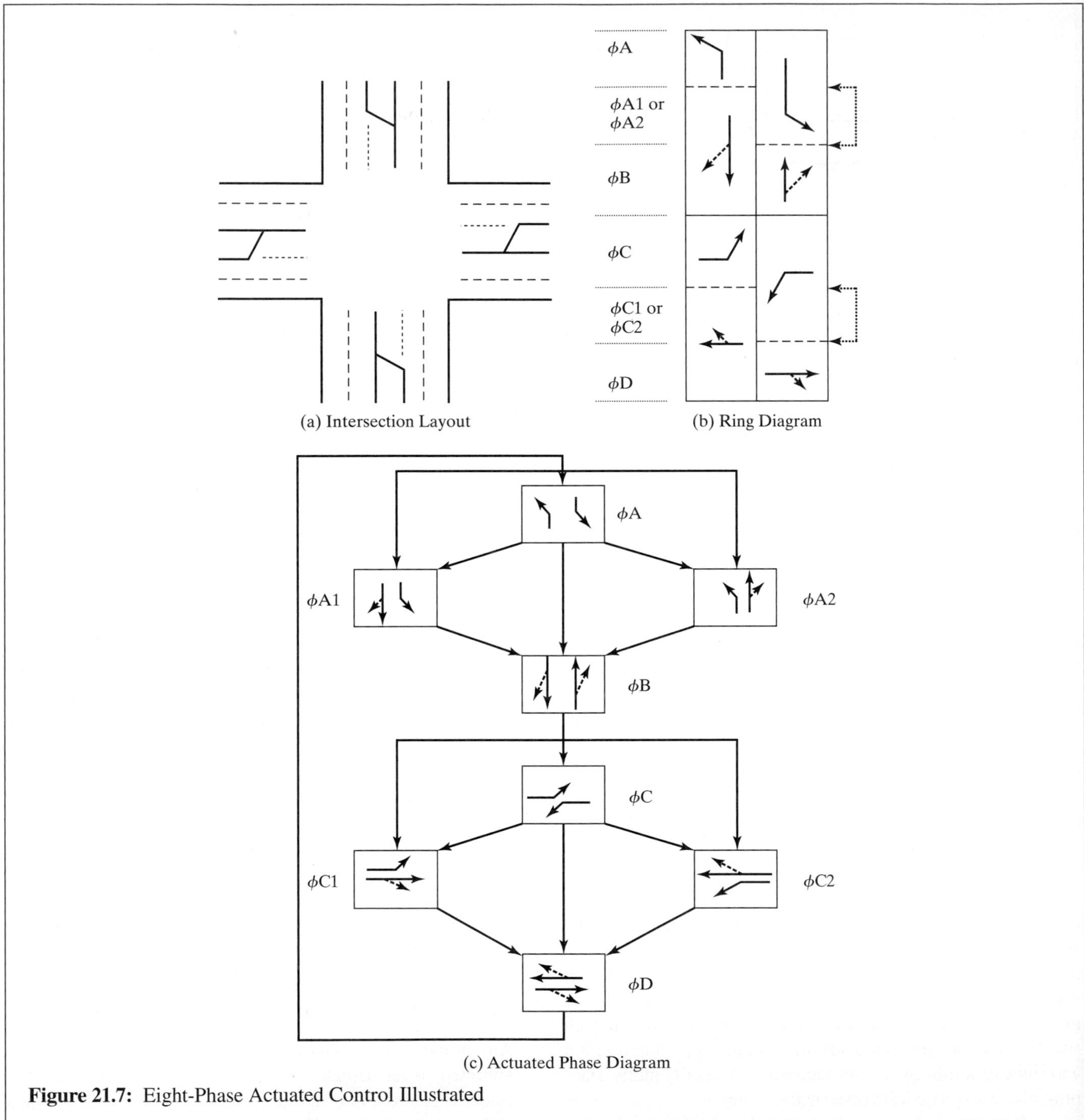

Figure 21.7: Eight-Phase Actuated Control Illustrated

In this case, exclusive left-turn phases and leading greens are provided for both streets, and both streets have exclusive left-turn lanes as shown.

This type of actuated signalization provides for complete flexibility in both the phase sequence and in the timing of each phase. Each street may start its green phases in one of three ways, depending on demand:

- An exclusive left-turn phase in both directions if left-turn demand is present in both directions

- A leading green phase (in the appropriate direction) if only one left-turn demand is present
- A combined through and right-turn phase in both directions if no left-turn demand is present in either direction

If the first option is selected, the next phase may be a leading green if one direction still has left-turn demand when the other has none, or a combined through and right-turn phase if both left-turn demands are simultaneously satisfied during the exclusive left-turn phase.

The ring diagram assumes that a full sequence requiring both the exclusive left-turn phase and the leading green phase (for one direction) are needed. The partial phase boundaries are shown as dashed lines because the relative position of these may switch from cycle to cycle depending on which left-turn demand flow is greater. If the entire sequence is needed, there are *four* phase transitions in either ring, making this (as a maximum) a *four-phase* signal plan. Thus, even though the controller defines eight potential phases, during any given cycle, a maximum of four phases may be activated.

As we noted previously, actuated control is generally used where signalized intersections are relatively isolated, although modern signal systems now coordinate actuated controllers. The type of flexibility provided by eight-phase actuated control is most effective where left-turn demands vary significantly over the course of the day.

The Exclusive Pedestrian Phase

The exclusive pedestrian phase was a unique approach to the control of situations in which pedestrian flows are a significant, or even dominant, movement at a traffic signal. Originally developed by New York City traffic engineer Henry Barnes in the 1960s for Manhattan, this type of phasing is often referred to as the "Barnes Dance."

Figure 21.8 illustrates this phasing. During the exclusive pedestrian phase, pedestrians are permitted to cross the intersection in any direction, including diagonally. All vehicular movements are stopped during the exclusive pedestrian phase. The exclusive pedestrian phase is virtually never used where more than two vehicular phases are needed.

The exclusive pedestrian phase has several drawbacks. The primary problem is that the entire pedestrian phase must be treated as lost time in terms of the vehicular signalization. Delays to vehicles are substantially increased because of this, and vehicular capacity is significantly reduced.

The exclusive pedestrian phase never worked well in the city of its birth. Where extremely heavy pedestrian flows exist, such as in Manhattan, the issue of clearing them out of the intersection at the close of the pedestrian phase is a major enforcement problem. In New York, pedestrians occupied intersections for far longer periods than intended, and the negative impacts on vehicular movement were intolerable.

The exclusive pedestrian phase works best in small rural or suburban centers, where vehicular flows are not extremely high and where the volume of pedestrians is not likely to present a clearance problem at the end of the pedestrian phase. In such cases, it can provide additional safety for pedestrians in environments where drivers are not used to negotiating conflicts between vehicles and pedestrians.

Signalization of T-Intersections

T-intersections present unique problems along with the opportunity for unique solutions using the combination of geometric design and imaginative signal phasing. This is particularly true where the one opposed left turn that exists at a T-intersection requires a protected phase.

Figure 21.9 illustrates such a situation along with several candidate solutions. In Figure 21.9 (a), there are no turning lanes provided. In such a case, providing the WB left turn with a protected phase requires that each of the three approach legs have its own signal phase. Although achieving the required protected phasing for the opposed left turn, such phasing is not very efficient in that each movement uses only one of three phases. Delays to all vehicles tend to be longer than they would be if more efficient phasing could be implemented.

If an exclusive left-turn lane is provided for the WB left-turn movement and if separate lanes for left and right turns are provided on the stem of the T, a more efficient phasing can be implemented. In this plan, the intersection geometry is used to allow several vehicular movements to use two of the three phases, including some overlaps between right turns from one street and selected movements from the other. This is illustrated in Figure 21.9 (b).

If a left-turn lane for the WB left turn can be combined with a channelizing island separating the WB through movement from all other vehicle paths, a signalization can be adopted in which the WB through movement is never stopped. Figure 21.9 (c) illustrates this approach. Note that this particular approach can be used only where there are no pedestrians present or where an overpass or underpass is provided for those crossing the E-W artery.

In each of the cases shown in Figure 21.9, a three-phase signal plan is used. Using geometry, however, additional movements can be added to each of the signal phases, improving the overall efficiency of the signalization. As the signal plan becomes more efficient, delays to drivers and passengers will be reduced, and the capacity for each movement will be increased.

Note that this example starts with the assumption that the WB left turn needs to be protected. If this were not true, a simple two-phase signal could have been used.

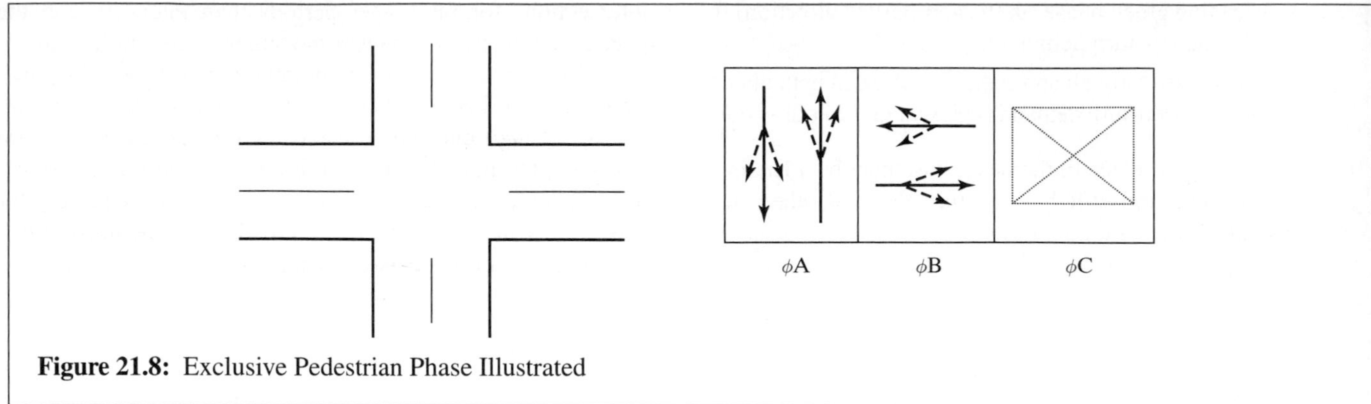

Figure 21.8: Exclusive Pedestrian Phase Illustrated

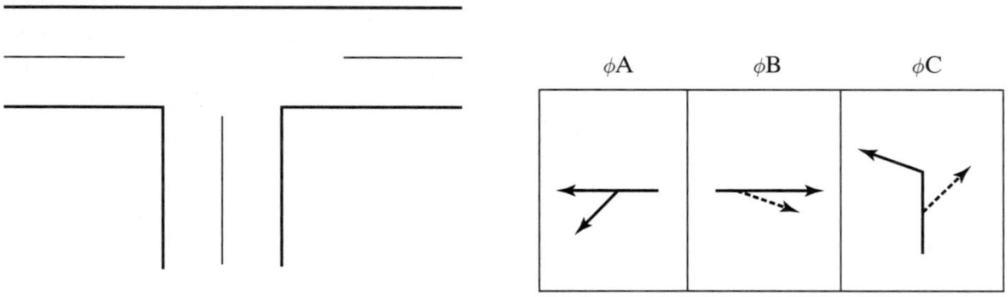

ϕA ϕB ϕC

(a) T-Intersection, No LT Lane, Protected Phasing

ϕA ϕB ϕC

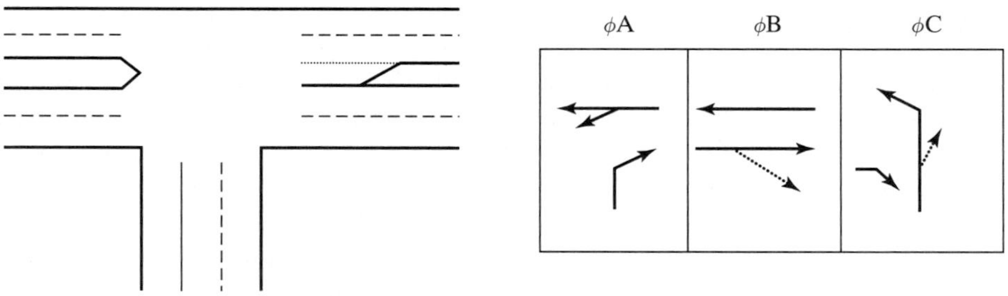

(b) T-Intersection, LT Lane, Protected Phasing

ϕA ϕB ϕC

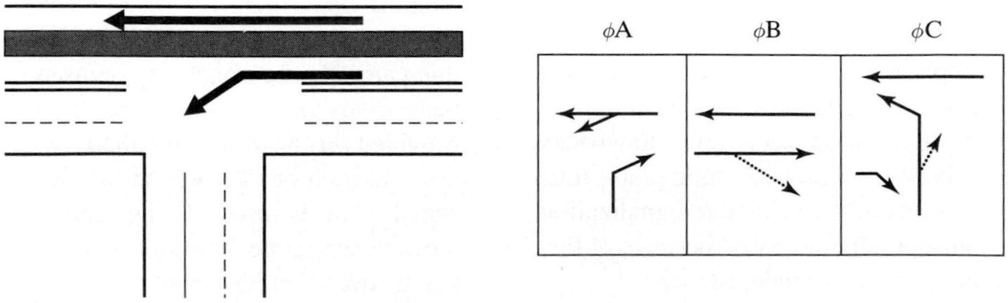

(c) T-Intersection, Channelized Through Movement

Figure 21.9: Signalization Options at a T-intersections

Five- and Six-Leg Intersections

Five- and six-leg intersections are a traffic engineer's worst nightmare. Although somewhat rare, these intersections do occur with sufficient frequency to present major problems in signal networks. Figure 21.10 illustrates a five-leg intersection formed when an off-ramp from a limited access facility is fed directly into a signalized intersection.

In the example shown, a four-phase signal phase plan is needed to provide a protected left-turn phase for the E-W artery. Had the N-S artery required a protected left-turn phase as well (an exclusive LT lane would have to be provided), then a five-phase signalization could have resulted. Addition of a fifth, and, potentially, even a sixth phase creates inordinate amounts of lost time, increases delay, and reduces capacity to critical approaches and lane groups.

Wherever possible, design alternatives should be considered to eliminate five- and six-leg intersections. In the case illustrated in Figure 21.10, for example, redesign of the ramp to create another separate intersection should be considered. The ramp could be connected to either of the intersecting arteries in a T-intersection. The distance from the new intersection to the main intersection would be a critical feature and should be arranged to avoid queuing that would block egress from the ramp. It may be necessary to signalize the new intersection as well.

In Manhattan (New York City), Broadway created a major problem in traffic control. The street system in most of Manhattan is a perfect grid, with the distance between N-S avenues (uptown/downtown) a uniform 800 feet, and the distance between E-W streets (crosstown) a uniform 400 feet. Such a regular grid, particularly when combined with a one-way street system (initiated in the early 1960s), is relatively easy to signalize. Broadway, however, runs diagonally through the grid, creating a series of major multileg intersections involving three major intersecting arteries. Some of these "major" intersections include Times Square and Herald Square, and all involve major vehicular and major pedestrian flows. To take advantage of the signalization benefits of a one-way, uniform grid street system, through flow on Broadway is banned at most of these intersections. This has effectively turned Broadway into a local street, with little through traffic. Through traffic is forced back onto the grid. Channelization is provided that forces vehicles

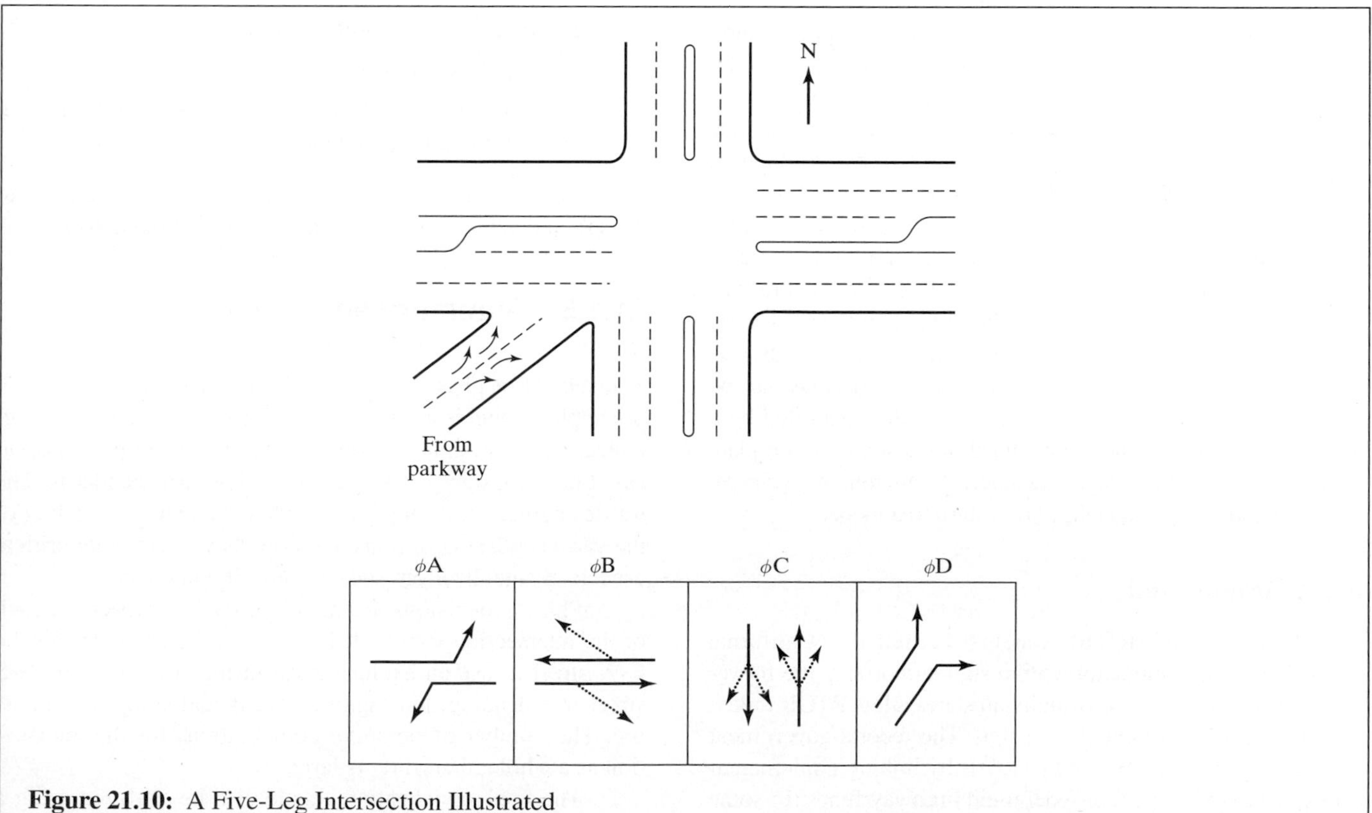

Figure 21.10: A Five-Leg Intersection Illustrated

approaching on Broadway to join either the avenue or the street, eliminating the need for multiphase signals. In addition, channelizing islands have been used to create unique pedestrian environments at these intersections.

Right-Turn Phasing

Although the use of protected left-turn phasing is common, the overwhelming majority of signalized intersections handle right turns on a permitted basis, mostly from shared lanes. Protected right-turn phasing is used only where the number of pedestrians is extremely high. Modern studies show that a pedestrian flow of 1,700 peds/h in a crosswalk can effectively block all right turns on green. Such a pedestrian flow is, however, extremely rare, and exists only in major urban city centers. Although use of a protected right-turn phase in such circumstances may help motorists, it may worsen pedestrian congestion on the street corner and on approaching sidewalks. In extreme cases, it is often useful to examine the feasibility of pedestrian overpasses or underpasses. These would generally be coordinated with barriers preventing pedestrians from entering the street at the corner. It should be noted, however, that pedestrian overpasses and/or underpasses are inconvenient for pedestrians and may pose security risks, primarily at night.

Compound right-turn phasing is usually implemented only in conjunction with an exclusive left-turn phase on the intersecting street. For example, NB and SB right turns may be without pedestrian interference during an EB and WB exclusive left-turn phase. Permitted right turns would then continue during the NB and SB through phase.

Exclusive right-turn lanes are useful where heavy right-turn movements exist, particularly where right-turn-on-red is permitted. Such lanes can be easily created on streets where curb parking is permitted. Parking may be prohibited within several hundred feet of the STOP line; the curb lane may then be used as an exclusive right-turn lane. Channelized right turns may also be provided. Channelized right turns are generally controlled by a YIELD sign and need not be included in the signalization plan. Chapter 19 contains a more detailed discussion of exclusive right-turn lanes and channelized right-turn treatments.

Right-Turn-on-Red

"Right-turn-on-red" (RTOR) was first permitted in California in 1937 only in conjunction with a sign authorizing the movement [*4*]. In recent years, virtually all states allow RTOR unless it is specifically prohibited by a sign. The federal government encouraged this approach in the 1970s by linking implementation of RTOR to receipt of federal-aid highway funds. In some urban areas, like New York City, right-turn-on-red is still generally prohibited. Signs indicating this general prohibition must be posted on all roadways entering the area. All RTOR laws require that the motorist stop before executing the right-turn movement on red.

When implemented using a shared right-turn through lane, the utility of RTOR is affected by the proportion of through vehicles using the lane. When a through vehicle reaches the STOP line, it blocks subsequent right-turners from using RTOR. Thus provision of an exclusive right-turn lane greatly enhances the effectiveness of RTOR.

The major issues regarding RTOR continue to be (1) the delay savings to right-turning vehicles, and (2) the increased accident risk such movements cause. An ITE practice [*5*] states that the delay to an average right-turning vehicle is reduced by 9% in central business districts (CBDs), 31% in other urban areas, and 39% in rural areas. Another early study on the safety of RTOR [*6*] found that only 0.61% of all intersection accidents involved RTOR vehicles and that these accidents tended to be less severe than other intersection accidents.

Because there are potential safety issues involving RTOR, its use and application should be carefully considered. The primary reasons for prohibiting RTOR include:

1. Restricted sight distance for right-turning motorist
2. High speed of conflicting through vehicles
3. High flow rates of conflicting through vehicles
4. High pedestrian flows in crosswalk directly in front of right-turning vehicles

Any of these conditions would make it difficult for drivers to discern and avoid conflicts during the RTOR maneuver.

21.1.5 Summary and Conclusion

The subject of phasing along with the selection of an appropriate phase plan is a critical part of effective intersection signalization. Although general criteria have been presented to assist in the design process, there are few firm standards. The traffic engineer must apply a knowledge and understanding of the various phasing options and how they affect other critical aspects of signalization, such as capacity and delay.

Phasing decisions are made for each approach on each of the intersection streets. It is possible, for example, for the E-W street to use an exclusive left-turn phase while the N-S street uses leading and lagging greens and compound phasing. The number of potential combinations for the intersection as a whole, therefore, is large.

The final signalization should also be analyzed using a comprehensive signalized intersection model (such as the 2000

Highway Capacity Manual) or simulation (such as CORSIM/NETSIM). This allows for fine tuning of the signalization on a trial-and-error basis and for a wider range of alternatives to be quickly assessed.

21.2 Determining Vehicular Signal Requirements

Once a candidate phase plan has been established, it is possible to establish the "timing" of the signal that would most effectively accommodate the vehicular demands present.

21.2.1 Change and Clearance Intervals

The terms "change" and "clearance" interval are used in a variety of ways in the literature. They refer to the *yellow* and *all-red* indications, respectively, that mark the transition from GREEN to RED in each signal phase. The *all-red* interval is a period during which all signal faces show a RED indication. The MUTCD specifically prohibits the use of a *yellow* indication to mark the transition from RED to GREEN, a practice common in many European countries.

Although the MUTCD does not strictly require *yellow* and/or *all-red* intervals, the Institute of Transportation Engineers (ITE) recommends that both be used at all signals. In most states, it is legal to *enter* an intersection on *yellow*. Therefore, the function of these critical intervals is as follows:

- *Change interval (yellow)*. This interval allows a vehicle that is one safe stopping distance away from the STOP line when the GREEN is withdrawn to continue at the approach speed and enter the intersection legally on *yellow*. "Entering the intersection" is interpreted to be the front wheels crossing over the intersection curb line.
- *Clearance interval (all-red)*. Assuming that a vehicle has just entered the intersection legally on *yellow*, the *all-red* must provide sufficient time for the vehicle to cross the intersection and clear its back bumper past the far curb line (or crosswalk line) before conflicting vehicles are given the GREEN.

The ITE recommends the following methodology for determining the length of the *yellow* or change interval [7]:

$$y = t + \frac{1.47\,S_{85}}{2a + (64.4 * 0.01G)} \tag{21-2}$$

where: y = length of the *yellow* interval, s
t = driver reaction time, s
S_{85} = 85th percentile speed of approaching vehicles, or speed limit, as appropriate, mi/h
a = deceleration rate of vehicles, ft/s^2
G = grade of approach, %
64.4 = twice the acceleration rate due to gravity, which is 32.2 ft/s^2

This equation was derived as the time required for a vehicle to traverse one safe stopping distance at its approach speed. Commonly used values for key parameters include a deceleration rate of 10.0 ft/s^2 and a driver reaction time of 1.0 seconds.

The ITE also recommends the following policy for determining the length of *all-red* clearance intervals [7]:

For cases in which there is no pedestrian traffic:

$$ar = \frac{w + L}{1.47\,S_{15}} \tag{21-3a}$$

For cases in which significant pedestrian traffic exists:

$$ar = \frac{P + L}{1.47\,S_{15}} \tag{21-3b}$$

For cases in which some pedestrian traffic exists:

$$ar = \max\left[\left(\frac{W + L}{1.47\,S_{15}}\right), \left(\frac{P}{1.47\,S_{15}}\right)\right] \tag{21-3c}$$

where: ar = length of the all-red phase, seconds
W = distance from the departure STOP line to the far side of the farthest conflicting traffic lane, feet
P = distance from the departure STOP line to the far side of the farthest conflicting crosswalk, feet
L = length of a standard vehicle, usually taken to be 18 to 20 feet
S_{15} = 15th percentile speed of approaching traffic, or speed limit, as appropriate, mi/h

The difference between the three equations involves pedestrian activity levels and the decision to clear vehicles beyond the line of potential conflicting vehicle paths and/or conflicting pedestrian paths before releasing the conflicting flows. Equation 21-3c, which addresses the most frequently occurring situations—some, but not significant pedestrian flows—is a compromise. If the pedestrian clearance distance, P, is used, the length of the vehicle is not added (i.e., the timing would provide for the *front* bumper of a vehicle to reach the far crosswalk line before releasing the conflicting vehicular and pedestrian flows).

To provide for optimal safety, the equations for *yellow* and *all-red* intervals use different speeds: the 85th percentile and the 15th percentile, respectively. Because speed appears in the numerator of the *yellow* determination and in the denominator of the *all-red* determination, accommodating the majority of motorists safely requires the use of different percentiles. If only the average approach speed is known, the percentile speeds may be estimated as:

$$S_{15} = S - 5$$
$$S_{85} = S + 5 \tag{21-4}$$

where: S_{85} = 85th percentile speed, mi/h
S_{85} = 85th percentile speed`, mi/h
S = average speed, mi/h

Where approach speeds are not measured and the speed limit is used, both the *yellow* and *all-red* intervals will be determined using the same value of speed. This, however, is not a desirable practice.

Use of these ITE policies to determine *yellow* and *all-red* intervals assures that drivers will not be presented with a "dilemma zone," which occurs when the combined length of the change and clearance intervals is not sufficient to allow a motorist who cannot safely stop when the *yellow* is initiated to cross through the intersection and out of conflicting vehicular and/or pedestrian paths before those flows are released. Where *yellow* and *all-red* phases are mistimed and a dilemma zone is created, agencies face possible liability for accidents that occur as a result.

Note that some states have a somewhat different law concerning the yellow interval: In these states it is legal to enter the intersection on yellow only if the driver can successfully cross and clear the intersection before the yellow signal expires. This is much more difficult for drivers because they don't know when the yellow will expire. In states using this law, the all-red phase is sometimes not included, although this is not a recommended practice. If the all-red phase was eliminated, then the yellow would have to be extended to include the total time for both.

Consider the following example: Compute the appropriate change and clearance intervals for a signalized intersection approach with the following characteristics:

- Average approach speed = 35 mi/h
- Grade = −2.5%
- Distance from STOP line to far side of the most distant lane = 48 ft
- Distance from STOP line to far side of the most distant cross-walk = 60 ft
- Standard vehicle length = 20 ft
- Reaction time = 1.0 s
- Deceleration rate = 10 ft/s^2
- Some pedestrians present

To apply Equations 21-2 and 21-3, estimates of the 15th and 85th percentile speeds are needed. Using Equation 21-4:

$$S_{85} = 35 + 5 = 40 \text{ mi/h}$$
$$S_{15} = 35 - 5 = 30 \text{ mi/h}$$

Using Equation 21-2, the length of the change or *yellow* interval should be:

$$y = 1.0 + \frac{1.47 * 40}{[2 * 10] + [64.4 * 0.01 * (-2.5)]}$$
$$= 1.0 + \frac{58.8}{20 - 1.61} = 4.2 \text{ s}$$

Equation 21-3c is used to compute the length of the clearance or *all-red* phase because there are some, but not significant, pedestrian flows present. The length of the clearance interval is the maximum of:

$$ar = \frac{48 + 20}{1.47 * 30} = \frac{68}{44.1} = 1.5 \text{ s}$$
$$ar = \frac{60}{1.47 * 30} = \frac{60}{44.1} = 1.4 \text{ s}$$

In this case, 1.5 seconds would be applied.

21.2.2 Determining Lost Times

The 2000 (and forthcoming 2010) edition of the *Highway Capacity Manual* [*8*] indicates that lost times vary with the length of the *yellow* and *all-red* phases in the signal timing. Thus it is no longer appropriate to use a constant default value for lost times as was historically done in most signal timing methodologies. The HCM now recommends the use of the following default values for this determination:

- Start-up lost time, ℓ_1 = 2.0 s/phase
- Motorist use of *yellow* and *all-red*, e = 2.0 s/phase

Using these default values, lost time per phase and lost time per cycle may be estimated as follows:

$$\ell_2 = Y - e \tag{21-5}$$
$$Y = y + ar \tag{21-6}$$
$$t_L = \ell_1 + \ell_2 \tag{21-7}$$

where: ℓ_1 = start-up lost time, s/phase
ℓ_2 = clearance lost time, s/phase
t_L = total lost time, s/phase
y = length of *yellow* change interval, s
ar = length of *all red* clearance interval, s
Y = total length of change and clearance intervals, s

In the example of the previous section, the *yellow* interval was computed as 4.2 seconds, and the *all-red* interval was found to be 1.5 seconds. Using the recommended default values for ℓ_1 and e, respectively, lost times would be computed as:

$$Y = 4.2 + 1.5 = 5.7\,\text{s}$$
$$\ell_2 = 5.7 - 2.0 = 3.7\,\text{s}$$
$$t_L = 2.0 + 3.7 = 5.7\,\text{s}$$

Note that when the HCM-recommended default values for ℓ_1 and e (both 2.0 s) are used, the lost time per phase, t_L is always equal to the sum of the *yellow* and *all-red* intervals, Y. Because the lost time for each phase may differ, based on different *yellow* and *all-red* intervals, the total lost time per cycle is merely the sum of lost times in each phase, or:

$$L = \sum_{i}^{n} t_{Li} \tag{21-8}$$

where: L = total lost time per cycle, s
t_{Li} = lost time for phase i, s
n = number of discrete phases in cycle

21.2.3 Determining the Sum of Critical-Lane Volumes

To estimate an appropriate cycle length and to split the cycle into appropriate green times for each phase, it is necessary to find the *critical-lane volume* for each discrete phase or portion of the cycle.

As discussed in Chapter 20, the *critical-lane volume* is the per-lane volume that controls the required length of a particular phase. For example, in the case of a simple two-phase signal, on a given phase the EB and WB flows move simultaneously. One of these per-lane volumes represents the most intense demand, and that is the one that will determine the appropriate length of the phase.

Making this determination is complicated by two factors:

- Simple volumes cannot be simply compared. Trucks require more time than passenger cars, left and right turns require more time than through vehicles, vehicles on a downgrade approach require less time than vehicles on a level or upgrade approach. Thus *intensity* of demand is not measured accurately by simple volume.
- Where phase plans involve overlapping elements, the ring diagram must be carefully examined to determine which flows constitute *critical-lane volumes*.

Ideally, demand volumes would be converted to equivalents based on all of the traffic and roadway factors that might affect intensity. For initial signal timing, however, this is too complex a process. Demand volumes can, however, be converted to reflect the influence of the most significant factors affecting intensity: left and right turns. This is accomplished by converting all demand volumes to *equivalent through vehicle units* (tvu's). Through vehicle equivalents for left and right turns are shown in Tables 21.1 and 21.2, respectively.

These values are actually a simplification of a more complex approach in the *Highway Capacity Manual* analysis model for signalized intersections, and they form an appropriate basis

Table 21.1: Through-Vehicle Equivalents for Left-Turning Vehicles, E_{LT}

Opposing Flow V_o (veh/h)	Number of Opposing Lanes, N_o		
	1	2	3
0	1.1	1.1	1.1
200	2.5	2.0	1.8
400	5.0	3.0	2.5
600	10.0*	5.0	4.0
800	13.0*	8.0	6.0
1,000	15.0*	13.0*	10.0*
≥1,200	15.0*	15.0*	15.0*
E_{LT} for all *protected* left turns = 1.05			

*The LT capacity is only available through "sneakers."

Table 21.2: Through-Vehicle Equivalents for Right-Turning Vehicles, E_{RT}

Pedestrian Volume in Conflicting Crosswalk, (peds/h)	Equivalent
None (0)	1.18
Low (50)	1.21
Moderate (200)	1.32
High (400)	1.52
Extreme (800)	2.14

for signal timing and design. In using these tables, the following should be noted:

- Opposing volume, V_o, includes only the through volume on the opposing approach, in veh/h.
- Interpolation in Table 21.1 for opposing volume is appropriate, but values should be rounded to the nearest tenth.
- For right turns, the "conflicting crosswalk" is the crosswalk through which right-turning vehicles must pass.
- Pedestrian volumes indicated in Table 21.2 represent typical situations in moderate-sized communities. Pedestrian volumes in large cities, like New York, Chicago, or Boston, may be much higher, and the relative terms used (low, moderate, high, extreme) are not well correlated to such situations.
- Interpolation in Table 21.2 is not recommended.

Once appropriate values for E_{LT} and E_{RT} have been selected, all right- and left-turn volumes must be converted to units of "through-vehicle equivalents." Subsequently, the demand intensity *per lane* is found for each approach or lane group.

$$V_{LTE} = V_{LT} * E_{LT}$$
$$V_{RTE} = V_{RT} * E_{RT} \qquad (21\text{-}9)$$

where: V_{LTE} = left-turn volume in through-vehicle equivalents, tvu/h
V_{RTE} = right-turn volume in through-vehicle equivalents, tvu/h

Other variables are as previously defined.

These equivalents are added to through vehicles that may be present in a given approach or lane group to find the total equivalent volume and equivalent volume per lane in each approach or lane group:

$$V_{EQ} = V_{LTE} + V_{TH} + V_{RTE}$$
$$V_{EQL} = V_{EQ}/N \qquad (21\text{-}10)$$

where: V_{EQ} = total volume in a lane group or approach, tvu/h
V_{EQL} = total volume per lane in a lane group or approach, tvu/h/ln
N = number of lanes

Finding the critical-lane volumes for the signal phase plan requires determining the critical path through the plan (i.e., the path that controls the signal timing). This is done by finding the path through the signal phase plan that results in the highest possible sum of critical-lane volumes. Because most signal plans involve two "rings," alternative paths must deal with two potential rings for each discrete portion of the phase plan. It must also be noted that the critical path may "switch" rings at any full phase boundary (i.e., a phase boundary that cuts through both rings). This process is best understood through examples. Figure 21.11 shows a ring diagram for a signalization with overlapping phases. Lane volumes, V_{EQL}, are shown for each movement in the phase diagram.

To find the critical path, the controlling (maximum) equivalent volumes must be found for each portion of the cycle, working between full-phase transition boundaries. For the combined Phase A in Figure 21.11, the volumes that control the total length of A1, A2, and A3 are on Ring 1 or Ring 2. As shown, the maximum total comes from Ring 2 and

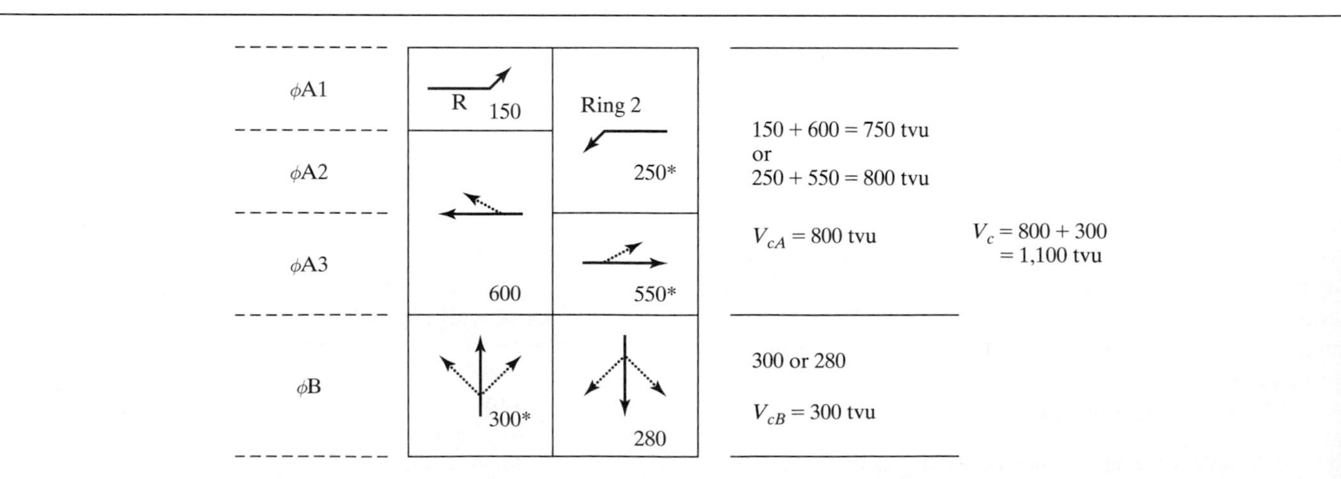

Figure 21.11: Determining Critical Lane Volume Illustrated

yields a total critical-lane volume of 800 tvu's. For Phase B, the choice is much simpler because there are no overlapping phases. Thus the Ring 1 total of 300 tvu's is identified as critical. The critical path through the cycle is now indicated by asterisks, and the sum of critical-lane volumes is 800 + 300 = 1,100 tvus. In essence, if the intersection is thought of in terms of a number of vehicles in single lanes seeking to move through a single common conflicting point, the signal, in this case, must have a timing that is sufficient to handle 1,100 tvu's through this point. The determination of critical-lane volumes is further illustrated in the complete signal-timing examples included in the last section of this chapter.

21.2.4 Determining the Desired Cycle Length

In Chapter 20, an equation describing the maximum sum of critical-lane volumes that could be handled by a signal was manipulated to find a desirable cycle length. That equation is used to find the desired cycle length, based on tvu volumes, and a default value for saturation flow rate. The default saturation flow rate, 1,615 tvu's per hour of green, assumes typical conditions of lane width, heavy-vehicle presence, grades, parking, pedestrian volumes, local buses, area type, and lane utilization. Common default values for saturation flow rate range from 1,500 to 1,700 in the literature, but these commonly also account for typical left-turn and right-turn percentages as well. The method presented here makes these adjustments by converting demand to equivalent through-vehicle units.

When the default value for saturation flow rate is inserted into the relationship, the desired cycle length is computed as:

$$C_{des} = \frac{L}{1 - \left[\dfrac{V_c}{1615 * PHF * (v/c)}\right]} \qquad (21\text{-}11)$$

where: C_{des} = desirable cycle length, s
L = total lost time per cycle, s/cycle
PHF = peak-hour factor
v/c = target v/c ratio for the critical movements in the intersection

Use of the peak-hour factor ensures that the signal timing is appropriate for the peak 15 minutes of the design hour. Target v/c ratios are generally in the range of 0.85 to 0.95. Very low values of v/c increase delays because vehicles are forced to wait while an unused green phase times out. Values of $v/c > 0.95$ indicate conditions in which frequent individual phase or cycle failures are possible, thereby increasing delay.

Consider the example illustrated previously in Figure 21.11. The sum of the critical-lane volumes for this case was shown to be 1,100 veh/h. What is the desirable cycle length for this three-phase signal if the total lost time per cycle is 4 s/phase × 3 phases/cycle = 12 s/cycle, the peak-hour factor is 0.92, and the target v/c ratio is 0.90? Using Equation 21-11:

$$C_{des} = \frac{12}{1 - \left[\dfrac{1100}{1615 * 0.9 * 0.90}\right]} = 67.6\,\text{s}$$

If this were a pretimed signal, timing dials (or modules) are available in 5-second increments between cycle lengths of 30 and 90 seconds, and in 10-second increments between 90 and 120 seconds. Thus a 70-second cycle would be adopted in this case. For actuated controllers, cycle lengths vary, although this equation might be used to obtain a very rough estimate of the expected average cycle length.

21.2.5 Splitting the Green

Once the cycle length is determined, the available *effective green time* in the cycle must be divided among the various signal phases. The available effective green time in the cycle is found by deducting the lost time per cycle from the cycle length:

$$g_{TOT} = C - L \qquad (21\text{-}12)$$

where: g_{TOT} = total effective green time in the cycle, s.
C, L as previously defined

The total effective green time is then allocated to the various phases or subphases of the signal plan in proportion to the critical lane volumes for each phase or subphase:

$$g_i = g_{TOT} * \left(\frac{V_{ci}}{V_c}\right) \qquad (21\text{-}13)$$

where: g_i = effective green time for Phase i, s
g_{TOT} = total effective green time in the cycle, s
V_{ci} = critical lane volume for Phase or Subphase i, veh/h
V_c = sum of the critical-lane volumes, veh/h

Returning to the example of Figure 21.11, the situation is complicated somewhat by the presence of overlapping phases. For the critical path, the following critical-lane volumes were obtained:

- 250 veh/h/ln for the *sum* of Phases A1 and A2
- 550 veh/h/ln for Phase A3
- 300 veh/h/ln for Phase B

Remembering that the desired cycle length of 70 seconds contains 12 seconds of lost time, the total effective green time may be computed using Equation 21-12:

$$g_{TOT} = 70 - 12 = 58\,\text{s}$$

Using Equation 21-13 and the critical-lane volumes just noted, the effective green times for the signal are estimated as:

$$g_{A1+A2} = 58 * \left(\frac{252}{1100}\right) = 13.2\,\text{s}$$

$$g_{A3} = 58 * \left(\frac{550}{1100}\right) = 29.0\,\text{s}$$

$$g_{B} = 58 * \left(\frac{300}{1100}\right) = 15.8\,\text{s}$$

The sum of these times (13.2 + 29.0 + 15.8) must equal 58.0 seconds, and it does. Together with the 12.0 seconds of lost time in the cycle, the 70-second cycle length is now fully allocated.

Because of the overlapping phases illustrated in this example, the signal timing is still not complete. The split between phases A1 and A2 must still be addressed. This can be done only by considering the noncritical Ring 1 for Phase A because this ring contains the transition between these two sub-phases. The total length of Phase A is 13.2 + 29.0 = 42.2 s On the noncritical ring (Ring 1), critical-lane volumes are 150 for Phase A1 and 600 for the sum of Phases A2 and A3. Using these critical-lane volumes:

$$g_{A1} = 42.2 * \left(\frac{150}{150 + 600}\right) = 8.4\,\text{s}$$

By implication, g_{A2} is now computed as the total length of Phase A, 42.2 s, minus the effective green times for Phases A1 and A3, both of which have now been determined (8.4 seconds and 29.0 seconds, respectively). Thus:

$$g_{A2} = 42.2 - 8.4 - 29.0 = 4.8\,\text{s}$$

The signal timing is now complete except for the conversion of effective green times to actual green times:

$$G_i = g_i - Y_i + t_{Li} \qquad (21\text{-}14)$$

where: G_i = actual green time for Phase i, s
g_i = effective green time for Phase i, s
Y_i = total of yellow and all-red intervals for Phase i, s
t_{Li} = total lost time for Phase i, s

Because information on the timing of yellow and all-red phases was not provided for the example, this step cannot be completed. Full signal-timing examples in the last section of this chapter will fully illustrate determination of actual green times.

As a general rule, very short phases should be avoided. In this case, the overlapping Phase A2 has an effective green time of only 4.8 seconds. When converted to actual green, this value would either stay the same or decrease. In either case, the short overlap period may not provide sufficient efficiency to warrant the potential confusion of drivers. The short Phase A2, in this case, may be one argument in favor of simplifying the phase plan by using a common exclusive left-turn phase.

21.3 Determining Pedestrian Signal Requirements

To this point in the process, the signal design has considered vehicular requirements. Pedestrians, however, must also be accommodated by the signal timing. Problems arise because pedestrian requirements and vehicular requirements are often quite different. Consider the intersection of a wide major arterial and a small local collector. Vehicle demand on the major arterial is more intense than on the small collector, and the green split for vehicles would generally result in the arterial receiving a long green and the collector a relatively short green.

This, unfortunately, is exactly the opposite of what pedestrians would require. During the short collector green, pedestrians are crossing the wide arterial. During the long arterial green, pedestrians are crossing the narrower collector. In summary, pedestrians require a longer green during the shorter vehicular green, and a shorter green during the longer vehicular green.

The *Highway Capacity Manual* [8] suggests the following minimum green-time requirements for pedestrians:

$$G_p = 3.2 + \left(2.7 * \frac{N_{ped}}{W_E}\right) + \left(\frac{L}{S_p}\right)$$

$$for\ W_E > 10\,\text{ft}$$

$$G_p = 3.2 + (0.27 * N_{ped}) + \left(\frac{L}{S_p}\right)$$

$$for\ W_E \leq 10\,\text{ft} \qquad (21\text{-}15)$$

where: G_p = minimum pedestrian crossing time, s
L = length of the crosswalk, ft
S_p = Average walking speed of pedestrians, ft/s
N_{ped} = number of pedestrians crossing per phase in a single crosswalk, peds
W_E = width of crosswalk, ft

In Equation 21-15, 3.2 seconds is allocated as a minimal start-up time for pedestrians. A pedestrian just starting to cross the street at the end of 3.2 seconds requires an additional (L/S_p) seconds to cross safely. The second term of the equation allocates additional start-up time based on the volume of pedestrians that need to cross the street. In effect, this equation provides that the minimum pedestrian green (WALK) indication (where pedestrian signals are employed) would be the sum of the first and second terms of the equation or:

$$WALK_{\min} = 3.2 + \left(2.7 * \frac{N_{ped}}{W_E}\right) for\ W_E < 10\,\text{ft}$$

$$WALK_{\min} = 3.2 + (0.27 * N_{ped})$$

$$for\ W_E \leq 10\,\text{ft} \qquad (21\text{-}16)$$

The flashing "up-raised hand" (DON'T WALK) signal (which is the pedestrian clearance interval) is most often (L/S_p) measured from the end of the vehicular all-red phase.

The WALK interval may be longer than the minimum required by pedestrians, if the vehicular green is longer than needed.

The total length of the WALK + Flashing DON'T WALK intervals can be considered in a number of different ways. ITE practice allows that pedestrians may be in the crosswalk during the green, yellow, and all-red intervals. Some jurisdictions, however, do not want to permit pedestrians to be in the crosswalk during the all-red, and others do not want pedestrians in the crosswalk during the yellow either.

For a signal timing to be viable for pedestrians, the minimum pedestrian crossing requirement, G_p in each phase must be compared with the time the relevant agency wishes pedestrians to be in the crosswalk:

$$G_p \leq G + y + ar, \quad or$$

$$G_p \leq G + y, \quad or$$

$$G_p \leq G \qquad (21\text{-}17)$$

If the chosen condition is not met, pedestrians are not safely accommodated, and changes must be made to provide for their needs.

Where the minimum pedestrian condition is not met in a given phase, two approaches may be taken:

1. A pedestrian actuator may be provided. In this case, when pushed, the *next* green phase is lengthened to provide $G_p = G + y + ar\ (or\ G + y, or\ G)$. The additional green time is subtracted from other phases (in a pretimed signal) to maintain the cycle length. When pedestrian actuators are provided, pedestrian signals *must* be used.
2. Retime the signal to provide the minimum pedestrian need in all cycles. This must be done in a manner that also maintains the vehicular balance of green times and results in a longer cycle length.

The first approach has limited utility. Where pedestrians are present in most cycles, it is reasonable to assume that the actuator will always be pushed, thus destroying the planned vehicular signal timing. In such cases, the approach should be to retime the signal to satisfy both vehicular and pedestrian needs in every cycle. Pedestrian actuators are useful in cases where pedestrians are relatively rare or where actuated signal controllers are used.

In the second case, the task is to provide the minimum pedestrian crossing time while maintaining the balance of effective green needed to accommodate vehicles.

Consider the case of the vehicular signal timing for a two-phase signal shown in Table 21.3. Minimum pedestrian needs are also shown for comparison.

In this case, Phase A serves a major arterial and thus has the longer vehicular green but the shorter pedestrian requirement. Phase B serves a minor cross-street but has the longer pedestrian requirement. Pedestrian requirements must be compared with the vehicular signal timing, using Equation 21-17. In this case, we apply the most liberal policy, which allows pedestrians to be in the crosswalk during G, y, and ar.

Table 21.3: Sample Signal Timing

Phase	Green Time G (s)	Yellow + All Red Y (s)	Lost Time t_L (s)	Pedestrian Requirement G_p (s)
A	40.0	5.0	4.0	20.0
B	15.0	5.0	4.0	30.0

$$G_{pA} = 20.0 \geq G_A + Y_A = 40.0 + 5.0 = 45.0\,\text{s OK}$$

$$G_{pB} = 30.0 \geq G_B + T_B = 15.0 + 5.0 = 20.0\,\text{s NG}$$

The effective green times for Phases A and B may be computed as:

$$g = G + Y - t_L$$

$$g_A = 40.0 + 5.0 - 4.0 = 41.0\,\text{s}$$

$$g_B = 15.0 + 5.0 - 4.0 = 16.0\,\text{s}$$

The signal must be retimed to result in a $G + Y$ for Phase B of at least 30.0 seconds while maintaining the relative balance of effective green time needed by vehicles in both phases (i.e., a ratio of 41.0 to 16.0). For Phase B to have a $G + Y$ of 30.0 seconds, the effective green time would have to be increased to:

$$g_B = 30.0 - 4.0 = 26.0\,\text{s}$$

To maintain the original ratio of vehicular green time, the effective green time for Phase A must also be increased:

$$\frac{g_A}{26.0} = \frac{41.0}{16.0}$$

$$g_A = \frac{41.0 * 26.0}{16.0} = 66.6\,\text{s}$$

The actual green times would become:

$$G_A = 66.6 - 5.0 + 4.0 = 65.6\,\text{s}$$

$$G_B = 26.0 - 5.0 + 4.0 = 25.0\,\text{s}$$

This yields a cycle length of $65.6 + 5.0 + 25.0 + 5.0 = 100.6$ s. If this intersection were under pretimed control, a 110-second cycle would be needed, and would be re-split to maintain the original proportion of effective green times. The provision of a timing that safely accommodates both pedestrians and vehicles results in an increase in the cycle length from 65 seconds to 110 seconds. The downside of this retiming would be an increase in delay to motorists and passengers.

Figure 21.12 further illustrates the relationship between vehicular and pedestrian phases for three basic cases.

a. $G_p = G + Y$
b. $G_p < G + Y$, and
c. $G_p > G + Y$.

The example once again uses the most liberal pedestrian policy, which allows pedestrians in the crosswalk during G, y, and ar.

In Case 1, where $G_p = G + Y$, the minimum WALK period is given with a pedestrian clearance interval of L/S_p. In this case, as in all cases, the vehicular RED indication coincides with the pedestrian DON'T WALK interval. In Case 2, where the vehicular signal is more than adequate for pedestrians, the WALK interval is longer than the minimum, essentially whatever time can be given after providing the pedestrian clearance interval of L/S_p. In Case 3, the vehicular green is not sufficient for pedestrians, so a permanent DON'T WALK is present. In this case, pedestrian push-button actuators *must* be provided. When pushed, the next vehicular green phase will be lengthened to provide for $G_p = G + Y$ (Case 1). Note that pedestrian signals are not required in all cases. They should, however, be provided whenever pedestrian safety might be compromised without them.

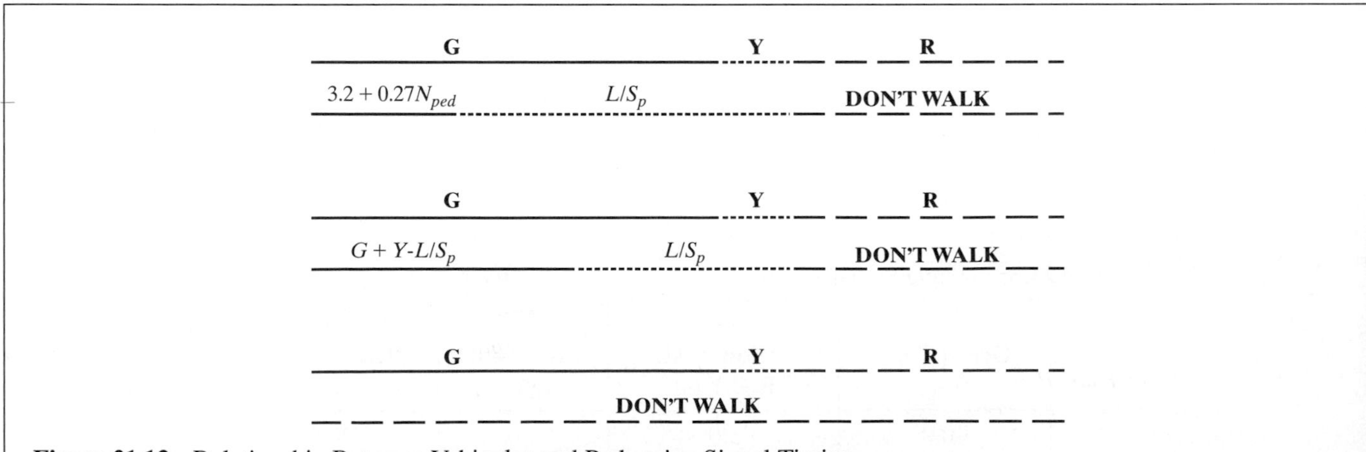

Figure 21.12: Relationship Between Vehicular and Pedestrian Signal Timing

21.4 Compound Signal Timing

Although it is recommended that most initial signal timings avoid compound phasing (protected + permitted or permitted + protected), the methodology of this chapter can be easily adapted to do so, if desired. To estimate a compound phasing, the analyst will have to predetermine *how many* of the subject left turns will be made in the permitted portion of the phase, and *how many* will be made in the protected portion of the phase. Once this is done, timing can be estimated by adapting the methodology of this chapter as follows:

- Treat the protected and permitted portions of the phase as if they were separate phases.
- In converting volumes to tvu's, use different equivalents (Table 21.1) as appropriate for each portion of the phase.
- Estimate the cycle length (C) and green splits (g) treating the protected and permitted portions of the phase separately.
- Remember that there will be "yellow" between the green arrow and the green ball as the phase transitions from protected to permitted (or vice versa). This yellow counts as green time for left turns.

21.5 Sample Signal Timing Applications

The procedures presented in Sections 21.1 through 21.3 will be illustrated in a series of signal-timing applications. The following steps should be followed:

1. Develop a reasonable phase plan in accordance with the principles discussed in Section 21.1. Use Equation 21-1 or local agency guidelines to make an initial determination of whether left-turn movements need to be protected. Do not include compound phasing in preliminary signal timing; this may be tried as part of a more comprehensive intersection analysis later.
2. Convert all left-turn and right-turn movements to equivalent through vehicle units (tvu's) using the equivalents of Tables 21.1 and 21.2, respectively.
3. Draw a ring diagram of the proposed phase plan, inserting lane volumes (in tvu's) for each set of movements. Determine the critical path through the signal phasing as well as the sum of the critical-lane volumes (V_c) for the critical path.
4. Determine *yellow* and *all-red* intervals for each signal phase.
5. Determine lost times per cycle using Equations 21-5 through 21-7.
6. Determine the desirable cycle length, C, using Equation 21-11. For pretimed signals, round up to reflect available controller cycle lengths. An appropriate *PHF* and reasonable target v/c ratio should be used.
7. Allocate the available effective green time within the cycle in proportion to the critical lane volumes for each portion of the phase plan.
8. Check pedestrian requirements and adjust signal timing as needed.

Example 21-1: Signal-Timing Case 1: A Simple Two-Phase Signal

Consider the intersection layout and demand volumes shown in Figure 21.13. It shows the intersection of two streets with one lane in each direction and relatively low turning volumes. Moderate pedestrian activity is present, and the *PHF* and target v/c ratio are specified.

Solution:

Step 1: *Develop a Phase Plan* Given that there is only one lane for each approach, it is not possible to even consider including protected left turns in the phase plan. However, a check of the criteria of Equation 21-1 and Figure 21.1 (not illustrated here) show that no protected left turns are required for this case:

- EB: $V_{LT} = 10 < 20$
  xprod $= 10 * 315/1 = 3{,}150 < 50{,}000$
- WB: $V_{LT} = 12 < 200$
  xprod $= 12 * 420/1 = 5{,}040 < 50{,}000$
- NB: $V_{LT} = 10 < 200$
  xprod $= 10 * 400/1 = 4{,}000 < 50{,}000$
- SB: $V_{LT} = 10 < 200$
  xprod $< 10 * 375/1 = 3{,}750 < 50{,}000$

A simple two-phase signal, therefore, will be adopted for this intersection.

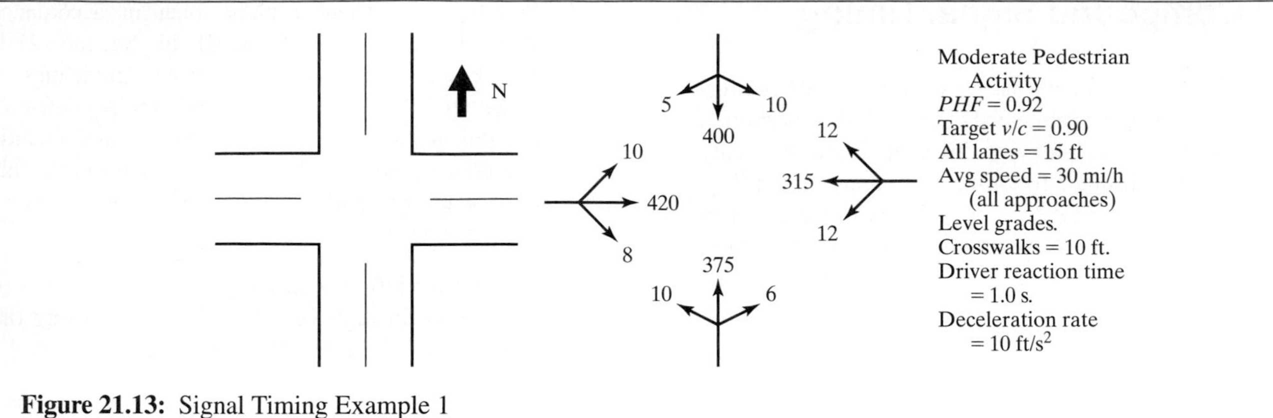

Figure 21.13: Signal Timing Example 1

Step 2: *Convert Volumes to Through-Vehicle Equivalents* The conversion of volumes to tvu's is illustrated in Table 21.4. Equivalent values are taken from Tables 21.1 and 21.2, and they are interpolated for intermediate values of opposing volume. Note that all through vehicles are equivalent to 1.0 tvu.

Step 3: *Determine Critical-Lane Volumes* The critical path through the signal phase plan is illustrated in Figure 21.14. As a two-phase signal, this is a relatively simple determination. For Phase A, either the EB or WB approach is critical. Because the EB approach has the higher lane volume, 470 tvu/h, this is the critical movement for Phase A. For Phase B, either the NB or SB approach is critical; SB has the higher lane volume (454 tvu/h), so this is the critical movement for Phase B. The sum of the critical-lane volumes, therefore, is 470 + 454 = 924 tuv/h.

Step 4: *Determine Yellow and All-Red Intervals* *Yellow* and *all-red* intervals are found using Equations 21-2 and 21-3. The average approach speed for all approaches is 30 mi/h. Thus the $S_{85} = 30 + 5 = 35$ mi/h, and the $S_{15} = 30 - 5 = 25$ mi/h. Because moderate numbers of pedestrians are present, the all-red interval will be computed using Equation 21-3b, which allows vehicles to clear beyond the far crosswalk line. The distance to be crossed during the *all-red* clearance interval is the sum of two 15-foot lanes and a 10-foot crosswalk, or $P = 15 + 15 + 10 = 40$ ft. Then:

$$y = t + \frac{1.47 S_{85}}{2a + (6.44 * 0.01G)}$$

$$= 1.0 + \frac{1.47 * 35}{(2 * 10) + (0)}$$

$$= 3.6 \text{ s}$$

$$ar = \frac{P + L}{1.47 S_{15}} = \frac{40 + 20}{1.47 * 25} = 1.6 \text{ s}$$

Table 21.4: Computation of Through Vehicle Equivalent Volumes for Signal Timing 1

Approach	Movement	Volume (Veh/h)	Equivalent Tab 21.1, 21.2	Volume (tvu/h)	Lane Group Volume (tvu/h/ln)	Vol/Lane (tvu/h/ln)
EB	L	10	3.90	39		
	T	420	1.00	420	470	470
	R	8	1.32	11		
WB	L	12	5.50	66		
	T	315	1.00	315	397	397
	R	12	1.32	16		
NB	L	10	5.00	50		
	T	375	1.00	375	433	433
	R	6	1.32	8		
SB	L	10	4.70	47		
	T	400	1.00	400	454	454
	R	5	1.32	7		

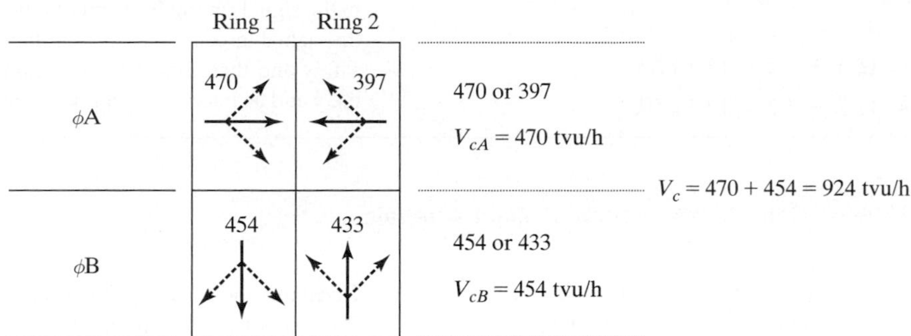

Figure 21.14: Determination of Critical Lane Volumes: Signal Timing Case 1

Because both streets have the same width, crosswalk width, and approach speed, the values of y and ar are the same for both Phases A and B of the signal.

Step 5: *Determination of Lost Times* Lost times are found using Equations 21-5 through 21-7. In this case, the recommended 2.0-second default values for start-up lost time (ℓ_1) and extension of effective green into yellow and all-red (e) are used:

$$Y = y + ar = 3.6 + 1.6 = 5.2\,\text{s}$$
$$\ell_2 = Y - e = 5.2 - 2.0 = 3.2\,\text{s}$$
$$t_L = \ell_1 + \ell_2 = 2.0 + 3.2 = 5.2\,\text{s}$$

Because both phases have the same value, the total lost time per cycle, L, is $5.2 + 5.2 = 10.4$s Note that in all cases where the recommended default values for ℓ_1 (2.0 s) and e (2.0 s) are used, lost time per phase (t_L) is the same numerical value as the sum of the yellow and all red intervals (Y).

Step 6: *Determine the Desirable Cycle Length* Equation 21-11 is used to determine the desirable cycle length:

$$C_{des} = \frac{L}{1 - \left(\dfrac{V_c}{1615 * PHF * v/c}\right)}$$

$$C_{des} = \frac{10.4}{1 - \left(\dfrac{924}{1615 * 0.92 * 0.90}\right)} = \frac{10.4}{0.31} = 33.5\,\text{s}$$

Assuming this is a pretimed controller, a desirable cycle length of 35 seconds or 40 seconds would be used. For the purposes of this signal timing case, the minimum value of 35 seconds will be used.

Step 7: *Allocate Effective Green to Each Phase* Given a 35-second cycle length with 10.4 seconds of lost time per cycle, the amount of effective green time to be allocated is $35.0 - 10.4 = 24.6$s. The allocation is done using Equation 21-13:

$$g_A = g_{TOT} * \left(\frac{V_{cA}}{V_c}\right) = 24.6 * \left(\frac{470}{924}\right) = 12.5\,\text{s}$$

$$g_B = g_{TOT} * \left(\frac{V_{cB}}{V_c}\right) = 24.6 * \left(\frac{454}{924}\right) = 12.1\,\text{s}$$

The cycle length may be checked as the total of effective green times plus the lost time per cycle, or $12.5 + 12.1 + 10.4 = 35.0$ s Effective green times may be converted to actual green times using Equation 21-14:

$$G_A = g_A - Y_A + t_{LA} = 12.5 - 5.2 + 5.2 = 12.5\,\text{s}$$
$$G_B = g_B - Y_B + t_{LB} = 12.1 - 5.2 + 5.2 = 12.1\,\text{s}$$

Again, note that when default values for start-up lost time (2.0 seconds) and extension of effective green into yellow and all-red (2.0 seconds) are used, the actual green time is numerically the same as effective green time.

Step 8: *Check Pedestrian Requirements* Equation 21-15 is used to compute the minimum pedestrian green requirement for each phase. Because both streets have equal width and equal crosswalk widths and because pedestrian traffic is "moderate" in all crosswalks, the requirements will be the same for each phase in this case. From Table 21.2, the default pedestrian volume for "moderate" activity is 200 peds/h. The number of pedestrians per cycle (N_{ped}) is based on the number of cycles per hour (3,600/35 = 102.9, say 103 cycles/h). The number of pedestrians per cycle is then 200/103 = 1.94, say 2 peds/cycle. Then:

$$G_{pA,B} = 3.2 + (0.27N_{ped}) + \left(\frac{L}{S_p}\right)$$

$$G_{p,B} = 3.2 + (0.27 * 2) + \left(\frac{30}{4.0}\right) = 11.2\,\text{s}$$

For this signal to be safe for pedestrians:

$$G_p < G + Y$$

$$G_p = 11.2 < 12.5 + 5.2 = 17.7 \text{ s } OK$$

$$G_p = 11.2 < 12.1 + 5.2 = 17.3 \text{ s } OK$$

The signal safely accommodates all pedestrians. No changes in the signal timing for vehicular needs are required. Note that the actual green times are sufficient to handle pedestrians safely and that allowing pedestrians in the crosswalk during the y and ar intervals is not necessary.

Example 21-2: Signal Timing Case 2: Intersection of Major Arterials

Figure 21.15 illustrates the intersection of two four-lane arterials with significant demand volumes and exclusive left-turn lanes provided on each approach.

Step 1: *Develop a Phase Plan* Each left-turn movement should be checked against the criteria of Equation 21-1 to determine whether or not it needs to be protected. The ITE criteria of Figure 21.1 is also checked but is not shown here.

- EB: $V_{LT} = 35 < 200$
  xprod = 35 * (500/2) = 8,750 < 50,000
  Figure 21.1 criteria not met.
  No protection needed.
- WB: $V_{LT} = 25 < 200$
  xprod = 25 * (610/2) = 22,875 < 50,000
  Figure 21.1 criteria not met.
  No protection needed.
- NB: $V_{LT} = 250 > 200$
  Protection needed.
- SB: $V_{LT} = 220 > 200$
  Protection needed.

Given that the NB and SB left turns require a protected phase, the next issue is how to provide it. The two opposing left-turn volumes, 220 veh/h (NB) and 250 veh/h (SB), are not numerically very different. Therefore, there appears to be little reason to separate the NB and SB protected phases. An exclusive left-turn phase will be used on the N-S arterial. A single phase using permitted left turns will be used on the E-W arterial.

Step 2: *Convert Volumes to Through Vehicle Equivalents* Through-vehicle equivalents are obtained from Tables 21.1 and 21.2 for left and right turns, respectively. The computations are illustrated in Table 21.5.

Note that exclusive LT lanes must be established as separate lane groups, with their demand volumes separately computed, as shown in Table 21.5. The equivalent for all protected left turns (Table 21.1) is 1.05.

Step 3: *Determine Critical Lane Volumes* As noted in Step 1, the signal phase plan includes an exclusive LT phase for the N-S artery and a single phase with permitted left turns for the E-W artery. Figure 21.16 illustrates this and the determination of critical-lane volumes.

Phase A is the exclusive N-S LT phase. The heaviest movement in the phase is 263 tvu/h for the SB left turn. In

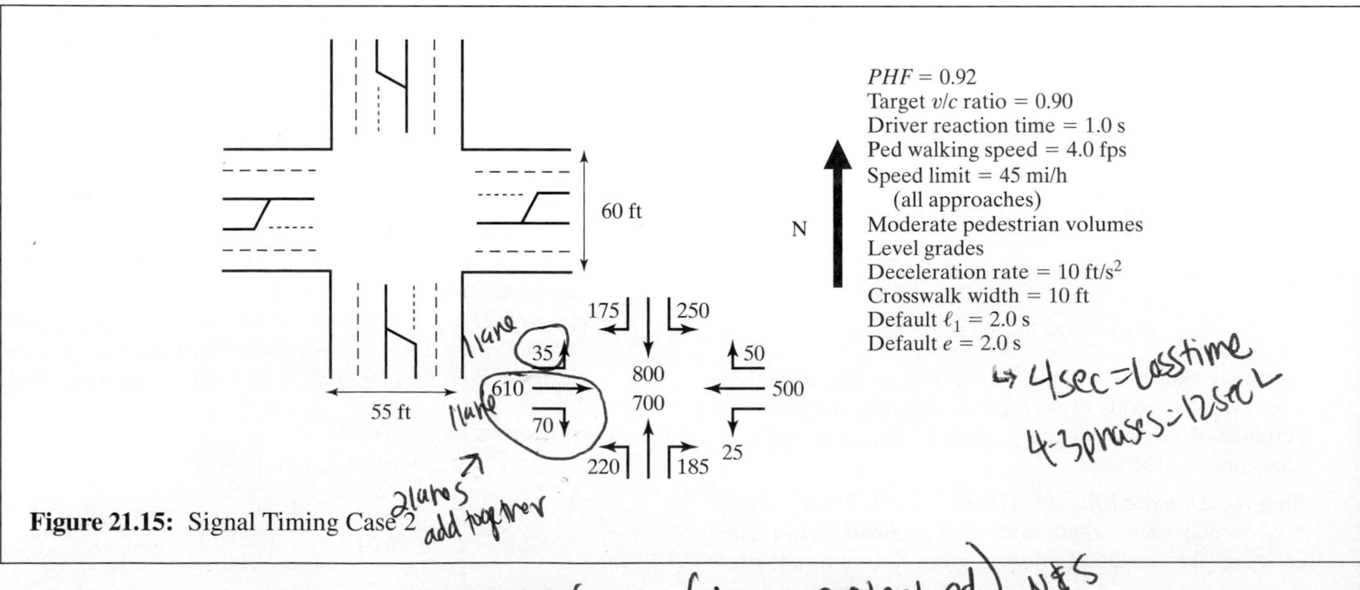

Figure 21.15: Signal Timing Case 2

Table 21.5: Computation of Through Vehicle Equivalent Volumes for Signal Timing 2

Approach	Movement	Volume (Veh/h)	Equivalent Tab 21-1,2	Volume (tvu/h)	Lane Group Volume (tvu/h)	Vol/Lane (tvu/h/ln)
EB	L	35	4.00*	140	140	140
	T	610	1.00	610	702	351
	R	70	1.32	92		
WB	L	25	5.15*	128	128	128
	T	500	1.00	500	566	283
	R	50	1.32	66		
NB	L	220	1.05	231	231	231
	T	700	1.00	700	944	472
	R	185	1.32	244		
SB	L	250	1.05	263	263	263
	T	800	1.00	800	1031	516
	R	175	1.32	231		

*Interpolated by opposing volume.

Phase B, the heavier movement is the SB through and right turn, with 516 tvu/h. In Phase C, both E-W left-turn lane groups and through/right-turn lane groups move at the same time. The heaviest movement is the EB TH/RT lanes, with 351 tvu/h. The sum of critical-lane volumes, V_c therefore, is $263 + 516 + 351 = 1{,}130$ tvu/h.

Note that each "ring" handles two sets of movements in Phase C. This is possible, of course, because it is the same signal face that controls all movements in a given direction. The left-turn lane volume cannot be averaged with the through/right-turn movement because lane-use restrictions are involved. All left turns must be in the left-turn lane; none may be in the through/right-turn lanes.

Step 4: *Determine Yellow and All-Red Intervals* Equation 21-2 is used to determine the length of the *yellow* interval; Equation 21-3b is used to determine the length of the *all-red* interval. Because a speed limit—45 mi/h—is given rather than a measured average approach speed, there will be no differentiation between the S_{85} and S_{15} The speed limits on both arteries are the same, so the *yellow* intervals for all three phases will also be the same:

$$y_{A,B,C} = 1.0 + \frac{1.47 * 45}{(2 * 10) + (0)} = 4.3 \text{ s}$$

The *all-red* intervals will reflect the need to clear the full width of the street plus the width of the far crosswalk. The width of the N–S street is 55 feet, and the width of the E-W street is 60 feet. The width of a crosswalk is 10 feet. During the N-S left-turn phase, it will be assumed that a vehicle must clear the entire width of the E-W artery. Thus, for Phase A,

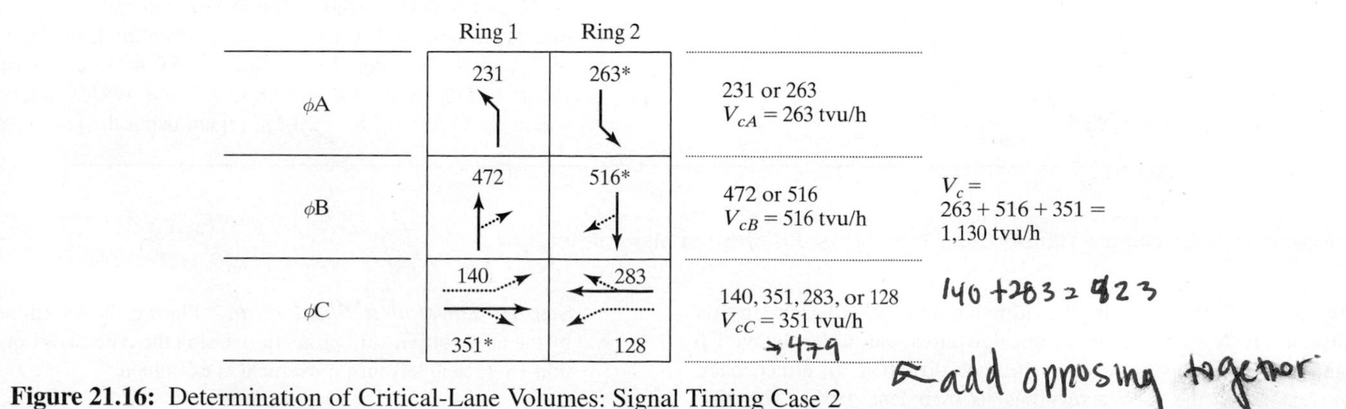

Figure 21.16: Determination of Critical-Lane Volumes: Signal Timing Case 2

the width to be cleared (*P*) is 60 + 10 = 70 feet; for Phase B, it is also 60 + 10 = 70 feet; for Phase C, the distance to be cleared is 55 + 10 = 65 feet. Thus:

$$ar_{A,B} = \frac{70 + 20}{1.47 * 45} = 1.4 \text{ s}$$

$$ar_C = \frac{65 + 20}{1.47 * 45} = 1.3 \text{ s}$$

where 20 feet is the assumed length of a typical vehicle.

Step 5: *Determination of Lost Times* Remembering that where the default values for ℓ_1 and e are both 2.0 seconds that the lost time per phase, t_L, is the same as the sum of the yellow plus all-red intervals, Y:

$$Y_{A,B} = t_{LA,B} = 4.3 + 1.4 = 5.7 \text{ s}$$

$$Y_C = t_{LC} = 4.3 + 1.3 = 5.6 \text{ s}$$

Based on this, the total lost time per cycle, L, is 5.7 + 5.7 + 5.6 = 17.0 seconds, noting that $Y_{A,B}$ occurs *twice*, at the end of *both* phases B and C.

Step 6: *Determine the Desirable Cycle Length* The desirable cycle length is found using Equation 21-11:

$$C_{des} = \frac{17}{1 - \left(\frac{1{,}130}{1{,}615 * 0.92 * 0.90}\right)} = 109.7 \text{ s}$$

Assuming that this is a pretimed signal controller, a cycle length of 110 seconds would be selected.

Step 7: *Allocate Effective Green to Each Phase* In a cycle length of 110 seconds, with 17 seconds of lost time per cycle, the amount of effective green time that must be allocated to the three phases is 110 − 17 = 93 s. Using Equation 21-13, the effective green time is allocated in proportion to the phase critical-lane volumes:

$$g_A = 93 * \left(\frac{263}{1{,}130}\right) = 21.6 \text{ s}$$

$$g_B = 93 * \left(\frac{516}{1{,}130}\right) = 42.5 \text{ s}$$

$$g_C = 93 * \left(\frac{351}{1{,}130}\right) = 28.9 \text{ s}$$

The cycle length is now checked to ensure that the sum of all effective green times and the lost time equals 110 seconds: 21.6 + 42.5 + 28.9 + 17.0 = 110 OK. Note that when the default values for ℓ_1 and e (both 2.0 seconds) are used, actual green times, G, equal effective green times, g.

Step 8: *Check Pedestrian Requirements* Pedestrian requirements are estimated using Equation 21-15. In this case, note that pedestrians will be permitted to cross the E-W artery only during Phase B. Pedestrians will cross the N-S artery during Phase C. The number of pedestrians per cycle for all crosswalks is the default pedestrian volume for "moderate" activity, 200 peds/h, divided by the number of cycles in an hour (3600/110 = 32.7 cycles/h). Thus $N_{ped} = 200/32.7 = 6.1$ peds/cycle. Required pedestrian green times are:

$$G_{pB} = 3.2 + (0.27 * 6.1) + \left(\frac{60}{4.0}\right)$$
$$= 3.2 + 1.6 + 15.0 = 19.8 \text{ s}$$

$$G_{pC} = 3.2 + (0.27 * 6.1) + \left(\frac{55}{4.0}\right)$$
$$= 3.2 + 1.6 + 13.8 = 18.6 \text{s}$$

The minimum requirements are compared to the sum of the green, yellow, and all-red times provided for vehicles:

$$G_{pB} = 19.8 \text{ s} < G_B + Y_B = 42.5 + 5.7$$
$$= 48.2 \text{ s } OK$$

$$G_{pC} = 18.6 \text{ s} < G_C + Y_C = 28.9 + 5.6$$
$$= 34.5 \text{ s } OK$$

Therefore, no changes to the vehicular signal timing are required to accommodate pedestrians safely. Note that pedestrians are more than accommodated by the vehicular greens, so it is not necessary to allow pedestrians in the crosswalk during y and ar intervals.

For major arterial crossings, pedestrian signals would normally be provided. During Phase A, all pedestrian signals would indicate "DON'T WALK." During Phase B, the pedestrian clearance interval (the flashing DON'T WALK) would be L/S_p or 60/4.0 = 15.0 s. The WALK interval is whatever time is left in $G + Y$, (or G + y, or G) counting from the end of *Y (or y or G)*: using G + Y, 48.2 − 15.0 = 33.2 s. During Phase *C*, L/S_p is 55/4.0 = 13.8 s, and the WALK interval would be 34.5 − 13.8 = 20.7 s. (again using the end of Y).

Example 21-3: Signal-Timing Case 3: Another Junction of Major Arterials

Figure 21.17 illustrates another junction of major arterials. In this case, the E-W artery has three through lanes, plus an exclusive LT lane and an exclusive RT lane in each direction. In effect, each movement on the E-W artery has its own lane group. The N-S artery has two lanes in each direction, with no exclusive LT or RT lanes. There are no pedestrians present at this intersection.

Step 1: *Develop a Phase Plan* Phasing is determined by the need for left-turn protection. Using the criteria of Equation 18-1, each left turn movement is examined.

- *EB:* V_{LT} = 300 veh/h > 200 veh/h
  Protected phase needed.

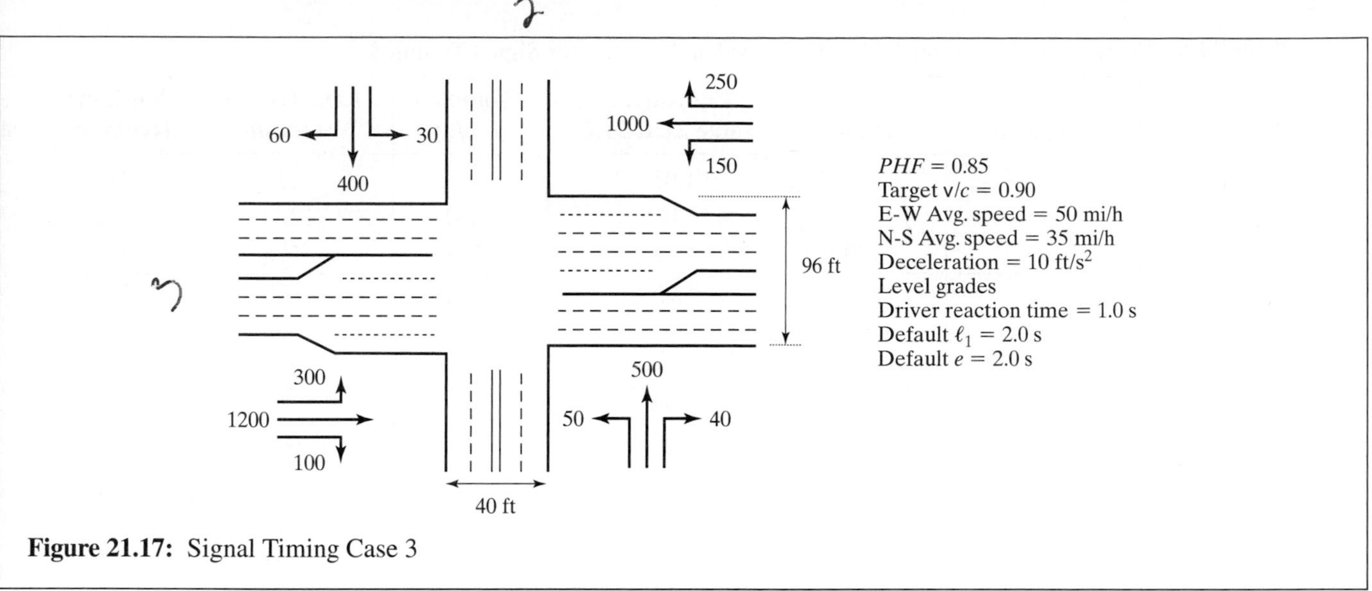

Figure 21.17: Signal Timing Case 3

- *WB:* V_{LT} = 150 veh/h < 200 veh/h
  xprod = 150 * (1200/3)
  = 60,000 > 50,000
  Protected phase needed.
- *NB:* V_{LT} = 50 veh/h < 200 veh/h
  xprod = 50 * (400/2)
  = 10,000 < 50,000
  Protected phase not needed.
- *SB:* V_{LT} = 30 veh/h < 200 veh/h
  xprod = 30 * (500/2)
  = 7,500 < 50,000
  Protected phase not needed.

The results are fortunate. Had protected phasing been required for the NB and SB approaches, the lack of an exclusive LT lane on these approaches would have caused a problem.

The E-W approaches have LT lanes, and protected left turns are needed on both approaches. Because the LT volumes EB and WB are very different (300 veh/h versus 150 veh/h), a phase plan that splits the protected LT phases would be advisable. A NEMA phase plan, using an exclusive LT phase followed by a leading green for the EB direction, will be employed for the E-W artery.

Step 2: *Convert Volumes to Through-Vehicle Equivalents* Tables 21.1 and 21.2 are used to find through-vehicle equivalents for left- and right-turn volumes, respectively. Conversion computations are illustrated in Table 21.6. Note that the EB and WB approaches have a separate lane group for each movement, whereas the NB and SB approaches have a single lane group serving all movements from shared lanes.

Step 3: *Determine Critical-Lane Volumes* Figure 21.18 shows a ring diagram for the phase plan discussed in Step 1 and illustrates the selection of the critical-lane volumes.

The phasing involves overlaps. For the combined Phase A, the critical path is down Ring 1, which has a sum of critical-lane volumes of 649 tvu/h. For Phase B, the choice is simpler because there are no overlapping phases. Ring 2, serving the NB approach, has a critical-lane volume of 349 tvu/h. The sum of all critical-lane volumes (V_c) is 649 + 349 = 998 tvu/h.

Note also that overlapping phases have a unique characteristic. In this example, for overlapping Phase A, the largest left-turn movement is EB and the largest through movement is EB as well. Because of this, the overlapping phase plan will yield a smaller sum of critical lane volumes than one using an exclusive left-turn phase for both left-turn movements. Had the largest left-turn and through movements been from opposing approaches, the sum of critical-lane volumes would be the same for the overlapping sequence and for a single exclusive LT phase. In other words, little is gained by using overlapping phases where a left turn and its opposing through (through plus right turn) movement are the larger movements.

Step 4: *Determine Yellow and All-Red Intervals* Equation 21-2 is used to determine the appropriate length of the *yellow* change intervals. Note that the signal design is a *three*-phase signal and there are three transitions in the cycle. Because of the overlapping sequence, the transition at the end of the protected EB/WB left turns occurs at different times on Ring 1 and Ring 2. For simplicity, it is assumed that left-turning vehicles from the EB and WB approaches cross the entire width of the N-S artery. *All-red* intervals are determined using Equation 21-3a because there are no pedestrians present.

Table 21.6: Computation of Through Vehicle Equivalent Volumes for Signal Timing 3

Approach	Movement	Volume (Veh/h)	Equivalent Table 21.1, 21.2	Volume (tvu/h)	Lane Group Vol (tvu/h)	Vol/Lane (tvu/h/ln)
EB	L	300	1.05	315	315	315
	T	1200	1.00	1200	1200	400
	R	100	1.18	118	118	118
WB	L	150	1.05	158	158	158
	T	1000	1.00	1000	1000	334
	R	250	1.18	295	295	295
NB	L	50	3.00	150		
	T	500	1.00	500	697	349
	R	40	1.18	47		
SB	L	30	4.00*	120		
	T	400	1.00	400	591	296
	R	60	1.18	71		

*Interpolated by opposing volume.

Percentile speeds are estimated from the measured average approach speeds given:

$$S_{85EW} = 50 + 5 = 55 \text{ mi/h}$$
$$S_{15EW} = 50 - 5 = 45 \text{ mi/h}$$
$$S_{85NS} = 35 + 5 = 40 \text{ mi/h}$$
$$S_{15NS} = 35 - 5 = 30 \text{ mi/h}$$

Then:

$$y_{A1,A2,A3} = 1.0 + \frac{1.47 * 55}{(2 * 10) + (0)} = 5.0 \text{ s}$$

$$y_B = 1.0 + \frac{1.47 * 40}{(2 * 10) + (0)} = 3.9 \text{ s}$$

$$ar_{A1,A2,A3} = \frac{40 + 20}{1.47 * 45} = 0.9 \text{ s}$$

$$ar_B = \frac{96 + 20}{1.47 * 30} = 2.6 \text{ s}$$

where 20 feet is the assumed average length of a typical vehicle.

Step 5: *Determination of Lost Times* Because the problem statement specifies the default values of 2.0 s each for start-up lost time and extension of effective green into yellow

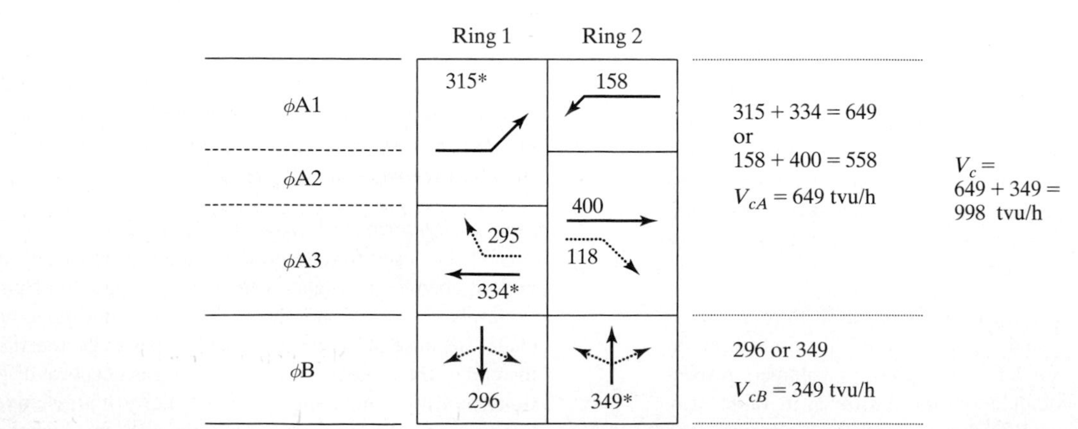

Figure 21.18: Determination of Critical-Lane Volumes: Signal Timing Case 3

and all-red intervals, the total lost time in each phase, t_L is equal to the sum of the yellow and all red intervals, Y. Thus:

$$t_{LA1/A2} = Y_{A1/A2} = 5.0 + 0.9 = 5.\text{ s}$$
$$t_{LA3} = Y_{A3} = 5.0 + 0.9 = 5.9\text{ s}$$
$$t_{LB} = Y_B = 3.9 + 2.6 = 6.5\text{ s}$$

Note from Figure 21.18 that the first phase transition occurs at the end of Phase A1, but only on Ring 2. A similar transition occurs at the end of Phase A2, but only on Ring 1. The two other transitions, at the end of Phases A3 and B, occur on both rings. Thus the total lost time per cycle, L is 5.9 + 5.9 + 6.5 = 18.3 s and the phase plan represents a three-phase signal.

Step 6: *Determine the Desirable Cycle Length* The desirable cycle length is found using Equation 21-11:

$$C_{des} = \frac{18.3}{1 - \left[\dfrac{998}{1615 * 0.85 * 0.90}\right]} = \frac{18.3}{0.192} = 95.3\text{ s}$$

Assuming this is a pretimed controller, a cycle length of 100 seconds would be selected.

Step 7: *Allocate Effective Green to Each Phase* A signal cycle of 100 seconds with 18.3 seconds of lost time has 100.0 − 18.3 = 81.7 s of effective green time to allocate in accordance with Equation 21-13. Note that in allocating green to the critical path, Phases A1 and A2 are treated as a single segment. Subsequently, the location of the Ring 2 transition between Phases A1 and A2 will have to be established.

$$g_{A1+A2} = 81.7 * \left(\frac{315}{998}\right) = 25.8\text{ s}$$
$$g_{A3} = 81.7 * \left(\frac{334}{998}\right) = 27.3\text{ s}$$
$$g_B = 81.7 * \left(\frac{349}{998}\right) = 28.6\text{ s}$$

The specific lengths of Phases A1 and A2 are determined by fixing the Ring 2 transition between them. This requires consideration of the noncritical path through combined Phase A, which occurs on Ring 2. The total length of combined Phase A is the sum of g_{A1+A_2} and g_{A3}, or 25.8 + 27.3 = 53.1 s. The Ring 2 transition is based on the relative values of the lane volumes for Phase A1 and the combined Phase A2/A3, or:

$$g_{A1} = 53.1 * \left(\frac{158}{158 + 400}\right) = 15.0\text{ s}$$

By implication, Phase A2 is the total length of combined Phase A minus the length of Phase A1 and Phase A3, or:

$$g_{A2} = 53.1 - 15.0 - 27.3 = 10.8\text{ s}$$

Now, the signal has been completely timed for vehicular needs. With the assumption of default values for $\ell_1(2.0\text{ s})$ and e (2.0 s), actual green times are equal to effective green times (numerically, although they do not occur simultaneously):

$$G_{A1} = 15.0\text{ s}$$
$$G_{A2} = 10.8\text{ s}$$
$$Y_{A1/A2} = 5.9\text{ s}$$
$$G_{A3} = 27.3\text{ s}$$
$$Y_{A3} = 5.9\text{ s}$$
$$G_B = 28.6\text{ s}$$
$$Y_B = 6.5\text{ s}$$
$$C = 100.0\text{ s}$$

There is no Step 8 in this case because there are no pedestrians at this intersection and, therefore, no pedestrian requirements to be checked.

Example 21-4: Signal-Timing Case 4: A T-Intersection

Figure 21.19 illustrates a typical T-intersection, with exclusive lanes for various movements as shown. Note that there is only one opposed left turn in the WB direction.

Step 1: *Develop a Phase Plan* In this case, there is only one opposed left turn to check for the need of a protected phase. As the WB left turn > 200 veh/h, it should be provided with a protected left-turn phase. There is no EB or SB left turn, and the NB left turn is unopposed. The standard way of providing for the necessary phasing would be to use a leading WB green with no lagging EB green.

Step 2: *Convert Volumes to Through-Vehicle Equivalents* Table 21.7 shows the conversion of volumes to through vehicle equivalents, using the equivalent values given in Tables 21.1 and 21.2 for left and right turns, respectively.

Note that the NB left turn is treated as an opposed turn with V_o = 0 veh/h. There are different approaches that have been used to address left turns that are unopposed due to one-way streets and T-intersections, reasons other than the presence of a protected left-turn phase. Such a movement could also be treated as any protected left turn and an equivalent of

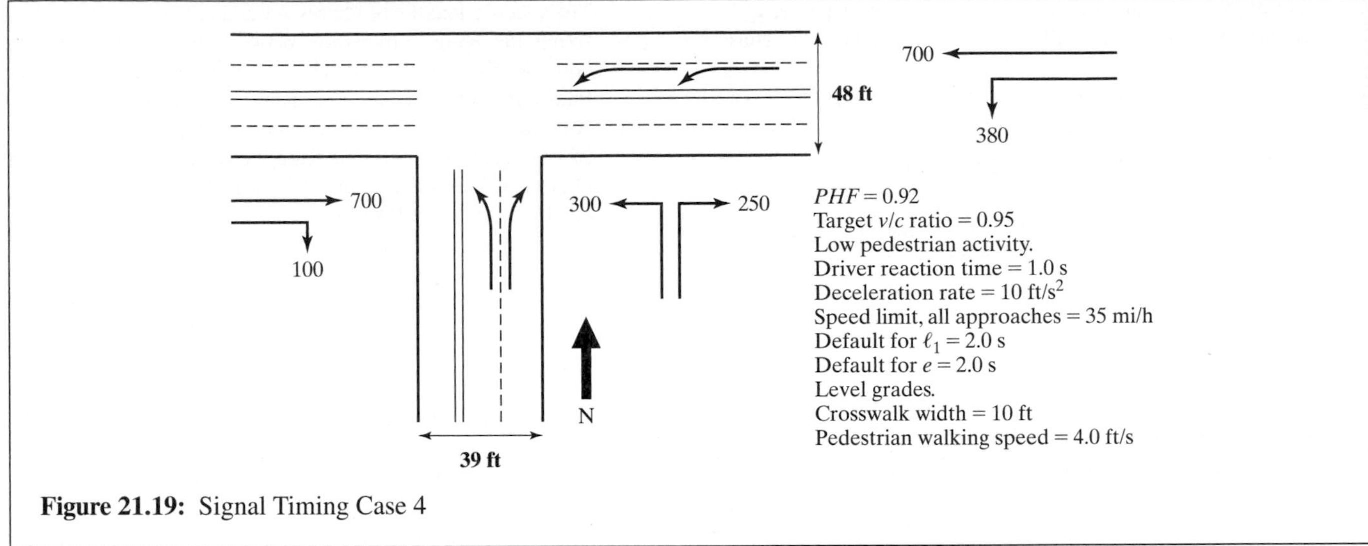

Figure 21.19: Signal Timing Case 4

1.05 applied. In some cases, particularly unopposed left turns from a one-way street, the movement is treated as a right turn, using the appropriate factor based on pedestrian interference.

Step 3: *Determine Critical-Lane Volumes* Figure 21.20 shows the ring diagram for the phasing described in Step 1 and illustrates the determination of the sum of critical-lane volumes.

In this case, the selection of the critical path through combined Phase A is interesting. Ring 1 goes through two phases; Ring 2 goes through only one. In this case, the critical path goes through Ring 1 and has a total of three phases. Had the Phase A critical path been through Ring 2, the signal would have only two critical phases. In such cases, the highest critical-lane volume total *does not alone determine the critical path*. Because one path has an additional phase and, therefore, an additional set of lost times, it could possibly be critical even if it has the lower total critical-lane volume. In such a case, the cycle length would be computed using *either* path, and the one yielding the largest desirable cycle length would be critical. In this case, the path yielding three phases has the highest sum of critical-lane volumes, so only one cycle length will have to be computed.

Step 4: *Determine Yellow and All-Red Intervals* Both *yellow* and *all-red* intervals for both streets will be computed using Equations 21-2 and 21-3a (low pedestrian activity) and the speed limit of 35 mi/h for both streets. Because a measured average speed was not given, the 85th and 15th percentile speeds cannot be differentiated. For Phases A1 and A2, it will be assumed that both the left-turn and through movements from the E-W street cross the entire 39-foot width of the N-S street. Similarly, in Phase B, it will be assumed that both movements cross the entire 48-foot width of the E-W street. Then:

$$y_{A1,A2,B} = 1.0 + \frac{1.47 * 35}{(2 * 10) + (0)} = 3.6 \text{ s}$$

$$ar_{A1,A2} = \frac{39 + 20}{1.47 * 35} = 1.1 \text{ s}$$

$$ar_B = \frac{48 + 20}{1.47 * 3.5} = 1.3 \text{ s}$$

Table 21.7: Computation of Through Vehicle Equivalent Volumes for Signal Timing 3

Approach	Movement	Volume (veh/h)	Equivalent Tab 21.1,2	Volume (tvu/h)	Lane Group Vol (tvu/h)	Vol/Lane (tvu/h/ln)
EB	T	700	1.00	700	821	411
	R	100	1.21	121		
WB	L	380	1.05	399	399	399
	T	700	1.00	700	700	700
NB	L	300	1.10	330	330	330
	R	250	1.21	303	303	303

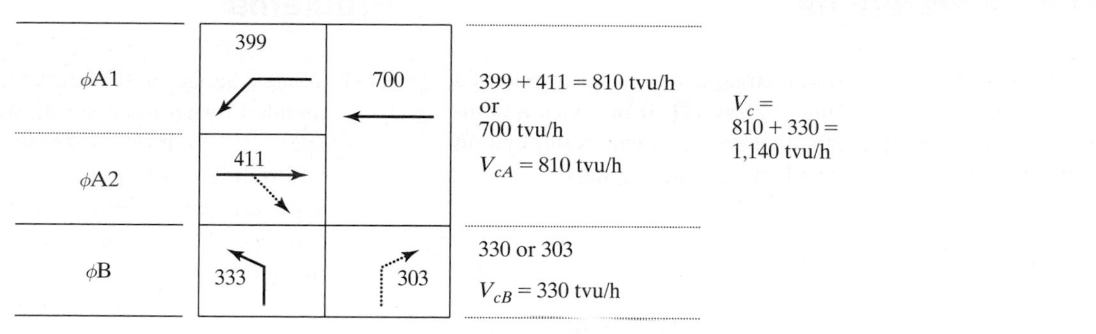

Figure 21.20: Determination of Critical Lane Volumes: Signal Timing Case 4

Step 5: *Determination of Lost Times* Once again, 2.0-second default values are used for start-up lost time (ℓ_1) and extension of effective green into yellow and all-red (e), so that the total lost time for each phase is equal to the sum of the yellow plus all-red intervals:

$$Y_{A1} = t_{LA1} = 3.6 + 1.1 = 4.7 \text{ s}$$
$$Y_{A2} = t_{LA2} = 3.6 + 1.1 = 4.7 \text{ s}$$
$$Y_B = t_{LB} = 3.6 + 1.3 = 4.9 \text{ s}$$

The total lost time per cycle, therefore, is 4.7 + 4.7 + 4.9 = 14.3 seconds.

Step 6: *Determine the Desirable Cycle Length* Equation 21-11 is once again used to determine the desirable cycle length, using the sum of critical-lane volumes, 1,140 tvu/h:

$$C_{des} = \frac{14.3}{1 - \left(\dfrac{1{,}140}{1{,}615 * 0.92 * 0.950}\right)} = \frac{14.3}{0.192} = 74.5 \text{ s}$$

For a pretimed controller, a cycle length of 75 s would be implemented.

Step 7: *Allocate Effective Green to Each Phase* The available effective green time for this signal is 75.0 − 14.3 = 60.7 s. It is allocated in proportion to the critical-lane volumes for each phase:

$$g_{A1} = 60.7 * \left(\frac{399}{1140}\right) = 21.2 \text{ s}$$

$$g_{A2} = 60.7 * \left(\frac{411}{1140}\right) = 21.9 \text{ s}$$

$$g_B = 60.7 * \left(\frac{330}{1140}\right) = 17.6 \text{ s}$$

Because the usual defaults for ℓ_1 and e are used, actual green times are numerically equal to effective green times.

Step 8: *Check Pedestrian Requirements* Although there is low pedestrian activity at this intersection, pedestrians must still be safely accommodated by the signal phasing. It will be assumed that pedestrians cross the N-S street only during Phase A2 and that pedestrians crossing the E-W street will use Phase B. The number of pedestrians per cycle in each crosswalk is based on the default volume for "low" activity—50 peds/h (Table 21.2)—and the number of cycles per hour − 3,600/75 = 48 Then, N_{ped} in each crosswalk would be 50/48 = 1.0 ped/cycle Equation 21-15 is used to compute minimum pedestrian requirements:

$$G_{pA2} = 3.2 + (0.27 * 1) + \left(\frac{39}{4.0}\right) = 13.2 \text{ s}$$

$$G_{pB} = 3.2 + (0.27 * 1) + \left(\frac{48}{4.0}\right) = 15.5 \text{ s}$$

These requirements must be checked against the vehicular green, yellow, and all-red intervals (or the green and yellow, or just the green, depending on local policy):

$$G_{pA2} = 13.2 \text{ s} < G_{A2} + Y_{A2} = 21.9 + 4.7 = 26.6 \text{ s OK}$$

$$G_{pB} = 15.5 \text{ s} < G_B + Y_B = 17.6 + 4.9 = 22.5 \text{ s OK}$$

Pedestrians are safely accommodated by the vehicular signalization, and no changes are required. It is noted that pedestrians could be entirely accommodated by the green and that it is not necessary in this case to allow pedestrians in the crosswalk during the y and ar intervals.

21.6 References

The Federal Highway Administration has recently released the *Manual of Traffic Signal Timing* [9]. It provides a complete, up-to-date review of signal timing practices throughout the United States for pretimed and actuated signals.

References

1. Pusey, R., and Butzer, G., "Traffic Control Signals," *Traffic Engineering Handbook*, 5th Edition (Pline, J., ed.), Institute of Transportation Engineers, Washington DC, 2000.
2. Asante, S., Ardekani, S., and Williams, J., "Selection Criteria for Left-Turn Phasing and Indication Sequence," *Transportation Research Record 1421*, Transportation Research Board, Washington DC, 1993.
3. *Manual of Uniform Traffic Control Devices*, Draft, Federal Highway Administration, U.S. Department of Transportation, Washington DC, 2007.
4. McGee, H., and Warren, D., "Right Turn on Red," *Public Roads*, Federal Highway Administration, U.S. Department of Transportation, Washington DC, June 1976.
5. "Driver Behavior at RTOR Locations," ITE Technical Committee 4M-20, *ITE Journal*, Institute of Transportation Engineers, Washington DC, April 1992.
6. McGee, H., "Accident Experience with Right Turn on Red," *Transportation Research Record 644*, Transportation Research Board, Washington DC, 1977.
7. "Recommended Practice: Determining Vehicle Change Intervals," ITE Technical Committee 4A-16, *ITE Journal*, Institute of Transportation Engineers, Washington DC, May 1985.
8. *Highway Capacity Manual*, 4th Edition, Transportation Research Board, Washington DC, 2000.
9. Koonce, P., Rodergerdts, L., Lee, K., Quayle, S., Beaird, S., Braud, C., Bonneson, J., Tarnoff, P., and Urbanik, T., *Manual of Traffic Signal Timing*, Final Report, FHWA Contract No. DTFH61-98-C-00075, Kittelson and Associates, Inc., Portland, OR, June 2008. Available online.

Problems

21-1. What change and clearance intervals are recommended for an intersection with an average approach speed of 35 mi/h, a grade of −2%, a cross-street width of 50 feet, and 10-foot crosswalks? Assume a standard vehicle length of 20 feet, a driver reaction time of 1.0 seconds, and significant pedestrian movements.

21-2. An analysis of pedestrian needs at a signalized intersection is undertaken. Important parameters concerning pedestrian needs and the existing vehicular signal timing are given in the table here. Are pedestrians safely accommodated by this signal timing? If not, what signal timing should be implemented? Assume that the standard default values for start-up lost time and extension of effective green into yellow and all-red (2.0 seconds each) are in effect.

Phase	G (s)	Y (s)	G_p (s)
A	18.0	4.5	30.0
B	60.0	4.0	15.0

21-3–21-7. Develop a signal design and timing for the intersections shown in Figures 21.21 to 21.25. In each case, accommodate both vehicular and pedestrian needs. Where necessary to make assumptions on key values, state these explicitly. If a successful signal timing *requires* geometric changes, indicate these with an appropriate drawing.

In general, the following values should be used for all problems:

- All volumes are in veh/h
- Pedestrian walking speed = 4.0 ft/s
- Vehicle deceleration speed = 10.0 ft/s^2
- Driver reaction time = 1.0 s
- Length of typical vehicle = 20 ft
- Level grades unless otherwise indicated

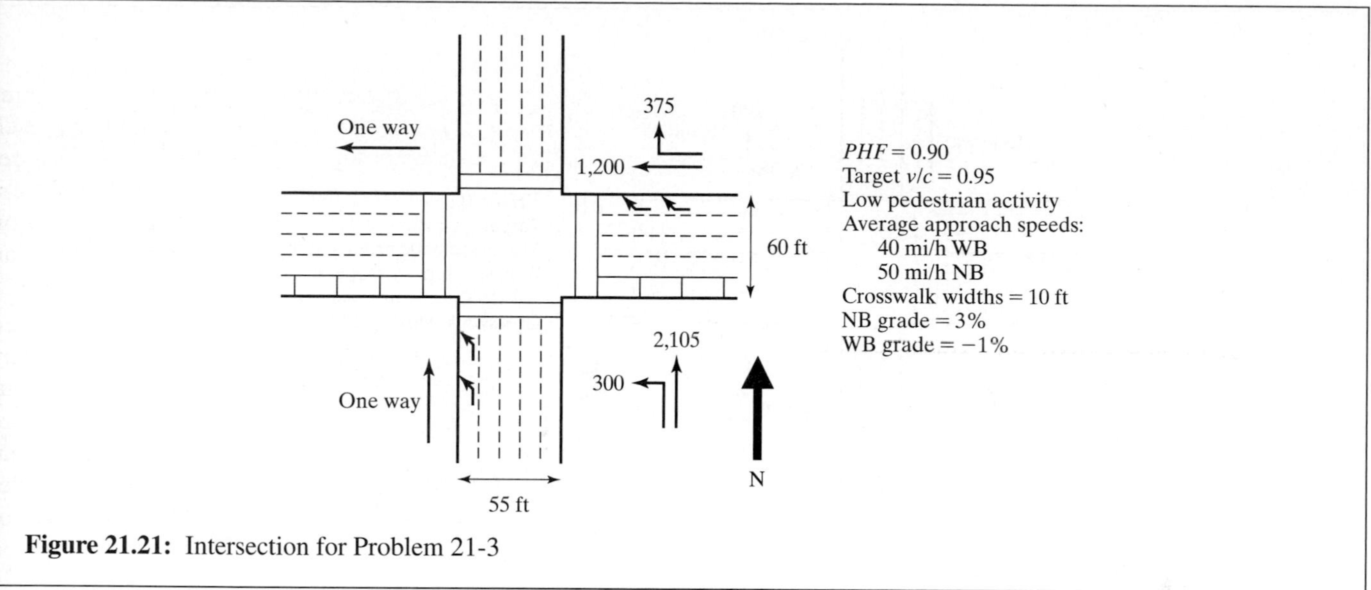

Figure 21.21: Intersection for Problem 21-3

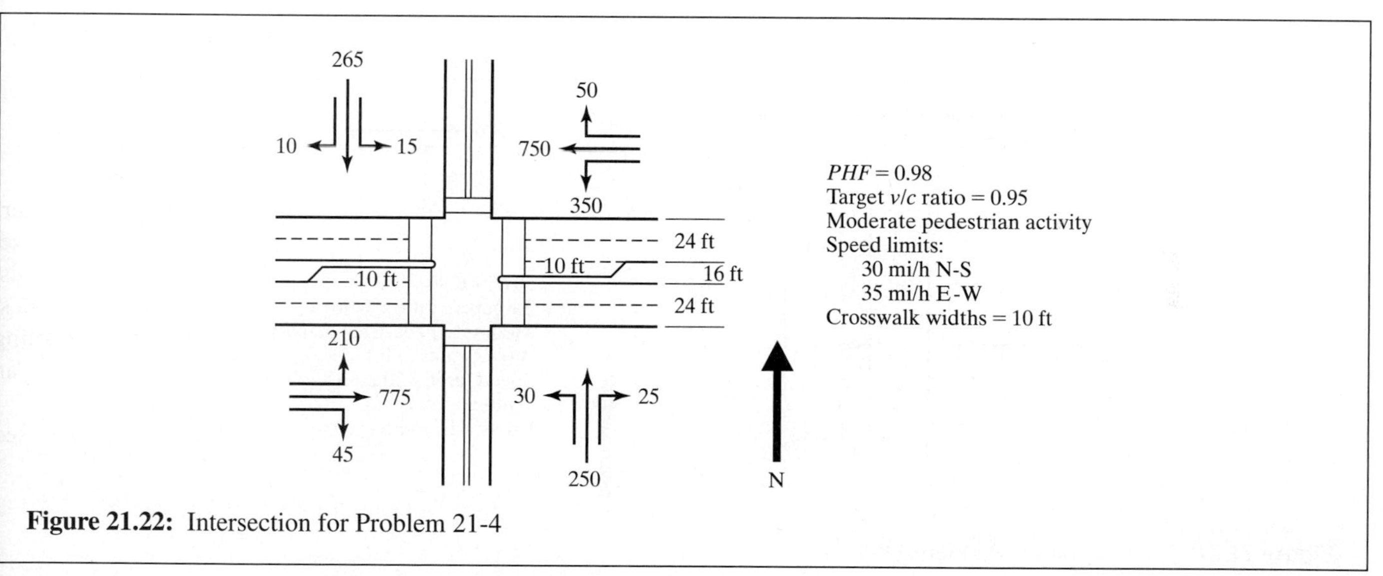

Figure 21.22: Intersection for Problem 21-4

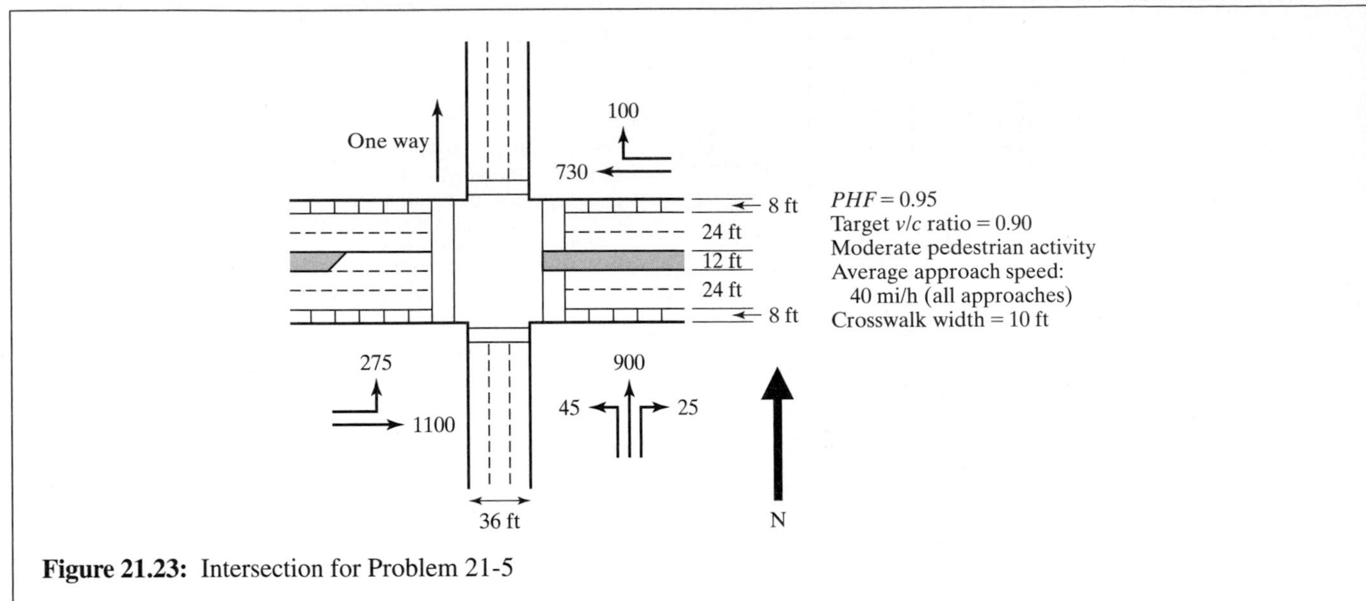

Figure 21.23: Intersection for Problem 21-5

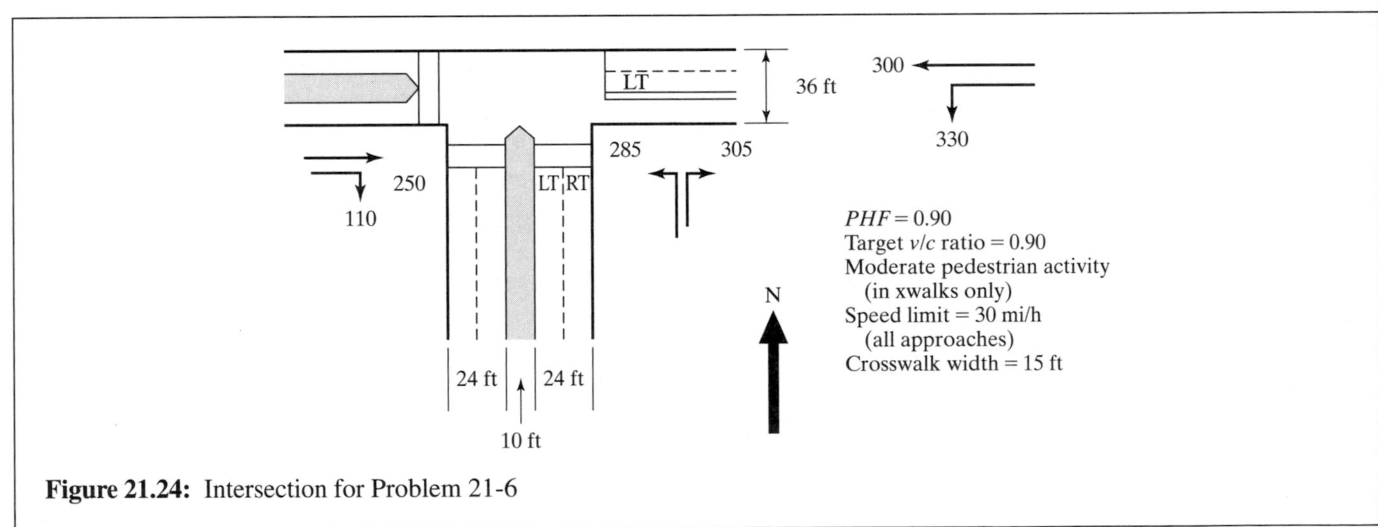

Figure 21.24: Intersection for Problem 21-6

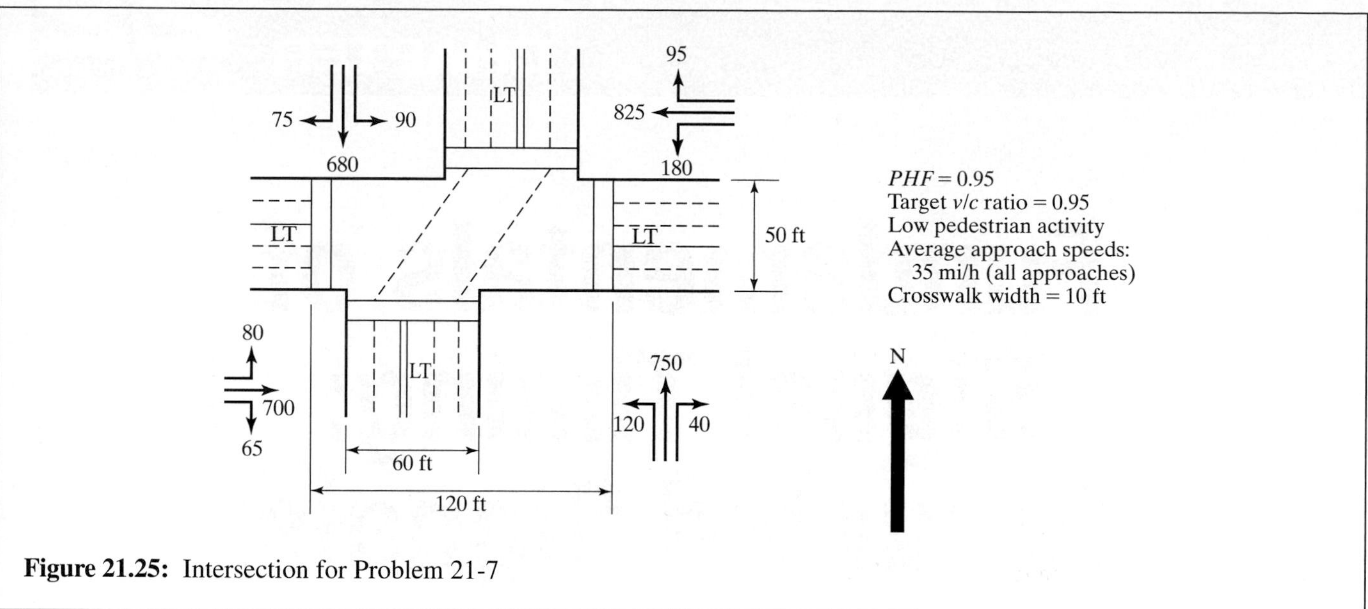

Figure 21.25: Intersection for Problem 21-7

CHAPTER 22

Fundamentals of Signal Timing: Actuated Signals

When pretimed signal controllers are employed, the phase sequence, cycle length, and all interval times are uniform and constant from cycle to cycle. Pretimed controllers can provide for several predetermined time periods during which different pretimings may be applied. During any one period, however, each signal cycle is an exact replica of every other signal cycle.

Actuated control uses information on current demands and operations, obtained from detectors within the intersection, to alter one or more aspects of the signal timing on a cycle-by-cycle basis. Actuated controllers may be programmed to accommodate:

- Variable phase sequences (e.g., optional protected LT phases)
- Variable green times for each phase
- Variable cycle length, caused by variable green times

Such variability allows the signal to allocate green time based on current demands and operations. Pretimed signals are timed to accommodate average demand flows during a peak 15-minute demand period. Even within that period, however, demands vary on a cycle-by-cycle basis. Thus, it is, at least conceptually, more efficient to have signal timing vary in the same way.

Consider the situation illustrated in Figure 22.1. Five consecutive cycles are shown, including the capacity and demand during each. Note that over the five cycles shown, the signal has the capacity to discharge 50 vehicles and that total demand during the five cycles is also 50 vehicles. Thus, over the five cycles shown, total demand is equal to total capacity.

Actual operations over the five cycles, however, result in a queue of unserved vehicles with pretimed operation. In the first cycle, 10 vehicles arrive and 10 vehicles are discharged. In the second, 6 vehicles arrive and 6 are discharged. In the third cycle, 8 vehicles arrive and 8 are discharged. Note that from the second and third cycles, there is unused capacity for an additional 6 vehicles. In cycle 4, 12 vehicles arrive and only 10 are discharged, leaving a queue of 2 unserved vehicles. In cycle 5, 14 vehicles arrive and only 10 are discharged, leaving an additional 4 unserved vehicles. Thus, at the end of the five cycles, there is an unserved queue of 6 vehicles. This occurs despite the fact that over the entire period, the demand is equal to the capacity.

The difficulty with pretimed operation is that the unused capacity of six vehicles in cycles 2 and 3 may not be

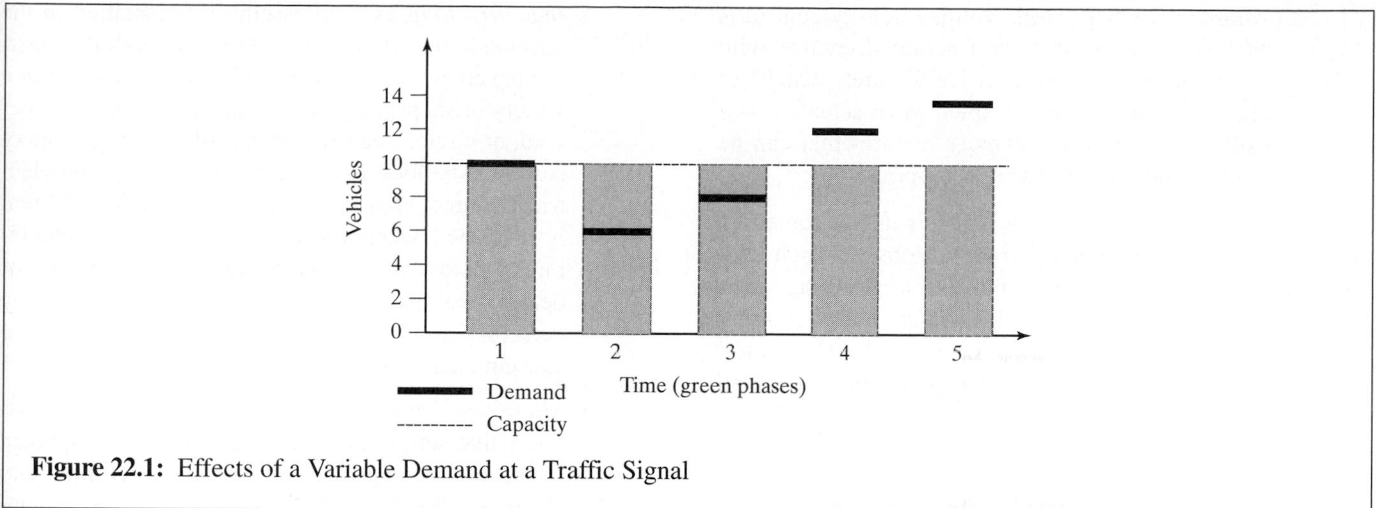

Figure 22.1: Effects of a Variable Demand at a Traffic Signal

used by excess vehicles arriving in cycles 4 and 5. If the signal had been a properly timed actuated signal, the green in cycles 2 and 3 could have been terminated when no demand was present and additional green time could have been added to cycles 4 and 5 to accommodate a higher number of vehicles. The ability of the signal timing to respond to short-term variations in arrival demand makes the overall signal operation more efficient. Even if the total amount of green time allocated over the five cycles illustrated did not change, the ability to "save" unused green time from cycles 2 and 3 to increase green time in cycles 4 and 5 would significantly reduce delay and avoid or reduce a residual queue of unserved vehicles at the end of the five-cycle period.

Another major benefit of actuated signal timing is that a single programmed timing pattern can flex to handle varying demand periods throughout the day, including peak and off-peak periods and changes in the balance of movements.

If the advantages of allowing signal timing to vary on a cycle-by-cycle basis are significant, why aren't all signalized intersections actuated? The principal issue is coordination of signal systems. To effectively coordinate a network of signals to provide for progressive movement of vehicles through the system, all signals must operate on a uniform cycle length. Thus, where signals must be interconnected for progressive movement, the cycle length cannot be permitted to vary at different intersections. This removes the principal benefit of actuated control in such circumstances, the ability to vary the cycle length. The additional cost of actuated signals and the required detection systems are also a consideration.

Actuated signal control is often used at isolated signalized intersections, usually a minimum of 2.0 miles from the nearest adjacent signal. Over the past two decades, however, the use of actuated signal controllers in a coordinated signal systems has greatly increased. In such systems, the cycle length must be kept constant, but it can be changed at intervals as short as 15 minutes, and the allocation of green time within the cycle may change on a cycle-by-cycle basis.

22.1 Types of Actuated Control

There are three basic types of actuated control, each using signal controllers that are somewhat different in their design:

1. *Semi-actuated control.* This form of control is used where a small side street intersects with a major arterial or collector. This type of control should be considered whenever Warrant 1B is the principal reason justifying signalization. Semi-actuated signals are almost always two phase, with all turns being made on a permitted basis. Detectors are placed only on the side street. The green is on the major street at all times unless a "call" on the side street is noted. The number and duration of side-street greens is limited by the signal timing and can be restricted to times that do not interfere with progressive signal-timing patterns along the collector or arterial.
2. *Full actuated control.* In full actuated operation, all lanes of all approaches are monitored by detectors. The phase sequence, green allocations, and cycle length are all subject to variation. This form of control is effective for both two-phase and multiphase operations and can accommodate optional phases.

3. *Volume-density control.* Volume-density control is basically the same as full actuated control with additional demand-responsive features, which are discussed later in this chapter. Most actuated controllers have volume-density features that can be implemented if their use is appropriate.

Computer-controlled signal systems do not constitute actuated control at individual intersections. In such systems, the computer plays the role of a large master controller, establishing and maintaining offsets for progression throughout a network or series of arterials. Computer-controlled systems can involve either pretimed or actuated signal controllers.

22.2 Detectors and Detection

The hardware for detection of vehicles is advancing rapidly. Pressure-plate detectors, popular in the 1970s and 1980s, are rarely used in modern traffic engineering. Most detectors rely on creating or observing changes in magnetic or electromagnetic fields, which occurs when a metallic object (a vehicle) passes through such a field. The *Traffic Engineering Handbook* [1] contains a useful summary of these detectors, which include:

- *Inductive loop.* A loop assembly is installed in the pavement, usually by saw-cutting through the existing pavement. The loop is laid into the saw cut in a variety of shapes, including square, rectangle, trapezoid, or circle. The saw cut is refilled with an epoxy sealant. The loop is connected to a low-grade electrical source, creating an electromagnetic field that is disturbed whenever a metallic object (vehicle) moves across it. This is the most common type of detector in use today. Figure 22.2 shows a loop detector installation with the epoxy-covered loop saw cut clearly visible.
- *Microloop.* This is a small cylindrical passive transducer that senses changes in the vertical component of the earth's magnetic field and converts them into electronically discernible signals. The sensor is cylindrical, about 2.5 inches in length and 0.75 inches in diameter. The probe is placed in a hole drilled in the roadway surface.
- *Magnetic.* These detectors measure changes in the concentration of lines of flux in the earth's magnetic field and convert such changes to an electronically discernible signal. The sensor unit contains a small coil of wire that is placed below the roadway surface.

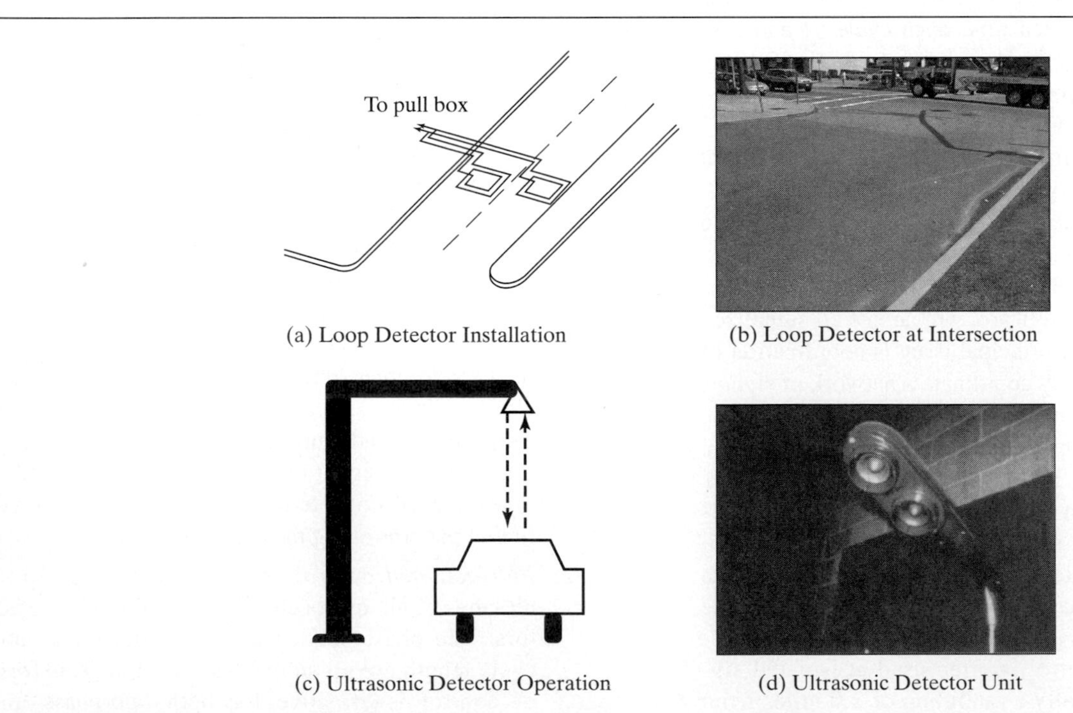

(a) Loop Detector Installation

(b) Loop Detector at Intersection

(c) Ultrasonic Detector Operation

(d) Ultrasonic Detector Unit

Figure 22.2: Loop and Ultrasonic Detectors Illustrated

For all of the magnetic class of detectors, one or more detectors must be used in each lane of each approach. A disadvantage is that all must be placed in or below the pavement. In areas where pavement condition is a serious issue, these detectors could become damaged or inoperable.

Another class of detector uses *sonic* or *ultrasonic* waves that can be emitted from an overhead or elevated roadside location. Such detectors rely on the echoes from reflected waves (ultrasonic) or on the Doppler principle of changes in reflected frequency when waves reflect back from a moving object (sonic). The emitted wave spreads in a cone-like shape and can, therefore, cover more than one lane with a single detector unit, depending on its exact placement.

Figure 22.2 also illustrates an ultrasonic detector. In the case of an overhead ultrasonic detector, the unit is calibrated for the time it takes a reflected wave to return to the sensor when it is reflected from the pavement. When a vehicle intercepts the emitted wave, it is reflected in a shorter time.

Other detector types are in use, such as radar, and even older pressure-plate systems. The vast majority of detectors are of the types described previously.

A rapidly emerging technology is optical video imaging, in which real-time video of an intersection approach or other traffic location is combined with computerized pattern-recognition software. Virtual detectors are defined within the video screen, and software is programmed to note changes in pixel intensity at the virtual detector location. Now in common use for data and remote observation, such detection systems only recently have been employed to operate signals in real time.

Of greater importance than the specific detection device(s) used is the *type* of detection. Two types of detection influence the design and timing of actuated controllers:

- *Point detection (Pulse Mode).* In this type of detection, only the fact that the detector has been "disturbed" is noted. The detector is installed at a "point," even though the detector unit itself may involve a short length. The detector operates in the "pulse" mode in that each vehicle passing through the detector field creates a short pulse of 0.10 to 0.15 seconds. The controller channel connected to a point detector operating in the pulse mode must be set to the "locked" feature. The controller essentially retains the observed pulse until the required green time has been allocated to serve the observed demand.
- *Presence detection.* In this type of detection, a length of an approach lane is included in the detection zone. The controller channel connected to such a detector is set to the "unlocked" feature. A "call" for service is retained until the vehicle is detected to have left the detection zone. At any given time, the number of vehicles stored in the detection zone is known. Most presence detection is accomplished using a single detector, of 20 feet or more. For very long detection zones, series of shorter loop detectors may be used.

The timing of an actuated signal is very much influenced by the type of detection in place. The use of presence detectors has greatly increased in recent years. Point and presence detectors are both in common use throughout the United States.

22.3 Actuated Control Features and Operation

Actuated signal controllers are manufactured in accordance with one of two standards. The most common is that of the National Electronic Manufacturer's Association (NEMA). NEMA standards specify all features, functions, and timing intervals, and timing software is provided as a built-in feature of the hardware (often referred to as "firmware"). The second set of standards is for the Type 170/270 class of controllers, used primarily by the California Department of Transportation and the New York State Department of Transportation. Type 170/270 controllers do not come with built-in software, which is generally available through third-party vendors. Although NEMA software cannot be modified by an agency, Type 170/270 software can be modified. U.S. manufacturers of signal controllers include Control Technologies, Eagle, Econolite, Kentronics, Naztec, and others. Most manufacturers maintain current Web sites, and we urge you to consult them for the most up-to-date descriptions of hardware, software, and functions.

22.3.1 Actuated Controller Features

Regardless of the controller type, virtually all actuated controllers offer the same basic functions, although the methodology for implementing them may vary by type and manufacturer. For each actuated phase, the following basic features must be set on the controller:

1. *Minimum Green Time (G_{min}).* Each actuated phase has a minimum green time, which serves as the smallest amount of green time that may be allocated to a phase when it is initiated, s.
2. *Passage Time (PT).* This time actually serves three different purposes: (a) It represents the maximum gap between actuations at a single detector required to retain the green. (b) It is the amount of time added

to the green phase when an additional actuation is received within the unit extension, *U*. (c) It must be of sufficient length to allow a vehicle to travel from the detector to the STOP line.

3. *Maximum Green Time (G_{max}).* Each phase has a maximum green time that limits the length of a green phase, even if there are continued actuations that would normally retain the green. The "maximum green time" begins when there is a "call" (or detector actuation) on a competing phase.

4. *Recall Settings.* Each actuated phase has a number of recall settings. The recall settings determine what happens to the signal when there is no demand.

5. *Yellow and All-Red Intervals.* Yellow and all-red intervals provide for safe transition from "green" to "red." They are fixed times and are not subject to variation, even in an actuated controller. They are found in the same manner as for pretimed signals (refer to Chapter 21).

6. *Pedestrian WALK ("Walking Man"), Clearance ("Flashing Up-raised Hand"), and DON'T WALK ("Up-raised Hand") intervals.* Pedestrian intervals must also be set. With actuated signals, however, the total length of the GREEN is not known. Thus pedestrian intervals are set in accordance with the minimum green time for each phase. Pedestrian pushbuttons are often, but not always, needed to ensure adequate crossing times.

Volume-density features add several other functions. They are generally used at intersections with high approach speeds (≥ 45 mi/h), and in conjunction with presence detectors. In addition to the normal features of any actuated controller, the volume-density features include two important functions:

1. *Variable Minimum Green.* Because presence detectors are capable of "remembering" the number of queued vehicles, the minimum green time may be varied to reflect the number of queued vehicles that must be served on the next "green" interval.

2. *Gap Reduction:* Using standard functions, the passage time is a constant value. Volume-density features allow the minimum gap required to retain the green to be reduced over time. Doing this makes it more difficult to retain the green on a particular phase as the phase gets longer. Implementing the gap-reduction feature usually involves identifying four different measures:

   a. Initial passage time: PT_1 (s) (maximum value)
   b. Final passage time, PT_2 (s) (minimum value)
   c. Time into the green that gap reduction begins, t_1 (s)
   d. Time into the green that gap reduction ends, t_2 (s)

3. Time t_1 begins when a "call" on a competing phase is noted.

Some controllers contain additional features that may be implemented. Those noted here, however, are common to virtually all controllers and controller types.

22.3.2 Actuated Controller Operation

Figure 22.3 illustrates the operation of an actuated phase based on the three critical settings: minimum green, maximum green, and the passage time.

When the green is initiated for a phase, it will be *at least* as long as the minimum green period, G_{min}. The controller divides the minimum green into an initial portion and a portion equal to one passage time. If an additional "call" is received during the initial portion of the minimum green, no time is added to the phase because there is sufficient time within the minimum green to cross the STOP line (yellow and all-red intervals take care of clearing the intersection). If a "call" is received during the last *PT* seconds of the minimum green, *PT* seconds of green are added to the phase. Thereafter, every time an additional "call" is received during a unit extension of *PT* seconds, an additional period of *PT* seconds is added to the green.

Note that the additional periods of *PT* seconds are added *from the time of the actuation or "call."* They are *not* added to the end of the previous unit extension because this would accumulate unused green times within each unit extension and include them in the total "green" period.

The "green" is terminated in one of two ways:

1. A unit extension of *PT* seconds expires without an additional actuation. Such a termination is commonly referred as a "gap out."

2. The maximum green is reached. Such a termination is referred to as a "max out." The maximum green begins timing out when a "call" on a competing phase is noted. During the most congested periods of flow, however, it may be assumed that demand exists more or less continuously on all phases. The maximum green, therefore, begins timing out at the beginning of the green period in such a situation.

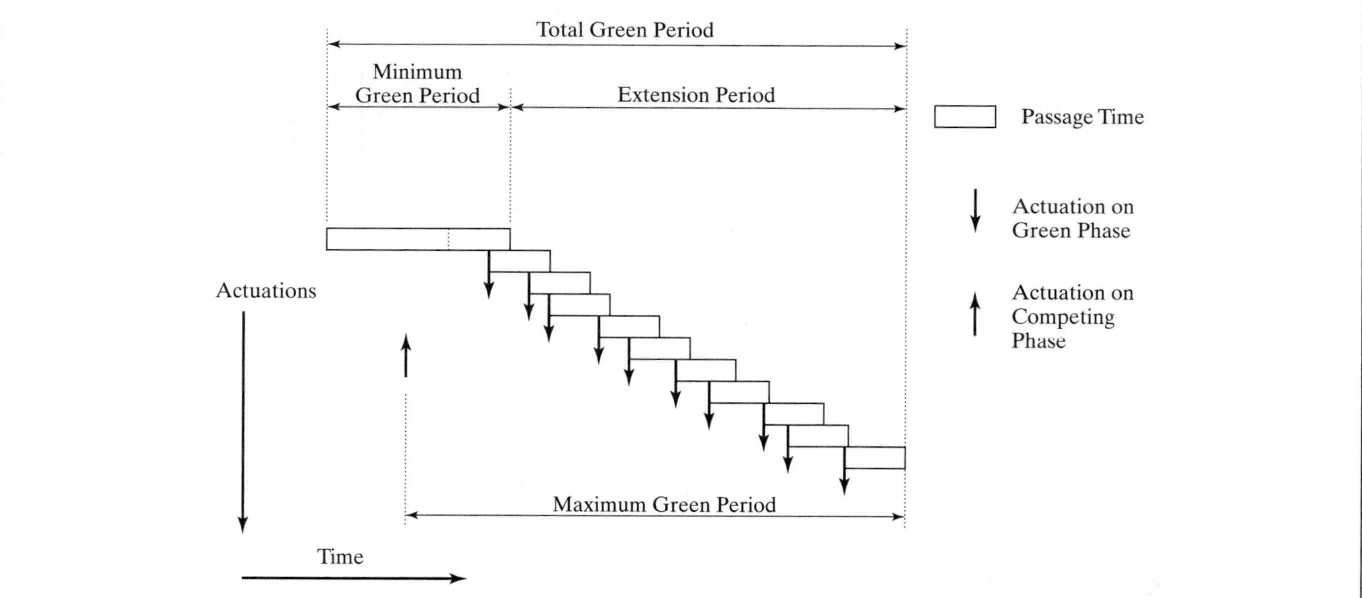

Figure 22.3: Operation of an Actuated Phase

(*Source:* Used with permission of Institute of Transportation Engineers, *Traffic Detector Handbook*, 2nd Edition, JHK & Associates, Tucson, AZ, p. 66.)

Assuming that demand exists continuously on all phases, the green period would be limited to a range of G_{min} to G_{max}. During periods of light flow, with no demand on a competing phase, the length of any green period can be unlimited, depending on the setting of the recall functions.

In most situations, parallel lanes on an approach operate in parallel with each other. For example, in a three-lane approach, there will be three detectors (one for each lane). If *any* of the three lanes receives an additional "call" within *PT* seconds, the green will be extended. Where multiple detectors are connected in series, using a single lead-in cable, gaps may reflect a lead vehicle crossing one detector and a following vehicle crossing another. Although this type of operation is less desirable, it is less expensive to install and therefore is used frequently.

Figure 22.4 illustrates the operation of the "gap-reduction" feature on actuated signal controllers. Note the four critical times that must be set on the controller. Depending on the manufacturer and model selected, there are a number of different protocols for implementing these four times, including:

- *BY-EVERY option.* Specify the amount of time by which the allowable gap is reduced after a specified amount of time. For example, for every 1.5 seconds of extension (after time x), reduce the allowable gap by 0.2 seconds.
- *EVERY SECOND option.* Specify the amount of time by which the allowable gap is reduced each second (after time x). For example, for every second of extension, reduce the allowable gap by 0.1 second.
- *TIME TO REDUCE option.* Specify a maximum and minimum allowable gap, and specify how long it will take to reduce from the maximum to the minimum (after time x). For example, the allowable gap will be linearly reduced from 3.5 seconds to 1.5 seconds over a period of 15 seconds.

22.4 Actuated Signal Timing and Design

In an actuated signal design, the traffic engineer does not provide an exact signal timing. Rather, a phase plan is established, and minima and maxima are set, along with programmed rules for determining the green period between limiting values based on vehicle actuations on detectors.

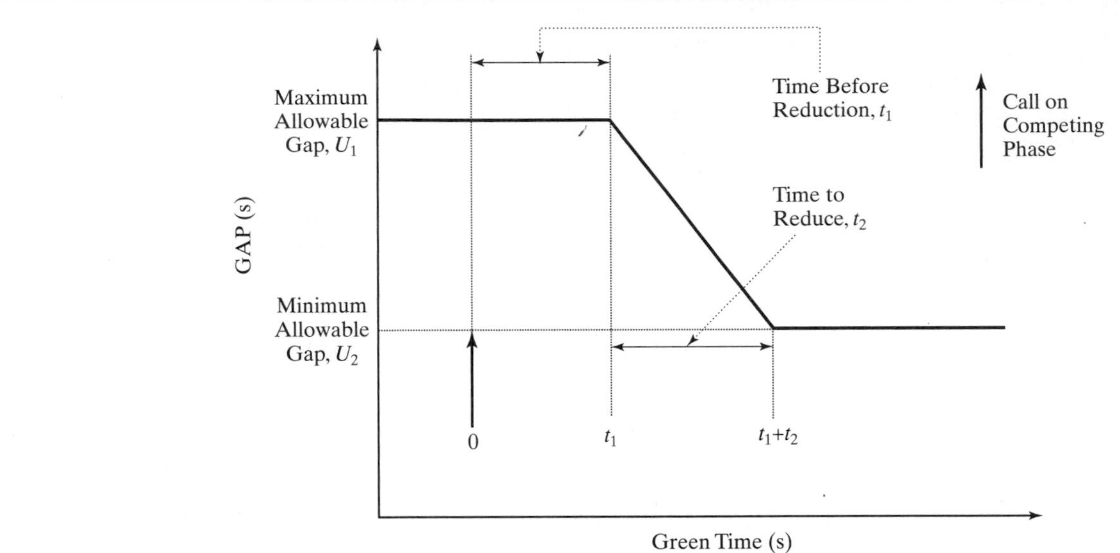

Figure 22.4: Gap Reduction Feature on Volume-Density Controllers

(*Source:* Used with Permission of Institute of Transportation Engineers, *Traffic Detector Handbook*, 2nd Edition, JHK & Associates, Tucson AZ, p. 68.)

22.4.1 Phase Plans

Phase plans are established using the same types of considerations as for pretimed signals (see Chapter 21). The primary difference is the flexibility in phase sequencing offered by actuated controllers.

Protected left-turn phases may be installed at lower left-turn flow rates because these phases may be skipped during any cycle in which no left-turn demand is present. There are no precise guidelines for minimum left-turn demands and/or cross-products, so the engineer has considerable flexibility in determining an optimum phase plan.

22.4.2 Minimum Green Times

Minimum green times must be set for each phase in an actuated signalization, including the nonactuated phase of a semi-actuated controller. The minimum green timing on an actuated phase is based on the type and location of detectors.

Point Detection

Point detectors only provide an indication that a "call" has been received on the subject phase. The number of calls experienced and/or serviced is not retained. Thus, if a point detector is located *d* feet from the STOP line, it must be assumed that a queue of vehicles fully occupies the distance *d*. The minimum green time, therefore, must be long enough to clear a queue of vehicles fully occupying the distance *d*, or:

$$G_{\min} = \ell_1 + 2.0 * Int\left[\frac{d}{25}\right] \qquad (22\text{-}1)$$

where G_{min} = minimum green time, s
ℓ_1 = start-up lost time, s
d = distance between detector and STOP line, ft
25 = assumed head-to-head spacing between vehicles in queue, ft

The integer function requires that the value of $d/25$ be rounded to the next highest integer value. In essence, it requires that a vehicle straddling the detector be serviced within the minimum green period. Various agencies set the value of ℓ_1 based on local policy. Values between 2.0 and 4.0 seconds are most often used. The *Traffic Signal Timing Manual* [2] recommends a value of 3.0 seconds be used.

Presence Detection

Where presence detectors are in use, the minimum green time can be variable, based on the number of vehicles sensed in the queue when the green is initiated. In general:

$$G_{min} = \ell_1 + 2n \qquad (22\text{-}2)$$

where ℓ_1 = start-up lost time, s

n = number of vehicles stored in the detection area, vehs

This is true, however, only if the front edge of the detector rests on (or very near—within 2 feet) of the STOP line. If the front edge of the detector is further away from the STOP line, the minimum green must assume that the distance between the front edge of the detector and the STOP line is full. Equation 22-1 is used, with d equal to the distance between the front edge of the detector and the STOP line.

22.4.3 Passage Time

As we noted previously, the passage time serves three different purposes. In terms of signal operation, it serves as both the minimum allowable gap to retain a green signal and as the amount of green time added when an additional actuation is detected within the minimum allowable gap.

The passage time is selected with three criteria in mind:

- The passage time should be long enough such that a subsequent vehicle operating in dense traffic at a safe headway will be able to retain a green signal (assuming the maximum green has not yet been reached).
- The passage time should not be so long that straggling vehicles may retain the green or that excessive time is added to the green (beyond what one vehicle reasonably requires to cross the STOP line on green).
- The passage time should not be so long that it allows the green to be extended to the maximum on a regular basis.

Presence Detection

Figure 22.5 illustrates the relationship between key variables at a presence detector. The key variable is the *maximum allowable headway* (MAH) or gap that will retain the green for a detector in a single lane. The illustration can be used to derive the following equation:

$$PT = MAH - \frac{L_v + L_d}{1.47\ S_a} \qquad (22\text{-}3)$$

where: PT = passage time, s,

MAH = maximum allowable headway, s,

S_a = average approach speed, mi/h,

L_v = length of vehicle, ft (use default value of 20 ft), and

L_d = length of the detection zone, ft.

In Figure 22.5, the passage time, PT, is the time that the detector is unoccupied. This is the setting on the controller that will determine whether the green is retained or not. Thus, if the maximum allowable headway is set, the passage time will be found by subtracting the amount of time it takes a vehicle traveling at the average approach speed to traverse both the length of the vehicle and the length of the detector.

Maximum allowable headways used are generally in the range of 2.0 to 4.5 seconds. Larger values tend to result in high delays. General practice is to use MAH = 3.0 seconds where the gap reduction feature is not in use, and 4.0 (for the maximum MAH) when the gap reduction feature is in use.

Point Detection

For point detection, the length of the detector is essentially "0" feet. Because a crossing vehicle registers only a pulse of 0.10 seconds to 0.15 seconds in duration, the length of the

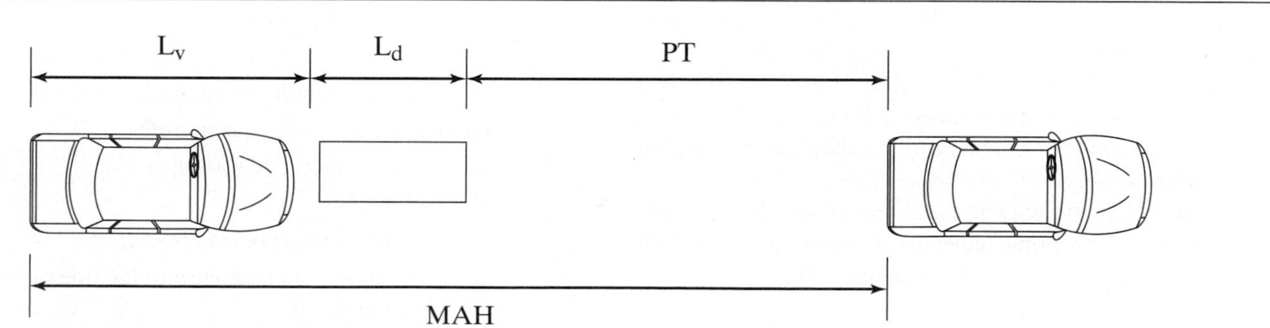

Figure 22.5: Relationship Between Passage Time, Detector and Vehicle Length, and Maximum Allowable Gap

(*Source:* Koonce, P., et al., *Traffic Signal Timing Manual*, Final Report, FHWA Contract No. DTFH61-98-C-00075, Kittelson and Associates Inc., Portland, OR, June 2008, Figure 5-4, p. 5-5.)

vehicle is irrelevant. Thus, for point detection, *PT* is equal to the maximum allowable headway, *MAH*.

Minimum Passage Time

Regardless of the type of detector being used, the passage time has a minimum value based on the location of the point detector, or the front edge of a presence detector. The passage time must be at least large enough to allow a vehicle traveling at the 15th percentile approach speed to traverse the distance between the detector (or front edge of the detector) to the STOP line, or:

$$PT_{min} = \frac{d}{1.47\ S_{15}} \tag{22-4}$$

where S_{15} is the 15th percentile approach speed, which may be estimated as the average approach speed minus 5 mi/h.

22.4.4 Detector Location

The minimum green time and the detector location are mathematically linked. Where presence detectors exist, and the front edge of the detector is within 2 feet of the STOP line, Equation 22-2 describes the relationship, which produces a variable minimum green time based on the number of vehicles stored within the detection area.

Where point detectors are used, or where presence detectors are more than 2 feet from the STOP line, the point detector (or front edge of the presence detector) is located to produce a preselected minimum green time.

Because many actuated signals are at locations where demands are quite low during off-peak periods, there is often the desire to keep minimum green times as low as possible, thus minimizing the waiting period for a vehicle on a competing phase when there is no demand on the subject phase. A practical minimum limit on the minimum green time is the assumed start-up lost time, $\ell_1 + 2.0$ s. This is the amount of time needed to process a single vehicle; it ranges between 4.0 and 6.0 seconds, depending on the assumed start-up lost time.

When this strategy is used, Equation 22-1 is used to compute the appropriate detector location for the selected minimum green. Consider the following situation:

> A minimum green on an approach to an actuated signal is to be set at 6.0 seconds, with an assumed start-up lost time of 4.0 seconds. How far may the detector be located from the STOP line?

From Equation 22-1:

$$G_{min} = 6.0 = 4.0 + 2.0\ Int\left[\frac{d}{25}\right]$$

$$Int\left[\frac{d}{25}\right] = \frac{6.0 - 4.0}{2.0} = 1.0$$

Due to the integer function, the detector may be located anywhere between 0.1 and 25.0 feet from the STOP line. Note that where presence detectors are used, the location refers to the *front* of the detector.

There are practical limitations on the placement of detectors that must be observed: The detector(s) must be placed such that no vehicle can arrive at the STOP line without having crossed the detector. In practical terms, this means that no detector can be placed where a vehicle can enter the traffic stream from driveway or curb parking space located between the detector and the STOP line. In many urban and suburban settings, this requires that the detector be located quite close to the STOP line.

Presence detectors are more flexible, in that they can detect vehicles entering the detection area from the side. Thus it is only the location of the *front* of the area detector that is limited as described previously.

22.4.5 Yellow and All-Red Intervals

Yellow and all-red intervals are determined in the same fashion as for pretimed signals:

$$y = t + \frac{1.47 * S_{85}}{2a + 64.4 * 0.01G} \tag{22-5}$$

$$ar = \frac{w + L}{1.47\ S_{15}} \quad \text{or} \quad \frac{P + L}{1.47\ S_{15}} \tag{22-6}$$

where y = yellow interval, s
ar = all red interval, s
S_{85} = 85th percentile speed, mi/h
S_{15} = 15th percentile speed, mi/h
a = deceleration rate (10 ft/s^2)
G = grade(%)
w = width of street being crossed, ft
P = distance from near curb to far side of far crosswalk, ft

As in the case of pretimed signals, yellow and all-red times must be known to determine the total lost time in the cycle, L, which is needed to determine maximum green times.

The relationships between yellow and all red times and lost times are repeated here for convenience:

$$L = \sum_i t_{Li}$$

$$t_{Li} = \ell_{1i} + \ell_{2i}$$

$$\ell_{2i} = Y_i - e_i$$

$$Y_i = y_i + ar_i \tag{22-7}$$

where L = total lost time in the cycle, s/cycle

t_{Li} = total lost time for Phase i, s

ℓ_{1i} = start-up lost time for Phase i, s (measured value, or 2.0 s default value)

ℓ_{2i} = clearance lost time for Phase i, s

e_i = encroachment of effective green into yellow and all-red periods for Phase i, s (measured value, or 2.0 s default value)

Y_i = sum of yellow and all red intervals for Phase i, s

y_i = yellow interval for Phase i, s

ar_i = all red interval for Phase i, s

Note that when the default values for ℓ_1 and e are used, the total lost time per cycle, L, is equal to the sum of the yellow and all-red phases associated with critical movements in the cycle, and that effective green, g, is equal to actual green, G.

22.4.6 Maximum Green Times and the Critical Cycle

The "critical cycle" for a full actuated signal is one in which each phase reaches its maximum green time. For semi-actuated signals, the "critical cycle" involves the maximum green time for the side street and the minimum green time for the major street, which has no detectors and no maximum green time.

Maximum green times for actuated phases and/or the minimum green time for the major street with semi-actuated signalization are found by determining a cycle length and initial green split based on average demands during the peak analysis period. The method is the same as that used for determining cycle lengths and green times for a pretimed signal:

$$C_i = \frac{L}{1 - \left[\dfrac{V_c}{1{,}615 * PHF * (v/c)}\right]} \tag{22-8}$$

where C_i = initial cycle length, s

V_c = sum of critical lane volumes, veh/h

PHF = peak hour factor

v/c = desired v/c ratio to be achieved

Because the objective in actuated signalization is to have little unused green time during peak periods, the v/c ratio chosen in this determination is taken to be 0.95 or higher in most applications.

Knowing the cycle length, green times are then determined as:

$$g_i = (C - L)*\left(\frac{V_{ci}}{V_c}\right) \tag{22-9}$$

where g_i = effective green time for Phase i, s

V_{ci} = critical lane volume for Phase i, veh/h

All other variables are as previously defined.

These computations result in a cycle length and green times that would accommodate the average cycle demands in the peak 15 minutes of the analysis hour. They are not, however, sufficient to handle perturbations occurring during the peak 15-minute demand period when individual cycle demands exceed the capacity of the cycle. Thus, to provide enough flexibility in the controller to adequately service peak cycle-by-cycle demands during the analysis period, green times determined from Equation 22-9 are multiplied by a factor of between 1.25 and 1.50. The results would then become the maximum green times for each phase and/or the minimum green time for a major street at a semi-actuated signal.

The "critical cycle length" is then equal to the sum of the actual maximum green times (and/or the minimum green time for a major street at a semi-actuated location) plus yellow and all-red transitions.

$$C_c = \sum_i (G_i + Y_i) \tag{22-10}$$

where C_c = critical cycle length, s

G_i = actual maximum green time for actuated Phase i, or actual minimum green time for the major street at a semi-actuated signal, s

Y_i = sum of yellow and all red intervals for Phase i, s

The timing of an actuated signal involves a number of practical considerations that may override the results of the computations as described. Particularly at a semi-actuated signal location with low side-street demands, the maximum green, G_{max}, may compute to a value that is less than the minimum green, G_{min}. Although a rarer occurrence, this could happen on a given phase at full actuated location as well, particularly where protected left-turn phases are involved. In such cases, the G_{max} is judgmentally set as $G_{min} + 2.0\,n$, where n is the maximum number of vehicles to be served during a single green phase. The value of n is usually approximately set as 1.5 times the average number of vehicles expected per cycle (an iterative concept because the cycle length would be needed to determine the value of n). However, to maintain an appropriate balance between all phases, values of G_{max} for other phases must then be adjusted to maintain a ratio equal to the balance of critical-lane volumes for each phase.

22.4.7 Pedestrian Requirements for Actuated Signals

As for pretimed signals, pedestrians require the following amount of time to safely cross a street:

$$G_p = 3.2 + 0.27\ N_{peds} + \left(\frac{L}{S_p}\right) \qquad (22\text{-}11)$$

where G_p = minimum pedestrian green time, s
L = length of crosswalk, ft
S_p = walking speed of pedestrians, default value 4.0 ft/s
N_{peds} = average number of pedestrians crossing the street in one crosswalk per cycle, peds

Where pedestrians are permitted to be in the crosswalk during the vehicular green time plus the yellow and all-red intervals, safe operation occurs only when:

$$G_p \leq G + y + ar \qquad (22\text{-}12)$$

for a given phase. Depending on local policy, pedestrian presence in either the yellow (y) and/or all red (ar) may not be permitted. In such cases, either ar or $y + ar$ is removed from Equation 22-12 as appropriate.

At actuated signal locations, however, there are several considerations in addition to those at pretimed locations:

1. The value of N_{peds} varies, as the cycle length is variable. For actuated signals, the value of N_{peds} is determined from the critical cycle length. This results in the *maximum feasible* number of pedestrians that might occur during a single cycle.
2. Because the length of green times also varies from cycle to cycle, safety can be assured only when Equation 22-12 is used with $G = G_{min}$ for each cycle.

With pretimed signals, when safe crossing is not assured, either the cycle length must be increased to accommodate both pedestrians and vehicles or pedestrian-actuated pushbutton and pedestrian signals must be installed.

With actuated signals, safe crossing based on minimum green times is most often *not* provided. Increasing minimum greens to accommodate pedestrians during every cycle is not an option because that would create inefficiencies for vehicles that the actuated signal was installed to avoid. Thus, whenever the minimum green *does not* provide safe crossing, a pedestrian pushbutton should be installed with pedestrian signals.

In such cases, the pedestrian signal rests on a DON'T WALK indication. When the pedestrian pushbutton is actuated, *on the next green phase* the minimum green time is increased to:

$$G_{\min,ped} = 3.2 + 0.27\ N_{peds} + \left(\frac{L}{S_p}\right) - y - ar \quad or$$

$$G_{\min,ped} = 3.2 + 0.27\ N_{peds} + \left(\frac{L}{S_p}\right) - y \quad or$$

$$G_{\min,ped} = 3.2 + 0.27\ N_{peds} + \left(\frac{L}{S_p}\right) \qquad (22\text{-}13)$$

as appropriate to local policy. When the pedestrian minimum green is called for, the WALK interval (Walking Man) is as follows:

$$WALK = 3.2 + 0.27\ N_{peds} \qquad (22\text{-}14)$$

and the pedestrian clearance interval (Flashing Up-raised Hand) is:

$$Up\text{-}raised\ Hand_{Flashing} = \frac{L}{S_p} \qquad (22\text{-}15)$$

All terms are as previously defined.

22.4.8 Dual-Entry Feature

The dual-entry feature, when engaged, calls phases that can time concurrently, even when only one of the phases is called. For example, if the NB and SB through phases generally run concurrently, then a call for service on either one would initiate both phases. When the feature is not engaged, only the phase that is "called" will run. Common practice is to engage this feature for pairs of phases that would commonly be expected to run simultaneously.

22.4.9 Recall Features

Recall settings determine what the signal controller will do when there is no demand present on one or more approach lane groups. A recall setting can be "off" or "on" for each phase; when "on," there are several different recall options:

- Minimum Recall: When engaged, the minimum recall feature places a "call" on the designated phase, even if there is no demand present. If there is no demand anywhere in the intersection, this would force the signal to initiate a green on the designated phase for a duration of at least the minimum green. Common practice is to engage the minimum recall feature on major-street through movements.

- Maximum Recall: When engaged, the feature causes a continuous "call" on the designated phase, even if there is no demand present. It forces every designated green phase to extend to its maximum green. This feature is not often used but is appropriate if vehicle detection is out of service or if a fixed-time operation is desired for some period of time.
- Pedestrian Recall: When engaged, the controller places a continuous call for service on the designated phase. It forces the minimum pedestrian green to be implemented during every green phase. It is most often used where pedestrian actuation buttons are out of service or during time periods when large numbers of pedestrians are present.
- Soft Recall: When engaged, the controller places a "call" on the designated phase when there are no competing calls present. This is most commonly used to for major-street through phases during light demand periods, to ensure that the signal rests on the main through movements, particularly at noncoordinated signals.

The minimum recall is the most frequently used recall setting. If all phases were set to minimum recall when there was no demand present, the signal would cycle through its phases using minimum green times. If all recall features were "off" during such a period, the green would stay on the last phase to have received a call, and would not move until a call on another phase was received. In the common usage, the recall would be set only on the major-street through movement, guaranteeing that the green would reside on the major-street through phases in the absence of any demand.

22.5 Examples in Actuated Signal Design and Timing

Timing of an actuated signal is less definitive than for pretimed signals and calls for more judgment to be exercised by the engineer. In any given instance, it is possible that several different signal timings and designs would work acceptably. Some of the considerations involved are best illustrated by example.

Example 22-1: A Semi-Actuated Signal Timing

Figure 22.6 shows an intersection that will be signalized using a semi-actuated controller. For convenience, the demand volumes shown have already been converted to through vehicle units (tvu's). This conversion is the same as for pretimed signals (see Chapter 21).

Step 1: *Phasing* Because this is a semi-actuated signal, there are only two phases, as follows:

- Phase 1—All First Avenue movements (minor street)
- Phase 2—All Main Street movements (major street)

Step 2: *Minimum Green Time and Detector Location* For a semi-actuated signal, only the side-street phase is actuated and only side-street approaches have detectors. Point detectors will be used. For semi-actuated signals, the objective is generally to provide only the amount of green time necessary to clear side-street vehicles, with as little unused green time as possible. Therefore, the minimum green time for First Avenue should be as low as possible. Using a start-up lost time of 3.0 seconds, the minimum green time that could be allocated would be 5.0 seconds. If G_{min1} is set at 5.0 seconds, then the detector placement is determined by solving for d in Equation 22-1:

$$G_{\min} = 5.0 = 3.0 + 2.0\,Int\left(\frac{d}{25}\right)$$

$$Int\left(\frac{d}{25}\right) = \frac{5.0 - 3.0}{2.0} = 1.0$$

The detector would be placed anywhere between 0.1 and 25.0 feet from the STOP line. It must be placed such that no vehicle can enter the approach without traversing the detector.

Step 3: *Passage Time* For point detectors, the passage time is equal to be the maximum allowable headway (*MAH*). The recommended value is 3.0 seconds. This must be greater than the passage time from the detector to the STOP line, assuming the maximum setback of 25.0 feet, or:

$$PT_{\min} = \frac{d}{1.47 * S_{15}} = \frac{25}{1.47 * (25 - 5)} = 0.85\,s < 3.0\,s \quad OK$$

The 3.0-second unit extension is safe and will be implemented.

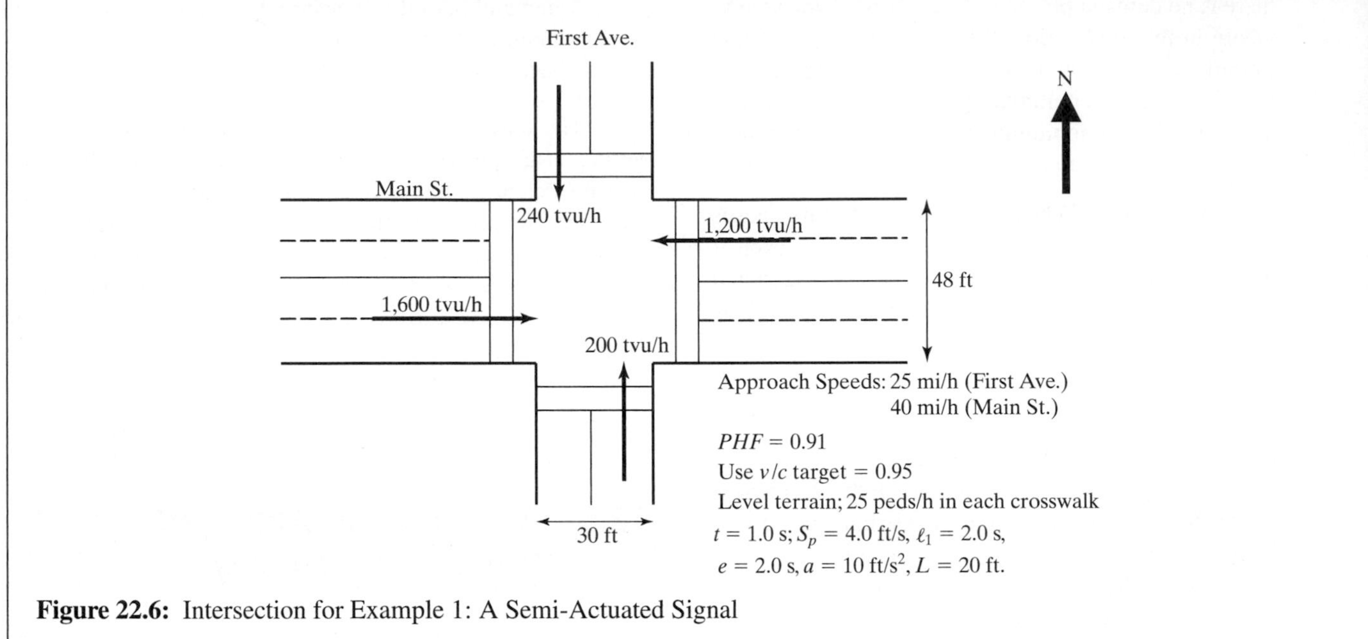

Figure 22.6: Intersection for Example 1: A Semi-Actuated Signal

Step 4: *Sum of Critical-Lane Volumes* All of the demand volumes of Figure 22.6 have already been converted into through vehicle equivalents (tvu's). The SB movement has a higher volume than the NB movement (both approaches have one lane). Thus the critical-lane volume for Phase 1 is 240 tvu/h. The EB volume of 1,600 tvu/h is critical for Phase 2 but is divided into two lanes. Thus the critical-lane volume for Phase 2 is 1,600/2 = 800 tvu/h. The sum of the critical-lane volumes, V_c, is 240 + 800 = 1,040 tvu/h.

Step 5: *Yellow and All-Red Times: Lost Time Per Cycle* To determine other signal timing parameters, an initial cycle length must be selected. This requires, however, that all lost times within the cycle be known, which requires that the yellow and all-red intervals be established. Yellow intervals for each phase are estimated using Equation 22-5; all-red intervals are estimated using Equation 22-6. Average approach speeds are given for Main Street and First Avenue. The 85th percentile speed may be estimated as 5 mi/h more than the average; the 15th percentile speed is estimated as 5 mi/h less than the average.

$$y = t + \frac{1.47\ S_{85}}{2a + 64.4\ (0.01\ G)}$$

$$y_1 = 1.0 + \frac{1.47(25 + 5)}{2 * 10 + 64.4\ (0.01 * 0)} = 3.2 \text{ s}$$

$$y_2 = 1.0 + \frac{1.47(40 + 5)}{2 * 10 + 64.4\ (0.01 * 0)} = 4.3 \text{ s}$$

$$ar = \frac{w + L}{1.47\ S_{15}}$$

$$ar_1 = \frac{48 + 20}{1.47(25 - 5)} = 2.3 \text{ s}$$

$$ar_2 = \frac{30 + 20}{1.47(40 - 5)} = 1.0 \text{ s}$$

With default values of 2.0 s used for both ℓ_1 and e, the lost time per cycle is equal to the sum of the yellow and all-red times in the cycle, or:

$$L = 3.2 + 2.3 + 4.3 + 1.0 = 10.8 \text{ s/cycle}$$

Step 6: *Maximum Green (Phase 1) and Minimum Green (Phase 2)* As a semi-actuated signal, the critical cycle is composed of the maximum green for the side street (First Avenue, Phase 1), the minimum green for the major street (Main Street, Phase 2), and the yellow and all-red intervals from each. The initial cycle length is estimated using Equation 22-8:

$$C_i = \frac{L}{1 - \left[\dfrac{V_c}{1{,}615 * PHF * (v/c)}\right]}$$

$$= \frac{10.8}{1 - \left[\dfrac{1{,}040}{1{,}615 * 0.92 * 0.95}\right]} = 41.1\ s$$

For a semi-actuated signal, this value does not have to be rounded. Green splits based on this cycle length are determined using Equation 22-9:

$$g_i = (C - L)*\left(\frac{V_{ci}}{V_c}\right)$$

$$g_1 = (41.1 - 10.8)*\left(\frac{240}{1{,}040}\right) = 7.0\ s$$

$$g_2 = (41.1 - 10.8)*\left(\frac{800}{1{,}040}\right) = 23.3\ s$$

Effective green times and actual green times are equal, given the default values of 2.0 seconds for both l_1 and e. Standard practice establishes the maximum green for the minor street and the minimum green for the major street as 1.50 times the above values, or:

$$G_{max1} = 1.50 * 7.0 = 10.5 \text{ s}$$

$$G_{min2} = 1.50 * 23.3 = 35.0 \text{ s}$$

The G_{max1} of 10.5 seconds compares favorably with the G_{min1} of 5.0 seconds established earlier, so no adjustment of this timing is necessary. The critical cycle length is the sum of $G_{max1} + G_{min2} + Y_1 + Y_2$, or:

$$C_c = 10.5 + 35.0 + 3.2 + 2.3 + 4.3 + 1.0$$
$$= 56.3 \text{ s}$$

Step 7: *Pedestrian Requirements* Pedestrians cross the minor street during Phase 2. Thus the pedestrian crossing requirement must be compared to the minimum green plus yellow and all-red provided in Phase 2. For the purposes of this computation, N_{peds}, which is the same for all crosswalks, will be based on the critical cycle length of 56.3 seconds.

Then:

$$N_{peds} = \frac{25}{(3{,}600/56.3)} = 0.39 \text{ peds/cycle}$$

$$G_{p2} = 3.2 + \frac{30}{4.0} + 0.27(0.39) = 10.8 \text{ s}$$

$$G_{p2} = 10.8 \leq G_{min2} + Y_2$$
$$= 35.0 + 4.3 + 1.0 = 40.3 \text{ s}$$

The minimum green provides more than enough time for safe crossings of the minor street during Phase 2. No pedestrian pushbutton is needed; pedestrian signals are optional.

Pedestrians cross the major street during Phase 1, which has a minimum green time of 5.0 seconds. Checking for pedestrian safety:

$$G_{p1} = 3.2 + \frac{48}{4.0} + 0.27(0.39) = 15.3 \text{ s}$$

$$G_{p1} = 15.3 \leq G_{min1} + Y_1$$
$$= 4.0 + 3.2 + 2.3 = 9.5 \text{ s}$$

Pedestrians *are not* safely accommodated by G_{min1}. Thus, for pedestrians crossing the major street, a pedestrian pushbutton must be provided, and pedestrian signals are mandatory. When pushed, the next green phase will provide a minimum green time of:

$$G_{min1,ped} = 15.3 - 3.2 - 2.3 = 9.8 \text{ s}$$

The pedestrian walk and clearance intervals would be as follows:

$$WALK_1 = 3.2 + 0.27(0.38) = 3.3 \text{ s}$$

$$\textit{Up-raised Hand}_{flashing1} = \frac{48}{4.0} = 12.0 \text{ s}$$

Example 22-2: A Variation on Example 22-1

What would change if the side-street critical-demand volume was only 85 veh/h, instead of the 240 veh/h of Example 22-1? None of the details of detector placement would change, nor would the length of the yellow and all-red phases. Thus the following values have already been determined:

- $G_{min} = 5.0$ s
- $Y_1 = 3.2 + 2.3 = 5.5$ s
- $Y_2 = 4.3 + 1.0 = 5.3$ s
- $L = 10.8$ s/cycle

The only significant change is in the critical lane volumes. For the minor street (First Avenue, Phase 1), the critical volume is now 85 veh/h. The major street (Main Street, Phase 2) critical-lane volume is unchanged: 1,600/2 = 800 veh/h. The sum of critical-lane volumes, therefore, is 800 + 85 = 885 veh/h. This will change the initial cycle length and the values of G_{max1} and G_{min2} that result from it:

$$C_i = \frac{10.8}{1 - \left[\frac{885}{1{,}615 * 0.92 * 0.95}\right]} = 29.0\ s$$

$$G_1 = (29.0 - 10.8) * \left(\frac{85}{885}\right) = 1.7\ s$$

$$G_2 = (29.0 - 10.8) * \left(\frac{85}{885}\right) = 16.5\ s$$

$$G_{max1} = 1.7 * 1.5 = 2.6\ s$$

$$G_{min2} = 16.5 * 1.5 = 24.8\ s$$

This timing is not reasonable because G_{max1} (2.6 s) is less than G_{min1} (5.0 s). An alternative way of establishing a reasonable G_{max1} must be found. The average number of expected vehicles per cycle arriving on the side street depends on the critical cycle length, which is not yet known. If a 60-second cycle length is assumed as a rough estimate, then 85/60 = 1.4 veh/cycle would be expected. If this is multiplied by 1.5, the result is 2.1 veh/cycle. This suggests that the maximum green time should accommodate two or three vehicles. The minimum green time of 5.0 seconds accommodates one vehicle. To guarantee that three vehicles could move through on a single green phase, G_{max1} would be increased by two unit extensions, or 5.0 + (2 * 3) = 11.0 seconds. A guarantee of two vehicles would require G_{max1} to be set at 5.0 + (1 * 3) = 8.0 seconds. Either choice is defendable, but 8.0 s will be used in the remainder of this illustration. The G_{min2} must now be increased to maintain the original balance of times (G_{max1} = 2.6 s; G_{min2} = 24.8 s). Then:

$$\frac{24.8}{2.6} = \frac{G_{min2}}{8.0}$$

$$G_{min2} = \frac{24.8 * 8.0}{2.6} = 76.3\ s$$

The critical cycle length now becomes:

$$C_c = 8.0 + 5.5 + 76.3 + 5.3 = 95.1\ s$$

The initial assumption of a 60-second cycle to determine the number of vehicles arriving in one cycle could be iterated, but given the approximate nature of these computations, that would not be necessary. Because the larger result would *increase* the number arriving per cycle, it might suggest that G_{max1} be set at 11.0 seconds rather than 8.0 seconds, but even this is a judgment call. The pedestrian safety checks could be redone to reflect small changes in N_{peds}, but this value would result in a change in the order of 0.10 and is also not necessary.

With a smaller side-street demand, the balance of green times is significantly changed, with the major street receiving a longer G_{min} and the side street receiving a smaller G_{max} than in the original solution.

Example 22-3: A Full-Actuated Signal

An isolated suburban intersection of two major arterials is to be signalized using a full actuated controller. Area detection is to be used, and there are no driveways or other potential entry points for vehicles within 300 feet of the STOP line on all approaches. The intersection is shown in Figure 22.7, and all volumes have already been converted to tvu's for convenience. Left-turn slots of 250 feet

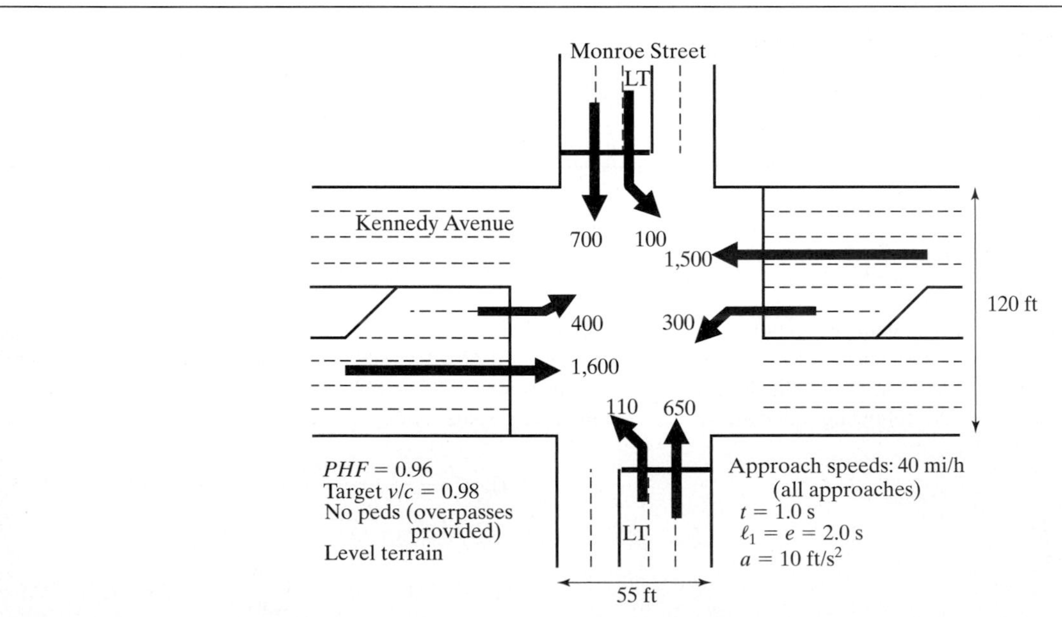

Figure 22.7: Intersection for Example 3: Full Actuated Control

in length are provided for each approach. The tvu conversions assume that a protected left-turn phase will be provided for all approaches. For all approaches and LT lanes, 50-feet presence detectors are provided. In each case, the front edge of the detector is located 2 feet from the STOP line.

Step 1: *Phasing* The problem statement indicates that protected left-turn phasing will be implemented on all approaches. Note that Kennedy Avenue has double left-turn lanes in each direction and that Monroe Street has a single left-turn lane in each direction.

At a heavily used intersection such as this, quad-eight phasing would be desirable. Each street would have an exclusive LT phase followed by a leading green in the direction of heavier LT flow and a TH/RT phase. As indicated in Chapter 21, such phasing provides much flexibility in that LT phasing is always optional and can be skipped in any cycle in which no LT demand is noted. The resulting signalization has a maximum of four phases in any given cycle and a minimum of two. It is treated as a *four-phase* signal because this option leads to the maximum lost times.

Quad-eight phasing involves overlaps (again, see Chapter 21) that would be taken into account if this were a pretimed signal. As an actuated signal, the worst-case cycle, however, would occur when there are no overlap periods. This would occur when the LT flow in opposing directions are equal. Thus, the signal timing will be considered as if this were a simple four-phase operation without overlaps. The controller, however, will allow one protected LT to be terminated before the opposing protected LT, creating a leading green phase. The four phases are:

- Phase 1—Protected LT for Kennedy Avenue
- Phase 2—TH/RT for Kennedy Avenue
- Phase 3—Protected LT for Monroe Street
- Phase 4—TH/RT for Monroe Street

Step 2: *Passage Times* As no gap reduction will be in use at this intersection, the maximum allowable headway, *MAH* is 3.0 seconds. This will be the same for all approaches, including left-turn phases. Then, using Equation 22-3:

$$PT = MAH - \frac{L_v + L_d}{1.47\, S_a}$$

$$PT = 3.0 - \frac{50 - 20}{1.47 * 40} = 2.5\, s$$

This value will be used for all approaches and LT lanes.

Step 3: *Minimum Green Times and Detector Placement* The detector design has been specified. The minimum green time for area detection is variable, based on the number of vehicles sensed within the detection area when the green is initiated. The value can vary from the time needed to service one waiting vehicle to the time needed to service *Int* (50/20) = 3.0 seconds. Using Equation 22-2, the range of minimum green times can be established for each approach. In this case, all values will be equal because the approach speeds are the same for all approaches and the detector location is common to every approach, including the LT lanes.

$$G_{min} = \ell_1 + 2n$$

$$G_{min,\, low} = 3.0 + (2 * 1) = 5.0\, s$$

$$G_{min,\, high} = 3.0 + (2 * 3) = 9.0\, s$$

Step 4: *Critical-Lane Volumes* Because the volumes given have already been converted to tvu's, critical-lane volumes for each phase are easily identified:

- Phase 1 (Kennedy Ave, LT) 400/2 = 200 tvu/h
- Phase 2 (Kennedy Ave, TH/RT) 1,600/4 = 400 tvu/h
- Phase 3 (Monroe St, LT) 110/1 = 110 tvu/h
- Phase 4 (Monroe St, TH/RT) 700/2 = 350 tvu/h
- V_c = 1,060 tvu/h

Step 5: *Yellow and All-Red Times: Lost Times* Yellow times are found using Equation 21-5; all-red times are found using Equation 20-5. With a 40-mi/h average approach speed for all movements, the S_{85} may be estimated as 40 + 5 = 45 mi/h, and the S_{15} may be estimated as 40 − 5 = 35 mi/h. Then:

$$y_{all} = 1.0 + \frac{1.47 * 45}{2 * 10 + 64.4(0.01 * 0)} = 4.3\, s$$

$$ar_{1,2} = \frac{55 + 20}{1.47 * 35} = 1.5\, s$$

$$ar_{3,4} = \frac{120 + 20}{1.47 * 35} = 2.7\, s$$

$$Y_{1,2} = 4.3 + 1.5 = 5.8\, s$$

$$Y_{3,4} = 4.3 + 2.7 = 7.0\, s$$

There are four phases in the worst-case cycle. Because the standard defaults are used for ℓ_1 and e, the total lost time is equal to the sum of the yellow and all-red intervals in the cycle:

$$L = 2 * 5.8 + 2 * 7.0 = 25.6\, s$$

Step 6: *Maximum Green Times and the Critical Cycle* The initial cycle length for determining maximum green times is:

$$C_i = \frac{25.6}{1 - \left[\frac{1{,}060}{1{,}615 * 0.96 * 0.98}\right]} = 84.8\, s$$

Green times are found as:

$$G_1 = (84.8 - 25.6) * \left(\frac{200}{1{,}060}\right) = 11.2\,s$$

$$G_2 = (84.8 - 25.6) * \left(\frac{400}{1{,}060}\right) = 22.3\,s$$

$$G_3 = (84.8 - 25.6) * \left(\frac{110}{1{,}060}\right) = 6.1\,s$$

$$G_4 = (84.8 - 25.6) * \left(\frac{350}{1{,}060}\right) = 19.5\,s$$

$$G_{\max 1} = 11.2 * 1.5 = 16.8\,s$$

$$G_{\max 2} = 22.3 * 1.5 = 33.5\,s$$

$$G_{\max 3} = 6.1 * 1.5 = 9.2\,s$$

$$G_{\max 4} = 19.5 * 1.5 = 29.3\,s$$

With area detection, the minimum green for all lane groups, including LT lanes, can be as high as 9.0 seconds. This limits the flexibility of LT Phase 3. Increasing the maximum greens beyond the computed values, however, will lead to an excessively long critical cycle length.

Thus it is recommended that the LT lanes use *point* detectors, placed so that the G_{min} for Phases 1 and 3 is a constant 5.0 seconds. The above G_{max} results will work in this scenario. It is, therefore, not recommended that any of these times be arbitrarily increased.

The critical cycle length becomes:

$$C_c = 16.8 + 5.8 + 33.5 + 5.8 + 9.2 + 7.0 + 29.3 + 7.0 = 114.4\text{ s}$$

Overpasses are provided for pedestrians, so no at-grade crossings are permitted, and no pedestrian checks are required for this signalization.

References

1. Pusey, R.S., and Butzer, G.L., "Traffic Control Signals," *Traffic Engineering Handbook*, 5th Edition, Institute of Transportation Engineers, Washington DC, 1999, pp. 453–528.
2. Koonce, P., et al., *Traffic Signal Timing Handbook*, Final Report, FHWA Contract No. DTFH61-98-C-00075, Kittelson and Associates Inc, Portland OR, June 2008.

Problems

Unless otherwise noted, use the following default values for each of the following actuated signal timing problems:

- Driver reaction time (t) = 1.0 s
- Vehicle deceleration rate = 10 ft/s^2
- Length of a vehicle = 20 ft
- Level terrain
- Low pedestrian activity at all locations (50 peds/h each cross walk)
- PHF = 0.90
- Target v/c ratio for actuated signals = 0.95
- Lane widths = 12 ft
- Crosswalk widths = 10 ft
- Pedestrian crossing speed = 4.0 ft/s
- All volumes in veh/h

If any other assumptions are necessary, specifically indicate them as part of the answer.

22-1. A semi-actuated signal is to be installed and timed for the location shown in Figure 22.8. Because of light side-street demand, a short minimum green of 6.0 seconds is desired. Point detectors will be used. For the conditions shown:

(a) How far from the STOP line should the side-street detectors be located?

(b) Recommend a passage time.

(c) Compute yellow and all-red times.

(d) Recommend a maximum side-street green and a minimum main street green.

(e) What is the critical cycle length?

(f) Are pedestrian signals and/or pushbuttons required for crossing the main street? The side street?

22-2. A full actuated controller must be retimed at the intersection shown in Figure 22.9. Detector locations are fixed from a previous installation and cannot be moved. For the conditions shown:

(a) Recommend a suitable phase plan for the signal.

(b) What minimum greens should be used?

(c) Recommend a passage time.

(d) Compute yellow and all-red times.

(e) Recommend maximum green times for each phase.

(f) What is the critical cycle length?

(g) Are pedestrian signals and/or pushbuttons needed for any phases?

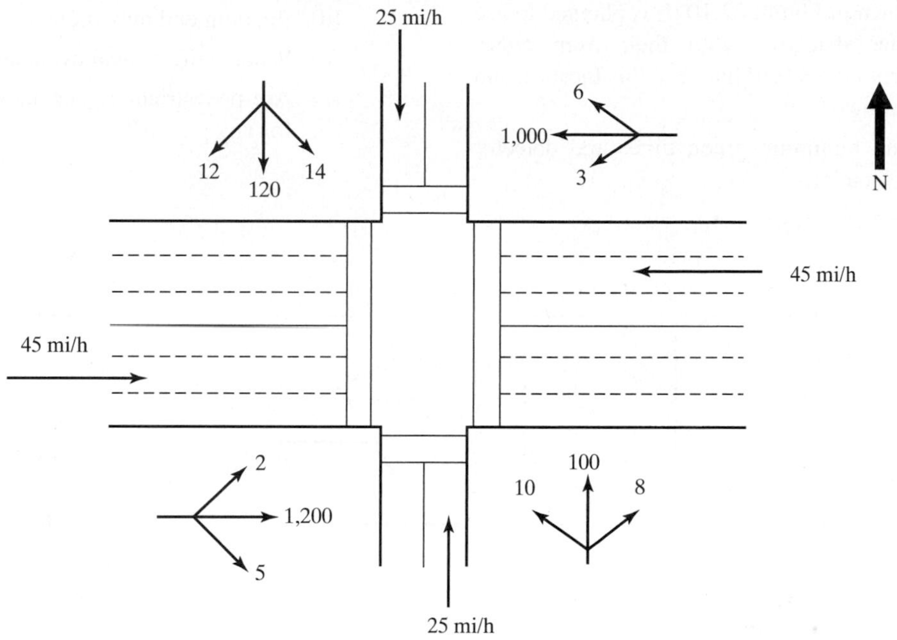

25 mi/h
12
120
14
6
1,000
3
N
45 mi/h
45 mi/h
2
1,200
5
100
10
8
25 mi/h

Figure 22.8: Intersection for Problem 1

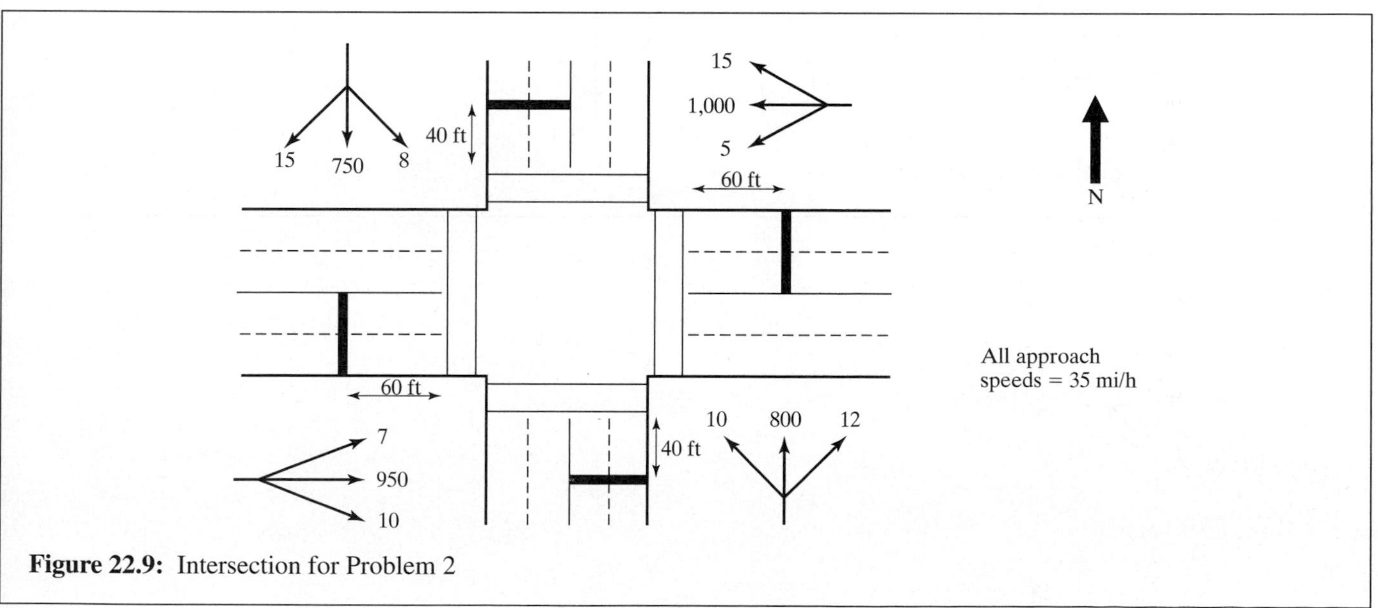

Figure 22.9: Intersection for Problem 2

22-3. A full actuated signal is to be installed at the major intersection shown in Figure 22.10. It is planned to use 60-foot presence detectors with their front edges located 2 feet from the STOP line. For this location and the conditions shown:

(a) Recommend minimum green times and detector placement locations.

(b) Recommend a passage time.

(c) Compute yellow and all-red times.

(d) Recommend maximum green times for each phase.

(e) What is the critical cycle length?

(f) Are pedestrian signals and/or pushbuttons needed for any phases?

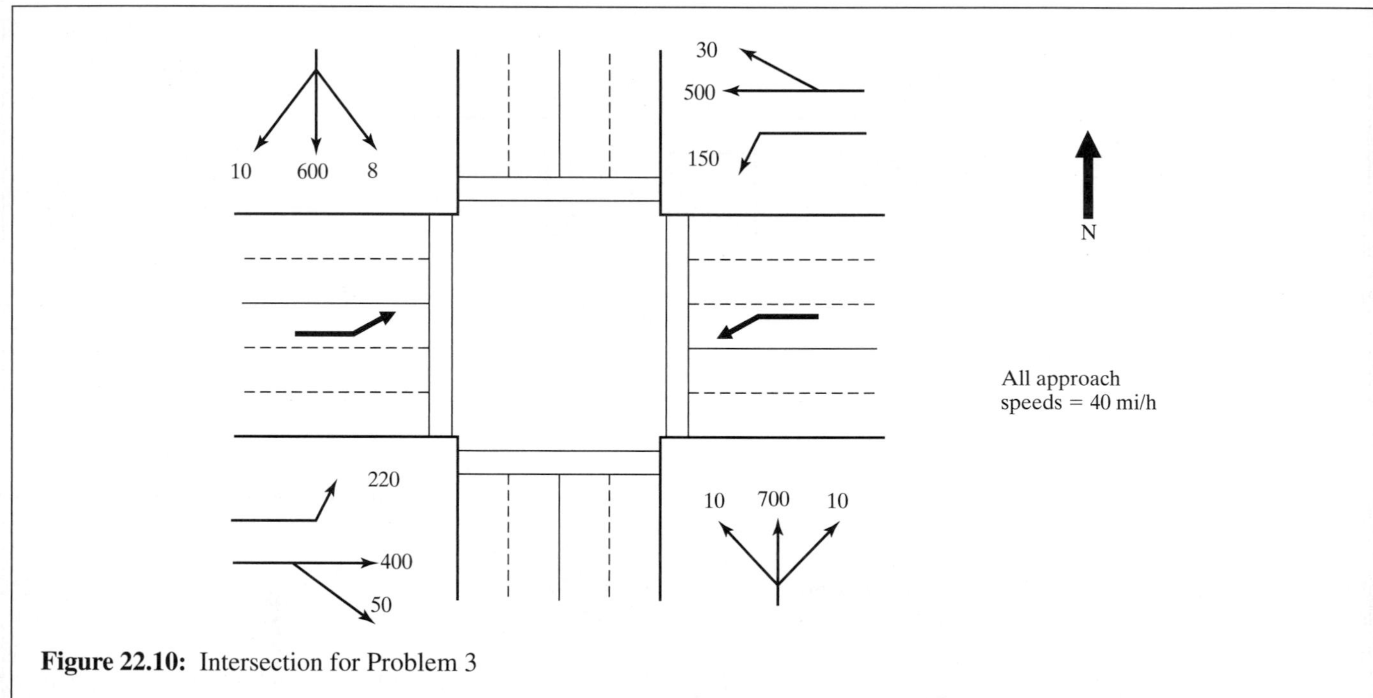

Figure 22.10: Intersection for Problem 3

CHAPTER 23

Critical Movement Analysis of Signalized Intersections

In Chapter 21, a detailed methodology for establishing optimal timing of isolated pretimed signals was presented. In this chapter, a relatively simple methodology for analysis of a signalized intersection is discussed; in Chapter 24, the complex model of the Highway Capacity Manual [*1*] is presented and illustrated.

Signal timing and analysis of signalized intersections are simply inverse applications of what should be the same model. In signal timing, demand flow rates are used with a known geometric design to establish the green times needed to accommodate the stated demands. In analysis, the signal timing, demand flows, and geometry are known, and the capacity (and/or level of service) of each approach or lane group is estimated. In essence, signal timing applications set green times to provide the necessary capacity. In analysis, the known signal timing is used to estimate the existing capacity.

Unfortunately, signal timing and signalized intersection analysis models are rarely the same and are often quite different from each other. The signalized intersection methodology of the 1965 Highway Capacity Manual [*2*], for example, used a methodology that was completely inconsistent with the signal timing methodologies of the day. This often presented the traffic engineer with an inexplicable conundrum: A signal timing designed to provide equivalent service to all major movements could produce wildly different levels of service for these same movements.

Today, the difference is in the level of complexity included in the models used. The HCM model for signalized intersection analysis is extremely complex and includes many iterative elements. It is generally recognized that the methodology cannot be implemented by hand. Software is the only practical means to implement the methodology. Signal timing must be more straightforward to be effective. Therefore, signal timing methodologies, although based on the same fundamental concepts as the HCM analysis model, incorporate many defaults and simplifications to allow for a relatively rapid and understandable approach to setting an appropriate timing pattern.

In 1980, the Transportation Research Board issued a set of preliminary capacity and level of service analysis methodologies (Interim Materials on Highway Capacity) in advance of the anticipated 1985 Highway Capacity Manual [*3*]. It included a relatively straightforward method for analysis of signalized intersections called "Critical

Movement Analysis" [*4*]. The methodology could be easily implemented by hand in a reasonable amount of time, and based level-of-service (LOS) determinations on the "sum of critical-lane flows," a concept discussed in Chapter 20. Although general delay estimates for each LOS for the intersection as a whole were provided by verbal description, there was no attempt to estimate average delays for individual movements or approaches.

The earliest work on critical-lane analysis of signalized intersections goes back to the original concepts of saturation headways and lost time developed by Bruce Greenshields. Its application to analysis originally appeared in 1961 in a paper by Capelle and Pinelle [*5*]. Messer and Fambro [*6*] produced a definitive methodology in 1977 that was adapted for inclusion in the Interim Materials. By the time the 1985 Highway Capacity Manual was published, however, Messer had built the basic procedure into a far more complex analysis methodology. Subsequent revisions have served to further complicate the approach.

As a result, there are traffic agencies that still (at this writing) use the methodology of the Interim Materials to analyze signalized intersections, including the California Department of Transportation. Because of the great complexity of the HCM approach, there are frequent calls for a return to a simpler approach similar to that of the Interim Materials.

23.1 The TRB Circular 212 Methodology

The methodology of TRB Circular 212 (Interim Materials on Highway Capacity) provides two levels of analysis: planning and operations/design. The first is done entirely in units of mixed veh/h with few adjustments. Average or "typical" conditions are assumed. The planning approach is as close to a "back of the envelope" approach as can be accomplished. The most complex signalized intersections could be analyzed in minutes. The operations and design model provides more detailed adjustments for heavy-vehicle presence, local bus presence, lane width, parking conditions, and other prevailing conditions. Even with these adjustments included, the most complex intersection could be analyzed by hand in less than 15 minutes.

Unlike the HCM, critical-lane or critical movement analysis applies all adjustment factors to the demand volume or flow rate. Thus volumes in veh/h are inflated by adjustment factors to reflect "through passenger car units (tpcus)," as is at least partially done for signal timing.

This chapter presents a design and operations methodology based on the concept of the Interim Materials. It has been updated, however, to reflect more recently calibrated adjustment factors, and for consistency with the signal timing methodology presented in Chapter 21. A delay prediction module has been added to the operations/design methodology to allow for better level of service estimates.

Although the operations/design model is not the same as that of the Interim Materials, the conceptual approach is the same: Only the numbers have been changed to reflect more modern conditions. This chapter does not present the planning level analysis of the Interin Materials, which should be consulted directly for a detailed description of that approach.

23.2 A Critical Movement Approach to Signalized Intersection Analysis

The methodology presented in this chapter is conceptually the same as the operations and design methodology of the 1980 Interim Procedure. It has, however, been updated to be more consistent with the signal timing methodology of this text and to reflect adjustment factors of the HCM, where it is appropriate to do so. A module for estimation of delay is added, based on theoretical models (as opposed to the more complex models of the HCM).

Unlike the HCM methodology, this critical-lane analysis approach makes all adjustments to demand volumes, as opposed to capacity or saturation flow rate. In the HCM methodology, saturation flow rates are *reduced* to reflect nonideal prevailing conditions. In this methodology, demand volumes are *increased* to reflect them.

23.2.1 Steps 1 Through 3

Steps 1 through 3 of the operations and design methodology are straightforward (1) Identify the lane geometry and use, (2) identify hourly demand volumes (in veh/h under prevailing conditions), and (3) specify the signal timing (a complete ring diagram is required).

In addition to the hourly demand volumes of Step 2, the prevailing conditions affecting each should be described in as much detail as possible. At a minimum, the following characteristics of each movement should be specified:

- Proportion of heavy vehicles
- Proportion of local buses
- Lane widths
- Approach grade
- Parking conditions on approach
- Pedestrian interference levels (either specify pedestrian volume in each crosswalk, or generally describe by level)

A "heavy vehicle" is any vehicle with more than four wheels on the ground, except for local buses, which are handled separately. In terms of defining buses, a "through bus" is one that is not making a stop to pick up or discharge passengers within the confines of the intersection (either near side or far side); a "local bus" is one that does make a stop to pick up or discharge passengers within the confines of the intersection (near side or far side). Thus a "through bus" is treated as a heavy vehicle.

The signal timing specified as part of Step 3 should include the green, yellow, and all red intervals. For the purpose of the operations and design approach, vehicles are assumed to move during the "effective green time," which for each phase may be computed as:

$$g = G + y + ar - \ell_1 - \ell_2 \qquad (23\text{-}1)$$

where: g = effective green time, s
G = actual green time, s
y = actual yellow time, s
ar = actual all-red time, s
ℓ_1 = start-up lost time, s
ℓ_2 = clearance lost time, s

For the vast majority of cases, the assumption that $(y + ar) = (\ell_1 + \ell_2)$ may be made. In this case, $g = G$. In cases where this assumption is known to be inaccurate, Equation 23-3 is used to estimate effective green time for each phase.

23.2.2 Step 4: Convert Demand Volumes to Equivalent Passenger-Car Volumes

Each movement demand volume is converted to a volume expressed in equivalent passenger-car units, using the following equation:

$$V_{pc} = (V P_{HV} E_{HV}) + (V P_{LB} E_{LB}) + V(1 - P_{HV} - P_{LB}) \qquad (23\text{-}2)$$

where: V_{pc} = volume in passenger-car equivalents, pc/h
V = volume in vehicles per hour under prevailing conditions, veh/h
P_{HV} = proportion of heavy vehicles (trucks, through buses, and RV's) in the movement, decimal
P_{LB} = proportion of local buses in the movement, decimal
E_{HV} = passenger-car equivalent of one heavy vehicle in movement
E_{LB} = passenger-car equivalent of one local bus in movement

The passenger-car equivalent for all heavy vehicles is 2.0. Passenger-car equivalents for local buses are more complex and shown in Table 23.1.

Local bus equivalents vary with a number of underlying conditions, including the percentage of local buses in the traffic stream, the total volume in the affected traffic stream, and

Table 23.1: Passenger-car Equivalents for Local Buses at Signalized Intersections

% Local Buses	No. of Lanes	Volume in Lane Group, veh/h				
		200	**400**	**600**	**800**	**≥1,000**
Buses Stopping in Travel Lane						
≤2%	1	1.8	2.7	3.5	4.4	5.3
	2	1.4	1.8	2.2	2.7	3.1
	≥3	1.3	1.5	1.8	2.1	2.4
5%	1	1.8	2.7	3.7	4.8	6.0
	2	1.4	1.8	2.3	2.7	3.2
	≥3	1.3	1.5	1.8	2.1	2.4
≥10%	1	1.9	2.9	4.2	5.7	7.7
	2	1.4	1.9	2.4	2.9	3.5
	≥3	1.3	1.6	1.9	2.2	2.5
Buses Stopping Off-Line or in Parking Lane						
≤2%	1	1.3	1.7	2.0	2.4	2.7
	2	1.2	1.3	1.5	2.7	1.8
	≥3	1.1	1.2	1.3	2.1	1.6
5%	1	1.3	1.7	2.1	2.4	2.8
	2	1.2	1.3	1.5	1.7	1.9
	≥3	1.1	1.2	1.3	1.5	1.6
≥10%	1	1.3	1.7	2.1	2.5	3.0
	2	1.2	1.3	1.5	1.7	1.9
	≥3	1.1	1.2	1.3	1.5	1.6

the number of lanes in the affected lane group. Exclusive LT lanes are generally not included as "affected lanes."

The type of bus stop also affects these equivalents. If the bus stops in a travel lane, the impact of the blockage is more severe than when the bus stops in a parking lane or in an offline bus bay. In general, local buses disrupt traffic primarily in the right-most lane of the affected lane group. Thus the overall impact of buses is reduced as the number of lanes in the lane group increases.

It is noted that the equivalents of Table 23.1 are generally based on the implied equivalents of the HCM signalized intersection analysis procedure. It is assumed that a bus stopping in a travel lane effectively blocks it for 14.4 seconds. A bus stopping in an offline position is assumed to block the right-most lane for 6.0 seconds as it enters and leaves the offline bus bay or parking lane.

23.2.3 Step 5: Convert Passenger-Car Equivalents to Through-Car Equivalents

After completing Step 4, all movement volumes have been converted to passenger-car equivalents. Now, left-turning vehicles and right-turning passenger-car volumes must be converted to equivalent through-car units.

Through-car equivalents for left-turning passenger cars are given in Table 23.2. These are the same equivalents used in timing traffic signals (Chapter 21). All protected left turns have an equivalent of 1.05, whereas the equivalents for permitted left turns depend on the opposing volume through which they turn and the number of lanes on the opposing through lane group.

Table 23.3 shows the through-car equivalents for right-turning passenger cars. These depend on the pedestrian volume in the conflicting crosswalk.

Table 23.2: Through-Car Equivalents for Left-Turning Vehicles, E_{LT}

Opposing Flow	Number of Opposing Lanes, N_o		
V_o (veh/h)	1	2	3
0	1.1	1.1	1.1
200	2.5	2.0	1.8
400	5.0	3.0	2.5
600	10.0*	5.0	4.0
800	13.0*	8.0	6.0
1,000	15.0*	13.0*	10.0*
≥1,200	15.0*	15.0*	15.0*

E_{LT} for all *protected* left turns = 1.05.
*The LT capacity is only available through "sneakers."

Table 23.3: Through-Car Equivalents for Right-Turning Vehicles, E_{RT}

Pedestrian Volume in Conflicting Crosswalk (peds/h)	Equivalent
None (0)	1.18
Low (50)	1.21
Moderate (200)	1.32
High (400)	1.52
Extreme (800)	2.14

It is appropriate to interpolate for intermediate values in Table 23.2, but this is not suggested for the values in Table 23.2, which are more approximate. The values in these tables are much simplified versions of adjustments in the HCM 2000. They eliminate the need to iterate solutions and greatly simplify the consideration of left and right turns.

Special Case: Compound LT Phases (Protected + Permitted and Permitted + Protected)

It is difficult to analyze compound phases in any detail in a simplified analysis approach such as CMA. In the HCM methodology (Chapter 24), protected and permitted portions of the phase are initially treated as two separate phases. This, in turn, requires that the left-turn demand flow be split between the two portions of the compound phase.

Two potential approaches for such an allocation are to (1) assume the full capacity of the initial portion of the phase is fully consumed with all remaining left turns assigned to the second portion of the compound phase, and (2) assume left-turn demand is assigned to the two portions of the phase in proportion to the capacities of two portions of the phase. Both are iterative because the initial assignment will influence the total tcu/h, which when allocated most often will violate the assumed initial distribution.

A simple methodology must find some more simplistic approach. The approach recommended here is that the entire left-turn demand flow be converted to tcu/h using a single equivalent that is an average of the equivalents for the protected portion of the phase and the permitted portion of the phase. It is further recommended that the average be approximately weighted using the relative green times in the compound phase. Thus the left-turn equivalent, E_{LT}, would be estimated as:

$$E_{LTC} = \frac{\left(E_{LTPT} * g_{LTPT}\right) + \left(E_{LTPM} * g_{LTPM}\right)}{g_{LTPT} + g_{LTPM}} \tag{23-3}$$

where: E_{LTC} = left-turn equivalent for compound LT phasing

E_{LTPT} = left-turn equivalent for protected portion of the compound LT phase

E_{LTPM} = left-turn equivalent for permitted portion of the compound LT phase

g_{LTPT} = effective green time for protected portion of the compound LT phase, s

g_{LTPM} = effective green time for permitted portion of the compound LT phase, s

This average equivalent is then used to convert the full left-turn flow to tcu/h. For compound phases, this approach is highly approximate but keeps the analysis in the range of a simple and straightforward methodology. Where more detailed analysis is required, the HCM methodology would be applied directly. Note that the effective green time for the *first* portion of the compound phase *includes* the yellow and all red intervals between the two portions.

23.2.4 Step 6: Converting Through-Car Equivalents Under Prevailing Conditions to Through-Car Equivalents Under Ideal Conditions

At this point in the methodology, all movement volumes have been converted to volumes in through-car equivalent units (tcu/h). They still represent, however, a number of prevailing conditions noted previously. Each movement volume must now be converted to a flow rate (for a 15-minute period) under equivalent ideal conditions. This conversion is computed as:

$$v = \frac{V_{tcu}}{PHF * f_w * f_g * f_p * f_{LU}} \quad (23\text{-}4)$$

where: v = movement flow rate under equivalent ideal conditions, tcu/h

V_{tcu} = movement volume, tcu/h, reflecting some prevailing conditions

PHF = peak-hour factor

f_w = adjustment factor for lane width (see Table 23.4)

f_g = adjustment factor for grade (see Table 23.5)

f_p = adjustment factor for parking (see Table 23.6)

f_{LU} = adjustment factor for lane utilization (see Table 23.7)

Table 23.4: Adjustment Factors for Lane Width

Lane Width (ft)	**9**	**10**	**11**	**12**	**13**	**14**	**15**
f_w	0.90	0.93	0.97	1.00	1.03	1.07	1.10

Table 23.5: Adjustment Factors for Grade

Grade (%)	**−6**	**−4**	**−2**	**0**	**2**	**4**	**6**
f_g	1.03	1.02	1.01	1.00	0.99	0.98	0.97

Table 23.6: Adjustment Factors for Parking Conditions

Parking Activity	**Number of Lanes in Lane Group**	**Adjustment Factor f_p**
High	1	0.825
	2	0.913
	3	0.942
Medium	1	0.850
	2	0.925
	3	0.950
Low	1	0.875
	2	0.938
	3	0.958
No parking lane	All	1.000

Table 23.7: Adjustment Factors for Lane Utilization

Lane Group Movements	**No. of Lanes in Lane Group**	**Adjustment Factor f_{LU}**
Through or shared	1	1.000
	2	0.952
	≥3	0.908
Exclusive left turn	1	1.000
	≥2	0.971
Exclusive right turn	1	1.000
	≥2	0.885

(*Source: Highway Capacity Manual,* 3rd Edition, Transportation Research Board, Washington DC, 2000, modified from Exhibit 10-23, p. 10–26.)

The peak-hour factor (*PHF*) converts a full-hour volume to a flow rate within the worst 15 minutes of the hour under consideration (usually one of the peak hours of the day). The adjustment factors account for nonstandard conditions of lane width (12-foot lanes are the base condition), grade (level grade is the base condition), parking activity adjacent to the lane group (no parking is the base condition), and lane use (uniform use of lanes is the base condition).

Lane widths less than 9 feet are not recommended; in fact, lane widths less than 11 feet are problematic where any appreciable number of heavy vehicles are present. A lane width of 16 feet or more will be used by drivers as *two* eight-foot lanes under the pressure of heavy traffic.

The parking adjustments of Table 23.6 focus on curb parking spaces within 250 feet of the stop line of the intersection approach. If there is no curb parking on the last 250 feet of the approach, there is no adjustment to be made. The extent of the adjustment is related to the number of parking maneuvers made (expressed as an equivalent hourly rate) and the number of lanes in the affected lane group. The number of movements (each parking maneuver and each unparking maneuver is separately counted) is generally characterized as "low" (0 to 7 mvts/h), "medium" (8 to 13 mvts/h), and "high" (more than 13 mvts/h) for the purposes of this methodology. Because parking maneuvers mainly affect the right-most lane (adjacent to the parking lane), the impact is dissipated by having multiple lanes for moving vehicles.

The lane utilization adjustment recognizes that where multiple lanes are present, vehicles will not divide themselves equally among all available lanes. The factor increases the equivalent ideal flow rate such that when the adjusted demand flow rate is divided equally into all available lanes of the lane group, the resulting average will represent the largest expected lane flow within the lane group.

23.2.5 Assigning Lane Flow Rates

Each movement volume is now expressed in through-car units under equivalent ideal conditions. It is now time to assign lane flows within each lane group. The following rules govern this assignment:

1. Where a separate LT lane (or lanes) exist, assign all LT tcus to this lane group. If more than one lane exists, divide the tcu/h equally among the lanes.
2. Where a separate RT lane (or lanes) exist, assign all RT tcus to this lane group. If more than one lane exists, divide the tcu/h equally among the lanes.
3. For all mixed lane groups (i.e., LT/TH/RT, LT/TH, or TH/RT) divide the total tcu/h equally among the lanes, *except that* (a) all LT tcus must be in the left-hand lane, and (b) all RT tcus must be in the right-hand lane.

23.2.6 Finding Critical-Lane Flow Rates for Each Signal Phase

Simple Case: No Overlapping or Compound Phases

The process of finding the critical-lane flow rate for each signal phase is relatively straightforward where there are no overlapping or compound phases. All lane flows moving during a given phase are examined. The critical-lane flow is the lane with the highest flow rate in adjusted tcu/h. When all critical-lane flows are identified, they are added to determine the sum of critical-lane flow rates. There will be one critical-lane flow rate for each phase of the signal.

Overlapping Phases

Finding the critical-lane flow rates for a signalization with overlapping phases must use the same technique described in the signal timing methodology of Chapter 21. A ring diagram of the signal timing must be drawn, with the appropriate critical-lane flow rates inserted into the proper parts of the diagram. An example is shown in Figure 23.1.

The ring diagram of Figure 23.1 depicts a standard NEMA phase sequence in which exclusive LT phases are followed by leading green phases. Each "box" of the ring

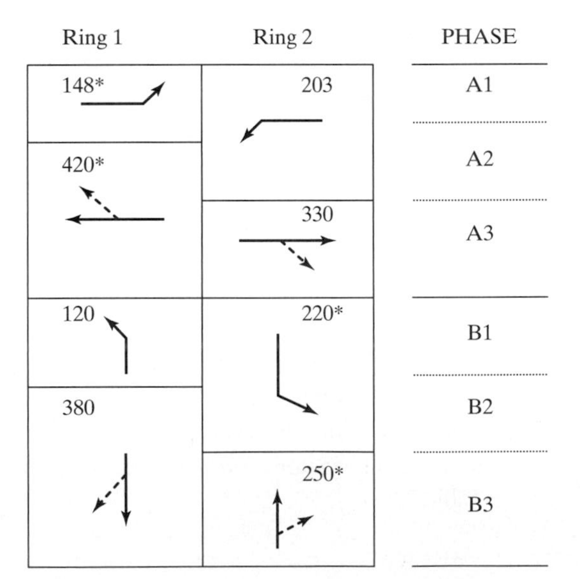

Figure 23.1: Sample Ring Diagram for Overlapping Phases

diagram shows the critical-lane flow rate (under equivalent ideal conditions) for the movements shown.

The "critical path" through the ring diagram is that path that results in the *highest* sum of critical-lane flow rates. It is found as follows:

- For Phases A1 to A3, the critical path can run down Ring 1 or Ring 2. The Ring 1 sum of critical-lane flow rates is 148 + 420 = 568 tcu/h. The Ring 2 sum is 203 + 330 = 533 tcu/h. The critical path runs through Ring 1 and the sum of the critical-lane flow rates for Phases A1 through A3 is **568 tcu/h.**
- For phases B1 to B3, the same choice is available. The Ring 1 sum of critical-lane flow rates is 120 + 380 = 500 tcu/h. The Ring 2 sum is 220 + 250 = 570 tcu/h. Ring 2 has the largest sum and therefore represents the critical path. The sum of critical-lane flow rates for Phase 2 is **570 tcu/h.**

The critical path through the signal timing is indicated by asterisks, and the sum of critical-lane flow rates for this example is 568 + 570 = 1,138 tcu/h.

Examination of the ring diagram identifies the critical path through the signal timing, and it also depicts the critical-lane flow rate for each phase and lane group in the intersection.

Compound Phases

When a protected + permitted or permitted + protected LT phase exists, it will be necessary to divide the demand between the two portions of the phase to determine the sum of critical-lane flows. The total LT flow in these cases has been converted to tcu/h using an average E_{LT} reflecting both portions of the phase. The converted flow can be divided in the proportion to the length of the effective green times for each portion of the phase. Note that a "yellow" transition between a leading protected phase to a permitted phase would be counted as effective green in this distribution.

23.2.7 Capacity and *v/c* Ratios

At this point in the analysis, the capacity of each critical lane must be estimated and compared to the demand flow rate using the *v/c* ratio.

Because all demand flow rates are expressed in tcu/h under equivalent ideal conditions, capacities must be stated in the same units. The recommended ideal saturation flow rate used to make these estimates is 1,900 tcu/h. The capacity of a critical lane within a lane group is:

$$c_i = 1900\left({}^{g_i}\!/_{C}\right) \tag{23-5}$$

where: c_i = capacity of critical lane in lane group i, tcu/h/ln
g_i = effective green time for lane group i, s
C = cycle length, s

In the case of *compound phasing,* the same equation is used, employing the effective green time for the entire compound phase (including the yellow and all red between the two portions of the phase). This is possible because the demand flow rates have been adjusted to reflect the different values of E_{LT} that apply in each portion of the phase. Capacity is stated in terms of tcu/h under ideal conditions.

The maximum sum of critical-lane flow rates that can be accommodated by the signalized intersection as a whole is:

$$c_{SUM} = 1900\left(\frac{\sum_{i=1-n} g_i}{C}\right) \tag{23-6}$$

where: c_{SUM} = maximum sum of critical-lane flow rates, tcu/h
g_i = effective green time for critical phase i, s
n = number of phases
C = cycle length, s

Once capacities for all critical lanes are determined, a *v/c* ratio can be computed for each:

$$X_i = {}^{v_i}\!/_{c_i} \tag{23-7}$$

where: X_i = *v/c* ratio for critical lane i
v_i = demand flow rate in critical lane i, tcu/h
c_i = capacity of critical lane i, tcu/h

In the case of *compound phasing,* a single value of X is computed based on the sum of the component phase demand flows, v, and the capacity of the compound phase, taken as a whole.

A "critical v/c ratio" for the entire intersection may also be computed as:

$$X_c = \sum_{i=1-n} v_{ci} \Big/ c_{SUM} \tag{23-8}$$

where: X_i = critical *v/c* ratio for intersection
v_{ci} = critical-lane flow rate i (in critical signal path), tcu/h
c_{SUM} = sum of critical-lane flow rates, tcu/h

Note that the *v/c* ratios produced in this step are analogous to those produced by the HCM 2000 methodology. The only difference is that several critical computations (notably the impact of turning movements) have been greatly simplified to avoid multiple iterations and to allow for a manual solution if desired.

At this point, each critical lane has been identified, and the sufficiency of the capacity provided to handle it (*v/c* ratio) has been assessed. A *v/c* ratio in excess of 1.0 signifies that the capacity is insufficient to accommodate the stated demand. A *v/c* ratio of 1.00 or less indicates that the capacity provided is sufficient to accommodate the demand. Although these computations are all based on a single *critical lane* within the intersection, when the *v/c* ratio is 1.00 or less, this means that all other lanes moving during that phase also have sufficient capacity.

The critical *v/c* ratio, X_c, is an important variable. If X_c is less than 1.00, it signifies that the existing signal cycle length, phase sequence, and the physical configuration of the intersection and its approaches are sufficient to handle all of the stated demand flows. If X_c is more than 1.00, the situation cannot be rectified without doing one or more of the following:

- Increasing the cycle length,
- Making the phase sequence more efficient (generally means adding protected or compound left-turn phasing), and/or
- Adding lanes to critical approaches and/or changing lane use restrictions for greater efficiency.

It is possible, however, that X_c is less than 1.00, but that some individual lane groups have X_i values that exceed 1.00. In this case, the difficulty can be resolved by reallocating green time within the existing phase sequence, cycle length, and geometric design. To do this, the methodology for signal timing found in Chapter 21 can be used, or (alternatively) the methodology for readjusting signal timing found in Chapter 24. The latter uses values of X_i directly, as opposed to critical-lane flow rates.

23.2.8 Delay and Level of Service

The planning approach estimated LOS on the basis of the sum of critical-lane volumes. Although this is sufficient for the most approximate of analyses, it contains a basic flaw. The presumption is that lower sums of critical-lane volumes produce higher quality operations. Although not exactly the same, use of the sum of critical-lane volumes is analogous to using *v/c* ratios (X_i) to determine level of service, and it implies that lower *v/c* ratios yield better operations.

The difficulty is that for signal timing, this is essentially not correct. In the analysis of uninterrupted flow facilities, lower *v/c* ratios lead directly to lower densities and higher speeds (i.e., higher quality operations). The critical difference between interrupted and uninterrupted flow, however, is in the determinants of capacity. For uninterrupted flow, capacity is based solely on the number of lanes present and the specifics of geometry (for a given traffic composition). For interrupted flow, it is based on the number of lanes present, the specifics of geometry, *and* the timing of the signal. In uninterrupted flow, we strive for low *v/c* ratios to allow for growth in traffic over time and to ensure that the road we build has sufficient space (lanes) to serve traffic for 20 to 30 years. When the time element is considered, however, the bottom line is a constant: There are 3,600, and only 3,600 seconds in an hour. There will never be more; there will never be less. Thus, although it is logical to provide excess capacity in terms of space, it is quite illogical to provide significant amounts of it in a signal timing.

If delay is used as the principal measure of LOS at intersections, very low *v/c* ratios often produce relatively high delays because vehicles are waiting on a RED signal while there are no vehicles present to use the GREEN on the other street. Unused green time is a *cause* of delay at a signalized intersection, not a cure for it. In general, *v/c* ratios at signalized intersections will be relatively high when optimal delay is achieved—often in the range of 0.80 – 0.95. Large amounts of unused green time occur when *v/c* ratios are low, and they generally indicate that the cycle length and green times are too long. Cycle lengths that are too long increase delay. On the other side of the coin, cycle lengths that are too short, pushing the v/c ratio close to 1.00, also increase delay—this time because of individual cycle failures that are likely to occur.

This methodology suggests that levels of service be directly related to estimated delay, just as in the more complex HCM methodology. It requires, therefore, that delays be estimated for each lane group in the intersection. To avoid the complexity of the HCM methodology, a more simplistic approach is taken using theoretical equations (see Chapter 20), with a simplified adjustment for progression quality.

The delay for a lane group is computed as:

$$d_i = d_{1i} * PF + d_{2i} \tag{23-9}$$

where: d_i = approach delay for lane group i, s/veh
d_{1i} = uniform delay for lane group i, s/veh
d_{2i} = overflow plus random delay for lane group i, s/veh
PF = progression adjustment factor

This equation is analogous to that used in HCM 2000, except for the following:

- It predicts approach delay, not control delay. The difference is slight. Approach delay is the sum of stopped delay plus acceleration and deceleration delay. Control delay is time-in-queue delay plus acceleration and deceleration delay. The difference is the time spent in queue while moving.

- It assumes there is no unserved queue present at the beginning of the analysis period. The HCM method adds a third element of delay when such queues are present.
- The progression factor recommended in this methodology is a much simplified version of those factors than used in the HCM.
- The equation for uniform delay is the same as in the HCM 2000. The equation for overflow delay is based on Akcelik's theoretical approach [7] as modified in the HCM 2000. Some simplifications of terms are introduced.

The uniform delay term, d_1, is computed using Webster's original delay equation:

$$d_{1i} = \frac{0.5\, C\left[1 - \left({}^{g_i}/_C\right)\right]^2}{1 - \left[\min\left(1, X_i\right) * \left({}^{g_i}/_C\right)\right]} \tag{23-10}$$

where all terms are as previously defined. The "*min*" function is used to emphasize that for uniform delay, the value of X_i cannot be higher than 1.00.

The overflow delay term is computed as follows:

$$d_{2i} = 225\left[\left(X_i - 1\right) + \sqrt{\left(X_i - 1\right)^2 + \frac{16\, X_i}{c_i N_i}}\right] \tag{23-11}$$

where N_i is the number of lanes in lane group *i*. This is essentially the HCM formula for the combination of overflow and random delay. Several things have been modified for this methodology:

- The analysis period is automatically set to 15 minutes (0.25 hour).
- Capacities c_i are computed per lane in this methodology. They are multiplied by N_i to provide a total lane group capacity, as needed for the HCM equation.
- From the HCM equation, adjustment factor k is set to 0.50, and adjustment factor I to 1.00.

Recommended progression factors (*PF*) are shown in Table 23.8. These are taken from the HCM 2000 but are very much simplified for this methodology. Instead of six arrival types specified in the HCM, this methodology uses only three terms:

- *Good progression:* Platoons arrive within the green phase, and the majority of vehicles move through the green without stopping.

Table 23.8: Progression Adjustment Factors for Lane Group Delay

Quality of Progression	Progression Factor (*PF*)	
	Pretimed Signals	**Actuated Signals**
Good	0.70	0.60
Random	1.00	0.85
Poor	1.25	1.06

- *Random arrivals:* The signal is isolated, or uncoordinated with adjacent signals; the proportion of arrivals on red and green approximates the proportions of red and green in the cycle.
- *Poor progression:* Platoons arrive when the signal is red, and/or the majority of vehicles must stop before proceeding through the intersection.

These definitions are somewhat imprecise, and they rely on the ability of the traffic engineer to make appropriate judgments on observing the arrival pattern or studying the time-space diagram(s) for a future situation.

The HCM varies this factor with g/C. To simplify for this methodology, values for a g/C of approximately 0.50 are used. For actuated signals, an additional adjustment if 0.85 is also included to roughly estimate the greater efficiency of these in allocating green time.

Once lane-group delays have been estimated, Table 23.9 is used to determine the level of service. A separate level of service is assigned to each lane group in the analysis. Where there is more than one lane group on an approach (such as in the case of an exclusive LT lane group or exclusive RT lane group), delays may be aggregated to obtain an average for the entire approach and a LOS assigned to the approach. Aggregation of approaches to determine an overall intersection delay and level of service is

Table 23.9: Levels of Service for Signalized Intersection Lane Groups and Approaches

Level of Service	Delay (s/veh)
A	≤10
B	>10–20
C	>20–35
D	>35–55
E	>55–80
F	>80

(*Source:* Used with permission of Transportation Research Board, National Research Council, *Highway Capacity Manual,* 4th Edition, Washington DC, 2000, Exhibit 16-2, p. 16-2.)

not recommended because this could easily mask serious problems on a particular approach or lane group.

Equation 23-12 is used to compute an average delay for an approach with more than one lane group.

$$d_{app} = \frac{\sum v_i d_i}{\sum v_i} \tag{23-12}$$

where: d_{app} = approach delay, s/veh

d_i = delay for lane group i, s/veh

v_i = demand flow for lane group i, pc/h

23.2.9 A Worksheet for Critical Movement Analysis

Although the critical movement analysis method recommended in this section is far less complex than the HCM it is relatively detailed and includes many steps. Because of this a worksheet is provided to help guide computations. It is shown in Figure 23.2. Instructions for the use of this three-page worksheet are given here:

1. Enter all identifying information concerning the location and time period of analysis in the top portion of page 1.
2. In the space provided (A), sketch a diagram of the intersection with demand volumes shown. Indicate (in parentheses) the percentage of heavy vehicles and percentage of local buses in each volume shown.
3. In the space provided (B), sketch a ring diagram of the signal timing, with complete timing details shown.
4. In the space provided (C), enter pedestrian flows for each crosswalk in peds/h, or generally characterize pedestrian flows as none, low, moderate, high, or extreme.
5. In the space provided (C), for each approach, enter the lane width (feet) and grade (%). Characterize parking activity as high, medium, low, or no parking lane.
6. Part D of the spreadsheet is on page 2 and consists of converting demands in veh/h to pc/h expressed as a flow rate under equivalent ideal conditions:
   a. For each approach and movement, enter the *number* of passenger cars (PC), heavy vehicles (HV), and local buses (LB). In the next line, enter the equivalents for PC, HV, and LB. All PCs have an equivalent of 1.000 (already shown on the worksheet), and all HVs have an equivalent of 2.000 (already shown on the worksheet). The equivalent for LBs is obtained from Table 23.5.
   b. Multiply all demand volumes in veh/h by the appropriate equivalent to determine the volume in pc/h. From this point on, only the "Total" column for each movement is pursued.
   c. For each movement, enter the left-turn equivalent from Table 23.6 or the right-turn equivalent from Table 23.7. The equivalent for through vehicles is always 1.000. Where a compound left-turn phase exists, use Part (E) of the worksheet to estimate the weighted average equivalent for use.
   d. For each movement, multiply the volume in pc/h by the appropriate equivalent to obtain the volume in tcu/h.
   e. For each movement, enter the following adjustment factors: lane width (Table 23.4), grade (Table 23.5), parking (Table 23.6), lane utilization (Table 23.7), and the peak-hour factor (*PHF*). Divide the volume in tcu/h by all of these factors to estimate the demand flow in ideal tcu/h.
7. In the space provided (F), sketch the intersection, and assign lane flow rates as appropriate to each lane shown.
8. In the space provided (G) sketch the signal ring diagram, assigning the appropriate critical-lane flow rate to each cell of the diagram. Find the critical path through the ring diagram and determine the sum of critical-lane flows.
9. Go to page 3 of the worksheet (H). The worksheet allows for up to 12 lane groups (one for each movement). In most cases, there will be fewer because many movements will share lanes. Where left turns and/or right turns share lanes with through movements, draw an arrow in the LT and/or RT cells pointing to the through lanes, where the appropriate critical-lane flow rate will be shown. Where separate LT and/or RT lanes exist, show the critical flow rate for each separately in the appropriate column.
10. Compute and enter the maximum sum of critical-lane flow rates and the critical *v/c* ratio as shown (H).
11. For each lane group, compute delay (I) as shown, and determine the level of service for each lane group (Table 23.9). Where desired, aggregate multiple lane groups to obtain an overall delay and LOS for the approach.

Worksheet for CMA: Operations and Design Method Pg 1

LOCATION: ______________________

TIME PERIOD OF ANALYSIS: __________ **TO** __________

ANALYST: ______________________ **DATE:** __________

(A) Diagram of Site and Demand Volumes (Show % HV and % LB)

(B) Ring Diagram with Signal Timing

(C) Pedestrian Flows Conflicting With:

EBRT	______	WBRT	______
NBRT	______	SBRT	______

Other Characteristics:

	Pkg Act	Ln Width	Grade
EB			
WB			
NB			
SB			

(D) Conversion of Volumes to Equivalent Ideal Flow Rates in tcu's/h (See Pg 2 of Worksheet)

(E) Compound Phase LT Equivalents

$$E_{LTC} = \frac{(E_{LTPT} * g_{LTPT}) + (E_{LTPM} * g_{LTPM})}{g_{LTPT} + g_{LTPM}}$$

COMPOUND PHASE LT EQUIVALENTS

EB	WB	NB	SB

(F) Diagram with Lane Flows

(G) Ring Diagram with Critical Lane Flows

Sum of Critical Lane Flows: __________

Figure 23.2: Worksheet for Critical Movement Analysis—Operations and Design—Page 1

Worksheet for CMA: Operations and Design Method Pg 2

	Item	LEFT				THROUGH				RIGHT			
EB		**PC**	**HV**	**LB**	**TOTAL**	**PC**	**HV**	**LB**	**TOTAL**	**PC**	**HV**	**LB**	**TOTAL**
	veh/h												
	equiv (Tab.23.1)	1.000	2.000		×	1.000	2.000		×	1.000	2.000		×
	pc/h												
	E_{LT}/E_{RT} (Tab. 23.2/3)	×	×	×		×	×	×		×	×	×	
	tcu/h	×	×	×		×	×	×		×	×	×	
	f_w (Tab. 23.4)	×	×	×		×	×	×		×	×	×	
	f_g (Tab. 23.5)	×	×	×		×	×	×		×	×	×	
	f_p (Tab. 23.6)	×	×	×		×	×	×		×	×	×	
	f_{LU} (23.7)	×	×	×		×	×	×		×	×	×	
	PHF												
	ideal tcu/h	×	×	×		×	×	×		×	×	×	
WB		**PC**	**HV**	**LB**	**TOTAL**	**PC**	**HV**	**LB**	**TOTAL**	**PC**	**HV**	**LB**	**TOTAL**
	veh/h												
	equiv (E)	1.000	2.000		×	1.000	2.000		×	1.000	2.000		×
	pc/h												
	E_{LT}/E_{RT}	×	×	×		×	×	×		×	×	×	
	tcu/h	×	×	×		×	×	×		×	×	×	
	f_w	×	×	×		×	×	×		×	×	×	
	f_g	×	×	×		×	×	×		×	×	×	
	f_p	×	×	×		×	×	×		×	×	×	
	f_{LU}	×	×	×		×	×	×		×	×	×	
	PHF												
	ideal tcu/h	×	×	×		×	×	×		×	×	×	
NB		**PC**	**HV**	**LB**	**TOTAL**	**PC**	**HV**	**LB**	**TOTAL**	**PC**	**HV**	**LB**	**TOTAL**
	veh/h												
	equiv (E)	1.000	2.000		×	1.000	2.000		×	1.000	2.000		×
	pc/h												
	E_{LT}/E_{RT}	×	×	×		×	×	×		×	×	×	
	tcu/h	×	×	×		×	×	×		×	×	×	
	f_w	×	×	×		×	×	×		×	×	×	
	f_g	×	×	×		×	×	×		×	×	×	
	f_p	×	×	×		×	×	×		×	×	×	
	f_{LU}	×	×	×		×	×	×		×	×	×	
	PHF												
	ideal tcu/h	×	×	×		×	×	×		×	×	×	
SB		**PC**	**HV**	**LB**	**TOTAL**	**PC**	**HV**	**LB**	**TOTAL**	**PC**	**HV**	**LB**	**TOTAL**
	veh/h												
	equiv (E)	1.000	2.000		×	1.000	2.000		×	1.000	2.000		×
	pc/h						×	×					
	E_{LT}/E_{RT}	×	×	×		×	×	×		×	×	×	
	tcu/h	×	×	×		×	×	×		×	×	×	
	f_w	×	×	×		×	×	×		×	×	×	
	f_g	×	×	×		×	×	×		×	×	×	
	f_p	×	×	×		×	×	×		×	×	×	
	f_{LU}	×	×	×		×	×	×		×	×	×	
	PHF												
	ideal tcu/h	×	×	×		×	×	×		×	×	×	

Figure 23.2: *(Continued)* Worksheet for Critical Movement Analysis—Operations and Design—Page 2

Worksheet for CMA: Operations and Design Method

Pg 3

(H) Critical Lane Computations

LANE GROUP	EBLT	EB	EBRT	WBLT	WB	WBRT	NBLT	NB	NBRT	SBLT	SB	SBRT
Critical Lane Flow, v_i												
Capacity of Critical Lane, c_i												
v/c Ratio of Critical Lane, X_i												

Maximum Sum of Critical Lane Flows:

$$c_{SUM} = 1900\left(\frac{\sum_{i=1-n} g_i}{C}\right) \quad \_\_\_\_\_\_\_\_$$

Critical v/c Ratio:

$$X_c = \sum_{i=1-n} v_{ci} \Big/ c_{SUM} \quad \_\_\_\_\_\_\_\_$$

(I) Delay and Level of Service Computations

LANE GROUP	EBLT	EB	EBRT	WBLT	WB	WBRT	NBLT	NB	NBRT	SBLT	SB	SBRT
Input Variables:												
Cycle Length, C												
Effective Green, g												
v/c Ratio, X												
Number of Lanes, N												
Capacity, c												
Delay Computations:												
Uniform Delay, d_i												
Prog. Factor, PF (Tab. 23.12)												
Overflow Delay, d_2												
Total Delay, d												
Level of Service (Tab. 23.13)												
Approach Delay												
Approach LOS (Tab. 23.13)												

DELAY EQUATIONS:

$$d_i = d_{1i} * PF + d_{2i}$$

$$d_{1i} = \frac{Q5C\left[1 - \left(g_i/C\right)\right]^2}{1 - \left[min(1, X_i) * \left(g_i/C\right)\right]}$$

$$d_{2i} = 225\left[(X_i - 1) + \sqrt{(X_i - 1)^2 + \frac{16\,X_i}{c_i N_i}}\right]$$

Figure 23.2: *(Continued)* Worksheet for Critical Movement Analysis—Operations and Design—Page 3

23.2.10 Summary

The critical movement analysis (CMA) methodology recommended in this section provides the same level of results as does the HCM. Its major simplifications include (1) greatly simplifying the determination of left- and right-turn adjustments, (2) greatly simplifying the approach to analysis of compound (protected + permitted or permitted + protected phasing), and (3) greatly simplifying delay computations. Another significant difference between this method and the HCM is that many of the adjustments are applied to the volume side of the equation, not the saturation flow rate (and therefore, capacity) side of the equation. This, however, is a difference of form, not substance.

These simplifications will mean that the CMA approach is less sensitive to the details of these portions of the analysis. Many of the complex additions to the HCM methodology have not been tested against field data, so although additional sensitivity is provided, no clear evidence indicates that the results are more accurate than a simpler approach. It is also true that the recommended approach here was not tested against field data either. Often, however, analysis of signalized intersections is done to test the impacts of various proposed schemes and approaches. The results of interest are the changes that occur in delay (and therefore LOS) and in v/c ratios. It is believed that the recommended approach is sufficiently detailed to reflect the vast majority of signal timing designs and physical designs that might be compared using it.

23.3 Sample Problems Using Critical Movement Analysis

23.3.1 Sample Problem 1: A Relatively Simple Problem

Intersection Description

Figure 23.3 shows a relatively simple intersection that will be analyzed using the recommended CMA methodology. It is noted that this same problem is repeated in Chapter 24, using the HCM methodology, so that the results may be compared. Figure 23.4 shows the worksheet with computational results summarized.

Figure 23.3 shows the intersection of a two-way arterial (EW) with a one-way arterial (NB). The two-way arterial has two lanes in each direction, with a left-turn lane in the WB approach. The one-way street has three lanes. There is no parking on any of the approaches, but the EW arterial has a bus stop serving 20 buses per hour.

The information in Figure 23.3 is more detailed than would normally be necessary for a CMA solution. It is given so that the more detailed analysis using the HCM methodology (Chapter 24) can also be applied, and the results compared. Key information for this solution is the % heavy vehicles, local bus volumes, pedestrian flows, lane widths, grades, and the "arrival type." For this methodology, the arrival type cited is equivalent to a "random" progression quality.

The signalization is typical for a three-leg intersection with only one opposed left turn: a leading green for the WB approach, followed by a EW through/right turn phase, and a phase for the NB approach.

Placing Input Data on Worksheet

Part (A) through (C) of page 1 of the worksheet provides space to inscribe the input information onto the worksheet. Note that the intersection diagram is simplified because a more detailing drawing is available in Figure 23.3.

Lane Group Decisions

Before starting, it is always beneficial to establish the "lane groups" to be analyzed. In this case, identifying lane groups is relatively easy. First, there are only three approaches. There is no SB approach on the one-way street. All vehicles flow from shared lanes on the WB and NB approaches, so each approach constitutes a single lane group. Because an exclusive LT lane exists on the EB approach, it will be treated as a separate lane group, with through and right-turning vehicles sharing the two lanes of the primary EB lane group. At the end of the analysis, a weighted average of delays on the two EB lane groups can be computed to characterize the entire EB approach.

Entering Demand Volumes (veh/h)

Demand volumes are entered on the top line of Part (D), page 2, of the worksheet for each movement. Note that volume is entered separately for the three classes of vehicles for each movement on each approach. This is done to simplify the conversion computations that follow.

For example, the EB through movement is 1,100 veh/h. It contains 10% heavy vehicles and 20 buses per hour. Thus, there are $1{,}100 * 0.10 = 110$ heavy veh/h, 20 local buses/h (given), and $1{,}100 - 110 - 20 = 970$ pc/h.

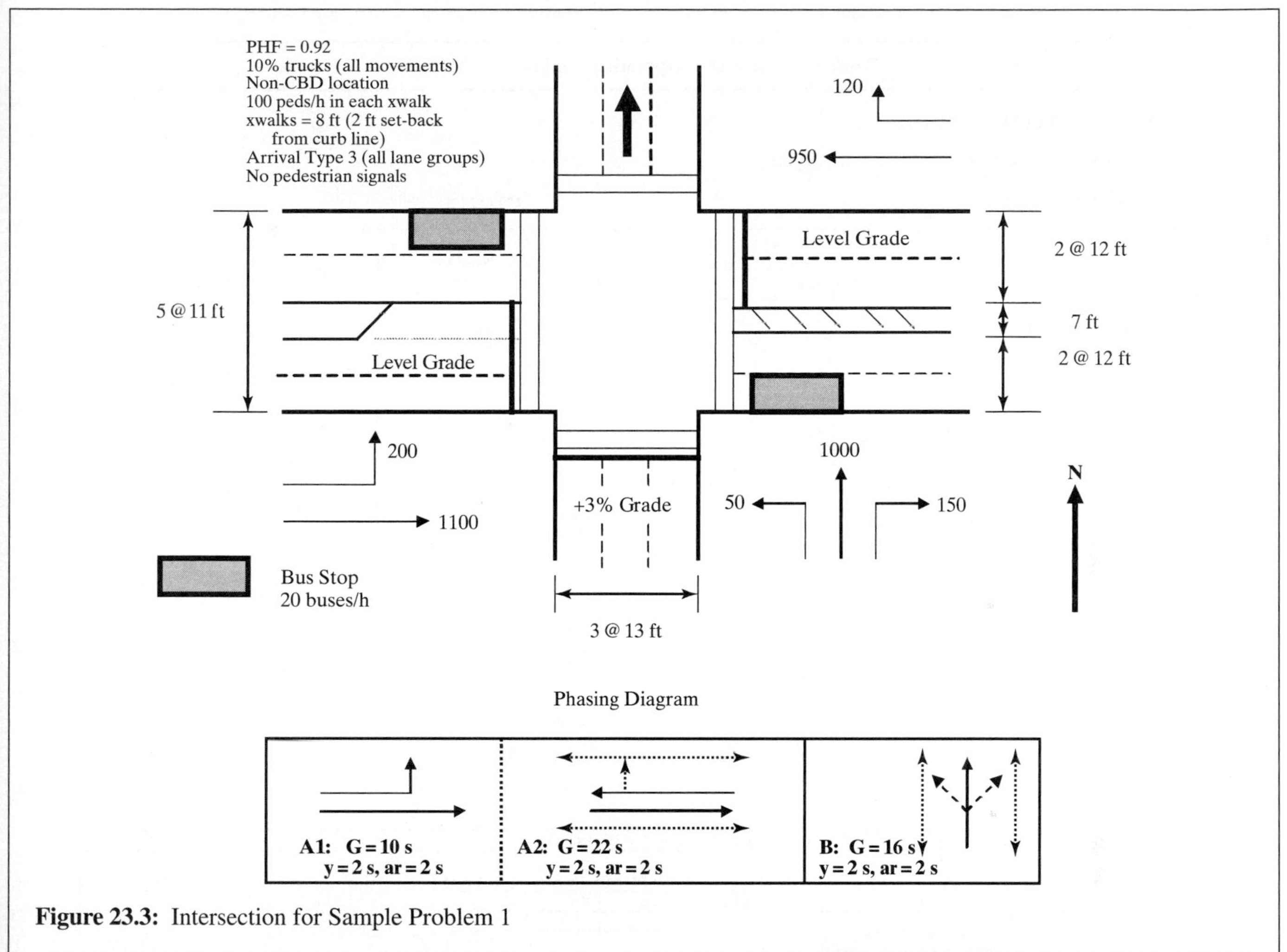

Figure 23.3: Intersection for Sample Problem 1

Conversions to Equivalent Passenger Cars (pc/h)

Equivalents are entered into the next line of Part (D). Passenger cars always have an equivalent of 1.000, heavy vehicles always have an equivalent of 2.000. The local bus equivalents are found in Table 23.1. The 20 buses/h in the EB direction operate in a two-lane group, with a lane group volume of 1,100 veh/h. The 20 buses represent $(20/1{,}100) \times 100 = 1.82\%$ of the lane-group demand. The buses stop in a travel lane. From Table 23.1, this is an equivalent of 3.1. The equivalent for WB local buses is found in a similar way from Table 23.1. Each equivalent is multiplied by the volume to obtain the passenger-car equivalents for each movement, which are then totaled. The results are entered into the third line (for each approach) of Part (D) of the worksheet. At this point, for each movement, only the total columns are used.

Conversions to Through-Car Equivalents (tcu/h)

Left-turn equivalents (Table 23.2) and right-turn equivalents (Table 23.3) are now entered. There are two left turns. The EB left turn is fully protected and has an equivalent of 1.05. The NB left turn is a special case. Because it is made from a one-way street, it is, by definition, unopposed. It is not "protected" in the sense that it has no separate phase. Moreover, left-turners must immediately go through a pedestrian crosswalk, just like right-turning vehicles. In this case, it is appropriate to treat it as another *right* turn, taking the equivalent from Table 23.3. All crosswalks have a pedestrian volume of 100 peds/h. No interpolation is permitted in this table, so the value for 50 peds/h is used (closest tabulated volume): The equivalent for all right turns, therefore, is 1.21.

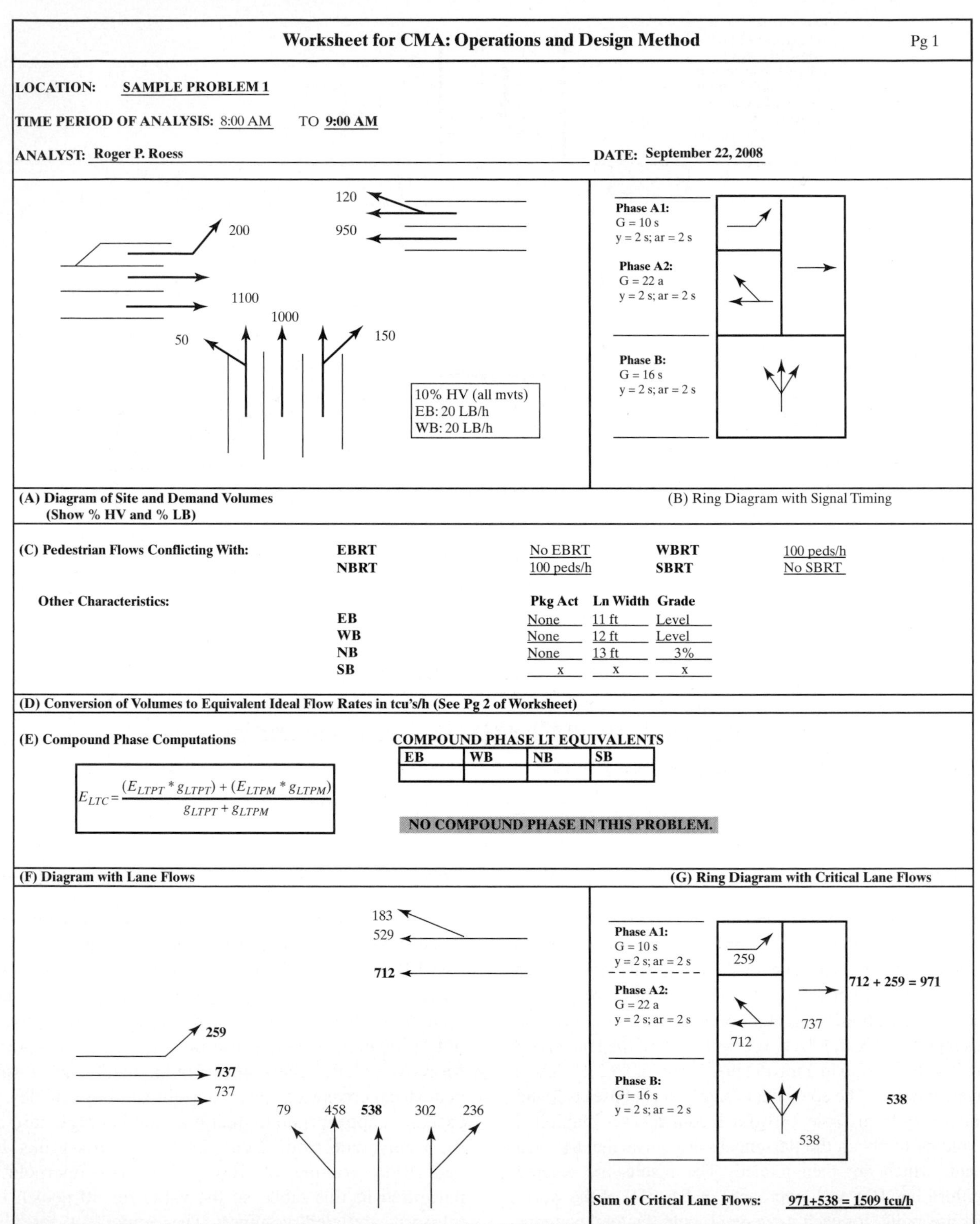

Figure 23.4: CMA Worksheet for Sample Problem 1—Page 1

Worksheet for CMA: Operations and Design Method													Pg 2
	Item	LEFT				THROUGH				RIGHT			
EB		**PC**	**HV**	**LB**	**TOTAL**	**PC**	**HV**	**LB**	**TOTAL**	**PC**	**HV**	**LB**	**TOTAL**
	veh/h	**180**	**20**	**0**	**200**	**970**	**110**	**20**	**1100**	**N/A**	**N/A**	**N/A**	**N/A**
	equiv (Tab.23.1)	1.000	2.000	N/A	×	1.000	2.000	**3.100**	×	1.000	2.000		×
	pc/h	**180**	**40**	**0**	**220**	**970**	**220**	**62**	**1252**				
	E_{LT}/E_{RT} (Tab. 23.2/3)	×	×	×	**1.05**	×	×	×	**1.00**	×	×	×	
	tcu/h	×	×	×	**231**	×	×	×	**1252**	×	×	×	
	f_w (Tab. 23.4)	×	×	×	**0.97**	×	×	×	**0.97**	×	×	×	
	f_g (Tab. 23.5)	×	×	×	**1.00**	×	×	×	**1.00**	×	×	×	
	f_p (Tab. 23.6)	×	×	×	**1.00**	×	×	×	**1.00**	×	×	×	
	f_{LU} (23.7)	×	×	×	**1.00**	×	×	×	**0.952**	×	×	×	
	PHF	×	×	×	**0.92**	×	×	×	**0.92**	×	×	×	
	ideal tcu/h	×	×	×	**259**	×	×	×	**1474**	×	×	×	
WB		**PC**	**HV**	**LB**	**TOTAL**	**PC**	**HV**	**LB**	**TOTAL**	**PC**	**HV**	**LB**	**TOTAL**
	veh/h	**N/A**	**N/A**	**N/A**	**N/A**	**835**	**95**	**20**	**950**	**108**	**12**	**N/A**	**120**
	equiv (E)	1.000	2.000		×	1.000	2.000	**3.100**	×	1.000	2.000		×
	pc/h					**835**	**190**	**62**	**1087**	**108**	**24**	**0**	**132**
	E_{LT}/E_{RT}	×	×	×		×	×	×	**1.00**	×	×	×	**1.21**
	tcu/h	×	×	×		×	×	×	**1087**	×	×	×	**160**
	f_w	×	×	×		×	×	×	**1.00**	×	×	×	**1.00**
	f_g	×	×	×		×	×	×	**1.00**	×	×	×	**1.00**
	f_p	×	×	×		×	×	×	**1.00**	×	×	×	**1.00**
	f_{LU}	×	×	×		×	×	×	**0.952**	×	×	×	**0.952**
	PHF	×	×	×		×	×	×	**0.92**	×	×	×	**0.92**
	ideal tcu/h	×	×	×		×	×	×	**1241**	×	×	×	**183**
NB		**PC**	**HV**	**LB**	**TOTAL**	**PC**	**HV**	**LB**	**TOTAL**	**PC**	**HV**	**LB**	**TOTAL**
	veh/h	**45**	**5**	**0**	**50**	**900**	**100**	**0**	**1000**	**135**	**15**	**0**	**150**
	equiv (E)	1.000	2.000	N/A	×	1.000	2.000	N/A	×	1.000	2.000	N/A	×
	pc/h	**45**	**10**	**0**	**55**	**900**	**200**	**0**	**1100**	**135**	**30**	**0**	**165**
	E_{LT}/E_{RT}	×	×	×	**1.21**	×	×	×	**1.00**	×	×	×	**1.21**
	tcu/h	×	×	×	**67**	×	×	×	**1100**	×	×	×	**200**
	f_w	×	×	×	**1.03**	×	×	×	**1.03**	×	×	×	**1.03**
	f_g	×	×	×	**0.985**	×	×	×	**0.985**	×	×	×	**0.985**
	f_p	×	×	×	**1.00**	×	×	×	**1.00**	×	×	×	**1.00**
	f_{LU}	×	×	×	**0.908**	×	×	×	**0.908**	×	×	×	**0.908**
	PHF	×	×	×	**0.92**	×	×	×	**0.92**	×	×	×	**0.92**
	ideal tcu/h	×	×	×	**79**	×	×	×	**1298**	×	×	×	**236**
SB		**PC**	**HV**	**LB**	**TOTAL**	**PC**	**HV**	**LB**	**TOTAL**	**PC**	**HV**	**LB**	**TOTAL**
	veh/h	N/A	N/A	N/A	N/A	N/A	N/A	N/A	N/A	N/A	N/A	N/A	N/A
	equiv (E)	1.000	2.000		×	1.000	2.000		×	1.000	2.000		×
	pc/h						×	×					
	E_{LT}/E_{RT}	×	×	×		×	×	×		×	×	×	
	tcu/h	×	×	×		×	×	×		×	×	×	
	f_w	×	×	×		×	×	×		×	×	×	
	f_g	×	×	×		×	×	×		×	×	×	
	f_p	×	×	×		×	×	×		×	×	×	
	f_{LU}	×	×	×		×	×	×		×	×	×	
	PHF	×	×	×		×	×	×		×	×	×	
	ideal tcu/h	×	×	×		×	×	×		×	×	×	

Figure 23.4: *(Continued)* CMA Worksheet for Sample Problem 1—Page 2

Worksheet for CMA: Operations and Design Method

Pg 3

(H) Critical Lane Computations

LANE GROUP	EBLT	EB	EBRT	WBLT	WB	WBRT	NBLT	NB	NBRT	SBLT	SB	SBRT
Critical Lane Flow, v_i	259	737	←		712	←	→	538	←		N/A	
Capacity of Critical Lane, c_i	317	1140			697			507				
v/c Ratio of Critical Lane, X_i	0.817	0.647			1.022			1.061				

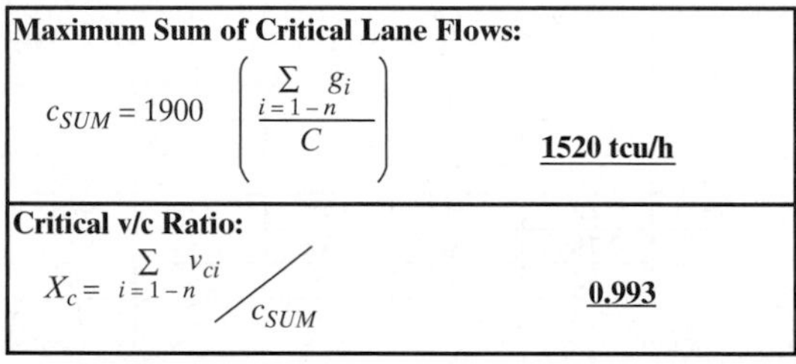

Maximum Sum of Critical Lane Flows:

$$c_{SUM} = 1900\left(\frac{\sum_{i=1-n} g_i}{C}\right)$$

1520 tcu/h

Critical v/c Ratio:

$$X_c = \sum_{i=1-n} v_{ci} \Big/ c_{SUM}$$

0.993

(I) Delay and Level of Service Computations

LANE GROUP	EBLT	EB	EBRT	WBLT	WB	WBRT	NBLT	NB	NBRT	SBLT	SB	SBRT
Input Variables:												
Cycle Length, C	60	60			60			60				
Effective Green, g	10	36			22			16				
v/c Ratio, X	0.817	0.647			1.022			1.061				
Number of Lanes, N	1	2			2			3				
Capacity, c	317	1140			697			507				
Delay Computations:												
Uniform Delay, d_i	24.1	7.8			19.0			22.0				
Prog. Factor, PF (Tab. 23.12)	1.00	1.00			1.00			1.00				
Overflow Delay, d_2	20.3	1.40			29.8			45.9				
Total Delay, d	44.9	9.3			48.8			67.9				
Level of Service (Tab. 23.13)	D	A			D			E				
Approach Delay		42.0										
Approach LOS (Tab. 23.13)		D										

DELAY EQUATIONS:

$$d_i = d_{1i} * PF + d_{2i}$$

$$d_{1i} = \frac{Q5C\left[1-\left(g_i/C\right)\right]^2}{1-\left[min(1, X_i)*\left(g_i/C\right)\right]}$$

$$d_{2i} = 225\left[(X_i - 1) + \sqrt{(X_i - 1)^2 + \frac{16\,X_i}{c_i N_i}}\right]$$

Figure 23.4: *(Continued)* CMA Worksheet for Sample Problem 1—Page 3

Conversions to Through-Car Equivalents Under Ideal Conditions (tcu/h)

The next several entries involve the *PHF* and adjustments for nonideal conditions in the following categories: lane width (Table 23.4), grade (Table 23.5), parking conditions (Table 23.6), and lane utilization (Table 23.7). All lane groups have no parking ($f_p = 1.00$). The EB approach has a lane width of 11 feet ($f_w = 0.97$), the WB approach has 12-foot lanes ($f_w = 1.00$), and the NB approach has 13-foot lanes ($f_w = 1.03$). The EBLT lane group has one lane ($f_{LU} = 1.00$), the EB TH and WB lane groups have two lanes each ($f_{LU} = 0.952$), and the NB lane group has three lanes ($f_{LU} = 0.908$).

The through-car equivalent volumes for each movement are now divided by the *PHF* and all adjustment factors, resulting in the final determination of demand flow rates in tcu/h under equivalent ideal conditions, for each movement.

Assigning tcu/h to Specific Lanes

Converted volumes must now be assigned to specific lanes and lane groups as follows:

- *EB:* The EBLT lane group handles all of the 259 left-turning tcu/h. The two through lanes of the EB approach handle 1474 tcu/h divided equally—737 tcu/h each.
- *WB:* The WB approach carries 183 tcu/h (right turn) and 1,241 tcu/h (through), for a total of 1424 tcu/h. Assuming this is uniformly split between the two lanes of the lane group, 712 tcu/h use each lane. In the left-most lane, this is 712 tcu/h of through vehicles only. The right-most lane carries all 183 right-turning tcu/h and the remaining 1,241 − 712 = 529 tcu/h through vehicles.
- *NB:* All NB flows operate from a single lane group. The lane group has a total flow rate of 79 + 1,298 + 236 = 1,613 tcu/h. This does not evenly divide into the three lanes available. One lane will carry 538 tcu/h; the other two will carry 537 tcu/h. The center lane can carry only through vehicles and is assigned 538 tcu/h. The right lane must accommodate all right-turning vehicles (236 tcu/h) and 537 − 236 = 302 through tcu/h. The left lane handles all 79 tcu/h of left-turning vehicles and 535 − 79 = 458 tvu/h.

The lane assignments are shown in the diagram in Part (F) of page 1 of the worksheet.

Determining Critical-Lane Flows and the Sum of Critical-Lane Flows

The highest single-lane flow moving in each signal phase is identified in bold typeface in Part (F) of the worksheet. These are then entered into the ring diagram. The highest sum of critical-lane flows passes through Phases A1, A2, and B on Ring 1 of the diagram. The sum of critical-lane flows is 1,509 tcu/h as shown.

Capacity and *v/c* Ratios

Part (H) on page 3 of the worksheet summarizes computations of capacity and *v/c* ratio for each lane group. The demand flow on each lane group is transferred from page 2 of the worksheet. There are four lane groups: EBLT, EB, WB, and NB. The capacity of each is computed using Equation 23-5. It will be assumed that actual green times are equal to effective green times. The EBLT phase is 10 seconds, the WB phase is 22 seconds, and the NB phase is 16 seconds. It must be remembered that the EB through movement is permitted during the EBLT phase, the WB phase, *and* the 4 seconds of yellow plus all red between them: 10 + 22 + 4 = 36 seconds. The cycle length is 60 seconds. Capacities are then computed as:

$$c = 1900\left({}^{g}\!/_{c}\right)$$

$$c_{EBLT} = 1900\left({}^{10}\!/_{60}\right) = 317 \text{ tcu/h/ln}$$

$$c_{EB} = 1900\left({}^{36}\!/_{60}\right) = 1{,}140 \text{ tcu/h/ln}$$

$$c_{WB} = 1900\left({}^{22}\!/_{60}\right) = 697 \text{ tcu/h/ln}$$

$$c_{NB} = 1900\left({}^{16}\!/_{60}\right) = 506 \text{ tcu/h}$$

The critical *v/c* ratio for the intersection is computed using Equation 23-6:

$$c_{SUM} = 1900\left(\frac{\sum g_i}{C}\right) = 1900\left(\frac{10 + 22 + 16}{60}\right)$$
$$= 1{,}520 \text{ tcu/h}$$

With both demand flows and capacities now expressed in tcu/h for critical lanes, *v/c* ratios can be directly computed:

$$X_{EBLT} = {}^{259}\!/_{317} = 0.817$$

$$X_{EB} = {}^{737}\!/_{1140} = 0.647$$

$$X_{WB} = {}^{712}\!/_{697} = 1.022$$

$$X_{NB} = {}^{538}/_{507} = 1.061$$

$$X_c = \frac{259 + 712 + 538}{1520} = 0.993$$

These results are significant. Of the critical lanes, two operate at *v/c* ratios in excess of 1.00, signifying that demand exceeds the available capacity (WB, NB). The critical *v/c* ratio, however, is under 1.00 (though barely so), indicating that a reallocation of green time might provide a workable signal timing (i.e., *without* lengthening the cycle length, providing a more efficient phase plan, or adding lanes to a lane group or lane groups). This situation is addressed subsequently.

Delay and Level of Service

Average delays per vehicle can now be computed using Equations 23-9, 23-10, and 23-11. Because the stated arrival pattern is random, the progression adjustment factor for all lane groups is 1.00. Total delays per vehicle are computed as:

$$d = d_1 PF + d_2$$

The first term, d_1, is uniform delay, and is computed using Equation 23-12:

$$d_1 = \frac{0.5\, C\left[1-\left({}^{g}/_{C}\right)\right]^2}{1-min\,[(1, X)] * \left({}^{g}/_{C}\right)}$$

$$d_{1EBLT} = \frac{0.50 * 60 * \left(1-{}^{10}/_{60}\right)^2}{1-\left(0.817 * {}^{10}/_{60}\right)} = 24.1 \text{ s/veh}$$

$$d_{1EB} = \frac{0.50 * 60 * \left(1-{}^{36}/_{60}\right)^2}{1-\left(0.647 * {}^{36}/_{60}\right)} = 7.8 \text{ s/veh}$$

$$d_{1WB} = \frac{0.50 * 60 * \left(1-{}^{22}/_{60}\right)^2}{1-\left(1 * {}^{22}/_{60}\right)} = 19.0 \text{ s/veh}$$

$$d_{1NB} = \frac{0.50 * 60 * \left(1-{}^{16}/_{60}\right)^2}{1-\left(1 * {}^{16}/_{60}\right)} = 22.0 \text{ s/veh}$$

The second term, d_2, is overflow delay, and is computed using Equation 23-11:

$$d_2 = 225\left[(X - 1) + \sqrt{(X - 1)^2 + \left(\frac{16\,X}{c\,N}\right)}\right]$$

$$d_{2EBLT} = 225\left[(0.816 - 1) + \sqrt{(0.816 - 1)^2 + \left(\frac{16 * 0.817}{317 * 1}\right)}\right] = 20.3 \text{ s/veh}$$

$$d_{2EB} = 225\left[(0.647 - 1) + \sqrt{(0.647 - 1)^2 + \left(\frac{16 * 0.647}{1140 * 2}\right)}\right] = 1.4 \text{ s/veh}$$

$$d_{2WB} = 225\left[(1.022 - 1) + \sqrt{(1.022 - 1) + \left(\frac{16 * 1.022}{697 * 2}\right)}\right] = 29.8 \text{ s/veh}$$

$$d_{2NB} = 225\left[(1.061 - 1) + \sqrt{(1.061 - 1)^2 + \left(\frac{16 * 1.061}{507 * 3}\right)}\right] = 45.9 \text{ s/veh}$$

Total delays for each lane group are:

$$d_{EBLT} = 24.1(1) + 20.3 = 44.4 \text{ s/veh}$$

$$d_{EB} = 7.8(1) + 1.4 = 9.3 \text{ s/veh}$$

$$d_{WB} = 19.0(1) + 29.8 = 48.8 \text{ s/veh}$$

$$d_{NB} = 22.0(1) + 45.9 = 67.9 \text{ s/veh}$$

From the criteria of Table 23.13, the levels of service are EBLT (LOS D), EB (LOS A), WB (LOS D), NB (LOS E).

Analysis of Results

Any analysis must consider *both* the LOS designation *and* the *v/c* ratios. Interestingly, none of the lane groups have a LOS of F, even though two of them have a *v/c* ratio in excess of 1.00. This is because of the 15-minute analysis period and the assumption of no preexisting unserved queue at the beginning of the analysis period. If a *v/c* ratio of greater than 1.00 persists for more than 15 minutes, the delay will quickly increase to values that would be classified as LOS F.

The critical *v/c* ratio is less than 1.00, suggesting that green time could be reallocated within the existing cycle length to achieve better results. The critical *v/c* ratio, however, is 0.993, meaning that such a signal retiming would still consume 99.3% of the available capacity. This is obviously higher than one would like.

Table 23.10: Comparison of CMA and HCM Analysis: Results for Sample Problem 1

Lane Group	*v/c* Ratios		Delays (s/veh)	
	CMA	HCM	CMA	HCM
EBLT	0.817	0.819	44.9	47.9
EB	0.647	0.653	9.3	9.7
WB	1.022	1.030	48.8	53.8
NB	1.061	1.059	67.9	64.8

Comparing to the HCM Model

As we noted previously, this sample problem is re-solved in Chapter 24 using the more complex HCM model. The resulting *v/c* ratios and delays are compared in Table 23.10.

In this case, both the *v/c* ratios and delays are quite comparable. In general, the CMA results show somewhat less delay than the HCM methodology. The differences are small, however, and none of the levels of service would have been different from those assigned as a result of CMA.

Improving the Signalization

Because the critical *v/c* ratio is less than 1.00, it is possible to reduce all of the *v/c* ratios to less than 1.00 by simply reallocating the green time. Using the approach of Chapter 21, effective green time would be reallocated in the ratio of critical-lane flow rates for each phase. The critical-lane flow rates estimated in this problem are as follows: Phase A1: 259 tcu/h, Phase A2: 712 tcu/h, and Phase B: 538 tcu/h. The signal has a 60-second cycle length, and it will be assumed that the seconds of "yellow" and "all red" intervals remains unchanged. It is also assumed (as was done in the analysis) that effective green equals actual green.

Then:

$$g_{TOT} = 60 - 12 = 48\,\text{s}$$

$$g_{A1} = 48 * \left({}^{259}/_{1509}\right) = 8.2\,\text{s}$$

$$g_{A2} = 48 * \left({}^{712}/_{1509}\right) = 22.7\,\text{s}$$

$$g_{B} = 48 * \left({}^{538}/_{1509}\right) = 17.1\,\text{s}$$

As might be expected, the retiming reduces the green time for Phase A1 slightly while increasing it in Phases A2 and B. Although this will produce an acceptable signal timing, it is barely so because the critical *v/c* ratio was 0.993, with virtually no unused green time.

It might be wise to consider an increase in the cycle length, as well as a reallocation of green time. A cycle length of 90 seconds would lead to the following signal timing:

$$g_{TOT} = 90 - 12 = 78\,\text{s}$$

$$g_{A1} = 78 * \left({}^{259}/_{1509}\right) = 13.4\,\text{s}$$

$$g_{A2} = 78 * \left({}^{712}/_{1509}\right) = 36.8\,\text{s}$$

$$g_{B} = 78 * \left({}^{538}/_{1509}\right) = 27.8\,\text{s}$$

The full worksheets are not shown here to conserve space, but Table 23.11 shows the resulting *v/c* ratios and delays for each lane group under these two signal timing options.

The two reallocations result in *v/c* ratios that are all below 1.00; that is, no lane group "fails," leading to unserved queues at the end of the 15-minute analysis period. With a 60-second cycle length, the delay for the EBLT lane group reaches more than 80 s/veh, or LOS F. High delays are always expected for short exclusive LT phases because stopped vehicles must wait a significant time to reacquire the green. Delays in the WB and NB lane groups is significantly reduced, even at a cycle length of 60 seconds because they no longer "fail."

The results improve for a 90-second cycle length. Delay to the EBLT lane group is reduced, and the LOS is improved to E. Delay in the WB and NB lane groups is further reduced, primarily because fewer isolated cycle failures are expected with the longer cycle length. This would probably be the recommended timing, assuming there are no system constraints on selecting a cycle length.

Table 23.11: Delay, *v/c* Ratio, and Delay Results from Reallocation of Green

Item	Lane Group			
	EBLT	EB	WB	NB
***Cycle Length* = 60 s**				
v/c	0.977	0.667	0.990	0.994
***d* (s/veh)**	88.50	10.40	43.00	50.70
LOS	F	B	D	D
***Cycle Length* = 90 s**				
v/c	0.916	0.644	0.916	0.917
***d* (s/veh)**	78.1	13.2	36.5	44.3
LOS	E	B	D	D

23.3.2 Sample Problem 2: Example with Compound Phasing

Intersection Description

Figure 23.5 is the worksheet for Sample Problem 2. A full description of the intersection and signal timing are found in sections (A), (B), and (C) of page 1. The intersection consists of two four-lane arterials, with left-turn lanes for the EB and WB approaches. The signalization includes a protected + permitted phase for the EB and WB left turns. The intersection may be considered to be isolated; that is, vehicles arrive randomly on all approaches.

Conversion of Volumes to pc/h

Page 2 of the worksheet shows all volume conversion computations. Each movement volume contains 10% heavy vehicles and no local buses. The volumes are segregated for conversion, with results rounded to the nearest whole vehicle. There are no equivalents to "look up" because all passenger cars have an equivalent of 1.00 and all heavy vehicles have an equivalent of 2.0.

Conversion of Volumes to tcu/h

In this step, left-turn and right-turn volumes are converted to through-car units. Through cars have an equivalent of 1.00. Right-turn equivalents are selected from Table 23.3 for 50 peds/h in each crosswalk. This equivalent is 1.21 and applies to all right turns on all approaches.

Left-turn equivalents are found in Table 23.3:

- NB left turns are permitted, and face an opposing flow of 650 veh/h in two lanes. Interpolating, the equivalent for NB left turns is 5.75.
- SB left turns are permitted, and face an opposing flow of 500 veh/h in two lanes. Interpolating, the equivalent for SB left turns is 4.0.
- The EB left turns have a protected + permitted compound phase. The equivalent is estimated as the weighted average of the equivalents for protected LTs and permitted LTs. The weighting is proportional to the effective green time in the protected and permitted portions of the phase. In the signalization shown, the protected portion has 8 seconds of green and 2 seconds of yellow (which counts as effective green), or 10 seconds. The permitted portion of the phase is 40 seconds. Protected LTs have an equivalent of 1.05. The EB left turn faces an opposing flow of 700 veh/h in two lanes. From Table 23.6, this equivalent is 6.5. Then:

$$E_{LTEB} = \frac{(10 * 1.05) + (40 * 6.5)}{(10 + 40)} = 5.41$$

- The WB left turns share the same compound phase as EB left turns. The equivalent for the protected portion of the phase is again 1.05. The WB left turns face an opposing flow of 800 veh/h in two lanes. From Table 23.8, the equivalent for this LT is 8.0. Then:

$$E_{LTWB} = \frac{(10 * 1.05) + (40 * 8.0)}{(10 + 40)} = 6.41$$

These equivalents are used to convert volumes in pc/h to tcu/h on the second page of the worksheet in Figure 23.5.

Converting Volumes in tcu/h to tcu/h Under Equivalent Ideal Conditions

The final step in converting demand volumes involves adjustments for the PHF and for lane width, grade, parking conditions, and lane utilization. None of the lane groups have parking, and all are on level grades. Thus the adjustment factors for these two conditions are both 1.00. Lane width adjustment factors are found in Table 23.4. All EB and WB lane groups have 12-foot lanes, and the adjustment factor for these is 1.00. The NB and SB lane groups, however, have 11-foot lanes, and an adjustment factor of 0.97 must be applied. The lane-utilization factor is found in Table 23.7. The EB and WB left-turn lane groups have one lane, for which the adjustment factor is 1.00. All other lane groups have two lanes, and the adjustment factor for these is 0.952. The PHF for all lane groups is given as 0.96.

Assigning Demand Flows to Lanes and Determining Critical Lanes

In section (F) on page 1 of the worksheet (Figure 23.5), lane volumes are assigned to individual lanes. The highest lane volumes for each signal phase are identified in bold type. Determining the critical-lane flows, however, requires that the EB and WB left-turn demands be split between the protected and permitted portions of the compound phase (in proportion to the length of the portions of the phase). Because the compound LT phase has 10 seconds of protected green (which

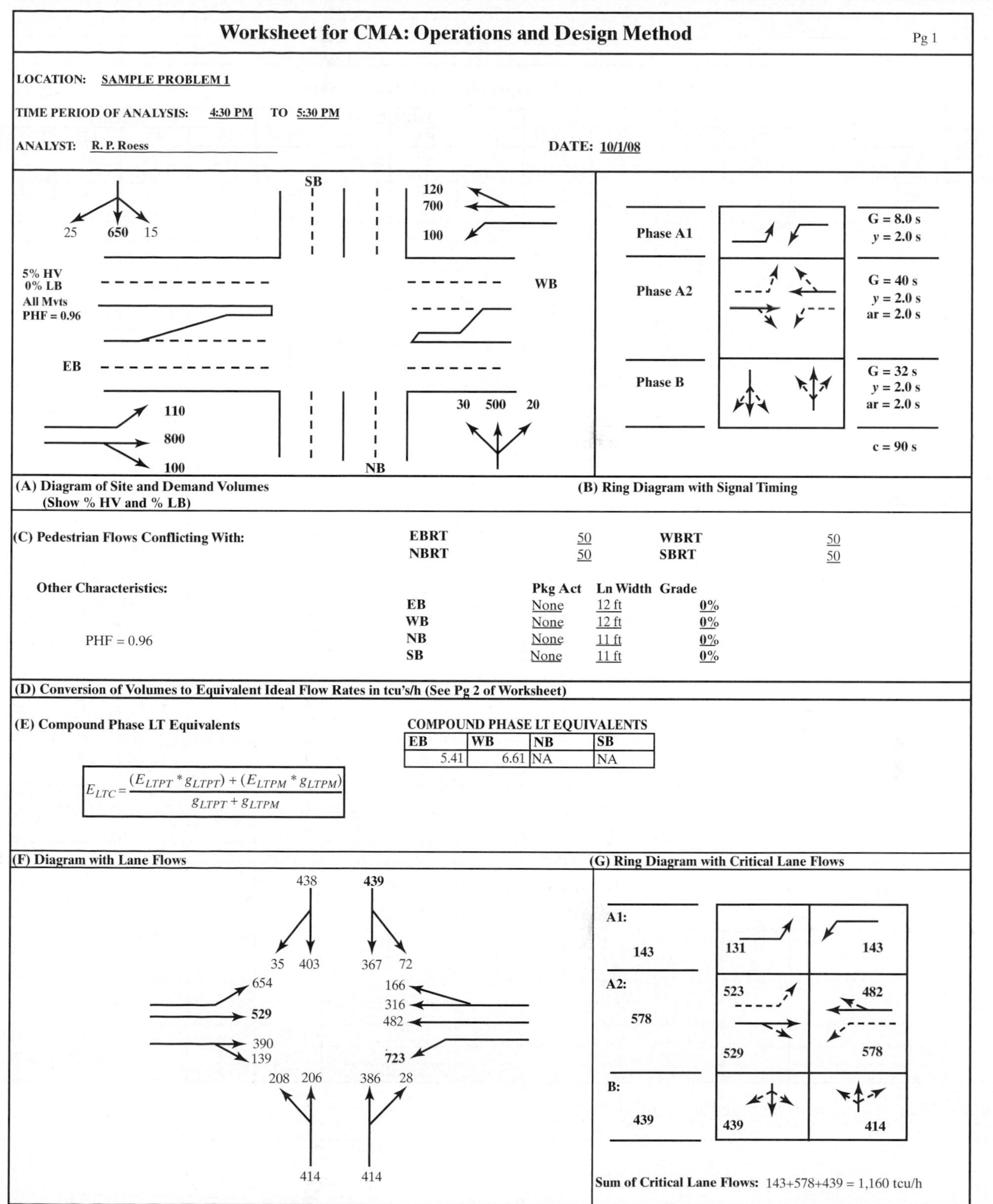

Figure 23.5: Sample Problem 2—Page 1

Worksheet for CMA: Operations and Design Method Pg 2

	Item	LEFT				THROUGH				RIGHT			
EB		**PC**	**HV**	**LB**	**TOTAL**	**PC**	**HV**	**LB**	**TOTAL**	**PC**	**HV**	**LB**	**TOTAL**
	veh/h	104	6	0	110	760	40	0	800	95	5	0	100
	equiv (Tab.23.1)	1.000	2.000		×	1.000	2.000		×	1.000	2.000		×
	pc/h	104	12	0	116	760	80	0	840	95	10	0	105
	E_{LT}/E_{RT} (Tab. 23.2/3)	×	×	×	5.41	×	×	×	1	×	×	×	1.21
	tcu/h	×	×	×	628	×	×	×	840	×	×	×	127
	f_w (Tab. 23.4)	×	×	×	1	×	×	×	1	×	×	×	1
	f_g (Tab. 23.5)	×	×	×	1	×	×	×	1	×	×	×	1
	f_p (Tab. 23.6)	×	×	×	1	×	×	×	1	×	×	×	1
	f_{LU} (23.7)	×	×	×	1	×	×	×	0.952	×	×	×	0.952
	PHF	×	×	×	0.96	×	×	×	0.96	×	×	×	0.96
	ideal tcu/h	×	×	×	**654**	×	×	×	**919**	×	×	×	**139**
WB		**PC**	**HV**	**LB**	**TOTAL**	**PC**	HV	**LB**	**TOTAL**	**PC**	**HV**	**LB**	**TOTAL**
	veh/h	95	5	0	100	665	**35**	0	700	114	**6**	0	120
	equiv (E)	1.000	2.000		×	1.000	2.000		×	1.000	2.000		×
	pc/h	95	10	0	105	665	70	0	735	114	12	0	126
	E_{LT}/E_{RT}	×	×	×	6.61	×	×	×	1	×	×	×	1.21
	tcu/h	×	×	×	694	×	×	×	1	×	×	×	152
	f_w	×	×	×	1	×	×	×	1	×	×	×	1
	f_g	×	×	×	1	×	×	×	1	×	×	×	1
	f_p	×	×	×	1	×	×	×	1	×	×	×	1
	f_{LU}	×	×	×	1	×	×	×	0.952	×	×	×	0.952
	PHF	×	×	×	0.96	×	×	×	0.96	×	×	×	0.96
	ideal tcu/h	×	×	×	**723**	×	×	×	**804**	×	×	×	**166**
NB		**PC**	**HV**	**LB**	**TOTAL**	**PC**	**HV**	**LB**	**TOTAL**	**PC**	**HV**	**LB**	**TOTAL**
	veh/h	28	2	0	30	475	25	0	500	19	1	0	20
	equiv (E)	1.000	2.000		×	1.000	2.000		×	1.000	2.000		×
	pc/h	28	4	0	32	475	50	0	525	19	2	0	21
	E_{LT}/E_{RT}	×	×	×	5.75	×	×	×	1	×	×	×	1.21
	tcu/h	×	×	×	184	×	×	×	525	×	×	×	25
	f_w	×	×	×	0.97	×	×	×	0.97	×	×	×	0.97
	f_g	×	×	×	1	×	×	×	1	×	×	×	1
	f_p	×	×	×	1	×	×	×	1	×	×	×	1
	f_{LU}	×	×	×	0.952	×	×	×	0.952	×	×	×	0.952
	PHF	×	×	×	0.96	×	×	×	0.96	×	×	×	0.96
	ideal tcu/h	×	×	×	**208**	×	×	×	**592**	×	×	×	**28**
SB		**PC**	**HV**	**LB**	**TOTAL**	**PC**	**HV**	**LB**	**TOTAL**	**PC**	**HV**	**LB**	**TOTAL**
	veh/h	14	1	0	15	617	33	0	650	24	1	0	25
	equiv (E)	1.000	2.000		×	1.000	2.000		×	1.000	2.000		×
	pc/h	14	2	0	16	617	66	0	683	24	2	0	26
	E_{LT}/E_{RT}	×	×	×	4	×	×	×	1	×	×	×	1.21
	tcu/h	×	×	×	64	×	×	×	683	×	×	×	34
	f_w	×	×	×	0.97	×	×	×	0.97	×	×	×	0.97
	f_g	×	×	×	1	×	×	×	1	×	×	×	1
	f_p	×	×	×	1	×	×	×	1	×	×	×	1
	f_{LU}	×	×	×	0.952	×	×	×	0.952	×	×	×	0.952
	PHF	×	×	×	0.96	×	×	×	0.96	×	×	×	0.96
	ideal tcu/h	×	×	×	**72**	×	×	×	**770**	×	×	×	**35**

Figure 23.5: *(Continued)* Sample Problem 2—Page2

Worksheet for CMA: Operations and Design Method

Pg 3

(H) Critical Lane Computations

LANE GROUP	EBLT	EB	EBRT	WBLT	WB	WBRT	NBLT	NB	NBRT	SBLT	SB	SBRT
Critical Lane Flow, v_i	694	529	←	723	482	←	→	414	←	→	439	←
Capacity of Critical Lane, c_i	1055	844		1055	844			676			676	
v/c Ratio of Critical Lane, X_i	0.62	0.627		0.685	0.571			0.612			0.649	

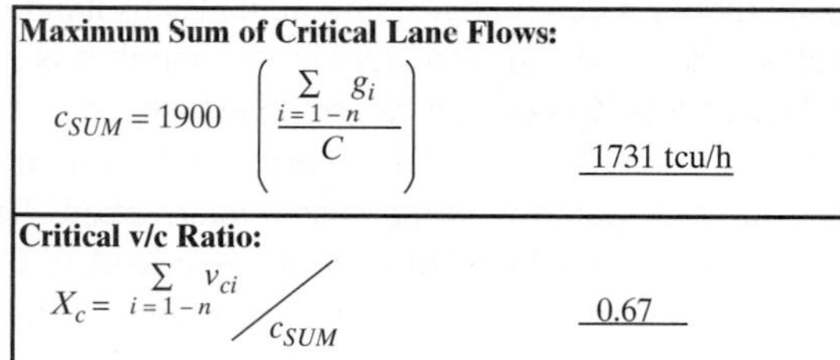

Maximum Sum of Critical Lane Flows:

$$c_{SUM} = 1900 \left(\frac{\sum_{i=1-n} g_i}{C} \right)$$

1731 tcu/h

Critical v/c Ratio:

$$X_c = \sum_{i=1-n} v_{ci} \Big/ c_{SUM}$$

0.67

(I) Delay and Level of Service Computations

LANE GROUP	EBLT	EB	EBRT	WBLT	WB	WBRT	NBLT	NB	NBRT	SBLT	SB	SBRT
Input Variables:												
Cycle Length, C	90	90		90	90			90			90	
Effective Green, g	50	40		50	40			32			32	
v/c Ratio, X	0.62	0.627		0.685	0.571			0.612			0.649	
Number of Lanes, N	1	2		1	2			2			2	
Capacity, c	1055	844		1055	844			676			676	
Delay Computations:												
Uniform Delay, d_i	13.6	19.3		14.4	18.6			23.9			24.3	
Prog. Factor, PF (Tab. 23.12)	1	1		1	1			1			1	
Overflow Delay, d_2	3.1	2		4.1	1.6			2.4			2.8	
Total Delay, d	16.7	21.3		18.5	20.2			26.2			27.1	
Level of Service (Tab. 23.13)	B	C		B	C			C			C	
Approach Delay		18.7			19.1							
Approach LOS (Tab. 23.13)		B			B							

DELAY EQUATIONS:

$$d_i = d_{1i} * PF + d_{2i}$$

$$d_{1i} = \frac{Q5C\left[1-\left(\frac{g_i}{C}\right)\right]^2}{1-\left[min(1, X_i)*\left(\frac{g_i}{C}\right)\right]}$$

$$d_{2i} = 225\left[(X_i-1)+\sqrt{(X_i-1)^2+\frac{16\,X_i}{c_i N_i}}\right]$$

Figure 23.5: *(Continued)* Sample Problem 2—Page3

includes the yellow), and 40 seconds of permitted green, the EB and WB left-turn flows are allocated as follows:

$$v_{EBLTprot} = 654 * \left({}^{10}/_{50}\right) = 131 \text{ tcu/h}$$

$$v_{EBLTperm} = 654 - 131 = 523 \text{ tcu/h}$$

$$v_{WBLTprot} = 723 * \left({}^{10}/_{50}\right) = 145 \text{ tcu/h}$$

$$v_{SBLTperm} = 723 - 145 = 578$$

In part (G) of the worksheet, these and all other demand flows are assigned to the appropriate portion of the signal ring diagram. Note that in Phase A2, there are four lane flows moving: the portion of the EB and WB left turns in the permitted phase, and the lane flows on the through and right-turn lane groups. The selection of critical-lane flows is shown on the worksheet. Of particular note is that the critical-lane flow rate for Phase A2 is in the WB left-turn lane, which is generally not a desirable condition.

Delay and Level of Service

Section (I) of the worksheet (page 3) shows the results of delay computations. These computations use Equations 23-9, 23-10, and 23-11. The intersection is isolated, so the progression factor, *PF,* is 1.00. The resulting levels of service are B for the left-turn lane groups and C for all other lane groups. It is noted that the EB and WB through-lane groups have delays that are just above the maximum boundary for LOS B. When the EB and WB lane groups are averaged, the aggregate EB and WB delays yield LOS B for the approaches.

In general, this intersection is operating well within acceptable limits. Most of the lane-group *v/c* ratios are relatively low. A shorter cycle length might be possible and might yield lower delays. The EB and WB protected LT phase might be proportionally increased by several seconds to eliminate the critical nature of the WB LT flow in the permitted portion of the phase.

23.4 Closing Comments

Critical-lane analysis has been used to evaluate the performance of signalized intersections for many years, and it has been the basis for signal timing for even more.

In this chapter, a very basic "back of the envelope" planning approach has been presented and was taken directly from TRB Circular 212, published in 1980. The planning approach is valid only for the most general of estimates and should not be used where a more detailed analysis of conditions is needed. The planning approach might be used, for example, to generally evaluate the basic viability of a future intersection configuration and signal timing. No decision on any specific design or signal timing would ever be made using the planning approach.

The operational analysis and design approach to critical movement analysis presented in this chapter is an updated model, taking into account some of the latest developments in signal timing methodologies and in the HCM methodology for analysis of signalized intersections. Although it is not a "back of the envelope" process, it can be done by hand in a reasonable amount of time. It is sensitive to many of the important factors affecting the operation of signalized intersections but is not as detailed as the HCM methodology, described in Chapter 24. It does yield delay estimates and can provide detailed insight into how key features of the intersection and signal timing affect overall performance.

References

1. *Highway Capacity Manual*, Transportation Research Board, Washington DC, 2000.
2. *Highway Capacity Manual*, Special Report 87, Transportation Research Board, Washington DC, 1965.
3. *Highway Capacity Manual*, Special Report 209, Transportation Research Board, Washington DC, 1985.
4. *Interim Materials on Highway Capacity*, Circular 212, Transportation Research Board, Washington DC, 1980.
5. Capelle, D.G., and Pinell, C., "Capacity Study of Signalized Diamond Interchanges," Highway Research Bulletin 291, Transportation Research Board, Washington DC, 1961.
6. Messer, C.J., and Fambro, D.B., "A New Critical Lane Analysis for Intersection Design," 56th Annual Meeting of the Transportation Research Board, Washington DC, January 1977.
7. Akcelik, R., "SIDRA for the Highway Capacity Manual, *Compendium of Papers,* 60th Annual Meeting of the Institute of Transportation Engineers, Washington DC, 1990.
8. Reilly, W., et al., "Signalized Intersection Capacity Study," Final Report, NCHRP Project 3-28(2), JHK & Associates, Tucson AZ, 1982.

Problems

23-1. The intersection shown here is to be analyzed using CMA. The following information in available for this intersection: (a) PHF = 0.90. (b) Progression quality = Good (EB), Poor (WB), and Random (NB). (c) 5% HV in all movements, no local buses. (d) no pedestrians present (overpasses are provided). For the intersection as shown:

(a) Analyze the intersection using the CMA Planning Approach.

(b) Using the Operations and Design Approach, determine the existing *v/c* ratio, delay, and level of service for each lane group in the intersection.

(c) Aggregate the lane group delays for each approach and determine the existing level of service for each.

(d) Evaluate how well the intersection is operating, based on these results. If necessary, suggest changes in the physical design or signal timing that will result in better operations. It is *not* necessary to reanalyze the modified intersection to demonstrate the effectiveness of the recommended improvements.

(e) Comment on the differences between the results of the Planning Approach and the Operations and Design Approach.

23-2. Analyze the intersection shown using the CMA Operations and Design Approach to determine the *v/c* ratio, delay, and LOS for each lane group and each approach. Assess the operation and recommend changes to the physical design or signalization that you think are needed to obtain acceptable operations. Use CMA to assess the effectiveness of your recommended improvements.

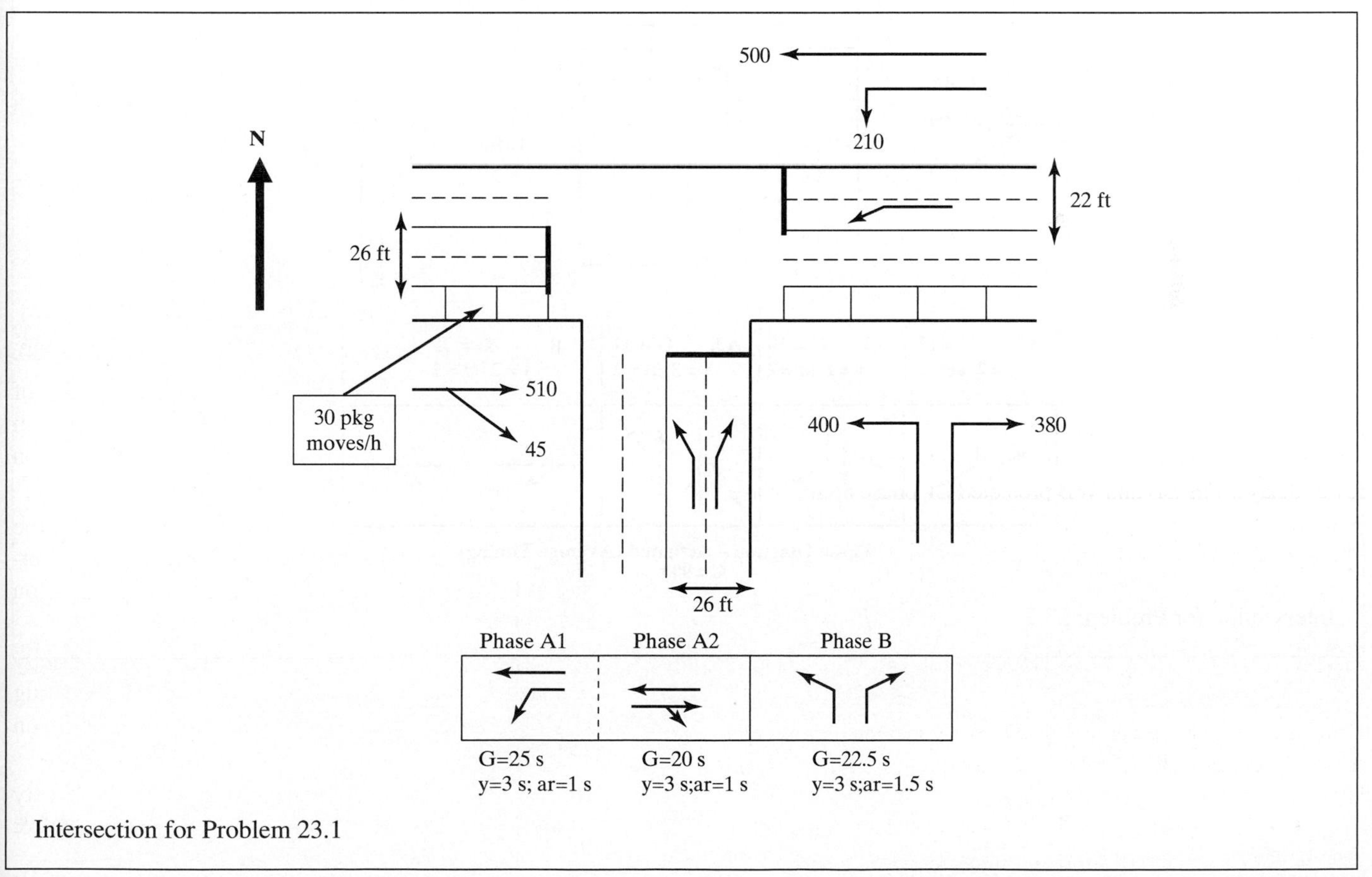

Intersection for Problem 23.1

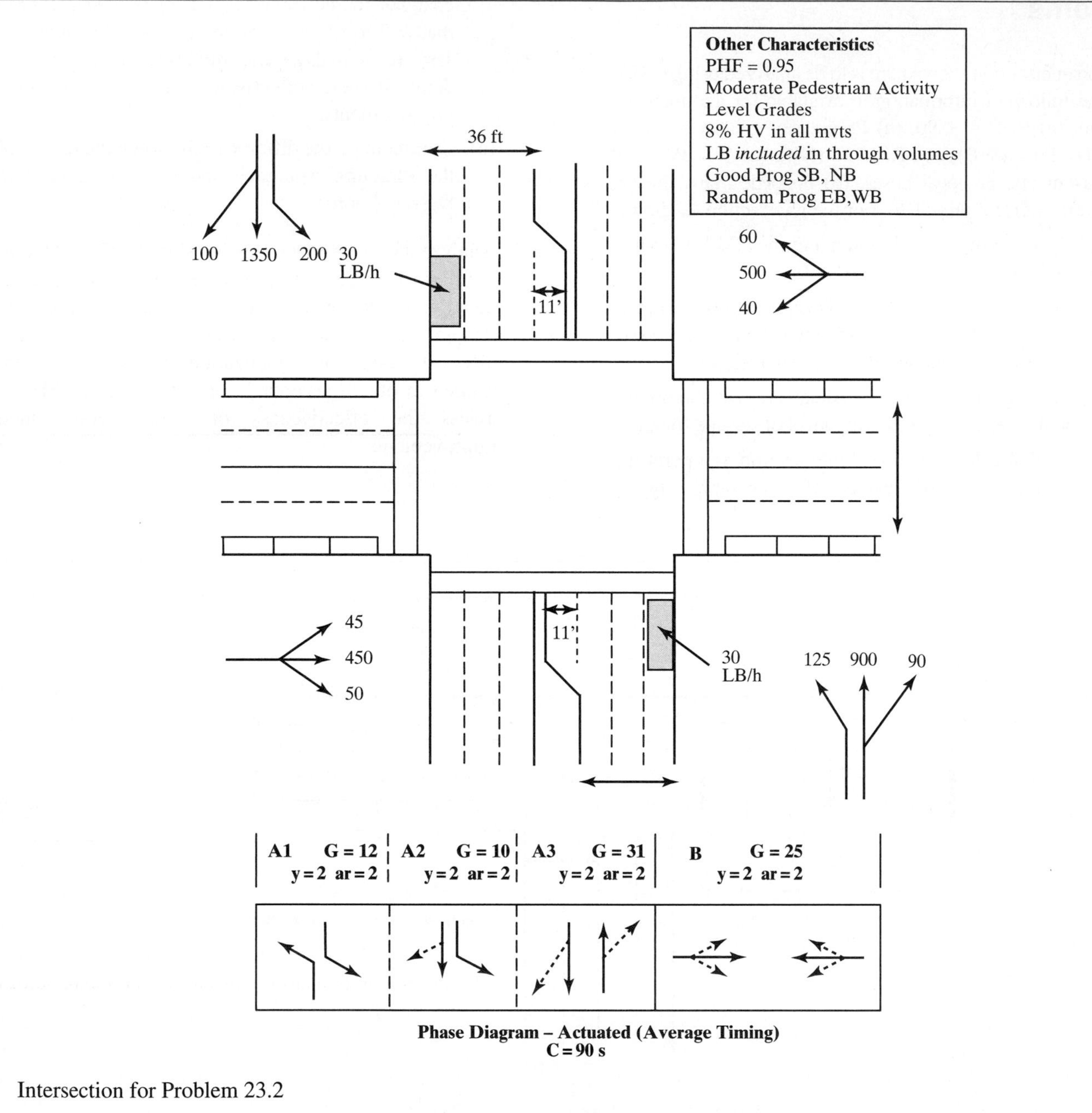

Intersection for Problem 23.2

CHAPTER 24

Analysis of Signalized Intersections

24.1 Introduction

The signalized intersection is the most complex location in any traffic system. In Chapter 20, we presented some simple models of critical operational characteristics. In Chapter 21, these were applied to create a simple signal-timing methodology. The signal-timing methodology, however, involved a number of simplifying assumptions, among which was a default value for saturation flow rate that reflected "typical" conditions. A complete analysis of any signalized intersection requires use of a more complex model that addresses all of the many variables affecting intersection operations as well as some of the more intricate interactions among component flows.

The most frequently used model for analysis of signalized intersections in the United States is the model contained in the *Highway Capacity Manual* (HCM) [*1*]. This model first appeared in the 1985 edition of the HCM and has been revised and updated in subsequent editions (1994, 1997, 2000, and 2010 forthcoming at this writing). The model has become increasingly complex, involving several iterative elements. It has become difficult, if not impossible, to do complete solutions using this model by hand. Its implementation, therefore, has been primarily through computer software that replicates the model.

The forthcoming 2010 edition of the HCM will contain three significant changes from previous editions:

1. The model has been set up to handle actuated signal analysis directly; previous editions handled actuated signals by requiring users to enter average signal parameters, such as cycle length and green splits, with analysis proceeding as if a pretimed signal was present.
2. The estimation of delay is now partially modelled using Incremental Queue Analysis (IQA). IQA allows a more detailed analysis of arriving and departing vehicle distributions.
3. The definition of lane groups has been altered. In general, this will lead to more lane groups being identified and separately analyzed as part of the methodology.

This text focuses on the analysis of pretimed signals because it is more straightforward to present basic modeling theory for these. The approach to actuated signals is quite complex and will require software for implementation. This text provides a general overview of what the actuated signal analysis methodology entails.

Other models are in use elsewhere. SIDRA [*2,3*] is a model and associated computer package developed for use in Australia by the Australian Road Research Board (ARRB). Some of its elements, particularly in delay estimation, have been adapted and applied in the HCM. A Canadian model also exists [*4*] and is the official model used throughout Canada. All of these models are "deterministic" analytic models. In deterministic models, the same input data produces the same result each time the model is applied.

A number of simulation models may also be used to analyze individual intersections, as well as networks. Some of the most frequently used are SimTraffic, VISSIM, AIMSUM, and CORSIM. Simulation often introduces stochastic elements. This means that the same input parameters do not lead to the same results each time the simulation is run. Because of this characteristic, most stochastic simulation models are run multiple times with a given set of input data, and average results of those runs are used in evaluating the intersection.

This chapter describes the overall concepts and some of the details of the HCM methodology. Some of the complexities are noted without going into full detail, and you are encouraged to consult the HCM directly for additional information on such subjects. This chapter is based on the latest research that is shaping the methodology for the 2010 edition of the HCM. At this writing, however, the final version of this methodology is not yet approved. You should consult HCM 2010 directly when it is published to ensure that you are using the latest version of this methodology.

24.2 Conceptual Framework for the HCM 2010 Methodology

Five fundamental concepts are used in the HCM 2010 signalized intersection analysis methodology that should be understood before considering any of the details of the model:

- The critical-lane group concept
- The v/s ratio as a measure of demand
- Capacity and saturation flow rate concepts
- Level-of-service (LOS) criteria and concepts
- Effective green time and lost-time concepts

Some of these are similar to what we have been discussing in other chapters; they are repeated here for reinforcement and because of their central importance in understanding the HCM model.

24.2.1 The Critical-Lane Group Concept

The signal-timing methodology of Chapter 21 relied on *critical lanes* and *critical-lane volumes*. In doing so, however, it was generally assumed that movements sharing a set of lanes on an approach would evenly divide among the available lanes, at least in terms of "through vehicle units" (tvu's). In the HCM model, the total demand in a set of lanes is used. Instead of identifying a set of "critical lanes," a set of "critical-lane groups" is identified.

Critical-lane analysis compares actual demand flows in a single lane with the saturation flow rate and capacity of that lane. Critical-lane *group* analysis compares actual flow with the saturation flow rate and capacity of a group of lanes operating in equilibrium. Figure 24.1 illustrates the difference.

Where several lanes operate in equilibrium (i.e., where there are no lane-use restrictions impeding driver selection of lanes), the *lane group* is treated as a single entity.

Not all methodologies do this. Both the Australian and Canadian models focus on individual lanes, taking into account unequal use of lanes. The HCM also accounts for unequal use of lanes through a process of adjustments to saturation flow rates.

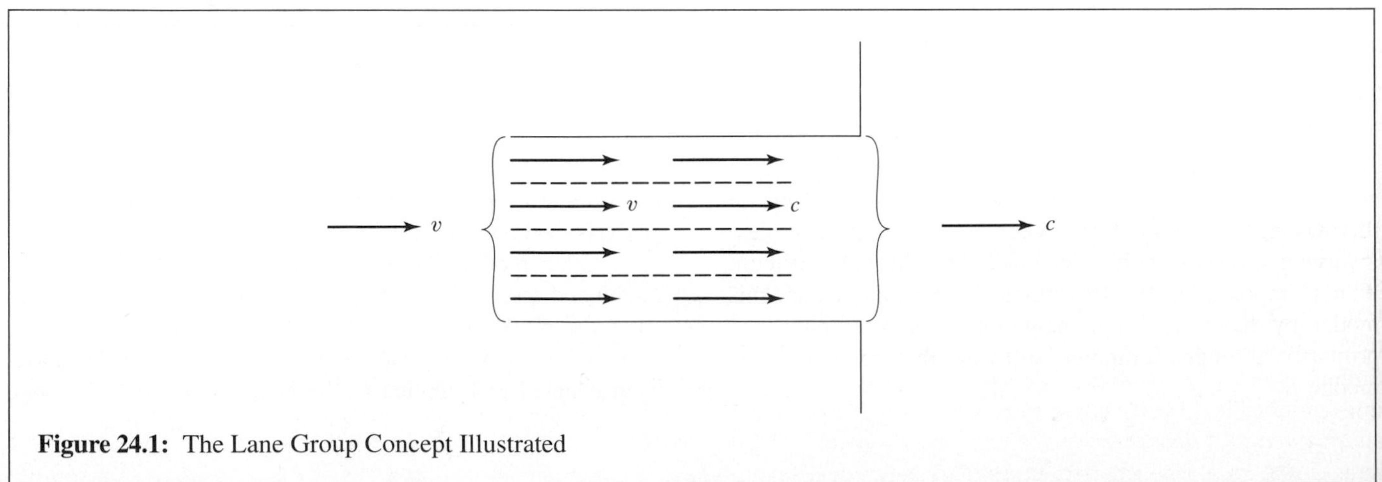

Figure 24.1: The Lane Group Concept Illustrated

24.2.2 The *v/s* Ratio as a Measure of Demand

In Chapter 21, a signal timing methodology was based on conversion of demand volumes to "through-vehicle equivalents." This allowed volumes with markedly different percentages of right- and left-turning vehicles to be directly compared in the determination of "critical lanes." It was assumed that all other conditions that might affect the equivalency of volumes (heavy vehicles, grades, parking conditions, etc.) were "typical."

In the HCM model, demand flow rates are not converted. They are stated as "veh/h" under prevailing conditions. Without conversion to some common base, therefore, flow rates cannot be compared directly with critical lanes or lane groups.

The HCM model includes adjustments for a wide variety of prevailing conditions, including the presence of left- and right-turning vehicles. All adjustments, however, are applied to the saturation flow rate, not to demand volumes. As a result, the methodology yields saturation flow rates and capacities that are defined in terms of prevailing conditions. These are then compared with demand volumes that reflect the same prevailing conditions.

To obtain a single parameter that will allow the intensity of demand in each lane group to be compared directly, the demand flow rate, *v* is divided by the saturation flow rate, *s*, to form the "flow ratio," *v/s* because the prevailing conditions in each lane group are reflected in both the flow rate and the saturation flow rate values, this dimensionless number may be used to represent the magnitude of the demand in each lane group.

24.2.3 Capacity and Saturation Flow Rate Concepts

The HCM model does not produce a value for the capacity of the intersection. Rather, each lane group is considered separately, and a capacity for each is estimated.

Why not simply add all of the lane group capacities to find the capacity of the intersection as a whole? Doing so would ignore the fact that traffic demand does not reach its peak on all approaches at the same time. Unless the demand split on each of the lane groups matched the split of capacities, it would be impossible to successfully accommodate a total demand equal to a capacity so defined. Further, signal timings may change during various periods of the day, yielding significantly different capacities on individual lane groups and, indeed, different sums. In effect, the "capacity" of the intersection as a whole is not a useful or relevant concept. The intent of signalization is to allocate sufficient time to various lane groups and movements to accommodate demand. Capacity is provided to specific movements to accommodate movement demands.

The concept of intersection capacity should not be confused with the "sum of critical-lane volumes" introduced in Chapters 20 and 21. The latter considers only the critical lanes and depends on a specified cycle length. The concepts of saturation flow rate, capacity, and *v/c* ratios are all interrelated in the HCM analysis model.

Saturation Flow Rates

In Chapters 20 and 21, we assumed the saturation headway or saturation flow rate reflecting prevailing conditions was known. A key part of the HCM model is a methodology for estimating the saturation flow rate of any lane group based on known prevailing traffic parameters. The algorithm takes this form:

$$s_i = s_o N \prod_i f_i \qquad (24\text{-}1)$$

where: s_i = saturation flow rate of lane group i under prevailing conditions, veh/hg

s_o = saturation flow rate per lane under base conditions, pc/hg/ln

N = number of lanes in the lane group

f_i = multiplicative adjustment factor for each prevailing condition i

The HCM now provides 11 adjustment factors covering a wide variety of potential prevailing conditions. Each adjustment factor involves a separate model, some of which are quite complex. These are described in detail later in the chapter.

Note that the algorithm includes multiplication of the base saturation flow rate by the number of lanes in the lane group, *N*. This produces a *total* saturation flow rate for the lane group in question.

Capacity of a Lane Group

The relationship between saturation flow rates and capacities is the same as that presented in Chapters 20 and 21. The saturation flow rate is an estimate of the capacity of a lane group if the signal were green 100% of the time. In fact, the signal is only effectively green for a portion of the time. Thus:

$$c_i = s_i (g_i/C) \qquad (24\text{-}2)$$

where: c_i = capacity of lane group i, veh/h
s_i = saturation flow rate of lane group i, veh/hg
g_i = effective green time for lane group i, s
C = cycle length, s

The *v/c* Ratio

In signal analysis, the *v/c* ratio is often referred to as the "degree of saturation" and given the symbol "*X*." This is convenient because the term "*v/c*" appears in many equations that can be more simply expressed using a single variable, *X*.

The *v/c* ratio, or degree of saturation, is a principal output measure from the analysis of a signalized intersection. It is a measure of the sufficiency of available capacity to handle existing or projected demands. Obviously, cases in which $v/c > 1.00$ indicate a shortage of capacity to handle the demand. Care must be taken, however, in analyzing such cases, depending on how the *v/c* value was determined.

Capacity, which is difficult to directly observe in the field, is most often estimated using Equation 24-2. Measured demands are usually a result of counts of *departure flows*. Departure flows are counted because it is easier to classify them by movement as they depart the intersection. True demand, however, must be based on *arrival flows*. There are several different scenarios in which a *v/c* ratio in excess of 1.00 can be achieved:

1. A departure count is compared to a capacity estimated using Equation 24-2. It is theoretically impossible to count a departure flow that is in excess of the true capacity of the lane group. In this case, obtaining a *v/c* ratio greater than 1.00 (assuming the departure counts are accurate) represents an *underestimate* of the capacity of the lane group. The estimated saturation flow rate resulting from the HCM model is lower than the actual value being achieved.
2. An arrival count is compared to an estimate of capacity using Equation 24-2. In this case, the arrival count (assuming it is accurate and complete) represents existing demand. A *v/c* ratio in excess of 1.00 indicates that queuing is likely to occur. If queues are, in fact, not observed, it is another indication that the capacity has been *underestimated* by the model. Note that arrival counts do not capture such demand elements as traffic diverted to other routes or repressed demand.
3. A forecast future demand is compared to an estimated capacity using Equation 24-2. In this case, the forecast demand is always an arrival demand flow, and a *v/c* ratio in excess of 1.00 indicates that queuing is likely to occur, based on the estimated value of capacity.

The key in all cases is that capacity is an *estimate* based on nationally observed norms and averages. In any given case, the actual capacity can be either higher or lower than the estimate. In fact, actual capacity has stochastic elements and will vary over time at any given location and over space at different locations.

Analytically, the *v/c* ratio for any given lane group is found directly by dividing the demand flow rate by the capacity. Another expression, however, can be derived by inserting Equation 24-2 for capacity:

$$X_i = \frac{v_i}{c_i} = \frac{(v/s)_i}{(g/C)_i} \tag{24-3}$$

where: X_i = degree of saturation (*v/c* ratio) for lane group i
v_i = demand flow rate for lane group i, veh/h
c_i = capacity for lane group i, veh/h
$(v/s)_i$ = flow ratio for lane group i
$(g/C)_i$ = green ratio for lane group i

Because demands are eventually expressed as *v/s* ratios in the HCM model, the latter form of the equation is sometimes convenient for use.

Although the HCM does not define a capacity for the entire intersection, HCM 2010 does define a *critical v/s ratio* for the intersection. It is defined as the sum of the flow ratios on critical-lane groups divided by the sum of the g/C ratios of critical-lane groups, or:

$$X_C = \frac{\sum_i (v/s)_{ci}}{\sum_i (g_{ci}/C)} \tag{24-4}$$

where: X_c = critical *v/c* ratio for the intersection
$(v/s)_{ci}$ = v/s ratio for critical-lane group i, veh/h
g_{ci} = effective green time for critical-lane group i, s
C = cycle length, s

The term $\Sigma(g_{ci}/C)$ is the total proportion of the cycle length that is effectively green for all critical-lane groups. Because the definition of a critical-lane group is that one and only one such lane group must be moving during all phases, the only time a critical movement is *not* moving is during the lost times of the cycle. Thus $\Sigma(g_{ci}/C)$ may also be expressed as:

$$\frac{C - L}{C}$$

where L is the total lost time per cycle. Inserting this into Equation 24-4 yields:

$$X_C = \frac{\sum_i (v/s)_{ci}}{\left(\frac{C-L}{C}\right)} = \sum_i (v/s)_{ci} * \left(\frac{C}{C-L}\right) \quad (24\text{-}5)$$

Because the value of X_c varies with cycle length, it is difficult to apply to future cases in which the exact signal timing may not be known. Thus, for analysis purposes, the 1997 edition of the HCM defined a value of X_c based on the maximum feasible cycle length, which results in the minimum feasible value of X_c. For pretimed signals, the maximum feasible cycle length is usually taken to be 120 seconds, but this is sometimes exceeded in special situations. For actuated signals, longer cycle lengths are not as rare, and 150 seconds is usually used as a practical maximum. Equation 24-5 then becomes:

$$X_{C\min} = \sum_i (v/s)_{ci} * \left(\frac{C_{\max}}{C_{\max} - L}\right) \quad (24\text{-}6)$$

where: X_{cmin} = minimum feasible v/c ratio

C_{max} = maximum feasible cycle length, s

The latter value is more useful in comparing future alternatives, particularly physical design scenarios. The cycle length is assumed to be the maximum and is, in effect, held constant for all cases compared. Use of the maximum cycle length gives a view of the "best" critical v/c ratio achievable through signal timing, given the physical design and the phase plan specified.

Note that subsequent editions of the HCM abandoned this concept. However, it remains a useful one for situations in which exact signal timings are not known.

The critical v/c ratio, X_c is an important indicator of capacity sufficiency in analysis. If X_c is ≤ 1.00 then the proposed physical design, cycle length, and phase plan are sufficient to handle all critical demands. This *does not* mean that all lane groups will operate at $X_i \leq 1.00$. It does, however, indicate that all critical-lane groups can achieve $X_i \leq 1.00$ by reallocating the green time within the existing cycle and phase plan. When $X_c > 1.00$ then sufficient capacity may be provided only by taking one or more of the following actions:

- Increasing the cycle length
- Devising a more efficient phase plan
- Adding a lane or lanes to one or more critical-lane groups

Increasing cycle length can add small amounts of capacity because the lost time per hour is diminished. Devising a more efficient phase plan generally means considering additional left-turn protection or making a fully protected left turn a protected plus permitted left turn. It may also mean consideration of more complex phasing such as leading and lagging greens and/or exclusive left-turn phases followed by a leading green in the direction of heaviest left-turn flow. Chapter 21 contains full discussion of various phasing options.

In many cases, significant capacity shortfalls can be remedied only by adding one or more lanes to critical-lane groups. This increases the saturation flow rate and capacity of these lane groups while the demand is constant.

24.2.4 Level-of-Service Concepts and Criteria

Level of service is defined in the HCM in terms of *total control delay* per vehicle in a lane group. "Total control delay" is basically *time in queue* delay, as defined in Chapter 20, plus acceleration-deceleration delay. LOS criteria are shown in Table 24.1.

HCM 2010 will add the concept that *any* lane group operating at a v/c ratio greater than 1.00 is also labeled as LOS F. In effect, any signalized intersection lane group that has an average delay greater than 80 s/veh *or* a $v/c > 1.00$ is operating at LOS F.

Because delay is difficult to measure in the field and because it cannot be measured for future situations, delay is estimated using analytic models, some of which were discussed in

Table 24.1: Level of Service Criteria for Signalized Intersections

Level of Service	Control Delay (s/veh)
A	≤10
B	>10–20
C	>20–35
D	>35–55
E	>55–80
F	>80 or v/c >1.00

(*Source:* Used with permission of Transportation Research Board, National Research Council, *Highway Capacity Manual*, 4th Edition, Washington DC, 2000, Exhibit 16-2, p. 16-2, as modified by vote of the HCQSC of TRB.)

Chapter 20. Delay is not a simple measure, however, and varies (in order of importance) with the following measures:

- Quality of progression
- Cycle length
- Green time
- *v*/*c* ratio

Because of this, LOS results must be carefully considered. It is possible, for example, to obtain a result in which delay is greater than 80 s/veh (LOS F) while the *v*/*c* ratio is less than 1.00. Thus, at a signalized intersection, LOS F does not necessarily imply there is a capacity deficiency. Such a result is relatively common for short phases (such as LT phases) in a long cycle length or where the green splits are grossly out of sync with demands.

The reverse, however, no longer leads to confusion. If $v/c > 1.00$ for a short time—one 15-minute interval, for example—delay could be less than 80 s/veh but now must still be labeled LOS F.

Understanding the results of a signalized intersection analysis requires consideration of *both* the LOS and the *v*/*c* ratio for each lane group. Only then can the results be understood in terms of the sufficiency of the capacity provided and of the acceptability of delays experienced by road users.

24.2.5 Effective Green Times and Lost Times

The relationship between effective green times and lost times is discussed in detail in Chapter 21. In terms of capacity analysis, any given movement has effective green time, g_i, and effective red time, r_i. Figure 24.2 illustrates how these values are related to actual green, yellow, and red times in the HCM.

Where there are overlapping phases, care must be taken in the application of lost times. When a movement continues into a subsequent phase, there is startup lost time but no clearance lost time is incurred between the phases but only at the end of the second phase.

The case of a leading and lagging green phase with protected plus permitted left turns is illustrated in Figure 24.3. Eastbound (EB) movements begin in Phase 1a and continue in Phase 1b. The startup lost time is only applied in Phase 1a, and clearance lost time is applied at the end of Phase 1b. Westbound (WB) movements, however, begin in Phase 1b and have

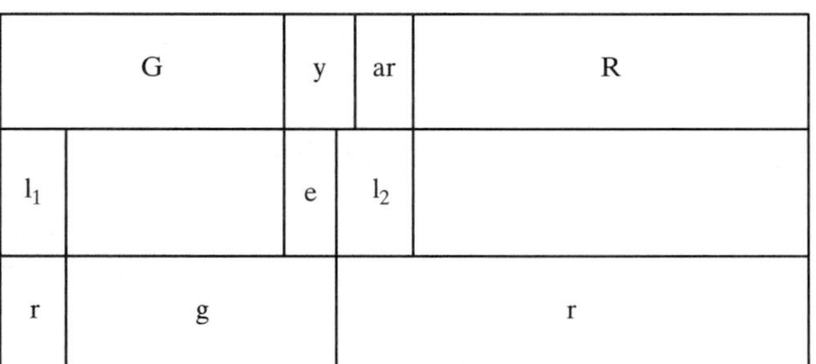

Figure 24.2: Effective Green Times and Lost Times in the HCM

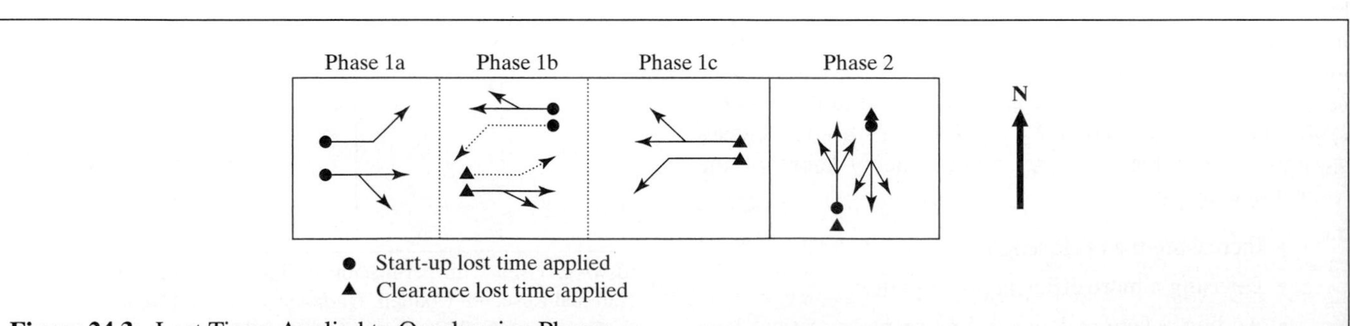

Figure 24.3: Lost Times Applied to Overlapping Phases

their startup lost time applied there. Thus, in Phase 1b, startup lost time for WB movements is effective green time for EB movements. Because no movements begin in Phase 1c, no startup lost times are assessed here, but there is clearance lost time applied for the WB movements. All NB/SB movements flow in Phase 2; their lost times are assessed in this phase. Essentially, three sets of lost times are applied over four subphases. Because effective green times affect capacity and delay, a systematic way of properly accounting for lost times must be followed.

Effective green and red times may be found as follows:

$$g_i = G_i + y_i + ar_i - \ell_1 - \ell_2$$
$$g_i = G_i + e - \ell_1$$
$$r_i = C - g_i \tag{24-7}$$

where: g_i = effective green time for Phase I, s
G_i = actual green time for Phase i, s
Y_i = $y_i + ar_i$, sum of yellow and all-red time for Phase i, s
l_1 = startup lost time, s
l_2 = clearance lost time, s
e = extension of effective green into yellow and all-red, s
r_i = effective red time for Phase i, s

24.3 The Basic Model

24.3.1 Model Structure

The basic structure of the HCM model for signalized intersections is relatively straightforward and includes many of the conceptual treatments presented in Chapter 20. The model becomes extremely complex, however, when permitted or compound left turns are involved. The complex interactions between permitted left turns and opposing flows must be fully incorporated into models. This results in algorithms with many variables and several iterative aspects. It also becomes very complex when actuated signals are analyzed. For simplicity and clarity of presentation, this chapter focuses on applications to pretimed signals.

In this section of the chapter, the building blocks of the HCM procedure are described and illustrated for basic cases involving only protected left turns for pretimed signals. In this way, the fundamental approach of the methodology can be presented without diversions into lengthy detail concerning permitted left turns and actuated signal complications. Subsequent sections of the chapter address these details. The structure of the HCM model is illustrated in Figure 24.4.

The methodology is modular, as depicted in Figure 24.4. The modules are described briefly here:

1. *Input module.* The input data include complete descriptions of traffic characteristics, roadway geometry, and signalization. For pretimed signals, the signal cycle and all of its intervals must be specified. For actuated signals, controller settings are specified, and the methodology directly incorporates these settings into critical determinations.
2. *Define movement groups and adjusted flow rates.* In this analysis module, movement groups for analysis are defined, and appropriate demand flow rates for each are estimated.
3. *Compute lane group flow and saturation flow rates.* In this module, analysis lane groups are identified, and demand flow rates are assigned to each. The saturation flow rate for each lane group is estimated.
4. *Input or compute phase duration.* For pretimed signals, timing parameters are specified as part of the input data; for actuated signals, phase durations are estimated in this module.
5. *Compute capacity.* The capacity of each lane group is estimated, and *v/c* ratios are computed.
6. *Compute performance measures.* Control delay for each lane group is estimated and the appropriate LOS assigned; aggregations of control delay for approaches and the overall intersection are computed. The queue storage ratio is estimated as an additional performance measure.

24.3.2 Analysis Time Periods

The basic time period for analysis recommended by the HCM remains a peak 15-minute period within the analysis hour, which is most often (but need not be) one of the peak hours of the day. Beginning with the HCM 2000, however, the HCM provides for some flexibility in this regard, recognizing that delay is particularly sensitive to the analysis period, especially when oversaturation exists. There are three basic time options for analysis:

1. The peak 15 minutes within the analysis hour
2. The full 60-minute analysis hour
3. Sequential 15-minute periods for an analysis period of one hour or greater

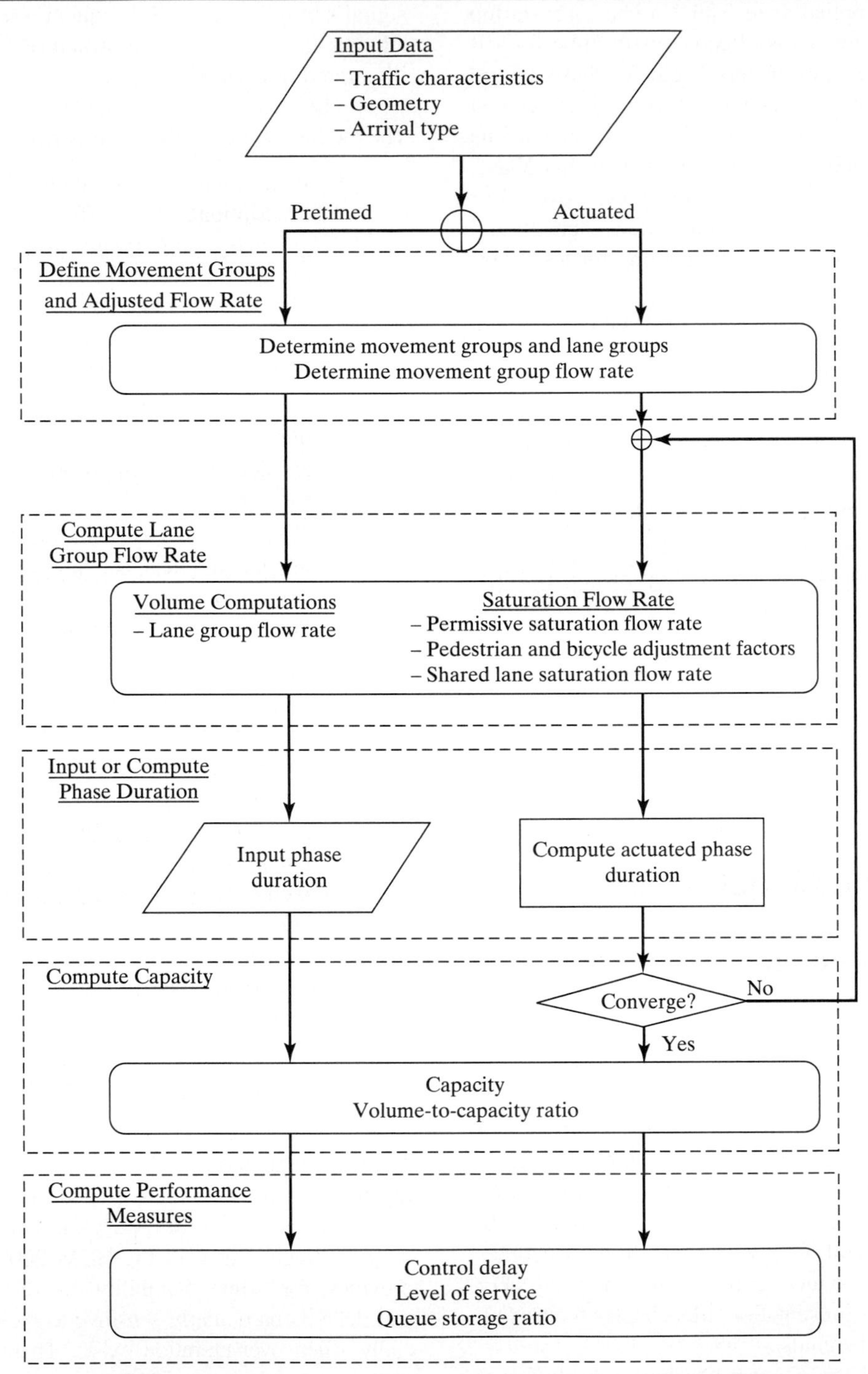

Figure 24.4: Flow Chart of the HCM 2010 Model for Signalized Intersections

(*Source:* NCHRP 3-92, Production of the 2010 Highway Capacity Manual, *Draft Chapter 18*, Exhibit 18-11, pp. 18–31.)

The first option is appropriate in cases where no oversaturation exists (i.e., no lane groups have $v/c > 1.00$). This focuses attention on the worst period within the analysis hour, where 15 minutes remains the shortest period during which stable flows are thought to exist. The second option allows for an analysis of average conditions over the full analysis hour. It could, however, mask shorter periods during which $v/c > 1.00$, even though the full hour has sufficient capacity.

The third option is the most comprehensive. It requires, however, that demand flows be measured or predicted in 15-minute time increments, which is often difficult. It allows, however, for the most accurate analysis of oversaturation conditions. The initial 15-minute period of analysis would be selected such that all lane groups operate with $v/c \leq 1.00$ and would end in a 15-minute period occurring after all queues have been dissipated. During each 15-minute period, residual queues of unserved vehicles would be estimated and would be used to estimate additional delay due to a queue existing at the start of an analysis period in the subsequent interval. In this way, the impact of residual queues on delay and LOS in each successive period can be estimated.

24.3.3 Input

The Input Module involves a full specification of all prevailing conditions existing on all approaches to the signalized intersection under study. As the model has become more complex in order to address a wider range of conditions, the input information required has also become more comprehensive.

Table 24.2 summarizes all of the input data needed to conduct a full analysis of a signalized intersection. Most of the variables included in Table 24.2 have been previously defined. Others require some additional definition or discussion. The HCM also provides recommendations for default values that may be used in cases where field data on a particular characteristic is not available. Caution should be exercised in using these because the accuracy of v/c, delay, and LOS predictions is influenced.

Geometric Conditions

Parking conditions must be specified for each movement group. For a typical two-way street, each approach either has curb parking or not. On a one-way street approach, parking may exist on the right and/or left side or on neither. For the purposes of the intersection, curb parking is noted only if it exists within 250 feet of the STOP line of the approach. Most of the other geometric conditions that must be specified are commonly used variables that have been defined elsewhere in this text.

The HCM provides default values for only two of the geometric variables in Table 24.2, as follows:

- *Lane width, W*: default value = 12 ft
- *Grade, G*: default values = 0% (level); 3% (moderate grades); 6% (steep grades)

Using a default for grade involves at least a general categorization of the grade from field observations.

Where wide lanes exist, some observation of their use should be made. Lanes of 16 to 20 feet often become two lanes under intense demand, particularly if it is a curb lane that could be used as a through lane plus a narrow RT lane. The analyst should try to characterize lanes as they would be used, not necessarily as they are striped. In most cases, however, these would be the same.

Traffic Conditions

A number of interesting variables are included in the list of traffic conditions to be specified.

Arrival Type "Arrival type" is used to describe the quality of progression for vehicles arriving on each approach. Arrival type has a major impact on delay predictions but does not have significant influence on other portions of the methodology. There are six defined arrival types, 1 through 6, with AT 1 representing the worst progression quality and AT 6 representing the best progression quality. Definitions are given in Table 24.3, and the verbal description defining each arrival type is shown in Table 24.4.

If the proportion of vehicles arriving on green is observed in the field, then the AT is computed as:

$$AT = 3\frac{P}{g/C} \tag{24-8}$$

where: P = proportion of vehicles arriving on green, decimal

g = effective green time for movement, s

C = cycle length for movement, s

The result may be rounded to the nearest integer.

If no field data are available, the descriptions shown in Table 24.5 may be used with knowledge of signal spacing to make a rough default estimate of arrival type.

Given the significant impact arrival type can have on delay estimates, it is important that a common arrival type be used when comparing different intersection designs and signal timings. High delays should not be simply dismissed or mitigated by assuming an improved progression quality.

Table 24.2: Data Requirements for Automobiles Signalized Intersection Analysis

Type of Condition	Parameter	Basis
Geometric Conditions	Number of Lanes, N	Movement Group
	Average Lane Width, W (ft)	Movement Group
	Number of Receiving Lanes	Approach
	Grade, G (%)	Approach
	Existence of LT or RT Lanes	Movement Group
	Length of Storage Bay for LT or RT lane (ft)	Movement Group
	Parking Conditions (Yes/No)	Movement Group
Traffic Conditions	Demand Volume (or flow rate) by Movement, V (veh/h)	Movement Group
	RTOR Flow Rate, (veh/h)	Approach
	Base Saturation Flow Rate, s_o (pc/hg/ln)	Movement Group
	Peak Hour Factor, PHF	Intersection
	Percent Heavy Vehicles, P_T (%)	Movement Group
	Pedestrian Flow in Conflicting Crosswalk, v_p (peds/h)	Approach
	Local Buses Stopping at Intersection, N_B (buses/h)	Approach
	Parking Activity, N_m (maneuvers/h)	Movement Group
	Arrival Type, AT	Movement Group
	Proportion of Vehicles Arriving on Green, P	Movement Group
	Speed Limit (mi/h)	Approach
Signalization Conditions (Pretimed Signals)	Cycle Length, C (s)	Movement Group
	Green Time, G (s)	Phase
	Yellow Plus All-Red Interval, Y (s)	Phase
	Pedestrian Push-Button (Yes/No)	Phase
	Minimum Pedestrian Green, G_p (s)	Phase
	Phase Plan	Intersection
Other	Analysis Period Duration (min)	Intersection
	Area Type	Intersection

Note: Movement = one value for each LT, TH, and RT movement.
Movement group = one value for each turn movement with exclusive turn lanes and one value for each through movement (inclusive of any turn movements in a shared lane).
Approach = one value or condition for the intersection approach.
Intersection = one value or condition for the intersection.
Phase = one value or condition for each signal phase.
(*Source:* Modified from NCHPR 3-92, Production of the 2010 Highway Capacity Manual, *Draft Chapter 18*, Exhibit 18-6, p. 18-8.)

Pedestrian Flows Pedestrian flows in conflicting crosswalks must be specified for a signalized intersection analysis. The "conflicting crosswalk" is the crosswalk that right-turning vehicles turn through. Pedestrian flows are best measured in the field, but the HCM 2000 provided the following default values:

- *CBD locations*: 400 peds/h per crosswalk
- *Other locations*: 50 peds/h per crosswalk

These defaults will not be included in HCM 2010, emphasizing the need to obtain accurate pedestrian counts whenever possible.

Parking Activity Parking activity is measured in terms of the number of parking maneuvers per hour into and out of parking spaces (N_m) located within 250 feet of the STOP line of the lane group or approach in question. Movements into and out of parking spaces have additional negative impacts on operations because the lane adjacent to the parking lane is

Table 24.3: Arrival Types Defined

Arrival Type	Description
1	Dense platoon containing over 80% of the lane group volume, arriving at the start of the red phase. This AT is representative of network links that may experience very poor progression quality as a result of conditions such as overall network optimization.
2	Moderately dense platoon arriving in the middle of the red phase, or dispersed platoon containing 40% to 80% of the lane group volume, arriving throughout the red phase. This AT is representative of unfavorable progression on two-way streets.
3	Random arrivals in which the main platoon contains less than 40% of the lane group volume. This AT is representative of operations at isolated and noninterconnected signalized intersections characterized by highly dispersed platoons. It may also be used to represent uncoordinated operation in which the benefits of progression are minimal.
4	Moderately dense platoon arriving in the middle of the green phase or dispersed platoon containing 40% to 80% of the lane group volume, arriving throughout the green phase. This AT is representative of favorable progression on a two-way street.
5	Dense to moderately dense platoon containing over 80% of the lane group volume, arriving at the start of the green phase. This AT is representative of highly favorable progression quality, which may occur on routes with low to moderate side-street entries and that receive high-priority treatment in the signal timing plan.
6	This arrival type is reserved for exceptional progression quality with near-ideal progression characteristics. It is representative of very dense platoons progressing over a number of closely spaced intersections with minimal or negligible side-street entries.

(*Source:* Used with Permission of Transportation Research Board, National Research Council, *Highway Capacity Manual*, 4th Edition, Washington DC, 2000, Exhibit 16-4, p. 16-4.)

Table 24.4: Relationship Between Platoon Ratio and Arrival Type

Arrival Type	Progression Quality
1	Very poor
2	Unfavorable
3	Random arrivals
4	Favorable
5	Highly favorable
6	Exceptional

(*Source:* NCHRP 3-92, Production of the 2010 Highway Capacity Manual, *Draft Chapter 18*, Exhibit 18-8, p. 18-11.)

disrupted for some finite amount of time each time such a maneuver takes place. This is in addition to the frictional impacts of traveling in the lane adjacent to the parking lane. Parking activity should be observed in the field but it is often not readily available. The HCM recommends the use of the default values shown in Table 24.6 in such cases.

Local Buses A local bus is defined as one that stops within the confines of the intersection, either on the near side or far side of the intersection, to pick up and/or discharge passengers. A local bus that does not stop at the intersection is included as a heavy vehicle in the heavy-vehicle percentage. Once again, this is a variable best measured in the field. If field measurements and/or bus schedules are not available, the HCM recommends the following default values be used (assuming there is a bus route on the street in question with a bus stop within the confines of the intersection):

- *CBD location*: 12 buses/hour
- *Other location*: 2 buses/hour

Other Default Values Other traffic conditions are well defined elsewhere in this text. The HCM does recommend default values for some of these variables if they are unavailable from field data or projections:

- *Base saturation flow rate*: 1,900 pc/hg/ln for cities with population $>$250,000; 1,750 pc/hg/ln otherwise.
- *Heavy-vehicle presence*: 3.0%
- *Peak-hour factor*: 0.92 for total entering volume $\geq$1,000 veh/h; 0.90 otherwise

Table 24.5: Progression Quality and Arrival Type

Arrival Type	Progression Quality	Signal Spacing (ft)	Conditions Under which Arrival Type Is Likely to Occur
1	Very poor	≤1,600	Coordinated operation on a two-way street where the subject direction does not receive good progression.
2	Unfavorable	>1,600–3,200	A less extreme version of arrival type 1.
3	Random arrivals	>3,200	Isolated signals or widely-spaced coordinated signals.
4	Favorable	>1,600–3,200	Coordinated operation on a two-way street where the subject direction receives good progression.
5	Highly favorable	≤1,600	Coordinated operation on a two-way street where the subject direction receives good progression.
6	Exceptional	≤800	Coordinated operation on a two-way street in dense networks and central business districts.

(*Source:* NCHRP 3-92, Production of the 2010 Highway Capacity Manual, *Draft Chapter 18*, Exhibit 18-32, pp. 18–89.)

Table 24.6: Recommended Default Values for Parking Activity

Street Type	Number of Parking Spaces within 250 ft of STOP Line	Parking Time Limit (*h*)	Maneuvers per Hour (N_m): Recommended Default Value
Two-Way	10	1 2	16 8
One-Way	20	1 2	32 16

(*Source:* Used with permission of Transportation Research Board, National Research Council, *Highway Capacity Manual*, 4th Edition, Washington DC, 2000, Exhibit 10-20, pp. 10–25.)

As we noted previously, use of default values should be avoided whenever local field data or projections are available. For each default value used in lieu of specific data for the intersection, the accuracy of predicted operating conditions becomes less reliable.

Signalization Conditions Virtually all of the signalization conditions that must be specified for an analysis are available from field observations of signal operation or from a signal design and timing analysis. The HCM provides a procedure for "quick" estimation of signal timing that involves many defaults and assumptions. The "quick" method can become relatively complex, despite the simplifying assumptions. The signal timing approach outlined in Chapter 21 of this text is less complex and can be used to obtain an initial signal timing for a capacity and LOS analysis. The analysis time period (*T*) has been discussed earlier in this chapter, and no further discussion is needed.

The analysis methodology uses an algorithm to determine the adequacy of pedestrian crossing times. It is the same as that recommended for use in signal timing in Chapter 21, or:

$$G_p = 3.2 + 0.27\, N_{peds} + \left(\frac{L}{S_p}\right) \quad \text{for } W_E \leq 10 \text{ ft}$$

$$G_p = 3.2 + 2.7\left(\frac{N_{peds}}{W_E}\right) + \left(\frac{L}{S_p}\right) \quad \text{for } W_E > 10 \text{ ft} \tag{24-9}$$

where: G_p = minimum green time for safe pedestrian crossings, s
L = length of the crosswalk, ft
S_p = walking speed of pedestrians, ft/s
N_{peds} = number of pedestrians crossing during one green interval, peds/cycle
W_E = width of the crosswalk, ft

The HCM checks green times against these minimum values and issues a warning statement if pedestrian crossings are unsafe under these guidelines. Analysis, however, may continue whether or not the minimum pedestrian green condition is met by the signalization or not.

Also note that many local and state agencies have their own policies on what constitutes safe crossings for pedestrians. These may always be applied as a substitute for Equation 24-9.

24.3.4 Movement Groups, Lane Groups, and Demand Volume Adjustment

One of the complexities being introduced in HCM 2010 is an extremely intricate process for assigning demand flow rates to analysis lane groups. This is necessary because the HCM 2010 will treat a single shared lane as a separate lane group, not as a part of a combined TH/LT or TH/RT lane. It is now necessary to estimate the exact composition of the shared lane.

Converting Demand Volumes to Demand Flow Rates

The process starts, however, with the determination of movement demand flow rates. It is most desirable to specify the demand flow rates by 15-minute analysis period as input from field data or projections. If, however, demand is specified as hourly volumes, then they must be converted to flow rates:

$$v = \frac{V}{PHF} \tag{24-10}$$

where: v = demand flow rate, veh/h

V = demand volume, veh/h

PHF = peak-hour factor

This computation assumes that all movements peak during the *same* 15-minute interval. This is almost always a false assumption. It does, however, represent a worst-case assumption. If an intersection works under this assumption, then the actual distribution of flow rates over the four 15-minute intervals of the demand volume hour will work as well or better than the analysis result indicates.

Establishing Analysis Lane Groups

HCM 2010 defines between one and five lane groups for analysis on each approach. Six different types of lane groups can exist on an intersection approach:

- A single-lane approach in which all three movements (LT, TH, and RT) are made from one lane.
- An exclusive LT lane or lanes; multiple LT lanes form a single lane group.
- An exclusive RT lane or lanes; multiple RT lanes form a single lane group.
- Exclusive TH lanes (no turns from these lanes) on an approach form a single lane group.
- A shared LT/TH lane; such a lane is treated as a separate lane group.
- A shared RT/TH lane; such a lane is treated as a separate lane group.

Figure 24.5 illustrates common combinations of lane groups on a signalized intersection approach.

Lanes that exclusively serve LT, TH, or RT movements are straightforward: they are simply established as analysis lane groups. Where shared lanes potentially exist, the problem is more difficult. In these cases, the mix of movements in shared lanes must be estimated. Depending on relative demand flow rates, a lane intended for shared use of LT and TH vehicles might, for example, actually operate as an exclusive LT lane; if it operates as a shared lane, the mix of LT and TH demand flow rates must be determined.

HCM 2010 presents an iterative analysis for this that optimizes the delay to motorists by assuming that each driver chooses a lane placement that minimizes their travel time through the intersection. This methodology, documented in Chapter 31 of HCM 2010 (which will only be available in electronic format, ideally online) cannot be implemented by hand and will be incorporated into the HCS package and other computer tools. This procedure is not detailed here.

The methodology, as for many variables, notes that field-measured values for lane group demand flow rates is preferred over the estimation procedure and allows lane group demand volumes or flow rates to be specified as input. The approximate "through-vehicle-equivalents" used in the signal timing methodology of Chapter 21 can be used to get a rough estimation of lane distribution by lane group by assuming that equivalent lane flow rates should be equalized across the approach.

24.3.5 Estimating the Saturation Flow Rate for Each Lane Group

Among the most complex computations in the HCM is the estimation of the saturation flow rate for each of the defined

Number of Lanes	Movements by Lanes	Movements Groups (MG)	Lane Groups (LG)
1	Left, thru., & right:	MG 1:	LG 1:
2	Exclusive left:	MG 1:	LG 1:
	Thru. & right:	MG 2:	LG 2:
2	Left & thru.:	MG 1:	LG 1:
	Thru. & right:		LG 2:
3	Exclusive left:	MG 1:	LG 1:
	Through:	MG 2:	LG 2:
	Thru. & right:		LG 3:

Figure 24.5: Common Movement and Lane Groups on a Signalized Intersection Approach

(*Source:* NCHRP 3-92, Production of the 2010 Highway Capacity Manual, *Draft Chapter 18*, Exhibit 18-12, pp. 18–33.)

lane groups in the intersection. The saturation flow rate is computed as:

$$s = s_o N f_w f_{HV} f_g f_p f_{bb} f_a f_{LU} f_{RT} f_{LT} f_{Rpb} f_{Lpb} \quad (21\text{-}11)$$

where: s = Saturation flow rate for the lane group, veh/h

s_o = base saturation flow rate, pc/hg/ln (1,900 pc/hg/ln, 1,750 pc/hg/ln, or locally calibrated value)

N = number of lanes in the lane group

f_i = adjustment factor for prevailing condition i (w = lane width; HV = heavy vehicles; g = grade; p = parking; bb = local bus blockage; a = area type; LU = lane use; RT = right turn; LT = left turn; Rpb = pedestrian/bicycle interference with right turns; and Lpb = pedestrian/bicycle interference with left turns)

Of these 11 adjustment factors, 8 were introduced with the original methodology in the 1985 HCM. The lane-use adjustment was moved from the volume side of the equation to the saturation flow rate side in the 1997 HCM. The pedestrian/bicycle interference factors were added in HCM 2000. All are retained in HCM 2010.

Eight of the adjustments are relatively straightforward. Pedestrian/bicycle adjustments are a little complex but are still tractable and can be done manually with a little care. The left-turn adjustment is straightforward unless permitted or compound left turns are involved, in which case the model becomes cumbersome and impractical to implement manually. In this section, left-turn adjustments are defined only for protected turns. A subsequent section discusses the more complex model for permitted left turns.

Adjustment for Lane Width

Although the standard lane width at signalized intersection is 12 feet, recent research [8] has indicated that as long as lane width is between 10 and 12.9 feet, there is no impact on saturation flow rate or capacity. Thus there are only three values for the lane width adjustment factor:

$f_w = 0.96$ Lane with <10 ft

$f_w = 1.00$ 10 ft $\leq$ lane width ≤ 12.9 feet

$f_w = 1.04$ Lane with >12.9 feet

Lane widths less than 10 feet are not recommended.

Adjustment for Heavy Vehicles

The adjustment for heavy vehicles is done separately from consideration of grade impacts. This separation, different from the approach taken for uninterrupted flow facilities, recognizes that approach grades affect the operation of all vehicles, not just heavy vehicles. A "heavy vehicle" is any vehicle with more than four wheels touching the ground during normal operations.

$$f_{HV} = \frac{1}{1 + P_{HV}(E_{HV} - 1)} \quad (24\text{-}12)$$

where: f_{HV} = heavy-vehicle adjustment factor

P_{HV} = proportion of heavy vehicles in the lane group demand flow

E_{HV} = passenger-car equivalent for a heavy vehicle

For signalized intersections, the value of E_{HV} is a constant: 2.00.

Heavy vehicles are not segmented into separate classes. Thus they include trucks, recreational vehicles, and buses not stopping within the confines of the intersection. Buses that do stop within the confines of the intersection are treated as a separate class of vehicles: local buses.

Adjustment for Grade

The adjustment for grade is found as:

$$f_g = 1 - \left(\frac{G}{200}\right) \quad (24\text{-}13)$$

where: f_g = Grade adjustment factor

G = Grade %

Remember that downgrades have a negative percentage, resulting in an adjustment in excess of 1.00. Upgrades have a positive percentage, resulting in an adjustment of less than 1.00.

Adjustment for Parking Conditions

The parking adjustment factor involves two variables: (1) parking conditions and movements, and (2) the number of lanes in the lane group. If there is no parking adjacent to the lane group, the factor is, by definition, 1.00. If there is parking adjacent to the lane group, the impact on the lane directly adjacent to the parking lane is a 10% loss of capacity due to the frictional impact of parked vehicles, plus 18 seconds of blockage for each movement into or out of a parking space within 250 feet of the STOP line. Thus, the impact on an adjacent lane is:

$$P = 0.90 - \left(\frac{18\ N_m}{3{,}600}\right)$$

where: P = adjustment factor applied only to the lane adjacent to parking lane

N_m = number of parking movements per hour into and out of parking spaces within 250 ft of the STOP line (mvts/h)

It is then assumed that the adjustment to additional lanes in the lane group is 1.00 (unaffected), or:

$$f_p = \frac{(N - 1) + P}{N}$$

where: N = number of lanes in the lane group

These two expressions are combined to yield the final equation for the parking adjustment factor:

$$f_p = \frac{N - 0.10 - \left(\frac{18\ N_m}{3{,}600}\right)}{N} \geq 0.05 \quad (24\text{-}14)$$

There are several external limitations on this equation:

- $0 \leq N_m \leq 180$; if $N_m > 180$, use 180 mvts/h
- $f_p(\text{min}) = 0.05$
- $f_p(\text{no parking}) = 1.00$

On a one-way street with parking on both sides, N_m is the total number of right- and left-parking lane maneuvers.

Adjustment for Local Bus Blockage

The local bus blockage factor accounts for the impact of local buses stopping to pick up and/or discharge passengers at a near-side or far-side bus stop within 250 feet of the near or far STOP line. Again, the primary impact is on the lane in which the bus stops (or the lane adjacent in cases where an offline bus stop is provided). It assumes that each bus blocks the lane for 14.4 seconds of green time. Thus:

$$B = 1.0 - \left(\frac{14.4 N_B}{3{,}600}\right)$$

where: B = adjustment factor applied only to the lane blocked by local buses

N_B = number of local buses per hour stopping

As in the case of the parking adjustment factor, the impact on other lanes in the lane group is assumed to be nil, so that an adjustment of 1.00 would be applied. Then:

$$f_{bb} = \frac{(N - 1) + B}{N}$$

where: N = number of lanes in the lane group

Combining these equations yields:

$$f_{bb} = \frac{N - \left(\frac{14.4 N_B}{3{,}600}\right)}{N} \geq 0.05 \qquad (24\text{-}15)$$

There are also several limitations on the use of this equation:

- $0 \leq N_B \leq 250$; if $N_B > 250$, use 250 b/h
- $f_{bb}(\text{min}) = 0.05$

If the bus stop involved is a terminal location and/or layover point, field studies may be necessary to determine how much green time each bus blocks. The value of 14.4 seconds in Equation 24-15 could then be replaced by a field-measured value in such cases.

Adjustment for Type of Area

As we noted previously, signalized intersection locations are characterized as "CBD" or "Other," with the adjustment based on this classification:

- *CBD location*: $f_a = 0.90$
- *Other location*: $f_a = 1.00$

This adjustment accounts for the generally more complex driving environment of central business districts and the extra caution that drivers often exercise in such environments. Judgment should be exercised in applying this adjustment. Not all central business districts have such complex environments that headways would be increased for that reason alone. Some non-CBD locations may have a combination of local environmental factors that would make the application of this adjustment advisable.

Adjustment for Lane Utilization

The adjustment for lane utilization accounts for unequal use of lanes by the approaching demand flow in a multilane group. Where demand volumes can be observed on a lane-by-lane basis, the adjustment factor may be directly computed as:

$$f_{LU} = \frac{v_g}{v_{g1} N} \qquad (24\text{-}16)$$

where: v_g = demand flow rate for the lane group, veh/h
v_{g1} = demand flow rate for the single lane with the highest volume, veh/h/ln
N = number of lanes in the lane group

When applied in this fashion, the factor adjusts the saturation flow rate downward so that the resulting *v*/*c* ratios and delays represent, in effect, conditions in the worst lane of the lane group. Although the HCM states that a lane utilization factor of 1.00 can be used "when uniform traffic distribution" can be assumed, the use of this factor is no longer optional, as was the case in the 1985 and 1994 editions of the manual.

Where the lane distribution cannot be measured in the field or predicted reliably, the default values shown in Table 24.7 may be used.

Table 24.7: Default Values for the Lane Utilization Adjustment Factor

Lane Group Movements	No. of Lanes in Lane Group	Traffic in Most Heavily Traveled Lane (%)	Lane Utilization Factor f_{LU}
Through or Shared	1	100.0	1.000
	2	52.5	0.952
	3[a]	36.7	0.908
Exclusive Left Turn	1	100.0	1.000
	2[a]	51.5	0.971
Exclusive Right Turn	1	100.0	1.000
	2[a]	56.5	0.885

[a]If lane group has more lanes, field observations are recommended; if not, use smallest value shown in this exhibit.

(*Source:* Used with permission of Transportation Research Board, National Research Council, *Highway Capacity Manual*, 4th Edition, Washington DC, 2000, Exhibit 10-23, pp. 10–26.)

Adjustment for Right Turns

The right-turn adjustment factor accounts for the fact that such vehicles have longer saturation headways than through vehicles because they are turning on a tight radius requiring reduced speed and greater caution. This factor no longer accounts for pedestrian interference with right-turning vehicles as editions of the HCM before 2000 did. With the introduction of new adjustments for pedestrian/bicycle interference, this element was removed from the right-turn adjustment factor. Right turns occur under three different scenarios:

- From an exclusive RT lane
- From a shared lane
- From a single-lane approach

For right turns from an exclusive RT lane, the adjustment factor for right turns (f_{RT}) is a constant 0.85. For shared-lane turns and single-lane approaches, HCM 2010 provides a complex iterative procedure for determining the saturation flow rate directly. This procedure is not detailed here.

Adjustment for Left Turns

If not for the existence of left turns, the HCM model for signalized intersections would be relatively straightforward. There are six basic situations in which left turns may be made:

- *Case* 1: Exclusive LT lane with protected phasing
- *Case* 2: Exclusive LT lane with permitted phasing
- *Case* 3: Exclusive LT lane with compound phasing
- *Case* 4: Shared lane with protected phasing
- *Case* 5: Shared lane with permitted phasing
- *Case* 6: Shared lane with compound phasing

All of these options are frequently encountered in the field, with the exception of Case 4, which exists primarily on one-way streets with no opposing flows.

As was the case for right turns, left-turning vehicles have a lower saturation flow rate than through vehicles due to the fact that they are executing a turning maneuver on a restricted radius. The reduction in saturation flow rate for left-turning vehicles is less than that for right-turning vehicles because the radius of curvature involved in the maneuver is greater. Right-turning vehicles have a sharper turn than left-turning vehicles. The left-turn adjustment factor (f_{LT}) for fully protected phases is 0.95.

For left turns made from shared lanes and/or with permitted or protected/permitted phasing, the calculations are far more complex. The process for estimating the adjustment factor in these cases is conceptually described later in this chapter.

Adjustments for Pedestrian and Bicycle Interference with Turning Vehicles

These two adjustment factors were added to the HCM in 2000 to account for the interference of both pedestrians and bicycles with right- and left-turning vehicles at a signalized intersection. As was the case for right-turn and left-turn adjustment factors, the methodology (explained here) is relatively simple for cases in which left-turns on one-way streets and/or right turns are made from exclusive lanes. The procedure becomes far more complex and iterative when shared lanes or left turns on a two-way street are involved. Consult the HCM 2010 directly for details of the shared-lane methodology.

Figure 24.6 illustrates the conflicts between turning vehicles, pedestrians, and bicycles.

Right-turning vehicles encounter both pedestrian and bicycle interference virtually immediately on starting their maneuver. Left-turning vehicles encounter pedestrian interference after moving through a gap in the opposing traffic flow. Because bicycles are legally required to be on the right side of the roadway, it is assumed that left turns experience no bicycle interference.

The basic modeling approach is to estimate the proportion of time that the pedestrian-vehicle and bicycle-vehicle are blocked to vehicles (because they are occupied by

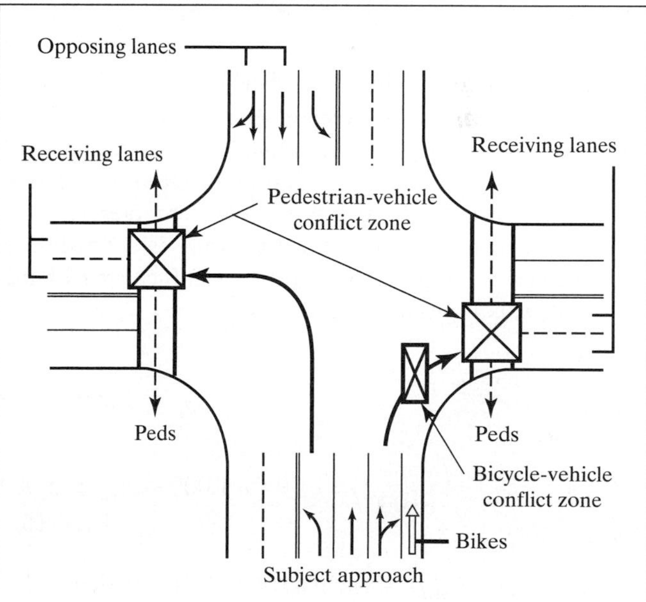

Figure 24.6: Pedestrian and Bicycle Interference with Turning Vehicles

(*Source:* NCHRP 3-92, Production of the 2010 Highway Capacity Manual, *Draft Chapter 18*, Exhibit 18-14, pp. 18–38.)

pedestrians and/or bicyclists, who have the right of way). Only when these conflict zones are unblocked can vehicles travel through them.

The following computational steps are followed:

Step 1: *Estimate Pedestrian Flow Rate During Green Phase (Left and Right Turns)* This is the actual pedestrian demand flow rate during the green phase. The rate is adjusted to reflect the fact that pedestrians are moving only during the green phase of the signal.

$$v_{pedg} = v_{ped}\left(\frac{C}{g_{ped}}\right) \leq 5{,}000 \qquad (24\text{-}17)$$

where: v_{pedg} = flow rate for pedestrians during the green phase, peds/hg

v_{ped} = demand flow rate for pedestrians during analysis period, peds/h

C = cycle length, s

g_{ped} = green phase for pedestrians, s

The value of g_p is taken to be the sum of the pedestrian walk and clearance intervals where they exist, or the length of the vehicular effective green where no pedestrian signals exist. Note that *two* values of this parameter are computed, one for pedestrians in the crosswalk in conflict with right turns, and one for pedestrians in conflict with left turns.

Step 2: *Estimate the Average Pedestrian Occupancy in the Conflict Zone (Left and Right Turns)* "Occupancy" measures represent the proportion of green time during which pedestrians and/or bicycles are present in a particular area for which the measure is defined. The occupancy of pedestrians in the conflict area of the crosswalk is given by Equation 24-18:

$$OCC_{pedg} = \frac{v_{pedg}}{2{,}000} \qquad \text{for } v_{pedg} \leq 1{,}000$$

$$OCC_{pedg} = 0.40 + \left(\frac{v_{pedg}}{10{,}000}\right) \leq 0.90 \text{ for } 1{,}000 < v_{pedg} \leq 5{,}000 \qquad (24\text{-}18)$$

where: OCC_{pedg} = occupancy of the pedestrian conflict area by pedestrians

All other variables as are as previously defined.

The first equation assumes that each pedestrian blocks the crosswalk conflict area for approximately 1.8 seconds, considering walking speeds and the likelihood of parallel crossings. At higher demand flows, the likelihood of parallel crossings is much higher, and each additional pedestrian blocks the crosswalk conflict area for another 0.36 seconds.

Step 3: *Estimate the Bicycle Flow Rate During the Green Phase (Right Turns Only)* The bicycle flow rate during the green phase is found in the same way that the pedestrian flow rate during green was estimated:

$$v_{bicg} = v_{bic}\left(\frac{C}{g}\right) \leq 1900 \qquad (24\text{-}19)$$

where: v_{bicg} = bicycle flow rate during the green phase, bic/hg

v_{bic} = demand flow rate for bicycles during the analysis period, bic/h

C = cycle length, s

g = effective green time for vehicular movement, s

Step 4: *Estimate the Average Bicycle Occupancy in the Conflict Zone (Right Turns Only)* Bicycle occupancy may then be estimated as:

$$OCC_{bicg} = 0.02 + \left(\frac{v_{bicg}}{2700}\right) \qquad (24\text{-}20)$$

where: OCC_{bicg} = occupancy of the conflict area by bicycles

All other variables are as previously defined.

Step 5: *Estimate the Conflict Zone Occupancy (Left and Right Turns)* The occupancies computed in Steps 2 and 3 treat each element (pedestrians and bicycles) separately. Further, it is assumed that turning vehicles are present to block during all portions of the green phase.

Equation 24-21 is used to estimate the conflict zone occupancy for right turns from exclusive lanes with no bicycles present and for left turns from an exclusive lane on a one-way street:

$$OCC_r = \left(\frac{g_{ped}}{g}\right) OCC_{pedg} \qquad (24\text{-}21)$$

where: OCC_r = occupancy of the conflict zone

For *right-turning vehicles*, where both pedestrian and bicycle flows exist, Equation 24-22 is used. Because the two interfering flows overlap, simply adding the two occupancy values would result in too great an adjustment. Allowing for the overlapping impact of pedestrians and bicycles on right-turning vehicles:

$$OCC_r = \left(\frac{g_{ped}}{g} OCC_{pedg}\right) + OCC_{bicg} - \left(\frac{g_{ped}}{g} OCC_{pedg} OCC_{bicg}\right) \quad (24\text{-}22)$$

All terms are as previously defined.

Step 6: *Estimate the Unblocked Portion of the Phase, A_{pbT} (Right and Left Turns)* Once the occupancies of the conflict zone have been established, the unblocked time for conflict zones (one for right turns, one for left turns) are computed as follows:

$$A_{pbT} = 1 - OCC_r \quad \text{if } N_{rec} = N_{turn}$$
$$A_{pbT} = 1 - 0.6OCC_r \quad \text{if } N_{rec} > N_{turn} \quad (24\text{-}23)$$

where: N_{rec} = number of receiving lanes (lanes into which right- or left-turning movement are made)

N_{turn} = number of turning lanes (lanes from which right- or left-turning movement are made)

All other variables are as previously defined.

Where the number of receiving lanes is equal to the number of turning lanes, drivers have virtually no ability to avoid the conflict area. Where the number of receiving lanes exceeds the number of turning lanes, drivers can maneuver around pedestrians and bicyclists to a limited extent.

Step 7: *Determine Adjustment Factors* The adjustment factors for interference of pedestrians and/or bicyclists with left-turn and right-turn movements can be determined as follows:

Adjustment Factor for Pedestrian and Bicycle Interference with Right Turns (f_{Rpb})

- If there are no conflicting pedestrians or bicyclists, $f_{Rpb} = 1.0$.
- If right turns are made as a protected phase, $f_{Rpb} = 1.0$.
- If right turns are made from an exclusive lane with permitted phasing, $f_{Rpb} = A_{pbT}$.
- If right turns are made from an exclusive lane with compound phasing, then $f_{Rpb} = A_{pbT}$ for the permitted portion of the phase and $f_{Rpb} = 1.0$ for the protected portion of the phase.
- If right turns are made from a shared lane, a complex procedure for determination of the shared lane-group saturation flow rate is presented in HCM 2010, Chapter 31.

Adjustment Factor for Pedestrian and Bicycle Interference with Left Turns (f_{Lpb})

- If there are no conflicting pedestrians, $f_{Lpb} = 1.0$.
- If left turns are made as a protected phase, $f_{Lpb} = 1.0$.
- If left turns are made from an exclusive lane on a one-way street with permitted phasing, $f_{Lpb} = A_{pbT}$.
- If left turns are made from an exclusive lane on a one-way street with compound phasing, then $f_{Lpb} = A_{pbT}$ for the permitted portion of the phase and $f_{Lpb} = 1.0$ for the protected portion of the phase.
- If left turns are made from a shared lane or on a two-way street, a complex procedure for determination of the shared lane-group saturation flow rate is presented in HCM 2010, Chapter 31.

Summary The estimation of the saturation flow rate for each defined analysis lane group can be a very complex and iterative procedure. This text outlines detailed methods for simple cases; more complex cases are treated in the HCM 2010 and can only be implemented through the use of software. It is important, however, to understand the basic relationships and concepts as applied to simple cases. The relationships and concepts do not fundamentally change for more complex situations—only the details of application.

24.3.6 Determine Lane Group Capacities and *v/c* Ratios

At this point in the analysis, lane groups have been established, and demand flow rates, *v*, for each lane group defined. Saturation flow rates, *s*, for each lane group have been estimated. As a result, both the demand and saturation flow rates for each lane group have been adjusted to reflect the same prevailing conditions. At this point, the ratio of *v* to *s* for each lane group can be computed and now be used as the variable indicating the relative demand intensity on each lane group.

Several important analytic steps may now be accomplished:

1. The *v/s* ratio for each lane group is computed.
2. Relative *v/s* ratios are used to identify the critical-lane groups in the phase plan; the sum of critical-lane group *v/s* ratios is computed.

3. Lane group capacities are computed (Equation 24-2).
4. Lane group v/c ratios are computed (Equation 24-3).
5. The critical v/c ratio for the intersection is computed (Equation 24-5).

The first is a simple manipulation of an earlier output to conform to the equations indicated, explained in the "concepts" (Section 24.2) portion of this chapter.

The remaining critical analysis that must be completed is the identification of critical-lane groups. In Chapter 21, critical lanes were identified by finding the critical path through the signal ring diagram that resulted in the highest sum of critical-lane volumes, V_c. The procedure here is exactly the same, except that instead of adding critical-lane volumes, v/s ratios are added. This process is best illustrated by an example, shown in Figure 24.7.

The illustration shows a signal with leading and lagging green phases on the E-W arterial and a single phase for the N-S arterial. This involves overlapping phases, which must be carefully considered.

The critical path through Phases A1 through A3 is determined by which ring has the highest sum of v/s ratios. In this case, as shown in Figure 24.8 the left ring has the highest total, yielding a sum of v/s ratios of 0.52. The critical path through Phase B is a straightforward comparison of the two rings, which have concurrent phases. The highest total again is on the left ring, with a v/s of 0.32. In this case, the critical path through the signal is entirely along the left ring, and the sum of critical-lane v/s ratios is $0.52 + 0.32 = 0.84$.

The meaning of this sum should be clearly understood. In this case, 84% of real time must be devoted to effective green if

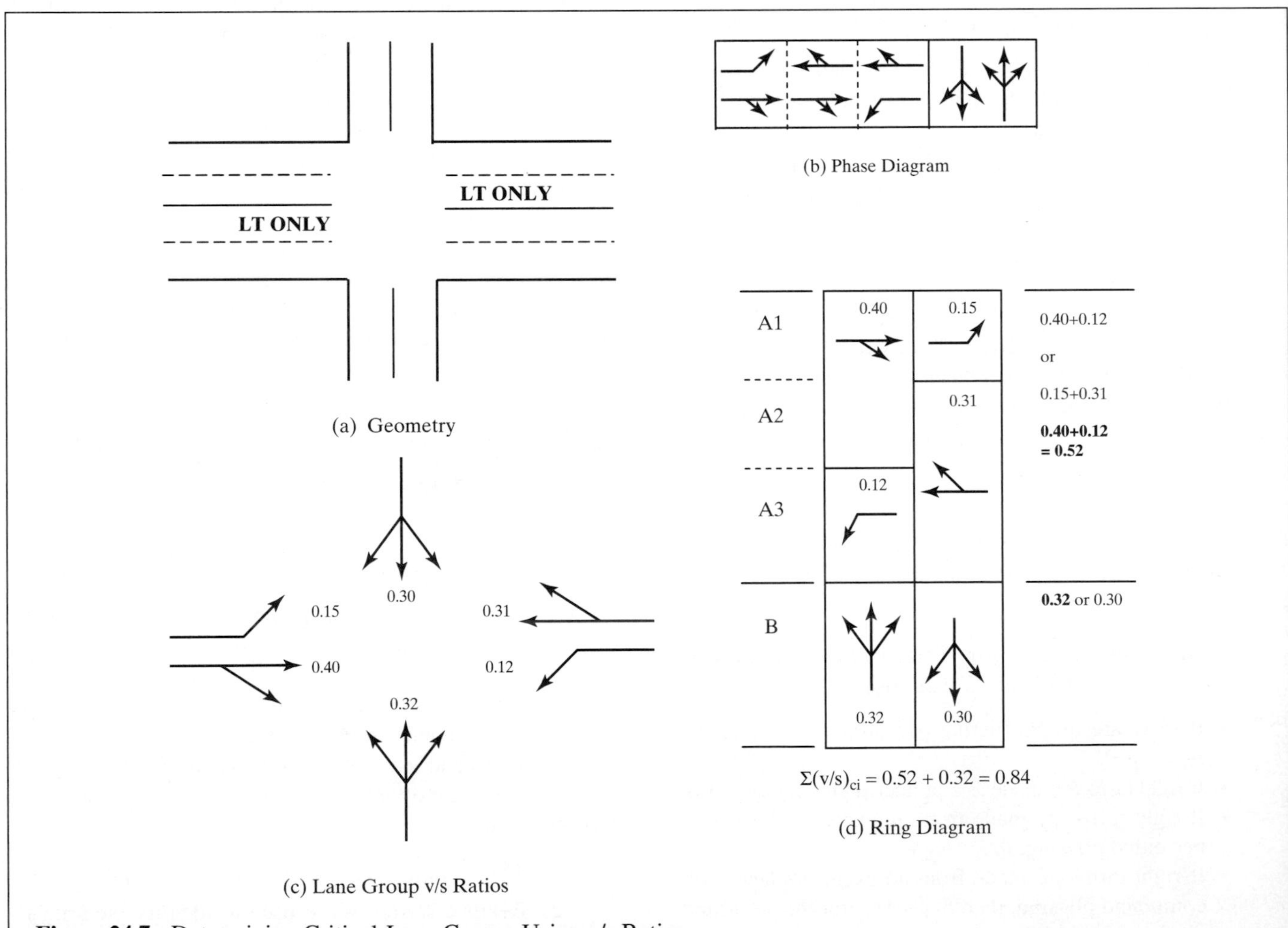

Figure 24.7: Determining Critical-Lane Groups Using v/s Ratios

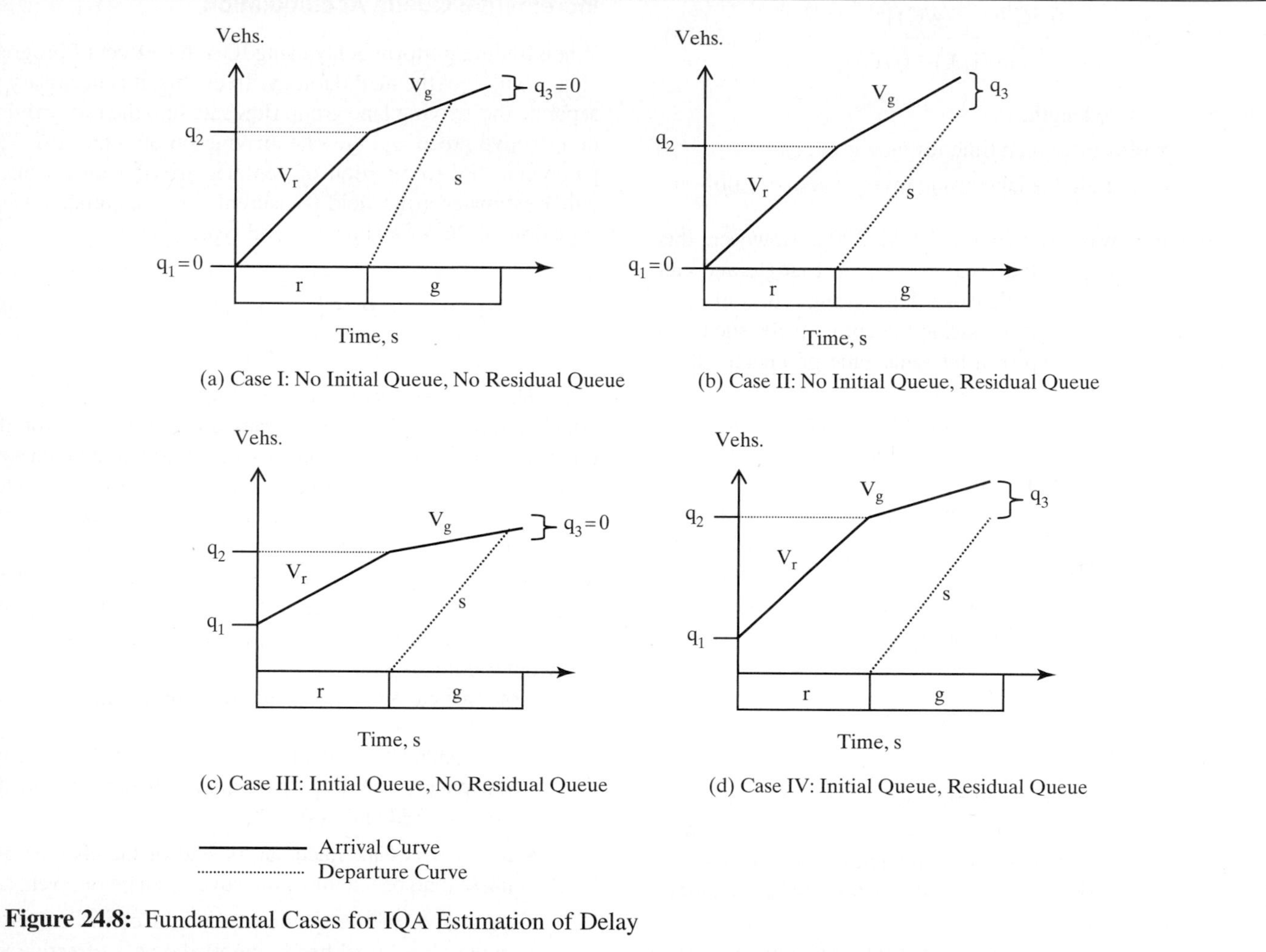

Figure 24.8: Fundamental Cases for IQA Estimation of Delay

all of the critical-lane group flows are to be accommodated. In effect, there must be $0.84 * 3600 = 3{,}024$ s of effective green time per hour to handle the demands indicated. Conversely, there is at most 16% of real time available to devote to lost times within the cycle. This fact is used later in the chapter to revise the signal timing if the analysis indicates that improvements are necessary.

24.3.7 Estimating Delay and Level of Service

LOS are based on delay, as discussed previously. Specific criteria were given in Table 24.1. In the analysis of capacity, values of the *v*/*c* ratio for each lane group will have been established. Using these results, and other signalization information, the delay for each lane group may be computed as:

$$d = d_1 + d_2 + d_3 \tag{24-24}$$

where: d = average control delay per vehicle, s/veh
d_1 = average uniform delay per vehicle, s/veh
d_2 = average incremental delay per vehicle, s/veh
d_3 = additional delay per vehicle due to a preexisting queue, s/veh

Uniform Delay

As discussed in Chapter 20, uniform delay can be obtained using Webster's uniform delay equation in the following form:

$$d_1 = \frac{0.5C\left[1 - (g/C)\right]^2}{1 - \left[\min(1, X) * (g/C)\right]} \tag{24-25}$$

where: C = cycle length, s
g = effective green time for lane group, s
X = v/c ratio for lane group (max value = 1.00)

This equation was used in the HCM 2000. However, this formulation calculates the area of a single triangle and can only accurately predict delay of a lane group with only one green time and one red time during the entire cycle, such as a through-only lane group or a left-only lane group served by protected phasing only. A better way to calculate uniform delay is to use a queue accumulation method known as Incremental Queue Accumulation (IQA) as documented in References 9 and 10. The IQA method does not limit the shape of the queue accumulation diagram to a simple triangle but rather can describe all types of complex phasing. The area of the resulting polygon is then found by breaking the polygon into component triangles and trapezoids for which the area can be found. This is the approach that will be taken in the 2010 HCM.

Effect of Progression

Webster's uniform-delay equation assumes that arrivals are uniform over time. In fact, arrivals are at best random and most often platooned as a result of coordinated signal systems. The quality of signal coordination or progression can have a monumental impact on delay.

Consider the following situation: An approach to a signalized intersection is allocated 30 seconds of effective green out of a 60-second cycle. A platoon of 15 vehicles at exactly 2.0 seconds headway is approaching the intersection. Note that the 15 vehicles will exactly consume the 30 seconds of effective green available (15*2.0 = 30). Thus, for this signal cycle, the v/c ratio is 1.0.

With perfect progression provided, the platoon arrives at the signal just as the light turns green. The 15 vehicles proceed through the intersection with no delay to any vehicle. In the worst possible case, however, the platoon arrives just as the light turns red. The entire platoon stops for 30 seconds, with every vehicle experiencing virtually the entire 30 seconds of delay. When the green is initiated, the platoon fully clears the intersection. In both cases, the v/c ratio is 1.0 for the cycle. The delay, however, could vary from 0 s/veh to almost 30 s/veh, dependent solely on when the platoon arrives (i.e., the quality of the progression).

Incremental Queue Accumulation

When finding uniform delay using IQA, the effect of progression is built into the methodology. To do this, it is necessary to separate the arriving lane group flow rate into the rate arriving on effective green and the rate arriving on effective red. The parameter "P" (proportion of vehicles arriving on green) is either estimated from field measurements or estimated using Equation 24-26 for a known arrival type.

$$P = \left(\frac{AT}{3}\right)\left(g/C\right) \tag{24-26}$$

The analysis begins by creating a delay polygon in which arrival and departure curves are constructed for the effective red portion of the phase and the effective green portion of the phase (in that order). The construction starts with effective red because that is the time a queue will be expected to form. Depending on whether or not there is an initial queue at the beginning of the effective red, and whether or not there is a residual unserved queue left at the end of the effective green phase, a number of polygon shapes are possible, as shown in Figure 24.8.

Note that each polygon deals with three queues:

- q_1 = size of queue at the beginning of the effective red phase, resulting from unserved demand on the previous red phase(s), veh,
- q_2 = size of the queue at the end of the effective red phase (and beginning of effective green phase), veh, and
- q_3 = size of the queue at the end of the effective green phase (and beginning of the next effective red phase), veh.

In a series of phases, q_3 at the end of the first computational cycle becomes q_1 for the beginning of the next.

If a signal is operating as desired, that is, without breakdowns, then in each cycle, the queue of vehicles that accumulates during the red phase (q_2) and arrive during the green phase will all depart during the green phase, leaving no residual queue (q_3) which would become an initial queue (q_1) for the next phase. This occurs in Figure 24.8 (a) and (c). In Figures 24.8 (b) and (d), however, the saturation flow rate is not sufficient for the departure curve to "catch up" with the arrival curve, and a residual queue remains at the end of the green. This becomes an initial queue for the next phase.

Note that all of the rates in Figure 24.8 are in veh/s because the time scale in this diagram is in seconds. In the equations that follow, these rates are expressed in their traditional units of veh/h.

The following steps are followed to determine the uniform delay using IQA:

1. The arrival rate during the effective red is given by Equation 24-27:

$$V_r = \frac{(1 - P)VC}{r} \tag{24-27}$$

where: V_r = arrival flow rate during effective red, veh/h
P = proportion of vehicles arriving on green
V = average arrival flow rate, veh/h
C = cycle length, s
r = effective red time, s

The proportion of vehicles arriving on green, P, can either be observed in the field or computed using Equation 24-8 for the known arrival type.

2. The queue at the end of the red time is found using Equation 24-28:

$$q_2 = q_1 + \left(\frac{v - s}{3600}\right) * \Delta t \tag{24-28}$$

$$q_2 \geq 0$$

where: q_1 = queue at the beginning of effective red, veh
q_2 = queue at the end of effective red, veh
v = average arrival rate; during red time $v = V_r$, veh/h
s = average saturation flow rate; during red time s = 0 veh/h
Δt = elapsed time; during red time, $\Delta t = r$, effective red time, s

3. The uniform delay during the effective red time is found using Equation 24-29:

$$d_r = \Delta t * \left(\frac{q_1 + q_2}{2}\right) \tag{24-29}$$

where all terms are as previously defined, and d_r = uniform delay during effective red, s. Note that this *does not include* delay to an initial queue or delay to vehicles in a residual queue.

4. The same steps must be repeated to find the uniform delay during effective green time. The starting point, however, is the queue at the end of the previous effective red time, or q_2. The arrival rate during the effective green time is found using Equation 24-30:

$$V_g = \frac{VP}{(g/C)} = \frac{VPC}{g} \tag{24-30}$$

where: V_g = average arrival rate during effective green, veh/h
g = effective green time, s.
All other variables as previously defined.

5. For an undersaturated condition, if you use Equation 24-28 to find the queue at the end of the effective green, the queue will be negative because there will be more departures than arrivals. Because of this, it is necessary to find the time when the queue is fully dissipated, that is, when q_3 (queue at the end of effective green) is equal to "0." To find the time Δt_2, where $q_3 = 0$, set the number of arrivals on green (n_a) equal to the number of departures. The number of arrivals during effective green is $V_g/3600 * \Delta t$, veh. The departure rate during effective green is the adjusted saturation flow rate, s, for the lane group. Thus the number of departures until the queue is zero is $s/3600 * \Delta t_2 = q_2 + n_a$. In other words, the number of departures during Δt_2 is set equal to the number of arrivals during Δt_2 plus the queue at the end of the effective red. Then:

$$q_2 + \left(\frac{V_g}{3600}\right) * \Delta t_2 = \left(\frac{s}{3600}\right) * \Delta t_2$$

And:

$$\Delta t_2 = \frac{3600\, q_2}{s - V_g} \tag{24-31}$$

Where a residual queue remains at the end of the green phase, Δt_2 is equal to the effective green time.

6. The delay during time Δt_2 is found using Equation 24-32:

$$d_g = \Delta t_2 * \left(\frac{q_2 + q_3}{2}\right) \tag{24-32}$$

where d_g = uniform delay during the green cycle, s, and all other terms are as previously defined. For undersaturated flow, q_3 = "0." Where there is a residual queue:

$$q_3 = q_1 + \left[\frac{(V_g * g) + (V_r * r) - (s * g)}{3600}\right] \tag{24-33}$$

where all terms are as previously defined. Arrival rates on green and red and the saturation flow rate are expressed in units of veh/h; effective green time and effective red time are in seconds.

7. Uniform delay is then the sum of the uniform delays incurred during the effective red and effective green phases divided by the number of vehicles experiencing the delay:

$$d_1 = \frac{(d_r + d_g)}{(q_2 + n_a)} \tag{24-34}$$

All terms are as previously defined.

For a more complicated phasing, these same steps are repeated for each component of the total polygon. The components of the polygon are determined by each part of a phase where the arrival rate and departure rate are constant. If either change, a separate component shape is formed and the previous steps are done for that component part. Arrival rates may change due to the effect of platooning. Departure rates will change due to the phase going from red to green, of course, but they also may change when the queue clears or the opposing queue clears. In cases of protected plus permitted (or permitted plus protected) phasing, the saturation flow rate is different for the permitted and protected portions of the phase. Therefore, although the shape of the polygons can get very complicated, the basic principles are relatively simple.

It is important to emphasize the IQA is used to estimate only uniform delay. It does not include additional delay due to oversaturation or additional delay due to the existence of an initial queue.

Incremental Delay

The incremental-delay equation is based on Akcelik's equation (see Chapter 20) and includes incremental delay from random arrivals as well as overflow delay when v/c ratio > 1.00. For a pretimed controller it is estimated as:

$$d_2 = 900\,T + \left[(X-1) + \sqrt{(X-1)^2 + \left(\frac{8kIX}{cT}\right)}\right] \tag{24-35}$$

where: T = analysis time period, h
X = v/c ratio for lane group
c = capacity of lane group, veh/h
k = adjustment factor for type of controller
I = upstream filtering/metering adjustment factor

For pretimed controllers, or unactuated movements in a semi-actuated controller, the k factor is always 0.50. The upstream filtering/metering adjustment factor is only used in arterial analysis. A value of 1.00 is assumed for all analyses of individual intersections.

For actuated controllers and phases, the relationship is more complex. For details, consult HCM 2010, Chapter 18 directly.

Initial Queue Delay

Equation 24-35 is used to find the delay due to an initial queue:

$$d_3 = \frac{3600}{vT}\left(t\,\frac{Q_b + Q_e - Q_{eo}}{2} + \frac{Q_e^2 - Q_{eo}^2}{2c} - \frac{Q_b^2}{2c}\right) \tag{24-36}$$

where: $Q_e = Q_b + t(v - c)$
Q_b = initial queue at beginning of analysis period T, veh
v = demand flow rate during analysis period T, veh/h
T = analysis period, h
t = duration of unmet demand in analysis period h
Q_{eo} = queue at the end of the first saturation analysis period when $v > c$ and $Q_b = 0$
$Q_{eo} = T(v - c)$ if $v \geq c$; then $t = T$
$Q_{eo} = 0$ if $v < c$; then $t = Q_b/(c - v) \leq T$

Assuming that an initial queue exists (i.e., $Q_b > 0$), an existing queue at the beginning of the analysis period results in additional delay beyond adjusted uniform and incremental delay (d_1 and d_2) that must be estimated. In addition, when there is an initial queue, the uniform delay calculation must be adjusted as follows:

$$d_1(adj) = d_s\frac{t}{T} + d_1\frac{T-t}{T} \tag{24-37}$$

where: $d_1(adj)$ = the new adjusted uniform delay
d_s = saturated delay (based on the IQA method setting $v = c$)
d_1 = previous uniform delay with given arrival rate, v

When oversaturation exists, it is best to analyze consecutive 15-minute analysis periods, beginning with the period before a queue appears and ending in the period after dissipation of the queue.

Aggregating Delay

Once appropriate values for d_1, d_2, and d_3 are determined, the total control delay per vehicle for each lane group is known. Levels of service for each lane group are assigned using the criteria of Table 24.1.

The HCM allows for lane group delays to be aggregated to approach delays and an overall intersection delay, as follows:

$$d_A = \frac{\sum_i d_i v_i}{\sum v_i}$$

$$d_I = \frac{\sum_A d_A v_A}{\sum_A v_A} \qquad (24\text{-}38)$$

where: d_i = total control delay per vehicle, lane group i, s/veh

d_A = total control delay per vehicle, approach A, s/veh

d_I = total control delay per vehicle for the intersection as a whole, s/veh

v_A = demand flow rate, approach A

v_i = demand flow rate, lane group i

Levels of service may be applied to individual lane groups and approaches using the criteria of Table 24.1. Note that average delays are weighted by the number of vehicles experiencing the delays. At this writing, it is not clear whether the HCM 2010 will permit aggregating LOS to the full intersection level. Many believe that this is a misleading measure, which can be used to mask critical problems in the intersection, particularly failures on individual lane groups or approaches.

24.3.8 Interpreting the Results of Signalized Intersection Analysis

At the completion of the HCM analysis procedure, the traffic engineer has the following results available for review:

- *v/c* ratios (*X*) for every lane group
- Critical *v/c* ratio (X_c) for the intersection as a whole
- Delays and levels of service for each lane group
- Delays and levels of service for each approach
- Delay for the overall intersection. HCM 2010 may or may not allow this delay to be used to obtain an overall intersection LOS.

All of these results must be considered to obtain a complete overview of predicted operating conditions in the signalized intersection and to get an idea of how to address any problems revealed by the analysis.

As noted earlier, the *v/c* ratio and delay values are not strongly linked, and a number of interesting combinations can arise. The *v/c* ratio for any lane group, however, represents an absolute prediction of the sufficiency of the capacity provided to that lane group. Further, the critical *v/c* ratio represents an absolute prediction of the total sufficiency of capacity in all critical-lane groups. The following scenarios may arise:

- *Scenario 1:* $X_c \leq 1.00$; *all* $X_i \leq 1.00$. These results indicate there are no capacity deficiencies in any lane group. If there are no initial queues, then there will be no residual queues in any lane group at the end of the analysis period. The analyst may wish to consider the balance of *X* values among the various lane groups, particularly the critical-lane groups. It is often a policy to provide balanced *X* ratios for all critical-lane groups. This is best accomplished when all critical-lane groups have $X_i \approx X_c$.
- *Scenario 2:* $X_c \leq 1.00$; *some* $X_i > 1.00$. As long as $X_c \leq 1.00$, all demands can be handled within the phase plan, cycle length, and physical design provided. All X_i values may be reduced to values less than 1.00 by reallocation of green time from lane groups with lower X_i values to those with $X_i > 1.00$. A suggested procedure for reallocation of green time is presented later in this chapter.
- *Scenario 3:* $X_c > 1.00$; *some or all* $X_i > 1.00$. In this case, sufficient capacity can be provided to all critical-lane groups only by changing the phase plan, cycle length, and/or physical design of the intersection. Improving the efficiency of the phase plan involves considering protected left-turn phasing where none exists or protected + permitted phasing where fully protected phasing exists. This may have big benefits, depending on the magnitude of left-turn demands. Increasing the cycle length will add small amounts of capacity. This may not be practical if the cycle length is already long or where the capacity deficiencies are significant. Adding lanes to critical-lane groups will have the biggest impact on capacity and may allow for more effective lane use allocations.

Delays must also be carefully considered but should be tempered by an understanding of local conditions. LOS designations are based on delay criteria, but acceptability of various delay levels may vary by location. Drivers in a small rural CBD will not accept the delay levels that drivers in a big city will.

As noted earlier, LOS F may exist where *v*/*c* ratios are less than 1.00. This situation may imply a poorly timed signal (retiming should be considered), or it may reflect a short protected turning phase in a relatively long cycle length. The latter may not be easily remedied; indeed, long delay to a relatively minor movement at a busy urban or suburban intersection is sometimes intentional.

Aggregate levels of service for approaches—and particularly for the intersection—may mask problems in one or more lane groups. Individual lane group delays and levels of service should always be reported and must be considered with aggregate measure. This is often a serious problem when consultants or other engineers report only the overall intersection LOS, as permitted through HCM 2000. The Highway Capacity and Quality of Service Committee of the Transportation Research Board is considering whether an aggregate intersection LOS should be permitted in future HCMs because of its potential to mask deficiencies in individual lane groups.

Where lane group delays vary widely, some reallocation of green time may help balance the situation. However, when changing the allocation of green time to achieve better balance in lane group delays, the impact of the reallocation on *v*/*c* ratios must be watched carefully.

24.4 A "Simple" Sample Problem

As long as applications do not involve permitted or compound left turns, shared lanes, or left-turns from a two-way street, a manual implementation of these models can be illustrated. This section provides a "simple" sample problem illustrating the application and interpretation of the HCM signalized intersection analysis procedure, incorporating many of the features that will be included in HCM 2010. In subsequent sections, some of the more complex elements of the model are presented.

The signalized intersection shown in Figure 24.9 involves a simple two-phase signal at an intersection of two one-way streets. All turns are made from a separate lane, there are no left turns from two-way streets, and all elements of the HCM procedure can be demonstrated in detail. The problem is an analysis of an existing situation. Improvements should be recommended if warranted.

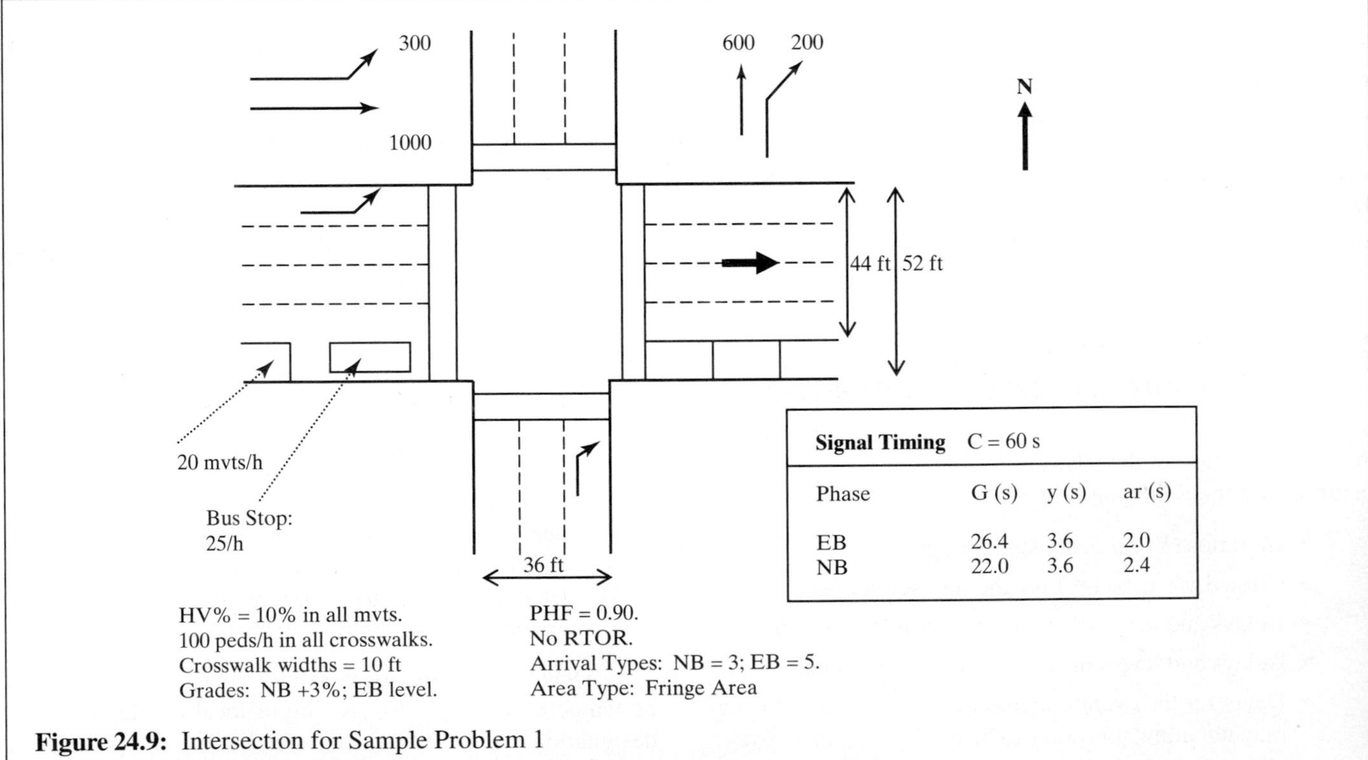

Figure 24.9: Intersection for Sample Problem 1

24.4.1 Input

All of the input variables needed for this analysis are specified in Figure 24.9. The minimum green times required for safe pedestrian crossings, however, must be computed. These are used to check actual signal timings for pedestrian safety. A warning would be issued if the green times were not sufficient to handle pedestrian safety requirements. Equation 24-9 applies:

$$G_p = 3.2 + 0.27\,N_{peds} + \left(\frac{L}{S_p}\right) \leq G + y + ar$$

where: G_p = minimum pedestrian green, s
L = length of the crosswalk, ft
S_p = crossing speed, default value = 4.0 ft/s
N_{peds} = peds/cycle in crosswalk,
peds/cycle = 100/(3,600/60) = 1.7 peds/cycle
G = vehicular actual green time, s
$y + ar$ = yellow plus all-red time, s

Then:

$$G_{pE} = 3.2 + (0.27 * 1.7) + \left(\frac{36}{4.0}\right) = 12.7\ s < 26.4 + 3.6 + 2.0 - 32.0\ s$$

$$G_{pN} = 3.2 + (0.27 * 1.7) + \left(\frac{52}{4.0}\right) = 28.0\ s < 22.0 + 3.6 + 2.5 = 28.1\ s$$

Pedestrians, therefore, are safely accommodated by the signal timing as shown. No warning is necessary.

Note that in some jurisdictions, pedestrians are only permitted in the crosswalk during the green and yellow intervals (*not* the all red); in others, they are permitted in the crosswalk only during the green interval (*not* the yellow or all red). Should either of these approaches be in effect, then the NB green *would not* be sufficient to accommodate pedestrian safely. A pedestrian pushbutton would have to be provided, or (more likely) a retiming would implemented to satisfy *both* pedestrian and vehicular needs.

24.4.2 Volume Adjustment

The intersection is fairly simple. There are four lane groups: two in each direction. The two legal turning movements have exclusive lanes. Flow rates are estimated using the *PHF*. These computations are summarized in Table 24.8.

24.4.3 Saturation Flow Rate Estimation

The saturation flow rate for each lane group is computed using Equation 24-11:

$$s = s_o\, N f_w\, f_{HV}\, f_g\, f_p\, f_{bb}\, f_a\, f_{LU}\, f_{RT}\, f_{LT}\, f_{Rpb}\, f_{Lpb}$$

where: s = saturation flow rate under prevailing conditions, veh/hg
s_o = saturation flow rate under ideal conditions, default value = 1,900 pc/hg/ln
N = number of lanes in the lane group
f_i = adjustment factor for prevailing condition "i"

where: w = lane width
HV = heavy vehicles
g = grade
p = parking
bb = local bus blockage
a = area type
LU = lane utilization
RT = right turn
LT = left turn
Rpb = ped/bike interference with right turns
Lpb = ped/bike interference with left turns

Table 24.8: Flow Rate Computations for Sample Problem

Movement	*V* (veh/h)	*PHF*	Flow Rate (veh/h)	Lane Group Flow Rate (veh/h)
EB-LT	300	0.90	333	333
EB-TH	1000	0.90	1111	1111
NB-TH	600	0.90	667	667
NB-RT	200	0.90	222	222

For the simple case posed in this sample problem, the adjustment factors are found in appropriate equations and/or tables from the chapter. These computations are summarized in Table 24.9.

The new adjustment factors for pedestrian and bicycle interference with right and left turns require multiple steps, each of which is illustrated in Table 24-10. Note there are no bicycles at this location, so only pedestrian interference is evaluated for the NB right-turn lane group and EB left-turn lane group.

The determination of these factors also requires that g_p, the pedestrian green be determined. This is generally taken to be the length of the pedestrian walk plus clearance intervals, which (for a pretimed signal) is generally equal to the effective green period, which is the same as the actual green period, G, using the standard default values for l_1 and e. Thus $g_{p,NB} = 22.0$ seconds and $g_{p,EB} = 26.4$ seconds.

The saturation flow rates for each lane group are now estimated as shown in Table 24.11.

Table 24.9: Computation of Adjustment Factors for Sample Problem 1

$f_w\,(all) = 1.00 \quad all\ lane\ widths\ 10\ ft - 12.9\ ft$
$f_{HV}\,(all) = \frac{1}{1 + P_{HV}(E_{HV} - 1)} = \frac{1}{1 + 0.10(2-1)} = 0.909\ (Eq\ 24-12)$
$f_{gEBTH/LT} = 1.00\ (no\ grade)$ $f_{gNBTH/RT} = 1 - \left(\frac{G}{200}\right) = 1 - \left(\frac{3}{200}\right) = 0.985\ (Eq\ 24-13)$
$f_{pEBTH} = \frac{N - 0.10 - \left(\frac{18N_m}{3600}\right)}{N} = \frac{3 - 0.10 - \left(\frac{18*20}{3600}\right)}{3} = 0.950\ (Eq\ 24-14)$ $f_p\,(all\ other\ lane\ groups) = 1.00\ (No\ Parking)$
$f_{bbEBTH} = \frac{N - \left(\frac{14.4*N_B}{3600}\right)}{N} = \frac{3 - \left(\frac{14.4*25}{3600}\right)}{3} = 0.967\ (Eq\ 24-15)$ $f_{bb}\,(all\ other\ lane\ groups) = 1.00\ (No\ Local\ Buses)$
$f_a\,(all) = 1.00\ (Fringe\ Area)$
$f_{LUEBTH} = 0.908\ (Table\ 24-7, 3\ Lanes)$ $f_{LUNBTH} = 0.952\ (Table\ 24-7,.2\ Lanes)$ $f_{LUEBLT/NBRT} = 1.00\ (Table\ 24-7, 1\ Lane)$
$f_{RTNBRT} = 0.85\ (Exclusive\ RT\ Lane)$ $f_{RT}\,(all\ other\ lane\ groups) = 1.00\ (No\ RTs)$
$f_{LTEBLT} = 0.95(Exclusive\ LT\ Lane)$ $f_{LT}\,(all\ other\ lane\ groups) = 1.00(No\ LTs)$

Table 24.10: Computation of Pedestrian Interference Factors

$v_{pedgNB} = v_{ped}\left(\frac{C}{g_{ped}}\right) = 100\left(\frac{60}{22}\right) = 273\ peds/hg\ (Eq\,24-17)$ $v_{pedgEB} = 100\left(\frac{60}{26.4}\right) = 227\ peds/hg$
$OCC_{pedNB} = \frac{v_{pedg}}{2000} = \frac{273}{2000} = 0.137\ (Eq\,24-18)$ $OCC_{pedEB} = \frac{227}{2000} = 0.114$
$OCC_{rNB} = \left(\frac{g_{ped}}{g}\right) OCC_{pedg} = \left(\frac{22}{22}\right) 0.137 = 0.137\ (Eq\,24-21)$ $OCC_{rEB} = \left(\frac{26.4}{26.4}\right) 0.114 = 0.114$
$A_{pbTNB} = 1 - 0.6\,OCC_r = 1 - (0.60 * 0.137) = 0.918 (Eq\,24-23)$ $A_{pbTEB} = 1 - (0.60 * 0.114) = 0.932$
$f_{pbRT}\ (NBRT) = A_{pbTNB} = 0.918$ $f_{pbRT}\ (all\ other\ lane\ groups) = 1.00 (No\ ped\ conflict)$ $f_{pbLT}\ (EBLT) = 0.932$ $f_{pbLT}\ (all\ other\ lane\ groups) = 1.00 (No\ ped\ conflict)$

24.4.4 Capacity Analysis

Adjusted demand flow rates and saturation flow rates may now be used to compute *v/s* ratios for the various lane groups, allowing the selection of critical-lane groups. In this case, the selection of critical-lane groups is fairly trivial because there is only one signal phase with two lane groups for each of two approaches. Thus, by definition, the two phases *must* be critical.

Capacity of a lane group is computed using Equation 24-2:

$$c_i = s_i * \left(\frac{g_i}{C}\right)$$

where: c_i = capacity of lane group i, veh/h
s_i = Saturation flow rate for lane group i, veh/hg
g_i = effective green time for lane group i, s
C = cycle length, s

It should be noted that if standard default values for startup lost time (ℓ_1) – 2.0 s, and for extension of effective green into yellow and all-red (e) – 2.0 s are used, the effective green time (g) and

Table 24.11: Estimation of Saturation Flow Rates Using Equation 24-11

Lane Group	s_o (pc/hg/ln)	N (lanes)	f_w	f_{HV}	f_g	f_p	f_{bb}	f_a	f_{LU}	f_{RT}	f_{LT}	f_{Rpb}	f_{Lpb}	s (veh/hg)
EB LT	1900	1	1.000	0.909	1.000	1.000	1.000	1.000	1.000	1.000	0.950	1.000	0.932	**1529**
EB TH	1900	3	1.000	0.909	1.000	0.985	0.967	1.000	0.908	1.000	1.000	1.000	1.000	**4481**
NB TH	1900	2	1.000	0.909	0.985	1.000	1.000	1.000	0.952	1.000	1.000	1.000	1.000	**3239**
NB RT	1900	1	1.000	0.909	0.985	1.000	1.000	1.000	1.000	0.850	1.000	0.918	1.000	**1327**

the actual green time (G) are numerically equivalent. Capacity analysis computations are summarized in Table 24.12.

The critical v/c ratio is computed using Equation 24-5:

$$X_C = \sum_i (v/s)_{ci} * \left(\frac{C}{C - L} \right)$$

where $\Sigma(v/s)_{ci}$ = sum of the critical-lane group v/s ratios, $0.248 + 0.206 = 0.454$

L = total lost time per cycle, s

Other terms as previously defined.

The total lost time per cycle is the sum of the start-up and clearance lost times for each phase, using the standard default values for ℓ_1 and e. Thus:

$$\ell_{1,EB} = 2.0 \text{ s}$$

$$\ell_{2,EB} = y + ar - 2.0 = 3.6 + 2.0 = 3.6 \text{ s}$$

$$\ell_{1,NB} = 2.0 \text{ s}$$

$$\ell_{2,NB} = 3.6 + 2.4 - 2.0 = 4.0 \text{ s}$$

$$L = 2.0 + 3.6 + 2.0 + 4.0 = 11.6 \text{ s}$$

Then:

$$X_c = 0.454 * \left(\frac{60}{60 - 11.6} \right) = 0.563$$

The capacity results indicate there is no capacity problem at this intersection.

24.4.5 Delay Estimation and Level of Service

The average control delay per vehicle for each lane group is computed using IQA for d_1(Equation 24-34), and Equations 24-35, and 24-36 for d_2 and d_3, respectively:

$$d = d_1 + d_2 + d_3$$

$$d_2 = 900\, T \left[(X - 1) + \sqrt{(X - 1)^2 + \frac{8kIX}{cT}} \right]$$

where: d = total control delay per vehicle, s/veh

d_1 = uniform delay, s/veh

d_2 = incremental delay (random + overflow), s/veh

d_3 = delay per vehicle due to preexisting queues, s/veh (no pre-existing queues in this case, $d_3 = 0.0$ s/veh)

T = analysis period, h, 0.25 h or 15 m in this case

X = v/c ratio for subject lane group

k = incremental delay factor for controller type 0.50 for all pretimed controllers

I = upstream filtering adjustment factor (always 1.0 for isolated intersection analysis)

Other variables are as previously defined.

Table 24.12: Capacity Analysis Results

Lane Group	v (veh/h)	s (veh/hg)	v/s	G (s)	C (s)	C (veh/h)	X (v/c)
EB LT	333	1529	0.218	26.4	60	673	0.495
EB TH	1111	4481	**0.248**	26.2	60	1972	0.577
NB TH	667	3239	**0.206**	22.0	60	1188	0.561
NB RT	222	1327	0.167	22.0	60	487	0.456

Note: Bold type indicates critical v/s ratio for each phase.

First-Term Delay (d_1)

To find d_1 for the EB LT lane group, first find:

v = 333 veh/h
$P = (AT/3) * (g/C) = (5/3) * (26.4/60) = 0.733$
V_r = Arrival rate on red = $(1-P)v(C/r)$
$= (1-0.733) * 333 * (60/33.6) = 159$ veh/h (Equation 24-27)

V_g = Arrival rate on green = $vP(C/g)$
$= 333 * 0.733 * (60/25.4) = 555$ veh/h (Equation 24-30)

1. Start at beginning of effective red, where the initial queue is assumed to be zero, $q_1 = 0$.
2. Find the queue at the end of the red time, $q_2 = q_1 + (v-s)/3600 * \Delta t \geq 0$, where v = average arrival rate, which during the red time = V_r, and $s = 0$, during red because there are no departures. Δt is the effective red time for the EB phase, which is equal to the cycle length minus the effective green time, or 60.0 − 26.4 = 33.6 seconds. Then, using Equation 24-28, $q_2 = 0 + (159-0)/3600 * 33.6 = 1.48$ veh.
3. Find the incremental delay during the effective red time using Equation 24-29: $d_r = \Delta t * (q_1 + q_2)/2 = 33.6 * (0 + 1.48)/2 = 25.9$ s/veh.
4. Assuming undersaturated flow, the time until the queue clears during the green phase, Δt_2, is found using Equation 24-31:

$$\Delta t_2 = \frac{3600 q_2}{s - V_g} = \frac{3600 * 1.48}{1529 - 555} = 5.47 \text{ s}$$

5. The delay during Δt_2 is found using Equation 24-34:

$$d_g = \Delta t_2 * \left(\frac{q_2 + q_3}{2}\right) = 5.47 * \left(\frac{1.48 + 0.00}{2}\right) = 4.0 \text{ s}$$

6. Uniform delay may now be computed using Equation 24-37:

$$d_1 = \frac{(d_r + d_g)}{(q_2 + n_a)} = \frac{25.9 + 4.0}{1.48 + 4.07} = 5.2 \text{ s/veh}$$

where n_a is the number of vehicles arriving on green $= 555 * (26.4/3600) = 4.07$.

Computations for the other three lane groups follow the same procedure and are summarized in Table 24.13.

Table 24.13: Computations for d_1 Summarized

Variable	Lane Group			
	EB LT	**EB TH**	**NB TH**	**NB RT**
v (veh/h)	333	1111	667	222
s (veh/hg) on red	0	0	0	0
s (veh/hg) on green	1529	4481	3239	1327
AT	5	5	3	3
C(s)	60	60	60	60
g(s)	26.4	26.4	22	22
r(s)	33.6	33.6	38	38
q_1 (veh)	0	0	0	0
p	0.733	0.733	0.367	0.367
V_r (veh/h)	159	529	667	222
V_g (veh/h)	555	1852	667	222
q_2 (veh)	1.48	4.94	7.04	2.34
q_3 (vehs)	0	0	0	0
Δt_2 (s)	5.47	6.76	9.85	7.63
n_a	4.07	13.58	4.08	1.36
d_r (s)	24.9	83.0	133.8	44.5
d_g (s)	4.0	16.7	34.7	8.9
d_1 (s/veh)	5.2	5.4	15.2	14.5

Second Term Delay (d_2)

The second-term delay for the EB LT lane group is computed as follows:

$$d_2 = 900T\left[(X-1) + \sqrt{(X-1)^2 + \frac{8kIX}{cT}}\right]$$

where: $T = 0.25$ h.
$X = v/c$ ratio $= 333/673 = 0.495$
$v = 333$ veh/h
c = capacity $= s * (g/C) = 1529 * (26.4/60) = 673$ veh/h
$k = 0.50$ (pretimed controller)
$I = 1.00$ (intersection analysis)

$$d_2 = 900 * 0.25\left[(0.495 - 1) + \sqrt{(0.495 - 1)^2 + \frac{8 * 0.50 * 1.00 * 0.495}{673 * 0.25}}\right]$$

$$d_2 = 225 * \left[-0.505 + \sqrt{0.255 + \frac{1.98}{168.25}} \right]$$

$$= 225 * \left[-0.505 + \sqrt{0.26677} \right] = 2.6 \text{ s/veh}$$

Computations for all of the lane groups are summarized in Table 24.14.

Third Term Delay (d_3)

Because there are no unserved queues at the end of any green phase, all values of d_3 are "0."

Total Delay and Level of Service

The total delay for each lane group is summarized in Table 24.15. Levels of service are assigned to each lane group according to the criteria of Table 24.1.

Delay may be aggregated for each approach, with the average delay weighted by the demand flow rate on each of the lane groups:

$$d_E = \frac{(333 * 7.8) + (1111 * 7.6)}{(333 + 1111)} = 7.65 \text{ s/veh } (LOS = A)$$

$$d_N = \frac{(667 * 17.1) + (222 * 17.6)}{(667 + 222)} = 17.22 \text{ s/veh } (LOS = B)$$

Although an aggregation for the overall intersection could also be completed, it is not a very meaningful number and not recommended.

Table 24.14: Computations for d_2 Summarized

Variable	Lane Group			
	EB LT	EB TH	NB TH	NB RT
v(veh/h)	333	1111	667	222
s(veh/hg)	1529	4481	3239	1327
g(s)	26.4	26.4	22	22
C(s)	60	60	60	60
k	0.5	0.5	0.5	0.5
i	1	1	1	1
T(h)	0.25	0.25	0.25	0.25
c(veh/h)	673	1972	1188	487
X	0.495	0.563	0.562	0.456
X–1	−0.505	−0.437	−0.438	−0.544
d_2(s/veh)	2.6	1.2	1.9	3.1

Table 24.15: Total Delay and Level of Service for Lane Groups of Sample Problem 24.1

Lane Group	D_1 (s/veh)	d_2 (s/veh)	d_3 (s/veh)	d (s/veh)	LOS
EB LT	5.2	2.6	0	7.8	A
EB TH	5.4	1.2	0	7.6	A
NB TH	15.2	1.9	0	17.1	B
NB RT	14.5	3.1	0	17.6	B

From Table 24.1, these delays are in the range of LOS A and B for both lane groups and the intersection as a whole, a very acceptable result.

24.4.6 Analysis

The delay and capacity results both indicate that the intersection is operating acceptably. Some improvement, however, might be possible through a retiming of the signal to achieve better balance between the two phases and to achieve a higher utilization of available green time.

The critical *v/c* ratio was determined to be 0.563. This is a low value, indicating that 43.7% of green time is unused. Optimal delays usually occur when *v/c* ratios are in the range of 0.80 to 0.95. The value 0.563 indicates that the cycle length is probably too long for the current demand volumes. The lane groups have similar *v/c* ratios, so the balance of green time is reasonable.

This was, in effect, a very simple situation with no complicating factors. The solution of the problem, however, is still lengthy and contains many steps and computations. For these reasons, most traffic engineers use a software package to implement this methodology. When more complicated situations arise, there is no choice but to use software.

24.4.7 What If There Is a d_3?

What would happen if the demand flow rate for the NB TH lane group were actually 2,000 veh/h? This would be greater than the capacity of the lane group, 1,972 veh/h. The answer would depend heavily on how long this situation persisted. For the sake of illustration, assume this condition exists for four consecutive 15-minute periods. There is no preexisting queue at the beginning of the first 15-minute interval.

Because $v > c$, there *will be* a queue at the end of the first 15-minute interval, and therefore a queue at the beginning of the *second* 15-minute interval (and the *third* and the *fourth*). Assuming the intervals are consecutively numbered 1 through 4, the queues at the beginning and end of each of the four periods can be found as:

$$\begin{aligned}
Q_e &= Q_b + t(V - c) \\
Q_{b1} &= 0.00\,(given) \\
Q_{e1} &= 0 + 0.25 * (2000 - 1972) = 7 \text{ vehs} = Q_{b2} \\
Q_{e2} &= 7.0 + 0.25 * (2000 - 1972) = 14\text{vehs} = Q_{b3} \\
Q_{e3} &= 14.0 + 0.25 * (2000 - 1972) = 21\text{vehs} = Q_{b4} \\
Q_{e4} &= 21.0 + 0.25 * (2000 - 1972) = 28\text{vehs}
\end{aligned}$$

Each of these periods will have the same d_1 and d_2 values.

The first-term delay, however, must be computed using Equation 24-37 because there is an unserved queue:

$$d_1(adj) = d_s \frac{t}{T} + d_1 \frac{T - t}{T}$$

In this case, the entire analysis period is oversaturated, and $t = T$. Then, $d_1\,(adj) = d_s$. This is computed as shown previously, but with the assumption that $v = s = 1972$ veh/h. When this is done (not shown here), the result is $d_1(adj) = 8.0$ s/veh.

The second-term delay is computed using Equation 24-35, but with a v/c ratio of 1.00 and $c = 1972$ veh/h:

$$d_2 = 900T\left[(X - 1) + \sqrt{(X - 1)^2 + \frac{8kI}{cT}}\right]$$

$$= 900 * 0.25$$

$$\left[(1 - 1) + \sqrt{(1 - 1)^2 + \frac{8 * 0.50 * 1 * 1}{1972 * 0.25}}\right]$$

$$d_2 = 225 * \left[0 + \sqrt{0 + \frac{4}{493}}\right]$$

$$= 225 * \left[0 + \sqrt{0.008112}\right] = 20.3 \text{ s/veh}$$

The third term delay, d_3, accounts for the delay due to the preexisting queues that are present in the second through fourth 15-minute analysis periods. This value will be different for each of these periods.

$$d_3 = \frac{3600}{vT}\left(t\frac{Q_b + Q_e - Q_{eo}}{2} + \frac{Q_e^2 - Q_{eo}^2}{2c} - \frac{Q_b^2}{2c}\right)$$

Values of Q_b and Q_e were computed previously for each period. Q_{eo} is the queue at the end of the first time period in which $v > c$, in this case, Q_{e1}. T and t are equal; both are 0.25 h. Demand flow rate, v, is 2000 veh/h. Then:

$$d_{31} = \frac{3600}{2000 * 0.25}\left[\left(0.25\frac{0 + 7 - 7}{2}\right) + \left(\frac{7^2 - 7^2}{2 * 1972}\right) + \left(\frac{0^2}{2 * 1972}\right)\right] = 0.0 \text{ s/veh}$$

$$d_{32} = \frac{3600}{2000 * 0.25}\left[\left(0.25\frac{7 + 14 - 7}{2}\right) + \left(\frac{14^2 - 7^2}{2 * 1972}\right) + \left(\frac{7^2}{2 * 1972}\right)\right] = 13.0 \text{ s/veh}$$

$$d_{33} = \frac{3600}{2000 * 0.25}\left[\left(0.25\frac{14 + 21 - 7}{2}\right) + \left(\frac{21^2 - 7^2}{2 * 1972}\right) + \left(\frac{14^2}{2 * 1972}\right)\right] = 26.3 \text{ s/veh}$$

$$d_{34} = \frac{3600}{200 * 0.25}\left[\left(0.25\frac{21 + 28 - 7}{2}\right) + \left(\frac{28^2 - 7^2}{2 * 1972}\right) + \left(\frac{21^2}{2 * 1972}\right)\right] = 39.9 \text{ s/veh}$$

To each of these delays, the first (8.0 s/veh) and second (20.3 s/veh) term delays must be added, so that:

d (first period) = 28.3 s/veh
d (second period) = 41.3 s/veh
d (third period) = 52.6 s/veh
d (fourth period) = 68.2 s/veh

This is considerably higher than the 7.6 s/veh of delay for the initial conditions described for this lane group. However, although the lane group in this case would be operating at LOS F ($v/c > 1.00$), even after four consecutive 15-minute periods, the delay values do not rise to the level that would be labeled LOS F in the absence of a failure. This is because after the four periods, the unserved queue is only 28 vehicles of 2,000 vehicles that have arrived during the four periods.

The length of the queue is approximately 28/3 = 9.33 veh/ln (about 280 to 320 feet) at the end of the fourth period. This may be a concern if the next upstream signal is close enough that a 280- to 320-foot queue would negatively impact saturation flow rates of vehicles attempting to leave the upstream signal.

It is also obvious that the longer this situation continues, the higher the delays will become. Delays would top 80 s/veh in a fifth period and would become extremely difficult should the condition continue into a sixth period or more.

24.5 Complexities

Previous sections of this chapter dealt with portions of the HCM model for analysis of signalized intersections that are at least somewhat straightforward and can be reasonably illustrated through manual applications. In this section of the chapter, some of the more intricate portions of the model are discussed. Some elements are not completely detailed. Consult the HCM directly for fuller descriptions.

The following aspects of the model are addressed:

- Permitted left turns
- Analysis of compound left-turn phasing
- Using analysis parameters to adjust signal timing
- Analysis of actuated signals

HCM 2010 introduces a number of new features for signalized intersection analysis. Most importantly, the methodology focuses on the analysis of actuated signals, with pretimed signals comprising an application in which signal parameters are specified instead of being estimated. Incremental queue analysis replaced Webster's delay equation for uniform delay and incorporates the effects of progression directly, eliminating the need for an adjustment factor to account for this.

In many cases, complex situations are not handled by discrete adjustment factors but by directly estimating saturation flow rate for the situation. Thus, in estimating the saturation flow rate for lane groups including permitted left turns, no left-turn adjustment factor is specifically identified but it is implicitly included in the models that are used. Because of this, the sections that follow focus on the conceptual approach, with some, but often incomplete, exposition of the specific analytic models used in each case.

24.5.1 Permitted Left Turns

The modeling of permitted left turns must account for the complex interactions between permitted left turns and the opposing flow of vehicles. These interactions involve several discrete time intervals within a green phase that must be separately addressed.

Figure 24.10 illustrates these portions of the green phase. It shows a subject approach with its opposing flow. When the green phase is initiated, vehicles on both approaches begin to move. Vehicles from the standing queue on the opposing approach move through the intersection *with no gaps*. Thus *no* left turn from the subject approach may proceed during the time it takes this opposing queue of vehicles to clear the intersection. If a left-turning vehicle arrives in the subject approach during this time, it must wait, *blocking the left-most lane*, until the opposing queue has cleared. After the opposing queue has cleared, left turns from the subject approach are made through gaps in the now unsaturated opposing flow. The rate at which they can be made as well as their impact on the operation of the subject approach depends on the number of left turns and the magnitude and lane distribution of the opposing flow.

Another fundamental concept is that left-turning vehicles have no impact on the operation of the subject approach *until the first left-turning vehicle arrives*. This is an intuitively obvious point, but it has often been ignored in previous methodologies.

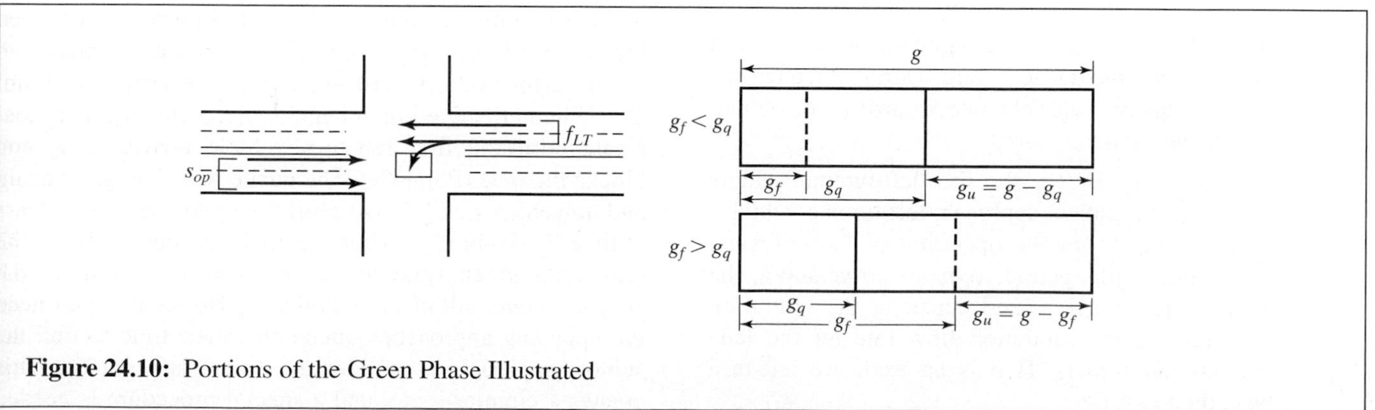

Figure 24.10: Portions of the Green Phase Illustrated

The three distinct portions of the green phase may be defined as follows:

- g_q = average amount of green time required for the opposing queue of standing vehicles to clear the intersection, s
- g_f = average amount of green time before the first left-turning vehicle arrives on the subject approach, s
- g_u = average amount of time *after the arrival of the first left-turning vehicle* that is not blocked by the clearance of the opposing queue, s

Figure 24.10 also illustrates the relationship between these key variables. The value of g_u depends on the relative values of g_f and g_q:

$$g_u = g - g_q \quad \text{for } g_q \geq g_f \qquad (24\text{-}39)$$

$$g_u = g - g_f \quad \text{for } g_q < g_f$$

where: g = total effective green for phase, s

Other variables as previously defined.

When defined in this fashion, g_u represents the actual time (per phase) that left turns actually filter through an unsaturated opposing flow.

Lastly, left turns can move during the clearance lost time as sneakers.

Basic Model Structure

Permissive left turns are analyzed using the IQA method. In the queue accumulation process, one needs to determine the arrival and departure rates during each portion of the green phase, using the preceding definitions of critical portions of the green phase. Thus the model for left-turns must consider what type of left-turn operations is taking place at various times within a given green phase.

- *Interval 0: r.* Starting at the beginning of effective red time, assume the queue is zero. During effective red time, the queue builds at a rate V_r, arrival rate on red. The departure rate is zero.
- *Interval 1: g_f.* Before the first left-turning vehicle arrives in the subject approach, left-turning vehicles have no impact on the operation of the left lane. Thus, during this period, vehicles arrive at Vg, the arrival rate on green, and depart at the saturation flow rate, s, the saturation flow rate for the lane without left turners. If it is an exclusive left-turn lane, then $g_f = 0$.
- *Interval 2: $(g_q - g_f)$.* If the first left-turning vehicle on the subject approach arrives before the opposing queue clears ($g_q > g_f$), the vehicle must wait, blocking the left lane during this interval. No vehicle can move in the left lane while the left-turner waits. Therefore, the saturation flow rate is 0.00. Where $g_f \geq g_q$, this time period does not exist. Where the opposing approach is a single-lane approach, some left turns can be made during this period, as discussed later. The arrival rate is V_g.
- *Interval 3: g_u.* This is the period during which left turns from the subject approach filter through an unsaturated opposing flow. During this period of time, the saturation flow rate must be adjusted by a left-turn adjustment factor between 0.00 and 1.00, to reflect the impedance of the opposing flow. The adjustment factor for this period is referred to as F_1. Arrival rate is V_g.

  When the *opposing approach* has a single lane, a unique situation for left-turners arises. Left-turning vehicles located within the opposing standing queue will create gaps in the opposing queue as it clears. Left-turners from the subject approach may make use of these gaps to execute their turns. Thus, when the opposing approach has only one lane, some left turns from the subject approach can be made during the time period $(g_q - g_f)$ and the saturation flow rate will have an adjustment factor for this period of some value between 0.00 and 1.00, designated as F_2.
- *Interval 4:* sneakers. During the ending or clearance lost time, l_2, left turns can move as sneakers. The number of sneakers depends on the proportion of left turns in the lane group, P_L. The saturation flow rate during this period is $3600 * (1 + P_L)/l_2$.

Consider the simple shared left-through lane with permissive-only phasing, as shown in Figure 24.11. When the phase is effectively red, vehicles continue to arrive, but the departure rate is zero and the queue grows (to Point Q_r). When effective green begins, the queue begins to dissipate until the first left-turn vehicle arrives at g_f and blocks the lane (Point Q_f). The queue will then grow again and no vehicles will depart until the opposing queue clears at time g_q (Point Q_q). After the opposing queue clears, the remaining green time left is g_u. During this time, the subject queue will clear at Point Q_p. Notice that you need the opposing approaches queue clearance time to find the subject approach queue clearance time at point Q_p. This creates a circular logic, and a special procedure is needed

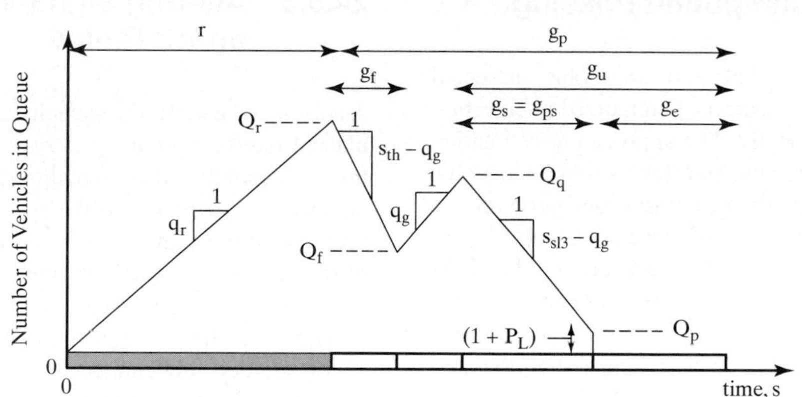

Figure 24.11: Queue Accumulation Polygon for a Shared Lane with Permitted Left Turns

(*Source:* NCHRP 3-92, Production of the 2010 Highway Capacity Manual, *Draft Chapter 31*, Exhibit 31-12.)

for this special case of both opposing approaches having shared left-through lanes with permissive phasing. Thus in order to find the area of the triangles and trapezoids that make up the polygon for a left-turn movement that has permissive left turns, you first need to determine the portions of the effective green phase: g_f, g_q, g_u; as well as the adjustment factors F_1 and F_2. Models for finding each of these variables are found in the HCM2010, but will not be given here.

Summary for Permitted Left Turns

The HCM 2010 does not actually produce a left-turn adjustment factor for permitted turns (the model is included in equations to estimate the saturation flow rate for a lane group with permitted left turns). If one were actually computed, it would be a composite based on three fundamental time periods, as described previously:

- During g_f, the time before the first left-turning vehicle arrives, left turns have no impact on discharge, and the left-turn adjustment factor is 1.00.
- During $g_{diff} = g_q - g_f \geq 0$, two situations are possible: If the opposing approach has only one lane, then the left-turn adjustment factor is F_2. If there are two or more lanes on the opposing approach, then this time is completely blocked to left-turning vehicles, and the effective left-turn adjustment factor is 0.
- During g_u, the left-turn adjustment factor is F_1.

Thus, the left-turn adjustment factor would be of the form:

$$f_{LT} = \frac{g_f}{g}(1.0) + \frac{g_{diff}}{g}(0.0) + \frac{g_u}{g}F_1 = \frac{g_f}{g} + \frac{g_u}{g}F_1 \tag{24-45}$$

for multilane opposing approaches. For single-lane opposing approaches, the adjustment factor would be in the form of:

$$\begin{aligned} f_{LT} &= \frac{g_f}{g}(1.0) + \frac{g_{diff}}{g}(F_2) + \frac{g_u}{g}(F_1) \\ &= \frac{g_f}{g} + \frac{g_{diff}}{g}F_2 + \frac{g_u}{g}F_1 \end{aligned} \tag{24-46}$$

All variables are as previously defined.

In the past, "sneakers" have been added as an additional term to Equations 24-45 and 24-46. If it is assumed that each successful sneaker adds 2.0 seconds of effective green to the actual effective green, then the following term could be added to Equations 24-45 or 24-26 to account for them:

$$g_{sn} = \frac{2.0(1 + P_L)}{g} \tag{24-47}$$

where g_{sn} is the additional effective green used by sneakers. It is assumed that there will always be at least one sneaker that can get through per phase. The maximum number that could successfully "sneak" is two vehicles. The probability of the second vehicle in line also being a left-turner is P_L.The term for sneakers can be added to Equations 24-45 and 24-46, or it could be treated as an effective minimum value for f_{LT}.

Of course, HCM 2010 never actually produces a left-turn adjustment factor for permitted turns, as we have noted.

24.5.2 Modeling Compound Phasing

Protected plus permitted and/or permitted plus protected phasing is the most complex aspect of signalized intersection operations to model analytically. The approach to estimating saturation flow rates, capacities, and delays, however, is the same as just shown but with a more complicated polygon.

This section describes some of these alterations in general, with some illustrations. We urge you to consult the HCM directly for full details of all aspects of compound phasing analysis.

In terms of saturation flow rates and capacities, the general approach taken in the HCM is straightforward. The protected and permitted portions of the phase are separated, with saturation flow rates and capacities computed separately for each portion of the phase. The appropriate green times are associated with each portion of the phase. The protected portion of the phase is analyzed as if it were a fully protected phase (i.e., $f_{LT} = 0.95$); the permitted portion of the phase is analyzed as if it were a fully permitted phase, using the left-turn model described in the previous section.

Although the separation of protected and permitted portions of the phase works well in the computation of saturation flow rates and capacities, the determination of *v/c* ratios is more difficult. In theory, what is needed is an algorithm for assigning the demand flow to the two portions of the phase. Unfortunately, the division of demand between protected and permitted portions of the phase depends on many factors, including the platoon arrival structure on the approach. The HCM takes a very simplistic view:

- Demand uses the full capacity of the *first* portion of the phase, regardless of whether it is protected or permitted.
- All demand unserved by the first portion of the phase is assigned to the second portion of the phase.

Thus the first portion of the phase has a maximum *v/c* ratio of 1.00. If all demand can be handled in the first portion of the phase, no demand is assigned to the second. In these cases, it might cause the engineer to question the need for compound phasing.

Where the first portion of the phase cannot accommodate the demand, all remaining flow is assigned to the second. The *v/c* ratio of the second portion of the phase, therefore, can range from 0.00 (when no demand is assigned) to a value > 1.00, when the total demand exceeds the total capacity of the compound phase.

The prediction of delay for compound phasing is even more complex than a simple permissive phasing, but the steps are the same, finding the area of the component polygons.

24.5.3 Altering Signal Timings Based on *v/s* Ratios

As discussed earlier, the capacity and/or delay results of a signalized intersection analysis may indicate the need to adjust the cycle length and/or to reallocate green time. Although the engineer could return to the methodology of Chapter 21, a retiming of the signal may be accomplished using the results of the analysis as well. The methodology outlined here may *not* be used if the phase plan is to be altered as well because it assumes that the phase plan on which the initial analysis was conducted does not change.

As a result of an analysis, *v/s* ratios for each lane group have been determined and the critical path through the phase plan has also been identified. Therefore, the sum of critical-lane group *v/s* ratios is known.

If the cycle length is to be altered, a desired value may be estimated using Equation 24-5, which defines X_c, the critical *v/c* ratio for the intersection. The equation is solved for the cycle length, C:

$$X_c = \sum_i (v/s)_{ci}\left(\frac{C}{C - L}\right)$$

$$C = \frac{LX_c}{X_c - \sum_i (v/s)_{ci}} \tag{24-48}$$

where all terms have been previously defined. In this application, the engineer would choose a target value of X_c. Just as in Chapter 21, the target value is usually somewhere in the range of 0.80 to 0.95. The optimal delays usually occur with *v/c* ratios in this range.

Once a cycle length has been selected, even if it is the same as the initial cycle length used in the analysis, green times may be allocated using Equation 24-3, which is solved for g_i:

$$X_i = \frac{(v/s)_i}{(g/C)_i}$$

$$g_i = (v/s)_i \frac{C}{X_i} \tag{24-49}$$

The most often applied strategy is to allocate the green such that all values of X_i for critical-lane groups are equal to X_c Other strategies are possible, however.

Consider the problem shown with analysis results summarized in Figure 24.12. Instead of demand flow rates, the *v/s* ratios determined for each lane group are shown on the

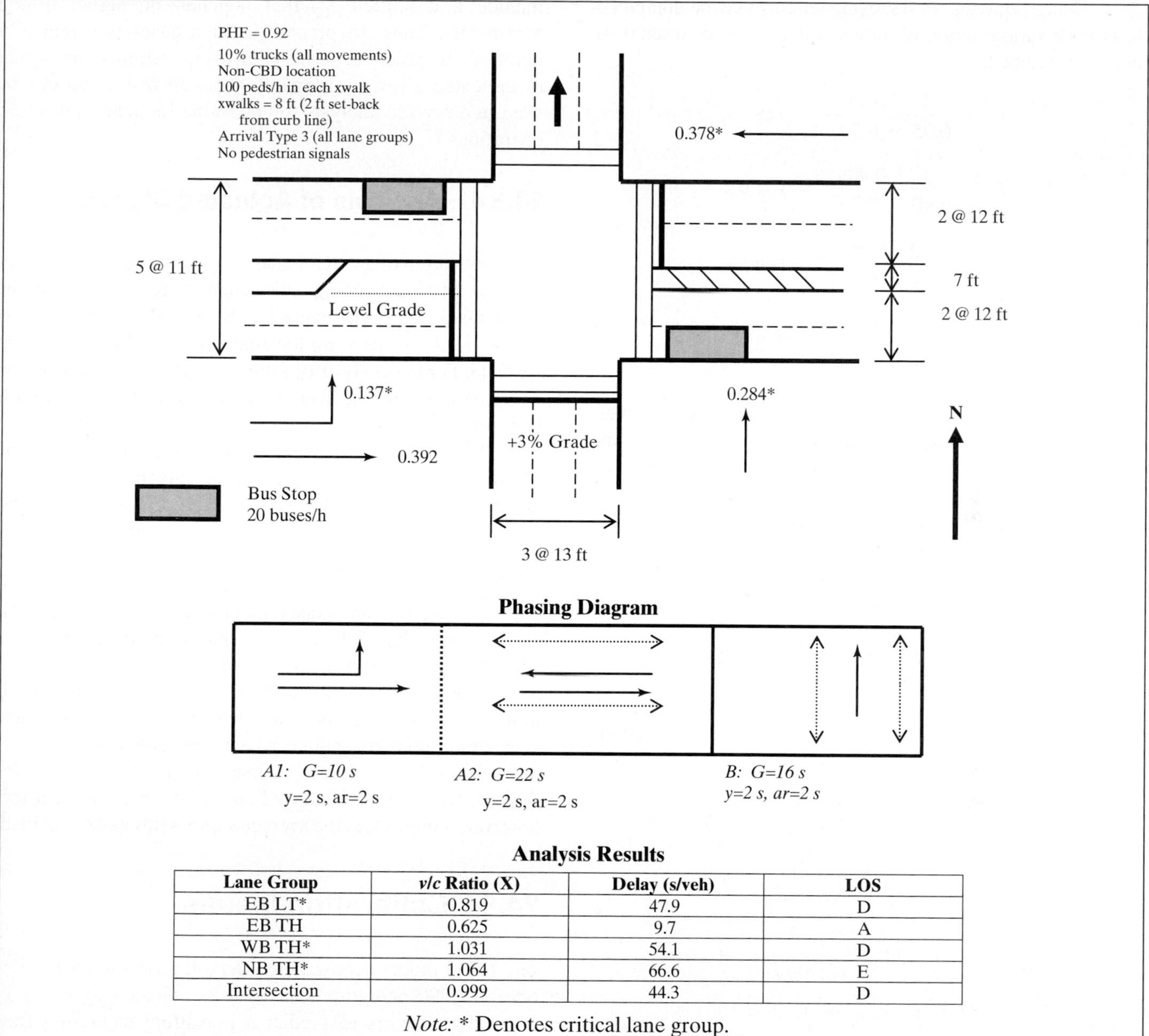

Analysis Results

Lane Group	*v/c* Ratio (X)	Delay (s/veh)	LOS
EB LT*	0.819	47.9	D
EB TH	0.625	9.7	A
WB TH*	1.031	54.1	D
NB TH*	1.064	66.6	E
Intersection	0.999	44.3	D

Note: * Denotes critical lane group.

Figure 24.12: Adjustment to Signal Timing Illustrated

diagram of the intersection. This is an intersection in a busy downtown area, and, except for the protected EB LT, no turns are permitted.

The figures show there are two distinct problems identified. The X_c value of 0.999 indicates that the cycle length is barely adequate to satisfy critical demands. An increase in cycle length appears to be justifiable in this case. Further, two lane groups, the WB and NB lane groups (both critical), have *v/c* ratios in excess of 1.00, indicating that a deficiency of capacity exists in these phases. Thus some reallocation of green time to these lane groups also appears to be appropriate.

Using Equation 24-48, cycle lengths may be computed to provide various target v/c ratios. Values of X_c between 0.80 and 0.95 will be tried:

$$C = \frac{12 * 0.95}{0.95 - 0.797} = 745 \text{ s}$$

$$C = \frac{12 * 0.90}{0.90 - 0.797} = 104.9 \text{ s}$$

$$C = \frac{12 * 0.85}{0.85 - 0.797} = 192.5 \text{ s}$$

$$C = \frac{12 * 0.80}{0.80 - 0.797} = 3{,}200 \text{ s}$$

It is impractical to provide for $X_c = 0.80$, or even $X_c = 0.85$, because the resulting cycle lengths required are excessive. Note that this computation *can* result in a negative number, which implies that *no* cycle length can satisfy the target X_c value. In this case, a cycle length between 80 seconds and 120 seconds appears to be reasonable. For the remainder of this illustration, a cycle length of 100 seconds is used. For a 100- second cycle length:

$$X_c = 0.797\left(\frac{100}{100 - 12}\right) = 0.906$$

Equation 24-43 is used to reallocate the green time within the new 100- second cycle length. A target X_i of 0.906 will be used for each critical-lane group:

$$g_{A1} = G_{A1} = 0.137 * \left(\frac{100}{0.906}\right) = 15.1 \text{ s}$$

$$g_{A2} = G_{A2} = 0.378 * \left(\frac{100}{0.906}\right) = 41.7 \text{ s}$$

$$g_B = G_B = 0.282 * \left(\frac{100}{0.906}\right) = 31.2 \text{ s}$$

$$15.1 + 41.7 + 31.2 + 12.0 = 100 \text{ OK}$$

The retimed signal will provide for v/c ratios of 0.906 in each critical-lane group and for a critical v/c ratio of 0.906 for the intersection. Delays could be recalculated based on these results to determine the impact of the retiming on delay and LOS, but these will surely be positive, given that two of the three critical-lane groups "failed" in the initial analysis, with $v/c > 1.00$.

In this example, there were no permitted or compound left turns. Thus the retiming suggested is exact. Where permitted or compound left turns exist, saturation flow rates include a permitted f_{LT} that depends on signal timing parameters. Thus the process in such cases is technically iterative. In practical terms, however, retiming the signal as indicated gives a reasonably accurate result that can be used in a revised analysis to determine its exact impact on operations.

24.5.4 Analysis of Actuated Signals

The application of an HCM analysis model requires the specification of signal timing. For actuated signals, the average phase times and cycle length for the period of analysis must be provided. For existing locations, this can be observed in the field. For the analysis of future locations, or consideration of signal retiming, the average phase and cycle lengths must be estimated.

The HCM 2010 provides a detailed model for the estimation of actuated signal timing, given controller and detector parameters as inputs. It is an algorithm that requires many iterations and is not possible to compute manually by hand.

Previous editions of the HCM recommended a rough estimation procedure using Equations 24-42 and 24-43 of the previous section with a high target v/c ratio in the range of 0.95 to 1.00. As a rough estimate, this is still a viable approach, when no software to implement the HCM's more detailed approach is available. It assumes, however, that controller settings and detector locations are optimal, which is not always the case.

The delay model, of course, specifically accounts for the positive impact of actuated control on operations. It too, however, requires that the average signal settings be specified.

24.6 Calibration Issues

The HCM model is based on a default base saturation flow rate of 1,900 pc/hg/ln. This value is adjusted by up to 11 adjustment factors to predict a prevailing saturation flow rate for a lane group. The HCM provides guidance on the measurement of the prevailing saturation flow rate, s. Although it allows for substituting a locally calibrated value of the base rate, s_o, it does not provide a means for doing so. It also does not provide a procedure for measuring lost times in the field.

It is also useful to quickly review how the calibration of adjustment factors of various types may be addressed, even if this is impractical in many cases. A study procedure for measuring delays in the field is detailed in Chapter 10.

24.6.1 Measuring Prevailing Saturation Flow Rates

As defined in Chapter 20, saturation flow rate is the maximum average rate at which vehicles in a standing queue may pass through green phase, after startup lost times have been dissipated. It is measured on a lane-by-lane basis through observations of headways as vehicles pass over the STOP line of the intersection approach. The first headway begins when the green is initiated and ends when the first vehicle in queue crosses the STOP line (front wheels). The second headway begins when the first vehicle (front wheels) crosses the STOP line and ends when the second vehicle in queue (front wheels) crosses the STOP line. Subsequent headways are similarly measured.

The HCM suggests that for most cases, the first four headways include an element of lost time and thus are not included in saturation flow rate observations. Saturation headways, therefore, begin with the fifth headway in queue and end when the last vehicle in the standing queue crosses the STOP line (again, front wheels). Subsequent headways do not necessarily represent saturation flow.

24.6.2 Measuring Base Saturation Flow Rates

The base saturation flow rate assumes a set of "ideal" conditions that include 12-foot lanes, no heavy vehicles, no turning vehicles, no local buses, level terrain, and non-CBD location, among others. It is usually impossible to find a location that has all of these conditions.

In calibrating a base saturation flow rate, a location is sought with near ideal physical conditions. An approach with three or more lanes is recommended because the middle lane can provide for observations without the influence of turning movements. Heavy vehicles cannot be avoided, but sites that have few heavy vehicles provide the best data. Even where data are observed under near ideal physical conditions, *all headways* observed after the first heavy vehicle must be discarded when considering the base rate.

24.6.3 Measuring Startup Lost Time

If the first four headways contain a component of start-up lost time, then these headways can be used to measure the startup lost time. If a saturation headway for the data has been established as h s/veh, then the lost time component in each of the first four headways is $(h_i - h)$, where h_i is the total observed headway for vehicles 1 to 4 in the queue. The startup lost time is the sum of these increments. Both saturation flow rate and startup lost time are observed for a given lane during each signal cycle. The calibrated value for use in analysis would be the average of these observations.

Startup lost time under base conditions can be observed as well by choosing a location and lane that conforms to the base conditions for geometrics with no turning vehicles and by eliminating consideration of any headways observed after the arrival of the first heavy vehicle.

24.6.4 An Example of Measuring Saturation Flow Rates and Startup Lost Times

The application of these principles is best illustrated through example. Table 24.17 shows data for six signal cycles of a center lane of a three-lane approach (no turning vehicles) that is geometrically ideal. In general, calibration would involve more cycles and several locations. To keep the illustration to a reasonable size, however, the limited data of Table 24.17 are used.

Note that saturation conditions are said to exist only between the fifth headway and the headway of the last vehicle present in the standing queue when the signal turns green. Only the headways occurring between these limits can be used to calibrate saturation flow rate. The first four headways in each queue will be used subsequently to establish the startup lost time.

The saturation headway for the lane in question is the average of all observed headways representing saturated conditions. As seen in Table 24.17, there are 41 observed saturation headways totaling 96.0 seconds.

From this data, the average saturation headway (under prevailing conditions) at this location is:

$$h = \frac{96.0}{41} = 2.34 \text{ s/veh}$$

From this, the saturation flow rate for this lane may be computed as:

$$s = \frac{3{,}600}{2.34} = 1{,}538 \text{ veh/hg/ln}$$

If a lane group had more than one lane, the saturation headways and flow rates would be separately measured for each lane. The saturation flow rate for the lane group is then the sum of the saturation flow rates for each lane.

Table 24.17: Example in Measuring Saturation Flow Rate and Startup Lost Time

Queue Position	Observed Headways (s) in Cycle No. ____						Sum of Sat Hdwys	No. of Sat Hdwys
	1	2	3	4	5	6		
1	3.5	2.9	3.9	4.2H	2.9	3.2	0.0	0
2	3.2	3.0	3.3	3.6	3.5H	3.0	0.0	0
3	2.6	2.3	2.4	3.2H	2.7	2.5	0.0	0
4	2.8H	2.2	2.4	2.5	2.1	2.9H	0.0	0
5	2.5	2.3	2.1	2.1	2.2	2.5	13.7	6
6	2.3	2.1	2.4	2.2	2.0	2.3	13.3	6
7	3.2H	2.0	2.4	2.4	2.2	2.3	14.5	6
8	2.5	1.9	2.2	2.3	2.4	2.0	13.3	6
9	4.5	2.9H	2.7H	1.9	2.2	2.4	12.1	5
10	6.0	2.5	2.4	2.3	2.7H	2.1	12.0	5
11		2.8H	4.0	2.2	2.4	2.0	9.4	4
12		2.5	7.0	2.9H	5.0	2.3	7.7	3
13		5.0		4.1		6.0	0.0	0
14		7.5					0.0	0
15							0.0	0
Sum							**96.0**	**41**

Notes: H = heavy vehicle.
Single underline: beginning of saturation headways.
Double underline: end of standing queue clearance; end of saturation headways.
Italics: saturation headway under base conditions.

Measuring the base saturation flow rate for this location involves eliminating the impact of heavy vehicles, assuming all other features of the lane conform to base conditions. Because the heavy vehicles may conceivably influence the behavior of any vehicle in queue behind it, the only headways that can be used for such a calibration are those before the arrival of the first heavy vehicle. Again, saturation headways begin only with the fifth headway. Looking at Table 24.18, only eight headways qualify as saturation headways occurring before the arrival of the first heavy vehicle:

- Headways 5 to 8 of Cycle 2
- Headways 5 to 8 of Cycle 3

Table 24.18: Calibration of Startup Lost Time from Table 24.17 Data

Position In Queue	Observed Headway (s) for Cycle No. ____						Avg *h* (s)	h_{avg}: 2.175 (s)
	1	2	3	4	5	6		
1	2.5	2.9	3.9	H	2.9	3.2	3.080	0.905
2	3.2	3.0	3.3	H	H	3.0	3.125	0.950
3	2.6	2.3	2.4	H	H	2.5	2.450	0.275
4	H	2.2	2.4	H	H	H	2.300	0.125
Sum								**2.255**

Notes: H = headway occurring after arrival of first heavy vehicle.

The sum of these eight headways is 17.4 seconds, and the base saturation headway and flow rate may be computed as:

$$h_o = \frac{17.4}{8} = 2.175 \text{ s/veh}$$

$$s_o = \frac{3{,}600}{2.175} = 1{,}655 \text{ pc/hg/ln}$$

Startup lost time is evaluated relative to the base saturation headway. It is calibrated using the first four headways in each queue because these contain a component of startup lost time in addition to the base saturation headway. Because the lost time is relative to base conditions, however, only headways occurring before the arrival of the first heavy vehicle can be used. The average headway for each of the first four positions in queue is determined from the remaining measurements. The component of startup lost time in each of the first four queue positions is then taken as $(h_i - h_o)$. This computation is shown in Table 24.18, which eliminates all headways occurring after the arrival of the first heavy vehicle. The startup lost time for this lane is 2.255 s/cycle.

Where more than one lane exists in the lane group, the startup lost time would be separately calibrated for each lane. The startup lost time for the lane group would be the *average* of these values.

Clearly, for actual calibration, more data would be needed and should involve a number of different sites. The theory and manipulation of the data to determine actual and base saturation flow rates, however, does not change with the amount of data available.

24.6.5 Calibrating Adjustment Factors

Of the 11 adjustment factors applied to the base saturation flow rate in the HCM model, some are quite complex and would require major research studies for local calibration. Included in this group are the left-turn and right-turn adjustment factors and the pedestrian/bicycle interference adjustment factors. A number of the adjustment factors are relatively straightforward and would not be difficult to calibrate locally, at least theoretically. It may always be difficult to find appropriate sites with the desired characteristics for calibration. Three adjustment factors involve only a single variable:

- Lane width (12-foot base condition)
- Grade (0% base condition)
- Area type (non-CBD base condition)

Two additional factors involve two variables:

- Parking (no parking base condition)
- Local bus blockage (no buses base condition)

The heavy-vehicle factor involves some very special considerations, and the lane utilization factor should be locally measured in any event and is found using Equation 24-16 or default values.

Calibration of all of these factors involves the controlled observation of saturation headways under conditions in which only one variable does not conform to base conditions. By definition, an adjustment factor converts a base saturation flow rate to one representing a specific prevailing condition, or:

$$s = s_o f_i \tag{24-50}$$

where f_i is the adjustment factor for condition i. Thus, by definition, the adjustment factor must be calibrated as:

$$f_i = \frac{s}{s_o} = \frac{(3{,}600/h)}{(3{,}600/h_o)} = \frac{h_o}{h} \tag{24-51}$$

All terms are as previously defined.

For example, to calibrate a set of lane-width adjustment factors, a number of saturation headways would have to be determined at sites representing different lane widths but where all other underlying characteristics conformed to base conditions. For example, if the following data were obtained for various lane widths, i:

$$h_{10} = 2.6 \text{ s/veh}$$
$$h_{11} = 2.4 \text{ s/veh}$$
$$h_{12} = 2.1 \text{ s/veh (base conditions)}$$
$$h_{13} = 2.0 \text{ s/veh}$$
$$h_{14} = 1.9 \text{ s/veh}$$

Adjustment factors for the various observed lane widths could then be calibrated using Equation 24-51:

$$f_{w10} = \frac{2.1}{2.6} = 0.808$$

$$f_{w11} = \frac{2.1}{2.4} = 0.875$$

$$f_{w12} = \frac{2.1}{2.1} = 1.000$$

$$f_{w13} = \frac{2.1}{2.0} = 1.050$$

$$f_{w14} = \frac{2.1}{1.9} = 1.105$$

Adjustment factors for lanes wider than 12 feet are greater than 1.000, indicating that saturation flow rates increase from the base value for wide lanes (> 12 ft). For lanes narrower than 12 feet, the adjustment factor is less than 1.000, as expected.

Similar types of calibration can be done for any of the simpler adjustments. If a substantial database of headway measurements can be achieved for any given factor, regression analysis may be used to determine an appropriate relationship that describes the factors.

Calibrating heavy-vehicle factors (or passenger-car equivalents for heavy vehicles) is a bit more complicated and best explained by example. Refer to the sample problem for calibration of prevailing and base saturation flow rates. In this case, all conditions conformed to the base, except for the presence of heavy vehicles. Of the 41 observed saturation headways, 6 were heavy vehicles, representing a population of $(6/41) * 100 = 14.63\%$.

The actual adjustment factor for this case is easily calibrated. The base saturation headway was calibrated to be 2.175 s/veh, and the prevailing saturation flow rate (representing all base conditions, except for heavy-vehicle presence) was 2.34 s/veh. The adjustment factor is:

$$f_{HV} = \frac{2.175}{2.34} = 0.929$$

This calibration, however, is only good for 14.63% heavy vehicles. Additional observations at times and locations with varying heavy-vehicle presence would be required to generate a more complete relationship.

There is another way to look at the situation that produces a more generic calibration. If all 41 headways had been passenger cars, the sum of the headways would have been $41 * 2.175 = 89.18$ s. In fact, the sum of the 41 headways was 96.0 seconds. Therefore, the six heavy vehicles caused $96.00 - 89.18 = 6.82$ s of additional time consumption due to their presence. If *all* of the additional time consumed is assigned to the six heavy vehicles, each heavy vehicle accounted for $6.82/6 = 1.137$ s of additional headway time. If the base saturation headway is 2.175 s/veh, the saturation headway for a heavy vehicle would be $2.175 + 1.137 = 3.312$ s/veh. Thus one heavy vehicle consumes as much headway time as $3.312/2.175 = 1.523$ passenger cars. This is, in effect, the passenger-car equivalent for this case, E_{HV}. This can be converted to an adjustment factor using Equation 24-14:

$$f_{LT} = \frac{1}{1 + 0.146(1.523 - 1)} = 0.929$$

This is the same as the original result. It allows, however, for calibration of adjustment factors for cases with varying heavy-vehicle presence, assuming that the value of E_{HV} is not affected by heavy-vehicle presence.

24.6.6 Normalizing Signalized Intersection Analysis

In many cases, it will be difficult or too expensive to calibrate individual factors involved in signalized intersection analysis. Nevertheless, in some cases, it will be clear that the results of HCM analysis are not correct for local conditions. This occurs when the results of analysis are compared to field measurements and obvious differences arise. It is possible to "normalize" the HCM procedure by observing departure volumes on fully saturated, signalized intersection approaches—conditions that connote capacity operation.

Consider the case of a three-lane intersection approach with a 30-second effective green phase in a 60-second cycle. Assume further that the product of all 11 adjustment factors that apply to the prevailing conditions is 0.80. Then:

$$s = s_o NF = 1{,}900 * 0.80 = 4{,}560 \text{ veh/hg}$$

$$c = 4{,}500 * \left({}^{30}\!/_{60}\right) = 2{,}280 \text{ veh/h}$$

This is the predicted capacity of the lane group using the HCM model. Despite this result, field observations measured a peak 15-minute departure flow rate from this lane group (under fully saturated conditions) of 2,400 veh/h.

The measured value represents a field calibration of the actual capacity of the lane group because it was observed under fully saturated conditions. Because it is more than the estimated value, the conclusion must be that the estimated value using the HCM model is too low. The difficulty is that it may be too low for many different reasons:

- The base saturation flow rate of 1,900 pc/hg/pl is too low.
- One or more adjustment factors is too low.
- The product of 11 adjustment factors is not an accurate prediction of the *combination* of prevailing conditions existing in the lane group.

All of this assumes that the measured value was accurately observed. The latter point is a significant difficulty with the methodology. Calibration studies for adjustment factors focus on isolated impacts of a single condition. Is the impact of 20% heavy vehicles in an 11-foot lane on a 5% upgrade the same as the product of the three appropriate adjustments, $f_{HV} * f_w * f_g$? This premise, particularly where there are 11 separately calibrated adjustments, has never been adequately tested using field data.

The local traffic engineer does not have the resources to check the accuracy of each factor involved in the HCM model, let alone the algorithms used to generate the estimate of capacity. But the value of the base saturation flow rate may be adjusted to reflect the field-measured value of capacity. The measured capacity value is first converted to an equivalent value of prevailing saturation flow rate for the lane group:

$$s = \frac{c}{(g/C)} = \frac{2{,}400}{0.50} = 4{,}800\,\text{veh/hg}$$

Using Equation 24-12, with the product of all adjustment factors of 0.80, the base saturation flow rate may be normalized:

$$S_o = \frac{s}{NF} = \frac{4{,}800}{3 * 0.80} = 2{,}000\,\text{pc/hg/ln}$$

This normalized value may now be used in subsequent analyses concerning the subject intersection. If several such "normalizing" studies at various locations reveal a common area-wide value, it may be more broadly applied.

It must be remembered, however, that this process does *not* mean that the actual base saturation flow rate is 2,000 pc/hg/ln. If this value were observed directly, it might be quite different. It reflects, however, an adjusted value that normalizes the entire model for a number of underlying local conditions that renders some base values used in the model inaccurate.

24.7 Summary

The HCM model for analysis of signalized intersections is complex and incorporates many submodels and many algorithms. These result from a relatively straightforward model concept to handle the myriad different conditions that could exist at a signalized intersection.

References

1. *Highway Capacity Manual*, 4th Edition, Transportation Research Board, National Research Council, Washington DC, 2000.
2. Akcelik, R., "SIDRA for the Highway Capacity Manual," *Compendium of Papers*, 60th Annual Meeting of the ITE, Institute of Transportation Engineers, Washington DC, 1990.
3. Akcelik, R., *SIDRA 4.1 User's Guide*, Australian Road Research Board, Australia, August 1995.
4. Teply, S., *Canadian Capacity Guide for Signalized Intersections*, 2nd Edition, Institute of Transportation Engineers, District 7—Canada, June 1995.
5. Reilly, W., et al., "Signalized Intersection Capacity Study," *Final Report*, NCHRP Project 3-28 (2), JHK & Associates, Tucson AZ, December 1982.
6. Roess, R., et al., "Levels of Service in Shared-Permissive Left-Turn Lane Groups at Signalized Intersections," *Final Report*, Transportation Research Institute, Polytechnic University, Brooklyn NY, 1989.
7. Bonneson, J. "Signalized Intersection: Procedure for Estimating Fully-Actuated Phase Duration," Working Paper #6.2, NCHRP Project 3-92, 2008.
8. Zegeer, J.D. *Field Validation of Intersection Capacity Factors*, Transportation Research Record 1091, Transportation Research Board. 1986.
9. Strong, D., Rouphail, N., and Courage, K., "New Calculation Method for Existing and Extended HCM Delay Estimation Procedures," Proceedings 85th Annual TRB meeting, Washington DC, January 2006.
10. Strong, D., and Rouphail, N., "Incorporating the Effects of Traffic Signal Progression into the Proposed Incremental Queue Accumulation (IQA) Method," Proceedings 85th Annual TRB Meeting, Washington DC, January 2006.

Problems

24-1. A one-way intersection approach of four lanes has the following characteristics:

- 60-s effective green time in a 100-s cycle
- Four 11-ft lanes
- 10% heavy vehicles
- 3% upgrade
- Parking on one side with 15 mvts/h within 250 ft of the stop line
- 20 local buses/h stopping to pick up and drop off passengers
- 8% right turns from an exclusive RT lane
- 12% left turns from an exclusive LT lane
- 100 peds/h in each crosswalk
- No bicycle traffic
- A CBD location
- No opposing approach

Estimate the saturation flow rate and capacity of the three lane groups on this approach.

24-2. The intersection shown in Figure 24-13 is to be analyzed using the HCM 2010 methodology. All computations are to be done by hand but may be checked using any appropriate software.

(a) Determine the existing *v/c* ratio and LOS for each lane group in the intersection.

(b) Determine the LOS for each approach and for the intersection as a whole.

(c) Make recommendations for improvements in the signal timing, if your results indicate this is needed.

The following additional information is available concerning the intersection for Problem 24-2: $PHF = 0.92$: Arrival types: 5 WB, 2 EB, 3 NB; $\%HV = 12\%$ in all movements; no pedestrians—overpasses provided.

24-3. The intersection of Grand Blvd. and Crescent Ave. is shown in Figure 24-14. It is a simple intersection of two one-way arterials in a busy downtown area.

(a) Determine the delays, levels of service, and *v/c* ratios for each approach, and for the intersection as a whole.

(b) If a signal retiming is indicated, propose an appropriate timing that would result in equal *v/c* ratios for the critical-lane groups.

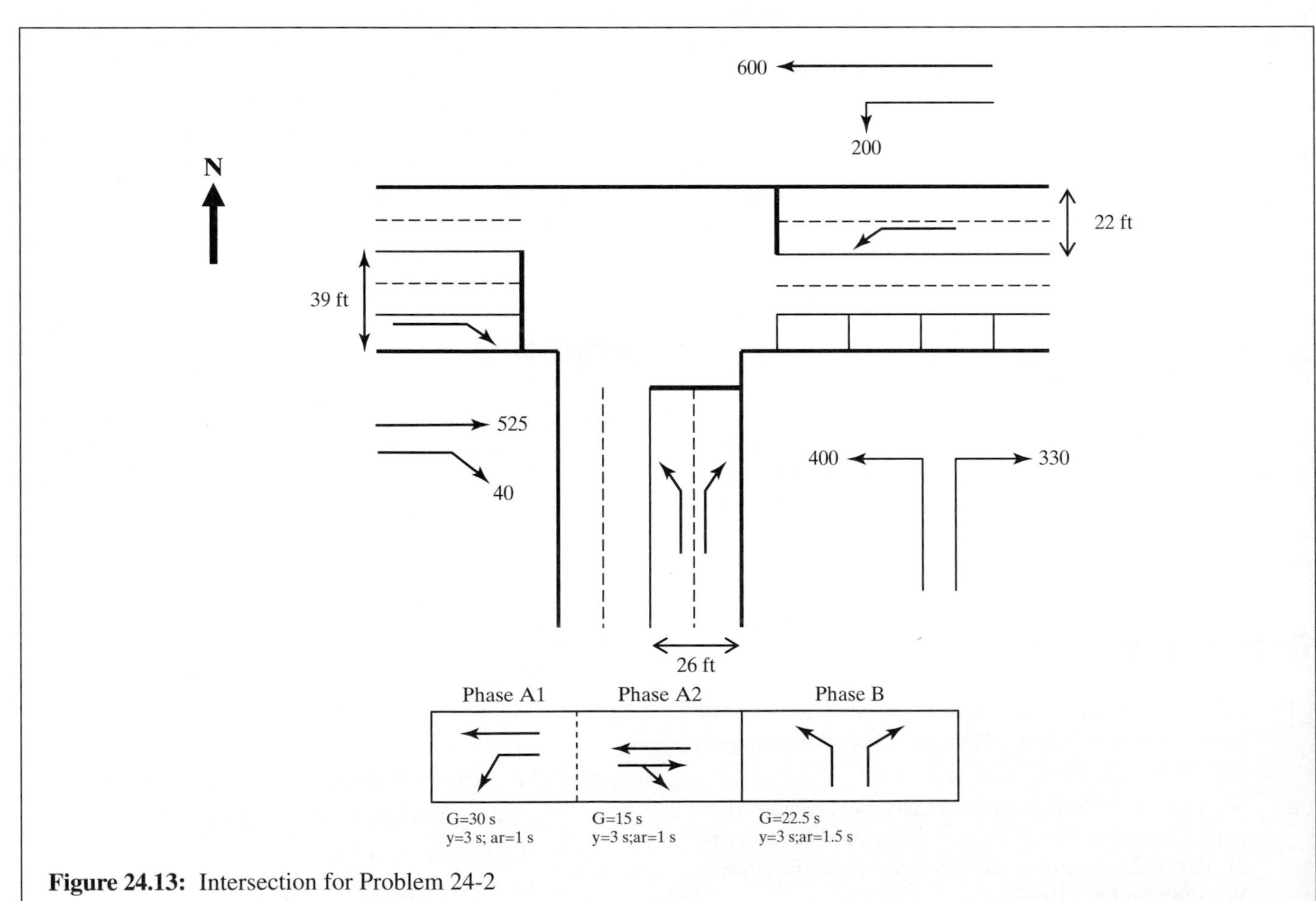

Figure 24.13: Intersection for Problem 24-2

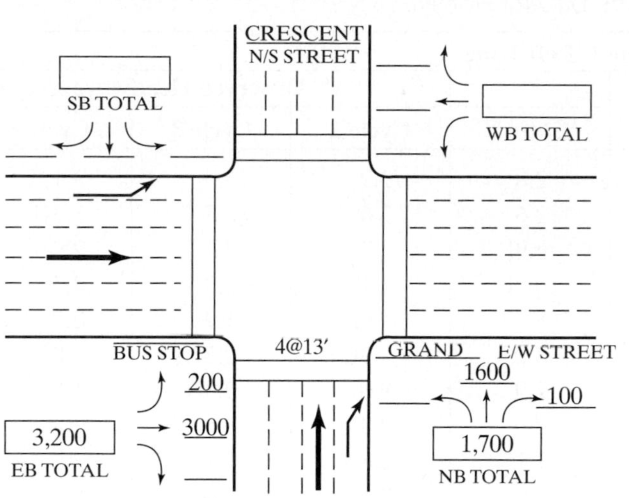

(a) Geometry and Demand

Lane Group	Grade (%)	Heavy Veh (%)	Local Buses/h, N_B	PHF	Conflicting Peds	Arrival Type
EB LT	0	15	0	0.90	0	5
EB TH	0	15	40	0.90	400	5
NB TH	0	5	0	0.90	400	3
NB RT	0	5	0	0.90	0	3

(b) Characteristics

Signal Phase	Green, G (s)	Yellow, y (s)	All Red, ar (s)
EB LT/TH	25	3	2
NB TH/RT	25	3	2

(c) Signal Timing

Figure 24.14: Intersection for Problem 24-3

24-4. Consider the following results of an HCM analysis of a WB lane group at a signalized intersection, as summarized here:

- $v = 800$ veh/h
- $c = 775$ veh/h
- $g/C = 0.40$
- $C = 90$ s
- Initial queue $= 0.0$
- $T = 0.25$ h
- $AT = 3$

(a) What total control delay is expected during the first 15-minute analysis interval that these conditions exist?

(b) If these conditions exist for two additional successive 15-minute periods, what would the total control delay be for those periods?

24-5. For the data shown in Table 24.19:

(a) Determine the prevailing saturation flow rate for the lane group illustrated in the data.

(b) Determine the base saturation flow rate for this lane group.

(c) Determine the startup lost time for this lane group.

24-6. Using the data in Table 24.19, calibrate the passenger-car equivalent for heavy vehicles for the lane group depicted and the heavy-vehicle adjustment factor. Demonstrate that they yield the same results.

Table 24.19: Data for Problem 24.5

Data for Lane 1, Left Lane

Veh. in Queue	Observed Headways (sec)				
	Cycle 1	Cycle 2	Cycle 3	Cycle 4	Cycle 5
1	2.8	2.9	3.0	3.1	2.7
2	2.6	2.6	2.5	3.5H	2.6
3	3.9L	2.3	2.2	2.9	2.5
4	10.2H	2.1	2.0	2.5	2.0
5	8.7	4.0L	1.9	2.2	1.9
6	3.0	9.9L	2.2	2.0	1.9
7	2.9	9.8	2.9H	1.9	3.6HL
8	5.0	3.3	2.6	1.8	9.0
9	7.1	2.8	2.1	7.0	4.0
10	9.0	2.2	4.0	8.0	4.9
11		1.9	5.0		9.0
12		5.5			
13		4.0			
14					
15					

Data for Lane 2, Center Lane

Veh. in Queue	Observed Headways (sec)				
	Cycle 1	Cycle 2	Cycle 3	Cycle 4	Cycle 5
1	2.8	2.9	2.9	2.7	2.9
2	2.7	2.5	2.5	2.6	2.3
3	2.3	2.2	2.1	2.3	2.1
4	2.1	2.0	2.0	1.9	2.1
5	2.8H	1.9	1.8	1.9	1.9
6	2.3	1.9	2.0	1.9	2.0
7	2.6H	2.0	2.1	1.8	2.4H
8	2.1	2.1	1.9	2.0	2.5H
9	1.9	4.5	1.8	1.9	6.0
10	5.0	4.4	2.1	2.0	9.0
11			5.6	7.1	
12			3.3		
13					
14					
15					

(Continued)

Table 24.19: Data for Problem 24.5 *(Continued)*

Data for Lane 3, Right Lane

Veh. in Queue	**Observed Headways (sec)**				
	Cycle 1	**Cycle 2**	**Cycle 3**	**Cycle 4**	**Cycle 5**
1	3.0	2.8	3.1	3.9R	2.8
2	2.5	2.5	2.7	2.8	2.6
3	2.1	2.1	2.8R	2.1	2.1
4	1.9	1.9	2.3	1.9	1.8
5	2.5H	2.0	3.2RH	1.9	1.9
6	2.3	2.1	2.5	1.8	1.9
7	2.4R	2.5R	2.3	1.7	2.1
8	2.2	2.1	2.0	3.7	1.9
9	4.4	1.9	1.8	5.0	1.9
10	6.0	3.5	2.6R		4.7
11		4.0	7.0		
12		5.0			
13		2.9			
14					
15					

Note: H = heavy vehicle; L = left turn; R = right turn.
Underline = last vehicle in standing queue.

24-7. The capacity of a signalized intersection lane group is estimated using the HCM methodology, with standard values as follows:

- $S_o = 1{,}900$ pc/hg/ln
- $N = 2$ lanes
- $F = 0.75$ (product of all adjustment factors)
- $g/C = 0.60$

If a capacity of 1,900 veh/h for this lane group was measured in the field, what normalized value of s_o should be used to adjust the HCM methodology to yield the correct estimate?

CHAPTER 25

Intelligent Transportation Systems in Support of Traffic Management and Control

It has been approximately 20 years since the first ISTEA (Interstate Surface Transportation Efficiency Act) legislation explicitly addressed the theme of "intelligent transportation systems" (ITS), in an effort that was then known as intelligent vehicle highways systems (IVHS).

Since that time, the concept has evolved to embrace a multimodal approach that (1) provides information to the traveling public on modal choices, operating status of the transportation system, and options for their decision making, and (2) provides the agencies responsible for the public infrastructure with standards and practices that are directed to interoperability of equipment and systems.

Based on the extensive public infrastructure and the emphasis on interoperability, there is a natural role for the public agencies responsible for the infrastructure and for professional and trade organizations that serve as forums, leaders, and technology accelerators.

At the same time, there are private sector initiatives that seek to serve the consumer (the traveler) with innovative products and services, *at a pace dictated by market forces and consumer desires.* In many ways, this is a concurrent but independent effort.

Consider the following:

- *Cell phones* are now pervasive and provide not only a means of communicating new information to the consumer, but also a source of travel times, traffic volumes, and even origin destinations and route selections. The data that can flow from these commercial providers can aid transportation planning, traffic operations and control, and real-time decision making by both consumers and agencies.[1]

- *In-vehicle navigation systems* are becoming a standard item, whether as original equipment or as a convenient (and relatively low-priced) after-market option. These systems are based on the global positioning system (GPS) and capable of accepting real-time updates on traffic conditions and delays from the service provider.
- *The technologies are merging* at a pace that may make even the preceding words dated during the life of this edition of this textbook. Cell phones have blended with navigation systems and personal data assistants (PDAs), so that consumers can expect to be routed to their chosen restaurant on a walking trip, see all choices, buy theater tickets, and check train schedules and delays—in real time.
- *The smart car* is emerging, driven by competitive forces, new safety standards, and consumer expectations. Collision warnings, adaptive speed control, backup sensors, and even automated parking are available. The "smart car" is also becoming the safe car, with intelligent air bags (side as well as front) and improved design. The path to the "automated car" is not clear but will emerge. However, it is easy to forecast the car (or companion device) that not only selects routes but identifies and perhaps reserves parking spaces.
- *Transportation modeling* is becoming more sophisticated because of the needs for dynamic assignment in real time (when incidents become known, people reroute, from their then-existing positions, not their starting point), and for receiving and using as input more detailed data than ever before (flows, travel times, selected paths, deduced origins and destinations, times loaded onto the network), and computing speed.
- *Bus transit* can interact with traffic signals, to provide bus priority systems (BPS) that aid on-time performance and reduce delay in these high-occupancy vehicles (as well as overall delay). GPS-based information provides performance and planning data.
- *Travel* can be made more efficient by electronic toll collection, integrated "one-card" systems, variable pricing, flexible designation of high-occupancy lanes (HOV) and high-occupancy toll (HOT) lanes.
- *Transportation Management Centers (TMCs)* can update control policies, traffic signal settings, metering, and even lane uses based on current information, including weather, traffic loads, and incidents. The TMC can assist in promulgating advisories and in coordinating with other regional TMCs. The same can be done for planned maintenance and for the maintenance and protection of traffic (MPT).
- *Responses to emergencies,* ranging from hurricanes to terrorist actions, can be enhanced by the ITS technologies and systems now available. Routes may need to be planned in real time, communicated to the public, and monitored for performance and incidents.
- *Management of scarce resources* can be achieved with ITS technologies, by encouraging travel in hours with lower demand by variable pricing.

This last item, of course, deals with the issue of *congestion pricing,* which is often a highly charged and politically sensitive subject. But given that resources are finite, the issue may well be arriving at a palatable and equitable formulation for congestion pricing, not focusing on whether or not it should exist. Indeed, the existing system of tolls, parking fees, transit and transportation vouchers, and such can be thought of as an *existing* congestion pricing formulation. But rather than focus on this issue extensively, for the purposes of this chapter and this text, it is sufficient to note that the enabling technologies for other formulations are in place due to ITS technologies.

25.1 ITS Standards

An excellent source of information on current ITS standards in the United States is the Research and Innovative Technology Administration (RITA) of the USDOT, at http://www.standards.its.dot.gov/.

The Web site notes that "The ITS Standards Program has teamed with standards development organizations and public agencies to accelerate the development of open, non-proprietary communications interface standards." To a very large extent, the key words are "open" and "non-proprietary."

[1]The issue of privacy is acknowledged and must be addressed. But the authors of this text see that as a issue that will be addressed and not a long-term impediment to the use of this data. The consumer's desire for the benefits of this data will be met in a way that respects the privacy of the individual because it is the same population that wants both the privacy and the benefits.

This site is the repository and path to such information as:

- Existing ITS standards
- Current development status
- Deployment resources for state and local ITS deployers
- Documents, such as guides, best practices, and lessons learned
- Technical assistance and training opportunities

The standards themselves are referred to by their National Transportation Communications for ITS Protocol (NTCIP) numbers, and they fit within an overall National ITS Architecture.

25.2 National ITS Architecture

For current information on the overall National ITS Architecture, visit the USDOT's Federal Highway Administration (FHWA) Web site at http://www.ops.fhwa.dot.gov/its_arch_imp/index.htm. This site contains the detailed FHWA rule and Federal Transit Administration (FTA) policy related to the overall architecture, as well as links to the detailed architecture itself, which can be found at http://www.iteris.com/itsarch/.

The architecture defines the functions required for ITS, the physical entities or subsystems where the functions reside, and the information and data flows that connect the functions and physical subsystems into an integrated system.

Table 25.1 shows the "interface classes" used in the architecture and the "standards application areas" associated with each. In the cited source, there are hyperlinks to the standards application areas.

25.3 ITS Organizations and Sources of Information

In the United States, the Intelligent Transportation Society of America (ITSA or, more commonly, "ITS America") is

Table 25.1: ITS Standards Application Areas

National ITS Architecture Interface Class	Standards Application Areas
Center to Center: This class of application areas includes interfaces between transportation management centers.	Data Archival Incident Management Rail Coordination Traffic Management Transit Management Traveler Information
Center to Field: This class of application areas includes interfaces between a management center and its field equipment (e.g., traffic monitoring, traffic control, environmental monitoring, driver information, security monitoring, and lighting control).	Data Collection/Monitoring Dynamic Message Signs Environmental Monitoring Lighting Management Ramp Metering Traffic Signals Vehicle Sensors Video Surveillance
Center to Vehicle/Traveler: This class of application areas includes interfaces between a center and the devices used by drivers or travelers. It includes interfaces with motorists and travelers for exchange of traveler and emergency information as well as interfaces between management centers and fleet vehicles to support vehicle fleet management.	Mayday Transit Vehicle Communications Traveler Information
Field to Field: This class of application areas includes interfaces between field equipment, such as between wayside equipment and signal equipment at a highway rail intersection.	Highway Rail Intersection (HRI)
Field to Vehicle: This class of application areas includes wireless communication interfaces between field equipment and vehicles on the road.	Probe Surveillance Signal Priority Toll/Fee Collection

(*Source:* http://www.standards.its.dot.gov/learn_Application.asp.)

the primary organization identified with ITS. It was founded at the time of the 1991 ISTEA legislation cited at the beginning of this chapter. It has a network of regional/state chapters or affiliates, generally identified by such names as ITS-NY. The ITS America Web site is at http://www.itsa.org and provides a wealth of information. Two important annual events are the ITS America Annual Meeting and the ITS World Congress. There are a number of comparable organizations throughout the world, such as ITS-Japan and ITS-Europe (ERTICO).

Important publications include *Traffic Technology International* and *ITS International,* information about which can be found at the Web sites http://www.ukipme.com/mag_traffic.htm and http://www.itsinternational.com/, respectively. The *Journal of the Transportation Research Board,* also known as the *Transportation Research Record,* is an important source of information on ITS research and related traffic and transportation research.

There is also the IACP Technology Clearinghouse, established in 1997 through a cooperative agreement between the International Association of Chiefs of Police (IACP), the Federal Highway Administration (FHWA), and the National Highway Traffic Safety Administration (NHTSA), which has its Web site located at http://www.iacptechnology.org/IntelligentTransportationSystems.html. It is the conduit for USDOT to address "address data collection, information management, and reporting requirements with law enforcement information technologists at all levels"; the Web site notes that "The Technology Clearinghouse is a forum to advance the interests of the FHWA on matters relating to conventional traffic enforcement, automated enforcement, data collection, commercial vehicle enforcement, and many other areas of critical mutual interest."

25.4 ITS-Related Commercial Routing and Delivery

Routing systems are of great importance to trucking and service vehicles. This is a large specialty market, with software available that can compute long-haul routing, urban routing, and provides answers that take into account the set of scheduled pickup and delivery points. Dynamic rerouting, as new pickups are added en route, is feasible.

Today, it is commonplace for package delivery services (FedEx, UPS, and others) to offer real-time package tracking to customers on their Web sites. Using bar-code scanning technology and wireless communication, packages are tracked in detail from origin to destination. At the delivery point, the driver uses a computer-based pad to record delivery time and often the receiver's signature. This information is available in virtually real time to the sender.

Clearly, the package delivery services found a differentiating service feature that has rapidly been adopted in a highly competitive industry. What was special a few years ago has now become the expected standard of service.

The same data allow the service providers to obtain a wealth of data on the productivity (and downtime) of their vehicles and drivers, and on the cost of delivery in various areas.

25.5 Sensing Traffic by Virtual and Other Detectors

There is indeed a future in which cell phones and other devices can provide such a wealth of data that the traffic engineer and manager must worry about being overwhelmed with data. They must plan for that day.

Indeed, tags such as E-ZPass (used by a number of states along the Northeast corridor) do already provide the opportunity to sample travel times and estimate volumes, and E-ZPass readers can be used for such purposes even where tolls are not collected.

But the day when these newer technologies can be the primary source of data has not yet arrived. Indeed, there are still a number of years in which other innovative and traditional sensors will be the primary source of traffic information. The mainstay of traffic detection for decades has been magnetic sensing of vehicles by a loop installed into the road surface. It is still important in many jurisdictions, but some have moved away from loops because of cuts made into the pavement during maintenance, weather issues such as frost heave, or relative cost.

These are three alternatives that are coming into greater use:

1. *Virtual detectors* generated in software, using a standard or infrared video camera to capture an image of the traffic and generate estimates of flow, speed, queue, and/or spatial occupancy. Refer to Figure 25.1 for an illustration. Using such a tool, the transportation professional can "locate" numerous "detectors" essentially by drawing them on top of the intersection image, and depending on the software to process the data.
2. *Side-fire microwave detectors,* used to identify flows and point occupancy in each lane (depending on location, in both directions), with the results summarized

Figure 25.1: Software-Based Virtual Detector, Using Video Camera

(*Source:* Courtesy of NYCDOT.)

by lane or by lane group, for user-specified amounts of time. Refer to Figure 25.2 for an illustration of the coverage of such a detector. The particular unit illustrated is an RTMS (Remote Traffic Microwave Sensor) detector by Electronic Integrated Systems Inc. It is capable of covering eight lanes and uses wireless connections to send the signals to its SPIDER network in a nearby control cabinet.

3. *Wireless detectors imbedded in pavement,* such as illustrated in Figure 25.3. Models are available for presence or count, and they are generally imbedded one per lane. The units are approximately 3 × 3 inches, 2 inches deep. Compared to loops, the manufacturer claims an easier install and less susceptibility to being broken than a loop with a much larger footprint. Figure 25.4 shows a travel time map generated from sets of such detectors, using software to identify the "signature" or profile of individual vehicles.

There are variations on these types, including a detector that uses a 360° video image for "area occupancy" detection.

The use of infrared imaging allows vehicles to be detected in a variety of weather conditions. Another variant is the use of sophisticated algorithms based on coverage of the underlying pavement image allows data to be collected from stationary traffic as well as moving traffic.

25.6 Traffic Control in an ITS Environment

Four of the remaining chapters in this textbook focus on (1) good practice and basic principles for coordinating traffic signals when the v/c is less than 1.00 (Chapter 26) and when the v/c is greater than 1.00 (i.e., oversaturated; Chapter 27), (2) arterial multimodal performance (Chapter 28), and (3) basic principles to consider in the planning and design of arterials and networks (Chapter 29). The final chapter focuses on assessing and mitigating the effects of development on the operation of the transportation system (Chapter 30).

Figure 25.2: Illustration of the Placement of an RTMS Detector

(*Source:* Image of RTMS detector courtesy of EIS, Inc.)

Although there are indeed long-term planning issues spanning years or decades, the underlying theme for much of this is the use of traffic *control* measures (cycle length, split, offset, number and arrangement of phases) to adapt the system to the present situation (demand, weather, incidents), achieving good performance and service for the users.[2]

Within the context of balanced use, the operational problem often reduces to viewing the situation as a feedback control loop and applying basic principles with computerized or other control. The situation as the authors see it is depicted in Figure 25.5, and can be summarized as follows (for clarity, it is referred to as a street system, but it could include freeways and/or several modes):

- The primary input to the system is determined by *the users,* who make decisions on the mode(s) to use, the time(s) to depart for the trip, and the route(s) to take.
- The users are of course influenced in their decisions by their knowledge of such other "inputs" to their decision process as
  - The status or condition of the network, as reflected in underlying capacities, weather, and incidents.
  - The settings of the traffic control devices.
  - Information available to the user from public sources (variable message signs, radio advisories, highway advisory radio [HAR], public radio) and subscription services (vehicle navigation systems, cell phone messages, other).

[2]Some of the longer term issues are addressed in Chapters 28 and 29, including good design of the facilities in the first instance; the use of roundabouts, traffic calming, and balanced uses; the multimodal nature of most facilities; and the need to design with that recognition.

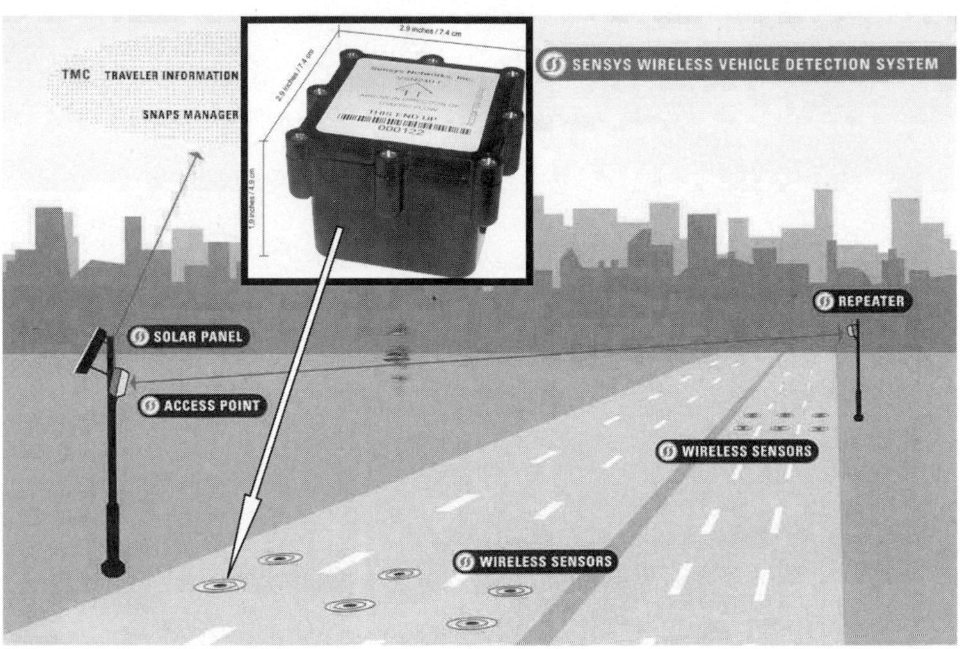

Figure 25.3: Wireless Detectors Imbedded into Road

(*Source:* Courtesy of Sensys Networks, Inc.)

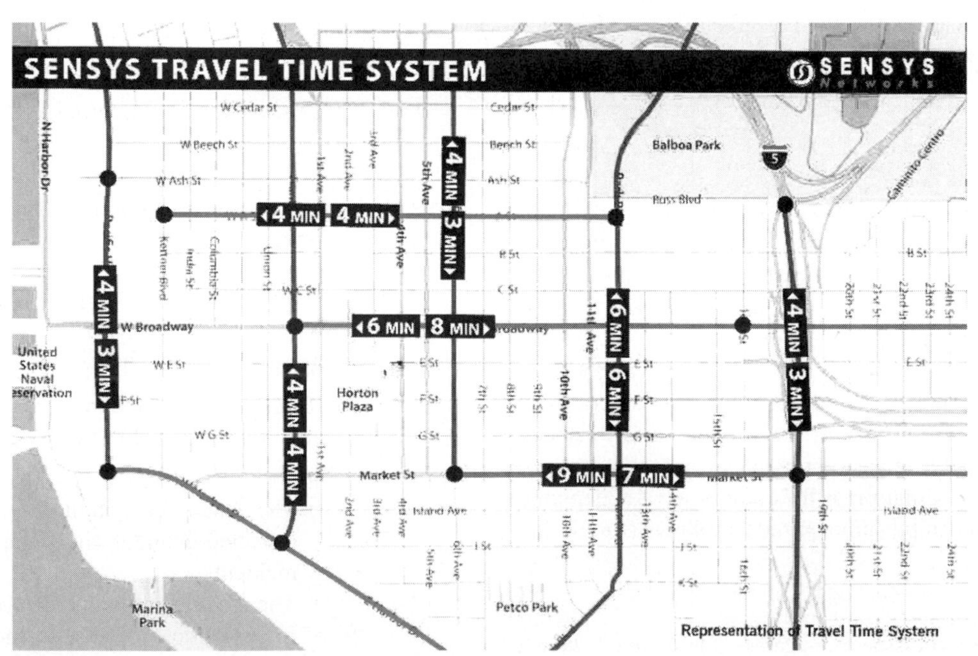

Figure 25.4: Travel Time Estimates from Sets of Figure 25-3 Detectors

(*Source:* Courtesy of Sensys Networks, Inc.)

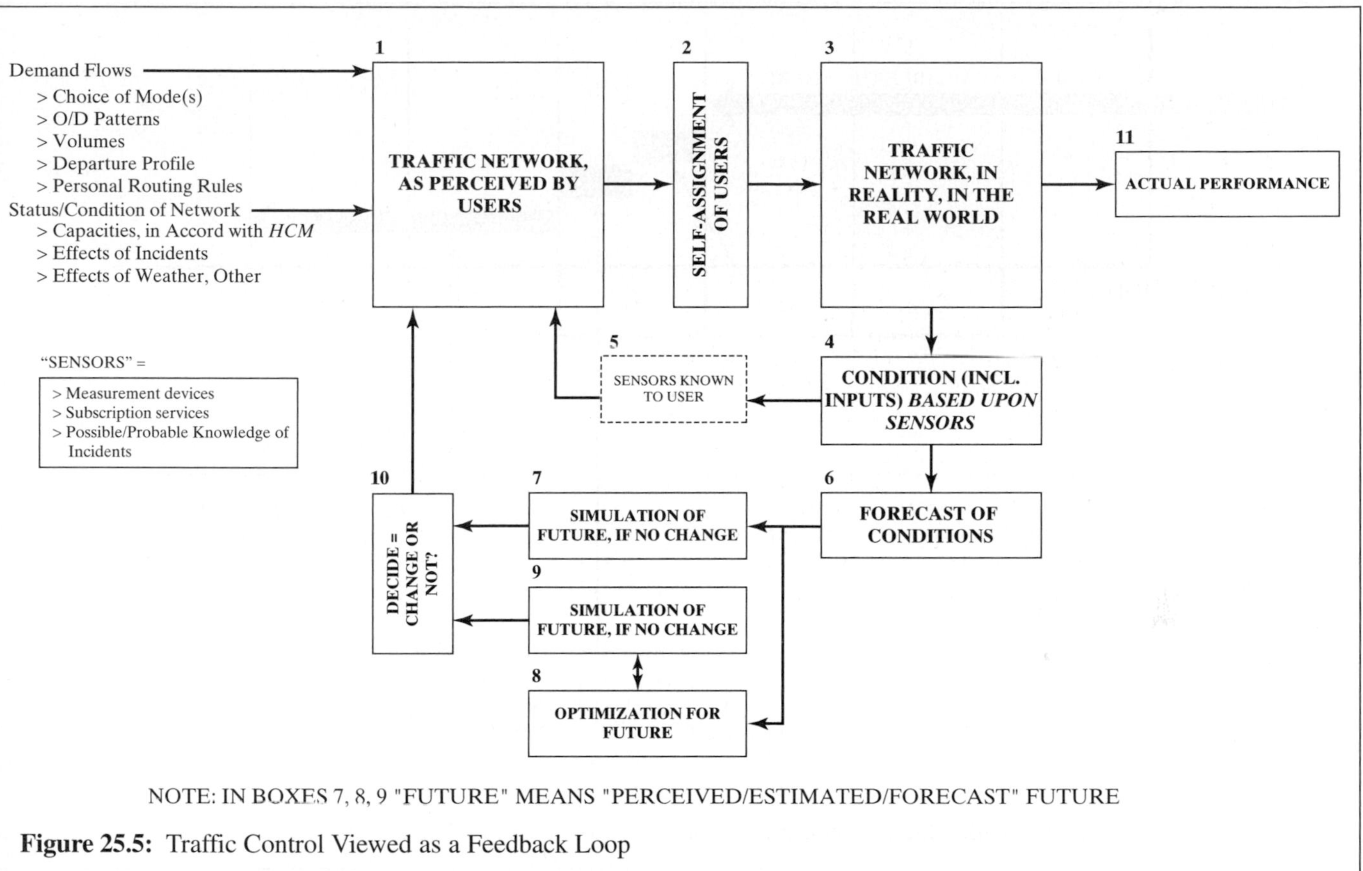

Figure 25.5: Traffic Control Viewed as a Feedback Loop

- All of this user knowledge, which many be imperfect, contributes to **Box 1** of Figure 25.5, namely the state of the network, *as perceived by users.* This is *not* necessarily the true state of the network.
- But it is this perception that leads to **Box 2,** namely the self-assignments that the users make to the network. It may also influence some of the user's initial conditions, such as departure time and mode.
- The self-assignments lead to **Box 3,** namely the actual loading of the network, with the consequent actual performance (**Box 11**) that may not be seen by anyone in real time but does exist.
- The next step in the process is **Box 4,** which reflects the conditions on the network *as observed by the many sensors that are available.*
    - For instance, public agencies may have a set of video cameras, side-fire microwave detectors, magnetic cans or loops in the pavement, and video for supplemental information. Some or all of this may be available on the Web, to users and to services that package or relay the information.
    - At the same time, subscription services may be using these and other sources (e.g., cell phone locations, travel times deduced, etc.) to update their advisories.
- The key phrase is "the sensors that are available," with the caveat "to the limit of their precision." Thus the knowledge of the system is *not perfect*. Indeed, the real challenge is how to make good decisions with imperfect information; more on this in the next section.
- So the public agency operating the signals takes the information from Box 4 and uses it in **Box 6** to *forecast* what is likely to be happening at some (reasonably close) future interval. At the same time, we have the users gaining knowledge from **Box 5,** which also works off Box 4 (although neither may have completely identically information because of sources used).
- The *user* will be making decisions based on perceptions from Box 5, influenced by their own personal routing

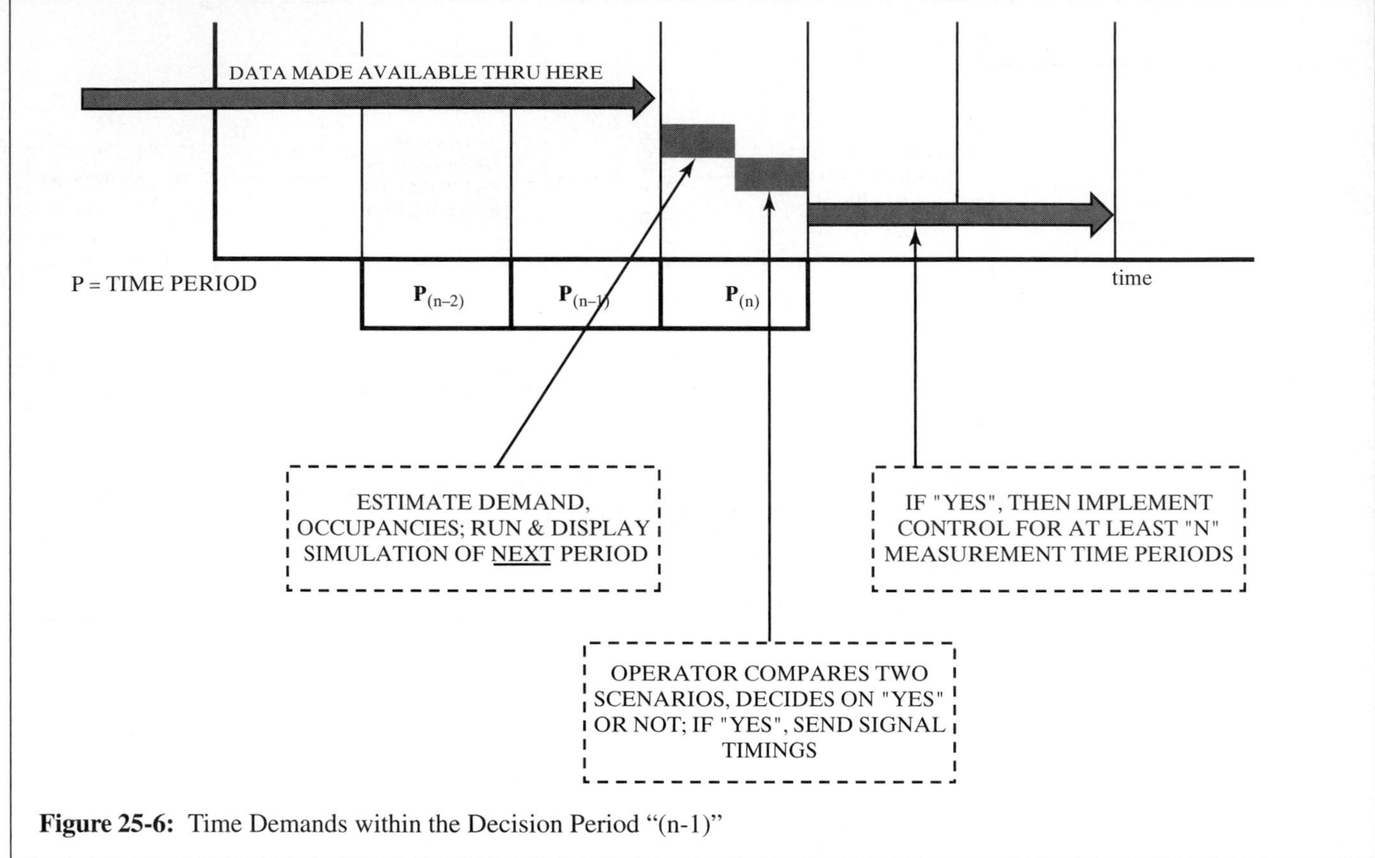

Figure 25-6: Time Demands within the Decision Period "(n-1)"

preferences and rules.[3] This will change upcoming conditions, even as the responsible agency tries to use Box 6 information to improve the operation.

- There are now two paths to take, so that one can assess what is expected with no changes (**Box 7**) vis-à-vis what is expected to happen if there *were* changes to the control settings (**Boxs 8 and 9**). Note that Box 8 is really an optimization algorithm, and that Boxes 8 and 9 are paired and might even have some iterative steps. Note also—and especially—that *speed* is needed in doing all that is required within Boxes 7, 8, and 9 in a timely fashion.
- **Box 10** is the decision element. It can be automated or can involve a human in the decision loop. If the latter, then there must be time *and* information for the human to make an informed decision.
- Whatever the decision, the original system (Box 1) is influenced by it, and the users experience the result and perhaps further adapt their own actions.

This then is the decision loop that goes on continually. At one extreme, there need be no changes; a simple "open loop" predetermined signal control plan can be used. At the other extreme, a highly responsive adaptive control can be used; more on this in the next section.

Notice that the discussion on Box 10 indicated that "time *and* information" is needed. Figure 25.6 shows the time aspect: If one is to consider changing the control for Interval or Period "n," everything must be done in Period "(n-1)," based on data available at the end of Period "(n-2)." Thus Period "(n-1)" has to include all the work shown in Boxes 6 through 9 of Figure 25.5 *and* allow for the decision making *and* allow for the communication and implementation of the new signal settings.

As to the "information" needs, Boxes 7 and 9 assume if there is a human operation in the loop, visualizations of the future situations must be simulated *and* displayed in a way

[3] At one time many years ago, a traffic manager might limit the information flow to Box 5 as a means of controlling options available to the user. Technology has made that approach largely irrelevant.

that will be effective in aiding the decision. There will also need to be some effective summary of the metrics, so that the visual is complemented by quantitative metrics, displayed as a table or bar charts.

25.7 How Fast Is Fast Enough?

In the preceding discussion, the promise was made "more on this in the next section." In one case, it was the issue that "the real challenge is how to make good decisions with imperfect information," specifically in the context of both limited observability (due to detector intensity and/or technology and/or cost) *and* the need to forecast into the future. In the other case, it was that there are a range of solutions, from a simple predetermined signal timing plan to "a highly responsive adaptive control." Indeed, the research literature and many dissertations have focused much effort on the latter problem.

Truly, if the daily pattern was totally repetitive and there were no weather problems (capacity tends to decrease in inclement weather), no incidents, and no special events, a single predetermined pattern would be just fine.

If there were advance knowledge of the major patterns within such a day, one could easily predetermine suitable plans for each pattern, and implement them by time of day. The same could be said of systematic changes due to day of week, or season, or even weather: Predetermined plans could be developed and implemented by time of day (with a check of the weather).

It is at this point that the lecturer usually says, "But things are not so predictable, they never are, and truly responsive control is necessary." Attention then focuses on the requirements to attain truly responsive control.

There are several major problems with this sequence:

1. The most glaring fault is that traffic patterns *are* indeed very regular, with the same peaks occurring at about the same times each day, on most days. Viewed from the perspective of when specific higher demand levels are reached, traffic looks even more regular, with those levels being reached at that same time each day plus or minus 10 minutes or so.
2. The next fault is that one assumes that the information coming from the detectors is perfectly valid. Generally, it is not, and it needs to be smoothed to account for anomalies. This smoothing means "longer," not "sooner."
3. The third fault is that one forgets that the information is *not* being used to make instantaneous decisions but rather to generate *forecasts* of what is (probably) going to exist at the (future) time that the decision is implemented. Refer to Figure 25.6.
4. The fourth fault is that one does not move from one signal plan to another immediately. Pedestrian phases must be respected, red and green times must not be too short (lest vehicle-vehicle accidents be induced), and so forth. It can take several cycle lengths to transition to a new plan, and the optimization (Box 8 of Figure 25.5) rarely takes this into account.
5. Lastly, a transition is usually "planned" simply to get from one set of signal timings to another, not to do so while also balancing traffic needs as the v/c hovers between 0.90 and 1.10. Thus the transition itself can precipitate a problem, and a series of them can almost assure a problem.

Now, it is perfectly true that unexpected things happen: Demand does not follow the usual pattern, a special event or even a holiday is overlooked, the weather is more severe than anticipated, and so forth.

We are *not* recommending blind adherence to predetermined patterns with no appreciation of actual conditions. But we do strongly assert, based on experience and data, that:

1. Traffic is indeed very regular when one takes into account day of week, season, and weather. Good plans can be implemented by advance offline planning and tend to work most of the time.
2. It is indeed useful to check what the sensors are saying, so that if there is a *very substantial* deviation from what is expected, a plan better suited to the observed profile can be implemented.[4]
3. At the same time, there are locations that have such complex or rapidly changing patterns that the underlying pattern is not discernible to the external observer. Such cases may occur near college campuses, which are driven by class schedules and special events that vary by the hour, over the week. Another such case may be malls that have sales or events that elude the external observer, or multiplex theaters that suddenly have hit movies on three screens, and so on. In these cases, an adaptive traffic plan may indeed be the most suitable.

[4]This approach is akin to selecting from a library of N signal timing plans, to match the best one to current conditions, except that we emphasize that there is a "standard" plan for the time slot that will almost surely be effective while allowing actions "by exception" when observations show a clear discrepancy.

But even in this last case, the detectors have to be well placed to maximize insight, and the changes have to occur at a pace consistent with the observability and the forecasting.

The authors do not question that advanced control policies, with operational sensors capable of delivering the precision needed on the variables driving the policy (e.g., queues, flows, turns) can deliver "more" than the approach put forth in Section 25.7. As an exercise for us and for you, we constructed the hypothesized curves in Figure 25.7 (it is also the basis for an assigned problem at the end of this chapter).

Figure 25.7 does not contain information on the cost of adding the features needed to move from one "stage" to the next, but we do believe that the cost to move beyond Stage 4 in the general case and Stage 5 in the special case will not prove cost effective or even feasible for most jurisdictions.

25.8 Emerging Issues

At the risk of being dated by events, the authors suggest that the following are major emerging issues that will draw your attention in your career:

1. Congestion pricing will grow as a priority, given that ITS technologies are creating the enabling infrastructure. Beyond the technological issues, the market, economic, and elasticity issues need much research.
2. There is a competition for who provides data, and with what quality and reliability. Manufacturers of cell phones and related devices will need to provide features and "applications," and will be driven by solely market forces. Agencies have a sensor infrastructure and have not yet come to terms with whether it is even responsible to depend on vendors that may disappear with the next "shake-out." But the overlap and inefficiencies will have to be addressed.
3. There has to be a next generation of models that specifically address time-varying demand profiles, dynamic route assignment, and elasticity in demand with cost and network status. These models have to be ready for a new data-rich environment, in terms of O/D patterns and route selections.
4. The national priority in the United States on antiterrorism and security, and the same priority in the international community, will set priorities for certain inspection and detection technologies and also influence the design of ITS systems. With the growing concern over security in the ace of terrorism, ITS technology will be extremely relevant, and the range of applications—including cargo and passenger inspection—will surely expand.

One issue that would be worthy of becoming an "emerging issue" is not on the list simply because the initiative is already

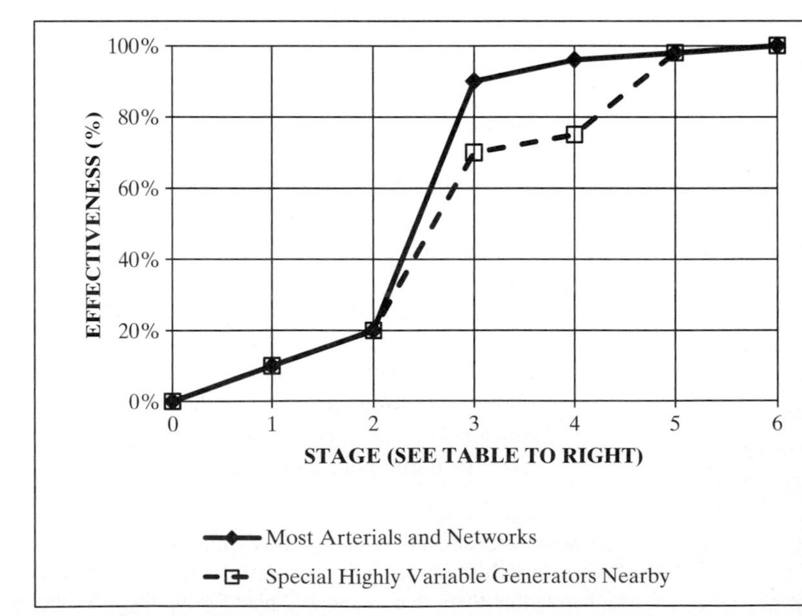

STAGE	DESCRIPTION
0	No systematic signal timing plans, matched to demand; no central control or coordination
1	Computer control of signals, giving ability to monitor equipment
2	Sensor network added, to collect data for off-line plans
3	Multi-dial control, or computer control with "N" signal timing plans, each matched to relevant demand periods
4	Ability to change plans "by exception" based upon sensor data
5	Ability to implement adaptive control at select locations
6	Highly responsive control with extensive detectorization

Figure 25.7: Hypothesized Relative Effectiveness of Features in Control System

underway: The nation's 911 systems are presently based on voice, and they were not built for multimedia communication in a wireless, mobile society. For more information on this initiative, refer to http://www.its.dot.gov/ng911/index.htm.

25.9 Summary

This chapter has in some ways raised more issues than it answers, and it skips some of details of specific ITS systems. This is done intentionally because (1) the field is moving rapidly and any "snapshot" of its present state is sure to be dated rapidly, perhaps even by the publication date of the text, and (2) the real issue is for you to be prepared to expand your view of providing transportation service in a highly competitive market in which computing, communications, and Web services are being used in novel ways.

Furthermore, the evolving roles of private and public sectors—in some ways, structure vis-à-vis market responsiveness—should draw your attention. Today's "right answer" can be swept away by what the enabling technologies make available.

And there is another fundamental issue for you to consider: Manufacturers need to devise products that are both more attractive and differentiated (at least in the short term, until the competition copies success). Transportation data and information is not an end in its own right—the traditional view in our profession—but rather it is a product enhancement or a service. Private sector forces may provide a data-rich environment for transportation professionals as a by-product of their own work *and at an innovative pace driven by that work and its market.* And this pace far exceeds the traditional pace of public sector planning and innovation and the orderly process of standardization.

Problems

25-1. Consider the feasibility of designing a Web site that offers the public the opportunity to have several service providers bid on para-transit trips. A user might indicate that he or she needs to get to the airport by a certain time and have flexibility in the length of travel time (or not) and in multiple pickups (or not). The user might also give advance notice of several days or only hours. Service providers would then bid for the particular trip or for a set of trips specified by the user. Consider how to market the service, what options it should provide the user, and why it might (or might not) be attractive to service providers. Write a paper not to exceed 10 pages on such a system.

25-2. Refer to Figure 25.7, and generate your own estimates of the effectiveness percentages for the "stages" shown. Refine the definition of the stages if you wish. Support the argument with facts and citations, in a paper not to exceed five pages.

25-3. Address the first "emerging issue" in Section 25.8, with particular attention to the research needed to advance the state of the art to where it must be to make informed decisions. This may require a literature search and thus be a major course project. But two timelines—the *desire* for informed congestion pricing implementations and the *advance planning/research* needed—may be inconsistent.

25-4. Address the second "emerging issue" in Section 25.8, with emphasis on how much data can be obtained and how it can be used in routing and assignment algorithms. At the same time, address the privacy issues and how personal privacy can realistically be assured. Prepare for (a) a class discussion of this issue, and/or (b) a 5- to 10-page paper on this issue, as specified by the course instructor.

25-5. Address the third "emerging issue" in Section 25.9, with emphasis on how closely existing advanced models for dynamic assignment truly address the issue. Prepare for (a) a class discussion of this issue, and/or (b) a 5- to 10-page paper on this issue, as specified by the course instructor.

25-6. Address the fourth "emerging issue" in Section 25.8, with emphasis on the most acute needs in face of the range of threats that must be considered. Take reasonable account of the reality that passive action—detection and then remedy—may not be wise or cost effective.

CHAPTER 26

Signal Coordination for Arterials and Networks: Undersaturated Conditions

26.1 Basic Principles of Signal Coordination

In situations where signals are close enough together so that vehicles arrive at the downstream intersection in platoons, it is necessary to coordinate their green times so that vehicles may move efficiently through the *set* of signals. It serves no purpose to have drivers held at one signal watching wasted green at a downstream signal, only to arrive there just as the signal turns red.

In some cases, two signals are so closely spaced that they should be considered one signal. In other cases, the signals are so far apart that they may be considered isolated intersections. However, vehicles released from a signal often maintain their grouping for well over 1,000 feet. Common practice is to coordinate signals less than a mile apart on major streets and highways.

26.1.1 A Key Requirement: Common Cycle Length

In coordinated systems, all signals must have the same cycle length. This is necessary to ensure that the beginning of green occurs at the same time relative to the green at the upstream and downstream intersections. There are some exceptions, where a critical intersection has such a high volume that it may require a double cycle length, but this is done rarely and only when no other solution is feasible.

26.1.2 The Time-Space Diagram and Ideal Offsets

The time-space diagram is a plot of signal indications as a function of time for two or more signals. The diagram is scaled with respect to distance, so that one may easily plot vehicle positions as a function of time. Figure 26.1 is a time-space diagram for two intersections. Standard conventions are generally used in such figures: A green signal indication is shown by a blank or simple line (——), yellow by a shaded line (////////), and red by a solid line (▬▬). In many cases, such diagrams show only effective green and effective red, as shown in Figure 26.1. This figure illustrates the path (trajectory) that a vehicle takes as time passes. At $t = t_1$, the first signal turns green. After some lag, the vehicle starts and moves down the street. It reaches the second intersection at some time $t = t_2$. Depending on the indication of that signal, it either continues or stops.

The difference between the two green initiation times (i.e., the difference between the time when the upstream intersection turns green and the downstream intersection turns green) is referred to as the *signal offset* or simply the *offset.* In Figure 26.1, the offset is defined as t_2 minus t_1. Offset is usually expressed as a positive number between zero and the cycle length. This definition is used throughout this and other chapters in this text.

Other definitions of offset are used in practice. For instance, offset is sometimes defined relative to one reference upstream signal, and sometimes it is defined relative to a standard zero. Some signal hardware uses "offset" defined in terms of red initiation, rather than green; other hardware uses the end of green as the reference point. Some hardware uses offset in seconds; other hardware uses offset as a percentage of the cycle length.

The "ideal offset" is defined as exactly the offset such that, as the first vehicle of a platoon just arrives at the downstream signal, the downstream signal turns green. It is usually assumed that the platoon was moving as it went through the upstream intersection. If so, the ideal offset is given by:

$$t_{Ideal} = L/S \qquad (26\text{-}1)$$

where: t_{Ideal} = ideal offset, s
L = distance between signalized intersections, ft
S = vehicle speed, ft/s

If the vehicle were stopped, and had to accelerate after some initial startup delay, the ideal offset could be represented by Equation 26-1 plus the startup time at the first intersection (which would usually add 2 to 4 seconds). In general, the startup time would only be included at the *first* of a series of

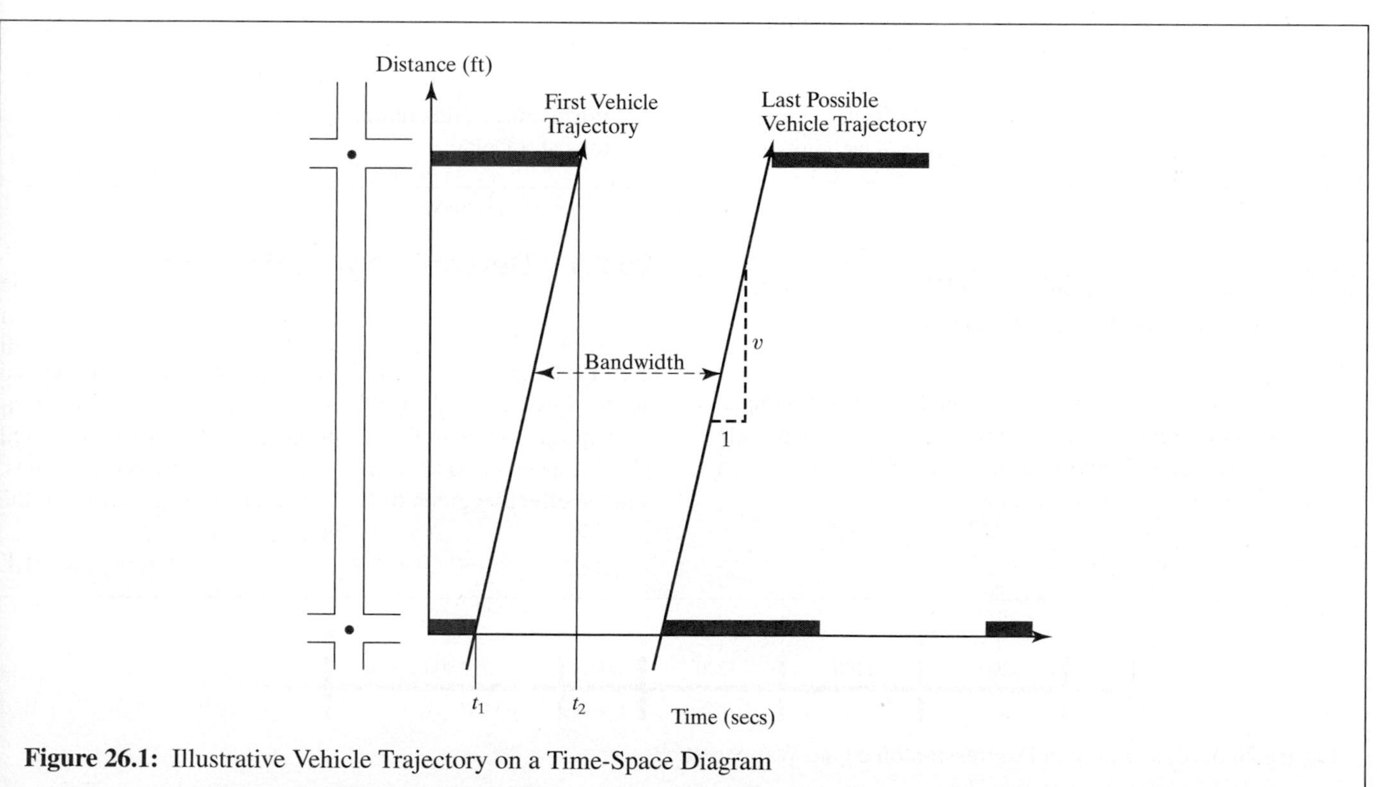

Figure 26.1: Illustrative Vehicle Trajectory on a Time-Space Diagram

signals to be coordinated, and often not at all. Usually, this will reflect the ideal offset desired for maximum bandwidth, minimum delay, and minimum stops. Even if the vehicle is stopped at the first intersection, it will be moving through most of the system.

Figure 26.1 also illustrates the concept of *bandwidth,* the amount of green time that can be used by a continuously moving platoon of vehicles through a series of intersections. In Figure 26-1, the bandwidth is the entire green time at both intersections because several key conditions exist:

- The green time at both intersections are the same.
- The ideal offset is illustrated.
- There are only two intersections.

In most cases, the bandwidth will be less, perhaps significantly so, than the full green time.

Figure 26.2 illustrates the effect of offset on stops and delay for a platoon of vehicles leaving one intersection and passing through another. In this example, a 25-second offset is ideal because it produces the minimum delay and the minimum number of stops. The effect of allowing a poor offset to exist is clearly indicated: Delay can climb to 30 seconds per vehicle, and the stops to 10 per cycle. Note that the penalty for deviating from the ideal offset is usually not equal in positive and negative deviations. An offset of (25 + 10) = 35 seconds causes much more harm than an offset of (25 − 10) = 15 seconds, although both are 10 seconds from the ideal offset. Figure 26.2 is illustrative because each situation would have similar but different characteristics.

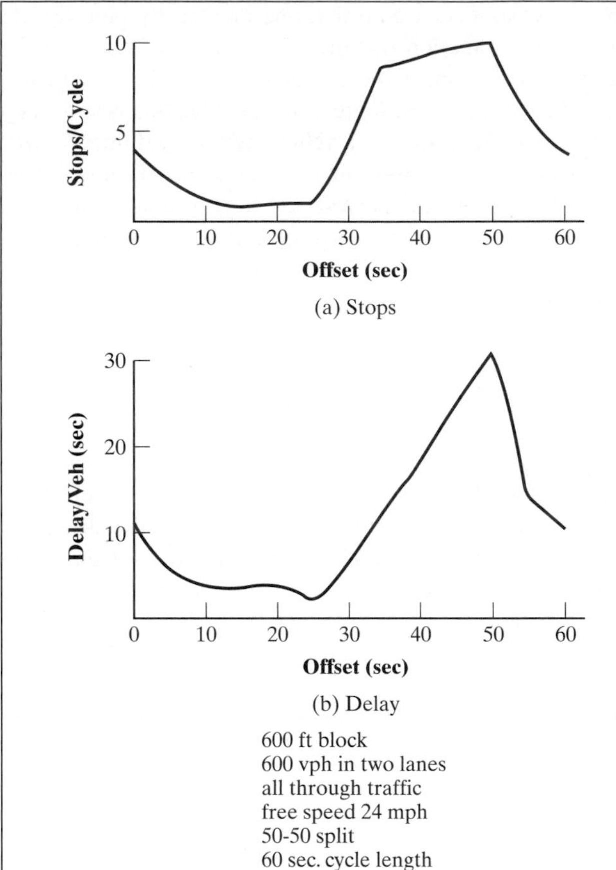

Figure 26.2: Illustration of the Effects of Offset on Stops and Delay

26.2 Signal Progression on One-Way Streets

Signal progression on a one-way street is relatively simple. For the purpose of this section, it will be assumed that a cycle length has been chosen and that the green allocation at each signal has been previously determined.

26.2.1 Determining Ideal Offsets

Consider the one-way arterial shown in Figure 26.3, with the link lengths indicated. Assuming no vehicles are queued at the signals, the ideal offsets can be determined if the platoon speed is known. For the purpose of illustration, a desired platoon speed of 60 ft/s is used. The cycle length is 60 seconds, and the effective green time at each intersection is 50% of the

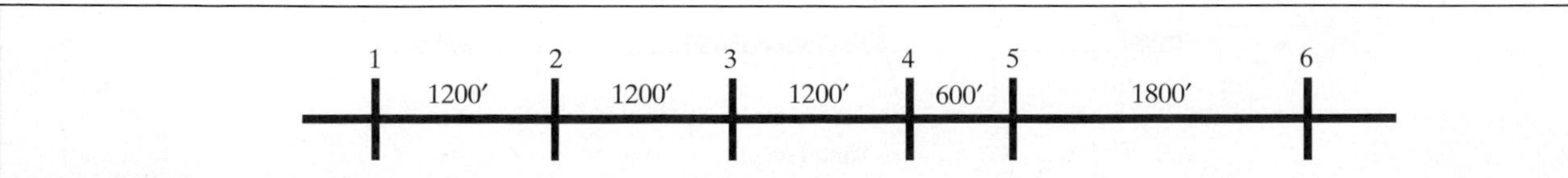

Figure 26.3: Case Study in Progression on a One-Way Street

Table 26.1: Ideal Offsets for Case Study

Signal	Relative to Signal	Ideal Offset
6	5	1,800/60 = 30 s
5	4	600/60 = 10 s
4	3	1,200/60 = 20 s
3	2	1,200/60 = 20 s
2	1	1,200/60 = 20 s

cycle length, or 30 seconds. Ideal offsets are computed using Equation 26-1 and are illustrated in Table 26.1.

Note that neither the cycle length nor the splits enter into the computation of ideal offsets. To see the pattern that results, the time-space diagram should be constructed according to the following rules:

1. The vertical should be scaled so as to accommodate the dimensions of the arterial, and the horizontal so as to accommodate at least three to four cycle lengths.
2. The beginning intersection (Number 1, in this case) should be scaled first, usually with main street green (MSG) initiation at $t = 0$, followed by periods of green and red (yellow may be shown for precision). See Point 1 in Figure 26.4.
3. The main street green (or other offset position, if MSG is not used) of the next downstream signal should be located next, relative to $t = 0$ and at the proper distance from the first intersection. With this point located (Point 2 in Figure 26-4), fill in the periods of effective green and red for this signal.
4. Repeat the procedure for all other intersections, working one at a time. Thus, for Signal 3, the offset is located at point 3, 20 seconds later than Point 2, and so on.

Figure 26.4 has some interesting features that can be explored with the aid of Figure 26.5.

First, if a vehicle (or platoon) were to travel at 60 fps, it would arrive at each of the signals just as they turn green; this is indicated by the solid trajectory lines in Figure 26.5. The solid trajectory line also represents the speed of the "green wave" visible to a stationary observer at Signal 1, looking downstream. The signals turn green in order, corresponding to the planned speed of the platoon, and give the visual effect of a wave of green opening before the driver. Third, note that there is a "window" of green in Figure 26.5, with its end indicated by the dotted trajectory line, which is also the trajectory of the *last* vehicle that could travel through the progression without

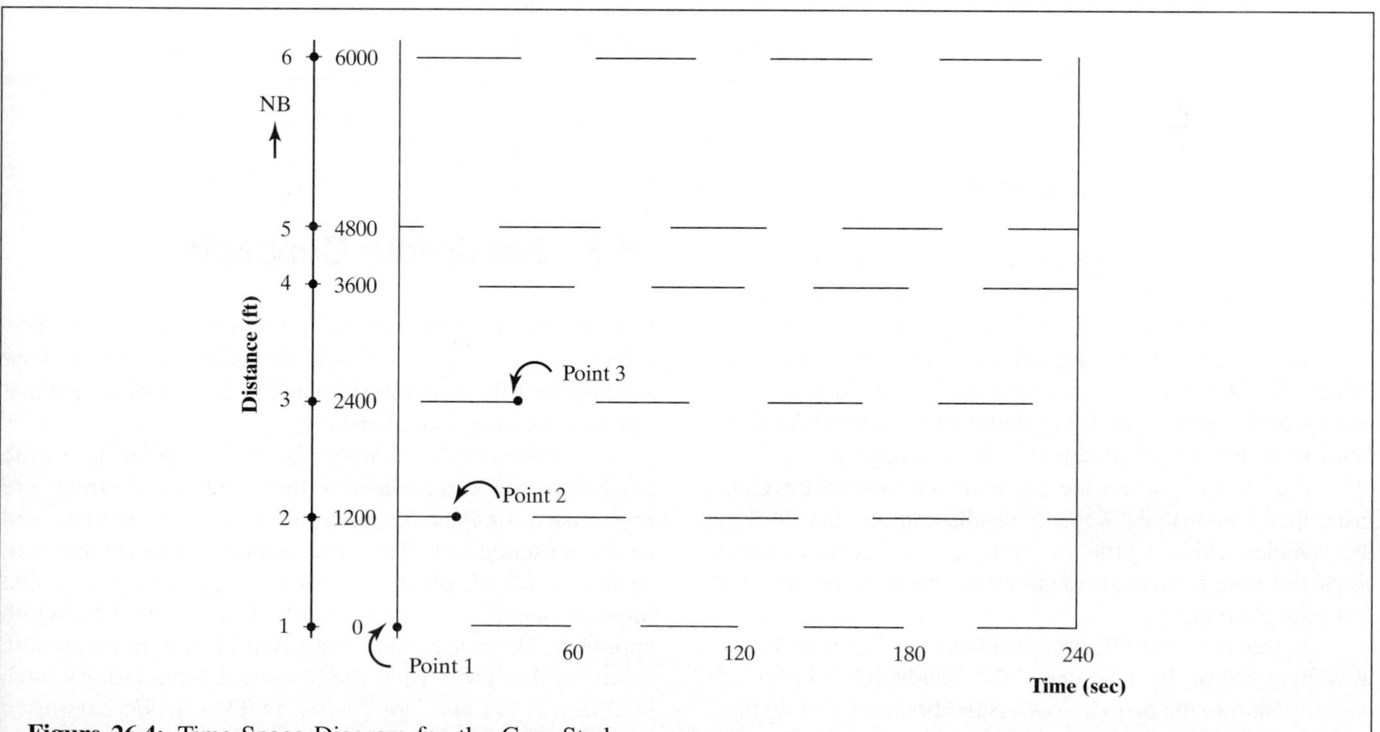

Figure 26.4: Time-Space Diagram for the Case Study

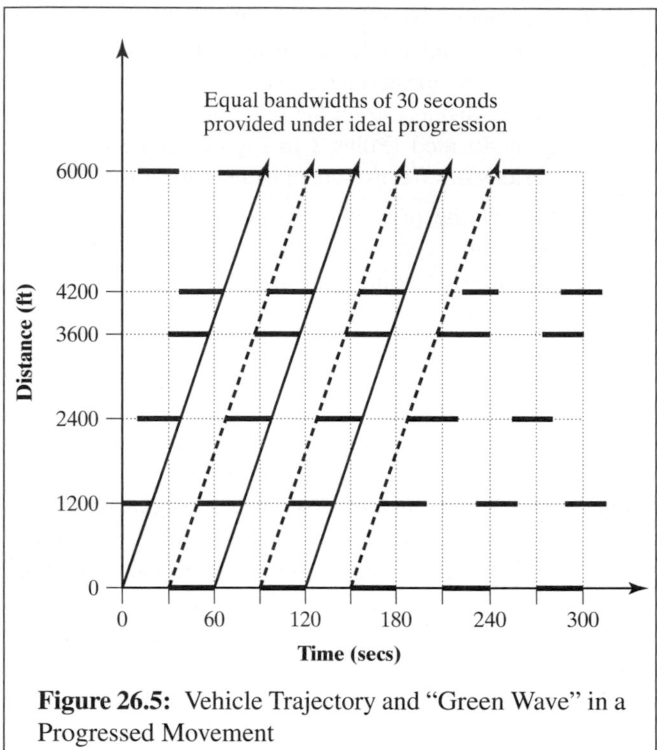

Figure 26.5: Vehicle Trajectory and "Green Wave" in a Progressed Movement

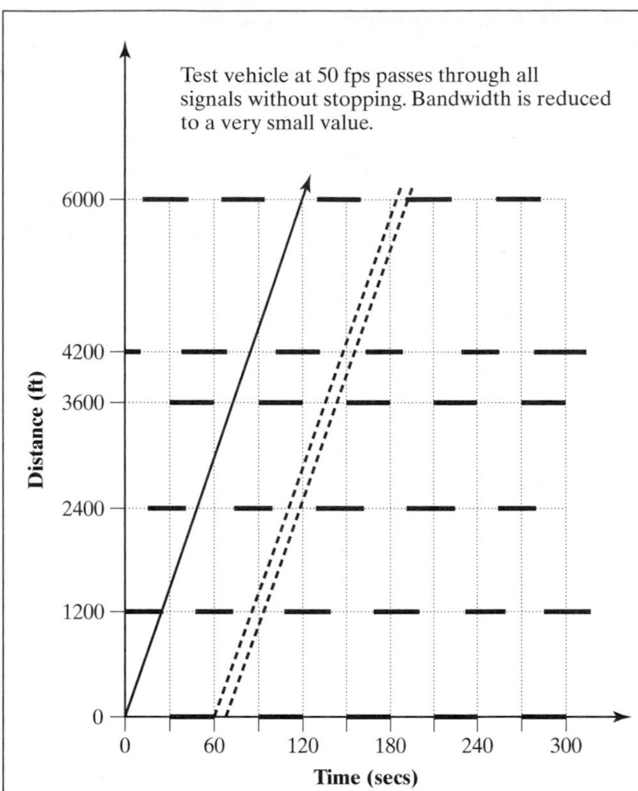

Figure 26.6: The Effect of a 50 ft/s Platoon Speed on Progression

stopping at 60 ft/s. This "window" is the bandwidth, as defined earlier. Again, in this case it equals the green time because all signals have the same green time and have ideal offsets.

26.2.2 Potential Problems

Consider what would happen if the actual speed of vehicle platoons in the case study was 50 ft/s, instead of the 60 ft/s anticipated. The green wave would still progress at 60 ft/s, but the platoon arrivals would lag behind it. The effect of this on bandwidth is enormous, as shown in Figure 26.6. Only a small window now exists for a platoon of vehicles to continuously flow through all six signals without stopping.

Figure 26.7 shows the effect of the vehicle traveling faster than anticipated (70 ft/s in this illustration). In this case, the vehicles arrive a little too early and are delayed; some stops will have to be made to allow the "green wave" to catch up to the platoon.

In this case, the effect on bandwidth is not as severe as in Figure 26.6. In this case, the bandwidth impact of *underestimating* the platoon speed is (60 ft/s instead of 70 ft/s) is not as severe as the consequences of *overestimating* the platoon speed (60 ft/s instead of 50 ft/s).

26.3 Bandwidth Concepts

Bandwidth is defined as the time difference between the first vehicle that can pass through the entire system without stopping and the last vehicle that can pass through without stopping, measured in seconds.

The bandwidth concept is very popular in traffic engineering practice because the windows of green are easy visual images for both working professionals and public presentations. The most significant shortcoming of designing offset plans to maximize bandwidths is that internal queues are often overlooked in the bandwidth approach. There are computer-based maximum bandwidth solutions that go beyond the historical formulations, such as PASSER [*1*] and Tru-Traffic TS/PP [*2*]. We have used the latter in professional practice and show an example in Section 26.7.

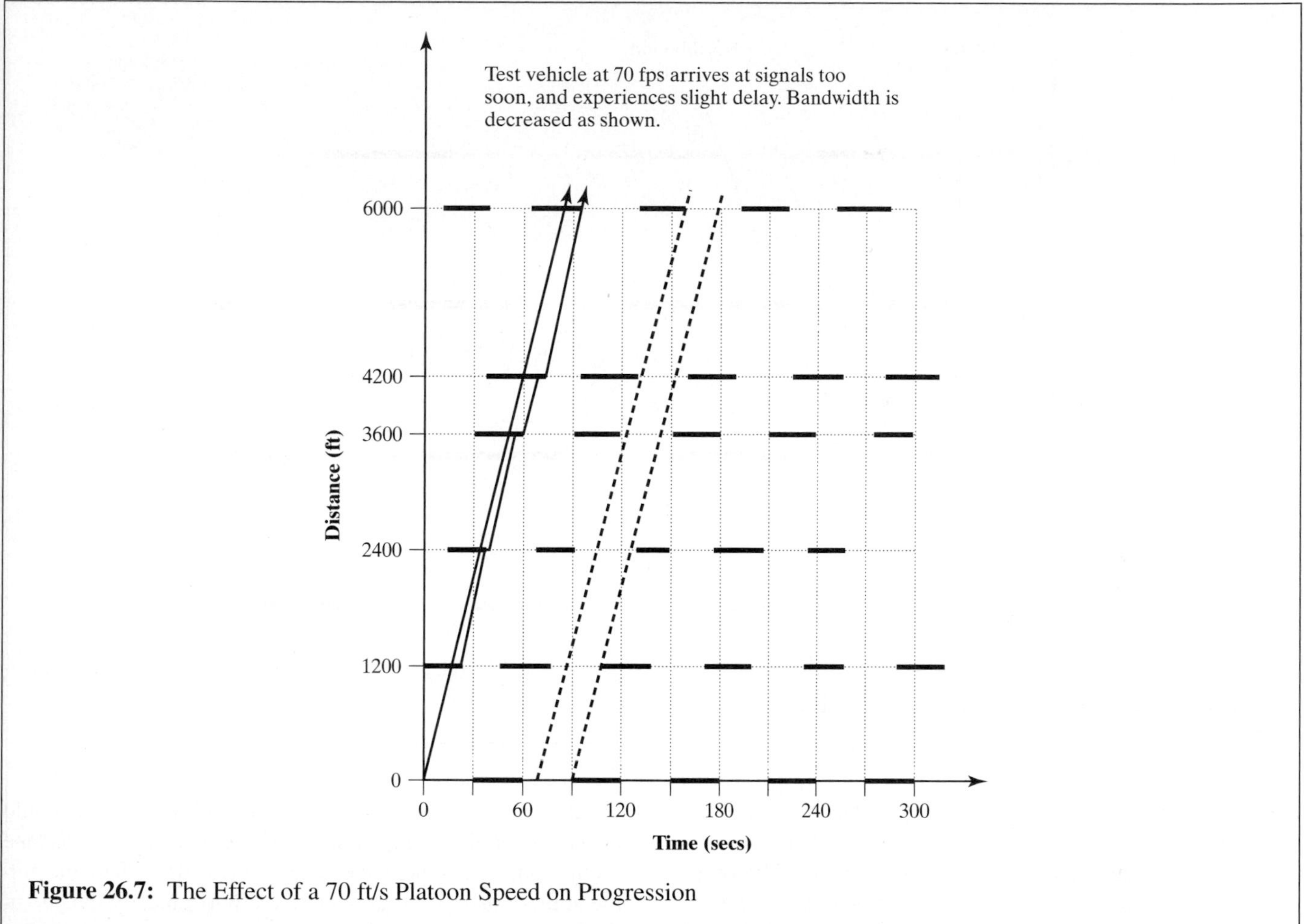

Figure 26.7: The Effect of a 70 ft/s Platoon Speed on Progression

26.3.1 Bandwidth Efficiency

The efficiency of a bandwidth is defined as the ratio of the bandwidth to the cycle length, expressed as a percentage:

$$EFF_{BW} = \left(\frac{BW}{C}\right) * 100\% \tag{26-2}$$

where: EFF_{BW} = bandwidth efficiency (%)
BW = bandwidth (s)
C = cycle length (s)

A bandwidth efficiency of 40% to 55% is considered good. The bandwidth is limited by the minimum green in the direction of interest.

Figure 26.8 illustrates the bandwidths for one signal-timing plan. The northbound efficiency can be estimated as (17/60) * 100% = 28.4%. The southbound bandwidth is obviously terrible; there is no bandwidth through the defined system. The northbound efficiency is only 28.4%. This system is badly in need of retiming, at least on the basis of the bandwidth objective. Just looking at the time-space diagram, one might imagine sliding the pattern at Signal 4 to the right and the pattern at Signal 1 to the left, allowing some coordination for the southbound vehicles.

26.3.2 Bandwidth Capacity

In terms of vehicles that can be put through the system of Figure 26.7 without stopping, the northbound bandwidth can carry 17/2.0 = 8.5 vehicles per lane per cycle in a nonstop path through the defined system, assuming that the saturation headway is 2.0 s/veh. Thus the northbound direction can handle 8.5 veh/cycle * 1 cycle/60 sec * 3,600 sec/hr = 510 veh/h/ln very efficiently if they are

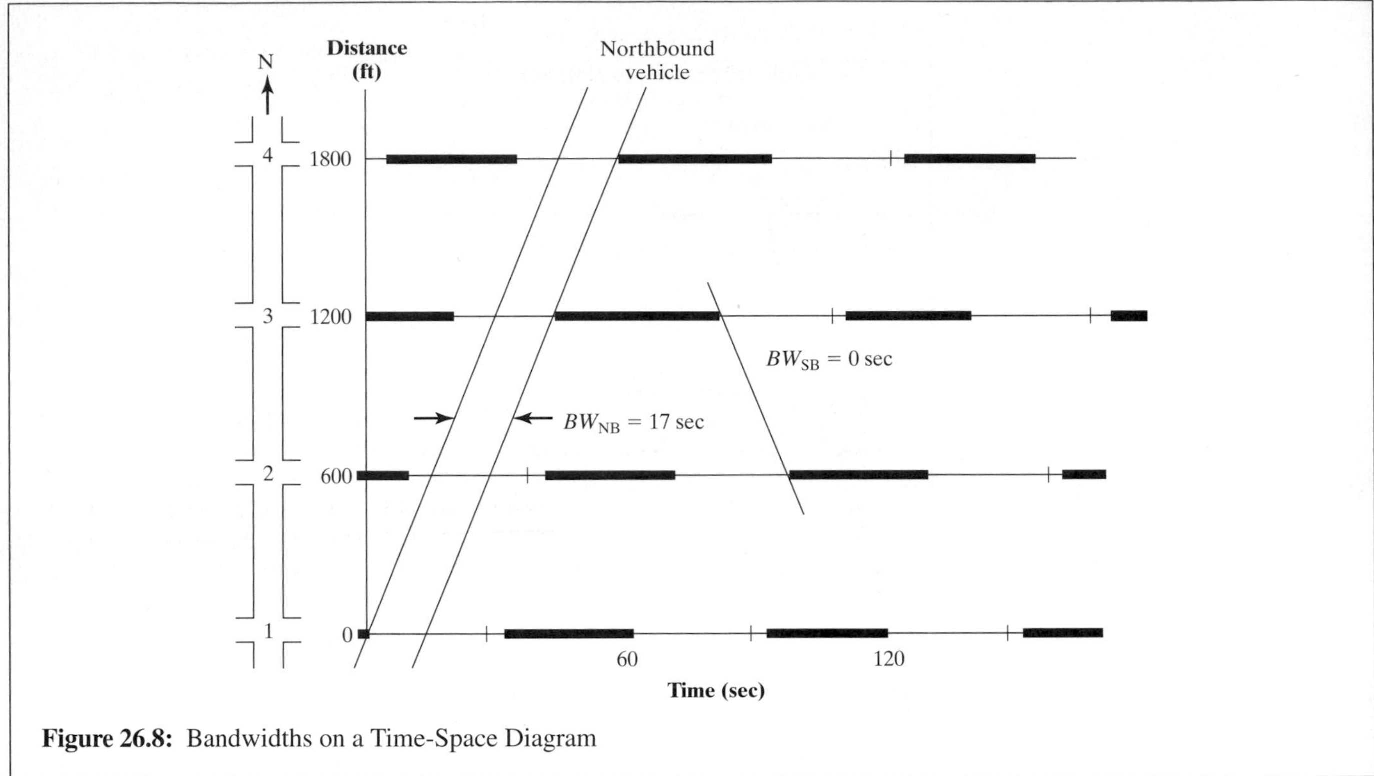

Figure 26.8: Bandwidths on a Time-Space Diagram

organized into eight-vehicle platoons when they travel through this system.

If the per lane demand volume is less than 510 vphpl and if the flows are well organized (and if there is no internal queue development), the system will operate well in the northbound direction, even though better timing plans might be obtained.

In general terms, the number of vehicles that can pass through a defined series of signals without stopping is called the *bandwidth capacity*. The illustrated computation can be described by the following equation:

$$c_{BW} = \frac{3600 * BW * NL}{C * h} \tag{26-3}$$

where: c_{BW} = bandwidth capacity, veh/h
BW = bandwidth, s
NL = number of through lanes in the indicated direction
C = cycle length, s
h = saturation headway, s

Equation 26-3 does not contain any factors to account for nonuniform lane utilization and is intended only to indicate some limit beyond which the offset plan will degrade, certainly resulting in stopping and internal queuing. It should also be noted that bandwidth capacity is *not* the same as lane group capacity. Where the bandwidth is less than the full green time, there is additional lane group capacity outside of the bandwidth.

26.4 The Effect of Queued Vehicles at Signals

To this point, it has been assumed there is no queue standing at the downstream intersection when the platoon (from the upstream signal arrives). This is generally not a reasonable assumption. Vehicles that enter the traffic stream between platoons will progress to the downstream signal, which will often be "red." They form a queue that partially blocks the progress of the arriving platoon. These vehicles may include stragglers from the last platoon, vehicles that turned into the block from unsignalized intersections or driveways, or vehicles that came out of parking lots or parking spots. The ideal offset must be adjusted to allow for these vehicles, so as to avoid unnec-essary stops. The situation without such an adjustment is depicted in Figure 26.9, where it can be seen

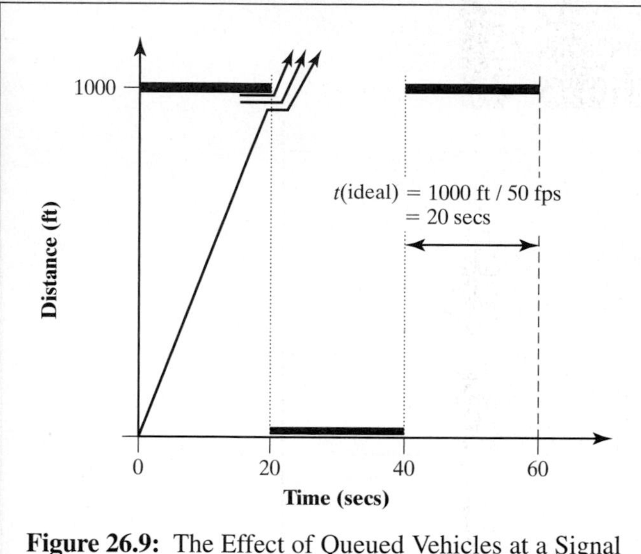

Figure 26.9: The Effect of Queued Vehicles at a Signal

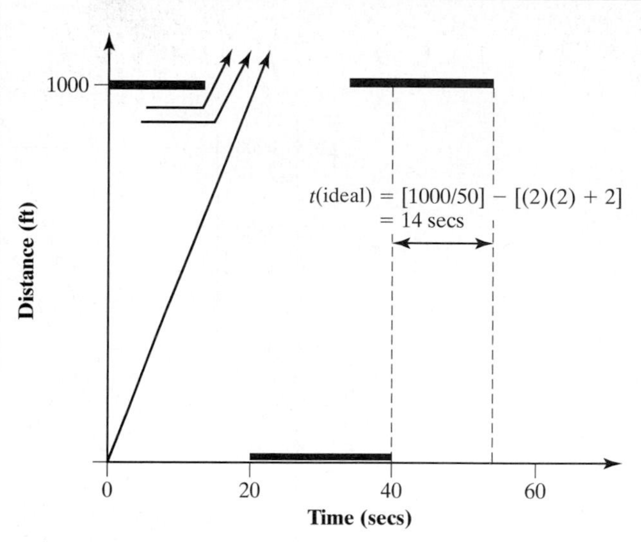

Figure 26.10: Adjustment in Offset to Accommodate Queued Vehicles

that the arriving platoon is delayed behind the queued vehicles as the queued vehicles begin to accelerate through the intersection.

To adjust for the queued vehicles, the ideal offset is adjusted as follows:

$$t_{adj} = \frac{L}{S} - (Qh + \ell_1) \tag{26-4}$$

where: t_{adj} = adjusted ideal offset, s
L = distance between signals, ft
S = speed, ft/s
Q = number of vehicles queued per lane, veh
h = discharge headway of queued vehicles, s/veh
ℓ_1 = start-up lost time, s

The lost time is counted only at the first downstream intersection, at most: If the vehicle(s) from the preceding intersection were themselves stationary, their startup causes a shift that automatically takes care of the startup at subsequent intersections.

Offsets can be adjusted to allow for queue clearance before the arrival of a platoon from the upstream intersection. Figure 26.10 shows the situation for use of the modified ideal offset equation.

Figure 26.11 shows the time-space diagram for the case study of Figure 26.4, given queues of two vehicles per lane in all links. Note that the arriving vehicle platoon has smooth flow, and the lead vehicle has 60 ft/s travel speed. The visual image of the "green wave," however, is much faster, due to the need to clear the queues in advance of the arriving platoon.

The "green wave" or the progression speed, as it is more properly called, is traveling at varying speeds as it moves down the arterial. The "green wave" will appear to move ahead of the platoon, clearing queued vehicles in advance of it. The progression speed can be computed for each link as shown in Table 26.2.

Note, however, that the bandwidth, and therefore the bandwidth capacity, is now much smaller. Thus, by clearing out the queue in advance of the platoon, more of the green time is used by queued vehicles, and less is available to the moving platoon.

The preceding discussion assumes the queue is known at each signal. In fact, this is not an easy number to know. However, if we know there is a queue and know its approximate size, the link offset can be set better than by pretending that no queue exists.

Consider the sources of the queued vehicles:

- Vehicles turning in from upstream side streets during their green (which is main street red)
- Vehicles leaving parking garages or spaces
- Stragglers from previous platoons.

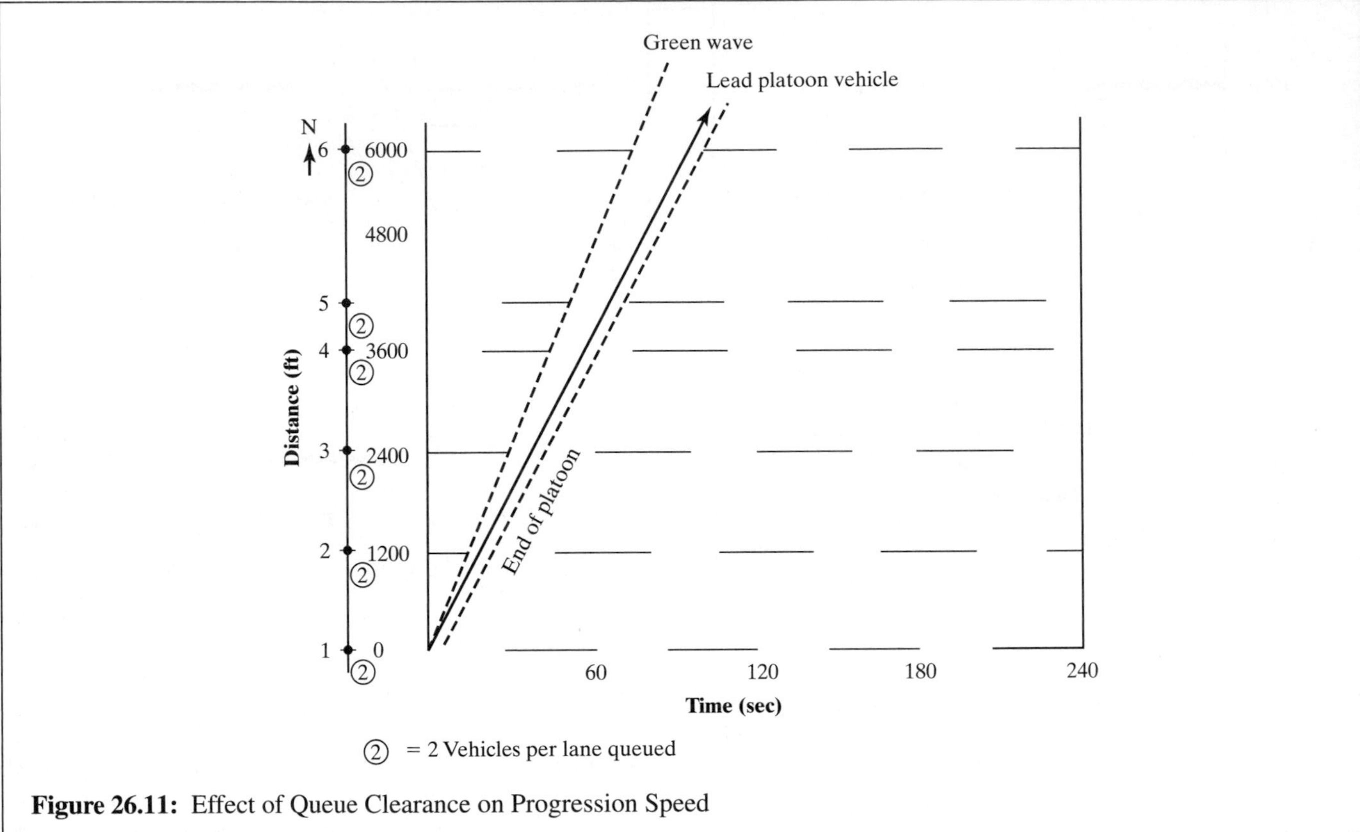

Figure 26.11: Effect of Queue Clearance on Progression Speed

There can be great cycle-to-cycle variation in the actual queue size, although the average queue size may be estimated. Even at that, queue estimation is a difficult and expensive task. Even the act of adjusting the offsets can influence the queue size. For instance, the arrival pattern of the vehicles from the side streets may be altered. Queue estimation is therefore a significant task in practical terms.

26.5 Signal Progression for Two-Way Streets and Networks

The task of progressing traffic on a one-way street has been relatively straightforward. To highlight the essence of the problem on a two-way street, assume the arterial shown in Figure 26.5 is a two-way street rather than a one-way street. Figure 26.12

Table 26.2: Progression Speeds in Figure 26-11

Link	Link Offset (s)	Speed of Progression (ft/s)
Signal 1 → 2	(1,200/60) − (4 + 2) = 14	1,200/14 = 85.7
Signal 2 → 3	(1,200/60) − (4) = 16	1,200/16 = 75
Signal 3 → 4	(1,200/60) − (4) = 16	1,200/16 = 75
Signal 4 → 5	(600/60) − (4) = 6	600/6 = 100
Signal 5 → 6	(1,800/60) − (4) = 26	1,800/26 = 69.2
Total Offset = 78 sec		

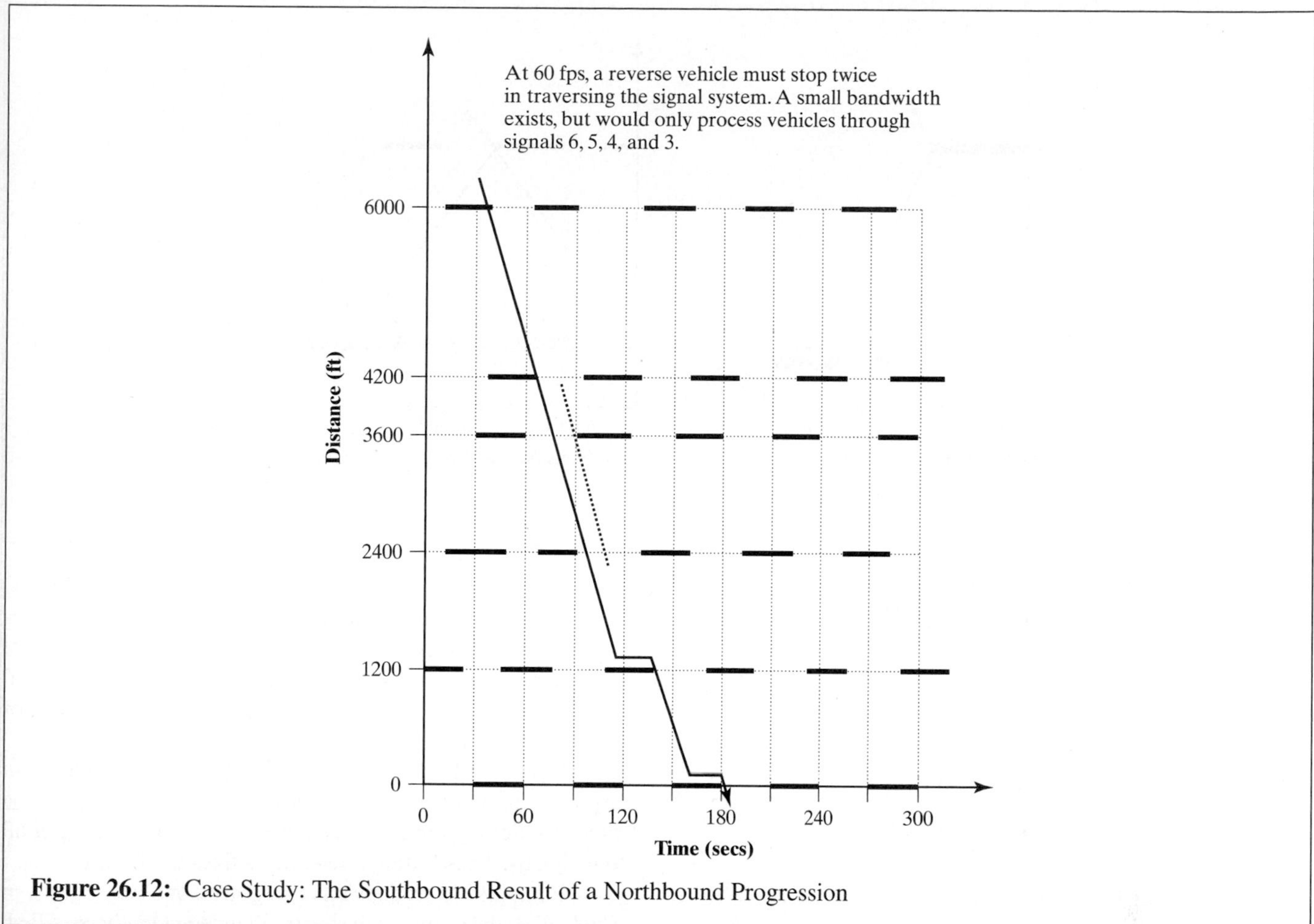

Figure 26.12: Case Study: The Southbound Result of a Northbound Progression

shows the trajectory of a *southbound* vehicle on this arterial. The vehicle is just fortunate enough not to be stopped until Signal 2, but is then stopped again for Signal 1, for a total of 2 stops and 40 seconds of delay. There is no bandwidth, meaning it is not possible to have a vehicle platoon pass along the arterial nonstop.

Of course, if the offsets or the travel times had been different, it might have been possible to have a southbound bandwidth through all six signals.

26.5.1 Offsets on a Two-Way Street

Note that if any offset were changed in Figure 26.12 to accommodate the southbound vehicles, then the northbound bandwidth would suffer. For instance, if the offset at Signal 2 were decreased by 20 seconds, then the pattern at that signal would shift to the left by 20 seconds, resulting in a "window" of green of only 10 seconds on the northbound, rather than the 30 seconds in the original display (Figure 26.5).

The fact that the offsets on a two-way street are interrelated presents one of the most fundamental problems of signal optimization. Note that inspection of a typical time-space diagram yields the obvious conclusion that the offsets in two directions add to one cycle length, shown in Figure 26.13 (a). However, for longer blocks, the offsets might add to two (or more) cycle lengths, shown in Figure 26.13 (b).

Figure 26.13 illustrates both actual offsets and travel times, which are not necessarily the same. Although the engineer might desire the ideal offset to be the same as the travel times, this is not always the case. Once the offset is specified in one direction, it is automatically set in the other.

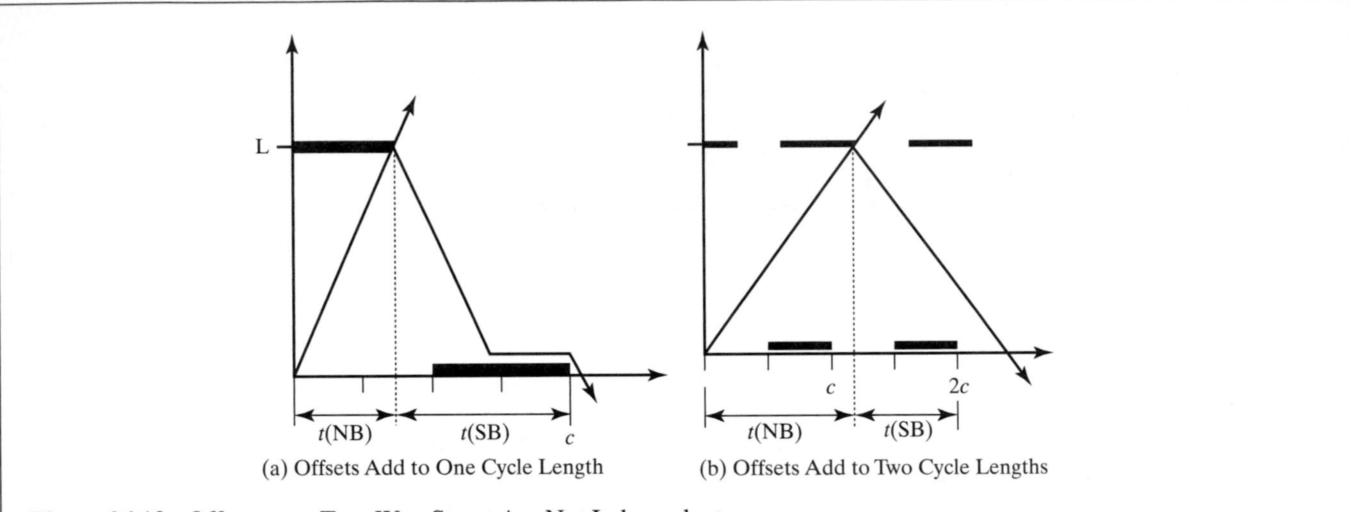

Figure 26.13: Offsets on a Two-Way Street Are Not Independent

The general expression for the two offsets in a link on a two-way street can be written as:

$$t_{1i} + t_{2i} = nC \tag{26.5}$$

where: t_{1i} = offset in direction 1 (link i), s
t_{2i} = offset in direction 2 (link i), s
n = integer value
C = cycle length, s

To have $n = 1$ (Figure 26.13a), $t_{1i} \leq C$; to have $n = 2$ (Figure 26-13b), $C < t_{1i} \leq 2C$.

Any actual offset can be expressed as the desired "ideal" offset, plus an "error" or "discrepancy" term:

$$t_{actual(i,j)} = t_{ideal(i,j)} + e_{ij} \tag{26.6}$$

where j represents the direction and i represents the link. In a number of signal optimization programs that are used for two-way arterials, the objective is to minimize some function of the discrepancies between the actual and ideal offsets.

26.5.2 Network Closure

The relative difficulty of finding progressions on a two-way street, compared to on a one-way street, might lead one to conclude that the best approach is to establish a system of one-way streets, to avoid the problem. A one-way street system has a number of advantages, not the least of which is elimination of left turns against opposing traffic. One-way streets simplify network signalization, but they do not eliminate closure problems, and they carry other practical disadvantages. See Chapter 29 for additional discussion of one-way streets.

Figure 26.14 illustrates network closure requirements. In any set of four signals, offsets may be set on three legs in one direction. Setting three offsets, however, fixes the timing of all four signals. Thus setting three offsets fixes the fourth.

Figure 26.15 extends this to a grid of one-way streets, in which all of the north-south streets are independently specified. The specification of one east-west street then "locks in" all other east-west offsets. Note that the key feature is that an open

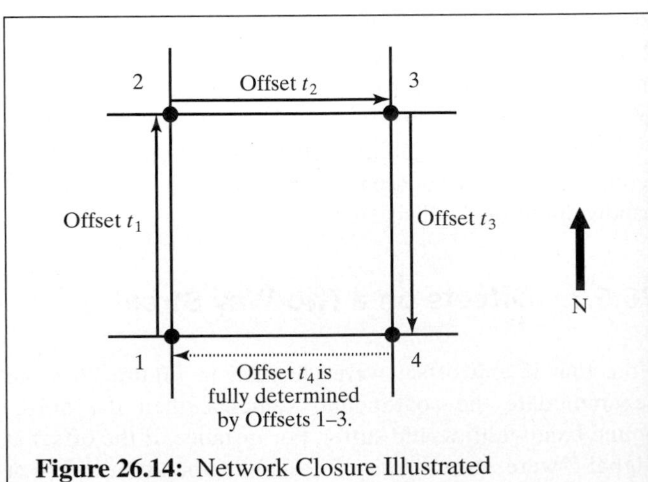

Figure 26.14: Network Closure Illustrated

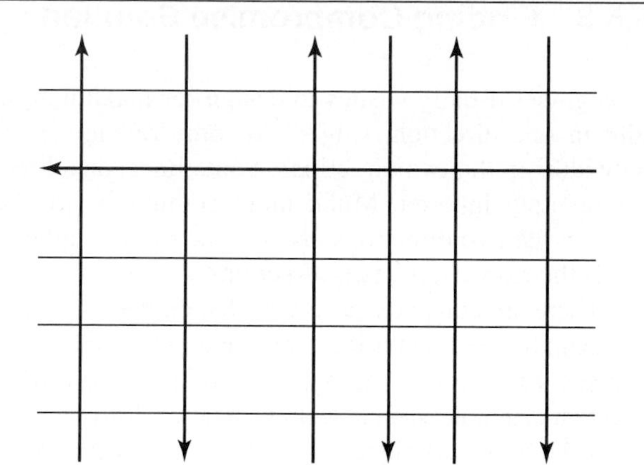

Figure 26.15: Impact of Closure on a Rectangular Street Grid

tree of one-way links can be completely independently set, and that it is the closing or "closure" of the open tree that presents constraints on some of the links.

To develop the constraint equation, refer to Figure 26.14 and walk through the following steps, keying to the green in all steps:

Step 1: Begin at Intersection 1 and consider the green initiation to be time $t = 0$.

Step 2: Move to Intersection 2, noting that the link offset t_1 specifies the time of green initiation at this intersection relative to its upstream neighbor. Thus green starts at Intersection 2 facing northbound at $t = 0 + t_1$.

Step 3: Recognizing that the westbound vehicles get released after the NS green is finished, green begins at Intersection 2 facing west at:

$$t = 0 + t_1 + g_{NS,2}$$

Step 4: Moving to intersection 3, the link offset in Link B specifies the time of green initiation at Intersection 3 relative to Intersection 2. Thus the green begins at Intersection 3, facing west at

$$t = 0 + t_1 + g_{NS,2} + t_2$$

Step 5: Similar to Step 3, the green begins at Intersection 3, but facing south, after the EW green is finished at time

$$t = 0 + t_1 + g_{NS,2} + t_2 + g_{EW,3}$$

Step 6: Moving to Intersection 4, the green begins in the southbound direction after the offset t_3 is added:

$$t = 0 + t_1 + g_{NS,2} + t_2 + g_{EW,3} + t_3$$

Step 7: Turning at Intersection 4, it is the NS green that is added to be at the start of green facing east.

$$t = 0 + t_1 + g_{NS,2} + t_2 + g_{EW,3} + t_3 + g_{NS,4}$$

Step 8: Moving to Intersection 1, it is t_4 that is relevant to be at the start of green facing east:

$$t = 0 + t_1 + g_{NS,2} + t_2 + g_{EW,3} + t_3 + g_{NS,4} + t_4$$

Step 9: Turning at Intersection 1, green will begin in the north direction after the EW green finishes:

$$t = 0 + t_1 + g_{NS,2} + t_2 + g_{EW,3} + t_3 + g_{NS,4} + t_4 + g_{EW,1}$$

This will bring us back to where we started. Thus this is either $t = 0$ or a multiple of the cycle length.

The following relationship results:

$$nC = 0 + t_A + g_{NS,2} + t_B + g_{EW,3} + t_C + g_{NS,4} + t_D + g_{EW,1} \quad (26\text{-}7)$$

where the only caution is that the g values should really include the change and clearance intervals.

Note that Equation 26-7 is a more general form of Equation 26-5, for the two-way arterial is a special case of a network. The interrelationships stated in Equation 26-7 are constraints on freely setting all offsets. In these equations one can trade off between green allocations and offsets. To get a better offset in Link 4, one can adjust the splits as well as the other offsets.

Although it is sometimes necessary to consider networks in their entirety, it is common traffic engineering practice to decompose networks into noninterlocking arterials whenever possible. Figure 26.16 illustrates this process.

Decomposition works well where a clear center of activity can be identified, and where few vehicles are expected to pass through the center without stopping (or starting) at or near the center. As the discontinuity in all progressions lies in and directly around the identified center, large volumes passing through can create significant problems in such a scheme.

In summary, if offsets are set in one direction on a two-way street, then the reverse direction is fixed. In a network, you can set any "open tree" of links, but links that close the tree already have their offsets specified.

You are advised to check the literature for the optimization programs in current use. At the present time, the dominant program in the United States seems to be Synchro [*3*]; several states specify its use in signal optimization and in traffic impact assessments and/or accept Synchro output for levels of

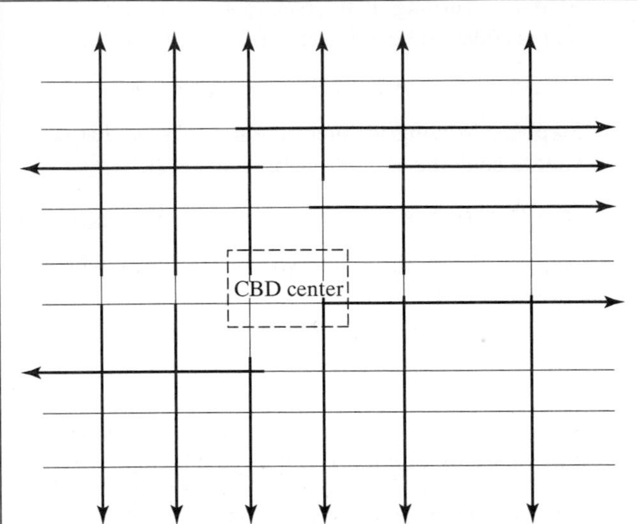

Figure 26.16: Decomposing a Network Into Noninterlocking Arterial Segments

service, equivalent to the Highway Capacity Manual (*HCM*) [*4*] for their purposes.[1] Section 26.7 contains an illustration of Synchro and its use.

TRANSYT [*5*] has been used over the years for arterials and networks and PASSER II [*2*] for certain arterials. The current version of the HCS+ software [*6*] for intersections is linked to TRANSYT-7F.

The programs just cited are for signal optimization, including determination of offsets. Another group of models exist that simulate and display the results, generally with both 2D and 3D visualizations. These models include VISSIM [*7*], AIMSUN [*8*], SimTraffic[2] [*3*], CORSIM [*9*], and PARAMICS [*10*].

With all of these models, it is important that the user make sure the default parameters reflect the reality of the jurisdiction or area in which they are being used. For instance, the discharge headway may be different than the 1.9 s/veh used in the *HCM,* work zone capacities may be handled differently than the *HCM* or local practice, and so forth.

[1]This is not to say that Synchro produces the identical answers as the *HCM* in all cases. We are simply reporting the state of the practice, namely that a number of states treat the Synchro outputs as having the same weight as if they came from the *HCM.*

[2]SimTraffic is available bundled with Synchro, and Synchro can directly feed SimTraffic. However, they are fundamentally different modeling approaches and can produce different estimates of level of service.

26.5.3 Finding Compromise Solutions

The engineer usually wishes to design for maximum bandwidth in one direction, subject to some relation between bandwidths in the two directions. Sometimes, one direction is completely ignored. Much more commonly, the bandwidths in the two directions are designed to be in the same ratio as the flows in the two directions.

There are computer programs that do the computations for maximum bandwidth that are commonly used by traffic engineers, as mentioned earlier. Thus it is not worthwhile to present an elaborate manual technique here. However, to get a feel for the basic technique and trade-offs, a small "by hand" example is shown.

Refer to Figure 26.17, which shows four signals and decent progression in both directions. For purposes of illustration, assume it is given that a signal with 50-50 split must be located midway between intersections 2 and 3. Figure 26.18 shows the possible effect of inserting the new signal into the system. It would appear there is no way to include this signal without destroying one or the other bandwidth, or cutting both in half.

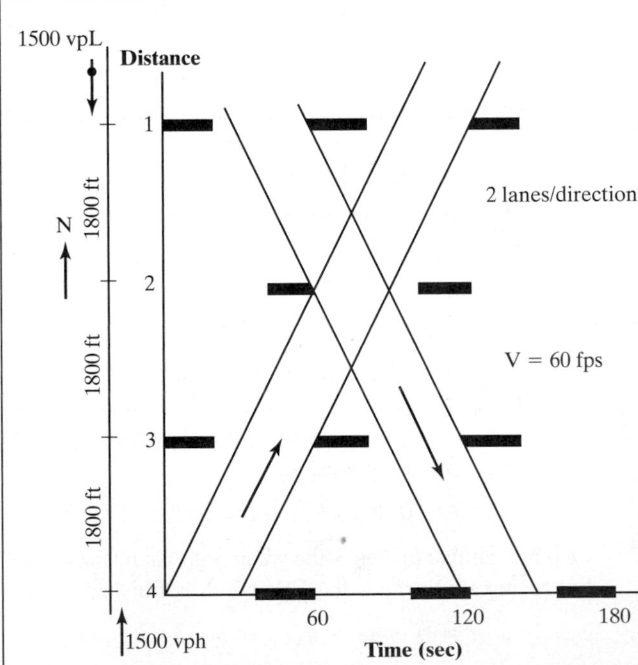

Figure 26.17: Case Study: Four Intersections with Good Two-Way Progression

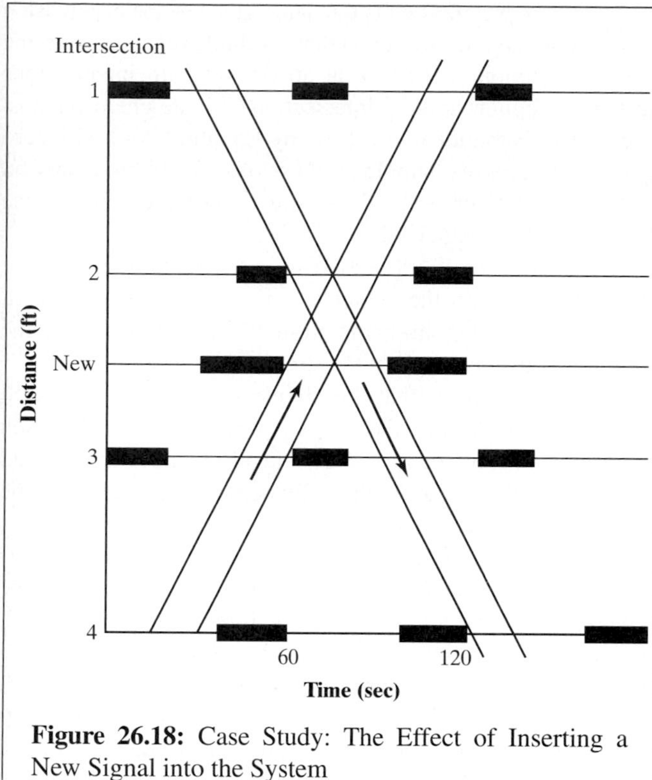

Figure 26.18: Case Study: The Effect of Inserting a New Signal into the System

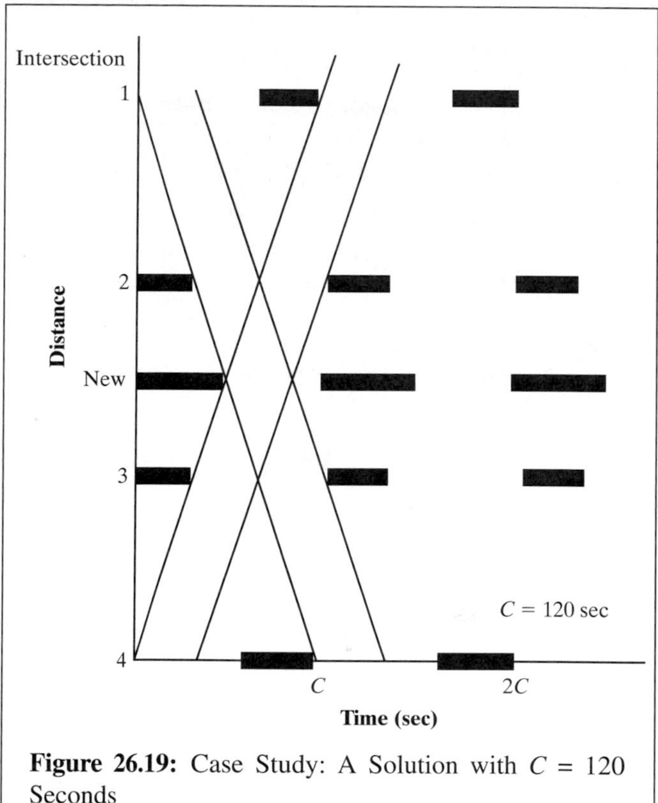

Figure 26.19: Case Study: A Solution with C = 120 Seconds

To solve this problem, the engineer must move the offsets around until a more satisfactory timing plan develops. A change in cycle length may even be required.

Note that the northbound vehicle takes 3600/60 = 60 s to travel from Intersection 4 to Intersection 2, or—given $C = 60$ seconds—one cycle length. If the cycle length had been $C = 120$ seconds, the vehicle would have arrived at Intersection 2 at $C/2$, or half the cycle length. If we try the 120-second cycle length, then a solution presents itself.

Figure 26.19 shows one solution to the problem, for $C = 120$ seconds, which has a 40-second bandwidth in both directions for an efficiency of 33%. The 40-second bandwidth can handle (40/2.0) = 20 vehicles per lane per cycle. Thus if the demand volume is greater than 3,600(40)(2)(2.0)(120) = 1,200 veh/h, then it will not be possible to process the vehicles nonstop through the system.

As indicated in the original information (see Figure 26.17), the northbound demand is 1500 veh/h. Thus there will be some difficulty in the form of excess vehicles in the platoon. They can enter the system but cannot pass Signal 2 nonstop. They will be "chopped off" the end of the platoon and be queued vehicles in the next cycle. They will be released in the early part of the cycle and arrive at Signal 1 at the beginning of red. Figure 26.20 illustrates this, showing that these vehicles then disturb the next northbound through platoon.

Note the Figure 26.20 illustrates the limitation of the bandwidth approach when internal queuing arises, disrupting the bandwidth. The figure also shows the southbound platoon pattern, suggesting that the demand of exactly 1,200 veh/h might give rise to minor problems of the same sort at Signals 3 and 4.

If one were to continue a trial-and-error attempt at a good solution, it should be noted that:

- If the green initiation at Intersection 1 comes earlier in order to help the main northbound platoon avoid the queued vehicles, the southbound platoon is released sooner and gets stopped or disrupted at Intersection 2.
- Likewise, shifting the green at Intersection 2 cannot help the northbound progression without harming the southbound progression.
- Nor can shifting the green at Intersection 3 help the southbound progression without harming the northbound progression.

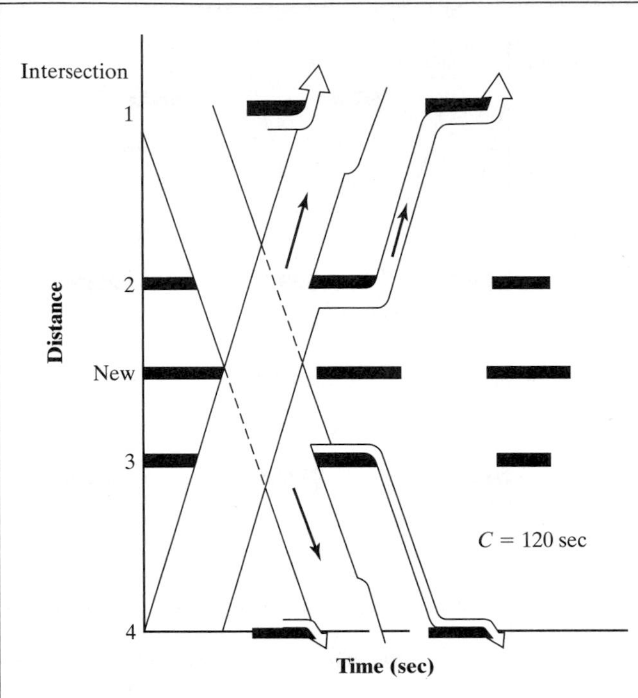

Figure 26.20: Case Study: Effect on Platoons with Demand Volume 1,500 Veh/h

- Some green can be taken from the side street and given to the main street.
- It is also possible that the engineer may decide to give the northbound platoon a more favorable bandwidth because of its larger demand volume.

This illustration showed insights that can be gained by simple inspection of a time-space diagram, using the concepts of bandwidth, efficiency, and an upper bound on demand volume that can be handled nonstop.

26.6 Common Types of Progression

26.6.1 Progression Terminology

The sole purpose of this section is to introduce some common terminology:

- Simple progression
- Forward progression
- Flexible progression
- Reverse progression

Simple progression is the name given to the progression in which all signals are set so that a vehicle released from the first intersection will arrive at all downstream intersections just as the signals at those intersections initiate green. That is, each offset is the ideal offset, set by Equation 26-4 with zero queue. Of necessity, simple progressions are effective only on one-way streets or on two-way streets on which the reverse flow is small or neglected.

Because the simple progression results in a green wave that advances with the vehicles, it is often called a *forward progression,* taking its name from the visual image of the advance of the green down the street.

It may happen that the simple progression is revised two or more times in a day, so as to conform to the direction of the major flow, or to the flow level (because the desired platoon speed can vary with traffic demand). In this case, the scheme may be referred to as a *flexible progression.*

Under certain circumstances, the internal queues are sufficiently large that the ideal offset is negative; that is, the downstream signal must turn green before the upstream signal, to allow sufficient time for the queue to start moving before the arrival of the platoon. Figure 26.21 has link lengths of 600 feet, platoon speeds of 60 ft/s, and internal queues averaging 7 vehicles per lane at each intersection. The visual image of such a pattern is of the green marching upstream, toward the drivers in the platoon. Thus it is referred to as a *reverse progression.*

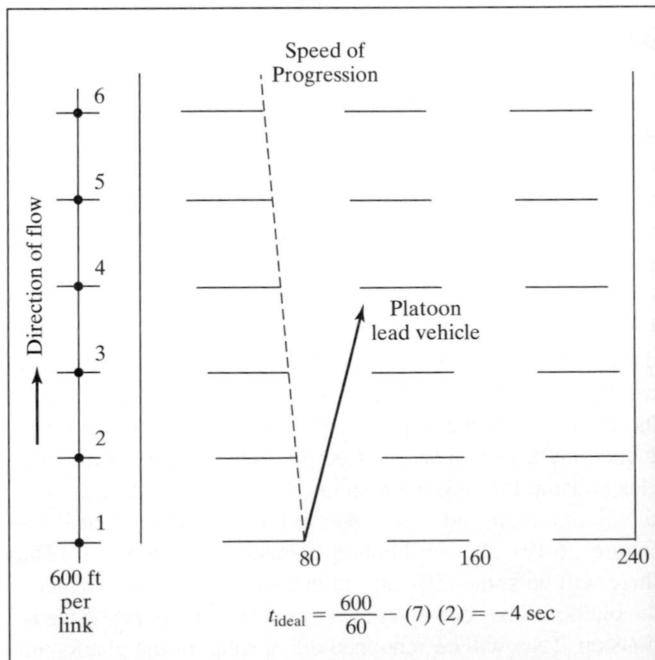

Figure 26.21: Illustration of a Reverse Progression

Figure 26.21 also illustrates one of the unfortunate realities of so many internal queued vehicles: The platoon's lead vehicle only gets to Signal 4 before encountering a red indication. As the platoon passes Signal 3, there are only 12 seconds of green to accommodate it, resulting in all vehicles beyond the sixth (i.e., 12/2 = 6) being cut off at Signal 3.

In the next several sections, common progression systems that can work extremely effectively on two-way arterials and streets are presented. As you will see, these systems rely on having uniform block lengths and an appropriate relationship among block length, progression speed, and cycle length. Because achieving one of these progressions has major benefits, the traffic engineer may wish to set the system cycle length based on progression requirements, introducing design improvements at intersections where the system cycle length would not provide sufficient capacity. Rather than increase the system cycle length to accommodate the needs of a single intersection, redesign of the intersection should be attempted to provide additional capacity at the desired system cycle length.

26.6.2 The Alternate Progression

For certain uniform block lengths, and all intersections with a 50-50 split of effective green time, it is possible to select a feasible cycle length such that:

$$\frac{C}{2} = \frac{L}{S} \qquad (26\text{-}8)$$

where: C = cycle length, s
L = block length, ft
S = platoon speed, ft/s

In this situation, the progression of Figure 26.22 can be obtained. There is no limit to the number of signals that may be included in the progression.

The name for this pattern is derived from the "alternate" appearance of the signal displays: As the observer at Signal 1 looks downstream, the signals alternate—red, green, red, green, and so forth.

The key to Equation 26-8 is that the ideal offset in either direction (with zero internal queues) is L/S. That is, the travel time to each platoon is exactly half the cycle length, so that the two travel times add up to the cycle length.

The efficiency of an alternate system is 50% in each direction because all of the green is used in each direction. The bandwidth capacity for an alternate progression is found using Equation 26-3, and noting that the bandwidth, BW, is

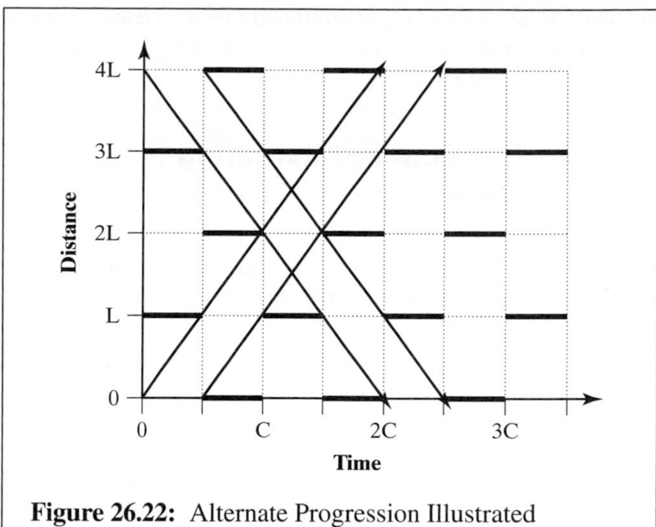

Figure 26.22: Alternate Progression Illustrated

equal to half the cycle length, C. If a saturation headway of 2.0 s/veh is assumed, then:

$$c_{BW} = \frac{3600 * BW * NL}{h * C} = \frac{3600 * 0.5C * NL}{2.0 * C} = 900NL$$

where all terms are as previously defined. This is an approximation based on the assumed saturation headway of 2.0 s/veh. The actual saturation headway may be determined more accurately using the *Highway Capacity Manual* procedure for intersection analysis.

Note that if the splits are not 50-50 at some signals, then (1) if they favor the main street, they simply represent excess green, suited for accommodating miscellaneous vehicles, and (2) if they favor the side street, they reduce the bandwidths.

As a practical matter, note the range of the block lengths for which alternate patterns might occur. Using Equation 26-8, appropriate block lengths are computed for platoon speeds of 30 and 50 mi/h (that is, 45 and 75 ft/s), and cycle lengths of 60 and 90 seconds. The results are shown in Table 26.3. These

Table 26.3: Come Illustrative Combinations for Alternate Progression

Cycle Length (s)	Platoon Speed (fps)	Matching Block Length (ft)
60	45	1,350
60	75	2,250
90	45	2,025
90	75	3,375

results are illustrative; other combinations are clearly possible as well. All of these signal spacings imply a high-type arterial, often in a suburban setting.

26.6.3 The Double-Alternating Progression

For certain uniform block lengths with 50-50 splits, it is not possible to satisfy Equation 26-8, but it is possible to select a feasible cycle length such that:

$$\frac{C}{4} = \frac{L}{S} \tag{26-9}$$

In this situation, the progression illustrated in Figure 26.23 can be obtained.

The key is that the ideal offset in either direction (with zero internal queues) over *two* blocks is one half of a cycle length, so that two such travel times (one in each direction) add up to a cycle length. There is no limit to the number of signals that can be involved in this system, just as there was no limit with the alternate system.

The name of the pattern is derived from the "double alternate" appearance of the signal displays; that is, as the observer at Signal 1 looks downstream, the signals alternate in pairs—green, green, red, red, green, green, red, red, and so forth.

The efficiency of the double alternate signal system is 25% in each direction because only half of the green is used in each direction. The upper limit on the bandwidth capacity

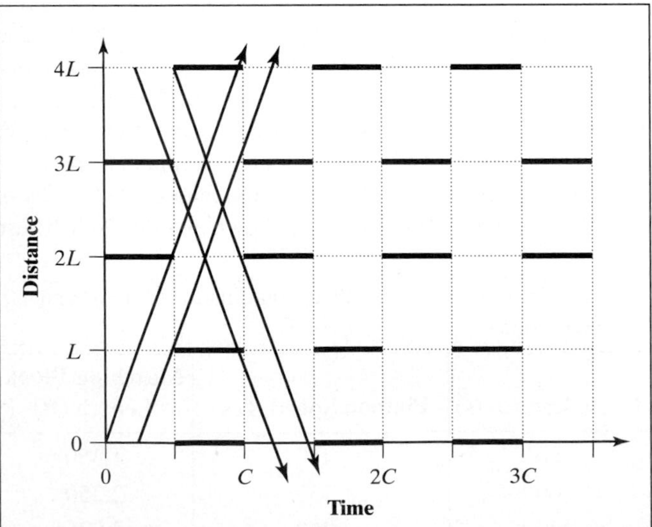

Figure 26.23: Double Alternate Progression Illustrated

Table 26.4: Illustrative Combinations for Double-Alternate Progression

Cycle Length (s)	Platoon Speed (fps)	Matching Block Length (ft)
60	45	675
60	75	1,125
90	45	1,012
90	75	1,688

may be approximated by assuming a 2.0 s/veh saturation headway and noting that the BW is a quarter of C.

As with the alternate system, if the splits are not 50-50 at some signals, then (1) if they favor the main street, they simply represent excess green, suited for accommodating miscellaneous vehicles, and (2) if they favor the side street, they reduce the bandwidths.

$$c_{BW} = \frac{3600 * BW * NL}{h * C} = \frac{3600 * 0.25C * NL}{2.0 * C} = 450NL$$

Table 26.4 shows some illustrative combinations of cycle length, platoon speed, and block lengths for which a double alternating progression would be appropriate. Other combinations, of course, are possible as well. Some of these signal spacings represent a high-type arterial. With the shorter cycle lengths, however, some urban facilities could also have the necessary block lengths.

26.6.4 The Simultaneous Progression

For very closely spaced signals, or for rather high vehicle speeds, it may be best to have all the signals turn green at the same time. This is called a simultaneous system because all the signals turn green simultaneously. Figure 26.24 illustrates a simultaneous progression.

The efficiency of a simultaneous system depends on the number of signals involved. For N signals:

$$EFF(\%) = \left[\frac{1}{2} - \frac{(N-1)*L}{S*C}\right]*100\% \tag{26-10}$$

For four signals with $L = 400$ feet, $C = 80$ seconds, and $S = 45$ ft/s, the efficiency is 16.7%. For the same number of signals with $L = 200$ feet, it is 33.3%.

Simultaneous systems are advantageous only under a limited number of special circumstances. The foremost of these special circumstances is very short block lengths. The simultaneous system has an additional advantage, however,

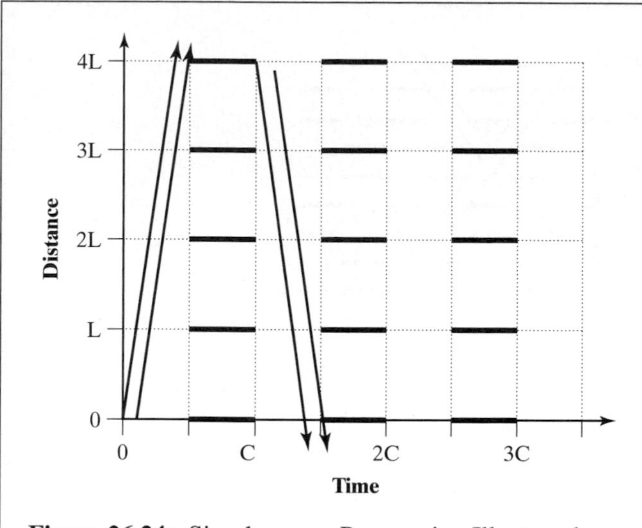

Figure 26.24: Simultaneous Progression Illustrated

that is not at all clear from a bandwidth analysis: Under very heavy flow conditions, it forestalls breakdown and spillback. This is so because (1) it allows for vehicle clearance time at the downstream intersection where queues inevitably exist during heavy flow, and (2) it cuts platoons off in a way that generally prevents blockage of intersections. This works to the advantage of cross traffic. Specific plans for controlling spillback under heavy traffic conditions are discussed in Chapter 27.

26.6.5 Insights from the Importance of Signal Spacing and Cycle Length

It is now clear that:

- All progressions have their roots in the desire for ideal offsets.
- For certain combinations of cycle length, block length, and platoon speed, some very satisfactory two-way progressions can be implemented.
- Other progressions can be designed to suit individual cases, using the concept of ideal offset and queue clearance, trial-and-error bandwidth based approaches, or computer-based algorithms.

A logical first step in approaching a system is simply to ride the system and inspect it. As you sit at one signal, do you see the downstream signal green, but with no vehicles being processed? Do you arrive at signals that have standing queues but were not timed to get them moving before your platoon arrived? Do you arrive on the red at some signals? Is the flow in the other direction significant, or is the traffic really a one-way pattern, even if the streets are two way?

It is very useful to sketch out how much of the system can be thought of as an "open tree" of one-way links. This can be done with a local map and an appreciation of the traffic flow patterns. A distinction should be made among:

- Streets that are one way.
- Streets that can be treated as one way, due to the actual or desired flow patterns.
- Streets that must be treated as two way.
- Larger grids in which streets (one way and two way) interact because they form unavoidable "closed trees" and are each important in that they cannot be ignored for the sake of establishing a "master grid" that is an open tree.
- Smaller grids in which issue is not coordination but rather local land access and circulation, so that they can be treated differently. Downtown grids may well fall into the latter category, at least in some cases.

The next most important issue is the cycle length dictated by the signal spacing and platoon speed. Attention must focus on the combination of cycle length, block length, and platoon speed, as shown earlier in this chapter.

Figure 26.25 shows the three progressions of the preceding sections—alternate, double alternate, and simultaneous—on the same scale. The basic "message" is that as the average signal spacing decreases, the type of progression best suited to the task changes.

Figure 26.26 illustrates a hypothetical arterial that comes from a low-density suburban environment with a larger signal spacing, into the outlying area of a city, and finally passes through one of the city's CBDs. As the arterial changes, the progression used may also be changed, to suit the dimensions.[3]

Note that the basic lesson here is that a system can sometimes be best handled by breaking it up into several smaller systems. This can be done with good effect on even smaller systems, such as 10 consecutive signals, of which a

[3]Of course, if the flow is highly directional—as may well be from the suburbs in the morning—then these suggestions are superseded by the simple expedient of treating the streets as one-way streets and imposing a simple forward progression, with queue clearance if needed.

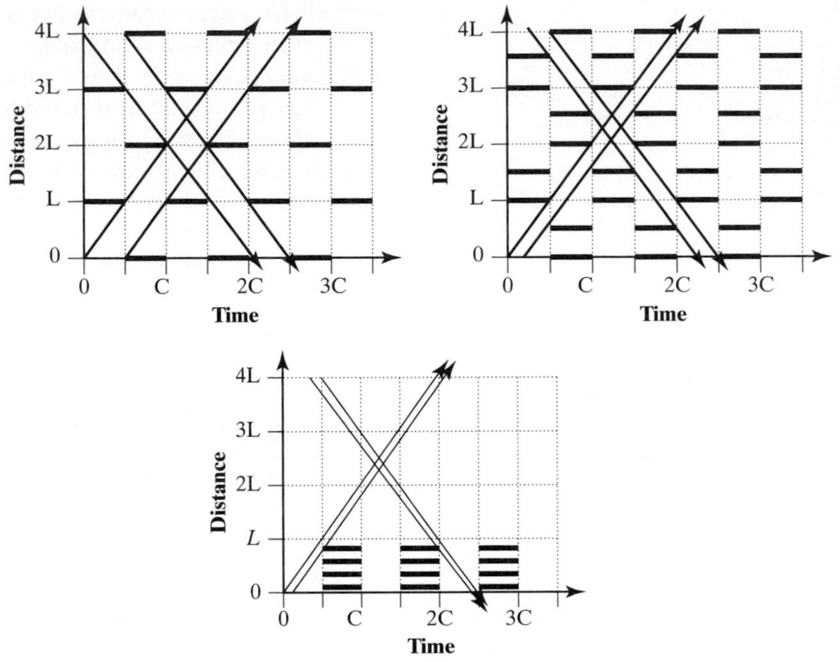

Figure 26.25: Comparison of Scales on Which Standard Patterns Are Used

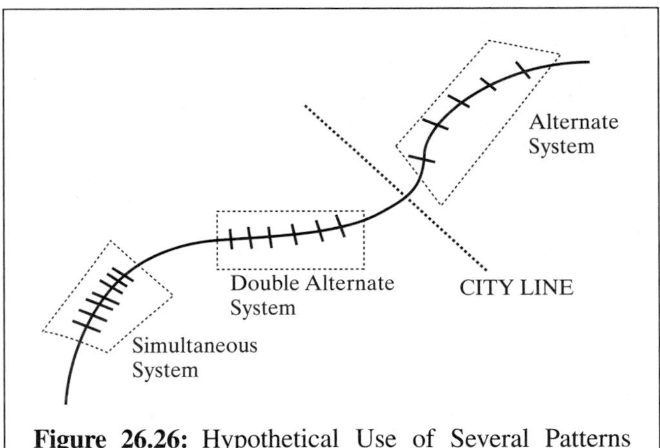

Figure 26.26: Hypothetical Use of Several Patterns Along the Same Arterial

contiguous 6 are spaced uniformly and the other 4 also uniformly, but at different block lengths. Note that to the extent that block lengths do no exist perfectly uniformly, these plans can serve as a basis from which adaptations can be made. Note also that the suitability of the cycle length has been significant. It is often amazing how often the cycle length is poorly set for system purposes.

26.7 Software for Doing Signal Progression

The purpose of this section is to *illustrate* the use of software that is commercially available to provide ease of computation in determining progressions. You cannot expect to become an effective user of either program from these brief illustrations. But it is common (and advisable) that such computer programs be used in at least the second course in a traffic engineering sequence. The user manuals and lecturer presentation, focused on case study or project applications, can provide the needed background. (In our own courses, we also depend on the students climbing the learning curve quickly as part of a project-based course.)

The first program that is illustrated, TruTraffic TS/PP, is especially interesting because it is based on the bandwidth approach that is considered "dated" by those who have used the programs that focus on delay-optimized signal settings that take into account internal queueing. But the simple fact is that bandwidth-based solutions are less data intensive, very suitable in many applications, and easy to manage.

The second program that is illustrated (Synchro) is the most commonly used delay- or stops-based optimization program in the United States. As already noted, a number of states accept its output as equivalent to *HCM* results, at least for sets of signals on an arterial or network. Synchro has an imbedded macro model of flow profiles that it uses to estimate delays and stops as it iteratively finds a solution that minimizes an overall "objective function" that is expressed in terms of stops and delays.

26.7.1 Bandwidth-Based Solutions

Bandwidth-based solutions remain an important tool for traffic engineers, particularly for off-peak periods when demand is relatively small and/or when flow is highly directional. Although prior examples in this chapter emphasized achieving windows of green along the entire arterial, bandwidth solutions also can be used to move platoons along relatively long arterials in a set of bandwidths, with the breaks between bandwidths occurring where it is logical or suitable to stop and re-form platoons—for instance, just upstream of a set of closely spaced signals, so that the platoons that might overflow the short block spacings are not stopped in that section of the arterial. Bandwidth solutions are also used effectively to discourage speeding, encourage adherence to the speed limit, and identify green that can be allocated to increased pedestrian walking times [*8*].

Figure 26.27 shows the output of TruTraffic TS/PP for the arterial addressed in Figure 26.4, considering it as a two-way arterial on which we wish to achieve equal bandwidths (16 seconds) in the two directions if at all possible.

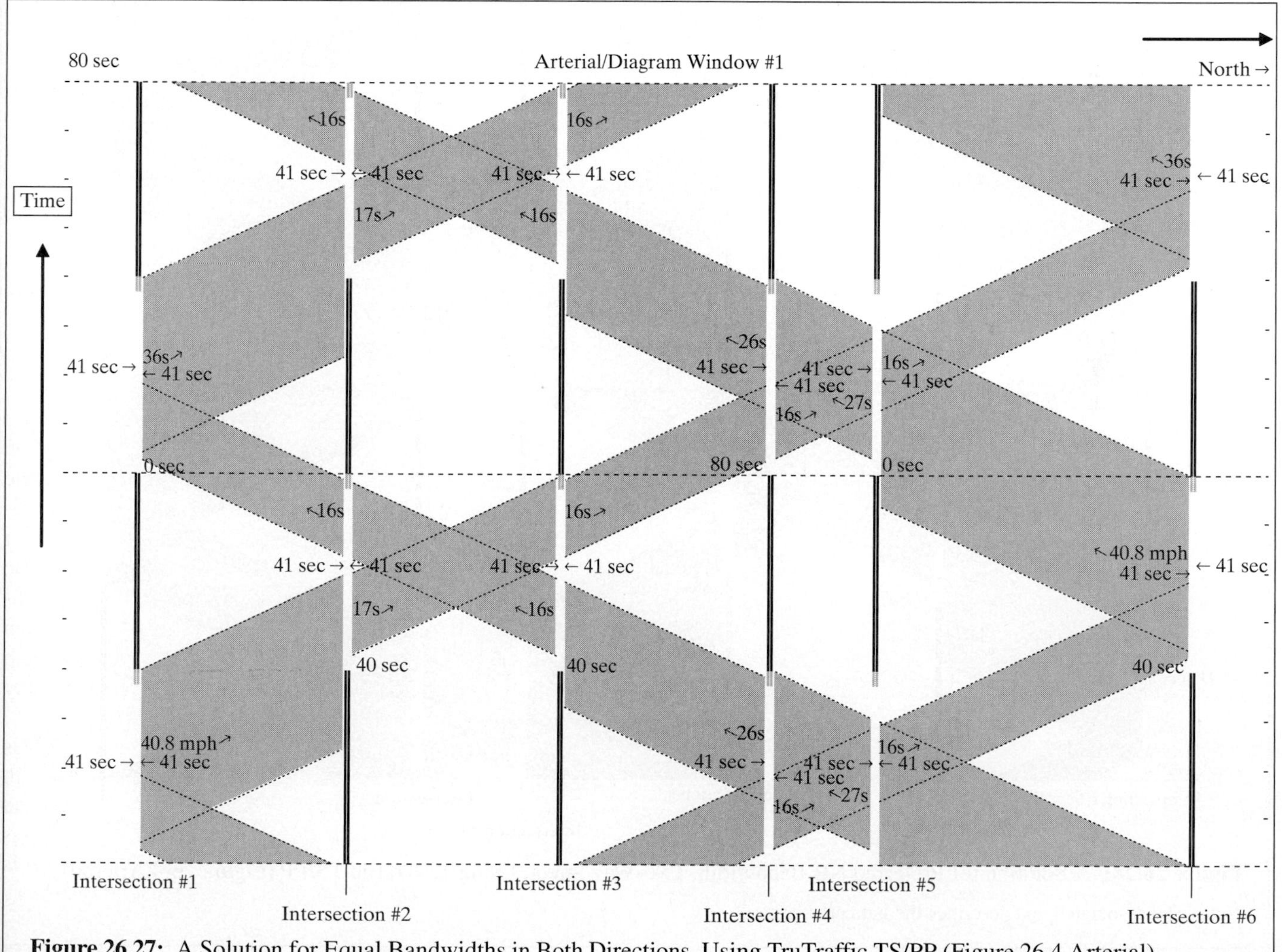

Figure 26.27: A Solution for Equal Bandwidths in Both Directions, Using TruTraffic TS/PP (Figure 26.4 Arterial)

Because of the way in which the output is printed (generally, on longer paper), the directions are somewhat reversed from the prior time-space diagrams. In this chapter arrows have been added, to emphasize the directions in which time and distance both increase. From this figure, note that:

- The bandwidths are shown in seconds, and partial bandwidths are shown when breaks are required. The "message" in such cases is that the platooned vehicles can travel only so many blocks without stopping;
- To visualize a vehicle going faster than the design speed of the bandwidth, it may be easiest for the reader to turn the display so that time is on the bottom, given the reversal in the figure.

Figure 26.28 shows the same illustration, except the target is to get a NB bandwidth of at least 20 seconds, with "as good as possible" in the SB.

The bandwidths shown in Figure 26.28 vary as opportunity presents itself, indicating that the band can be wider in some segments along the arterial than in others; this is shown in both

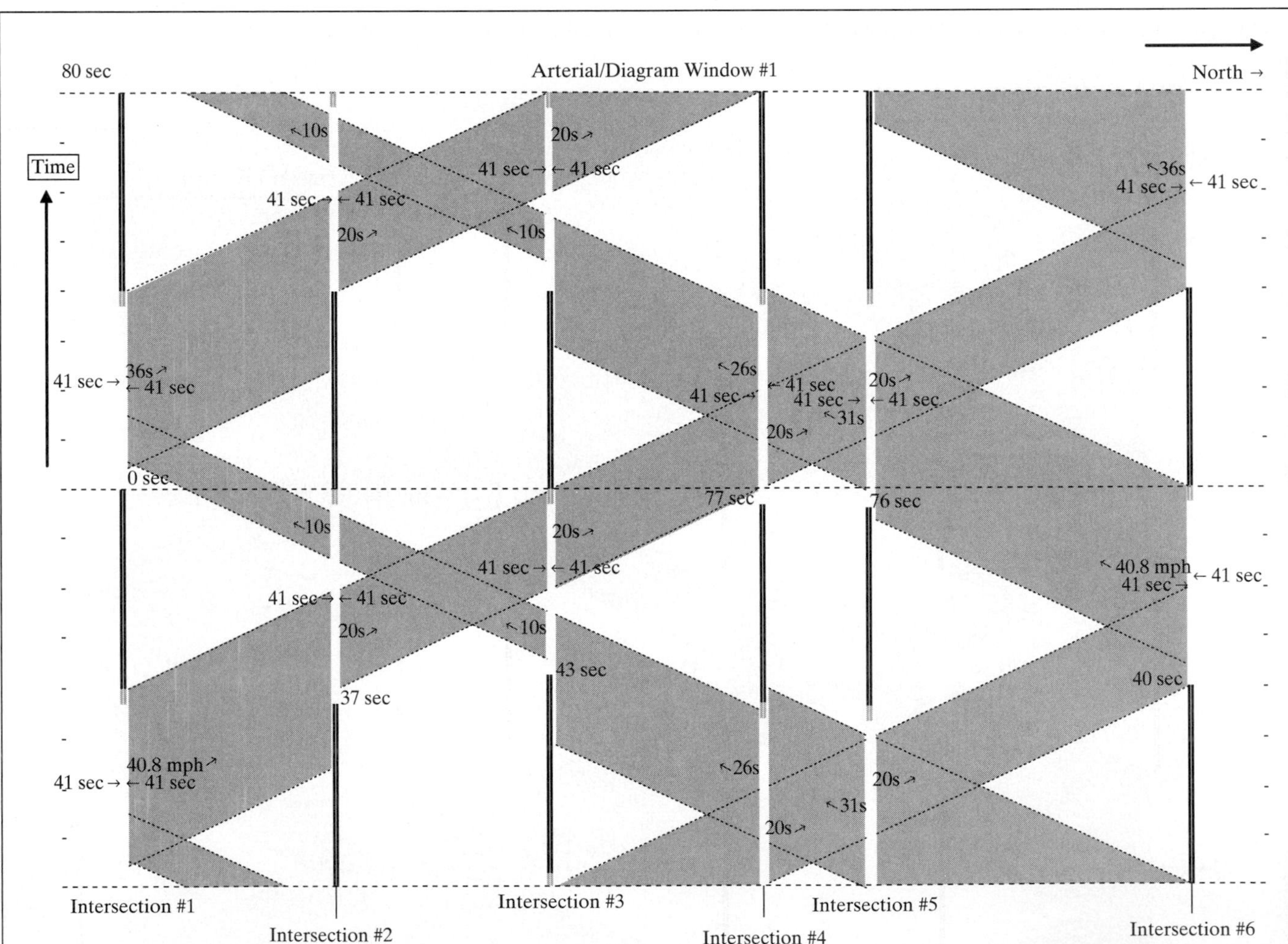

Figure 26.28: A Solution for Preferred NB Bandwidth, Two-Way Street, Using TruTraffic TS/PP (Figure 26-4 Arterial)

Note: The associated text describes the indicated items.

directions. TruTraffic also allows the user to turn this and other features "on" or "off," so that

a. One can emphasize a fixed bandwidth that can move the entire distance (or some large subsegment), rather than that shown in Figure 26.28; for instance, in the northbound direction, the band can be reduced to 20 seconds in Figure 26.28, emphasizing the best band that can get through with no one stopping.
b. One can also restart the bands at any intersection, designating a new start point.

If the bandwidth in "a" is capable of handling the existing traffic, this approach has meaning. Referring to Equation 26-3 and using $BW = 20$ seconds and $C = 80$ seconds, one can compute $c_{BW} = 3{,}600(20)(1)/(80)(2.1) = 429$ vph on a *per lane* basis.

The actual trajectories along the arterial depend on where the platoon started, turns in and out, and some dispersion, even when the demand is less than c_{BW}. In practice, one has to take into account that when northbound "early greens" are given to allow southbound bandwidth, northbound speeding may be allowed because drivers may perceive short-term gain by moving faster.

When the demand exceeds c_{BW} and TruTraffic or such tool is still used, more has to be taken into account. For instance, refer to Figure 26.29: When the northbound platoon fills most of the bandwidth from Intersection 1 to Intersection 2,

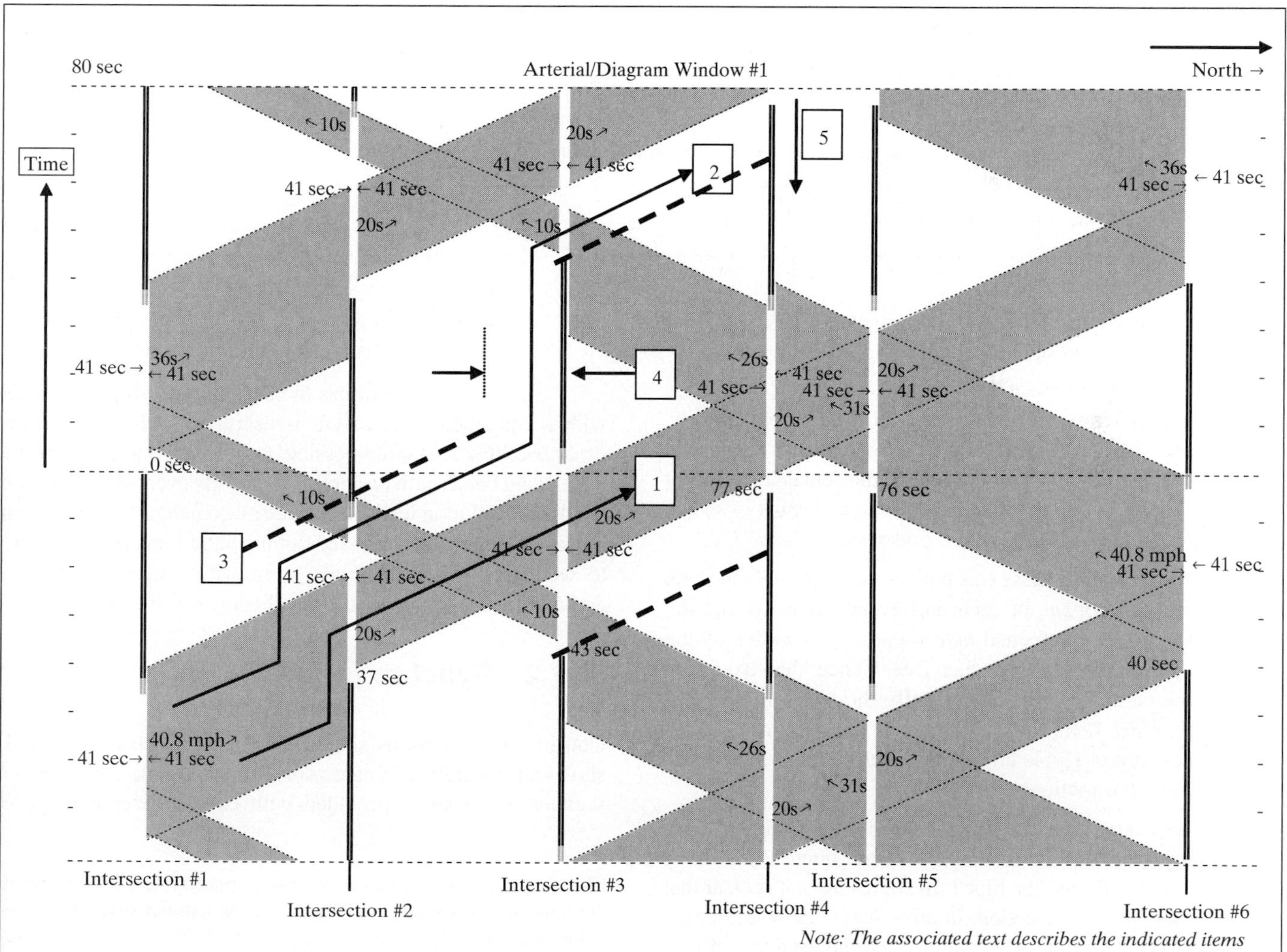

Figure 26.29: Some of the Considerations in Using TruTraffic to Design Bandwidths

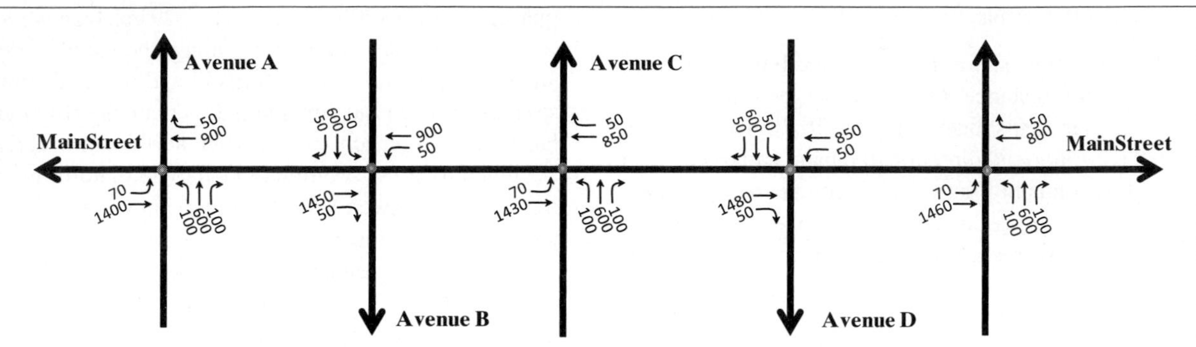

- All the intersections are spaced at 1500 feet
- 2 lanes per direction on Main St
- left turn bays on Main St, 250 ft, both directions
- RTOR prohibited everywhere
- 60 fps free flow speed, Main Street
- All avenues have 2 moving lanes, but are one-way streets
- PHF = 0.91
- Minimum ped crossing times = 17 sec across side streets, 30 sec Main St

Figure 26.30: Inputs for Illustrative Synchro Case

- The lead part of the platoon must stop at Intersection 2 and then continue as shown by ___.
- The second part of the platoon, designated by ___, is queued temporarily at Intersection 2 and then "clipped" and stopped at Intersection 3. It can extend to the dashed line shown by ___, depending on the demand. *Further, it then becomes the lead part of the next platoon, as shown by the dashed line between Intersections 3 and 4.*
- Notice that because this pattern is repetitive, the same reality happens in each and every cycle, so that the same thing happened *before* path __ as shown by the earlier (i.e., lower) dashed line. Hence, the vehicles in the path ___ were not actually the lead vehicles *and they are temporarily queued* at Intersection 4. This concern or ripple effect exists throughout the system, in both directions;
- At every intersection, but most notably ones close together, one must assure that the queues shown by ___ can fit into the block. In some cases, it is clear that they do, by inspection. In other cases, the width of the stopped section of the band has to be used to estimate the queue size, convert it to an expected length, and see if it approaches or exceeds the block length.[4]
- Given the knowledge of the presence of vehicles as shown by ___, one may wish to consider moving the green initiation at ___ as shown, although this has an adverse effect on the southbound traffic.

The TruTraffic tool has the capability to show the added widths described here, and it is useful in such cases as just described. It is also quite possible to use such a tool (really, the underlying bandwidth principle) effectively even when *apparent* bandwidth solution leads to such displacements.[5] At some point however, delay/stopped-based optimization tools may be easier to use and more appropriate, even with rough estimates of demand (i.e., without all the detail shown in the next section).

26.7.2 Synchro

Consider the inputs as shown in Figure 26.30. Figure 26.31 shows an illustrative Synchro solution for a cycle length of 100 seconds; the user is provided with choices over a range of

[4]The practical result is that very closely spaced intersections operate simultaneously, so that the platoon moves without stopping in that short distance.

[5]Notice that the demand does not have to exceed *or even approach* the capacity of the *signals*. The issue at hand is the demand that can be accommodated by c_{BW}. Hence solutions that involve narrow bands over considerable lengths might look good initially but can induce problems when demand exceeds c_{BW}.

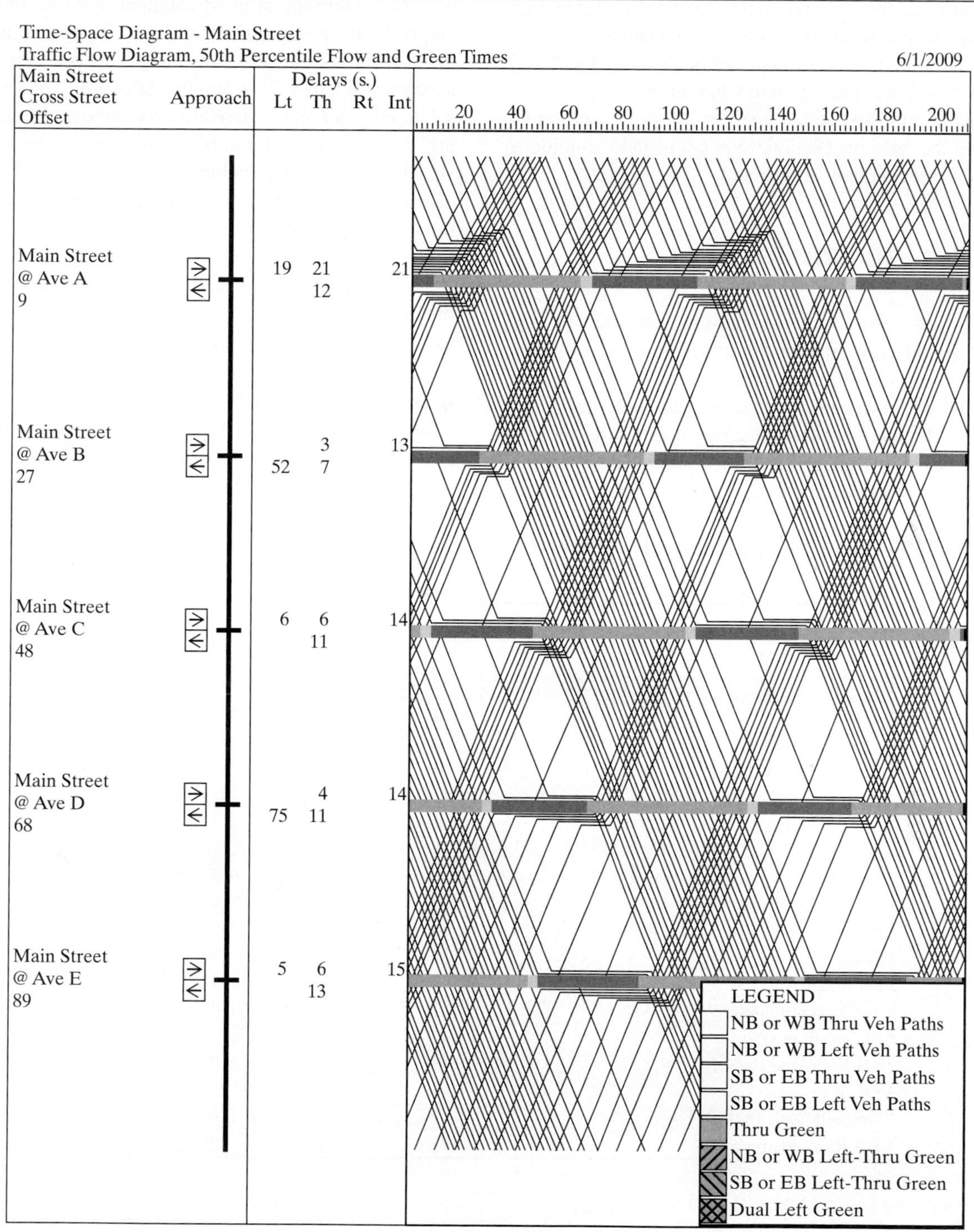

Figure 26.31: Sample Synchro Time-Space Diagram for Illustrative Case

user-specified cycle lengths and selects from them. Table 26.5 shows sections of the standard Synchro output tables.

Although this illustration will not be discussed in detail at this point, it will be used again in Chapter 30.

In practice, engineers often fine-tune the optimized signal offsets during the field installation, to adapt to field conditions (that is, how drivers actually arrive, etc.).

26.8 Closing Comments

This chapter has introduced the basic considerations and concepts of signal coordination for undersaturated flows on one-way and two-way arterials, and in networks.

Although the commercially available computer programs are extremely efficient and cost-effective tools for developing signal timing plans (including offsets, for coordination), you will find that thoughtful reflection on the basic principles will help not only at arriving at good solutions, but also in checking the outputs of the programs. For instance, if one is fortunate enough to have block lengths (L) and desired speeds (S) that happen to approximate a half cycle length ($L/S \sim C/2$), one can expect the solution to look like a standard alternative progression. If it does not, then a significant number of turn-ins or internally generated vehicles (e.g., from a midblock parking lot) may be skewing the offsets, so that queues are cleared and overall delay/stops minimized. But if this potential explanation is absent, another explanation might be that the user made an error with the inputs.

There are now excellent Web sites that can serve as the basis for keeping up to date and aware of the trends related to signal timing and coordination. For instance, the TRB Traffic Signal Systems Committee has its Web site at http://www.signalsystems.org.vt.edu/. FHWA's Office of Operations Web site on publications at http://www.ops.fhwa.dot.gov/publications/publications.htm is an excellent source of key material, much of which can be downloaded. The *Traffic Signal*

Table 26.5: Segments of Synchro Output, Related to Illustrative Case

Detailed Measures of Effectiveness

3. Main Street & Ave A

Direction	EB	WB	NB	All
Volume (vph)	1470	950	800	3220
Control Delay / Veh (s/v)	21	12	31	21
Queue Delay / Veh (s/v)	0	0	0	0
Total Delay / Veh (s/v)	21	12	31	21
Total Delay (hr)	8	3	7	18
Stops / Veh	0.74	0.58	0.83	0.72
Stops (#)	1085	555	665	2305
Average Speed (mph)	19	28	8	19
Total Travel Time (hr)	16	10	9	34
Distance Traveled (mi)	306	270	69	644
Fuel Consumed (gal)	28	18	14	60
Fuel Economy (mpg)	10.9	15.4	4.8	10.7
CO Emissions (kg)	1.97	1.23	1.01	4.20
NOx Emissions (kg)	0.38	0.24	0.20	0.82
VOC Emissions (kg)	0.46	0.28	0.23	0.97
Unserved Vehicles (#)	0	0	0	0
Vehicles in dilemma zone (#)	70	64	40	174

Arterial Level of Service: EB Main Street

Cross Street	Arterial Class	Flow Speed	Running Time	Signal Delay	Travel Time (s)	Dist (mi)	Arterial Speed	Arterial LOS
Ave A	II	41	23.9	20.6	44.5	0.21	16.8	E
Ave B	II	41	29.0	3.3	32.3	0.28	31.7	B
Ave C	II	41	29.0	6.2	35.2	0.28	29.1	B
Ave D	II	41	29.0	3.6	32.6	0.28	31.4	B
Ave E	II	41	29.0	5.7	34.7	0.28	29.5	B
Total	II		139.9	39.4	179.3	1.34	27.0	C

Table 26.5: Segments of Synchro Output, Related to Illustrative Case (*Continued*)

HCM Signalized Intersection Capacity Analysis
3. Main Street & Ave A 6/1/2009

Movement	EBL	EBT	EBR	WBL	WBT	WBR	NBL	NBT	NBR	SBL	SBT	SBR
Lane Configurations	↰	↑↑			↑↱			↰↑↱				
Ideal Flow (vphpl)	1900	1900	1900	1900	1900	1900	1900	1900	1900	1900	1900	1900
Total Lost time (s)	4.0	4.0			4.0			4.0				
Lane Util. Factor	1.00	0.95			0.95			0.95				
Frt	1.00	1.00			0.99			0.98				
Flt Protected	0.95	1.00			1.00			0.99				
Satd. Flow (prot)	1770	3539			3511			3451				
Flt Permitted	0.20	1.00			1.00			0.99				
Satd. Flow (perm)	371	3539			3511			3451				
Volume (vph)	70	1400	0	0	900	50	100	600	100	0	0	0
Peak-hour factor, PHF	0.91	0.91	0.91	0.91	0.91	0.91	0.91	0.91	0.91	0.91	0.91	0.91
Adj. Flow (vph)	77	1538	0	0	989	55	110	659	110	0	0	0
RTOR Reduction (vph)	0	0	0	0	0	0	0	0	0	0	0	0
Lane Group Flow (vph)	77	1538	0	0	1044	0	0	879	0	0	0	0
Turn Type	Perm						Perm					
Protected Phases		4			8			2				
Permitted Phases	4						2					
Actuated Green, G (s)	56.0	56.0			56.0			36.0				
Effective Green, g (s)	56.0	56.0			56.0			36.0				
Actuated g/C Ratio	0.56	0.56			0.56			0.36				
Clearance Time (s)	4.0	4.0			4.0			4.0				
Lane Grp Cap (vph)	208	1982			1966			1242				
v/s Ratio Prot		c0.43			0.30							
v/s Ratio Perm	0.21							0.25				
v/s Ratio	0.37	0.78			0.53			0.71				
Uniform Delay, d1	12.2	17.1			13.8			27.5				
Progression Factor	1.00	1.00			0.76			1.00				
Incremental Delay, d2	5.0	3.1			0.9			3.4				
Delay (s)	17.2	20.2			11.4			30.9				
Level of Service	B	C			B			C				
Approach Delay (s)		20.0			11.4			30.9			0.0	
Approach LOS		C			B			C			A	

Intersection Summary			
HCM Average Control Delay	20.2	HCM Level of Service	C
HCM Volume to Capacity ratio	0.75		
Actuated Cycle Length (s)	100.0	Sum of lost time (s)	8.0
Intersection Capacity Utilization	68.0%	ICU Level of Service	C
Analysis Period (min)	15		

c Critical Lane Group

Timing Manual is available at http://www.ops.fhwa.dot.gov/publications/fhwahop08024/index.htm.

References

1. *PASSER II-90 Microcomputer User's Guide,* distributed by the McTrans Center, Gainsville, FL, 1991.
2. Bullock, G., "Tru-Traffic TS/PP Version 7.0 Time-Space/Platoon-Progression Diagram Generator: Users Manual," April 2007, www.tsppd.com.
3. SYNCHRO Studio 7, Synchro plus SimTraffic and 3D Viewer User Guide, Trafficware Ltd., June 2006, www.trafficware.com.
4. *Highway Capacity Manual,* Transportation Research Board, Washington, DC, 2000.
5. TRANSYT-7F Release 11, Users Guide, distributed by the McTrans Center, University of Florida, 2008, http://mctrans.ce.ufl.edu/featured/TRANSYT-7F/.
6. HCS+ Release 5: Users Manual, distributed by the McTrans Center, University of Florida, 2008, http://mctrans.ce.ufl.edu/hcs/.

7. VISSIM 5.0 User's Guide, PTV, 2007, http://www.ptvamerica.com/vissim.html

8. AIMSUN Users Manual v6, TSS-Transport Simulation Systems, October 2008.

9. TSIS User's Guide, Version 6.0, CORSIM User Guide, distributed by the McTrans Center, University of Florida, 2008.

10. S-Paramics microscopic simulation software, distributed by SIAS Limited (http://www.sias.com/ng/sparamicshome/) and Quadstone Paramics distributed by Quadstone (http://www.paramics-online.com/).

11. Discussion with John Tipaldo, PhD PE, NYCDOT, May 2009, regarding off-peak use of bandwidth solutions.

Problems

26-1. Two signals are spaced at 1,000 feet on an urban arterial. It is desired to establish the offset between these two signals, considering only the primary flow in one direction. The desired progression speed is 40 mph. The cycle length is 60 seconds. Saturation headway may be taken as 2.0 sec/veh, and the startup lost time as 2.0 seconds.

(a) What is the ideal offset between the two intersections assuming that vehicles arriving at the upstream intersection are already in a progression (i.e., a moving platoon) at the initiation of the green?

(b) What is the ideal offset between the two intersections assuming that the upstream signal is the first in the progression (i.e., vehicles are starting from a standing queue)?

(c) What is the ideal offset assuming that an average queue of three vehicles per lane is expected at the downstream intersection at the initiation of the green? Assume the base conditions of part a.

(d) Consider the offset of part a. What is the resulting offset in the opposite (off-peak) direction? What impact will this have on traffic traveling in the opposite direction?

(e) Consider the offset of part a. If the progression speed was improperly estimated, and the actual desired speed of drivers was 45 mi/h, what impact would this have on the primary direction progression?

26-2. Consider the time-space diagram for this problem. For the signals shown:

(a) What is the NB progression speed?

(b) What is the NB bandwidth and bandwidth capacity? Assume a saturation headway of 2.0 s/veh.

(c) What is the bandwidth in the SB direction for the same desired speed as the NB progression speed? What is the SB bandwidth capacity for this situation?

(d) A new development introduces a major driveway that must be signalized between intersections 2 and 3. It requires 15 seconds of green out of the 60-second system cycle length. Assuming that you had complete flexibility as to the exact location of the new driveway, where would you place it? Why?

26-3. A downtown grid has equal block lengths of 750 feet along its primary arterial. It is desired to provide for a progression speed of 30 mi/h, providing equal service to traffic in both directions along the arterial.

(a) Would you suggest a alternate or a double-alternate progression scheme? Why?

(b) Assuming your answer to part a, what cycle length would you suggest? Why?

26-4. Refer to the figure below. Trace the lead NB vehicle through the system. Do the same for the lead SB vehicle. Use a platoon speed of 50 ft/s. Estimate the number of stops and the seconds of delay for each of these vehicles.

26-5. Refer to the figure below. Find the NB and the SB bandwidths (in seconds). Determine the efficiency of the system in each direction and the bandwidth capacity. There are three lanes in each direction. The progression speed is 50 ft/s.

26-6. (a) If vehicles are traveling at 60 fps on a suburban road, and the signals are 2,400 feet apart, what cycle length would you recommend? What offset would you recommend?

(b) If an unsignalized intersection is to be inserted at 600 feet from one of the signalized intersections, what would you recommend?

26-7. You have two intersections 3,000 feet apart and have achieved some success with a 50-50 split, 60-second cycle length, and simultaneous system.

(a) Draw a time-space diagram and analyze the reason for your success.

(b) A developer who owns the property fronting on the first 2000 feet of the subject distance plans a major employment center. She plans a major driveway and asks your advice on its location. What is your recommendation, and why?

26-8. (a) Consider four intersections, spaced by 500 feet. The platoon speed is 40 ft/s. Recommend a set of offsets for the eastbound direction, considering only the eastbound traffic.

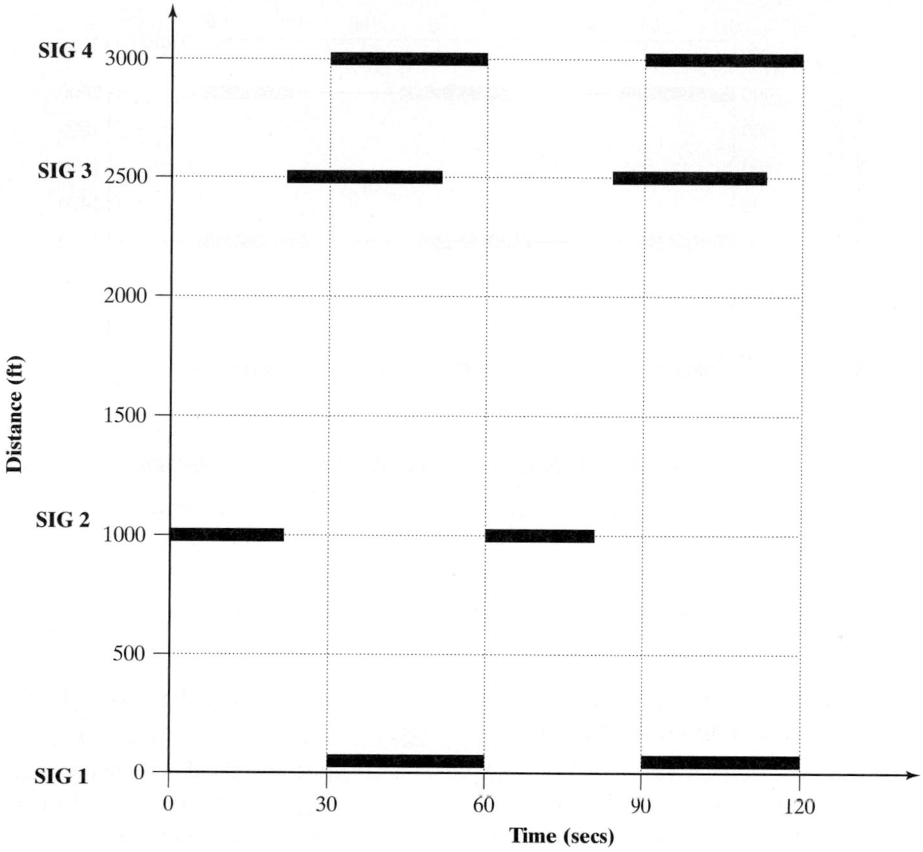

Time-Space Diagram for Problem 26-2

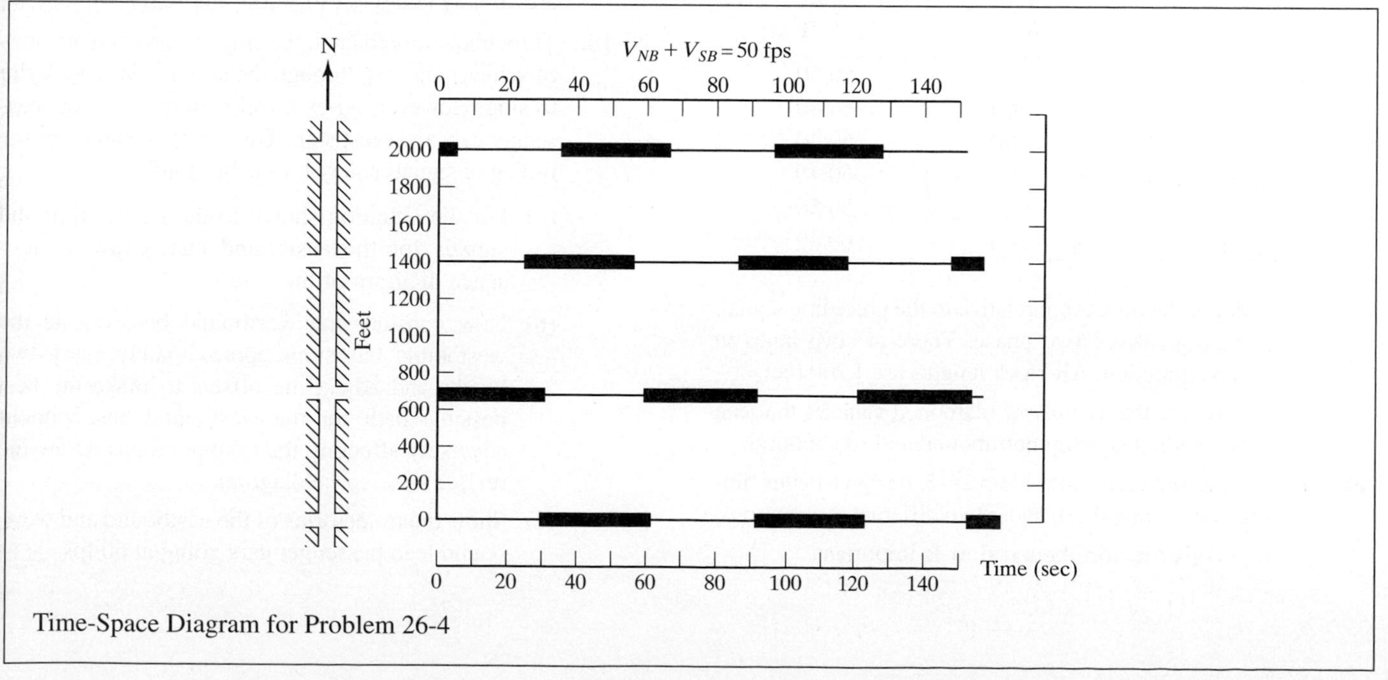

Time-Space Diagram for Problem 26-4

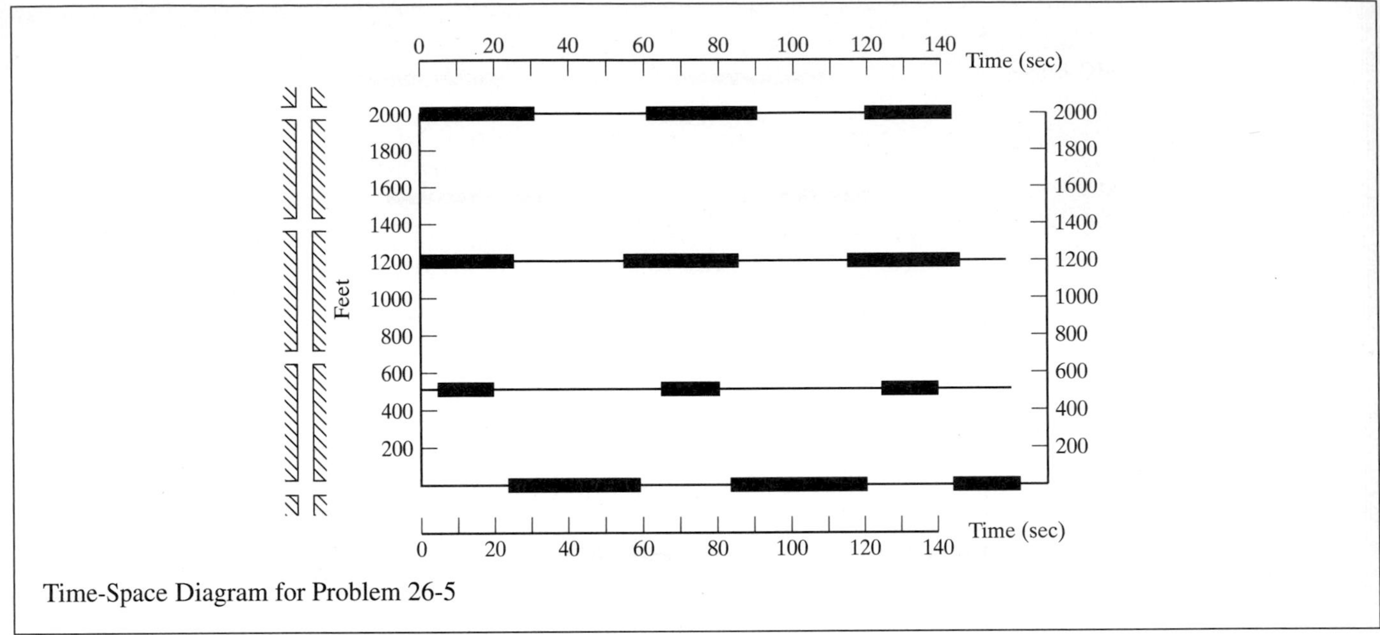

Time-Space Diagram for Problem 26-5

(b) If there are queues of three vehicles at each of the intersections, recommend a different set of offsets (if appropriate).

26-9. (a) Construct a time-space diagram for the following information and estimate the northbound bandwidth and efficiency for platoons going at 50 ft/s:

Signal No.	Offset (sec)	Cycle length	Split (MSG first)
6	16	60	50:50
5	16	60	60:40
4	28	60	60:40
3	28	60	60:40
2	24	60	50:50
1		60	60:40

All of the offsets are relative to the preceding signal. All signals are two phase. There are two lanes in each direction. All block lengths are 1,200 feet.

(b) Estimate the number of platooned vehicles that can be handled nonstop northbound and southbound.

26-10. For the situation in Problem 26-5, design a better timing plan (if possible), under two different assumptions:

(a) Only the northbound flow is important.

(b) The two directions are equally important.

26-11. Find the offset for a link of 1,500 ft, no standing queue at the downstream signal, and a platoon traveling at 40 ft/s. Re-solve if there is a standing queue of eight vehicles per lane.

26-12. Develop an arterial progression for the situation shown in the figure below. Use a desired platoon speed of 40 ft/s. For simplicity, the volumes shown are already corrected for turns and PHF.

26-13. Throughout this chapter, the emphasis was on platoons of vehicles moving through the system, with no desire to stop. However, buses travel slower than most passenger cars and must stop. This problem addresses the timing of signals solely for the bus traffic.

(a) For the situation shown in the figure, time the signals for the eastbound bus. Draw a time-space diagram of the solution.

(b) Now consider the westbound bus. Locate the westbound bus stops approximately every two blocks and adjust the offsets to make the best possible path for the westbound bus, without adversely affecting the eastbound bus. Draw the revised time-space diagram.

(c) Show the trajectories of the eastbound and westbound lead passenger cars going at 60 fps.

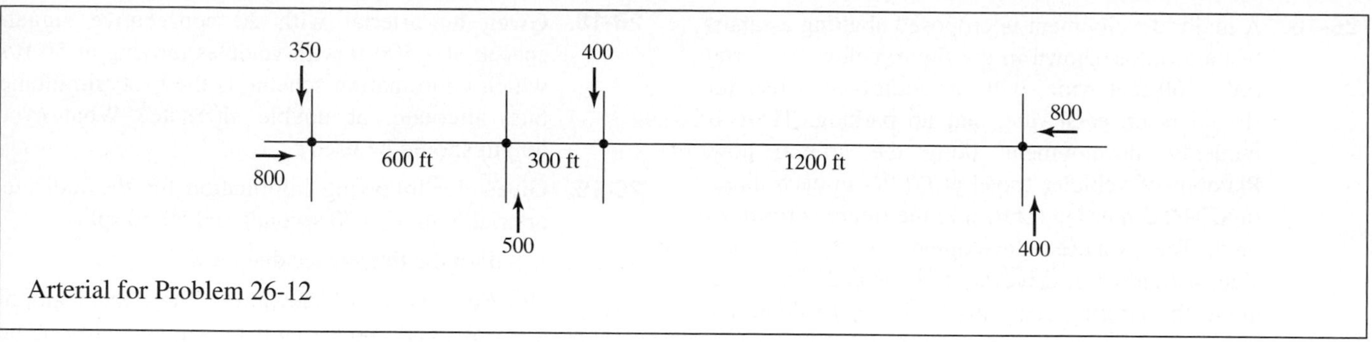

Arterial for Problem 26-12

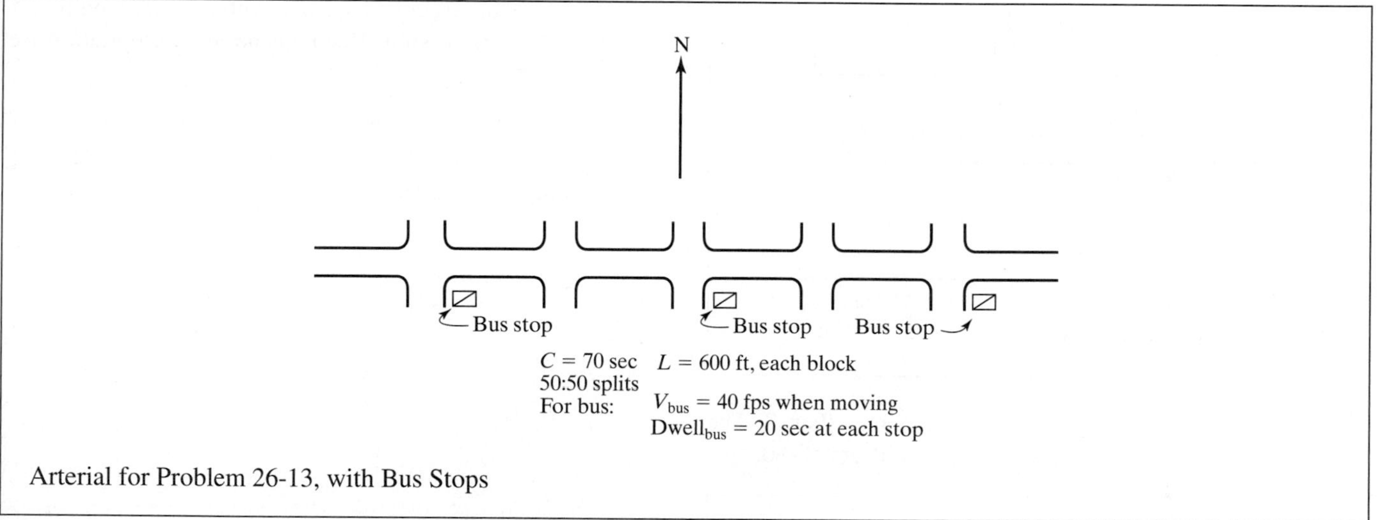

Arterial for Problem 26-13, with Bus Stops

26-14. Refer to the figure below. Second street is southbound with offsets of +15 seconds between successive signals. Third street is northbound with offsets of +10 seconds between successive signals. Avenue A is eastbound, with a +20-second offset of the signal at Second Street and Avenue A relative to the signal at Third Street and Avenue A. Given this information, find the offsets along Avenues B through J. The directions alternate, and all splits are 60-40, with the 60 on the main streets (2nd and 3rd Streets).

26-15. Given three intersections, spaced 600 feet apart, each with $C = 60$ seconds and 50-50 split, find an offset pattern that equalizes the bandwidth in the two directions. *Hint:* Set the first and the third relative to each other, and then do the best you can with the second intersection. This is a good way to start.

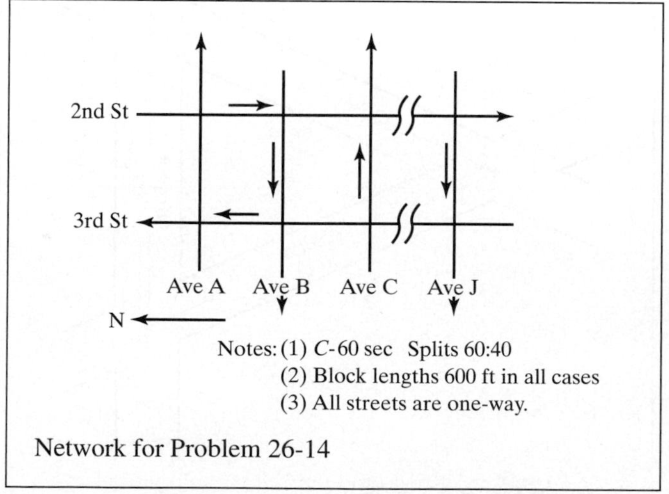

Network for Problem 26-14

26-16. A major development is proposed abutting a suburban arterial as shown in the figure below. The arterial is 60-feet wide, with an additional 5 feet for shoulders on each side, and no parking. There is moderate development along the arterial now. Platoons of vehicles travel at 60 ft/s in each direction. The center lane shown in the figure is for turns only. The proposed development is on the north side, with a major driveway to be added at 900 feet along the arterial, requiring a signal. Evaluate the impact of this development in detail. Be specific, and illustrate your points and recommendations.

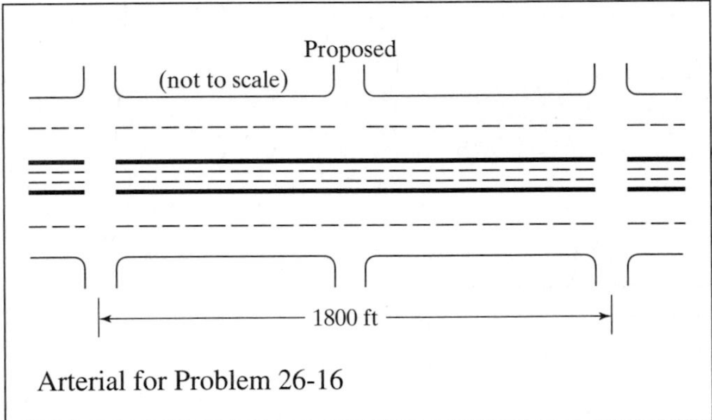

Arterial for Problem 26-16

26-17. Refer to the figure. Find the unknown offset X. The cycle length is 80 seconds. The splits are 50-50.

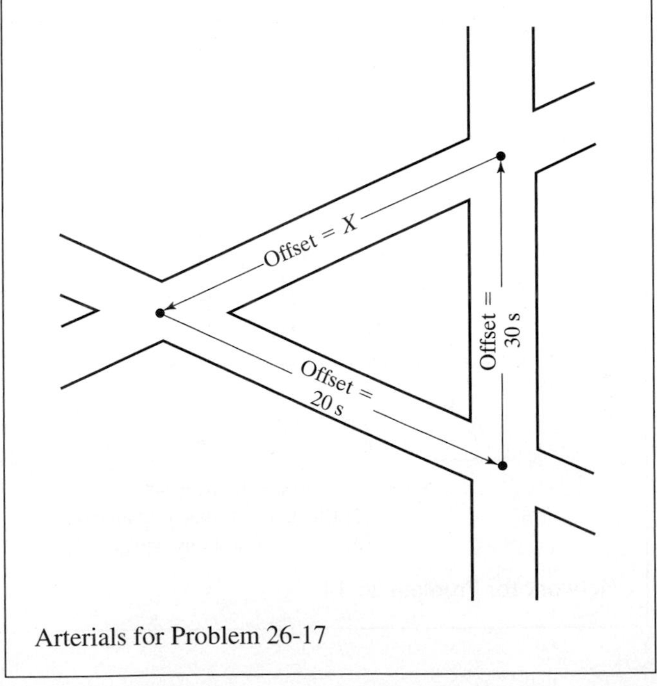

Arterials for Problem 26-17

26-18. Given an arterial with 20 consecutive signals, spaced at 1,500 ft with vehicles moving at 50 ft/s, which coordination scheme is the best: simultaneous, alternate, or double alternate? What cycle length should be used?

26-19. Given the following information for the indicated arterial, with $C = 70$ seconds and 50-50 splits:

(a) Plot the time-space diagram.

(b) Find the two bandwidths. Show them graphically and find the numeric values. If they do not exist, say so.

(c) An intersection is to be placed midway between Intersections 3 and 4, with $C = 70$ seconds and a 50-50 split. Recommend an appropriate offset.

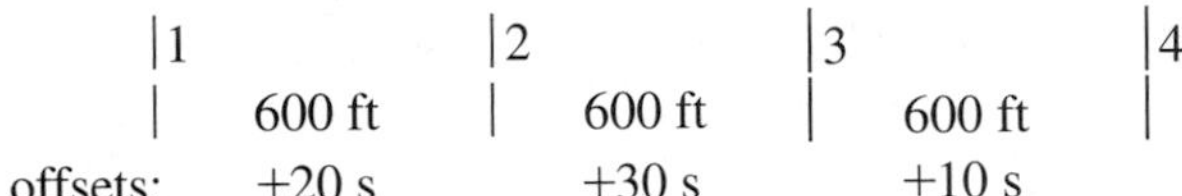

CHAPTER 27

Signal Coordination for Arterials and Networks: Oversaturated Conditions

It is well recognized that the oversaturated traffic environment is fundamentally different from the undersaturated environment. In undersaturated networks, capacity is adequate and queue lengths are generally well contained within individual approaches.

The oversaturated environment is characterized by an excess of demand relative to capacity ($v/c > 1.00$) and thus unstable queues that tend to expand over time. This tends to lead to situations in which queues spill back into the upstream intersections and block other flows. This in turn inhibits discharge on one or more approaches, in effect reducing capacity when it is most needed.

Control policies for oversaturated networks thus focus on maintaining and fully exploiting capacity to *maximize productivity* (vehicle throughput) of the system by controlling the inherent instability of queue growth.

27.1 System Objectives for Oversaturated Conditions

For undersaturated flow ($v/c < 1.00$) on arterials and in networks, the primary objective is to attain smooth flow as described in Chapter 26, often expressing this objective in terms of minimizing delay and stops. This primary objective tends to become the "prime directive" and even the obsession.

But when approaches, intersections, or systems become oversaturated, the mission and objective is *fundamentally different.* But many people do not change their mindset, and—even as this is written—most of the computational tools (i.e., signal optimization programs) do not address oversaturation directly.[1]

When networks are congested, the explicit objectives change to:

- *Maximize system throughput.* This is the primary objective. *I*t is achieved by (a) avoiding queue spill back, which blocks intersections and wastes green time; (b) avoiding starvation: the tardy arrival of traffic at the stop bar that wastes green time; and (c) managing queue formation to yield the highest service rate across the stop bar.
- *Fully utilize storage capacity (queue management).* This objective seeks to confine congested conditions to a limited area by managing queue formation in the context of a "feed forward" system.
- *Provide equitable service.* Allocate service to cross street traffic and to left-turners so that all travelers are serviced adequately and the imperative of traffic safety is observed.

Because intersection blockage can so degrade the network, its removal must be the keystone of the prime objective for the traffic engineer.

27.2 Root Causes of Congestion and Oversaturation

Reference [1] addressed the "root causes" of congestion and oversaturation, so that practitioners could put both the apparent problem *and* the physical limits on remedying it into perspective as they try to identify solutions.

Many of the root causes are imbedded in the fabric of our urban areas and the routes connecting them. Consider, for instance, the following list, extended from [1]:

- *Convergence of routes.* It is inevitable that congestion will arise as several routes converge on such natural features as river crossings (bridges, tunnels), paths out of valleys or between surrounding hills, paths to the waterfront, airports, or major attractors. Yet this convergence is imbedded in the history and evolution of our cities, sometimes over centuries or decades;
- *Crossings of major routes.* It tends to be unavoidable that two major arterials cross each other as they carry their respective through movements from the origins to destinations consistent with their design. It is also common that turn movements at these intersections of major routes tend to be significant as people move from one to the other. These intersections become "critical intersections" at which demand exceeds capacity, and there is no true "major" and "minor" street, rendering the adage of giving priority to the major street moot;
- *Natural features, historic sites, and special architecture.* It is common that topography (knolls, hills) have shaped the development of the area, that large parks or commons areas exist, and that monuments in town circles, large libraries or railroad stations, and, more recently, auto-free zones define where the flows are and the paths they must take from one critical intersection to another;
- *Street spacings.* They are also frequently defined by history and uses that predate motor vehicles. In some cases, they are defined by the original surveys (as in the western United States). It is truly serendipitous when street spacings fit the neat "alternate progression" needs implied by the $L/S = C/2$ relation of Chapter 26;
- *Unavailability of alternative routes.* The work in [1] uncovered the apparent paradox that medium-size cities and towns reported more perceived congestion that either their larger or smaller counterparts. A little probing revealed that small areas may have one main street with short peak "hours" and that large areas may have multiple routes between any origin and destination (subject to preceding points, of course) *but* that medium-size areas had significant traffic that still centered around one major arterial or street;
- *Natural variability.* The traditional view of both capacity and demand is that they are both known and invariant, at least for the period of the analysis.[2] But yet both are random around these expected values, leading to situations in which v/c fluctuates well above (and below) its nominal value.

[1]Specifically, they do not model or take into account the effects of spillback and intersection blockage. Reference [2] notes that "those methods seemed insensitive to even the existence of growing residual queues, let alone (consideration of them) in the optimization objective. This impression is supported by a cursory review of available tools. . . . No existing popular tools could be identified . . . that optimized specifically to maximize throughput or to manage queues."

[2]It is true that peaking is taken into account by the peak-hour factor (PHF), but this simply increases the value of the constant that is put into the steady-state equations that exist in the literature and key references.

This mixture of history, topography, convergence of routes, relative size, and natural variability creates an environment that virtually defines both the street system and the critical intersections. Added to this is the mixture of modes that are common in most urban areas (and suburban towns)—buses, paratransit, truck deliveries, and more recently package delivery services, bicycles, and a growing emphasis on pedestrian travel as a mode worthy of attention (and design) in its own right.[3]

27.3 Overall Approaches to Address Oversaturation

The *overall approach* stated in [1] as a set of logical steps was as follows:

- Address the root causes of congestion—first, foremost, and continually;
- Update the signalization, for poor signalization is frequently the cause of what looks like an incurable problem;
- If the problem persists, use novel signalization to minimize the impact and spatial extent of the extreme congestion;
- Provide more space, by use of turn bays and parking restrictions;
- Consider both prohibitions and enforcement realistically—is it effort likely to be effective or futile? Will it only transfer the problem to another location? Is it contrary to the fundamental use of the area?
- Take other available steps, such as right-turn-on-red, recognizing that the benefits will generally not be as significant as either signalization or more space;
- Develop site-specific evaluations where there are conflicting goals, such as providing local parking versus moving traffic, when the decision is ambiguous. Explicitly consider the solution in terms of economics.

The last category was intended (for instance) to focus the debate on the use of space, by quantifying the effects and tradeoffs—for example, use for good delivery, bus lane, parking, or through/turning traffic.

[3]The 2003 edition of the *MUTCD* required that pedestrian crossing times be computed based on the entire travelled path, rather than to center of the far lane. As of this writing, the 2010 edition of the *MUTCD* is likely to reduce the assumed pedestrian walking time from 4.0 ft/s to 3.5 ft/s, thereby increasing minimum phase durations.

The preceding list was constructed in [1] with some allowance for ease of implementation: It is generally easier to change signalization than to remove urban parking, it is generally easier to treat spot locations than entire arterials, and so forth.

Reference [2] expressed the approach in terms of *throughput strategies* (which it considered curative) and *queue management strategies* (which it considered palliative).

For curative *throughput strategies*, the experts consulted in the course of the Reference [2] work identified three categories or themes:

- Make the best use of the physical space available in the intersection.
- Make the best use of the green time in the cycle.
- Minimize the negative effects of other influences.

The specific techniques identified in [2] included:

- Work back from the downstream bottleneck;
- Run nearby intersections on a single controller, so that they stay in lock-step even during transitions;
- Improve the lane utilization;
- Run heavy left turns on a lag phase;
- Find and use the right cycle;
- Service heavy movements more than once in a cycle ("phase reservice");[4]
- Consider the effect of buses;
- Minimize the effect of pedestrian movements;
- Seek all possible available green time, even if exotic signal controller features must be used;
- Consider congested and uncongested movements separately;
- Prevent actuated short greens.

The more detailed advice in [2] on some of these includes (1) in general, it is good to set green times so that one bus per cycle can be serviced on relevant phases, when one considers bus dwell times; (2) pedestrian actuation buttons allow long phase minima to be avoided when the button is not actuated; (3) slow-starting trucks may lead to a short actuated phase, and gap settings have to consider this; (4) lanes get less productive over time, even with a standing queue, so that greens of up to 30 seconds are more productive than longer ones (indeed, it was noted that this is particularly true for through lanes next to left-turn bays, where movements into the turn bay actually leave gaps in the thru flow *if* the green time is longer than the bay length).

[4]This technique assumes there are some phases for which the v/c is less than 1.00, perhaps due to a combination of demand and minimum pedestrian requirements on phase duration.

For the palliative *queue management strategies,* the experts consulted in [2] identified techniques including:

- Reduce minor splits to encourage diversion (although it was noted that this generally does not work);
- Run nearby intersections on a single controller, so that they stay in lock-step even during transitions;
- Balance the queues for conflicting approaches;
- Prevent queues from spreading congestion;
- Meter traffic into the bottlenecks;
- Prevent downstream queues from backing up into the bottleneck intersections.

The approaches enumerated in [1] and [2] tend to be *operational* in nature, looking for short-term solutions to existing critical problems. There are indeed longer term remedies that can be addressed, mostly falling under the general heading of *transportation demand management.* These include:

- Assuring that alternative *higher occupancy modes* of transportation are available, such as bus routes, light rail, and even heavy rail. But in these alternatives, it is necessary to plan over the long term for *integrated* and *complete* systems, rather than fragmented routes;
- Encouraging *higher occupancy use* of automobiles by providing the support facilities (e.g., park and ride lots), lower tolls, special use lanes, and even incentives (e.g., parking or parking reimbursement);
- Encouraging *temporal shifts in demand* by incentives and disincentives (congestion pricing, for instance, can be viewed both ways), and by strategic planning with large employers.

To a large extent, these are excellent long-term measures that cannot and should not be ignored. The present chapter, however, focuses on shorter term operational issues—the result of inaction or inattention on the broader scale.

One long-term trend deserves special mention in this section. The historic model is of the "hub city" such as New York or Chicago in which the population surrounds the core in rings and travels to it in the journey to work. For several decades, census data have been showing a distinctly different trend, even in these prototypical areas: What would have been called "suburb-to-suburb" work trips are growing, and hub-bound travel is decreasing. The result can be depicted as a set of mini-cities within the same region, depending on true arterials and other facilities to link them. Concurrent with this, the directional patterns become more balanced, so that setting signals for "tidal flow" (80-20 directional splits) in the morning and evening becomes less typical.

27.4 Classification

NCHRP 194 [1] included the words "Three groups of terms must be defined" and then listed (1) congestion-related terms, (2) terms related to the types of oversaturation, and (3) characteristics distinguishing the productivity of an intersection and the perception of congestion. Table 27.1 shows the attempt of that era to define the congestion-related terms.

The question of classification of congestion has been addressed repeatedly in the literature. For instance, Reference [2] notes that [3] and [4] are two of the few works on metering, but that much work was not clearly related to actions that practitioners would take. Reference [2] also recommends a simplified and hierarchical classification, namely:

- Light traffic
- Moderate traffic
- Heavy traffic
- Oversaturated operations

It further notes that "Experts were not generally interested in defining the point of saturation in a precise or scientifically

Table 27.1: Early Definition of Congestion-Related Terms (c. 1978)

	Congested		
	Saturated		
Uncongested	**Stable**	**Unstable**	**Oversaturated**
No queue formation.	Queue formation, but not growing. Delay effects local.	Queue formation and growing. Delay effects still local. A transient state may be only of short duration.	Queue formation and growing to a point where upstream intersection performance is adversely affected.

(*Source:* Reference 1, p. 113.)

Table 27.2: Types of Oversaturation / Congestion

Types of Oversaturation	**Types of Congestion**
NCHRP 194 [1], 1978	**ITE Webinar, October 11, 2007 [5]**
Type I—The critical intersection (CI) has a smaller *g/C* ratio than does the upstream intersection. Type II—The CI and the upstream intersection have the same *g/C* ratio. However, the capacity of the CI is less than that of the upstream intersection because of factors other than the g/C ratio (such as turning movements and/or physical conditions). Type III—Heavy turn-in movements from the upstream cross street fill up the entire link or a significant part of it during a red phase on the arterial and cause spill back on the arterial. Type IV—This type of oversaturation of a CI results from the signal offset between the CI and its upstream intersection. Type V—This is defined as being a combination of two or more of Types I through IV oversaturation.	Type I—CI has less green time than upstream intersections (typical of CI at junction of crossing arterials). Type II—CI has lower capacity due to additional phases, geometrics, or backups from downstream intersection. Type III—CI has greater demand due to heavy turning at upstream intersection. Type IV—Congestion due to offset relationship with upstream intersection.

elegant way, but were rather interested in conditions that would justify a change in strategy or more detailed adjustment to operation on the ground." We agree with this classification and with the observation.

It was interesting that some of the thoughts from the cited earlier era (c. 1978) are still reflected in current practice and in professional development, including an ITE webinar [5] given in October 2007. Refer to Table 27.2 for an illustration.

These two events—separated by so many years—highlight the reality that engineers are coping with the same fundamental problems, often expressed in the same ways.

The needs of different modes sometimes reinforce each other in selecting a treatment. Whereas tradition often favors longer cycle lengths as demand grows, *shorter* cycle lengths (1) are advocated in [1] and in [5] in order to achieve shorter queues and more phases per hour, but they also (2) allow for lower pedestrian waiting time and better levels of service for pedestrians, an emerging priority in the multimodal environment. The case can also be made that *turn restrictions* aid more than one mode, and that *some bus measures* (e.g., stop location) can aid several modes.

27.5 Metering Plans

One short-term demand management strategy is *metering*. Three forms of metering can be applied within a congested traffic environment, characterized by demand exceeding supply (i.e., *v/c* deficiencies): internal, external, and release.

Internal metering refers to the use of control strategies within a congested network so as to influence the distribution of vehicles arriving at or departing from a critical location. The vehicles involved are stored on links defined to be part of the congested system under control, so as to eliminate or significantly limit the occurrence of either upstream or downstream intersection blockage.

Note that the metering concept does not explicitly minimize delay and stops, but rather it manages the queue formation in a manner that maximizes the productivity of the congested system.

Figures 27.1 and 27.2 show situations in which internal metering might be used: (1) controlling the volume being discharged at intersections upstream of a critical intersection (CI), thus creating a "moving storage" situation on the upstream links, (2) limiting the turn-in flow from cross streets, thus preserving the arterial for its through flow, and (3) metering in the face of a backup from "outside."

External metering refers to the control of the major access points to the defined system, so that inflow rates into the system are limited if the system is already too congested (or in danger of becoming so).

External metering is convenient conceptually because the storage problem belongs to "somebody else," outside the system. However, there may be limits to how much metering can be done without creating major problems in the "other" areas. Figure 27.3 shows a network with metering at the access points.

As a practical matter, there must be a limited number of major access points (such as river crossings, a downtown

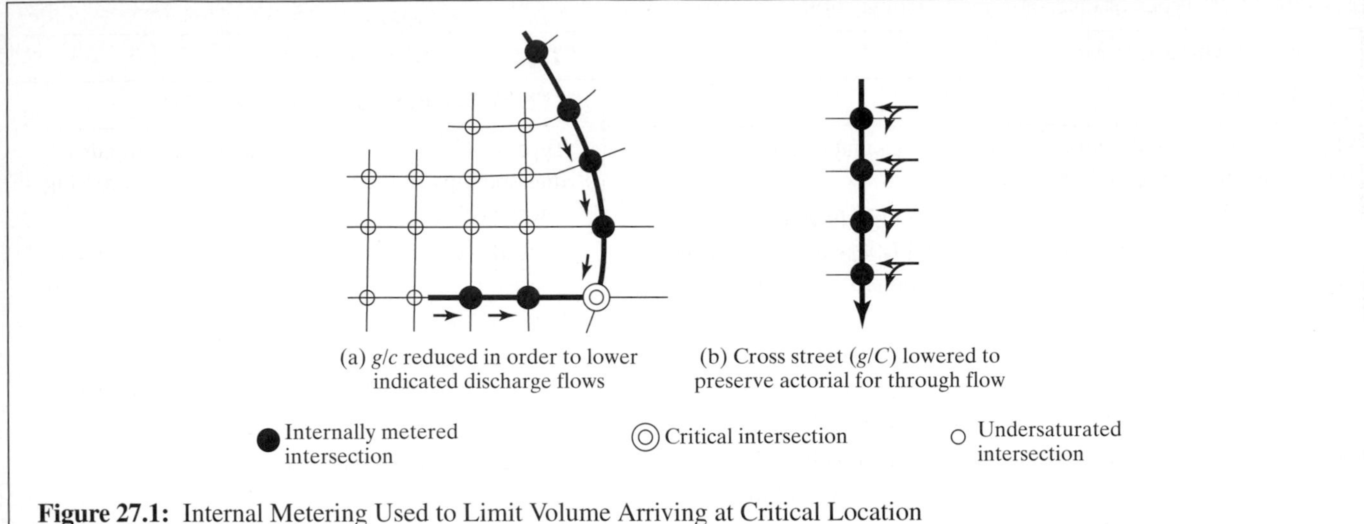

Figure 27.1: Internal Metering Used to Limit Volume Arriving at Critical Location

surrounded by water on three sides, a system that receives traffic from a limited number of radial arterials, etc.). Without effective control of access, the control points can potentially be bypassed by drivers selecting alternative routes.

Release metering refers to cases where vehicles are stored in such locations as parking garages and lots, from which their release can be controlled (at least in principle). The fact that they are stored "off street" also frees the traffic engineer of the need to worry about their storage and their spill-back potential.

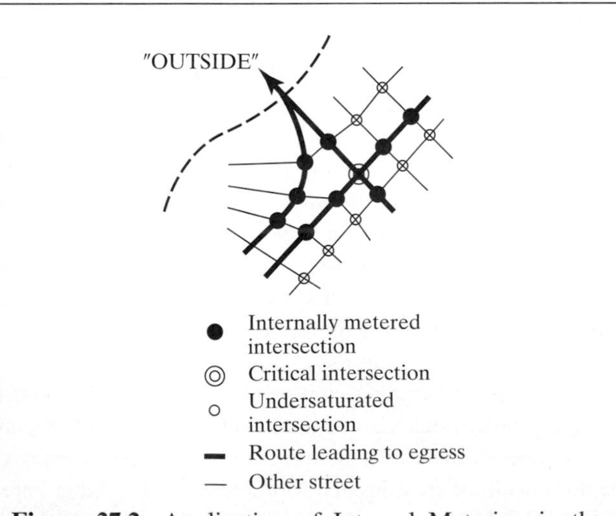

Figure 27.2: Application of Internal Metering in the Face of a Backup from "Outside"

Release metering can be used at shopping centers, mega-centers, major construction sites, and other concentrations. Although there are practical problems with public (and property-owner) acceptance, this could even be—and has been—a developer strategy to lower the facility's discharge rates so that adverse impacts are avoided.[5] Such strategies are of particular

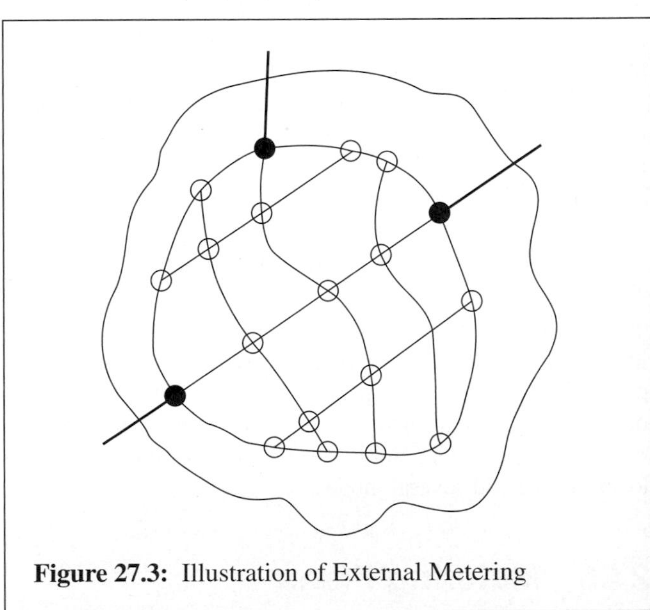

Figure 27.3: Illustration of External Metering

[5]Traffic impacts by developers leads to requirements for traffic mitigation, generally at the cost to the developer. The general subject is addressed in Chapter 30.

interest when the associated roadway system is distributing traffic to egress routes or along heavily congested arterials.

27.6 Signal Remedies

It is difficult to overstate how often the basic problem is poor signalization. Once the signalization is improved through reasonably short cycle lengths, proper offsets (including queue clearance), and proper splits, then many problems disappear. Sometimes, of course, there is just too much traffic. At such times *rapid adjustment to splits* to meet short-term relative demand changes (i.e., in competing directions), *equity offsets* to lower the likelihood of spill back and to aid cross flow, a *different concept of splits*, namely to manage the spread of congestion, and *phase reservice* may be appropriate if other options cannot be called on. These options may be used as distinct treatments or as part of a *metering plan* (addressed in the previous section).

27.6.1 Responsive/Adaptive Phase Duration Changes

Locations such as college entrances have short bursts of inflows followed by short bursts of outflows (both in the order of 15 to 25 minutes), directly related to their class schedules. Control in such cases must adapt to the rapidly changing demand, in order to avoid precipitating oversaturation that can promulgate and perpetuate itself.

27.6.2 Shorter Cycle Lengths

An earlier chapter demonstrated that increasing the cycle length does *not* substantially increase the capacity of the intersection (a change of +50% in cycle length could add +5% to 8% in capacity, under favorable conditions). But the favorable conditions include maintaining high discharge rates over long phases, and this does not occur in practice.

Rather, the emphasis should be on the related reality that as the cycle length increases, so do the lengths of stored queue and the length of the discharged platoons, which then arrive at downstream intersections that may have shorter cycle lengths and cannot be stored or processed easily.

Thus the likelihood of intersection blockage increases, with substantial adverse impacts on system capacity. This is particularly acute when short link lengths are involved.

Note that a critical flow of v_i veh/hr/lane nominally discharges v_i *C/3600* vehicles in a cycle. If each vehicle requires D feet of storage space, the length of the downstream link in a congested environment (assuming the downstream signal can process the queue in one cycle but that it will be forced to stop) would have to be:

$$L \geq \left(\frac{v_i C}{3600}\right) \text{D} \qquad (27\text{-}1)$$

where L is the *available* downstream space in feet. This "available" space may be the full link length or by some lower value, perhaps 150 feet less than the true length (to keep the queue away from the discharging intersection or to allow for turn-ins). The engineer will have to decide that, noting that [1] shows results that queues that occupy more than 85% of the link length actually inhibit the discharge rate at the upstream intersection.

Equation 27-1 may be rearranged as:

$$C \leq \left(\frac{L}{D}\right)\left(\frac{3600}{v_i}\right) \qquad (27\text{-}2)$$

Note that v_i in this case is the discharge volume per downstream lane, which may differ from the demand volume, particularly at the fringes of the "system" being considered. Refer to Figure 27.4 for an illustration of this relationship. Note that only rather high flows (≥800 veh/h/ln) and short blocks will create very severe limits on the cycle length. However, these are just the situations of most interest for conditions or extreme congestion. *Note that the discharge volume v_i depends on the upstream demand and (g/C) allocation,* and that this analysis really has to be carried along the arterial.

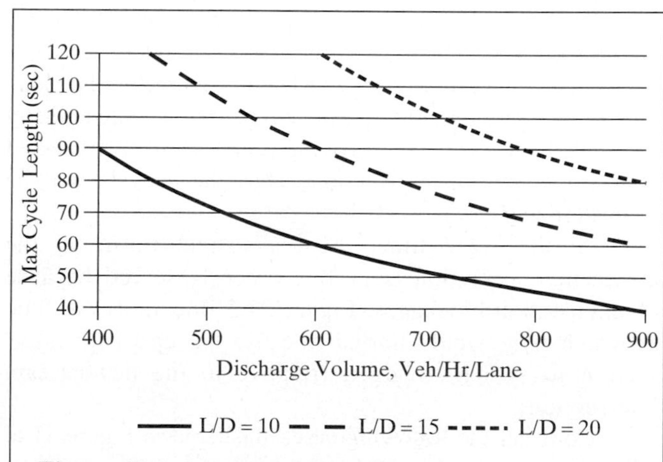

Figure 27.4: Maximum Cycle Length as a Function of Block Length[6]

[6]It is recognized that cycle lengths shorter than 60 seconds are rare, but they are shown to emphasize the shape of the curves.

In Figure 27.4, note that D equals approximately 25 feet, so that block lengths with *available* storage of 250 feet to 500 feet are represented (300 to 600 feet, if the 85% rule of thumb cited earlier is used).

The important lesson from Figure 27.4 is that shorter cycle lengths are not only good but *necessary* to manage the size of queues arriving in downstream links.

This analysis assumes that the downstream link can itself discharge the arriving queue in one cycle. To achieve this (for instance, at a critical intersection), it may be necessary to allow the downstream capability determine the upstream discharge, which would have to be achieved by reassigning green time there (to minor movements) or imposing an all-red (i.e., metering). Failing this, the downstream queues will grow and other measures will be needed to avoid spill back.

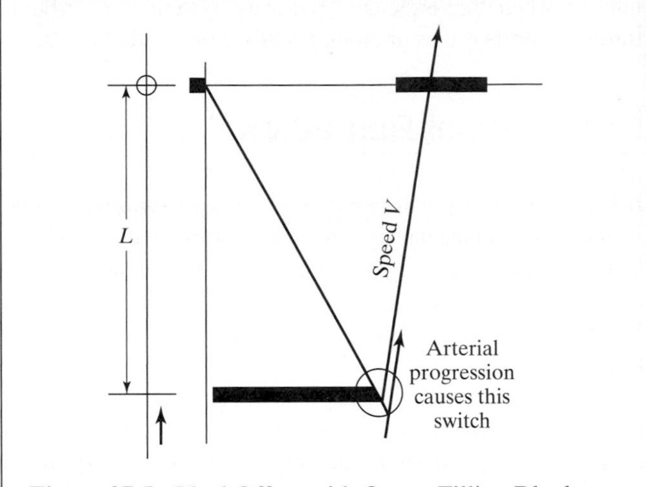

Figure 27.5: Ideal Offset with Queue Filling Block

27.6.3 Equity Offsets

Offsets on an arterial are usually set to move vehicles smoothly along the arterial, which is logical. If no queues exist on the arterial, the ideal offset is L/S, where L is the signal spacing in feet and v is the vehicle speed in feet per second. If a queue of Q vehicles exists, the ideal offset t_{ideal} is:

$$t_{ideal} = \frac{L}{S} - Q * h \qquad (27\text{-}3)$$

where h is the discharge headway of the queue, in seconds. Clearly, as Q increases, t_{ideal} decreases, going from a "forward" progression to a "simultaneous" progression to a "reverse" progression.

Unfortunately, as the queue length approaches the block length, such progressions lose meaning, for it is quite unlikely that both the queue *and* the arriving vehicles will be processed at the downstream intersection. Thus the arrivals will be stopped in any case.

At the same time, the cross-street traffic at the upstream intersection is probably poorly served because of intersection blockages. Figure 27.5 illustrates the time at which t_{ideal} would normally cause the upstream intersection to switch to green (relative to the downstream intersection).

Consider the following case, illustrated in Figure 27.6: Allow the congested arterial to have its green at the upstream intersection until its vehicles just begin to move. Then switch the signal so these vehicles flush out the intersection but no new vehicles continue to enter.

At the same time, this gives the cross-street traffic an opportunity to pass through a clear intersection. This concept, defined as *equity offset*, can be translated into this equation:

$$t_{equity} = g_i C - \frac{L}{S_{acc}} \qquad (27\text{-}4)$$

where g_i is the *upstream* main street (i.e., the congested intersection) green fraction and S_{acc} is the speed of the "acceleration wave" shown in Figure 27.7.

A typical value is 16 fps. Comparing Figure 27.5 and 27.7, it is clear that equity offset causes the upstream signal to go red just when "normal" offsets would have caused it to switch to green in this particular case. This is not surprising, for the purpose is different: Equity offsets are intended to be fair (i.e., equitable) to cross-street traffic.

Simulation tests using a microscopic simulation model have shown the value of using equity offsets: Congestion does not spread as fast as otherwise and may not infect the cross streets at all.

Figure 27.8 (a) shows a test network used to test the equity-offset concept. Link 2 is upstream of the critical intersection (CI). For the demands and signal splits shown it is likely to accumulate vehicles, with spill back into its upstream intersection likely. If this occurs, the discharge from Link 1 will be blocked and its queue will grow. In the extreme, congestion will spread.

The equity offset is computed as

$$t_{equity} = (0.60)(60) - \frac{600}{16} = -1.5\,s$$

using Equation 27-4. (At 25 feet per vehicle and a platoon speed of 50 ft/s, Equation 27-3 would have yielded

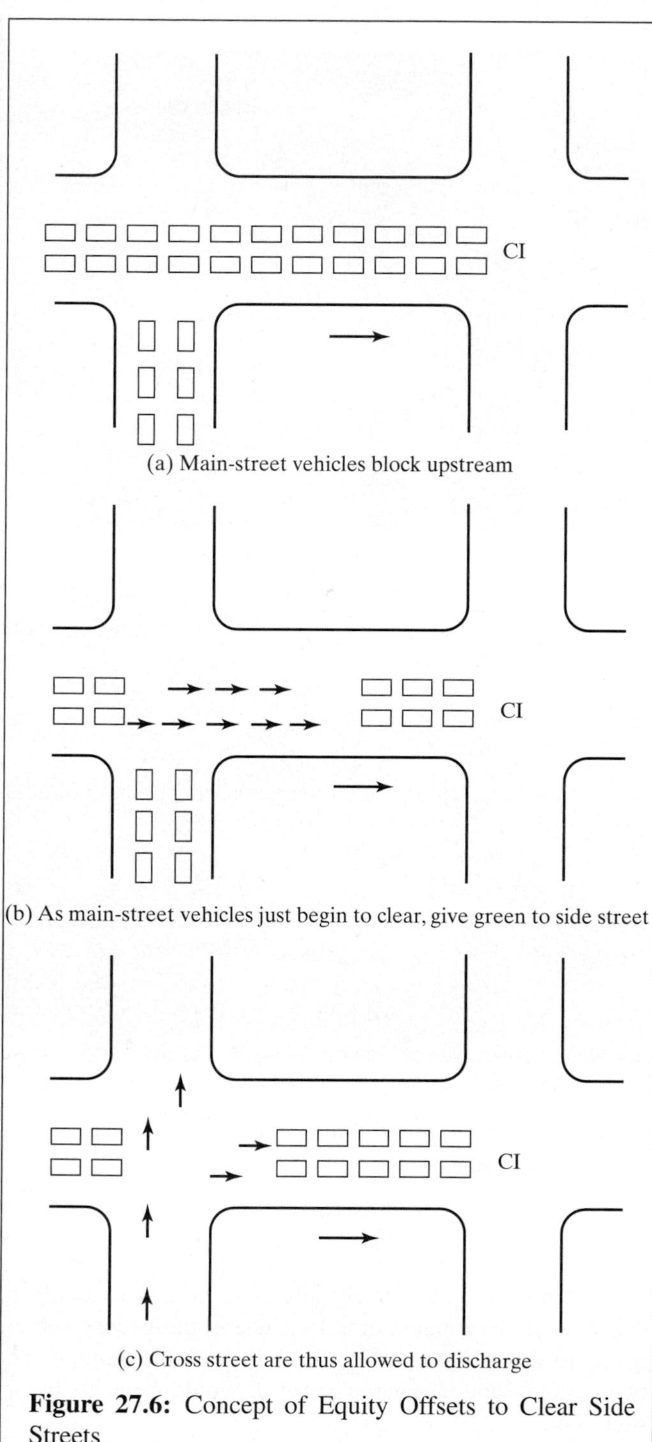

Figure 27.6: Concept of Equity Offsets to Clear Side Streets

t_{ideal} = (600/50) − (24)(2)= −36 seconds for progressed movement. Of course, progressed movement is a silly objective when 24 vehicles are queued for 30 seconds of green.)

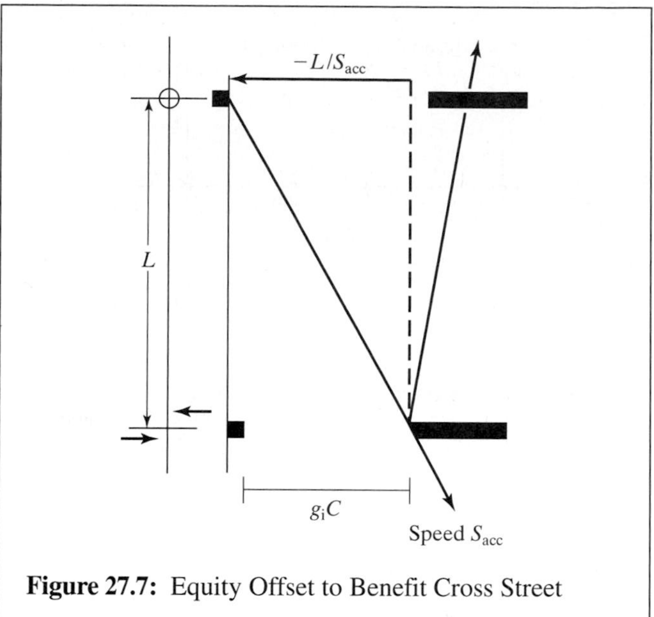

Figure 27.7: Equity Offset to Benefit Cross Street

Figure 27.8 (b) shows the side-street queue (i.e., the Link 1 queue) as a function of the main-street offset. Note that an offset of –36 seconds is the same as an offset of +24 seconds when C = 60 seconds, due to the periodic pattern of the offsets. Figure 27.8 (b) shows the best result for allowing the side street to clear when the equity offset (offset = −1.5 seconds) is in effect, and, in this case, the worst results when the queue-adjusted "ideal offset" (offset = 24 seconds) would have been in effect.

The preceding discussion assumes that the cross-street traffic does not turn into space opened on the congested arterial. If a significant number of cross-street vehicles do turn into the arterial, a modification in the offset is appropriate to assure that the upstream traffic on the congested arterial also has its fair share.

The equity offset concept has been used to keep side-street flows moving when an arterial backs up from a critical intersection (CI). It may also be used to keep an arterial functioning when the cross streets back up across the arterial from their critical intersections.

Figure 27.9 shows another illustration of the concept that offsets can be designed to relieve congestion and the likelihood of upstream intersection blockage.

27.6.4 Imbalanced Split

For congested flow, the standard rule of allocating the available green in proportion to the relative demands could be used, but it

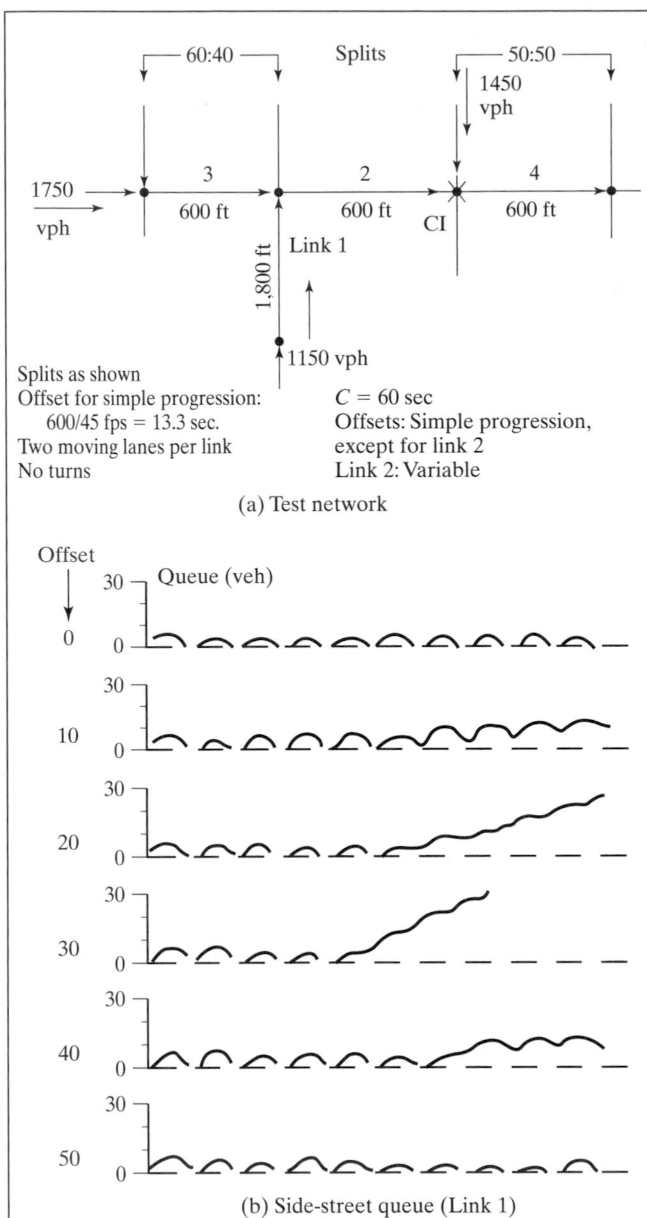

Figure 27.8: Equity Offsets Avoid Side-Street Congestion, Despite Spill Backs

(a) High Demand and "Traditional" Offsets, with Consequent Congestion

(b) The Same High Demand but an Offset Tailored to Congestion Relief

Figure 27.9: Another Illustration of Offsets Used to Avoid Spill Back

does not address an important problem. Consider the illustration of Figure 27.10. If the prime concern is to avoid impacting Route 347 and First Ave. (but with little concern for the minor streets in between, if any), it is not reasonable to use a 50:50 split.

Considering that the relative storage available is 750 feet in one direction and 3000 feet in the other, and we wish neither to be adversely affected, the impact could be delayed for the longest time by causing the excess-vehicle queue to grow in proportion to their available storage. The two critical-lane discharge flows f_i would have to be set such that:

$$\frac{d_1 - f_1}{d_2 - f_2} = \frac{L_1}{L_2} \tag{27-5}$$

and:

$$f_1 + f_2 = CAP \tag{27-6}$$

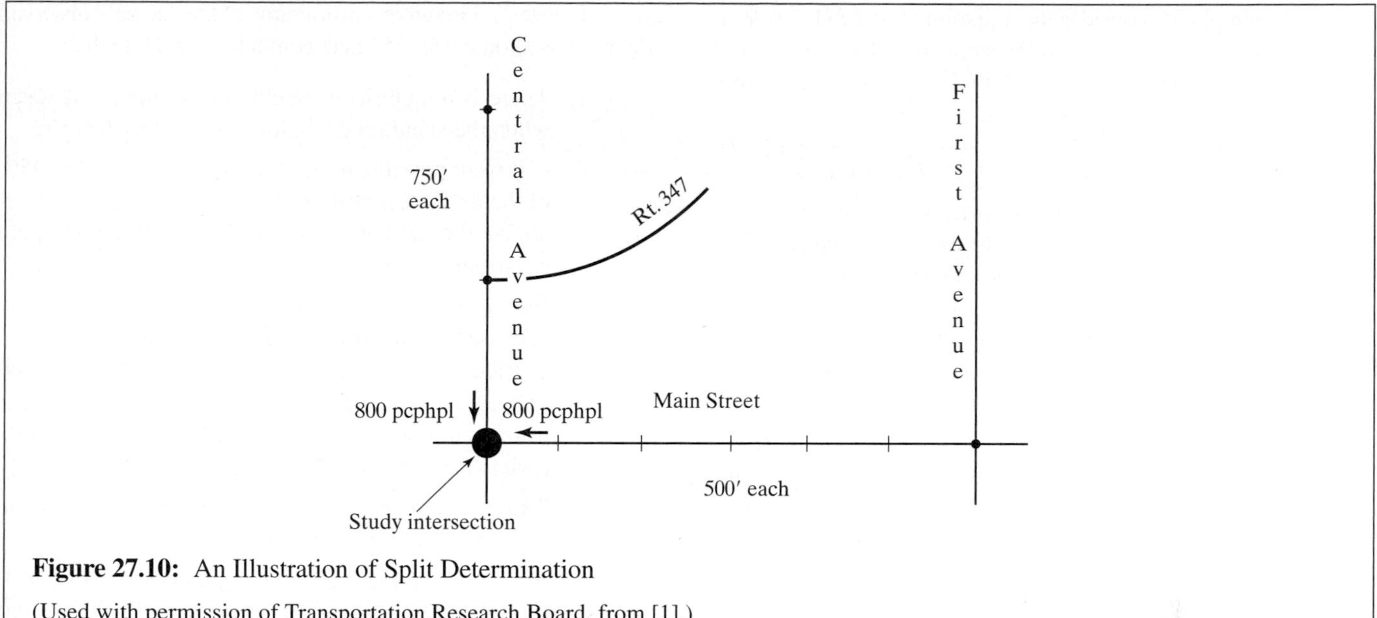

Figure 27.10: An Illustration of Split Determination

(Used with permission of Transportation Research Board, from [1].)

where d_i are the demands (veh/h/ln), L_i is the storage and CAP is the sum of the critical-lane flows (i.e., the capacity figure).

For the illustrative problem, using CAP = 1550 veh/h/ln, the preceding equations result in $f_1 = 954$ veh/h/ln and $f_2 = 759$ veh/h/ln, where direction 1 is the shorter distance. This is a 56:44 split.

Note that in the extreme, if only one direction has a cross route that should not be impacted, much of the green could be given to that direction (other than some minimum for other phases) in order to achieve that end.

27.6.5 Phase Reservice

The term *phase reservice* refers to servicing important phases more than once in a cycle, by going back to them, generally to the disbenefit of side street movements on other phases. The technique is for clearing queues on protected lefts and saturated approaches, but it generally requires that there are undersaturated phases at the intersection, so that one can "catch up" with servicing them on a future cycle if necessary.

Phase reservice can aid the basic objectives of maximizing throughput and queue management. It does require that both drivers and pedestrians become familiar with this sort of operation, so that all concerned are aware that the "normal" sequence of phases cannot be counted on.

27.6.6 Pedestrian Minima Provided Only Upon Request

In areas in which there are relatively little pedestrian traffic, satisfying the pedestrian minimum crossing times on all approaches may lead to phases that waste green time and to longer-than-necessary cycle lengths. In such cases, traffic engineers sometimes do consider invoking pedestrian minima only when there is an actual pedestrian actuation of a pedestrian button.

This of course requires that the intersection have functioning pedestrian buttons in place (as well as pedestrian signals) and that the pedestrians learn that the only way to assure the pedestrian crossing times is to *use* the buttons.[7]

27.7 Variations in Demand and Capacity

In current (and historic) methods, traffic demand and the capacity of the traffic signal are taken as fixed values, rather than intrinsically random.

[7]Although there is no factual basis to support the supposition, the authors do speculate that this environment might be an ideal application of the pedestrian "countdown" signals that are coming into use. The pedestrian is rewarded by the action of pushing the button having a definitive and highly visible effect, namely the pedestrian signal changing *and* providing information on time remaining.

For simplicity, consider the equation D = 1 / (1 – {*v*/*c*}), a very oversimplified version of the standard delay equation but sufficient for the illustration. In probability theory, it is clear that

$$E[D] \neq \frac{1}{1 - E[v/c]} \neq \frac{1}{1 - (E[v]/E[c])} \tag{27-7}$$

That is, the expected value of a quantity such as delay *cannot* be computed by simply putting the expected values of the input values into the right-hand side of a static equation.

For years, such an approach was *good enough*, considering (1) the amount of data available, (2) the cost of data, and (3) the need for results that were good enough for the applications at hand.

Still, it was always known there were problems—delay data had very high variability, particularly as the *v*/*c* ratio increases (plots of delay versus *v*/*c* have trends, but also high scatter of the data at high *v*/*c*). In addition, it was known that some data (maybe much data) represented the transition between two demand levels, rather than "steady-state" data from different equilibrium conditions. But the need for data—and the cost of it—often competed with such precision. And, besides, the results were *good enough* for the applications at hand.

27.7.1 An Illustration of the Effects of Demand and Capacity Variability on Delay

For present purposes, we consider the case in which the expected value of demand is 1700 veh/h and of capacity is 1900 veh/h. The computations of delay are illustrative, and we do not list values for all of the factors in the *HCM*. The computed *v*/*c* ratio is 0.89 in this case, and the computed delay is 51.0 s/veh.

Table 27.3 shows the same computation but with normally distributed values of demand and capacity. A total of 525 samples, or "observations," were generated in a spreadsheet for this illustration.

Figure 27.11 contains histograms of the "actual" observed delays, based on the spreadsheet computations. Note that:

1. There is a significant fluctuation in the delay, even when the standard deviation is as low as 30 vph;
2. If it were possible to use the rule of thumb that 95% of the delay observations fall within 2 standard deviations, the estimated range is 24.4 to 85.2 vph, quite a variation around an observed mean of 54.8 vph;
3. But it is *not* possible to apply this rule of thumb because it is only truly valid for the symmetric normal distribution. As Figure 27.11b shows, when $\sigma = 60$ vph, the delay distribution is *not* symmetric but rather has a long tail—*and* a noticeable set of *apparently* anomalous very high delays.
4. Thus the common rule of thumb would *underestimate* the spread in asymmetric cases.
5. As to the "apparent" anomaly cited in number 3, Figure 27.12 shows the range of *v*/*c* ratios that existed in Case 2. Although the average was close to the 0.89 computed from averages, the *range* is from about *v*/*c* = 0.80 up to over 1.00. These higher values in particular would lead to very long delays, if the intersection approach were long enough to receive and store the vehicles.

We did additional analyses that showed that even when $\sigma = 30$ s/veh, the asymmetric distribution occurs when {*v*/*c*} is higher. The long-tailed delay distribution of Figure 27.11.b can be expected routinely and must be taken into account.

27.7.2 Practical Implications

Clearly, high {*v*/*c*} ratios should be avoided, to avoid the large standard deviations demonstrated in even this simple illustration.

Table 27.3: An Illustrative Study

	demand	**capacity**	**est avg v/c**	**est std v/c**	**delay based upon expected values**	**related std**
mean=	1700	1900				
std=						
> Traditional	0	0	0.89	0.00	51.0	
> Case 1	30	30	0.90	0.02	54.8	15.2
> Case 2	60	60	0.90	0.04	70.5	73.8
					sec/veh	sec/veh

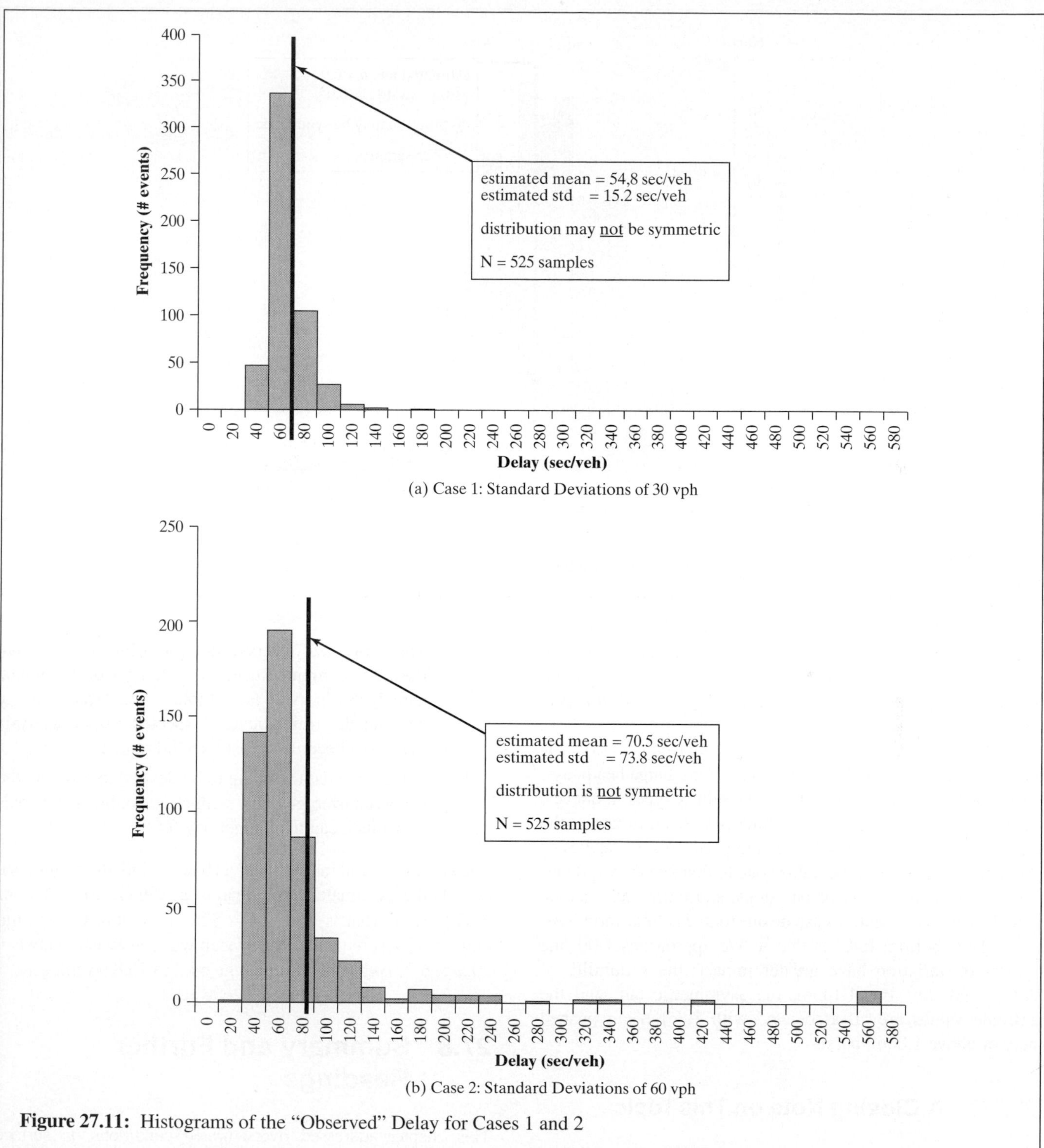

Figure 27.11: Histograms of the "Observed" Delay for Cases 1 and 2

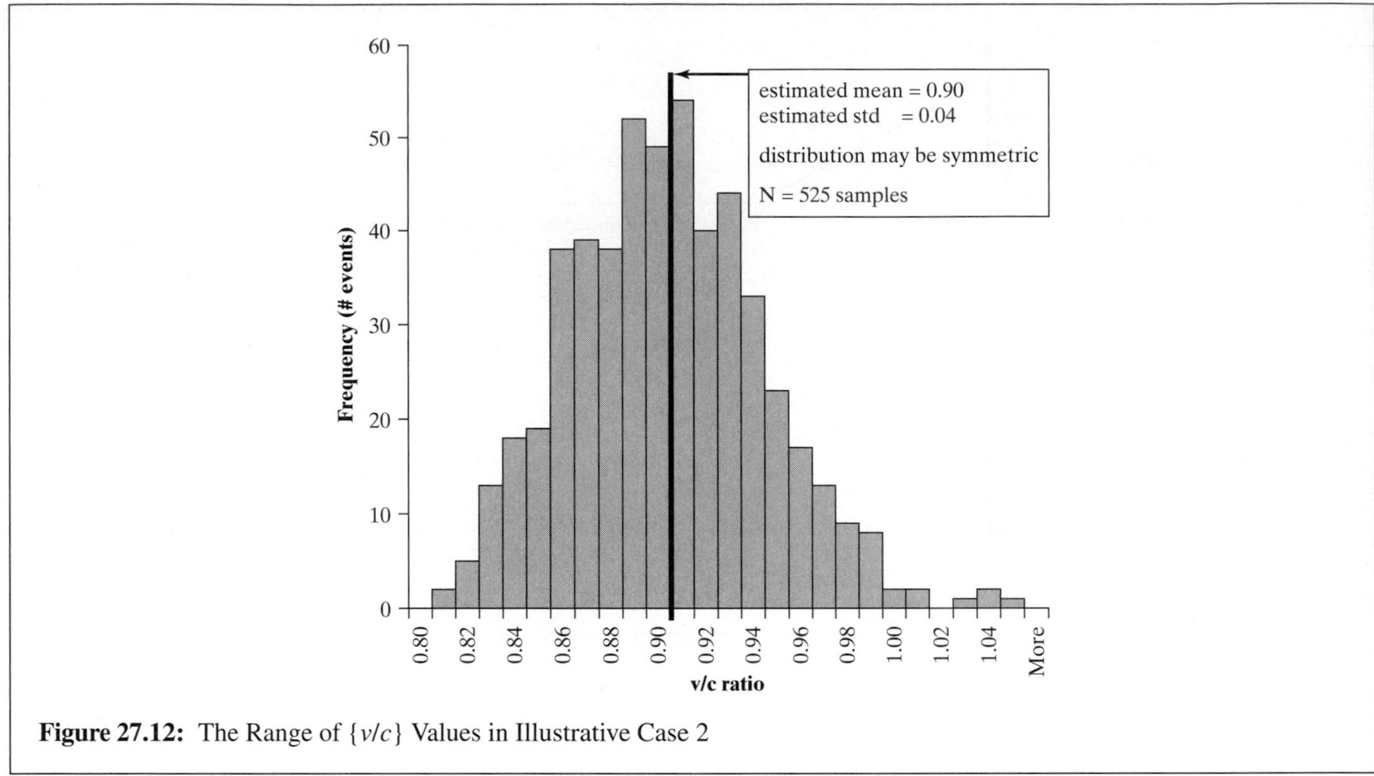

Figure 27.12: The Range of $\{v/c\}$ Values in Illustrative Case 2

But the subject of this chapter is the oversaturated condition, which occurs when v/c exceeds 1.00 systematically—or when randomness pushes it there, and the inability to "catch up" perpetuates it.

Capacity variations can be due to such factors as (a) "natural" variations due to the randomness of discharge headways, loss times, the percentage of different vehicle types in queues, the percentage of turns in a given time period, left-turners at the head of a queue in a shared lane, and even (b) singularities or "outliers" in some of these values due to double parkers, deliveries, buses hanging out of bus stops, and other such factors. Local "demand" variations can be due to true natural variations.

The "bottom line" is that as v/c approaches 1.00, the effects of variation have greater impact: the variability of delay (and thus travel time), the asymmetric tail, and the extreme values are all associated with $\{v/c\}$ being pushed near or above 1.00 in bursts.

27.7.3 A Closing Note on This Topic

Figure 27.11b results in an (estimated) average observed delay of 70.5 s/veh with the 525 observations, which raises two troublesome issues:

- The "observed" average is quite a bit different than the value computed using the averages of the inputs, namely 50.1 s/veh (see Table 27.3, "Traditional"), and the deterministic or "expected value" analysis may have been used past its valid range;
- How much data is going to be desired to estimate the observed average delay with good confidence, versus how much can the budget pay for?

Because the spread in Figure 27.11b is so high, the confidence bound on the estimate of the mean is $\pm\ 1.96\,(73.8)/\sqrt{525}$, or $\pm$ 6.3 s/veh. That is, after N = 525 observations, we could only be 95% sure that the true mean was somewhere between 70.5 $\pm$ 6.3 s/veh (i.e., in the range 64.2 to 76.8 s/veh).

27.8 Summary and Further Readings

This chapter addressed oversaturated conditions on surface streets. The problem of congestion and saturation is widespread and is not often approached consistently. Definite measures can be taken, but preventive action addressing the root causes must be given a high priority. Among the possible measures, those relating to signalization generally can have the greatest impact. The nonsignal remedies are in no way to be minimized, particularly those that provide space, whether for direct productivity increases or for removing impediments to the principal flow.

A wealth of information is available at the TRB Traffic Signal Systems Committee Web site [6], including the set of presentations at the Committee's January 2007 Workshop on "Operating Traffic Signal Systems in Oversaturated Conditions" and its July 2006 Mid-Year Meeting that focused on oversaturation. This material was also revisited in the context of this proposal. In the interest of space, the individual materials are not enumerated or commented on at this time. It is refreshing to see the number of case studies of actual applications that appear in this compendium of information.

To assist engineers and planners in applying HCM methodologies, the HCM 2000 includes default values for many of the more difficult-to-obtain input parameters and variables. Field measured default values for use in applications of the HCM were assembled in the research performed in NCHRP Report 599 [7], *Default Values for Highway Capacity and Level of Service Analyses*. Of the 63 default values used in the HCM 2000, that research found 19 values that had a high degree of impact on the service measures. For the HCM 2000 Urban Streets analysis, signal density was found to have a high impact. For signalized intersections, the peak-hour factor, length of analysis period, arrival type, adjusted saturation flow rate, lane width, percent heavy vehicles, and lane utilization were found to have high impact on the estimate of delay. These high-impact variables could also be considered for their impact on the variability in demand and capacity.

References

1. McShane, W.R., et al., "Traffic Control in Oversaturated Street Networks," National Cooperative Highway Research Program, (NCHRP) Report 194, Transportation Research Board, National Research Council, Washington, DC, 1978.
2. Denney, R.W., Head, L., Spencer, K., "Signal Timing Under Saturated Conditions," FHWA Office of Operations, FHWA-HOP-09-008, November 2008.
3. Lieberman, E., and Messer, C., "Internal Metering Policy for Oversaturated Networks," Final Report, NCHRP 3-38(4), 1992.
4. Lieberman, E.B., Chang, J., and Prassas, E.S., "Formulation of Real-Time Control Policy for Oversaturated Arterials," *Transportation Research Record 1727,* TRB, National Research Council, Washington DC, 2000.
5. "Signal Timing for Congested Conditions," ITE Professional Development Program, October 11, 2007 webinar, presented by Woody Hood, Maryland DOT.
6. TRB Traffic Signal Systems Committee (TRB Committee ANB25) Web site, http://www.signalsystems.org.vt.edu/
7. Zegeer, J., Vandehey, M., Blogg, M., Nyguen, K., and Ereti, M., *Default Values for Highway Capacity and Level of Service Analyses*, NCHRP Report 599, Transportation Research Board, National Research Council, Washington, DC, 2008.

CHAPTER 28

Analysis of Streets in a Multimodal Context

Planning and design of urban streets has changed drastically over the past two decades. Twenty years ago, the mode that led the design of an urban street was the automobile. Even urban streets that moved large numbers of buses every day were planned for automobiles, without much consideration for the buses except for providing a bus stop. Priorities have changed, however, because of federal funding acts promoting alternative modes of travel. In 1991, the Intermodal Surface Transportation Efficiency Act (ISTEA) was passed. This act totally overhauled the federal aid highway system. ISTEA and its successors (TEA21 and SAFETYLU) encouraged multimodality for the first time, and they set aside significant amounts of money for promoting alternative modes of transportation. Nowadays, arterial planning initiatives should lead toward an integrated urban environment that is safer overall, encourages multimodal use, and balances the use of streets. On an arterial, the usual modes recognized are automobile, transit, pedestrian, and bicycle. Automobiles are going to continue to be the major mode on arterials in the United States, but multimodal initiatives can often be done without costing needed vehicular capacities and can often make the movements of vehicles better defined and safer for all modes, including vehicles.

28.1 Arterial Planning Issues and Approaches

The current literature on arterial planning focuses on planning for "complete streets," that is, streets that are multimodal, that will lessen the impact of traffic on their community's streets to promote safer, more livable, and "sustainable" environments for all users [*1–3*]. Consider as an example a typical urban street with two lanes per direction serving automobiles, buses, bicycles, and pedestrians. There is parking on both sides and sidewalks for the pedestrians. In this "before" condition, drivers often have to do broken-field running, that is, drivers would move around left-turn queues in the "left-through" lane, and double-parked cars and buses in the "right-through" lane causing frequent lane changes and conflicts. Further, there is no well-defined bicycle path. In the "after" condition of a well-planned "complete-street" initiative, there are well-defined bicycle lanes, buffers between buses and cars, left-turn pockets, and single lanes for through movements. The relevance of this initiative is that it is a long-term infrastructure, with new goals to determine its usefulness. When making this type of change, some of the questions that should now be asked are:

1. What modes, if any, have really lost capacity that was used effectively?
2. What modes, if any, gained capacity or become feasible?
3. What is the new mix of demand levels (by mode) that are feasible with the new design?
4. What are the estimated safety improvements, based on (a) actual usage in the after condition and (b) potential usage in the after condition, both compared to like volumes (by mode) in the "before" design?
5. How does the project add to a future integrated system?

Figures 28.1 and 28.2 show a street before and after it was redesigned to be a complete street. In Figure 28.1, there are two lanes in each direction for traffic and a parking lane. There is no defined bicycle path for the cyclist who will then use the right-hand lane for traveling. In Figure 28.2, the automobiles are given only one lane for travel on the street; however, at the intersection there is a left-turn bay so that through vehicles are not trapped waiting behind a left-turn vehicle. There is also a defined bicycle path, which has the added benefit of discouraging cars from double parking on the street.

28.2 Multimodal Performance Assessment

There are standard methods for assessing the performance of each mode on an arterial and these are summarized here. The chapter does not present the details of each model but rather just discusses the variables that are important for each mode. Many of the variables are common to more than one mode, which allows an analyst to see how changing a variable will

Figure 28.1: Ninth Street in Brooklyn, New York, before Redesign as a Complete Street
(*Source:* http://gowanuslounge.blogspot.com/2007/04/dot-moving-ahead-with-ninth-street.html.)

Figure 28.2: Ninth Street in Brooklyn, New York, after Redesign as a Complete Street
(*Source:* http://gowanuslounge.blogspot.com/2007/04/dot-moving-ahead-with-ninth-street.html.)

affect each mode differently. Currently, the 2000 HCM [*5*] is not a multimodal document. This, however, will be changing when the next edition is published, based on research done under NCHRP 3-70, titled "Multimodal Level of Service Analysis for Urban Streets." Reference 6 contains the results of that research, which provides an integrated multimodal methodology for analyzing an urban arterial. This type of methodology provides a way for transportation professionals to incorporate smart-growth principles and context-sensitive design into their evaluations. A multimodal level-of-service (LOS) analysis simultaneously evaluates the quality of service of each of the four modes on the arterial: automobile, bicycle, pedestrian, and transit. By doing a multimodal analysis, the analyst can see how improving one mode may benefit or harm another mode. Each mode is assigned its own LOS rating, but there is no overall LOS that describes the arterial as a whole or weights the individual modes together. All four models produce LOS letter grades from their respective models. Each of the models output a numerical value that is converted to a letter grade as shown in Table 28.1.

Table 28.1: Level of Service Grades for Multimodal Analysis

Numerical Value Output from LOS Models	LOS Letter Grade
≤2.0	A
>2–≤2.75	B
>2.75–≤3.50	C
>3.50–≤4.25	D
>4.25–≤5.00	E

(*Source:* NCHRP Report 616 [*6*].)

28.2.1 Bicycle Level of Service

The bicycle LOS methodology was originally developed by Sprinkle Consulting, Inc. (SCI) [*7*] and was enhanced in the NCHRP 3-70 work [*6*]. The methodology is based on a quality of service rating tied to the comfort and convenience that bicyclists experience. This is very different than the method that is in the bicycle chapter, Chapter 19, of the 2000HCM [*5*], which has a LOS based on bicycle speed and another LOS based on the delay that bicycles experience crossing an intersection. The SCI approach gives a rating that is tied to the actual experience of the bicyclist. Some of the factors that go into the methodology and are important to a bicyclist's comfort on an urban street are the effective width of the outside through lane, the speed and volume of vehicles on the roadway, the percentage of trucks, is there a bike lane or no bike lane, and the pavement condition. The actual number of bicycles or the speed of the bicycles is not a factor, but what is measured is the bicyclists' comfort in sharing the roadway with vehicles. The model looks at the bicyclist both on the link between intersections (the segment LOS) and as they cross the intersection (intersection LOS) and then these are combined into a weighted average for the facility. The actual equations used to determine the point score that defines LOS can be found in Reference 6.

28.2.2 Pedestrian Level of Service

The pedestrian LOS methodology [*6*] takes a different philosophical approach to assessing LOS than that for the other modes. The Pedestrian LOS is defined as the worse of two different LOS methodologies. One method considers the density of pedestrians on the sidewalks and corners, this is the method in Chapter 18 of the 2000HCM. The second method takes into account the nondensity attributes that consider the comfort and convenience experienced by pedestrians traversing the roadway due to the environment rather than the number of pedestrians using the facility. The worse LOS rating from these two different approaches is then defined as the pedestrian LOS for the facility.

Chapter 18 of the 2000 HCM [*5*] provides LOS criteria for sidewalks based on the average spacing per pedestrian, which depends on the number of pedestrians and width of the sidewalk. Table 28.2 shows these LOS criteria. In large cities with very high volumes of pedestrians, the HCM model is usually the more significant because it looks at the crowding of the pedestrians.

Table 28.2: HCM Criteria for Pedestrian LOS for Sidewalks

Level of Service	Space per Pedestrian (sq ft)	Equivalent Maximum Flow Rate (peds/hr/ft)
A	>60	≤300
B	>40–60	>300–≤420
C	>24–40	>420–≤600
D	>15–24	>600–≤900
E	>8–15	>900–≤1,380
F	≤8	>1,380

(*Source:* Exhibit 93 NCHRP Report 616 [*6*].)

The second of the two pedestrian methods was developed by SCI [*8*] and enhanced in the NCHRP 3-70 research [*5*]. Similar in perspective to the SCI bicycle model, it is completely dependent on environmental factors. Thus it gives a rating as to the comfort pedestrians experience walking an urban street. This model combines three different LOS scores: one score for the segment LOS for the pedestrians, one score for intersection LOS for the pedestrians, and a third score that rates crossing difficulty midblock. Some of the factors that go into the pedestrian segment model are the vehicle volume and speed, the separation of the vehicles from the pedestrians, the existence of a sidewalk or not, and the width of the outside lane. Factors important for the pedestrian intersection LOS include the number of right-turn-on-red vehicles and the number of permitted left-turners, the number of lanes to be crossed, and the average delay that pedestrians must wait to cross the street. The midblock crossing factor considers the expected waiting time trying to find an acceptable gap to cross the street. Again, the number of pedestrians is not important in any of these models, but rather it is a measure of the comfort and convenience experienced by pedestrians traversing the roadway due to the environment rather than the number of pedestrians using the facility. The model is a regression equation using the scores of the three models, which results in the same range of point scores as the bicycle model, with the categories also the same as shown in Table 28.1 [*7*].

28.2.3 Bus Level of Service

The model for the bus LOS score is a function of two other scores. One score is based on the quality of service perceived by riders concerning ease of access to the bus. This is based on the pedestrian LOS score for the facility. The second score combines a score based on waiting time and travel time of the trip. The waiting time score is a function of the headway between buses, and the travel time score is a function of the perceived travel time of the service. A regression equation then uses these two scores to come up with a LOS point score as shown in Table 28.1.

28.2.4 Automobile Level of Service

The urban street LOS for automobiles is based on the average speed of the through vehicles over the facility length. Because speed is distance divided by time, the average travel speed on an urban street is the length of segment divided by the average travel time. The average travel time is composed of the running time (RT) on the segment plus the control delay time at the intersection (D). Thus average travel speed of the through vehicles on an urban segment is:

$$Spd = \frac{3600 * L}{5280(\text{RT} + \text{D})} \tag{28-1}$$

where: Spd = arterial average travel speed, mi/h
L = length of the arterial segment under study, ft
RT = running time on the segment, s
D = control delay at the signalized intersection incurred by the through lanes, s/veh. This value is found from the delay equations in Chapter 24.

An estimation of running time (RT) can be found in Reference 9 as follows:

$$RT = \frac{6 - l_1}{.0025L} + \frac{3600L}{5280S_f}f + \sum_{i=1}^{N_a} d_{lr} + d_{other} \tag{28-2}$$

where: RT = running time on the segment, s
l_1 = start up lost time, s
L = length of the segment under study, ft
S_f = segment free flow speed, mi/h
f = adjustment factor that accounts for traffic density, where

$$f = \frac{2}{1 + \left(1 - \frac{v}{52.8NS_f}\right)^{0.21}} \tag{28-3}$$

where: v = volume on the segment in vph
N = number of through lanes on the segment
d_{lr} = delay from left and right-turn vehicles on the segment, s/veh
N_a = number of access points on the segment
d_{other} = other delay along the segment (from parking, bicycles, or pedestrians), s/veh

The delay from left and right turn vehicles on the segment into mid-segment access points, d_{lr}, can be estimated from Table 28.3. This delay refers to the delay experienced by through vehicles when they find themselves behind a vehicle making a turn into a midblock access point. If a more accurate estimation is desired, refer to the research report in Reference 9.

Table 28.3: Delay from Left- and Right-Turning Vehicles on the Segment, d_{lr}

Mid-Segment Volume Vphpl	Through Vehicle Delay (s/veh/pt)		
	1 Lane	**2 Lanes**	**3 Lanes**
200	0.04	0.04	0.05
300	0.08	0.08	–0.09
400	0.12	0.15	0.15
500	0.18	0.25	0.15
600	0.27	0.41	0.15
700	0.39	0.72	0.15

(*Source:* NCHRP 3-79 Final Report [9].)

For a multimodal urban street, the LOS for automobiles is the probability of users rating the facility each LOS grade. For each segment of the facility, the probability of choosing each LOS grade is found; and then a weighted average for the segment is found, based on the percentage of users that weight the segment each LOS. Thus auto LOS is the sum of the probabilities from A to F that an individual will choose a specific LOS grade multiplied times the weighting factor for that grade.

To find the probability of a given LOS, say C for example, one finds the probability a user will rate the facility LOS C or worse as follows:

$$\Pr(LOS \leq C) = \frac{1}{1 + \exp\left(-\alpha_C - \sum_k \beta_k x_k\right)} \tag{28-4}$$

Where α_C is alpha value for LOS C, which is the maximum numerical value for LOS C, as shown in Table 28.4 for each LOS, and β_k is a calibration parameter for the individual attribute, *X*, which is being used to determine LOS on the facility.

Table 28.4: Alpha Values for Each Level of Service

Alpha Values for LOS	Value
A	1.1614
B	–0.6234
C	–1.7389
D	–2.7047
E	–3.8044

(*Source:* NCHRP Report 616 [6].)

Then to get the probability that an individual user would choose the given LOS:

$$\Pr(LOS = C) = \Pr(LOS \leq C) - \Pr(LOS \leq C - 1) \quad (28\text{-}5)$$

Where $\Pr(LOS \leq C-1)$ is the $\Pr(LOS \leq B)$. Thus the probability of choosing an individual LOS is the probability of a user choosing that LOS or worse minus the probability of the user choosing one better LOS grade or worse.

28.2.5 Florida Quality/Level of Service Handbook

The Florida Quality/Level of Service (Q/LOS) Handbook [*10*] provides a planning methodology for determining measures of effectiveness for auto, pedestrians, bicycles, and buses. The models used are the HCM 2000 model for autos, a planning version of the SCI methods for pedestrians and bicycles, and a modified version of the bus measures described earlier from Reference [11]. With the Florida Q/LOS handbook there is a software package called ARTPLAN [*12*] that performs the analyses and allows for easy comparisons of LOS of each of the modes. Thus it is very useful for alternative analyses of a designs effect on all modes. Tables 28.5 and 28.6 show the results of an ARTPLAN analysis for an arterial that was converted from a two-lane per direction facility to a roadway with one through lane per direction but with bike lanes and turn bays. The results show that although the speed for the vehicles does drop from approximately 27.5 mi/h to 23.7 mi/h, the LOS does not change but remains at LOS C. The bicycle LOS improves a letter grade from D to C. Most interesting is the only minor loss in capacity for the vehicles when reducing the through lanes from two to one. In the before case, the capacity is 1197 vph; in the after case the capacity is 756 vph in the one lane, but you have removed the left turners to a separate bay. So, if we add back the number of left turners to the capacity of this approach, we have an additional 93 vehicles moving through the intersection an hour or 849 vph. Thus by removing a lane, you are reducing the capacity by only 348 vph.

28.3 Summary

This chapter has covered information on analyzing arterials as multimodal facilities. This is an important trend in traffic engineering being emphasized by all levels of DOTs. There has been a substantial amount of research on multimodal quality of flow indices, much of it originally initiated by the Florida Department of Transportation (FDOT). Likewise, there has been considerable work done on the individual modes—vehicular, fixed-route transit, bicycle, and pedestrian. At the same time, this same work has raised fundamental issues on how to establish comparability across the modes, for the comprehension of users,

Table 28.5: "Before" Two-Mile Arterial Roadway Description and Results

Roadway Variables		**Traffic Variables**		**Control Variables**		**Multimodal Variables**	
Area type	Large Urbanized	AADT	14,500	No. signals	4	Bike lane	No
Class	II	K factor	0.09	Control type	Pretimed	Outside lane width	Typical
Posted speed	35	D factor	0.55	Cycle length	90 s	Pavement condition	Typical
No. through lanes	4	PHF	0.925	g/C	0.45	Sidewalk	Yes
Median type	Nonrestrictive	% Heavy vehicles	2	Arrival type	4	Sidewalk separation	Typical
Left-turn lanes	No	% lefts	12			Sidewalk protective barrier	Yes
Right-turn lanes	No	% rights	12			Bus frequency (buses/hour)	4
Adjusted saturation flow rate per lane	1333					Bus span of service	24
BEFORE RESULTS							
Auto speed (mi/h)	27.5	Auto LOS	C	Bike LOS	D	Ped LOS C	Bus LOS B

Table 28.6: "After" Arterial Roadway Description and Results

Roadway Variables		**Traffic Variables**		**Control Variables**		**Multimodal Variables**	
Area type	Large urbanized	AADT	14,500	No. signals	4	Bike lane	Yes
Class	II	K	0.09	Control type	Pretimed	Outside lane width	Typical
Posted speed	35	D	0.55	Cycle length	90	Pavement condition	Typical
No. through lanes	1	PHF	0.925	g/C	0.5	Sidewalk	Yes
Median type	Nonrestrictive	% Heavy vehicles	2	Arrival type	4	Sidewalk separation	Typical
Left-turn lanes	Yes	% lefts	12			Sidewalk protective barrier	Yes
Right-turn lanes	Yes	% rights	12			Bus frequency (buses/hr)	5
Adjusted Saturation flow rate per lane	1,680					Bus span of service	24
AFTER RESULTS							
Auto Speed (mi/h)	23.7	Auto LOS	C	Bike LOS C	Ped LOS C	Bus LOS	B

elected officials, and transportation professionals. The essence of the multimodal approach is to provide realistic choices to travelers and ideally to create more livable cities.

References

1. "The Effects of Traffic Calming Measures on Pedestrian and Motorist Behavior," FHWA Report No. FHWA-RD-00-194, August 2001.
2. Leonard, J., and Davis, J., "Urban Traffic Calming Treatments: Performance Measures and Design Conformance," *ITE Journal,* August 1997.
3. "Designing Streets for Pedestrian Safety," C. Hardej, NYCMTC, June 2007; http://nysmpos.org/conference/Presnetation/1.%20Safety%20Roundtable/Hardej.pdf.
4. "The Gowanus Lounge," http://gowanuslounge.blogspot.com/2007/04/dot-moving-ahead-with-ninth-street.html.
5. *Highway Capacity Manual,* Transportation Research Board, National Research Council, Washington DC, 2000.
6. *NCHRP Report 616, "Multimodal Level of Service Analysis for Urban Streets,"* Transportation Research Board, Washington DC, 2008.
7. Landis, B., et al., "Real-time Human Perceptions: Toward a Bicycle Level of Service, *Transportation Research Record 1578,* Transportation Research Board, Washington DC, 1997.
8. Landis, B., et al., "Modeling the Roadside Walking Environment: A Pedestrian Level of Service" *Transportation Research Record 1773,* Transportation Research Board, Washington DC, 2001.
9. "Predicting the Performance Measures on Urban Streets," Final Report, NCHRP Project 3-79, 2008.
10. Florida Department of Transportation, *Quality/Level of Service Handbook,* Tallahassee, FL, 2007.
11. *Transit Capacity and Quality of Service Manual,* TCRP Web Document 6, Transportation Research Board, Washington DC, 1999.
12. LOSPLAN Software, Florida Department of Transportation, 2007.

Problems

For the following problems, download the free ARTPLAN software from the Florida DOT Web site: http://www.dot.state.fl.us/planning/systems/sm/los/los_sw2M2.shtm.

28-1. Find the LOS for autos, bicycles, pedestrians, and buses on a four-lane (both directions) arterial, with no left-turn lanes, AADT = 14,000, k = 0.12, D = 0.55, 4 buses per hour. Do this for the case where no bike lane exists and then again with a bike lane.

28-2. Using the same inputs as in problem 28-1, change the four-lane arterial to a two-lane arterial with left-turn bays. Again do this for the case where no bike lane exists and then again with a bike lane.

28-3. How does the LOS for the arterials above change with and without a sidewalk?

28-4. How does the LOS change for each mode in problem 1, if you increase the AADT by 2,000 vpd up to 20,000 vpd?

CHAPTER 29

Planning, Design, and Operation of Streets and Arterials

By definition, an *arterial* is intended to serve *through* traffic; *collectors* and *local streets* serve other functions and constitute a hierarchy of roads serving a community. On an arterial, the most common modes are automobile, transit, pedestrian, and bicycle.

At a planning level, it is important to evaluate the impacts of a project on the quality of service provided to users of all modes. These are implemented in design and in operations using basic principles, initially and during rehabilitation or evolution (for instance, due to general growth or major generators). Chapter 26 addressed signal-based methods of assuring smooth flow on arterials and in networks, but we now focus on the facility.

Much of the operational task is *preserving the arterial function* (i.e., service to through traffic) in face of growth that imposes other uses on the facility. One leading example is a state arterial linking major population centers: As time passes, towns along the route have their own growth and use the state arterial as their "main street"; the arterial devolves into a major local street, and its design use is compromised; in the extreme, both state and local governments look for the construction of "bypass" arterials that route interurban traffic around the new or larger town, replacing the function of the original road.

At the same time, the largest mileage of traffic facilities consists of local streets, providing primarily access service in a variety of settings from densely populated urban areas to remote rural areas. Local street networks provide vital access service to abutting lands. Access is required by both goods and people and must be carefully managed to strike a balance between providing all of the needed service without creating too much congestion.

This chapter addresses some design principles and some methods of preserving arterial function. The latter methods often fall under the rubric of *access management.* The chapter also addresses concepts related to the *balanced use of streets* and the theme of *complete streets* and of *traffic calming.* The role of *roundabouts* in these concepts and in modern design is considered. *Network issues* such as one-way streets and pairs are addressed. Finally, a set of special cases (most related to special signal needs) are considered.

29.1 Kramer's Concept of an Ideal Suburban Arterial

In a paper that could serve as a classic position paper on the proper design and operation of an arterial, Kramer [*6*] enumerates two design principles that he recommends are *essential* to be adopted and followed at *all* intersections of a suburban arterial with public streets and private driveways, in addition to standard AASHTO, FHWA, and other applicable criteria:

Principle 1

Establish an absolute minimum percentage of green time that will be displayed to arterial thru traffic. At every location where all crossing or entering (design year) traffic volumes cannot be accommodated within the remaining percent of green time allocated to those movements, *a full or partial grade separation is mandated.*

All locations having traffic movements that will be permitted to fully cross both directions of the arterial at grade, or otherwise cause both directions of the arterial to stop simultaneously, *must be predetermined before reconstruction* and a series of signal cycle lengths preestablished that will provide for full-cycle and/or half-cycle offsets for these critical at-grade intersections over the design life of the facility.

Kramer observed that at-grade crossings of the arterial should involve indirect turning movements, and that this should be assured by raised median barriers.

Principle 2

The signal spacing dominates all other considerations, and the cycle length must be picked to be in harmony with the signal spacing (i.e., the geometry).

Kramer also identifies 10 characteristics of an ideal suburban arterial; refer to Table 29.1.

Moreover, he held that all of these characteristics can be provided at reasonable construction costs within the existing right-of-way for typical divided suburban arterials, recognizing that some spot acquisitions (taking of land) might be needed.

To a large extent, Kramer lays out a grand concept for the fresh design of a new arterial. But he goes beyond that, for as he indicates, the same principles apply to the redesign of a facility. And they are guidance for operational improvements and for new developments along existing facilities.

Can they apply to urban arterials? The authors would answer, "Yes, they can be excellent guidance" while recognizing that developed urban areas may well not have the right-of-way and cannot apply all of his guidance.

At the same time, many urban arterials in existing large cities had their roots in "suburban" or even rural levels of development, as the city expanded to encompass them.

Table 29.1: Kramer's 10 Characteristics of an Ideal Suburban Arterial

1. Its three or four lanes in each direction of travel would receive a minimum of two-thirds to three-fourths of the signal cycle as green time at all intersections encountered along its entire length.
2. Each direction of travel would be signalized for progressive movement so that traffic would simultaneously flow as smoothly in each direction as if it were two parallel one-way streets.
3. Through traffic would be protected (by signalization) from conflicting left turns from the opposing direction.
4. Direct left turns would be provided from the arterial at frequent intervals and would be protected by signalization from conflicting through traffic movements from the opposing direction.
5. The facility would accommodate all maneuvers of increased truck sizes and combinations allowable under federal legislation.
6. Pedestrians crossing this arterial would be provided protected signal phasing and be free from (lawful) conflict with any vehicular traffic crossing their path; and the spacing of pedestrian crossings would be so convenient as to discourage pedestrian crossing at unprotected locations.
7. The facility would also provide for transit operations that would not impede through traffic movement at any bus stop.
8. Transit bus operations would be enhanced by providing stops at all convenient locations in close proximity to protected pedestrian crossings.
9. The geometric design of the facility would accommodate the infusion of additional major traffic generators with minimal adverse effect to the road user; that is, through traffic could continue to receive a minimum of two-thirds to three-fourths of the signal cycle as arterial green time.
10. Signalization timing and offset programs for this arterial would be independently variable for each direction to take into account changes in traffic volumes, provide for special event (stadium) traffic, and accommodate an uninterrupted flow for emergency vehicles having on-board preemption equipment.

Sections of Queens Boulevard in New York City are an example of a road that has considerable width and buffers that serve local functions while allowing separated through flows.[1]

29.2 Principles Guiding Local Streets

Two principles dominate traffic engineering on local street networks:

1. The design of local street networks should reenforce their function and purpose as such.
2. Through traffic should be discouraged through a variety of traffic engineering measures.

These principles are addressed using a variety of measures.

In addition, a number of aspects of local streets must be considered as part of an overall management and operational approach:

1. Adequate allocation of curb space should be provided among the competing uses (parking, loading/unloading, transit/taxi).
2. Safety of motorists and pedestrians must be assured.
3. Blockage of local streets to through movement must be controlled and minimized.
4. Appropriate access to and through the local network must be provided for emergency vehicles.

All of these speak to logical concerns and focus on a commonsense proposition: Local streets must be designed, managed, controlled, and operated in a manner that enhances and preserves their intended function.

The three primary categories of local streets are (1) residential, (2) industrial, and (3) commercial. In a perfect world, all traffic on local street networks is either accessing a particular land use at a particular location or circulating while seeking a parking place or a location. There should be little "through" flow, that is, motorists traveling significant distances along local streets (i.e., more than two to three blocks).

[1]Although the history of the development in this case did induce problems by violating Kramer's second principle, namely the primacy of signal spacing. If the combination of block lengths and feasible cycle lengths is poor, the engineer (and public) will live with the challenges of making reasonable accommodations to shifting conditions.

29.3 Access Management

Access management has been one of the emerging themes in traffic engineering, with many states creating the explicit legislation and administrative infrastructure needed. This section summarizes some of the operational aspects related to access management, but you may refer to the excellent Web site, http://www.accessmanagement.info/ [*2*] of the TRB Access Management Committee for information on recent conferences, state Web sites, resource materials, and multimedia materials. It cites the *TRB Access Management Manual* [*3*], developed under the guidance of the committee.

The committee's Web site states that "Access management is much more than driveway regulation. It is the systematic control of the location, spacing, design and operation of driveways, median crossings, interchanges, and street connections."

Reference 4 provides guidance on managing access in a community, by enumerating 10 ways. Refer to Table 29.2.

29.3.1 Primary Operations Measures in Access Management

The primary operations measures that may be taken in access management are:

- Achieve proper signal spacing.
- Minimize conflicts by proper median treatments.
  - Two-way left-turn lanes (TWLTL)
  - Restricted/raised medians

Table 29.2: Ten Ways to Manage Access in a Community [*4*]

1. Lay the foundation for access management, by addressing access management in the local comprehensive plan.
2. Restrict the number of driveways per lot.
3. Locate the driveways away from intersections.
4. Connect parking lots and consolidate driveways.
5. Provide residential access.
6. Increase minimum lot frontage on major roads.
7. Promote a connected street system.
8. Encourage internal access to outparcels.
9. Regulate the location spacing and the design of driveways.
10. Coordinate with the State Department of Transportation.

- Minimize frictions by controlling driveway number, placement, and design.
- Separate and/or direct flows by use of back streets, side-street access, lanes divided from the through lanes for local service (frontage roads, in effect).

Much has been said about the first item, and it will not be addressed further in this section.

The several operational measures just cited do have overlaps. For instance, installing a raised median does determine what driveway movements are feasible. Likewise, separating or directing flows to other access points is in effect influencing the number—or existence—of driveways. The same is true of concentrating flows into fewer driveways.

29.3.2 Proper Median Treatments

Figure 29.1 shows an existing two-way left turn lane (TWLT) from a Google Earth view. TWLTLs are very popular in some jurisdictions because they (1) remove turning traffic from through lanes, (2) do not require radical changes in access to existing land uses, and (3) allow some movements such as lefts from a driveway to be made in two stages—one move to the TWLTL, followed by another merging into the traffic, so that two independent and smaller gaps are needed, not one combined and larger gap. Note the number of driveways along this section of arterial.

Some of the literature has focused on the reality that raised medians have lower midblock accident experiences than TWLTLs. This is somewhat obvious, if only because the number of conflicts is reduced when the median is raised; refer to Figure 29.2.

Reference 5 reports that at midblock locations, the accident rate per million vehicle miles of traffic is:

2.43 accidents/MVM for undivided
1.66 accidents/MVM for TWLTL
1.09 accidents/MVM for restrictive medians

Reference 6 also addresses such differences, including the statistical significance of its findings.

This suggests the clear advantage of installing raised medians. However, it does not tell the complete story, for in some cases the raised median is simply not a viable option due to needed service to land uses, lack of alternative paths to those land uses, and/or budget. The engineer may be faced with choosing an attainable 32% improvement over an unattainable 55% improvement.

29.3.3 Control Number, Placement, and Design of Driveways

Reference 7 is a 1996 TRB Circular on driveway and street intersection spacing and is especially relevant. It includes a

Figure 29.1: A Two-Way Left Turn Lane (TWLTL) on an Arterial

(*Source:* Image from Google Earth.)

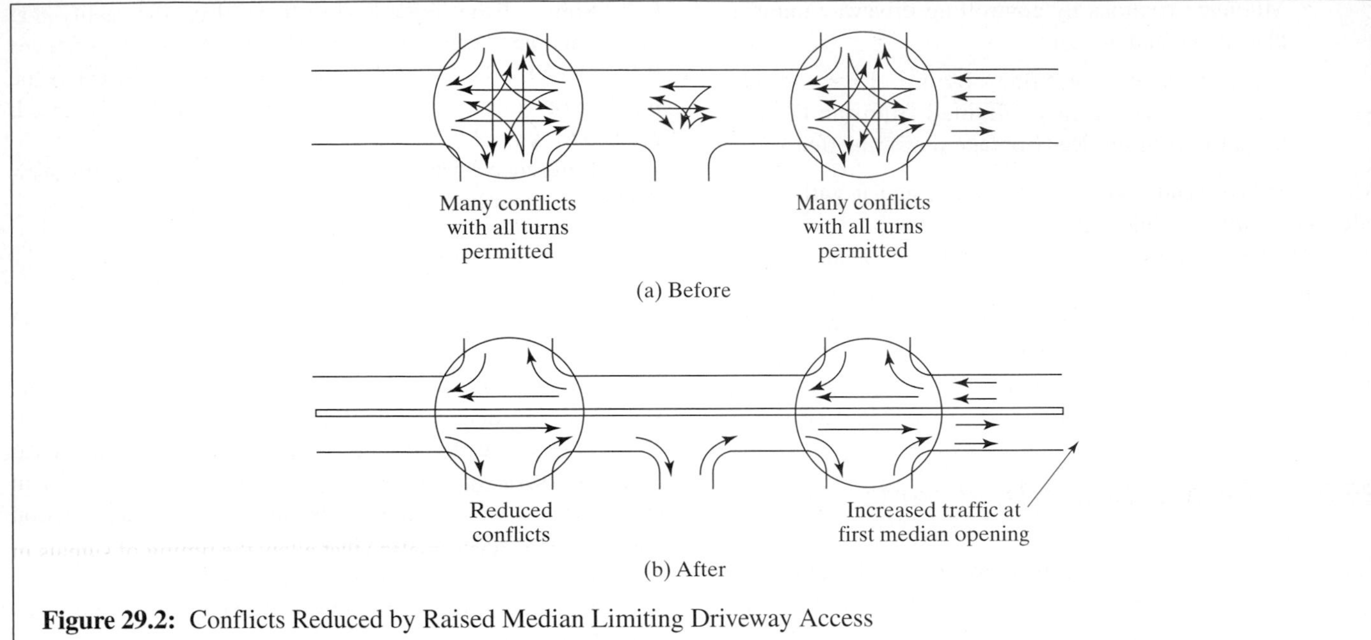

Figure 29.2: Conflicts Reduced by Raised Median Limiting Driveway Access

discussion of practices in various states and other jurisdictions. It also includes a discussion of minimum driveway spacings to avoid facing the driver with too many overlapping decisions. Refer to Figure 29.3 for one illustration of the problem.

Reference 8 used simulation to gain insight into the effects of driveways, including eliminating left turns at the driveway and having decel/accel lanes associated with the driveway. Figure 29.4 shows the left-turns-eliminated results on the westbound through traffic for the case indicated, with the initial case being 30% lefts and 70% rights and the second case being the same volume but 100% right turns (as assured by a raised median or signing, for instance). There was no significant effect on the eastbound through traffic, and it is not shown.

Reference 8 also found that the presence of deceleration lanes into the driveway had a significant (2 mi/h) benefit to the eastbound through traffic, but that an acceleration lane had

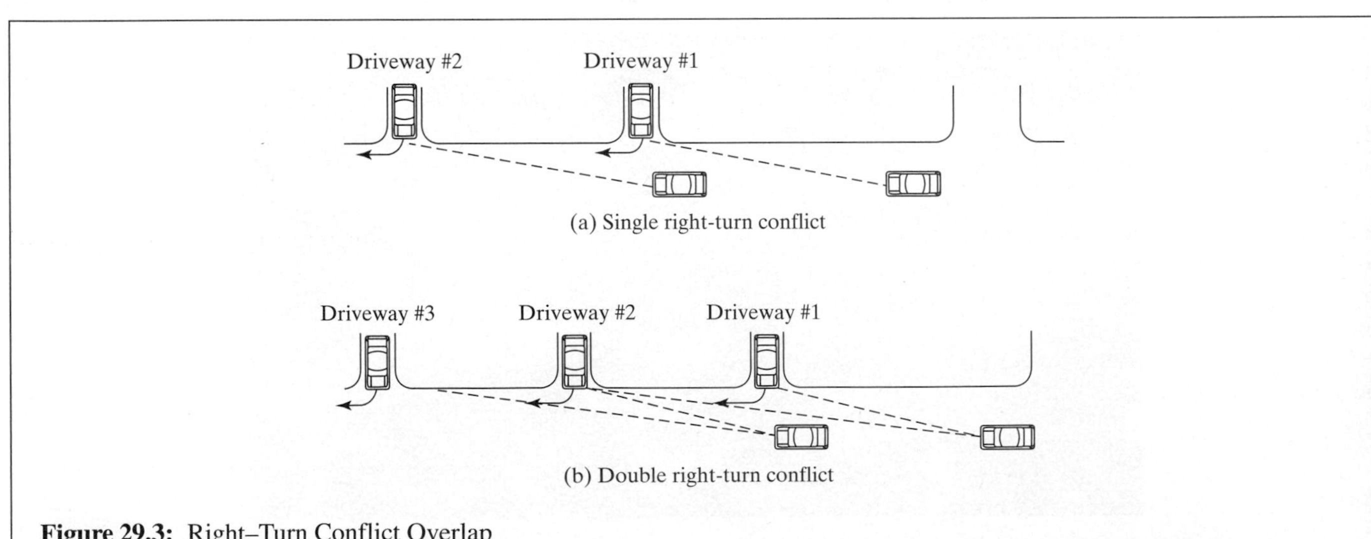

Figure 29.3: Right–Turn Conflict Overlap

(*Source:* "Driveway and Street Intersection Spacing", *Circular 456*, Transportation Research Board, National Research Council, Washington DC, March 1996.)

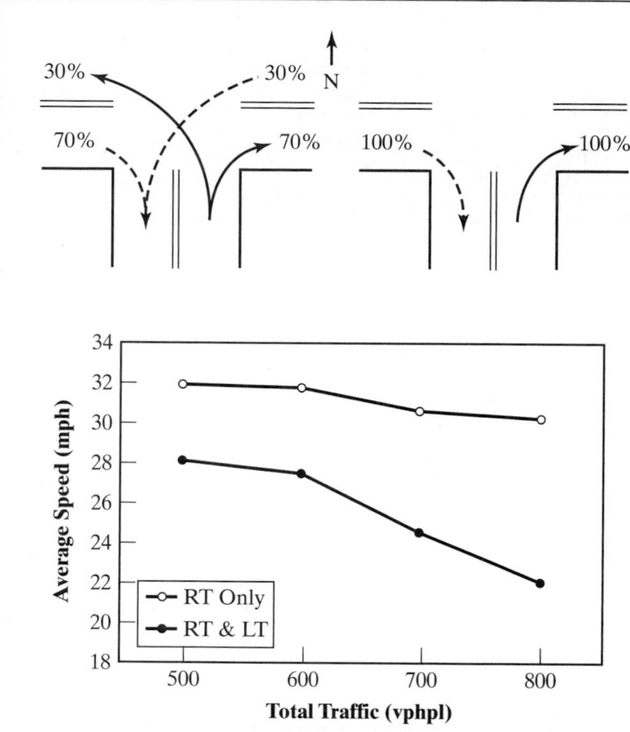

Figure 29.4: The Effect of Eliminating Left Turns at Driveways

(*Source:* McShane, W., et al., Insights into Access Management Details using TRAF-NETSIM, presented at the Second Annual Conference on Access Management, August, 1996.)

little benefit to eastbound through traffic (although it may have benefited the departing driveway traffic itself). On the other hand, the very existence of the driveway had a – 6 mi/h disbenefit due to the existence of the driveway, so only a third of the disruption was mitigated by the deceleration lane.

Clearly, *minimizing the number and intensity of driveways* is highly desirable. The wisdom of redirecting flows to back streets, side streets, and/or restricting local access to frontage roads seems effective: If it can be done, the number and intensity of driveways on principal arterials can be minimized.

However, there are two major impediments to this very desirable approach:

1. Do the alternative path and access exist?
2. Are the intersections properly designed to accept the additional load from the "minor" streets, which now include this redirected traffic?

Another impediment, of course, is having the legal and regulatory authority to require this approach. As already cited, the legislation must be in place and the comprehensive plan must address requirements. Reference 2 is a fine resource for the state of the practice in the various states and for model laws.

29.3.4 Separation of Functions

To assure that local activity, such as land access, parking, and short trips do not disrupt the arterial function, some arterials now have an "inner" and "outer" roadway, just as some freeways do; refer to Figure 29.5. This construction also ensures that the arterial function is "built into concrete" and defined for the future as well.

Earlier chapters in the text presented a number of concepts for limiting the negative impact of left turns on arterial operation, including jug-handles, continuous flow intersections, and various U-turn treatments, as well as unique intersection channelization plans that allow the timing of signals in two directions to be decoupled. Figure 29.6 shows an interesting design at a major shopping center: The eastbound lefts do *not* have a signal controlling them; rather, they move "in the shadow" of the two-phase signal to right in the figure, which has the SB lefts on one phase and the main street east-west through traffic on the other. The eastbound lefts also benefit from gaps in the westbound flow, even when it is moving. The SB right-turners leave from the driveway on the left in the figure, so that they do not conflict with the EB lefts turning in.

Note that the roadway inside the site must "sort them out," but this is done internal to the site and not at the expense of the through traffic on the arterial.

29.4 Balanced Streets and Complete Streets

A multimodal focus to transportation facilities is not merely a current emphasis but an explicit legislated mandate in federal and other legislation. For instance, FHWA provides design guidance that "bicycling and walking facilities will be incorporated into all transportation projects unless exceptional circumstances exist" [9].

Related to this is a growing emphasis on the need to "balance the street" with the service it provides to several modes, and to "complete the street" by broadening the focus and attention to modes other than the automobile.

Reference 10 notes that "A balanced street is one that works for all users. Achieving the right balance between transit and other uses is a delicate matter, as it is easy for buses, in particular, to overwhelm these other uses." The same reference

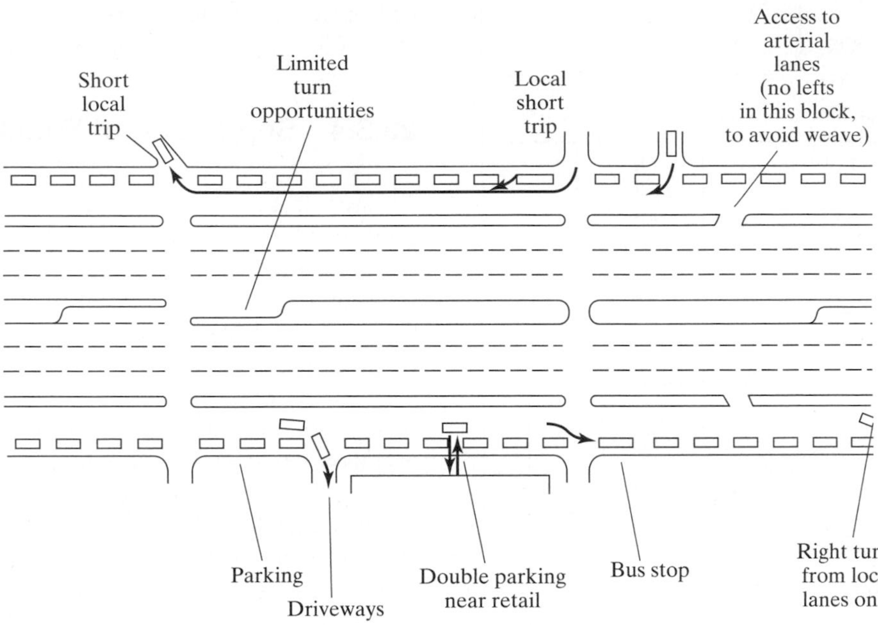

Short local trip

Limited turn opportunities

Local short trip

Access to arterial lanes (no lefts in this block, to avoid weave)

Parking

Driveways

Double parking near retail

Bus stop

Right turns from local lanes only

Figure 29.5: Separation of Local Functions from Arterial Traffic

Figure 29.6: An Interesting Driveway Design at a Shopping Center

(*Source:* Image from Google Earth.)

enumerates a number of strategies, including sufficiently wide sidewalks, amenities for pedestrians and transit riders, priority lanes for transit vehicles, and traffic-calming measures for automobiles.

Reference 11 is a Web site that provides a good source of current information and directs the visitor to a slideshow summary related to complete streets [*12*] that observes that "Complete streets are designed and operated

Table 29.3:
Complete Streets Policy in Delaware

STATE OF DELAWARE
OFFICE OF THE GOVERNOR

Executive Order Number Six

April 24, 2009

TO: HEADS OF ALL STATE DEPARTMENTS AND AGENCIES

RE: CREATING A COMPLETE STREETS POLICY

WHEREAS, walking is the most fundamental mode of physical transportation; and

WHEREAS, bicycling promotes healthier lifestyles; and

WHEREAS, walking and bicycling are simple fitness activities that can prevent disease, improve physical health and assist in fostering mental well-being; and

WHEREAS, by walking and bicycling you help to reduce greenhouse gas emission by reducing the time you spend in your car; and

WHEREAS, my administration, along with the Delaware Department of Transportation, promotes the walkability and bicycle friendliness of communities through principles such as context sensitive design, mobility-friendly design, mixed-use and infill developments; and

WHEREAS, the Delaware Department of Transportation has developed user friendly design standards for pedestrian, bicycle, as transit facilities; and

WHEREAS, the Delaware Department of Transportation has the opportunity to create and improve transportation facilities for all users by implementing these principles and standards through its projects; and

WHEREAS, the Advisory Council on Pedestrain Awareness and Walkability and the Delaware Bicycle Council serve as advisors to the Delaware Department of Transportation; and

WHEREAS, a complete streets Policy means deliberately planning, designing, building, and maintaining streets for all modes of transportation;

NOW, THEREFORE, I, JACK A. MARKELL, by virtue of the authority vested in me as Governor of the State of Delaware, do hereby declare and order the following:

1. The Delaware Department of Transportation ("DelDot") shall enhance its multi-modal initiative by creating a Complete Street Policy that will promote safe access for all users, including pedestrians, bicyclists, motorists and bus riders of all ages to be able to safely move along and across the streets of Delaware;
2. The Delaware Bicycle Council, the Advisory Council on Pedestrain Awareness and Walkability, and the Elderly & Disabled Transit Advisory Council shall assist DelDOT with this endeavor;
3. A Complete Streets Policy should:
    1. Solidify DelDOT's objective of creating a comprehensive, integrated, connected transportation network that allows users to choose between different modes of transportation;
    2. Establish that any time DelDOT builds or maintains a roadway or bridge, the agency must whenever possible accommodate other methods of transportation.
    3. Focus not just on individual roads, but changing the decision-making and design process so that all users are considered in planning, designing, building, operating and maintaining all roadways;
    4. Recognize that all streets are different and user needs should be balanced in order to ensure that the solution will enhance the community;
    5. Apply to both new and retrofit projects, including planning, design, maintenance, and operations for the entire right-of-way;
    6. Ensure that any exemption to the Complete Streets Policy is specific and documented with supporting data that indicates the basis for the decision;
    7. Direct the use of the latest and best design standards as they apply to bicycle, pedestrian, transit and highway facilities;
4. DelDOT, with the assistance of the advisory councils, shall create the Policy and deliver it the Governor for consideration no later than September 30, 2009.

(*Source:* Reference 13.)

so that they are safe, comfortable, and convenient for all users—pedestrians, bicyclists, motorists, and transit riders of all ages and abilities." It also cites the interesting statistics that of all metro-area trips, 50% are three miles or less, 28% are less than a mile, and 65% of all trips under a mile are taken by automobile.

Table 29.3 shows the directive to establish a Complete Streets Policy in Delaware [*13*], issued as this text goes to the publisher. It defines the essential elements of that policy.

29.5 Traffic Calming

Traffic calming techniques have been under development since the 1970s, with much of the early work done in Europe. New attitudes toward traffic and its management began to develop in the United States, with explicit direction by federal legislation in 1991 and 1998 (ISTEA and TEA-21, respectively) and later legislation.

In its most simple terms, traffic calming is about preserving the function of local streets. Although the initial focus of traffic calming activities was on residential local street systems, broadened interest has resulted in applications to other street networks as well. References 14 and 15 represent excellent treatments of the state of the art in traffic calming through 2000. For more recent information, see an FHWA Web site on traffic calming [*16*] that provides information on measures, programs, and other agencies. Other very useful Web sites include the home page of TrafficCalming.org [*17*], work in San Jose, California [*18*], and an online *Traffic Calming Handbook* by PennDOT [*19*].

As noted previously, *local streets* are intended to provide primarily land access service. When traffic begins to use local streets for through movements, this generally leads to volume and speed levels that are incompatible with this primary function. Traffic calming is a set of traffic engineering measures and devices that are intended to address these problems. These are the specific goals of traffic calming:

1. Reduce traffic volumes on local streets.
2. Reduce traffic speeds on local streets.
3. Reduce truck and other commercial traffic on local streets.
4. Reduce accidents on local streets.
5. Reduce negative environmental impacts of traffic, such as air and noise pollution and vibrations.
6. Provide a safer and more inviting environment for pedestrians and children.

Most of the objectives of traffic calming are best achieved through design. In fact, most traffic calming projects are generally retro fits to a local street system that was improperly designed in the first place.

29.5.1 Overview

Figure 29.7 provides an illustration of a traffic calming plan for an urban local street network serving a primarily residential neighborhood with local stores and merchants present. Note that although many different traffic calming treatments are incorporated into the overall plan, not all street links are directly involved. Nevertheless, there are only one or two paths left where a motorist could drive straight through the area on a local street. These were doubtless carefully chosen and probably represent collector-type facilities where local stores are located.

29.5.2 Illustrative Techniques

In both of the examples in Figure 29.8, through movement on the local street is *physically barred.* In (a), two dead-end streets are created with a midblock barrier. The barrier can be landscaped and may also provide for pedestrian refuge and crossings. In (b), a cul-de-sac is created by blocking the intersection access of the local street. This also creates the opportunity for creative landscaping and provides for uninterrupted pedestrian movement on one side of the intersection.

Figure 29.9 illustrates *diagonal diverters.* These devices require all vehicles to turn as they pass through the intersection, and they provide for pedestrian refuge and connectivity. Diagonal diverters can be used at many intersections in a grid street network, essentially providing for the type of nonstraight streets illustrated in the design of Figure 29.7. They correct the most difficult aspects of grid networks, which are straight, encourage through movements, and also encourage higher than desirable speeds.

The use of *chicanes* to both narrow a traveled way and to create a serpentine driving path in a straight right-of-way is illustrated in Figure 29.10. As with most other traffic calming devices, they are intended primarily to reduce speed, but they also discourage through movement and can reduce volumes as well. Continuous use of chicanes can create longitudinal areas in which enhanced environments for pedestrians, bicycles, and residents can be created.

Figure 29.11 illustrates two other techniques for traffic calming.

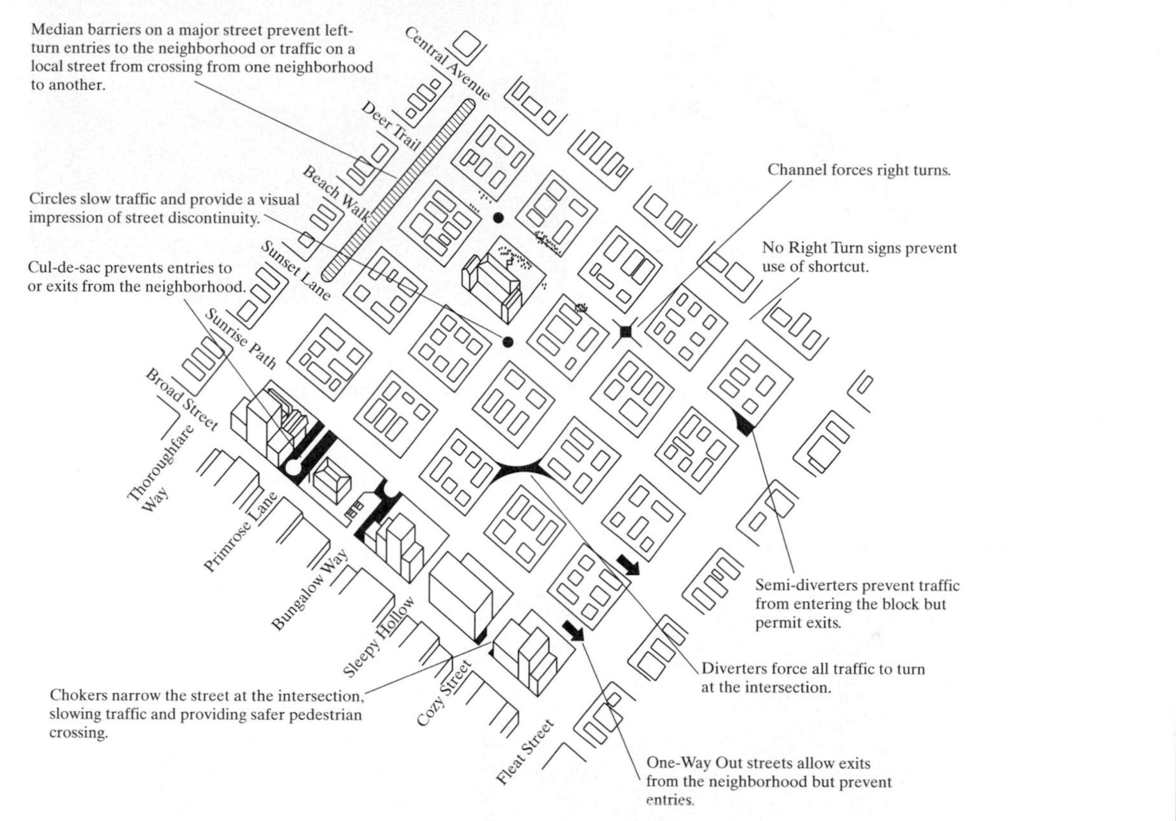

Figure 29.7: Illustration of Traffic Calming Devices Applied to a Neighborhood Grid

(*Source:* Smith, D.T., *State of the Art: Residential Traffic Management*, FHWA Research Report RD-80-092, December 1980.)

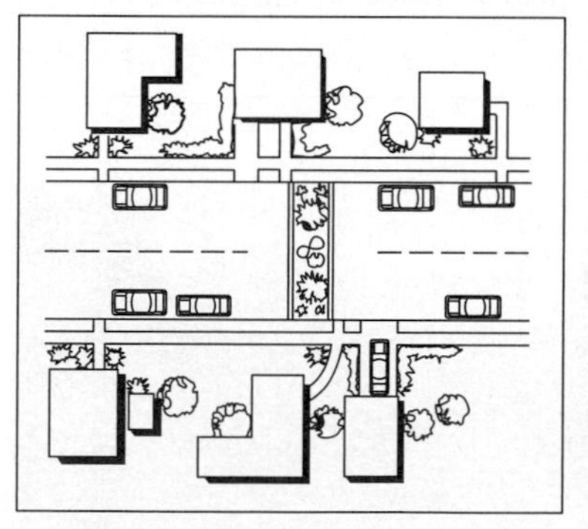

(a) Street blockage creating two dead ends

(b) Creation of cul-de-sac by blocking street access at intersection

Figure 29.8: Examples of Full Street Closures

(*Source:* Ewing, R., *Traffic Calming: State of the Practice*, Federal Highway Administration and the Institute of Transportation Engineers, Washington DC, August 1999.)

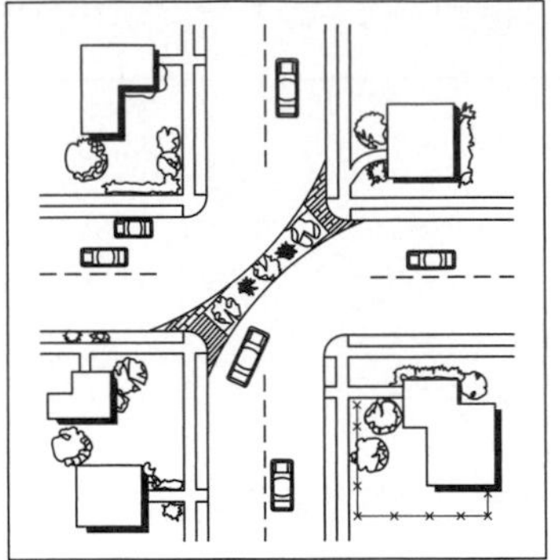

Figure 29.9: Semi-Diverters Blocking Through Movement at an Intersection

(*Source:* Ewing, R., *Traffic Calming: State of the Practice*, Federal Highway Administration and the Institute of Transportation Engineers, Washington DC, August 1999.)

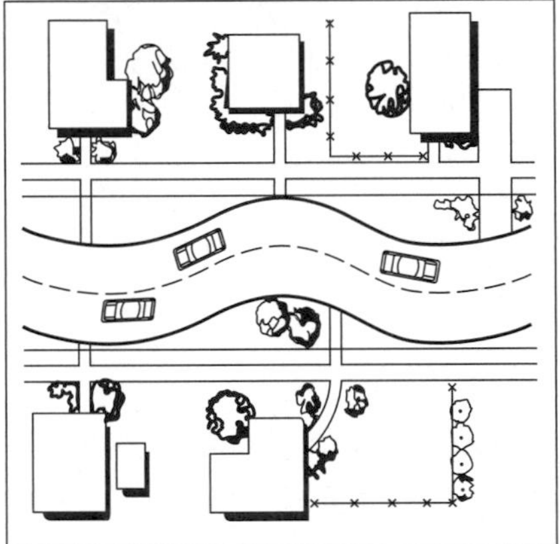

Figure 29.10: Use of Chicanes Illustrated

(*Source:* Ewing, R., *Traffic Calming: State of the Practice*, Federal Highway Administration and the Institute of Transportation Engineers, Washington DC, August 1999.)

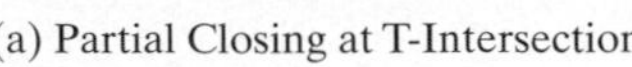

(a) Partial Closing at T-Intersection

(b) Partial Closing to Through Movement

Figure 29.11: Miscellaneous Traffic Calming Devices illustrated

(*Source:* Photos courtesy of City of Portland Department of Transportation.)

29.5.3 Impacts and Effectiveness of Traffic Calming Measures

On local street networks, the three most tangible objectives of traffic calming are reductions in speed, volumes, and accidents. Reference 15 contains detailed summaries of results of hundreds of traffic calming projects along with the reductions achieved in these three critical areas. Table 29.4 summarizes some of these results.

We caution you about the data of Table 29.4: Some of the sample sizes are relatively small, and all of the average results have relatively large standard deviations. Still, it is clear from the data that speed humps and speed tables are devices that are effective in all three areas, reducing speeds, volumes,

Table 29.4: Reported Impacts of Traffic Calming Measures

Speed Impacts			
Traffic Calming Measure	*Average Impact on 85th Percentile Speed*		
	Number of Samples	**Δ Speed (mi/h)**	**Percent Decline (%)**
12-ft Speed Humps	179	−7.60	22
14-ft Speed Humps	15	−7.70	23
22-ft Speed Tables	58	−6.60	18
Longer Speed Tables	10	−3.20	9
Raised Intersections	3	−0.30	1
Traffic Circles	45	−3.90	11
Narrowings	7	−2.60	4
One-Lane Slow Points	5	−4.80	14
Half Closures	16	−6.00	19
Diagonal Diverters	7	−1.40	4
Volume Impacts			
Traffic Calming Measure	*Average Impact on Volumes*		
	Number of Samples	**Δ Volume (veh/day)**	**Percent Decline (%)**
12-ft Speed Humps	143	−355	18
14-ft Speed Humps	15	−529	22
22-ft Speed Tables	46	−415	12
Traffic Circles	49	−293	5
Narrowings	11	−263	10
One-Lane Slow Points	5	−392	20
Full Closures	19	−691	44
Half Closures	53	−1611	42
Diagonal Diverters	27	−501	35
Other Vol. Controls	10	−1,167	31
Accidents/Year			
Traffic Calming Measure	*Average Impact on Accident Occurrence*		
	Number of Samples	**Δ Accidents (Acc/Yr.)**	**Percent Decline (%)**
12-ft Speed Humps	50	−0.33	13
14-ft Speed Humps	5	−1.73	40
22 ft Speed Tables	8	−3.05	45
Traffic Circles	130	−1.55	71

(*Source:* Compiled from Reference 15, Tables 5.1, 5.2, and 5.7.)

and accident occurrence. Volume reductions are greatest with techniques that partially or completely block through vehicles; this result, of course, is logical.

29.6 Roundabouts

The *modern roundabout* has some has some features in common with the older "traffic circle" design, but it is fundamentally different in scale and in key design features such as (1) approach roadways angled at a tangent to the center island, so as to direct flow smoothly, and (2) an emphasis on paths being developed to decrease lane changing. Figure 29.12 shows a rendering of a modern roundabout, including yield signs and markings. This is taken from the NYSDOT Web site on roundabouts [*21*].

NYSDOT has adopted a "Roundabouts First" Policy that requires that "when a project includes reconstructing or constructing new intersections, a roundabout alternative is to be analyzed to determine if it is a feasible solution based on site constraints, including ROW, environmental factors, and other design constraints" (Chapter 5, State Design Manual, which can be accessed through Reference 22), with some exceptions noted. But "When the analysis shows that a roundabout is a feasible alternative, it should be considered the Department's preferred alternative due to the proven substantial safety benefits and other operational benefits."

Figure 29.12: Rendering of a Modern Roundabout

(*Source:* Reference 21.)

A number of states have implemented roundabouts, and roundabouts are common in many other countries; indeed, the United States has been slower to encourage roundabouts. Two useful sources on the developing state of the practice, and related policy, are the Web sites cited in References 23 and 24.

Of course, part of the history in the reticence in the United States has been a systematic move away from large traffic circles, generally with approach roadways perpendicular to the interior circle. For the most part, the remaining large traffic circles in the United States include signature architecture or are focal points of a radial road system or have historic significance that transcends their traffic operational problems (e.g., Grand Army Plaza, Brooklyn, New York; Columbus Circle, New York City; DuPont Circle, Washington, DC).

At the same time, small *neighborhood* traffic circles have been used in traffic calming projects. Refer to Figure 29.13 (a).

29.7 Network Issues

A number of network issues deserve special attention. These include *one-way street pairs and systems* and *special use lanes.*

29.7.1 One-Way Street Systems

One-way street systems represent the ultimate solution to elimination of left-turn conflicts at intersections and the congestion that they may cause. It is hard to imagine, for example, the Manhattan street system operating as two-way streets, yet this was the case until the early 1960s, when NYC traffic engineer Henry Barnes introduced the one-way street concept. For high-density street networks with many signalized intersections, one-way streets are attractive because:

- Providing signal progression is a relatively simple task with no special geometric constraints and no multiphase signals to deal with.
- There are no opposing flows to create operational problems for left-turning vehicles.
- There are related benefits for safety and capacity in many cases.

Safety and capacity benefits should not be overlooked. The elimination of opposed left turns removes a major source

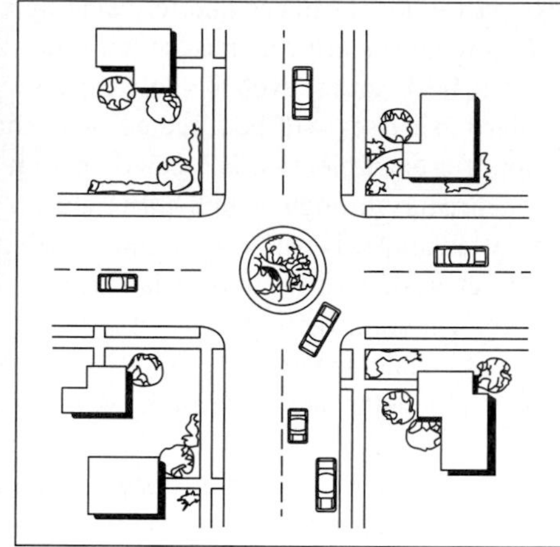

(a) Neighborhood Traffic Circle Illustrated

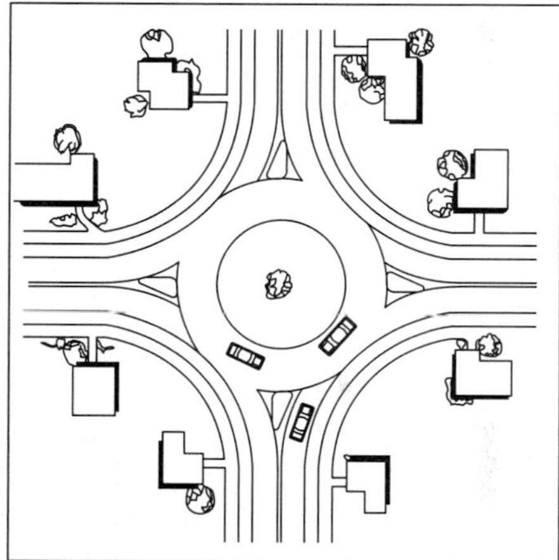

(b) Roundabouts

Figure 29.13: Neighborhood Traffic Circles and Roundabouts

(*Source:* Ewing, R., *Traffic Calming: State of the Practice*, Federal Highway Administration and the Institute of Transportation Engineers, Washington DC, August 1999.)

of intersection accidents. A one-way street may allow more efficient lane striping and additional capacity. A 50-foot street width would normally be striped for four lanes, two in each direction, as a two-way street. As a one-way street, it is possible to stripe for five lanes.

One study [25] reported that a conversion of a two-way street system to one-way operation resulted in:

- A 37% reduction in average trip time
- A 60% reduction in the number of stops
- A 38% reduction in accidents

Although these results are certainly illustrative, every case has unique features, and these percentage improvements will not be automatically achieved in every case.

A successful one-way street program, however, depends on having paired "couplets" within close proximity. A pair of streets with a similar cross section and number of lanes must be available, so that capacity and travel times in two directions can be balanced. It would also be difficult to operate a couplet where one street is a "main street," with many large trip generators, and the paired street is not.

Business owners often object to installing one-way street systems, fearing lost business due to changing circulation patterns. This would clearly be an issue in the pairing of a main street with a back street, even if their capacities were similar. Particular businesses will be affected more than others. A gas station, for example, that is located on a far corner of two one-way streets, through which turns are no longer made, is a typical example. Nevertheless, most studies show that one-way street systems in commercial areas generally results in enhanced economic activity.

The most noticeable negative impact of one-way street systems is that trip lengths are increased, as illustrated in Figure 29.14.

Another impact of one-way street systems affects transit routing. Bus routes that formerly operated on one street under two-way operation must now use two streets, one block apart, under one-way operation. This is more confusing to users and lengthens some pedestrian trips from bus stops to the desired destination. It becomes a particular problem when a major generator or generators are located on one of the two streets, such as a subway station, a major department store, and so on. In this case, a substantial pedestrian flow may be generated as transit users access the system.

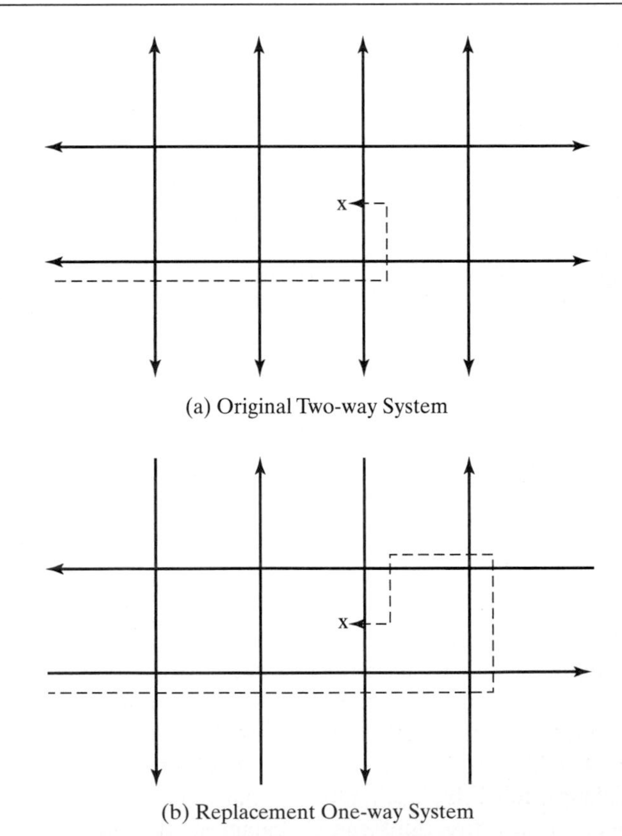

Figure 29.14: Increased Trip Lengths on a One-Way Street System Illustrated

Although there are clearly some disadvantages associated with one-way street systems, where widespread network congestion exists, and where reasonably paired couplets exist, it is often a simple way of accomplishing many traffic efficiencies. Table 29.5 provides a summary of the advantages and disadvantages of one-way street systems.

29.7.2 Special Use Lanes

Special use lanes on urban streets can be used to improve circulation and flow under a variety of circumstances. The most common forms for use on urban street networks are:

- Exclusive transit or HOV lanes
- Reversible lanes
- Two-way left-turn lanes

Exclusive transit lanes are provided where bus volumes are high, and where congestion due to regular traffic produces unreasonable delays to transit vehicles. Although a single transit lane can be provided along the right curb of a local street, where sufficient lanes are available, it is recommended that two lanes be provided, as illustrated in Figure 29.15. Because such lanes are provided when bus volumes are relatively high, a second bus lane is a virtual requirement to allow buses to pass each other, particularly where there are multiple routes using the facility that do not have the same bus stop locations. The provision of the second lane allows buses to "go around" each other as needed, without encroaching on mixed traffic lanes.

Such lanes expedite transit flow and act as an encouragement for travelers to use public transportation as opposed to driving. There two notable disadvantages of such lanes:

- Right-turning vehicles from mixed lanes must be allowed to use the bus lanes at intersections where right turns are permitted.
- Enforcement is difficult; no parking, standing, or stopping can be permitted in the bus lanes.

Often, transit-only lanes are heavily criticized by taxi operators. In some cases, disputes are avoided by allowing all high-occupancy vehicles (defined as two or more, or three or more persons per vehicle) to use the lanes. Taxis, however, may cause congestion within the lane by making curbside pickups or dropoffs at undesignated locations. Occupancy requirements are also difficult to enforce on urban streets.

Table 29.5: Advantages and Disadvantages of One-Way Street Systems

Advantages
Improved ability to coordinate traffic signals.
Removal of opposed left turns.
Related quality of flow benefits such as increased average speed and decreased delays.
Better quality of flow for bus transit; lower transit operating costs.
Left-turn lanes not needed.
More opportunity to maneuver around double-parked or slow-moving vehicles.
Ability to maintain curb parking longer than otherwise possible (due to capacity benefits).
Capacity Benefits:
Reduced left-turn pce's.
Fewer signal phases (at signalized intersections).
Reduced delay.
Better utilization of street width.
Safety Benefits:
Intersection LT conflicts removed.
Midblock LT conflicts removed.
Improved driver field of vision.
Disadvantages
Increased trip lengths for some/most/all vehicles, pedestrians, and transit routes.
Some businesses negatively affected.
Signal coordinated in grid still poses closure problem.
Transit route directions now separated by at least one block.
For transit routes, a 50% reduction in right-hand lanes; may create bus stop capacity problem.
Concern of businesses about potential negative impacts.
Fewer turning opportunities.
Additional signing needed to designate "one-way" designations, turn prohibitions, and restricted entry, as required by MUTCD.

Reversible lanes are used where highly directional traffic distributions exist and there are sufficient lanes on facilities to allow this. In such a plan, one or more lanes serves traffic in different directions according to the time of day. Reversible lanes are best controlled using overhead signals with green and red X's to signify which direction is permitted. Such signals should be supplemented by clear regulatory signing indicating the hours of operation in each direction. Reversible lanes cannot be used where intersections have exclusive LT lanes. The benefit of reversible lanes is that capacity in each direction can be varied in response to demand fluctuations. The most serious disadvantage is the cost to control such lanes and the confusion to drivers.

The two-way left-turn lane is provided on two-way urban streets where (1) there is an odd number of lanes, such as 3, 5, or 7, and (2) where midblock left turns are a significant cause of congestion. When employed, such a lane allows vehicles making a midblock left turn into or from a driveway to use the center lane as a refuge. In effect, vehicles waiting to make a left turn no longer block a through lane of traffic.

29.8 Special Cases

The following are four arterial management issues related to signalization:

- Transitions from one plan to another
- Coordinating multiphase signals
- Multiple and sub-multiple cycle lengths
- Diamond interchanges

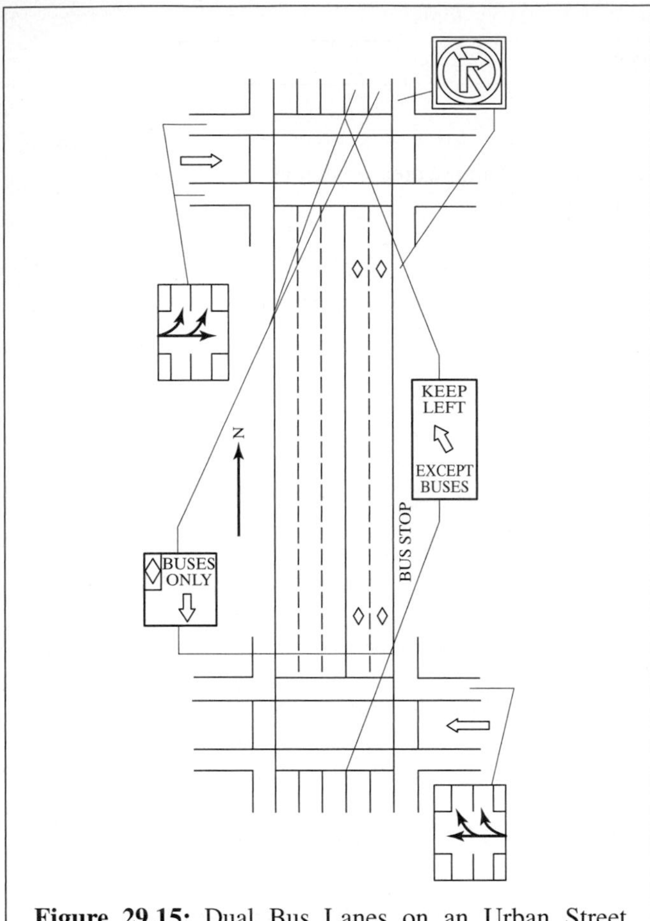

Figure 29.15: Dual Bus Lanes on an Urban Street Illustrated

29.8.1 Transitions from One Signal Plan to Another

Very little appears in the literature on finding a "best" or optimum way of moving from one plan to another in an orderly and efficient way. Nonetheless, it is an important problem; some engineers feel that the transition from one plan to another during peak loads is more disruptive than having the wrong plan in operation.

The fundamental problem is to get from "Plan A" to "Plan B" without allowing:

- Red displays so short as to leave pedestrians stranded in front of moving traffic
- Such short green displays that drivers get confused and have rear-end accidents as one stops but the other does not
- Excessive queues to build up during excessively long greens
- Some approaches to be "starved" for vehicles due to long red displays upstream, thus wasting their own green

Further, based on lessons learned in one demonstration system, Reference 26 reports that:

> No more than two signal cycles should be used to change an offset, and the offset should be changed by lengthening the cycle during transition if the new offset falls within 0 and 70 percent of the cycle length. The cycle should be shortened to reach the new offset if the offset falls within the last 30 percent (70–100 percent) of the cycle.

The most basic transition algorithm is the "extended main-street green" used in conventional hardware: At each signal, the old plan is kept in force until main-street green (MSG) is about to end, at which time MSG is extended until the time at which the new plan calls for its termination. Clearly, some phase durations will be rather long. However, the policy is simple, safe, and easily implemented.

Reference 27 reported on the test of six transition algorithms (some of which were boundary cases and not field-implementable algorithms) because of the logical importance of effective transitions in computer-controlled systems, which update their plans frequently. Over the range of situations simulated (volume increasing, decreasing, constant), the "extended main-street green" algorithm was no worse than any of the special designs.

29.8.2 Coordinating Multiphase Signals

Multiphase signals (more than two phases) are sometimes required by local policy, dictated by safety considerations (for instance, when lefts must cross a very wide opposite direction) or needed to reduce the pce of the left-turners in the face of an opposing flow.

When multiphase signals exist along a two-way arterial, they actually introduce another "degree of freedom" in attempts to get good progressions in two directions. Consider the time-space diagram of Figure 29.16, with the northbound progression set first, as shown. The usual challenge is to find a "best" southbound progression that does not disrupt this northbound success; the usual constraint is that the southbound and northbound greens must occur at the same time, so that perhaps only part of the "window" can be used for southbound platoon movement. This is illustrated in Figure 29.16.

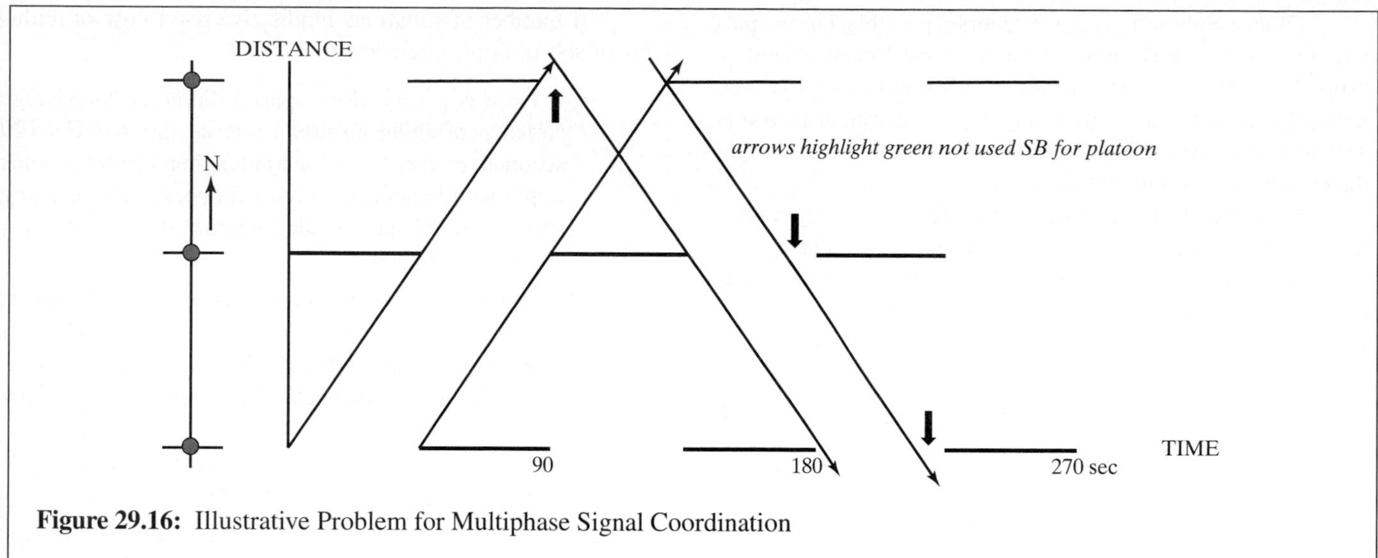

Figure 29.16: Illustrative Problem for Multiphase Signal Coordination

However, there is some new flexibility, introduced by the multiple phases. Refer to Figure 29.17, which shows that the SB through green can be "moved around" by the simple expedient of locating the protected left turns in various places. In this particular example, it gives the SB through a flexibility of ± 10 seconds relative to the (fixed) NB through-green initiation.

Observe that the SB window can be made wide by "pushing" the SB through at any intersection to an extreme in most cases. Further, the direction of the "push" generally alternates SB down the arterial (first one extreme and then the other) except for fortuitous spacings. Thus the actual selected settings will frequently alternate between "Pair 2" and "Pair 3" in Figure 29.17.

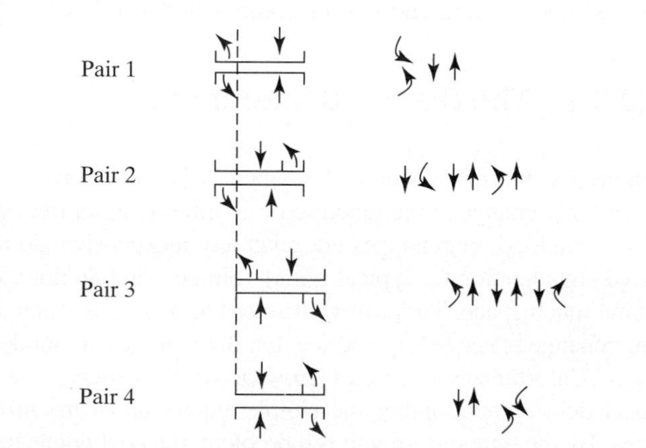

Figure 29.17: Candidate Phase Plans at Each Intersection of a Multiphase Arterial

Crowley [28] and Messer et al. [29] have both addressed the optimization of signal progressions along two-way multiphase arterials. Messer used his work as the basis for the PASSER program, which implemented this policy.

29.8.3 Multiple and Sub-Multiple Cycle Lengths

Coordination systems operate on the principle of moving platoons of vehicles efficiently through a number of signals. To do this, a common cycle length is almost always assumed.

Chapter 26 emphasized that the selection of cycle length is a system consideration. In addition, (1) delay is rather insensitive to cycle-length variation over a range of cycle lengths, (2) capacity does not increase significantly with increased cycle length, and (3) real net capacity is likely to decrease if large platoons are encouraged because of storage and spillback problems.

Nonetheless, there are situations in which multiple or sub-multiple cycle lengths can work to advantage or when other combinations are necessitated. As a matter of definition: If the system is at C = 60 seconds and one intersection is put at C = 120 seconds, it is a "multiple" of the system cycle length. If, however, the system is at C = 120 seconds and one intersection is put at C = 60 seconds, it is a "sub-multiple" of the system cycle length.

Other combinations are, of course, possible, but the pattern they induce will only repeat after the lowest period (a "supercycle length," so to speak) has passed. For a 60-second and a 90-second pair of cycle lengths, the common period is 180 seconds. For a 60-second and a 75-second cycle length, the common period is 300 seconds.

Figure 29.18 shows one of the effects of having a multiple cycle length in the system: The platoon discharged from the greater cycle length (and thus greater phase duration) moves into the downstream link, to be processed in two parts; with very good offset, this could be no problem, for much of the platoon is kept moving. However, for poor offsets, the entire platoon could become a queue, if only for a short time.

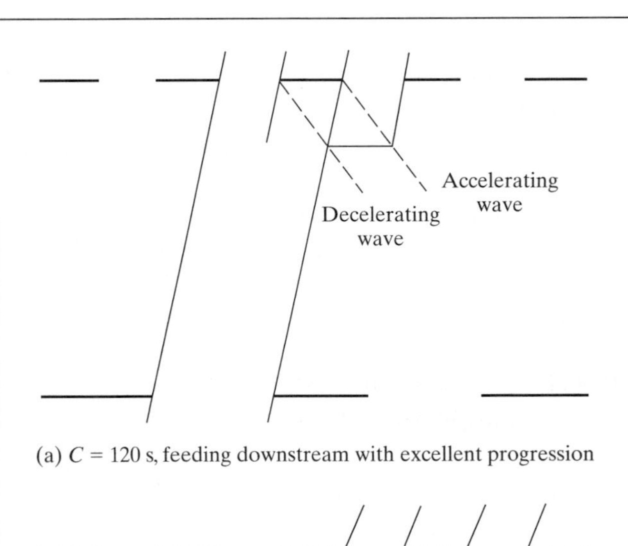

(a) C = 120 s, feeding downstream with excellent progression

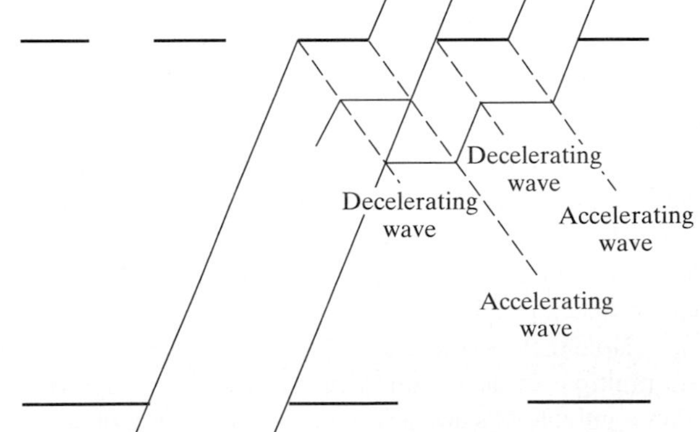

(b) C = 120 s, feeding downstream with unsatisfactory progression

Figure 29.18: One of the Effects of a Multiple Cycle Length in the System

A number of situations might give rise to use of multiple or sub-multiple cycle lengths:

- If there is a very close spaced diamond interchange somewhere along an urban arterial that has C = 120 seconds (or even C=90 seconds), then C = 60 seconds might well benefit the diamond operation by keeping platoons small and avoiding internal storage problems and even spillback.
- At a very wide intersection of two major arterials, at least one of which has a C = 60 seconds, the added lost time per phase (due to longer clearance intervals associated with the wide intersection) may present a problem, and it may be best to reduced the number of cycles per hour at this one intersection by increasing the cycle length to C=120 seconds—if storage permits.
- If turns dictate that multiphasing is required and if the geometry of the overall arterial indicates that the system cycle length should not be changed, then it may be best to allow this one intersection to have a different (greater) cycle length.
- At the intersection of two major arterials, each with its own system cycle length dictated by its own geometry, it may be necessary to accept a different cycle length at the common intersection.

Other examples could be constructed, but these serve for illustration, and are representative.

In considering the use of multiple or sub-multiple cycle lengths, attention must be paid to upstream and downstream storage, relative g/C ratios, and the length of the common period (the least common denominator). Some savings can be achieved in particular cases (e.g., Reference 30) but usually at the expense of a markedly more complicated analysis.

29.8.4 The Diamond Interchange

Figure 29.19 shows a sketch of a typical or "conventional" diamond interchange at the juncture of an arterial and a freeway. Such diamond interchanges are relatively inexpensive, do not need signalization for typical initial volumes, and do not consume much space. They are well suited to locations where an interchange is needed for service, but the volumes are modest.

Unfortunately, volumes grow at such locations, due to local development and/or the simple expansion of the urban area. By the time this growth is a problem, the contiguous land has often been developed and its cost is prohibitive (not to mention the practical and political problem with acquiring such land). Thus the option of a total redesign is often not open.

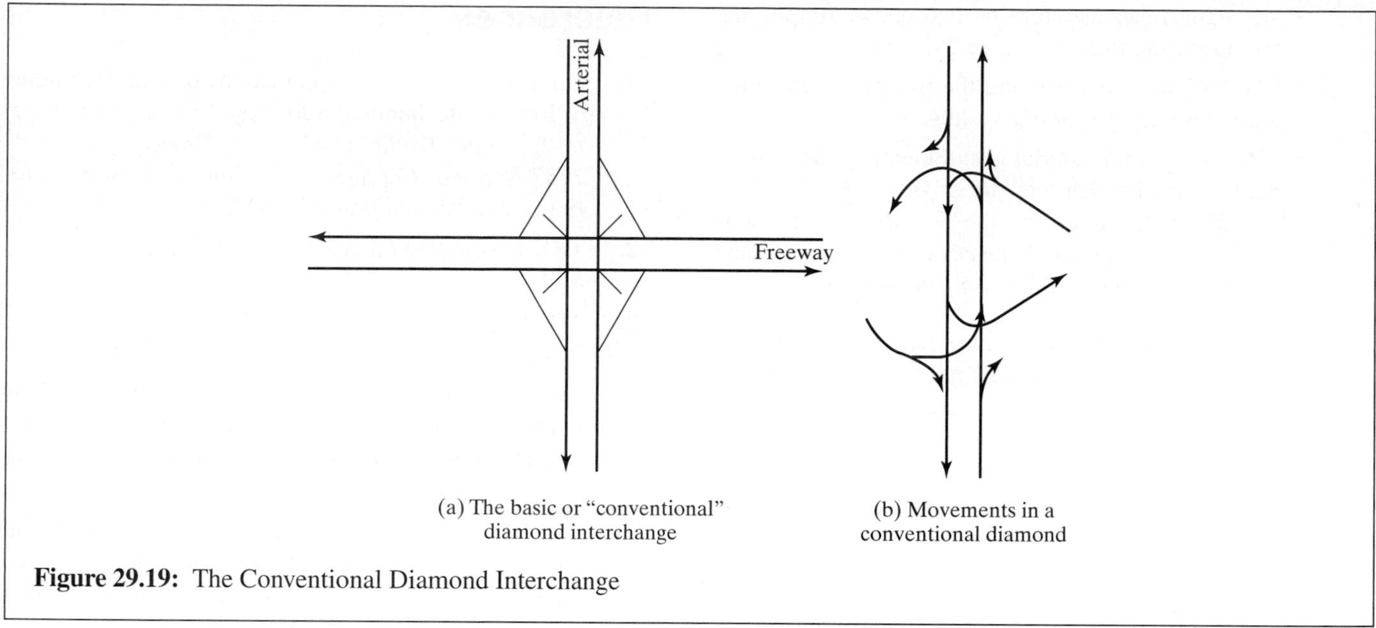

(a) The basic or "conventional" diamond interchange

(b) Movements in a conventional diamond

Figure 29.19: The Conventional Diamond Interchange

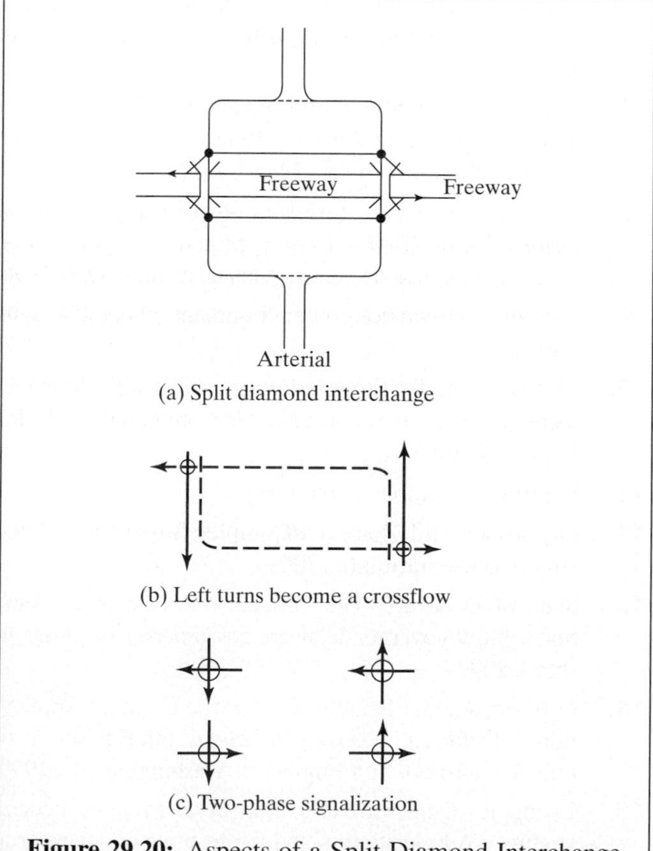

(a) Split diamond interchange

(b) Left turns become a crossflow

(c) Two-phase signalization

Figure 29.20: Aspects of a Split Diamond Interchange

Historically, much of the literature on diamond interchanges has been concerned with signal optimization (see, for instance, References 31 and 32). The reason is simple: With the space unavailable, signalization is one of the few hopes for coping with the problem.

The inherent problems of a conventional diamond interchange under heavy volumes are considerable: As illustrated in Figure 29.19 (b), there are numerous conflicting movements. Further, the bridge over the highway is relatively short (typically 300 feet) and must be used for storage of left-turners during certain phases. Last, volumes from the freeway tend to back up onto the freeway, substantially degrading its performance. All of this is exacerbated by periodic intersection blockages as queues develop due to the left turns from the arterial onto the freeway.

The true problem is that the initial design did not anticipate the traffic growth, and the initial economic analysis—if any—did not include "life cycle" costs reflecting traffic growth, delay costs, travel-time costs, accident costs, and land and construction capital costs [*33*].

One design alternative that can be considered is the "split diamond" illustrated in Figure 29.20 (a).

- The storage leading off the freeway is greater, for it includes the EW space between the NB and SB sections of the "split" arterial.
- There are no left-turn flows crossing or competing for green with opposing flows. Refer to Figure 29.20 (b).

- All signals can therefore be two phase, simplifying the operation. Refer to Figure 29.20 (c).
- The EW storage space and the two-phase operation allow for narrower bridge widths.
- If the split of the arterial is sufficiently wide, drivers may not realize that they are driving through a "diamond" configuration, for the two arterial directions could be considerable distances from each other, and the enclosed land could be fully developed.

Some signal optimization is in order, for the "closure" equations for optimum progressions must be tailored to the geometry and the traffic patterns. There are other diamond configurations and variations on each. However, the conventional and split serve to introduce the principal issues in this rather familiar—and troublesome—configuration.

One of the inescapable lessons of any consideration of the diamond interchange is that *the true solution is much more than an optimum signalization.* The literature is filled with better signalization schemes because the given condition in so many localities has always been, "We have a problem, but cannot provide more space: What can we do?" The answer is then, "Patch and cope, using signalization."

For you, however, the true solution—and lesson—is to avoid such ends by initial design, if at all possible. Cost-effective alternatives based on life-cycle costs are one method of enhancing the chance of "building in" a better solution.

29.9 Summary

During the several editions of this text, the state of the practice has evolved to a healthy emphasis on serving multiple modes of transportation on a shared facility, balancing the needs of the several user groups.

The chapter has focused on arterials and on local streets, and on related design principles and methods of preserving arterial function. The latter methods often fall under the rubric of *access management.* Other important themes, sure to recur in your career, are the *balanced use of streets* and *complete streets,* and *traffic calming.* The role of *roundabouts* in these concepts and in modern design is considered. *Network issues* such as one-way streets and pairs were addressed. Finally, a set of special cases (most related to special signal issues) were considered.

The references cited in the chapter have evolved also and now include a number of Web sites that can serve as excellent ways for you to supplement the text, keeping up to date on rules and practices and seeing new cases and installations.

References

1. Kramer, R.P., "New Combinations of Old Techniques to Rejuvenate Jammed Suburban Arterials," *Strategies to Alleviate Traffic Congestion, Proceedings of the 1987 National Conference*, Institute of Transportation Engineers, Washington DC, 1988.
2. TRB Committee on Access Management, http://www.accessmanagement.info/.
3. "TRB Access Management Manual," Transportation Research Board, ISBN: 0-309-07747-8, 2003.
4. "Ten Ways to Manage Roadway Access in Your Community," Center for Urban Transportation Research (CUTR), University of South Florida for Florida DOT, 1998.
5. Long, G.D., et al., "Safety Impacts of Selected Median and Access Design Features," report to Florida Department of Transportation, Transportation Research Center, University of Florida, Gainesville, FL, 1995.
6. Bowman, B.L., and Vecellio, R.L., "The Effect of Urban/Suburban Median Types on Both Vehicular and Pedestrian Safety," Transportation Research Record 1445, 1994.
7. "Driveway and Street Intersection Spacing," *Circular 456*, Transportation Research Board, National Research Council, Washington DC, March 1996.
8. McShane, W., et al., *Insights into Access Management Details Using TRAF-NETSIM*, presented at the second annual conference on Access Management, August 1996.
9. http://www.fhwa.dot.gov/environment/bikeped/design.htm.
10. "Transit-Friendly Streets: Design and Traffic Management Strategies to Support Livable Communities," TCRP Report 33, 1998.
11. http://www.completestreets.org/.
12. http://www.slideshare.net/CompleteStreets/complete-streets-presentation (c. 2009).
13. State of Delaware: The Official Web Site of the First State, http://governor.delaware.gov/orders/exec_order_6.shtml, 2009.
14. O'Brien, A., and Brindle, R., "Traffic Calming Applications," *Traffic Engineering Handbook*, 5th Edition, Institute of Transportation Engineers, Washington DC, 1999.
15. Ewing, R., *Traffic Calming: State of the Practice*, Federal Highway Administration and the Institute of Transportation Engineers, Washington DC, August 1999.

16. http://www.fhwa.dot.gov/environment/tcalm/.

17. http://www.trafficcalming.org/.

18. http://www.sanjoseca.gov/transportation/traffic_calming.htm.

19. http://www.dot.state.pa.us/Internet/Bureaus/pdBHSTE.nsf/infoTrafficCalming?readform

20. Smith, D.T., *State of the Art: Residential Traffic Management*, FHWA Research Report RD-80-092, December 1980.

21. https://www.nysdot.gov/portal/page/portal/main/roundabouts.

22. https://www.nysdot.gov/portal/page/portal/divisions/engineering/design/dqab/.

23. http://www.roundaboutsusa.com/design.html.

24. http://roundabout.kittelson.com/.

25. Karmeier, D., "Traffic Regulations," *Traffic and Transportation Engineering Handbook*, 2nd Edition, Institute of Transportation Engineers, Washington DC, 1982.

26. Bissell, H.H., and Cima, B.T., "Dallas Freeway Corridor Study," *Public Roads,* 45, no. 3, 1982.

27. Ross, Pal, "An evaluation of Network Signal Timing Transition Algorithms," *Transportation Engineering,* September 1977.

28. Crowley, K.W., "Arterial Signal Control," Ph.D. dissertation, Polytechnic Institute, Brooklyn, NY, 1972.

29. Messer, C.J., et al., *A Variable-Sequence Multiphase Progression Optimization Program, Highway Research Record 445*, Transportation Research Board, National Research Council, Washington DC, 1973.

30. Kreer, J.B., "When Mixed Cycle Length Signal Timing Reduces Delay," *Traffic Engineering,* March 1977.

31. Messer, C.J., et al., "Optimization of Pretimed Signalized Diamond Interchanges Using Passer III," *Transportation Research Report 644*, Transportation Research Board, National Research Council, Washington DC, 1977.

32. Messer, C.J., et al., "A Real-Time Frontage Road Progression Analysis and Control Strategy," *Transportation Research Report 503*, Transportation Research Board, National Research Council, Washington DC, 1974.

33. Oh, Y.T., "The Effectiveness for the Selections of Various Diamond Interchange Designs," Ph.D. dissertation, Polytechnic University, Brooklyn, NY, 1988.

Problems

29-1. The street network shown here serves a small downtown street network. The main street (shown as the thick black line) has six lanes, three in each direction. Most, but not all, of the major generators are located on this street. Two bus routes use this street as well. All other local streets shown have standard 40-feet street widths with metered parking on both sides. There is no parking on the main street. Intersection congestion has created significant delays throughout the network. A one-way street system and other measures are under consideration. Is a one-way street system an appropriate solution? Why or why not? If it is, propose a specific plan. Propose a control plan for the network that would alleviate some of the existing problems. State any assumptions made. The spacing between North-South streets is 600 feet. The spacing between East-West streets is 700 feet.

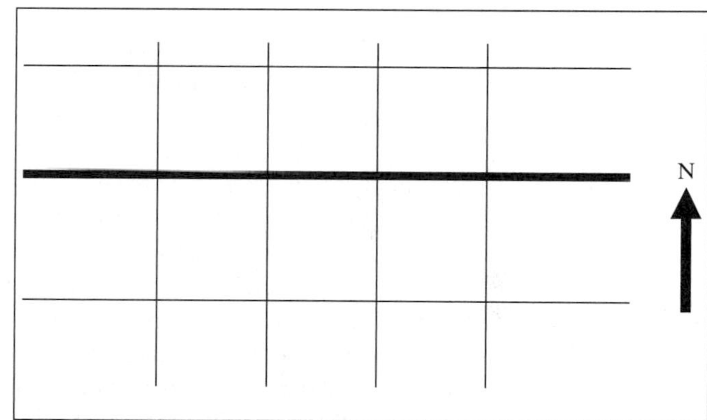

29-2. Class Project. Establish working groups of from three to four people. Each group will select a local street network exhibiting problems that might be addressed by traffic calming measures. The group will survey the network and analyze what specific traffic problems need to be addressed. The objectives to be achieved should be clearly stated. A plan of action should be proposed, with each group making a public presentation to the class. If possible, involve local officials or groups in the project and invite them to the presentations.

CHAPTER 30

Traffic Impact Analysis

In the United States, landmark legislation regarding the environmental impact of federal actions came into effect with the signing of the National Environmental Policy Act (NEPA) on January 1, 1970 [*1*].

NEPA's procedures apply to all agencies in the executive branch of the federal government and generally require an *environmental assessment* (EA) document that will result in a finding of no significant impact (FONSI) or an *environmental impact statement* (EIS) that includes a detailed process for its development, submission, review, and consequent decision making. The legislation also established the Council on Environmental Quality (CEQ). See Reference 2 for more information on CEQ and its role.

Over the years since 1970, the definition of a "major federal action" by an agency in the executive branch has come to include most things that the agency could prohibit or regulate [*3*]. This has come to the current state of the practice that a project is required to meet NEPA guidelines whenever a federal agency provides any portion of the financing for the project and sometimes when it simply reviews the project.

An environmental assessment (EA) or consequent EIS includes attention to a full range of potential environmental impacts, and it certainly includes those due to traffic. Indeed, the traffic impact work is generally an important input to the assessment of noise and pollution impacts (due to the related mobile source emissions).

The individual states have generally passed their own environmental legislation, extending the range of needed environmental impact assessments, following a process akin to the federal one. For instance, the State of New York has its State Environmental Quality Review Act (SEQR) [*4*].

Local governments generally have their own legislation and processes for actions taken at their own level. New York City has a CEQR process [*5*], and there is a full range of such legislation throughout the United States. It is imperative that a practicing professional be aware of the governing laws in a jurisdiction, including which level of government has purview on a given project *and* which agency will be the *lead agency* for the specific project or activity.

A traffic impact analysis (TIA) is a common element of both EA and EIS documents required by the relevant level of government, or can be required as a separate submittal by an agency that has jurisdiction. Despite its short form name, the TIA must have information on both impact and proposed mitigation.

Different jurisdictions have their own guidance on what constitutes a significant impact that requires mitigation be evaluated and addressed. In some cases, it is a certain change in v/c ratio at an intersection and/or level of service change on approaches, at intersections, and/or on arterials. Again, it is imperative that the practicing professional have knowledge of the specific requirements on state and local levels, and federal (if applicable).

Generally, all such legislation requires that the environmental impacts be identified and estimated using the current state of the art/state of the practice tools and methods, and that mitigation be investigated and proposed to the extent possible. It is *not* required that full mitigation be achieved, but rather that the impacts and effects be fully disclosed so that the relevant decision maker as established by law can make a fully informed decision on whether the project is permitted to proceed. Indeed, the challenges to a decision tend to be on whether the process was followed rigorously, whether proper methodologies were used, and whether there was full disclosure of impacts. Provided with proper information and following an orderly process, the law explicitly vests the decision authority in a specific agency or designated position, *and the decision itself is not a valid basis for litigation.*[1]

As a practical matter, a recommendation that there are impacts that cannot be fully mitigated will be the basis for lively discussion in the review process, and reviews tend to go smoother when full mitigation is feasible.

30.1 Scope of This Chapter

You should not expect to have mastered the ability to conduct full and complete traffic impact assessments after simply reading this chapter, or even this entire text.

Rather, this chapter is intended to focus your attention on how information from the preceding chapters must be brought to bear in executing a traffic impact assessment, and on how you must use this knowledge to create design concepts that can mitigate impacts.

One of the authors has taught a project-based course centered around a traffic impact assessment and has used the ITE *Transportation Impact for Site Development* [6] as a companion text for that course (it was a second course in a sequence, and also covered several chapters from this text and built on the chapters taught in the first course).

This chapter provides an overview of the process and techniques in the next two sections and then provides two case studies that can be used as course projects or as the basis for discussion. The chapter does *not* provide total solutions to either of the case studies, and this is intentional: At this point, the learning is best done by meeting the challenges in a project-based experience, interacting with the instructor. (Issues are identified, and some guidance provided, but a definitive "correct solution" is absent by design.)

[1]This last statement, indeed the entire paragraph, is drawn from both the law and the practice, but nuances can be better explained by the attorney on the team.

30.2 An Overview of the Process

This section focuses on the process as shown in Figure 30.1. There are variations on this (and more comprehensive versions, for specific localities), but it can serve the purpose of this chapter. The boxes are numbered for convenience and are referred to as "steps" in this section.

Step 1 is rather self-explanatory, but it is easy to encounter both clients and other professionals who have views and preferences that influence the traffic engineer's work, simply because each of us knows more of our specialty than all the others. In one enjoyable instance, an architect casually moved 500,000 square feet of one tower across the project to another tower because "it looked more balanced" *after* the detailed traffic circulation work was done. (To that point, it had been part of the "dominant tower" and "signature building.")

Step 2 is a working plan for the entire project, and specifically preparation for Step 4. These are the important issues:

- What do the local regulations and practices require, in terms of hours/days of data, analysis tools, required methods (e.g., HCM, CLV, Synchro), triggers for mitigation, and other?
- What do the local regulations and practices require, in terms of site development as it affects traffic? This may include setbacks, buffer zones, mandated allowances for parking and transit, mandated emphasis on traffic calming within the site, and other.
- What days and periods within the day(s) are justified by the project or by local regulations and practices?
  - AM, midday, PM are commonly required.
  - Weekend may be required for some developments.
- What analysis periods are required?
  - The most common are the existing condition, the future no-build (FNB), and future build (FB).
  - On some projects, the period of construction is so large and/or so long that the peak of the construction period must also be analyzed.
- What exactly is the base case for analysis? Is it the future no build (FNB) with existing signalization or the FNB with optimized signals, or other?
- What are the local growth rates to be used, and what if any planned and approved major developments are there? Is there local guidance (i.e., a guidebook or set of tables) on trip generation rates or is the ITE *Trip Generation* publication [7] sufficient?

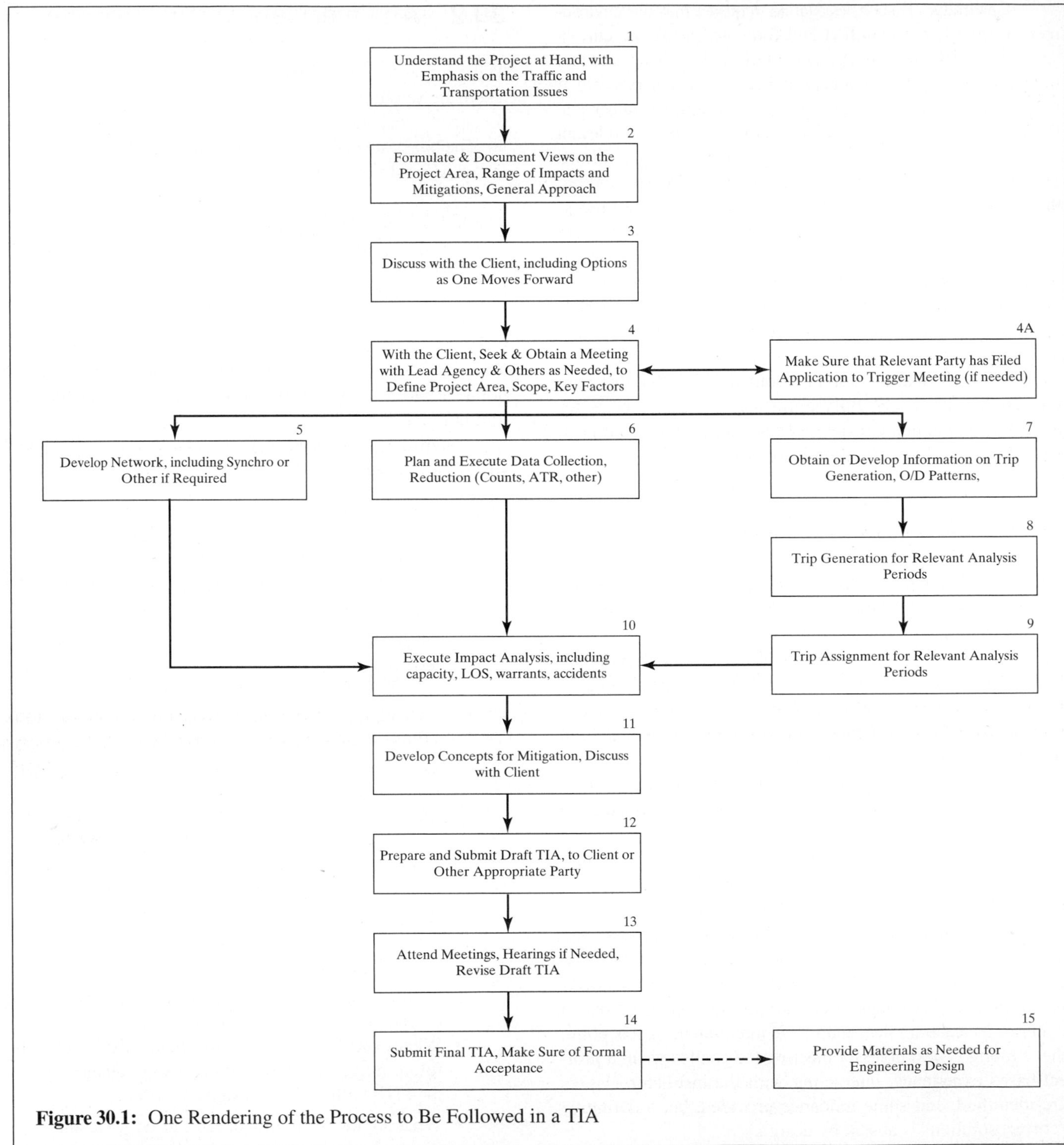

Figure 30.1: One Rendering of the Process to Be Followed in a TIA

- What are the relevant road system and transit facilities? How far will concentrated traffic flow before it disperses into background levels? What intersections and other key points are affected and need analysis?
- What data exists, and how in what form is it available (counts, ATR stations, accident data, other)?

That is not to say that all these questions will be answered in Step 2, but rather that the engineer must get a handle on each of them, particularly with regard to the extent of the project area and the intersections/facilities to be affected.

Step 3 is important, in that the client and/or their team (attorney, architect) must understand the difficulties that are likely to be encountered in the process, and the need for a reasonable project area. If it is obvious to the traffic professional that the impact area is larger than the nearby intersection, it will also be obvious to the professionals who are reviewing the TIA. Some clients may like it to be smaller (sometimes much smaller), but unless they have time to iterate with the reviewing agency, both time and cost will dictate that they be made aware of reality.

Of course, that is *not* to say that the defined project area (i.e., the impact area) *must* be large. In many projects only a small number of key intersections are involved before the traffic distributes into the background levels. Smaller is better, but reasonable is best.

Step 4, the meeting with the cognizant lead agency and other relevant parties (e.g., state, counties, towns), is a defining moment in the project. It is generally a formal step in the local process, needs an agenda, and must seek to arrive at a set of action items covering all points raised in Step 2, but especially these:

- Mutual agreement on the defined impact area (the "project area") for the analysis.
- Mutual agreement on the supporting data program (data to be collected, amounts, number of days, and so forth), and on the key intersections/facilities to be analyzed.
- Clear understanding of local requirements on growth rates to be used, approved projects that need to be considered beyond the background growth rates, and standard practice documents to be used (many jurisdictions have publications or memoranda specifying these, including tools and techniques to be used).

To avoid wasted effort and awkwardness later, it is best—let us say vital—that the lead agency sign off on the agreed items. This may take the form of a letter accepting the minutes of the meeting but in some cases is an e-mail acknowledging the discussion and accepting the minutes. Verbal approvals are not really useful, if only because personnel change over the course of a project and because people invariably have slightly different recollections.

The traffic professional must be aware that in some jurisdictions, such a formal meeting and agreement is simply not possible *unless the formal application to start the process has been filed* (Step 4A).[2] This application generally involves more issues than traffic (including timing issues known only to the client) and has to be filed by the appropriate party with the client's approval. The NYS SEQR is one such process.

After Step 4, the project tends to kick into high gear. A set of three major activities happen, somewhat concurrently:

1. The network is encoded into the analysis tool(s) to be used, whether they be spreadsheets or computer programs (Step 5); more is said on this in the next section.
2. The agreed data is collected and summarized (Step 6) and made ready for the analysis.
3. The references for trip generation rates that have been agreed, whether based on local practices or ITE [*7*] or information provided by the traffic professional and sourced, is documented (Step 7), used to establish number of trips generated (Step 8), and then distributed onto the network (Step 9) for each relevant time period (e.g., AM, midday, PM) at each relevant stage (e.g., Existing, FNB, FB).

The next step requires a good bit of careful work but is somewhat anticlimactic, given the above. Step 10 is the actual execution of the analysis that is the heart of the "impact" section of the TIA.

Step 11 is the most creative and demanding part of the entire exercise because design is a creative process as well as an orderly one: The traffic professionals must identify one or more mitigation plans that address the adverse impacts that become clear in the analysis work of Step 10.

The sets of solutions available in Step 11 (the mitigation plans) include the following:

- Retiming of signals, including different phases and cycle lengths, as well as different offset plans.
- Addition of signals as warranted by increased traffic or other factors.
- Addition of driveways for the project as needed, consistent with local access management policies and

[2] Some dialogue has been known to take place on a conversational basis, but it is not reasonable for the traffic professional or the client to expect these to be binding.

maintenance of arterial flow and function. The driveways may include such innovative designs as illustrated in Figure 29.5 (Chapter 29).

- Addition of lanes approaching and departing from intersections, to increase throughput on specific approaches.
- Addition of additional intersection(s) as needed or of additional lanes.
- Use of jug-handles and other solutions to reduce left-turn activity.
- Programs to increase average vehicle occupancy.
- Other traffic demand management (TDM) solutions, such as shifting work hours at the project site and/or sponsoring employee van pooling and/or transit check programs.

For solutions in the last two bulleted items, it is incumbent on the applicant (the client and their traffic professional) to make it clear how such policies and practices will *really* be put into effect. This will be expected by the TIA reviewers if the mitigation depends on them, and approval may be contingent on terms related to the proposed programs.

It is ***imperative*** in our view that the client be briefed in advance of the draft report about the mitigation options and the related first-cut (i.e., rough cut) estimate of costs. This is for a very practical reason: The client will probably have to pay for some or all of the mitigation and should know what the costs are likely to be.

At this stage (Step 11), there is likely to be some good discussions about the trade-offs among alternative mitigation approaches. These will often involve cost, ease of approval, and speed of the approval process (in some cases, the expression "time is money" is very apt because of the overall project costs and schedule). In some cases, other factors known best to the client will arise, such as work rules for employees, as a cost factor.

Step 12 is the formal preparation of the draft TIA, internal reviews for quality control, client review and comment, and submittal to the lead agency by the appropriate party. This draft may include a CD containing the data files and the input streams for any computer programs used, as well as sample animations.

Step 13 is the review and approval process, which will surely involve meetings convened by the lead agency and may well involve public presentations and hearing, receipt and documentation of comments, and revisions to the TIA.

Step 14 is the submittal of the final TIA document, either as part of another document or a stand-alone document (depending on the required process), leading (one would anticipate) to final, formal acceptance.

Step 15 is internal to the client or their team. The functional designs and traffic loads developed in the course of the TIA are an important input to the engineering design that must generally follow acceptance of the TIA by the cognizant agency.

A final note on time frames: This overall process is not instantaneous. All of the preceding steps can easily consume 6 to 12 months. Following approval of the mitigation plan in the TIA (for that is what acceptance means, as used here), the next steps are detailed engineering design, submittal of permit applications and related approvals, followed by construction. The construction period may be shut down in the winter months, and an Maintenance and Protection of Traffic (MPT) Plan is generally required as part of the permit process. For sizable projects with a reasonable amount of mitigation work, these extra steps can add up to 15 to 18 months *after* the TIA acceptance. It is possible that the total process may move more rapidly, but that needs to be assessed.

30.3 Tools, Methods, and Metrics

This text has presented information on the state of the art and the state of the practice in traffic engineering, with a strong emphasis on the levels of service as defined in the *HCM*. Chapter 23 presented the critical movement analysis (CMA) approach to intersection capacity and timing, as an alternative method advocated or used by some.

But you *must* appreciate that it is the local jurisdiction—usually at the state level—that determines the exact method to be used in that jurisdiction. And some details of design practice (including the acceptance of some design concepts) are sometimes delegated to the local district or regional offices, so that variation within a state can be expected. When roads are solely within the control of a county or town, their rules and procedures may prevail. Therefore, knowledge of local practice and rules is *essential* to the practicing traffic professional.

Fortunately, these rules are usually easily obtained, and they are posted on the official Web sites of the state or local jurisdiction. Equally fortunately, the review process usually involves a *lead agency* that coordinates the information and needs.

At the same time, it is sometimes natural for counties to have different concerns and priorities than states, or for one region to have more precise rules and practices than its neighbor. Most often, the goodwill and professionalism of all concerned overcome potential problems, but there are protocols and practices to respect.

To consider the range of practices with regard to just intersection and arterial evaluation, the following is informative (and based in fact):

- Some require impact to be expressed in terms of *level of service (LOS) changes* by intersection or by lane group, and cite the HCS+ software [*8*] as the expected tool.
- Some want *both LOS and v/c* ratio changes to be reported, with the HCS+ software for single intersections and Synchro [*9*] for sets of intersections and arterials.
- Others specifically mandate a *critical lane volume (CLV)* methodology provided by the state (e.g., Maryland). The "not to exceed" CLV is 1450 veh/h. The procedure is rooted in the method introduced in Reference 10 as interim materials to the *HCM*. Rather than being considered "dated," it is reemerging as an effective and efficient tool, and it is the logical foundation for the treatment in Chapter 23 of this text.
- Several states accept *Synchro LOS results* as if they were as equally valid (and exactly the same) as the *HCM* results.
- A number of states require "*Synchro visualizations*" of the traffic conditions, although the actual visualizations are produced by a separate tool (the SIMTraffic simulator [*11*]) that is sold as a companion to Synchro.[3]
- Some states focus on the *intersections rather than the arterial*, primarily by silence on the arterial impacts (i.e., average travel speeds and arterial LOS).
- At least one state had begun to focus on *arterials to the exclusion of intersections,* at least in the initial planning-level review. That has evolved to a more balanced view that includes arterial LOS *and* intersection LOS *and* intersection *v/c* ratio.

In terms of traffic visualizations, other tools are commercially available and have merit, including VISSIM [*12*] and AIMSUN [*13*].

Related to the discussion of the critical movement analysis, consider the values of "maximum sum of critical movement volumes" that can be accommodated for various conditions. The computations are done using a loss time per phase of 4.0 seconds and a discharge headway of 1.9 sec/veh

Table 30.1: Values of Maximum Sum of Critical Movement Volumes (veh/h), for Various Conditions, Including Cycle Length

h =	1.9 sec/veh,	0% trucks
pce_{TRUCK} =	2.0	5% trucks
t_{LOSS} =	4.0 sec/phase	

Cycle Length (sec)	Number of Phases		
	2	3	4
60	1564	1444	1323
70	1598	1495	1392
80	1624	1534	1444
90	1644	1564	1484
100	1660	1588	1516
110	1673	1608	1542
120	1684	1624	1564

$v/c = 0.90$

Cycle Length (sec)	Number of Phases		
	2	3	4
60	1408	1299	1191
70	1438	1346	1253
80	1462	1380	1299
90	1480	1408	1335
100	1494	1429	1364
110	1506	1447	1388
120	1516	1462	1408

for passenger cars (consistent with the HCM saturation flow rate of 1900 pcphpl). With 5% trucks, the discharge headway is changed to 2.0 sec/veh and used in Table 30.1.

The upper set of values in Table 30.1 is based on full utilization of the green by the vehicles; the lower set is based on 90% utilization (that is, a $v/c = 0.90$ on each phase). The lower set, of course, is 10% lower than the upper set, by simple arithmetic.

Some observations are in order, using the lower set:

- For a two-phase signal with a cycle length C = 80 seconds, the number shown (namely, 1,462 veh/h) is comparable to the CLV upper limit cited above (namely, 1,450 veh/h).
- Each additional phase decrease the value by about 5%, using C = 80 seconds as a reference condition.
- This is probably an overstatement because added phases tend to imply longer cycle lengths, so a decrease

[3]As noted in Chapter 26, the two tools sometimes produce radically different results, particularly when intersection spillback and blockage is involved.

of about 2.5% per added phase is a plausible rule of thumb.

- Within the lower set of values, starting with C = 80 seconds as the reference, each increase of 10 seconds of cycle length adds 1% to the displayed value. But one must remember discussions in this text that indicate the saturation flow becomes less efficient, so the nominal improvement is probably not significant.
- Using the same starting point, each decrease of 10 seconds in cycle length loses 2% on the displayed value. But the main purpose of doing so would be queue management in congestion, a different priority.

Finally, this little exercise with Table 30.1 is interesting, but not dominant, for two distinct reasons: (1) the traffic professional is governed by the formal procedure adopted locally, not this exercise, in applying corrections, (2) changes to "maximum sum of critical movement volumes" is *not* identical to *capacity* because the nominal losses cited here can be adjusted with shifts of green time (remember, v/c = 0.90 is used) to favor approaches with more lanes and thus more vehicles *and* because a measure such as added phases is generally taken to correct a problem that already degraded the base number.

30.4 Case Study 1: Driveway Location

This case study only addresses one fragment of a traffic impact analysis, namely the effect of driveway location on quality of flow along the arterial. The lessons to be learned by this exercise include:

1. The new driveway will add flows to both the NB and SB arterial flows, and do so "between main street platoons," thus making coordination less effective.
2. The new driveway will take vehicles from the passing platoons, thus leaving holes in them and making them less cohesive, also making coordination less effective.
3. The new driveway can totally disrupt the NB and/or SB green bandwidths if it is poorly placed within the block.

The case to be considered by the reader is the same basic case shown toward the end of Chapter 26, so the solution shown there in Figure 26.29 is available as a starting point.

You are expected to use the information in Table 30.2, which is the same as Table 26.5. A significant development is

Table 30.2: Inputs for Case Study 1

		Hourly Flows (vph)											
		NB			EB			SB			WB		
	Link Length (ft)	L	T	R	L	T	R	L	T	R	L	T	R
Main St & Avenue E		70	1460						**800**	50	100	**600**	100
	1500												
Main St & Avenue D			1480	50	50	**600**	50	50	850				
	1500												
Main St & Avenue C		70	1430						850	50	100	**600**	100
	1500												
Main St & Avenue B			1450	50	50	**600**	50	50	900				
	1500												
Main St & Avenue A		70	**1400**						900	50	100	**600**	100

2 lanes per direction on Main St
left turn bays on Main St, 250 ft, both directions
RTOR prohibited everywhere
60 fps free flow speed, Main Street
All avenues have 2 moving lanes, but are one-way streets
PHF = 0.91
Minimum ped crossing times = 17 sec across side streets, 30 sec Main St

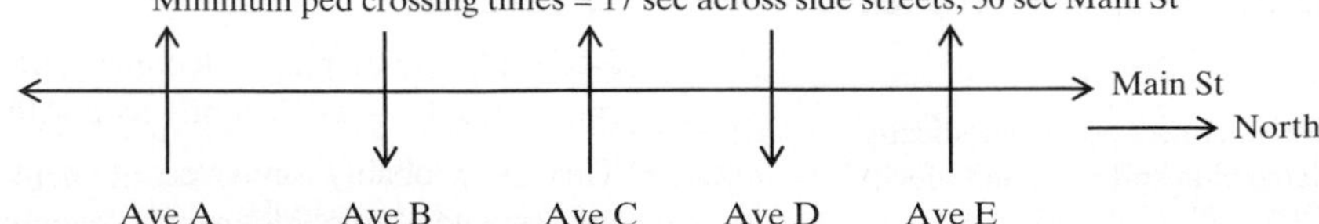

proposed for a site on the east side of Main Street, between Avenues B and C. Here are the specifics:

- The development will add "X" vehicles per hour to the Main Street traffic heading NB and "X" vehicles per hour to the Main Street traffic heading SB. Each of these flows will turn into the driveway(s) provided.
- At the same time, the development will generate "X" exiting vehicles per hour exiting and heading SB on Main Street and another "X" vehicles per hour exiting and heading NB on Main Street.
- The preferred driveway (pair) is to be the configuration shown in Figure 29.5. The signalized intersection is to be located "N" feet north of Avenue B.

Using Synchro and SIMTraffic, you are asked to analyze the impacts for four scenarios. It is not necessary to propose mitigation at this stage, unless the instructor makes that an addition part of the assignment.

You may choose to use VISSIM or AIMSUN for the visualizations; that is acceptable. The choice can be made based on tools available to the college or to you (limited student versions of some tools are available at low cost).

You should be aware that the emphasis is on *relative* impacts of the four scenarios. However, some of the tools may model some effects different than others, and the signal optimization program (i.e., Synchro) may consider some queueing effects differently than the simulation models. You may have to fine-tune the signal timing results. Also, some of the simulation models may give metrics such as arterial travel time more easily than others.

30.5 Case Study 2: Most Segments of a Traffic Impact Analysis

Case Study 2 is much more comprehensive than the first case study but is less intense than a full traffic impact study, in that (1) the trip assignment paths are specified in detail, (2) the project area is defined, (3) the data are provided and are balanced so that it is internally consistent,[4] (4) applicable local rules are provided, in terms of requirements.

[4]In general, traffic counts provide numbers that simply do not add up. This may be due to parking lots or generators within individual links, but it can also be due to simple random errors in the field work. In the latter case, and assuming reasonable variations, the counts are then balanced by the analyst to reflect a more realistic snapshot.

Some of you may believe that the specified requirements on buffer zones, parking spaces per unit of activity, and other elements are very restrictive. They were, however, assembled from real requirements in real locations within the United States. There are locations that have most if not all of these requirements.

The only major embellishment is the requirement that the site allow for both (1) transit access as if 20% of the trips were using public transportation, and (2) parking that recognizes that 95% of the trips will arrive by auto. Although this is unrealistic *in the short term* and might be viewed by some as a burden, we believe it could represent good long-term planning on the part of the local jurisdiction. For instance, it provides a critical part of the enabling infrastructure that will lead to a future transit use of 20% (other parts include a bus route system that is sufficiently complete to enable the trips and sufficiently frequent to make them attractive).[5]

Some information is *not* provided. Actual trip generation rates are not provided but can be obtained from Reference 7 or 8 or from a Web search that will give you the specific numbers used in certain jurisdictions. Spatial requirements for parking are not provided but can be found or estimated as suggested in Problem 30-3. Other needed information can be found in various sources or provided by the instructor.

30.5.1 The Project Area and the Existing Condition

Figure 30.2 shows the project area, including the two parcels that will be of interest in this case study. Table 30.3 provides details on the streets (number of lanes, etc.) and the available right-of-way.

Table 30.4 provides the hourly *volumes* for four periods that may be relevant to the project at hand. Other important information, such as the PHF, is included in Table 30.5.

Figure 30.3 shows the sources of *new* traffic attracted to the developments at the site(s) of interest. Trips return from the site(s) to these destinations. Note that the *magnitude* of the traffic is *not* specified, either as arriving traffic or departing traffic in each of the four analysis periods (AM, midday, PM, and weekend).

[5]There was a time when fully accessible transportation was questioned on a cost-effectiveness basis, given that the full set of requirements did not exist for a meaningful number of trips. Those requirements included curb cuts, accessible entrance, accessible rest rooms, legible signage, and then accessible buses and rail transit. Policy decisions to build this infrastructure systematically over decades has resulted in accessible *systems* in a number of locales.

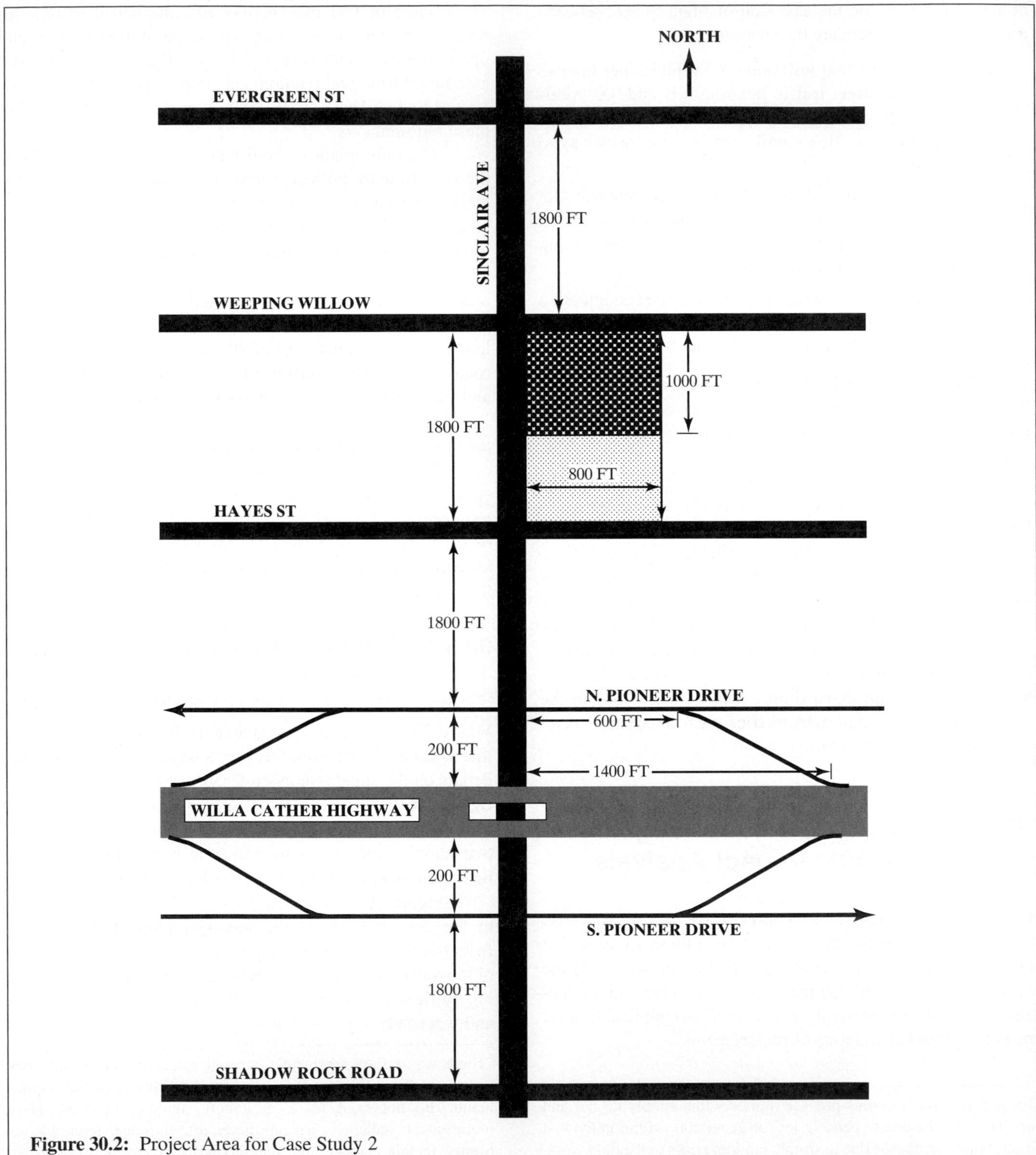

Figure 30.2: Project Area for Case Study 2

Table 30.3: Details of Streets Within the Project Area, Case Study 2

	R.O.W.	Lanes
SINCLAIR AVE	100ft	2 12-ft lanes each direction; 10-ft shoulders; grass median, including 12-ft left turn bay, 150 ft long plus taper
EVERGREEN ST	80ft	1 12-ft lane each direction; 10-ft shoulders; grass median, including 12-ft left turn bay, 150 ft long plus taper
WEEPING WILLOW	80ft	1 12-ft lane each direction; 10-ft shoulders; grass median, including 12-ft left turn bay, 150 ft long plus taper
HAYES ST	80ft	1 12-ft lane each direction; 10-ft shoulders; grass median, including 12-ft left turn bay, 150 ft long plus taper
N. PIONEER DR	60ft	2 12-ft lanes WB, highway exit added as 3rd 12-ft lane (on left), then dropped as highway entrance; minimal shoulders
S. PIONEER DR	60ft	2 12-ft lanes WB, highway exit added as 3rd 12-ft lane (on left), then dropped as highway entrance; minimal shoulders
SHADOW ROCK RD	80ft	1 12-ft lane each direction; 10-ft shoulders; grass median, including 12-ft left turn bay, 150 ft long plus taper
WILLA CATHER HWY	Note 1	Note 2

Note 1 = All land between highway & service roads is part of the right of way;
Highway is presently 2 12 ft lanes each direction, 10 ft shoulders, 40 ft grass median;
Highway bridges over sinclair ave are a pair, each with 50 ft of roadway.

Note 2 = All ramps are single lanes, tapered; measurements shown on figure are to the gore area or the R.O.W. boundary.

Signal phasing = Protected lefts required, no permissive lefts allowed; lefts can be lead or lag; rtor generally permitted.

Distances = All distances shown along sinclair ave are measured to the R.O.W. boundaries, and do not include the R.O.W. itself.

Discussion Point 1

You must find an appropriate source of trip generation rates for the uses proposed (see next subsection). This includes consideration of both entering and departing traffic in each time period that is relevant.

Note that Figure 30.3 applies only to new traffic; it may be true (depending on the land use at the site of interest) that some percentage of the traffic arriving at the site is drawn from existing traffic that passes the site. The practical implication is that the *existing* traffic that goes to the site does not add to the volumes for the purpose of impact assessment. It does, however, use parking that needs to be provided within the site, and this must be taken into account.

Discussion Point 2

Clearly, there is less impact (and therefore less mitigation) if a good percentage of the traffic using the site is diverted from traffic that would pass in any case. This point is made in some cases for gas stations and for breakfast shifts at fast-food operations. But for other uses, the percentage is probably small. You must obtain information for the specific uses and/or argue the case.

30.5.2 Proposed Use(s) of the Two Site(s)

Refer to Figure 30.2. The plan is to develop the northern property as commercial space, specifically a suburban shopping mall, and the southern property as a multiplex theater. The multiplex is to have eight theaters, of which four will have 400 seats and four will have 200 seats. The shopping mall is to be built out to the limits of the local code (more on this later).

If it suits this plan, the two properties can be combined and considered as one.

Discussion Point 3

At the risk of getting ahead of ourselves, the local code requirements in the next subsection will require buffer zones between the two sites if they remain on the records as two distinct parcels. This may use space that could be dedicated to parking or other uses. However, there may be other issues of interest to the client (i.e., the owner of the parcels), such as future flexibility allowed by keeping them distinct or even some tax implications (none of us are knowledgeable in this area).

Table 30.4: Traffic Counts in the Project Area, Existing Condition, Case Study 2

PFH = 0.85 %TRUCKS = 5.0%

ALL VOLUMES IN HOURLY VPH EXISTING CONDITION, AM PEAK	NB			SB			EB			WB		
	L	T	R	L	T	R	L	T	R	L	T	R
SINCLAIR AVE & EVERGREEN ST	120	960	50	60	650	50	60	270	40	80	360	40
SINCLAIR AVE & WEEPING WILLOW	40	970	60	60	660	50	80	280	60	80	320	80
SINCLAIR AVE & HAYES ST	60	990	60	60	690	80	40	320	40	40	280	40
SINCLAIR AVE & N. PIONEER DR	90	1010			630	100				60	600	100
SINCLAIR AVE & S. PIONEER DR		1040	120	100	590		60	550	100			
SINCLAIR AVE & SHADOW ROCK RD	90	1060	60	60	590	60	40	320	40	60	340	60

WILLA CATHER HWY			
HOURLY	COUNT STATION AT EAST END OF SKETCH	5200	EB
		4850	WB
	AT RAMPS TO N. PIONEER	300	
	FROM N. PIONEER	250	
	TO S. PIONEER	270	
	FROM S. PIONEER	300	

EXISTING CONDITION, MIDDAY WEEKDAY	NB			SB			EB			WB		
	L	T	R	L	T	R	L	T	R	L	T	R
SINCLAIR AVE & EVERGREEN ST	84	672	35	42	455	35	42	189	28	56	252	28
SINCLAIR AVE & WEEPING WILLOW	28	679	42	42	462	35	56	196	42	56	224	56
SINCLAIR AVE & HAYES ST	42	693	42	42	483	56	28	224	28	28	196	28
SINCLAIR AVE & N. PIONEER DR	63	707			441	70				42	420	70
SINCLAIRAVE & S. PIONEER DR		728	84	70	413		42	385	70			
SINCLAIR AVE & SHADOW ROCK RD	63	742	42	42	413	42	28	224	28	42	238	42

WILLA CATHER HWY			
HOURLY	COUNT STATION AT EAST END OF SKETCH	3640	EB
		3395	WB
	AT RAMPS TO N. PIONEER	210	
	FROM N. PIONEER	175	
	TO S. PIONEER	189	
	FROM S. PIONEER	210	

(Continued)

Table 30.4: Traffic Counts in the Project Area, Existing Condition, Case Study 2 (*Continued*)

ALL VOLUMES IN HOURLY VPH

PFH = 0.85 %TRUCKS = 5.0%

EXISTING CONDITION, PM PEAK	**NB**			**SB**			**EB**			**WB**		
	L	**T**	**R**	**L**	**T**	**R**	**L**	**T**	**R**	**L**	**T**	**R**
SINCLAIR AVE & EVERGREEN ST	40	660	80	40	960	60	50	360	120	50	270	60
SINCLAIR AVE & WEEPING WILLOW	60	670	80	80	1030	80	50	320	40	60	280	60
SINCLAIR AVE & HAYES ST	40	670	40	40	970	40	80	280	60	60	320	60
SINCLAIR AVE & N. PIONEER DR	100	650			1010	60				120	550	100
SINCLAIR AVE & S. PIONEER DR		650	60	100	1070		100	600	90			
SINCLAIR AVE & SHADOW ROCK RD	40	590	60	60	1060	40	60	340	90	60	320	60

WILLA CATHER HWY			
HOURLY	COUNT STATION AT EAST END OF SKETCH	4850	EB
		5200	WB
	AT RAMPS TO N. PIONEER	300	
	FROM N. PIONEER	270	
	TO S. PIONEER	250	
	FROM S. PIONEER	300	

EXISTING CONDITION, SATURDAY	**NB**			**SB**			**EB**			**WB**		
	L	**T**	**R**	**L**	**T**	**R**	**L**	**T**	**R**	**L**	**T**	**R**
SINCLAIR AVE & EVERGREEN ST	**120**	960	**50**	**60**	**650**	**50**	**60**	**270**	**40**	**80**	**360**	**40**
SINCLAIR AVE & WEEPING WILLOW	**40**	970	**60**	**60**	660	**50**	**80**	**280**	**60**	**80**	**320**	**80**
SINCLAIR AVE & HAYES ST	**60**	990	**60**	**60**	690	**80**	**40**	**320**	**40**	**40**	**280**	**40**
SINCLAIR AVE & N. PIONEER DR	**90**	1010			630	**100**				**60**	**300**	**100**
SINCLAIR AVE & S. PIONEER DR		1040	**120**	**100**	590		**60**	**250**	**100**			
SINCLAIR AVE & SHADOW ROCK RD	**90**	**1060**	**60**	**60**	590	**60**	**40**	**320**	**40**	**60**	**340**	**60**

WILLA CATHER HWY			
HOURLY	COUNT STATION AT EAST END OF SKETCH	**5200**	EB
		4850	WB
	AT RAMPS TO N. PIONEER	**300**	
	FROM N. PIONEER	**250**	
	TO S. PIONEER	**270**	
	FROM S. PIONEER	**300**	

Table 30.5: Requirements from Local Ordinances

1. Each property must be at least 5 acres in this zoning, and have at least 300 ft of frontage on the main road. Building heights may not exceed 35 feet. The building(s) cannot cover more than 30% of the property.
2. There must be 100 feet of buffer facing all roads and other properties, except that the buffer may be reduced to 50 feet if the use on the other property is nonresidential. If the use on the other property is residential,[6] a 6-foot barrier wall must be constructed of stone or heavy timber and of appearance consistent with like construction in the jurisdiction.
3. For the shopping mall and like uses, 5 parking spaces per 1000 square feet of gross floor area (GFA) must be provided.
4. For the motion picture theater(s), 1 parking space per 3 seats must be provided.
5. The design must allow for up to 20% of the trips arriving by bus transit, for anticipated mode shifts over coming decades.[7] However, in assigning trips to modes, no more than 5% of the anticipated trips can be assigned to transit, given current realities.
6. Access management principles must be strongly considered and used to the maximum extent feasible, consistent with state requirements.[8]
7. Adequate loading bays for goods delivery and pickup must be provided, generally separated from the visitors to the site(s).
8. Entrances/exits for goods vehicles need to accommodate large tractor-trailers. All entrances/exits must accommodate a hook-and-ladder fire truck.
9. Internal circulation at any site shall be such that safe pedestrian and transit uses are encouraged, and that traffic calming principles are used.
10. Mitigation plans are expected (a) if the *v/c* ratio on any approach at any existing intersection is increased by more than 0.03, (b) if the LOS on any approach at any existing intersection is decreased (made poorer) by one level of service, and (c) if the arterial LOS of service on any arterial is decreased (made poorer) by one level of service.
11. If additional intersection(s) is/are proposed, the intersection *v/c* and LOS at each such intersection must be comparable to nearby existing intersections *and* the arterial level of service cannot be decreased (made poorer).
12. The future no-build and future build traffic levels are to be taken as 10 years from the present (i.e., existing) condition.
13. A section of the impact analysis shall address the construction phase. This section may be qualitative and descriptive, *if* no significant impact is anticipated to peak hour traffic, but the means for achieving this shall be described.

30.5.3 Local Code & Local Ordinance Requirements

Table 30.5 contains requirements that are extracted from various local codes, so that you will have a reasonable set of design considerations for this case study.

30.5.4 Other Given Conditions

With regard to *basic traffic engineering*, the local practice dictates:

- Use a loss time of 4.0 seconds per phase.
- For saturation flow rates, use 1900 pcphpl for the through lanes.
- All left turns at signalized intersections are to be protected.
- RTOR at signalized intersections is prohibited, except if there is appropriate channelization in a separate lane, with at least YIELD sign control.

The *background traffic*, exclusive of these sites, is to be taken as 3.0% per year (and is to be compounded). No other planned or approved developments of note exist or are assumed to be incorporated in the background growth.

The local *topography* has no significant elevation changes.

With regard to *tools and references*, you are expected to use *at least* the following tools, references, and practices in the course of this case study:

1. Signal timing (including alternate phase plans) for both existing and all future conditions is to be done using Synchro.

[6]The adjacent properties to the east are zoned commercial in this case study.

[7]Assume the primary (sole) bus routes run north-south on the main arterial.

[8]Use the State of Florida guidelines or practices, if needed.

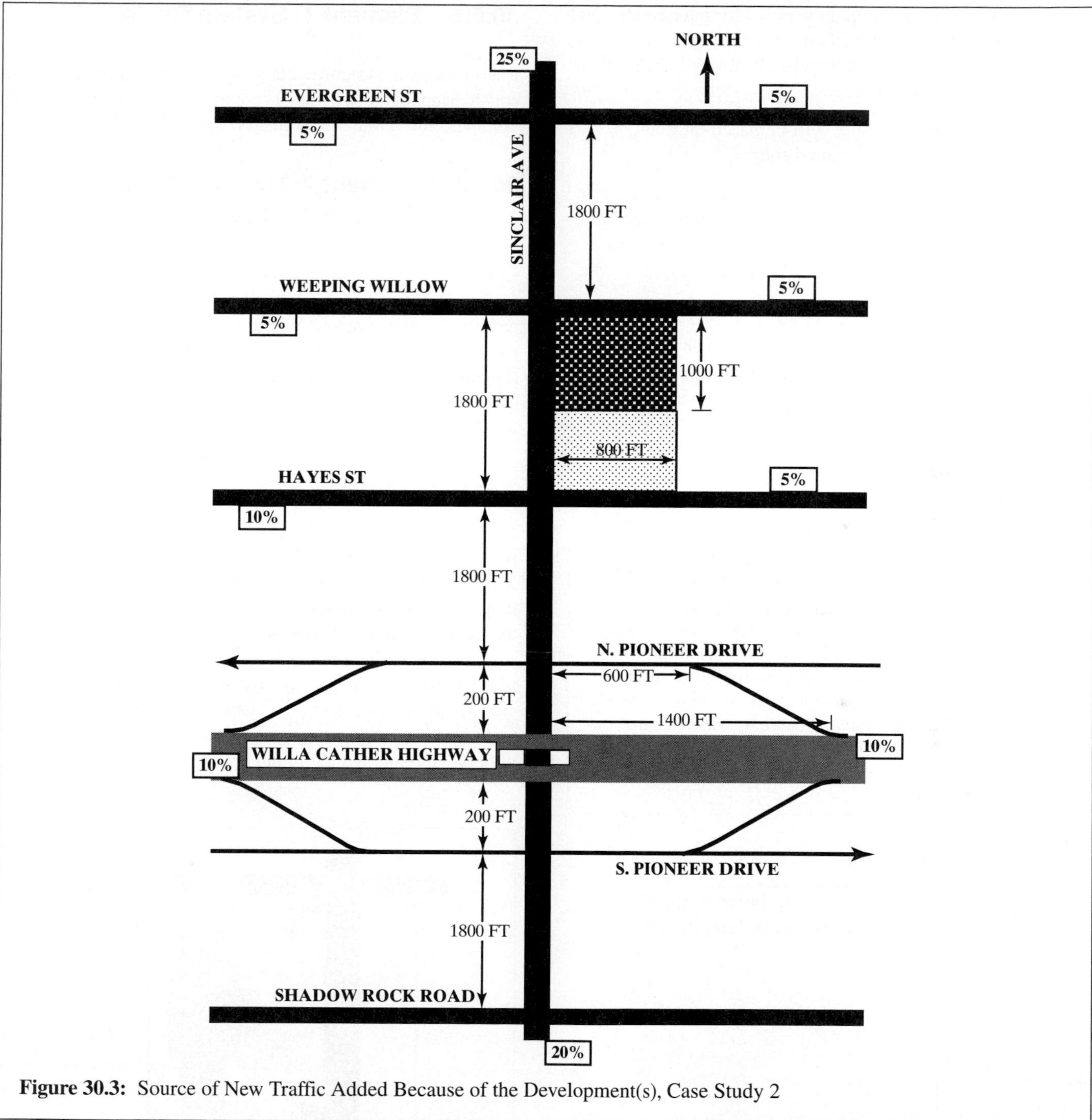

Figure 30.3: Source of New Traffic Added Because of the Development(s), Case Study 2

2. Given that the existing signal timings are not available (they usually are) *and* given that the local jurisdiction had scheduled signal retiming work in any case, "existing" signal timings are to be determined using Synchro optimization for the most suitable cycle length (more on this in the next subsection).

3. Simulations and visualizations are to be done with a tool such as SIMTraffic, VISSIM, or AIMSUN.

4. Levels of service and v/c ratios for intersections and arterials[9] may be obtained from Synchro results, unless the simulations/visualizations indicate there is a clear inconsistency.
5. Trip generation rates will be based on References 6 or 7 or other documented source.
6. Parking layout and internal circulation using relevant ITE or other documents, or the approach suggested in Problem 30-3.
7. For this particular case study, this textbook may be used as a reference document.

If the traffic professional uses HCS+ or other such tool for capacity analysis or signal timing, that traffic professional must submit a reasoned argument on why these results are more relevant than Synchro, recognizing that this particular jurisdiction prefers and is accustomed to receiving Synchro results.

Note: The remaining subsections divide the traffic impact and mitigation work into a set of "elements" so that they can be easily assigned in parts, generating discussion and learning as the course progresses. The schedule used by one of the authors is shown below, merely as a suggestion. Periodic presentations by the student groups are to be encouraged, and a final presentation of the impact and proposed mitigation is an essential experience. (Work submitted at the intermediate due dates was evaluated on the basis of the learning experience, but the final comprehensive report and presentation was viewed with a higher level of expectation.)

Element	Due Date, Relative to the Week the Case Study Was Begun
1	1
2	2
3	4
4	5
5	6
6	7
7	9
8	10

[9]Remember that arterial LOS is based on the average travel speed of the *through* vehicles.

30.5.5 Element 1: System Cycle

Recommend a system cycle length along Sinclair Avenue, considering signal spacing, a reasonable vehicle speed, traffic volumes, and number of signal phases.

30.5.6 Element 2: The Developer's Favorite Access Plan

Consider the following hypothetical situation: The developer is very interested in a design that combines the two parcels, centers the development, and uses only two major driveways for public access, including transit. Refer to Figure 30-3. (Details of parking lot and internal circulation are not shown; they would remain for the traffic professional to work out within the overall concept.)

Clearly, this creates two five-legged intersections that need a multiphase signal plan.

You are to evaluate the operational needs of these intersections and the required flow patterns given the arrival/departure suggested within Figure 30.4. Table 30.5 provides the existing flows, and the growth rate is known. For the added traffic, you may have to make some assumptions (just to get started) or use the trip generation references already cited.

Keep in mind that (1) the client really wants this "innovative" approach, and (2) you have to be the responsible professional, sometimes achieving what is desired by creative design and sometimes presenting a reality that serves the client well while disappointing their initial notions.

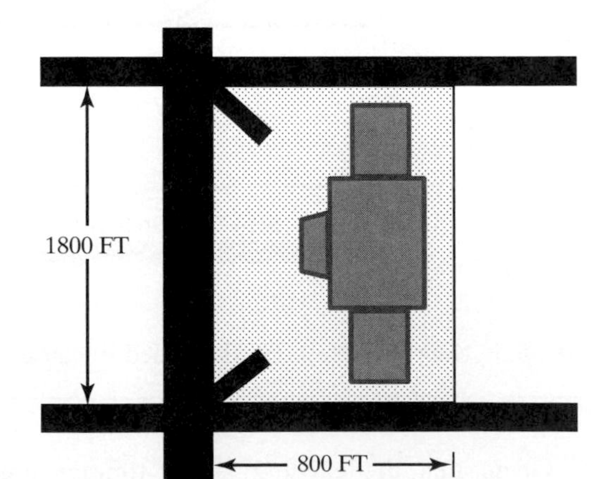

Figure 30.4: A Concept of the Development, from the Client

Discussion Point 5
Yes, the authors have revealed they are pessimistic about making the five-leg intersections work, given the flow rates and patterns. However, the obligation is for *you* to find a successful solution *or* to clearly explain to the client why the concept is not in their best interests.

Discussion Point 6
Remember that in *Element 1*, you recommended a system cycle length. That cycle length may be inconsistent with the values needed at the proposed five-leg intersections.

The end product of the work on Element 2 should be a PowerPoint presentation explaining your findings, in terms that both your peers and a nonspecialist can understand. Use no more than five slides.

30.5.7 Element 3: Existing Conditions, Capacity, and LOS Analyses

Use Synchro (and other tools, as needed) to estimate the existing and future no-build (FNB) conditions at each intersection, and for each lane group at each intersection, and for the overall arterial.

Summarize the results in the format of Table 30.6. (The future build [FB] column will be completed in a later element of this work.)

Table 30.6: Format to Be Used in Element 3 Work

	v/c			LOS		
	Existing	**FNB**	**FB w/o Mitigation**	**Existing**	**FNB**	**FB w/o Mitigation**
Intersection						

Intersection	Lane Group						

Arterial		Average Travel Speed (mph)			LOS		
	NB						
	SB						

Note: FB not done in Element 3.

One table is needed for each time period of interest. If it is less than the four periods for which data has been provided, indicate why the period is irrelevant to the analysis, given the anticipated uses.

Discussion Point 7

Before this analysis is begun, the instructor will have to specify the system cycle length to be used by all teams in the course. This can be assigned after the submission and discussion of the Element 1 work.

Anticipate what traffic improvements will be needed due solely to the "Future No-Build" traffic levels.

Discussion Point 8

It is logical that some operational problems will arise, due simply to the annual growth of traffic. In some cases, these would logically require some attention/action, independent of the added traffic due to the specific new development (i.e., the two sites at hand). It is good to know this, in discussions with the local jurisdiction.

Given that you were not provided with existing signal timing, you may want to use the critical movement analysis as a reasonable method of determining the initial signal timing and phasing. This can be refined by using Synchro.

Discussion Point 9

This work is part of this student assignment, but existing signal plans generally exist and are available from the local jurisdiction or state.

Discussion Point 10

In the same spirit of a limited student assignment, the application of the *traffic signal warrants* from the state's MUTCD[10,11] is not part of this assignment. In some applications, there are unsignalized intersections that may need to be signalized, due to a combination of background growth and the project at hand. There may also be new intersections due to driveways, and these will have to be evaluated for signalization. The instructor may wish to require a warrant analysis for at least these new prospective intersections.

[10]Remember that the simple satisfaction of one or more traffic signal warrants does *not* mandate the installation of a signal. Rather, the sense of the MUTCD is that a signal should not be installed *unless* one or more warrants are satisfied, and the satisfaction of warrants is simply the trigger that allows this evaluation (an engineering study, involving many factors) to proceed.

[11]Some states specify that certain warrants are not used determining whether a signal is justified (e.g., the peak hour volume warrant).

Discussion Point 11

Again, in the same spirit of a limited student assignment, this particular case study does not require you to acquire and analyze the accident experience in the project area.

30.5.8 Element 4: Trip Generation

You have already been guided to the sources from which to obtain trip generation *rates*. These will often be expressed in terms of trips per hour for the peak hour and other hours. The "peak hour" for the development's traffic may however *not be the same as the general peak hour*. This depends on the proposed use (e.g., supermarket, shopping center, multiplex cinema). The traffic professional must take care that the terminology used in the trip generation source refers to the traditional peak hour or the peak activity hour (and day) of the specific use.

Likewise, the construction phase—if it has a noticeable impact on traffic levels at all—may actually define the peak hour for analysis.[12]

In this Element 4, you are expected to generate estimates of the number and routes of the trips generated *based on the desired maximum build-out of the site(s)*.

Discussion Point 12

The emphasis in the preceding sentence is intentional. Given local rules as stated in Table 30-6, there are clearly limits imposed by available square footage of the "footprint" of the development. This does not necessarily mean that the rest of the site can support this maximum development. But it is a starting point.[13]

Discussion Point 13

The local rule on building height may tempt the architect or the traffic professional to think in terms of a two-story shopping mall, with stores on both levels. However, if one is inclined to go down this path, it is important to discuss how many *successful* and *attractive* malls in the region are two-story designs. At the time of this writing, many malls are characterized by large open spaces, common areas, and generally one-story operation. That is

[12]For many projects, the construction phase workers will arrive and depart in off-peak hours (relative to the existing peaks), and not create intense loads. In some cases, materials delivery may need to be noted and considered.

[13]The traffic professional must recognize that the developer and perhaps the architect may well focus on the maximum that can be done. It falls to the traffic professional to point out that the support functions also mandated by local rules (e.g., required parking spaces, internal circulation, space for transit and for goods deliveries) actually impose constraints that limit the size of the development, by the arithmetic imposed.

not to say a two-story design cannot be found that would succeed, but rather that it should be a discussion point in class. Even with such a design, there are still likely to be significant open spaces and common areas that use up part of the footprint, leaving less gross floor area (GFA) than twice the footprint.

Estimate the required parking, consider the overall properties, and start forming opinions on whether the available space can accommodate the parking requirement.

Discussion Point 14

This is an exercise in arithmetic, but an important one. The underlying issues were discussed earlier and in related footnotes. Remember that local rules express parking requirements in terms of spaces per 1,000 square feet of GFA or spaces per seats (in the theaters). GFA includes common areas, hallways, and such.

It is recognized that the parking layout is going to be a bit of a challenge for you, unless you are drawing on other knowledge (i.e., other sources than this textbook). Fortunately, such sources exist, in print or on the Web. Real-world examples based on actual developments are also available from Google Earth or other tools.

Discussion Point 15

An area of lively discussion is sometimes the average vehicle occupancy for the different uses at the site. This is rendered moot by the local codes. At the same time, the numbers used in the codes implicitly assume average vehicle occupancies *and* percentage trips by auto. As the years pass, the age of these implicit assumptions may be a basis for discussion, if more recent data prove inconsistent.

Determine whether combining the sites provides any advantages.

Discussion Point 16

This point has already been raised, but this is a logical point to remind you of it.

Likewise, take into account the special local requirement on building now to accommodate future transit usage of 20% of the trips.

Discussion Point 17

Note that this is a mandate and related to the long-term planning of the local jurisdiction. The spirit and intent is that this transit service be accommodated on-site, in attractive and efficient areas. Simply depending on bus stops on the local arterial will *not* suffice in this submittal.

30.5.9 Element 5: Determine the Size of the Development, Trips Generated, and Internal Circulation

Continue the work begun in Element 4, with special attention to the needs of internal circulation, parking, transit, safe pedestrian travel, and space for goods vehicles. If the overall requirements dictate a smaller build-out than the maximum *nominally* allowed by the zoning, be prepared to address this.

Of necessity, internal circulation will depend on driveway location(s) and on any special design features on the arterial. Although these are nominally Element 6 of the project, there is an overlap.

30.5.10 Element 6: Driveway Locations and Special Arterial and Intersection Design Features

Taking into account the mandate for access management to the extent feasible, develop recommendations on driveway locations, special arterial features, and intersections (present or proposed). Consider the "lessons learned" in Case Study 1 with regard to driveway locations.

Remember the needs of goods delivery, which is addressed in the local rules. Consider the possibility (likelihood) of separate driveways for these deliveries, and remember that any "behind" the building has to allow sufficient space to turn and/or maneuver large trucks (some truck bay designs may affect this also).

30.5.11 Element 7: Mitigation Measures

Return to Table 30.6. Complete the "Future Build" (FB) sections, assuming no mitigation. Highlight all entries that are determined to have an impact, in accord with local rules.

Develop ideas on mitigation in detail. The teams working on the project will need to be creative while recognizing that improvements cost money and will probably be paid by the client.

A guiding principle sometimes overlooked at this stage is that the mitigation and related design is not simply intended to meet the minimum requirements of local rules, but also to assure that the development operates smoothly into the future and is/remains attractive to the public and the occupants of the businesses. It is sometimes useful for the traffic professional to have this discussion with the client and their team, and to develop a minimal "Plan A" and a "Plan B" so the client can see the costs and benefits of any enhancements the traffic professional believes can serve the overall success.

Add columns to Table 30-6 for Plans A and B. Prepare presentation materials. Be ready to engage in discussion on the plan(s) recommended.

Discussion Point 18
Whereas Plans A and B may be for internal discussion, with only one included in the Draft Final Report of Element 8, in this student assignment, both plans will be included in the Draft Final Report.

Discussion Point 19
Because this assignment will culminate in the presentation of the Draft Final Report and not include the usual next round of agency review and hearings, culminating in a Final Report submittal, the word "Draft" is dropped in Element 8.

Discussion Point 20
You are reminded of the list of possible mitigation actions enumerated in Step 11 of Section 30.2, and invited to add to the list as needed for this specific project.

30.5.12 Element 8: Final Report and Presentation

Each group will have 20 minutes to present their findings and recommendations. Business attire is required. The group need not assign everyone a speaking role (although there is merit to that), but the instructor may (i.e., will) direct questions to any group member, so *all* group members have to be fully prepared.

A final report not exceeding 30 pages is required 24 hours prior to the class, sent by e-mail to the instructor in PDF format. The PowerPoint slides for the presentation have to be sent at the same time.

30.6 Summary

This chapter introduced you to the topic of traffic impact studies, including an overview of the process and emphasizing the need for creative design in meeting mitigation needs associated with a significant development. (Minor projects may result in a finding or no significant impact, although some estimates of future traffic load and a permitting process are still involved.)

Case Study 2 was used by one of the authors as the basis for a second course in traffic engineering. Reference 6 was specified as a required companion text, in this mode. Early lectures covered other chapters of this textbook and Case Study 1. A presentation and a work session on computer tools was included. When Case Study 2 was begun, the class time was devoted to additional chapters of this textbook, discussion of each Element when it was initiated and when it was due, and ad hoc discussions based on information requests from the students.

Case Study 2 can also be used as the basis of a few lectures in a course that is not project based, with emphasis on the discussion points enumerated throughout.

References

1. http://www.epa.gov/oecaerth/basics/nepa.html.
2. http://nepa.gov/nepa/nepanet.htm.
3. http://en.wikipedia.org/wiki/National_Environmental_Policy_Act.
4. http://www.dec.ny.gov/permits/357.html.
5. http://www.nyc.gov/html/oec/html/ceqr/ceqrfaq.shtml.
6. "Transportation Impact Analyses for Site Development: An ITE Proposed Recommended Practice," Institute of Transportation Engineers, Washington DC, 2005.
7. *Trip Generation, 8th edition: An ITE Informational Report*, Institute of Transportation Engineers, Washington DC, 2008.
8. HCS+ Release 5: Users Manual, distributed by the McTrans Center, University of Florida, 2008. http://mctrans.ce.ufl.edu/hcs/
9. SYNCHRO Studio 7, Synchro plus SimTraffic and 3D Viewer User Guide, Trafficware Ltd., June 2006. www.trafficware.com
10. Transportation Research Circular 212, Interim Materials on Highway Capacity, January 1980, TRB, National Academy of Sciences, Washington, DC.
11. VISSIM 5.0 User's Guide, PTV, 2007.http://www.ptvamerica.com/vissim.html
12. AIMSUN Users Manual v6, TSS-Transport Simulation Systems, October 2008.http://www.aimsun.com

Problems

Note: The instructor may prefer that all problems in this chapter be done as group or team assignments, with the team not to exceed three members (perhaps four in a very large class).

30-1. Do the analysis and impact assessment as specified in Case Study 1.

30-2. For the higher value of "X" and the better value of "N" in Case Study 1 as determined in Problem 30-1, recommend any additional mitigation measures appropriate and provide the supporting analysis.

30-3. For Case Study 1 and the higher value of "X," lay out a functional design of the parking lot, allowing for both the entering and departing values of "X" and recognizing that they may be competing for internal roadway use and for parking spaces over the hour of analysis.

If you are not provided with information on the parking requirements and do not have access to a reference providing that information, you should either (a) consult sources on the Web, using a search engine, or (b) use a tool such as Google Earth to "visit" a known suburban parking lot, and estimate the space per parked vehicle (taking into account space needed to travel to the parking spaces and any separation between or at the end of rows of cars).

30-4. For Case Study 2, execute and submit Element 1 in accord with the schedule in the chapter or the instructor's specification. Submit an analysis for the group, not to exceed three pages.

30-5. For Case Study 2, execute and submit Element 2 in accord with the schedule in the chapter or the instructor's specification.

30-6. For Case Study 2, execute and submit Element 3 in accord with the schedule in the chapter or the instructor's specification.

30-7. For Case Study 2, execute and submit Element 4 in accord with the schedule in the chapter or the instructor's specification.

30-8. For Case Study 2, execute and submit Element 5 in accord with the schedule in the chapter or the instructor's specification.

30-9. For Case Study 2, execute and submit Element 6 in accord with the schedule in the chapter or the instructor's specification.

30-10. For Case Study 2, execute and submit Element 7 in accord with the schedule in the chapter or the instructor's specification.

30-11. For Case Study 2, execute and submit Element 8 in accord with the schedule in the chapter or the instructor's specification.

Index